Keys to Nearctic Fauna

Thorp and Covich's Freshwater Invertebrates

A Global Series of Books on the Identification,
Ecology, and General Biology of Inland Water Invertebrates
by Experts from Around the World

Fourth Edition
Series Editor: James H. Thorp

Volume I: Ecology and General Biology
Edited by James H. Thorp and D. Christopher Rogers
Published 2015

Volume II: Keys to Nearctic Fauna
Edited by James H. Thorp and D. Christopher Rogers
Published 2016

Volume III: Keys to Palaearctic Fauna
Edited by D. Christopher Rogers and James H. Thorp
Expected Publication Date: 2017

Volumes in Preparation and Under Contract
Keys to Neotropical and Antarctic Fauna
Keys to Neotropical Hexapoda
Keys to Fauna of the Australian Bioregion

Possible Future Volumes of the Fourth Edition
Keys to Oriental and Oceana Fauna
Keys to Oriental and Oceana Hexapoda
Keys to Palaearctic Hexapoda
Keys to Afrotropical Fauna
Keys to Afrotropical Hexapoda

Related Publications
Ecology and Classification of North American Freshwater Invertebrates
Edited by J.H. Thorp and A.P. Covich
First (1991), Second (2001), and Third (2010) Editions
Field Guide to Freshwater Invertebrates of North America
by J.H. Thorp and D.C. Rogers

Keys to Nearctic Fauna

Thorp and Covich's Freshwater Invertebrates - Volume II

Fourth Edition

Edited by

James H. Thorp
D. Christopher Rogers

AMSTERDAM • BOSTON • HEIDELBERG • LONDON • NEW YORK • OXFORD • PARIS
SAN DIEGO • SAN FRANCISCO • SINGAPORE • SYDNEY • TOKYO
Academic Press is an imprint of Elsevier

Academic Press is an imprint of Elsevier
125 London Wall, London EC2Y 5AS, UK
525 B Street, Suite 1800, San Diego, CA 92101-4495, USA
225 Wyman Street, Waltham, MA 02451, USA
The Boulevard, Langford Lane, Kidlington, Oxford OX5 1GB, UK

ISBN: 978-0-12-385028-7

British Library Cataloguing-in-Publication Data
A catalogue record for this book is available from the British Library

Library of Congress Cataloging-in-Publication Data
A catalog record for this book is available from the Library of Congress

For information on all Academic Press publications
visit our website at http://store.elsevier.com/

Publisher: Cathleen Sether
Acquisition Editor: Laura Kelleher
Editorial Project Manager: Rowena Prasad
Production Project Manager: Julia Haynes
Designer: Greg Harris

Typeset by TNQ Books and Journals
www.tnq.co.in

Dedications from the Editors

To Henry B. Ward, George C. Whipple, W. Thomas Edmondson, and Robert W. Pennak—pioneers who blazed a publishing trail with books on the ecology and identification of North American freshwater invertebrates.

To Alan P. Covich, a longtime friend and valued colleague, who not only helped develop the first three editions but also made possible the fourth edition's improved taxonomy and worldwide coverage by introducing the current editors to each other.

James H. Thorp and D. Christopher Rogers

Contents

Contributors to Volume II

Fernando Álvarez [Chapter 16] Departamento de Zoología, Instituto de Biología, U.N.A.M., Circuito exterior s/n, Ciudad Universitaria, Copilco, Coyoacán, A.P. 70-153, México, Distrito Federal. C.P. 04510, México; email: falvarez@servidor.unam.mx

Bonnie A. Bain [Chapter 12] Department of Biological Sciences, Southern Utah University, Cedar City, Utah 84720, USA; email: bain@uss.edu

Ilse Bartsch [Chapter 16] Forschungsinstitut Senckenberg, c/o DESY, Gebaeude 3, Raum 316, Notkestr. 85, 22607, Hamburg, Germany; email: bartsch@meeresforschung.de

Valerie Behan-Pelletier [Chapter 16] Agriculture and Agri-Food Canada, K.W. Neatby Building, 960 Carling Avenue, Ottawa, Ontario K1A 0C6, Canada; email: valerie.behan-pelletier@agr.gc.ca

Matthew G. Bolek [Chapter 10] Department of Zoology, Oklahoma State University, 501 Life Sciences West, Stillwater, Oklahoma 74078, USA; email: bolek@okstate.edu

Ralph O. Brinkhurst [Chapter 12] 205 Cameron Court, Hermitage, Tennessee 37076, USA

Francisco Brusa [Chapter 5] División Zoologia Invertebrados, Museo de La Plata, FCNyM-UNLP, 1900 La Plata, Argentina; email: fbrusa@fcnym.unlp.edu.ar

Richard D. Campbell [Chapter 4] Department of Developmental and Cell Biology, University of California, Irvine, CA, USA; post mail: 2561 Irvine Ave., Costa Mesa, California, 92627 USA; email: rcampbel@uci.edu

Joo-lae Cho [Chapter 16] Invertebrate Research Division, National Institute of Biological Resources, Environmental Research Complex, Gyoungseo-dong, Incheon, 404-170, South Korea; email: Joolae@Korea.kr

David R. Cook [Chapter 16] 7725 North Foothill Drive South, Paradise Valley, Arizona 85253, USA; email: watermites@msn.com

Kevin S. Cummings [Chapter 11] Illinois Natural History Survey, Center for Biodiversity, 607 East Peabody Drive, Champaign, Illinois 61820, USA; email: ksc@inhs.uiuc.edu

Cristina Damborenea [Chapter 5] División Zoología Invertebrados, Museo de La Plata, FCNyM-UNLP, Paseo del Bosque, 1900 La Plata, Argentina; email: cdambor@fcnym.unlp.edu.ar

R. Edward DeWalt [Chapter 16] Illinois Natural History Survey, Center for Biodiversity, 607 East Peabody Drive, Champaign, Illinois 61820, USA; email: edewalt@inhs.illinois.edu

Genoveva F. Esteban [Chapter 2] Conservation Ecology and Environmental Sciences Group, Faculty of Science and Technology, Bournemouth University, Dorset, United Kingdom; email: gesteban@bournemouth.ac.uk

James W. Fetzner Jr. [Chapter 16] Biodiversity Services Facility, Section of Invertebrate Zoology, Carnegie Museum of Natural History, 4400 Forbes Avenue, Pittsburgh, Pennsylvania 15213-4080, USA; email: FetznerJ@CarnegieMNH.org

Bland J. Finlay [Chapter 2] School of Biological and Chemical Sciences, Queen Mary University of London, The River Laboratory, Wareham, Dorset, BH20 6BB, United Kingdom; email: b.j.finlay@qmul.ac.uk

Stuart R. Gelder [Chapter 12] Department of Science and Math, University of Maine at Presque Isle, Presque Isle, Maine 04769, USA; email: stuart.gelder@umpi.edu

Fredric R. Govedich [Chapter 12] Department of Biological Sciences, Southern Utah University, 351 West University Blvd, Cedar City, Utah 84720, USA; email: govedich@suu.edu

Daniel L. Graf [Chapter 11] The Academy of Natural Sciences, 1900 Benjamin Franklin Parkway, Philadelphia, Pennsylvania 19103, USA; email: grad@acnatsci.org

Roberto Guidetti [Chapter 15] Department of Biology, University of Modena and Reggio Emilia, via Campi 213/D, 41125, Modena, Italy; email: roberto.guidetti@unimore.it

Ben Hanelt [Chapter 10] Department of Biology, University of New Mexico, 163 Castetter Hall, Albuquerque, New Mexico 87131, USA; email: bhanelt@unm.edu

Brenda J. Hann [Chapter 16] Department of Biological Sciences, W463 Duff Roblin, University of Manitoba, Winnipeg, Manitoba R3T 2N2, Canada; email: hann@cc.umanitoba.ca

Tom Hansknecht [Chapter 16] Barry A. Vittor and Associates, Inc., 8060 Cottage Hill Rd., Mobile, Alabama 36695, USA; email: bvataxa@bvaenviro.com

David J. Horne [Chapter 16] School of Geography, Queen Mary University of London, Mile End Road, London E1 4NS, United Kingdom; email: d.j.horne@qmul.ac.uk

Julian J. Lewis [Chapter 16] Lewis & Associates LLC, 17903 State Road 60, Borden, Indiana 47106-8608, USA; email: lewisbioconsult@aol.com

Lawrence L. Lovell [Chapter 12] Research Associate, Polychaetous Annelids, Research & Collections, Natural History Museum of Los Angeles County, 900 Exposition Blvd., Los Angeles, California 90007, USA; email: lllpolytax@gmail.com

Tobias Kånneby [Chapter 7] Department of Zoology, Swedish Museum of Natural History, 10405, Stockholm, Sweden; email: tobias.kanneby@nrm.se

Renata Manconi [Chapter 3] Dipartimento di Scienze della Natura e del Territorio (DIPNET), Università di Sassari, Muroni 25, I-07100, Sassari, Italy; email: r.manconi@uniss.it

William E. Moser [Chapter 12] Smithsonian Institution, National Museum of Natural History, Department of Invertebrate Zoology, Museum Support Center, 4210 Silver Hill Road, Suitland, Maryland 20746, USA; email: moserw@si.edu

Diane R. Nelson [Chapter 15] Department of Biological Sciences, East Tennessee State University, Johnson City, Tennessee 37614-1710, USA; email: janddnelson@yahoo.com

Carolina Noreña [Chapter 5] Departamento Biodiversidad y Biología Evolutiva, Museo Nacional de Ciencias Naturales (CSIC), Madrid, España; email: norena@mncn.csic.es

Roy A. Norton [Chapter 16] SUNY College of Environmental Science and Forestry, 134 Illick Hall, 1 Forestry Drive, Syracuse, New York 13210, USA; email: ranorton@esf.edu

Alejandro Oceguera-Figueroa [Chapter 12] Laboratorio de Helmintologiá, Instituto de Biologiá, Universidad Nacional Autoñoma de México, Tercer circuito s/n, Ciudad Universitaria, Copilco, Coyoacán. A.P. 70-153, Distrito Federal, C. P. 04510, México; email: aoceguera@ib.unam.mx

Anna J. Phillips [Chapter 12] Smithsonian Institution, National Museum of Natural History, Department of Invertebrate Zoology, 10th and Constitution Ave, NW, Washington, DC 20560-0163, USA; email: phillipsaj@si.edu

George O. Poinar Jr. [Chapter 9] Department of Zoology, Oregon State University, Corvallis, Oregon 97331, USA; email: poinarg@science.oregonstate.edu

Wayne Price [Chapter 16] Department of Biology, University of Tampa, 401 W. Kennedy Blvd., Tampa, Florida 33606, USA; email: wprice@ut.edu

Roberto Pronzato [Chapter 3] Dipartimento di Scienze della Terra, dell'Ambiente e della Vita (DISTAV), Università di Genova, Area Scientifico-Disciplinare 05 (Scienze biologiche), Settore BIO/05, Genova, Italy; email: pronzato@dipteris.unige.it

Lorena Rebecchi [Chapter 15] Department of Biology, University of Modena and Reggio Emilia, via Campi 213/D, 41125, Modena, Italy; email: lorena.rebecchi@unimore.it

Janet W. Reid [Chapter 16] Virginia Museum of Natural History, 1001 Douglas Avenue, Martinsville, Virginia 24112, USA; email: jwrassociates@sitestar.net

Vincent H. Resh [Chapter 16] Department of Environmental Science, Policy, and Management, University of California, 305 Wellman Hall, Berkeley, California 94720, USA; email: resh@berkeley.edu

Dennis J. Richardson [Chapter 12] School of Biological Sciences, Quinnipiac University, 275 Mt. Carmel Avenue, Hamden, CT 06518, USA; email: Dennis.Richardson@quinnipiac.edu

D. Christopher Rogers [Chapters 1, 11, 16] Kansas Biological Survey and Biodiversity Institute, Higuchi Hall, University of Kansas, 2101 Constant Avenue, Lawrence, Kansas 66047, USA; email: branchiopod@gmail.com

S.S.S. Sarma [Chapter 8] Laboratorio de Zoología Acuática, Unidad de Morfología y Función, Facultad de Estudios Superiores, Universidad Nacional Autónoma de México, Av. de lo Barrios, no. 1, Los Reyes, Tlalnepantla, Edo. de Méx. C.P. 54090, México; email: sssarma@gmail.com

Andreas Schmidt-Rhaesa [Chapter 10] Zoological Museum, University Hamburg, Martin Luther-King. Platz 3, 20146 Hamburg, Germany; email: andreas.schmidt-rhaesa@uni-hamburg.de

Hendrik Segers [Chapter 8] School of Freshwater Biology, Belgian Biodiversity Platform, Royal Belgian Institute of Natural Sciences, Vautierstraat 29, B-1000, Brussels, Belgium; email: Hendrik.Segers@naturalsciences.be

Alison J. Smith [Chapter 16] Department of Geology, Kent State University, Kent, Ohio 44242, USA; email: alisonjs@kent.edu

Ian M. Smith [Chapter 16] Systematic Acarology, Environmental Health Program, Agriculture and Agri-Food Canada, K.W. Neatby Building, 960 Carling Ave., Ottawa, Ontario K1A 0C6, Canada; email: smithi@agr.gc.ca

T.W. Snell [Chapter 8] School of Biology, Georgia Institute of Technology, 310 Ferst Drive, Atlanta, Georgia 30332, USA; email: terry.snell@biology.gatech.edu

Malin Strand [Chapter 6] The Swedish Species Information Centre, Swedish University of Agricultural Sciences, Uppsala, Sweden; email: malin.strand@slu.edu

Per Sundberg [Chapter 6] Department of Zoology, University of Gothenburg, P.O. Box 463, SE-405 30 Gothenburg, Sweden; email: P.Sundberg@zool.gu.se

Christopher A. Taylor [Chapter 16] Curator of Fishes and Crustaceans, Prairie Research Institute, Illinois Natural History Survey, University of Illinois at Urbana-Champaign, 1816 S. Oak, Champaign, Illinois 61820, USA; email: ctaylor@inhs.illinois.edu

Roger F. Thoma [Chapter 16] Midwest Biodiversity Institute, 4673 Northwest Parkway, Hilliard, Ohio 43026, USA; email: cambarus1@mac.com

James H. Thorp [Chapters 1, 11, 12] Kansas Biological Survey and Department of Ecology and Evolutionary Biology, University of Kansas, 2101 Constant Avenue, Lawrence, Kansas 66047, USA; email: thorp@ku.edu

Robert J. Van Syoc [Chapter 16] California Academy of Sciences, Department of Invertebrate Zoology and Geology, 55 Music Concourse Drive, San Francisco, California 94118, USA; email: Bvansyoc@calacademy.org

L. Cristina de Villalobos [Chapter 10] Facultad de Ciencias Naturales y Museo, Departamento de Invertebrados, Paseo del Bosque S/N 1900 La Plata, Argentina; email: villalo@fcnym.unlp.edu.ar

Robert L. Wallace [Chapter 8] Department of Biology, Ripon College, 300 Seward Street, Ripon, Wisconsin 54791, USA; email: wallacer@ripon.edu

Elizabeth J. Walsh [Chapter 8] Department of Biological Science, University of Texas at El Paso, 500 W. University Avenue, El Paso, Texas 79968, USA; email: ewalsh@utep.edu

Alan Warren [Chapter 2] Department of Life Sciences, Natural History Museum, Cromwell Road, London SW7 5BD, United Kingdom; email: a.warren@nhm.ac.uk

Timothy S. Wood [Chapters 13, 14] Department of Biological Sciences, Wright State University, 3640 Colonel Glen Highway, Dayton, Ohio 45435, USA; email: tim.wood@wright.edu

Fernanda Zanca [Chapter 10] Facultad de Ciencias Naturales y Museo, Departamento de Invertebrados, Paseo del Bosque S/N 1900 La Plata, Argentina; email: fmzanca@fcnym.unlp.edu.ar

About the Editors

Dr. James H. Thorp has been a Professor in the Department of Ecology and Evolutionary Biology at the University of Kansas (Lawrence, KS, USA) and a Senior Scientist in the Kansas Biological Survey since 2001. Prior to returning to his alma mater, Prof. Thorp was a Distinguished Professor and Dean at Clarkson University, Department Chair and Professor at the University of Louisville, Associate Professor and Director of the Calder Ecology Center of Fordham University, Visiting Associate Professor at Cornell, and Research Ecologist at the University of Georgia's Savannah River Ecology Laboratory. He received his Baccalaureate from the University of Kansas (KU) and both Masters and PhD degrees from North Carolina State. Those degrees focused on zoology, ecology, and marine biology with an emphasis on the ecology of freshwater and marine invertebrates. Dr. Thorp has been on the editorial board of three freshwater journals and is a former President of the International Society for River Science. He teaches freshwater, marine, and general ecological courses at KU, and his master's and doctoral graduate students work on various aspects of the ecology of organisms, communities, and ecosystems in rivers, reservoirs, and wetlands. Prof. Thorp's research interests and background are highly diverse and span the gamut from organismal biology to community, ecosystem, and macrosystem ecology. He works on both fundamental and applied research topics using descriptive, experimental, and modeling approaches in the field and lab. While his research emphasizes aquatic invertebrates, he also studies fish ecology, especially as related to food webs. He has published more than hundred refereed journal articles, books, and chapters, including three single-volume editions of *Ecology and Classification of North American Freshwater Invertebrates* (edited by J.H. Thorp and A.P. Covich) and the first volume (Ecology and General Biology) in the current fourth edition of *Thorp and Covich's Freshwater Invertebrates*.

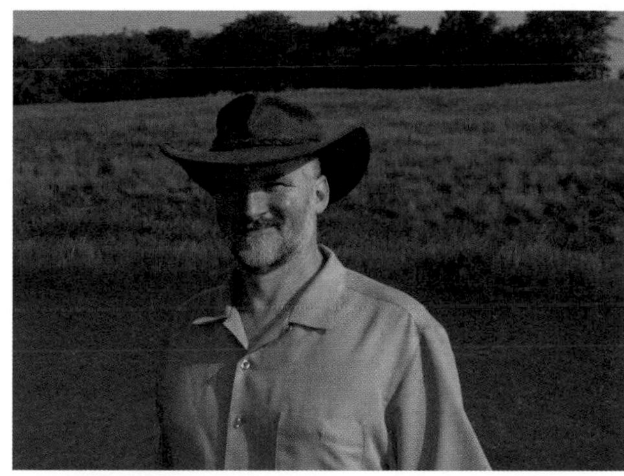

Dr. D. Christopher Rogers is a research zoologist at the University of Kansas with the Kansas Biological Survey and is affiliated with the Biodiversity Institute. He received his PhD degree from the University of New England in Armidale, NSW, Australia. Christopher specializes in freshwater crustaceans (particularly Branchiopoda and Decapoda) and the invertebrate fauna of seasonally astatic wetlands on a global scale. He has numerous peer reviewed publications in crustacean taxonomy and invertebrate ecology, as well as published popular and scientific field guides and identification manuals to freshwater invertebrates. Christopher is an Associate Editor for the *Journal of Crustacean Biology* and a founding member of the Southwest Association of Freshwater Invertebrate Taxonomists. He has been involved in aquatic invertebrate conservation efforts all over the world.

Preface to the Fourth Edition

Those readers familiar with the first three editions of our invertebrate book (*Ecology and Classification of North American Freshwater Invertebrates*, edited by J.H. Thorp and A.P. Covich) will note that the fourth edition has expanded from a North American focus to worldwide coverage of inland water invertebrates. We gave our book series on inland water invertebrates the name *Thorp and Covich's Freshwater Invertebrates* to: (1) associate present with past editions, unite current volumes, and link to future editions; (2) establish a connection between the ecological and general biology coverage in Volume I with the taxonomic keys in the remaining volumes; and (3) give credit to Professor Alan Covich for his work on the first three editions. For the sake of brevity, we refer to the current edition as T&C IV. Whether the fifth edition of T&C will ever appear is certainly problematic, but who knows! At present we are considering producing up to 11 volumes in the fourth edition.

While I am the sole editor of the book series at this point, Christopher has been a major and highly valued partner in developing ideas for the fourth edition and is thus far an editor on the first three volumes (senior editor on the third). He will also play a major role in many of the remaining volumes because of his diverse and global knowledge of freshwater invertebrates, especially in the area of taxonomy. As we made significant progress on the first three volumes, we began contacting some potential coeditors and authors to develop volumes for other zoogeographic regions and negotiations with a few of those volumes are now underway. However, we are still seeking experts in fields of invertebrate taxonomy for various zoogeographic regions to serve as highly dependable coeditors, especially those who both work and live in the zoogeographic regions covered by the various future volumes.

Our concept for T&C IV included producing one book (Volume I, published in late 2014 with a 2015 copyright date) with 6 chapters on general environmental issues applicable to many invertebrates, followed by 35 chapters devoted to individual taxa at various levels (order to phylum, or even multiple phyla in the case of the protozoa). Volume I was designed both as an independent book on ecology and general biology of various invertebrate taxa and as a companion volume for users of the keys in the regional taxonomic volumes, thereby reducing the amount of information duplicated in the taxonomic volumes. The perhaps 10 taxonomic volumes to be published in the next decade or so will contain both keys for identifying invertebrates in specific zoogeographic regions and descriptions of detailed anatomical features needed to employ those keys.

While the vast majority of authors in T&C editions I–III were from the United States or Canada, we attempted in T&C IV to attract authors from many additional countries in six continents. Although we largely succeeded in this goal, we expect the fifth edition of T&C—if it is ever published—to continue increasing the proportion of authors from outside North America as our books become better known internationally.

Our goals for T&C IV are to improve the state of taxonomic and ecological knowledge of inland water invertebrates, help protect our aquatic biodiversity, and encourage more students to devote their careers to working with these fascinating organisms. These goals are especially important because the verified and probable losses of species in wetlands, ponds, lakes, creeks, and rivers around the globe exceed those in most terrestrial habitats.

James H. Thorp

Preface to Volume II

This is the second volume of the fourth edition of *Thorp and Covich's Freshwater Invertebrates* (T&C IV) and the first to focus almost exclusively on taxonomy. Information on the ecology and general biology of the groups can be found in Volume I (Ecology and General Biology, edited by Thorp & Rogers, 2015), the companion text for the current and all remaining books in this series. All taxonomic volumes (other than those focused exclusively on Hexapoda) are expected to consist of an introductory chapter, a chapter on protozoa (multiple kingdoms), and 14 chapters on individual phyla from Cnidaria to Arthropoda. Some of the chapters are very small (e.g., Chapter 14 on Entoprocta), whereas others are huge, especially Chapter 16 on Arthropoda.

A typical chapter includes a short introduction, a brief discussion of limits to identification of taxa in that chapter, important information on terminology and morphology that is needed to use the keys, techniques for preparing and preserving material for identification (also covered in Volume I), the taxonomic keys, and a few references. In the large chapters on Mollusca (11), Annelida (12), and Arthropoda (16), different individuals have contributed separate sections, and thus there are multiple sections on introduction through keys and references. While this may confuse some readers, it has allowed us to gain contributions from an increased number of experts around the world.

The multilevel keys are formatted to enable users to work easily at the level of their taxonomic expertise and the needs of their project. For that reason, we separated keys by major taxonomic divisions. For example, a student in a college course might work through one or more of the initial crustacean keys to determine the family in which a freshwater shrimp belongs. In contrast, someone working on an environmental monitoring project might need to identify a crayfish or crab to genus or even species, and thus would use the relevant, detailed keys that require more background experience. We also designed the keys, where possible, to proceed from a general to a specific character within a couplet.

We have asked authors to include only taxa that are recognized internationally by publication in reputable scientific journals that follow the International Code of Zoological Nomenclature. Thus, no taxa that have merely been proposed should be included even if they have been identified by the world's expert on that group. "Common" species are not designated because a common species in one area may not be common in another, and this designation can lead to overly frequent and false identifications. Authors have been encouraged to end the keys at the point where further identification without genetic analysis is not practical or when it is clear that too many of the extant fauna have yet to be described in scientific publications.

Users of these keys need to realize that taxonomy is a growing and vibrant field in which new taxa are being described and previously accepted relationships reevaluated. For some users, this volume may be sufficient for their needs, but for others, a companion text listing known species in a smaller geographic region may also be helpful.

This edition is strongly focused on species found in fresh through saline inland waters, with a nonexclusive emphasis on surface waters, thereby reflecting the bias of existing scientific literature. Again, most estuarine and parasitic species are not covered in this book, but we do discuss species whose life cycle includes a free-living stage (e.g., Nematomorpha) and species that live in hard freshwaters through to brackish waters even though they may be normally associated with estuarine or marine habitats in some parts of their life cycles (e.g., some shrimp and crabs).

It is our hope that scientists and students from around the world will benefit from this volume. Suggestions for improving future volumes are welcome.

Editors
James H. Thorp
D. Christopher Rogers

Acknowledgments for Volume II

Many people contributed to this volume in addition to the chapter authors and those acknowledged in individual chapters. We greatly appreciate all our colleagues who have contributed information, figures, or reviews to Volume II, and also thank those who provided similar services for the earlier editions, upon which the present book partially relies. We are again grateful to the highly competent people at Academic Press/Elsevier who helped in many aspects of the book's production from the original concept to the final marketing. In particular, we appreciate our association with Elsevier editors and production team including Candace Janco, Rowena Prasad, Laura Kelleher, and the entire United States and overseas production teams, especially Julia Haynes.

James H. Thorp
D. Christopher Rogers

Chapter 1

Introduction[1]

James H. Thorp
Kansas Biological Survey and Department of Ecology and Evolutionary Biology, University of Kansas, Lawrence, KS, USA

D. Christopher Rogers
Kansas Biological Survey and Biodiversity Institute, University of Kansas, Lawrence, KS, USA

Chapter Outline

INTRODUCTION TO THIS VOLUME AND CHAPTER 1

This is the second volume in the fourth edition of *Thorp and Covich's Freshwater Invertebrates*. Unlike the first three editions of *Ecology and Classification of North American Freshwater Invertebrates* (edited by Thorp and Covich in 1991, 2001, and 2010), the fourth edition has been split into multiple texts, with Volume I (Thorp & Rogers, 2015) providing global coverage of the ecology, general biology, phylogeny, and collection techniques for inland water invertebrates. Subsequent volumes provide keys to identify fauna in specific zoogeographic regions. This division of volumes enabled us to produce reasonable sized volumes at relatively moderate prices instead of publishing one massive, high priced tome. While some labs may have multiple copies of the "Keys to Fauna" in their region, we also recommend that they have at least one copy of Volume I, in order to obtain useful background information on each invertebrate group.

The current chapter is organized into an introduction, a section explaining the organization of most taxonomic chapters, and a key to larger taxonomic groups. This chapter's key is designed to help the reader locate the most pertinent chapter (important probably only for students and beginning taxonomists) and begin identifying organisms in their samples. Readers will note that chapters within and among volumes vary in specificity of their taxonomic keys. This reflects both the likely percent of the fauna that has been named and how easily taxa can be separated by alpha taxonomic methods and associated keys.

COMPONENTS OF TAXONOMIC CHAPTERS

This volume is an identification manual to the inland water invertebrates of the Nearctic Region where we present information needed to diagnose and determine these organisms to various taxonomic levels. Other information concerning ecology, morphology, physiology, phylogeny, and both collecting and culturing techniques can be found in Volume I of this series. Each of the remaining 15 chapters in the current volume is limited to a single phylum, except Chapter 2's coverage of multiple phyla of unicellular protists. Chapter 2 is designed for readers who only need general information about protists. We have attempted to include the following five sections in those chapters: (1) a brief introduction to the broader taxon; (2) a description of identification limitations for each taxon; (3) details of pertinent terminology and morphology; (4) information on preparing and preserving specimens for identification; and (5) taxonomic keys (separated by level of identification). A restricted number of especially pertinent references are given in each chapter following appropriate taxonomic sections. Readers can find a much more extensive list of references to their group in

1. This chapter was written to be a useful starting point for taxonomic volumes (II, III, etc.) in all zoogeographic regions. Consequently, there will be only minor differences among volumes.

Thorp and Covich's Freshwater Invertebrates. http://dx.doi.org/10.1016/B978-0-12-385028-7.00001-9

Volume I (Chapters 3 and 7–41) along with more details on collecting, preparation, and preserving major taxa. Figures in each chapter are limited to those needed for effective use of the keys. For additional anatomical information, including figures, see the relevant chapter in Volume I.

HOW TO USE THIS VOLUME

There is an old maxim that says "keys are written by people who do not need them for people who cannot use them." We have made every effort to make these keys as user friendly as publication limitations would permit.

Each section begins with a basic introduction to the morphology and terminology used in diagnosing the taxa of that section. Limitations to the current state of taxonomic knowledge are also presented so that the reader may gauge the reliability of the information presented. Only the established, peer reviewed scientific literature was used to define the taxonomic categories and epithets included. All names, as far as we are aware, conform to the International Code of Zoological Nomenclature (ICZN). All nomina and taxonomic arrangements used, as well as the rejection of old names was based on peer reviewed scientific literature. Names from unpublished manuscripts, dissertations, "in house" designations, or records that have not been validated are not acceptable. Provisional names and species designated "taxon 1" or "species 1" were not used unless they were previously recognized and accepted in the peer reviewed scientific literature (Richards & Rogers, 2011). No new species descriptions or previously unpublished taxonomic arrangements are presented.

The keys are dichotomus (no triplets or quadruplets are used) and are hierarchical. Thus, for a given group, the first keys are to the highest taxonomic category. The second set of keys is to the next level, the third set to the level below that one, and so on, down to the lowest justifiable taxonomic level based on current knowledge of that group. This level is different for different groups depending upon the state of resolution in the scientific literature. Organisms not identifiable beyond a particular taxonomic level are left at that level.

Properly prepared keys typically employ specific, primary, diagnostic characters. Older keys often use different characters than the more recent keys. This shift in primary characters results from systematists and taxonomists testing the importance of characters. The ultimate goal of the systematist is to ensure that the interpretation of which characters are important will converge with biological reality. To a non-taxonomist, this process may seem merely to be "lumping and splitting," rather than the result of employing the scientific method to reveal natural relationships.

Surprisingly, many users do not know how to interpret a dichotomus key, making the fundamental assumption that a correct identification answer is always present in the key. This assumption generally takes one of the following three forms:

1. *All species are identifiable using a given key*. Many new species have yet to be described, let alone discovered. Generalized geographic ranges are provided for most taxa presented herein, yet species ranges shrink, swell, and change elevation constantly, particularly as weather and climate patterns shift. Species disperse, colonize, and suffer stochastic local extinctions. In addition to these natural processes, some species are introduced intentionally or accidentally by humans, and sometimes their establishment allows other species to invade as well.
2. *All variation is accounted for in the key*. As stated above, identification keys use specific, primary, diagnostic characters. Problems in identification are compounded by taxa that: (a) have different character states at different times; (b) only have diagnostic characters at certain life stages or in certain genders; and/or (c) have severely truncated morphology (often due to lack of sexual selection) and lack morphological characters to separate the species. Furthermore, new variation within taxa is continually developing, and thus, one cannot assume that species are immutable or develop tools predicting those changes.
3. *The key is a sufficient identification tool in and of itself*. A key is just a tool. The fact that one has a bolt that needs removing and a wrench of the correct size does not mean that the bolt can be loosened. Similarly, identification keys are tools to aid in taxon identification. They are primarily tools to eliminate incorrect taxa from the range of possible choices, narrowing the field to the names that may be applicable. Keys are the process of elimination. The possibility that the specimen to be identified is new, a hybrid, anomalous, or a recent invasive colonist is always a possible answer. This is fundamental to using any identification key.

Once one arrives at a name or group of possible names for a specimen in hand, the specimen should then be compared against descriptions, distribution maps, and figures of that and other taxa in that group. The descriptions, figures, and maps are other tools to be used in identification. Direct comparison of the specimen at hand with identified museum material or using molecular comparisons is also sometimes necessary for a correct identification.

Species are not immutable, fixed in location and form. They change constantly and will continue to do so, confounding keys and any other identification method, such as trait tables, character matrices, or even genetic analyses. This is why biology is far behind physics in the development of unified theories: biology is far more complex than physics, as it involves more interacting parts and processes.

KEY TO KINGDOMS AND PHYLA IN THIS VOLUME

A major change in the identification keys for our fourth edition has been to include multiple keys per chapter that generally start with a class level key and proceed to finer and finer divisions. These allow users to work at their levels of interest, need, and skill without having to wade through extraneous taxa not in the direct line to the taxon of interest.

The following key was derived in part from Chapter 1 in Volume I of the fourth edition. It is meant to allow you to move to the next level of keys, which will be in individual chapters.

Freshwater Invertebrate Kingdoms and Phyla

1 Multicellular, heterotrophic organisms as individuals or colonies (sometimes with symbiotic autotrophs) .... kingdom Animalia 2

1' Unicellular (or acellular) organisms present as individuals or colonies with nuclei irregularly arranged; heterotrophic and/or autotrophic; multiple phyla within the autotrophic protozoa phyla .... kingdom Protista [Chapter 2]

2(1) Radially symmetric or radially asymmetric organisms living individually or in colonies 3

2' Individuals bilaterally symmetric 4

3(2) Surface not porus; oral tentacles always present around a closeable mouth; colonial or single, mostly single polyp forms (primarily hydra) or rarely medusoid form (freshwater jellyfish); adults with a single central body cavity opening to the exterior and surrounded by cellular endoderm, acellular mesoglea, and cellular ectoderm .... phylum Cnidaria [Chapter 4]

3' Surface porus; colonial; tentacles absent; no closable orifices; without discrete organs; cellular-level (or incipient tissue-level) construction; variable, non-distinct colony shapes, including encrusting, rounded, or digitiform growth forms; skeleton of individual siliceous spicules and a collagen matrix; internal water canal system; may contain symbiotic algae; the sponges .... phylum Porifera [Chapter 3]

4(2) Oral region with numerous tentacles or cilia distributed around the mouth; organism never with eversible jaws and never vermiform as adult 5

4 Oral region with two or no tentacles, or tentacles behind the mouth 7

5(4) Oral region with tentacles, organisms in gelatinoids or branching colonies 6

5' Oral region ringed with cilia, muscular pharynx (mastax) with complex set of jaws; single free swimming, or semi-sessile living singly or in small colonies; wheel animals, or rotifers .... phylum Rotifera [Chapter 8]

6(5) Oral tentacles (the lophophore) in a "U" or "horseshoe" shape around mouth; anus opens outside of lophophore; colonial animals, often in massive colonies attached to hard surfaces; true bryozoans .... phylum Ectoprocta (Bryozoa) [Chapter 13]

6' Both mouth and anus open within lophophore; individual (non-colonial) animals with a calyx containing a single whorl of 8–16 ciliated tentacles .... phylum Entoprocta [Chapter 14]

7(4) Not with the combination of characteristics described below 8

7' Small (50–800 μm), spindle- or tenpin-shaped, ventrally flattened with a more or less distinct head bearing sensory cilia; cuticle usually ornamented with spines or scales of various shapes; posterior of body often formed into a furca with distal adhesive tubes; gastrotrichs (pseudocoelomates) .... phylum Gastrotricha [Chapter 7]

8(7) Anterior mouth and posterior anus present 9

8' Flattened or cylindrical, acoelomate worms with only one, ventral digestive tract opening; sometimes with evident head; turbellarian flatworms (commonly called planaria, a non-specific, and usually incorrect name) .... phylum Platyhelminthes [Chapter 5]

9(8) Vermiform or not, eversible oral proboscis not present, although eversible jaws or other mouthparts may occur 10

9' Long, flattened, unsegmented worms with an eversible proboscis; ribbon worms .... phylum Nemertea [Chapter 6]

10(9) Body not enclosed in a single, spiraled shell or in a hinged, bivalved shell; or if a bivalved shell is present, then animal has jointed legs 11

10' Soft-bodied coelomates whose viscera is covered (in freshwater species) by a single or dual (hinged), hard calcareous shell; with a ventral muscular foot; fleshy mantle covers internal organs; snails, clams, and mussels .... phylum Mollusca [Chapter 11]

11(10) Segmented legs absent in all life stages; if jaws are present, then body with at least 20 segments 12

11' Adults and most larval stages with legs; if larvae without legs or prolegs (some insects), then cephalic region with paired mandibles, or eversible head, always with less than 15 body segments 14

12(11) Organism vermiform, not segmented 13

12' Organism vermiform or not, body segmented .... phylum Annelida [Chapter 12]

13(12) Body cylindrical, usually tapering at both ends; cuticle without cilia, often with striations, punctuations, minute bristles, etc.; 1 cm long (except family Mermithidae, <6 cm); nematodes, roundworms .... phylum Nemata [Chapter 9]

13'	Body with anterior tip normally obtusely rounded or blunt, posterior tip may be bi- or trilobed; cuticle opaque to dark brown or black, and epicuticle usually crisscrossed by minute grooves; length several cm to 1 m, width 0.25–3 mm; only adults with free-living stage; hairworms or horsehair worms	phylum Nematomorpha [Chapter 10]
14(11)	Four pairs of clawed, non-jointed legs; water bears	phylum Tardigrada [Chapter 15]
14'	Adults and most larvae with jointed legs, or legs lacking, or more or less than four pairs	phylum Arthropoda [Chapter 16]

REFERENCES

Richards, A.B. & D.C. Rogers. 2011. Southwest Association of Freshwater Invertebrate Taxonomists (SAFIT) list of freshwater macroinvertebrate taxa from California and adjacent states including standard taxonomic effort levels. 266 pp.

Thorp, J.H. & D.C. Rogers (eds.). 2015. Ecology and General Biology. Volume I of Thorp and Covich's Freshwater Invertebrates, Fourth Edition. Academic Press, Elsevier, Boston, MA.

Chapter 2

Protozoa

Alan Warren
Department of Life Sciences, Natural History Museum, London, UK

Genoveva F. Esteban
Bournemouth University, Faculty of Science and Technology, Dorset, UK

Bland J. Finlay
School of Biological and Chemical Sciences, Queen Mary University of London, The River Laboratory, Wareham, Dorset, UK

PROTOZOA

Chapter Outline

INTRODUCTION

During the last 20 years, studies on the systematics and evolution of unicellular eukaryotes (algae, protozoa, and lower fungi) have been in a state of great activity. Over this period, many taxonomic boundaries, including those between the algae and protozoa, have been broken down and new relationships established (Cavalier-Smith, 2010; Adl et al., 2012). As a result, the constituent organisms are grouped together by some workers as protists, reviving the term originally coined by Haekel (1866), or as protoctists (Margulis et al., 1989), although many systematists believe that such groups have no evolutionary or systematic validity. By contrast, other workers have proposed systems that retain the Kingdom Protozoa, albeit with much modified definitions and boundaries (Cavalier-Smith, 2010). Nevertheless, the terms algae and protozoa are still useful in a functional or ecological sense, defining (primarily) photoautotrophic and heterotrophic protists, respectively.

Protozoa *sensu lato*, which means first animals, are a diverse assemblage that comprises a number of separate lineages representing almost all the major eukaryote clades, including alveolates, stramenopiles, amoebozoans, opisthokonts, rhizarians, and excavates (Cavalier-Smith, 2010; Adl et al., 2012). Protozoa typically measure 5 to 1000 μm in size, and most are visible only with the aid of a microscope. There is considerable morphological and physiological diversity within the group. Because actively feeding protozoa need water, all free-living (non-parasitic) protozoa are essentially aquatic, living in freshwater (including soil), brackish, and marine environments.

LIMITATIONS

There are a number of factors that pose significant limitations to the taxonomy of protozoa. These include: (1) the lack of adequate methods for the fixation and long-term preservation of specimens for much of the ca. 350-year history of the discipline of protozoology; (2) an absence of type specimens for most species; (3) a lack of sufficient morphological features for species circumscription; (4) inadequate species descriptions for reliable identification; (5) high rates of synonymy; (6) insufficient numbers of trained taxonomists; (7) undersampling and a large unknown species diversity; and (8) technical difficulties in culturing many

Thorp and Covich's Freshwater Invertebrates. http://dx.doi.org/10.1016/B978-0-12-385028-7.00002-0

species, which is sometimes a prerequisite for adequate characterization.

It is often difficult and time-consuming to identify protozoa to the level of species. In many cases, unambiguous identification requires specialized staining techniques or the use of electron microscopy. The taxonomic grouping used in our key is an amalgamation of publications by specialists on the different groups (e.g., Lee et al., 2000; Lynn, 2008; Bass et al., 2009; Cavalier-Smith, 2010; Smirnov et al., 2011; Adl et al., 2012). While the taxonomy of many groups is based on a combination of cell morphology, ultrastructural features, and molecular data, this key is designed to make possible the identification of many protozoa to the family level using light microscopy alone. Although observation of living organisms is important for identification, the key should still be useful for many fixed samples. The illustrations used as examples here are of one or more species considered typical of a genus.

Although this key primarily deals with free-living protozoa, some ciliates that are commensal or parasitic, e.g., certain groups of suctorians and oligohymenophoreans, are also included.

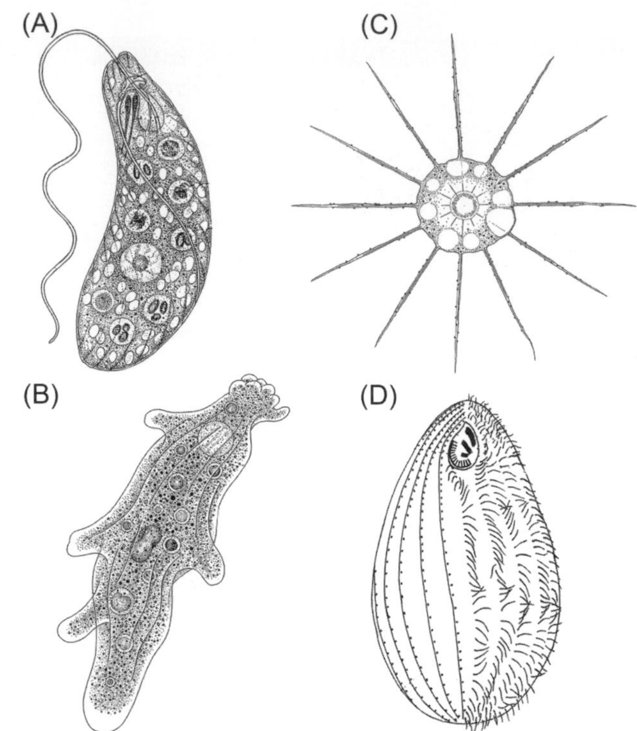

FIGURE 2.1 Examples of the main functional groups of protozoa. (A) *Peranema trichophorum*—a flagellate; (B) *Amoebae proteus*—an amoeba; (C) *Actinophrys sol*—a heliozoan; (D) *Tetrahymena* sp.—a ciliate. *After Vickerman & Cox (1967) A, B; Siemensa (1991) C; Curds (1982) D.*

TERMINOLOGY AND MORPHOLOGY

Traditionally, free-living protozoa have been divided into three main groups according to their morphology and means of locomotion: flagellates, amoebae (including heliozoans), and ciliates (Fig. 2.1). Of these, only the ciliates are a truly natural, monophyletic group, the flagellates and amoebae being polyphyletic and include groups that may be only distantly related. Nevertheless, from a practical viewpoint, it is still sometimes useful to refer to these groupings because isolation, cultivation, and identification methods used are often the same within each group.

Flagellates

Flagellates are characterized by the possession of one or more flagella, which are long, tapering, hair-like appendages that act as organelles of locomotion and feeding (Fig. 2.1 A). In free-living taxa, as opposed to parasitic species, the number of flagella is limited; *Paramastix* has two rows of 8–12 flagella, but most others have 1–4 (usually 2). Typically, where two flagella are present, one may project forward, and the other trails behind. Often, the organism's flagella are longer than its body. There are several groups of heterotrophic flagellates in freshwater: choanoflagellates, kinetoplastids, diplomonads, and bicoecids. These are raised to phyla by some authors, while bicoecids are occasionally put with chrysophytes. Some amoeboid forms, such as cercomonads and the Schizopyrenida, or amoeboflagellates, also have flagella but are treated here with the amoebae.

Other groups of flagellates contain mostly or entirely autotrophic forms with chloroplasts. However, many of the pigmented, autotrophic taxa are also capable of phagotrophy, producing an overall condition called mixotrophy (Sanders, 1991; Esteban et al., 2010), and also among these groups are some wholly heterotrophic species. The groups with many mixotrophic or heterotrophic taxa include cryptophytes, chrysophytes, dinoflagellates, and euglenoids, and are usually considered phyla. Pigmentation and chloroplast morphology are important taxonomic characters for some of these groups.

Choanoflagellates, or collared flagellates, are distinctive for the collar that surrounds the single flagellum (Fig. 2.2 B–H). They bear a strong resemblance to sponge choanocytes. Most choanoflagellates attach to the substrate or are colonial, and many have an external, loose-fitting covering or lorica, although this may be difficult to see with the light microscope.

Bicoecids (Fig. 2.2 I) resemble choanoflagellates, although they lack a collar. Like choanoflagellates, they are enclosed in a lorica and have a flagellum that is used to create a feeding current. A second flagellum lies along the cell and continues posteriorly to become an attachment to the base of the lorica.

Kinetoplastids (Fig. 2.2 J, N–P) are known mostly as parasites, especially *Trypanosoma* and its relatives,

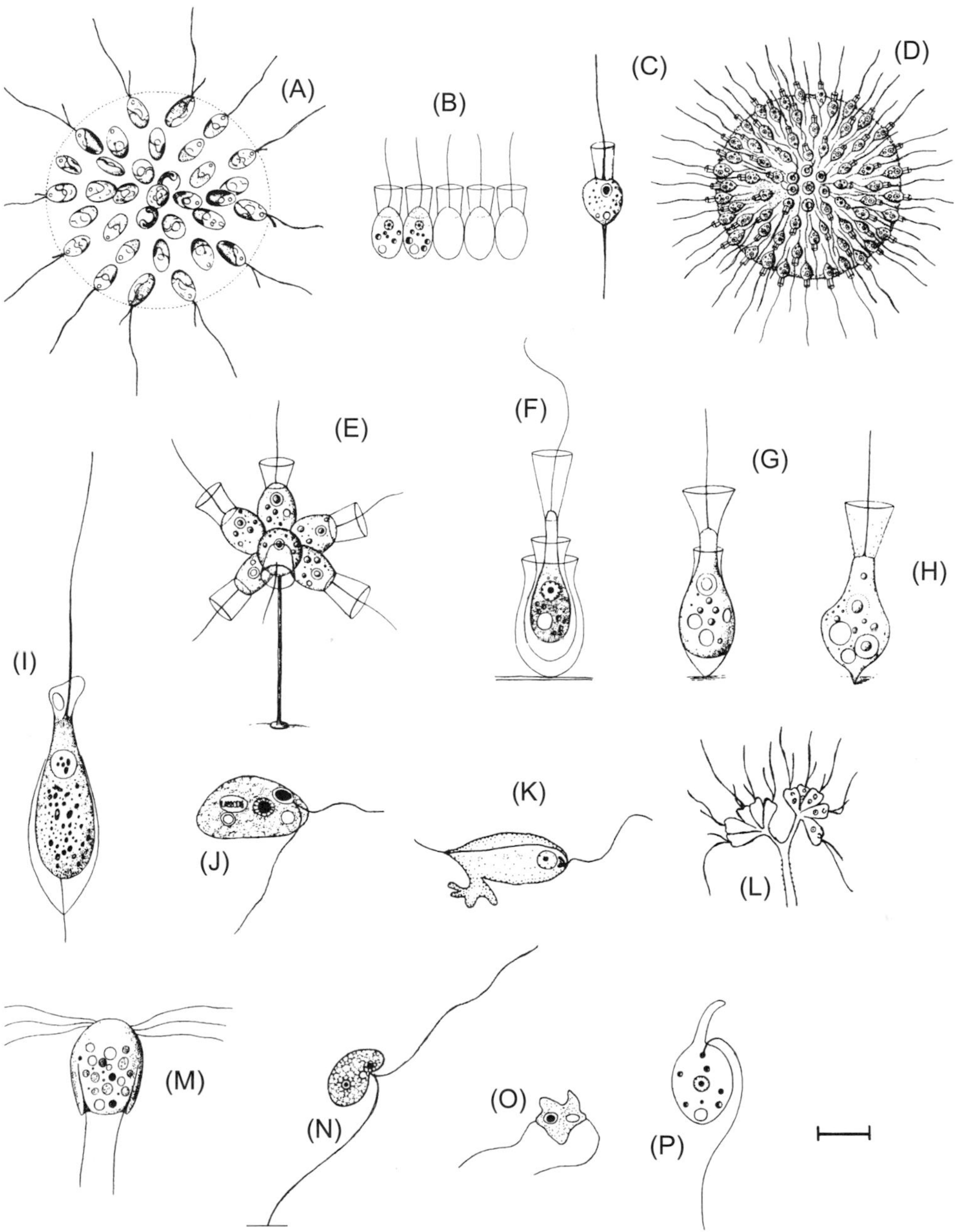

FIGURE 2.2 (A) *Uroglena americana* (mixotrophic); (B) *Desmarella moniliformis*; (C, D) *Sphaeroeca volvox*, individual and colony; (E) *Codosiga botrys*; (F) *Diploeca plactita*; (G) *Salpingoeca fusiformis*; (H) *Monosiga ovata*; (I) *Bioeca lacustris*; (J) *Bodo caudatus*; (K) *Cercomonas* sp.; (L) *Cephalomonas cyclopum*; (M) *Hexamita inflata*; (N, O) *Pleuromonas jaculans*, attached and amoeboflagellate forms; (P) *Rhynchomonas nasuta*. Scale 2.5 μm for P; 5 μm for F, G, H, I, K, L; 10 μm for A, B, C, J, M, N, O; 20 μm for E; and 30 μm for D. *After: Bourelly (1968) L; Calaway & Lackey (1962) N, O, P; Lackey (1959) B, F; Lee et al. (1985) K; Pascher (1913) C, D, E, G, H, I, J, M.*

but many members of the suborder Bodina live in freshwater (Vickerman, 1976). The best-known genus is *Bodo*, which, like other bodonids, has two flagella (Fig. 2.2 J) one of which trails, while the other extends ahead.

The cryptomonads include many common heterotrophs and autotrophs and a few mixotrophs. The two flagella are unequal in length and arise from a subapical invagination commonly referred to as a "gullet," although it does not appear to be the site of ingestion in heterotrophic forms. The pellicle is covered with plates, although these also are not generally visible.

The dinoflagellates (Fig. 2.3 A–C) form a very large and unique group, which is probably more important in marine than freshwater environments. Their unique arrangement of flagella, one spiraling around the cell in a groove (girdle) and a second distally directed in another groove (sulcus), makes them distinctive. Again, heterotrophy and mixotrophy are common. A covering of plates

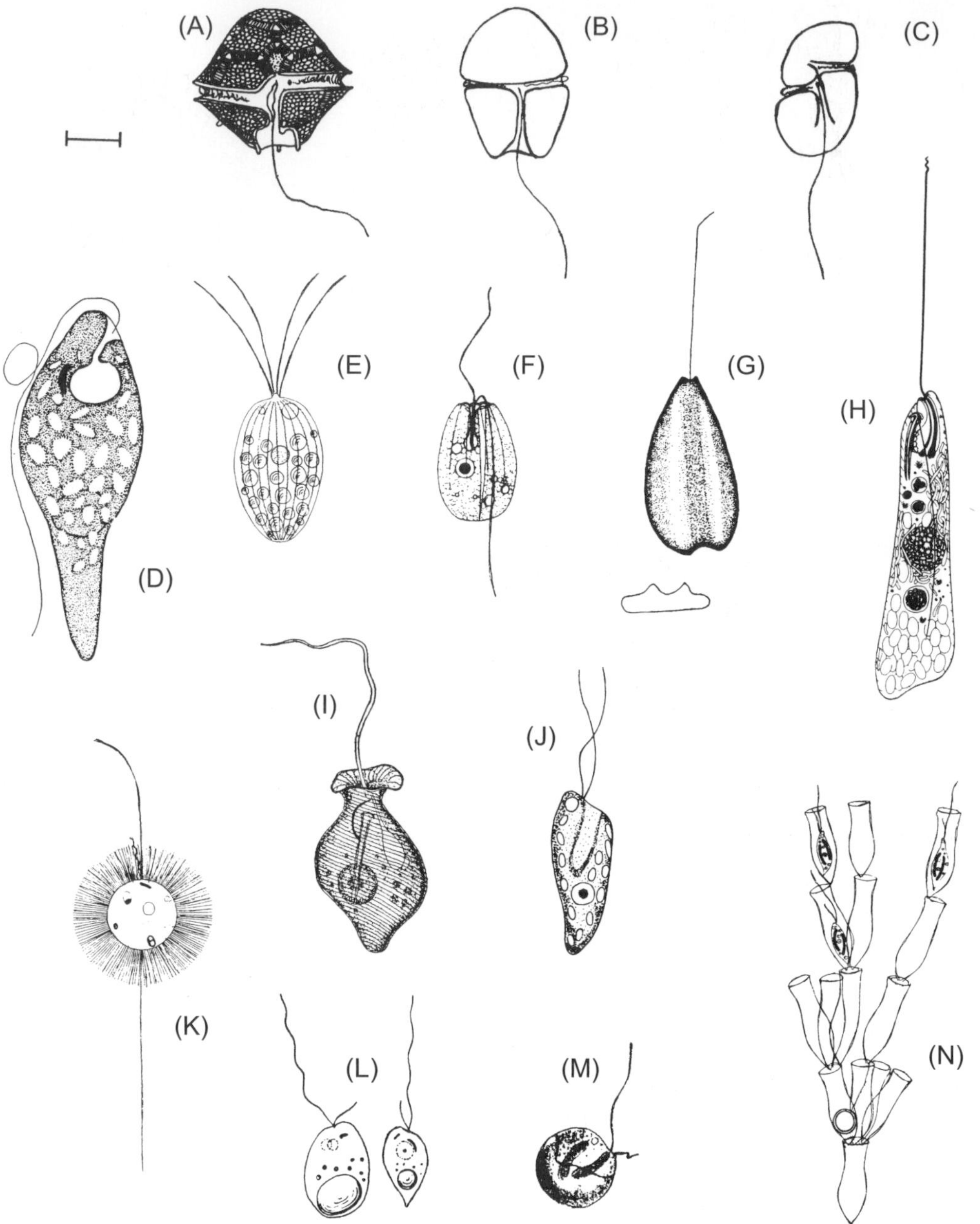

FIGURE 2.3 (A) *Peridinium*; (B) *Gymnodinium*; (C) *Gyrodinium*; (D) *Khawkinea halli*; (E) *Polytomella citri*; (F) *Entosiphon sulcatum*; (G) *Petalomonas abcissa*; (H) *Peranema trichophorum*; (I) *Urceolus;* (J) *Chilomonas paramecium*; (K) *Paraphysomonas vestita*; (L) *Spumella* (*Monas*) *vivipara*, two cell shapes; (M) *Ochromonas variabilisa*; (N) *Dinobryon sertularia* (mixotrophic). Scale 5 µm for E; 10 µm for A, B, C, G, I, J, K, L, M; and 20 µm for D, F, H, N. *After: Bourelly (1968) L; Calaway & Lackey (1962) E, F, J, N; Eddy (1930) A; Jahn & McKibben (1937) D; Leedale (1985) H; Pascher (1913) M; Lemmerman (1914) K; Shawhan & Jahn (1947) G; Smith (1950) I.*

may or may not be present (hence the terms armored and naked dinoflagellates).

Chrysophytes are generally small, and they prey on bacteria. They have two unequal flagella, one long and directed anteriorly, the other short and directed laterally (Fig. 2.3 K–M). They are naked or covered in fine siliceous scales (Esteban et al., 2012), which are not always visible with light microscopy; many are amoeboid. Their carbohydrate storage product, chrysolaminarin, occurs in liquid globules and may be useful in recognizing the members of this group. Chrysophytes contain both colorless heterotrophs and pigmented mixotrophs.

Euglenids are generally large flagellates with two flagella, although in many taxa, only one flagellum emerges from the gullet (Fig. 2.3 D). Several heterotrophic species creep over the substrate with the second flagellum trailing and hidden beneath the cell (Fig. 2.3 F–H), as in some bodonids. The euglenids are currently assigned to the supergroup Excavata (Adl et al., 2012).

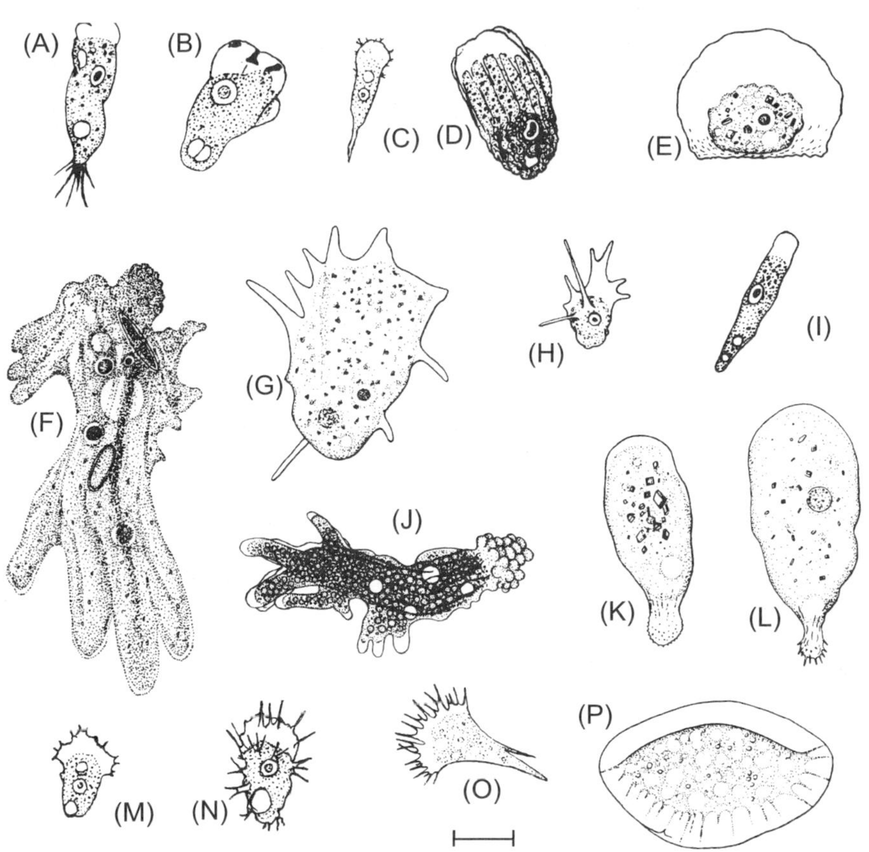

FIGURE 2.4 (A) *Vahlkampfia avaria*; (B) *Naegleria*; (C) *Stachyamoeba lipophora*; (D) *Thecamoeba sphaeronucleolus*; (E) *Vanella miroides*; (F) *Amoeba proteus*; (G) *Mayorella bigomma*; (H) *Vexillifera telemathalassa*; (I) *Hartmannella vermiformis*; (J) *Chaos illinoisense*; (K) *Saccamoeba lucens*; (L) *Trichamoeba cloaca*; (M) *Echinamoeba exudans*; (N) *Acanthamoeba*; (O) *Filamoeba nolandi*; (P) *Hylodiscus rubicundus*. Scale 10 μm for A, B, C, E, I, M; 15 μm for H, N, O, P; 30 μm for D, G, K, L; 50 μm for F; and 100 μm for J. *After: Bovee (1985) A, B, C, D, H, I, J, M, N, O, P; Kudo (1966) F; Page (1988) E, G, K, L, P.*

Amoebae

The primary characteristic of amoebae is their possession of pseudopodia, retractile processes that serve as organelles of locomotion and feeding (Fig. 2.1 B). There is considerable diversity of structure in the amoebae, particularly in the character of any shell or skeletal material that may be present, and in the type of pseudopodium, for example, broadly lobed, needle-like, or reticulate. Amoebae range in size from only a few micrometers to 2 mm in diameter. Although many lack a fixed external morphology, the characteristic morphologies shown by the various taxa are surprisingly distinctive, even if difficult to quantify (Fig. 2.4). By using also the number, size, and structure of organelles and characteristics of tests (where present), identification is not as difficult for living specimens as might be imagined. The morphology of amoebae is plastic. Many adopt a stellate morphology if suspended in water, but few are truly planktonic; rather, they live on surfaces or in sediments. In most, for example, *Amoeba* (Figs. 2.1 B and 2.4 F), the cytoplasm is divided into an inner granular endoplasm and an outer hyaline ectoplasm, or hyaloplasm, with a characteristic thickness and distribution around the cell. Locomotion may be achieved by extending many pseudopodia simultaneously, as in *Amoeba* (Figs. 2.1 B and 2.4 F), or by moving as a single mass on a broad front (2.4 E, P), or as a cylinder (limax amoebae, Fig. 2.4 I, K, L). Not only do pseudopodia have characteristic shapes, but the tail end or uroid may be distinctive (Fig. 2.4 J, L), and the cell surface may be distinctly sculptured, as in *Thecamoeba* (Fig. 2.4 D). The classification of the naked, lobose amoebae was recently revised by Smirnov et al. (2011).

Other groups of amoebae, notably the testate amoebae, possess shells (or tests) that may be proteinaceous, agglutinate, siliceous or calcareous in composition (Figs. 2.5 A–Q and 2.6 B–K). These are generally vase-shaped, with a single opening through which pseudopodia emerge. Many are terrestrial, but benthic forms are common, and a few are planktonic. Identification of testate amoebae is mainly based on shell characters, i.e., size, shape, and composition.

Heliozoans

Heliozoans and pseudoheliozoans are roughly spherical amoebae with many stiff projections called axopodia radiating outward from the cell surface (Figs. 2.1 C, 2.7, and 2.8 D, E, I, J, L). The axopodia give heliozoans their characteristic sun-like appearance for which they

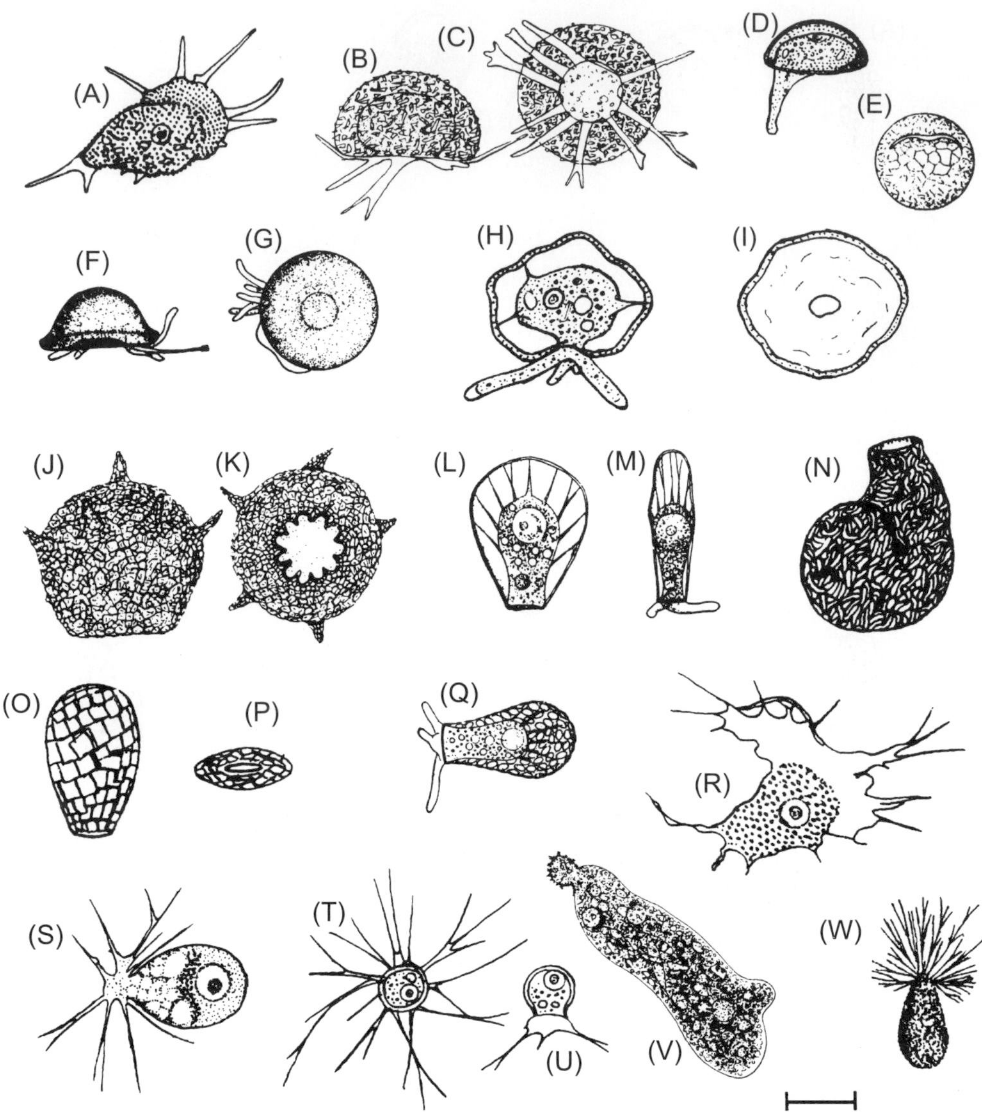

FIGURE 2.5 (A) *Cochliopodium bilimbosum*; (B, C) *Phryganella nidulus*, side and oral views; (D) *Pyxidicula operculata*; (E) *Plagiopyxis callida*; (F, G) *Arcella vulgaris*, side and dorsal views; (H, I) *Penardochlamys arcelloides*, side and oral views; (J, K) *Difflugia corona*, side and oral views; (L, M) *Hyalosphenia cuneata*; (N) *Lesquereusia spiralis*; (O, P) *Quadrulella symmetrica*; (Q) *Nebela collaris*; (R) *Penardia granulose*; (S) *Chlamydophrys minor*; (T, U) *Lecythium hyalinum*, dorsal and side views; (V) *Pelomyxa palustris*; (W) *Pseudo difflugia gracilis*. Scale 10 μm for C, D, R, S; 30 μm for H, I, L, M, T, U, W; 45 μm for G, N, O, P; 60 μm for Q; 90 μm for B, E, F, J, K; and 500 μm for V. *After: Bovee (1985) A, B, C, H, I, J, K, N, O, P, R, T, U; Deflandre (1959) D, E, F, G, L, M, Q, S, W; Kudo (1966) V.*

are named, and are variously used for capturing food, sensation, movement, and attachment. Axopodia are strengthened by a microtubular array called an axoneme or stereoplasm. The term axoneme is also used to describe the microtubular core of cilia and flagella, but this does not imply homology, and the origin and ultrastructure of axonemes is diverse (Yabuki et al., 2012). Most heliozoans lack the skeleton that is so characteristic of their marine counterparts such as Radiolaria and Acantharia, although some are covered in siliceous or organic scales (Fig. 2.7 F, H), and some have a perforated shell or capsule (order Desmothoracida, Fig. 2.7 A). Although heliozoans are frequently planktonic, they are found primarily on or near the benthos. Some heliozoans traverse the bottom with a unique tumbling motion, resulting from controlled changes in the length of the axopodia. Many sessile forms with stalks are known. In sessile forms, cell division is likely to be unequal, producing a dispersal stage that may be flagellated or amoeboid.

Ciliates

The ciliates (phylum Ciliophora) form a natural group distinguishable from other protozoa by a number of specialized features, including the possession of cilia, which are short hair-like processes, at some stage in their life cycle,

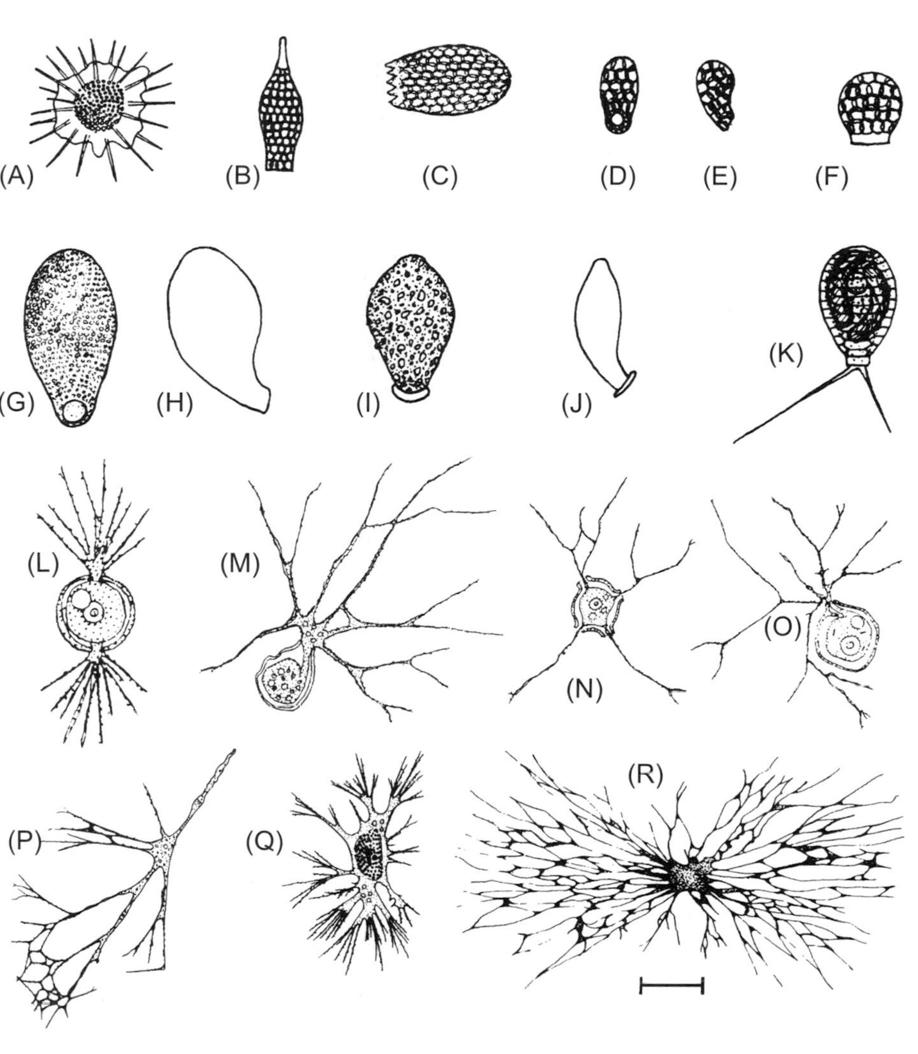

FIGURE 2.6 (A) *Vampyrella lateritia*; (B) *Paraeuglypha reticulata*; (C) *Euglypha tuberculata*; (D, E) *Trinema enchelys*, oral and side views; (F) *Sphenoderia lenta*; (G, H) *Cyphoderia ampulla*; (I, J) *Campascus triqueter*; (K) *Paulinella chromatophora*; (L) *Diplophyrys archeri*; (M) *Liekerkuehnia wagnerella*; (N) *Microcometes paludosa*; (O) *Microgromia haeckeliana*; (P) *Biomyxa vegans*; (Q) *Chlamydomyxa montana*; (R) *Reticulomyxa filosa*. Scale 10 μm for L, N, O; 15 μm for K; 25 μm for A, B, C; 40 μm for D, E, F, G, H; 50 μm for I, J, M, Q; 80 μm for P; and 10,000 μm for R. *After: Bovee (1985) B, D, E, F, G, H, I, J, K, L, M, N, O, P, Q, R; Deflandre (1959) A, C.*

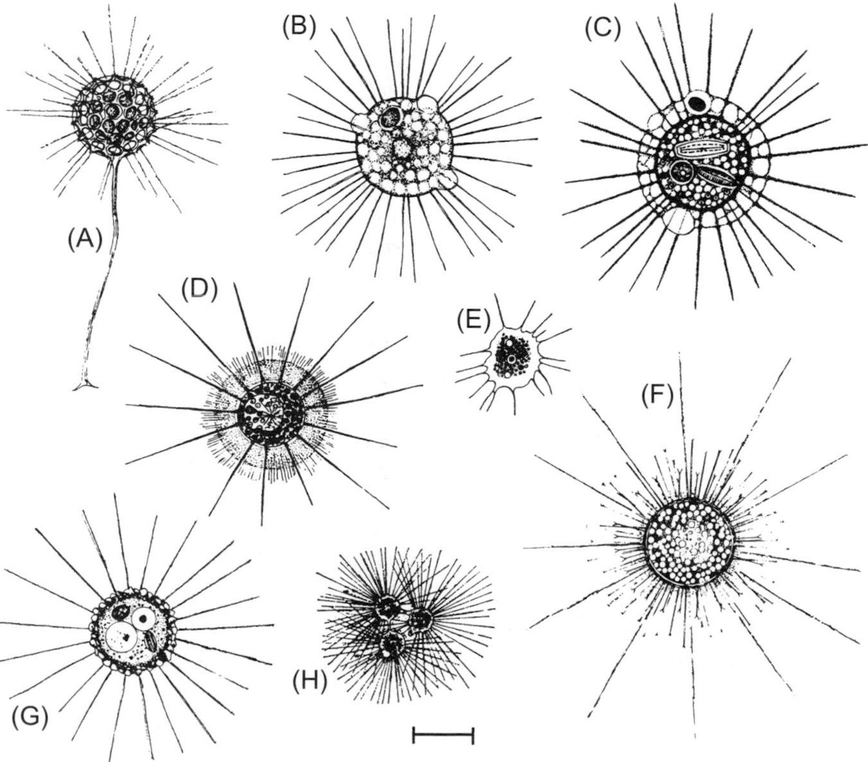

FIGURE 2.7 (A) *Clathrulina elegans*; (B) *Actinophrys sol*; (C) *Actinosphaerium eichhorni*; (D) *Heterophrys myriopoda*; (E) *Ciliophrys infusorium*; (F) *Acanthocystis turfacea*; (G) *Lithocolla globosa*; (H) *Raphidiophrys elegans*. Scale 15 μm for E; 30 μm for B, D, G; 50 μm for A; 75 μm for F, H; and 160 μm for C. *After: Deflandre (1959) H; Kudo (1966) B, C, D, E; Rainer (1968) A, F, G.*

PROTOZOA

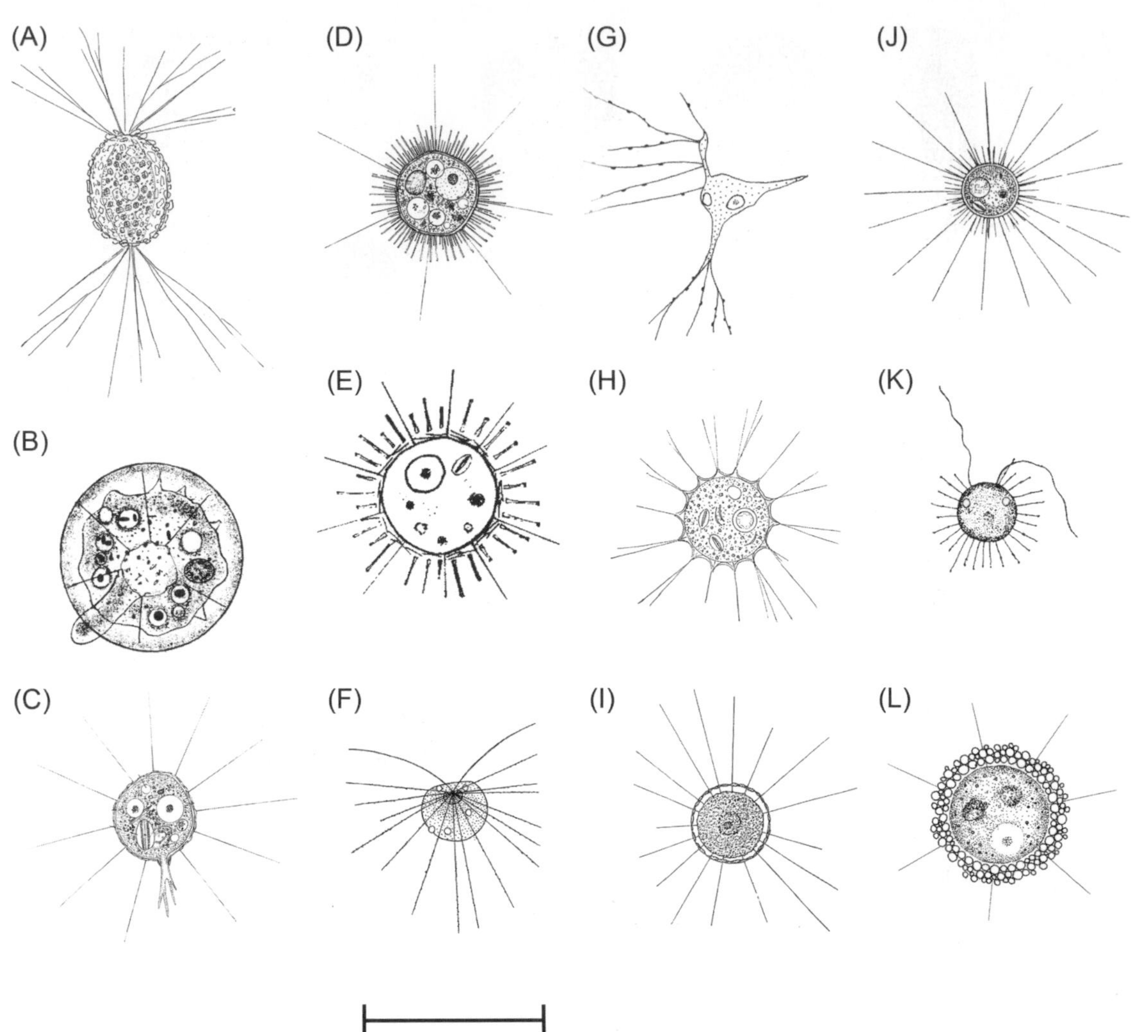

FIGURE 2.8 (A) *Amphitrema stenostoma*; (B) *Microchlamys patella*; (C, I) *Pinaciophora fluviatilis*; (D) *Rabdiophrys anulifera*; (E) *Rabdiaster pertzovi*; (F) *Heliomorpha depressa*; (G) *Limnofila mynlikovi*; (H) *Clathrella foreli*; (I) *Pinaciophora fluviatilis*; (J) *Acanthoperla ludibunda*; (K) *Acinetactis mirabilis*; (L) *Pompholyxophrys punicea*. Scale = 200 μm C, I; 100 μm D; 50 μm A, B, G, H, J, K, L; 25 μm E, F. *After Greef (1869) I; Lemmermann (1914) K; Mikrjukov (1999) J; Mikrjukov (2001) E; Mikrjukov & Mylnikov (1995) (called* Penardia cometa*) G; Penard (1902) A, B; Penard (1905) H; Rainer (1968) C, D; Schoutenden (1907) F; Siemensma (1991) L.*

the presence of two types of nuclei, and a unique form of sexual reproduction called conjugation. A representative ciliate is shown in Fig. 2.1 D. The body surface is covered with cilia, which are mostly aligned in rows called kineties. The pattern of kineties is interrupted in the region of the mouth where there may be specialized oral cilia used for feeding. The cilia may be reduced in number, especially in sessile forms, or organized into larger compound ciliary organelles, such as cirri. The only large group that does not always possess cilia is the Suctoria; these are sessile predators whose dispersal stages are, however, ciliated. This distinctive group is easily recognized by its feeding tentacles. The novice should take care not to confuse small, ciliated animals with ciliates; the size range of ciliates overlaps that of several metazoan groups, such as turbellarians, rotifers, and gastrotrichs. Some ciliates are mixotrophic due to the presence of endosymbiotic algae, or by sequestering chloroplasts from ingested algae that are kept functional in the ciliate cytoplasm (Esteban et al., 2010).

The ciliates are divisible into 12 classes (Adl et al., 2012). Members of the class Karyorelictea are thought primitive for the group, with numerous non-dividing macronuclei that are not highly polyploid. They are largely benthic, the best-known freshwater example being *Loxodes* (Fig. 2.9 J). Compound ciliary organelles associated with the cytostome are prominent in the classes Heterotrichea and Spirotrichea. Large heterotrichs, such as *Stentor* and *Spirostomum* (Fig. 2.10 A–F), are familiar as teaching material. Spirotrichs are abundant in many freshwater habitats, from plankton (choreotrichs and oligotrichs, Fig. 2.11 S–W) to the benthos (e.g., many stichotrichs and hypotrichs).

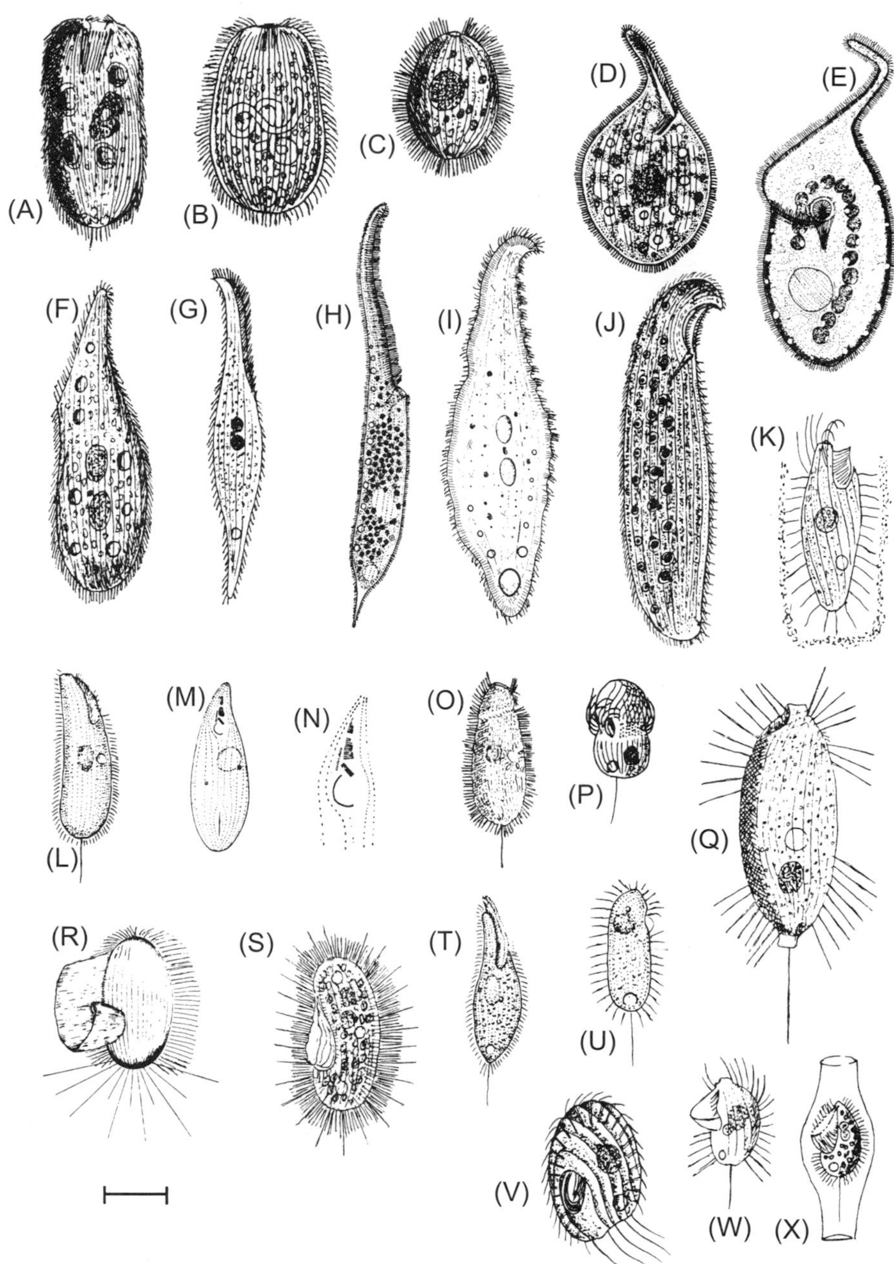

FIGURE 2.9 (A) *Prorodon teres*; (B) *Pseudoprorodon ellipticus*; (C) *Holophyra simplex*; (D) *Trachelius ovum*; (E) *Paradileptus robustus*; (F) *Amphileptus claparedi*; (G) *Litonotus fascicola*; (H) *Dileptus anser*; (I) *Loxophyllum helus*; (J) *Loxodes magnus*; (K) *Cyrtolophosis mucicola*; (L, M, N) *Philasterides armata*, live, silver-stained, and oral detail of silver-stained specimen; (O) *Loxocephalus plagius*; (P) *Urozona bütschlii*; (Q) *Balanonema biceps*; (R) *Pleuronema coronatum*; (S) *Histiobalantium natans*; (T) *Cohnilembus pusillus*; (U) *Uronema griseolum*; (V) *Cinetochilum margaritaceum*; (W) *Cyclidum glaucoma*; (X) *Calyptotricha pleuronemodies*. Scale 10 μm for K, Q; 15 μm for P, V; 20 μm for T, U, W, X; 25 μm for G, H, L, M; 30 μm for C, I, S; 40 μm for B, R; 50 μm for F; 60 μm for A, O; and 75 μm for D, E, J. *After: Corliss (1979) R; Dragesco (1966a) I; Grolière (1980) M, N; Kahl (1930–1935) A, B, C, F, G, J, K, O, P, Q, S, V, W, X; Kudo (1966) I; Noland (1959) L, T, U.*

Stichotrichs and hypotrichs (Figs. 2.11 A–H, N–Q; and 2.12 X, Y) are mostly dorsoventrally flattened crawlers with compound ciliary structures called cirri.

The Nassophorea are named for their basket-like nasse or cyrtos supporting the cytopharynx (Fig. 2.12 V, W, Z). The armophoreans were formerly placed in the Heterotrichea but are now recognized as a separate class, Armophorea, established on the basis of small subunit (SSU) rRNA gene sequence data. Armophoreans are found only in anoxic habitats, benthic, pelagic, or as endosymbionts in the digestive systems, mainly of invertebrates. Armophoreans are free-swimming, typically small to medium-size, with multiple adoral polykinetids and a somatic ciliature that is typically holotrichous but sometimes reduced (Fig. 2.11 K, R).

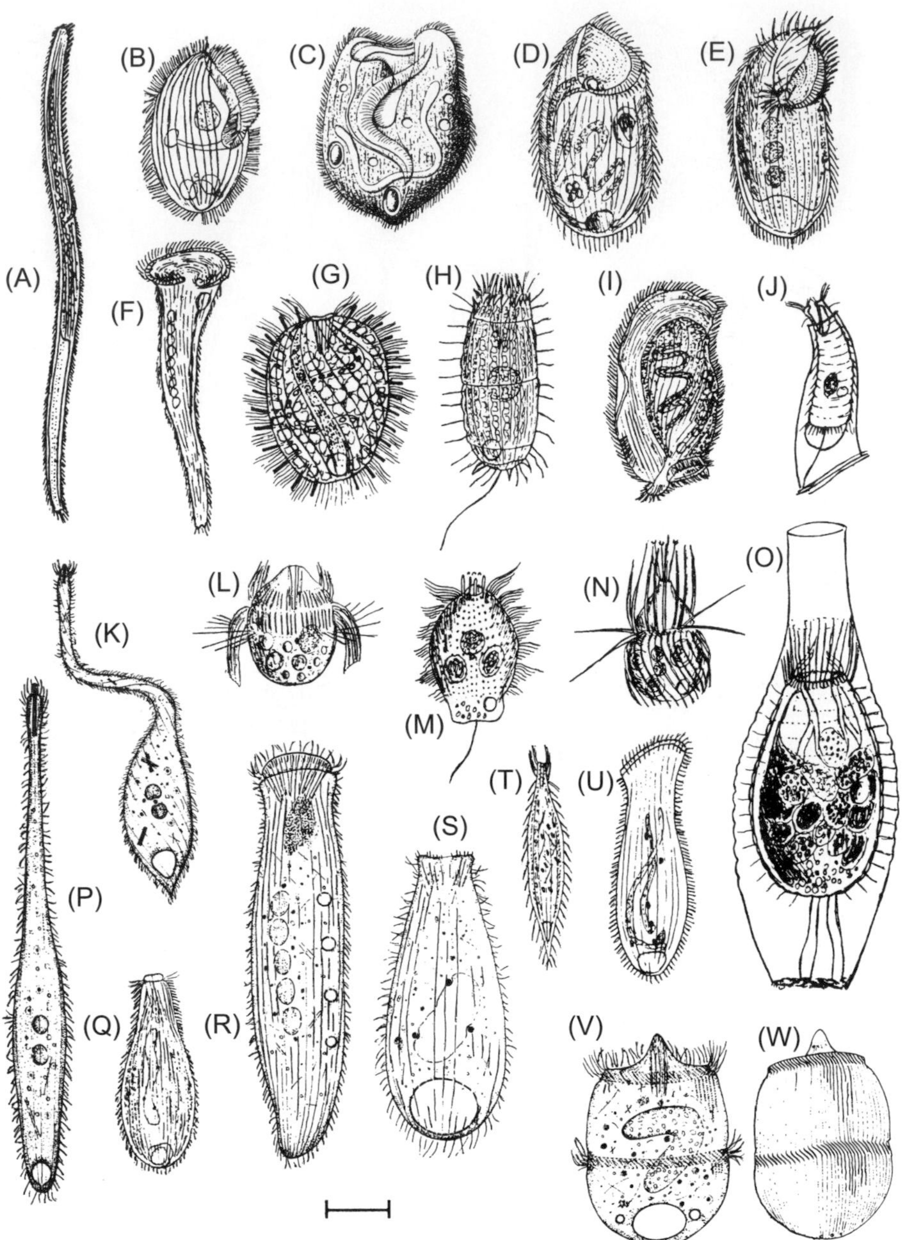

FIGURE 2.10 (A) *Spirostomum minus*; (B) *Blepharisma lateritium*; (C) *Bursaria truncatella*; (D) *Climacostomum virens*; (E) *Condylostoma tardum*; (F) *Stentor polymorphus*, half extended; (G) *Actinobolina radians*; (H) *Coleps hirtus*; (I) *Bryophyllum lieberkühni*; (J) *Metacystis recurva*; (K) *Lacrymaria olor*; (L) *Askenasia volvox*; (M) *Urotricha farcta*; (N) *Mesodinium pulex*; (O) *Vasicola ciliata*; (P) *Trachelophyllum apiculatum*; (Q) *Enchelyodon elegans*; (R) *Homalozoon vermiculare*; (S) *Enchelys simplex*; (T) *Chaenea teres*; (U) *Spathidium spathula*; (V, W) *Didinium nasutum*, live and silver-stained. Scale 10 μm for M, N; 20 μm for H, J, L, P, S; 30 μm for G, O, U; 40 μm for B, K, T; 60 μm for E, Q, R; 80 μm for D, V, W; 100 μm for A, F, I; and 200 μm for C. *After: Dragesco (1966a) K, S, V, W; Dragesco (1966b) P, R; Kahl (1930–1935) A, B, D, E, F, G, H, I, J, L, M, N, O, Q, T, U; Kent (1882) C.*

Classes Prostomatea (Fig. 2.10 J, O) and Litostomatea (Figs. 2.9 D, E, H; and 2.13 J, M) are largely predators, often of other ciliates. Prostomes generally have apical cytostomes, while many litostomes have subapical, sometimes slit-like cytostomes. The mouth is encircled by a crown of cilia from whose bases (kinetosomes) arise the rhabdos, a cylinder of microtubules surrounding and supporting the cytopharynx. Toxicysts are found in most species and are used to subdue active prey. Toxicysts may be found around the cytostome, on a proboscis, on tentacles, or elsewhere on the body. A number of short, specialized kineties (rows of kinetosomes) are often found near the anterior. This brosse (brush) probably assists in prey recognition.

Class Phyllopharyngea contains the distinctive Suctoria (Figs. 2.13 B, F, I; 2.14; 2.15 A–C; and 2.16 B, C, J, L), sessile or free-floating predators of other ciliates. Suctoria are unusual in that most have several "sticky" feeding tentacles rather than a single mouth. Suctoria reproduce by unequal

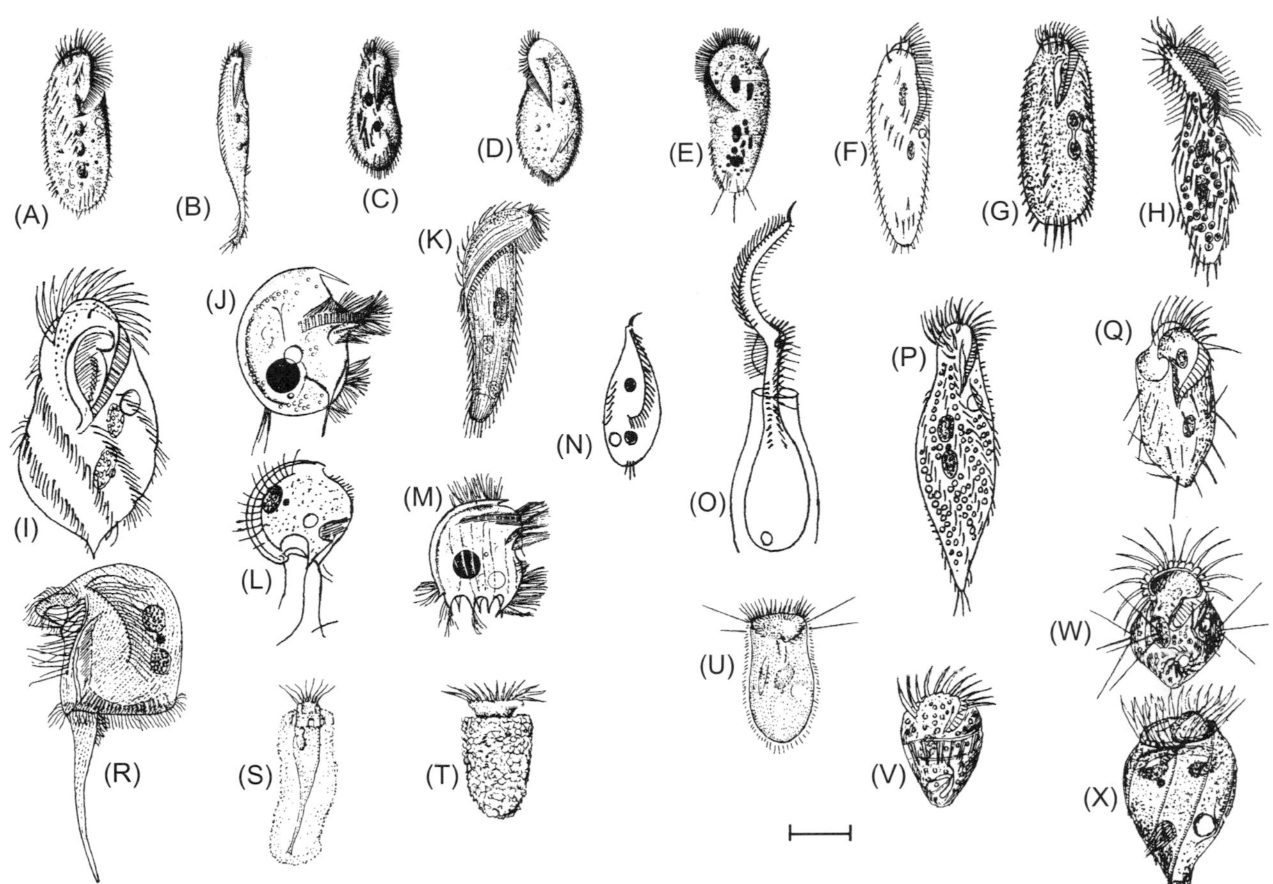

FIGURE 2.11 (A) *Gastrostyla steini*; (B) *Uroleptus piscis*; (C) *Oxytricha fallax*; (D) *Urostyla grandis* (dorsal view); (E) *Stylonychia mytilus* (dorsal view); (F) *Gonostomum affine*; (G) *Tetrastyla oblonga*(called Amphisiella oblonga); (H) *Stichotricha aculeata*; (I) *Hypotrichidium conicum*; (J) *Discomorphella pectinata*; (K) *Metopus es*; (L) *Myelostoma flagellatum*; (M) *Saprodinium dentatum*; (N,O) *Chaetospira mülleri*, contracted and extended forms; (P) *Strongylidium crassum*; (Q) *Psilotricha acuminata*; (R) *Caenomorpha medusula*; (S) *Tintinnidium fluviatile*; (T) *Tintinnopsis cylindricum*; (U) *Strombidinopsis setigera*; (V) *Strombidium viride*; (W) *Halteria grandinella*; (X) *Strobilidium gyrans*. Scale 15 μm for L; 25 μm for H, W, X; 30 μm for F, I, J, P, Q, R, T; 40 μm for A, G, K, M, N, O, S, U, V; 60 μm for B; 80 μm for C, E; and 140 μm for D. *After: Jankowski (1964a,b) J, M; Kahl (1930–1935) F, G, H, I, K, L, N, O, P, Q, R, V, W, X; Kent (1882) A, B, C, D, E; Noland (1959) S, T, U.*

binary fission (budding), which yields a ciliated dispersal stage or "swarmer." Other groups within the Phyllopharyngea include the Cyrtophoria, which contains surface-associated algivores such as *Chilodonella* (Fig. 2.17 T), plus a diverse array of epizooic and free-living forms such as chonotrichians and rhynchodians (Gong et al., 2009).

Colpodeans (Figs. 2.16 F, G, M; 2.17 K, L, N, P, S; and 2.18 G) are not common in freshwater environments, most being terrestrial bacterivores. They are more likely to be encountered in small, temporary waters. Plagiopylea is a riboclass whose monophyly, like the class Armophorea, is based only on the evidence of sequences of the SSU rRNA gene. Also like the armophoreans, plagiopyleans are considered to be anaerobic or microaerophilic and include groups not formerly thought to be phylogenetically related, e.g., the "classic" plagyopyleans (Fig. 2.17 M), which were formerly placed in the Colpodea and resemble colpodids in form, and the odontostomes (Fig. 2.11 J, M). Recently, another anoxic ciliate lineage, which was initially known only from marine environmental rRNA sequence data, has been characterized, based on which the class Cariacotrichea was established (Orsi et al., 2011).

Members of the Oligohymenophorea are mostly microphagous, and this class is named for the compound ciliary organelles that are found in a buccal cavity surrounding the cytostome. The most common pattern (in subclasses Hymenostomatia, Scuticociliatia, and Peniculia; Figs. 2.9 L–X; 2.15 H, I; and 2.17 A–J) is three polykinetids on the left side of the buccal cavity and an undulating membrane on the right. The net result is three brushes, the polykinetids, working against a curved wall, the undulating membrane, to deliver small particles to the cytostome. The large subclass Peritrichia (Figs. 2.12 A–U, 2.13 H, and 2.18 I) contains sessile bacterivores in which the buccal cavity is deepened as an infundibulum, and the polykinetids wind down it to the cytostome after encircling a prominent peristome. Somatic ciliature is absent in most species. Many are attached to the substrate by a stalk, as in the common *Vorticella* (Fig. 2.12 K),

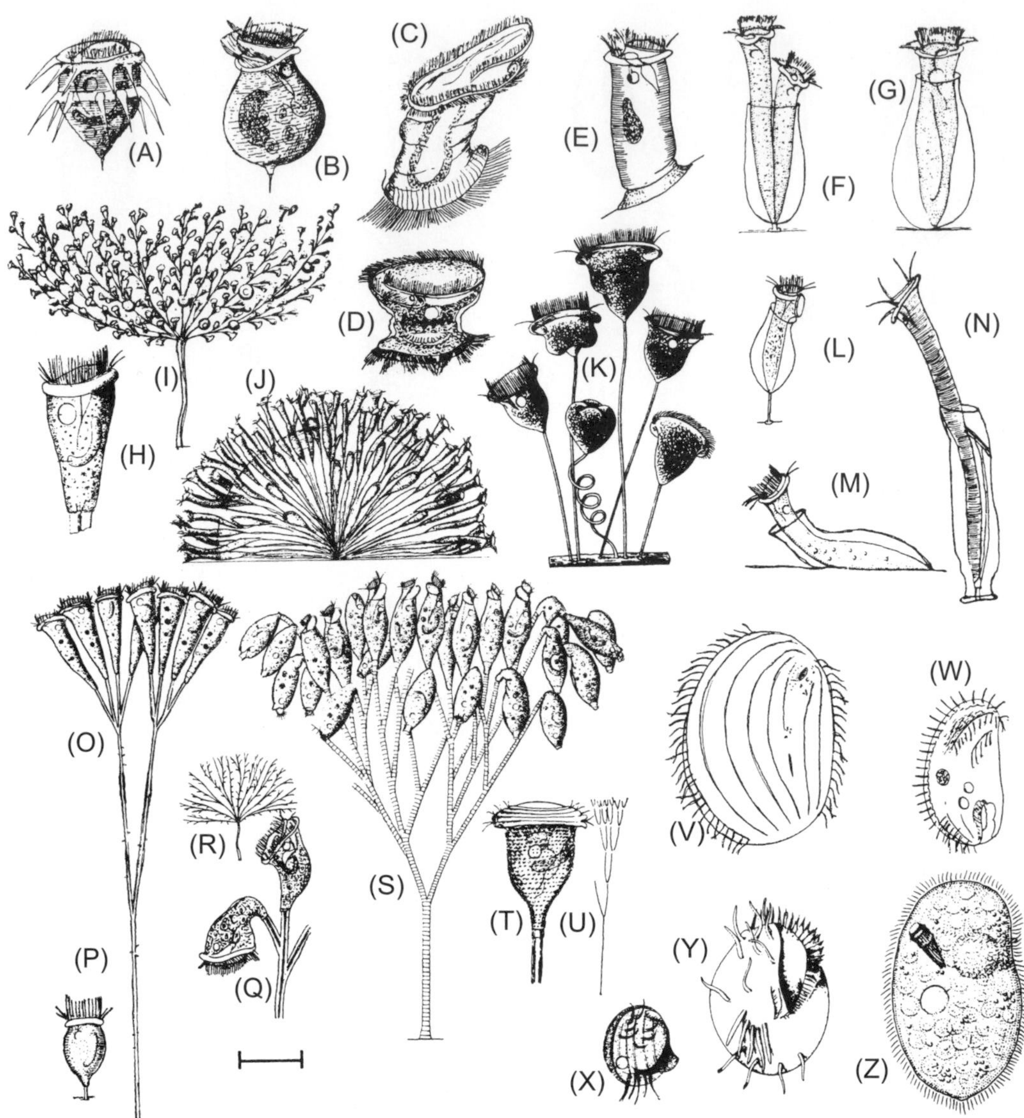

FIGURE 2.12 (A) *Hastatella radians*; (B) *Astylozoon faurei*; (C) *Urceolaria mitra*; (D) *Trichodina pediculis*; (E) *Scyphidia physarum*; (F) *Cothurnia imberbis*; (G) *Vaginicola ingenita*; (H, I) *Zoothamnium arbuscula*, individual and colony; (J) *Ophrydium eichhorni*; (K) *Vorticella campanula*; (L) *Pyxicola affinis*; (M) *Platycola decumbens (*called *Platycola longicollis)*; (N) *Thuricola folliculata*; (O) *Epistylis plicatilis*; (P) *Rhabdostyla pyriformis*; (Q, R) *Carchesium polypinum*, individual and colony; (S) *Opercularia nutans*; (T, U) *Campanella umbellaria*, individual and colony; (V) *Pseudomicrothorax agilis*; (W) *Microthorax pusillus*; (X) *Aspidisca costata*; (Y) *Euplotes patella*; (Z) *Nassula ornata*. Scale 15 μm for V, W; 20 μm for A, B, G, P; 25 μm for D, E, H, F, X; 30 μm for C, Z; 40 μm for L, M, S, Y; 50 μm for O; 75 μm for K, N, Q, U; and 200 μm for I, J. *After: Corliss (1979) V, Y; Kahl (1930–1935) A, B, C, D, E, H, L, N, Q, R, T, U, W; Kent (1882) I, J, K, O, S, X; Noland (1959) F, G, M, P.*

and a few are secondarily free-swimming. Peritrichs may be either solitary or colonial.

MATERIAL PREPARATION AND PRESERVATION

To identify certain species of protozoa, it may be necessary to cultivate them. This involves isolating them from other (contaminant) organisms and then growing them in a culture medium. In general, all initial manipulations and transfers should be performed where possible in media with pH and osmotic potential similar to those at the site of isolation. Likewise, it is optimal that suitable temperature, light, and oxygen tension regimes should also be maintained throughout the isolation and culturing processes. Numerous methods for the isolation and cultivation of protozoa have been reported, and these have been reviewed or summarized on a number of occasions (Finlay et al., 1988; Kirsop & Doyle, 1991; Nerad, 1993; Lee & Soldo, 1992; Tompkins et al., 1995; Day et al, 2007).

Methods to collect protozoa are described in Volume I's chapter on protozoa, but below we describe more detailed methods for isolating, culturing, and preserving selected groups.

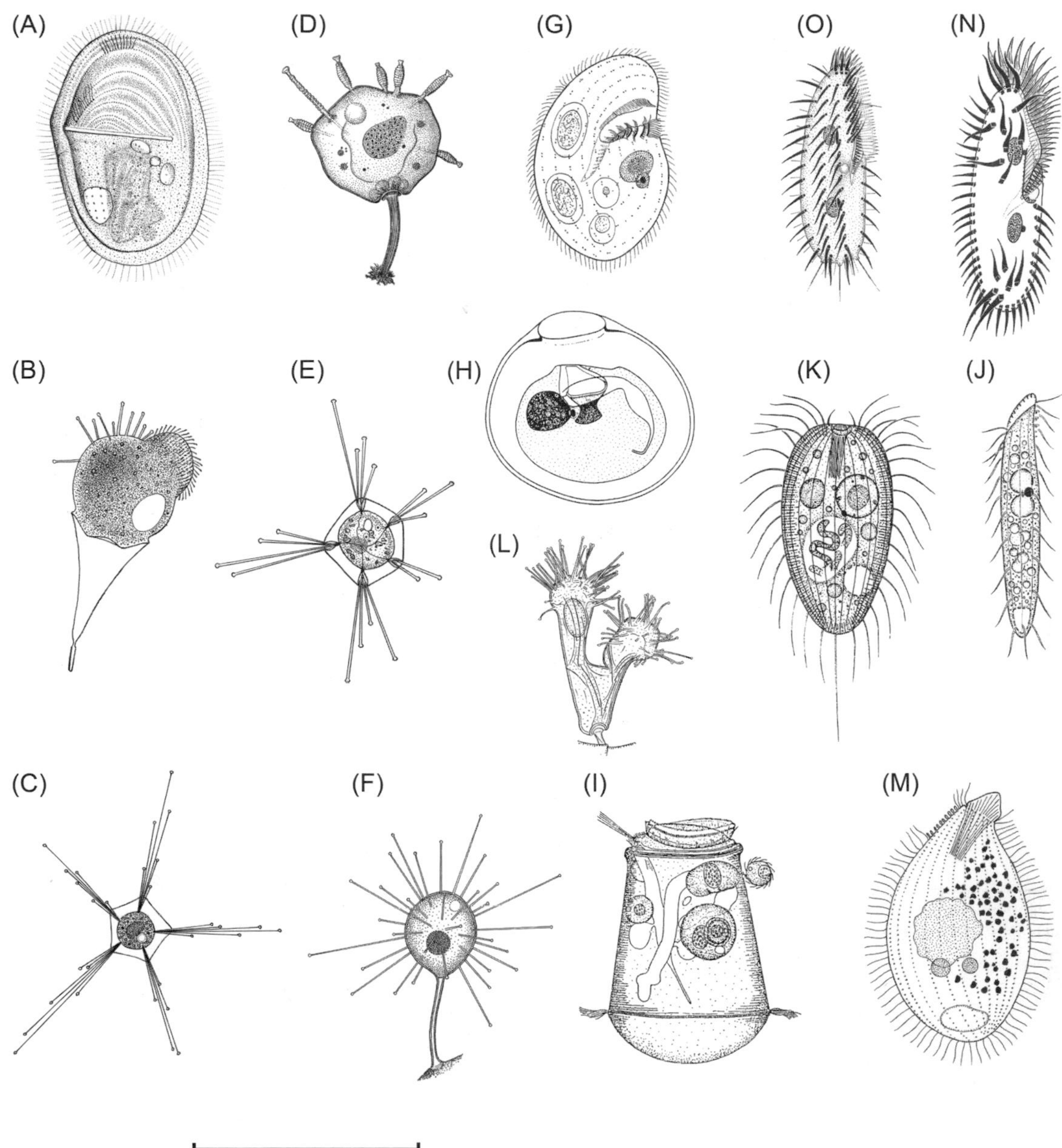

FIGURE 2.13 (A) *Gastronauta* sp; (B) *Paracineta patula*; (C) *Metacineta micraster* var. *pentagonalis* (called *M. pentagonalis* in Nozawa 1939); (D) *Choanophrya infundibulifera*; (E) *Solenophrya micraster*; (F) *Prodiscophrya collini*; (G) *Bryometopus pseudochilodon*; (H) *Usconophrys aperta*; (I) *Endosphaera engelmanni* in cytoplasm of *Opisthonecta henneguyi*; (J) *Apertospathula armata*; (K) *Apsikrata gracilis*; (L) *Lecanophryella paraleptastaci*; (M) *Lagynophrya fusidens*; (N) *Trachelostyla ciliophorum*; (O) *Wallackia schiffmanni*. Scale = 200 µm C, I; 100 µm B, E, O; 50 µm A, G, H, J, K, L, M, N; 25 µm D, F. *After Clamp (1991) H; Curds (1982) A, B, C, D, E, F, M; Curds et al. (1983) G, N, O; Dovgal (1985) L; Foissner & Xu (2006) J; Foissner (1984) K; Matthes (1971) I.*

Isolation

There is a wide variety of methods of isolation, and these can broadly be classified into three categories: enrichment methods, dilution methods, and physical methods. Enrichment is the inoculation of a field sample into an equal or greater volume of suitable medium and incubation under favorable conditions. By inoculation of parallel cultures in a range of media, different organisms will be selected. For bacterivorous protozoa, the simplest way to enrich a sample is to add boiled grains of barley, wheat or rice, which will promote the growth of bacteria and thereby produce a food source for the protozoa. Some commonly used enrichment methods are described in Finlay et al. (1988) and Lee & Soldo (1992). The development of bacterivores may also encourage the growth of carnivorous protozoa that will feed upon them. Dilution methods are most effective for use on preponderantly uniprotozoan samples. Material is sequentially diluted in an appropriate medium and incubated under favorable conditions. The greatest dilution in which growth occurs is likely to be uniprotozoan and usually isolates

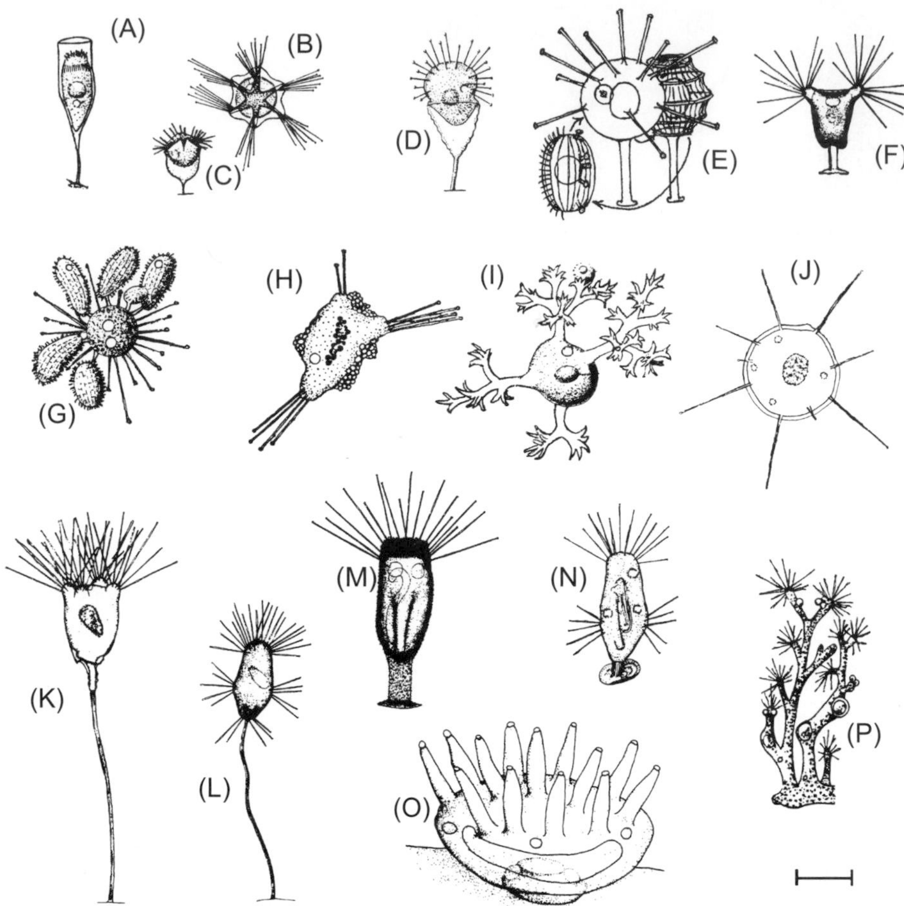

FIGURE 2.14 (A) *Thecacineta cothurniodes*; (B, C) *Metacineta mystacina*, top and side views; (D) *Paracineta crenata*; (E) *Podophrya fixa*, showing trophont, encysted form, and swarmer; (F) *Acineta limnetis*; (G) *Sphaerophyra magna*; (H) *Trichophyra epsitylidis*; (I) *Dendrocometes paradoxus*; (J) *Heliophrya reideri*; (K) *Tokophrya quadripartita*; (L) *Multifasciculatum elegans*; (M) *Squalorophyra macrostyla*; (N) *Discophrya elongata*; (O) *Stylocometes digitalis*; (P) *Dendrosoma radians*. Scale 15 μm for E, H, J, O; 30 μm for A, D, F, G; 50 μm for I, L, M, N; 75 μm for B, K, 150 μm for C; and 2000 μm for P. *After: Corliss (1979) P; Goodrich & Jahn (1943) F, K, L, M; Kent (1882) G, I; Matthes (1954) J, O; Noland (1959) A, B, C, D, N; Small and Lynn (2000) E, H.*

the most abundant species in a sample. Details of various dilution methods are described in Cowling (1991) and Finlay et al. (2000). Once isolated, it may be important to reduce the volume of liquid in which the cell is contained, thereby initiating a quorum-sensing mechanism. Physical methods involve the selection of individual protozoan cells and their transfer into a growth medium. Micropipetting with thin capillary pipettes, working under a dissecting microscope, can be used for a wide variety of protozoa, particularly those that are relatively large and/or slow. Other methods of isolation include silicone oil plating, flow cytometry, agar plating, and electromigration. Silicone oil plating involves the isolation of clone-founding cells within microdroplets formed from vortex-mixed oil/culture emulsions (Soldo & Brickson, 1980). Flow cytometry is an automated means of discriminatory cell sorting and isolation on the basis of various cell attributes including size and density. It is particularly useful for cells that contain pigments that give a fluorescent signal and has also been applied successfully to isolate protozoa using their fluorescent food vacuole contents (Keenan et al., 1978). Agar plating methods rely on discerning colony growth of isolated clones on agar surfaces or within agar, and are particularly useful for amoebae and some flagellates. Usually one or two drops of sample are placed onto a non-nutrient agar plate that has been streaked with a suitable food organism, and then incubated. Amoebae then migrate across the agar surface away from site of inoculation, thereby isolating themselves from other organisms in the sample. The amoebae may then be picked off and subcultured (Lee & Soldo, 1992; Day et al., 2007). Electromigration is a method for obtaining concentrated suspensions of ciliated and flagellated protozoa relatively free of bacteria and other organisms. It works on the principle that many ciliates and flagellates orient themselves in a direct current and migrate toward the cathode (Schmidt, 1982). It is particularly useful for the isolation of organisms from mud and sediment samples.

Cultivation

To maintain cultures of protozoa long term, it is necessary to provide a medium that suits each species and a supply of

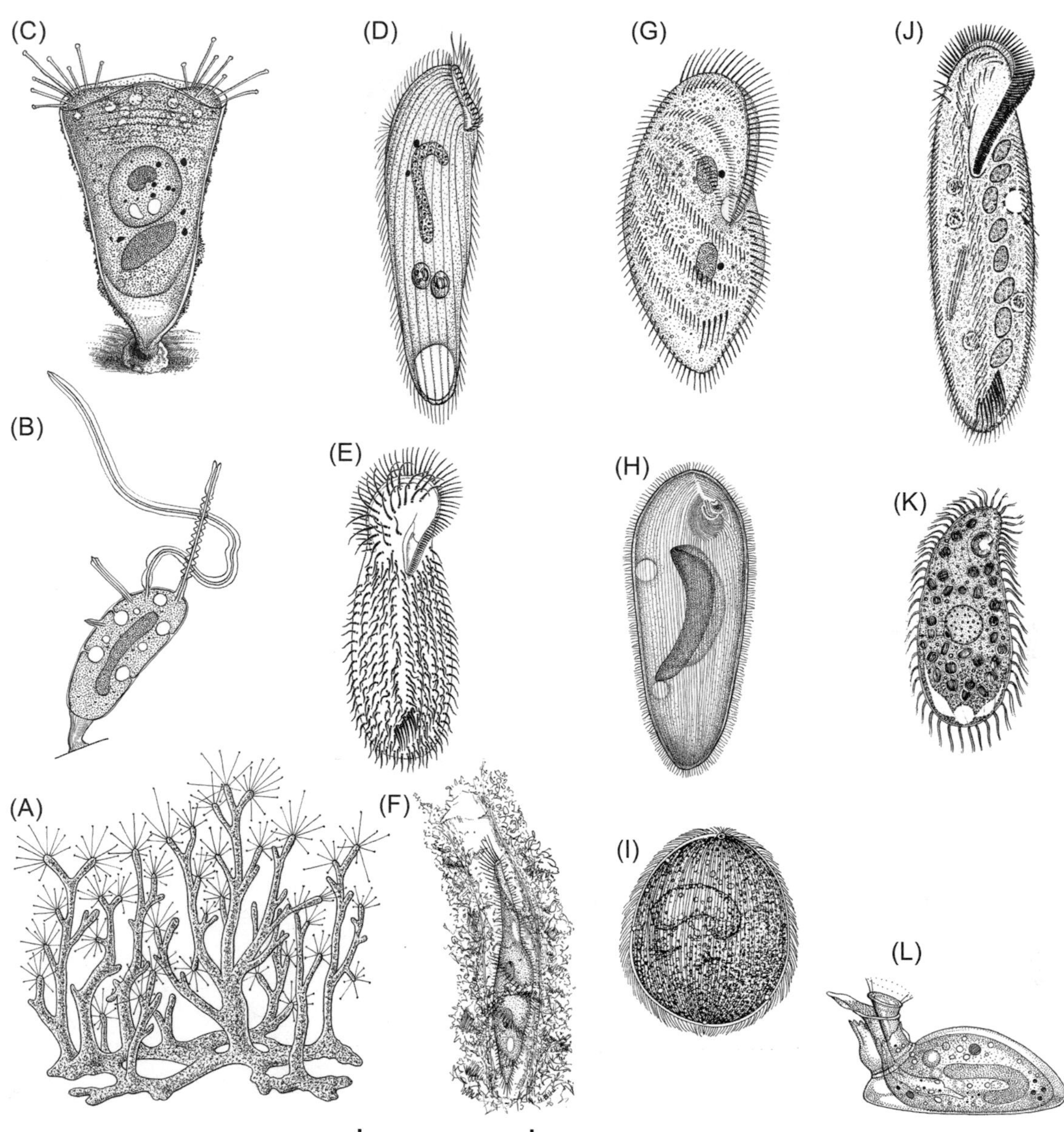

FIGURE 2.15 (A) *Dendrosoma radians*; (B) *Rhynchophrya palpans*; (C) *Periacineta linguifera*; (D) *Rheichenowella nigricans*; (E) *Pseudourostyla levis*; (F) *Stichotricha* sp.; (G) *Kerona pediculus*; (H) *Ophryoglena rhabdocaryon*; (I) *Ichthiophthirius multifiliis*; (J) *Pseudokeronopsis similis*; (K) *Parabryophrya penardi*; (L) *Lagenophrys nassa*. Scale = 200 μm B, C, I; 100 μm E, K; 50 μm A, G, H, J, L; 25 μm D, F. *After Corliss (1979) I; Curds (1982) A, B, C, D, E, F; Curds et al. (1983) G, H, L; Shi et al. (2007) J; Foissner (1985) K.*

appropriate food. Various publications provide comprehensive information or refer to media preparations for protozoa (Kirsop & Doyle, 1991; Lee & Soldo, 1992; Nerad, 1993; Tompkins et al., 1995; Finlay et al., 2000; Day et al., 2007). However, certain isolation techniques, growth media, and culture conditions suit a wide range of organisms. Some of these are discussed below.

The methods for the cultivation of flagellates and ciliates are often identical or similar, so these two groups will be dealt with together and the amoebae separately. The choice of culture medium will depend largely upon what the protozoan feeds. Many flagellates and ciliates eat bacteria, and in these cases, non-selective media, designed for the growth of bacterial populations, may be used. For many species, isolates may be cultured in the presence of mixed bacterial flora that coexisted with the target organism in its original habitat. Alternatively, selective cultures may be obtained by incubating the protozoa in an inorganic salt solution along with an appropriate food organism. In some cases, non-pathogenic laboratory cultures of bacteria may be used, whereas in others it may be necessary to isolate bacteria from the original sample and use one or more of these strains as the selected food organisms.

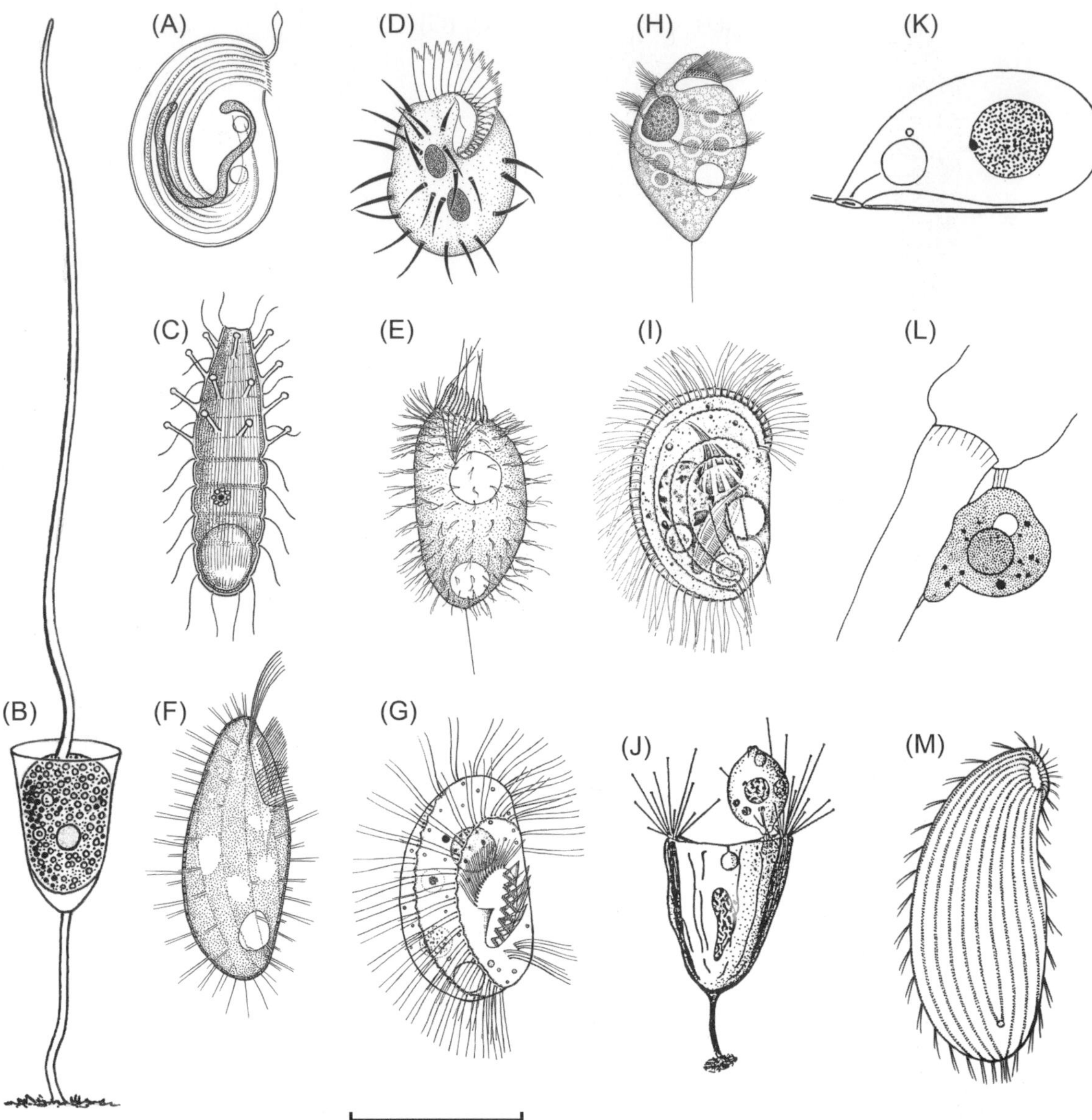

FIGURE 2.16 (A) *Lophophorina capronata*; (B) *Acinetopsis elegans*; (C) *Encelyomorpha vermicularis*; (D) *Psilotricha viridis*; (E) *Plagiocampa ovata*; (F) *Cyrtolophosos mucicola*; (G) *Kreyella minuta*; (H) *Trimyema compressum*; (I) *Pseudochlamydonella rheophyla*; (J) *Pseudogemma pachystyla* (trophont on *Acineta tuberosa*); (K) *Trypanococcus rotiferorum*; (L) *Manuelophrya parasitica*; (M) *Woodfruffia spumacola*. Scale = 40 μm B, D, H, J, L, M; 30 μm A, E; 20 μm C, K; 15 μm G, F, I. *After Batisse (1968) J; Batisse (1994) K; Curds (1982) A, B, C, F, G, H, M; Curds et al. (1983) D, E; Foissner (1993) I; Matthes (1988) L.*

For omnivores and carnivores, an examination of the contents of the food vacuoles may give an indication of the preferred food. The organism can then be incubated in the presence of its natural prey. In some cases, it may be necessary to carry out replicated feeding experiments using a range of food organisms to determine which will support the growth of the isolate.

Culture media may be categorized into four main types: plant infusions, soil extract-based media, inorganic salt solutions, and specific (organically rich) media. Plant infusions are commonly used for bacterivorous flagellates and ciliates. The principle is that organic compounds leach out of plant material, and these support bacterial growth. The most commonly used are lettuce, hay, powdered cereal leaf, and grains of rice, wheat, and barley. Soil extract media are similar to plant infusions in the sense that organic compounds that will support the growth of bacteria are extracted from the soil. Thus, these media may also be used for the cultivation of bacterivorous species. Inorganic salt solutions provide a balanced medium for the growth of many protozoa. However, they contain negligible quantities of organic matter, so the addition of food organisms, or a carbon source, is essential. Such media are commonly used for the cultivation of carnivorous species. Specific, defined, media may be used for producing axenic cultures. Such media invariably contain

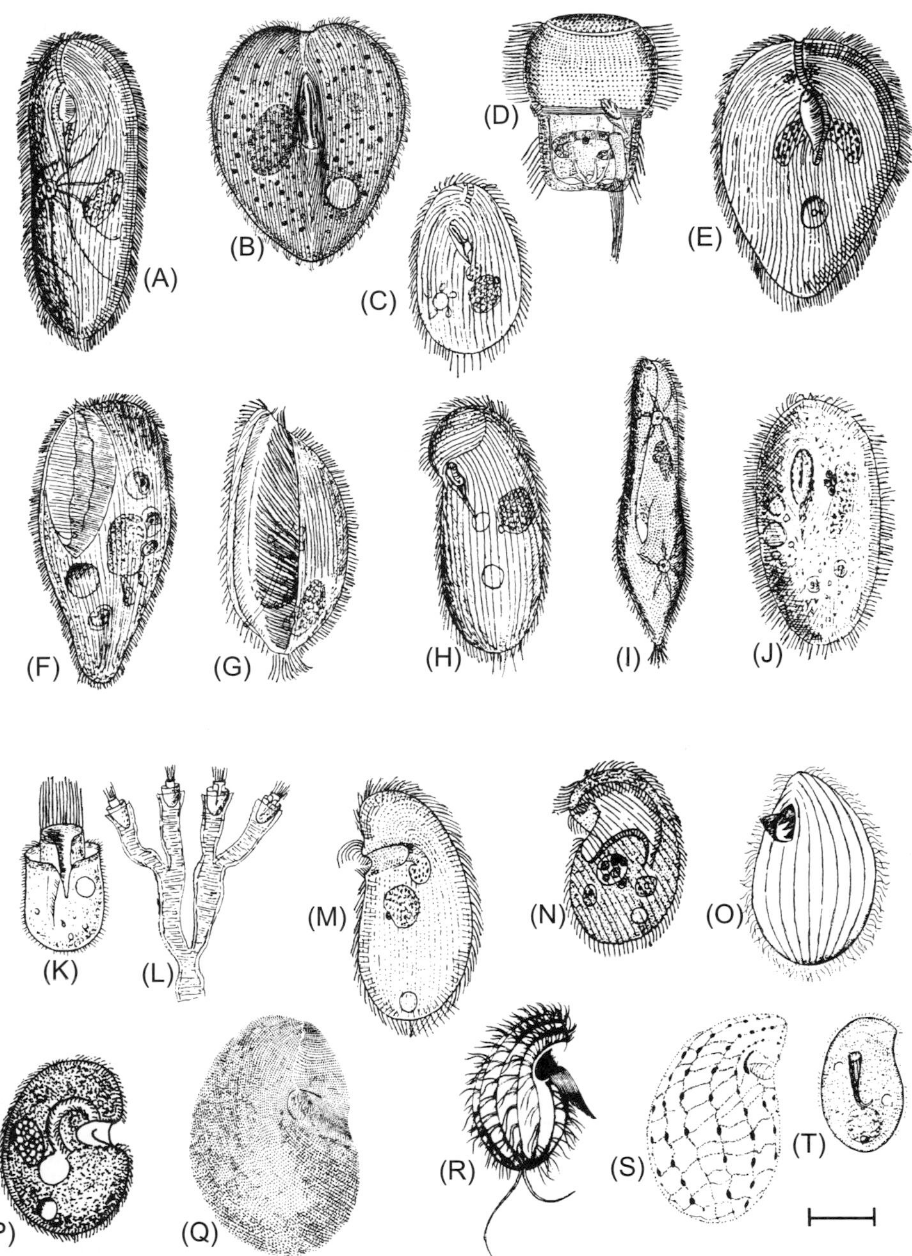

FIGURE 2.17 (A) *Frontonia leucas*; (B) *Stokesia vernalis*; (C) *Glaucoma scintillans*; (D) *Urocentrum turbo*; (E) *Parastokesia bütschlii* (called *Disematostoma bütschlii*); (F) *Turaniella vitrea*; (G) *Lembadion magnum*; (H) *Colpidium colpoda*; (I) *Paramecium caudatum*; (J) *Clathrostoma viminale*; (K, L) *Maryna socialis*, individual and colony; (M) *Plagiopyla nasuta*; (N) *Bresslaua vorax*; (O) *Tetrahymena pyriformis*; (P, Q) *Tillina magna*, live and line drawing of silver-stained specimen; (R, S) *Colpoda steini*, live and silver-stained; (T) *Chilodonella uncinata*. Scale 15 μm for G, O, R; 25 μm for C, H, S, T; 30 μm for D, F; 40 μm for B, E, J, M; 60 μm for I, N; 75 μm for A, K, Q; 100 μm for P; and 300 μm for P. *After: Corliss (1979) O, R; Dragesco (1966b) B; Kahl (1930–1935) A, C, D, E, F, G, H, I, J, K, L, M, P; Kudo (1966) N; Lynn (1976) S; Lynn (1977) Q; Noland (1959) T.*

high concentrations of dissolved organic compounds, usually derived from animal sources. Optimal maintenance conditions (e.g., temperature, pH, oxygen tension) and the frequency of subculturing may be highly variable among different species and should usually take account of the conditions in the natural habitat and the feeding strategy of the isolate. For example, algivorous species should be kept in conditions of illumination that allow an adequate algal food supply to be maintained, whereas bacterivorous and carnivorous species may best be cultivated in the dark in order to control algal contaminants.

Amoebae may be cultured in liquid media, on agar, or in biphasic media. Generally, the larger forms are grown in liquid culture, and the smaller forms are grown on agar. In most cases, amoebae can be cultivated on a non-nutrient medium (liquid or agar) with a suitable bacterial food organism, although the medium can be enriched to stimulate growth of the bacterial

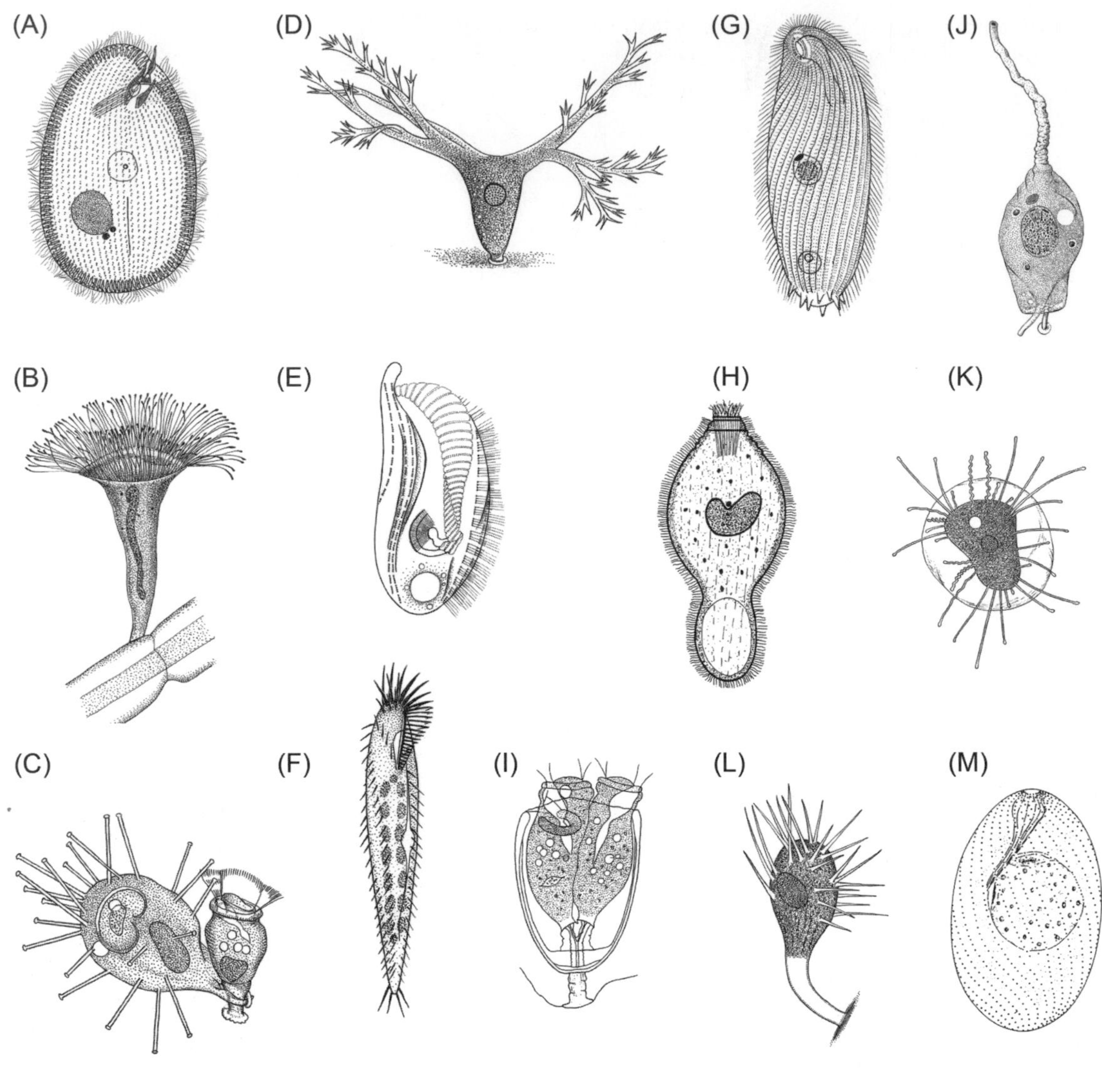

FIGURE 2.18 (A) *Furgasonia blochmanni*; (B) *Spelaeophrya troglocardis*; (C) *Erastophrya chattoni* (growing on the peritrich *Apiosoma*); (D) *Cometodendron eretum*; (E) *Phacodinium metchnicoffi*; (F) *Uroleptoides kihni*; (G) *Cirrophrya haptica*; (H) *Lagynus elegans*; (I) *Rovinjella sphaeromae*; (J) *Rhyncheta cyclopum*; (K) *Mucophrya pelagica*; (L) *Echinophrya horrida*; (M) *Pseudoholophrya terricola*. Scale = 200 μm C, I; 100 μm B, E; 50 μm A, G, H, J, K, L; 25 μm D, F, M. *After Berger et al. (1984) M; Curds (1982) A, B, C, D, E, F, J, K, L; Curds et al. (1983) G; Dovgal (1985) L; Matthes (1972) I; Sola et al. (1990) H.*

food. Cultures of bacterivores should normally be cultivated in the dark, whereas algivorous amoebae, or those with algal symbionts, will require a light-dark cycle. Testate amoebae are often maintained on cereal leaf agar overlaid with a cereal leaf infusion. Details of media formulations and culture methods may be found in (Lee & Soldo, 1992; Nerad, 1993; Tompkins et al., 1995; Finlay et al., 2000; Day et al., 2007).

Preservation

Wherever possible, protozoa should be observed *in vivo* to determine their behavior and certain features that may only be visible in live cells. Nevertheless, it may also be necessary to employ preservation methods, for example: (1) if it is not possible to observe the sample for a long period after collection; (2) to observe certain features not visible in live specimens; or (3) to maintain a reference collection of the organisms or permanent record of the sample. A useful general fixative is Lugol's Iodine, 1% volume/volume, or higher in saline or hard water (Taylor & Heynen, 1987). Mercuric chloride has been used extensively but should probably be discontinued for safety and environmental reasons. These fixatives do not lend themselves to the identification of ciliates, nor to the detection of chromatophores in small flagellates. Filtration methods may aid with both of these problems. The quantitative protargol method or QPS (Montagnes & Lynn, 1987a,b; Skibbe, 1994) produces permanent, quantitative, stained preparations for identification of ciliates and flagellates, although it requires that samples be fixed in a concentrated Bouin's fixative. Various types of silver-staining techniques, which highlight ciliary patterns, have been used in the identification of ciliates (Lee et al., 1985; Foissner, 1991). Other fixatives, such as glutaraldehyde and/or osmium tetroxide (OsO_4), are used if cells are to be examined by electron microscopy. A comprehensive account many of the main commonly used methods for the collection, isolation, cultivation, and preservation of protozoa is given in Lee & Soldo (1992).

ACKNOWLEDGMENTS

The contents of this chapter are based heavily on the chapter on Protozoa by Bill Taylor and Bob Sanders in the third edition of Ecology and Classification of North American Freshwater Invertebrates (Thorp & Covich, 2010), which we have expanded and updated where necessary. We would therefore like to thank Bill and Bob for allowing us to use so much of their material, and in doing so, acknowledge their important contribution to this chapter.

KEYS TO PROTOZOA

Key to Major Functional Groups of Protozoa

The phylum Cercozoa includes organisms that are amoeboid, flagellate, or both. This group is therefore included in the keys to both these major functional groups.

1 Flagella never or rarely present; primary organelles of locomotion and feeding are either pseudopodia or cilia 2

1' Commonly 1–4 flagella (Fig. 2.1 A), 16–24 in one free-living genus (*Paramastix*) .... **Flagellated protozoa [p. 23]**

2(1) Main organelles of locomotion and feeding are pseudopodia as in Fig. 2.1 B, C. Rarely, flagella may be present as well (e.g., Figs. 2.5 S–W and 2.6 A–I, K, P) .... **Amoeboid protozoa [p. 24]**

2' Simple cilia or compound ciliary organelles characteristic (but see Fig. 2.14 A–P) and present in at least one part of life cycle; subpellicular infraciliature present even when cilia are not; two types of nuclei (macronucleus and micronucleus) with rare exceptions (Fig. 2.1 D); ciliated protozoa .... phylum **Ciliophora [p. 27]**

Key to Major Groups of Flagellated Protozoa

1 Without a median groove 2

1' With median groove (annulus and sulcus); two flagella, one extending transversely around the cell (Fig. 2.3 A–C) .... phylum Dinoflagellata

[Note: This group contains many or mostly photosynthetic members; several genera, however, have one or more species that lack chloroplasts and are thus heterotrophic, or contain chloroplasts but also exhibit heterotrophic nutrition.]

2(1) Without pharyngeal rods or paramylon as a reserve material 3

2' Often large flagellates with pharyngeal rods (sometimes difficult to see) or containing the reserve material paramylon; one or usually two flagella arise from within an anterior invagination (reservoir); contractile vacuole associated with reservoir; elongate; body shape often plastic when living (Fig. 2.3 D, F–I) .... phylum Euglenida

[Note: This group contains many or mostly photosynthetic members; several genera, however, have one or more species that lack chloroplasts and are thus heterotrophic, or contain chloroplasts but also exhibit heterotrophic nutrition.]

3(2) Cell with 1–4 flagella (commonly two), usually of unequal length 4

3' Cell usually with two or four equal flagella; plastids two-membraned and containing chlorophyll a and b and starch as a carbohydrate storage product (Fig. 2.3 E) .... phylum Chlorophyta

[Note: This group contains many or mostly photosynthetic members; several genera, however, have one or more species that lack chloroplasts and are thus heterotrophic, or contain chloroplasts but also exhibit heterotrophic nutrition.]

4(3) Cells without deep gullet 5

4' Small cells with a deep, subapical gullet; two nearly equal length flagella arise from gullet (Fig. 2.3 J) .... phylum Cryptophyta

[Note: This group contains many or mostly photosynthetic members; several genera, however, have one or more species that lack chloroplasts and are thus heterotrophic, or contain chloroplasts but also exhibit heterotrophic nutrition.]

5(4) Cells lacking the characteristics described below 6

5' Typically with two unequal flagella; long flagellum usually directed forward during swimming, with short flagellum directed backward if emergent; compact Golgi body frequently visible anterior to nucleus of larger cells; chrysolaminarin vesicle(s) often fill posterior of cell; contractile vacuole usually present in extreme anterior end of cell; many species capable of simultaneous photosynthesis and phagocytosis (Figs. 2.3 K–N and 2.2 A) .... phylum Chrysophyta

[Note: This group contains many or mostly photosynthetic members; several genera, however, have one or more species that lack chloroplasts and are thus heterotrophic, or contain chloroplasts but also exhibit heterotrophic nutrition.]

6(5) Without collar, or with indistinct collar 7

6' Single anterior flagellum encircled laterally by a tentacular, funnel-shaped collar; solitary or colonial; with or without theca (Fig. 2.2 B–H) .... class Choanoflagellida

7(6) Without protoplasmic collar 8

7' Like Choanoflagellida above, but with second trailing flagella attached to base of lorica; protoplasmic collar; bicoecid flagellates (affinities uncertain) (Fig. 2.2 I) .... class Bicoecea

8(7) Ventral cytostome or flagellar pocket, with associated flagella; either kinetoplast or flagellar mastigont system present; ingestion not by pseudopodia 9

8' Flagellates with two unequal length flagella, one trailing; no kinetoplast or flagellar mastigont system; ingestion by pseudopodia (Fig. 2.2 K) .... phylum Cercozoa

9(8) With one or usually two flagella arising from a flagellar pocket (depression); characteristically elongate or bean-shaped. Unique organelle, the kinetoplast, usually associated with the flagella (Fig. 2.2 J, L, N–P) .... class Kinetoplastida

9' Cells with one or two nucleus-flagella complexes (karyomastigonts) each with 1–4 flagella; no Golgi apparatus; when two karyomastigonts, mirrored symmetry of nuclei and flagella; diplomonads (Fig. 2.2 M) .... order Diplomonadida

PROTOZOA

Key to Major Groups of Amoeboid Protozoa

1 Pseudopodia either blunt and hyaline, or filiform and sometimes branching or anastomosing, or thread-like and often branching and anastomosing, or stiff and radiating 2

1' Large, cylindrical or ovoid multinucleate amoeba; bacterial symbionts; usually containing mineral particles; non-motile flagella-like extensions; oxygen-poor habitats (Fig. 2.5 V) .... phylum Amoebozoa, class Archamoebea *Pelomyxa*

2(1) Pseudopodia hyaline, usually blunt, eruptive but filiform process may occur (Figs. 2.4, 2.5 A–Q, and 2.8 B); flagellated stages may be present 3

2' Pseudopodia not blunt and eruptive 4

3(2) Small, usually <65 μm long, with eruptive pseudopodia, flagellated stage common in one group .... class **Heterolobosea**, order **Schizopyrenida [p. 24]**

3' No flagellated stages .... class **Lobosea [p. 24]**

4(2) Pseudopodia not as filopodia 5

4' Filiform pseudopodia (filopodia) sometimes branching or anastomosing, granular or hyaline, many with flagella at some stage in their life cycle; some with test (Figs. 2.5 S–W and 2.6 A–I, K, P) .... phylum **Cercozoa [p. 25]**

5(4) Pseudopodia thread-like and delicate with finely granular appearance, often branching and anastomosing to form complex reticulum (reticulopodia); test often present (Figs. 2.6 M, O, Q, R and 2.8 A) .... traditionally grouped as phylum **Granuloreticulosea [p. 26]**

5' Long slender axopodia or filopodia radiating 3-dimensionally from cell; with or without skeletal elements (Figs. 2.1 C, 2.7, 2.8 D, I–J, L) .... **Heliozoans and Pseudoheliozoans [p. 26]**

Protozoa: Amoeboid Protozoa: Heterolobosea: Schizopyrenida: Families

1 Amoeboid form usually cylindrical, often monopodial; usually uninucleate; nucleolus divides to form polar masses in mitosis; temporary flagellate stage common (Fig. 2.4 A, B) .... Vahlkampfiidae

1' Amoeba flattened or limax; often multinucleate, nucleolus disintegrates during mitosis, no flagellate stage known (Fig. 2.4 C) .... Gruberellidae

Protozoa: Amoeboid Protozoa: Lobosea: Families

1 Lacking external test, traditionally grouped as subclass Gymnamoebia 2

1' Incompletely enclosed in a test or other flexible cuticle of microscales traditionally grouped as subclass Testacealobosia 10

2(1) Cylindrical or flattened; flattened forms with regular outline; no trailing uroidal filaments, with rare exceptions; not strikingly eruptive .... order Euamoebida 3

2' Usually flattened; frequent changes in shape typical; sometimes eruptive; subpseudopodia usually present, often furcate 8

3(2) Without subpseudopodia 4

3' With subpseudopodia 6

4(3) Cell body flattened 5

4' Cell body subcylindrical 7

5(4) Cell usually oblong; crescent-shaped hyaline margin at anterior end; pellicle-like layer, with dorsum often wrinkled and/or ridged; usually uninucleate; no cytoplasmic crystals (Fig. 2.4 D) .... Thecamoebidae

5' Body usually fan-shaped, oval, or spoon-shaped, with hyaline margin occupying up to half of length (Fig. 2.4 E) .... Vannellidae

6(3) Subpseudopodia hyaline, blunt, digitiform, usually from anterior hyaline margin; uninucleate; nucleolar material in central body (Fig. 2.4 G) .... Paramoebidae

6' Few slender, conical, or linear subpseudopodia, from anterior hyaline margin or cell surface; uninucleate (Fig. 2.4 H) .... Vexilliferidae

7(4) Most species polypodial; length usually more than 75 µm; uni- or multinucleate; numerous cytoplasmic crystals (Fig. 2.4 F, J, L) .... Amoebidae

7' Cell monopodial, pseudopods rare; uninucleate with central nucleolus; cytoplasmic crystals in some; cysts common (Fig. 2.4 I, K) .... Hartmannellidae

8(2) Cell flattened, triangular, trapezoid or irregular in outline 9

8' Cell regularly discoid, flattened ovoid, or fan-shaped; usually broader than wide; postcentral granular mass, usually surrounded, sometimes completely, by hyaline border with short subpseudopodia (Fig. 2.4 P) .... Hyalodiscidae

9(8) Cell flattened, broad and irregular in outline, though sometimes elongate during locomotion; slender tapering subpseudopodia, sometimes furcate, produced from broad, hyaline lobopodium; often with small lipid globules; uninucleate (Fig. 2.4 N) .... Acanthamoebidae

9' Several to many fine, sometimes furcate subpseudopodia, finer than in Acanthamoebidae (Fig. 2.4 M, O) .... Echinamoebidae

10(1) Test more or less rigid with distinct aperture .... order Arcellinida 11

10' Discoid or sometimes globose amoeba incompletely closed in a flexible tectum, no well-defined aperture (Fig. 2.5 A) .... Cochliopodiidae

11(10) Pseudopodia digitate and finely granular 12

11' Pseudopodia conical, clear, sometimes anastomosing; test with siliceous material embedded or attached (Fig. 2.5 B, C) .... Phryganellidae

12(11) Test membranous or chitinoid, pliable or rigid; no plates or scales, but may have attached debris 13

12' Test chitinoid or not, rigid, with embedded and/or attached plates, scales, siliceous granules .... suborder Difflugina 16

13(12) Test round; aperture ventral .... suborder Arcellina 14

13' Test oval to flask-shaped; non-areolar, clear; aperture terminal (Fig. 2.5 L, M) .... Hyalospheniidae

14(13) Test flexible to semi-rigid; finely or not areolate 15

14' Test rigid, areolar, smooth; aperture ventral, round (Fig. 2.5 D, F, G) .... Arcellidae

15(14) Test not areolate; cytoplasm not enclosed in a separate membrane sac (Fig. 2.5 H, I) .... Microcoryciidae

15' Test finely areolate; cytoplasm enclosed in a separate membrane sac (Fig. 2.8 B) .... Microchlamyiidae

16(12) Aperture round, broadly oval or wavy (Fig. 2.5 J, K, N) .... Difflugiidae

16' Aperture slit-like or narrow oval 17

17(16) Aperture terminal 18

17' Aperture anterioventral, invaginated, slit-like with overhanging lip (Fig. 2.5 E) .... Plagiopyxidae

18(17) Test particles rectangular (Fig. 2.5 O, P) .... Paraquadrulidae

18' Test particles not rectangular (Fig. 2.5 Q) .... Nebelidae

Protozoa: Amoeboid Protozoa: Cercozoa: Subphyla

1 Filopodia very fine, branching or unbranching, often with obvious granules (extrusomes); many biciliate; with or without a test .... **Filosa [p. 26]**

1' Filopodia or reticulopodia non-granular; cilia and test absent .... **Endomyxa [p. 26]**

PROTOZOA

Protozoa: Amoeboid Protozoa: Cercozoa: Filosa: Families

1 Without distinct test, sometimes with scales 2

1' With test 3

2(1) Filopodia extremely slender, branching, regularly granular; if present, cilia not visible with light microscope; cell typically small, often globular (Fig. 2.8 G) .... Limnofilidae

2' Filopodia numerous, extremely long and branching; two long cilia sometimes present; may be found in organic mud of hot springs .... Mesofilidae

3(1) Test without scales; may have spines and/or attached debris 4

3' Test with secreted, siliceous scales arranged in definitive patterns 5

4(3) Test round, thin; may have spines or spicules (Fig. 2.5 S–U) .... Chlamydophryidae

4' Test rigid and agglutinated; cell does not fill test (Fig. 2.5 W) .... Pseudodifflugiidae

5(3) Scales not as below 6

5' Scales round to elliptical, thin, overlapping, adjacent or scattered (Fig. 2.6 B–F) .... Euglyphidae

6(5) Scales circular or oval; test usually with aperture at end of a neck bent to one side (Fig. 2.6 G–I) .... Cyphoderiidae

6' Scales long, with long axes perpendicular to aperture; aperture at end of short neck that is not bent; test ovoid, <45 μm (Fig. 2.6 K) .... Paulinellidae

Protozoa: Amoeboid Protozoa: Cercozoa: Endomyxa: Families

1 Medium to large (30 μm to >1000 μm); filopodia non-anastomosing and more or less radiate (Fig. 2.6 A) .... Vampyrellidae

1' Small (<50 μm); filopodia lack microtubules and tend to project more from one face of the cell than from others (Figs. 2.5 R and 2.6 P) .... Biomyxidae

Protozoa: Amoeboid Protozoa: Granuloreticulosea: Classes and Families

1 With a test, traditionally grouped as class Monothalamea 2

1' Without a test, traditionally grouped as class Athalamea 3

2(1) Test with one aperture 4

2' Test with more than one aperture (Fig. 2.8 A) .... Amphitremidae

3(1) Test flattened on one side (Fig. 2.6 O) .... Microgromiidae

3' Test not flattened on one side (Fig. 2.6 M) .... Lieberkuehnidae

4(2) Multinucleate and highly reticulate plasmodia (Fig. 2.6 R) .... Reticulomyxidae

4' With or without anastomosing pseudopodia; one or a few nuclei; body mass more or less round (Fig. 2.6 Q) .... Chlamydomyxidae

Protozoa: Amoeboid Protozoa: Heliozoa and Pseudoheliozoa

[Note: some forms with pseudopodia are members of other protist groups, for example *Diplophrys* (Fig. 2.6 L) which is a labyrinthulid.]

1 With axopodia radiating from centrosomes; axonemes visible with light microscope 2

1' With thin, radiating filopodia; axonemes absent .... pseudoheliozoans (cercozoan subphylum Filosa) 6

2(1) Skeleton absent 3

2' With skeleton of siliceous or organic plates and/or spicules; sometimes stalked; axopodia and stalk highly contractile (Fig. 2.7 D, F, H) .... phylum Heliozoa, order Centrohelida

3(2) With two cilia 4

3' Cilia absent 5

4(3) Axopodia with regularly spaced, complex extrusomes; centrosome embedded in depression of centrally located nucleus (Fig. 2.8 F) .... phylum Cercozoa, family Heliomorphidae

4' Axopodia bear moving granules; centrosome not in depression of nucleus which is offset (Fig. 2.8 K) .... family Acinetactidae

5(3) No centroplast or axoplast; axopods granule-studded and thicker at bases; large central nucleus surrounded by lacunar ectoplasm or several nuclei at periphery of central area with vesicular ectoplasm (Fig. 2.7 B, C).. order Actinophryida

5' Microtubule organizing center of dense plaques from the nuclear membrane or from centroplast; may be confused with Actinophryida without knowledge of fine structure (origin and pattern of axopod microtubules) (Fig. 2.7 E).. order Ciliophyrida

6(1) Test with mineralized perles, or plate scales .... orders Perlofilida and Rotosphaerida 7

6' Cell enclosed by a latticed organic capsule (skeleton), a rigid theca, or by irregular particles adhering to the outer surface 12

7(6) Test of mineralised spherical perles; outer layer with apically pointed silica scales (Fig. 2.8 J) .... order Perlofilida, family Acanthoperlidae

7' Test of plate scales .... order Rotospherida and perlofilid family Pompholyxophryidae 8

8(7) Test typically with two-tier plate scales; flagella absent .... order Rotospherida 9

8' Body spherical, coated with a single layer of siliceous perles (Fig. 2.8 L) .... family Pompholyxophryidae

9(8) Test with columnar radial spines, flared out both basally and apically 10

9' Test without columnar spines 11

10(9) Outer plate scales with single, large hole (Fig. 2.8 D).. family Rabdiophryidae

10' Outer plate scales without holes (Fig. 2.8 E) .... family Rabdiasteridae

11(9) With scalloped test of regular, unperforated cup-shaped scales; filopodia in groups of 1–3 (Fig. 2.8 H) .... family Cathrellidae

11' Test not scalloped; filopodia extremely thin, not in groups (Fig. 2.8 C) .... family Piaciophoridae

12(6) Cell enclosed in latticed organic capsule (skeleton); generally stalked; no centroplast; cell body spherical in adults (Fig. 2.7 A) .... order Desmothoracida

12' Cell enclosed in rigid theca with many tine pores through which filopodia emerge; filopodia with long, thin extrusomes (Fig. 2.7 G)......... .... subclass Testosia, family Lithocollidae

Key to Classes and Subclasses of Ciliophora

1 Cilia present in trophont (active) stages; retractile suctorial tentacles only in family Actinobolinidae (class Litostomatea, order Haptorida) 2

1' Suctorial tentacles present (absent in one endocommensal group); no true cytostome or cytopharynx; adults (trophonts) usually sessile, many species ectosymbiotic, some planktonic species; cilia absent except in free-swimming dispersal larval stages; with or without lorica ....class Phyllopharyngea, subclass **Suctoria [p. 28]**

[Note: Patterns of division and release of the larvae form the basis for dividing the subclass into orders. Other characters are useful in identification, but knowledge of the full life cycle is often required before suctorians can be confidently assigned to a genus.]

2(1) Conspicuous buccal ciliature at apical pole; buccal ciliature winds clockwise toward the center when viewed from oral end; somatic ciliature reduced or absent; mobile or sessile; solitary, gregarious or colonial; some species loricate; oral region can contract and withdraw in most species; with either scopula or complex adhesive disc (holdfast) at aboral end of cell .... class Oligohymenophorea, subclasses Mobilia and Peritrichia 3

2' Not as above: without scopula or complex adhesive disc at aboral end of cell 4

3(2) Trophont mobile; symbiotic on other organisms; with complex adhesive disc (holdfast) at aboral end of cell .... subclass **Mobilia [p. 30]**

3' Trophont only rarely motile, usually attached to substratum, which may be inanimate objects or other organisms, via a stalk, lorica or scopula; many gregarious or colonial species .... subclass **Peritrichia [p. 30]**

4(2) Oral area not bordered by an adoral zone of membranelles; no ventral cirri 5

4' Oral area usually bordered by a well-developed adoral zone of membranelles (AZM) consisting of more than three membranelles (polykineties); with or without ventral cirri 11

5(4) Cytostome at end of a buccal cavity or vestibulum with cilia or paroral membrane(s) associated 6

5' Cytostome at or near surface; buccal cavity, if present, without cilia or paroral membranes 7

6(5) Alveoli well-developed and revealed as a prominent argyrome, typically reticulate; body shape variable, many reniform and flattened....... .... class **Colpodea [p. 31]**

6' Alveoli not well developed, no prominent argyrome; body typically ovoid to elongate ovoid elongate ovoid .... class **Oligohymenophorea [p. 31]**

7(5) Cytostome lateral or ventral 8

7' Cytostome at or near anterior end, or continuing down side as a slit 9

PROTOZOA

8(7) Circular mouth located midventrally; no proboscis; large cytostome ... 10

8' Cytostome a barely visible lateral slit on the convex side of a tapering front end, or a lateral opening at the base of an anterior proboscis (Fig. 2.9 J; see also family Spathidiidae Fig. 2.10 I, U) ... class Karyorelictea, family Loxodidae

9(7) Oral ciliature as simple kinetids; with dorsal brosse or brush formed of specialized dikinetids bearing clavate cilia ... class **Litostomatea [p. 33]**

9' With circumoral ciliation composed of dikinetids; without dorsal brosse ... class **Prostomatea [p. 34]**

10(8) Body nearly ellipsoid, rounded in cross section, sometimes flattened ventrally; medium to large (some 100 µm); densely ciliated all over; oral depression present ... class **Nassophorea [p. 34]**

10' Body usually flattened; ventrum ciliated, dorsum bare or with a few cilia; anterior preoral arcs of right ventral ciliary rows continuous with more posterior parts ... class Phyllopharyngea, subclass **Cyrtophoria [p. 34]**

11(4) Somatic ventral ciliature as polykinetids (cirri) in groups or files on ventral surface ... 12

11' Somatic ciliature as monokinetids or dikinetids, but not cirri ... 14

12(11) Body typically oval to rectangular, sometime ellipsoidal, rigid and dorsoventrally flattened, often heavily ribbed; ventral cirri often conspicuous; pellicular alveoli well developed ... 13

12' Body elongate, often flexible; ventral cirri often inconspicuous; pellicular alveoli poorly developed ... class Spirotrichea, subclass **Stichotrichia [p. 35]**

13(12) Body ovoid with prominent longitudinal ridges on each side; AZM stretching almost entire body length terminating near posterior pole; ventral ciliature comprising delicate cirri in widely spaced rows (Fig. 2.18 E) ... class Spirotrichea, subclass Protocruziidia, family Phacodiniidae

13' Body ovoid to rectangular or ellipsoidal; AZM not usually stretching to posterior pole ... class Spirotrichea, subclass **Hypotrichia [p. 35]**

14(11) Somatic cilia sparse or absent ... 15

14' Body surface densely ciliated ... 16

15(14) Body flattened, rigid, often with spines; body ciliature present as short, generally obvious rows; oral ciliature relatively inconspicuous; mainly anaerobic ... class **Plagiopylea [p. 36]**

15' Membranelles numerous in complete, or almost complete, circle at oral end; body generally conical or bell-shaped; mostly planktonic ... class Spirotrichea, subclasses **Choreotrichia** and **Oligotrichia [p. 36]**

16(14) Body small to medium, usually twisted to left; oral region spiraled with paramembranelles; with hydrogenosomes instead of mitochondria; in richly organic sediments with low, or no, oxygen ... class **Armophorea [p. 36]**

16' Body medium to large, often elongate and contractile; somatic ciliation holotrichous; left oral polykinetids conspicuous, typically paramembranelles encircling the anterior end clockwise before plunging into the oral cavity; one or more parorals on right side ... class **Heterotrichea [p. 36]**

Protozoa: Ciliophora: Suctoria: Orders

1 Budding begins in a pouch ... 2

1' Budding and cytokinesis on surface of trophont ... **Exogenida [p. 28]**

2(1) Exogenous budding occurring in a brood pouch; swarmers become free-swimming in pouch before emerging through birth pore; swarmer small and ciliated ... **Endogenida [p. 29]**

2' Cytokinesis of swarmer completed exogenously, after the emergence of everted bud on cell surface; swarmer often ellipsoidal, flattened ... **Evaginogenida [p. 29]**

Protozoa: Ciliophora: Suctoria: Exogenida: Families

1 Trophont sac-like or spherical, basally attached to bottom of lorica near junction with stalk ... 2

1' Trophont not sac-like, not basally attached to lorica, or without lorica ... 3

2(1) Tentacles capitate, grouped apically in single fascicle or row; swarmers ovoid with somatic kineties in U-shape around body (Fig. 2.13 B) ... Paracinetidae

2' Tentacles clavate, in a group on narrow, rounded distal end of body; swarmers flattened or vermiform, ciliated on one margin (Fig. 2.14 A) ... Thecacnetidae

3(1) Without a rod-like tentacle or protuberance of stylotheca ... 4

3' Ectoparasitic, attached to host (usually peritrich ciliates) by rod-like tentacle or protuberance of stylotheca (Fig. 2.16 L) ... Manuelophryidae

4(3) Trophont medium to large, cylindrical, conical or trumpet-shaped; ectocommensal on crustaceans (Fig. 2.18 B) .... Spelaeophryidae

4' Trophont small to medium, spheroid, pyriform or goblet-shaped 5

5(4) Lorica with several radial slits in distal half through which tentacles project; sometimes stalked; free-living or ectocommensal on invertebrates or other ciliates (Fig. 2.13 C) .... Metacinetidae

5' Typically aloricate and stalked 6

6(5) Trophont goblet-shaped or laterally flattened; attached to antennules of harpacticoid copepods (Fig. 2.13 L) .... Lecanophryidae

6' Trophont spheroid or pyriform; planktonic or sessile, often attached to other ciliates as parasites (Fig. 2.14 E, G) .... Podophryidae

Protozoa: Ciliophora: Suctoria: Endogenida: Families

1 With stalk 2

1' Without stalk 5

2(1) With lorica 3

2' Without lorica 4

3(2) Trophonts laterally flattened, trapezoid, triangular or discoid; lorica often triangular; one type of tentacle arranged in two, rarely three, fascicles or rows (Fig. 2.14 F) .... Acinetidae

3' Trophont trapezoid, laterally flattened; loricate; tentacles of two types, i.e., agile prehensile and regular feeding ones; ectosymbionts on plants and invertebrates (Fig. 2.16 B) .... Acinetopsidae

4(2) Trophont globular to ellipsoidal; tentacles funnel-like; attached to cyclopoid crustaceans (Fig. 2.13 D) .... Choanophryidae

4' Trophont ovoid, cylindrical or triangular, often flattened; tentacles capitate; free-living (Fig. 2.14 K, L) .... Tokophryidae

5(1) Without lorica or, if present, lorica mucoid 6

5' Trophont small, spheroid to ovoid; tentacles capitate; with lorica that is attached to substrate by basal surface; in periphyton or plankton (Fig. 2.13 E) .... Solenophryidae

6(5) Tentacles present 7

6' Tentacles absent; trophont ovoid to spheroid; endoparasitic in cells and tissues of hosts such as other ciliates and invertebrates (Fig. 2.13 I) .... Endosphaeridae

7(6) Attached to substratum by basal body surface or protuberance of body 8

7' Attached to substratum by tentacles or cinctum 10

8(7) Trophont ovoid, pyriform, truncate or branching 9

8' Trophont flattened; some species in mucoid lorica; tentacles capitate or rod-like; ectocommensals on invertebrates and vertebrates including the gills of fishes (Fig. 2.18 K) .... Trichophryidae

9(8) Trophont small, pyriform to ovoid; tentacles agile, very flexible; ectoparasites on crustaceans (Fig. 2.18 J) .... Rhynchetidae

9' Trophont medium to large, pyriform to truncate to branching; tentacles capitate not conspicuously flexible; some free-living, some endosymbionts some ectosymbionts on turtles or the gills of crustaceans (Fig. 2.15 A) .... Dendrosomatidae

10(7) Trophonts small to medium, ovoid to irregular; attached to host by arm-like cinctum; hypocommensals on peritrich ectosymbionts of fishes (Fig. 2.18 C) .... Erastophryidae

10' Trophonts small, globular to ellipsoid; with lorica; tentacles rod-like, one to several serving both for feeding and attachment; parasites of other ciliates (e.g., folliculinids and suctorians) (Fig. 2.16 J) .... Pseudogemmidae

Protozoa: Ciliophora: Suctoria: Evaginogenida: Families

1 Tentacles present; not parasites of rotifers 2

1' Tentacles absent; trophont small and sac-like; stalk absent; swarmer ellipsoidal, flattened, with several longitudinal kineties; parasites of tissues of rotifers (Fig. 2.16 K) .... Trypanococcidae

2(1) Tentacles ramified 3

2' Tentacles not ramified 4

3(2) Trophont vase-like, branched, lifted off substrate by basal protuberance; ectocommensals on gammarid crustaceans (Fig. 2.18 D) .... Cometodendridae

3' Trophont hemispherical or disc-shaped, unbranched, basal protuberance absent; ectocommensals on gammarid amphipods (Fig. 2.14 I, O) .... Dendrocometidae

PROTOZOA

4(2) Tentacles capitate 5

4' Tentacles not capitate 6

5(4) Trophont small, spheroid; stalked; tentacles evenly distributed over body; macronucleus globular; in periphyton (Fig. 2.13 F) .... Prodiscophryidae

5' Trophont small to medium, discoid, sometimes sac-like; sometimes stalked; tentacles in fascicles or evenly distributed over body; macronucleus ellipsoid, ribbon-like or ramified; in periphyton or ectocommensal on arthropods (Fig. 2.14 M, N)........ Discophryidae

6(4) Tentacles rod-like, neither solitary nor in fascicles 7

6' Tentacles clavate or knobbed, either solitary or in fascicles........ 8

7(6) Trophont small, ovoid to spheroid; without stalk or lorica; tentacles on one side of body only; macronucleus globular; hydrogenosomes present; in anaerobic or microaerophilic habitats (Fig. 2.16 C) .... Enchelyomorphidae

7' Trophont small to medium, ovoid, discoid or sac-like, spread over substratum; sometimes stalked; tentacles rod-like, evenly distributed or arranged in rows; macronucleus elongate ellipsoid; ectocommensal on isopod and amphipod crustaceans (Fig. 2.18 L) .... Stylocometidae

8(6) Stalk absent........ 9

8' Stalk present; trophont small, elongate, laterally flattened; tentacles agile and contractile; macronucleus ribbon-like; ectoparasites of discophryid suctorians (Fig. 2.15 B) .... Rhynchophryidae

9(8) Trophonts discoid, often flattened, attached to substratum by adhesive disc; lorica absent; tentacles knobbed, extensible, solitary or in fascicle; in periphyton or ectocommensal on invertebrates (Fig. 2.14 J)........ Heliophryidae

9' Trophont laterally flattened; with lorica or stylotheca; tentacles clavate, in fascicles; in periphyton or endocommensal on invertebrates (Fig. 2.15 C)........ Periacinetidae

Protozoa: Ciliophora: Oligohymenophorea: Mobilia: Families

1 Denticles of holdfast with hooks and spines; ectosymbionts on a wide range of vertebrate and invertebrates hosts e.g., *Hydra*, fishes, amphibians (Fig. 2.12 D) .... Trichodinidae

1' Denticles of holdfast simple, toothed; elongated macronucleus, ectosymbionts on turbellarians (Fig. 2.12 C); .... Urceolariidae

Protozoa: Ciliophora: Oligohymenophorea: Peritrichia: Families

1 Attached to substratum via a stalk, lorica or scopula; one genus (*Planeticovorticella*) unattached 2

1' Free-swimming, without stalk or lorica; swims with oral end forward; with or without aboral ring of cilia (Fig. 2.12 A, B) .... Astylozoidae

2(1) Stalked or, if stalkless, with lorica 3

2' Without stalk or lorica, attached to substratum via scopula; solitary; epibionts of invertebrates or on the gills of fishes (Fig. 2.12 E)........ Scyphidiidae

3(2) With contractile stalk, although *Ophrydium* has a non-contractile stalk (Fig. 2.12 J); solitary or colonial 4

3' If stalked, stalk non-contractile; solitary or colonial 5

4(3) Stalk contracts in a spiral fashion; in colonial forms, stalk myoneme discontinuous so each zooid or branch contracts independently (Fig. 2.12 K, Q, R) .... Vorticellidae

4' Stalk contracts in a zig-zag or other fashion, but not spirally; in colonial forms stalk myoneme continuous so colony contracts in union (Fig. 2.12 H, I) .... Zoothamniidae

5(3) With lorica 6

5' Without lorica (although lorica may be present in some species; if in doubt follow both parts of key) 9

6(5) Lorica ovoid or hemispheroid, flattened; if present, peristomial lip rigid 7

6' Lorica conical to cylindrical, generally slender; peristomial lip retractable 8

7(6) Lorica aperture closeable by lip-like folds; peristomial lip absent; symbionts of invertebrates or attached to aquatic plants (Fig. 2.15 L) .... Lagenophryidae

7' Lorica aperture not closeable; peristomial lip present; attached to isopod crustaceans (Fig. 2.13 H)........ Usconophryidae

8(6) Stalk in two parts, a contractile proximal part within the lorica, and a non-contractile distal part outside the lorica; attached to crustaceans (Fig. 2.18 I) .... Rovinjellidae

8' If present, stalk non-contractile; lorica generally slender; zooid typically extends well beyond opening of lorica; attached to plants, inanimate objects or as symphorionts (Fig. 2.12 F, G, L–N)........ Vaginicolidae

9(5) Peristomial disc raised on a short neck; peristomial lip absent; some species with a highly developed theca; stalked; commonly as epibionts of arthropods but also on inanimate objects (Fig. 2.12 S) .... Opercularidae

9' Peristomial disc not raised on a short neck; peristomial lip present; stalked; solitary or colonial; symphorionts on a wide range of hosts, or on inanimate objects (Fig. 2.12 O, P, T, U) .... Epistylididae

Protozoa: Ciliophora: Colpodea: Families

1 Body ovoid; oral region large relative to body size; argyrome 'kreyellid', i.e., highly reticulated and densely subdivided 2

1' Body shape variable; argyrome 'colpodid' or 'playophorid, i.e., not highly reticulated or densely subdivided........ 3

2(1) Body length exceeds 40–50 μm; somatic ciliation holotrichous forming a conspicuous postoral suture (Fig. 2.13 G) .... Bryometopidae

2' Body flattened, length less than 40–50 μm; somatic ciliation reduced or absent on dorsal and left sides (Fig. 2.16 G) .... Kreyellidae

3(1) Body asymmetric, often reniform or ovoid with lobes; paroral as few to many rows, often disordered 4

3' Body elongate ovoid or reniform; paroral as a single row of dikinetids 6

4(3) Oral region not conspicuously large relative to body size; aperture opening round to reniform; vestibulum funnel-shaped, extends from ventral to dorsal side 5

4' Oral apparatus very large; oral opening conspicuously key-hole-shaped, occupies anterior area and extends to mid-region of cell; body large to very large (Fig. 2.10 C)........ Bursariidae

5(4) Body typically reniform; oral apparatus in anterior half of body; division in a cyst (Fig. 2.17 N, P–S) .... Colpodidae

5' Body ovoid with large preoral and small postoral lobes; oral apparatus in posterior half of body; often enclosed in gelatinous lorica but some species only rarely build, or readily desert, the lorica (Fig. 2.17 K, L) .... Marynidae

6(3) Body broad-to-narrow ovoid; oral region equatorial to apical; paroral as a file of kineties on right, few to many polykineties on left; micronucleus in perinuclear space of macronucleus........ 7

6' Body elongate ovoid; oral opening distinctly subapical, shallow or funnel-shaped; micro-and macronuclei with separate nuclear membranes; typically found in temporary ponds (Fig. 2.15 K) .... Bryophryidae

7(6) Body small to large; oral region in anterior 1/3 of body; somatic ciliation holotrichous........ 8

7' Body small, ovoid; oral region equatorial, with distinct cytopharyngeal basket; somatic kineties only on right (= ventral) surface (Fig. 2.16 I) .... Pseudochlamydonellidae

8(7) Body broad or elongate ovoid, never in a lorica; paroral not segmented; several to many left polykinetids 9

8' Body narrow ovoid, sometimes in a gelatinous lorica; paroral in two segments, anterior of which is conspicuously ciliated, posterior inconspicuous; 3–5 left polykinetids (Fig. 2.9 K) .... Cyrtolophosididae

9(8) Body broadly ovoid; oral region subapical on right side, slanted, with distinctive adoral zone that is longer than the paroral (Fig. 2.16 M) .... Woodruffiidae

9' Body elongate ovoid; oral region near truncated anterior margin; length of adoral zone about equal to paroral (Fig. 2.18 G) .... Platyophryidae

Protozoa: Ciliophora: Oligohymenophorea: Subclasses

[Note: identification of these subclasses and the taxa within them is difficult without special preparation techniques, such as silver staining.]

1 Somatic kinetids mostly as dikinetids; postoral suture may be present 2

1' Somatic kinetids as monokinetids; with preoral suture but no postoral suture; mouth in anterior quarter of cell .... **Hymenostomatia [p. 31]**

2(1) Somatic ciliation typically dense with preoral and postoral sutures; oral polykinetids typically elongate and parallel to long axis of oral cavity; extrusomes mostly as trichocysts; scutica absent .... **Peniculia [p. 32]**

2' Somatic ciliation sometimes sparse; oral region variable in shape; extrusomes mostly as mucocysts; scutica present........ **Scuticociliatia [p. 32]**

Protozoa: Ciliophora: Hymenostomatia: Families

1 Body small to large, elongate ovoid to spherical; somatic ciliation very dense; histophagous on invertebrates or parasites of fishes; with organelle of Lieberkühn (watchglass organelle) 2

1' Body usually small to medium, sometimes large, ovoid or pyriform; primarily bacterivorous but some histophagous or parasitic; without organelle of Lieberkühn 3

2(1) Theront elongate ovoid, tomont spherical; caudal cilium present in theront; parasitic invading the gills and integuments of fishes causing white spot disease (Fig. 2.15 I) .... Ichthyophthiriidae

2' Elongate ovoid; posterior end of polykinetid 2 enlarged and beats like a brush; histophagous on invertebrates or parasitic on molluscs (Fig. 2.15 H) .... Ophryoglenidae

3(1) Body ovoid, ellipsoid, cylindrical or pyriform; right ventral kineties do not twist abruptly to run parallel to anterior suture (except in *Glaucomella*) 4

3' Body elongate ovoid, sometimes tapering posteriorly; right ventral kineties curving left, twisting abruptly anterior of the oral region, to run parallel to the anterior suture (Fig. 2.17 F, H) .... Turaniellidae

4(3) Body pyriform to cylindrical; bases of membranelles of uniform width; paroral ciliated along its entire length; mouth roughly triangular; most species free-living but some as facultative or obligate parasites in a wide range of hosts (Figs. 2.1 D and 2.17 O)........ Tetrahymenidae

4' Body ovoid to ellipsoid; oral cavity relatively large; posterior portion of paroral unciliated; with two conspicuous oral membranelle-like "lips" that seem to vibrate; free-living, sometimes carnivorous (Fig. 2.17 C) .... Glaucomidae

Protozoa: Ciliophora: Peniculia: Families

1 Oral region conspicuously large, covering most of ventral surface 2

1' Oral region typically covers <50% of ventral surface 3

2(1) Body distinctly heart- or cone-shaped with flattened ventral surface and humped dorsal surface; mouth V-shaped; left oral cilia as widely spaced polykinetids (Fig. 2.17 B) .... Stokesiidae

2' Body broadly ovoid; cilia of paroral membrane long, left oral cilia appearing as one long polykinetid (Fig. 2.17 G) .... Lembadionidae

3(1) Body ovoid or elongate ovoid; somatic ciliation holotrichous 4

3' Body short, cylindrical, with larger rounded anterior half; somatic ciliation as a distinct equatorial girdle; caudal cilia forming a conspicuous tuft used for temporary attachment (Fig. 2.17 D) .... Urocentridae

4(3) Mouth area sunken in, with oral groove leading to it, or with ciliated pharyngeal tube, or both 5

4' Mouth area subapical and elongate; prebuccal area shallow or absent; usually with many ophrokineties to right of buccal region; nematodesmata prominent to side and rear of mouth; contractile vacuoles with long collecting canals (Fig. 2.17 A, E) .... Frontoniidae

5(4) Mouth within an oral vestibule; nematodesmata form a ring or "basket" around mouth (Fig. 2.17 J) .... Clathrostomatidae

5' Medium to large (150 μm in some species); body foot-shaped or ellipsoidal, with pointed or rounded caudal end; long, broad, ciliated oral groove leads into buccal cavity; two contractile vacuoles; one species with algal symbionts (Fig. 2.17 I) .... Parameciidae

Protozoa: Ciliophora: Scuticociliatia: Families

1 With linear oral furrow or groove leading from anterior end to mouth bordered on right by paroral membrane 2

1' Without linear preoral groove bordered by membranes, but generally with membranes inside oral cavity 7

2(1) Paroral membrane along furrow *not* double 3

2' With an apparent double paroral membrane on the right side of the furrow (actually the paroral membrane plus a row of somatic cilia) (Fig. 2.9 T) .... Cohnilembidae

3(2) Bottom of preoral furrow not ciliated, paroral membrane curves around rear of mouth 4

3' Preoral furrow long, shallow, ciliated with three ciliary fields; generally a single caudal cilium; body long, ovoid (Fig. 2.9 L–N) .... Philasteridae

4(3) Does not produce a lorica 5

4' Dwells in lorica that is open at both ends; resembles *Pleuronema* or *Cyclidium* (Fig. 2.9 X) .... Calyptotrichidae

5(4) Medium to large; somatic ciliation often dense; long, stiff cilia present either caudally or interspersed between regular cilia 6

5' Very small to small (15–60 μm long); somatic ciliation sparse; single, conspicuous distinct caudal cilium (Fig. 2.9 W) .... Cyclidiidae

6(5) Paroral membrane a prominent velum extending from anterior to well past equator of cell; one to many stiff, long caudal cilia (Fig. 2.9 R) .... Pleuronematidae

6' Somatic ciliation dense with long, stiff, bristle-like cilia interspersed between regular cilia over body surface; paroral membrane less prominent than above (Fig. 2.9 S) .... Histiobalantiidae

7(1) With 1–2 girdles of cilia 8

7' Cilia not in girdles, usually in longitudinal rows 9

8(7) Oral cavity equatorial; distinct constriction in the ciliated girdle; ends bare except for caudal cilium (Fig. 2.9 P) .... Urozonidae

8' Oral cavity deep, equatorial; two distinct ciliary girdles; ciliary tuft on posterior end (Fig. 2.17 D) .... Urocentridae

9(7) Mouth at, or anterior to, middle of cell 10

9' Oral area large, toward posterior half of cell (midventral); body usually flattened; cilia denser ventrally (Fig. 2.9 V) .... Cinetochilidae

10(9) Small, <50 μm; anterior pole flat, unciliated (Fig. 2.9 U) .... Uronematidae

10' Somatic ciliation even; body long-ovoid; oral area small, closer to anterior end (Fig. 2.9 O, Q) .... Loxocephalidae

Protozoa: Ciliophora: Litostomatea: Orders

1 Oral region typically circular surrounded by circumoral dikinetids 2

1' Oral region elongated with slit-like cytostome and oral kinetids as left and right components that extend along the edge of the laterally flattened body; somatic ciliation on both sides of body, typically more dense on the right side .... **Pleurostomatida [p. 33]**

2(1) Oral region apical and domed, without nematodesmata or microtubules of rhabdos; somatic cilia bristle-like, of at least two types, arranged in girdles around the body; body globular to sub-spheroid .... Cyclotrichiida (Fig. 2.10 L, N); one family Mesodiniidae

2' Oral region with microtubules or nematodesmata of rhabdos; somatic ciliation holotrichous, but restricted to girdles in didiniids .... **Haptorida [p. 33]**

Protozoa: Ciliophora: Litostomatea: Pleurostomatida: Families

1 Right somatic kineties converge forming a spica in anterior mid-region; oral region with one left and one right perioral kineties (Fig. 2.9 F) .... Amphileptidae

1' Right somatic kineties terminate along rightmost perioral kinety, thus spica absent; oral region with one left and two right perioral kineties (Fig. 2.9 G, I) .... Litonotidae

Protozoa: Ciliophora: Litostomatea: Haptorida: Families

1 Somatic cilia not arranged in girdles around body 2

1' Somatic cilia arranged in one or two girdles that encircle the body (Fig. 2.10 V, W) .... Didiniidae

2(1) Apex fan-shaped to varying degrees; mouth a long slit, beginning at anterior end; slit may extend down side of cell, or at lasso-shaped oral bulge 3

2' Mouth at anterior end, rounded or only slightly elongated 4

3(2) Body flattened with obliquely truncate anterior end; oral region usually elongate dorsoventrally with slit-like cytostome; circumoral dikinetids as proliferated anterior fragments of somatic kineties (Fig. 2.10 I, U) .... Spathidiidae

3' Body elongate ovoid; oral region apical or subapical forming a lasso-shaped bulge surrounded by an unclosed ring of circumoral dikinetids (Fig. 2.13 J) .... Apertospathulidae

4(2) Without tentacles 5

4' Body ovoid; with tentacles extending from the cell in all directions when at rest, but retracted and hardly visible when swimming (Fig. 2.10 G) .... Actinobolinidae

5(4) Anterior tapering to a ciliated neck; neck set off by groove and a circle of longer cilia; contractile (Fig. 2.10 K) .... Lacrymariidae

5' Not as above 6

6(5) Oral region apical; body lacks a dorsal proboscis 7

6' Oral region distant from apical region and lies at base of dorsal proboscis (Fig. 2.9 D, E, H) .... Tracheliidae

7(6) Oral region apical but not forming a dome 8

7' Oral region simple dome, sometimes pointed 9

8(7) Body ovoid to flask-shaped, usually shorter than four times width; oral region flat, often at end of extensible neck; 2–4 brosse kineties (Fig. 2.10 Q–S) .... Enchelyidae

8' Body small, ovoid to elongate; without an extensible neck; brosse kineties of many rows in which clavate dikinetids alternate with typical somatic monokinetids (Fig. 2.18 M) .... Pseudoholophryidae

9(7) Long-ovoid or flask-shaped; oral region circular to elliptical; circumoral dikinetids at end of, and never exceeding the number of, somatic kineties (Fig. 2.10 P) .... Trachelophyllidae

9' Ovoid to elongate; oral region apical with oral dikinetids evenly surrounding cytostome; somatic ciliation often more dense in anterior half (Fig. 2.10 T) .... Acropisthiidae

Protozoa: Ciliophora: Prostomatea: Orders

1 Body ovoid to cylindroid; somatic ciliation holotrichous, sometimes reduced in posterior half and lacks obvious radial symmetry; brosse present; toxicysts present in oral region .... **Prorodontida [p. 34]**

1' Body cylindroid; somatic ciliation holotrichous with clear radial symmetry; brosse and toxicysts absent .... **Prostomatida [p. 34]**

Protozoa: Ciliophora: Prostomatea: Prorodontida: Families

1 Body without armor plates 2

1' Translucent $CaCO_3$ plates (armor) in cortex; typically barrel-shaped; body frequently spiny, often with prominent anterior and caudal thorns; long caudal cilium common; brosse present, but inconspicuous (Fig. 2.10 H) .... Colepidae

2(1) Posterior region of body ciliated 3

2' Posterior 1/3–1/5th of body unciliated, except for one or more long caudal cilia; other somatic ciliation evenly distributed (Fig. 2.10 M) .... Urotrichidae

3(2) Brosse inconspicuous, comprising 3 or 4 units either between perioral and circumoral ciliature, or on posterior right of oral area; body small, ovoid or pyriform 4

3' Brosse as an extension of the unclosed circumoral ciliature or as several to many kinetofragments 5

4(3) Body pyriform; somatic ciliation as a girdle encircling cell apex; brosse as 3 or 4 inconspicuous rows between perioral and circumoral ciliature (Fig. 2.18 H) .... Lagynidae

4' Body ovoid; somatic ciliation holotrichous; oral region with extensible lappets; brosse as 3 units on posterior right of oral area (Fig. 2.16 E) .... Plagiocampidae

5(3) Brosse short, extending backward from apical mouth on one side; body ellipsoid (Fig. 2.9 A, B) .... Prorodontidae

5' Brosse as several to many kinetofragments; body ovoid to cylindroid; oral region apical to subapical, surrounded by circumoral dikinetids; similar to and easily confused with *Prorodon* (Fig. 2.9 C) .... Holophryidae

Protozoa: Ciliophora: Prostomatea: Prostomatida: Families

1 Free-swimming but living in a pseudochitinous lorica; one or more caudal cilia; pantenes conspicuous (Fig. 2.10 J, O) .... Metacystidae

1' Without a lorica; pantenes present but not conspicuous; somatic kineties mostly bipolar, oral region apical, surrounded by simple dikinetids (Fig. 2.13 K) .... Apsiktratidae

Protozoa: Ciliophora: Nassophorea: Families

1 Somatic ciliature relatively sparse with few, widely separated kineties, kinetosomes without proximal and distal cartwheel; cyrtos small; somatic extrusomes as trichocysts .... order Microthoracida 2

1' Somatic ciliature usually dense with closely spaced kineties, kinetosomes with proximal and distal cartwheel; cyrtos large; frange with polykinetid extending left from postoral region, sometimes to dorsal surface .... order Nassulida (Fig. 2.12 Z) .... Nassulidae

2(1) Cytostome in rear half of body; small cyrtos hidden by ventral pellicular fold (Fig. 2.12 W) .... Microthoracidae

2' Cytostome opens laterally in anterior one-third of body, long tubular cyrtos; body nearly oval (Fig. 2.12 V) .... Leptopharyngidae

Protozoa: Ciliophora: Phyllopharyngea: Families

1 Dorsal and ventral surfaces join with a simple margin, without a 'railway-track-like' groove 2

1' Dorsal and ventral surfaces separated by a 'railway-track-like' groove; typically >3 oral kineties (Fig. 2.16 A) .... Lynchellidae

2(1) Body ovoid and flattened; oral opening as large transverse groove; oral ciliature as a single kinety encircling oral opening (Fig. 2.13 A) .... Gastronautidae

2' Body width 2/3 length, with prominent beak-like protuberance in anterior-left; oral ciliature as one preoral and two circumoral kineties (Fig. 2.17 T) .... Chilodonellidae

Protozoa: Ciliophora: Spirochotrichea: Stichotrichia: Orders

1 Ventral cirri arranged in longitudinal files, either linear or zig-zag 2

1' Ventral cirri heavy and conspicuous, typically arranged in localized groups; body outline often oval to elliptical, sometimes elongate .... **Sporadotrichida [p. 35]**

2(1) Frontoventral cirri arranged in longitudinal or zig-zag files .... **Urostylida [p. 35]**

2' Frontoventral cirri arranged in one or more linear files (not zig-zag); body outline typically elongate, sometimes very drawn out posteriorly .... **Stichotrichida [p. 35]**

Protozoa: Ciliophora: Spirotrichea: Stichotrichia: Sporadotrichida: Families

1 Small to large, elongate, sometimes tailed; two or more macronuclear nodules 2

1' Small, spheroid to sub-ovoid; somatic ciliature reduced to a few long (>10 μm), (often stiff) bristle-like cirri arranged circumferentially around the equator effecting a jumping motion; single macronucleus (Fig. 2.11 W) .... Halteriidae

2(1) Ventral cirri in anterior and posterior regions but not in mid-body; two to many macronuclear nodules (Fig. 2.13 N) .... Trachelostylidae

3' Ventral cirri large and distinctive, typically 18 in number, scattered over body including mid-body; typically two macronuclear nodules (Fig. 2.11 A, C, E) .... Oxytrichidae

Protozoa: Ciliophora: Spirotrichea: Stichotrichia: Urostylida: Families

1 Midventral complex with a single zig-zag file of paired cirri and with or without marginal rows 2

1' Body medium to large, elongate ovoid; frontoventral cirral zig-zag with 1-6 files; multiple left and right marginal cirral rows (Fig. 2.15 E) .... Pseudourostylidae

2(1) Frontal cirri in a bicorona; midventral complex comprises midventral cirral pairs only; one marginal row on each side; caudal cirri absent (Fig. 2.15 J) .... Pseudokeronopsidae

2' Many frontal cirri arranged in a bicorona, tricorona or multicorona; midventral complex often with midventral rows; often >1 marginal row on each side; caudal cirri often present (Fig. 2.11 B, D) .... Urostylidae

Protozoa: Ciliophora: Spirotrichea: Stichotrichia: Stichotrichida: Families

1 Two or more ventral cirral files; when present, marginal rows often not extending entire body length 2

1' Ventral cirri in a single file; marginal rows typically extending entire body length (Fig. 2.18 F) .... Amphisiellidae

2(1) Ventral files not ending on dorsal surface; dorsal files not winding helically around body 3

2' Ventral cirral files curved or spiraling around body, some ending on dorsal surface; dorsal files in a strip that winds helically around body (Fig. 2.15 F) .... Spirofilidae

3(2) Transverse and caudal cirri present 4

3' Transverse and caudal cirri absent; medium to large, elongate ovoid; typically >2 midventral files (Fig. 2.13 O) .... Kahliellidae

4(3) Ventral cirral files in several oblique rows, first row curved along anterior border; ventral cirri generally in several oblique rows between right and left marginal rows (Fig. 2.15 G) .... Keronidae

4' Ventral cirri long and sparse; marginal cirri strongly reduced; oval to elliptical, sometimes with spiny posterior extensions and/or zoochlorellae (Fig. 2.16 D) .... Psilotrichidae

Protozoa: Ciliophora: Spirotrichea: Hypotrichia: Families

1 Body dorsoventrally flattened, sometimes ridged; ventral ciliation as cirri arranged in localized groups 2

1' Body ellipsoidal; somatic ciliation as delicate cirri in files; oral cavity supported by a basket of nematodesmata (Fig. 2.15 D) .... Reichenowellidae

2(1) AZM reduced, located centrally and inconspicuous; no left marginal or caudal cirri; transverse cirri conspicuous (Fig. 2.12 X) .... Aspidiscidae

PROTOZOA

2' AZM well developed and conspicuous in anterior half of body; marginal and caudal cirri present (Fig. 2.12 Y) .... Euplotidae

Protozoa: Ciliophora: Plagiopylea: Families

1 Body round to discoidal, helmet-shaped or box-shaped, laterally compressed, with armor-like cuirass and often short spines 2

1' Body reniform and flattened, with oral region at indented part; oral cavity deep and transverse; somatic ciliation holotrichous and dense; common in anaerobic habitats (Fig. 2.17 M) .... Plagiopylidae

2(1) Body discoidal; somatic ciliature sparse; oral cavity in posterior half 3

2' Body box- or helmet-shaped, generally with short posterior spines; somatic ciliature relatively dense, in short rows at front and rear; oral cavity equatorial (Fig. 2.11 M) .... Epalxellidae

3(2) Body smooth in outline apart from two prominent anterior spines and one posterior spine; band of cilia on ridge overhanging buccal cavity (Fig. 2.11 J) .... Discomorphellidae

3' Body smooth in outline without spines; somatic ciliature very sparse with longer, cirrus-like cilia at posterior end (Fig. 2.11 L) .... Myelostomatidae

Protozoa: Ciliophora: Spirotrichea: Choreotrichia and Oligotrichia: Families

1 Oral ciliature forms a closed circle .... subclass Choreotrichia 2

1' Oral cilature forms an open anterior collar with an anterioventral "lapel"; typically rounded or gently pointed posteriorly; somatic kineties reduced in number, forming girdles and spirals .... subclass Oligotrichia (Fig. 2.11 V) .... Strombidiidae

2(1) Loricate, attached to inner wall of lorica 3

2' Not loricate 4

3(2) Delicate, gelatinous or mucoid tubular lorica, with attached debris, often translucent (Fig. 2.11 S) .... Tintinnididae

3' Lorica rigid, agglomerate of mineral grains and diatom fragments (Fig. 2.11 T) .... Codonellidae

4(2) Rows of somatic ciliature equally distributed around body, extending length of body; body cilia may be long (Fig. 2.11 U) .... Strombidinopsidae

4' One or more usually short rows of body cilia; body ciliature short (Fig. 2.11 X) .... Strobilidiidae

Protozoa: Ciliophora: Armophorea: Families

1 Body slightly twisted to left; long caudal spine commonly present; somatic kineties as small kineties or tufts (Fig. 2.11 R) .... Caenomorphidae

1' Body conspicuously contorted, anterior part twisted to left; somatic ciliation holotrichous (Fig. 2.11 K) .... Metopidae

Protozoa: Ciliophora: Heterotrichea: Families

1 Body not trumpet-shaped 2

1' Body trumpet-shaped; highly contractile; oral ciliature spirals clockwise around flared anterior end; often pigmented and/or with algal endosymbionts; sometimes attached to substratum (Fig. 2.10 F) .... Stentoridae

2(1) Body not laterally compressed 3

2' Body laterally compressed, pyriform to ellipsoid; often pigmented pink, red, or purple; long, narrow peristome on left margin; non-contractile (Fig. 2.10 B) .... Blepharismidae

3(2) Body ovoid, ellipsoidal or elongate; oral region expansive; oral ciliature prominent 4

3' Body often elongate, cylindrical and very contractile oral cavity shallow; peristomial area long, narrow, oral ciliature inconspicuous (Fig. 2.10 A) .... Spirostomidae

4(3) Body ovoid, often anteriorly pointed; buccal cavity occupying much of anterior part of body; peristomial field ciliated, bordered by adoral polykinetids; paroral membrane inconspicuous (Fig. 2.10 D) .... Climacostomidae

4' Body ellipsoidal to very elongate, contractile; paroral membrane prominent; peristomial field absent (Fig. 2.10 E) .... Condylostomatidae

REFERENCES

Adl, S.M., A.G.B. Simpson, C.E. Lane, J. Lukeš, D. Bass, S.S. Bowser, M.W. Brown, F. Burki, M. Dunthorn, V. Hampl, A. Heiss, M. Hoppenrath, E. Lara, L. le Gall, D.H. Lynn, H. McManus, E.A.D. Mitchell, S.E. Mozley-Stanridge, L.W. Parfrey, J. Pawlowski, S. Rueckert, L. Shadwick, C.L. Schoch, A. Smirnov & F.W. Spiegel. 2012. The revised classification of the eukaryotes. Journal of Eukaryotic Microbiology 59: 429–493.

Bass, D., A.T. Howe, A.P. Mylnikov, K. Vickerman, E.E. Chao, J.E. Smallbone, J. Snell, C. Cabral & T. Cavalier-Smith. 2009. Phylogeny and classification of Cercomonadida (Protozoa, Cercozoa): *Cercomonas, Eocercomonas, Paracercomonas*, and *Cavernomonas* gen. nov. Protist 160: 483–521.

Cavalier-Smith, T. 2010. Kingdoms Protozoa and Chromista and the eozoan root of the eukaryotic tree. Biology Letters 6: 342–354.

Cowling, A.J. 1991. Chapter 27: Free-living heterotrophic flagellates: methods of isolation and maintenance, including sources of strains in culture. Pages 477–491 *in*: D.J. Patterson and J. Larson (eds), The Biology of Free-Living Heterotrophic Flagellates. Systematics Association Special Volume No. 45, Clarendon Press, Oxford, UK.

Day, J.G., U. Achilles-Day, S. Brown & A. Warren. 2007. Chapter 7: Cultivation of algae and protozoa. Pages 79–92 *in*: C.J. Hurst et al. (eds), Manual of Environmental Microbiology. 3rd ed. ASM Press, Washington, USA.

Esteban, G.F., T. Fenchel & B.J. Finlay, 2010. Mixotrophy in ciliates. Protist 161: 621–641.

Esteban, G.E., B.J. Finlay, K.J. Clarke. 2012. Priest Pot in the English Lake District: a showcase of microbial diversity. Freshwater Biology 57: 321–330.

Finlay, B.J., A. Rogerson & A.J. Cowling. 1988. A Beginners Guide to the Collection, Isolation, Cultivation and Identification of Freshwater Protozoa. Culture Collection of Algae and Protozoa, Freshwater Biological Association, Ambleside, UK. 78 pp.

Finlay, B.J., H.I.J. Black, S. Brown, K.J. Clarke, G.F. Esteban, R.M. Hindle, J.L. Olmo, A. Rollett & K. Vickerman. 2000. Estimating the growth of the soil protozoan community. Protist 151: 69–80.

Foissner, W. 1991. Basic light and scanning electron-microscopic methods for taxonomic studies of ciliated protozoa. European Journal of Protistology 27: 313–330.

Gong, J., T. Stoeck, Z. Yi, M. Miao, Q. Zhang, D.McL. Roberts, A. Warren & W. Song. 2009. Small subunit rRNA phylogenies show that the class Nassophorea is not monophyletic (Phylum Ciliophora). Journal of Eukaryotic Microbiology 56: 339–347.

Keenan, K., E. Erlich, K.H. Donnelly, M.B. Basel, S.H. Hutner, R. Kassoff & S.A. Crawford. 1978. Particle-based axenic media for tetrahymenids. Journal of Protozoology 25: 385–387.

Kirsop, B. & A. Doyle (eds). 1991. Maintenance of Microorganisms and Cultured Cells. Academic Press Ltd., London, UK. 288 pp.

Lackey, J.B. 1959. Chapter 8: Zooflagellates. Pages 190–231 *in*: W.T. Edmondson (ed), Freshwater Biology. Wiley, New York, USA.

Lee, J.J., S.H. Hutner & E.C. Bovee (eds). 1985. An Illustrated Guide to the Protozoa. Society of Protozoologists. Lawrence, Kansas, USA. 629 pp.

Lee, J.J., G.F. Leedale & P. Bradbury (eds). 2000. An Illustrated Guide to the Protozoa. 2nd edition. Society of Protozoologists. Lawrence, Kansas, USA. 1432 pp.

Lee, J.J. & A.T. Soldo (eds). 1992. Protocols in Protozoology. Society of Protozoologists, Lawrence, Kansas, USA.

Lynn, D.H. 2008. The Ciliated Protozoa: Characterization, Classification and Guide to the Literature. 3rd edition. Springer, Dordrecht, The Netherlands. 605 pp.

Margulis, L., J.O. Corliss, M. Melkonian & D.J. Chapman (eds). 1989. Handbook of Protoctista. Jones and Bartlett, Boston, USA. 914 pp.

Montagnes, D.J.S. & D.H. Lynn. 1987a. A quantitative protargol stain (QPS) for ciliates: a method description and test of its quantitative nature. Marine Microbial Food Webs 2: 83–93.

Montagnes, D.J.S. & D.H. Lynn. 1987b. Agar embedding on cellulose filters: an improved method of mounting protists for protargol and Chatton-Lwoff staining. Transactions of the American Microscopical Society 106: 183–186.

Nerad, T.A. (ed.) 1993. ATCC Catalogue of Protists. 18th edition. ATCC, Rockville, USA. 100 pp.

Orsi, W., V. Edgcomb, J. Faria, W. Foissner, W.H. Fowle, T. Hohmann, P. Suarez, C. Taylor, G.T. Taylor, P. Vd'acny & S.S. Epstein. 2011. Class Cariacotrichea, a novel ciliate taxon from the anoxic Cariaco Basin, Venezuela. International Journal of Systematic and Evolutionary Microbiology 62: 1435–1433.

Sanders, R.W. 1991. Mixotrophic protists in marine and freshwater ecosystems. Journal of Protozoology 38: 76–80.

Schmidt, H.J. 1982. New methods for cultivating, harvesting and purifying mass cultures of the hypotrich ciliate *Euplotes aediculatus*. Journal of Protozoology 29: 132–135.

Skibbe, O. 1994. An improved quantitative protargol stain for ciliates and other planktonic protists. Archiv für Hydrobiologie 130: 339–348.

Smirnov, A.V., E. Chao, E.S. Nassonova & T. Cavalier-Smith. 2011. A revised classification of naked lobose amoebae (Amoebozoa: Lobosa). Protist 162: 545–570.

Soldo, A. & S.A. Brickson. 1980. A simple method for plating and cloning ciliates and other protozoa. Journal of Protozoology 27: 328–331.

Taylor, W.D. & M.L. Heynen. 1987. Seasonal and vertical distribution of Ciliophora in Lake Ontario. Canadian Journal of Fisheries and Aquatic Sciences 44: 2185–2191.

Thorp, J.H. & A.P. Covich (eds). 2010. Ecology and Classification of North American Freshwater Invertebrates. 3rd edition. Academic Press. Boston, USA. 1021 pp.

Tompkins, J., M.M. DeVille, J.G. Day & M.F. Turner. 1995. Culture Collection of Algae and Protozoa Catalogue of Strains. CCAP, Ambleside, UK. 203 pp.

Vickerman, K. 1976. The diversity of kinetoplastid flagellates. Pages 1–34 *in*: W.H.R. Lumsden and D.A. Evans (eds), Biology of Kinetoplastida. Academic Press, New York, USA.

Yabuki, A., E.E. Choi, K.I. Ishida & T. Cavalier-Smith. 2012. *Microheliella maris* (Microhelida ord. n.), an ultrastructurally highly distinctive new axopodial protist species and genus, and the unity of phylum Heliozoa. Protist 163: 356–388.

Chapter 3

Phylum Porifera

Renata Manconi

Dipartimento di Scienze della Natura e del Territorio (DIPNET), Università di Sassari, Sassari, Italy

Roberto Pronzato

Dipartimento di Scienze della Terra, dell'Ambiente e della Vita (DISTAV), Università di Genova, Genova, Italy

Chapter Outline

INTRODUCTION

The World Porifera Database (continuously updated on line) (Van Soest et al., 2013) reports ~8400 valid sponge species to date. Only few taxa, grouped in the order Spongillida Manconi & Pronzato, 2002, live in freshwater and brackish habitats (see Manconi & Pronzato, 2015). Although taxonomic richness values are continuously under revision, there are currently 7 families, 47 genera, and 236 species distributed among six continents (all but Antarctica). The Nearctic biogeographic region hosts 2 families, 14 genera, and 30 species (Fig. 3.1; Table 3.1; Appendix 3.1), 17 of which are endemic (>50%). The highest number of shared species with other biogeographic regions is with the Palaearctic (13 species) followed by Neotropical (10), Oriental (6), and Afrotropical and Australian (5). No species are shared with the Pacific Oceanic Island Region.

Despite the extreme rarity of some species of Spongillida, no species are currently protected by laws or international conventions in the Nearctic. No reliable data exist, due to the scarcity of studies, on the conservation status of freshwater sponges despite their key functional role in inland water ecosystems as both benthic active filter-feeders and as natural mesocosms for other taxa.

LIMITATIONS

The Nearctic freshwater sponges are well known and readily identified to species level. Fundamental studies and monographs (Potts, 1887; Penney, 1960; Penney & Racek, 1968; Smith, 2001; Reiswig et al., 2010) provide exhaustive data on the inland Porifera fauna of the region, although further investigations in inadequately explored habitat/areas should yield new taxa.

Spongillida families presently occurring in the Nearctic Region are the Metaniidae with two species and the Spongillidae with 28 species. However, middle Eocene fossil remains of the family Potamolepidae were recently reported from the Giraffe kimberlite maar in northern Canada (Pisera et al., 2013), and recent surveys in the poorly known Tennessee and southern Appalachians streams and rivers yelded the discovery of a rich and diversified sponge fauna,

Thorp and Covich's Freshwater Invertebrates. http://dx.doi.org/10.1016/B978-0-12-385028-7.00003-2

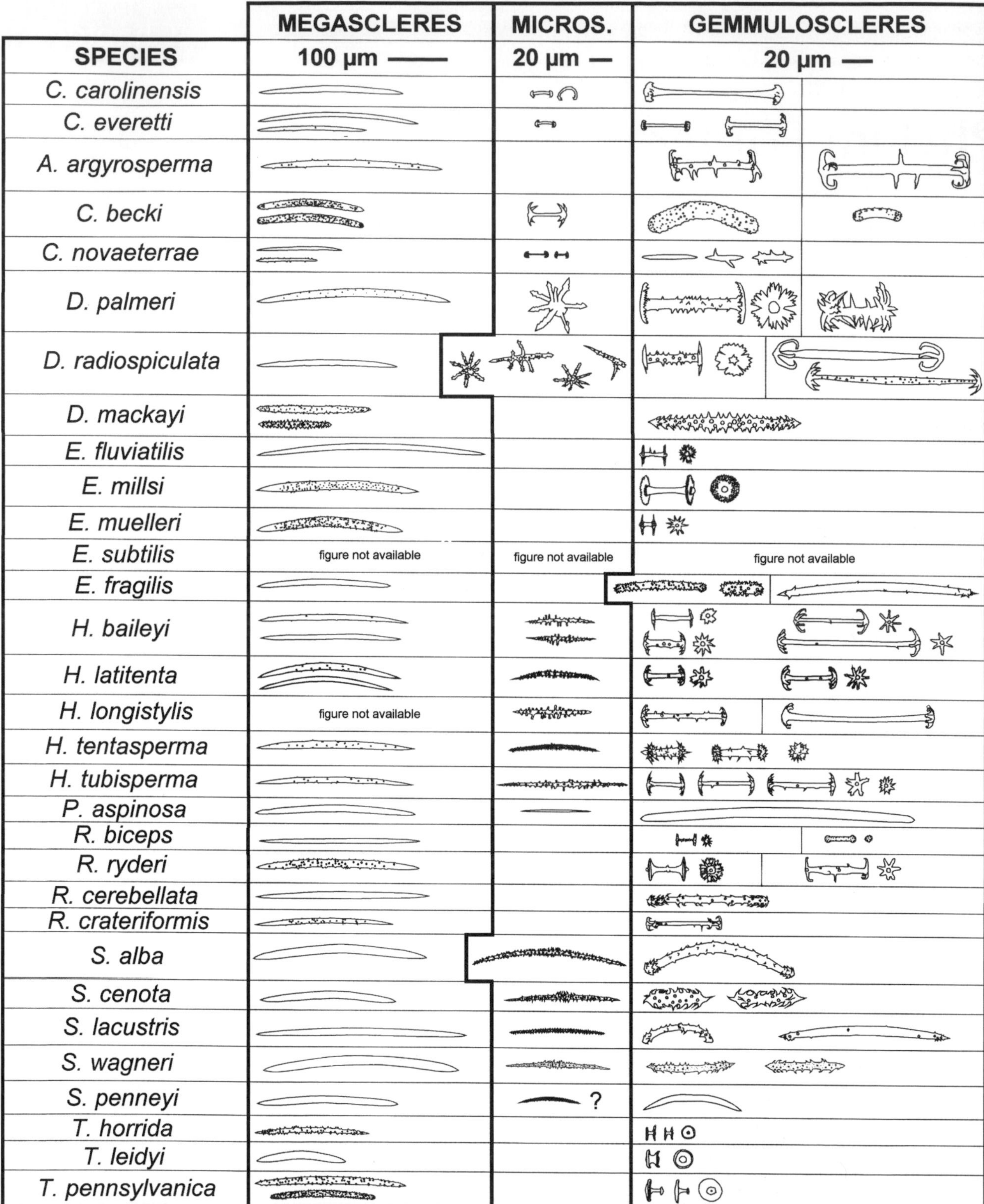

FIGURE 3.1 All Nearctic species of sponges. Synthetic representation of the spicular complements of Nearctic freshwater sponge species. Drawings were made from the most authoritative sources available. Dimension scales are most precise as possible. *Modified from Reiswig et al., 2010.*

TABLE 3.1 Checklist and Geographic Range of Nearctic Freshwater Sponge Species

SPONGILLIDA Manconi & Pronzato, 2002	
METANIIDAE Volkmer-Ribeiro, 1986	
Corvomeyenia Weltner, 1913	NA-NT
**C. carolinensis* Harrison 1971	NA
**C. everetti* (Mills, 1884)	NA
SPONGILLIDAE Gray, 1867	
Anheteromeyenia Schröder, 1927	NA-NT
**A. argyrosperma* (Potts, 1880)	NA
Corvospongilla Annandale, 1911	PA-NA-NT-AT-OL
**C. becki* Poirrier, 1978	NA
**C. novaeterrae* (Potts, 1886)	NA
Dosilia Gray, 1867	OL-AT-NA-NT
D. palmeri (Potts, 1885)	NA-NT
**D. radiospiculata* (Mills, 1888)	NA
Duosclera Reiswig & Ricciardi, 1993	NA
**D. mackayi* (Carter, 1885)	NA
Ephydatia Lamouroux, 1816	PA-NA-AT-OL-AU-NT-PAC
E. fluviatilis (Linnaeus, 1759)	PA-NA-AT-OL-AU
**E. millsi* (Potts, 1887)	NA
E. muelleri (Lieberkuhn, 1855)	PA-NA
**E. subtilis* Weltner, 1895	NA
Eunapius Gray, 1867	PA-AT-OL-AU-NA-NT
E. fragilis (Leidy, 1851)	PA-NA-AT-NT-OL-AU
Heteromeyenia Potts, 1881	NA-PA-NT-AU-PAC
H. baileyi (Bowerbank, 1863)	NA-PA-NT-PAC
**H. latitenta* (Potts, 1881)	NA
**H. longistylis* Mills, 1884	NA
**H. tentasperma* (Potts, 1880)	NA
**H. tubisperma* (Potts, 1881)	NA
Pottsiela Volkmer-Ribeiro et al., 2010	NA-PA
P. aspinosa Potts, 1880	NA-PA
Racekiela Bass & Volkmer-Ribeiro, 1998	PA-NA-NT
**R. biceps* (Lindenschmidt, 1950)	NA
R. ryderi (Potts, 1882)	PA-NA-NT
Radiospongilla (Penney & Racek, 1968)	AU-PAC-NT-PA-AT-OL-NA
R. cerebellata (Bowerbank, 1863)	AU-AT-OL-PA-NA
R. crateriformis (Potts, 1882)	NA-NT-PA-AU-OL
Spongilla Lamarck, 1816	PA-NA-OL-AT-AU-NT

Continued

PHYLUM PORIFERA

TABLE 3.1 Checklist and Geographic Range of Nearctic Freshwater Sponge Species—cont'd

S. alba Carter, 1849	OL-AT-AU-NT-PA-NA
S. cenota Penney & Racek, 1968	NT-NA
S. lacustris (Linnaeus, 1759)	PA-NA
**S. wagneri* Potts, 1889	NA
Stratospongilla Annandale, 1909	OL-AT-PA-AU-NA
**S. penneyi* (Harrison, 1979)	NA
Trochospongilla Vejdowsky, 1883	PA-NA-NT-OL-AU-AT
T. horrida (Weltner, 1893)	PA-NA
T. leidyi (Bowerbank, 1863)	NA-NT
**T. pennsylvanica* (Potts, 1882)	NA

*Nearctic endemics. OL = Oriental Region; NA = Nearctic Region; PA = Palaearctic Region; AU = Australian Region; AT = Afrotropical Region; NT = Neotropical Region; PAC = Pacific Islands Region.

including a new genus and a new species of Potamolepidae (Copeland et al., 2015).

A few Nearctic Spongillida species (e.g., *Radiospongilla cerebellata, R. crateriformis, Spongilla alba, S. lacustris,* and *Ephydatia fluviatilis*) are widespread, but the majority of taxa show a restricted geographic range. Smith (2001) reported that several species have become locally extint, mainly due to the pollution.

The following keys focus on the distribution of sponges within the Nearctic bioregion but include some overlap with the Palaearctic (Beringia, Greenland) and the Neotropical (Baja Peninsula at Tropic of Cancer, Mexican Lowlands, Southern Florida Peninsula). We also refer to species distributions reported in Bânârescu (1992) and Manconi & Pronzato (2007, 2008, 2009).

TERMINOLOGY AND MORPHOLOGY

Nearctic freshwater sponges range in size from <1 mm to large patches up to a meter in diameter. In most cases they are encrusting, but massive and ramified specimens are also known. The color in life varies from white to brown to green in different light intensities. Structural consistency varies according with the amount of spongin and siliceous spicules in the skeleton. The skeleton consists in a fibrous network of spongin and siliceous spicules (spicular complement) ranging from large (megascleres) to small (microscleres, when present). Freshwater sponges produce gemmules as resistant bodies armed by special spicules (gemmuloscleres) to survive potential stressing seasonal conditions (e.g., drying or icing). Gemmules can also play a potential role of passive dispersal bodies by means of water flow, wind, or animal carriers (e.g., birds or migrant mammals). Colonization by sponges can occur on a wide array of natural (living or mineral) and artificial substrates.

Terminology of diagnostic traits and spicules follows Manconi & Pronzato (2002, p. 924) and Pronzato & Manconi (2002). Alveolate skeleton: network of fibers with rounded meshes; reticulate skeleton: network of fibers with primary (ascending) fibers connected by transversal secondary fibers; oxeas (skeleton and gemmules): monaxial spicules with pointed tips; strongyles (skeleton and gemmules): monaxial spicules with rounded tips; gemmuloscleres (gemmular spicules): spicules of various kind; birotule: gemmulosclere with rotules at both tips; rotule: disk-shaped tips; pseudobirotule: gemmulosclere with pseudorotules at both tips; pseudorotule: tip as hooks; pseudobirotule: gemmulosclere with distal pseudorotules; tubelliform gemmulosclere: trumpet-like spicule; parmuliform: gemmulosclere with a single rotule supporting a short, acute, conical stem; botryoidal: gemmulosclere with distal cluster of concavities at the shaft tips.

MATERIAL PREPARATION AND PRESERVATION

Sponge specimens can be kept long-term either dried or preserved in ethanol. For standard methods to characterize diagnostic traits of the spicular complement and of gemmules, see Chapter 8 (Manconi & Pronzato, 2015) in the first volume of this series. Species identification is not an easy task, and light microscopy and/or scanning electron microscopy are needed.

Spongillida species shows species-specific mineral (silica) spicules: megascleres and microscleres from the sponge skeleton and gemmuloscleres from gemmules.

In most cases, the gemmuloscleres and gemmular architecture supply essential key diagnostic traits for identification.

ACKNOWLEDGMENTS

This work was supported in part by grants from the Regione Autonoma Sardegna (RAS) Progetto CRP-60215-2012 LR7-2007 'Promozione della ricerca scientifica e dell'innovazione tecnologica in Sardegna', EU-7FP BAMMBO n. 265896, Italian MIUR-PRIN, Fondazione Banco di Sardegna, University of Sassari and University of Genoa.

KEYS TO SPONGILLIDA OF THE NEARCTIC REGION

See Fig. 3.1 for illustrations of megascleres, microscleres, and gemmuloscleres for all taxa keyed below. Additional figures are noted in the individual species accounts in Appendix 3.1.

We continue to use the generic name *Trochospongilla* despite Pinheiro & Nicacio (2012) proposed to resurrect the generic name *Tubella* to replace *Trochospongilla*. This change is incorrect because of in contrast with the Art. 23.2 (very old custom) and Art. 23.9.1 (use of the name by several authors) of the ICZN (International Code of Zoological Nomenclature).

Spongillida: Families

1 Gemmuloscleres as only pseudobirotules to tubelliform, parmuliform almost exclusively radially arranged; skeletal network alveolate-reticulate with scanty spongin; megascleres (oxeas/strongyles) smooth to variably ornamented; microscleres usually present .... Metaniidae, one genus: **Corvomeyenia**

1' Gemmuloscleres as oxeas, strongyles, birotules, pseudobirotules, botryoidal, club-like radially to tangentially arranged; skeletal network pauci-spicular, irregularly reticulate; megascleres as oxeas and strongyles smooth to variably ornamented; microscleres usually present .... **Spongillidae [p. 43]**

Spongillida: Metaniidae: *Corvomeyenia:* Species

1 Microscleres mostly straight pseudobirotules .... *Corvomeyenia everetti* (Mills, 1884)

[Canada: Nova Scotia. USA: Delaware, Massachusetts, Wisconsin]

1' Microscleres, mostly bent pseudobirotules .... *Corvomeyenia carolinensis* Harrison, 1971

[USA: South Carolina, Connecticut]

Spongillida: Spongillidae: Genera

1 Birotules and pseudobirotules absent 2

1' Birotules and pseudobirotules present 7

2(1) Microscleres and/or gemmuloscleres present (not pseudobirotules and/or birotules) 3

2' Microscleres and gemmuloscleres absent; gemmular theca armed by gemmuloscleres very similar to short megascleres; microscleres absent; megascleres as spiny oxeas of two size classes .... *Duosclera mackayi* (Carter, 1885)

[Canada: eastern provinces. USA: eastern states]

3(2) Megascleres as smooth oxeas 4

3' Megascleres as spiny oxeas; microscleres apparently absent; gemmuloscleres as spiny strongyles radially arranged in the gemmular theca .... **Radiospongilla [p. 44]**

4(3) Microscleres present 5

4' Microscleres absent; gemmuloscleres as oxeas/strongyles tangentially arranged at the surface to embedded in the gemmular theca .... *Eunapius fragilis* (Leidy, 1851)

[Nearctic: east of continental divide. Cosmopolitan]

5(4) Microscleres as spiny oxeas to absent 6

5' Microscleres and gemmuloscleres smooth to very sparsely microspiny .... *Pottsiela aspinosa* (Potts, 1880)

[E-Canada, USA, China]

6(5) Microscleres as spiny oxeas to absent; gemmuloscleres as oxeas abruptly pointed, irregularly tangential to the outer layer of gemmular theca .... *Stratospongilla penneyi* (Harrison, 1979)

[USA: Florida]

6' Microscleres as slender spiny oxeas; gemmuloscleres, when present, spiny oxeas/strongyles often very curved tangentially to irregularly arranged .... **Spongilla [p. 44]**

7(1) Birotules with rotules bearing spiny to indented edges, usually associated with pseudobirotules 8

7' Birotules with rotules bearing smooth edges, pseudobirotules absent; microscleres absent; megascleres as densely spiny oxeas; gemmuloscleres as smooth short birotules radially arranged in the gemmular theca .... **Trochospongilla [p. 44]**

8(7) Birotules and/or pseudobirotules present, exclusively as gemmuloscleres 9

8' Birotules and/or pseudobirotules present, exclusively as microscleres; micropseudobirotules as microscleres; gemmules free or sessile; gemmuloscleres as spiny oxeas and/or strongyles tangential to the gemmular theca; megascleres smooth/microtubercled .... **Corvospongilla [p. 45]**

9(8) Birotules and/or pseudobirotules gemmuloscleres belonging to two morphs or two size classes (birotules/pseudobirotules, pseudobirotules/pseudobirotules) 10

9' Birotules and/or pseudobirotules gemmuloscleres of a single morph 11

10(9) Gemmuloscleres birotules and/or exclusively pseudobirotules radially arranged; microscleres usually acanthoxeas; megascleres acanthoxeas .... **Heteromeyenia [p. 45]**

10' Gemmuloscleres birotules (short shaft) and pseudobirotules or modified birotules (long shaft) usually radially arranged; microscleres absent; megascleres oxeas .... **Racekiela [p. 45]**

11(9) Gemmuloscleres birotules 12

11' Gemmuloscleres pseudobirotules; microscleres absent; megascleres spiny oxeas .... *Anheteromeyenia argyrosperma* (Potts, 1880)

[Canada: Nova Scotia; USA: New England and Great Lakes regions]

12 (11) Birotules with short shaft (usually less than the rotule diameter); gemmuloscleres birotules with short shaft and rotules of identical diameter; microscleres absent; megascleres microspiny and/or smooth oxeas .... **Ephydatia [p. 45]**

12' Birotules with long spiny shaft (2–3 times the rotule diameter); gemmuloscleres birotules and/or pseudobirotules with spiny shaft; microscleres from spiny oxeas to typical aster-like with spiny rays; megascleres smooth to spiny oxeas .... **Dosilia [p. 45]**

Spongillida: Spongillidae: *Radiospongilla:* Species

1 Megascleres with sparse small spines .... *Radiospongilla crateriformis* (Potts, 1882)

[Eastern North America from Canada to Mexico. Neotropical, Palaearctic, Oriental and Australian Regions]

1' Megascleres entirely smooth .... *Radiospongilla cerebellata* (Bowerbank, 1863)

[USA: Texas. Oriental and Afrotropical Regions]

Spongillida: Spongillidae: *Spongilla:* Species

1 Microscleres as oxeas with conspicuously denser and longer stout spines in the central region 2

1' Microscleres entirely spiny 3

2(1) Gemmuloscleres as spiny oxeas (65–86 × 9–13 μm) extremely stout and stubby .... *Spongilla cenota* (Penney & Racek, 1968)

[USA: Florida. Mexico: Yucatan. Neotropics]

2' Gemmuloscleres slightly bent oxeas (48–75 × 6–8 μm) with spines more dense towards the tips .... *Spongilla wagneri* (Potts, 1889)

[USA]

3(1) Gemmulosclere slightly/strongly bent oxeas/strongyles (18–130 × 1–10 μm), usually with strong, bent spines more dense at the tips .... *Spongilla lacustris* (Linnaeus, 1759)

[Cosmopolitan]

3' Gemmuloscleres strongyles (78–130 × 5–10 μm) with spines more dense at the tips forming distinct club-like heads, but not pseudorotules .... *Spongilla alba* (Carter, 1849)

[Southeastern coastal regions. Cosmopolitan]

Spongillida: Spongillidae: *Trochospongilla:* Species

1 Megascleres spinose 2

1' Megascleres smooth or with very fine spines .... *Trochospongilla leidyi* (Bowerbank, 1863)

[USA: eastern states to Texas]

2(1) Gemmuloscleres with rotules diameters of two size classes .... *Trochospongilla pennsylvanica* (Potts, 1882)

[Canada: Newfoundland, Nova Scotia. USA: eastern states.]

2' Gemmuloscleres with rotules of nearly equal diameters .... *Trochospongilla horrida* (Weltner, 1893)
[Holarctic]

Spongillida: Spongillidae: *Corvospongilla:* Species

1 Megascleres as stout spiny oxeas, gemmuloscleres as short spiny strongyles .... *Corvospongilla becki* (Poirrier, 1978)
[USA: Louisiana]

1' Megascleres as oxeas, gemmuloscleres as short oxeas with a few large spines .... *Corvospongilla novaeterrae* (Potts, 1886)
[Canada: Newfoundland, Nova Scotia. USA: Connecticut]

Spongillida: Spongillidae: *Heteromeyenia:* Species

1 Gemmular foramen tubular with typical long terminal cirrus filaments 2

1' Gemmular foramen without terminal cirrus filaments .... *Heteromeyenia baileyi* (Bowerbank, 1863) (including *H. longistylis*)
[Nearctic: east of continental divide. Palaearctic and Neotropical Regions]

2(1) Gemmular foramen with 3/7 cirrus filaments 3

2' Gemmular foramen with flat disk bearing 1/2 very long cirrus filaments .... *Heteromeyenia latitenta* (Potts, 1881a)
[USA: Indiana, Illinois, New York, Ohio, Pennsylvania, Wisconsin. Mexico: northern states]

3(2) Gemmular foramen very long tube, 0.5–0.9 times the gemmular diameter .... *Heteromeyenia tubisperma* (Potts, 1881b)
[USA: Eastern states and California (?)]

3' Gemmular foramen short tube <0.4 times of gemmular diameter, very long cirrus filaments .... *Heteromeyenia tentasperma* (Potts, 1880)
[USA: New England to Ohio]

Spongillida: Spongillidae: *Racekiela:* Species

1 Megascleres as spiny oxeas, microscleres absent, gemmuloscleres as birotules/pseudobirotules .... *Racekiela ryderi* (Potts, 1882)
[Canada: Nova Scotia. USA: eastern states to Texas. Neotropical Region: Belize]

1' Megascleres as smooth oxeas, microscleres absent, gemmuloscleres as birotules and biceps .... *Racekiela biceps* (Lindenschmidt, 1950)
[USA: Michigan]

Spongillida: Spongillidae: *Ephydatia:* Species

1 Rotules of gemmuloscleres with indented margins 2

1' Rotules of gemmuloscleres with typical lacinulate margins .... *Ephydatia millsi* (Potts, 1887)
[USA: Florida]

2(1) Mean gemmulosclere length (>20 μm) with rotules with more than 20 teeth not deeply incised 3

2' Mean gemmulosclere length less than 20 μm with rotules bearing fewer than 12 teeth deeply incised into long rays, gemmulosclere length less than or equal to rotule diameter .... *Ephydatia muelleri* (Lieberkühn, 1856)
[Cosmopolitan]

3(2) Mean diameter of rotules <11 μm .... *Ephydatia subtilis* Weltner, 1895
[USA: Florida]

3' Mean diameter of rotules >11 μm with sometimes large spines .... *Ephydatia fluviatilis* (Linnaeus, 1759)
[Holarctic and Afrotropical Regions]

Spongillida: Spongillidae: *Dosilia:* Species

1 Gemmuloscleres as birotules in two categories nearly equal in length (55–85 μm), differing in the shaft spinosity .... *Dosilia palmeri* (Potts, 1885b)
[USA: Arizona, Florida, Louisiana, New Mexico, Texas. Mexico: widespread]

1' Gemmuloscleres as short birotules and long pseudobirotules .... *Dosilia radiospiculata* (Mills, 1888)
[Canada: southern regions. USA: central and western states. Mexico: widespread]

REFERENCES

Annandale, N. 1909. Beiträge zur Kenntnis der Fauna von Süd-Afrika. Ergebnisse einer Reise von Prof. Max Weber im Jahre 1894. IX. Freshwater sponges. Zoologische Jahrbücher. Abteilung für Systematik, Geographie und Biologie der Tiere 27(6): 559–568.

Annandale, N. 1911. Freshwater sponges, hydroids and Polyzoa. Porifera. Pages 27–126, 241–245 *in*: Shipley AE (ed.), Fauna of British India, Including Ceylon and Burma. Taylor & Francis, London.

Annesley, J., J. Jass & D. Watermolen. 2008. Wisconsin freshwater sponge species documented by scanning electron microscopy. Journal of Freshwater Ecology 23(2): 263–272.

Annesley, J., J. Jass & R. Henderson. 2011. Definition of species-specific traits for freshwater sponges (Porifera: Spongillidae) in the genus *Dosilia*. Proceedings of the Biological Society of Washington 124(2): 53–61.

Barnes, D.K. & T.E. Lauer. 2003. Distribution of freshwater sponges and bryozoans in Northwest Indiana. Proceedings of the Indiana Academy of Science 112(1): 29–35.

Banarescu, P. 1992. Zoogeography of freshwaters. 2. Distribution and dispersal of freshwater animals in North America and Eurasia. Wiesbaden, Aula, pp. 513–1091.

Bowerbank, J.S. 1863. A Monograph of the Spongillidae. Proceedings of the Zoological Society of London 1863: 440–472, pl. XXXVIII.

Candido, J.L., C. Volkmer-Ribeiro, F.L.S. Filho, B.J. Turcq, T. Desjardins & A. Chauvel. 2010. Microsclere variations of *Dosilia pydanieli* (Porifera, Spongillidae) in Caracaranã Lake (Roraima-Brazil). Palaeoenvironmental implication. Biociencias 8(2): 77–92.

Carter, H.J. 1849. A descriptive Account of the freshwater sponges (genus *Spongilla*) in the Island of Bombay, with observations on their structure and development. Annals and Magazine of Natural History, Ser. 2, 4(20): 81–100, pls III-V.

Carter, H.J. 1881. History and classification of the known species of *Spongilla*. The Annals and Magazine of Natural History Series 5, 7(38): 77–107.

Carter, H.J. 1885. Note on *Spongilla fragilis*, Leidy, and a new Species of *Spongilla* from Nova Scotia. Annals and Magazine of Natural History, Ser. 5, 15(85): 18–20.

Copeland, J., R. Pronzato & R. Manconi. 2015. Discovery of living Potamolepidae (Porifera: Spongillina) from Nearctic freshwater with description of a new genus. Zootaxa 3957(1): 37–48.

De Santo, E.M. & P.E. Fell. 1996. Distribution and ecology of freshwater sponges in Connecticut. Hydrobiologia 341: 81–89.

Eshleman, S.K. 1949 (1950). A key to Florida's freshwater sponges with descriptive notes. Quarterly Journal of the Florida Academy of Science 12(1): 35–44.

Ezcurra de Drago, I.D. 1975. El genero *Ephydatia* Lamouroux (Porifera, Spongillidae) sistematica y distribución. Physis (B) 34(89): 157–174.

Ezcurra de Drago, I.D. 1976. *Ephydatia mülleri* (Lieberkühn) in Africa, and the systematic position of *Ephydatia japonica* (Hilgendorf) (Porifera, Spongillidae). Arnoldia Rhodesia 35(7): 1–7.

Gray, J.E. 1867. Notes on the Arrangement of Sponges, with the Descriptions of some New Genera. Proceedings of the Zoological Society of London 1867(2): 492–558, pls XXVII-XXVIII.

Harrison, F.W. 1971. A taxonomical investigation of the genus *Corvomeyenia* Weltner (Spongillidae) with an introduction of *Corvomeyenia carolinensis* sp. nov. Hydrobiologia 38(1): 123–140.

Harrison, F.W. 1974. Porifera. Pages 29–66 *in*: C.W. Hart and S.L.H. Fuller (eds), Pollution Ecology of Fresh-water Invertebrates, Academic Press, New York.

Harrison, F.W. 1979. The taxonomic and ecological status of the environmentally restricted spongillid species of North America. V. *Ephydatia subtilis* (Weltner) and *Stratospongilla penneyi* sp. nov. Hydrobiologia 65: 99–105.

Hilgendorf, F. 1882. *Spongilla fluviatilis* Lieberkühn var. *Japonica* vor. Sitzungsberichte der Gesellschaft naturforschender Freunde zu Berlin 1882(2): 26.

Jewell, M.E. 1935. An ecological study of the fresh-water sponges of northern Wisconsin. Ecological Monograph 5: 461–504.

Jewell, M.E. 1939. An ecological study of the freshwater sponges of Wisconsin. II. The Influence of Calcium. Ecology 20: 11–28.

Leidy, J. 1851. On *Spongilla*. Proceedings of the Academy of Natural Sciences of Philadelphia 5: 278.

Lieberkühn, N. 1856. Zusätze zur Entwicklungsgeschichte der Spongillen. Müller Archiv 1856: 496–514.

Lindenschmidt, M.J. 1950. A new species of freshwater sponge. Transactions of the American Microscopical Society 69(2): 214–216.

Linnaeus, C. 1759. Systema Naturae per regna tria naturæ, secundum classes, ordines, genera, species, cum characteribus, differentiis, synonymis, locis. Tomus II. Editio decima, reformata. Holmiæ, Laurentii Salvii: 825–1348.

MacKay, A.H. 1889. Freshwater sponges of Canada and Newfoundland. Proceedings and Transactions of the Royal Society of Canada 7: 85–95.

Manconi, R. 2008. The genus *Ephydatia* (Spongillina: Spongillidae) in Africa: a case of Mediterranean *vs.* southern Africa disjunct distribution. Biogeographia 29: 19–28.

Manconi, R. & R. Pronzato. 2000. Rediscovery of the type material of *Spongilla lacustris* (L., 1759) from the Linnean Herbarium. Italian Journal of Zoology 67(1): 89–92.

Manconi, R. & R. Pronzato. 2002. Suborder Spongillina subord. nov.: Freshwater sponges. Pages 921–1020 *in* J.N.A. Hooper and R.W.M. Van Soest (eds), Systema Porifera. A guide to the classification of sponges. Vol. 1, Kluwer Academic/Plenum Publishers, New York.

Manconi, R. & R. Pronzato. 2004. The genus *Corvospongilla* Annandale (Haplosclerida, Spongillina, Spongillidae) with description of a new species from eastern Mesopotamia, Iraq. Archiv für Hydrobiologie, Suppl., Monographic Studies 151: 161–189.

Manconi, R. & R. Pronzato. 2007. Gemmules as a key structure for the adaptive radiation of freshwater sponges: A morpho-functional and biogeographical study. Pages 61–77 *in*: M.R. Custódio, G. Lôbo-Hajdu, E. Hajdu and M. Muricy (eds), Porifera Research: Biodiversity, Innovation, Sustainability, Serie Livros 28, Museu Nacional, Rio de Janeiro.

Manconi, R. & R. Pronzato. 2008. Global diversity of sponges (Porifera: Spongillina) in freshwater. Hydrobiologia 595: 27–33.

Manconi, R. & R. Pronzato. 2009. Atlas of African freshwater sponges. Studies in Afrotropical Zoology, Royal Museum for Central Africa, Tervuren, vol. 295, 214 pp.

Manconi, R. & R. Pronzato. 2011. Suborder Spongillina (freshwater sponges). Pages 341–366 *in*: M. Pansini, R. Manconi, R. Pronzato (eds), Porifera I. Calcarea, Demospongiae (*partim*), Hexactinellida, Homoscleromorpha. Fauna d'Italia, vol. XLVI, Calderini-Il Sole24Ore, Bologna.

Manconi, R. & R. Pronzato. 2015 Phylum Porifera. Pages 133–157 in: J. Thorp & D.C. Rogers (eds), Ecology and General Biology: Thorp and Covich's Freshwater Invertebrates. 4th Edition, Vol. 1. Academic Press London. http://dx.doi.org/10.1016/b978-0-12-385026-3.00008-5.

Mills, H. 1884. Thoughts on the Spongidae, with reference to the American sponges of the freshwater group with some accounts of them in detail. Proceedings of the American Society of Microscopists 1884: 131–147.

Mills, H. 1888. A new freshwater sponge. *Heteromeyenia radiospiculata* n. sp. Annals and Magazine of Natural History, Ser. 6, 1(4): 313–314.

Neidhoefer, J.R. 1940. The fresh-water sponges of Wisconsin. Transactions of the Wisconsin Academy of Sciences Arts and Letters 32: 177–197.

Old, M. 1931. Taxonomy and distribution of the fresh-water sponges (Spongillidae) of Michigan. Michigan Academy of Science, Arts and Letters 15: 439–477.

Old, M.C. 1936. Yucatan freshwater sponges. Pages 29–32 *in*: A.S. Pearse, E.P. Creaser and F.G. Hall (eds) The cenotes of Yucatan: a zoological and hydrographic survey. Carnegie Institution Publication 457.

Pennak, R.W. 1989. Fresh-water Invertebrates of the United States. 3rd ed. John Wiley, New York.

Penney, J.T. 1960. Distribution and bibliography (1892-1957) of the freshwater sponges. University of South Carolina Publication, Ser. 3, 3: 1–97.

Penney, J.T. & A.A. Racek. 1968. Comprehensive revision of a worldwide collection of freshwater sponges (Porifera: Spongillidae). Bulletin of the United States National Museum 272: 1–184.

Pinheiro, U. & G. Nicacio. 2012. Resurrection and redefinition of the genus Tubella (Porifera: Spongillidae) with a worldwide list of valid species. Zootaxa 3269: 65–68.

Pisera A., P.A. Siver & A.P. Wolfe 2013. A first account of freshwater sponges (Demospongiae, Spongillina, Potamolepidae) from the Middle Eocene: Biogeographic and Paleoclimatic implications. Journal of Paleontology 87(3): 373–378.

Poirrier, M.A. 1972. Additional records of Texas fresh-water sponges (Spongillidae) with the first record of *Radiospongilla cerebellata* (Bowerbank, 1863) from the Western Hemisphere. Southwestern Naturalist 16: 434–435.

Poirrier, M.A. 1978 *Corvospongilla becki* n. sp., a new freshwater sponge from Louisiana. Transactions of the American Microscopical Society 97(2): 240–243.

Potts, E. 1880. On freshwater sponges. Proceedings of the Academy of Natural Sciences of Philadelphia 1880: 356–357.

Potts, E. 1881a. Some new genera of freshwater sponges. Proceedings of the Academy of Natural Sciences of Philadelphia 1881: 149–150.

Potts, E. 1881b. A new form of fresh-water sponge. Proceedings of the Academy of Natural Sciences of Philadelphia 1881(2): 176.

Potts, E. 1882. Three more fresh-water sponges. Proceedings of the Academy of Natural Sciences of Philadelphia 1882(1): 12–14.

Potts, E. 1884. Some modifications observed in the form of sponge spicules. Proceedings of the Academy of Natural Sciences of Philadelphia: 184–185.

Potts, E. 1885a. A new freshwater sponge from Nova-Scotia. Proceedings of the Academy of Natural Sciences of Philadelphia 1885: 28–29.

Potts, E. 1885b. Freshwater sponges from Mexico. Proceedings of the United States National Museum 8: 587–589.

Potts, E. 1886. Freshwater sponges from Newfoundland: a new Species. Proceedings of the Academy of Natural Sciences of Philadelphia 1886: 227–230.

Potts, E. 1888 (1887). Contributions towards a synopsis of the American forms of fresh-water Sponges with descriptions of those named by other authors and from all parts of the world. Proceedings of the Academy of Natural Sciences of Philadelphia 39(1887): 158–279, pls V-XII.

Potts, E. 1889 Report upon some fresh-water sponges collected in Florida by Jos. Wilcox, Esq. Transactions of Wagner Free Institute of Sciences Philadelphia 2: 5–7.

Pronzato, R. & R. Manconi. 2002. Atlas of European fresh-water sponges. Annali del Museo Civico di Storia naturale di Ferrara 4: 3–64.

Reiswig, H.M. & A. Ricciardi. 1993a. Re-examination of *Corvospongilla novaeterrae* (Potts) (Porifera, Spongillidae), an environmentally restricted freshwater sponge from eastern Canada. Canadian Journal of Zoology 71: 1954–1962.

Reiswig, H.M. & A. Ricciardi. 1993b. Resolution of the taxonomic status of the freshwater sponges *Eunapius mackayi*, *E. igloviformis*, and *Spongilla johanseni* (Porifera: Spongillidae). Transactions of the American Microscopical Society 112(4): 262–279.

Reiswig, H.M., T.M. Frost & A. Ricciardi. 2010. Porifera. Pages 91–123 *in*: J.H. Thorp and A.P. Covich (eds), Ecology and Classification of North American Freshwater Invertebrates, vol. 4, Elsevier Inc.

Ricciardi, A. & H.M. Reiswig 1993. Freshwater sponges (Porifera, Spongillidae) of eastern Canada: taxonomy, distribution, and ecology. Canadian Journal of Zoology 71: 665–682.

Rioja, E. 1940. Estudios hidrobiológicos. I. Estudio critico sobre las esponjas del lago de Xochimilco. Anales del Instituto de Biologia 11(1): 173–189.

Saller, U. 1990a. A redescription of the freshwater sponge *Trochospongilla horrida* (Porifera, Spongillidae). Abhandlungen und Verhandlungen des Naturwissenschaftlichen Vereins in Hamburg (NF) 31/32: 163–174.

Saller, U. 1990b. The formation of the gemmule shells in the freshwater sponge *Trochospongilla horrida* (Porifera, Spongillidae). Zoologische Jahrbücher Abteilung für Anatomie und Ontologie der Tiere 120: 239–250.

Schröder, K. 1927. Über die Gattungen *Carterius* Petr, *Astromeyenia* Annandale und *Heteromeyenia* Potts (Porifera: Spongillidae). Spongilliden-Studien III. Zoologischer Anzeiger 73(5–8): 101–112.

Smith, D.G., 2001. Pennak's freshwater invertebrates of the United States: Porifera to Crustacea. 2-Porifera, Wiley & Sons, New York, 41–58.

Smith, F. 1918. A new species of *Spongilla* from Oneida Lake, New York. Technical Publications New York State University Coll. For. Cornell University 9: 239–243.

Smith, F. 1921. Distribution of the freshwater sponges of North America. Illinois Department Natural History Survey 14: 10–22.

Van Soest, R.W.M., N. Boury-Esnault, J.N.A. Hooper, K. Rützler, N.J. de Voogd, B. Alvarez de Glasby, E. Hajdu, A.B. Pisera, R. Manconi, C. Schoenberg, D. Janussen, K.R. Tabachnick, M. Klautau, B. Picton, M. Kelly & J. Vacelet. 2013. World Porifera Database (WPD). Available online at http://www.marinespecies.org/porifera.

Volkmer-Ribeiro, C., V. De Souza Machado, K. Fürstenau-Oliveira & F.V. Soares. 2010. New genus of freshwater sponges with a new species from Amazonian waters (Porifera, Demospongiae). Revista de Ciências Ambientais 4(1): 47–64.

Weltner, W. 1893. Über die Autorenbezeichnung von *Spongilla erinaceus*. Berichte der Gesellschaft für Naturforschender Freunde Berlin 1893: 7–13.

Weltner, W. 1895. Spongillidenstudien III. Katalog und Verbreitung der bekannten Süsswasserschwämme. Archiv für Naturgeschichte 61(1): 114–144.

Weltner, W. 1913. Süsswasserschwämme (Spongillidae) der Deutschen Zentralafrika - Expedition 1907-1908. Wissenschaftliche Ergebnisse der Deutschen Antarktischen Expedition 1938–39, Hamburg 12: 475–485.

APPENDIX 3.1 - TAXONOMIC ACCOUNTS OF NEARCTIC PORIFERA

Family Metaniidae Volkmer-Ribeiro, 1986

Corvomeyenia Weltner, 1913

Corvomeyenia everetti (Mills, 1884)

Meyenia everetti Mills, 1884: 146

Ephydatia everetti Jewell, 1935, 1939; Neidhoefer, 1940

Corvomeyenia everetti Penney & Racek, 1968; Ricciardi & Reiswig, 1993; Manconi & Pronzato, 2002, 2007; Reiswig et al., 2010

Figure 3.2

Description. Growth form encrusting to with slender finger-like projections, small, very thin, delicate. **Consistency** very soft, fragile. **Color** emerald green *in vivo*. **Surface** conulose, with hispidation due to the apical spicules at the tips of ascending fibers. **Oscules** not conspicuous. **Inhalant apertures** scattered. **Ectosomal skeleton** armed by microscleres (micropseudobirotules) in the well-developed dermal membrane. **Choanosomal skeleton** irregular, reticulate network of vague ascending spicular primary fibers joined by secondaries, with scattered microscleres. **Spongin** scanty, except for the gemmular theca. **Basal spongin plate** not recorded. **Megascleres** oxeas (143–(218)–285 × 6–9 µm), fusiform, slender, slightly bent, entirely smooth. Rare oxeas sparsely spiny sometime present. **Microscleres** pseudobirotules (14–(18)–26 × 2 µm) with slender, smooth, slightly bent (33%) to straight shaft, with distal umbonate pseudorotules (3–(5)–7 µm in diameter) of 3–8 hooks distinctly recurve. **Gemmules** axial in finger-like projections, large, subspherical (480–530 µm in diameter Penney & Racek, 1968; 710–902 µm in diameter Ricciardi & Reiswig, 1993) armed by a cage of skeletal spicules and dense, radial gemmuloscleres not embedded in the pneumatic layer. **Foramen** with collar, short. **Gemmular theca** trilayered well developed. **Outer layer** of compact spongin with few protruding longer gemmuloscleres. **Pneumatic layer** of small, regular spongin chambers, well-developed and supported by distal tips of gemmuloscleres not embedded in it. **Inner layer** of compact spongin with sublayers, not adhering to the pneumatic layer. **Gemmuloscleres** pseudobirotules notably variable in length (33–(59)–78 × 3–5 µm) with slender, straight to slightly curved, smooth shafts, with distal pseudorotules (10–(20)–26 µm in diameter) bearing 5–8 recurve hooks.

Habitat. Lentic (pond, spring) in clear water from near sea level (Pictou County) to ca. 700 m ASL in the type locality. On submerged aquatic plants and weeds in shallow water. Associated with zoochlorellae.

Geographic distribution. Endemic to the Nearctic Region. Known until now from the eastern half of Canada and eastern United States (Delaware, Massachusetts, Nova Scotia, and Wisconsin).

Remarks. This species is characterized by a peculiar architecture of the gemmular theca as highlighted by SEM cross sections (radial gemmuloscleres not embedded in the pneumatic layer) diverging from all the other freshwater sponges. See remarks on *C. carolinensis*. The etymology of the specific epithet refers to the type locality name the Gilder Pond on Mount Everett.

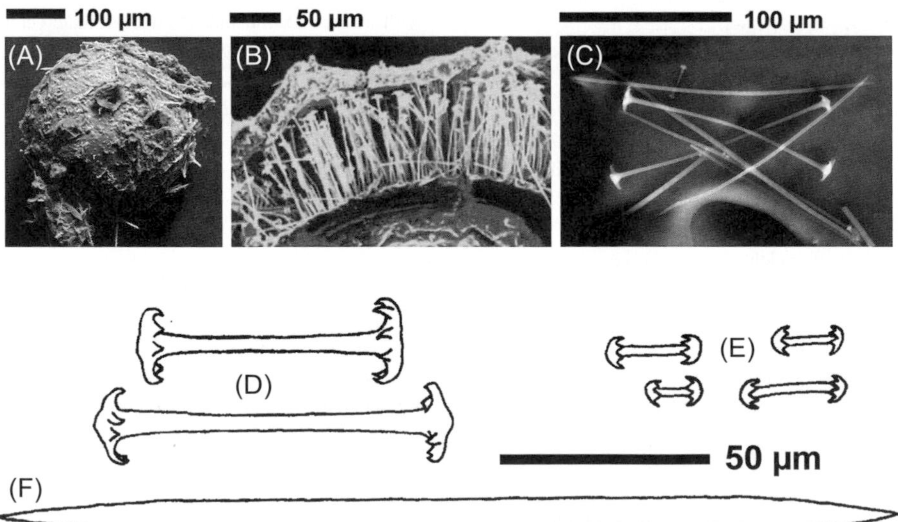

FIGURE 3.2 *Corvomeyenia everetti*: (A) entire gemmule with evident foramen (center, top) and irregular surface of outer layer; (B) gemmular theca trilayered (from top to bottom): thin outer layer, pneumatic layer, radial gemmuloscleres and inner layer (cross section) with radially arranged gemmuloscleres not embedded in pneumatic layer; (C) spicular complement (megascleres, microscleres, gemmuloscleres); (D) gemmuloscleres (pseudobirotules); (E) microscleres (birotules); (F) megascleres (oxea). *Figures A and C modified from Annesley et al., 2008; B modified from Volkmer-Ribeiro et al., 2005; D–F modified from Harrison, 1971.*

PHYLUM PORIFERA

Corvomeyenia carolinensis Harrison, 1971

Corvomeyenia carolinensis Harrison, 1971: 129; De Santo & Fell, 1996; Manconi & Pronzato, 2002, 2007; Reiswig et al., 2010

Figure 3.3

Description. Growth form encrusting, occasionally with outgrowths. **Consistency** fragile. **Color** green *in vivo* due to symbiosis with zoochlorellae. **Surface** conulose, with conules up to 3 mm in height, supporting the dermal membrane. **Oscules** elevated (3–6 mm in diameter, 6–12 mm in height). **Inhalant apertures** scattered. **Ectosomal skeleton** of microscleres and without special architecture and tips of ascending fibers. **Choanosomal skeleton** as a network of ascending paucispicular primary fibers joined by secondaries, with scattered microscleres. **Spongin** scanty, except for the gemmular theca. **Basal spongin plate** not recorded. **Megascleres** oxeas (194–(244)–280 × 5–(9)–10 μm) slender, straight to slightly curved, entirely smooth. **Microscleres** small micropseudobirotules with straight to strongly curved (84%) up to have rotules in contact, smooth shafts (15–(20)–25 μm in length) bearing pseudorotules (4–(6)–7 μm in diameter) with 4–6 recurve hooks. **Gemmules** subspherical (748–1012 μm in diameter). **Foramen** simple. **Gemmular theca** trilayered. **Outer layer** of compact spongin with protruding radially embedded gemmuloscleres. **Pneumatic layer** armed by radial gemmuloscleres. **Inner layer** of compact spongin with sublayers. **Gemmuloscleres**, also scattered freely in the sponge body, long pseudobirotules with straight to slightly bent, smooth shafts (60–(118)–158 μm in length) and with pseudorotules (13–(17)–22 μm in diameter) bearing 5–8 recurve hooks.

Habitat. Lentic. Shallow water, on submerged vegetation. Perennial throughout entire year. Gemmulation occurs at the beginning of hotter weather in early June.

Geographic distribution. Endemic to the Nearctic Region. Restricted until now to the type locality Adams Pond (five miles southeast of Columbia, Richmond County) in South Carolina and one lake in Connecticut.

Remarks. The extremely bent microbirotule microscleres are typical of *C. carolinensis* although Reiswig et al. (2010) report 'Distinction between *C. carolinensis* and *C. everetti* is questionable; both were reported from a single lake in Connecticut and differ mainly in the proportion of curved microscleres.' The etymology of the specific epithet refers to South Carolina.

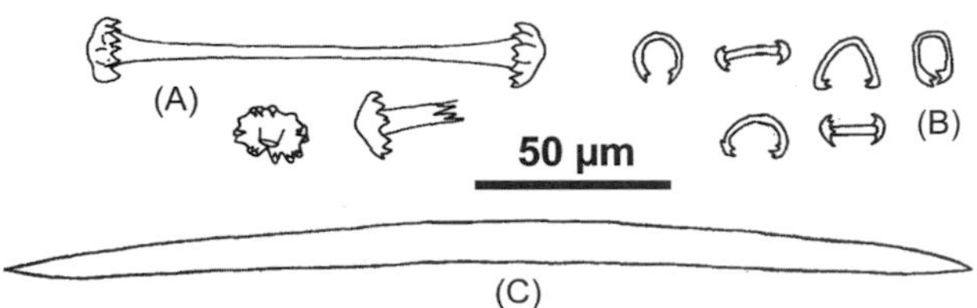

FIGURE 3.3 *Corvomeyenia carolinensis*: (A) gemmuloscleres (pseudobirotules) and details of their pseudorotules; (B) birotules (microscleres) from straight to extremely bent up to ring-shaped; (C) oxea (megasclere). *Figures A–C modified from Harrison, 1971.*

Family Spongillidae Gray, 1867

Duosclera Reiswig & Ricciardi, 1993b

Duosclera mackayi (Carter, 1885)

Spongilla mackayi Carter, 1885: 19

Spongilla igloviformis Potts, 1887; Jewell, 1935, 1939; Neidhoefer, 1940; Eshleman, 1950

Eunapius igloviformis Penney & Racek, 1968; Reiswig & Ricciardi, 1993b

Eunapius mackayi Reiswig & Ricciardi, 1993b

Duosclera mackayi Reiswig & Ricciardi, 1993b; Manconi & Pronzato, 2002, 2007; Reiswig et al., 2010

Figure 3.4

Description. Growth form encrusting. **Consistency** fragile both *in vivo* and dry condition. **Color** green. **Surface** with slight hispidation of more or less erected spicules. **Oscules** not conspicuous. **Inhalant apertures** scattered. **Ectosomal skeleton** without special architecture; tips of ascending fibers support the dermal membrane. **Choanosomal skeleton** irregular, vague network, with few ascending paucispicular primary fibers joined by secondaries. **Spongin** scanty, except for the gemmular theca. **Basal spongin plate** not recorded. **Megascleres** oxeas sharply pointed of two size classes, long oxeas (177–(200)–302 × 7–(12)–18 μm) relatively scarce in the skeleton, straight to slightly curved, entirely spiny by scattered, coarse, blunt, procurved (directed distally) spines; short oxeas (79–(156)–267 × 2–(8)–20 μm, excluding spines) abundantly scattered in the choanosome, densely covered with acute, stout spines frequently with microspines at the tips, from straight in the central portion to long, pointed, recurve at the tips. **Microscleres** absent. **Gemmules** subspherical (180–800 μm in diameter) usually in subglobular clusters sharing a central space, occasionally in carpets; gemmular clusters armed by a shared cage of intercrossing gemmuloscleres of various size. **Foramen** simple with a small collar; in gemmular groups foramina always oriented inwards the central space or when in carpets towards the substrate. **Gemmular theca** trilayered, armed by gemmuloscleres from tangentially to radially arranged except near the foraminal aperture. **Outer layer** of compact spongin with superficial ornamentation as polygonal ridges (pneumatic layer) particularly evident around the foramen where spicules are absent. **Pneumatic layer** from scarcely developed to absent. **Inner layer** of compact spongin with sublayers. **Gemmuloscleres** oxeas (165 × 10 μm) of the same size and shape as the short class of megascleres, which are always present also when gemmules are absent.

Habitat. Lentic. Symbiotic with green algae. Associated with *C. everetti* and *T. pennsylvanica*.

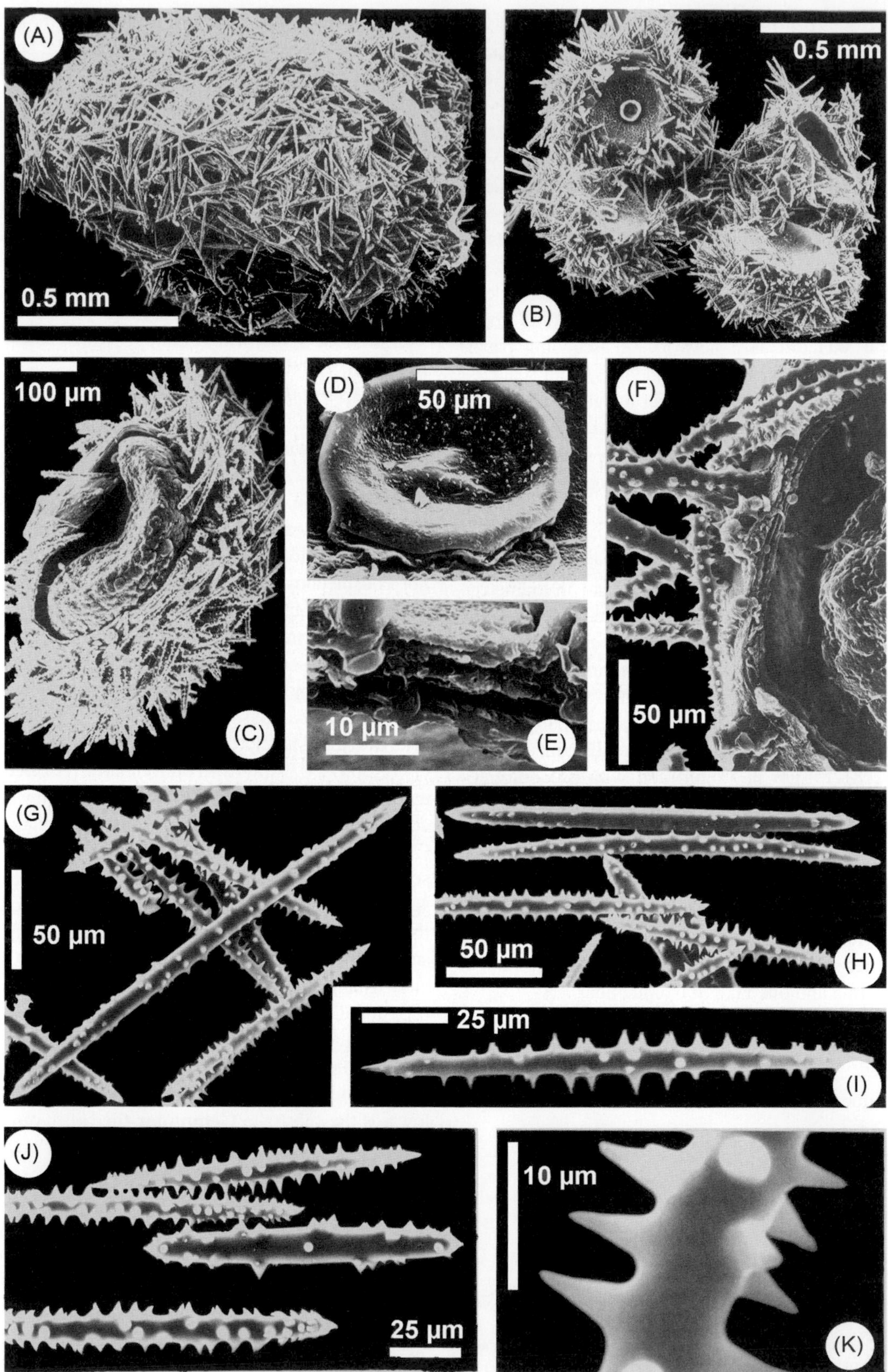

FIGURE 3.4 *Duosclera mackayi*: (A) gemmules in a group with surface armed by a shared irregular layer of tangential to radial spicules; (B) gemmules of a group with evident foramina surrounded by a scarcely developed pneumatic layer as polygonal ornamentations; (C) gemmular theca (cross-section); (D) foramen simple with a short collar at the surface of a gemmule; (E, F) gemmular theca with sublayered inner layer of compact spongin (cross section); (G, H) spiny oxeas (gemmuloscleres and megascleres); (I) megasclere; (J) gemmuloscleres; (K) close-up of gemmulosclere spiny surface. Gemmuloscleres and megascleres diverge only for a few morphological traits.

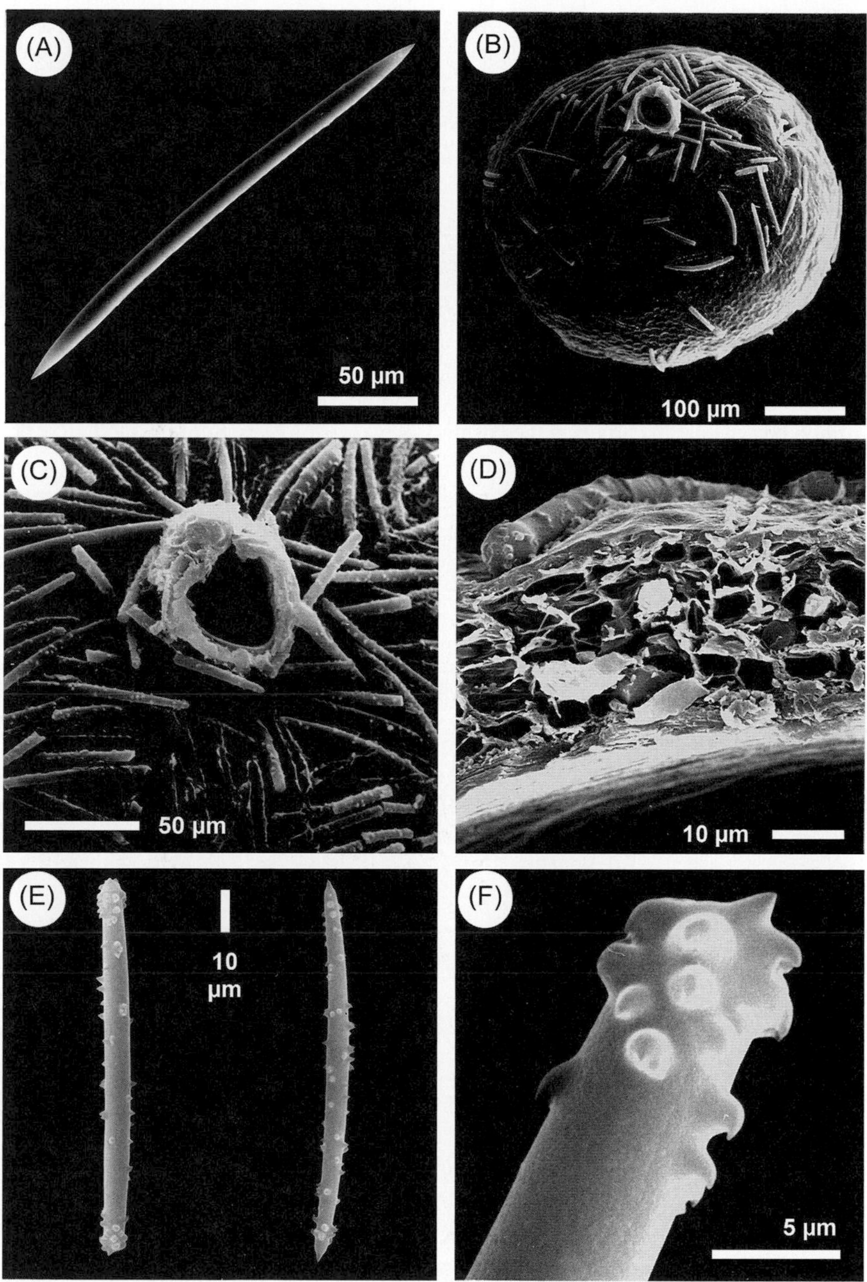

FIGURE 3.5 *Eunapius fragilis*: (A) oxea with smooth surface (megasclere); (B) gemmule with evident foramen and gemmuloscleres tangentially arranged on the outer layer of compact spongin; (C) foramen with collar surrounded by tangential gemmuloscleres; (D) gemmular theca with outer layer (left), pneumatic layer of chambered spongin and inner layer of compact multilayered spongin with (cross section); (E) strongyle and oxea (gemmuloscleres); (F) strongyle tip (gemmulosclere).

Geographic distribution. Restricted to the Nearctic Region. Known until now from the United States and Canada (Florida, Georgia, Louisiana, Massachusetts, Michigan, New Brunswick, Newfoundland, New Jersey, Nova Scotia, New York, Quebec, and Wisconsin).

Remarks. The genus is monotypic. *D. mackayi* is the type species of the genus (Reiswig & Ricciardi, 1993b). Orientation of gemmules can help distinguish *D. mackayi* from *E. fragilis*. The etymology of the specific epithet refers to the name of the type locality, MacKay's Lake (Pictou, Nova Scotia).

Eunapius Gray, 1867

Eunapius fragilis (Leidy, 1851)

Spongilla fragilis Leidy, 1851: 278

Eunapius fragilis Penney & Racek, 1968; Ricciardi & Reiswig, 1993; Pronzato & Manconi, 2002; Manconi & Pronzato, 2002, 2007, 2009; Reiswig et al., 2010

Figure 3.5

Description. Growth form encrusting, variably thick. **Consistency** notably soft and fragile both *in vivo* and dry condition. **Color** whitish to greenish. **Surface** slight hispid due to more or less erected spicules. **Oscules** not conspicuous *in vivo*, scattered in a network of subdermal canals. **Inhalant apertures** scattered. **Ectosomal skeleton** without special architecture, tips of ascending fibers support the dermal membrane. **Choanosomal skeleton** irregularly reticulate network with ascending paucispicular fibers joined by secondaries. **Spongin** scanty, except for the gemmular theca and the basal spongin plate. **Basal spongin plate** developed. **Megascleres** smooth oxeas (180–270×4–15 μm Penney & Racek, 1968; 165–(189)–271 μm in length Ricciardi & Reiswig, 1993). **Microscleres** absent. **Gemmules** subspherical (350–450 μm Penney & Racek, 1968; 300–400 μm Ricciardi & Reiswig, 1993) enclosed in a common brown coat in carpets at the basal portion in encrusting specimens, or in groups (2–5) scattered in the skeletal network in thicker specimens. Gemmules groups are always directed outward or, in layers on pavements, away from the substrate. **Foramen** tube-like. **Gemmular theca** trilayered. **Outer layer** of compact spongin thin. **Pneumatic layer**, shared by carpets or group of gemmules, of more or less rounded spongin chambers, with 1–4 layers of gemmuloscleres tangentially embedded. Inner layer of compact laminar spongin with sublayers. **Gemmuloscleres** irregularly tangential and more or less embedded into the theca, straight to slightly curved strongyles to oxeas (75–140×3–8 μm Penney & Racek, 1968; 32–(57)–140 μm Ricciardi & Reiswig, 1993) covered with conspicuous spines which are often more dense near the tips.

Habitat. Wide range of lentic and lotic habitats.

Geographic distribution. Recorded in the Nearctic Region from the United States and Canada (Colorado, Connecticut, Florida, Illinois, Indiana, Iowa, Kansas, Kentucky, Maine, Michigan, Minnesota, Montana, Newfoundland, New Jersey, New Scotland, New York, Ohio, Pennsylvania, Texas, Wisconsin, and Wyoming). Apparently widespread from all zoogeographic regions except Antarctica. Although considered usually truly cosmopolitan, it seems probable that it is a species complex, as suggested also by the high variability of morphological traits.

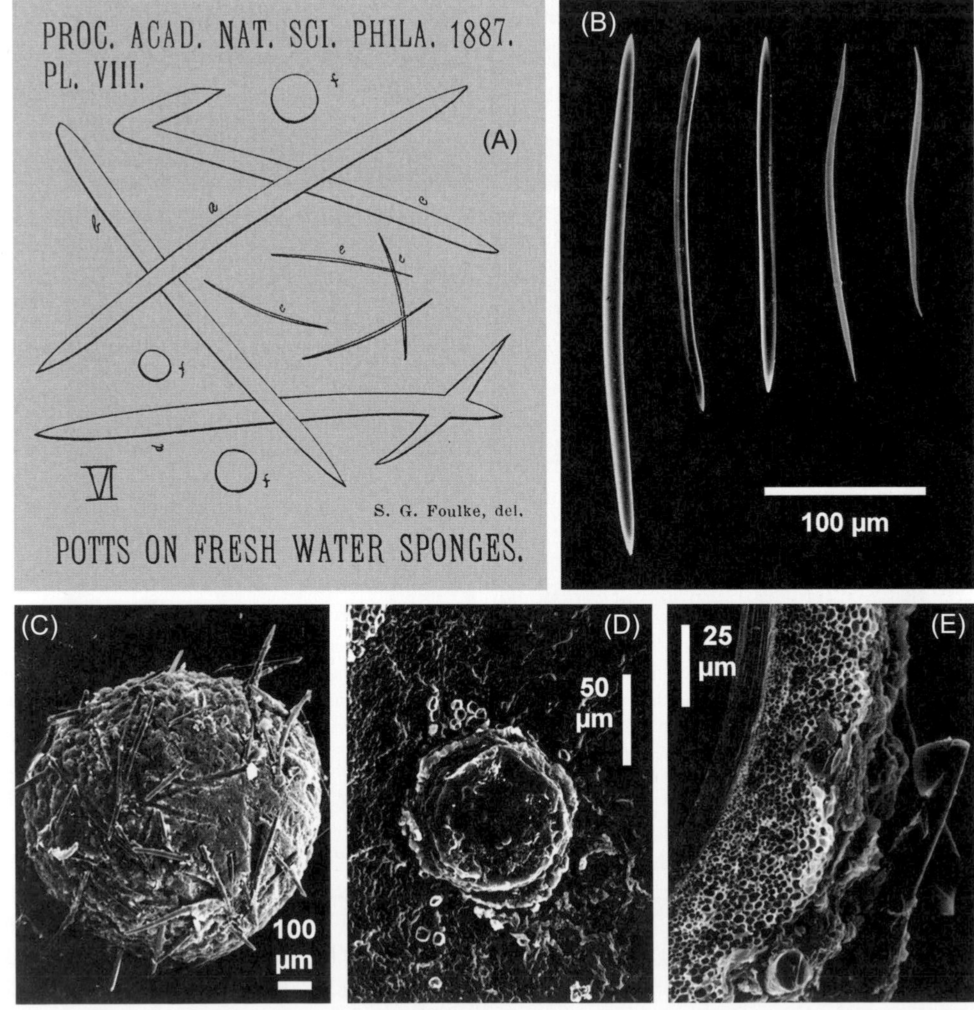

FIGURE 3.6 *Pottsiela aspinosa*: (A) spicular complement with also aberrant morphs; (B) oxeas entirely smooth from straight to sinuous; (C) gemmule with irregularly arranged gemmuloscleres; (D) foramen with short collar; (E) cross section of the gemmular theca with inner layer of compact spongin (left) and pneumatic layer. *Figure A modified from Potts, 1887.*

Remarks. *E. fragilis* diverges from a similar species *D. mackayi* for the orientation of gemmules. In the former, the foraminal aperture is always directed outward from a cluster or upward from a pavement layer, while those in *D. mackayi* are always oriented inward or towards a substrate (Reiswig et al., 2010). The etymology of the specific epithet refers to the fragile consistency of this sponge.

Pottsiela Volkmer-Ribeiro, de Souza Machado, Fürstenau-Olivera & Vieira Soares, 2010

Pottsiela aspinosa (Potts, 1880)

Spongilla aspinosa Potts, 1880: 357; Penney & Racek, 1968; Barnes & Lauer, 2003; Manconi & Pronzato, 2002, 2007; Reiswig et al., 2010

Pottsiela aspinosa Volkmer-Ribeiro, de Souza Machado, Fürstenau-Olivera & Vieira Soares, 2010

Figure 3.6

Description. Growth form encrusting, irregular, thin to branch with long, slender cylindrical branches. **Consistency** soft *in vivo*. **Color** bright green *in vivo*. **Surface** smooth. **Oscules small**, not conspicuous. **Inhalant apertures** scattered not conspicuous. **Ectosomal skeleton** armed by microscleres in the dermal membrane. **Choanosomal skeleton** as a network of paucispicular ascending primary fibers joined by secondaries, and scattered microscleres. **Spongin** quite abundant in the fibers and the gemmular theca. **Basal spongin plate** not recorded. **Megascleres** oxeas (155–(274)–338 µm in length), slender, entirely smooth, abruptly pointed. **Microscleres**, from rare to abundant in the dermal membrane, straight to bent, smooth, gradually pointed, slender microxeas (30–42 × 1–1.5 µm Penney & Racek, 1968; 21–(50)–78 µm in length Reiswig et al., 2010). **Gemmules** rare, subspherical (450–700 µm in diameter) in clusters armed by gemmuloscleres resembling small megascleres. **Foramen** slightly elevated with a shallow simple collar. **Gemmular theca** trilayered. **Outer layer** of compact spongin with gemmuloscleres frequently tangentially embedded. **Pneumatic layer** of spongin thin, chambered. **Inner layer** of compact spongin with sublayers. **Gemmuloscleres** smooth oxeas (129–(274)–306 µm in length) with very abruptly pointed or blunt tips.

Habitat. Lentic, on submerged logs and timbers in clear water at a few meters of depth below the surface or on sphagnum, grass, weeds near the surface. Sometimes perennial.

Geographic distribution. Known until now only from the Nearctic Region, reported from the United States (Florida, Indiana, Michigan, New Jersey, and Virginia) and Canada. Also doubtfully recorded from China.

Remarks. The etymology of the specific epithet refers to the smooth surface of all spicular types.

Stratospongilla Annandale, 1909

Stratospongilla penneyi (Harrison, 1979)

Stratospongilla penneyi Harrison, 1979: 100; Manconi & Pronzato, 2007; Reiswig et al., 2010

Figure 3.7

Description. Growth form encrusting. **Consistency** somewhat fleshy and elastic. **Color** lead gray *in vivo*. **Surface** irregular and hispid. **Oscules** not conspicuous. **Inhalant apertures** scattered. **Ectosomal skeleton** with microscleres in the dermal membrane, without special architecture. **Choanosomal skeleton** irregular network of ascending paucispicular primary fibers joined by secondaries, and scattered microscleres. **Spongin** scanty, except for the gemmular theca. **Basal spongin plate** not recorded. **Megascleres** oxeas (215–(243)–296 × 6–(10)–13 µm in length) slightly curved, abruptly pointed, smooth to very delicately microspiny. **Microscleres** oxeas (38–(52)–75 × 1–(2)–3 µm in length) bent, very slender, entirely covered by very small spines. **Gemmules** scattered in the skeletal network, armed by a tangential cage of megascleres. **Foramen** slightly elevated, circular, with short collar. **Gemmular theca** trilayered armed by tangential gemmuloscleres. **Outer layer** of compact spongin with tangentially partially embedded gemmuloscleres. **Pneumatic layer** of spongin chambered with small, more or less rounded meshes, with tangential gemmuloscleres irregularly partially embedded. **Inner layer** of compact spongin with sublayers. **Gemmuloscleres** oxeas (48–(84)–123 × 4–(5)–7 µm in length) curve or slightly to distinctly bent at each distal third, smooth to delicately microspined, with sharply pointed tips.

Habitat. Lentic water, on aquatic plants.

Geographic distribution. Endemic to the Nearctic Region. Known until now only from the type locality, a canal connecting Lake Gentry and Cypress Lake, at overpass of Route 523, Kissimmee River basin, Osceola County, in southern Florida.

Remarks. The validity and taxonomic status of *S. penneyi* need further investigations after the revision of the genus *Stratospongilla* to which this species was previously assigned (Manconi & Pronzato, 2002). The main problematic trait in *S. penneyi* is the reported presence of skeletal microscleres, whereas the genus *Stratospongilla* lacks skeletal microscleres, but gemmules bear strongyles tangentially arranged in the outer layer plus acanthoxeas radially arranged in the intermediate layer of the theca (cf. Manconi & Pronzato, 2002, Fig. 74, p. 963). This problem can be solved defining the presence of microscleres in the dermal membrane and choanosomal skeleton and also by SEM investigations of the gemmular architecture by cross sections to ascertain the generic status of *S. penneyi*. The absence of reports of this species in the last 80 years, despite extensive collecting, make more problematic also the status of this species from a conservation point of view, it is indeed doubtful if it is extant. New sampling campaign not only in the hot season can however highlight its presence (Harrison, 1979). The etymology of the specific epithet refers to the American spongologist J.T. Penney.

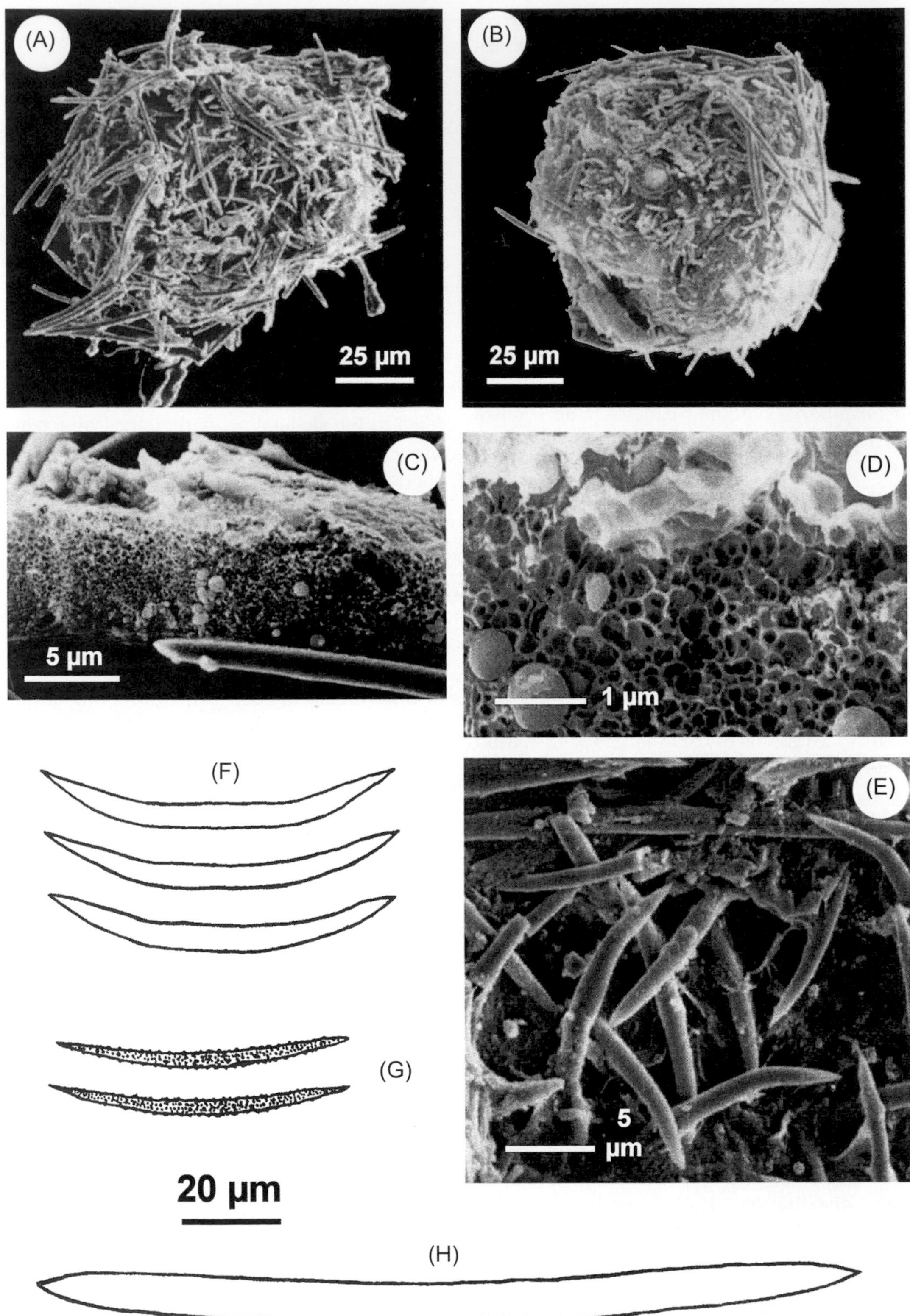

FIGURE 3.7 *Stratospongilla penneyi*: (A, B) gemmules with central foramen (B) and irregularly arranged gemmuloscleres and megascleres at the surface of gemmular theca; (C, D) different magnifications of the gemmular theca with pneumatic layer of chambered spongin; (E) smooth oxeas (gemmuloscleres) tangentially arranged on the surface of gemmular theca; (F) oxeas (gemmuloscleres); (G) microscleres; (H) oxea (megascleres). *Modified from Harrison, 1979.*

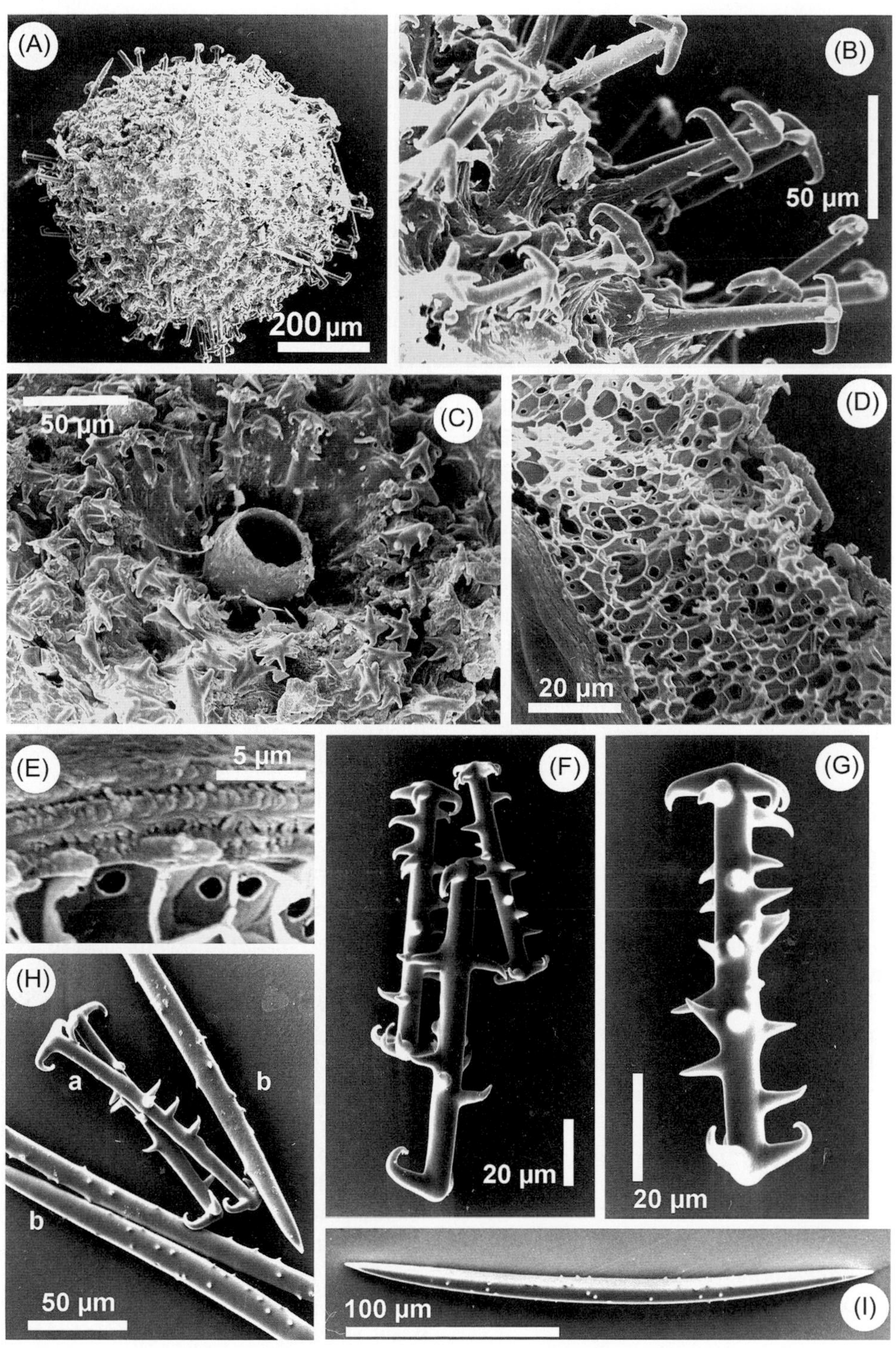

FIGURE 3.8 *Anheteromeyenia argyrosperma*: (A) gemmule with surface armed by radially arranged gemmuloscleres; (B) close-up of gemmular outer layer with protruding distal tips of gemmuloscleres; (C) foramen with short collar at the gemmular surface; (D) trilayered gemmular theca with thin outer layer of compact spongin, thick pneumatic layer of chambered spongin, and inner layer (cross section, from right to left); (E) gemmular theca with sub-layered inner layer of compact spongin (cross section, detail of D); (F and G) birotules of two different size classes, with spiny shaft and grouped hooks at the tips (gemmuloscleres); (H) spicular complement (gemmuloscleres and megascleres); (I) spiny oxeas (megascleres).

Anheteromeyenia Schröder, 1927

Anheteromeyenia argyrosperma (Potts, 1880)

Spongilla argyrosperma Potts, 1880: 357
Heteromeyenia argyrosperma Potts, 1887: 239
Anheteromeyenia argyrosperma Penney & Racek, 1968; Manconi & Pronzato, 2002, 2007; Reiswig et al., 2010
Figure 3.8

Description. Growth form encrusting as small to minute cushions. **Consistency** soft both *in vivo* and dry condition. **Color** gray or green *in vivo* if in symbiosis with zoochlorellae. **Surface** with a slight hispidation due to more or less emerging spicules. **Oscules** not conspicuous *in vivo*. **Inhalant apertures** scattered. **Ectosomal skeleton** without special architecture; tips of ascending fibers support the dermal membrane. **Choanosomal skeleton** irregular network, with ascending pauci-spicular fibers. **Spongin** in a small amount, except for the gemmular theca and the basal spongin plate. **Basal spongin plate** developed. **Megascleres** slender oxeas (240–(284)–304 × 13–17 µm), with abruptly pointed tips, sparsely covered on the entire shaft with small, sharply pointed spines. **Microscleres** absent. **Gemmules** silvery white, subspherical (400–450 µm in diameter). **Foramen** well evident, distinctly tubular; porus tube slender and narrowing towards the tip, without marginal cirri. **Gemmular theca** trilayered armed by a cage of megascleres variably developed and gemmuloscleres radially arranged. **Outer layer** of thin compact spongin with protruding gemmuloscleres. **Pneumatic layer** of spongin well-developed, thick with small, rounded chambers, with gemmuloscleres strictly radially embedded in the theca with distal tips from not emerging to considerably protruding from the outer layer. **Inner layer** of compact spongin with sublayers. **Gemmuloscleres** straight birotules of two different size classes similar in form, long (65–(81)–89 × 6–8 µm) and short (110–(130)–160 × 7–10 µm), with entirely spiny shaft and large acute spines conspicuously recurve particularly at the tips as claw-like hooks.

Habitat. On submerged sticks, stones.

Geographic distribution. Endemic to the Nearctic Region. Restricted until now to the eastern half of North America from Florida to Canada (Connecticut, Illinois, Indiana, Michigan, New Jersey, Nova Scotia, Pennsylvania, and Wisconsin).

Remarks. *A. argyrosperma* is the type species of the genus. The etymology of the specific epithet *argyrosperma* refers to the silvery white color of the gemmules.

Radiospongilla Penney & Racek, 1968

Radiospongilla crateriformis (Potts, 1882)

Spongilla crateriformis Potts, 1882: 12

Radiospongilla crateriformis Penney & Racek, 1968; Manconi & Pronzato, 2002, 2007; Reiswig et al., 2010

Figure 3.9

Description. Growth form cushion-like. **Consistency** soft *in vivo*. **Color** flesh to light green *in vivo*. **Surface** even. **Oscules numerous**, not conspicuous. **Inhalant apertures** scattered. **Ectosomal skeleton** without special architecture. **Choanosomal skeleton** irregular ascending pauci-spicular primary fibers joined by secondaries. **Spongin** scanty. **Basal spongin plate** not recorded. **Megascleres** oxeas (240–(278)–300 × 9–11 µm) slender, slightly curved, sharply pointed, sparsely covered by very small spines except at their tips. **Microscleres** when recorded similar to gemmuloscleres. **Gemmules** (370–450 µm in diameter Penney & Racek, 1968; 261–520 µm in diameter Reiswig et al, 2010) with gemmuloscleres radially arranged in a monolayer except for those displaced and slanting around the foramen to form a crater-like depression. **Foramen** tubular, short, within the gemmular theca. **Gemmular theca** trilayered. **Outer layer** of compact spongin present or absent, when present armed by distal rotules of gemmuloscleres. **Pneumatic layer** as a dense network of spongin fibers forming small, irregular meshes, with a monolayer of gemmuloscleres radially embedded. **Inner layer** of compact spongin with sublayers. **Gemmuloscleres** strongyles (60–(71)–80 × 3–5 µm) slender, spiny by small conical spines only at the tips or all over, but always more abundant at ends where larger, slightly recurve spines are sufficiently dense to form pseudorotules.

Habitat. Not reported.

Geographic distribution. Known in the Nearctic Region from southern Canada, Mexico and the United States (Alabama, District of Columbia, Illinois, Indiana, Maryland, Michigan, New York, Ohio, Pennsylvania, Texas, and Wisconsin). Also known from the West Indies (Cuba), Suriname, Brazil, China, Japan, Southeast Asia, and Australia.

Remarks. The presence of true microscleres in the entire genus needs further investigations. The etymology of the specific epithet refers to the foraminal crater-like depression.

Radiospongilla cerebellata (Bowerbank, 1863)

Spongilla cerebellata Bowerbank, 1863: 465

Radiospongilla cerebellata Penney & Racek, 1968; Poirrier, 1972; Manconi & Pronzato, 2002; Reiswig et al., 2010

Figure 3.10

Description. Growth form encrusting. **Consistency** soft. **Color** green to dark tan *in vivo*. **Surface** corrugate (brain-like). **Oscules** conspicuous. **Inhalant apertures** scattered. **Ectosomal skeleton** with immature gemmuloscleres in the dermal membrane without special architecture. **Choanosomal skeleton** as irregular network with ascending pauci-spicular primary fibers joined by secondaries. **Spongin** scanty, except for the gemmular theca. **Basal spongin plate** not recorded. **Megascleres** straight to slightly curved, smooth oxeas (240–330 × 10–12 µm). **Microscleres** apparently absent, although immature gemmuloscleres may be abundant in the dermal membrane. **Gemmules** subspherical (420–590 µm in diameter) single, scattered. **Foramen** tubular, slender, straight. **Gemmular theca** trilayered with gemmuloscleres arranged, in intact gemmules, in two layers arranged tangentially on the outer layer and radially on the inner layer. **Outer layer** of compact spongin. **Pneumatic layer** of fibrous spongin scarcely

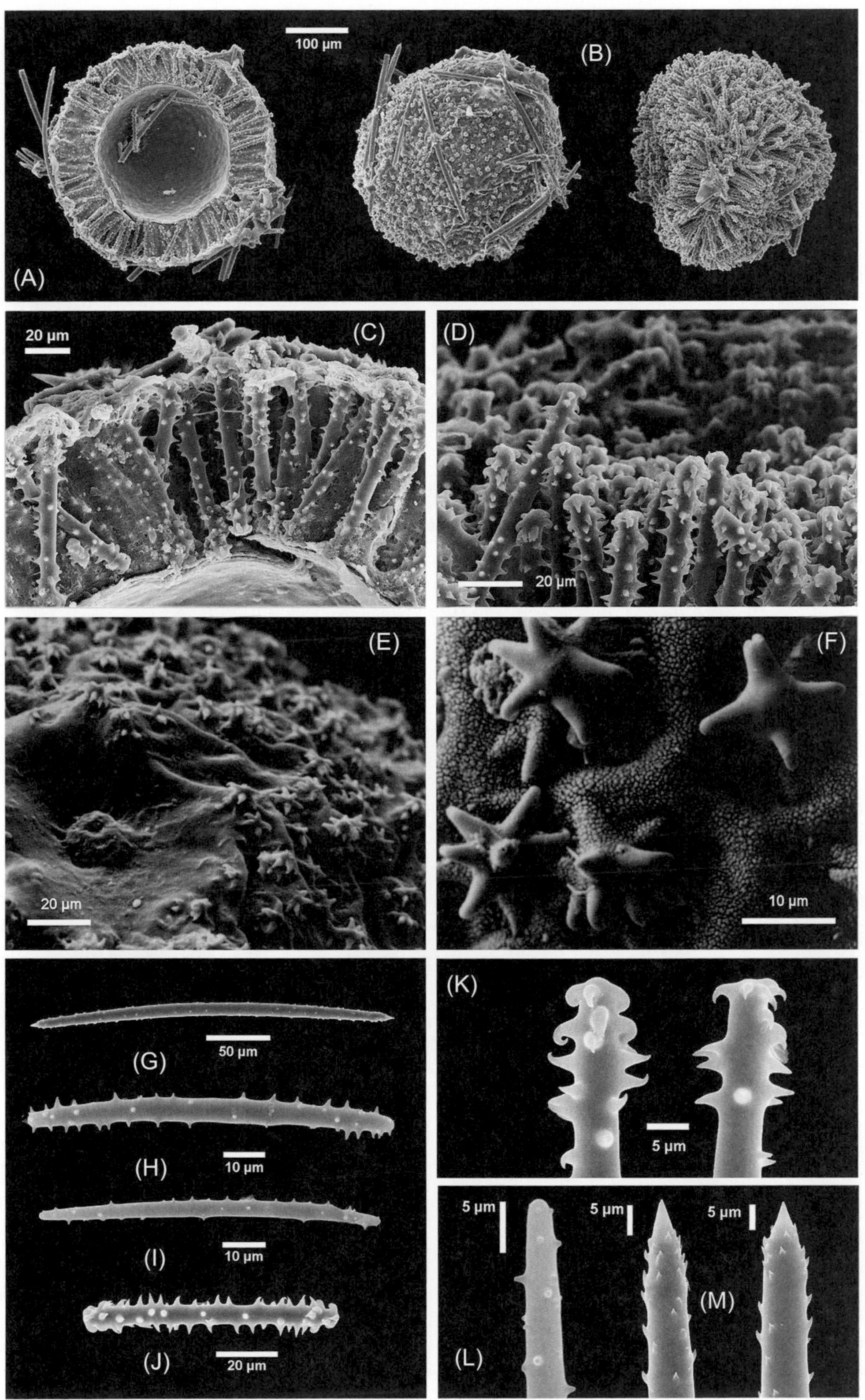

FIGURE 3.9 *Radiospongillia crateriformis*: (A) gemmule (cross section); (B) entire gemmule with outer layer and few tangential (oxeas) megascleres (left) and with radial gemmuloscleres not covered by outer layer (right); (C) gemmular theca with scarcely developed fibrous pneumatic layer and inner layer of compact spongin; (D) detail of gemmular surface without layer; (E) foramen of a gemmule with outer layer and emerging tips of gemmuloscleres; (F) detail of gemmular surface with tips of gemmuloscleres emerging from outer layer; (G) megasclere; (H, I) microscleres; (J) gemmulosclere; (K) gemmulosclere tips; (L) microsclere tip; (M) Megasclere tips.

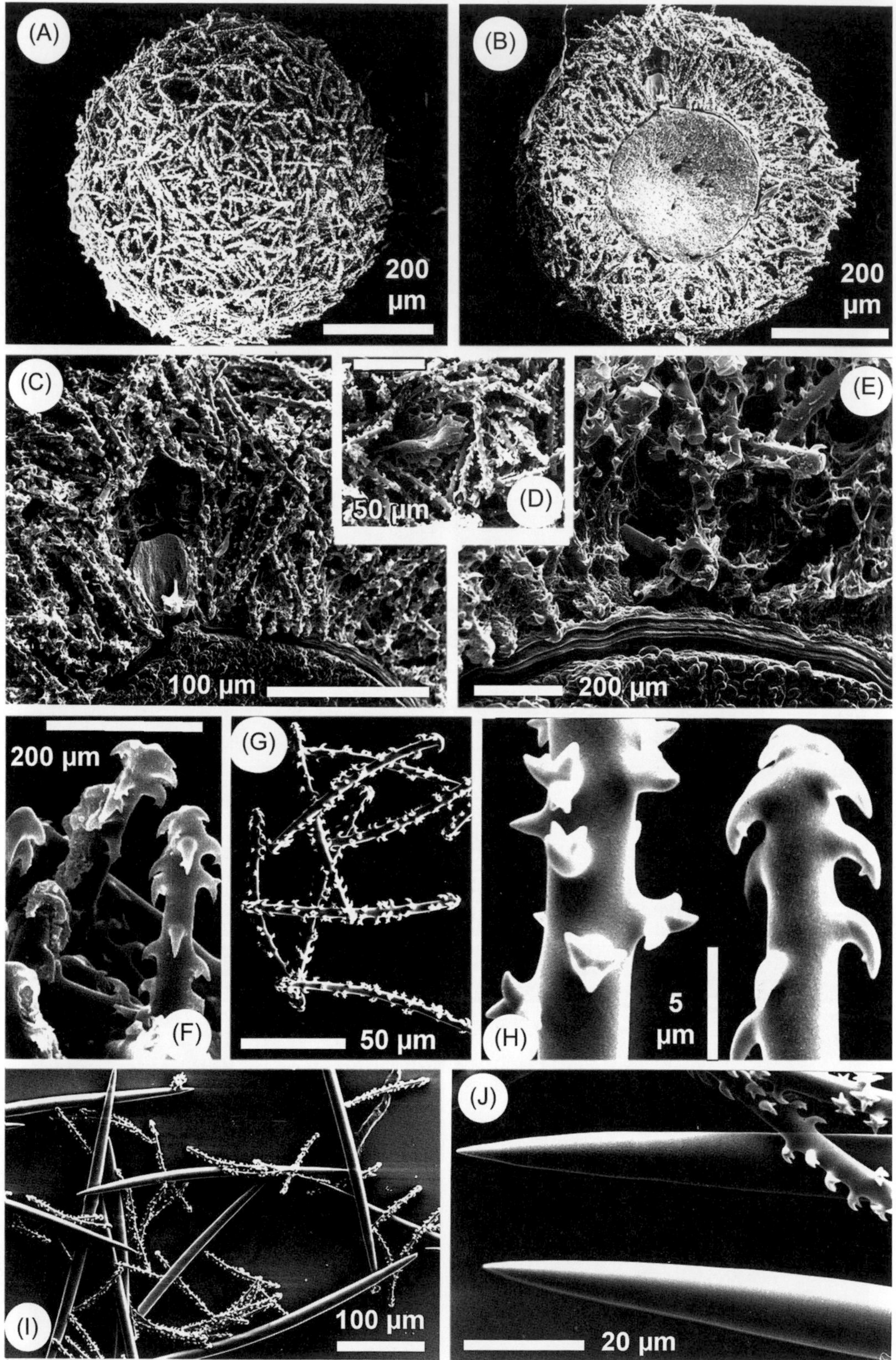

FIGURE 3.10 *Radiospongilla cerebellata*: (A) gemmule armed by gemmuloscleres tangentially arranged on the outer layer of compact spongin, and evident foraminal aperture (top); (B) gemmular theca with tangential and radial gemmuloscleres (cross section); (C, D) gemmular foramen (cross section); (E) detail of gemmular theca with fibrous pneumatic layer scarcely developed and inner layer of compact sublaminar spongin (cross section); (F) gemmuloscleres irregularly arranged on the gemmule surface; (G) gemmuloscleres; (H) gemmulosclere shaft and tip; (I) spicular complement (megascleres and gemmuloscleres); (J) megasclere tips.

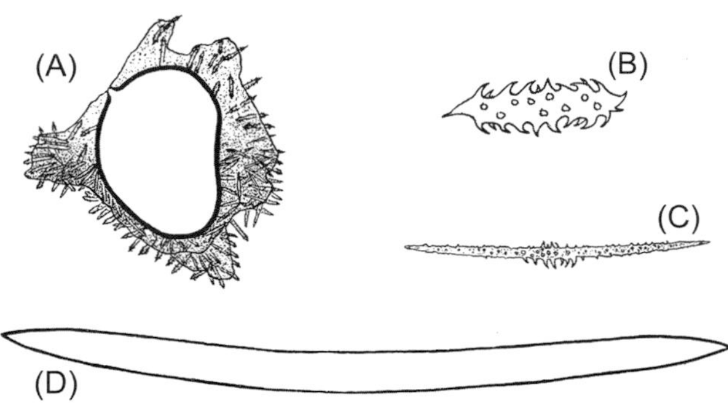

FIGURE 3.11 *Spongilla cenota*: (A) gemmule with lobate theca and embedded gemmuloscleres (cross section); (B) gemmulosclere; (C) spiny oxea (microsclere); (D) smooth oxea (megascleres). *Modified from Penney & Racek, 1968.*

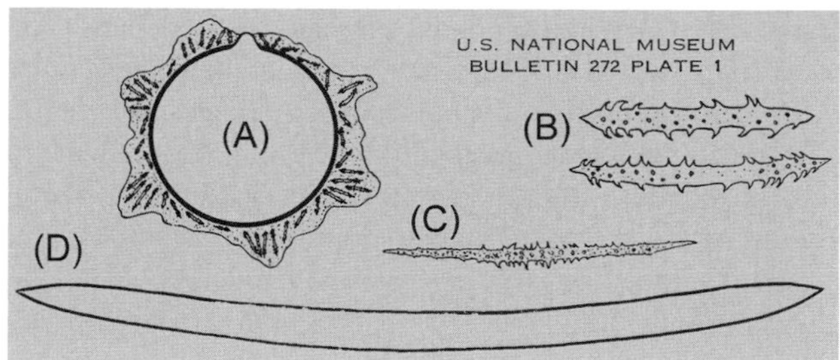

FIGURE 3.12 *Spongilla wagneri*: (A) gemmule with irregular theca with lobes and gemmuloscleres tangentially to radially embedded in the pneumatic layer (cross section); (B) oxeas, stout, with spines (gemmulosclere); (C) oxeas, slender with spines (microsclere); (D) smooth oxea (megasclere). *Modified from Penney & Racek, 1968.*

developed. **Inner layer** of compact spongin with sublayers. **Gemmuloscleres** strongyles usually distinctly curved, rarely straight, 72–110 µm in length, covered with abundant spines that are pronouncedly recurve toward the tips.

Habitat. Lotic, on submerged small logs and rocks in ca. 30–91 cm of depth. Active sponges in the hottest season of the Texas arid climate.

Remarks. The record of this Oriental species as an enclave in the arid western USA is problematic, although Poirrier (1972) reported, without description and figures of diagnostic traits, that the Texas specimens match the description by Penney & Racek (1968). We consider with doubt the presence of *R. cerebellata* in the Nearctic due to the extremely disjoint geographic pattern of this typically tropical-subtropical species. The presence of true microscleres in the entire genus needs further investigations. The specific epithet refers to the cerebellum-like growth form typical of this species.

Geographic distribution. Known until now, with doubt, in the Nearctic Region only from Texas (Pedernales River, Johnson City) but widely distributed in tropical and subtropical Asia and Africa.

Spongilla Lamarck, 1816

Spongilla cenota Penney & Racek, 1968

Spongilla lacustris Auctorum (in Old, 1936: 26; Rioja, 1940)

Spongilla cenota Penney & Racek, 1968: 18; Manconi & Pronzato, 2002, 2007; Reiswig et al., 2010

Figure 3.11

Description. Growth form irregularly massive cushions. **Consistency** firm, rather brittle in dry condition. **Color** light gray to green *in vivo*, whitish in dry condition. **Surface** smooth. **Oscules** small and not conspicuous. **Inhalant apertures** scattered. **Ectosomal skeleton** of abundant microscleres in the dermal membrane. **Choanosomal skeleton** dense network of paucispicular ascending primaries spicular fibers joined by secondaries, and microscleres. **Spongin** scanty, except for the gemmular theca. **Basal spongin plate** not recorded. **Megascleres** oxeas (310–410 × 14–22 µm) stout, entirely smooth, abruptly pointed. **Microscleres** oxeas (68–123 × 2–3 µm) slender, slightly bent, entirely ornamented by spines and tubercles small at the tips and large erect spines with frequent knob-like tips in the central portion of the shaft. **Gemmules** subspherical (480–650 µm in diameter) with irregular lobes scattered in the skeletal meshes. **Foramen** slightly elevated, simple. **Gemmular theca** trilayered. **Outer layer** of compact spongin partially covering gemmuloscleres. **Pneumatic layer** of chambered spongin very irregular in thickness, armed by gemmuloscleres ranging from irregularly and tangentially embedded in thinner parts of the pneuma to entirely more or less radially embedded in the larger lobes. **Inner layer** of compact spongin with sublayers. **Gemmuloscleres** oxeas (65–86 × 9–13 µm) very stout, slightly bent to straight, entirely covered with stout, sharp, recurve spines.

Habitat. Lotic/lentic.

Geographic distribution. Endemic to the Nearctic Region. Known until now only from Yucatan (Mexico), Costa Rica, and Florida. The type locality is the Cenote Xtoloc, near Chichen Itza (Yucatan).

Remarks. The etymology of the specific epithet refers to 'cenote,' a Mexican name for inland karstic formations corresponding to the 'blue holes' in the sea.

Spongilla wagneri Potts, 1889

Spongilla wagneri Potts, 1889: 7; Penney & Racek, 1968; Manconi & Pronzato, 2002, 2007; Reiswig et al., 2010

Figure 3.12

Description. Growth form encrusting, thin. **Consistency** rather brittle *in vivo*. **Color** whitish *in vivo*. **Surface** greatly roughened by microscleres extremely abundant in the dermal membrane. **Oscules** not conspicuous *in vivo*. **Inhalant apertures** scattered. **Ectosomal skeleton** as extremely abundant microscleres in the dermal membrane. **Choanosomal skeleton** irregular with a vague network of spicular primary and secondary fibers, and abundant scattered microscleres. **Spongin** scanty, except for the gemmular theca. **Basal spongin plate** not recorded. **Megascleres** stout, long oxeas entirely smooth (144–270×7–12μm). **Microscleres** slender, slightly curved oxeas (49–62×2–4μm), entirely spiny with erect, stout spines longer in the central region. **Gemmules** (470–610μm in diameter) abundant, particularly at the sponge base, irregularly subspherical, single. **Foramen** simple, slightly elevated. **Gemmular theca** trilayered with thick lobes. **Outer layer** of compact spongin thin, covering gemmuloscleres from irregularly to radially (in thicker lobes) arranged. **Pneumatic layer** unusually developed and irregular in thickness, with small, more or less rounded meshes. **Inner layer** of compact spongin with sublayers. **Gemmuloscleres** abundant, radially arranged in the lobes of the theca, slightly curved oxeas (48–75×6–8μm) with stout, sharp, recurve spines more dense towards the tips of shaft to form often distinct heads but not pseudorotules.

Habitat. Lentic, in slightly to strongly brackish water. Encrusting on *Lepas* sp. and *Serpula* sp., with gemmules attached to their skeleton.

Geographic distribution. Endemic to the Nearctic Region. Southeastern coastal regions of the United States (Florida, Louisiana, and South Carolina).

Remarks. We consider, in agreement with Penney & Racek (1968), *S. wagneri* as valid species for its typical morphotraits, although Reiswig et al. (2010) considered it junior synonym di *S. alba* (see remarks). The etymology of the specific epithet refers to the name of the collector Wagner.

PHYLUM PORIFERA

Spongilla lacustris (Linnaeus, 1759)

Spongilla lacustris Linnaeus, 1759: 1348; Penney & Racek, 1968; Manconi & Pronzato, 2000, 2002, 2007; Pronzato & Manconi, 2002; Reiswig et al., 2010

Figure 3.13

Description. Growth form from encrusting or massive to erected and branched or arborescent up to 40–50cm in height. **Color** whitish, yellow-orange, greenish or emerald green. **Consistency** soft, fragile. **Surface** hispid. **Oscules** scattered. **Inhalant apertures** scattered. **Ectosomal skeleton** with no special architecture, as tangential microscleres and spicular brushes at apices of primary fibers to support the dermal membrane. **Choanosomal skeleton** as an irregular network from isotropic in encrusting portions to anisotropic in finger-like outgrowths or branches; primary fibers pauci- to multi-spicular joined by pauci-spicular secondary fibers. **Spongin** quite abundant. **Basal spongin plate** not recorded. **Megascleres** oxeas (90–350×2–18μm Penney & Racek, 1968; 158–(254)–362μm in length Reiswig et al., 2010) entirely smooth, fusiform with tips from gently to sharply pointed, slightly spiny if associated to gemmules. **Microscleres** abundant to rare, scattered fusiform oxeas (25–178×2–8μm Penney & Racek, 1968; 30–(61)–130μm in length Reiswig et al., 2010) with dense spines on the entire length and microspinosities on spines, giving them an asterose shape. **Gemmules** (98–789μm) from subspherical to suboval in dense clusters or irregularly scattered in the skeletal network. **Gemmular cage** of megascleres sometimes present in unarmored gemmules. **Gemmules** of two types can be found sometimes in the same specimen: unarmored without gemmuloscleres or armored with gemmuloscleres tangential to partially embedded in the gemmular theca. **Gemmular theca** from thick-walled when trilayered to thin-walled when monolayered with a variable number of sublayers of compact spongin (3–7), corresponding to the inner layer of the thick walled gemmule. **Outer layer** and **pneumatic layer** well developed in thick-walled gemmules. **Foramen** slightly elevated with or without a simple collar or plate-like collar, generally single set in thick-walled gemmules; multiple foramina (1–6) common in thin-walled gemmules, but sometimes also present in thick-walled gemmules. **Gemmuloscleres** present in thick-walled gemmules or usually absent in thin-walled gemmules; when present, slightly to strongly bent oxeas to strongyles (80–130×3–10μm Penney & Racek, 1968; 21–130×1–10μm Manconi & Pronzato, 2002; 18–(32)–130μm in length Reiswig et al., 2010) with curved to straight spines more dense at the tips.

Habitat. Lotic and lentic, from coastal brackish waters to alpine lakes, from semi-arid zones to permafrost areas at the Northern Polar Circle. On a wide variety of substrata such as woodpiles, floating wood, shells, pebbles, boulders and rocks, metallic and concrete piles, and brick walls.

Geographic distribution. Palaearctic and Nearctic regions. One of the most common species throughout the United States and Canada (Alaska, Colorado, Connecticut, District of Columbia, Illinois, Iowa, Kansas, Maryland, Massachusetts, Minnesota, Montana, New Jersey, New York, Michigan, Ohio, Pennsylvania, Texas, Washington, Wisconsin, Wyoming, Newfoundland, and Nova Scotia).

Remarks. Type species of the genus and of the family Spongillidae. The etymology of the specific epithet refers to the lentic habitat (lake) from which Linnaeus collected this sponge. *Spongia lacustris* was described by Linnaeus (1759) as "*repens, fragilis, ramis teretibus obtusis*" ("creeping, fragile, with cylindrical branches showing swellings at their ends") in the second volume of the *Systema Naturae*, and its type material was recently rediscovered in the *Linnean Herbarium* (Manconi & Pronzato, 2000).

Spongilla alba Carter, 1849

Spongilla alba Carter, 1849: 83; Penney & Racek, 1968; Manconi & Pronzato, 2002, 2007; Reiswig et al., 2010

Not *S. wagneri* Potts, 1889

Figure 3.14

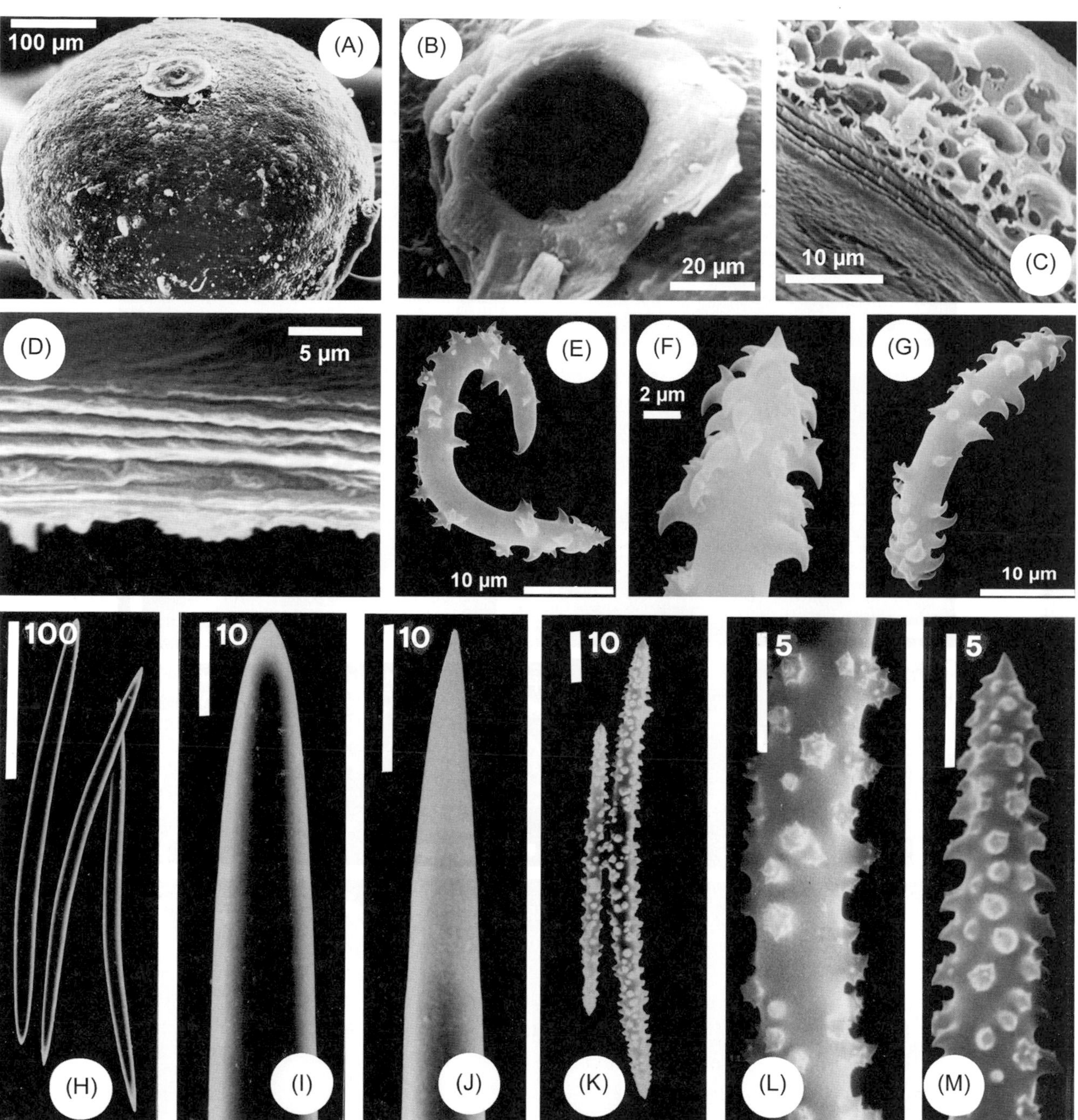

FIGURE 3.13 *Spongilla lacustris*: (A) gemmule with smooth outer layer of compact spongin, and evident foramen with collar; (B) gemmular foramen simple, with collar; (C) close up of the thick-walled gemmule with pneumatic layer; (D) close up of thin-walled gemmule with only multilayered inner layer of compact spongin; (E, G) oxeas to strongyles with large, bent spines (gemmuloscleres); (F) gemmulosclere tip; (H) smooth oxeas (megascleres); (I, J) oxeas tips (megascleres); (K) spiny oxeas (microscleres); (L) microsclere with shaft ornate by spines and rosettes of microspines; (M) microsclere tip.

Description. Growth form massive or encrusting as large patches. **Consistency** rather firm but brittle *in vivo*. **Color** pale gray to off-white *in vivo*. **Surface** smooth, slightly lobose, with abundant microscleres. **Oscules** large, not conspicuous *in vivo*. **Inhalant apertures** scattered. **Ectosomal skeleton** as extremely abundant microscleres tangentially arranged in the dermal membrane. **Choanosomal skeleton** as a dense network of ascending primary spicular fibers joined by secondaries, and scattered microscleres. **Spongin** scanty, except for the gemmular theca. **Basal spongin plate** not recorded. **Megascleres** stout, oxeas entirely smooth (256–420 × 12–22 µm). **Microscleres** oxeas (75–124 × 2–3 µm) slender, slightly curved, entirely spiny with stout spines. **Gemmules** abundant, scattered in the skeleton, subspherical, single (450–600 µm in diameter). **Foramen** simple, slightly elevated, with a shallow

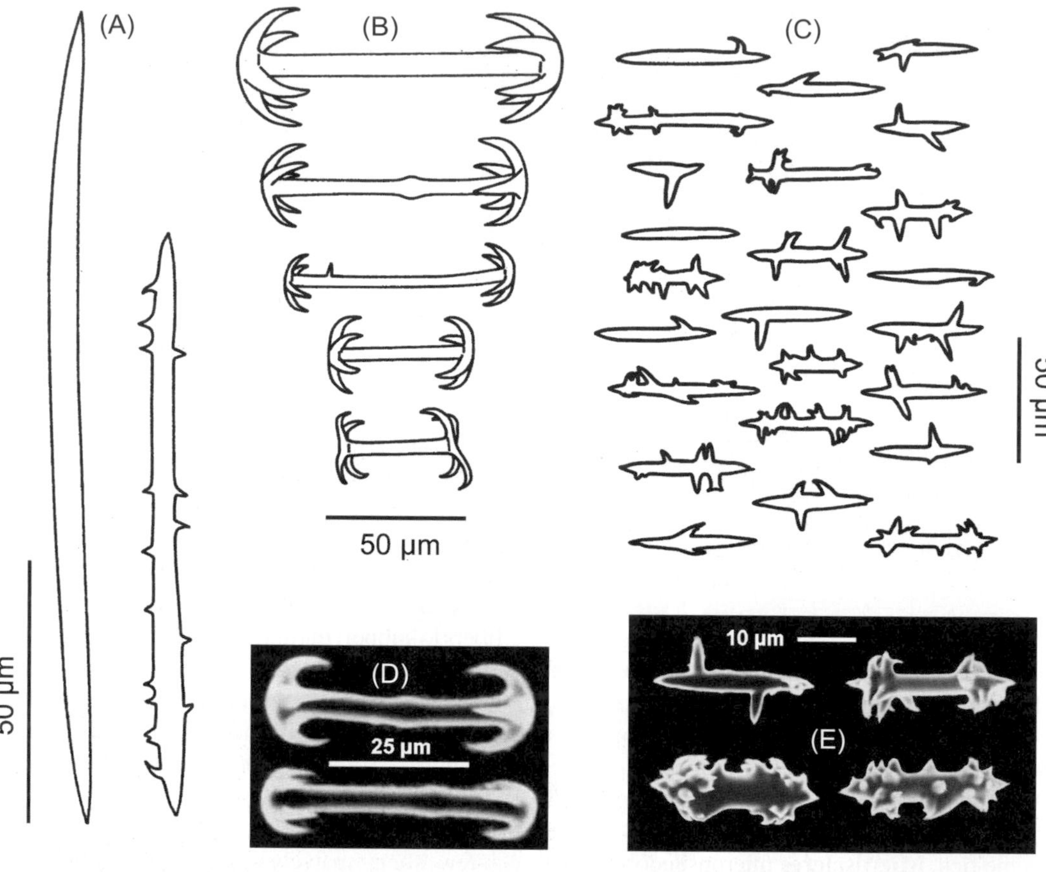

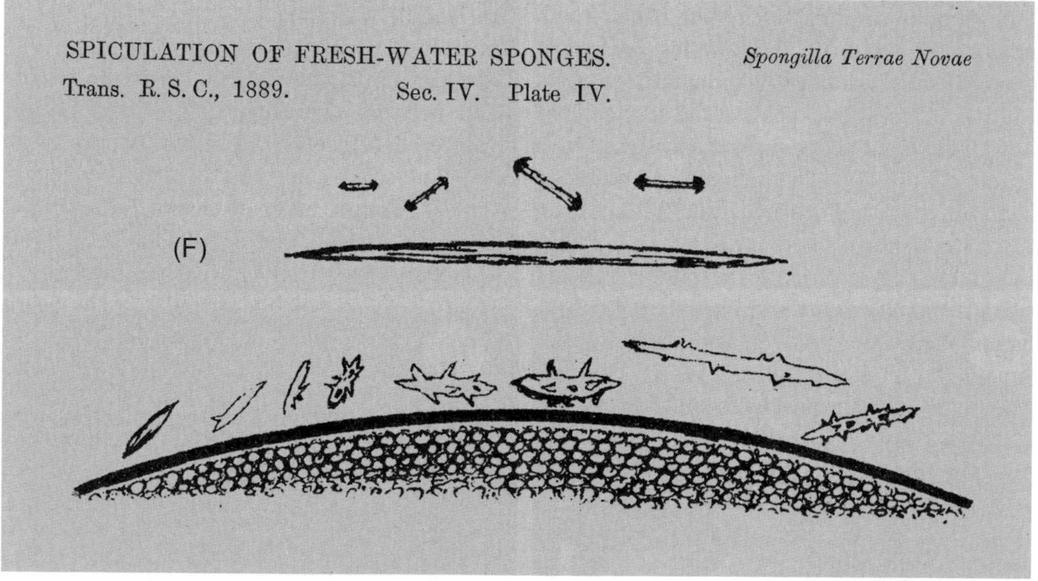

FIGURE 3.19 *Corvospongilla novaeterrae*: (A) oxeas from smooth to spiny (megascleres); (B, D) micropseudobirotules (microscleres); (C, E) gemmuloscleres oxeas with a few large spines; (F) spicular complement (microscleres, megasclere, gemmuloscleres) and gemmular theca with pneumatic layer of chambered spongin (cross section). *Figures A–C modified from Reiswig & Ricciardi, 1993b; D and E modified from Reiswig et al., 2010; F modified from MacKay, 1889.*

Heteromeyenia Potts, 1881a

Heteromeyenia baileyi (Bowerbank, 1863)

Spongilla baileyi Bowerbank, 1863: 451

Heteromeyenia repens Potts, 1887

Heteromeyenia baileyi Schröder, 1927; Penney & Racek, 1968; Manconi & Pronzato, 2002, 2007; Reiswig et al., 2010

Figure 3.20

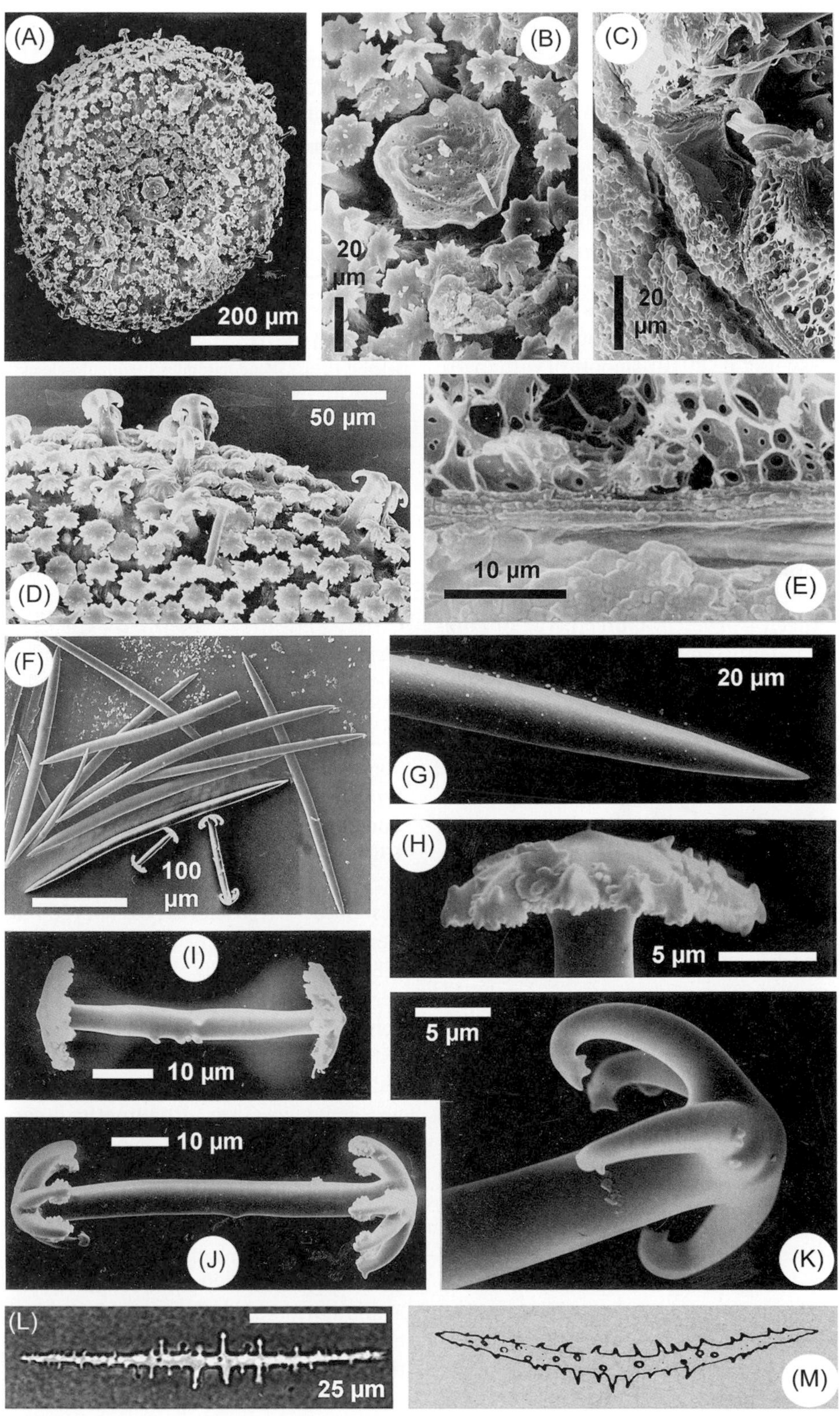

FIGURE 3.20 *Heteromeyenia baileyi*: (A) gemmule with evident foramen (center) and tips of gemmuloscleres emerging from the outer layer of compact spongin; (B) foramen surrounded by distal rotules of gemmuloscleres; (C) gemmular theca with foraminal tube in the pneumatic layer of chambered spongin (cross section); (D) gemmular outer layer with protruding tips of gemmuloscleres; (E) multilayered inner layer and pneumatic layer of the trilayered gemmular theca; (F) spicular complement oxeas (megascleres) and two types of gemmuloscleres (birotule and pseudobirotule); (G) oxea tip (megasclere); (H, I) birotule (short gemmuloscleres); (J, K) pseudobirotule (long gemmuloscleres); (L, M) spiny oxeas (microscleres). *Figure L modified from Reiswig et al., 2010; M modified from Penney & Racek, 1968.*

Description. Growth form encrusting, thin. **Consistency** fragile both *in vivo* and dry condition. **Color** bright green. **Surface** hispid due to more or less erected spicules at the tips of ascending fibers. **Oscules** not conspicuous. **Inhalant apertures** scattered. **Ectosomal skeleton** armed by microscleres and tips of ascending fibers support the dermal membrane, without special architecture. **Choanosomal skeleton** irregular network of ascending paucispicular fibers joined by secondaries, and scattered microscleres. **Spongin** scanty, except for the gemmular theca. **Basal spongin plate** not reported. **Megascleres** slender oxeas (255–315 × 11–10 µm Penney & Racek, 1968; 216–(247)–320 µm in length Reiswig et al., 2010), smooth or with sparse microspines except near the sharply pointed tips. **Microscleres** oxeas (75–85 × 2–3 µm Penney & Racek, 1968; 53–(67)–85 µm in length Reiswig et al., 2010) delicate and sharply pointed, slightly curved to almost straight, entirely spiny, with larger perpendicular spines bearing apical rosettes of microspines to knobs in the central region of the shaft. **Gemmules** subspherical, smooth, single (450–480 µm in diameter). **Foramen** without terminal cirrus projections. **Gemmular theca** trilayered. **Outer layer** of compact spongin with radially embedded gemmuloscleres. **Pneumatic layer** of spongin fibers forming more or less rounded meshes, with a monolayer of gemmuloscleres radially embedded. **Inner layer** of compact spongin with sublayers. **Gemmuloscleres** birotules of two types. Type A, birotules of two size classes, shorter (38–(51)–60 µm in length) with flat, serrate rotules (13–(18)–22 µm in diameter), and longer birotules very similar to the shorter, but with a wider size range (67–160 µm in length). Type B, long pseudobirotules (49–(70)–85 µm in length) with pseudorotules (18–(22)–28 µm in diameter) of long recurve hooks often with ornamented tips (from rosettes of microspines to knobs).

Habitat. Lotic/lentic.

Geographic distribution. Nearctic, Palaearctic, and Neotropical Regions. Widely distributed in Canada and throughout eastern United States from New York to Louisiana and to Texas. Type locality a stream of Canterbury Road, West Point, New York. A few reports from Europe and South America (Argentina) need confirmation.

Remarks. *H. baileyi* is the type species of the genus. The long spines on the central portion of microscleres are diagnostic, useful in distinguishing it from other species in the genus. Their maximum length is on average greater than the width of the microscleres. Reiswig et al. (2010) report 'The presence of two types of gemmuloscleres suggests that two distinct populations may be presently included under this species name; they may require separation at the species level when detailed studies are carried out.' To clarify this problematic point, it is necessary a revision in depth of gemmular architecture by scanning electron microscopy to check if both gemmuloscleres types are present in the same gemmule or if two types of gemmules exist in the same specimen. The etymology of the specific epithet refers to the name of the American naturalist and microscopist Prof. J.W. Bailey.

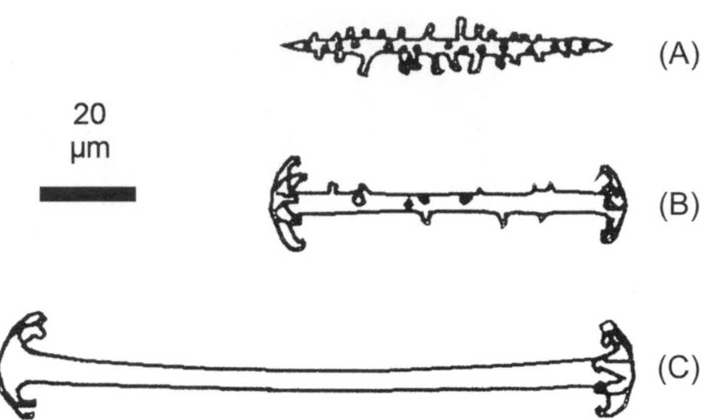

FIGURE 3.21 *Heteromeyenia longistylis*: (A) spiny oxea (microsclere); (B, C) pseudobirotules (gemmuloscleres) of two length classes. Megascleres were never figured. *Figures modified from Reiswig et al., 2010.*

Heteromeyenia longistylis Mills, 1884

Heteromeyenia longistylis Mills, 1884: 146; Penney & Racek, 1968; Reiswig et al., 2010

Figure 3.21

Description. Growth form small. **Consistency** fragile, apparently strengthened by gemmules. **Color** not recorded. **Surface** not recorded. **Oscules** not conspicuous. **Inhalant apertures** scattered. **Ectosomal skeleton** not described. **Choanosomal skeleton** not described. **Spongin** scanty, except for the gemmular theca. **Basal spongin plate** not described. **Megascleres** oxeas (259–330 µm in length) slender, very sparsely spined. **Microscleres** oxeas small (58–68 µm in length) fusiform, entirely spined with longer spines in the central portion bearing rosettes of microspines. **Gemmules** (480 µm in diameter). **Foramen** tubular with cirrus projections. **Gemmular theca** trilayered with radial gemmuloscleres. **Outer layer** with protruding longer gemmuloscleres. **Pneumatic layer** of spongin with gemmuloscleres radially embedded. **Inner layer** of compact spongin. **Gemmuloscleres** pseudobirotules in two size classes; shorter pseudobirotules (73–76 µm in length) with straight shaft bearing few scattered spines and slightly umbonate pseudorotules (20 µm in diameter); longer pseudobirotules (125–129 µm in length) bearing curved, smooth shafts and more hemispheric pseudorotules (23 µm in diameter) with marginal strongly curved hooks with ornamentations at the tips.

Habitat. Not recorded.

Geographic distribution. Endemic to the Nearctic Region. Known until now only from Pennsylvania.

Remarks. This rarely mentioned species is extremely poorly known and its taxonomic status is problematic. Megascleres have never been figured, and few data are

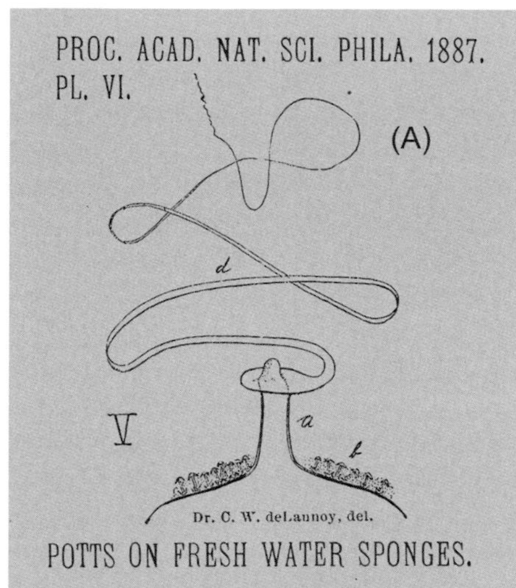

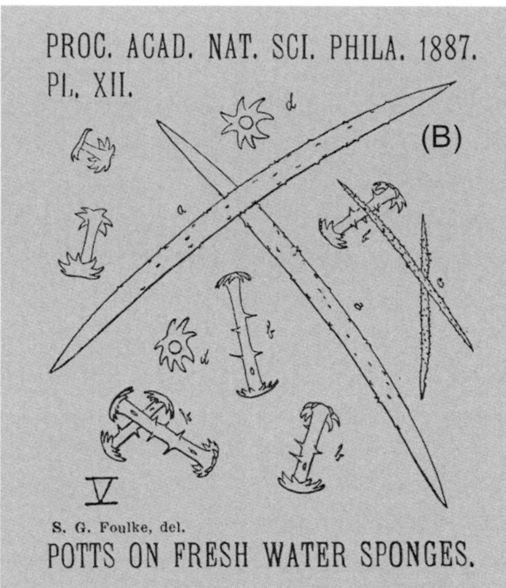

FIGURE 3.22 *Heteromeyenia latitenta*: (A) gemmule with funnel-shaped foramen bearing single long flat cirrus typical of the species, and gemmular theca (cross section) with a monolayer of gemmuloscleres radially arranged in the pneumatic layer; (B) spicular complement of gemmuloscleres, spiny oxeas (megascleres), and spiny oxeas (microscleres). *Figures modified from Potts, 1887.*

known of its gemmular foramen. The etymology of the specific epithet refers to the extremely long shafts of the gemmuloscleres. Reiswig et al. (2010) report 'It is almost certainly a junior synonym of *H. baileyi* as the two species cannot be separated by available characters.... Since, however, the taxonomic status of *H. longistylis* has not been formally resolved, it is included here in the valid species list.' Candido et al. (2010) also report that 'A study of samples collected throughout the area of distribution of the *D. radiospiculata* and *Heteromeyenia longistylis* may show that both are in reality the same species, what would take *Dosilia radiospiculata* (Mills, 1888) to the condition of junior synonym of *H. longistylis* (Mills, 1884).' We consider *H. longistylis* as valid until further detailed studies will be carried out.

Heteromeyenia latitenta (Potts, 1881a)

Carterella latitenta Potts, 1881a: 176

Heteromeyenia latitenta Penney & Racek, 1968; Manconi & Pronzato, 2002, 2007; Reiswig et al., 2010

Figure 3.22

Description. Growth form cushion-like, small. **Consistency** soft *in vivo*. **Color** green. **Surface** with distinct hispidation of more or less erected spicules. **Oscules** not conspicuous. **Inhalant apertures** scattered. **Ectosomal skeleton** without special architecture; tips of ascending fibers support the dermal membrane. **Choanosomal skeleton** irregular network of ascending fibers joined by secondaries, and scattered microscleres. **Spongin** scanty, except for the gemmular theca. **Basal spongin plate** not recorded. **Megascleres** oxeas (265–285 × 8–11 μm), straight, smooth to sparsely microspined except towards the sharply pointed tips. **Microscleres** oxeas (85–100 × 2–3 μm), slender, entirely spined with large spines in the central region. **Gemmules** subspherical (410–480 μm in diameter). **Foramen** tubular, long, with one or two very long, flat, ribbon-like cirrus projections originating from a flat disk. **Gemmular theca** with radially embedded gemmuloscleres. **Outer layer** with protruding radial gemmuloscleres. **Pneumatic layer** of spongin well developed with a monolayer of gemmuloscleres radially embedded. **Inner layer** not recorded. **Gemmuloscleres** birotules of one widely overlapping length group or two distinct length groups (50–55 to 60–78 μm in length) with stout shafts bearing numerous stout and pointed spines, and umbonate rotules (16–18 μm diameter) with margins bearing numerous conspicuous, recurve teeth.

Habitat. Lotic on rocks in rapidly running water.

Geographic distribution. Endemic to the Nearctic Region (type locality Chester Creek). Known until now exclusively from northeastern United States (New York, Pennsylvania, Ohio, Indiana, Wisconsin, and Illinois) and Mexico.

Remarks. Reiswig et al. (2010) report birotules of one widely overlapping length group or two distinct length groups, whereas Penney & Racek (1968) report two length groups. Detailed analysis of gemmular architecture and gemmuloscleres is needed to ensure correct genus allocation of this species. The etymology of the specific epithet refers to the shape of the several and notably long gemmular cirrus projections.

Heteromeyenia tubisperma (Potts, 1881b)

Carterella tubisperma Potts, 1881b: 150

Carterius tubisperma Potts, 1881b

Heteromeyenia tubisperma Schröder, 1927; Penney & Racek, 1968; Manconi & Pronzato, 2002, 2007; Annesley et al., 2008; Reiswig et al., 2010

Figure 3.23

Description. Growth form massive, encrusting. **Consistency** soft. **Color** brown to green. **Surface** uneven to

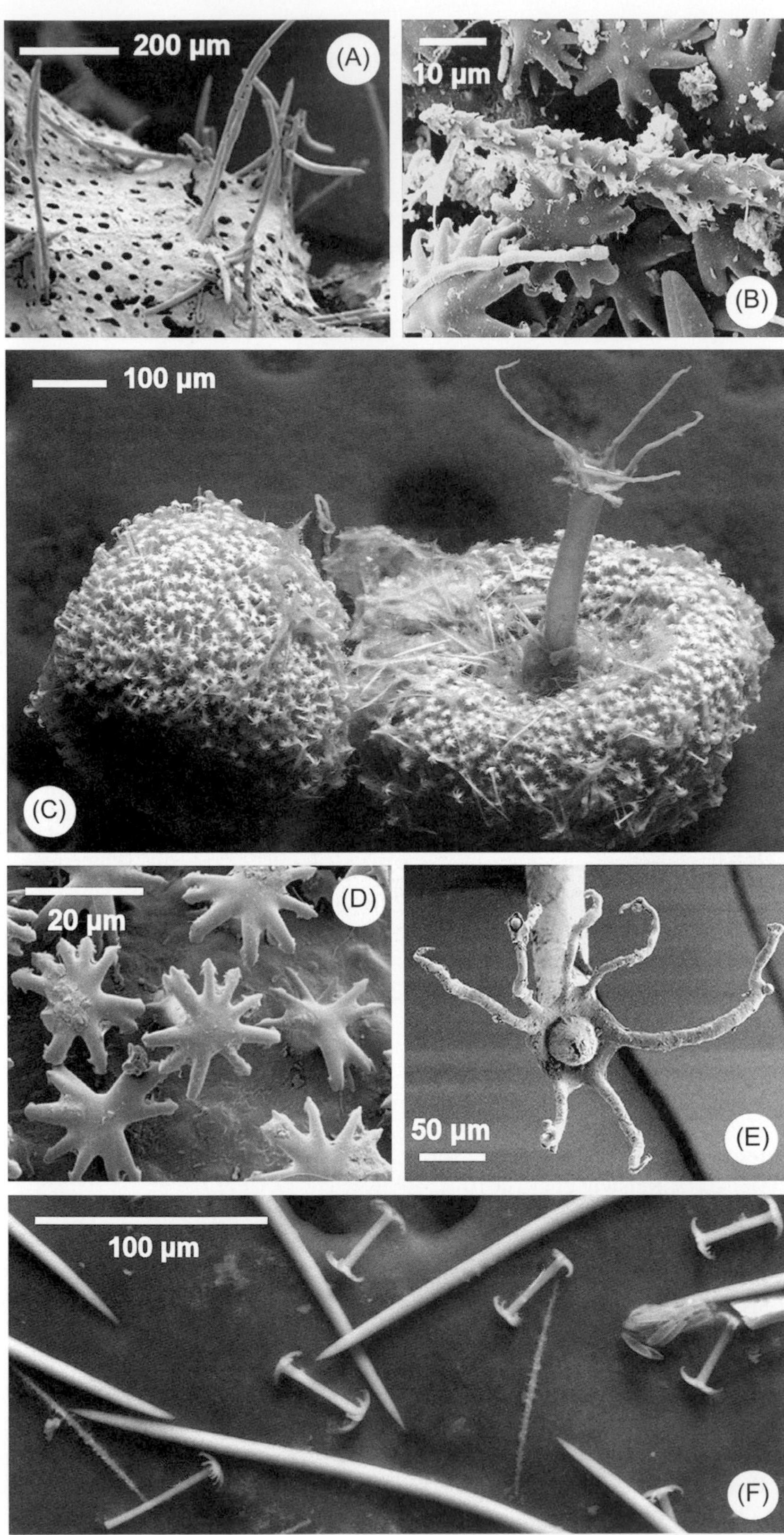

FIGURE 3.23 *Heteromeyenia tubisperma*: (A) sponge surface with cribrose dermal membrane supported by emerging megascleres at the tips of skeletal ascending fibers; (B) close-up of the gemmular surface armed by rotules of two types of gemmuloscleres, and a microsclere (center); (C) gemmules with funnel-shaped foramen bearing long cirri typical of the species; (D) gemmular surface armed by rotules of gemmuloscleres; (E) foramen with characteristic long cirri; (F) spicular complement with oxeas (megascleres), spiny slender oxeas (microscleres) and gemmuloscleres. *Figures modified from Annesley et al., 2008.*

slightly papillose. **Oscules** not conspicuous. **Inhalant apertures** scattered. **Ectosomal skeleton** without special architecture, tips of ascending fibers support the dermal membrane. **Choanosomal skeleton** irregular network, with ascending paucispicular primary fibers joined by secondaries, and scattered microscleres. **Spongin** scanty, except for the gemmular theca. **Basal spongin plate** not recorded. **Megascleres** oxeas (190–230×7–10 μm Penney & Racek, 1968; 190–(290)–337 μm in length Reiswig et al., 2010) slender, sharply pointed, smooth to sparsely microspiny. **Microscleres** oxeas (85–90×2–3 μm Penney & Racek, 1968; 73–(100)–118 μm in length Reiswig et al., 2010) slender, entirely spiny with spines small and recurve near the tips and larger, straight, and with knobs the central portion distinctly. **Gemmules** subspherical, single (500–550 μm in diameter) armed by radial megascleres. **Foramen** distinctly tubular, slender, very long (0.5–0.9 times the gemmule diameter) with 5–10 cylindrical cirrus projections in mature gemmules. **Gemmular theca** trilayered. **Outer layer** of compact spongin with protruding longer gemmuloscleres. **Pneumatic layer** of spongin with small, more or less rounded meshes, with a monolayer of gemmuloscleres radially embedded. **Inner layer** of compact spongin with sublayers. **Gemmuloscleres** birotules of one size class (33–(44)–70 μm in length) with stout shafts bearing a small number of acute spines and with both rotules of equal 12–(19)–25 μm diameters, consisting of an arrangement of lateral spines.

Habitat. Lotic in running water.

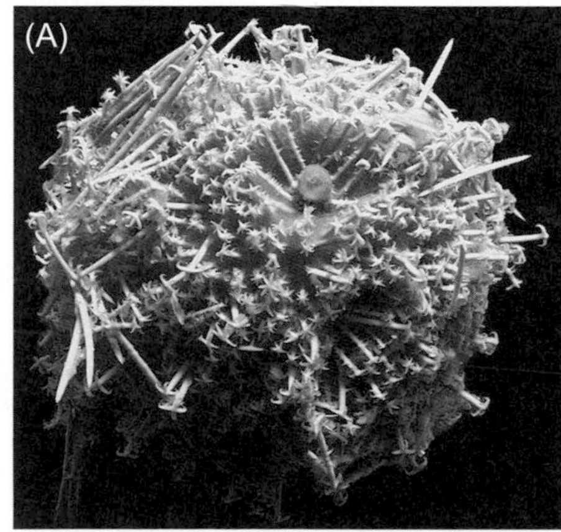

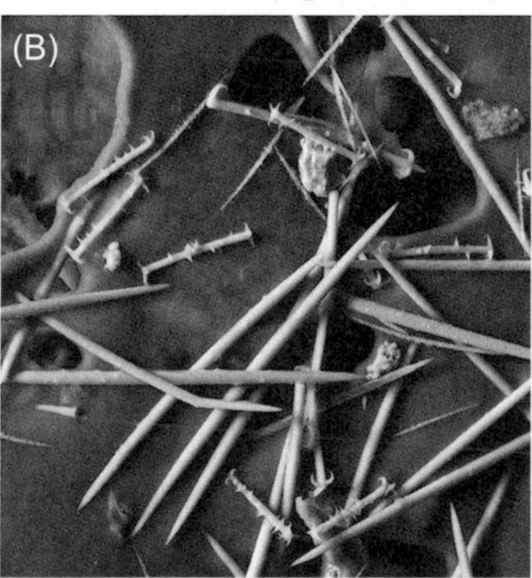

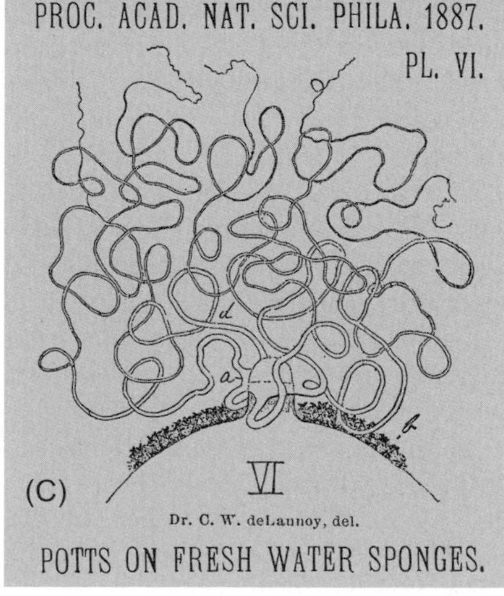

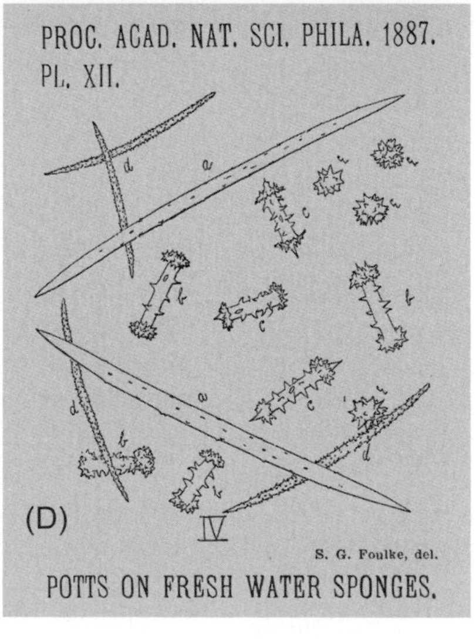

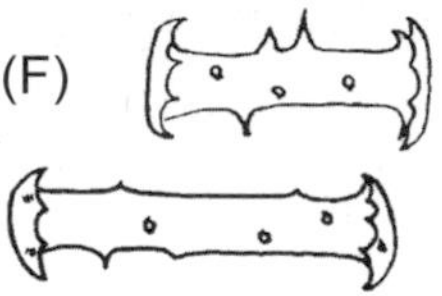

FIGURE 3.24 *Heteromeyenia tentasperma*: (A) gemmule armed by long gemmuloscleres protruding from outer layer, and evident tubular foramen lacking long cirri; (B) spicular complement (megascleres, microscleres, gemmuloscleres); (C) gemmular theca with gemmuloscleres radially arranged in the pneumatic layer and foramen with long cirri typical of the species; (D) spicular complement (megascleres, microscleres, gemmuloscleres); (E) spiny oxeas (microsclere); (F) gemmuloscleres. *Figures A and B modified from Annesley et al., 2008; C and D modified from Potts, 1887; E and F modified from Penney & Racek, 1968.*

Geographic distribution. Endemic to the Nearctic Region. Known until now from the eastern half of North America (California, Florida, Illinois, Indiana, Iowa, Kansas, Massachusetts, Michigan, New York, Ohio, and Wisconsin).

Remarks. Further detailed studies on this species are necessary to ascertain the existence of one or two size classes of gemmuloscleres as reported by Reiswig et al. (2010) and Penney & Racek (1968) respectively. *H. tentasperma* can be distinguished from *H. baileyi*, which has shorter microscleres with longer spines. The etymology of the specific epithet refers to the morphology of the gemmular foramen as a long tube.

Heteromeyenia tentasperma (Potts, 1880)

Spongilla tentasperma Potts 1880: 331

Spongilla tenosperma Potts 1880: 356

Carterius tenosperma Potts, 1887

Heteromeyenia tentasperma Penney & Racek, 1968; Manconi & Pronzato, 2002, 2007; Annesley et al., 2008; Reiswig et al., 2010

Figure 3.24

Description. Growth form encrusting to shallow cushion-like. **Consistency** not reliably recorded. **Color** yellowish to green. **Surface** hispid. **Oscules** not conspicuous. **Inhalant apertures** scattered. **Ectosomal skeleton** without special architecture, tips of ascending fibers support the dermal membrane. **Choanosomal skeleton** irregular network of ascending paucispicular primary fibers joined by secondaries, and scattered microscleres. **Spongin** scanty, except for the gemmular theca. **Basal spongin plate** not recorded. **Megascleres** oxeas (260–280×7–10 µm in length) very slender, sharply pointed, with sparse microspines. **Microscleres** oxeas (75–80×2–3 µm in length) slender with sparse microspines. **Gemmules** subspherical (420–450 µm in diameter) light yellow to brown, with gemmuloscleres radially arranged. **Foramen** distinctly tubular, short with 3–6 long and irregular cirrus projections resembling tentacles. **Gemmular theca** trilayered. **Outer layer** of compact spongin with emerging longer gemmuloscleres. **Pneumatic layer** of spongin well-developed with shorter gemmuloscleres entirely embedded. **Inner layer** of compact spongin. **Gemmuloscleres** pseudobirotules of two length groups (50–55 and 65–72 µm) with stout shafts bearing a small number of acute spines and with both burr-like pseudorotules of equal (15–18 µm) diameters and consisting of an arrangement of lateral spines.

Habitat. Lotic in running water. On aquatic plants and rocks. Symbiotic with algae.

Geographic distribution. Endemic to the Nearctic Region. Known until now from United States (New Jersey, New York, Ohio, and Pennsylvania).

Remarks. The etymology of the specific epithet refers to the extremely long tentacle-like cirrus projections, that contribute, together longer gemmuloscleres, to cluster gemmules after degeneration of the sponge.

Racekiela Bass & Volkmer-Ribeiro, 1998

Racekiela ryderi (Potts, 1882)

Heteromeyenia ryderi Potts, 1882: 13; Jewell, 1935, 1939; Neidhoefer, 1940; Ricciardi & reiswig, 1993; Manconi & Pronzato, 2002

Anheteromeyenia pictovensis Potts, 1885a

Anheteromeyenia ryderi Penney & Racek, 1968

Racekiela ryderi Manconi & Pronzato, 2007; Reiswig et al., 2010

Figure 3.25

Description. Growth form massive from cushion to dome-shaped, with lobes, several inches in diameter. **Consistency** fragile both *in vivo* and dry condition. **Color** light green *in vivo*. **Surface** more or less lobed, with radial exhalant canals. **Oscules** with five-six radial canals. **Inhalant apertures** scattered. **Ectosomal skeleton** without special architecture. **Choanosomal skeleton** irregular, delicate network of paucispicular ascending fibers joined by secondaries. Slender, small, smooth oxeas presumably belong to larvae. **Spongin** scanty, except for the gemmular theca. **Basal spongin plate** not recorded. **Megascleres** oxeas (190–220×3–19 µm Penney & Racek, 1968; 141–(220)–279 Reiswig et al., 2010) variably fusiform, gradually pointed, entirely spiny except at the tips with broadly conical spines often projected towards the tips. **Microscleres** absent. **Gemmules** spherical, single (320–350 µm in diameter Penney & Racek, 1968; 300–400 µm in diameter Ricciardi & Reiswig, 1993) scattered in the skeletal network. **Foramen** short, simple on a conical elevation. **Gemmular theca** trilayered with radially embedded gemmuloscleres. **Outer layer** of compact spongin with protruding gemmuloscleres. **Pneumatic layer** of spongin chambers small, more or less rounded meshes. **Inner layer** of compact spongin with sublayers, in contact with proximal rotules. **Gemmuloscleres** birotules long of two distinct size categories but equal in rotules diameter, and, with clear differences in their shapes; short birotules (30–40 µm Penney & Racek, 1968; 28–(34)–41 µm Reiswig et al., 2010) with shafts, with only one or a few straight spines, rapidly enlarging towards the flat, often microspined, rounded rotules with a margin, sometimes incurved, lacinulate to crenulate bearing a large number of small teeth and a diameter nearly as great as the length of the shaft; long birotules (50–75 µm Penney & Racek, 1968; 45–(64)–75 µm Reiswig et al., 2010) robust, with numerous recurve spines on their shaft and with strongly recurve hooks on their tips to form a pseudorotule with umbone, sometimes spiny at the apex.

Habitat. Lotic on timber and stones in shallow water. Megascleres extremely variable from habitat to habitat. Sometimes evergreen in winter under ice, with few gemmules (*H. pictovensis* from Nova Scotia).

Geographic distribution. Known until now from the Nearctic Region (type locality Cobb's Creek, tributary of Delaware River, near Philadelphia) from Canada and United States (Connecticut, Florida, Indiana, Iowa,

PHYLUM PORIFERA

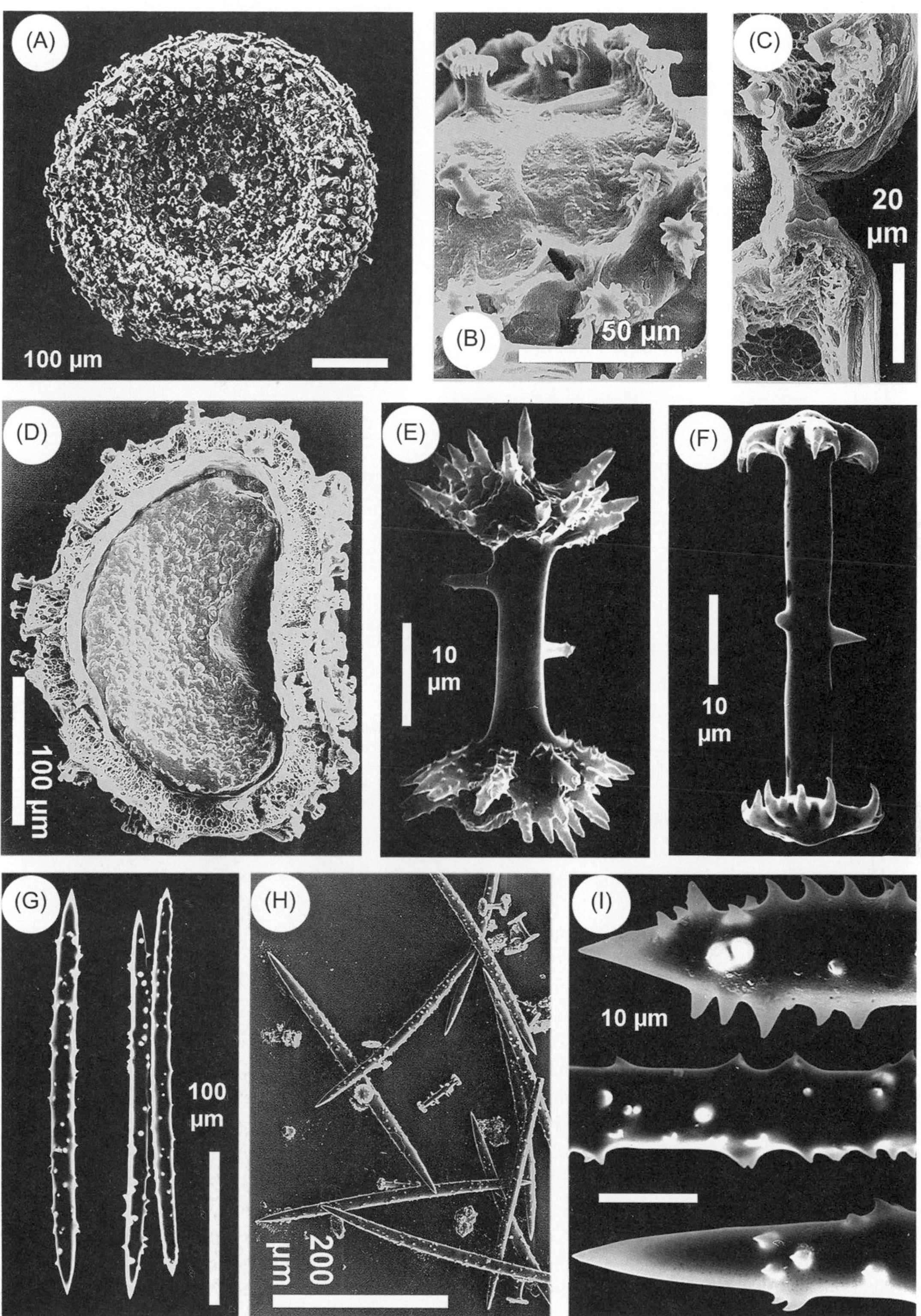

FIGURE 3.25 *Racekiela ryderi*: (A) gemmule armed by gemmuloscleres with evident foramen (center); (B) close-up of the gemmular outer layer of compact spongin with gemmuloscleres; (C) gemmular foramen with pneumatic layer (cross section); (D) gemmular theca with gemmuloscleres radially embedded in the pneumatic layer of chambered spongin (cross section); (E, F) different forms of gemmuloscleres; (G) spiny oxeas (megascleres); (H) spicular complement (megascleres and gemmuloscleres); (I) tips and shaft of megascleres.

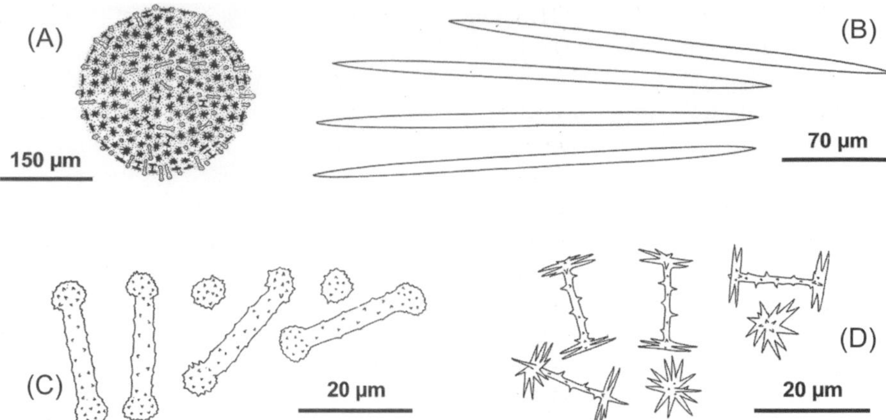

FIGURE 3.26 *Racekiela biceps*: (A) gemmule with two types of gemmuloscleres from radially to tangentially arranged; (B) oxeas (megascleres); (C, D) gemmuloscleres of two morphs, biceps (left) and birotules (right). *Figures modified from Lindenschmidt, 1950.*

Louisiana, Massachussetts, Michigan, New Hampshire, New Jersey, New Scotland, New York, Pennsylvania, Texas, and Virginia). Amphiatlantic in the N-Hemisphere. Also recorded from Belize.

Remarks. Type species of the genus. The etymology of the specific epithet refers to the collector Mr. J.A. Ryder.

Racekiela biceps (Lindenschmidt, 1950)

Anheteromeyenia biceps Lindenschmidt, 1950: 214; Penney & Racek, 1968; Manconi & Pronzato, 2002; Reiswig et al., 2010

Racekiela biceps Manconi & Pronzato, 2007

Figure 3.26

Description. Growth form cushion-like. **Consistency** rather firm *in vivo* and dry condition. **Color** yellow to green. **Surface** even. **Oscules** not conspicuous. **Inhalant apertures** scattered. **Ectosomal skeleton** without special architecture. **Choanosomal skeleton** irregular network of ascending primary fibers joined by secondaries. **Spongin** scanty, except for the gemmular theca. **Basal spongin plate** not recorded. **Megascleres** oxeas (260–310×15–17 μm Penney & Racek, 1968; 255–325 μm in length Reiswig et al., 2010) slender, smooth to spiny on the entire shaft with microspines except at tips. **Microscleres** absent. **Gemmules** subspherical (350–380 μm in diameter) abundant in the basal portion. **Foramen** simple, without tube, slightly elevated. Gemmular theca trilayered. Outer layer of compact spongin with emerging tips of gemmuloscleres. Pneumatic layer small, of more or less rounded meshes with radially, sometime slanting, embedded gemmuloscleres. Inner layer of compact spongin with sublayers. **Gemmuloscleres** of two types; short birotules (23–26×2 μm Penney & Racek, 1968; 17–22 μm in length Reiswig et al., 2010) abundant with notably slender shaft and flat, deeply serrate rotules (20–22 μm in diameter); long biceps, modified birotules (30–42×4–5 μm Penney & Racek, 1968; 24–30 μm in length Reiswig et al., 2010) entirely spiny, with stout shaft and knob-like rotules bearing coarse and blunt spines.

Habitat. Lotic water (inlet and outlet of a lake).

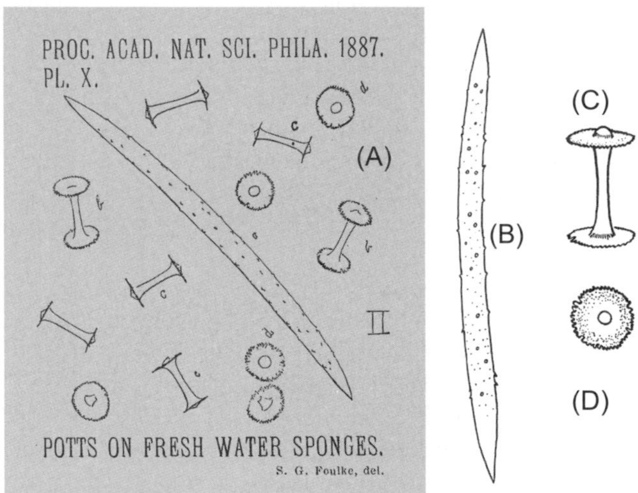

FIGURE 3.27 *Ephydatia millsi* (A) spicular complement (gemmuloscleres and megasclere); (B) oxea (megasclere); (C) birotule (gemmuloscleres); (D) rotule of a gemmulosclere. *Figure A modified from Potts, 1887; B modified from Penney & Racek, 1968.*

Geographic distribution. Endemic to the Nearctic Region. Known until now only from creeks near Douglas Lake, Cheboygan Co., Michigan.

Remarks. We consider *R. biceps* a valid species, although other authors report that 'This geographically, very restricted species is likely a variant of one of the more common species suggested that it is an ecomorph of *E. muelleri*, but their supporting evidence was unconvincing. The more likely possibility that this is an aberrant form of *E. fluviatilis* has not yet been explored (Reiswig et al., 2010).' The etymology of the specific epithet means 'two heads' in Latin and refers to the morphology of longer gemmuloscleres.

Ephydatia Lamouroux, 1816

Ephydatia millsi (Potts, 1887)

Meyenia millsii Potts, 1887: 225

Ephydatia millsii Smith, 1921; Penney & Racek, 1968; Manconi & Pronzato, 2002, 2007; Reiswig et al., 2010

Figure 3.27

Description. Growth form encrusting to massive, moderately thin, and flat. **Consistency** soft, texture loose. **Color** not recorded. **Surface** slightly hispid. **Oscules** inconspicuous. **Inhalant apertures** not recorded. **Ectosomal skeleton** with no special architecture; tips of ascending fibers support the dermal membrane. **Choanosomal skeleton** irregularly reticulate with ascending primary fibers joined by secondaries. **Spongin** scanty, except for the gemmular theca. **Basal spongin plate** not recorded. **Megascleres** slender oxeas (180–270×9–12 μm) nearly straight, spiny, with numerous microspines and few, scattered, low conical spines except at the abruptly pointed tips. **Microscleres** absent. **Gemmules** (300–360 μm in diameter) abundant, small, subspherical, smooth, single. **Foramen** simple, slightly elevated. **Gemmular theca** with radially embedded gemmuloscleres in a dense monolayer. **Pneumatic layer** feebly developed irregular chambers. **Inner layer** not described. **Gemmuloscleres** abundant birotules of one size class with smooth shafts (36–48 μm in length) increasing in thickness near the flat, umbonate, circular rotules of equal diameters (23–28 μm) with only very small incisions at their lacinulate margins, outer surface granulate.

Habitat. Lentic.

Geographic distribution. Endemic to the Nearctic Region. Known until now exclusively from the type locality Sherwood Lake near DeLand, Florida.

Remarks. Reiswig et al. (2010) report that this species may eventually prove to be an ecomorph of one of the more widely distributed *Ephydatia* species. Gemmuloscleres, however, resembles the small class of birotules in *Racekiela ryderi*. The etymology of the specific epithet refers to the name of the collector Mr. Henry Mills, therefore the epithet is *millsi* and not *millsii*.

Ephydatia muelleri (Lieberkühn, 1856)

Spongilla mülleri Lieberkühn, 1856: 510

Spongilla heterosclerifera Smith, 1918

Ephydatia mülleri Smith, 1921; Penney & Racek, 1968; Ezcurra de Drago, 1975, 1976

Ephydatia muelleri Manconi & Pronzato, 2002, 2007, 2009; Pronzato & Manconi, 2002; Reiswig et al., 2010

Figure 3.28

Description. Growth form encrusting to thin cushion. **Color** gray, yellowish, light brown, or green. **Consistency** compact and quite hard. **Oscules** scattered. **Surface** slightly hispid. **Ectosomal skeleton** with no special architecture, tips of ascending fibers support the dermal membrane. **Choanosomal skeleton** anisotropic with primary ascending fibers joined by vague secondaries. **Spongin** scanty except for gemmular theca. **Megascleres** acanthoxeas (171–(245)–350×9–20 μm) straight to slightly curved, covered by small spines except at the tips, rarely entirely smooth. **Microscleres** absent. **Gemmules** subspherical (350–450 μm in diameter) quite abundant and scattered singly in the choanosomal skeleton or in clusters at the sponge base. **Foramen** slightly elevated with flattened collar. **Gemmular theca** trilayered. **Outer layer** of compact spongin, usually scarcely developed, totally or partially covering emerging distal rotules of radially embedded gemmuloscleres. **Pneumatic layer** of chambered spongin with small, more or less rounded meshes and 1–4 layers of gemmuloscleres radially embedded. **Inner layer** of sublayered compact spongin. **Gemmuloscleres** birotules one class (8–(17)–28 μm in length) with short, thick, smooth shaft (5–10 μm) and rotules with indented margins (12–20 μm in diameter) with flat, irregularly shaped rotules of equal, 8–(15)–27 μm diameters and usually with fewer than 12 teeth deeply incised into long rays.

Habitat. In lentic and running waters (rivers, lakes, pools, caves) on rocks, boulders, wood, and artificial substrata.

Geographic distribution. Widely scattered in the Nearctic Region (Colorado, Illinois, Iowa, Massachusetts, Michigan, Montana, Newfoundland, New York, Nova Scotia, Pennsylvania, and Virginia) from cold to warm climates with a preference for temperate regions (Western Palaearctic, Nearctic, Mesoamerica and Asia) but also recorded from Sub-Saharan Africa (see Manconi, 2008).

Remarks. The status of this species is problematic, and several varieties were described. Morphological variations in the spicular complement and gemmular architecture suggest the possible existence of a species complex, supported also by the wide geographic range. *E. muelleri* diverges from *E. fluviatilis* for gemmulosclere length less than or equal to rotule diameter in *E. muelleri* (Ricciardi & Reiswig, 1993). *Ephydatia cooperensis* (Peterson and Addis) described, only from three lakes in the northern rocky mountains of western Montana, as the only member of a new genus, *Clypeatula*, now considered a synonym of *Ephydatia*, is a sister species of *E. fluviatilis*, as suggested by a recent molecular investigation. The etymology of the specific epithet refers to the name of J. Müller, the director of the Müller Archiv on which the original description of this species was published.

Ephydatia subtilis Weltner, 1895

Ephydatia subtilis Weltner, 1895: 141; Harrison, 1979; Reiswig et al., 2010

Description. Growth form encrusting, very thin. **Consistency** fragile. **Color, surface, inhalant apertures** and **oscules** not recorded. **Ectosomal skeleton** without special architecture, tips of ascending fibers support the dermal membrane. **Choanosomal skeleton** irregular reticulate network of ascending primary fibers joined by secondaries. **Spongin** scanty, except for the gemmular theca and the basal spongin plate. **Basal spongin plate** developed. **Megascleres** notably slender oxeas (158×2.6 μm) with sparse short spines. **Microscleres** absent. **Gemmules** subspherical, single (176 μm in diameter). **Foramen** simple. **Gemmular**

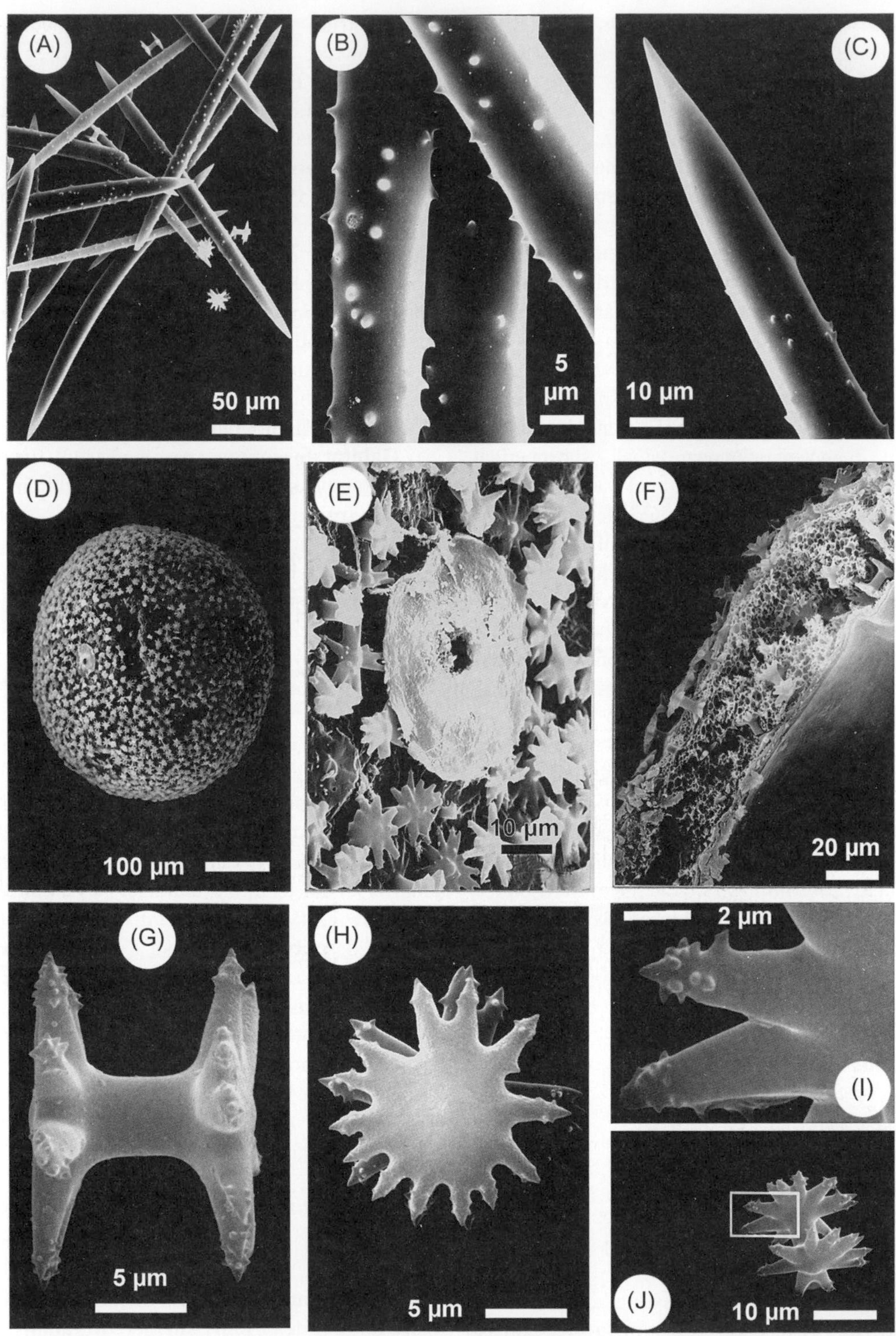

FIGURE 3.28 *Ephydatia muelleri*: (A) spicular complement (megascleres, gemmuloscleres); (B) spiny shafts of oxeas (megascleres); (C) oxea with smooth tip (megascleres); (D) entire gemmule with evident foramen (left) and distal rotules of gemmuloscleres in the gemmular outer layer; (E) foramen with collar at the gemmular surface; (F) gemmular theca with layers of gemmuloscleres embedded in pneumatic layer of chambered spongin and multi-layered inner layer (cross section); (G–J) different views of birotules (gemmuloscleres).

theca trilayered. Outer layer of compact spongin thin with a monolayer of radial gemmuloscleres. **Pneumatic layer** of spongin scarcely developed, thin. Inner layer not described. **Gemmuloscleres** delicate, slender birotules of variable size (average 23 μm in length) with smooth shaft (1.7 μm in thickness) and rotules (9.5 μm in diameter) deeply incised with 10–20 blunt rays.

Habitat. Lotic/lentic, on submerged sticks.

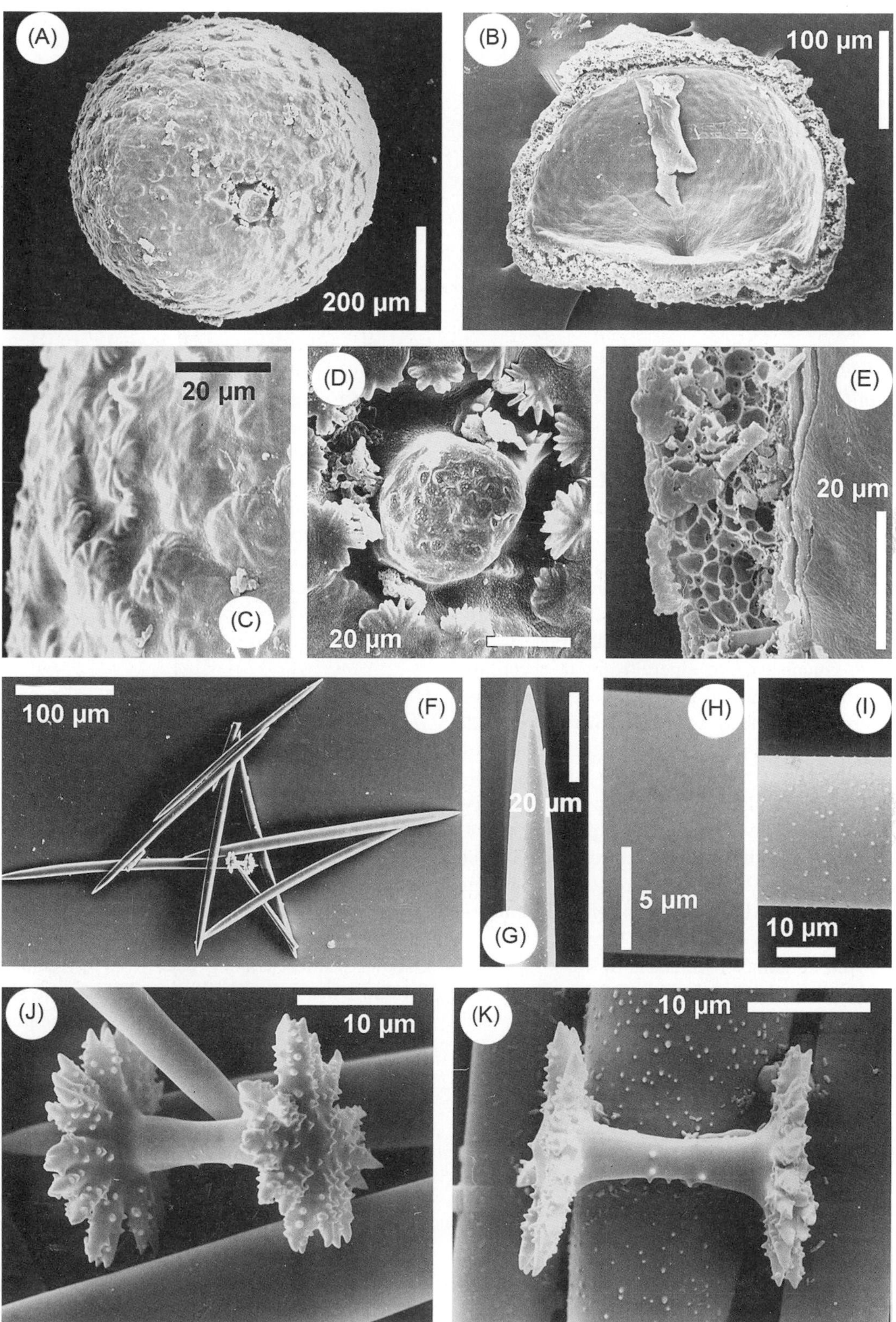

FIGURE 3.29 *Ephydatia fluviatilis*: (A) gemmule with evident foramen at the surface armed by tips of birotules covered by outer layer of compact spongin; (B) gemmule with birotules radially arranged in the pneumatic layer (cross section); (C) close-up of the gemmular outer layer covering gemmuloscleres tips; (D) foramen simple surrounded by distal rotules of gemmuloscleres; (E) gemmular theca with chambered pneumatic layer and inner layer (right) of compact multilayered spongin; (F) oxeas (megascleres) and a single birotule (gemmuloscleres); (G) tip of oxea (megascleres); (H, I) smooth (H) and tubercled (I) surface of oxeas (megascleres); (J, K) birotules (gemmuloscleres) among oxeas (megascleres).

Geographic distribution. Endemic to the Nearctic Region. Known until now only from the type locality Lake Kissimmee (Lake Tohopekaliga) in Florida.

Remarks. This often ignored species, known only from Weltner's (1895) original description, has never been figured. Although spicules are very similar to some forms of *E. fluviatilis*, rotules of the gemmuloscleres are much smaller than those of that species. Attempts to obtain new specimens from the type location failed (Harrison, 1974, 1979). The uncertain status of this species needs the analysis of the type specimen in the Humboldt Museum, Berlin. The etymology of the specific epithet refers to the delicate spicules.

Ephydatia fluviatilis (Linnaeus, 1759)

Spongia fluviatilis Linnaeus, 1759: 1348

Meyenia robusta Potts, 1887

Ephydatia fluviatilis Penney & Racek, 1968; Ezcurra de Drago, 1975; Ricciardi & Reiswig, 1993; Manconi & Pronzato, 2002, 2007, 2009, 2011; Pronzato & Manconi, 2002; Reiswig et al., 2010

Figure 3.29

Description. Growth form from encrusting, to bulbous or massive, with ridges or rounded lobes. **Consistency** firm but fragile in life, extremely brittle if dry. **Color** whitish, brown, greenish up to emerald green, according with the exposition to light and presence/absence of symbiotic zoochlorellae. **Oscules** scattered. **Surface** hispid. **Ectosomal skeleton** with no special architecture, tips of ascending primary fibers emerging at the surface. **Choanosomal skeleton** reticulate with paucispicular primary ascending fibers joined by secondaries. **Spongin** scanty, except for the gemmular theca. **Megascleres** slightly bent to rarely straight, oxeas (210–(343)–439 × 6–19 µm) from smooth to microspiny. **Microscleres** absent. **Gemmules** subspherical (350–450 µm in diameter Penney & Racek, 1968; 400–600 µm in diameter Ricciardi & Reiswig, 1993) abundant to rare, from scattered to grouped in carpets at the sponge basal portion. **Foramen** simple slightly elevated with a small collar. **Gemmular theca** trilayered. **Outer layer** well-developed, covering or not the distal rotule of gemmuloscleres. **Pneumatic layer** variably developed with irregular rounded chambers, and gemmuloscleres radially embedded in one layer. **Inner layer** of sublayered compact spongin in contact with the proximal rotule of birotules. **Gemmuloscleres** birotules of one class (20–(23)–30 × 13–(18)–24 µm) with a slender to stout, smooth to spiny shaft with a few large spines; rotules of equal diameter, flat, microspiny, with incised irregular margins bearing not less than 20 teeth that are not deeply incised.

Habitat. Lentic and lotic from springs to coastal brackish waters to inland salt lakes and caves on a variety of substrata (pebbles, rocks, shells, wood, glass, plastic, metallic objects, cement). Patches of dormant gemmules during dormancy remain adherent to dry hard substrata or can spread in the water body. Frequently associated with the invasive zebra mussel *Dreissena polymorpha*.

Geographic distribution. *E. fluviatilis* is one of the most common species widely distributed in the Nearctic and throughout the Holoarctic Region with scattered findings in the Afrotropical Region. Some reports of *E. fluviatilis* in the Nearctic Region probably referred to *E. muelleri* are considered with doubt. Apparently cosmopolitan with more frequent occurrence in temperate than in tropical zones (see Manconi, 2008; Manconi & Pronzato, 2009, 2011).

Remarks. *E. fluviatilis* diverges from *E. muelleri* for gemmuloscleres length always greater than rotule diameter in *E. fluviatilis* (Ricciardi & Reiswig, 1993). The status of this species is problematic. *Ephydatia japonica* Hilgendorf, 1882 and *Ephydatia robusta* Penney & Racek, 1968 from the Nearctic Region were considered junior synonym of *E. fluviatilis* by Reiswig et al. (2010). The wide geographic range and variability in both life cycle timing (aestivant vs. hibernant vs. perennial) and diagnostic morphotraits suggest that *E. fluviatilis* represents a complex of species. The etymology of the specific epithet refers to the lotic habitat (river) from where this species was firstly described by Linnaeus (1759).

Dosilia Gray, 1867

Dosilia palmeri (Potts, 1885b)

Meyenia plumosa var. *palmeri* Potts, 1885b: 587

Dosilia palmeri Penney & Racek, 1968; Manconi & Pronzato, 2002, 2007; Candido et al., 2010; Reiswig et al., 2010; Annesley et al., 2011

Figure 3.30

Description. Growth form massive, spherical, of large size. **Consistency** soft, very fragile, brittle. **Color** dark brown in dry condition. **Surface** with spicular dermal membrane, and hispidation due to the tips of ascending fibers. **Oscules** scattered along a network of subdermal canals. **Inhalant apertures** scattered. **Ectosomal skeleton** with microscleres and tips of ascending fibers supporting the dermal membrane without special architecture. **Choanosomal skeleton** reticulate, anisotropic network of ascending, paucispicular primary fibers joined by irregular secondaries, and with scattered microscleres. **Spongin** scanty, except for the gemmular theca and the basal spongin plate. **Basal spongin plate** not recorded. **Megascleres** slender microspined oxeas abruptly pointed (370–450 × 14–20 µm), slightly bent to nearly straight, with sparse small spines in the central portion. **Microscleres** from dominant aster-like (star-like with 8–12 smooth rays) of extremely variable size, to less frequent oxeas with few long rays. **Gemmules** scattered mainly towards the sponge basal portion, spherical to ovoid (450–690 µm in diameter) armed by radially arranged gemmuloscleres. **Foramen** as a short, straight tube not surpassing the outer layer of the theca. **Gemmular theca** trilayered with outer layer of compact spongin armed

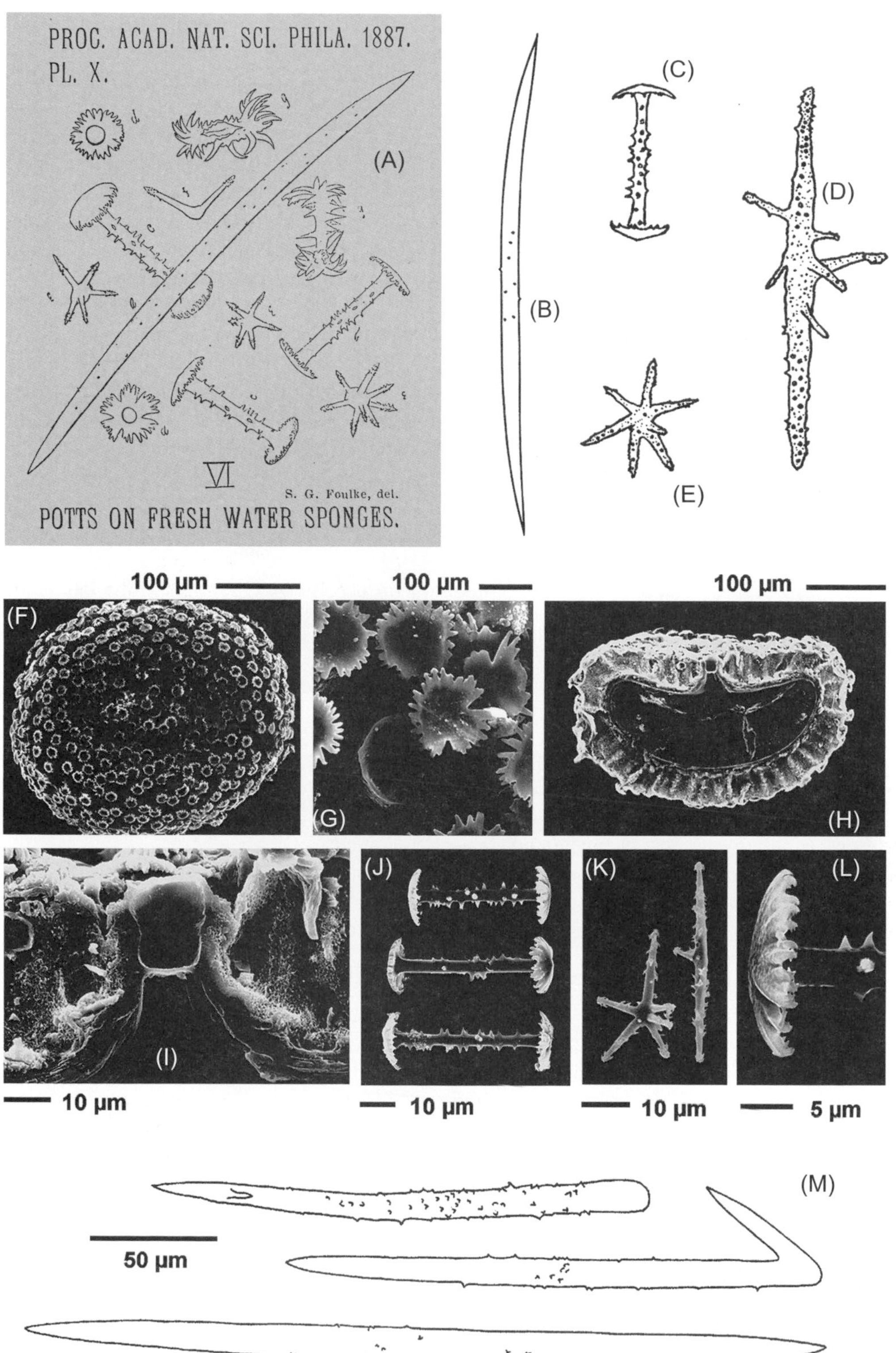

FIGURE 3.30 *Dosilia palmeri*: (A) spicular complement (gemmuloscleres, microscleres, megascleres); (B) spiny oxea (megasclere); (C) spiny birotule (gemmulosclere); (D, E) spiny microscleres from aster-like (E) to monaxial with long spines at the middle (D). The "g" type spicule of Potts are not represented by Penney & Racek (1968); the "D" type spicule of Penney & Racek (1968) is not reported by Potts; (F) gemmule with surface armed by gemmuloscleres; (G) close up of the gemmule surface showing the foramen and the rotules of gemmuloscleres; (H) gemmular theca with radial gemmuloscleres embedded in pneumatic layer (cross section); (I) gemmular foramen (cross section); (J) birotules (gemmuloscleres); (K) aster-like to oxeas (microscleres); (L) rotule of a gemmulosclere; (M) oxeas with spines and malformations (megascleres). *Figure A modified from Potts, 1887; B–E modified from Penney & Racek, 1968; F–L modified from Cândido et al., 2010.*

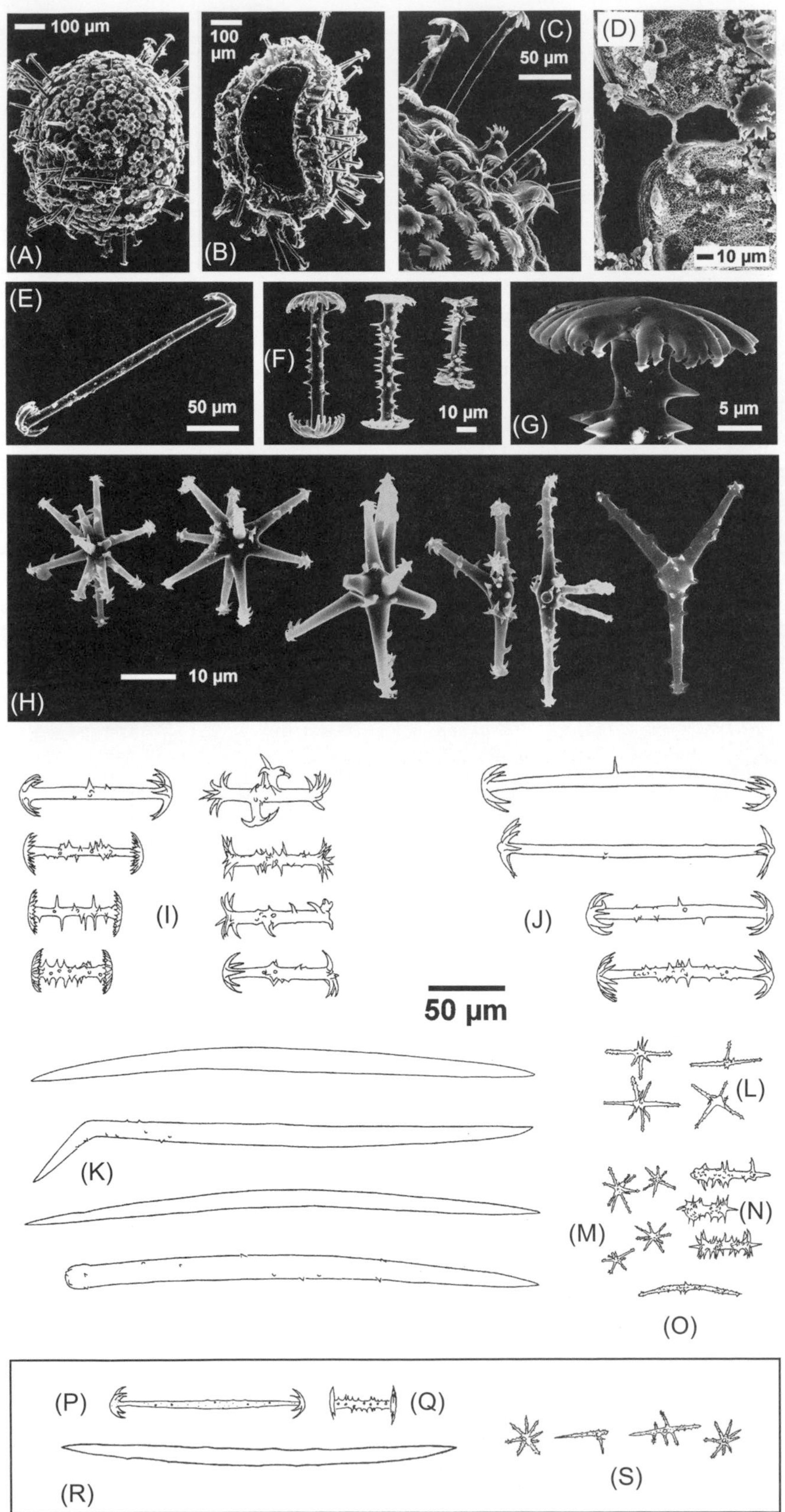

FIGURE 3.31 *Dosilia radiospiculata*: (A) gemmule with surface armed by gemmuloscleres; (B) gemmular theca with radial gemmuloscleres (cross section); (C) close-up of the gemmular outer layer with long and short gemmuloscleres; (D) gemmular foramen (cross section); (E) pseudobirotule (longer gemmulosclere); (F) birotules (shorter gemmuloscleres); (G) close-up of the rotule and shaft of a short gemmulosclere; (H) microscleres; (I) shorter gemmuloscleres of different morphs; (J) longer gemmuloscleres of different morphs; (K) megascleres; (L, M, N, O) microscleres of different morphs; (P, Q) gemmuloscleres; (R) megasclere; (S) microscleres. Penney & Racek (1968) report a few morphs of spicules (box at the bottom). *Figures A–O modified from Cândido et al., 2010; P–S modified from Penney & Racek, 1968.*

by distal rotules, sometimes protruding, of radially arranged gemmuloscleres. **Pneumatic layer** of spongin fibers with more or less rounded, small meshes, armed by radially arranged gemmuloscleres. **Inner layer** of compact spongin with sublayers. **Gemmuloscleres** birotules (55–85 μm in length) with strong spines on their central shaft (5–6 μm in diameter) and with equal sized rotules (23–28 μm in diameter) bearing numerous blunt recurve teeth resembling hooks of pseudobirotules.

Habitat. Lotic in floodplain areas in arid climate, on small trees along banks. Long-term aestivation (9–10 months); active phase in May–June characterized by a rapid growth. Sponges frequently found floating in water tanks about one month after the beginning of rainy season.

Geographic distribution. Nearctic and Neotropical (Mesoamerica) Region. Known until now from southern United States (Arizona, Florida, Louisiana, New Mexico, and Texas) and Mexico, and possibly other locations in Central America.

Remarks. The etymology of the specific epithet refers to the name of the collector Dr. Edward Palmer along the banks of Colorado River (type locality, near Lerdo, Sonora, New Mexico).

Dosilia radiospiculata (Mills, 1888)

Heteromeyenia radiospiculata Mills, 1888: 313

Dosilia radiospiculata Penney & Racek, 1968; Manconi & Pronzato, 2002, 2007; Reiswig et al., 2010; Annesley et al., 2011

Figure 3.31

Description. Growth form massive, bulbous. **Consistency** moderately soft, very brittle. **Color** not recorded. **Surface** nodular. **Oscules** not conspicuous *in vivo*, scattered in a network of subdermal canals. **Inhalant apertures** scattered. **Ectosomal skeleton** without special architecture, tips of ascending fibers support the dermal membrane. **Choanosomal skeleton** reticulate network, with ascending primary fibers joined by irregularly arranged secondaries, and with scattered microscleres. **Spongin** scanty, except for the gemmular theca. **Basal spongin plate** not recorded. **Megascleres** slender oxeas (290–400 × 14–23 μm), gradually pointed, slightly bent, from entirely smooth to entirely spiny by microspines. **Microscleres** aster-like (star-like), extremely variable in size, moderately abundant also near the gemmules, with a distinct central nodule bearing 6–10 microspined rays of extremely variable length with recurve spines to form, occasionally, a pseudorotule at the tips. Rare spiny oxeas with long straight rays in the central portion also present. **Gemmules** scattered mainly towards the sponge basal portion, subspherical (540–610 μm in diameter) with radially embedded gemmuloscleres. **Foramen** as a short, straight tube not surpassing the outer layer of the theca. **Gemmular theca** trilayered with gemmuloscleres radially embedded. **Outer layer** of compact spongin with longer gemmuloscleres (pseudobirotules) emerging from the surface with notably long distal portions. **Pneumatic layer** of spongin with more or less rounded small chambers bearing entirely embedded shorter gemmuloscleres. **Inner layer** of compact spongin with sublayers. **Gemmuloscleres** of two types, pseudobirotules and birotules, of distinctly different size classes; pseudobirotules long (120–230 μm) with bent, entirely smooth or with short spines fusiform shafts (16–20 μm in thickness), and apical pseudorotules of few, long, strongly recurve hooks; birotules short (45–82 μm) with spiny shaft (8–19 μm in thickness) bearing umbonate rotules (22–26 μm in diameter) with irregularly indented margins to bearing recurve teeth resembling hooks of pseudobirotules.

Habitat. Lotic in floodplains subjected to water level variations. On vegetation and artificial substrata (dam walls).

Geographic distribution. Known until now from southern Canada, United States (Arizona, California, Illinois, Louisiana, Ohio, and Texas) and Mexico. Type locality near Cincinnati along the Ohio River (Ohio).

Remarks. The etymology of the specific epithet refers to the aster-like (star-like) microscleres. Data in the literature focus on the problematic status of North American species belonging to the genus *Dosilia*. Penney & Racek (1968) report that '…it will be necessary to change its present specific name, since it represents a junior homonym of *D. plumosa* (Carter). The *nomen novum Dosilia heterogena* is herewith proposed.' Focusing on this point Candido et al. (2010) report 'The analysis of the type material of *Heteromeyenia plumosa* Weltner, 1895 (ZMB 1477) permitted the detection of the similarity, in size and shape, of the spicular components, with those of *Dosilia radiospiculata* (Mills, 1888), fully conforming to the description of the latter. The area of distribution is also the same. With this it is possible to definitively exclude the proposal of a *nomen novum Dosilia heterogena* put forward by Penney & Racek (1968) for *Heteromeyenia plumosa* of Weltner, placing it as a synonym of *Dosilia radiospiculata*.' Candido et al. (2010) report also that 'A study of samples collected throughout the area of distribution of the *D. radiospiculata* and *Heteromeyenia longistylis* may show that both are in reality the same species, what would take *Dosilia radiospiculata* (Mills, 1888) to the condition of junior synonym of *H. longistylis* (Mills, 1884).' We consider *D. radiospiculata* as valid species, and we maintain here the name *D. radiospiculata* because the name *D. heterogena* has never been used after 1968.

Chapter 4

Phylum Cnidaria

Richard D. Campbell
Department of Developmental and Cell Biology, University of California, Irvine, CA, USA

Chapter Outline

INTRODUCTION

A few cnidarians have adapted to freshwater. The most abundant, widespread, and well-studied of these are the hydras, members of the genus *Hydra*. Hydra are single polyps, generally 1–15 mm in length, with a whorl of tentacles at the top of a column and an adherent base at the bottom (Fig. 4.1 A). The next most widespread type is *Cordylophora caspia* Pallas 1771, a colonial species with polyps arising from stolons adherent to the substrate or from upright stalks (Fig. 4.1 B). Polyps bear a single cluster of tentacles. The third type of freshwater cnidarian is a morphologically diverse group referred to as the genera *Craspedacusta* and *Calpasoma*. These are minute polyps, some of which include a larger medusa in their lifecycle (Fig. 4.1 C, D). Cnidarians bear nematocysts (stinging cells) in their tentacles and on the bodies, and these are key elements in identifying species.

These three types of freshwater cnidarians belong to the class Hydrozoa but are only distantly related to one another. *Hydra* and *Cordylophora* represent long-separated branches within the Hydroidolina (Cartwright et al., 2008), while *Craspedacusta* and *Calpasoma* represent the Trachylina.

Hydras occur throughout the world (Jankowski et al., 2008) and are most diverse in North America, with most species being widely distributed. The far northern and southern portions of our region have restricted faunas. Hydras are present on Greenland, but other islands such as the Aleutians and Bermuda have not been studied. Also, we know little of their distributions in the arid regions of western North America and Mexico.

LIMITATIONS

The taxonomy of hydra has been clouded by the dearth and variability of morphological characters and by a tendency to describe new species with insufficient attention to variability. All hydras were originally included in the single genus *Hydra* Linnaeus, 1758. Schulze (1917) added two genera, *Pelmatohydra* and *Chlorohydra*, and these are sometimes still used. But, all species are generally now included in the single genus *Hydra*, which is informally divided into four species groups: vulgaris, oligactis, viridissima, and braueri (Campbell, 1987). Each group is a distinct clade defined by DNA sequencing (Martinez et al., 2010). Due to the many synonymous descriptions, hydras have been notoriously confusing to identify. Actually, there are few distinct species, with less than a dozen in the Nearctic.

Most species of hydra are easily distinguished by their nematocysts. However, hydras of the vulgaris group have always confused taxonomists and cannot be distinguished reliably. These species are: *H. americana* Hyman, 1929, *H. cauliculata* Hyman 1938, *H. rutgersensis* Forrest, 1963, *H. carnea* Agassiz, 1850, and *H. littoralis* Hyman 1931. The last two species are frequently cited in collecting and research. All of these hydras were at one time known as *H. vulgaris*, and this name is again being used increasingly for them. Thus until taxonomists clarify this situation, it is reasonable to consider all hydra of the vulgaris group as *H. vulgaris* Pallas 1766.

Cordylophora caspia Pallas, 1771 is sometimes considered as a cluster of similar species (see Chapter 9 in Volume I). Part of its polymorphism is associated with its distribution over a wide range of salinity habitats, from freshwater to seawater.

Thorp and Covich's Freshwater Invertebrates. http://dx.doi.org/10.1016/B978-0-12-385028-7.00004-4

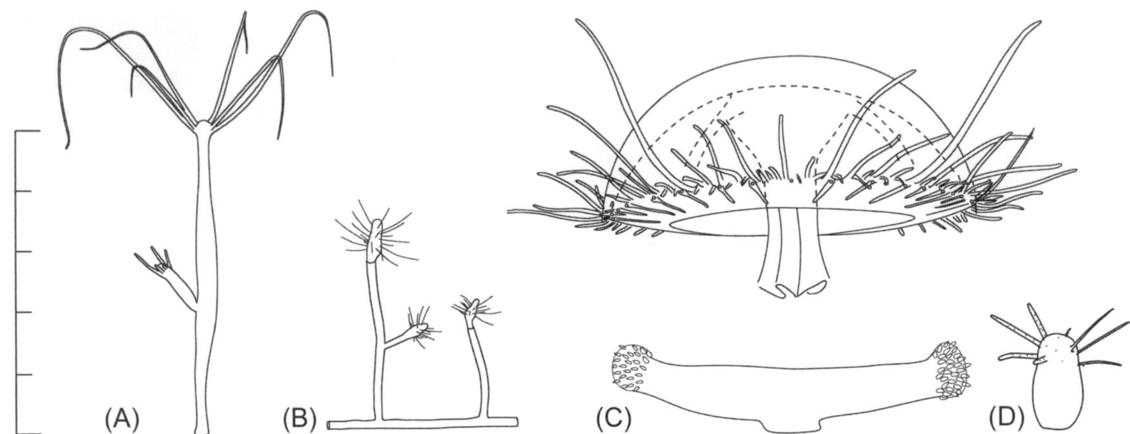

FIGURE 4.1 (A) Hydra; (B) a portion of colony of *Cordylophora caspia*; colonies can consist of thousands of hydranths; (C) *Craspedacusta sowerbii* medusa and polyps (modified from Buchert (1960) and Dejdar (1935)); (D) *Calpasoma dactyloptera*. Scale (mm) is for hydra, *Cordylophora*, and the medusa (which can grow to 20 mm diameter). The *Craspedacusta* polyp is under 1 mm in length and *Calpasoma* is about 0.3 mm high.

The minute, tentacle-less polyp *Craspedacusta sowerbii* Lankester, 1880 is the only local freshwater cnidarian to produce medusae (which were originally known as *Microhydra ryderi* Potts, 1885). Somewhat similar polyps, often with single-celled tentacles, might also represent this species, but they do not produce medusae and are usually referred to the species *Calpasoma dactyloptera* Fuhrmann, 1939.

Recent analyses of DNA sequences of cnidarians set out the relationships between these three groups of freshwater hydrozoans and their extensive marine relatives (Collins et al., 2006, 2008; Cartwright et al., 2008). Apparently, they evolved from marine species in three separate events and represent quite different body plans. Hence, our freshwater fauna, although sparse, is diverse.

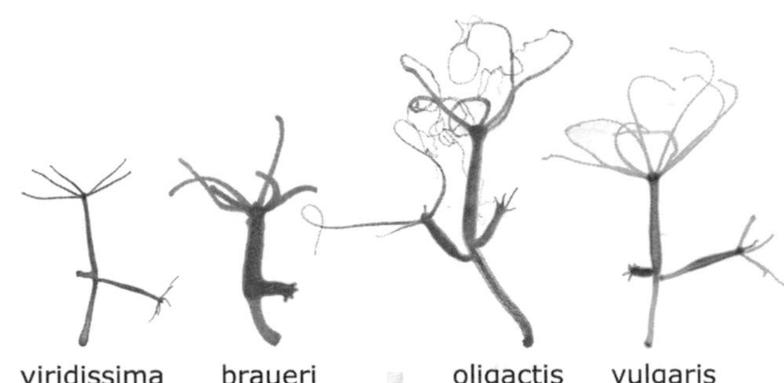

FIGURE 4.2 Typical morphology of hydra of the four groups of species.

TERMINOLOGY AND MORPHOLOGY

Freshwater cnidarians have simple, variable morphology without hard skeletons, so taxonomic characters are scarce. The overall morphological differences between the three major groups (*Hydra*, *Cordylophora*, and *Craspedacusta*) are obvious (Fig. 4.1).

The main characters often used in identifying hydra are overall body shape. The length of tentacles and body column and slenderness of the stalk may point to the species group of a hydra (Fig. 4.2), but these are characteristic only if the animal is growing vigorously with buds. Most recently collected hydra cannot be identified from their morphology. The early embryo, which forms externally in hydra, secretes a chitinous theca. The form and decoration of the theca are unique for each of the four groups of species and are similar for all species within a group (Fig. 4.3). Unfortunately, most collected and cultured specimens are not in a sexual condition.

The most important character in taxonomy of hydra is the structure of the nematocysts. Almost all species can be identified by examining nematocysts in a flattened tentacle

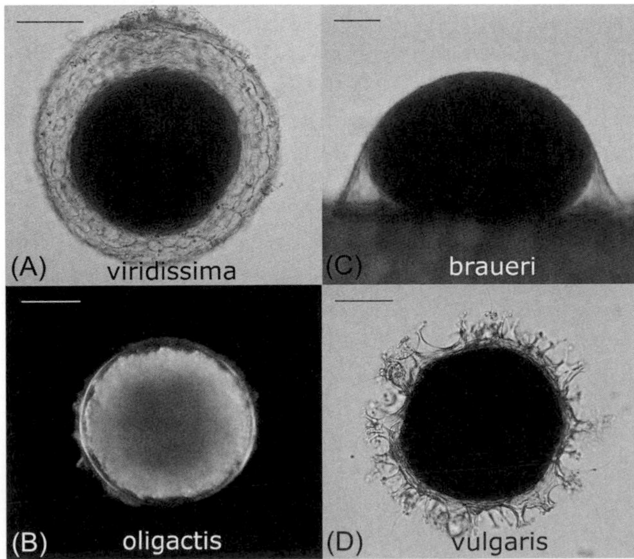

FIGURE 4.3 Typical embryothecae of the four groups of hydra species. Species shown are: (A) *H. viridissima*; (B) *H. oligactis* (photograph courtesy of Annalise Nawrocki); (C) *H. utahensis*; and (D) *H. carnea*. Scales are 100 µm.

under high magnification with a light microscope. The keys that follow are heavily based on nematocysts. Although setting up an oil-immersion squash of a tentacle may be an obstacle for a beginner or one working with other tools, it is the only reliable way to identify specimens, and once mastered, the analysis is simple.

MATERIAL PREPARATION AND PRESERVATION

Preservation

Polyps fixed by immersion in 70%–100% alcohol keep their morphology and can be used later for DNA sequencing. Before fixation, allow polyps to attach to the culture dish and expand. Remove excess water. Flooding small animals quickly with alcohol prevents their contraction. For larger hydra, allow them to relax and extend, and then control the degree of contraction (or stretching) by the rate of flooding with alcohol. Detach fixed polyps from the dish with a rapid sideways stroke of a needle against their point of attachment. *Cordylophora* are generally firmly attached to a substrate; either the substrate surface must be kept or else the parts of the colony above the substrate can be cut off. Store fixed material in alcohol or as microscope mounts by staining with Borax Carmine or other colorant, dehydrating, and covering with mounting medium and adding a cover glass.

Examining Nematocysts

Living nematocysts are the primary taxonomic characters for hydra. Fixation destroys their usefulness. Cut a whole tentacle from the body using a single, swift cut with a microscalpel or cut off a whole whorl of tentacles. Place the tissue under a cover glass supported by a small drop of water. Nematocysts are abundant in the tentacles. They are ovoid in shape with the axes pointing radially outward from the tentacle surface. You must flatten the tissue to rotate the nematocysts into the horizontal plane. Flatten the tissue by drawing off the water slowly, using a small piece of absorbent paper at the side of the cover glass. Watch the process using an oil immersion, 100× lens until the nematocysts are horizontal but not yet squashed.

Nematocysts are capsules with an internal tubule coiled in a characteristic pattern. There are four types of nematocysts in hydra, and each one must be recognized. First, notice the largest, and most obvious, pyriform nematocyst called a stenotele (Fig. 4.4 A). Then identify the smallest but most abundant capsule called a desmoneme (Fig. 4.4 D). It has a single coil of tubule inside an asymmetrical pyriform capsule. The two remaining nematocyst types called isorhizas are of intermediate size. The holotrichous isorhiza (Fig. 4.4 B) is the larger of the two and can be recognized by its three prominent transverse coils of tubule at one end of the capsule, except in *H. oligactis*, where the tubule is plied back and forth longitudinally (Fig. 4.5 D). The holotrichous isorhiza is the most important nematocyst for taxonomy. It is also the least numerous, and in rare cases, you may have to look at several tentacles to find them. This nematocyst type is more abundant at the base of the tentacles than towards the tip, and is usually numerous around the mouth. The fourth type of nematocyst, the atrichous isorhiza (Fig. 4.4 C), is between

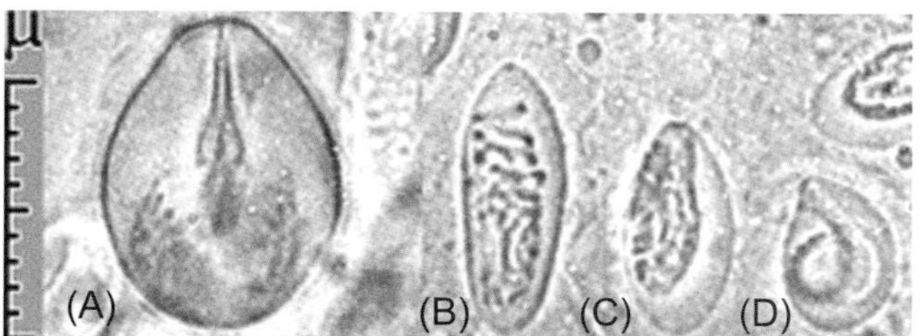

FIGURE 4.4 Four types of nematocysts in hydra: (A) Stenotele, (B) holotrichous isorhiza, (C) atrichous isorhiza, and (D) desmoneme. This shows nematocysts of *H. canadensis*, in the oligactis group.

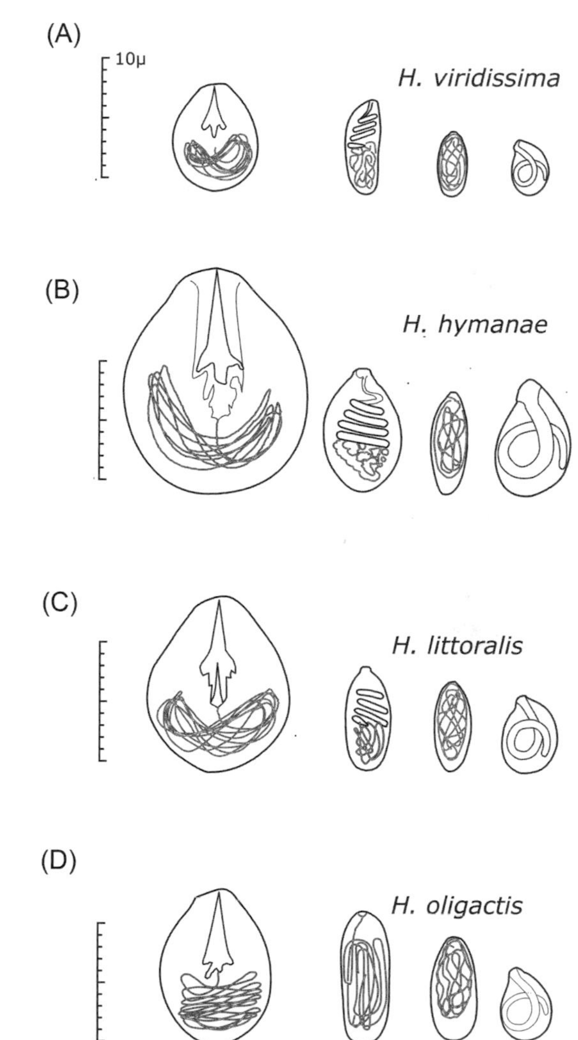

FIGURE 4.5 Typical nematocysts of the four groups of hydra species. (A) viridissima group, (B) braueri group, (C) vulgaris group, (D) oligactis group. Scale is 10 μm.

PHYLUM CNIDARIA

the holotrichous isorhiza and the desmoneme in size, and its tubule is so slender, indistinct, and tightly wound that its detailed coiling pattern is not visible. The atrichous isorhiza is not important in taxonomy, but it must be distinguished from the holotrichous isorhiza.

Nematocysts are most simply viewed using a brightfield microscope with the condenser lowered slightly to obtain better contrast. Differential Interference Contrast microscopes show the structures particularly well, but phase microscopes do not produce clear images.

KEYS TO FRESHWATER CNIDARIA

Class Hydrozoa: Families

1 Medusae or solitary polyps, sometimes with a few others fused or developing as buds 2

1' Colonial polyps, with individuals arising from a stolon attached to a substratum and from uprights .... Oceaniidae, 1 species: *Cordylophora caspia* Pallas, 1771

2(1) Polyps >1 mm in length .... **Hydridae [p. 88]**

2' Medusae, or polyps <1 mm in length .... **Olindiidae [p. 89]**

Cnidaria: Hydrozoa: Hydridae: *Hydra:* Species Groups

1 Polyps not green; stenotele nematocysts >11 μ in length; embryotheca either not tiled or not roughly spherical 2

1' Polyps green; stenotele nematocysts 9–11 μ in length; surface of embryotheca roughly spherical and cobbled with tiles (Fig. 4.3 A) .... viridissima group

2(1) Holotrichous isorhiza nematocysts narrowly oval or slipper-shaped, less than half as wide as long (Fig. 4.4 B); embryotheca not flattened 3

2' Holotrichous isorhiza nematocysts broadly lemon-shaped, at least half as wide as long (Fig. 4.5 B); embryotheca flattened against a substratum (Fig. 4.3 C) .... braueri group

3(2) Stenotele nematocysts appear all the same length (within 1 μm); embryotheca smooth and thin (Fig. 4.3 B) .... oligactis group

3' Stenotele nematocysts varied in length, typically about 13–18 μ; embryotheca thick with numerous radial spines (Fig. 4.3 D) (one or several confused species. See section on limitations above) .... vulgaris group

Hydra Species Groups: Viridissima: *Hydra:* Species

Probably all green hydras in our area represent *H. viridissima*. *H. hadleyi* is based on the character of an extra chamber in the embryotheca, but this is now known to be present in *H. viridissima*.

1 Embryotheca with one large empty chamber proximal to the embryo .... *Hydra hadleyi* (Forrest, 1959)

1' Embryotheca with only a narrow chamber proximal to embryo .... *Hydra viridissima* Pallas, 1766

Hydra Species Group: Oligactis: *Hydra:* Species

1. Internal tubule of the holotrichous isorhiza nematocysts with about 3 prominent transverse coils (Fig. 4.4 B) .... *Hydra canadensis* Rowan, 1930

1' Internal tubule of the holotrichous isorhiza nematocysts plied longitudinally throughout (Fig. 4.5 D) .... *Hydra oligactis* Pallas, 1766

Hydra Species Group: Braueri: *Hydra:* Species

1 Desmoneme nematocyst length <7 μm, shorter than the holotrichous isorhiza 2

1' Desmoneme nematocyst length >7 μm, approximately as long as the holotrichous isorhiza nematocyst (Fig. 4.5 B) .... *Hydra hymanae* Hadley & Forrest, 1949

[Canada; northern USA]

2(1) Budding polyps <3 mm in length 3

2' Budding polyps >3 mm in length .... *Hydra utahensis* Hyman, 1931
[Canada; northern USA]

3(2) Embryotheca sometimes spherical .... *Hydra lirosoma* Campbell, 1987
[USA: Georgia]

3' Embryotheca always flattened .... *Hydra minima* Forrest, 1963
[USA: New Jersey]

Cnidaria: Hydrozoa: Olindiidae: Species

1 Medusae, or polyps without tentacles .... *Craspedacusta sowerbii* Lankester, 1880

1' Polyps with tentacles consisting of a single cell each .... *Calpasoma dactyloptera* Fuhrmann, 1939

REFERENCES

Campbell, R.D. 1987. A new species of *Hydra* (Cnidaria: Hydrozoa) from North America with comments on species clusters within the genus. Zoological Journal of the Linnean Society of London 91: 253–263.

Cartwright, P., N.M. Evans, C.W. Dunn, A.C. Marques, M.P. Miglietta, P. Schuchert & A.G. Collins. 2008. Phylogenetics of Hydroidolina (Hydrozoa: Cnidaria). Journal of the Marine Biological Association of the United Kingdom 88: 1663–1672.

Collins, A.G., B. Bentlage, A. Lindner, D. Lindsay, S.H.D. Haddock, G. Jarms, J.L.Norenburg, T. Jankowski & P. Cartwright. 2008. Phylogenetics of Trachylina (Cnidaria: Hydrozoa) with new insights on the evolution of some problematical taxa. Journal of the Marine Biological Association of the United Kingdom 88: 1673–1685.

Jankowski, T., A.G. Collins & R. Campbell. 2008. Global diversity of inland water cnidarians. Hydrobiologia 595: 35–40.

Martinez, D.E., A.R. Iñiguez, K.M. Percell, J.B. Willner, J. Signorovitch & R.D. Campbell. 2010. Phylogeny and biogeography of *Hydra* (Cnidaria: Hydridae) using mitochondrial and nuclear DNA sequences. Molecular Phylogenetics and Evolution 57: 403–410.

Schulze, P. 1917. Neue Beiträge zu einer Monographie der Gattung Hydra. Archiv für Biontologie 4: 31–119.

Chapter 5

Phylum Platyhelminthes

Carolina Noreña
Departamento Biodiversidad y Biología Evolutiva, Museo Nacional de Ciencias Naturales (CSIC), Madrid, Spain

Cristina Damborenea, Francisco Brusa
División Zoología Invertebrados, Museo de La Plata, CONICET, La Plata, Argentina

Chapter Outline

INTRODUCTION

The flatworms of the phylum Platyhelminthes comprise free-living ("Turbellaria") and obligate parasitic organisms (Monogenea, Digenea, Aspidogastrea, and Cestoda, today grouped in Neodermata). "Turbellaria" includes an amazing variety of forms, but built in a similar way. All have the following characteristics: bilateral symmetry, organs embedded in a solid cellular matrix (the parenchyma), a sac-like gut without an anus, a nervous system with an anterior "brain" and lateral nerve chords, and internal fluids that are regulated by protonephridia. Most are cross-fertilizing hermaphrodites, with morphologically diverse structures like sclerotic parts of the male copulatory organ that forms stylets of different shape and complexity, and with endolecithal or ectolecithal eggs. Variation also can be found in the location and shape of the pharynx along the main body-axis.

Most "Turbellaria" are small with a body size of around 1 mm (microturbellaria), but others, such as Tricladida and Polycladida, have a body-size of 5–10 cm (macroturbellaria). They show a worldwide distribution and inhabit most kinds of freshwater, brackish, and marine habitats.

Based on their simple body structure, the "Turbellaria" has been considered one of the most basal bilateria (Littlewood & Bray, 2001). This hypothesis is supported by the results obtained from molecular phylogenetic analysis (Ruiz-Trillo et al., 1999, 2004; Baguñà & Riutort, 2004a). "Turbellaria" is a class constituted by a very heterogeneous group of orders. Besides the body plan, all orders share a free-living life style, similar features of the body wall (simple epithelium, muscle network, absence of cuticle) and regeneration based on stem cell-like neoblasts (Tyler & Hooge, 2004). These characters justified the grouping of three main clades: Acoelomorpha, Catenulida, and Rhabditophora (Tyler & Hooge, 2004). Recently, the Acoelomorpha (Acoela and Nemertodermatida) were placed outside the Platyhelminthes, with the status of a new phylum, because the singular molecular composition of the mitochondrial and nuclear genes and morphological characters such as duet-spiral cleavage, bi-spiral segmentation and an epithelium with an own root system network (Ruiz-Trillo et al., 2004; Baguñà & Riutort, 2004a,b).

LIMITATIONS

The keys were built with users in mind who are not trained in the characteristics of species of freshwater "Turbellaria." They use characters that can be easily seen, essentially in whole mounts of living animals by inexperienced observers and applied at higher systematic levels. In lower levels, however, identification requires more detailed observations. Once the identification is reached, it must be checked against the original description of the taxon or later descriptions in monographs such as Ferguson (1954) and the website created by Shärer (2014) for Macrostomida.

Thorp and Covich's Freshwater Invertebrates. http://dx.doi.org/10.1016/B978-0-12-385028-7.00005-6

Readers should also consult the following texts: Gilbert (1938) and Houben et al. (2014) for *Phaenocora*, Kenk (1974, 1989) for Tricladida, Luther (1955) for Dalyelliidae, Luther (1960) for Catenulida, Macrostomida, Lecithoepitheliata, Prolecithophora, and Proseriata, Luther (1963) for Typhloplanida, Nuttycombe & Waters (1938) for Stenostomidae (Catenulida), and Nuttycombe (1956) for Catenulidae. More general information on turbellarians is available in Cannon (1986), Tyler et al. (2006–2015), and Kolasa & Tyler (2010). Table 5.1 lists the freshwater turbellarian species known in the Nearctic Region.

TERMINOLOGY AND MORPHOLOGY

The following terminology is intended to assist in understanding the morphology and the use of turbellarian keys included in the chapter. It is based primarily on Cannon (1986), Rieger et al. (1991), and Richter et al. (2010). Adenodactyl: auxiliary glandular bulb in male reproductive system of some groups, mainly Tricladida. Adhesive glands: cell glands found in different parts of the body, which secrete adhesive substances. Ascus: glandular pocket near the gonopores. Ciliated pits: sensory structures (e.g., anterior end of Stenostomidae). Copulatory organ: the terminal part of the male reproductive system to deliver the sperm into the body of a partner at copulation. It is often bulbous (copulatory bulb), containing the sperm duct and elements of the prostate glands. It is a penis (penis papilla) if it is capable of protrusion but not armed with hard structures. It may be enclosed in an inner chamber (penis pocket) the walls of which can form a conical prominence that may serve to guide the penis (a penis sheath). The penis is said to be armed when a "cuticular" or "sclerotic" structure is present. A stylet is a single sclerotic tube or channel of variable shape; a cirrus is an eversible male duct; often spiny. Excretophore: large vacuolated cells of the intestine epithelium. Handles: structures in the basal part of the stylet. Light refracting bodies: unpigmented refracting bodies found in species of Stenostomidae. These bodies are formed by a variable number of spherical granules; some species have fewer than five and others more than 15. The light refracting bodies are frequently associated with the posterior cerebral lobes, or sometimes with the anterior lobes. Generally only one pair is found, although in some cases, more pairs are present. Neoblast: pluripotent cells responsible for the normal replacement of cells, cell substitution during regeneration, and stem cells for the gonads. Pharynx: the glandulo-muscular structure between mouth and intestine inwards. It is used to capture prey and may be of considerable taxonomic value. The main terms for practical use are: simplex (simple)—a ciliated mouth tube; plicatus (folded/plicate)—an annular fold which is variable in length, lacks a delimiting septum, and has a large protrusion capacity; bulbosus (bulbous)—a glandulo-muscular bulb delimited by a more or less muscular septum, only the end of the pharynx may be protruded; it could be rosulatus (rosulate)—bulbous and usually globular and dorso-ventrally oriented; or doliiformis (doliiform)—bulbous and usually barrel-shaped and horizontally (fronto-caudally) oriented; or variabilis (variable)—bulbous, but with variable shape and a weakly differentiated septum (see Volume I, Chapter 10). Proboscis: anterior protrusible organ, which is a notable feature of the rhabdocoel group Kalyptorhynchia. Prostatic vesicle: part of the male reproductive system. Structure containing gland cells and its secretion is combined with the sperm during their release. Prostomium: body region anterior to mouth; usually containing the brain. Protonephridium: organ containing one or more flame cells and connecting ducts, which presumably functions in an osmoregulatory fashion. Rhabdite: rod-shaped bodies found in the epidermal cells or below the ectoderm. They are characteristic for Platyhelminthes. Statocyst: an orientation organ of the nervous system; it contains a space enclosing a granule (the statolith) and is surrounded by sensory tissue. Vitellaria: gland cells that secrete nutrient material which is included into egg capsules. Zooid: individuals formed by paratomy.

MATERIAL PREPARATION AND PRESERVATION

In vivo observations are very important for the study of turbellarians because certain structures are only visible in living specimens. Once collected, the specimens will need to undergo different group-specific treatments for subsequent identification and preservation. For large planarians, external characteristics are easily observable under stereomicroscope. Specimens should then be fixed (in Steinman's, 10% formaldehyde or Bouin fixative) for later histological analysis. For microturbellarians, the specimens must first be observed live and their size, shape, and color noted. After placing the individual on a slide with a drop of water and either Vaseline or plasticine smeared on the coverslip edges, slowly squash it with the coverslip. The water excess can be removed with filter paper. Diagnostic structures of the internal organs of the transparent microturbellarians can be observed with this method. Some specimens must be fixed and preserved in polyvinyl lactophenol, which makes them more transparent, thus allowing the observation of sclerotized diagnostic structures (e.g., spines, stylets, and the cirrus). Microturbellarians can be fixed in a variety of conventional fixation fluids (e.g., Bouin, 5% formaldehyde) prior to being processed for histological analysis of the internal structures of diagnostic importance.

With few exceptions, the characters used in the following keys are viewable by transparency (squash methodology).

TABLE 5.1 List of Neartic Species of Turbellarians

Catenulida
Family Catenulidae
Catenula confusa Nuttycombe, 1956
Catenula lemnae Duges, 1832
Catenula leptocephala Nuttycombe, 1956
Catenula sekerai Beauchamp, 1919
Catenula turgida (Zacharias, 1902)
Catenula virginia Kepner & Carter, 1930
Family Chordariidae
Chordarium europaeum Schwank, 1980
Family Stenostomidae
Myostenostomum gigerium (Kepner & Carter, 1931)
Rhynchoscolex platypus Marcus, 1945
Rhynchoscolex simplex Leidy, 1851
Stenostomum anatirostrum Marcus, 1945
Stenostomum anops Nuttycombe & Waters, 1938
Stenostomum arevaloi Gieysztor, 1931
Stenostomum beauchampi Papi, 1967
Stenostomum bicaudatum Kennel, 1888
Stenostomum brevipharyngium Kepner & Carter, 1931
Stenostomum ciliatum Kepner & Carter, 1931
Stenostomum cryptops Nuttycombe & Waters, 1935
Stenostomum glandulosum Kepner & Carter, 1931
Stenostomum grande Child, 1902
Stenostomum kepneri Nuttycombe & Waters, 1938
Stenostomum leucops (Duges, 1828)
Stenostomum mandibulatum Kepner & Carter, 1931
Stenostomum membranosum Kepner & Carter, 1931
Stenostomum occultum Kolasa, 1971
Stenostomum pegephilum Nuttycombe & Waters, 1938
Stenostomum predatorium Kepner & Carter, 1931
Stenostomum pseudoacetabulum Nuttycombe & Waters, 1935
Stenostomum saliens Kepner & Carter, 1931
Stenostomum simplex Kepner & Carter, 1931
Stenostomum sphagnetorum Papi in Luther, 1960
Stenostomum temporaneum Kolasa, 1981
Stenostomum tuberculosum Nuttycombe & Waters, 1938
Stenostomum uronephrium Nuttycombe, 1931
Stenostomum ventronephrium Nuttycombe, 1932
Stenostomum virginianum Nuttycombe, 1931
Macrostomida
Family Macrostomidae
Macrostomum acutum Ax, 2008
Macrostomum appendiculatum Fabricius, 1826
Macrostomum bellebaruchae Ax, 2008
Macrostomum bicurvistyla Armonies & Hellwig, 1987
Macrostomum carolinense Ferguson, 1940
Macrostomu collistylum Ferguson, 1939
Macrostomum curvistylum Ferguson, 1939
Macrostomum curvituba Luther, 1947
Macrostomum frigorophilum Ferguson, 1940
Macrostomum gilberti Ferguson, 1939
Macrostomum glochistylum Ferguson, 1939
Macrostomum granulophorum Ferguson, 1940
Macrostomum hystricinum Beklemischev, 1951
Macrostomum lewisi Ferguson, 1939
**Macrostomum norfolkensis* Jones & Ferguson, 1940
Macrostomum ontarioense Ferguson, 1943
Macrostomum orthostylum Braun, 1885
Macrostomum phillipsi Ferguson & Stirewalt, 1938
Macrostomum recurvostylum Ferguson, 1940
Macrostomum reynoldsi Ferguson, 1939
Macrostomum riedeli Ferguson, 1939
Macrostomum ruebushi Ferguson, 1940
Macrostomum tenuicauda Luther, 1947
**Macrostomum thermophilum* Riedel, 1932
Macrostomum schmitti Hayes & Ferguson, 1940
Macrostomum sensitivum Silliman, 1884
Macrostomum shenandoahense Ferguson, 1940
Macrostomum tennesseensis Ferguson, 1939
Macrostomum truncatum Ferguson, 1940
Macrostomum tuba Graff, 1882
Macrostomum vejdovskyi Ferguson, 1940
Macrostomum virginianum Ferguson, 1937
Family Microstomidae
Microstomum bispiralis Stirewalt, 1937

Continued

TABLE 5.1 List of Neartic Species of Turbellarians—cont'd

Lecithoepitheliata
Family Prorhynchidae
Geocentrophora applanata (Kennel, 1888)
Geocentrophora baltica (Kennel, 1883)
Geocentrophora cavernicola Carpenter, 1970
Geocentrophora marcusi Darlington, 1959
Geocentrophora sphyrocephala de Man, 1876
Prorhynchus stagnalis Schultze, 1851
Prolecithophora
Family Plagiostomidae
Hydrolimax bruneus Girard, 1891
Hydrolimax grisea Haldeman, 1843
**Plagiostomum planum* Silliman, 1885
Bothrioplanida
Family Bothrioplanidae
Bothrioplana semperi Braun, 1881
Proseriata
Family Coelogynoporidae
Coelogynopora falcaria Ax & Sopott-Ehlers, 1979
Family Monocelididae
Monocelopsis carolinensis Ax, 2008
Family Otomesostomidae
Otomesostoma auditivum (Du Plessis, 1874)
Rhabdocoela
Dalyellioida
Family Provorticidae
Provortex virginiensis Ruebush & Hayes, 1939
Vejdovskya pellucida (Schultze, 1851)
Family Dalyelliidae
Castrella cylindrica Riedel, 1932
Castrella graffi Hayes, 1945
Castrella groenlandica Riedel, 1932
Castrella pinguis (Silliman, 1884)
Castrella truncata (Abidgaard, 1789)
Dalyellia alba Higley, 1918
Dalyellia idahoensis Knapp, 1954
Dalyellia viridis (Shaw, 1791)
Fulinskiella bardeaui (Steinböck, 1926)
Gieysztoria blodgetti (Silliman, 1884)
Gieysztoria choctaw Van Steenkiste, Gobert & Artois, 2011
Gieysztoria cuspidata (Schmidt, 1861)
Gieysztoria dodgei (Graff, 1911)
Gieysztoria eastmani (Graff, 1911)
Gieysztoria minima (Riedel, 1932)
Gieysztoria ornata (Hofsten, 1907)
Gieysztoria pavimentata (Beklemischew, 1926)
Gieysztoria pseudoboldgetti Luther, 1955
Gieysztoria triangulata (Robeson, 1931)
Microdalyellia abursalis (Ruebush, 1937)
Microdalyellia armigera (Schmidt, 1861)
Microdalyellia circulobursalis (Ruebush, 1937)
Microdalyellia deses (Riedel, 1932)
Microdalyellia fairchildi (Graff, 1911)
Microdalyellia gilesi Jones & Hayes, 1941
Microdalyellia groenlandica (Riedel, 1932)
Microdalyellia mohicana (Graff, 1911)
Microdalyellia rheesi (Graff, 1911)
Microdalyellia rochesteriana (Graff, 1911)
Microdalyellia rossi (Graff, 1911)
Microdalyellia ruebushi Luther, 1955
Microdalyellia schockaerti Willems, Artois, Jocque, & Brendonck, 2007
Microdalyellia sillimani (Graff, 1911)
Microdalyellia tennesseensis (Ruebush & Hayes, 1939)
Microdalyellia virginiana (Ruebush, 1937)
Pseudodalyellia alabamensis Van Steenkiste, Gobert & Artois, 2011
Typhloplanoida
Family Typhloplanidae
Acrochordonoposthia conica Reisinger, 1924
Adenoplea nanus Sayre & Wergen, 1994
Amphibolella segnis Findenegg, 1924
Ascophora elegantissima Findenegg, 1924
Bryoplana xerophila Van Steenkiste, Davison & Artois, 2010
Bothromesostoma personatum (Schmidt, 1848)
Castrada affinis Hofsten, 1907
Castrada borealis Steinböck, 1931

TABLE 5.1 List of Neartic Species of Turbellarians—cont'd

Castrada hofmanni Braun, 1885	*Tetracelis marmorosa* (Müller, 1773)
Castrada inermis Hofsten, 1911	*Typhloplana minima* (Fuhrmann, 1894)
Castrada libidinosa Hofsten, 1916	*Typhloplana viridata* (Abildgaard, 1789)
Castrada luteola Hofsten, 1907	*Typhloplanella halleziana* (Vejdovsky, 1880)
Castrada lutheri Kepner, Stirewalt & Ferguson, 1939	**Kalyptorhynchia**
Krumbachia minuta Ruebush, 1938	**Family Polycistididae**
Krumbachia virginiana (Kepner & Carter, 1931)	*Gyratrix hermaphroditus* Graff, 1831
Limnoruanis romanae Kolasa, 1977	*Opisthocystis goettei* (Bresslau, 1906)
Mesocastrada fuhrmanni Volz, 1898	**Tricladida**
Mesostoma angulare Higley, 1918	**Familia Dendrocoelidae**
Mesostoma arctica Hyman, 1938	*Procotyla fluviatilis* Leidy, 1857
Mesostoma californicum Hyman, 1957	*Procotyla typhlops* Kenk, 1935
Mesostoma columbianum Hyman, 1939	*Dendrocoelopsis alaskensis* Kenk, 1953
Mesostoma craci Schmidt, 1858	*Dendrocoelopsis americana* (Hyman, 1939)
Mesostoma curvipenis Hyman, 1955	*Dendrocoelopsis hymanae* Kawakatsu, 1968
Mesostoma ehrenbergii (Focke, 1836)	*Dendrocoelopsis piriformis* Kenk, 1953
Mesostoma georgianum Darlington, 1959	*Dendrocoelopsis vaginata* Hyman, 1935
Mesostoma macropenis Hyman, 1939	**Familia Dugesiidae**
Mesostoma macroprostatum Hyman, 1939	*Cura foremanii* (Girard, 1852)
Mesostoma platygastricum Hofsten, 1924	*Girardia arizonensis* Kenk, 1975
Mesostoma vernale Hyman, 1955	*Girardia dorotocephala* (Woodworth, 1897)
Microcalyptorhynchus virginianus Kepner & Ruebush, 1935	*Girardia jenkinsae* (Benazzi & Gourbault, 1977)
Olisthanella coeca (Silliman, 1885)	*Girardia tigrina* (Girard, 1850)
Opistomum pallidum Schmidt, 1848	*Schmidtea polychroa* (Schmidt, 1861)
**Phaenocora agassizi* Graff, 1911	**Familia Kenkiidae**
Phaenocora aglobulata Houben, van Steenkiste & Artois, 2014	*Kenkia glandulosa* (Hyman, 1956)
Phaenocora falciodenticulata Gilbert, 1938	*Kenkia lewisi* (Kenk, 1975)
Phaenocora gilberti Houben, van Steenkiste & Artois, 2014	*Kenkia rhynchida* Hyman, 1937
Phaenocora highlandense Gilbert, 1935	*Sphalloplana (Speophila) buchanani* (Hyman, 1937)
Phaenocora kepneri Gilbert, 1935	*Sphalloplana (Speophila) chandleri* Kenk, 1977
Phaenocora lutheri Gilbert, 1937	*Sphalloplana (Speophila) hoffmasteri* (Hyman, 1954)
**Phaenocora megacephala* (Higley, 1918)	*Sphalloplana (Speophila) hubrichti* (Hyman, 1945)
Phaenocora sulfophila (Gilbert, 1938)	*Sphalloplana (Speophila) hypogeal* Kenk, 1984
Phaenocora virginiana Gilbert, 1935	*Sphalloplana (Speophila) pricei* (Hyman, 1937)
Prorhynchella minuta Ruebush, 1939	*Sphalloplana (Speophila) weingartneri* Kenk, 1970
Protoascus wisconsinensis Hayes, 1941	*Sphalloplana (Speophila) californica* Kenk, 1977
Rhynchomesostoma rostratum (Müller, 1773)	*Sphalloplana (Speophila) consimilis* Kenk, 1977
Strongylostoma elongatum Hofsten, 1907	*Sphalloplana (Speophila) culveri* Kenk, 1977
Strongylostoma gonocephalum (Silliman, 1884)	*Sphalloplana (Speophila) evaginata* Kenk, 1977
Styloplanella strongylostomoides Findenegg, 1924	*Sphalloplana (Speophila) georgiana* Hyman, 1954

Continued

TABLE 5.1 List of Neartic Species of Turbellarians—cont'd

Sphalloplana (Speophila) percoeca (Packard, 1879)	*Phagocata nordeni* Kenk, 1977
Sphalloplana (Speophila) subtilis Kenk, 1977	*Phagocata oregonensis* Hyman, 1963
Sphalloplana (Polypharyngea) mohri Hyman, 1938	*Phagocata notorchis* Kenk, 1987
Familia Planariidae	*Phagocata procera* Kenk, 1984
Hymanella retenuova Castle, 1941	*Phagocata pygmaea* Kenk, 1987
Paraplanaria dactyligera (Kenk, 1935)	*Phagocata spuria* Kenk, 1987
Paraplanaria occulta (Kenk, 1969)	*Phagocata tahoena* Kawakatsu, 1968
Phagocata angusta Kenk, 1977	*Phagocata velata* (Stringer, 1909)
Phagocata bulbosa Kenk, 1970	*Phagocata vernalis* (Kenk, 1944)
Phagocata bursaperforata Darlington, 1959	*Phagocata virilis* Kenk, 1977
Phagocata carolinensis Kenk, 1979	*Phagocata woodworthi* Hyman, 1937
Phagocata fawcetti Ball & Gourbault, 1975	**Polycelis coronata coronata* (Girard, 1891)
Phagocata gracilis (Haldeman, 1840)	*Polycelis coronata monticola* Kenk & Hampton, 1982
Phagocata hamptonae Kenk, 1982	*Polycelis coronata brevipenis* Kenk, 1972
Phagocata holleri Kenk, 1979	*Seidlia remota* (Smith, 1988)
Phagocata morgani morgani (Stevens & Boring, 1906)	*Seidlia sierrensis* (Kenk, 1973)
Phagocata morgani polycelis (Kenk, 1935)	
Phagocata nivea Kenk, 1953	

The arrangement of orders and families follows the taxonomic listing of Tyler et al. (2006–2015).
*Indicates a doubtful species.

KEYS TO PLATYHELMINTHES

Platyhelminthes: Orders

1	Pharynx simple 2
1'	Pharynx plicatus or bulbous 4
2(1)	Intestine ill defined; statocyst with one statolith (Acoela) or more (Nemertodermatida) .... **Acoela [p. 97]**
2'	Intestine well defined; freshwater; planktonic and benthonic 3
3(2)	Protonephridium unpaired; excretory duct central and excretory pore caudal .... **Catenulida [p. 97]**
3'	Protonephridium paired .... **Macrostomida [p. 97]**
4(1)	Pharynx plicatus 5
4'	Pharynx bulbous 6
5(4)	Pharynx plicatus directed ventrally; intestine tubular; with statocyst or prominent paired ciliated pits; testes and ovaries dispersed .... **Proseriata [p. 97]**
5'	Pharynx plicatus directed backwards; intestine triradiate; without statocyst or prominent paired ciliated groves; with paired central eyes or numerous marginal eyes .... **Tricladida [p. 97]**
6(4)	Oral pore and genital pore separated 7
6'	Oral pore and male pore joined; frontal. Male copulatory organ with stylet; female gonads (ovaries and yolk glands) joined; diffuse .... Lecithoepitheliata, one family **Prorhynchidae [p. 102]**
7(6)	Pharynx bulbous *variabilis* (with variable shape); oral pore frontal; male pore caudal; without female copulatory organs; female gonads opens into the masculine atrium .... Prolecithophora, one family **Plagiostomidae [p. 103]**
7'	Pharynx bulbous *rosulatus* (dorsoventral orientated) or *doliiformis* (frontocaudally); with male and female copulatory organs; male and female pore ventral or caudal; female gonads (ovaries and yolk glands) separated .... **Rhabdocoela [p. 105]**

Platyhelminthes: Acoela: Families

1 Body shape elongated, thin; statocyst with one statolith; colorless or slighty pigmented; without pharynx, small rounded eyes well separated; oral pore at the middle of the ventral side .... Mecynostomidae, one genus *Limnoposthia* [Palaearctic]

1' Body shape pedunculate with a small tail and enfolded sides; statocyst with one statolith; gray or whitish pigmented; pharynx simple; orange or yellowish small crescent-shaped eyes behind the pharynx; oral pore at the anterior end .... Convolutidae, one genus *Oligochoerus* [Palaearctic]

Platyhelminthes: Catenulida: Families

1 Brain compact; oval; with or without statocyst; ciliated furrow separate the anterior end (*prostomium*) from the posterior region 2

1' Brain lobed; with anterior and posterior lobes; without statocyst; without ciliated furrow .... **Stenostomidae [p. 97]**

2(1) Oral pore near the *prostomium*; often forming chains of numerous and compact zooids (Figs. 5.1 A, B and 5.8 C) .... Catenulidae, one genus: *Catenula*

2' Oral pore away from the *prostomium*; often forming chains of not more than two or four zooids (Figs. 5.1 C and 5.8 A, B). .... Chordariidae, one species: *Chordarium europaeum* Schwank, 1980

Platyhelminthes: Catenulida: Stenostomidae: Genera

1 With ciliated pits; *prostomium* of variable length; generally with paired light refracting bodies associated to the posterior brain lobes; forming zooids 2

1' Without ciliated pits; with a long *prostomium*; without light refracting bodies; without zooids (Figs. 5.1 D and 5.8 D) ...... *Rhynchoscolex*

2(1) Without muscular belt between pharynx and intestine (Fig. 5.1 E) .... *Stenostomum*

2' With muscular belt between pharynx and intestine (Fig. 5.8 E, F) .... *Myostenostomum gigerium* (Kepner & Carter, 1931)

Platyhelminthes: Macrostomida

1 Body shape oval; with two eyes; without ciliated pits; adhesive posterior end; paired testes and ovaries; stylet tube or funnel-like; sexual reproduction. (Fig. 5.1 F) .... Macrostomidae, one genus: *Macrostomum*

1' Body shape elongated (often forming zooids); anterior and posterior end pointed; without eyes; with ciliated pits; ovaries unpaired; stylet turned; corkscrew-like; with sexual and asexual (preferred) reproduction (Fig. 5.1 G) .... Microstomidae, one genus: *Microstomum*

Platyhelminthes: Proseriata

1 With two pairs of ciliated pits; ovaries behind the pharynx 2

1' Without ciliated pits; ovaries before the pharynx 3

2(1) Without statocyst; pharynx tubular directed backwards; intestinal branches at pharynx level; common gonopore; posterior adhesive glands (Fig. 5.2 C) .... Bothrioplanidae, one species: *Bothrioplana semperi* Braun, 1881

2' With statocyst; pharynx directed ventral; without intestinal branches; female and male gonopore; without posterior adhesive glands (Fig. 5.2 E) .... Otomesostomidae, one species: *Otomesostoma auditivum* (Du Plessis, 1874)

3(1) Pharynx tubular directed backwards; female and male gonopores separate; statocyst anterior to the brain; without sensory pits (Fig. 5.2 F) .... Monocelididae sp., one species: *Monocelopsis carolinensis* Ax, 2008

3' Pharynx globular (dorsoventrad orientated); common gonopore; statocyst before the brain; with sensory pits (Fig. 5.2 D) .... Coelogynoporidae, one species: *Coelogynopora falcaria* Ax & Sopott-Ehlers, 1979

Platyhelminthes: Tricladida: Families

1 Usually unpigmented; with anterior glandulomuscular organ 2

1' Usually pigmented; without anterior glandulomuscular organ 3

2(1) Testes scattered; with two eyes; with adenodactyls .... **Dendrocoelidae [p. 98]**

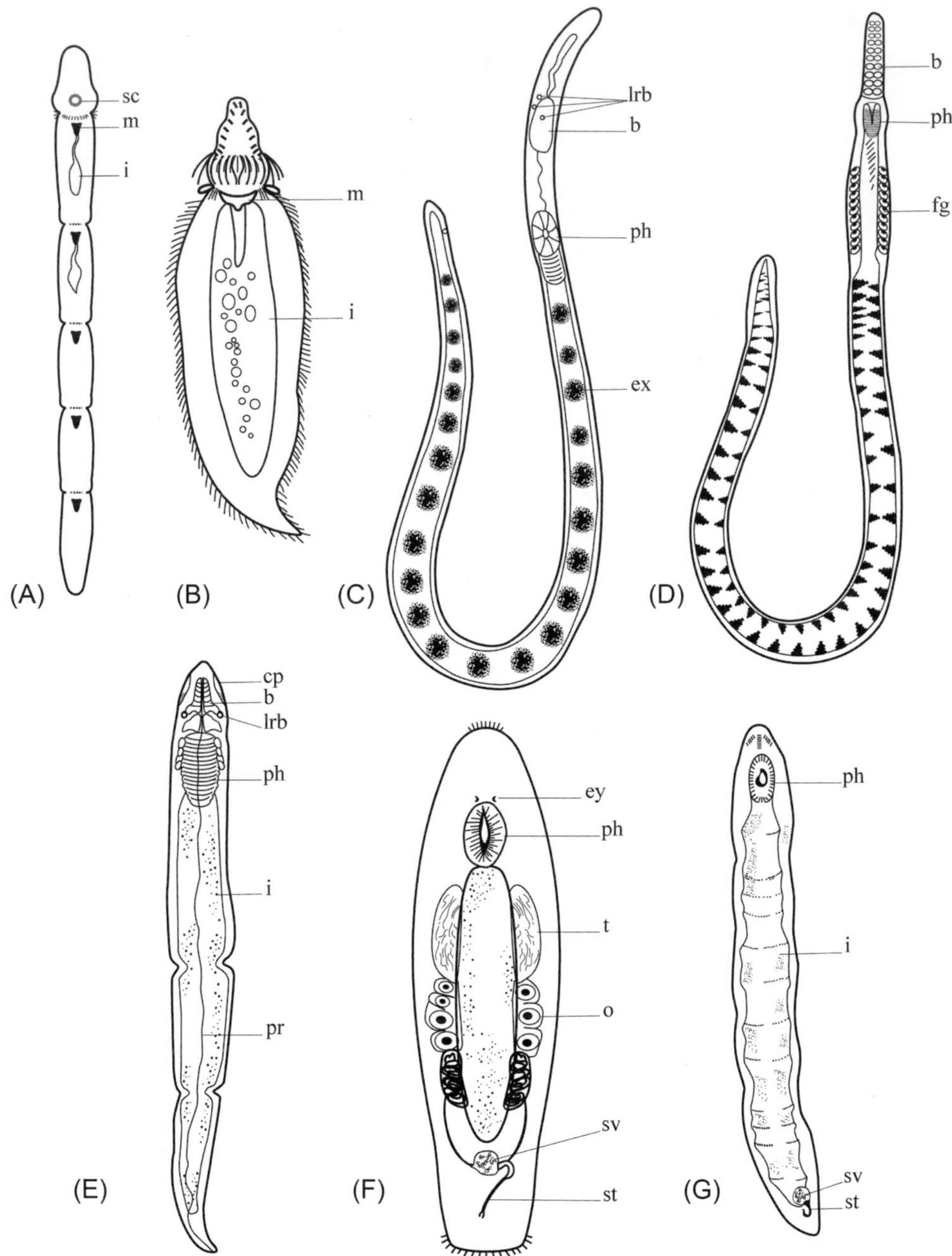

FIGURE 5.1 Schematic representation of some genera of Catenulida and Macrostomida from the freshwater environments. (A) *Catenula* sp. (Catenulida, Catenulidae) length: 1–5 mm; (B) *Catenula turgida* (Catenulida, Catenulidae) length 0.2–0.36 mm; (C) *Chordarium europaeum* (Catenulida, Chordariidae) length: 0.6–1 mm; (D) *Rhynchoscolex* sp. (Catenulida, Stenostomidae) length: 4–6 mm; (E) *Stenostomum* sp. (Catenulida, Stenostomidae) length: 0.5–4 mm; (F) *Macrostomum* sp. (Macrostomida, Macrostomidae) length: 1–3 mm; (G) *Microstomum* sp. (Macrostomida, Microstomidae) length: 0.4–4 mm. Abbreviations: b, brain; cp, ciliated pits; ex, excretophore; ey, eyes; fg, female gonads; i, intestine; lrb, light refracting bodies; m, mouth; o, ovary; ph, pharynx; pr, protonephridial duct; sc, statocyst; st, stylet; sv, seminal vesicle; t, testes.

2'	Testes before the pharynx; without eyes and without adenodactyls	**Kenkiidae [p. 99]**
3(1)	Head triangular; one pairs of eyes; white auricle and eyefields	**Dugesiidae [p. 99]**
3'	Head rounded, no triangular; two or numerous marginal eyes, sometimes absent; without white auricles or eyefields	**Planariidae [p. 100]**

Platyhelminthes: Tricladida: Dendrocoelidae: Genera

1	Penis bulb elongated (Fig. 5.10 A)	*Procotyla*
1'	Penis bulb rounded (Fig. 5.10 B)	*Dendrocoelopsis*

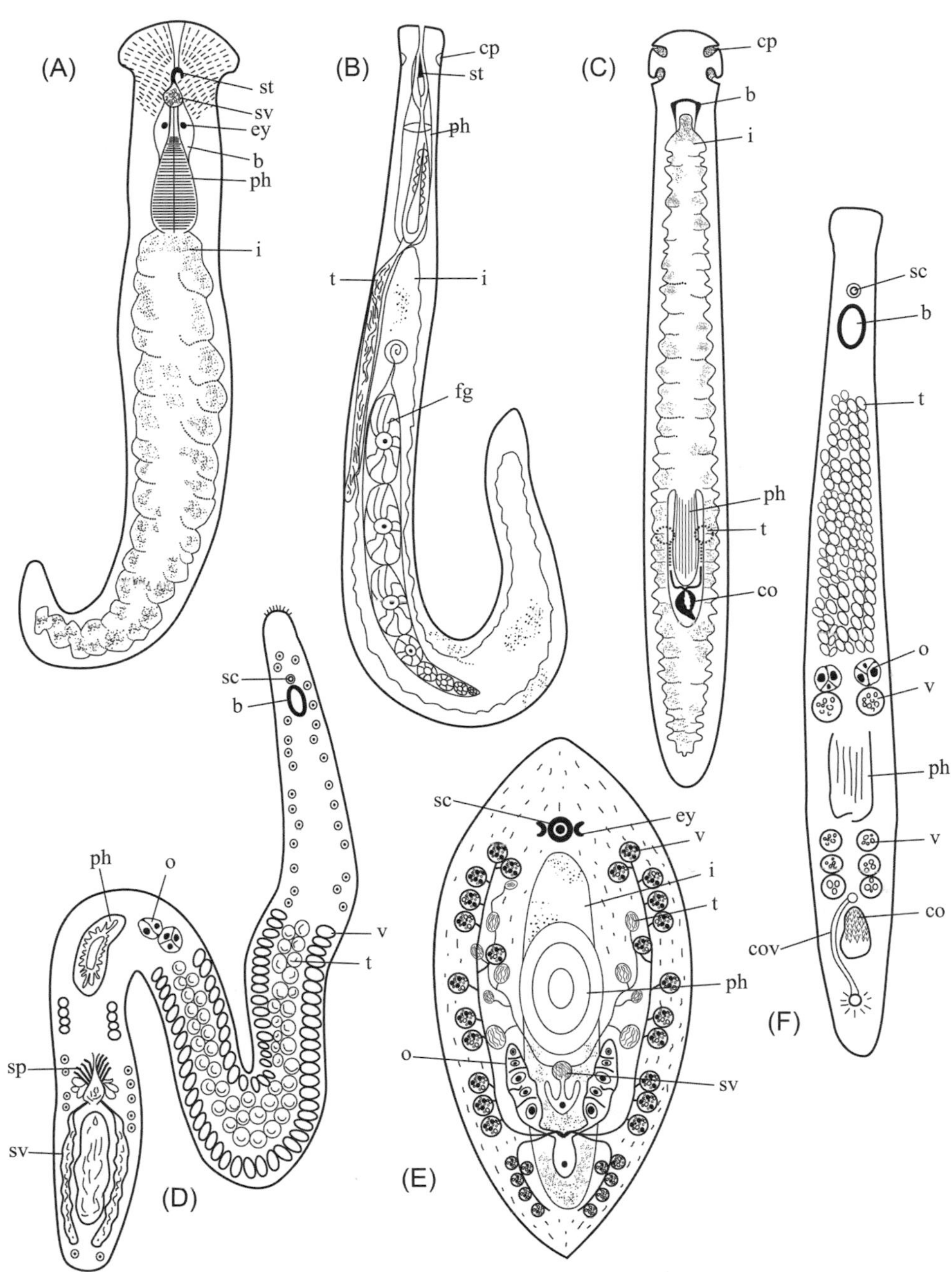

FIGURE 5.2 Schematic representation of freshwater genera and some freshwater species of Lecithoepitheliata and Proseriata. (A) *Geocentrophora* sp. (Lecithoepitheliata, Prorhynchidae) length: 1–4 mm; (B) *Prorhynchus stagnalis* (Lecithoepitheliata, Prorhynchidae) length: 2–7 mm; (C) *Bothrioplana semperi* (Proseriata, Bothrioplanidae) length: 1–3 mm; (D) *Coelogynopora* sp. (Proseriata, Coelogynoporidae) length: 5–10 mm; (E) *Otomesostoma auditivum* (Proseriata, Otomesostomidae) length: 3–5 mm; (F) *Monocelopsis* sp. (Proseriata, Monocelididae) length 1–1.5 mm. Abbreviations: b, brain; co, copulatory organ; cov, common oviduct; cp, ciliated pits; ey, eyes; fg, female gonads; i, intestine; o, ovary; ph, pharynx; sc, statocyst; sp, spines; st, stylet; sv, seminal vesicle; t, testes; v, vitellaria.

Platyhelminthes: Tricladida: Kenkiidae: Genera

1 Body elongated; flat; with a well-developed postpharyngeal section (Fig. 5.11 A) .... *Sphalloplana*

1' Body oval elongated-shaped; with reduced postpharyngeal section .... *Kenkia*

Platyhelminthes: Tricladida: Dugesiidae: Genera

1 Head triangular; pharynx pigmented; testes pre and post pharyngeal 2

1' Head bluntly triangular; pharynx unpigmented; testes pre pharyngeal; male copulatory organ small; penis papilla digitiform; bursa sac not developed (Fig. 5.10 C) .... *Cura foremanii* (Girard, 1852)

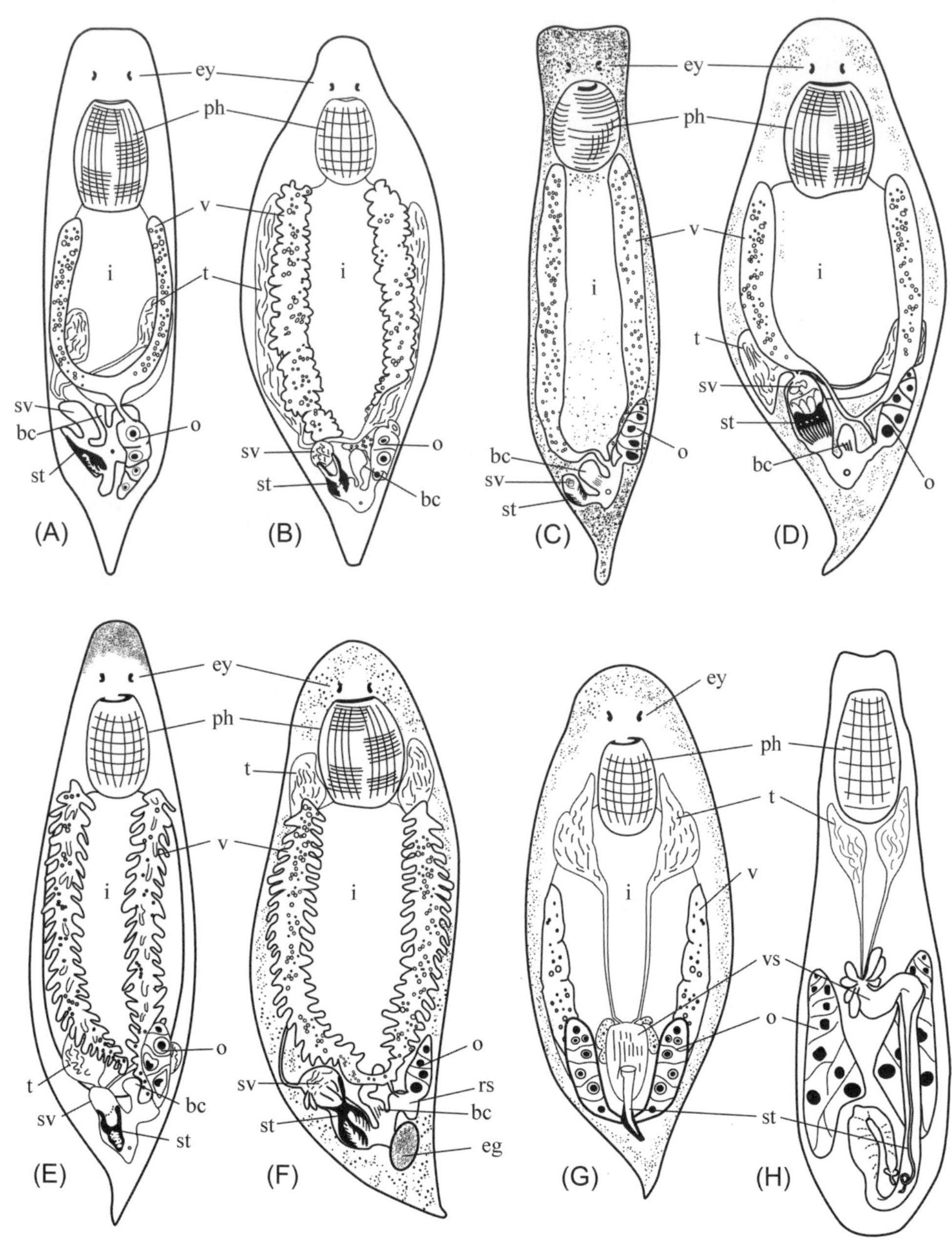

FIGURE 5.3 Schematic representation of characteristic genera of Dalyellioida (Rhabdocoela). (A) *Castrella* sp. (Dalyelliidae) length: 0.5–1 mm; (B) *Dalyellia* sp. (Dalyelliidae) length: 1.5–5 mm; (C) *Fulinskiella* sp. (Dalyellidae) length: 0.7–1 mm; (D) *Gieysztoria* sp. (Dalyelliidae) length: 0.7–1.35 mm; (E) *Microdalyellia* sp. (Dalyelliidae) length: 0.5–1.5 mm; (F) *Pseudodalyellia* sp. (Dalyelliidae) length: 0.6–1 mm; (G) *Provortex* sp. (Provorticidae) length: 0.5–1 mm; (H) *Vejdovskya* sp. (Provorticidae) length: 0.3–1 mm. Abbreviations: bc, bursa copulatrix; eg, eggs; ey, eyes; i, intestine; o, ovary; ph, pharynx; rs, receptaculum seminis; st, stylet; sv, seminal vesicle; t, testes; v, vitellaria.

2(1)	Head low triangular; testes forming dorsal clusters (Fig. 5.10 E)	*Schmidtea polychroa* (Schmidt, 1861)
2'	Head high triangular; testes numerous; usually ventral and scattered (Fig. 5.10 D)	*Girardia*

Platyhelminthes: Tricladida: Planariidae: Genera

1	Numerous eyes	2
1'	One pair of eyes or absent	3

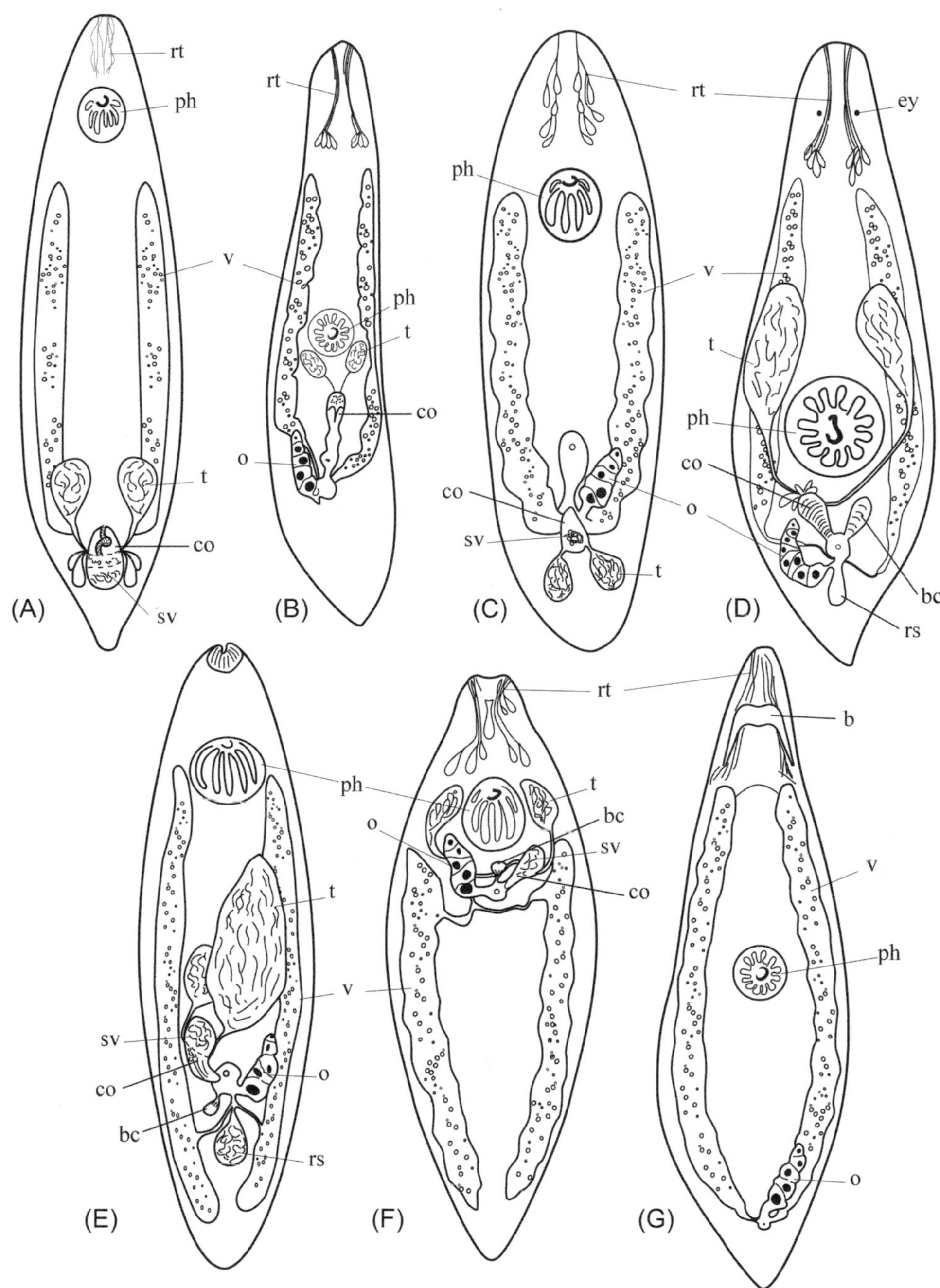

FIGURE 5.4 Schematic representation of characteristic genera of some subfamilies of Typhloplanidae (Typhloplanoida, Rhabdocoela). (A) *Acrochordonoposthia* sp. (Protoplanellinae) length: 0.9–1.5 mm; (B) *Amphibolella* sp. (Protoplanellinae) length: 1.4–1.8 mm; (C) *Bryoplana* sp. (Protoplanellinae) length: 0.4–0.5 mm; (D) *Krumbachia* sp. (Protoplanellinae) length: 1.7–2.8 mm; (E) *Microcalyptorhynchus* sp. (Protoplanellinae) length: 0.8–1 mm; (F) *Prorhynchella* sp. (Protoplanellinae) length: 0.6–1 mm; (G) *Olisthanella* sp. (Olisthanellinae) length: 1–1.5 mm. Abbreviations: b, brain; bc, bursa copulatrix; co, copulatory organ; ey, eyes; o, ovary; ph, pharynx; rs, receptaculum seminis; rt, rhabdite tracks; sv, seminal vesicle; t, testes; v, vitellaria.

2(1)	Eyes located along the anterior edge of the head (Fig. 5.11 D)	*Polycelis*
2’	Eyes located in two groups on both sides of the head (Fig. 5.11 E)	*Seidlia*
3(1)	Penis papilla long	4
3’	Penis papilla remarkable short; large male atrium	*Hymanella retenuova* Castle, 1941

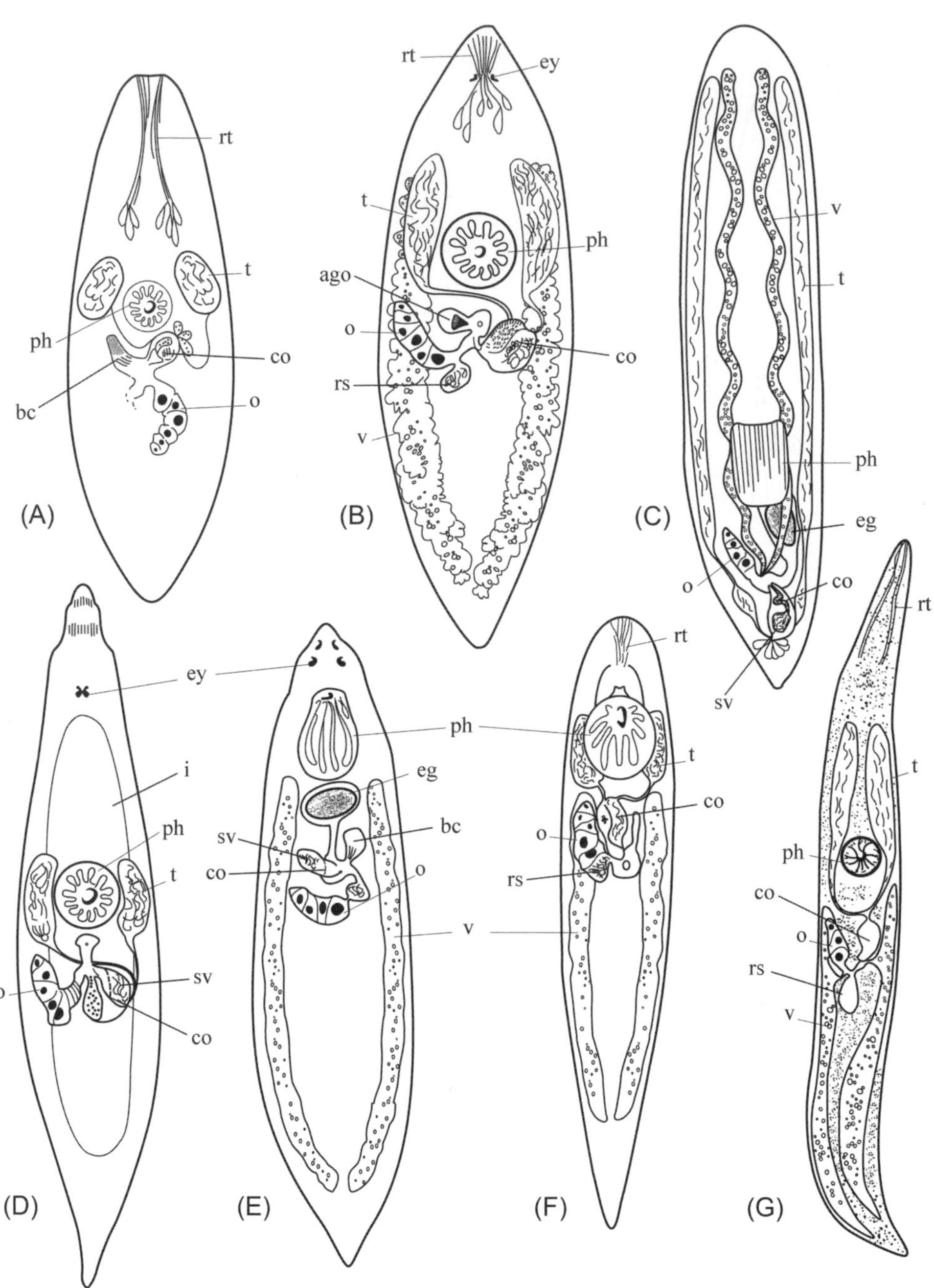

FIGURE 5.5 Schematic representation of characteristic genera of some subfamilies of Typhloplanidae (Typhloplanoida, Rhabdocoela). (A) *Castrada* sp. (Rhynchomesostominae) length: 0.5–1.5 mm; (B) *Mesocastrada* sp. (Rhynchomesostominae) length: 1–2.5 mm; (C) *Opistomum* sp. (Opistominae) length: 2–4.5 mm; (D) *Rhynchomesostoma* sp. (Rhynchomesostominae) length: 1–3.5 mm; (E) *Tetracelis* sp. (Rhynchomesostominae) length: 1–1.5 mm; (F) *Protoascus* sp. (Ascophorinae) length: 0.4–1.5 mm; (G) *Ascophora* sp. (Ascophorinae) length: 1–3 mm. Abbreviations: ago, atrial glandular organ; bc, bursa copulatrix; co, copulatory organ; eg, eggs; ey, eyes; i, intestine; o, ovary; ph, pharynx; rs, receptaculum seminis; rt, rhabdite tracks; sv, seminal vesicle; t, testes; v, vitellaria.

4(3) Adenodacty and bursal canal lead in the vicinity of gonopore (Fig. 5.11 B) .... *Paraplanaria*

4' Without adenodactyl (Fig. 5.11 C) .... *Phagocata*

Platyhelminthes: Lecithoepitheliata: Prorhynchidae: Genera

1 Anterior end elongated; sometimes light truncated; no fan expanded; without eyes; male stylet straight (Fig. 5.2 B) .... *Prorhynchus stagnalis* Schultze, 1851

1' Anterior end fan-like; with eyes except in some cave forms; male stylet curved; claw-like (Fig. 5.2 A) .... *Geocentrophora*

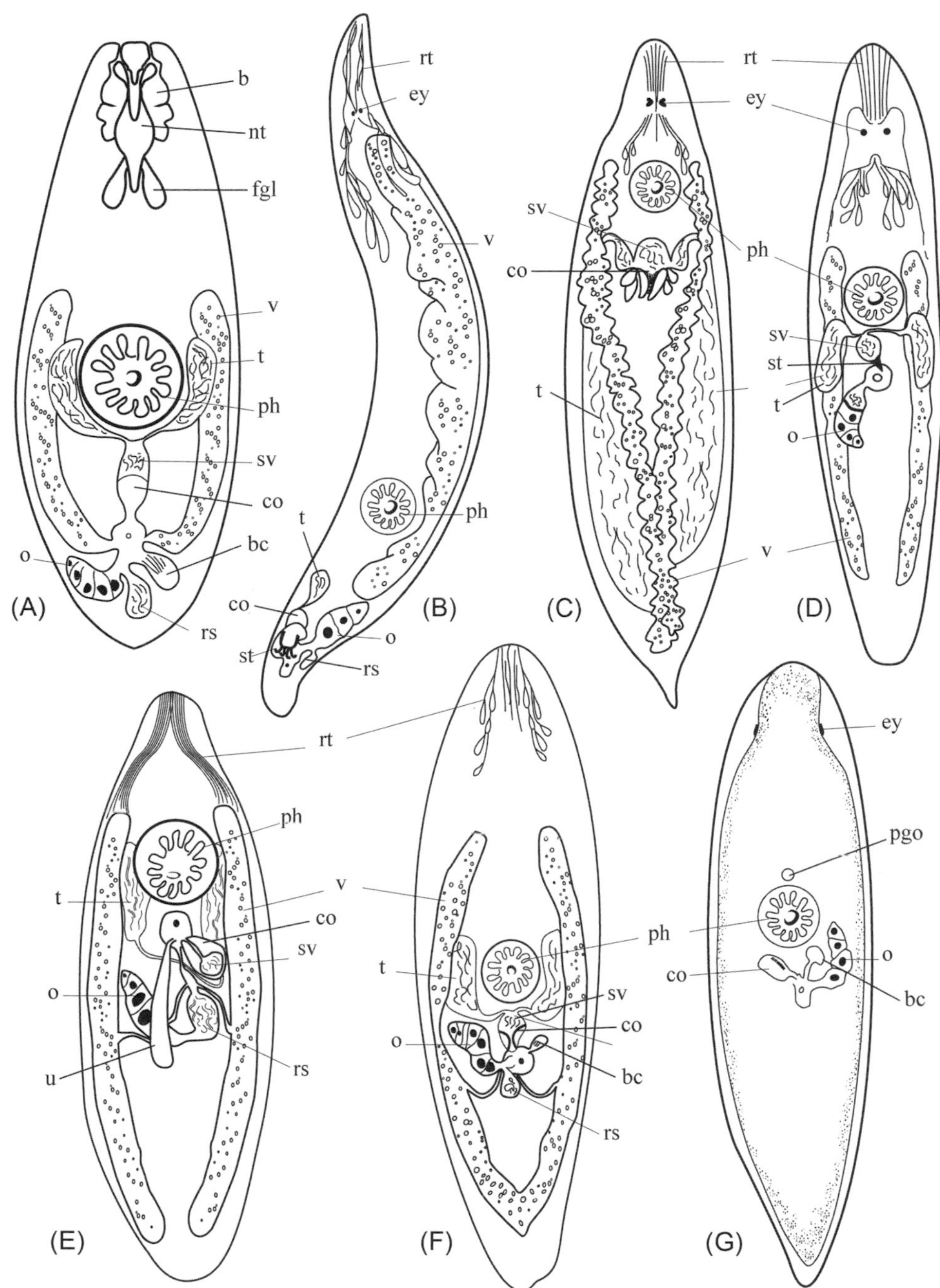

FIGURE 5.6 Schematic representation of characteristic genera of some subfamilies of Typhloplanidae (Typhloplanoida, Rhabdocoela). (A) *Adenoplea* sp. (Typhloplaninae) length: 0.42–0.6 mm; (B) *Limnoruanis* sp. (Typhloplaninae) length: 0.4–0.5 mm; (C) *Strongylostoma* sp. (Typhloplaninae) length: 0.75–1.5 mm; (D) *Styloplanella* sp. (Typhloplaninae) length: 0.8–1.3 mm; (E) *Typhloplana* sp. (Typhloplaninae) length: 0.7–1 mm; (F) *Typhloplanella* sp. (Typhloplaninae) length: 1.5–2.5 mm; (G) *Bothromesostoma* sp. (Mesostominae) length: 3–5 mm. Abbreviations: b, brain; bc, bursa copulatrix; co, copulatroy organ; ey, eyes; fgl, frontal glands; nt, nerve trunks; o, ovary; pgo, prepharyngeal glandular organ; ph, pharynx; rs, receptaculum seminis; rt, rhabdite tracks; st, stylet; sv, seminal vesicle; t, testes; u, uterus; v, vitellaria.

Platyhelminthes: Prolecithophora: Plagiostomidae: Genera

1 Female gonads forming two well defined characteristic "bodies" along the main body axis; ovaries follicular; prostatic vesicle very large; unarmed penis (Fig. 5.7 C) .... *Hydrolimax*

1' Female gonads on both sides of the intestine; ovaries compact; prostatic vesicle small; penis sometimes armed or reinforced .... *Plagiostomum planum* Silliman, 1885

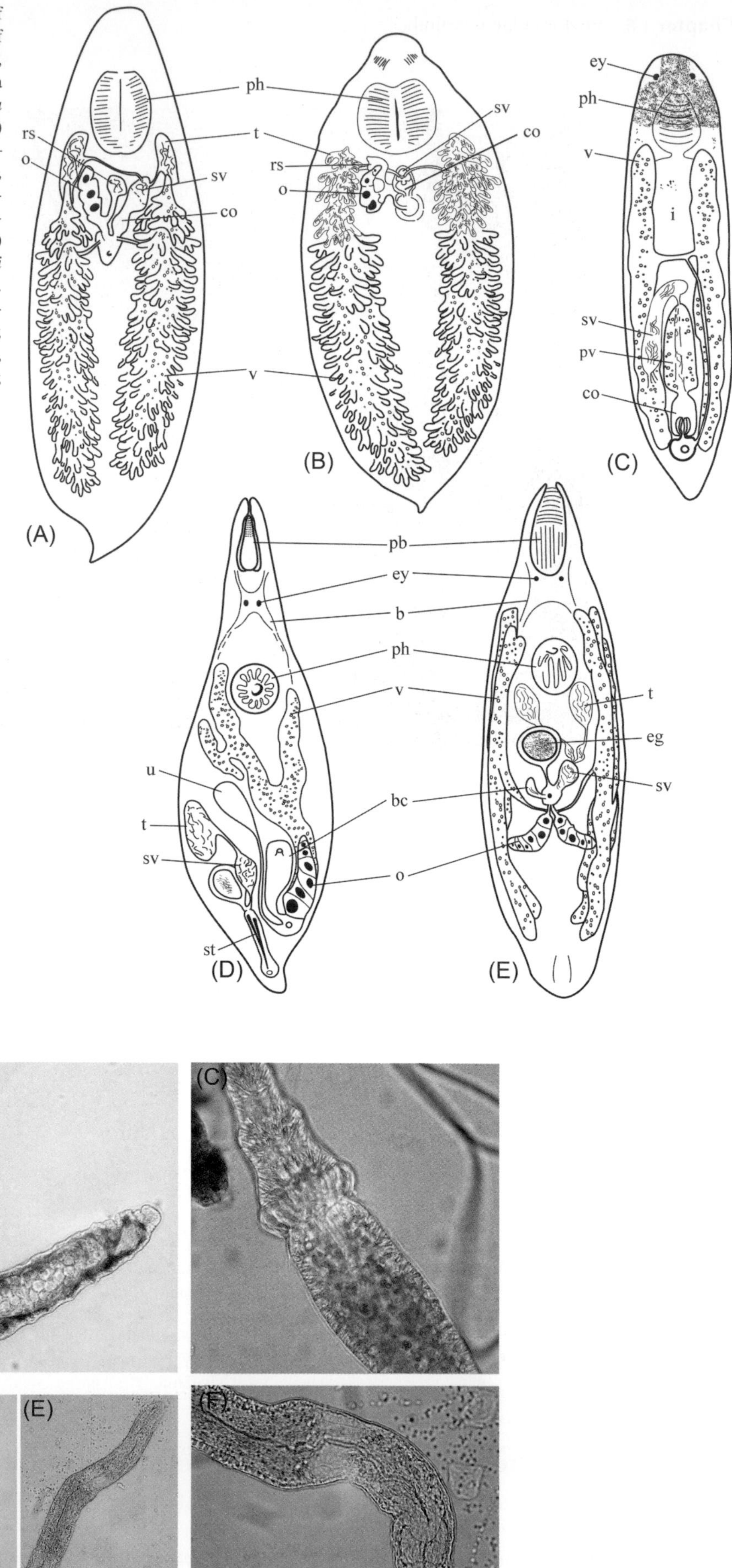

FIGURE 5.7 Schematic representation of characteristic genera of some subfamilies of Typhloplanidae (Typhloplanoida, Rhabdocoela), and of freshwater genera of Kalyptorhynchia (Rhabdocoela) and Prolecithophora. (A) *Phaenocora* sp. (Phaenocorinae) length: 2–6 mm; (B) *Phaenocora sulfophila* (Phaenocorinae) length: 3.5–4.8 mm; (C) *Hydrolimax grisea* (Plagiostomidae, Prolecithophora) with a partial sample of their pigmentation; length: 10–15 mm; (D) *Gyratrix hermaphroditus* (Polycistididae Kalyptorhynchia) length: 0.5–12 mm; (E) *Opisthocystis goettei* (Polycistididae, Kalyptorhynchia) length: 2.5–3 mm. Abbreviations: b, brain; bc, bursa copulatrix; co, copulatory organ; eg, eggs; ey, eyes; i, intestine; o, ovary; pb, proboscide; ph, pharynx; pv, prostatic vesicle; rs, receptaculum seminis; st, stylet; sv, seminal vesicle; t, testes; u, uterus; v, vitellaria.

FIGURE 5.8 Microphotographs of live microturbellarians. (A) General view of *Chordarium* sp.; (B) detail of the frontal region of *Chordarium* sp.; (C) anterior region of *Catenula turgida*; (D) anterior region of *Rhynchoscolex* sp.; (E) *Myostenostomum* sp.; (F) detail of the muscular intestinal zone of *Myostenostomum* sp.

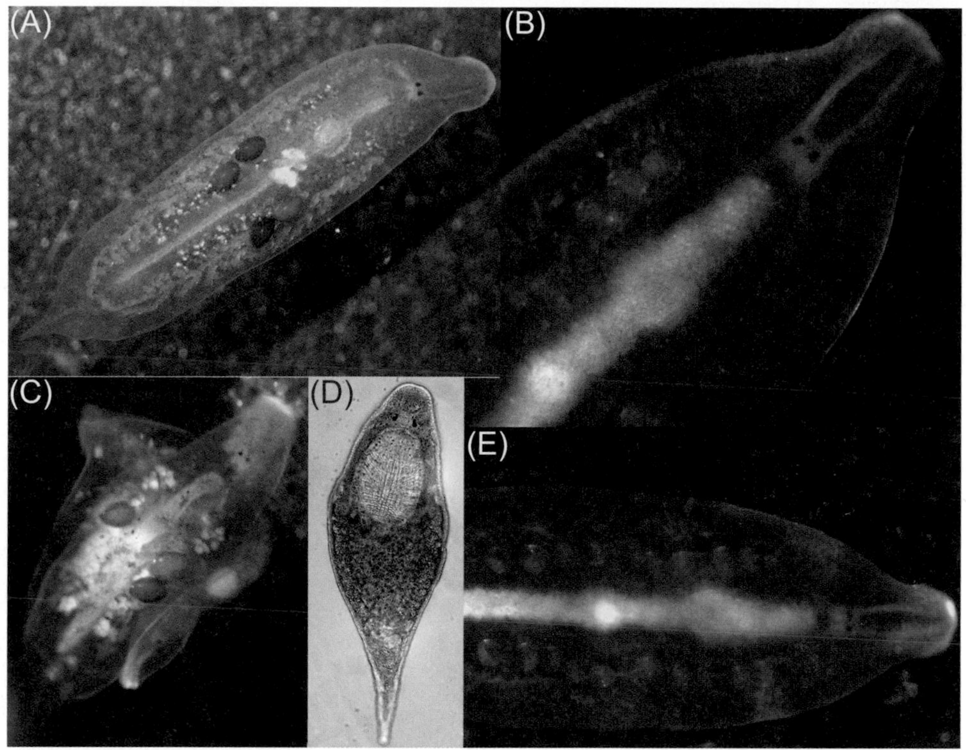

FIGURE 5.9 Microphotographs of live microturbellarians. (A) mature *Mesostoma* ehrenbergii with several dormant eggs in the uterus; (B) anterior region of *Mesostoma* ehrenbergii with several subitaneous eggs in the uterus; (C) *Mesostoma ehrenbergii* while mating; (D) *Gieysztoria* sp. with the developed ovary and vitellaria; (E) *Mesostoma ehrenbergii* with several eggs in the uterus and with some Cladocera in the intestine.

Platyhelminthes: Rhabdocoela: Families

1	Pharynx bulbous *rosulatus* (dorsoventral orientated)	2
1'	Pharynx bulbous *doliiformis* (fronto-caudally orientated)	3
2(1)	Without true proboscis	**Typhloplanidae [p. 105]**
2'	With true proboscis	**Polycistididae [p. 108]**
3(1)	Ovary single; male copulatory organ with a spiny stylet	**Dalyellidae [p. 108]**
3'	Ovary paired; male copulatory organ with a straight tubular stylet (bucket-like) without spines	**Provorticidae [p. 108]**

Platyhelminthes: Rhabdocoela: Typhloplanidae: Genera

1	Pharynx ventral directed	2
1'	Pharynx with other orientation	22
2(1)	Testes dorsal to the vitellaria	3
2'	Testes ventral to the vitellaria	7
3(2)	Testes sac-like, in front of the pharynx	4
3'	Testes mostly follicular, behind or on both sides of the pharynx, pharynx in the mid-body (second third) with two eyes	6
4(3)	Without eyes; with ascus	5
4'	Without eyes (sometimes with pigment patches); without ascus; male and female organ posterior; at the last third of the body (Fig. 5.4 G)	*Olisthanella coeca* (Silliman, 1885)
5(4)	Pharynx at the anterior end (first third of the body); female and male copulatory organ in the second third (Fig. 5.5 F)	*Protoascus wisconsinensis* Hayes, 1941
5'	Pharynx in the middle of the body (second third); female and male copulatory organs close behind the pharynx (Fig. 5.5 G)	*Ascophora elegantissima* Findenegg, 1924
6(3)	Dense dark brownish pigmentation on the dorsal side; with anterior pigment loss eyes fields; eyes masked; with a prepharyngeal glandular organ (Fig. 5.6 G)	*Bothromesostoma personatum* (Schmidt, 1848)
6'	Different tonalities (transparent; beige; caramel) and pattern (with spots or with thin stripes) of the dorsal pigmentation, eyes visible, without pre-pharyngeal glandular organ (Fig. 5.9 A–C, E)	*Mesostoma*

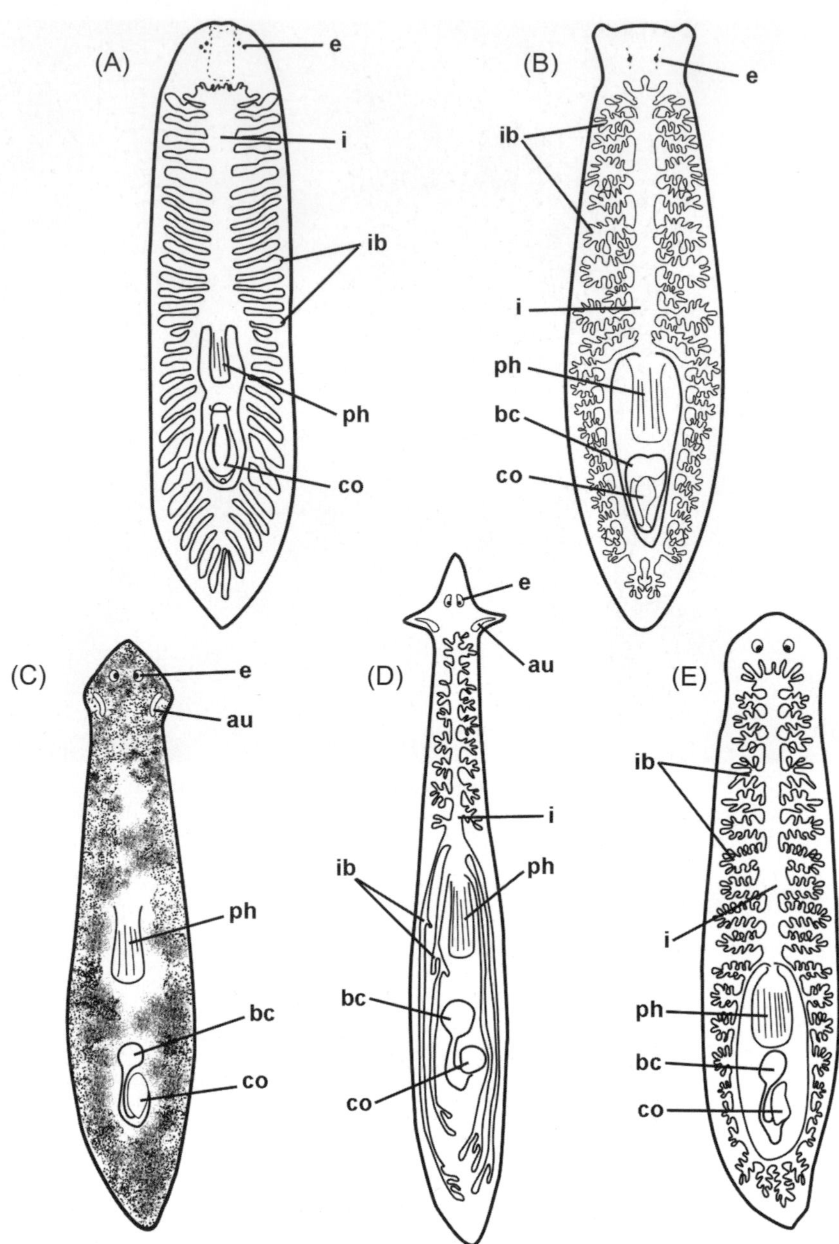

FIGURE 5.10 Schematic representation of some characteristic genera of Dendrocoelidae (Tricladida) and Dugesiidae (Tricladida). (A) *Procotyla* sp. (Dendrocoelidae) length: 10–12mm; (B) *Dendrocoelopsis* sp. (Dendrocoelidae) length: 14–22mm; (C) *Cura* sp. (Dugesiidae) length: 7–15mm; (D) *Girardia* sp. (Dugesiidae) length: 6–30mm; (E) *Schmidtea* sp. (Dugesiidae) length: ±20mm. Abbreviations: au, auricles; bc, bursa copulatrix; co, copulatory organ; e, eyes; i, intestine; ib, intestinal branches; ph, pharynx; vd, vas deferens.

7(2)	Pores of the protonephridium open near the oral pore or in the genital pore	8
7'	Pores of the protonephridium open on body surface	17
8(7)	Pores of the protonephridium open in the buccal cavity or near the pharynx	9
8'	Pores of the protonephridium open in the genital atrium	14
9(8)	Copulatory organ at the distal end of the body	10
9'	Copulatory organ at the middle of the body, close to the pharynx	11
10(9)	Male organ with a stylet funnel-like and smooth; body oval; without eyes and rounded (light truncated) anterior part (Fig. 5.6 A)	*Adenoplea nanus* Sayre & Wergen, 1994
10'	Male organ with a stylet crown-shaped with four curved spines; body elongated; with two eyes and an elongated anterior part (Fig. 5.6 B)	*Limnoruanis romanae* Kolasa, 1977

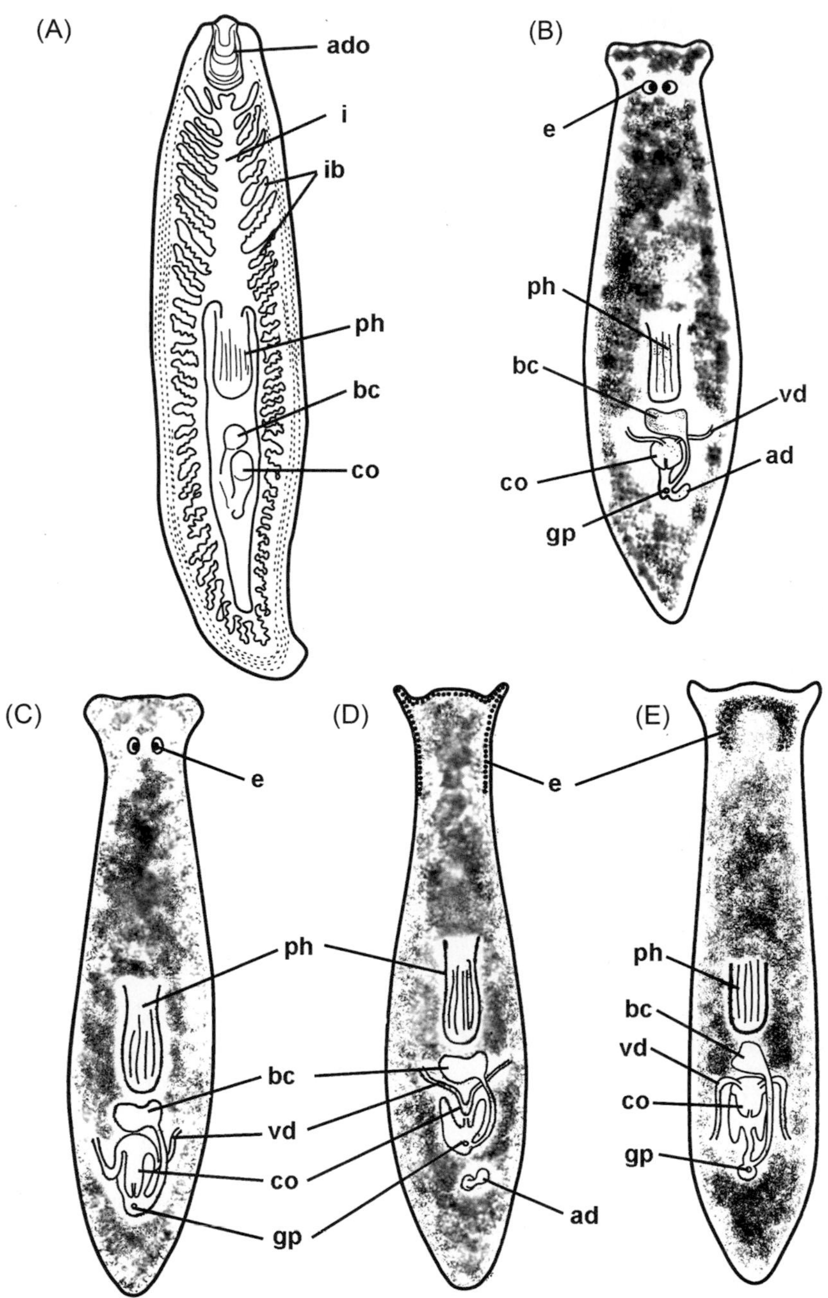

FIGURE 5.11 Schematic representation of some characteristic genera of Kenkiidae (Tricladida) and Planariidae (Tricladida) length: 7–15 mm. (A) *Sphalloplana* sp. (Kenkiidae) length: 5–20 mm; (B) *Paraplanaria* sp. (Planariidae) length: 9–13 mm; (C) *Phagocata* sp. (Planariidae) length: 10–14 mm; (D) *Polycelis* sp. (Planariidae) length: 7–12 mm; (E) *Seidlia* sp. (Planariidae) length: 11–17 mm. Abbreviations: ad, adenodactyl; ado, adhesive organ; bc, bursa copulatrix; co, copulatory organ; e, eyes; gp, gonopore; i, intestine; ib, intestinal branches; ph, pharynx; vd, vas deferens.

11(9)	Caudal testes	12
11'	Frontal testes, behind and before the pharynx	13
12(11)	Ejaculatory duct developed as spiny cirrus; testes stretching at the posterior end; with bursa copulatrix; with eyes (Fig. 5.6 C)	*Strongylostoma*
12'	Ejaculatory duct with cuticular lining, without spines; testes near the pharynx; vitellaria at the posterior end; without bursa copulatrix; without eyes (Fig. 5.6 E)	*Typhloplana*
13(11)	Without eyes, whitish, opaque, sometimes light pigmented; ejaculatory duct with cuticular lining; with bursa copulatrix (Fig. 5.6 F)	*Typhloplanella halleziana* (Vejdovsky, 1880)
13'	With eyes; transparent; with tube-like stylet (Fig. 5.6 D)	*Styloplanella strongylostomoides* Findenegg, 1924

14(8) With eyes; male organ with numerous spines in the bursa copulatrix, in the blind sacs or in the male atrium 15

14' Without eyes; copulatory organs in the middle, or between the first and the second third of the body. Male organ with numerous spines in blind sacs or in the bursa copulatrix (Fig. 5.5 A) .... *Castrada*

15(14) With two eyes 16

15' With four eyes (Fig. 5.5 E) .... *Tetracelis marmorosa* (Müller, 1773)

16(15) With a retractable anterior end (telescope-like) (Fig. 5.5 D) .... *Rhynchomesostoma rostratum* (Müller, 1773)

16' Without retractable anterior end (Fig. 5.5 B) .... *Mesocastrada fuhrmanni* Volz, 1898

17(7) Pharynx in the second third of the body 18

17' Pharynx anterior; in the first third of the body 19

18(17) Testes before the pharynx; copulatory organs posterior to the pharynx; on the last third of the body (Fig. 5.4 D) .... *Krumbachia*

18' Small testes behind the pharynx; copulatory organs close behind the pharynx on the second third of the body (Fig. 5.4 B) .... *Amphibolella segnis* Findenegg, 1924

19(17) Male copulatory organ without spines or cirrus 20

19' Male copulatory organ with a spiny cirrus (Fig. 5.4 A) .... *Acrochordonoposthia conica* Reisinger, 1924

20(19) Male and female copulatory organs on the last third of the body 21

20' Male and female organs close behind the pharynx; on the first third of the body (Fig. 5.4 F) .... *Prorhynchella minuta* Ruebush, 1939

21(20) Testes paired; anterior, voluminous and with uneven development; without eyes; anterior end with a small anterior projection like a proboscis (Fig. 5.4 E) .... *Microcalyptorhynchus virginianus* Kepner & Ruebush, 1935

21' Paired small posterior testes with equal development (Fig. 5.4 C).......... *Bryoplana xerophila* Van Steenkiste, Davison, & Artois, 2010

22(1) Pharynx forward directed; feathery vitellaria; male copulatory organ sometimes with spines .... *Phaenocora* (Fig. 5.7 A, B)

22' Pharynx backwards directed; vitellaria smooth; male copulatory organ cirrus-like; with spines (Fig. 5.5 C) .... *Opistomum pallidum* Schmidt, 1848

Platyhelminthes: Rhabdocoela: Polycistididae: Genera

1 Copulatory organs in the second third of the body, behind and near the pharynx; paired ovaries; short crooked funnel-like stylet; with excretory cup (Fig. 5.7 E).......... *Opisthocystis goettei* (Bresslau, 1906)

1' Copulatory organs in the last third of the body, behind and clear separate from the pharynx; posteriorly with a long fork-like stylet; with eyes (Fig. 5.7 D) .... *Gyratrix hermaphroditus* Graff, 1831

Platyhelminthes: Rhabdocoela: Dalyellidae: Genera

1 Stylet with two handles and two spiny branches 2

1' Stylet with one handle or without handles 5

2(1) Stylet with two handles of similar size 3

2' Stylet with two handles of different size (Fig. 5.3 F) .... *Pseudodalyellia alabamensis* Van Steenkiste, Gobert & Artois, 2011

3(2) Stylet compose by two joined handles, branches and spines 4

3' Stylet compose by two independent handles, branches and spines (Fig. 5.3 C).......... *Fulinskiella bardeaui* (Steinböck, 1926)

4(3) Stylet with narrow handles, usually longer than the branches, branches with different kind of spines, with a middle channel; with only one egg; without zoochlorellae (Fig. 5.3 E) .... *Microdalyellia*

4' Stylet with width handles, usually shorter than the branches, branches with robust spines, without a middle channel, several eggs stored in the parenchyma, sometimes with zoochlorellae (Fig. 5.3 B) .... *Dalyellia*

5(1) Stylet with one handle and two spiny branches (Fig. 5.3 A) .... *Castrella*

5' Stylet without handles, with a wide belt and different kind of spines (Figs. 5.3 D and 5.9 D) .... *Gieysztoria*

Platyhelminthes: Rhabdocoela: Provorticidae: Genera

1 Tubular stylet long; very fine and needle-like (Fig. 5.3 H) .... *Vejdovskya pellucida* (Schultze, 1851)

1' Tubular stylet short, width, truncated cone (Fig. 5.3 G).......... *Provortex virginiensis* Ruebush & Hayes, 1939

ACKNOWLEDGMENTS

We would like to thank Dr Thorp for inviting us to participate in this chapter, Dr Ana Maria Leal-Zanchet for allowing us to use the photographs of the catenulids (Fig. 5.8) and *Gieysztoria* (Fig. 5.9), and Sebastian Castorino for providing assistance in the vectorization of the drawings. This study received support from the I+D Project grant CGL 2010-15786/BOS and CGL2011-29916, which are financed by Spanish Ministry of Economy; from PIP 2010-390 and PIP 2013-2015-635(CONICET), and N11/728 FCNyM (UNLP).

REFERENCES

Baguñà, J. & M. Riutort. 2004a. Molecular phylogeny of the Platyhelminthes. Canadian Journal of Zoology 82: 168–193.

Baguñà, J. & M. Riutort. 2004b. The dawn of bilaterian animals: the case of acoelomorph flatworms. BioEssays 26: 1046–1057.

Cannon, L.R.G. 1986. Turbellaria of the world. A guide to families & genera. Queensland Museum, Brisbane. 186 pp.

Ferguson, F.F. 1954. Monograph of the Macrostomine worms of Turbellaria. Transactions of the American Microscopical Society 73: 137–164.

Gilbert, C.M. 1938. A remarkable North American species of the genus *Phaenocora*. Zeitschrift fuer Morphologie und Ökologie der Tiere 33: 53–71.

Houben, A.M., N. Van Steenkiste & T.J. Artois. 2014. Revision of *Phaenocora* Ehrenberg, 1836 (Rhabditophora, Typhloplanidae, Phaenocorinae) with the description of two new species. Zootaxa 3889: 301–354.

Kenk, R. 1974. Chapter 2: History of the study of Turbellaria in North America. Pages 17–22 *in*: N.W. Rise and M.P. Morse. (eds.), Biology of the Turbellaria. McGraw-Hill Book Company, New York, NY.

Kenk, R. 1989. Revised list of the North American freshwater planarians (Platyhelminthes: Tricladida: Paludicola). Smithsonian Contributions to Zoology 476: 1–10.

Kolasa, J. & S. Tyler. 2010. Chapter 6: Flatworms: Turbellarians and Nemertea. Pages 143–161 *in*: J.H. Thorp and A.P. Covich (eds.), Ecology and classification of North American freshwater invertebrates (Third Edition), Academic Press, San Diego, CA.

Littlewood, D.T.J. & R.A. Bray (eds.) 2001. Interrelationships of the Platyhelminthes. Taylor & Francis, London, New York.

Luther, A. 1955. Die Dalyelliiden (Turbellaria Neorhabdocoela) Eine Monographie. Acta Zoologica Fennica 87: 1–337.

Luther, A. 1960. Die Turbellarien Ostfennoskandiens. I. Acoela, Catenulida, Macrostomida, Lecithoepitheliata, Prolecithophora, und Proseriata. Fauna Fennica 7: 1–155.

Luther, A. 1963. Die Turbellarien Ostfennoskandiens. IV. Neorhabdocoela 2. Typhloplanoida: Typhloplanidae, Solenopharyngidae und Carcharodopharyngidae. Fauna Fennica 16: 1–161.

Nuttycombe, J.W. 1956. The Catenulida of the Eastern United States. American Midland Naturalist 55: 419–433.

Nuttycombe, J.W. & A.J. Waters 1938. The American species of the genus *Stenostomum*. Proceedings of the American Philosophical Society 79: 213–301.

Richter, S., R. Loese, G. Purschke, A. Schmidt-Rhaesa, G. Scholtz, T. Stach, L. Vogt, A. Wanninger, G. Brenneis, C. Döring, S. Faller, M. Fritsch, P. Grobe, C.M. Heuer, S. Kaul, O.S. Møller, C.H.G. Müller, V. Rieger, B.H. Rothe, M.E.J. Stegner & S. Harzsch. 2010. Invertebrate neurophylogeny: suggested terms and definitions for a neuroanatomical glossary. Frontiers in Zoology 7:1–49.

Rieger, R.M., S. Tyler, J.P.S. Smith III, & G. Rieger. 1991. Chapter 2: Turbellaria. Pages 7–140 *in*: F.W. Harrison and B.J. Bogitsh (eds.), Microscopic Anatomy of Invertebrates. Vol. 3: Platyhelminthes and Nemertinea. Wiley-Liss, New York, NY.

Ruiz-Trillo, I., M. Riutort, D.T.J., Littlewood, E.A. Herniou & J. Baguñà. 1999. Acoel flatworms: Earliest extant bilaterian metazoans, not members of Platyhelminthes. Science 283: 1919–1923.

Ruiz-Trillo, I., M. Riutort, H.M. Fourcade, J. Baguñà & J.L. Boore. 2004. Mitochondrial genome data support the basal position of Acoelomorpha and the polyphyly of the Platyhelminthes. Molecular Phylogenetics and Evolution 33: 321–332.

Schärer, L. 2014. Macrostomorpha taxonomy and phylogeny. http://macrostomorpha.info. Consulted 2014-12-04.

Tyler, S. & M. Hooge. 2004. Comparative morphology of the body wall in flatworms (Platyhelminthes). Canadian Journal of Zoology 82: 194–210.

Tyler, S., S. Schilling, M. Hooge, & L.F. Bush (comp.) 2006–2015. Turbellarian taxonomic database. Version 1.7 http://turbellaria.umaine.edu. Consulted 2014-12-04.

Chapter 6

Phylum Nemertea

Malin Strand
The Swedish Species Information Centre, Swedish University of Agricultural Sciences, Uppsala, Sweden

Per Sundberg
Department of Zoology, University of Gothenburg, Gothenburg, Sweden

PHYLUM NEMERTEA

Chapter Outline

INTRODUCTION

There are around 1275 species of nemerteans (phylum Nemertea) described (Kajihara et al., 2008), with the vast majority being marine, and only 22 species are currently known from freshwater habitats. Of these, only two species are recorded from the Nearctic region: one from Lake Huron in Canada, and the other from a stream in Chester County, Pennsylvania. One of the species (*Prostoma canadiensis* Gibson & Moore 1978) was later recorded from Holland (Moore & Gibson, 1985). The other species, *Prostoma asensoriatum* (Montgomery, 1896), is only recorded from the type locality, and thus endemic for the region. The species are both benthic, like all other known freshwater species. Being rare and often overlooked, the ecological impact of these species is unknown.

LIMITATIONS

More freshwater nemerteans have been named and described from the region, besides the two listed above. Coe (1901), for example, combined a number of previous names into *Tetrastemma rubrum* (later transferred to *Prostoma*). However, the descriptions of these species are inadequate and too vague to be acceptable, and Gibson & Moore (1976) disregarded *P. rubrum* (and others) leaving the *P. canadiensis* and *P. asensoriatum* as the only valid names for nemerteans in this region. The original description of *Prostoma asensoriatum* closely resembles *P. eilhardi*, but offers a number of anatomical differences. Although some characters listed are too variable to be used for species diagnostics (Moore & Gibson, 1976), others appear to be a reliable species indicator. We thus conclude that there are currently two valid species known from the region, but at the same time emphasize that the real number of species is uncertain and probably higher. We base this conclusion on experience from other regions, and especially marine habitats, where new species are commonly found whenever a taxonomist familiar with the group samples in an area.

The two Nearctic nemerteans are both small (up to 4 cm), slender, and orange-reddish, and both are difficult to distinguish using external characters. The number of eyes (up to 8 arranged in 2–4 pairs) varies intraspecifically and also by age, with a positive correlation between the number and age/size. Although it is stated that *P. canadiensis* is slightly shorter than *P. asensoriatum*, this is obviously a character that varies with age and degree of contraction. Nemerteans are very contractile, and this also affects the interpretation of anatomical characters like extension of organs, positions, and volumes. It is often stated in the literature that nemerteans can, and have to be, identified from internal characters. However, as shown by Moore & Gibson (1976), and also argued in other papers, many characters are for various reasons impossible to use in this context. Furthermore, internal characters can only (with few exceptions) be interpreted from histological sections. In most cases, these sections are too time-consuming/expensive to provide, and the

Thorp and Covich's Freshwater Invertebrates. http://dx.doi.org/10.1016/B978-0-12-385028-7.00006-8

technique also requires special skills and equipment, which are lacking in many situations. We therefore conclude that use of internal characters is rare in reality, despite what being stated in many publications as a necessary requirement for proper identification of nemertean species.

TERMINOLOGY AND MORPHOLOGY

A nemertean is recognized by its smooth gliding way of locomotion over the ground, using the cilia covering the body, and the fact that it contracts easily if disturbed. Nemerteans are often confused with flatworms, moving in the same way and also sometimes of similar body shape. The unique character identifying Nemertea as a monophyletic taxon is the eversible proboscis.

It will be difficult to distinguish the two species from external characters, although it is stated in the description of *P. asensoriatum* that the color is bright orange, while *P. canadiensis* is described as "bright and … shade of pink". The species can, however, be distinguished by the internal character presence/absence of frontal organ, which we judge as a valid species identifier. There may be cases when it is possible to observe this character in live specimens by the opening in the snout, but it will require considerable experience working with this and use of a dissecting microscope. A firm species identification would, however, require histological preparation, and in most situations, it still may not be possible to key the species below the level of genus. There are currently no published molecular markers for either species, but these may prove especially valuable in the future.

MATERIAL PREPARATION AND PRESERVATION

Nemerteans should always be studied alive whenever possible. The characters to be observed are presence/absence of eyes, number and pattern if present, color, position of mouth, and (in this case), the presence of a frontal organ. For histological sections, it is important to properly anesthetize the animal before fixation to minimize contractions (which will distort anatomical characters). MS 222 or $MgCl_2$ is recommended. A common fixative for nemerteans is Bouin's fluid/paraformaldehyde. For histological studies, fixed animals are embedded in paraffin wax (56 °C). Sections are in general 6 microns thick and stained by the Mallory trichrome method. Ethanol is required for later DNA extractions, but is less appropriate for histology. Another good fixative for RNA and DNA studies is RNAlater, which is easy to carry around (not being toxic nor flammable) and is especially suited for fieldwork. Store the animals at −20 °C or lower if their DNA is to be analyzed.

KEYS TO NEMERTEA

The currently two known species from the Nearctic region can be conclusively identified by the presence/absence of a frontal organ (Fig. 6.1), such as:

1 Frontal organ present .... *Prostoma canadiensis* Gibson & Moore 1978

1' Frontal organ absent .... *Prostoma asensoriatum* (Montgomery 1896)

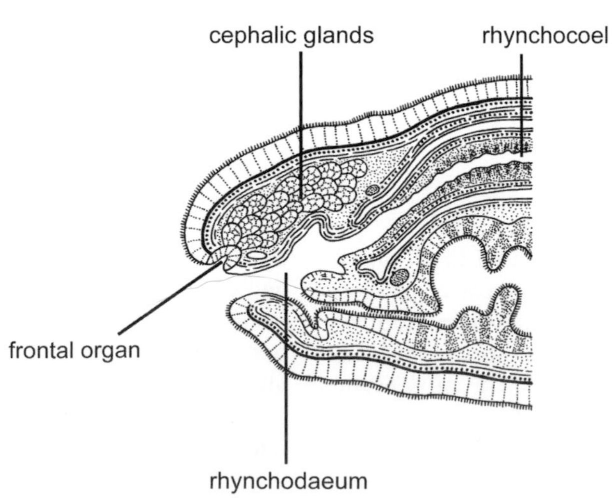

FIGURE 6.1 The placement of the frontal organ.

REFERENCES

Coe, W. R. 1901. Papers from the Harriman Alaska expedition. XX. The nemertean. Proceedings of the Washington Academy of Sciences 3: 1–110.

Kajihara, H., A. V. Chernyshev, S-C. Sun, P. Sundberg & F. B. Crandall. 2008. Checklist of Nemertean Genera and Species published between 1995 and 2007. Species Diversity 13: 245–274.

Moore, J. & R. Gibson. 1985. The evolution and comparative physiology of terrestrial and freshwater nemerteans. Biological Reviews 60: 257–312.

Sundberg, P. & R. Gibson. 2008. Global diversity of nemerteans (Nemertea) in freshwater. Hydrobiologia 595: 61–66.

Chapter 7

Phylum Gastrotricha

Tobias Kånneby
Department of Zoology, Swedish Museum of Natural History, Stockholm, Sweden

Chapter Outline

INTRODUCTION

Gastrotricha are small vermiform or tenpin-shaped acoelomate animals. The group is common to most aquatic environments and constitutes an important part of the meiofauna. To date, roughly 850 species have been described from all over the world. Freshwater gastrotrichs are, with very few exceptions, classified within the order Chaetonotida. They are among the smallest metazoans and can have a total body length of only 60 µm. However, most freshwater species attain larger body sizes, and some can reach lengths of up to 800 µm. Gastrotrichs are widely distributed in a variety of freshwater habitats. They are most common in epibenthic or periphytic habitats but may also be encountered in the interstitial. Bogs with *Sphagnum* spp. or small still nutrient rich ponds with *Lemna* spp. are especially rich and diverse in gastrotrichs.

The following works are helpful in species identification: Brunson (1950, 1959), who keyed and illustrated species then known from North American freshwaters; Robbins (1965, 1973), who gave additional information and drawings of North American freshwater species; d'Hondt (1971), who gave a key to the species of *Lepidodermella*; Kisielewski (1981), who made a critical evaluation of morphological characters that must be measured to identify a species; Kisielewski (1986), who gave a recent treatment of *Aspidiophorus*; Schwank (1990), who provided illustrated keys (in German) for all known freshwater species worldwide; Kisielewski (1991), who described many Neotropical species and addressed several important issues in gastrotrich systematics; Balsamo & Todaro (2002), who provided a key to the freshwater genera of the world; Kånneby et al. (2009) who gave an illustrated key to *Ichthydium*; and Kånneby et al. (2013), who provided the first molecular phylogeny of Chaetonotidae to evaluate the homology of cuticular characters for generic and subgeneric classification.

The identification key below includes all known genera of freshwater chaetonotidans worldwide. For monotypic genera, the species is always presented, although the genus may not have been reported from the Nearctic region. The key covers all Nearctic freshwater species published, as of December 2013. For more recent publications on the Nearctic freshwater gastrotrich fauna, the reader should refer to: (1) Schwank & Kånneby (2014), which redescribes seven species, from Ontario, Canada, previously considered *nomina nuda*; and (2) Kånneby & Wicksten (2014), which present the first record of a freshwater Macrodasyida, *Redudasys* sp., from the Northern hemisphere, found in Texas, USA. The identification key to species may sometimes seem too detailed; however, in this way new records or species for the region can be discerned. Rough distributions are given for each species as reported in the literature, but most species probably have a much wider distribution because of the few studies focusing on Nearctic freshwater Gastrotricha.

LIMITATIONS

It is relatively easy to identify most Nearctic freshwater gastrotrichs to genus and very difficult to identify them to species. Great patience is needed, since their minute, transparent, and soft bodies make them very hard to work with. Moreover, most of the freshwater gastrotrichs of the Nearctic are undoubtedly not described.

For species identification, animals need to be studied alive. Material preserved as whole mounts deteriorate rapidly over time, and important diagnostic characters may be extremely

PHYLUM GASTROTRICHA

Thorp and Covich's Freshwater Invertebrates. http://dx.doi.org/10.1016/B978-0-12-385028-7.00007-X

hard or impossible to discern. When studying the live animal, a good light microscope is essential (preferably with differential interference contrast (DIC) or phase contrast). I also recommend using a digital camera (with video capability to capture multiple focal planes) to record the animal.

The difficulty to work with the group have made gastrotrichs somewhat neglected by invertebrate zoologists and the group is in need of taxonomic revisions. The extent of cryptic species is poorly studied within the group, and it is possible that the true number of species is underestimated (see Kånneby et al., 2012). Current classification is, to a great extent, based on morphology, and many groups have recently been rendered non-monophyletic based on molecular data (Kånneby et al., 2013).

TERMINOLOGY AND MORPHOLOGY

In general the body of a freshwater gastrotrich is tenpin-shaped and can be divided into four parts (Fig. 7.1): (1) the anterior lobed head with cephalic plates and cephalic sensory ciliary tufts and/or tentacles; (2) the neck that separates the head from the trunk; (3) the trunk which constitutes the bulk of the body; and (4) the furca carrying 0–4, but usually 2 adhesive tubes. Exceptions to this generalization are the macrodasyidan gastrotrichs, with a vermiform body and more than four adhesive tubes (Fig. 7.2), and the semi-planktonic groups (e.g., Dasydytidae and Neogosseidae), with a sack-shaped body that lacks the posterior furca (Fig. 7.3 L–R, T, U).

The head can be rounded (one-lobed), three-lobed, or five-lobed (Fig. 7.4 A–C). The lobes are covered by cephalic plates, which in a species with a five-lobed head consist of a cephalion (Fig. 7.1.3), a pair of epipleura (Fig. 7.1.2), and a pair of hypopleura (Fig. 7.1.1). Cephalic sensory ciliary tufts usually originate from between the lateral or ventrolateral borders of the cephalic plates (Figs. 7.1.16, 7.4 A–E, and 7.6 F). The width of the neck region varies a lot between species, e.g., certain species may have a very constricted neck, whereas others almost completely lack a neck constriction. In the neck scale, columns usually converge to later diverge in the trunk region. In most species, a pair of anterior sensory bristles, inserted between the scales or on small papillae, is also present in the neck region (Fig. 7.1.6). The trunk constitutes most of the body and is usually also the widest part. In the posterior trunk region, a pair of posterior sensory bristles is often present; the bristles are usually inserted on specialized double-keeled scales (Fig. 7.1.12). The furca typically consists of two furcal branches, each with a cone-shaped basal part that bears the distal adhesive tube (Fig. 7.1).

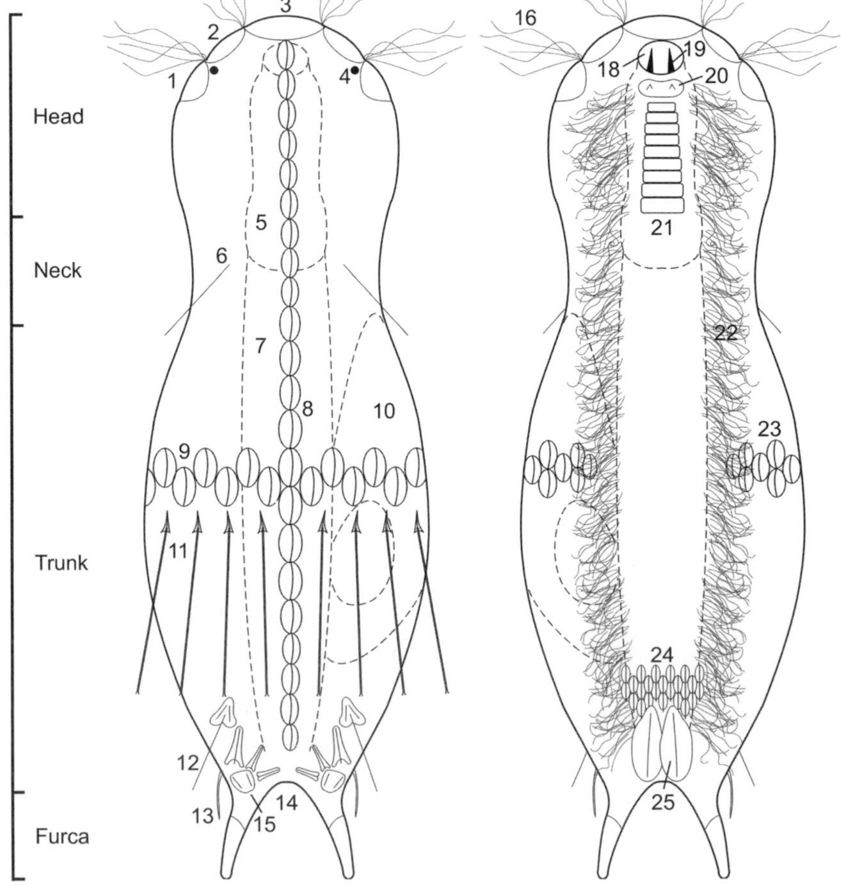

FIGURE 7.1 Schematic representation of a hypothesized Chaetonotid gastrotrich showing important diagnostic characters and the different body regions. Dorsal view to the left and ventral view to the right. Broken lines represent internal structures or structures on the opposite side of the body: (1) Hypopleura; (2) Epipleuria; (3) Cephalion; (4) Ocellar granule; (5) Pharynx with weak posterior swelling; (6) Anterior dorsal sensory bristle; (7) Intestine; (8) Dorsal column of 22 keeled scales; (9) Dorsal row of 13–14 alternating keeled scales; (10) Egg; (11) Spine girdle with bifurcated spines; (12) Posterior sensory bristle anchored by double-keeled scale; (13) Parafurcal spine; (14) Caudal incision; (15) Rounded double-keeled scale typical for certain species of *Chaetonotus* (*Hystricochaetonotus*); (16) Posterior sensory ciliary tuft; (17) Anterior sensory ciliary tuft; (18) Mouth; (19) Pharyngeal tooth; (20) Hypostomium; (21) Transverse ventral scale bars of the interciliary area typical of *Lepidodermella squamata* and *Chaetonotus maximus*; (22) Ventral ciliation; (23) Ventrolateral row of alternating keeled scales; (24) Ventral interciliary scales; (25) Ventral terminal keeled scales. *Redrawn from various sources.*

PHYLUM GASTROTRICHA

It is usually straight; but in certain species, it is forcipate (Fig. 7.4 I). In *Polymerurus*, the branches of the furca are annulate, appearing segmented (Fig. 7.4 G).

The body is covered by a cuticle that is often developed into various types of cuticular structures like scales, spines, and plates. These cuticular structures vary between species and are often diagnostic when identifying and describing species. An important character when determining a species is the number of scales. Columns of scales are parallel to the longitudinal body axis and counted from one side of the animal to the other. Rows of scales are perpendicular to the longitudinal body axis and are counted from anterior to posterior. The number of rows is usually higher than the number of columns. In Fig. 7.1, one dorsal column with 22 scales is depicted, hence the number of rows is 22. Moreover, in Fig. 7.1, one dorsal row with 13–14 alternating scales is depicted; hence the number of columns is 13–14. The number of columns is always counted where the body is widest (unless otherwise stated), and the number of rows is counted down the median of the animal. The dorsal number of columns is the columns seen on the dorsal side of the animal, while the total number of columns also includes the ventrolateral columns. In Fig. 7.1, there are 10 ventrolateral columns (5 on either side of the animal); hence the total number of columns is 23–24. The shape (outline) of individual scales is also an important character. A common type of scale is the three-lobed scale; it comes in different flavors ranging from rounded or weakly three-lobed (Fig. 7.5 L) via three-lobed (Fig. 7.5 A, B) to strongly three-lobed (Fig. 7.5 C, I). There are also rounded to oval scales (Fig. 7.5 F, G, R), five-lobed scales (Fig. 7.5 K, M), rhomboidal scales (Fig. 7.5 S), and polygonal scales (Fig. 7.5 U). Scales may be smooth (e.g., *Lepidodermella*) (Fig. 7.5 U), keeled (e.g., *Heterolepidoderma*) (Fig. 7.5 V), pedunculate (e.g., *Aspidiophorus*) (Fig. 7.5 T), or they can bear spines (e.g. *Chaetonotus*). The spine can be simple (Fig. 7.5 A), dentate (Fig. 7.5 B), bidentate (Fig. 7.5 Y), or carry lateral denticles and a bifurcate tip (Fig. 7.5 H). Certain species of *Ichthydium* and *Arenotus* lack scales and spines altogether and are said to have a naked cuticle.

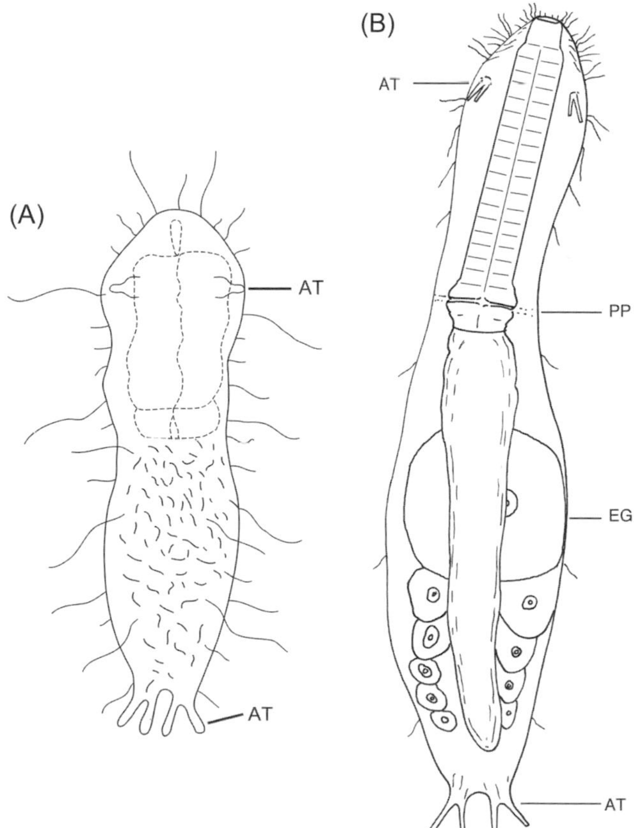

FIGURE 7.2 Freshwater macrodasyidan gastrotrichs: (A) *Marinellina flagellata*; (B) *Redudasys fornerise*. AT = Adhesive tube, EG = Egg, PP = Pharyngeal pore. *Based on Ruttner-Kolisko (1955); Tirjaková (1998); Todaro et al. (2012).*

The most striking feature of the ventral side is the ventral ciliation, which is used for locomotion. This ciliation often consists of two longitudinal bands going from the anterior to posterior (Figs. 7.1.22 and 7.6 E), but it can also consist of tufts. The area between the ciliary bands is called the interciliary area. It may be naked but is more often covered by scales, which can be smooth, keeled, spined, etc. (Figs. 7.1.24, 7.5 X, and 7.6 E). At the posterior end of the interciliary area, a pair of ventral terminal scales are often present (Fig. 7.1.25).

The mouth is situated terminally or subterminally in the anterior end (Fig. 7.1.18). It is connected to the pharynx, which is a muscular tube with (Fig. 7.4 B, C) or without terminal swellings (bulbs) (Fig. 7.4 A). The intestine is usually straight and may sometimes have a differentiated anterior part (Fig. 7.1.7).

MATERIAL PREPARATION AND PRESERVATION

Gastrotrichs can be collected from almost all aquatic environments, but are most common in still, nutrient rich waters. Material is collected with a plankton net with a mesh size of 25 μm or by collecting sediment, aquatic plants, and mosses by hand. In the laboratory, the samples can be kept in small aerated aquaria. They yield the highest diversity and densities of gastrotrichs just after collection. Subsamples are sucked up from the aquaria with a large pipette and put in petri dishes. The subsample can be treated with an isotonic solution of magnesium chloride (1% for freshwater); this anesthetizes the animals and prevents them from adhering to surfaces with their adhesive tubes.

Scan the subsamples under a dissecting microscope, preferably in transmitted light. Once an animal is located, transfer it with a micropipette to a clean microscope slide, and then mount the animal alive in a small amount of water. To prevent squashing of the animal, add a small amount of modeling clay under the edges of the cover slip. In order

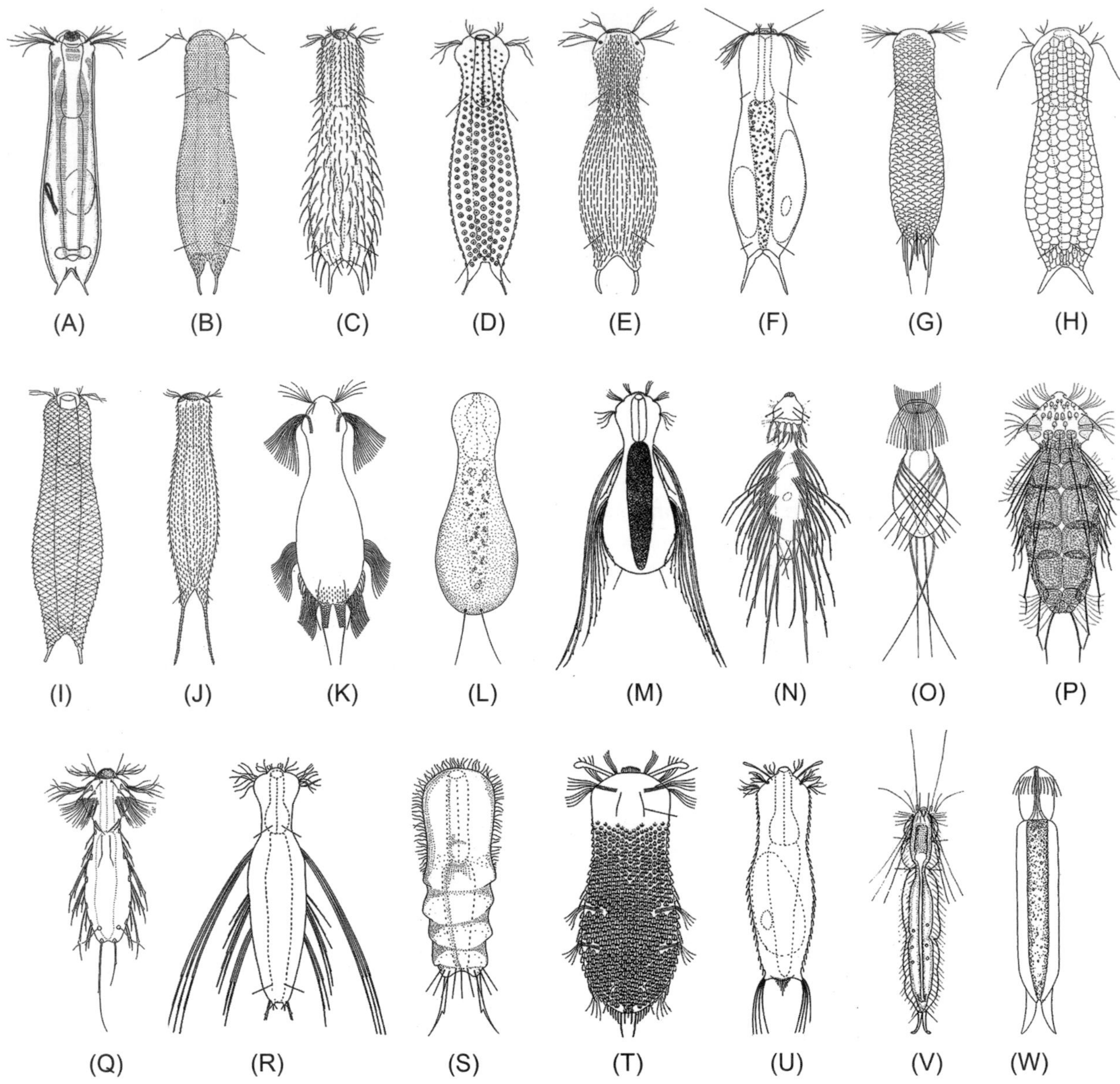

FIGURE 7.3 Freshwater chaetonotidan genera representing Chaetonotidae (A–K), Dasydytidae (L–R), Dichaeturidae (S), Neogosseidae (T, U), and Proichthydidae (V, W): (A) *Arenotus*; (B) *Aspidiophorus*; (C) *Chaetonotus*; (D) *Fluxiderma*; (E) *Heterolepidoderma*; (F) *Ichthydium*; (G) *Lepidochaetus*; (H) *Lepidodermella*; (I) *Rhomballichthys*; (J) *Polymerurus*; (K) *Undula*; (L) *Anacanthoderma*; (M) *Chitonodytes*; (N) *Dasydytes*; (O) *Haltidytes*; (P) *Ornamentula*; (Q) *Setopus*; (R) *Stylochaeta*; (S) *Dichaetura*; (T) *Kijanebalola*; (U) *Neogossea*; (V) *Proichthydioides*; (W) *Proichthydium*. *Adapted and modified from Strayer et al. (2009).*

to observe important characters, the animal should be oriented in a dorso-ventral fashion. Proper orientation can be achieved by adding small amounts of water to the sides of the cover slip or by removing water with a filter paper. This procedure creates currents under the cover slip, which should alter the animal slightly. If the animal gets stuck on either the surface of the slide or the cover slip, a gentle tapping with a pair of forceps on the cover slip can help.

When properly mounted, examine the animal under a light microscope equipped with DIC or phase contrast. Documentation should include pictures/recordings of the habitus and taxonomically important characters, such as distribution and shape of scales, spines, adhesive tubes, cephalic plates, pharynx, etc.

After documentation, the animal can be recovered from the slide for further treatment. The recovering procedure is tricky, but by adding water, the cover slip can be made to float and subsequently gently lifted with a fine insect needle. The animal is now free and can be sucked up with a micropipette and transferred to a vial containing the

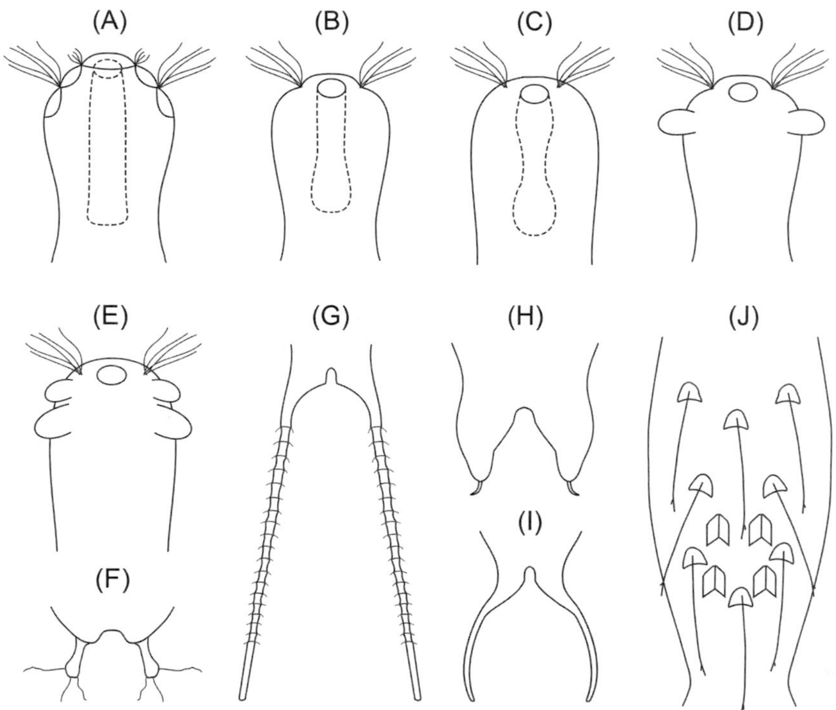

FIGURE 7.4 Schematic representation of different head-shapes (A–E), pharynx-shapes (A–C), caudal ends/furcas (F–I) and spine distribution (J): (A) Five-lobed head with two pairs of sensory ciliary tufts, and pharynx of equal thickness along its length; (B) Three-lobed head with one pair of sensory ciliary tufts, and pharynx with posterior swelling; (C) One-lobed or rounded head with one pair of sensory ciliary tufts, and pharynx with anterior and posterior swellings; (D) Three-lobed head where the posterior lobes are modified into dorsal flaps; (E) Five-lobed head where the middle and posterior lobes are modified into dorsal flaps; (F) The peg-like protuberances typical of *Stylochaeta*; (G) The long ringed (appearing segmented) furca typical of *Polymerurus*; (H) Furca with reduced/absent adhesive tubes typical of *Chaetonotus* (*Wolterecka*); (I) Forcipate furca present in for example *Ichthydium* and *Chaetonotus*; (J) Distribution of the conspicuous rows of longer thicker spines and spineless scales of many *Chaetonotus* (*Hystricochaetonotus*) spp. *Redrawn from Brunson (1950) and Schwank (1990).*

medium of choice. For whole mounts, the animal can be fixed in 10% buffered formalin and then dehydrated through a series of ethanol and glycerol to pure glycerol. Before mounting, a few drops of formalin can be added to the slide to prevent bacterial deterioration (Kånneby et al., 2009). Animals can also be fixed in 2.5% glutaraldehyde, exposed to osmium tetraoxide, dehydrated through an ethanol series, and mounted in epon. For molecular studies, animals are stored in 95% ethanol in −18 to 20 °C until further treatment.

KEYS TO GASTROTRICHA

Gastrotricha: Orders

1 Animal with at least three pairs of adhesive tubes (one anterior and two posterior) and a pair of pharyngeal pores (absent in Lepidodasyidae) (Fig. 7.2) .... Macrodasyida

[very rare; an almost entirely marine group, represented by only two nominal species in freshwaters; the genus *Redudasys* was recently reported from Texas, USA]

1' Animal lacking adhesive tubes (Fig. 7.3 L–R, T–U) or with one pair (very rarely two pairs) of adhesive tubes posteriorly (Fig. 7.3 A–J), and no pharyngeal pores .... **Chaetonotida [p. 119]**

[common and widespread in fresh-water]

Gastrotricha: Chaetonotida: Families

1 Posterior end of body usually with furca and adhesive tubes; body usually tenpin-shaped or strap-shaped (Fig. 7.3 A–K, S, V, W) 2

1' Posterior end of body without furca or adhesive tubes, although sometimes bearing spines or pegs; body usually sac- or tenpin-shaped (Fig. 7.3 L–R, T–U) 4

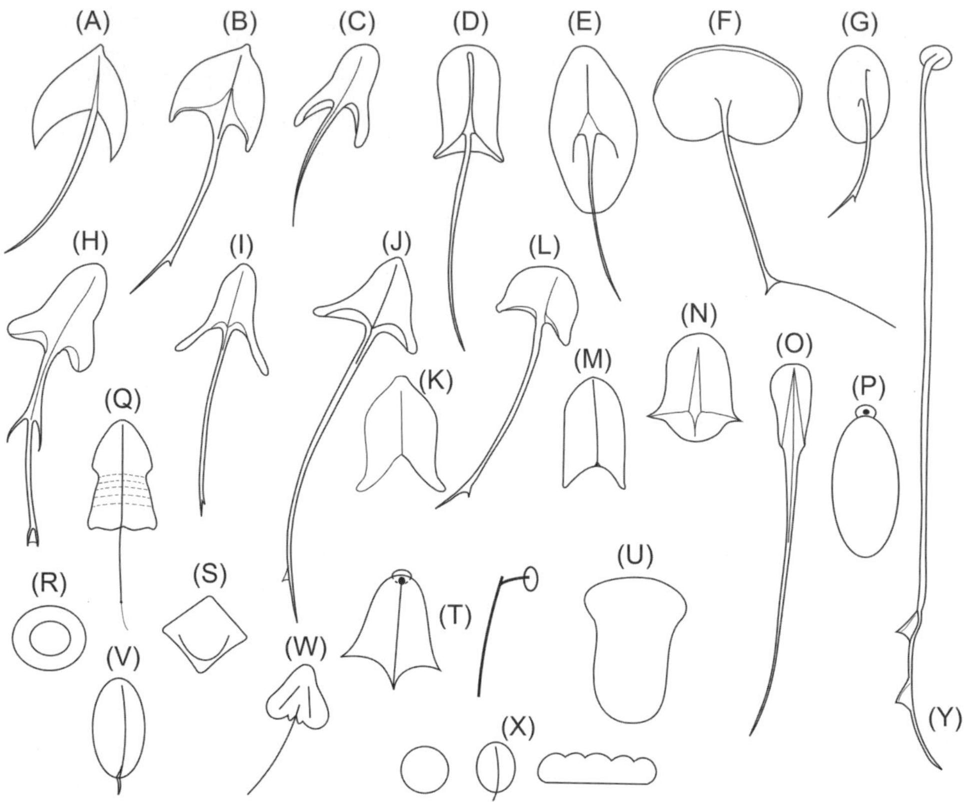

FIGURE 7.5 Schematic representation of scales and spines in some freshwater Chaetonotida: (A) Dorsal three-lobed scale with simple gently curved spine of *Chaetonotus* (*C.*) *maximus*; (B) Dorsal three-lobed scale with dentate spine of *C.* (*C.*) *similis*; (C) Dorsal strongly three-lobed scale with simple spine of *C.* (*C.*) *larus*; (D) Dorsal elongated three-lobed scale with simple spine of *C.* (*C.*) *hirsutus*; (E) Dorsal longitudinally oval scale with simple spine of *C.* (*Primochaetus*) *mutinensis*; (F) Dorsal transversely oval to weakly five-lobed scale with dentate spine bent at almost right angle of *C.* (*P.*) *heideri*; (G) Dorsal longitudinally oval scale with dentate spine of *C.* (*P.*) *chuni*; (H) Dorsal three-lobed scale with spine with two diametrically opposite lateral denticles and bifurcated tip of *C.* (*Schizochaetonotus*) *schultzei*; (I) Dorsal strongly three-lobed scale with dentate spine of *C.* (*Hystricochaetonotus*) *hystrix*; (J) Dorsal three-lobed scale with long thick dentate spine of *Hystricochaetonotus* sp.; (K) Dorsal five-lobed spineless scale of *C.* (*H.*) *octonarius*; (L) Dorsal semi-circular to weakly three-lobed scale with dentate spine of *C.* (*H.*) *octonarius*; (M) Dorsal weakly five-lobed spineless scale of *Hystricochaetonotus* sp.; (N) Dorsal polygonal spineless scale of *C.* (*Zonochaeta*) *succinctus*; (O) Dorsal girdle scale with simple spine of *C.* (*Z.*) *succinctus*; (P) Dorsal pedunculate oval scale of *Polymerurus rhomboides*; (Q) Dorsal scale with constriction and transverse lines thickenings, with thin dentate spine of *C.* (*Captochaetus*) *robustus*; (R) Round scale of *Fluxiderma*; (S) Rhomboidal scale of *Rhomballichthys*; (T) Pedunculate scale of *Aspidiophorus* sp., dorsal view to the left and lateral view to the right; (U) Dorsal smooth scale of *Lepidodermella squamata*; (V) Dorsal oval keeled scale with short simple spine of *Heterolepidoderma ocellatum*; (W) Double keeled three-lobed scale anchoring posterior sensory bristles; (X) Three types of ventral interciliary scales: round smooth scale to the left, rounded to weakly oval keeled scale with short simple spine in the middle, and transverse scale plate with jagged anterior edge present in the pharynx region of *C.* (*P.*) *macrolepidotus ophiogaster*; (Y) Small rounded rudimentary scale with strong bidentate spine of *Stylochaeta fusiformis*. *Redrawn from Schwank (1990) and references therein.*

2(1) Furca doubly branched; scales and spines sparse or absent (Fig. 7.3 S) .... Dichaeturidae, one genus: *Dichaetura*

[Palaearctic. Rare, not yet reported from the Nearctic]

2' Furca singly branched; scales and spines present or absent; common and widespread 3

3(2) Furca branches heavy, sickle-shaped, curved, and tapered, not distinctly divided into a cone-shaped basal part and a distal duct; head with long cilia that are not arranged in tufts; head plates absent (Fig. 7.3 V, W) .... **Proichthydiidae [p. 121]**

[Neotropical and Palaearctic. Rare, not yet reported from the Nearctic]

3' Furca branches usually with a cone-shaped base and a distal adhesive duct; body often with numerous spines, scales or spined scales; head with cilia arranged in tufts; cephalic plates present (Fig. 7.3 A–K) .... **Chaetonotidae [p. 121]**

[common and widespread]

4(1) Head with clavate tentacles (Fig. 7.3 T, U) .... **Neogosseidae [p. 128]**

[rare; semiplanktonic or pelagic]

4' Head without clavate tentacles (Fig. 7.3 L–R); .... **Dasydytidae [p. 129]**

[rare; semiplanktonic or pelagic]

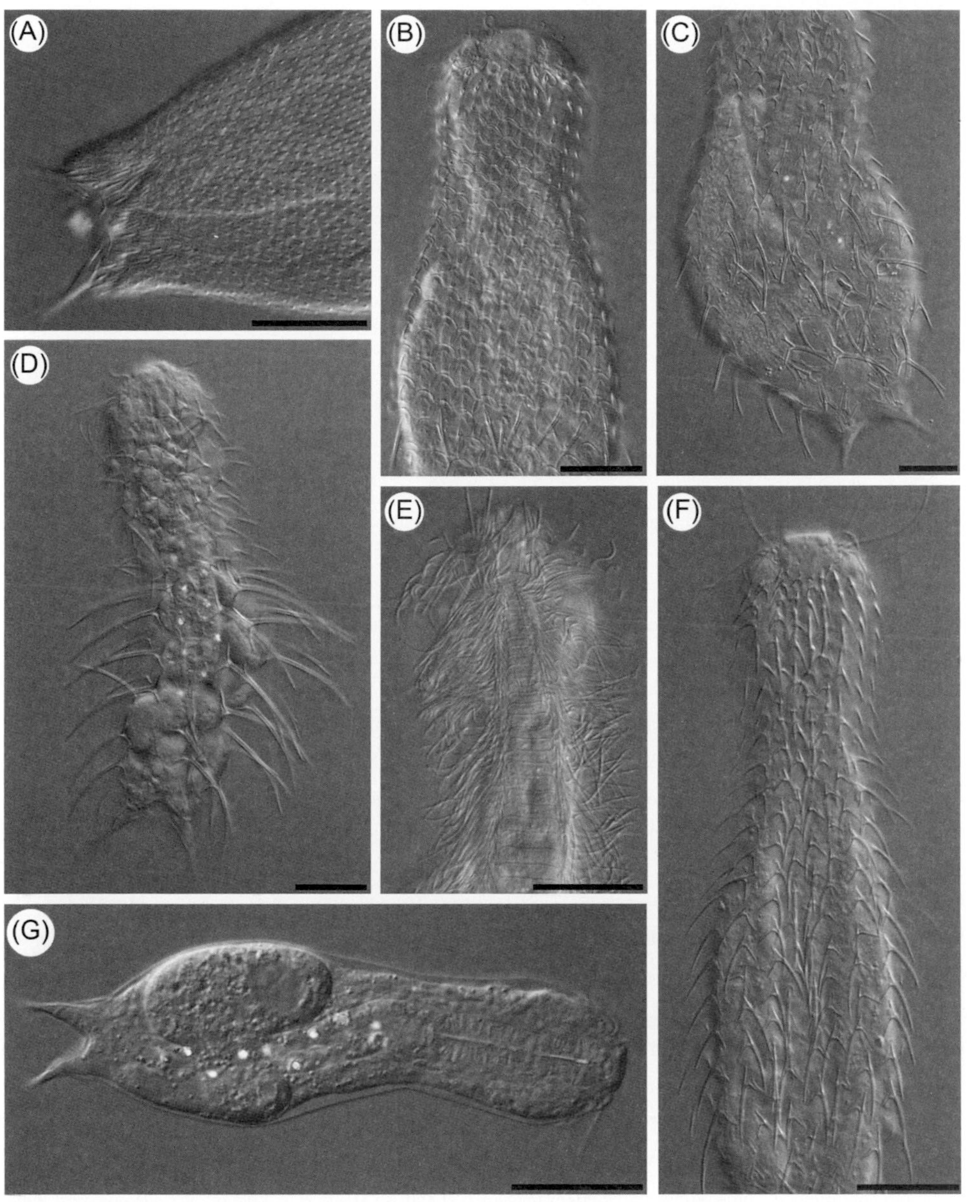

FIGURE 7.6 Photographs of live specimens: (A) *Aspidiophorus ophiodermus*, dorsal view showing pedunculate scales and keeled non-pedunculate scales in posteriormost trunk region; (B) *Chaetonotus* (*Primochaetus*) *acanthodes*, dorsal view showing distribution of scales and the girdle; (C) *C.* (*Hystricochaetonotus*) *hystrix*, dosal view showing distribution of strongly three-lobed scales with dentate spines; (D) *C.* (*H.*) *macrochaetus*, dorsal view showing drastic increase in spine length; (E) *C.* (*Chaetonotus*) *microchaetus*, ventral view of anterior portion of body showing transverse scale plates in the pharynx region; (F) *Chaetonotus* sp., dorsal view showing distribution of three-lobed scales with simple spines; (G) *Ichthydium podura*, habitus showing the smooth cuticle. Scale bars: A-B, 20 μm; C, 10 μm; D-G, 20 μm. *Photos: T. Kånneby.*

Gastrotricha: Chaetonotida: Proichthydiidae: Genera

1 Head bearing a row of cilia shorter than the head (Fig. 7.3 W); .... *Proichthydium coronatum* Cordero, 1918

[Neotropical. Rare, not yet reported from Nearctic]

1' Cephalic cilia row much longer than the head (Fig. 7.3 V); .... *Proichthydioides remanei* Sudzuki, 1971

[Palaearctic. Rare, not yet reported from Nearctic]

Gastrotricha: Chaetonotida: Chaetonotidae: Genera

1 Furca branches not annulate (not appearing segmented) and furca amounts to less than 20 percent of the total body length 2

1' Furca branches annulate (appearing segmented) (Figs. 7.3 J and 7.4 G), except for one uncommon European species, and amounts to at least 20–25 percent of total body length; body often large and without a distinct neck .... ***Polymerurus* [p. 122]**

[common]

2(1) Body dorsal surface without spines or scales, sometimes a few scales or spines at the furca bases and/or scales anchoring posterior sensory bristles (Fig. 7.3 A, F) 3

2' Body dorsal surface with numerous scales, spines, or spined scales 4

3(2) Cuticle very thick and smooth, distinct from the epidermis, entirely without scales (Fig. 7.3 A); mouth with large mouth ring and strong pharyngeal teeth (Fig. 7.1.19) .... *Arenotus strixinoi* Kisielewski, 1987

[Neotropical. Rare monotypic genus not yet reported from Nearctic waters]

3' Cuticle not especially thick, occasionally with cuticular folds and sometimes with scales near the base of the furca or the bases of posterior sensory bristles (Figs. 7.5 W and 7.6 G), or with minute cuticular structures; mouth ring small and without pharyngeal teeth (Fig. 7.3 F) .... ***Ichthydium* [p. 123]**

[common and widespread]

4(2) Spines or spined scales present and often numerous (Fig. 7.3 C, G) 5

4' Spines absent (occasionally a few spines are present at the base of the furca) 6

5(4) Ventral interciliary scales different in shape from dorsal scales; spines of various types (Figs. 7.3 C and 7.6 B–D, F); .... ***Chaetonotus* [p. 123]**

[common and widespread]

5' Ventral interciliary scales similar in shape to dorsal scales; posterior part of body with several long spines that reach beyond the end of the furca (Fig. 7.3 G) .... *Lepidochaetus zelinkai* (Grünspan, 1908)

[Canada: New Brunswick, Ontario]

6(5) Furca with adhesive tubes; body strap-shaped or tenpin-shaped; without groups of long cilia on head and posterior body 7

6' Furca without adhesive tubes; body markedly tenpin-like, with groups of long cilia on the head and posterior part of the body (Fig. 7.3 K); .... *Undula paraënsis* Kisielewski, 1991

[Neotropical. Rare monotypic, semiplanktonic genus not yet reported from Nearctic]

7(6) Body with dorsal scales not pedunculate 8

7' Body with dorsal surface covered with pedunculate scales (Figs. 7.3 B and 7.5 T), in some rare species scales of posterior dorsal surface non-pedunculate (Fig. 7.6 A) .... ***Aspidiophorus* [p. 127]**

[common]

8(7) Scales not keeled 9

8' Scales elongate to suboval, with longitudinal keels (Figs. 7.3 E and 7.5 V) .... ***Heterolepidoderma* [p. 128]**

[common and widespread]

9(8) Scales numerous, flat and polygonal 10

9' Few scales, circular; rare (Figs. 7.3 D and 7.5 R) .... *Fluxiderma concinnum* Roszczak, 1935

[USA: New Hampshire, New Jersey, Maine?]

10(9) Scales clearly rhomboid (Figs. 7.3 I and 7.5 S) .... *Rhomaballichthys*

[Palaearctic. Rare, not yet reported from Nearctic]

10' Scales smooth and rounded, some with squared edges (Figs. 7.3 H and 7.5 U) .... ***Lepidodermella* [p. 128]**

[Common and widespread]

Gastrotricha: Chaetonotida: Chaetonotidae: *Polymerurus*: Species

1 Cuticle with pedunculate or spined scales 2

1' Cuticle without scales and spines but with numerous small pointed excresences; furca almost as long as half of the total body length .... *Polymerurus callosus* Brunson, 1950

[USA: Arizona, Illinois, Indiana, Michigan]

2(1) Cuticle with elongated oval, pedunculate scales (Fig. 7.5 P); annulate part of furca without hairs/bristles .... *Polymerurus rhomboides* (Stokes, 1887)

[Canada: Ontario. USA: Illinois, New Jersey, Virginia]

2' Cuticle with spined scales; spines reach their greatest length at the posterior body end; annulate part of furca with or without hairs/bristles (Fig. 7.4 G) .... *Polymerurus nodicaudus* (Voigt, 1901)

[Canada: Ontario. USA: Indiana, New Jersey, North Dakota, Virginia]

Gastrotricha: Chaetonotida: Chaetonotidae: *Ichthydium*: Species

1 Head lobes developed as dorsal flaps (Fig. 7.4 D, E) 2

1' Head lobes not developed as dorsal flaps 3

2(1) Head with one pair of dorsal flaps (Fig. 7.4 D) .... *Ichthydium auritum* Brunson, 1950

[USA: Indiana, Michigan]

2' Head with two pairs of dorsal flaps (Fig. 7.4 E); adhesive tubes approximately 1/3 of the total body length and with pointed tips .... *Ichhydium macropharyngistum* Brunson, 1949

[USA: Michigan]

3(1) Head wider than body, single lobed and rounded to oval or rounded rectangular (Fig. 7.4 C) 4

3' Head of equal width or narrower than body, three- or five-lobed (Fig. 7.4 A, B) 5

4(3) Head rounded to oval; furca extremely short, only 1/25 of the total body length; pharynx with thick medial swelling .... *Ichthydium brachykolon* Brunson, 1949

[USA: Michigan]

4' Head rounded rectangular; furca forcipate (Fig. 7.4 I) and of normal size; pharynx without swellings .... *Ichthydium cephalobares* Brunson, 1949

[USA: Michigan]

5(3) Furca distinctly forcipate (Fig. 7.4 I) 6

5' Furca straight or weakly forcipate with hollow caudal cutting 7

6(5) Head distinctly five-lobed; total body length less than 100 μm; pharynx pear-shaped .... *Ichthydium minimum* Brunson, 1950

[USA: Michigan]

6' Head weakly five-lobed, but may appear rounded; total body length more than 100 μm; adhesive tubes fine; posterior sensory bristles anchored by papillae .... *Ichthydium forficula* Remane, 1927

[Canada: Ontario. USA: Michigan]

7(5) Head distinctly five-lobed; furca weakly forcipate with hollow caudal cutting or base of furca branches slightly enlarged 8

7' Head weakly three-lobed; furca straight; body plump; cuticle with minute cuticular ridges (Fig. 7.6 G) .... *Ichthydium podura* (Müller, 1773)

[Canada: Ontario? USA: New Jersey]

8(7) Lateral edges of cuticle accordion-like; body covering with 35–40 transverse cuticular ridges .... *Ichthydium sulcatum* (Stokes, 1887)

[Canada: Ontario. USA: Indiana, Michigan, New Jersey, Virginia?]

8' Lateral edges of body smooth; base of furca branches slightly enlarged; body covering apparently without transverse cuticular ridges .... *Ichthydium leptum* Brunson, 1949

[USA: Michigan]

Gastrotricha: Chaetonotida: Chaetonotidae: Chaetonotus: Subgenera

Members of the subgenera *Chaetonotus* (*Wolterecka*) and *Chaetonotus* (*Schizochaetonotus*) have been reported from Nearctic waters; however, the specimens were not determined to species or described (see Green, 1986 and Weiss, 2001).

1 Furca not reduced; adhesive tubes developed, never rudimentary 2

1' Furca more or less reduced, with rudimentary adhesive tubes (Fig. 7.4 H) and strong hooked spines dorsally; benthic; rare .... *Chaetonotus* (*Wolterecka*)

[rare]

2(1) Cuticle normally developed, uncolored and more or less transparent 3

2' Cuticle thick, three-layered, divided into: outer irregular granular layer, middle layer of rhomboidal scales with simple spines, and inner layer of basal cuticle; cuticle of distinct orange-brown color .... *Chaetonotus* (*Tristratachaetus*) *rhombosquamatus* Kolicka et al., 2013

[Palaearctic. Rare, not yet reported from the Nearctic]

3(2) Lateral denticles of spines present or absent, if present never inserted diametrically opposite 4

PHYLUM GASTROTRICHA

3' Scales three-lobed; spines with two diametrically opposite lateral denticles (Fig. 7.5 H), sometimes a distal denticle also present; benthic and periphytic; one freshwater species .... *Chaetonotus* (*Schizochaetonotus*) *schultzei* Metschnikoff, 1865

[USA: Arizona]

4(3) Trunk dorsally and laterally almost always without a transverse row (girdle) of long simple-, dentate- or bifurcated spines; if present, girdle-like spines simple and anchoring scales similar to the others of dorsal surface (Fig. 7.6 B) 5

4' Trunk dorsally and laterally with a transverse row of long simple- dentate- or bifurcated spines (Fig. 7.1.11); scales anchoring girdle spines different in size and shape from others of dorsal surface; at least one pair of long, usually thin parafurcal spines present (Fig. 7.1.13); .... ***Chaetonotus (Zonochaeta)* [p. 124]**

[benthic and periphytic; relatively rare]

5(4) Scales rounded- to five-lobed, if three-lobed without longitudinal keel; lateral denticle/s of spines present or absent; scales and spines sometimes reduced in posteriormost dorsal trunk region 6

5' Small species, 60–190 μm in total body length; scales three-lobed with distinct longitudinal keel, sometimes rounded three-lobed or more or less pentagonal (Figs. 7.5 K–M and 7.6 C, D); spines usually with one (sometimes two or seldom absent) lateral denticle; many taxa with reduced dorsal spines (and scales) and/or a conspicuous group of spines in trunk region (Fig. 7.4 J); benthic, periphytic and interstitial .... ***Chaetonotus (Hystricochaetonotus)* [p. 124]**

[common and widespread]

6(5) Total body length usually <370 μm; hypostomium present or absent, if present without deep transverse furrow; pharynx usually widens toward its posterior end, or with terminal swellings 7

6' Total body length 210–625 μm; hypostomium very well developed, with deep transverse furrow; dorsal trunk scales rounded to five-lobed (occasionally with constriction) (Fig. 7.5 Q); pharynx equal in width from anterior to posterior; benthic and periphytic. .... ***Chaetonotus (Captochaetus)* [p. 126]**

[rare]

7(6) Scales without keels, anterior edges of scales extroverted or flat and posterior edge with no or shallow rounded incision (Fig. 7.5 E–G); spines, anchored near posterior edge of scale, thick and straight, rarely curved basally; interciliary scales usually of same shape as dorsal ones; pharynx usually with terminal swellings; benthic and periphytic .... ***Chaetonotus (Primochaetus)* [p. 126]**

[common]

7' Anterior edges of scales not extroverted; posterior edge of scale with incision; scales with longitudinal keel along at least half of their length (Fig. 7.5 A–D); posteriormost trunk region occasionally with reduced spines (and scales); interciliary scales often different in size, shape and distribution from dorsal ones; benthic, periphytic and interstitial .... ***Chaetonotus (Chaetonotus)* [p. 127]**

[common and widespread]

PHYLUM GASTROTRICHA

Gastrotricha: Chaetonotida: Chaetonotidae: *Chaetonotus (Zonochaeta)*: Species

1 Larger species, >100 μm in total body length; girdle with more than 5 spines 2

1' Small species, <100 μm in total body length; 5 dentate spines in girdle, situated at the posterior end of pharynx; dorsal scales keeled .... *Chaetonotus* (*Z.*) *trichostichodes* Brunson, 1950

[USA: Michigan]

2(1) Head and neck spines poorly developed, not bluntly pointed; girdle spines anchored by scales 3

2' Head and neck spines bluntly pointed; 8 alternating simple spines, not anchored by scales, in girdle; one pair of dorsal spines and one ventral spine between furcal branches .... *Chaetonotus* (*Z.*) *palustris* Anderson & Robbins, 1980

[USA: Illinois]

3(2) Scales anterior to girdle polygonal, smaller and more numerous than scales posterior to girdle (Fig. 7.5 N, O); 9 simple, non-bifurcated girdle spines; interciliary area with 18–20 columns of small oval keeled scales, and up to 14 smooth transverse scale plates in pharynx region .... *Chaetonotus* (*Z.*) *succinctus* Voigt, 1902

[USA: New Jersey, Virginia]

3' Dorsal scales anterior and posterior to girdle of approximately same size and shape; 7–11 (maximum 20), bifurcated girdle spines (Fig. 7.1.11); interciliary area with 11–13 columns of small elongate keeled scales .... *Chaetonotus* (*Z.*) *bisacer* Greuter, 1917

[Canada: Ontario. USA: Indiana, New Jersey, Virginia]

Gastrotricha: Chaetonotida: Chaetonotidae: *Chaetonotus (Hystricochaetonotus)*: Species

1 Dorsal trunk region with rows or groups of longer thicker spines (Fig. 7.4 J); other spines usually short, fine and simple; scales sometimes more or less reduced, cuticle seldom completely naked 2

1' Dorsal trunk region without rows or groups of longer spines; a drastic or gradual increase in spine length in anterior- or mid-trunk region present (Fig. 7.6 C, D); scales never reduced 12

2(1) Dorsal trunk region with more than 10 thicker longer dentate spines distributed in 4–5 curved rows 3

2' Dorsal trunk region with less than 10 thicker longer simple, dentate or bidentate spines distributed in 1–3 rows 4

3(2) Dorsal trunk region with 13 long thick dentate spines distributed in 5 curved rows anchored directly on cuticle without scales; lateral body sides with short simple spines; head and neck region with 12–13 columns of simple spines .... *Chaetonotus* (*H.*) *enormis* Stokes, 1887

[USA: New Jersey]

3' Dorsal trunk region with 20 relatively delicate longer (16–24 μm) dentate spines distributed in 4 curved rows anchored by small three-lobed scales; head and neck region with 11 columns of simple spines (4–7 μm); lateral body sides with short simple spines; all spines anchored by reduced scales .... *Chaetonotus* (*H.*) *acanthophorus* Stokes, 1887

[USA: Arizona, Illinois, Indiana, Maine?, Michigan, New Hampshire, New Jersey]

4(2) Dorsal body surface completely or partly naked or with regions covered by thin reduced scales 5

4' Dorsal body surface covered by fully developed scales with spines; scales anchoring longer thicker spines very seldom reduced 10

5(4) Cuticle completely devoid of scales and spines except for 5–8 long thick spines in dorsal trunk region 6

5' Cuticle at least with lateral scales with short fine spines and 4–9 long thick spines in dorsal trunk region 7

6(5) Eight dentate spines, 43–48 μm in length and overshooting tip of furcal branches, originating from a triangular area in mid-trunk region .... *Chaetonotus* (*H.*) *trichodrymodes* Brunson, 1950

[USA: Michigan]

6' Five dentate spines, 13–19 μm in length in mid-trunk region .... Chaetonotus (*H.*) *quintospinosus* Greuter, 1917

[USA: Illinois]

7(5) Four to nine long thicker dentate spines in trunk region, if anchored in anterior trunk region not reaching the tip of furcal branches 8

7' Seven to eight long thicker dentate spines distributed in 2 rows in anterior trunk region; the 4–5 spines of the posterior row very long (40–45 μm, in European populations 70–90 μm) reaching the tip of the furcal branches; lateral scales with short fine simple spines always present; neck region sometimes with fine spines .... *Chaetonotus* (*H.*) *longispinosus* Stokes, 1887

[Canada: Ontario. USA: Illinois, New Jersey, Virginia]

8(7) Four to nine long thicker dentate spines anchored by three- to five lobed scales distributed in two to four rows in dorsal trunk region; grade of scale reduction variable 9

8' Eight long thicker dentate spines anchored by semi-circular to weakly three-lobed scales (Fig. 7.5 L) distributed in three columns (3, 2, 3 spines respectively) in dorsal trunk region; several keeled weakly three-lobed to five-lobed scales without spines interspersed between the eight thicker spines (Figs. 7.4 J and 7.5 K, M); two pairs of elongate three-lobed scales present just anterior to base of furcal branches .... *Chaetonotus* (*H.*) *octonarius* Stokes, 1887

[USA: Illinois, Michigan, New Hampshire, New Jersey]

9(8) Four to eight longer (12–25 μm) thicker dentate spines anchored by three- to five-lobed scales distributed in 2–3 alternating rows in dorsal trunk region; several large keeled five-lobed scales interspersed between 4–8 longer thicker spines (Fig. 7.4 J); lateral body sides with small three-lobed scales with fine simple spines; similar scales sometimes also in head and neck region .... *Chaetonotus* (*H.*) *spinulosus* Stokes, 1887

9' Seven to nine longer (18–26 μm) thicker straight dentate spines anchored by three- to five-lobed scales distributed in three to four rows in dorsal trunk region; grade of scale reduction variable, head and neck regions seldom fully covered by scales; scales more or less three-lobed with short fine simple spine; furcal bases either with round double keeled scales surrounded by 3–4 pairs of elongate keeled three-lobed scales (Fig. 7.1.15) or without round double keeled scales but with 3 pairs of elongated keeled three-lobed scales; interciliary area naked or with 7–10 columns of round scales; keeled ventral terminal scales present or absent .... *Chaetonotus* (*H.*) *aemilianus* Balsamo, 1978

[Canada: Ontario]

10(4) Four to five longer thicker dentate spines anchored by round or five-lobed scales distributed in two rows in mid-trunk region; other spines of dorsal surface anchored by round or five-lobed scales with short simple or dentate spine 11

10' Seven longer thicker bidentate spines in three rows in dorsal mid-trunk region; scales with short rigid simple spines distributed in 6–8 columns with 8–10 scales in each .... *Chaetonotus* (*H.*) *anomalus* Brunson, 1950

[USA: Michigan]

11(10) Total body length 110 μm; five longer thicker dentate spines anchored by five-lobed scales distributed in 2 alternating rows in mid-trunk region; anterior dorsal surface with 7–10 columns of three- to five-lobed scales with short fine simple or dentate spines; posterior trunk region with keeled five-lobed scales without spines; interciliary field with keeled terminal scales and small round scales .... *Chaetonotus* (*H.*) *ferrarius* Schwank, 1990

[Canada: Ontario]

PHYLUM GASTROTRICHA

11' Total body length 196 μm; four thicker longer dentate spines distributed in two columns and anchored by round scales in dorsal trunk region; other dorsal and lateral scales anchored by rounded scales .... *Chaetonotus* (*H.*) *spinifer* Stokes, 1887

[USA: New Jersey]

12(1) Dorsal spines gradually increase in length from anterior to posterior (Fig. 7.6 C); posteriormost trunk region with keeled scales lacking spines and/or with very short fine simple spine; bases of furcal branches with double keeled round scale surrounded by three elongated three-lobed scales (Fig. 7.1.15); hypostomium absent .... *Chaetonotus* (*H.*) *hystrix* Metschnikoff, 1865

[Canada: New Brunswick, Ontario]

12' Dorsal spines drastically increase in length and thickness in mid-trunk region (Fig. 7.6 D); bases of furcal branches without double keeled round scale, elongated three-lobed scales present; hypostomium sometimes present .... *Chaetonotus* (*H.*) *macrochaetus* Zelinka, 1889

[Canada: Ontario. USA: Illinois]

Gastrotricha: Chaetonotida: Chaetonotidae: *Chaetonotus (Captochaetus)*: Species

1 Dorsal scales without constriction, three- to five-lobed 2

1' Dorsal scales with constriction approximately 2/5 from anterior scale edge, scale surface posterior to constriction with transverse lines (thickenings) (Fig. 7.5 Q); spines with or without denticle; mouth very large, 30 μm in width, with approximately 15 cuticular ridges .... *Chaetonotus* (*C.*) *robustus* Davison, 1938

[USA: New Jersey, New York]

2(1) Dorsal scales three-lobed with dentate spines, in 14–16 columns and 20–25 rows; mouth with 10–15 cuticular ridges; intestinal wall of a pale sky-blue color .... *Chaetonotus* (*C.*) *gastrocyaneus* Brunson, 1950

[USA: Indiana, Michigan]

2' Dorsal trunk scales five lobed, head and neck scales rounded; scales bear simple spines; pharynx with cuticular teeth and ridges .... *Chaetonotus* (*C.*) *simrothi* Voigt, 1909

[Canada: Ontario]

Gastrotricha: Chaetonotida: Chaetonotidae: *Chaetonotus (Primochaetus)*: Species

1 Spine girdle present (Fig. 7.1.11) 2

1' Spine girdle absent 3

2(1) Furca forcipate (Fig. 7.4 I); scales oval with small posterior incision and keel drawn out into a short spine; girdle with 6 simple spines (8–12 μm); posteriorly a pair of parafurcal spines (17–18 μm) (Fig. 7.1.13); anterior interciliary area with 20–23 transverse plates with anterior jagged edge (Fig. 7.5 X), posterior interciliary area with large smooth (occasionally keeled) strongly overlapping scales .... *Chaetonotus* (*P.*) *macrolepidotus ophiogaster* Remane, 1927

[Canada: Ontario]

2' Furca straight; scales roundly three-lobed, strongly overlapping, giving the impression of a double anterior edge (edges of surrounding scales) (Fig. 7.6 B); girdle with 7–12 simple spines; dorsal spines always present anterior to girdle but sometimes absent posterior to girdle; posteriorly 2 pairs of parafurcal spines; interciliary area with rounded to oval keeled scales .... *Chaetonotus* (*P.*) *acanthodes* Stokes, 1887

[USA: New Jersey]

3(1) Spines dentate 4

3' Spines simple, hair-like and gently curved; dorsal scales longitudinally oval to weakly five-lobed (Fig. 7.5 E); posteriormost dorsal trunk region partly with smaller scales widely spaced apart; interciliary area mostly with round smooth scales .... *Chaetonotus* (*P.*) *mutinensis* Balsamo, 1978

[Canada: Ontario]

4(3) Scales thin, transversely oval to weakly five-lobed; spines dentate, very long and bent at almost right angles approximately halfway to 2/3 from body surface (Fig. 7.5 F); 7–9 columns and 20–23 rows of dorsal scales; interciliary area with oval keeled scales .... *Chaetonotus* (*P.*) *heideri* Brehm, 1917

[Canada: Ontario. USA: Ohio]

4' Scales longitudinally oval with thick curved dentate spines (Fig. 7.5 G); 9 columns and approximately 16 rows of dorsal scales; posteriormost dorsal trunk surface naked except for a pair of median dentate spines reaching the tip of the furcal branches; interciliary area terminally with 4 short spines, rest of interciliary area with numerous columns of small scales with short spines .... *Chaetonotus* (*P.*) *chuni* Voigt, 1901

[USA: Washington]

Gastrotricha: Chaetonotida: Chaetonotidae: *Chaetonotus (Chaetonotus)*: Species

Chaetonotus (*C.*) *tachyneusticus* Brunson, 1948, p. 350–351 is purposely left out from the key because of the poor description.

1 Head five-lobed (Fig. 7.4 A) to rounded (Fig. 7.4 C); spines anchored by scales 2

1' Head three-lobed with one pair of sensory ciliary tufts (Fig. 7.4 B); scales absent, spines anchored directly on cuticle by thickening; 8–13 alternating columns and 25–30 rows of simple gently curved spines .... *Chaetonotus* (*C.*) *formosus* Stokes, 1887

[USA: Michigan, New Jersey]

2(1) Total number of dorsal scales distributed in more than 6–8 columns or more than 8–9 rows 3

2' Head indistinctly five-lobed with two pairs of sensory ciliary tufts (Fig. 7.4 A); short simple spines increase marginally in length from anterior to posterior (3–7 μm) and distributed in 6–8 columns and 8–9 rows; shape of scales not described .... *Chaetonotus* (*C.*) *vulgaris* Brunson, 1950

[USA: Kansas, Michigan]

3(2) Head indistinctly five-lobed with two pairs of sensory ciliary tufts; dorsal trunk scales strongly three-lobed with simple gently curved spines (Figs. 7.5 C and 7.6 F); 5–9 dorsal columns and 12–20 dorsal rows of scales........ 4

3' Head distinctly five-lobed with two pairs of sensory ciliary tufts (Fig. 7.5 A); dorsal trunk scales three-lobed or rounded with simple or dentate spines; 10–25 columns and 20–65 rows of dorsal scales........ 5

4(3) Seven to nine columns and 13–20 rows of dorsal scales; interciliary scales rounded, smooth in anterior portion of body and keeled in posterior portion of body (Fig. 7.5 X) .... *Chaetonotus* (*C.*) *larus* (Müller, 1773)

[USA: Maine, New Jersey]

4' Approximately 8 columns and 12–15 rows of dorsal scales; interciliary scales absent .... *Chaetonotus* (*C.*) *aculeatus* Robbins, 1965

[USA: Illinois]

5(3) Dorsal trunk scales three-lobed with posterior incision; spines simple or dentate, not angular; refractive granules of head plates absent 6

5' Dorsal trunk scales rounded with strong posterior incision; spines simple, thick and angular; 11–13 columns and 20–25 rows of dorsal scales; cuticular plates of head with refractive granules; pharynx large and muscular .... *Chaetonotus* (*C.*) *brevispinosus* Zelinka, 1889

[Canada: Ontario. USA: New Hampshire, Ohio]

6(5) Posteriormost dorsal trunk region with small and very short spined scales 7

6' Posteriormost dorsal trunk region similar to rest of dorsal trunk region........ 9

7(6) Scales with dentate gently curved spine (Fig. 7.5 B); interciliary area below pharynx with oval to rounded smooth or keeled scales 8

7' Scales with simple gently curved spines (Figs. 7.5 A and 7.6 F) in 11–14 columns and 23–28 dorsal rows; interciliary area below pharynx with approximately 9 transverse cuticular plates (Figs. 7.1.21 and 7.6 E) .... *Chaetonotus* (*C.*) *maximus* Ehrenberg, 1838

[Canada: Ontario. USA: New Jersey, North Dakota]

8(7) Fourteen to seventeen columns and 22–27 rows of dorsal scales .... *Chaetonotus* (*C.*) *similis* Zelinka, 1889

[Canada: Ontario. USA: Michigan?, New Hampshire, New Jersey]

8' Eighteen to nineteen columns of dorsal scales in head and neck region and 10–11 columns of dorsal scales in trunk region; 33–40 rows of dorsal scales .... *Chaetonotus* (*C.*) *hoanicus* Schwank, 1990

[Canada: Ontario]

9(6) Twenty to twenty five columns and 58–65 rows of very small dorsal scales with short spines; caudally 5–6 pairs of longer (8.5–18 μm) spines; interciliary area with 9–12 columns of elongated oval to rounded keeled scales .... *Chaetonotus* (*C.*) *polyspinosus* Greuter, 1917

[Canada: Ontario]

9' Thirteen to fifteen columns and 20–26 rows of dorsal elongated three-lobed scales with thin hair-like gently curved spines (Fig. 7.5 D); posteriorly a pair of thicker parafurcal spines (up to 50 μm) that just overshoots the furcal branches .... *Chaetonotus* (*C.*) *hirsutus* Marcolongo, 1910

[Canada: Ontario]

Gastrotricha: Chaetonotida: Chaetonotidae: *Aspidiophorus*: Species

1 Body large and stout; dorsal surface with 25 columns and 40–45 rows of rhomboidal pedunculate scales (Fig. 7.5 T); each scale with lateral and medial keels; pharynx large, with three cuticular teeth .... *Aspidiophorus paradoxus* (Voigt, 1902)

[Canada: Ontario. USA: New Jersey]

1' Body elongate and thin; dorsal surface with 25–30 columns and 60–80 rows of smooth elongate pedunculate scales; posteriormost dorsal surface with both pedunculate and non-pedunculate keeled scales (Fig. 7.6 A) .... *Aspidiophorus schlitzensis* Schwank, 1990

[Canada: New Brunswick, Ontario]

Gastrotricha: Chaetonotida: Chaetonotidae: *Heterolepidoderma*: Species

1 Furca of normal size and shape (Fig. 7.1), with adhesive tubes contributing to approximately half of the total furca length...................... 2

1' Furca short and thick with conspicuously short (1/4 of the total furca length) notched adhesive tubes .... *Heterolepidoderma brevitubulatum* Kisielewski, 1981

[Canada: Ontario]

2(1) Total number of dorsal keeled scales high (>20 columns; >32 rows); keels not drawn out into a short simple spine................................. 3

2' Total number of dorsal keeled scales lower than above (15–20 columns; 20–25 rows); columns straight; scales with keels drawn out into a short simple spine (Fig. 7.5 V); ocellar granules present (Fig. 7.1.4) (sometimes absent in certain populations/individuals) .... *Heterolepidoderma ocellatum* (Metschnikoff, 1865)

[Canada: Ontario. USA: Illinois]

3(2) Body relatively long and slender; scales hexagonal or elongate oval in shape; pharynx without distinct swellings, but can widen posteriorly 4

3' Body stout and short; scales numerous, round to suboval; ocellar granules absent; pharynx with anterior and posterior swelling .... *Heterolepidoderma illinoisensis* Robbins, 1965

[USA: Illinois]

4(3) Head five-lobed; 20–25 dorsal columns of delicate hexagonal scales; keels only developed in middle trunk region, sometimes absent altogether; interciliary field naked; ocellar granules present or absent .... *Heterolepidoderma gracile* Remane, 1927

[Canada: Ontario. USA: Illinois]

4' Head three-lobed; approximately 25 columns of elongate oval scales; interciliary field with a pair of ventral terminal scales and 6–11 columns of scales; ocellar granules absent .... *Heterolepidoderma majus* Remane, 1927

[Canada: Ontario]

Gastrotricha: Chaetonotida: Chaetonotidae: *Lepidodermella*: Species

1 Head rounded or five-lobed with two pairs of sensory ciliary tufts.. 2

1' Head distinctly three-lobed with one pair of sensory ciliary tufts (Fig. 7.4 B) .... *Lepidodermella triloba* (Brunson, 1950)

[USA: Indiana, Michigan]

2(1) Seven to nine dorsal columns of smooth scales (Fig. 7.5 U); interciliary area with smooth transverse cuticular scale plates in pharynx region (Figs. 7.1.21 and 7.6 E) .... *Lepidodermella squamata* (Dujardin, 1841)

[Canada: Ontario. USA: Illinois, Indiana, Michigan, New Hampshire, New Jersey, New York, Ohio, Virginia, Washington]

2' Thirteen to seventeen dorsal columns of smooth pentagonal to hexagonal scales; interciliary area with >20 columns of fine keels .... *Lepidodermella zelinkai* (Konsuloff, 1913)

[Canada: Ontario]

Gastrotricha: Chaetonotida: Neogosseidae: Genera

Members of the genus *Kijanebalola* have been reported from Nearctic waters; however, the specimens were not determined to species or described (see Krivanek & Krivanek, 1958).

1 Posterior end of body with two groups of long spines (Fig. 7.3 U); elements of mouth ring jointed .... **Neogossea [p. 128]**

[rare]

1' Posterior end of body with single medial group of spines (Fig. 7.3 T); elements of mouth ring unjointed .... *Kijanebalola*

[USA: Louisiana]

Gastrotricha: Chaetonotida: Neogosseidae: *Neogossea*: Species

1 Several pairs of longer spines in mid-dorsal and mid-lateral trunk region; 6 pairs of caudal spines .... *Neogossea sexiseta* Krivanek & Krivanek, 1958

[USA: Louisiana]

1' Pairs of longer spines absent from mid-dorsal and mid-lateral trunk region; 8–12 pairs of caudal spines .... *Neogossea fasciculata* (Daday, 1905)
[USA: Louisiana]

Gastrotricha: Chaetonotida: Dasydytidae: Genera

1 Posterior end of body without peg-like protuberances 2

1' Posterior end of body with pair of peg-like protuberances (Figs. 7.3 R, 7.4 F, and 7.5 Y) .... ***Stylochaeta* [p. 129]**
[Rare]

2(1) Scales absent or small and inconspicuous; in one South American species scales are large and smooth 3

2' Body enclosed in a "lorica" of large, thick, ornamented scales (Fig. 7.3 P) .... *Ornamentula paraënsis* Kisielewski, 1991
[Neotropical. Rare, not yet reported from the Nearctic]

3(2) Head distinctly wider than neck; body with long lateral spines, some mobile; pharynx with one bulb or no bulb 4

3' Head much narrower than body and scarcely wider than neck; lateral spines absent or identical to dorsal spines, the latter may be reduced to a posterior pair only; pharynx with two bulbs (Fig. 7.3 L) .... *Anacanthoderma*
[Palaearctic. Rare, not yet reported from the Nearctic]

4(3) Lateral spines without denticles or with denticles not directed inwards towards the body 5

4' Body with 2–3 pairs of lateral spine bundles; spines with 1–2 lateral denticles directed inward towards the body (Fig. 7.3 M) .... *Chitonodytes*
[Palaearctic. Rare, not yet reported from the Nearctic]

5(4) Posterior end of body usually with caudal spines 6

5' Posterior end of body rounded, without caudal spines (Fig. 7.3 O) .... ***Haltidytes* [p. 129]**
[rare]

6(5) Lateral spines with 1–3 lateral denticles and often terminally bifurcated; spines of uniform thickness from their base to the last lateral denticle; pharynx usually with a distinct posterior bulb (Fig. 7.3 N) .... ***Dasydytes* [p. 130]**
[rare]

6' Lateral spines tapered, with at most 1 weak lateral denticle and never terminally bifurcated; pharynx without posterior bulb (Fig. 7.3 Q) .... *Setopus bisetosus* (Thompson, 1891)
[USA: New Jersey]

Gastrotricha: Chaetonotida: Dasydytidae: *Stylochaeta*: Species

1 Peg-like protuberances (styli) with 3 pairs of equally long hairs (Fig. 7.4 F); three pairs of shorter ventrolateral bifurcated spines anchored by three-lobed scales, anterior pair anchored in anteriormost group of long ventral spines; middle pair anchored between anterior and middle group of long ventral spines and posterior pair anchored in middle group of long ventral spines .... *Stylochaeta fusiformis* (Spencer, 1890)
[USA: New Jersey]

1' Peg-like protuberances with 2 pairs of short (5 µm) and 1 pair of long (20–30 µm) hairs; ventrolateral bifurcated spines anchored by three-lobed scales absent .... *Stylochaeta scirtetica* Brunson, 1950
[Canada: Ontario. USA: Arizona, Louisiana, Michigan, New Jersey]

Gastrotricha: Chaetonotida: Dasydytidae: *Haltidytes*: Species

1 One pair of ventral rigid spines 2

1' Three pairs of ventral rigid spines with thick bases; 5 pairs of dorsal movable spines .... *Haltidytes crassus* (Greuter, 1917)
[Canada: Ontario]

2(1) One pair of rigid straight caudal spines; 4 pairs of ventrolateral movable spines that goes around and cross each other distally .... *Haltidytes saltitans* (Stokes, 1887)
[USA: New Jersey]

2' Eight pairs of ventral spines of which 6 pairs are movable .... *Haltidytes ooeides* Brunson, 1950
[USA: Arizona, Michigan]

Gastrotricha: Chaetonotida: Dasydytidae: *Dasydytes*: Species

1 Two rows of dentate spines present in neck region; two pairs of shorter spines on head; dorsal spines bidentate, not bifurcated .... *Dasydytes monile* Horlick, 1975

[USA: Illinois]

1' Spines in neck region do not form two rows, one pair of longer spines on head; dorsal spines dentate with bifurcated tips .... *Dasydytes goniathrix* Gosse, 1851

[USA: Indiana]

REFERENCES

Balsamo, M. 1983. Gastrotrichi (Gastrotricha). Guide per il Riconoscimento delle Specie Animali delle Acque Interne Italiane 20. Consiglio Nazionale delle Richerce.

Balsamo, M. & M.A. Todaro. 2002. Gastrotricha. Pages 45–61 in: S.D. Rundle, A.L. Robertson & J.M Schmid-Araya (eds.). Freshwater meiofauna: biology and ecology. Backhuys Publishers, Leiden.

Brunson, R.B. 1950. An introduction to the taxonomy of the Gastrotricha with a study of eighteen species from Michigan. Transactions of the American Microscopical Society 69:325–352.

Brunson, R.B. 1959. Gastrotricha. Pages 406–419 *in*: W.T. Edmondson (ed.). Fresh-water biology. Second edition. John Wiley and Sons, New York.

Green, J. 1986. Associations of zooplankton in six crater lakes in Arizona, Mexico and New Mexico. Journal of Zoology, London (A) 208: 135–159.

Greuter, A. 1917. Beiträge zur Systematik der Gastrotrichen in der Schweiz. Revue Suisse de Zoologie 25:35–76.

d'Hondt, J.L. 1971. Note sur les quelques Gastrotriches Chaetonotidae. Bulletin de la Société, Zoologique de France 96:215–235.

Horlick, R. 1975. *Dasydytes monile*, a new species of gastrotrich from Illinois. Transactions of the Illinois State Academy of Sciences 68:61–64.

Kånneby, T., M.A. Todaro & U. Jondelius. 2009. One new species and records of *Ichthydium* Ehrenberg, 1830 (Gastrotricha: Chaetonotida) from Sweden with a key to the genus. Zootaxa 2278:26–46.

Kånneby, T., M.A. Todaro & U. Jondelius. 2012. A phylogenetic approach to species delimitation in freshwater Gastrotricha from Sweden. Hydrobiologia 683:185–202.

Kånneby, T., M.A. Todaro & U. Jondelius. 2013. Phylogeny of Chaetonotidae and other Paucitubulatina (Gastrotricha: Chaetonotida) and the colonization of aquatic ecosystems. Zoologica Scripta 42:88–105.

Kånneby, T. & M.K. Wicksten. 2014. First record of the enigmatic genus *Redudasys* Kisielewski, 1987 (Gastrotricha: Macrodasyida) from the Northern hemisphere. Zoosystema 36(4): 1–12.

Kisielewski, J. 1981. Gastrotricha from raised and transitional peat bogs in Poland. Monografie Fauny Polski 11:1–143.

Kisielewski, J. 1986. Taxonomic notes on freshwater gastrotrichs of the genus *Aspidiophorus* Voigt (Gastrotricha: Chaetonotidae), with descriptions of four new species. Fragmenta Faunistica 30:139–156.

Kisielewski, J. 1987. Two new interesting genera of Gastrotricha (Macrodasyida and Chaetonotida) from the Brazilian freshwater psammon. Hydrobiologia 153:23–30.

Kisielewski, J. 1991. Inland-water Gastrotricha from Brazil. Annales Zoologici 43, Supplement 2:1–168.

Krivanek, R.C. & J.O. Krivanek. 1958. Taxonomic studies on the Gastrotricha of Louisiana. ASB Bulletin 5: 12.

Remane, A. 1935–1936. Gastrotricha und Kinorhyncha. Klassen und Ordnungen des Tierreichs, Band 4, Abteilung 2, Buch 1, Teil 2, Lieferungen 1-2: 1-242, 373–385.

Robbins, C.E. 1965. Two new species of Gastrotricha (Aschelminthes) from Illinois. Transactions of the American Microscopical Society 84:260–263.

Robbins, C.E. 1973. Gastrotricha from Illinois. Transcations of the Illinois State Academy of Science 66: 124–126.

Ruttner-Kolisko, A. 1955. *Rheomorpha neiswestnovae* und *Marinellina flagellata*, zwei phylogenetisch intressante Wurmtypen aus dem Süsswasserpsammon. Österreichische Zoologische Zeitschrift 6:55–69.

Schwank, P. 1990. Gastrotricha. Pages 1–252 in: J. Schwoerbel & P. Zwick (eds.). Süsswasserfauna von Mitteleuropa, Band 3. Gustav Fischer Verlag, Stuttgart.

Schwank, P. & T. Kånneby. 2014. Contribution to the freshwater gastrotrich fauna of wetland areas of southwestern Ontario (Canada) with redescriptions of seven species and a check-list for North America. Zootaxa 3811(4): 463–490.

Strayer, D., Hummon, W.D., Hochberg, R. 2009. Gastrotricha. In: Thorp, J.H., Covich, A.P. (Eds.), Ecology and Classification of North American Freshwater Invertebrates, third ed. Elsevier Inc., New York, pp. 163–172.

Sudzuki, M. 1971. Die das Kapillarwasser des Lückensystems Bewohnenden Gastrotrichen Japans. I. Zoological Magazine (Tokyo) 80: 256–257.

Tirjaková, E. 1998. Zaujímavý nález Brušnobrvca (Gastrotricha) zčel'ade Dichaeuturidae. Folia Faun. Slovaca 3, 19–21.

Todaro, M.A., Dal Zotto, M., Jondelius, U., Hochberg, R., Hummon, W.D., Kånneby, T., Rocha, C.E.F. 2012. Gastrotricha: a marine sister for a freshwater puzzle. PLoS ONE 7 (2), e31740.

Weiss, M.J. 2001. Widespread hermaphroditism in freshwater gastrotrichs. Invertebrate Biology 120: 308–341.

Chapter 8

Phylum Rotifera

Robert L. Wallace

Department of Biology, Ripon College, Ripon, WI, USA

T.W. Snell

School of Biology, Georgia Institute of Technology, Atlanta, GA, USA

E.J. Walsh

Department of Biological Science, University of Texas at El Paso, El Paso, TX, USA

S.S.S. Sarma

Universidad Naćional Autónoma de México Campus Iztacala, México

Hendrik Segers

Royal Belgian Institute of Natural Sciences, Brussels, Belgium

Chapter Outline

INTRODUCTION

Classification schemes differ slightly in how they regard the four groups of rotifers: Seisonidea, Bdelloidea, Monogononta, and Acanthocephala, the latter being an obligatorily parasitic taxon previously treated as a distinct phylum (Garey et al., 1996; Garey et al., 1998; Mark Welch & Meselson, 2000; Fontaneto & Jondelius, 2011). The peculiar problem is that outside of the phylogenetic literature, acanthocephalans and rotifers are still treated separately. Basically, each group has ignored much of the research done by the other. However, as evidenced in the works cited above, there is ample support to unite the two taxa within a single classification. First, they both share an important, unique, synapomorphic characteristic. Within the syncytial integument of both lies a curious layer of two proteins called the intracytoplasmic lamina. These proteins provide a degree of stiffness to body wall depending on its thickness. More importantly, molecular evidence supports uniting these taxa; indeed, researchers are now arguing that acanthocephalans ought to be subsumed within Rotifera (Sørensen & Giribet, 2006; Segers, 2007; Wey-Fabrizius et al., 2014). Thus, a new classification is emerging that asserts that acanthocephalans and seisonids are sister groups within Pararotatoria (Min & Park, 2009; Fontaneto & Jondelius, 2011; Wey-Fabrizius et al., 2014). Pararotatoria and bdelloids then comprise Hemirotifera, with Hemirotifera and Monogononta comprising the Syndermata. Syndermata is then a sister taxon to the Gnathostomulida, thus creating a group of small, jawed animals, the Gnathifera (Sørensen, 2001; Leasi et al., 2012). The close affinity among these taxa has led some to propose that the Gnathifera be treated as a distinct phylum (Shiel et al., 2009). However, given the scope of this work, we ignore the problem of where to place the acanthocephalans and gnathostomulids, but look forward to additional study that will provide a better synthesis of these taxa.

Thorp and Covich's Freshwater Invertebrates. http://dx.doi.org/10.1016/B978-0-12-385028-7.00008-1

At one time rotifers were divided into two classes: those with two gonads (seisonids and bdelloids) were considered to be orders within class Digononta, leaving the Monogononta as a separate class (Pennak, 1989). In this key, two classes, Pararotatoria and Eurotatoria, are recognized (Melone & Ricci, 1995; Smith, 2001; Segers, 2002; Wallace et al., 2006; Wallace & Smith, 2009; Wallace & Snell, 2010). Older classification schema that considers rotifers as a class within the phylum Aschelminthes should be abandoned.

Systematics of Rotifer Classes

Class Pararotatoria, Family Seisonidea

This taxon comprises only two genera (*Paraseison* and *Seison*), each with two species. These are large (2–3 mm) dioecious, marine rotifers that live on the gills of crustaceans (Ricci, 1993; Sørensen et al., 2005; Leasi et al., 2011; Leasi et al., 2012). Sometimes described as aberrant, the corona of seisonids is very reduced and not used in locomotion. All species have paired gonads and a functional gut in both sexes. The jaws or trophi are described as being fulcrate. Females have ovaries without vitellaria. Sexes are of similar size and morphology.

Class Eurotatoria, Subclass Bdelloidea

This important subclass comprises 19 genera and, depending on the author, approximately 350–460 species (Ricci, 1987; Segers, 2002, 2007, 2008). A complicating factor in determining an accurate estimation of the number of species is that cryptic speciation (hidden biological diversity) appears to be widespread in bdelloids (Fontaneto et al., 2007a, 2009, 2011). Bdelloids possess a somewhat uniform body plan (Donner, 1965; Melone & Ricci, 1995) and are exclusively females, reproducing by parthenogenesis. Bdelloids are characterized by having paired ovaries with vitellaria, more than two pedal glands, and ramate trophi (Melone et al., 1998; Wallace & Ricci, 2002; Wallace & Snell, 2010). Nearly all bdelloids are microphagous with a corona of either two trochal discs or a modified ciliated field. Bdelloids often have a vermiform body with a pseudo-segmentation consisting of annuli that permits shortening and lengthening of the body by telescoping. Current taxonomy recognizes three orders: Adinetida (family Adinetidae), Philodinida (families Habrotrochidae and Philodinidae), and Philodinavida (family Philodinavidae) (Melone & Ricci, 1995).

Bdelloids generally are not caught in plankton tows, although they may be found in waters with dense vegetation. On the other hand, they often occur in sediments, among plant debris, or crawling on the surfaces of aquatic plants. Some forms inhabit the capillary water films of soils (Pourriot, 1979; Devetter, 2010) or covering mosses (Burger, 1948; Peters et al., 1993; Ricci & Caprioli, 2005; Bielańska-Grajner et al., 2011); this habitat has been referred to as the limnoterrestrial (Devetter, 2008; Segers, 2008). Bdelloids also are present in communities that develop in tree holes (Devetter, 2008) and the pitfall traps of the carnivorous plant *Sarracenia purpurea* Linnaeus, 1753 (Bateman, 1987; Błędzki & Ellison, 1998). Many species are capable of becoming desiccated and then rehydrated (Ricci & Fontaneto, 2009; Wilson, 2011). Unfortunately, there has been no systematic review of Nearctic bdelloids, but there are several works that should be consulted for more information (Bartoš, 1951; Donner, 1965; Pourriot, 1979; Koste & Shiel, 1986; Melone et al., 1998; Ricci & Melone, 2000).

Class Eurotatoria, Subclass Monogononta

The monogononts comprises the largest group of rotifers with >1500 species in more than 100 genera. While most are free-living, taking a benthic, free swimming, or sessile existence, a few genera are parasitic. Nearly all rotifers are found in inland waters, both fresh and alkaline, but some appear to be exclusive to the marine habitat. In many species, males have never been observed, but all monogononts are assumed to be dioecious. As the name implies, these rotifers possess a single gonad. Males are often structurally reduced with a vestigial gut that functions only in energy storage (Ricci & Melone, 1998). They have a shorter lifespan than females and are usually present in the plankton for only a few days or weeks each year. Monogononts are microphagous or raptorial, but a few are parasitic. The corona is more varied than in other rotifers; the corona may range from broad to narrow disks or possess ear-like lobes or be vase-shaped with reduced ciliation and long setae that are used in prey capture. This key recognizes two superorders: Pseudotrocha (order Ploima) and Gneisotrocha (orders Collothecaceae and Flosculariaceae).

Rotifers of the Nearctic

Here we recognize Phylum Rotifera to comprise 35 families with >130 genera. Of these, >100 genera and >900 species have been reported from inland waters of the Nearctic (NEA). Because of the relatively little taxonomic work done in the Nearctic, we feel that this assessment is a gross underestimation of the true species, richness.

LIMITATIONS

Except for the incomplete *Guide Series* (Segers, 1995a,b; De Smet, 1996; De Smet & Pourriot, 1997; Nogrady & Segers, 2002; Wallace et al., 2006), there has been no attempt to construct a comprehensive key to Rotifer. There are a variety of keys to rotifers, including those by Edmondson (1959) and Smith (2001), which cover some of the North American fauna to the level of genus, and Stemberger (1979) for the Laurentian Great Lakes, which provide a species-level key. Although specialized and covering other biogeographical realms, valuable keys can be found for Australian waters (Koste & Shiel, 1986, 1987, 1989a,b, 1990a,b, 1991; Shiel & Koste,1992, 1993; Shiel, 1995), the British Isles (Pontin, 1978), Italy (Braioni & Gelmini, 1983), and planktonic rotifers in general (Ruttner-Kolisko, 1974). Unfortunately, no key is comprehensive enough to cover all of the variations in size and morphology that are sometimes found within a species. Of all the efforts, Koste's (1978) revision of Voigt for the rotifers of central Europe comes the closest; although over 35 years old, it is still important. We urge caution in using keys that are available in electronic form (e.g., Internet or on CD-ROMs). In most instances, these are far too simplistic, often reflecting a suite of species from a rather narrow region; in no way should they be considered thorough. Nevertheless, Jersabek et al. (2014) and Jersabek & Leitner (2013) are two very valuable online resources.

A further limitation stems from the fact that too few species have photographic documentation. Thus, as with other keys to the Rotifera, we caution that the illustrations used in this one are not necessarily from specimens collected from the Nearctic, and we have had to rely on illustrations from the older literature. When the figures used to demonstrate the general form of a taxon are from the Nearctic, they are designed with the abbreviation NEA. For this work, we estimated the number of species in each genus using the assessment of Segers (2007) and Jersabek & Leitner (2013), including subspecific variants (subspecies), but not those species noted as *species inquirenda*—species requiring additional investigation before being accepted as a valid taxon. We also did not count species that appear to be strict halophiles (only inhabiting marine and brackish waters), recognizing that they could still be present in alkaline, inland waters.

Finally, it should be noted that emerging molecular data supports the concept of cryptic speciation in rotifers. Thus, there may be 3–14 times as many species within any currently recognized morphospecies (Gómez, 2005; Fontaneto et al., 2007a,b; Schröder and Walsh, 2007; Garcia-Morales and Elias-Gutierrez, 2013).

One of the fundamental differences among the higher taxonomic levels in the phylum Rotifera is the structure of the trophi that reside in a muscular pharynx called the mastax. Nine different types of trophi are recognized, along with various transitional forms (Koste, 1978; Koste & Shiel, 1987). Although the following key is designed with the non-specialist in mind, commensurate with the central importance of the trophi, we use both the structure of the trophi and other obvious characters of anatomy and morphology as principal points of separation of the taxa. Therefore, it is important to note that you may have to destroy one or more specimens to get a good look at the trophi. Consequently, because other morphological features also are critical to identification, it is important that you reserve several specimens to study whole, preferably alive. Taking photomicrographs of the live, intact animal before preservation or hydrolysis is always a prudent practice.

The obligatory marine taxon, class Pararotatoria (family Seisonidae) is not included herein. The following key is based on several important works (Koste, 1978; Segers, 2004; Wallace et al., 2006). The species records annotated herein for the Nearctic are noted in brackets; monospecific genera are designated as such. All data were derived from (Segers, 2007).

TERMINOLOGY AND MORPHOLOGY

Most rotifer trophi are composed of seven hardened elements or sclerites (Fig. 8.1; also Fig. 13.19 in Volume I). Three of these, the manubrium (manubria), ramus (rami), and uncus (unci) are paired, while the fulcrum is unpaired (see also Fig. 13.10 in Volume I). The central most elements are the rami that may contain one or more chambers. The fulcrum, which is absent in bdelloids, is located caudally between the rami and articulates with them. Together, the fulcrum and paired rami comprise the incus. Located left and right of the rami are the manubrium and uncus; together they comprise the malleus (plural, mallei). These elements are all held together and are moved by muscular action. Besides the discussion in Volume I of this series, additional information on rotifer trophi may be found in the following works: (Donner, 1956; Edmondson, 1959; Koste, 1978; Sørensen, 2006). The nine different types of trophi were established based on the size, shape, and positioning of the seven elements described above, as well as the presence of other accessory parts and variations. The "Rotifer trophi web page" (Anon, 2014) also provides some excellent scanning electron photomicrographs. Of the nine types, the fulcrate trophi of seisonids is not considered in this key. The other eight types, which are important to this key (cardate, forcipate, incudate, malleate, malleoramate, ramate, uncinate, and virgate), are briefly discussed below. They are also described in detail in Chapter 13 of Volume I.

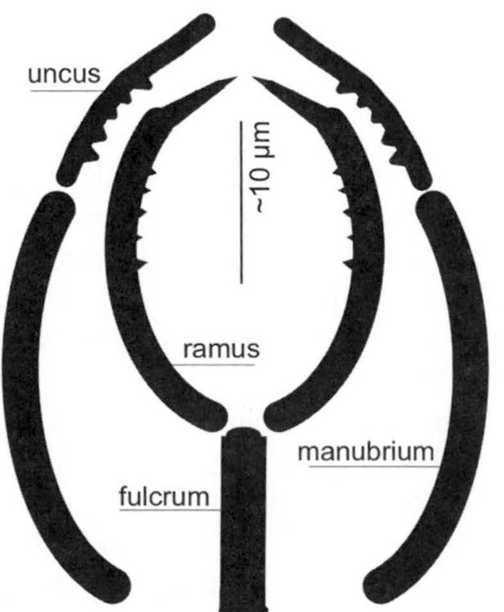

FIGURE 8.1 Schematic view of the basic elements of rotifer trophi. (See also Figs. 13.10 and 11 in Volume I for a complete overview of the trophi.)

Other important terms are defined below.

- Alula—Wing-shaped projections present at the base of each ramus.
- Basal chamber—Found in some trophi, a chamber or depression in the ramus with a window-like opening (basifenestra) that opens on the dorsal side.
- Cardate trophi—Trophi with a wide fulcrum and rami in the shape of a lyre (U-shaped).
- Cirri—Intermediate length (cilia-like) structures that are present in the coronal region.
- Forcipate trophi—Trophi in the form of forceps that can be extruded from the mouth (found in some creeping and parasitic species; no planktonic and semiplanktonic); compare to Incudate.
- Cryptic species—More than one population from the same or different habitats that with currently recognized morphological metrics are difficult or impossible to diagnose as species, but that may be differentiated by their genetic signatures. Together these form a species complex (e.g., *Brachionus plicatilis* species complex).
- Gelatinous tube—A thick secretion from glands in the body that form a gelatinous mass that may remain transparent or become occluded with bacteria, algae, and/or detritus.
- Hemiforcipate trophi—The specialized jaws of family Asciaporrectidae as described by De Smet (2006).
- Hypopharyngeal muscle—Powerful muscle in the mastax that helps in creating a strong pumping action in some species.
- Incudate trophi—The trophi of family Asplanchnidae (planktonic and semiplanktonic) that possess large, pincer-like rami, but reduced manubria and unci.
- Illoricate—A term used for rotifers with a thin body wall: one that is not thickened by the proteins of the intracytoplasmic lamina (compare loricate).
- Infundibulum—A cup-shaped corona found in Order Collothecacea.
- Intramalleus—An accessory part found in the trophi of some species; a connecting piece between the uncus and the manubrium.
- Loricate—A term used for rotifers with a thick body wall or lorica: one that is thickened by the proteins of the intracytoplasmic lamina; compare to illoricate.
- Malleate trophi—Trophi with a short fulcrum, but the manubrium and rami are strong; unci with 4–7 teeth; compare to Malleoramate.
- Malleoramate trophi—Rami with strong teeth and unci with many slender teeth; compare to Malleate.
- Modulus—A specialized organ located at the anterioventral end of some *Floscularia* that makes small, round, or bullet-shaped pellets (sometimes called pseudofecal pellets) that are composed of bacteria and tiny bits of detritus. The animal uses the pellets to construct a tall, turret-like structure as found in many medieval castles. These pellets should not be confused with the fecal masses that a few species embed within a more extensive gelatinous tube.
- Morphospecies—A species designated and clearly separated from other species based on recognized unequivocal morphological features.
- Oral plates—An accessory part found in some trophi; plates with tooth-like structures associated with the rami.
- Oviferon—A structure in *Sinantherina* and *Pentatrocha* that holds their parthenogenetic eggs.
- Preunci—An accessory part found in some trophi; a separate set of teeth present under the teeth in the uncus.
- Pseudunci—An accessory part of large hooked structures found in the trophi of family Birgeidae.
- Ramate—The trophi of bdelloid rotifers possessing wide, curved to half moon-shaped unci, but lacking the fulcrum.
- Rostrum—A small to prominent, beak-like structure that projects from the anterior end of rotifers, especially in some bdelloids.
- Setae—Very long cilia that line the edge of the corona (infundibulum) in *Collotheca* and the elongate arms (lobes) of *Stephanoceros*.
- Spurs—Short to long protrusions that project dorsally at the terminus of the foot. Note that the foot itself may be much shorter than the toes (compare toes).
- Subbasal chamber—A chamber or depression in the ramus with a window-like opening (subbasifenestra) that opens on the ventral side.
- Subunci—An accessory part found in some trophi; a narrow rim of small fused bumps or ridges that are attached to the uncus near the teeth.
- Sulcus—A hollow depression where the lorica is not overly thickened, usually between two regions or plates where the lorica is thickened (see loricate).
- Toes—Short to long protrusions that project from the terminus of the foot in many rotifers (compare spurs).
- Trochal pedicels—Paired parts of the bdelloid corona elevated by a flexible stalk above the rest of the body.
- Uncinate—The trophi of collothecid rotifers, possessing well-developed rami, but weak fulcrum and manubria; 3–5 teeth in the unci.
- Virgate—Trophi with an action that functions in pumping; often asymmetrical with most elements, having a tendency towards being slender; fulcrum and manubrium long.

MATERIAL PREPARATION AND PRESERVATION

Whenever possible, specimens should be examined alive and then studied after preservation to appreciate the consequence of fixation on body shape. This is especially important for bdelloids and illoricate monogononts, which will shrivel into a mass of tissue that is usually impossible to identify. While the trophi of these specimens may be extracted and used identification, the details of the body are often needed as well. Solutions of sugar (40% sucrose) and 4% formalin (Haney & Hall, 1973) or ethanol (Black & Dodson, 2003), which are employed in the preservation of cladocerans, do not prevent contraction in rotifers. However, in loricate taxa, such as *Lecane*, formalin fixation is important. In this case a so-called, hard fixation, with higher concentrations of formalin (>4%) will cause the tissues to contract revealing the details of the surface of the lorica, which is critical to the identification to the level of species (Segers, 1995a). Visualization of preserved rotifers may be enhanced by the use of the stain Rose Bengal (Rublee & Partusch-Talley, 1995; Rublee, 1998). While molecular techniques are being used to identify cryptic species in rotifers, its use has not yet been applied to the routine identification of species.

As of yet, there is no single anesthetic that has proved to be useful for a wide range of species. For example, carbonated water (club soda), 1% solution of $MgCl_2$, and various other narcotizing agents have been tried with mixed success. For a review of these techniques, consult the following publications: Edmondson (1959), Nogrady & Rowe (1993), Wallace et al. (2006), as well as Chapter 13 in Volume I. Two older techniques that kill the specimen in a more-or-less natural condition are the hot-water fixation technique of Edmondson (1959) and the formalin drop technique. However, both require much practice to master. In the later procedure, minute drops of formalin (<5 μL) are added over a long period of time (hours) to a small volume of water (~1 mL) in which a specimen is kept. For additional information on preservation, storage, and mounting, consult Wallace et al. (2006) and Jersabek et al. (2010).

Identification of rotifers usually requires examination of the live animal, which is then followed by close study of their trophi. However, for some taxa, such as those of family Asplanchnidae, identification to genus (*Asplanchna, Asplanchnopus*, and *Harringia*) is quite simple and requires only rudimentary understanding of rotifer morphology. On the other hand, while one can easily become familiar with the body form of subclass Bdelloidea or the monogeneric family Lecanidae (*Lecane*), identification to species is challenging. In many cases, it will be necessary to extract the trophi using bleach (sodium hypochlorite), which dissolves the animal's soft tissues. Edmondson's (1959) description of the extraction employs a depression slide to begin the process, but that necessitates moving the trophi from the depression slide to a flat slide after the soft tissues have disappeared. With practice this method works well, and it has the added advantage to being able to wash the bleach away while the trophi are in the well of the slide. Alternatively, the animal can be placed on a flat slide in a drop of water. To this, a small drop of bleach is added and a cover glass is carefully applied to the preparation. The animal is observed as the bleach dissolves its tissues, but observers must be diligent, as this can happen quite suddenly. Confounding this process is the fact that bubbles usually form, obscuring one's view. This technique also works without adding the cover glass. In that case, the trophi must be located, and then some of the bleach solution is removed with a micropipette, followed by the slow addition of clean water. That process is then repeated several times to wash away the bleach, which otherwise will continue its hydrolysis on the trophi. At this point, a cover glass may be gently added and the trophi examined. This slide will last until the liquid evaporates from under the cover glass. For a longer lasting preparation, the fluid must be replaced with a glycerol+4% formalin solution. To do this, prepare a series of 4% formalin solutions in increasing concentrations of glycerol: 10, 20, 30 ... 100%. Using the same technique used to remove the bleach water, the liquid is slowly replaced (over an hour or more) with the graded 4% formalin–glycerol series. Such preparations will last for days, if not longer. However, this preparation can be made to last longer by sealing the cover glass to the slide by applying a line of melted Vaspar (a 50:50 mixture of melted paraffin wax and petroleum jelly) to the four edges of the cover glass.

Lastly, it must be remembered that the trophi are tiny (<50 μm), delicate structures composed of 6–7 elements (see above) that have a precise three-dimensional arrangement.

KEY TO FRESHWATER ROTIFERS (CLASS EUROTATORIA)

Rotifer Subclasses

1 Rotifers with ramate trophi (Fig. 13.11 t in Volume I) and paired ovaries; usually not planktonic in open waters; movement along surfaces often in a leech-like fashion .... **Bdelloidea [p. 136]**

[Although this step obviously is very important, special care is generally not necessary to resolve this point. The ramate trophi of bdelloids, which lack the fulcrum and have the largest unci teeth in the middle, are usually identifiable in whole animals without resorting to isolating the trophi from the animal (Ricci & Melone, 2000; Fontaneto et al., 2007c). If the type of trophi and number of ovaries cannot be determined, there are other clues that may be useful for live organisms. Upon contacting a substratum, many bdelloids will crawl on the surface in a manner reminiscent of a leech, hence the etymon of the name (Greek, *bdella*, leech). Further, some bdelloids possess a corona that has the appearance of two separate wheels elevated on short stalks (pedicles), while only a few monogononts of the Flosculariidae give this impression, and most of those are permanently sessile.]

1' Rotifers with a single ovary; trophi other than ramate .... **Monogononta [p. 142]**

Rotifera: Bdelloidea Families

Important keys to the bdelloids include the following: (Bartoš, 1951; Edmondson, 1959; Donner, 1965; Koste & Shiel, 1986; Ricci & Melone, 2000; Kutikova, 2005; Wallace et al., 2006). Estimates of the number of species reported here are probably too low. This is no doubt due to the lack of taxonomists currently working with bdelloids in the Nearctic. Four genera not reported from this region are included here for completeness and to call attention to the need for additional research. The monospecific marine genus, *Zelinkiella* (Philodinidae), is not included herein.

1 Stomach with lumen 2

1' Stomach without lumen, as a syncytial mass of food vacuoles that gives the gut a frothy appearance .... **Habrotrochidae [p. 136]**

[The food vacuoles must not to be confused with oil droplets that are often present in other bdelloids or with the symbiotic algae in the stomach wall of Ituridae (*Itura*; Subclass Monogononta: Order Ploima).]

2(1) Trophi located deeper, not near the mouth, and are not exsertile 3

2' Trophi located near the mouth and are exsertile (protruded during feeding) .... **Philodinavidae [p. 137]**

3(2) Corona with paired trochal disks on raised pedicles. .... **Philodinidae [p. 137]**

3' Corona lacking trochal disks (no pedicles); ventrally a pair of flat, ciliated fields separated by a longitudinal groove ...... **Adinetidae [p. 138]**

Rotifera: Bdelloidea: Habrotrochidae: Genera

[3 genera; all reported in NEA]

1 Ring on trochal pedicels absent 2

1' Shelf-like ring present on the trochal pedicels (Fig. 8.2), but may not extend corona to feed; 6–10 teeth on each uncus; 200–500 μm; habitat mostly in soil water [2 NEA of 15 species] .... *Otostephanos* spp.

2(1) Each trochus partially covered by a wide dorsal, membrane-like hood seen when the animal is feeding (Fig. 8.3); 3–9 teeth on each uncus; ~200 μm; habitat mostly soil and on mosses [1 NEA of 12 species] .... *Scepanotrocha rubra* Bryce, 1910

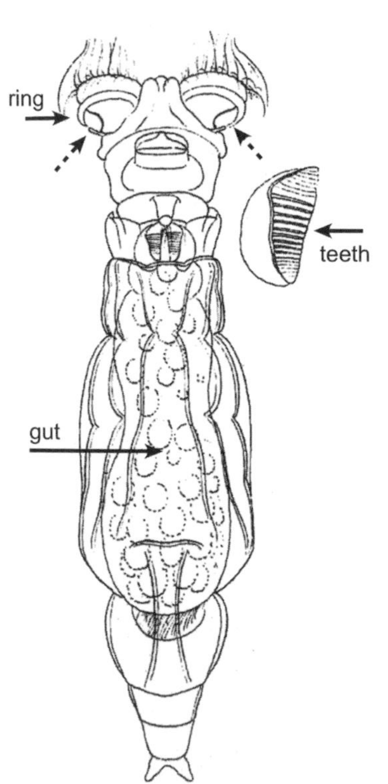

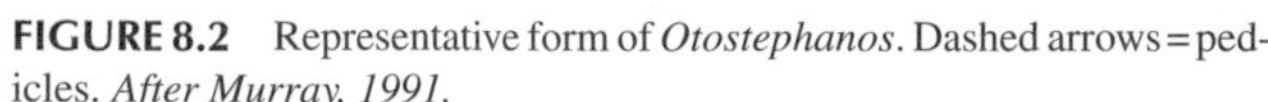
FIGURE 8.2 Representative form of *Otostephanos*. Dashed arrows = pedicles. *After Murray, 1991.*

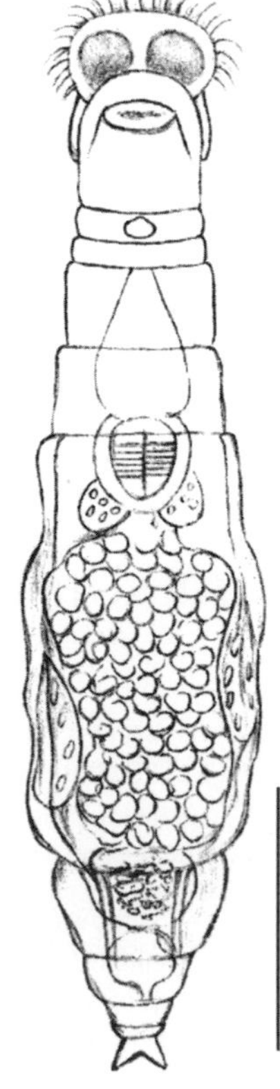

FIGURE 8.3 *Scepanotrocha rubra*. Bar = 50 μm. *After Bryce, 1910.*

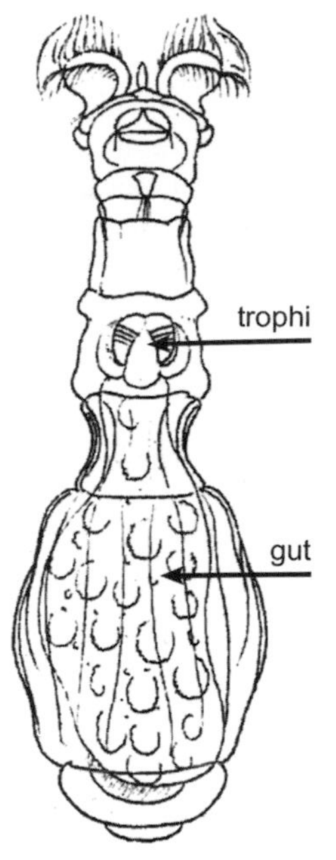

FIGURE 8.4 Representative form of *Habrotrocha*. *After Murray 1911.*

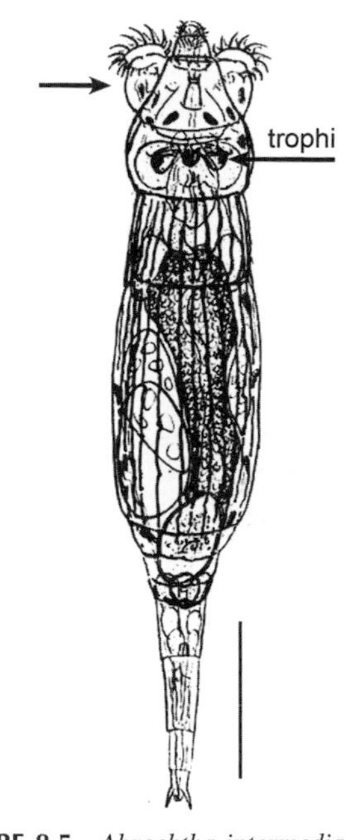

FIGURE 8.5 *Abrochtha intermedia*. Arrows: corona: bilobed-disk on short pedicels and trophi. Bar = 100 μm. *After de Beauchamp 1909.*

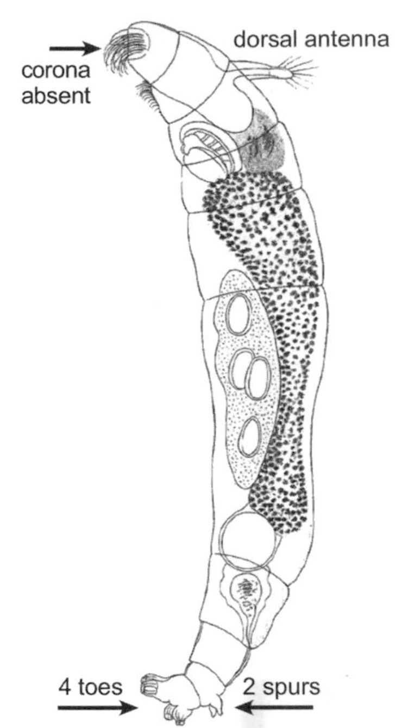

FIGURE 8.6 *Philodinavus paradoxus*. Animal 200–300 μm. *After Murray 1905.*

2' Hood extending over trochus is absent (Fig. 8.4); ≥10 teeth on each uncus of varying width; 150–400 μm; habitat aquatic sediments, soils, and mosses [22 NEA of 140 species] .... *Habrotrocha* spp.

Rotifera: Bdelloidea: Philodinavidae: Genera

[3 genera; 2 reported in NEA]

1 Corona reduced, with only a hint of being bilobed or not bilobed 2

1' Corona a poorly developed bilobed-disk on short pedicels, lacking upper and lower lips (Fig. 8.5); ~500 μm; ephemeral habitats [4 NEA of 5 species] .... *Abrochtha* spp.

2(1) Corona possessing two small, ciliated disks on the ventral side; ~130–300 μm, habitat includes mosses and algae. [2 species; neither reported in the Nearctic] .... *Henoceros* spp.

2' Corona with even greater reduction (no wheel-like structure present); cilia are restricted to a small region around the mouth (Fig. 8.6); ~200–300 μm; habitat includes mosses in well-oxygenated streams [1 NEA of 2 species] .... *Philodinavus paradoxus* (Murray, 1905)

Rotifera: Bdelloidea: Philodinidae: Genera

[11 genera; 10 reported in NEA]

1 Toes absent, but with an attachment disk on the terminal foot joint 2

1' Toes present on foot: numbering 2, 3, or 4 3

[One must take great care to distinguish toes (extensions of the pedal glands) from spurs (stiff, conical appendages lacking pedal bland ducts) (Ricci & Melone, 2000).]

2(1) Spurs on caudal region of foot exceptionally small (only mere bumps); obligatorily epizoic on freshwater crab (*Potamon*) (Van Damme & Segers, 2004); >500 μm; [2 species, not reported from the Nearctic] .... *Anomopus* spp.

2' Spurs on caudal region of foot not exceptionally small; corona wide; toes absent; (Fig. 8.7); usually ≥3 up to 10 teeth; 200–600 μm; on mosses, some epizoic; lives in the water film of soils and mosses [11 NEA of 49 species] .... *Mniobia* spp.

3(1) More than two toes on foot 4

3' Two toes on foot; eyespots absent; ca. 400 μm; [monospecific, not reported from the Nearctic] .... *Didymodactylos carnosus* Milne, 1916

4(3) Three toes on foot 5

PHYLUM ROTIFERA

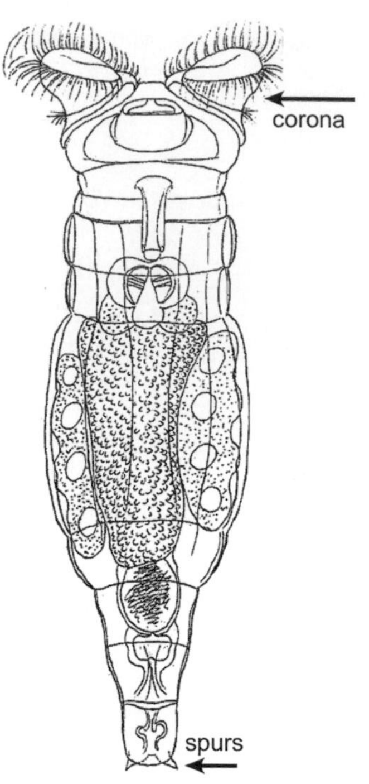

FIGURE 8.7 Representative form of *Mniobia*. *After Murray 1905.*

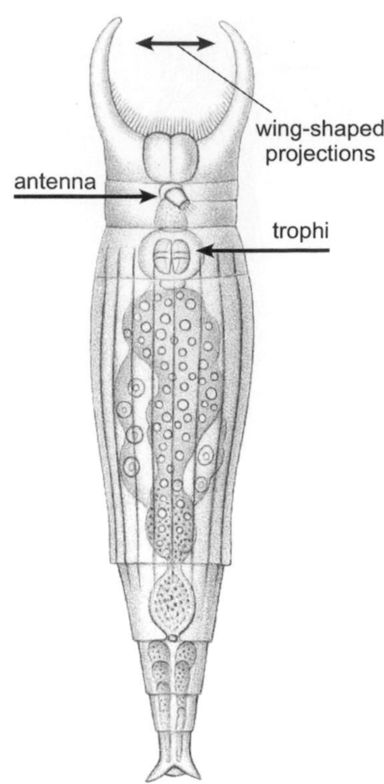

FIGURE 8.8 *Ceratotrocha cornigera*. Animals 150–350 μm. *After Weber 1898.*

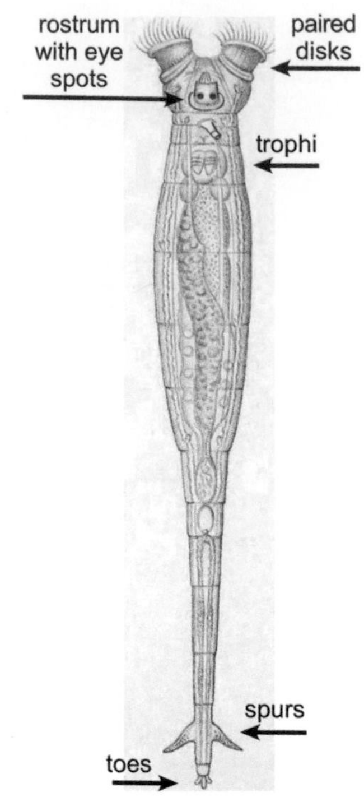

FIGURE 8.9 Representative form of *Rotaria*. *After Weber 1898.*

4'	Four toes on foot 7
5(4)	Edges of corona without horns or wing-shaped projections 6
5'	Edges of corona expanded into prominent, elongate, horn- or wing-shaped projections; 2–3 teeth on each uncus; (Fig. 8.8); 150–350 μm; habitat is the interstitial water films of soils [1 NEA of 4 species] .... *Ceratotrocha cornigera* (Bryce, 1893)
6(5)	Rostrum (eyes often present), spurs, toes, and dorsal antenna prominent; 3 toes and 2 long spurs; (Fig. 8.9); uncus with 2, rarely 3 teeth; ≤400–1600 μm; ovoviviparous; free swimming; habitat mostly water bodies [16 NEA of 29 species] .... *Rotaria* spp.
6'	Rostrum, spurs, toes (3 short), and dorsal antenna not prominent; (Fig. 8.10); surface may be sculptured; unci with 2, 3, rarely 5 teeth; 200–500 μm; oviparous; habitats include aquatic sediments and the interstitial film of soils and mosses [18 NEA of 104 species] .... *Macrotrachela* spp.
7(4)	Integument of trunk stiff (somewhat inflexible) and sculptured 8
7	Integument of trunk thin and flexible 9
8(7)	Spurs long; many species possess eyespots (Fig. 8.11); stiff spines usually present on body; foot long, often with long spurs; 300–500 μm; ovoviviparous; usually living in aquatic sediments [5 NEA of 37 species] .... *Dissotrocha* spp.
8'	Spurs short; eyespots absent (Fig. 8.12); body stiff and armor-like; foot short with 4 toes; 200–400 μm; oviparous; interstitial film on mosses [3 NEA of 14 species] .... *Pleuretra* spp.
9(7)	Spurs very long, flat and wide; foot ≥1/2 of total animal length (Fig. 8.13); 250–650 μm; oviparous or ovoviviparous; often epizoic on crustaceans [4 NEA of 5 species] .... *Embata* spp.
9'	Spurs short or, if long, not flat and wide; foot ≤1/2 of total animals length (Fig. 8.14); unci usually with 2 or more teeth; 100–800 μm; 1 viviparous species (not NEA); ponds and lakes and in the interstitial water film of soils and mosses [17 NEA of 53 species] .... *Philodina* spp.

Rotifera: Bdelloidea: Adinetidae: Genera

[2 genera; 1 reported in NEA]

1	Body fusiform, with distinct foot possessing two spurs (Fig. 8.15); corona developed as a ventral ciliate field, possessing a rake that it uses to collect food; trophi small, unci with 2 teeth; body dorsoventrally compressed; foot well develop; does not swim; 200–700 μm; habitat includes aquatic sediments and the interstitial film of mosses, lichens, and soils [8 NEA of 19 species] .... *Adineta* spp.
1'	Body vermiform possessing an indistinct foot lacking spurs, but with 10–12 papillae located near the terminus of foot; 175–250 μm. [2 species; not reported in the Nearctic] .... *Bradyscella* spp.

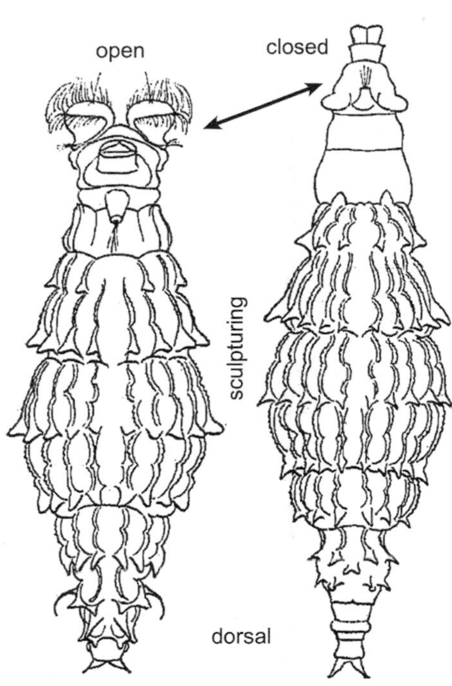

FIGURE 8.10 Representative form of *Macrotrachela. After Murray 1911.*

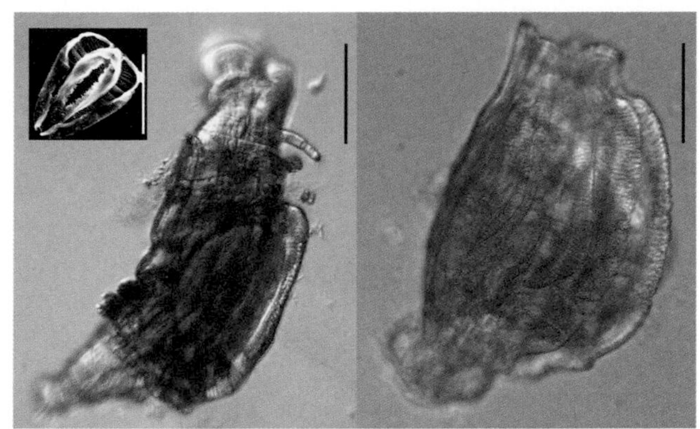

FIGURE 8.12 Representative form of *Pleuretra*. Left with trophi inset. Bars = 50 μm. (NEA)

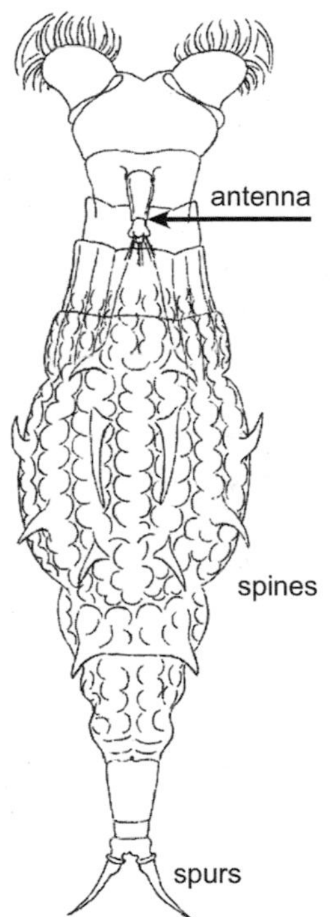

FIGURE 8.11 Representative form of *Dissotrocha. After Murray 1908.*

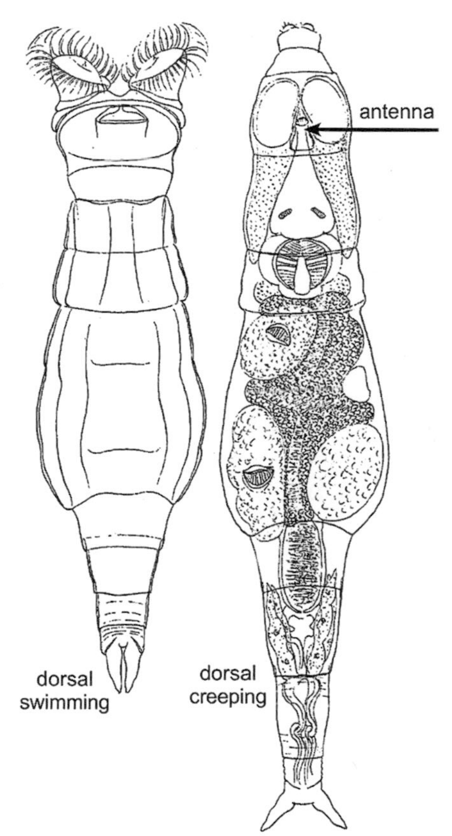

FIGURE 8.13 Representative form of *Embata. After Murray 1905.*

Rotifera: Monogonata: Orders

1 Trophi other than uncinate; corona with typical ciliated lobes or fields, or ciliated bands 2

1' Trophi uncinate (Fig. 13.10 h in Volume I); corona lacking typical ciliated regions, but is open as a small to large cup-shaped structure (infundibulum) with or without elongate setae (sometimes with cilia as well) around the margin (Fig. 13.8 in Volume I) .... **Collothecacea [p. 140]**

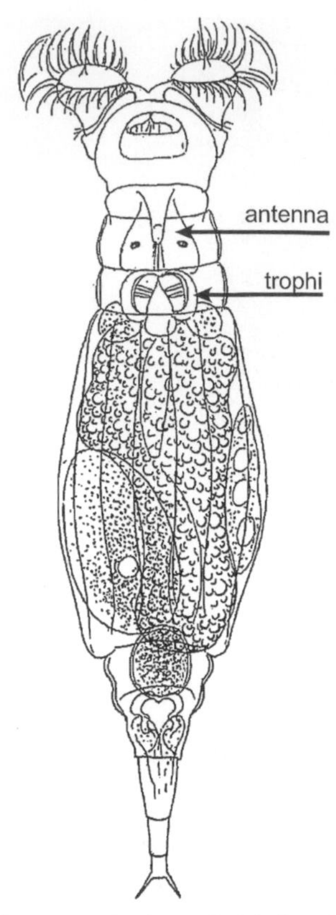

FIGURE 8.14 Representative form of *Philodina. After Murray 1906.*

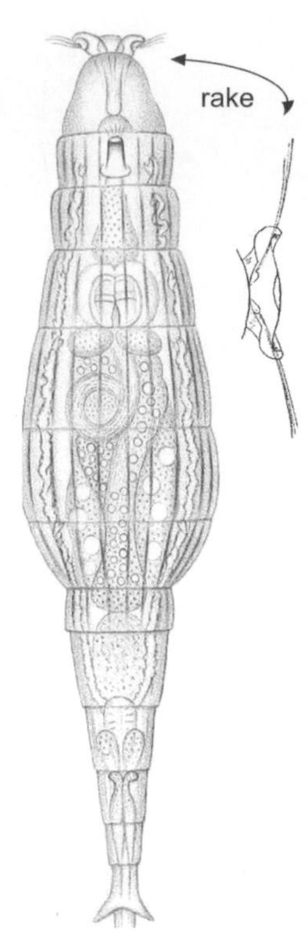

FIGURE 8.15 Representative form of *Adineta. After Weber 1998; insert in Janson 1893.*

2(1) Rotifers possessing malleoramate trophi (Fig. 13.11 i in Volume I) .... **Floscularíaceae [p. 142]**

[Malleoramate trophi possess a fulcrum, crescent-shaped manubria, and unci with a few clavate to numerous fine teeth nearly completely overlying the rami (compare to description in the key to Ploima); teeth close to the fulcrum are usually larger than those more distant. In live animals, the nearly constant movement of the trophi in a grinding or pounding action is characteristic.]

2' Rotifers with trophi other than malleoramate .... **Ploima [p. 148]**

Rotifera: Monogonata: Collothecacea: Families

Important keys to the Collothecacea include the following: (Bērziņš, 1951; Meksuwan et al., 2013).

1 Coronal margin of adults usually with long setae (sometimes absent) and/or ciliated; living in transparent gelatinous tubes; mostly sessile, but some planktonic in open waters .... **Collothecidae [p. 140]**

1' Coronal margin of adults without setae or cilia; with or without gelatinous tube .... **Atrochidae [p. 142]**

Rotifera: Monogonata: Collothecacea: Collothecidae: Genera

[2 genera; both reported in NEA]

1 Corona with very elongate lobes resembling arms, approximate as long as trunk; arms either stout with robust setae arranged in whorls and held at right angles to arms or slender with long slender setae not arranged in whorls; ≤1500 μm; sessile. (Fig. 8.16) [2 NEA of 2 species] .... *Stephanoceros* spp.

[In light of the fact that some members of the genus *Collotheca* also possess elongate lobes, Meksuwan et al. (2013) have recommended that the status of both genera be re evaluated.]

1' Corona lobes long, short, or absent; setae never in whorls; clear tube may be embedded with algae or covered with debris (Fig. 8.17); ~100–2500 μm. [19 NEA of ~50 species] .... *Collotheca* spp.

[While most are sessile, 4 species are planktonic (*C. libera, C. mutabilis, C. pelagica, C. polyphema*); 1 is both (*C. ornata*), and 1 is benthic (*C. crateriformis*). Nearly all produce a clear gelatinous tube; none are colonial, but may colonize substrata forming dense groupings (Wallace & Edmondson, 1986) (cf. colonial forming species of Flosculariidae).]

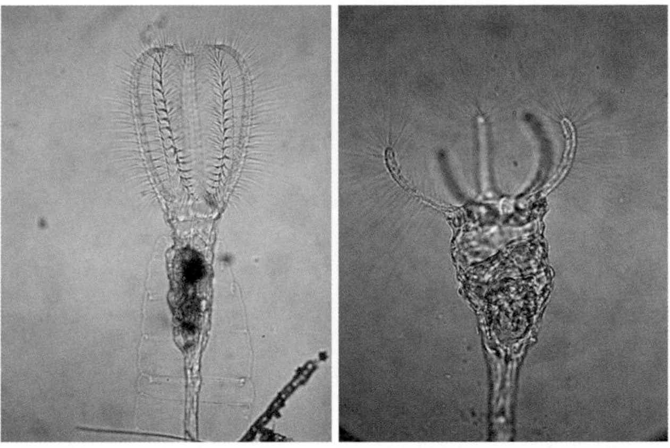

FIGURE 8.16 Genus *Stephanocerous: S. fimbriatus* (left) and *S. millsii* (right). Animals ca. 1000 µm. (NEA) *Photomicrographs courtesy of Adele Hochberg, University Massachusetts at Lowell USA.*

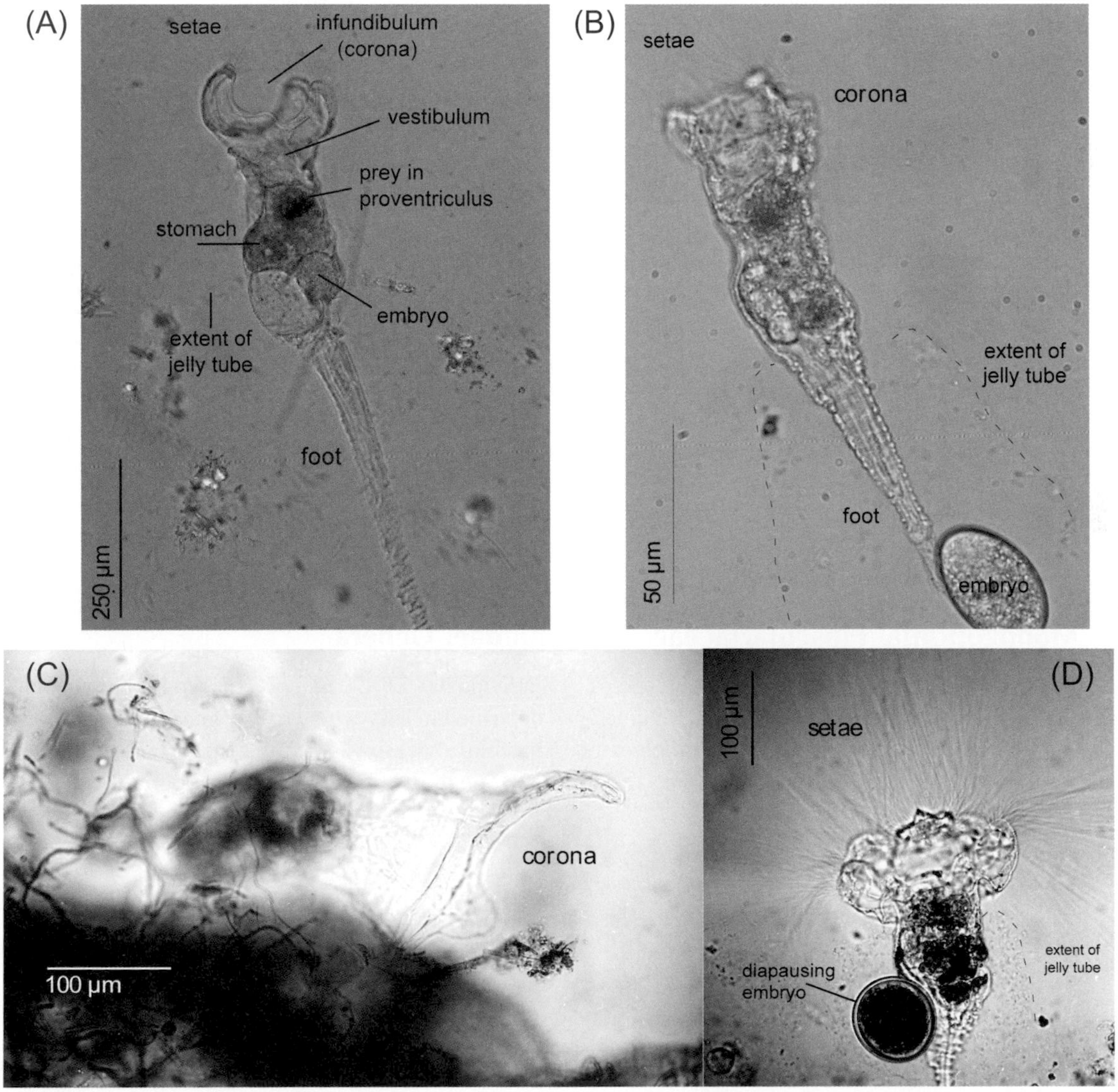

FIGURE 8.17 Representatives of genus *Collotheca*. (A) *Collotheca trilobata*, a sessile species; (B) *Collotheca libera* with one embryo, a planktonic species; (C) *Collotheca ferox*, a sessile species; (D) *Collotheca* sp., with one diapausing embryo. Scale bars as illustrated. (NEA)

Rotifera: Monogonata: Collothecacea: Atrochidae: Genera

[3 genera; all reported in NEA]

1 Coronal not large and bowl-shaped; sessile or mobile; trunk elongate; foot not a short stalk 2

1' Corona large and bowl-shaped, as wide as trunk of animal; trunk (body) as a slightly flattened sphere; mouth (infundibulum) positioned parallel to substratum; (≤1 mm); sessile, attached to substratum by a short, flexible centrally positioned, ventral foot; (Fig. 8.18) .... *Cupelopagis vorax* (Leidy, 1857)

2(1) Corona with a long dorsal lobe; transparent gelatinous tube; foot relatively short; ≤1500 µm; sessile, often present in colonies of *Sinantherina socialis* (Flosculariidae) on whose young it feeds (Fig. 8.19) .... *Acyclus inquietus* Leidy, 1882

[An undescribed congener, reported from Thailand (Meksuwan et al., 2013), should be explored further.]

2' Corona lacking the long dorsal lobe, but possessing horny process on dorsal side; tube lacking; foot short and hemispherical in shape; 1200–1500 µm; mobile. (Fig. 8.20) .... *Atrochus tentaculatus* Wierzejski, 1893

Rotifera: Monogononta: Flosculariaceae: Families

Important keys and other taxonomic papers on the Flosculariaceae include the following: (Edmondson, 1949; Segers, 1997; Segers & Wallace, 2001; Segers & Shiel, 2008; Segers et al., 2010). The monospecific genus *Pentatrocha* is not included herein, as it has not been reported from the Nearctic (Segers & Shiel, 2008; Meksuwan et al., 2011).

1 Body lacking hollow, arm-like, setose appendages; planktonic or sessile 2

1' Conical body with 6 arm-like, setose appendages that are outgrowths of the body wall; appendages inserted with powerful swimming muscles (Fig. 8.21); ~100–400 µm; exclusively planktonic. [1 genus; 11 NEA of 21 species] .... Hexarthridae, one genus: *Hexarthra* spp.

[Although illoricate, *Hexarthra* preserve well in formalin without serious contraction.]

2(1) Planktonic or not; solitary or colonial; lacking signature trophi described below and conical body of the Conochilidae 3

2' Planktonic, conical, illoricate animals within a loose gelatinous matrix; gut is U-shaped with anus exiting above the jelly margin when the animal is extended; foot unsegmented; trophi unique possessing asymmetrical unci, right unci teeth with intercalating projections; proximal projection of right manubrium well-developed, connecting to both ramus and proximal tooth of uncus plate. .... **Conochilidae [p. 142]**

3(2) Rotifers lacking a large, circular to lobate, auriciform corona 4

3' Rotifers typically with elongate bodies and large, circular (slightly elliptical) to lobate, cordiform, or auriciform corona; solitary or colonial; with or without a tube or gelatinous matrix; mostly sessile, but some free swimming. .... **Flosculariidae [p. 144]**

4(3) Without movable anterior spines (bristles) below the corona; body spherically-shaped (a round ball), with a corona as a circular band of cilia around the equator or towards one end or slightly elongate (oviform), with corona a slight extension of the body; **OR** with two movable anterior spines (bristles) below the corona, of varying lengths often much longer than the body and one (rarely two) rigid caudal spines; foot absent. .... **Trochosphaeridae [p. 146]**

4' Rotifers never spherical in shape; without long spines (bristles) below the corona. .... **Testudinellidae [p. 147]**

Rotifera: Monogononta: Flosculariaceae: Conochilidae: Genera

This family (Segers & Wallace, 2001) comprises <12 species forming autorecruitive colonies (Chapter 13, Volume I) of an adult and 1–3 recently hatched young or colonies of 5 to 100 individuals; occasionally enormous colonies (>400 individuals) of *C. unicornis* occur. Subitaneous eggs are deposited in jelly where they remain for a short period before hatching. Meksuwan et al. (2015) discusses the phylogenetic position of the family. [2 genera; both in NEA]

1 Corona horseshoe-shaped, possessing a ventral gap in the ciliation 2

1' Coronal other; colonies of 30–60 adults and juveniles; colonies ~1 mm in diameter; (in NEA only reported from rice paddies in USA (LA) ... . *Conochilopsis causeyae* (Vidrine, McLaughlin & Willis, 1985)

[Consult Vidrine et al., 1985; Gilbert & Burns, 2000; Segers & Wallace, 2001 for additional information.]

2(1) Corona slanted ventrally to body axis; antennae located inside corona field; diapausing egg smooth; colonies more or less spherical. (Fig. 8.22) [2 NEA of 2 species] .... *Conochilus* (Subgenus *Conochilus*) spp.

[Two important species are recognized *C. hippocrepis* Schrank, 1803 with ≤100 animals (2–4 mm in diameter) and *C. unicornis* Rousselet, 1892 with 5–35 animals (500–600 µm in diameter); however, enormous colonies of >400 individuals (>4 mm in diameter) have been reported from deep lakes (Wallace et al., 2015).]

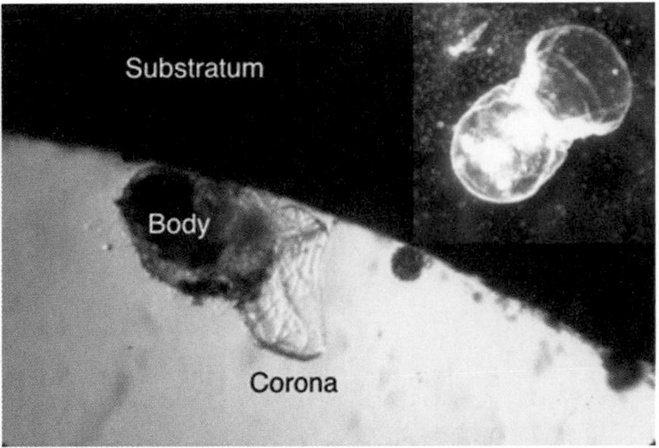

FIGURE 8.18 *Cupelopagis vorax*. Side view: animal attached to the undersurface of a hydrophyte; insert: from above, animal attached in a Petri dish. Body, without corona ~250 μm. (NEA)

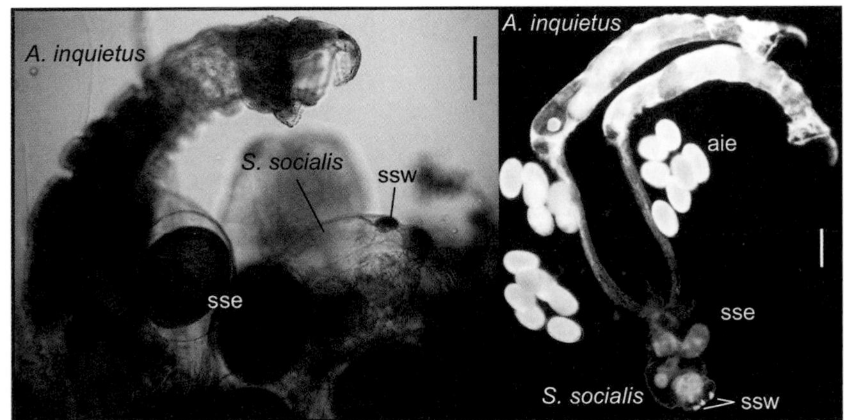

FIGURE 8.19 *Acyclus inquietus*. Animal within a colony of *S. socialis* (left); two adult *Acyclus* attached to a bit of substrate along with one *S. socialis* (right). Abbreviations: sse = *S. socialis* embryos; ssw = *S. socialis* warts; aie = *A. inquietus* embryos. Bars ~100 μm. (NEA) *Photomicrographs courtesy of Adele Hochberg, University Massachusetts at Lowell.*

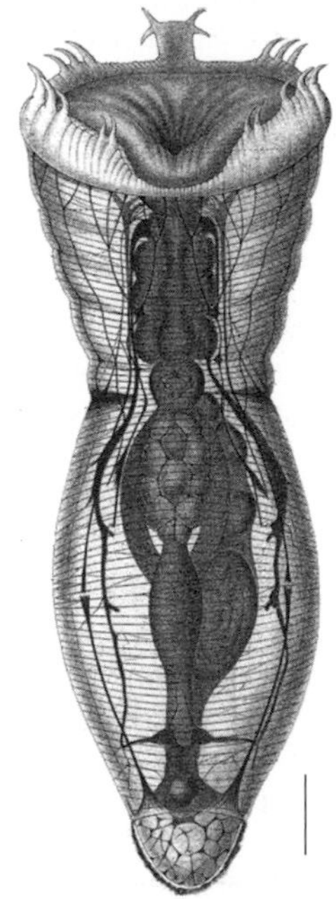

FIGURE 8.20 *Atrochus tentaculus*. Ventral view. Bar = 100 μm. *After Wierzejski, 1893.*

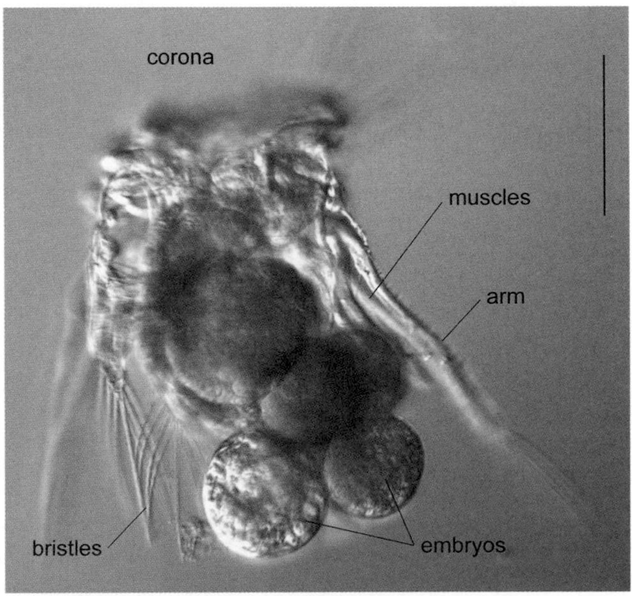

FIGURE 8.21 *Hexarthra*. Bar = 100 μm. (NEA)

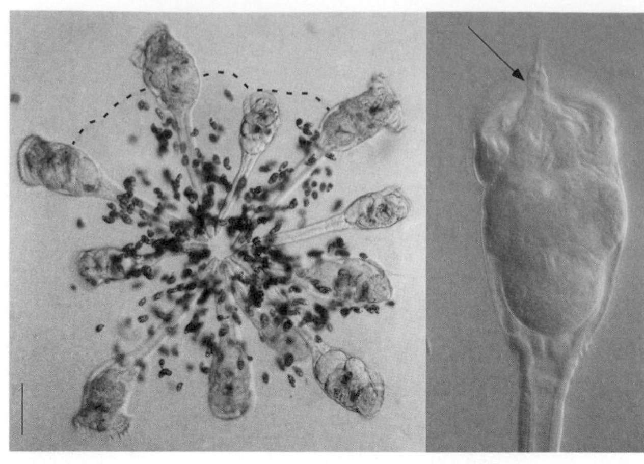

FIGURE 8.22 *Conochilus.* Small colony of *C. unicornis* (left). Dashed lines partially outline the margin of the gelatinous matrix. Dark ovoid objects are euglenoid algae colonizing the gelatinous matrix. Individual *C. unicornis* (right). Arrow = fused antennae within the corona. Bar = 100 µm. (NEA)

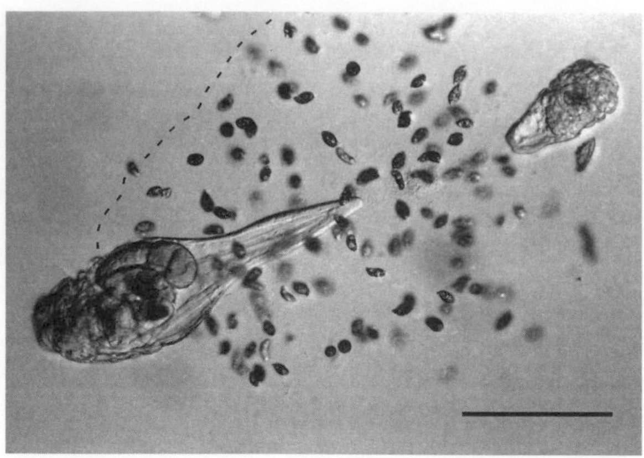

FIGURE 8.23 *Conochiloides*: an adult with one juvenile. Dashed lines partially outline the margin of the gelatinous matrix. Dark ovoid objects are euglenoid algae colonizing the gelatinous matrix. Bar = 250 µm. (NEA)

2' Corona perpendicular to body axis; antennae located dorsally outside of the coronal field; diapausing eggs are ornamented with spirals; colonies not spherical, typically an adult with 0–3 juveniles; colony's gelatinous matrix usually colonized by euglenoid algae (Fig. 8.23) [4 NEA of 4 species] .... *Conochilus* (Subgenus *Conochiloides*) spp.

[Genus *Conochiloides* has been subsumed within *Conochilus* (Ruttner-Kolisko, 1974; Koste, 1978; Segers & Wallace, 2001).]

Rotifera: Monogononta: Flosculariaceae: Flosculariidae: Genera

[9 genera; 8 reported in NEA]

1 Dorsal antennae (paired), not conspicuously long but not visible when animal contracts 2

1' Single, centrally placed, dorsal antenna, conspicuously long (longer than body width), visible when animal is contracted, ≤500 µm. (Fig. 8.24) [monospecific] .... *Beauchampia crucigera* (Dutrochet, 1812)

2(1) Corona with one or more pairs of distinct ear-like lateral lobes 3

2' Corona without distinct ear-like lobes (corona circular, subcircular, or at most, lobes indistinct to cordiform) 6

3 Corona with more than one pair of distinct lateral lobes on each side of the animal; tube a gelatinous matrix with or without pellets or debris 4

3' Corona with one pair of distinct lateral lobes on each side of the animal; animals usually solitary (occasionally small colonies, usually ≤12) in separate straight or slightly curved tubes, each formed as a distinct pipe of hardened secretions (Fig. 8.25); ≤1000 µm. [2 species with ringed tubes; 3 species with tubes as a granular matrix-(stucco-like)]; 3 NEA of 5 species] .... *Limnias* spp.

4(3) Corona very large with more than 4 lobes 5

4' Corona large with two pairs of distinct lateral lobes (= 4 lobes), one pair above the other; animals often colonial (Fig. 8.26); ≤1500 µm. [7 NEA of 9 species] .... *Floscularia* spp.

[Some species make pellets employing a specialized organ (modulus) located near the mouth on the ventral side; these are used to construct a pellet tube in which they reside (e.g., *F. conifera, F. ringens*) (Edmondson, 1945; Fontaneto et al., 2003). One species (*F. janus*) embeds fecal pellets in a gelatinous tube (cf. *Ptygura*); in *F. melicerta* the modulus produces a gelatinous tube in layers without pellets, but debris may accumulate within the jelly. Consult Segers (1997) for additional information.]

5(4) Corona with four pairs of distinct lateral lobes (= 8 lobes), one pair smaller; solitary in a gelatinous tube (Fig. 8.27); 200–2000 µm. [monospecific] .... *Octotrocha speciosa* Thorpe, 1893

[While the report from the Nearctic by Edmondson (1959) was called into question by Segers et al. (2010), specimens of this species have been observed in samples from Wisconsin (E.J. Walsh, personal observation; Fig. 8.27) and by Sarma & Elías-Gutiérrez (1998) from Mexico. For comments and additional illustrations, consult Segers & Shiel (2008), Segers et al. (2010), Meksuwan et al. (2011).]

5' Corona with 5 promanent and 2 small lobes; solitary or colonial in a gelatinous tube; ≤1300 µm. Consult (Meksuwan et al., 2011) for figures and additional details. [monospecific] .... *Lacinularoides coloniensis* (Colledge, 1918)

6(2) Corona cordiform (reniform), subquadrangular, or occasionally round; animals usually forming spherical to ellipsoid colonies with many individuals, occasionally solitary; tube, when present, a continuous gelatinous matrix 7

FIGURE 8.24 *Beauchampia*. Bar = 100 µm. (NEA)

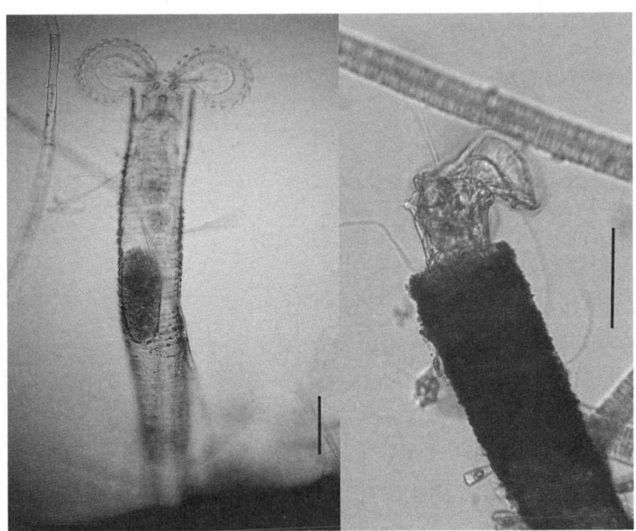

FIGURE 8.25 *Limnias*: *L. melicerta* (left) and *L. ceratophylli* (right). Bars = 100 µm. (NEA)

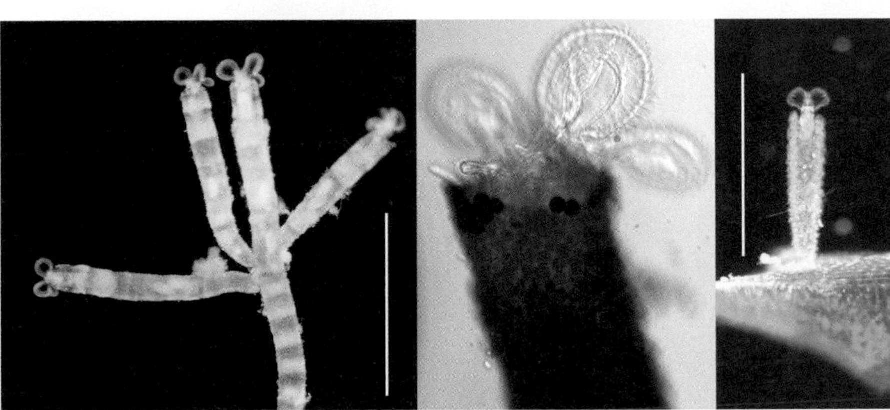

FIGURE 8.26 *Floscularia*: *F. conifera*, small colony (left); *F. ringens*, anterior (center); *F. janus* (right). Bars = 100 µm. (NEA) *Center photomicrograph courtesy of E.J. Walsh, University of Texas at El Paso.*

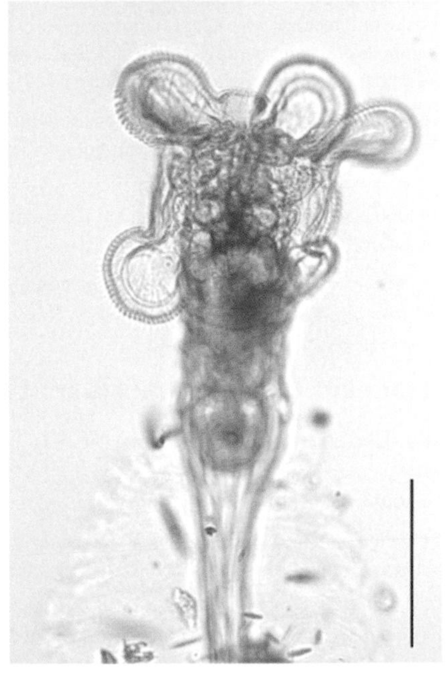

FIGURE 8.27 *Octotrocha speciosa*. Solitary individual in a gelatinous tube. Bar = 250 µm. (NEA) *Photomicrograph courtesy of E.J. Walsh, University of Texas at El Paso.*

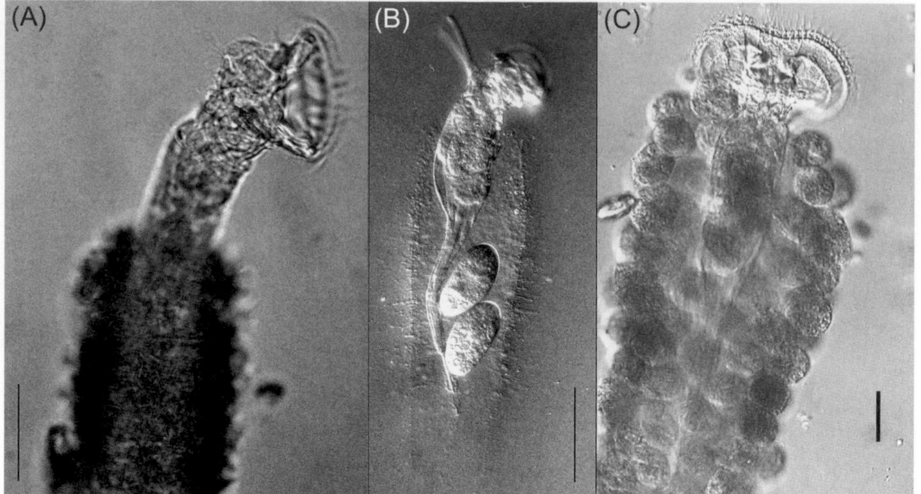

FIGURE 8.28 *Ptygura*: *P. beauchampi* (sessile), *P. libera* (planktonic, with two embryos), and *P. pilula* (sessile, with fecal pellets embedded in a gelatinous tube.) Bars = 50 μm. (NEA)

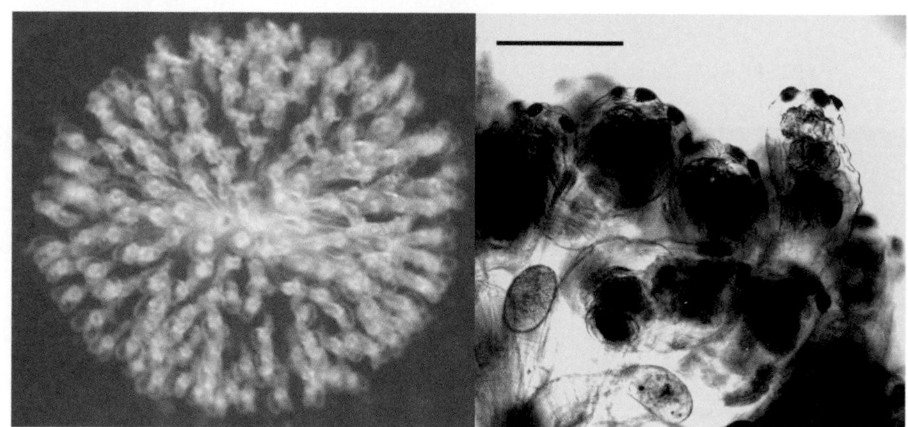

FIGURE 8.29 *Sinantherina*. Planktonic colony (left) ≥1 mm diameter and *S. socialis* (right). Bar = 200 μm. (NEA)

6' Corona round to slightly elliptical, or weakly-lobed; animals usually solitary in separate, gelatinous tubes that may be covered by or infiltrated with debris (Fig. 8.28); 150–1000 μm. [19 NEA of 31 species] .... *Ptygura* spp.

[One common species (*P. pilula*) embeds ovoid, fecal pellets in its gelatinous tube (cf. *Floscularia janus*); another, *P. (Floscularia) noodti*, (not been reported from the Nearctic), embeds elongate (vermiform) fecal pellets within its gelatinous tube. Some species of *Ptygura* may form intra- or interspecific colonies with other sessile Flosculariidae (Wallace, 1987).]

7(6) Possessing a specialized egg bearing structure (oviferon, ovifer) below anus to which eggs attach; solitary or colonial; sessile or planktonic; gelatinous matrix is absent (minimal?); individuals: 400–2000 μm, colonies: ≤4 mm. (Fig. 8.29) [4 NEA of 5 species] .... *Sinantherina* spp.

[*Pentatrocha gigantea* Segers & Shiel, 2008 also possesses an oviferon, but it has not been identified from the Nearctic. However, Segers & Shiel (2008) suggest potential confusion of this species with *Octotrocha speciosa* (see above). Both genera deserve additional study.]

7' Lacking an oviferon; usually colonial; sessile or planktonic; possessing gelatinous matrix; individuals: ≤2000 μm, colonies: ≤5 mm. (Fig. 8.30) [2 NEA of 6 species] .... *Lacinularia* spp.

Rotifera: Monogononta: Flosculariaceae: Trochosphaeridae: Genera

Consult Nogrady & Segers (2002) for details. [3 genera; all reported in NEA]

1 Body spherical-shaped (a round ball) or slightly elongate (oviform) 2

1' Rotifers not as above; possessing two movable anterior setae (bristles or spines) of varying lengths below the corona, often much longer than the body and one (rarely two) rigid caudal spines; foot absent; (Fig. 8.31); excluding setae ~40–325 μm. [8 NEA of 15 species] .... *Filinia* spp.

[*Filinia* (formerly Family Filiniidae) has been subsumed within Family Trochosphaeridae (Nogrady and Segers, 2002).]

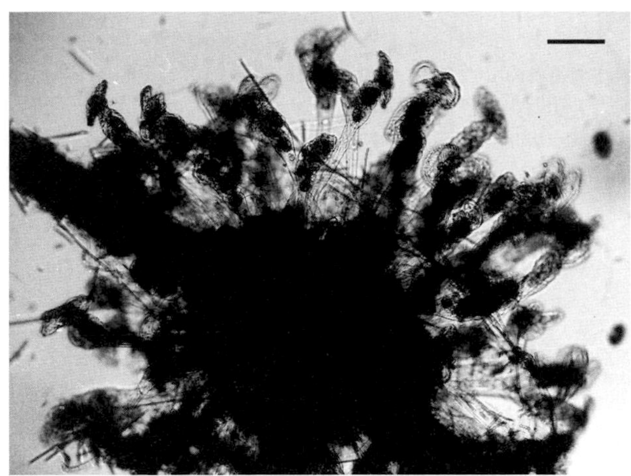

FIGURE 8.30 *Lacinularia*. Bar = 250 µm. (NEA)

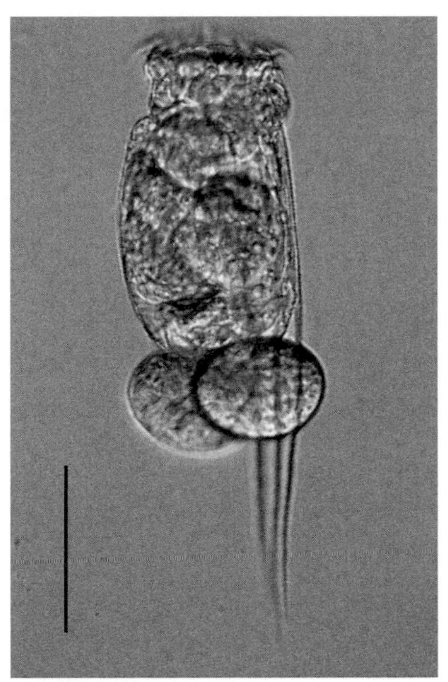

FIGURE 8.31 *Filinia*. Bar = 100 µm. *Photomicrograph courtesy of M.V. Sørensen, University of Copenhagen.*

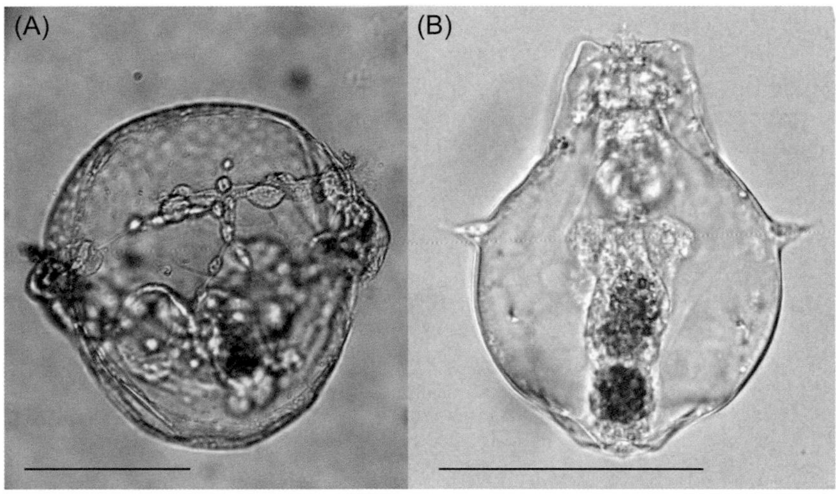

FIGURE 8.32 Family Trochosphaeridae. (A) *Trochosphaera*. (B) *Horaella*. Bars = 100 µm. (NEA) *Photomicrographs courtesy of S.S.S. Sarma, Universidad Naćional Autónoma de México Campus Iztacala, México.*

2(1) Body spherically-shaped, corona as a circular band of cilia around the equator or towards one end; 320–1100 µm; viviparous; (Fig. 8.32 A). [2 NEA of 2 species] .... *Trochosphaera* spp.

2' Body not spherically-shaped, slightly elongate (oviform); corona a slight extension of the body; ~125 × 160 µm; oviparous; (Fig. 8.32 B). [1 NEA of 2 species] .... *Horaella thomassoni* Koste, 1973

Rotifera: Monogononta: Flosculariaceae: Testudinellidae: Genera

[3 genera, 2 in NEA]

1 Lorica greatly flattened dorsoventrally with dorsal and ventral plates fused along the lateral margin (Fig. 8.33 A); foot annulated and retractile; ≤250 µm. [18 NEA of 38 species] .... *Testudinella* spp.

1' Lorica not greatly flattened, rather possess 4 nearly equal lobes in cross section; 1 pair frontal eyespots; foot absent; (Fig. 8.33 B); ≤120 µm. [3 NEA of 3 species] .... *Pompholyx* spp.

PHYLUM ROTIFERA

Rotifera: Monogononta: Ploima: Families

This key does not include two familes: family Clariaidae (*Claria*), which comprises a single species of endoparasitic rotifers of terrestrial Oligochaeta (Segers, 2008), and the monospecific Cotylegaleatidae (*Cotylegaleata*) (De Smet, 2007). Neither has been reported from the Nearctic. Also, the separation of two curious congenerics (*Pourriotia*, formerly *Proales*, Proalidae, currently Notommatidae), may be made based on the fact that both are parasitic in galls of the algal genus *Vaucheria*.

1 Rotifers with forcipate (Fig. 13.11 j in Volume I) or modified (protrusible) virgate trophi 2

1' Rotifers with trophi other than above 4

2(1) Not inhabiting shells of live testate amoeba 3

2' Inhabiting shells of live testate amoeba (genus *Diffugia*); hemiforcipate trophi described by De Smet (2006). [1 genus; 2 NEA of 3 species] .... Asciaporrectidae, one genus: *Asciaporrecta* spp.

3(2) Stomach filled with zoochlorellae; gastric glands absent, but with cecae directed towards anterior end; trophi not protrusible. Consult (De Smet & Pourriot, 1997) for details. [1 genus; 4 NEA of 6 species] .... Ituridae, one genus: *Itura* spp.

3' Not as above .... **Dicranophoridae [p. 150]**

4(1) Mostly littoral rotifers possessing cardate trophi (Fig. 13.11 u in Volume I) having a sucking action or trophi highly modified and stomach with zoochlorellae 5

4' Rotifers with trophi other than cardate 6

5(4) Manubria of the trophi with hooks that may be determined in lateral view of the animal (Fig. 13.11 u in Volume I), body fusiform, 300–500 μm. (Fig. 8.34). Consult Nogrady and Segers (2002) for details. [1 genus; 11 NEA of 13 species] .... Lindiidae, one genus: *Lindia* spp.

5' Highly modified trophi, possessing pseudunci, stomach with zoochlorellae, gastric glands absent, a rare littoral, ~275 μm. (Fig. 8.35) [monospecific] .... Birgeidae, one species: *Birgea enantia* Harring & Myers, 1922

6(4) Rotifers not possessing incudate or modified incudate trophi 7

6' Illoricate, saccate rotifers, with incudate or modified incudate trophi, (Fig. 13.11 g, r in volume I) .... **Asplanchnidae [p. 152]**

7(6) Rotifers possessing malleate trophi 8

[In malleate trophi (Fig. 13.11 a–c, e, f, p in volume I), each uncus has only 4–7 teeth (unlike the condition found in malleoramate trophi). The fulcrum may be short (malleate) or long (submalleate)]

7' Rotifers possessing virgate trophi 15

[Virgate trophi are often asymmetrical (Fig. 13.11 d, k–o, q, s in Volume I), with a long fulcrum and manubria and generally small rami.]

8(7) Loricate rotifers: body wall thickened and firm 9

8' Illoricate rotifers: body wall not thickened or firm 13

[Interpreting whether the body wall is thickened (loricate) or not (illoricate) can be difficult, and it does take some experience to judge this characteristic in forms that do not easily fall one way or the other. To get an indication as to how firm the lorica is, follow this procedure. When working with fresh material, apply gentle pressure from the point of a pencil or probe onto the cover glass while observing the specimen. If the body puffs out under pressure and returns to its original shape when the pressure is released, the specimen is illoricate. Loricate forms will exhibit much less flexibility of the body wall: i.e., they do not puff out. NB: If too much pressure is applied to the cover glass, the animal may be crushed, becoming irreversibly damaged, perhaps rendering them unidentifiable. In preserved materials, illoricate forms tend to shrivel up, while the body wall of loricate rotifers will retain its shape even if the inner organs separate from the body wall, collapsing into a central mass of tissue. Some members of the Brachionidae that are weakly loricate will key to the Epiphanidae.]

9(8) Lorica possessing furrows, grooves, or sulci; or with a dorsal, semicircular head shield covering the corona; or with a very strongly developed lorica and a dorsal transverse ridge 10

9' Lorica lacking furrows, grooves, sulci, or dorsal head shield; without a strongly developed lorica and dorsal transverse ridge 14

10(9) Lorica lacking medial, ventral furrow or notch, and lacking a head shield 11

10' Lorica with medial, ventral furrow (but no lateral furrow or grooves) extending the full length of the animal or with a ventral notch in which the foot lies, or with a dorsal, semicircular head shield covering the corona (includes former Colurellidae) .... **Lepadellidae [p. 152]**

11(10) Lorica not as below 12

11' Lorica with dorsal, medial sulcus (double keel) or a single dorsal keel .... **Mytilinidae [p. 154]**

12(11) Foot usually projecting from between dorsal and ventral plates at the posterior end of the lorica; dorsal and ventral plates separated by a deep furrow or groove .... **Euchlanidae [p. 154]**

12' Foot projecting through a foramen (hole) in ventral plate at the posterior end of the lorica; dorsal and ventral plates connected by a weak furrow (groove); ≤200 μm (excluding toes) (Fig. 8.36) [1 genus, 105 NEA of ~200 species] .... Lecanidae, one genus: *Lecane* spp.

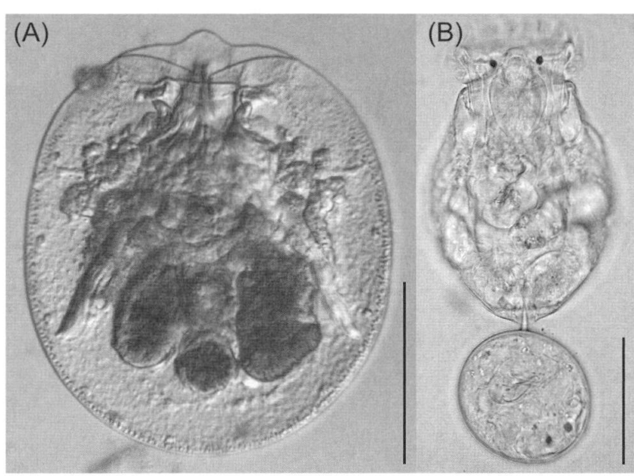

FIGURE 8.33 Family Testudinellidae: (A) *Testudinella* and (B) *Pompholyx*. Bars = 50 µm. (NEA) *Photomicrograph of* Pompholyx *courtesy of C. Jersabek, University of Salzburg, Austria.*

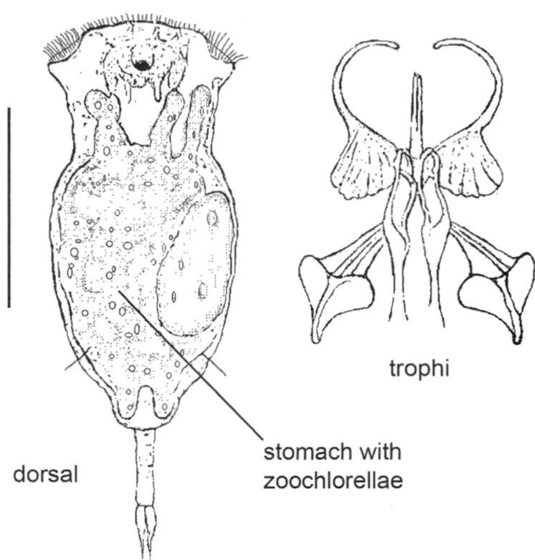

FIGURE 8.35 Family Birgeidae: *Birgea*. Whole animal (left); trophy (right). Bar = 100 µm. (NEA) *After Harring & Myers, 1922.*

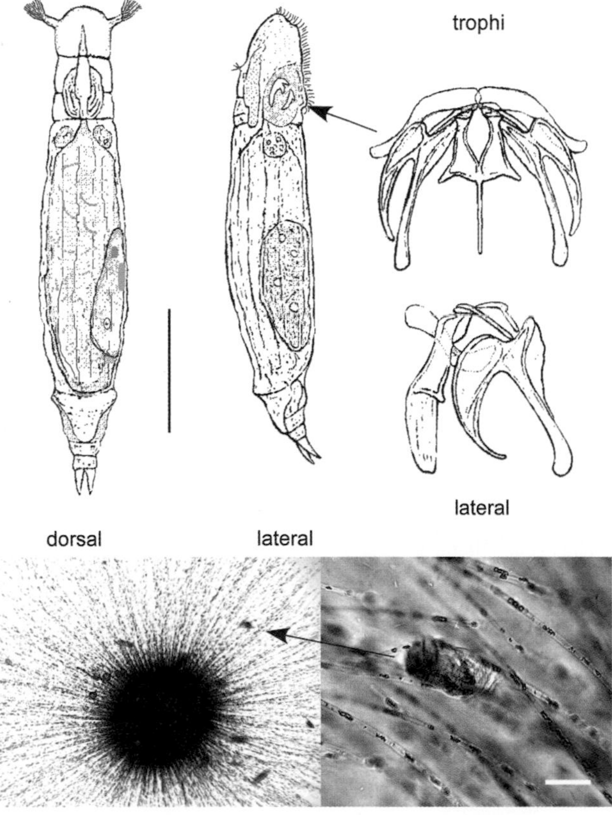

FIGURE 8.34 Family Lindiidae: *Lindia*. Top panel: left, whole animal; right, trophy. (NEA) (After Harring & Myers, 1922.) Lower panel: left, animals in a colony of the bluegreen bacteria, *Gloeotrichia*; right, close up. Bar ~100 µm

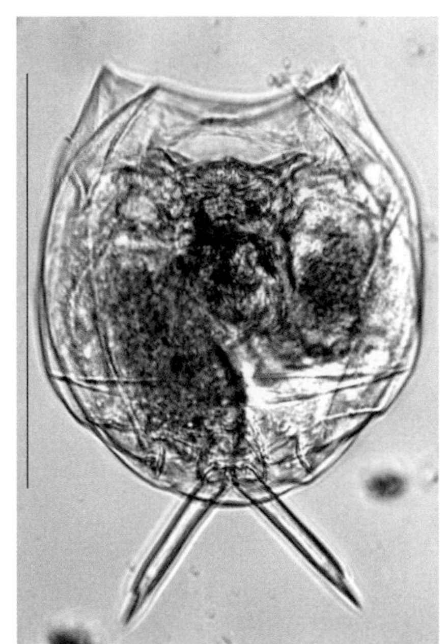

FIGURE 8.36 Family Lecanidae: *Lecane*. Bar = 100 µm. (NEA)

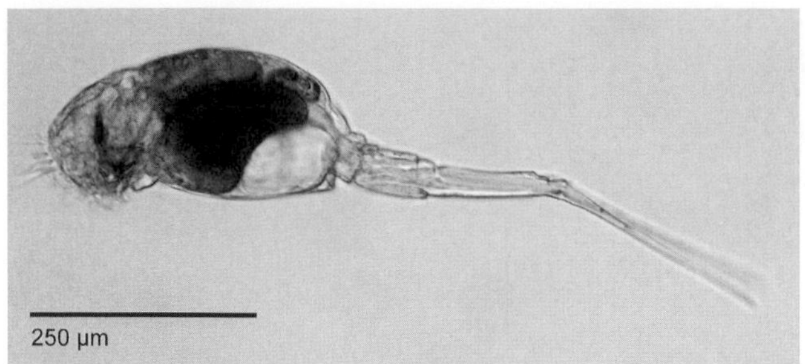

FIGURE 8.37 Family Scaridiidae: *Scaridium*. Bar=250 μm. *Photomicrograph courtesy of M.V. Sørensen, University of Copenhagen.*

[The genera *Hemimonostyla* (toes partially fused) and *Monostyla* (toes completely fused) has been subsumed within *Lecane* (with two toes that may be partially fused at the base) (Koste, 1978; Segers, 1995a). Use of these older names should be abandoned.]

13(8,) Mouth set in a funnel-shaped buccal field .... **Epiphanidae [p. 156]**

13' Mouth set in an oblique, ciliated field on ventral side; body swollen, vermiform, or fusiform .... **Proalidae [p. 156]**

14(9,) Lorica extending beyond body to head, foot, and toes .... **Trichotriidae [p. 158]**

14' Lorica not extending beyond body .... **Brachionidae [p. 158]**

15(7,) Body not twisted as a partial helix and/or trophi not asymmetrical 16

15' Body twisted as a partial helix (asymmetrical) and/or trophi asymmetrical **OR** small, saccate animals, present within the colonial alga *Volvox* where they feed from the inside on the algal cells. .... **Trichocercidae [p. 160]**

16(15) Not as below 17

16' Unci tips pointed outwards, capable of protruding through mouth; body elongate; head with two pairs of lateral lobes; corona with stiff, with preoral setae; foot with prominent muscles, foot and toes longer then rest of body; ≤450 μm. (Fig. 8.37). Consult Nogrady et al. (1995) for details. [1 genus, 3 NEA of 7 species] .... Scaridiidae, one genus: *Scaridium* spp.

17(16) Not as below 18

17' Body spindle-shaped; long, knobbed, lateral antennae located near the base of the foot and with long setae; stomach separated from intestine by a circle of bulbous glands; ≤1000 μm (acid waters with a pH < 6.0). (Fig. 8.38) [monospecific] .... Tetrasiphonidae, one species: *Tetrasiphon hydrocora* Ehrenberg, 1840

18(17) Not as below 19

18' Purple plates positioned anterior to the mastax; corona wide, flat, and somewhat circular; foot jointed and long, about 1/2 of the total length of the animal, 1 toe; uncommon, mostly littoral, may be present in plankton samples); consult Nogrady & Segers (2002) for details, 100–200 μm. (Fig. 8.39) [monospecific] .... Microcodidae, one species: *Microcodon clavus* Ehrenberg, 1830

19(18) Not as below 20

19' Small, ovate to saccate; often colored and somewhat laterally compressed; planktonic; consult Nogrady and Segers (2002) for details. .... **Gastropodidae [p. 160]**

20(19) Illoricate; corona as a circumapical ciliated band with conspicuous auriciform structures (auricles) possessing swimming cilia; apical field with four prominent setae (sensory bristles), foot rudimentary or present (retractable); **OR** illoricate; corona flat with a circumapical band of swimming cilia; possessing four groups of movable, flattened, xiphoid to plumiliform appendages (blades, paddles, fins); foot absent; **OR** loricate, sculptured having ridges, grooves, or areolations; corona a simple ciliated band with two digitiform appendages; foot annulated with 2 toes; **OR** loricate, covered by circular plaques; corona with protuberances next to mouth; foot with two sections. Consult Nogrady & Segers (2002) for details .... **Synchaetidae [p. 162]**

20' Animals not as above .... **Notommatidae [p. 162]**

Rotifera: Monogononta: Ploima: Dicranophoridae: Genera

Consult De Smet & Pourriot (1997) for details. The monospecific genus *Glaciera*, which has only been reported from high altitude lakes in the Austrian Alps, is not included in this key, but we recommend additional survey work in other high altitude lakes. According to De Smet (2003, 2006) "… similarities of several … features of the trophi of Asciaporrectidae, Dicranophoridae, and Ituridae suggest a sister group relationship." [19 genera; 17 reported in NEA]

1 Trophi forcipate, rami not bent as below; toes not laterally compressed 2

1' Trophi modified virgate with rami bent dorsally at 90° midway along their length (partly protrusible and adapted for grasping); epipharynx present; two laterally compressed toes; corona at an oblique angle to body; eyespot absent; illoricate; aquatic mosses and rocks [monospecific] .... *Dorria dalecarlica* Myers, 1933

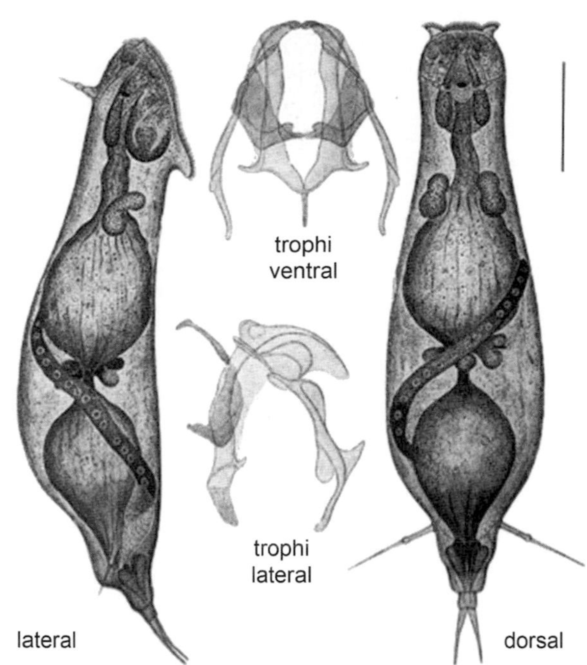

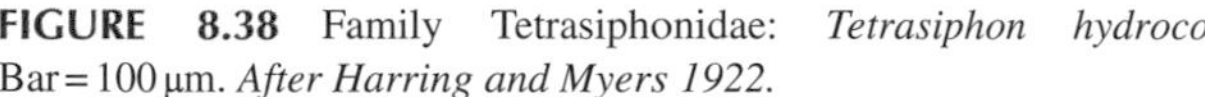

FIGURE 8.38 Family Tetrasiphonidae: *Tetrasiphon hydrocora.* Bar = 100 µm. *After Harring and Myers 1922.*

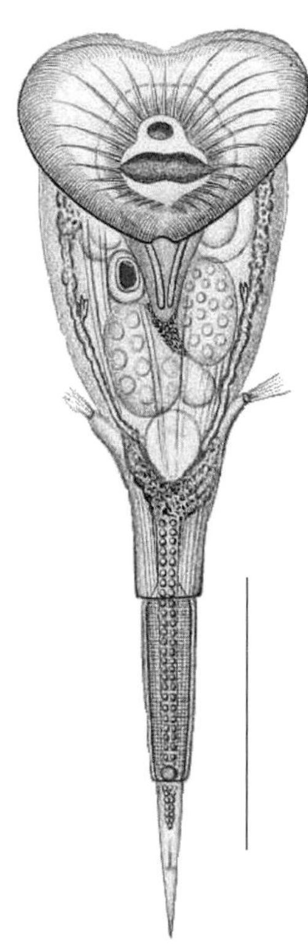

FIGURE 8.39 Family Microcodidae: *Microcodon clavus.* Bar = 50 µm. *After Weber 1898.*

[This species has been found among the moss *Fontinalis dalecarlica* and in mountain streams. Additional survey work is warranted.]

2(1) Corona present although sometimes reduced; mostly free-living, but some epizootic or endoparasitic of oligochaetes and slugs 3

2' Corona absent; parasitic (endoparasitic) in terrestial oligochaetes, only occasionally free of their host. [4 species, not reported from the Nearctic] .... *Balatro* spp.

[Additional survey work is warranted.]

3(2) Trophi symmetrical or slightly asymmetrical; teeth on rami often unpaired; unci composed of tooth and shaft 4

3' Trophi symmetrical or if asymmetrical; unci T-shaped or with 3 branches 16

4(3) Incus not Y-shaped; all elements of trophi not rod-like 5

4' Trophi long and slender; incus Y-shaped; rami and fulcrum long and straight; all elements of trophi as long slender rods [monospecific] .... *Streptognatha lepta* Harring & Myers, 1928

5(4) Ramus basal and subbasal chambers not positioned as below; fulcrum plate-shaped somewhat rod shape 6

5' Ramus with inner subbasal chamber, with basal chambers situated laterally on external edge of subbasal ones; fulcrum a small triangle or elongated triangle [1 NEA of 4 species] .... *Dicranophoroides caudatus* (Ehrenberg, 1834)

6(5) Unci hinged at the posterior end to ramus and manubrium as a 3-way joint 7

6' Unci hinged midway or near tip to middle or tip of ramus, not as a 3-way joint 9

7(6) Unci not fused; body wall strongly annulated 8

7' Unci fused to rami at their posterior ends; body wall strongly annulated [1 NEA of 2 species] .... *Parencentrum plicatum* (Eyferth, 1878)

[*Parencentrum* are reported to be more abundant in cold waters. One species of *Encentrum* possessing an annulated body might be confused (see couplet 13')]

8(7) Midway on outer margin of rami possessing a posteriorly directed projection that ends in a joint connecting uncus and manubrium to ramus; 4 eyespots without pigment [2 NEA of 4 species] .... *Myersinella* spp.

8' Outer margin of rami lacking projections as above, unci and manubrium hinged to corners of rami and bent at right angles near their midpoint [3 NEA of 6 species] .... *Erignatha* spp.

9(6,) Trophi usually with intramalleus linking manubrium and uncus 10

9' Trophi usually lacking intramalleus 14

[A few *Encentrum* species lack the intramalleus; a few *Dicranophorus* species possess the Intramalleus.]

10(9,) Trunk spines lacking; intramalleus not as a strong inwardly projecting spine 11

10' Trunk possessing 3–4 pairs of lateral protuberances (hooks) on trunk; intramalleus as a strong spine projecting inwardly; ≤450 µm. (Fig. 8.40) [monospecific; rare in Mexico, Sarma, per. obs.] .... *Kostea wockei* (Koste, 1961)

11 Body not as below; foot long or short 12

11' Body a broad, dorsoventrally flattened oval; foot short; [1 species; not reported from the Nearctic] .... *Wigrella depressa* Wiszniewski, 1932

12(11) Body not as below; usually spindle or cylindrical in shape, manubria not curved as below 13

12' Body rounded, bulbous, or saccate and arched dorsally; manubria of trophi curved (C-shaped) [monospecific, Lake Baikal endemic] .... *Infatana pomazkovae* Kutikova, 1985

13(12) Long foot, >1/5 total body length, toes end in rounded tips [3 NEA of 8 species] .... *Wierzejskiella* spp.

13' Short foot, <1/5 total body length, toes end in pointed tips [22 NEA of ~100 species] .... *Encentrum* spp.

14(9) Body not pyriform and furrows lacking; teeth on preunci or subunci lacking; body not covered with detritus 15

14' Body pyriform with deep furrows both length and crosswise; teeth on preunci or subunci present; body wall often covered with detritus; sediment surface and the psammolittoral; apparently collected from the Colorado River (Jersabek & Leitner 2013) [monospecific] .... *Donneria sudzukii* (Donner, 1968)

[*Paradicranophorus* (6 species, not reported from the Nearctic) will key here.]

15(14) Large corona; row of teeth usually present on rami (rarely absent) (Fig. 8.10 j in Volume I); usually with 2 eyespots (frontal); toes long; not endoparasitic [38 NEA of 53 species] .... *Dicranophorus* spp.

15' Small corona; teeth absent on rami; toes minute or absent; eyespots lacking; usually endoparasitic in oligochaetes and slugs [2 NEA of 7 species] .... *Albertia* spp.

16(3) One ramus possessing a large, aliform (quadrate) alula; rudimentary pointed toes, developed as a continuous extension of the foot, lacking a distinct junction (i.e., a yoke-shaped tail appendage) [monospecific] .... *Pedipartia gracilis* (Myers, 1936)

16' Alula, if present, rounded, triangular, or hook-shaped; toes not as above (foot to toe junction present); carnivorous; psammic and among algae and mosses [13 NEA of 17 species] .... *Aspelta* spp.

Rotifera: Monogononta: Ploima: Asplanchnidae: Genera

Consult (Nogrady & Segers, 2002) for details. [3 genera; all in NEA]

1 Animals possessing a foot 2

1' Animals lacking a foot (1 species with a rudimentary foot); intestine and anus absent; 200–2000 µm. (Fig. 8.41). [7 NEA of 9 species] .... *Asplanchna* spp.

[*Asplanchna herrickii* de Guerne, 1888 does possess an undeveloped (vestigial) pedal gland and rudimentary toes.]

2(1) Animals with complete gut (anus present); 400–700 µm. (Fig. 8.42) [2 NEA of 2 species] .... *Harringia* spp.

2' Animals with incomplete gut (both intestine and anus absent). (Fig. 8.43) [2 NEA of 4 species] .... *Asplanchnopus* spp.

Rotifera: Monogononta: Ploima: Lepadellidae: Genera

A rare, monospecific genus (*Diplois*), which has been reported from *Sphagnum* bogs in the Palearctic will key here, if its lateral furrows are missed: see Euchlanidae. [4 genera; 4 in NEA]

1 Head shield poorly developed, retractile 2

1' Animals possessing a dorsal semicircular head shield covering the corona; 80–225 µm. (Fig. 8.44) [5 NEA of 9 species] .... *Squatinella* spp.

2(1) Lorica laterally flattened, possessing at least a medial ventral furrow; with or without a sulcus 3

2' Lorica dorsoventrally flattened, with a deep ventral notch from which the foot extends; ~55–200 µm. (Fig. 8.45) [40 NEA of ~120 species] .... *Lepadella* spp.

3(2) Lorica with a longitudinal dorsal sulcus; foot + toes > ½ total animal length; terminal foot segment longer than toes; ~100–160 µm. (Fig. 8.46 [2 NEA of 2 species] .... *Paracolurella* spp.

3' Lorica lacking dorsal sulcus (continuous); foot + toes < ½ total animal length; terminal foot segment longer than toes; ~50–120 µm. (Fig. 8.47) [17 NEA of 23 species] .... *Colurella* spp.

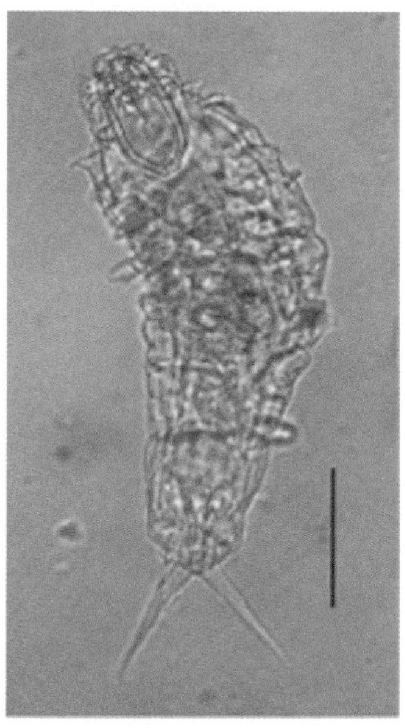

FIGURE 8.40 *Kostea wockei*. Bar ca. 100 µm. (NEA)

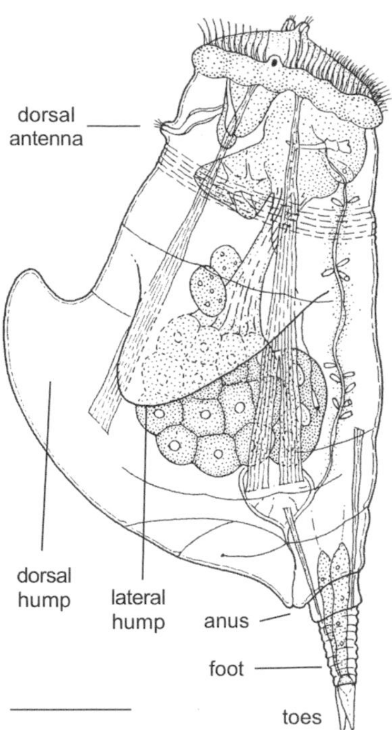

FIGURE 8.42 Family Asplanchnidae: *Harringia*, right side view. (Bar = 100 µm) (NEA) *After* American Museum Novitates *700, with permission.*

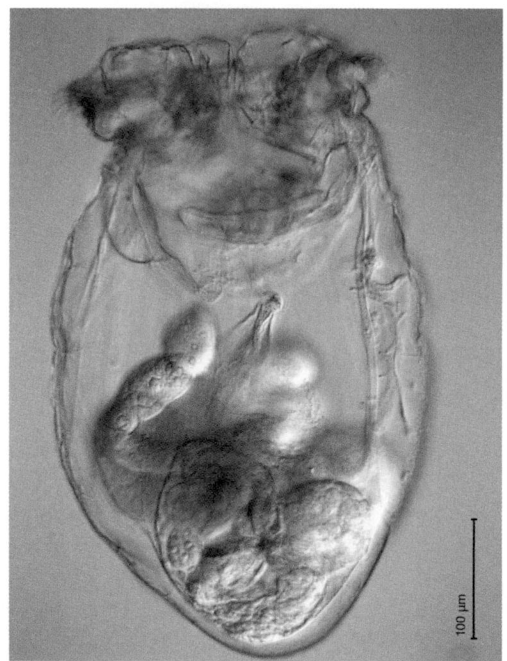

FIGURE 8.41 Family Asplanchnidae: *Asplanchna*. Bar = 100 µm. (NEA)

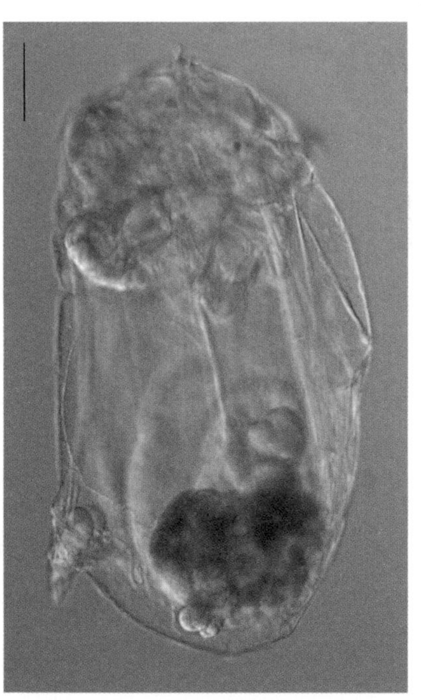

FIGURE 8.43 Family Asplanchnidae: *Asplanchnopus*. Bar = 100 µm. (NEA)

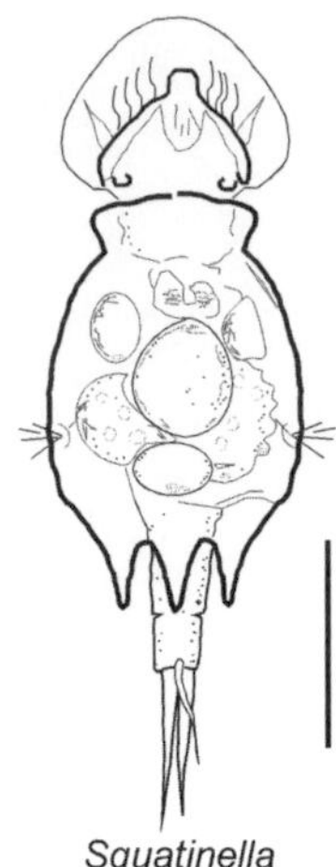

FIGURE 8.44 Family Lepadellidae: *Squatinella*. Schematic representation from life. Bar ~50 µm. (NEA)

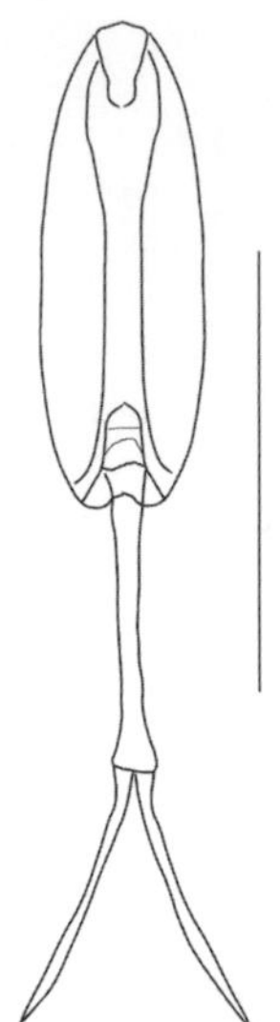

FIGURE 8.46 Family Lepadellidae: *Paracolurella*. Schematic representation from life (ventral view). Bar ~100 µm. (NEA)

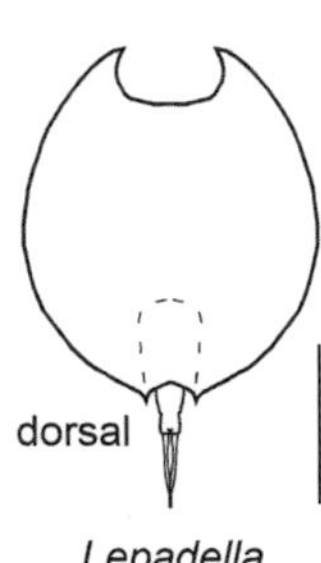

FIGURE 8.45 Family Lepadellidae: *Lepadella*. Schematic representation from life. Bar = ~50 µm. (NEA)

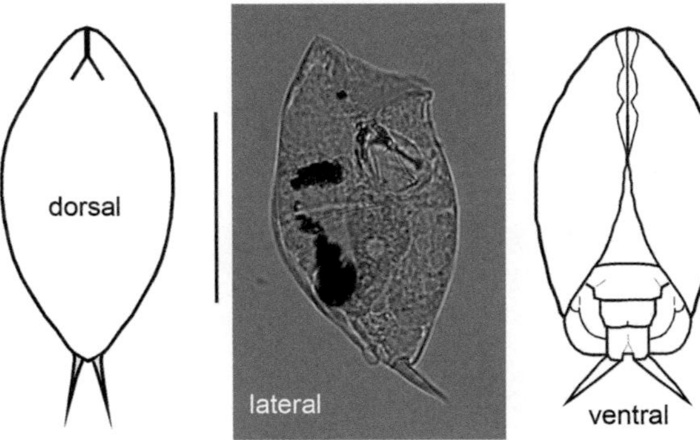

FIGURE 8.47 Family Lepadellidae: *Colurella*. Schematic representation from life. Bar ~100 µm. (NEA)

Rotifera: Monogononta: Ploima: Mytilinidae: Genera

[2 genera; both reported in NEA]

1 Animals with a dorsal longitudinal sulcus (in cross section appears double keeled); 170–380 µm. (Fig. 8.48) [7 NEA of ~15 species] .... *Mytilina* spp.

1' Animals with a dorsal ridge, but lacking dorsal double keel; 100–250 µm. (Fig. 8.49) [3 NEA of 11 species] .... *Lophocharis* spp.

Rotifera: Monogononta: Ploima: Euchlanidae: Genera

The monospecific genus, Pseudoeuchlanis Dhanapathi, 1978, from India is not covered here. [6 genera, 4 in NEA]

1 Lorica not pyriform without narrow neck, rather possessing 2–3 plates separated by grooves (sulci); toes not equal in length to body 2

1' Lorica pyriform with narrow neck, dorsal side swelling into a bulging hump, toes long, about equal or longer then the body, superficially resembling *Scaridium*, step 58), 420–760 µm. (Fig. 8.50) .... *Beauchampiella eudactylota* (Gosse, 1886)

2(1) Lorica of 2 plates, 1 dorsal and 1 ventral 3

2' Lorica of 3 plates, 1 ventral, 2 dorsal separated by sulci [monospecific; not recorded from the Nearctic] .... *Diplois daviesiae* Gosse, 1886

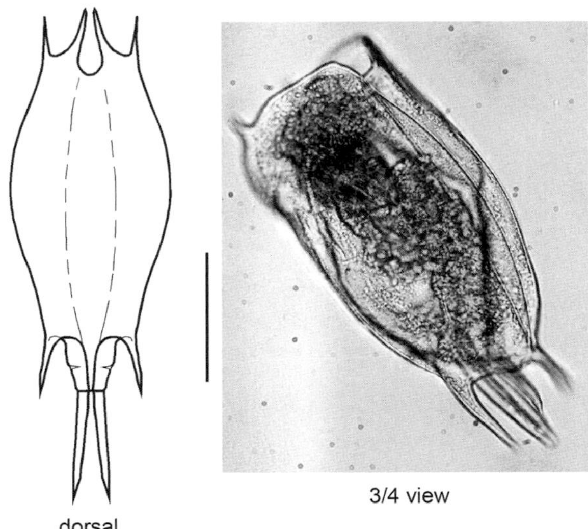

FIGURE 8.48 Family Mytilinidae: *Mytilina*. Bar = 100 µm. (NEA)

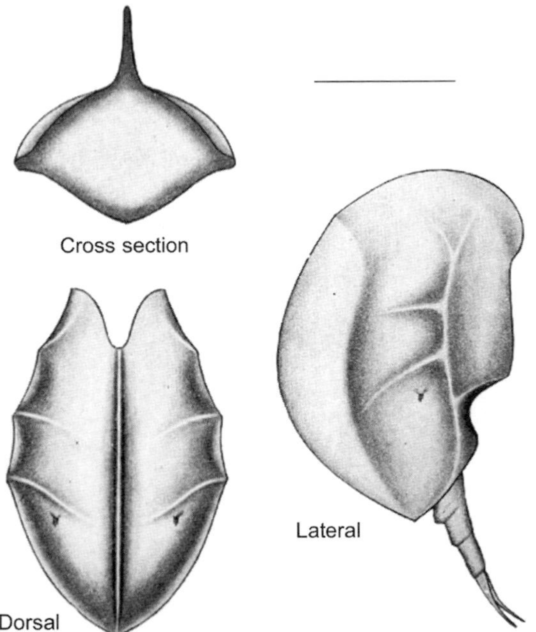

FIGURE 8.49 Family Mytilinidae: *Lophocharis*. Bar = 100 µm. *After Harring, 1916.*

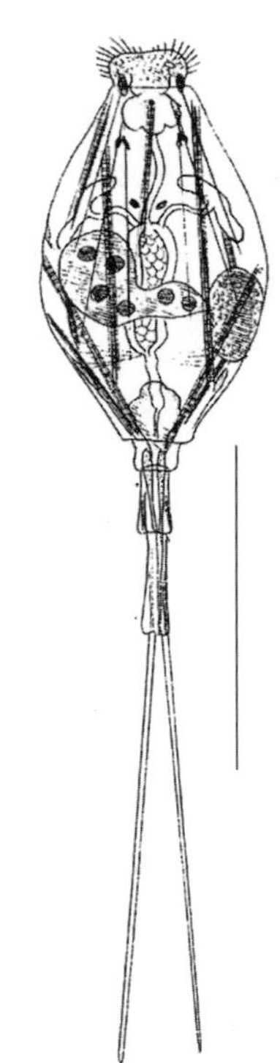

FIGURE 8.50 Family Euchlanidae: *Beauchampiella*. Bar ~250 µm. (NEA) *After Herrick, 1885.*

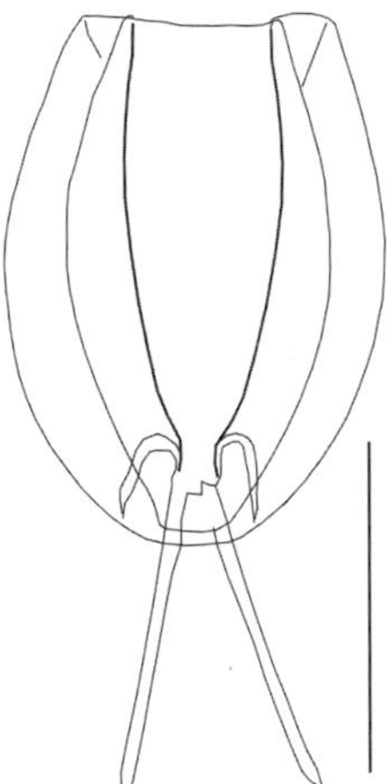

FIGURE 8.51 Family Euchlanidae: *Dipleuchlanis propatula* (dorsal). (Schematic representation from life.) Bar = 100 µm. (NEA)

[This taxon is in need of redescription, as it has been considered to be species inquirenda. However, it is currently recognized in the LAN.]

3Dorsal plate not flat, as least slightly arched; dorsal plate equal to, or wider than, ventral plate 4

3' Dorsal plate flat, narrower than the slightly arched ventral plate, plates separated by a sulcus around the margin (cf. *Euchlanis* and *Tripleuchlanis*), toes long ca. 3/4th body length, ~125–250 µm (including toes). (Fig. 8.51) [1 NEA of 3 species] .... *Dipleuchlanis propatula* (Gosse, 1886)

4 Dorsal plate arched (may be strongly), and indented posteriorly, ventral plate flat, plates separated by a sulcus around margin, foot short with two segments and two ensiform toes apically acute, ca. 120–400 µm. (Fig. 8.52) (cf. *Dipleuchlanis* and *Tripleuchlanis*) [15 NEA of 22 species] .... *Euchlanis* spp.

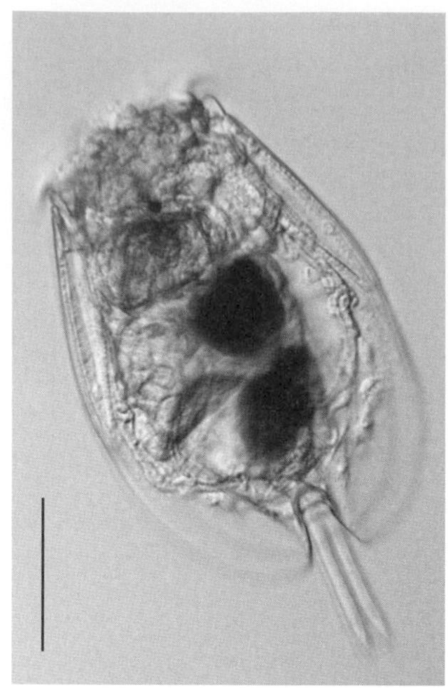

FIGURE 8.52 Family Euchlanidae: *Euchlanis*. Bar = 100 µm. (NEA)

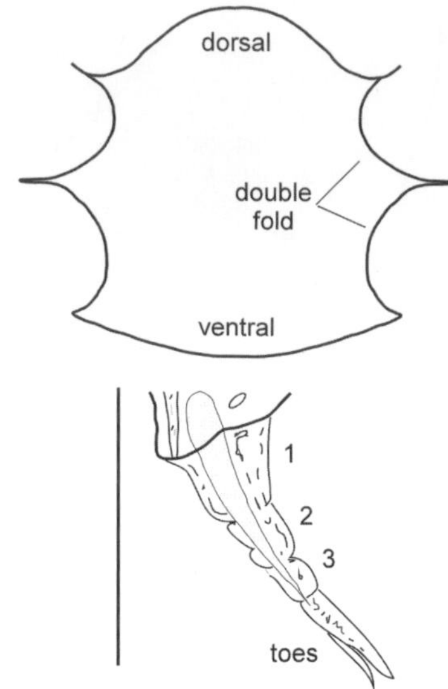

FIGURE 8.53 Family Euchlanidae: *Tripleuchlanis plicata*. Cross section (upper); foot, side view (lower). Bar = 100 µm. (NEA) *After Herrick 1885.*

4' Dorsal and ventral plates slightly arched, the wide marginal sulcus separating plates is divided by a projecting flange producing an extra fold (pleated, accordion-like), foot with three segments, toes shorter than in *Euchlanis* (cf. *Dipleuchlanis* and *Euchlanis*), ca. 250 µm. (Fig. 8.53) [monospecific] .... *Tripleuchlanis plicata* (Levander, 1894)

Rotifera: Monogononta: Ploima: Epiphanidae: Genera

[5 genera; all reported in NEA]

1 Body conical, cylindrical, or saccate, lacking prominent placations or folds 2

1' Body somewhat rectangular (square) or cylindrical and plicate 4

2(1) Proboscis lacking 3

2' Possessing a prominent dorsal proboscis extending from the corona, 2 eyespots, 200–250 µm. (Fig. 8.54). Consult (Melone, 2001) for a review of this species. [1 NEA of 4 species] .... *Rhinoglena frontalis* Ehrenberg, 1853

3(2) Body arched dorsally, in lateral view S-shaped, tapering to foot with longish toes; 200–350 µm; littoral (Fig. 8.55; trophi: Fig. 13.10 in Volume I) [monospecific] .... *Cyrtonia tuba* (Ehrenberg, 1834)

3' Body not arched and not S-shaped in lateral view; toes short; 200–500 µm. (Fig. 8.56) [5 NEA of 7 species] .... *Epiphanes* spp.

4(1) Body subquadrate in dorsal view, with a segmented conical foot, possessing a prominent dorsal spur on the last foot joint; dorsal side of the body with a few nearly parallel ridges originating dorsoventrally in the body wall and running laterally to near the corona; 170–250 µm. (Fig. 8.57): [1 NEA of 2 species] .... *Mikrocodides chlaena* (Gosse, 1886)

4' Body cylindrical (elongate), lacking a prominent dorsal spur on the last foot joint, possessing prominent transverse folds (plicate) (compare to *Taphrocampa*), posterior end with short cylindrical structures for egg attachment, 120–135 µm. (Fig. 8.58) [1 NEA of 3 species] .... *Proalides tentaculatus* de Beauchamp, 1907

Rotifera: Monogononta: Ploima: Proalidae: Genera

Consult De Smet (1996) for details. [4 genera; all reported in NEA]

1 Corona lacking cirri; head lacking styli; rostrum lacking or weakly developed 2

1' Corona possessing long cirri, styli directed posteriorly on both sides of the head, head with rostrum. [2 NEA of 2 species] .... *Bryceella* spp.

2(1) Corona not reduced, cilia not uniformly short, mouth not set posteriorly (set within the buccal field), head separated from trunk, foot not small 3

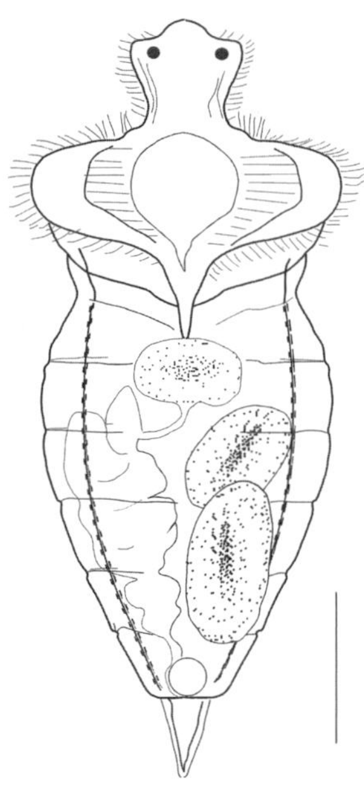

FIGURE 8.54 Family Epiphanidae: *Rhinoglena frontalis*. Ventral view. Bar = 50 µm. *After Hudson 1869.*

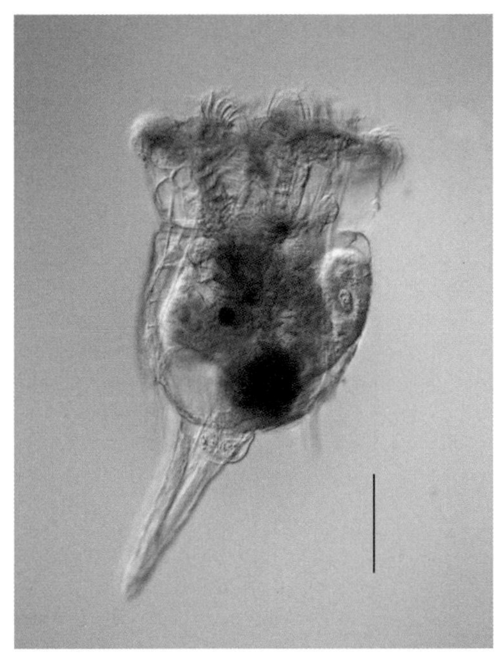

FIGURE 8.56 Family Epiphanidae: *Epiphanes*. Bar = 100 µm. (NEA)

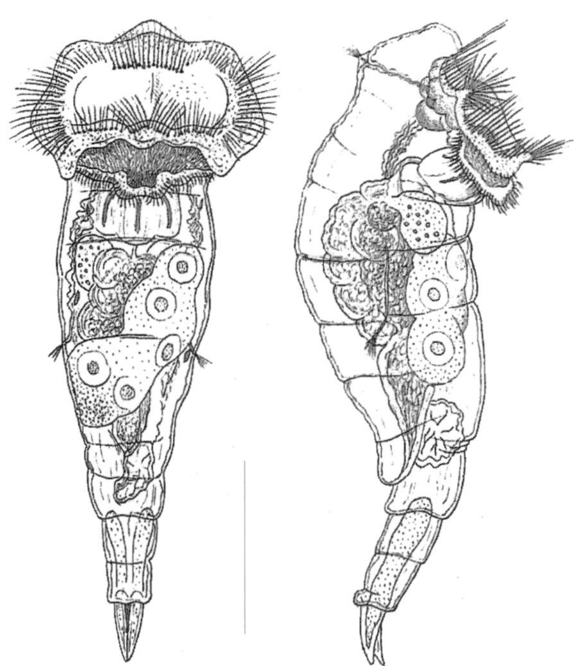

FIGURE 8.55 Family Epiphanidae: *Cyrtonia tuba*. Ventral (left); left side (right). Bar = 100 µm. *After Hood 1895.*

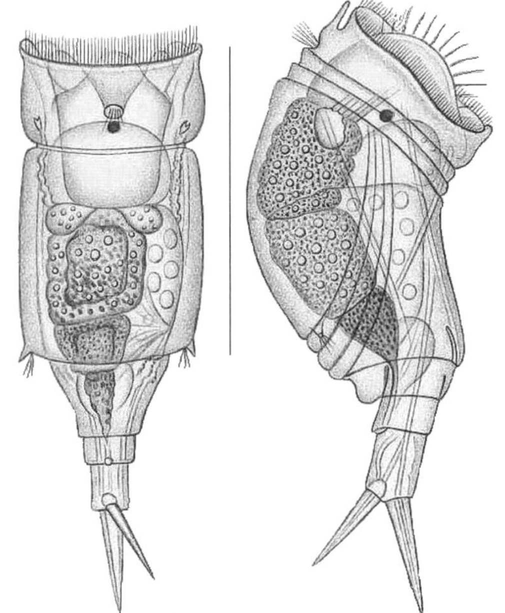

FIGURE 8.57 Family Epiphanidae: *Mikrocodides*. Arrow = spur. Dorsal (left); side (right). Bar = 100 µm. *After Weber 1998.*

2' Simple reduced corona, nearly frontal, covered with short cilia, mostly set posteriorly, head not well separated from trunk, foot small, toes small. [1 NEA of 3 species] .... *Wulfertia ornata* Donner, 1943

3(2) Tail or foot possessing a papilla with setae or spine, dorsal antennae as a papilla or projection. Consult De Smet (1996) for details. [6 NEA of 6 species] .... *Proalinopsis* spp.

3' Tail or foot lacking papilla, dorsal antennae not as above [25 NEA of ~35 species] .... *Proales* spp.

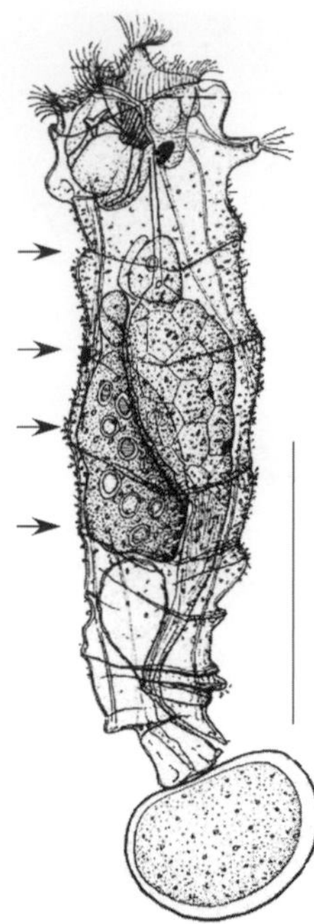

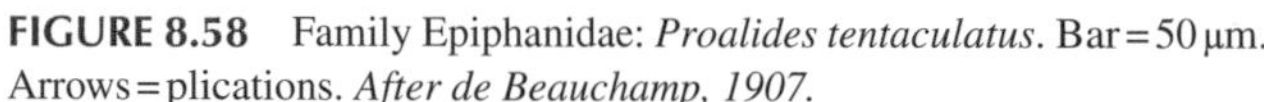

FIGURE 8.58 Family Epiphanidae: *Proalides tentaculatus*. Bar = 50 μm. Arrows = plications. *After de Beauchamp, 1907.*

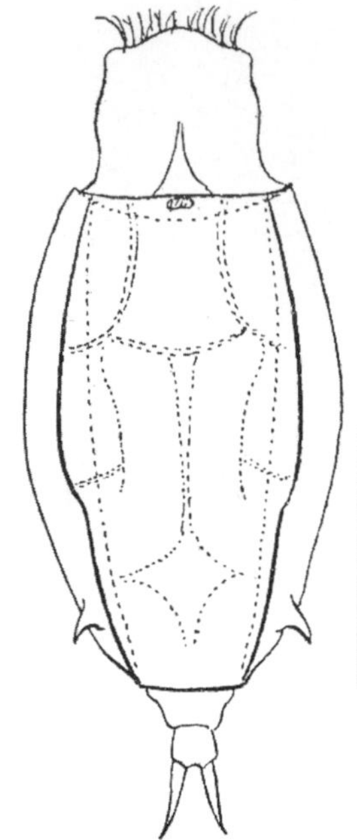

FIGURE 8.59 Family Trichotriidae: *Wolga*. Bar = 100 μm. *After Murray, 1913.*

[*Proales* is a very difficult genus. *Proales daphnicola* Thompson, 1892 is planktonic and may attach to cladocerans. *Proales werneckii* (Ehrenberg, 1834), which is parasitic in galls that it induces in filaments of the alga *Vaucheria*, has been moved into the genus *Pourriotia*, Notommatidae. That genus has a second species, *P. carcharodonta*, which also is parasitic in *Vaucheria* (De Smet, 2009a,b).]

Rotifera: Monogononta: Ploima: Trichotriidae: Genera

Genus *Pulchritia* (comprising two species) is not included herein as it has not been reported from the Nearctic (Luo & Segers, 2013). [4 genera, 3 in NEA]

1	Animals with numerous spines on body or with a pair of heavy spines on the dorsal side of the first foot joint 2
1'	Animals lacking numerous spines and lacking spines on the dorsal side of the first foot joint; ~300 μm. (Fig. 8.59) [monospecific] .... *Wolga spinifera* (Western, 1894)
2(1)	Animals with numerous, long, bilaterally placed spines; 90–250 μm. (Fig. 8.60) [4 NEA of 11 species] .... *Macrochaetus* spp.
2'	Animals without numerous bilaterally placed spines, but with a pair of heavy spines on the dorsal side of the foot, 200–400 μm. (Fig. 8.61) [6 NEA of 10 species] .... *Trichotria* spp.

Rotifera: Monogononta: Ploima: Brachionidae: Genera

[7 genera; all reported in NEA]

1	Foot present 2
1'	Foot absent 4
2(1)	Foot jointed, not with morphology or movement as below 3
2'	Foot ringed and capable of movement like a flexible hose, ca. 100–400 μm. (Fig. 8.62) [23 NEA of ~75 species] .... *Brachionus* spp.
3(2)	Body in dorsal view more or less rectangular, ~200 μm. (Fig. 8.63) [1 NEA of 4 species] .... *Plationus patulus* (Müller, 1786)
3'	Body in dorsal aspect more or less ovoid, ~165–350 μm. (Fig. 8.64) [2 NEA of 5 species] .... *Platyias* spp.
4(1)	Dorsal plate lacking prominent facets 5
4'	Dorsal plate of lorica possessing facets, ~150–450 μm. (Fig. 8.65) [22 NEA of 50 species] .... *Keratella* spp.

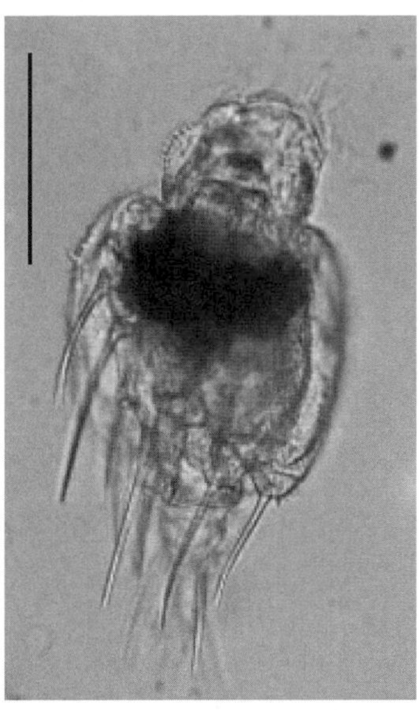

FIGURE 8.60 Family Trichotriidae: *Macrochaetus*. Bar = 100 μm. (NEA) *Photomicrograph courtesy of John Maccagno.*

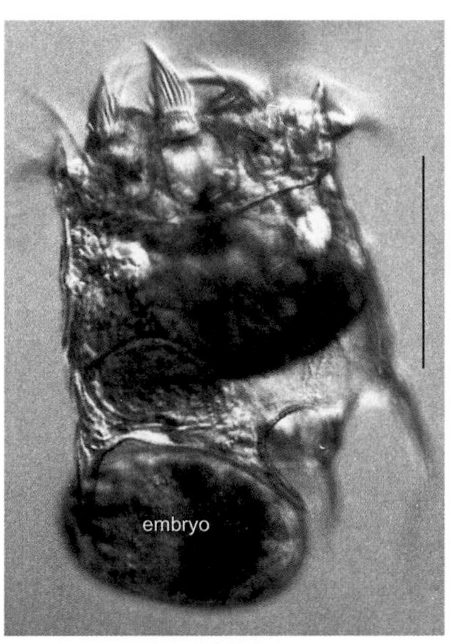

FIGURE 8.63 Family Brachionidae: *Plationus*. Bar = 100 μm. (NEA)

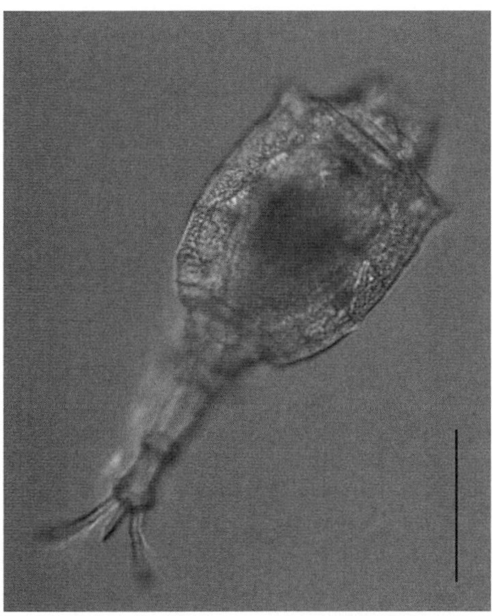

FIGURE 8.61 Family Trichotriidae: *Trichotria*. Bar = 100 μm. (NEA) *Photomicrograph courtesy of E.J. Walsh, University of Texas at El Paso.*

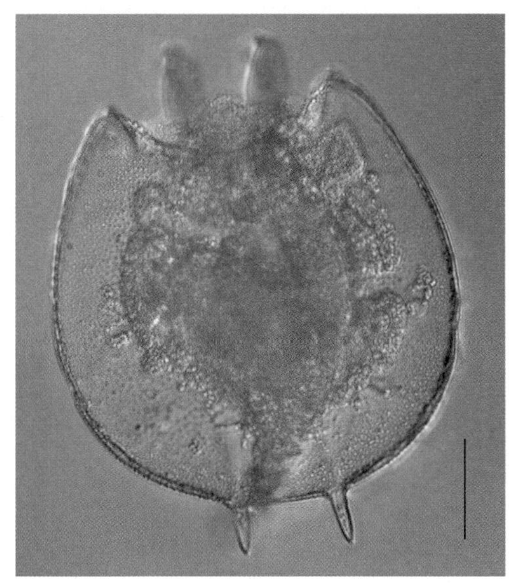

FIGURE 8.64 Family Brachionidae: *Platyias*. Bar = 50 μm. (NEA)

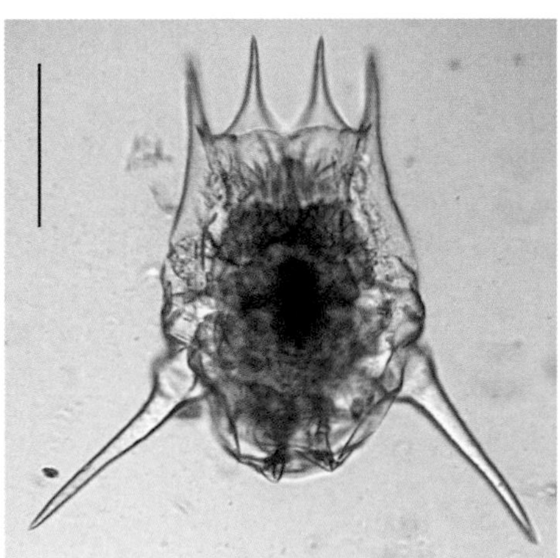

FIGURE 8.62 Family Brachionidae: *Brachionus*. Bar = 100 μm. (NEA)

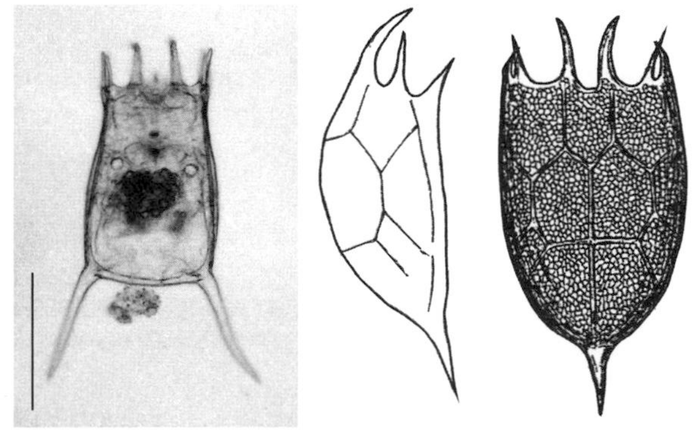

FIGURE 8.65 Family Brachionidae: *Keratella*. Bar = 100 μm. *After Lauterborn, 1900.*

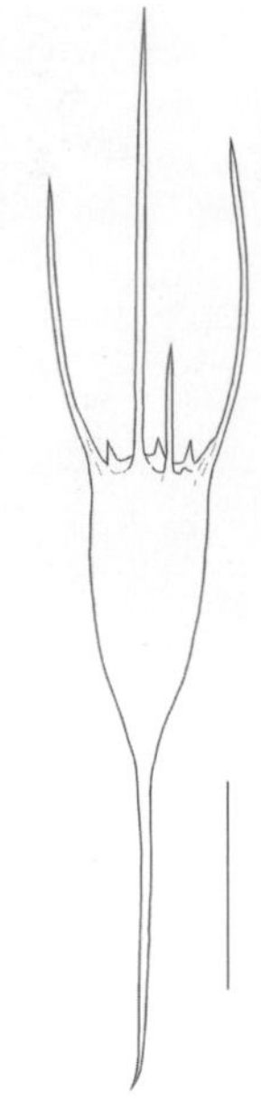

FIGURE 8.66 Family Brachionidae: *Kellicottia*. Bar = 100 μm. (NEA) (Line drawing from life.)

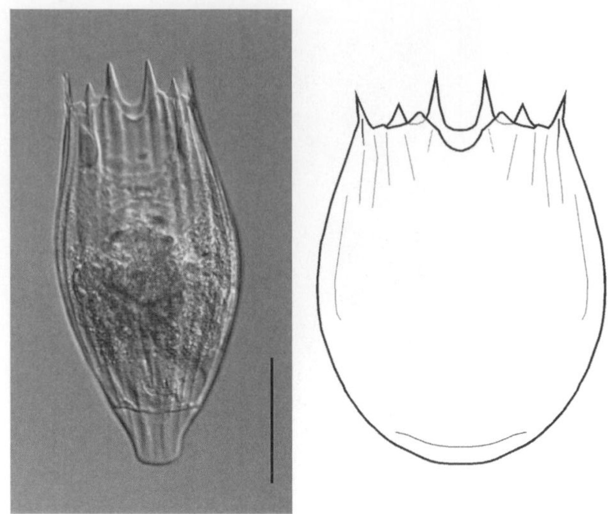

FIGURE 8.67 Family Brachionidae: *Notholca*. Bars = 50 μm. (NEA) (Line drawing from life.)

5(4)	Possessing symmetrical anterior spines or lacking spines	6
5'	Possessing asymmetrical anterior spines (unequal length), single long, slender posterior spine, ca. 100 μm, excluding spines. (Fig. 8.66) [2 NEA of 2 species]	*Kellicottia* spp.
6(5)	Six anterior spines, dorsal plate with longitudinal striations, anterior margin of ventral plate with scalloping, ≤200 μm. (Fig. 8.67) [13 NEA of 44 species]	*Notholca* spp.
6'	Lacking anterior spines, dorsal plate arched, ventral plate nearly flat, plates connected by narrow sulci of flexible cuticle, ~100 μm. (Fig. 8.68) [3 NEA of 11 species]	*Anuraeopsis* spp.

Rotifera: Monogononta: Ploima: Trichocercidae: Genera

[3 genera; all reported in NEA]

1	Body short, not twisted as a partial helix, foot and toes normal	2
1'	Body elongate, loricate, asymmetrical; twisted as a partial helix; toes mostly unequal, spine-shaped; trophi asymmetrical, ~100–500 μm. (Fig. 8.69) [51 NEA of ~70 species]	*Trichocerca* spp.
2(1)	Body showing in cross section 3–4 lobes (2 lateral, 1 dorsal, 1 ventral [inconspicuous], trophi asymmetrical, common in psammon and within of *Sphagnum*, length ca. 100 μm. (Fig. 8.70) [1 NEA of 2 species]	*Elosa worrallii* Lord, 1891
2'	Body not as above; ~100 μm; lives within and feeding on *Volvox* (Fig. 8.71) [monospecific]	*Ascomorphella volvocicola* (Plate, 1886)

Rotifera: Monogononta: Ploima: Gastropodidae: Genera

[2 genera; both reported in NEA]

1	Body ovate to saccate; 1–4 dark bodies in stomach (defecation bodies), foot absent, ~75–150 μm. (Fig. 8.72) [4 NEA of 9 species]	*Ascomorpha* spp.
1'	Body laterally compressed, dark bodies in gut absent, foot present, originating near middle of the ventral side and extending at approximately right angles to the body, ca. 80–400 μm. (Fig. 8.73) [3 NEA of 3 species]	*Gastropus* spp.

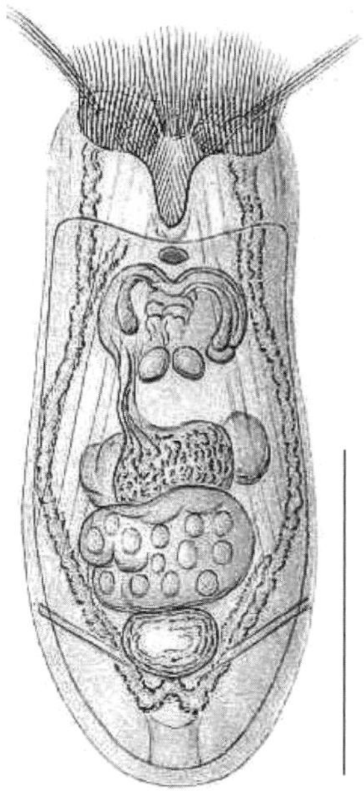

FIGURE 8.68 Family Brachionidae: *Anuraeopsis*. Bars = 50 µm. *After Gosse, 1851.*

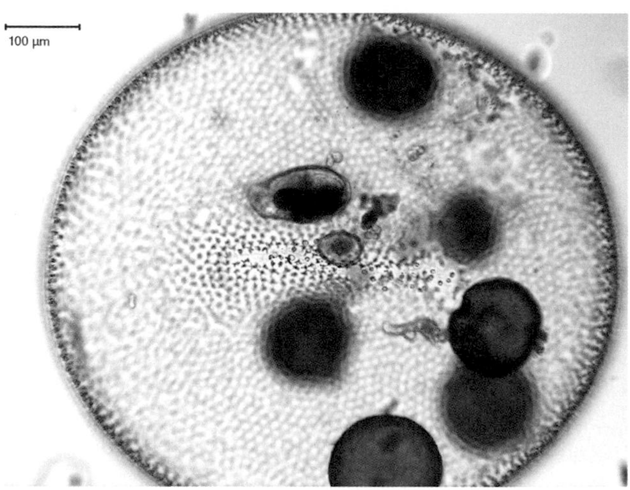

FIGURE 8.71 Family Family Trichocercidae: *Ascomorphella volvocicola* preying on cells in the colonial alga *Volvox*. Seen within this colony is an embryo (center) and just above it a female *A. volvocicola* (ca. 120 µm). Also present are at least six daughter colonies. This *Volvox* colony has sustained minor damage, as evidenced by the line of missing cells along the central portion of the equatorial region and in the upper right quadrant of the parent colony. Some unusual clumps of cells also are seen, one nearby the rotifer and another below and to the right. (NEA).

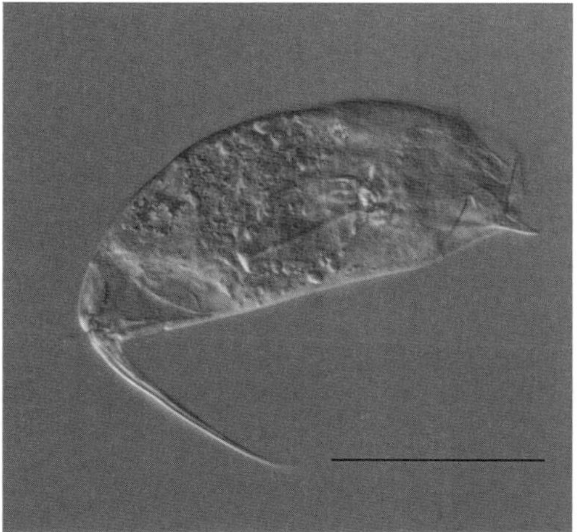

FIGURE 8.69 Family Trichocercidae: *Trichocera*. Bar = 50 µm. (NEA) *Photomicrograph courtesy of E.J. Walsh, University of Texas at El Paso.*

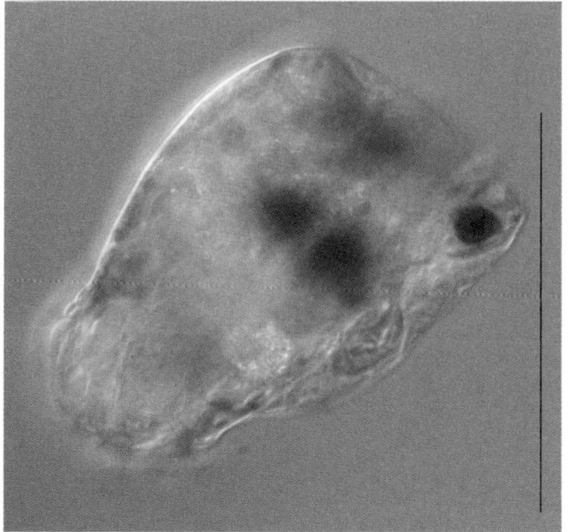

FIGURE 8.72 Family Gastropodidae: *Ascomorpha*. Schematic and photomicrograph. Bar = 100 µm. (NEA) *Photomicrograph courtesy of E.J. Walsh, University of Texas at El Paso, TX.*

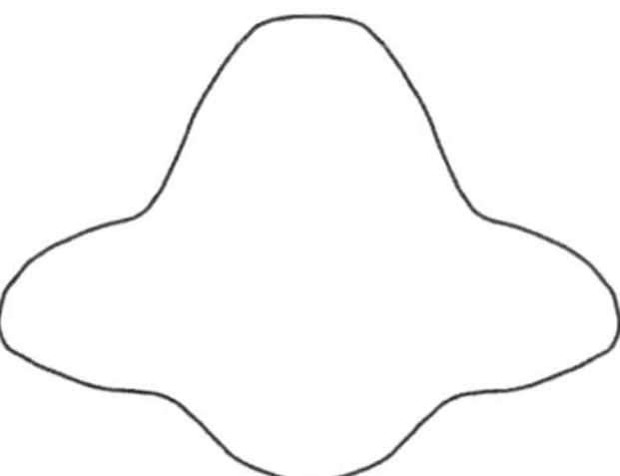

FIGURE 8.70 Family Trichocercidae: *Elosa worrallii*, cross sectional view. Bar = 50 µm. *After Lord, 1891.*

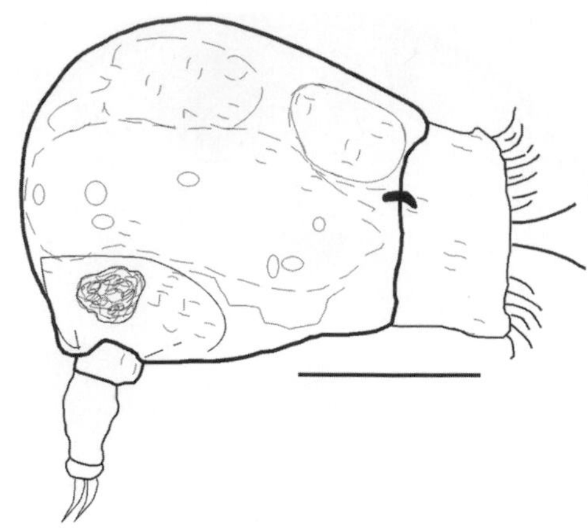

FIGURE 8.73 Family Gastropodidae: *Gastropus*. Bar = 100 µm. (NEA) *Drawn from a photomicrograph.*

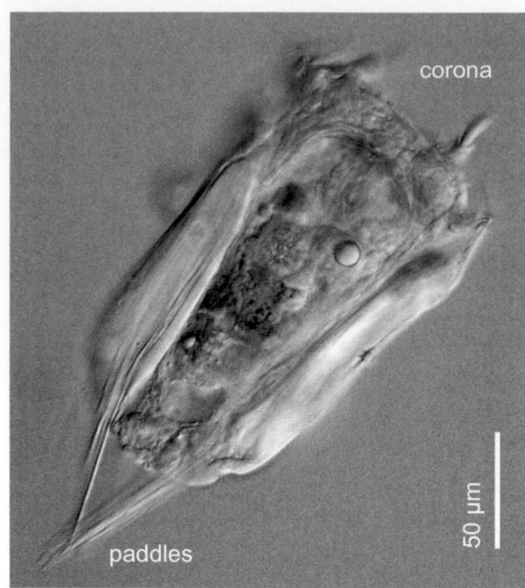

FIGURE 8.74 Family Synchaetidae: *Polyarthra*. Bar = 50 µm. *Photomicrograph courtesy of M.V. Sørensen, University of Copenhagen.*

Rotifera: Monogononta: Ploima: Synchaetidae: Genera

[4 genera; all reported in NEA]

1	Body of animal illoricate	2
1'	Body of animal loricate	3
2(1)	Animals with flat, blade- or paddle-shaped appendages (feather-shaped) positioned at the anterior, ca. 100–225 µm. (Fig. 8.74) [8 NEA of 11 species]	*Polyarthra* spp.
2'	Animals lacking paddles, head possessing sensory bristles (setae) with a lateral pair of auricles, ca. 70–600 µm. (Fig. 8.75) [15 NEA of ~35 species]	*Synchaeta* spp.
3(1)	Head shield present, body ornamented with grooves and ridges or reticulate and corona with digitiform palps, long flexible foot, ca. 130–500 µm. (Fig. 8.76) [3 NEA of 8 species]	*Ploesoma* spp.
3'	Head shield absent, grooves and ridges reduced or absent; lorica covered by tiny circular plaques; corona without palps, but on either side of the mouth two retractile, cylindrical protuberances, foot short, 150–210 µm. (Fig. 8.77) [monospecific]	*Pseudoploesoma formosum* (Myers, 1934)

Rotifera: Monogononta: Ploima: Notommatidae: Genera

The Notommatidae has been defined by Nogrady et al. (1995) as "a taxonomically unsatisfactory assemblage of diverse taxa, in need of revision by modern methodology." Given the dearth of competent rotifer taxonomists in the world, we do not see this situation changing in the near future. The following portion of this key covering the Notommatidae is based on Nogrady & Pourriot's remarkable effort in the reference cited above.

1	Not parasitic in the galls of *Vaucheria* (Xanthophyceae)	2
1'	Parasitic in galls of *Vaucheria*; ≤200 µm. Consult (Wallace et al., 2001; De Smet, 2009a,b) for details. (Fig. 8.78) [1 NEA of 2 species]	*Pourriotia werneckii* (Ehrenberg, 1834)
2(1)	Corona not as below	3
2'	Corona a circumapical band of cilia; mouth positioned deeply within a cylindrical snout developed as a suction ring; body slender, fusiform possessing a small conical foot; ectoparasitic on freshwater oligochaetes and leeches. Consult Nogrady et al. (1995) for details. [1 NEA of 3 species]	*Drilophaga judayi* Harring & Myers, 1922
3(2)	Vitellarium-shaped as a ribbon or band	4
3'	Vitellarium bean- or oval-shaped	6
4(3)	Nuclei within vitellarium arranged linearly	5
4'	Nuclei within vitellarium arranged randomly. Consult Nogrady et al. (1995) for details. [monospecific]	*Enteroplea lacustris* Ehrenberg, 1830

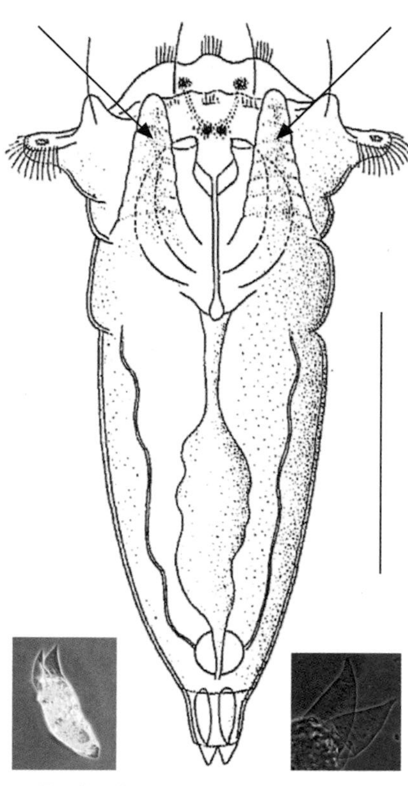

FIGURE 8.75 Family Synchaetidae: *Synchaeta*. (Line drawing of *Synchaeta bicornis*, showing two unusual prominences in this species; inserts preserved samples. Bar = 50 µm. (NEA). *Line drawing after Smith 1904; inserts courtesy of David Fischer, Cary Institute of Ecosystem Studies, Millbrook, NY.*

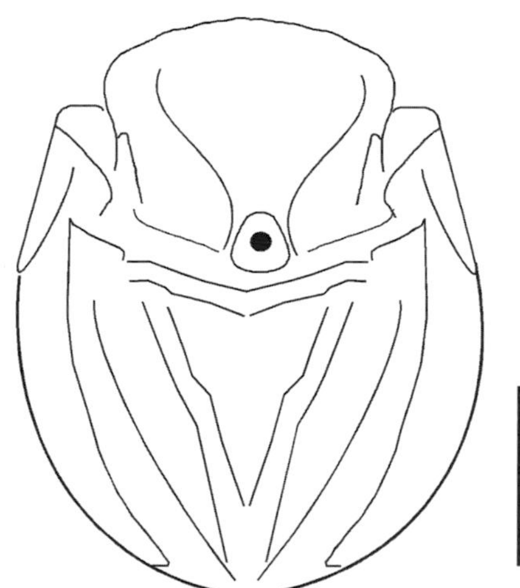

FIGURE 8.76 Family Synchaetidae: *Ploesoma*. Dorsal view drawn from life. Bar = 100 µm. (NEA)

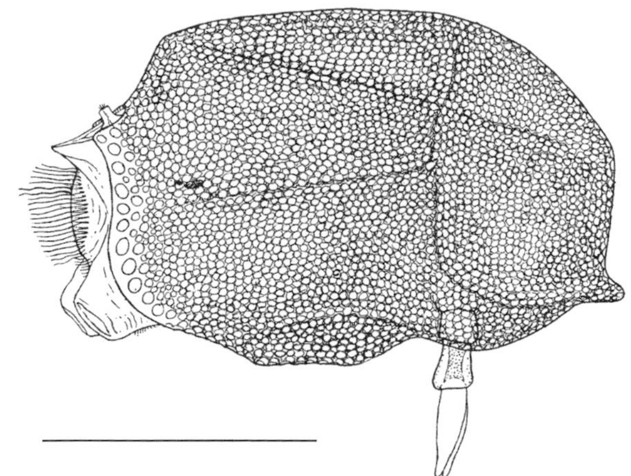

FIGURE 8.77 Family Synchaetidae: *Pseudoploesoma*. Bar = 100 µm. (NEA) *Line drawing after* American Museum Novitates *700, with permission.*

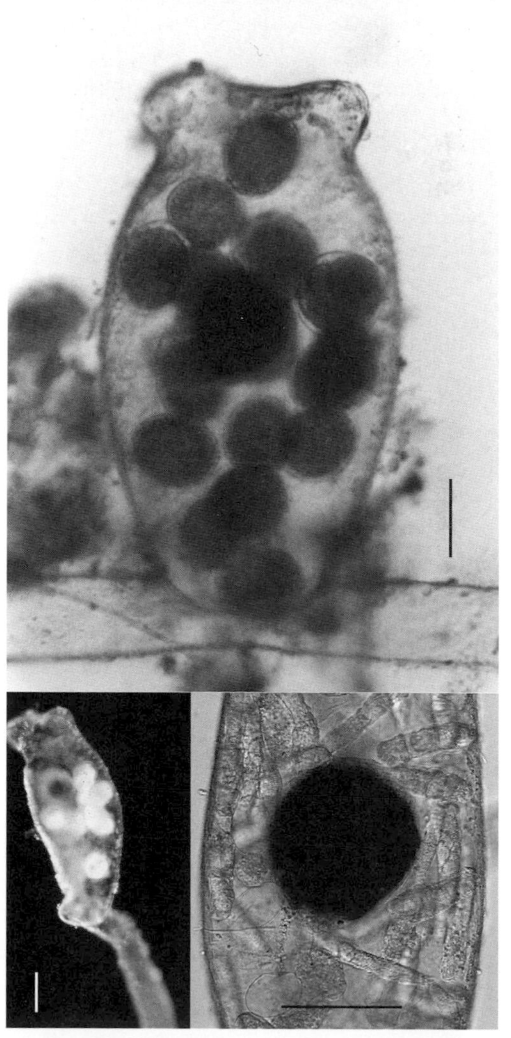

FIGURE 8.78 *Pourriotia (Proales) werneckii*. Bars = 100 µm. (NEA) (Photomicrographs of adult of *P. werneckii* (dark body) with numerous eggs within a gall of *Vaucheria* sp. (upper); parasitized gall (lower left); adult female *P. werneckii* with hatchlings (lower right). *Courtesy of D. Ott, University of Akron, Ohio.*

5(4) Head squarish, not as wide as body; no eyespots; conical foot tapers from body to 2 tiny toes; (genus not well described). Consult Nogrady et al. (1995) for details. [monospecific; not reported from the Nearctic] .... *Pseudoharringia similis* Fadeew, 1925

5' Head triangular to trapezoidal in shape, wider than body; eyespots on widely separated, small prominences; foot separated from body; toes long. Consult Nogrady et al. (1995) for details. [monospecific; not reported from the Nearctic] .... *Sphyrias lofuana* (Rousselet, 1910)

6(3,) Toes much shorter than body length 7

6' Toes relatively long, mostly longer than body, and unequal in length (except in one species). Consult Nogrady et al. (1995) for details. [14 NEA of 17 species] .... *Monommata* spp.

7(6) Body not as below; two toes 8

7' Soft, fusiform body, possessing 2 lateral humps (retractable); body tapering to conical foot (wrinkled); single conical toe (probably fusion of 2); animals always colored (rust to reddish purple). Consult Nogrady et al. (1995) for details. [monospecific] .... *Tylotrocha monopus* (Jennings, 1894)

8(7) Spine not present as below 9

8' Small or prominent spine present on the distal end of body or on the distal end of foot 10

9(8) Spine on distal end of body. Consult Nogrady et al. (1995) for details [1 NEA of 2 species] .... *Dorystoma caudata* (Bilfinger, 1894)

9' Spine on distal end of foot. Consult Nogrady et al. (1995) for details [monospecific] .... *Rousseletia corniculata* Harring, 1913

10(8,) Body illoricate and not as below 11

10' Body generally weakly loricate (occasionally illoricate); mostly cylindrical, but shape varies from stubby to fusiform; possessing 3 to 5 plates separated by sulci (furrows) (Fig. 8.79). Consult Nogrady et al. (1995) for details [79 NEA of ~156 species] ...... *Cephalodella* spp.

[Two genera *Metadiaschiza* and *Paracephalodella*, neither of which has been reported from Nearctic waters, are included in the key of Nogrady & Pourriot provided in Nogrady et al. (1995), but are now subsumed within *Cephalodella* (Segers, 2002, 2007). For a recent account of the trophi types in this genus, consult Fischer and Alrichs (2011).]

11(10) Body wall without spines 12

11' Body wall possesses rows of small spines. Consult Nogrady et al. (1995) for details [monospecific] .... *Pleurotrochopsis multispinosa* (Fadeew, 1925)

12(11) Not as below 13

12' Dorsal side of body with deep plicate folds (accordion-like). (Fig. 8.80); consult Nogrady et al. (1995) for details. [3 NEA of 4 species] .... *Taphrocampa* spp.

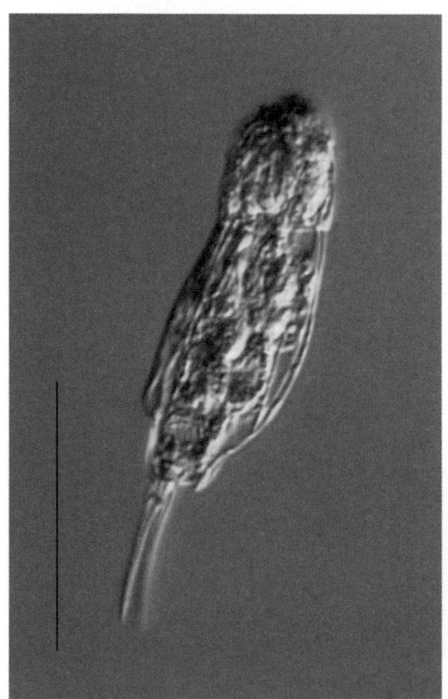

FIGURE 8.79 Family Notommatidae: *Cephalodella*. Bar = 100 µm. (NEA) *Photomicrograph courtesy of E.J. Walsh, University of Texas at El Paso, TX.*

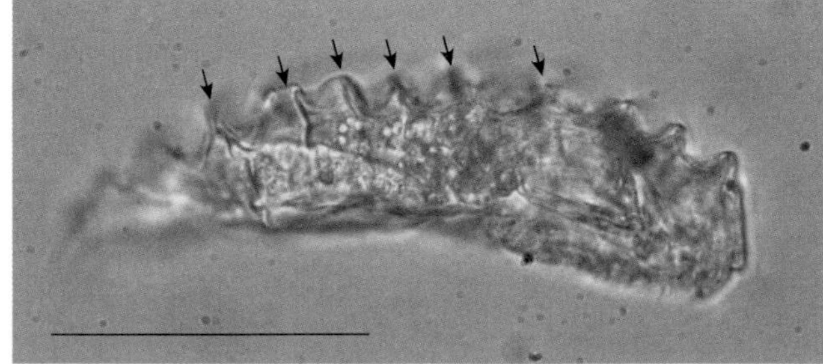

FIGURE 8.80 Family Notommatidae: *Taphrocampa*. Arrows indicate plication (folds). Bar = 100 µm. (NEA) *Photomicrograph courtesy of E.J. Walsh, University of Texas at El Paso, TX.*

13(12)	Animals possessing two frontal eyespots	14
13'	Animals lacking frontal eyespots	15
14(13)	Possessing 1 salivary gland; consult Nogrady et al. (1995) for details [6 NEA of 7 species]	*Eothinia* spp.
14'	Possessing 2 salivary glands; consult Nogrady et al. (1995) for details [6 NEA of 7 species]	*Eosphora* spp.
15(13,)	Auricles absent; anterior corona	16
15'	Auricles present (ciliated); ventral corona; consult Nogrady et al. (1995) for details [36 NEA of 49 species]	*Notommata* spp.
16(15)	Salivary glands are well-formed and symmetrical	17
16'	Salivary glands are rudimentary or asymmetrical; consult Nogrady et al. (1995) for details [5 NEA of 7 species]	*Resticula* spp.
17(16)	Rami triangular (strong, robust); consult Nogrady et al. (1995) for details [4 NEA of 7 species]	*Pleurotrocha* spp.
17'	Rami lyrate (curved, thin and slender); consult Nogrady et al. (1995) for details [4 NEA of 7 species]	*Pleurata* spp.

REFERENCES

Anon. 2014. The Rotifer Trophi Web Page. http://www.rotifera.hausdernatur.at/Rotifer_data/trophi/. (assessed 2014.06.23). Hosted by Rotifer World Catalog. Haus der Natur, Salzburg.

Bartoš, E. 1951. The Czechoslovak Rotatoria of the order Bdelloidea. Vestník Ceskoslovenské Zoologické Spolecnosti 15:241–500.

Bateman, L. E. 1987. A bdelloid rotifer living as an inquiline in leaves of the pitcher plant, *Sarracenia purpurea*. Hydrobiologia 1 47:129–133.

Bērziņš, B. 1951. On the Collothecacean Rotatoria with special reference to the species found in the Aneboda district, Sweden. Arkiv for Zoologi 1:565–592.

Bielańska-Grajner, I., A. Cudak & T. Mieczan. 2011. Epiphytic rotifer abundance and diversity in moss patches in bogs and fens in the Polesie National Park (Eastern Poland). International Review of Hydrobiology 96:29–38.

Black, A. R. & S. I. Dodson. 2003. Ethanol: a better preservation technique for *Daphnia*. Limnology and Oceanography: Methods 1:45–50.

Błędzki, L. A. & A. M. Ellison. 1998. Population growth and production of *Habrotrocha rosa* Donner (Rotifera: Bdelloidea) and its contribution to the nutrient supply of its host, the northern pitcher plant, *Sarracenia purpurea* L. (Sarraceniaceae). Hydrobiologia 385: 193–200.

Braioni, M. G. & D. Gelmini. 1983. Guide per il riconoscimento delle specie Animali delle acque interne Italiane. 23. Rotiferi Monogononti. Consiglio Nazionale delle Ricerche, Verona. p180.

Burger, A. 1948. Studies on the moss dwelling bdelloids (Rotifera) of eastern Massachusetts. Transactions of the American Microscopical Society 67:111–142.

De Smet, W. H. (edit.). 1996. Rotifera. Volume 4: The Proalidae (Monogononta). SPB Academic Publishing bv, The Hague.

De Smet, W. H. 2003. *Paradicranophorus sinus* sp. nov. (Dicranophoridae, Monogononta) a new rotifer from Belgium, with remarks on some other species of the genus *Paradicranophorus* Wiszniewski, 1929 and description of *Donneria* gen. nov. Belgium Journal of Zoology 133:181–188.

De Smet, W. H. 2006. Asciaporrectidae, a new family of Rotifera (Monogononta: Ploima) with description of *Asciaporrecta arcellicola* gen. et sp. nov. and *A. difflugicola* gen. et sp. nov. inhabiting shells of testate amoebae (Protozoa). Zootaxa 1339:31–49.

De Smet, W. H. 2007. Cotylegaleatidae, a new family of Ploima (Rotifera: Monogononta), for *Cotylegaleata perplexa* gen. et sp. nov., from freshwater benthos of Belgium. Zootaxa 1425:35–43.

De Smet, W. H. 2009a. *Pourriotia carcharodonta*, a new genus and species of monogonont rotifer from subantarctic Îles Kerguelen (Terres Australes et Antarctiques Françaises). Annales de Limnologie - International Journal of Limnology 39:273–280.

De Smet, W. H. 2009b. Pourriotia carcharodonta, a rotifer parasitic on *Vaucheria* (Xanthophyceae) causing taxonomic problems. Bulletin de la Société zoologique de France 134:195–202.

De Smet, W. H. & R. Pourriot (edit.). 1997. Rotifera. Volume 5: The Dicranophoridae (Monogononta) and: The Ituridae (Monogononta). SPB Academic Publishing, The Hague.

Devetter, M. 2008. Clearance rates of the bdelloid rotifer, *Habrotrocha thienemanni*, a tree-hole inhabitant. Aquatic Ecology 43:85–89.

Devetter, M. 2010. A method for efficient extraction of rotifers (Rotifera) from soils. Pedobiologia 53:115–118.

Donner, J. 1956. Rädertierre (Rotatorien). W. Ketter & Co. Stuttgart. English translation and adaptation by H.G.S. Wright (1966). Frederick Warne & Co., Ltd, New York.

Donner, J. 1965. Ordnung Bdelloidea (Rotatoria, Rädertiere). Akademie-Verlag, Berlin.

Edmondson, W. T. 1945. Ecological studies of sessile Rotatoria, Part II. Dynamics of populations and social structure. Ecological Monographs 15:141–172.

Edmondson, W. T. 1949. A formula key to the Rotatorian genus *Ptygura*. Transactions of the American Microscopical Society 68:127–135.

Edmondson, W. T. 1959. Rotifera. Pages 420–494 *in* W. T. Edmondson, editor. Freshwater Biology, 2nd ed. John Wiley & Sons, Inc., New York.

Fischer, C., and W. H. Alrichs. 2011. Revisiting the *Cephalodella* trophi types. Hydrobiologia 662:205–209.

Fontaneto, D., C. Boschetti & C. Ricci. 2007a. Cryptic diversification in ancient asexuals: evidence from the bdelloid rotifer *Philodina flaviceps*. Journal of Evolutionary Biology 21:580–587.

Fontaneto, D., I. Giordani, G. Melone & M. Serra. 2007b. Disentangling the morphological stasis in two rotifer species of the *Brachionus plicatilis* species complex. Hydrobiologia 583:297–307.

Fontaneto, D., E. A. Herniou, T. G. Barraclough, C. Ricci & G. Melone. 2007c. On the reality and recognisability of asexual organisms: morphological analysis of the masticatory apparatus of bdelloid rotifers. Zoologica Scripta 36:361–370.

Fontaneto, D., N. Iakovenko, I. Eyres, M. Kaya, M. Wyman & T. G. Barraclough. 2011. Cryptic diversity in the genus *Adineta* Hudson & Gosse, 1886 (Rotifera: Bdelloidea: Adinetidae): a DNA taxonomy approach. Hydrobiologia 662:27–33.

Fontaneto, D., and U. Jondelius. 2011. Broad taxonomic sampling of mitochondrial cytochrome c oxidase subunit I does not solve the relationships between Rotifera and Acanthocephala. Zoologischer Anzeiger 250:80–85.

Fontaneto, D., M. Kaya, E. A. Herniou & T. G. Barraclough. 2009. Extreme levels of hidden diversity in microscopic animals (Rotifera) revealed by DNA taxonomy. Molecular Phylogenetics and Evolution 53:182–189.

Fontaneto, D., G. Melone & R. L. Wallace. 2003. Morphology of *Floscularia ringens* (Rotifera, Monogononta) from egg to adult. Invertebrate Biology 122:231–240.

Garcia-Morales, A. E. & M. Elias-Gutierrez. 2013. DNA barcoding of freshwater Rotifera in Mexico: Evidence of cryptic speciation in common rotifers. Molecular Ecology Resources.

Garey, J. R., T. J. Near, M. R. Nonnemacher & S. A. Steven A. Nadler. 1996. Molecular evidence for Acanthocephala as a subtaxon of Rotifera. Journal of Molecular Evolution 43:287–292.

Garey, J. R., A. Schmidt-Rhaesa, T. J. Near & S. A. Nadler. 1998. The evolutionary relationships of rotifers and acanthocephalans. Hydrobiologia 387/388:83–91.

Gilbert, J. J. & C. W. Burns. 2000. Day and night vertical distributions of *Conochilus* and other zooplankton in a New Zealand reservior. Verhandlungen Internationale Vereinigung Limnologie 27:1909–1914.

Gómez, A. 2005. Molecular ecology of rotifers: from population differentiation to speciation. Hydrobiologia 546:83–99.

Haney, J. F. & D. J. Hall. 1973. Sugar-coated *Daphnia*: A preservation technique for Cladocera. Limnology and Oceanography 18:331–333.

Jersabek, C. D., E. Bolortsetseg & H. L. Taylor. 2010. Mongolian rotifers on microscope slides: Instructions to permanent specimen mounts from expedition material. Mongolian Journal of Biological Sciences 8:51–57.

Jersabek, C. D. & M. F. Leitner. 2013. The Rotifer World Catalog. World Wide Web electronic publication. http://www.rotifera.hausdernatur.at/, (accessed 2015.01.27).

Jersabek, C. D., H. Segers & P. J. Morris. 2014. An illustrated online catalog of the Rotifera in the Academy of Natural Sciences of Philadelphia (version 1.0: 2003-April-8). [WWW database] URL http://rotifer.ansp.org/rotifer.php.

Koste, W. 1978. *Rotatoria. Die Rädertiere Mitteleuropas.* 2 volumes. Gebrüder Borntraeger, Stuttgart.

Koste, W. & R. J. Shiel. 1986. Rotifera from Australian inland waters. I. Bdelloidea (Rotifera: Digononta). Australian Journal of Marine and Freshwater Research 37:765–792.

Koste, W. & R. J. Shiel. 1987. Rotifera from Australian inland waters. II. Epiphanidae and Brachionidae (Rotifera: Monogononta). Invertebrate Taxonomy 7:949–1021.

Koste, W. & R. J. Shiel. 1989a. Rotifera from Australian inland waters IV. Colurellidae (Rotifera: Monogononta). Transactions of the Royal Society of South Australia 113:119–143.

Koste, W. & R. J. Shiel. 1989b. Rotifera from Australian inland waters. III. Euchlanidae, Mytilinidae and Trichotriidae (Rotifera: Monogononta). Transactions of the Royal Society of South Australia 113:85–114.

Koste, W. & R. J. Shiel. 1990a. Rotifera from Australiam inland waters. VI. Proalidae, Lindiidae (Rotifera: Monogononta). Transactions of the Royal Society of South Australia 114:129–143.

Koste, W. & R. J. Shiel. 1990b. Rotifera from Australian inland waters V. Lecanidae (Rotifera: Monogononta). Transactions of the Royal Society of South Australia 114:1–36.

Koste, W. & R. J. Shiel. 1991. Rotifera from Australian inland waters. VII. Notommatidae (Rotifera: Monogononta). Transactions of the Royal Society of South Australia 115:111–159.

Kutikova, L. A. 2005. [The bdelloid rotifers of the fauna of Russia]. KMK Scientific Press Ltd. 314 p., Moscow. (In Russian).

Leasi, F., R. C. Neves, K. Worsaae & M. V. Sørensen. 2012. Musculature of *Seison nebaliae* Grube, 1861 and Paraseison annulatus (Claus, 1876) revealed with CLSM: a comparative study of the gnathiferan key taxon Seisonacea (Rotifera). Zoomorphology 131:185–195.

Leasi, F., G. W. Rouse & M. V. Sørensen. 2011. A new species of *Paraseison* (Rotifera: Seisonacea) from the coast of California, USA. Journal of the Marine Biological Association of the United Kingdom 92:959–965.

Luo, Y. & H. Segers. 2013. On *Pulchritia* new genus, with a reappraisal of the genera of Trichotriidae (Rotifera, Monogononta). Zookeys 342:1–12.

Mark Welch, D. & M. Meselson. 2000. Evidence for the evolution of bdelloid rotifers without sexual reproduction or genetic exchange. Science 288:1211–1215.

Meksuwan, P., P. Pholpunthin & H. Segers. 2011. Diversity of sessile rotifers (Gnesiotrocha, Monogononta, Rotifera) in Thale Noi Lake, Thailand. Zootaxa 2997:1–18.

Meksuwan, P., P. Pholpunthin & H. Segers. 2013. The Collothecidae (Rotifera, Collothecacea) of Thailand, with the description of a new species and an illustrated key to the Southeast Asian fauna. Zookeys 315:1–16.

Melone, G. 2001. *Rhinoglena frontalis* (Rotifera, Monogononta): a scanning electron microscopic study. Hydrobiologia 466/467:291–296.

Melone, G. & C. Ricci. 1995. Rotatory apparatus in Bdelloids. Hydrobiologia 313/314:91–98.

Melone, G., C. Ricci & H. Segers. 1998. The trophi of Bdelloidea (Rotifera): a comparative study across the class. Canadian Journal of Zoology 76:1755–1765.

Min, G.-S. & J.-K. Park. 2009. Eurotatorian paraphyly: Revisiting phylogenetic relationships based on the complete mitochondrial genome sequence of *Rotaria rotatoria* (Bdelloidea: Rotifera: Syndermata). BMC Genomics 10:533.

Nogrady, T., R. Pourriot & H. Segers, editors. 1995. Rotifera. Volume 3: The Notommatidae and: The Scaridiidae. SPB Academic Publishing, The Hague.

Nogrady, T. & T. L. A. Rowe. 1993. Comparative laboratory studies of narcosis in Brachionus plicatilis. Hydrobiologia 255/256:51–56.

Nogrady, T. & H. Segers (edit.). 2002. Rotifera. Volume 6: Asplanchnidae, Gastropodidae, Lindiidae, Microcodidae, Synchaetidae, Trochosphaeridae and *Filinia*. SPB Academic Publishers bv, The Hague.

Pennak, R. W. 1989. Fresh-Water Invertebrates of the United States: Protozoa to Mollusca. John Wiley & Sons, Inc., New York.

Peters, U., W. Koste & W. Westheide. 1993. A quantitative method to extract moss-dwelling rotifers. Hydrobiologia 255/256:339–341.

Pontin, R. M. 1978. *A Key to British Freshwater Planktonic Rotifera.* Freshwater Biological Association, Ambleside.

Pourriot, R. 1979. Rotiféres du sol. Revue d'Ecologie et de Biologie du Sol 16:279–312.

Ricci, C. 1987. Ecology of bdelloids: how to be successful. Hydrobiologia 147:117–127.

Ricci, C. 1993. Old and new data on Seisonidea (Rotifera). Hydrobiologia 255/256:495–511.

Ricci, C. & M. Caprioli. 2005. Anhydrobiosis in bdelloid species, populations and individuals. Intergrative and Comparative Biology 45:759–763.

Ricci, C. & D. Fontaneto. 2009. The importance of being a bdelloid: ecological and evolutionary consequences of dormancy. Italian Journal of Zoology 76:240–249.

Ricci, C. & G. Melone. 1998. Dwarf males in monogonont rotifers. Aquatic Ecology 32:361–365.

Ricci, C. & G. Melone. 2000. Key to the identification of the genera of bdelloid rotifers. Hydrobiologia 418:73–80.

Rublee, P. A. 1998. Rotifers in arctic North America with particular reference to their role in microplankton community structure and response to ecosystem perturbations in Alaskan Arctic LTER lakes. Hydrobiologia 387/388:153–160.

Rublee, P. A. & A. Partusch-Talley. 1995. Microfaunal response to fertilization of an arctic tundra stream. Freshwater Biology 34:81–90.

Ruttner-Kolisko, A. 1974. Planktonic rotifers: biology and taxonomy. Die Binnengewässer (Supplement) 26:1–146.

Sarma, S. S. S. & M. Elías-Gutiérrez. 1998. Rotifer diversity in a central Mexican pond. Hydrobiologia 387/388:47–54.

Schröder, T. & E. J. Walsh. 2007. Cryptic speciation in the cosmopolitan *Epiphanes senta* complex (Monogononta, Rotifera) with the description of new species. Hydrobiologia 593:129–140.

Segers, H. (edit.). 1995a. Rotifera. Volume 2: The Lecanidae (Monogononta). SPB Academic Publishing bv, The Hague.

Segers, H. (edit.). 1995b. Rotifera. Volume 3: The Scaridiidae (Monogononta). SPB Academic Publishing bv, Amsterdam.

Segers, H. 1997. Contribution to a revision of *Floscularia* Cuvier, 1798 (Rotifera: Monogononta): notes on some Neotropical taxa. Hydrobiologia 354:165–175.

Segers, H. 2002. The nomenclature of the Rotifera: annotated checklist of valid family and genus-group names. Journal of Natural History 36:631–640.

Segers, H. 2004. Rotifera: Monogononta. Pages 112–126 *in* C. Yule, M. & H. S. Yong (edit.), Freshwater invertebrates of the Malaysian region. Academy of Sciences Malaysia and Monash University, Kuala Lumpur, Malaysia.

Segers, H. 2007. Annotated checklist of the rotifers (Phylum Rotifera), with notes on nomenclature, taxonomy and distribution. Zootaxa 1564:1–104.

Segers, H. 2008. Global diversity of rotifers (Rotifera) in freshwater. Hydrobiologia 595:49–59.

Segers, H., P. Meksuwan & L.-o. Sanoamuang. 2010. New records of sessile rotifers (Phylum Rotifera: Flosculariacea, Collothecacea) from Southeast Asia. Belgium Journal of Zoology 140:235–240.

Segers, H. & R. J. Shiel. 2008. Diversity of cryptic Metazoa in Australian freshwaters: a new genus and two new species of sessile rotifer (Rotifera, Monogononta, Gnesiotrocha, Flosculariidae). Zootaxa 1750:19–31.

Segers, H. & R. L. Wallace. 2001. Phylogeny and classification of the Conochilidae (Rotifera, Monogononta, Flosculariacea). Zoologica Scripta 30:37–48.

Shiel, R. J. 1995. A guide to identification of rotifers, cladocerans and copepods from Australian inland waters. The Murray-Darling Freshwater Research Centre, Albury, Australia.

Shiel, R. J. & W. Koste. 1992. Rotifera from Australian inland waters VIII. Trichocercidae (Monogononta). Transactions of the Royal Society of South Australia 116:1–27.

Shiel, R. J. & W. Koste. 1993. Rotifera from Australian inland waters. IX. Gastropodiae, Synchaetidae, Asplanchnidae (Rotifera: Monogononta). Transactions of the Royal Society of South Australia 117:111–139.

Shiel, R. J., L. Smales, W. Sterrer, I. C. Duggan, S. Pichelin & J. D. Green. 2009. Phylum Gnathifera lesser jaw worms, rotifers, thorny-headed worms. Pages 137–158 *in* D. Gordon (edit.), New Zealand Inventory of Biodiversity, Volume 1. Kingdom Animalia ↓ Radiata, Lophotrochozoa, Deuterostomia. Canterbury University Press, Canterbury, NZ.

Smith, D. G. 2001. Pennak's Freshwater Invertebrates of the United States: Porifera to Crustacea, 4th ed. John Wiley & Sons, New Your.

Sørensen, M. V. 2001. On the phylogeny and jaw evolution in Gnathifera. University of Copenhagen, Copenhagen.

Sørensen, M. V. 2006. On the rotifer fauna of Disko Island, Greenland, with notes on selected species from a stagnant freshwater lake. Zootaxa 1241:37–49.

Sørensen, M. V. & G. Giribet. 2006. A modern approach to rotiferan phylogeny: combining morphological and molecular data. Molecular Phylogenetics and Evolution 40:585–608.

Sørensen, M. V., H. Segers & P. Funch. 2005. On a new *Seison* Grube, 1861 from coastal waters of Kenya, with a reappraisal of the classification of the Seisonida (Rotifera). Zoological Studies 44:34–43.

Stemberger, R. S. 1979. A guide to rotifers of the Laurentian Great Lakes. US Environmental Protection Agency, Cincinnati, OH. National Technical Information Service (PB80-101280), Springfield, VA.

Van Damme, K. & H. Segers. 2004. *Anomopus telphusae* Piovanelli, 1903, an epizoic bdelloid (Rotifera: Bdelloidea) on the Socotran endemic crab *Socotrapotamon socotrensis* (Hilgendorf, 1883). Fauna of Saudi Arabia 20:169–175.

Vidrine, M. F., R. E. McLaughlin & O. R. Willis. 1985. Free-swimming colonial rotifers (Monogononta: Flosculariacea: Flosculariidae) in Southwestern Louisiana rice fields. Freshwater Invertebrate Biology 4:187–193.

Wallace, R. L. 1987. Coloniality in the phylum Rotifera. Hydrobiologia 147:141–155.

Wallace, R. L. & W. T. Edmondson. 1986. Mechanism and adaptive significance of substrate selection by a sessile rotifer. Ecology 67:314–323.

Wallace, R. L., D. W. Ott, S. L. Stiles & C. K. Oldham-Ott. 2001. Bed and Breakfast: the parasitic life of *Proales werneckii* (Ploimida: Proalidae) within the alga *Vaucheria* (Xanthophyceae: Vaucheriales). Hydrobiologia 446/447:129–137.

Wallace, R. L. & C. Ricci. 2002. Rotifera. Pages 15–44 *in* S. D. Rundle, A. L. Robertson & J. M. Schmid-Araya (edit.), Freshwater Meiofauna: Biology and Ecology. Backhuys Publishers, Leiden.

Wallace, R. L. & H. A. Smith. 2009. Rotifera. Pages 689–703 *in* G. E. Likens (edit.), Encyclopedia of Inland Waters Elsevier, Oxford.

Wallace, R. L., T. Snell & H. A. Smith. 2015. Rotifer: Ecology and General Biology. Pages 225–271 *in* J. H. Thorp and D. C. Rogers, editors. Thorp and Covich's Freshwater Invertebrates. Elsevier, Waltham, MA.

Wallace, R. L. & T. W. Snell. 2010. Rotifera. Pages 173–235 *in* J. Thorp and A. Covich (edit.), Ecology and Classification of North American Freshwater Invertebrates. Elsevier, Inc, Amsterdam.

Wallace, R. L., T. W. Snell, C. Ricci & T. Nogrady. 2006. Rotifera. Volume 1: Biology, Ecology and Systematics (2nd edition). Backhuys Publishers, Leiden.

Wey-Fabrizius, A. R., H. Herlyn, B. Rieger, D. Rosenkranz, A. Witek, D. B. Mark Welch, I. Ebersberger & T. Hankeln. 2014. Transcriptome data reveal Syndermatan relationships and suggest the evolution of endoparasitism in Acanthocephala via an epizoic stage. PLoS ONE 9:e88618.

Wilson, C. G. 2011. Desiccation-tolerance in bdelloid rotifers facilitates spatiotemporal escape from multiple species of parasitic fungi. Biological Journal of the Linnean Society 104:564–574.

Chapter 9

Phylum Nemata

George O. Poinar Jr.
Department of Integrative Biology, Oregon State University, Corvallis, OR, USA

Chapter Outline

INTRODUCTION

Nematodes, as members of the phylum Nemata (or Nematoda), are one of the most abundant groups of invertebrates on earth and rival the Arthropoda in biodiversity and species abundance. While some 20,000 nematode species have been described, estimates for species diversity range from 100,000 to 10 million (Poinar, 2011). Their structure, physiology, diverse reproductive patterns, and adaptability have resulted in their invasion of more habitats on earth than any other multicellular group of animals. Their basic body plan is highly canalized as a result of selection pressure over millions of years. While this spells success for nematodes, it makes identification difficult, because fine details are often needed to differentiate between even families. Their diverse reproductive methods are also a key to their success. Aside from the standard amphimictic or bisexual condition, nematodes can undergo several types of autotoky (reproduction without males), including hermaphroditism, pseudogamy, and parthenogenesis. Some nematodes practice heterogamy (with sexual and autotokous life cycles) and switch from amphimixis to hermaphroditism during alternating life cycles. Unfortunately, very little is known about the biology and reproductive patterns of freshwater nematodes.

Nematodes are an extremely basal clade that has no close affinity to any extant invertebrate group, and the lineage from which they evolved is probably extinct. Nematodes have recently been included in a group of protostome animals called the Ecdysozoa based on phylogenetic trees constructed with 18S ribosomal RNA genes. This group also contains the arthropods and is characterized by invertebrates that shed their exoskeleton. This assignment is highly speculative, and others found that it was not supported. In fact, Blair et al. (2002) concluded that the arthropod members of the Ecdysozoa are more closely related to humans than to nematodes. Aquatic nematodes were originally placed in the now outdated morphological group known as the "aphasmidians" in contrast to the terrestrial "phasmidians." With the present classification, representatives of at least ten orders of nematodes occur in freshwater (Eyualem-Abebe et al., 2006). This great diversity shows how significant nematodes are in the general ecology of freshwater habitats.

LIMITATIONS

A compound microscope with high magnification lenses is needed to observe many of the diagnostic characters because most freshwater nematodes are so small that their distinguishing features are difficult to see. Only the mermithids (parasites of terrestrial and aquatic invertebrates, especially of insects) are large enough for the important characters to be seen without a dissecting microscope (Fig. 9.1). Over half of the described mermithid genera parasitize the immature and adult stages of freshwater insects. The free-living stages (postparasitic juvenile, adults, eggs, and infective juveniles) occur in freshwater habitats. Only the second and third stage juveniles occur inside invertebrate hosts. The eggs and infective stage juveniles are microscopic and normally not collected, but the postparasitic juveniles and adults are fairly large and easily detected.

Because the key to mermithids is based on adult characters, postparasitic juveniles emerging from hosts should be held in water until they molt to the adult stage. Even with many of the smaller, free-living freshwater nematodes, generic identification is only possible when the adults of

Thorp and Covich's Freshwater Invertebrates. http://dx.doi.org/10.1016/B978-0-12-385028-7.00009-3

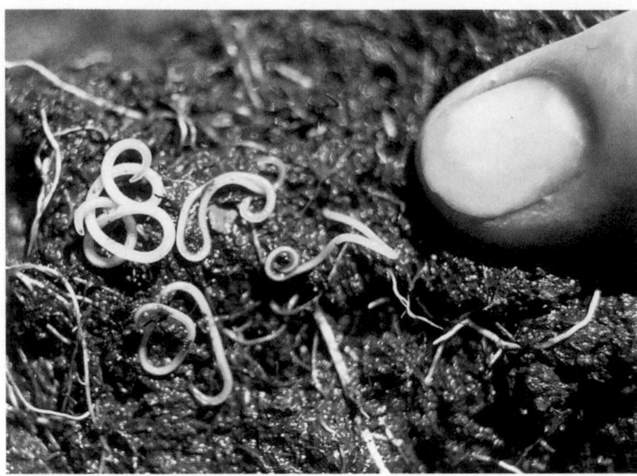

FIGURE 9.1 A long, white, smooth body and non-functional mouth are characters of free-living postparasitic and adult mermithid (Mermithidae) nematodes.

both sexes are available. Only genera of nematodes most likely to be encountered in freshwater habitats are included in the key. The study of freshwater nematodes is still in its infancy, and the assignment of many genera to the families listed below will certain be greatly modified after genomic studies are completed.

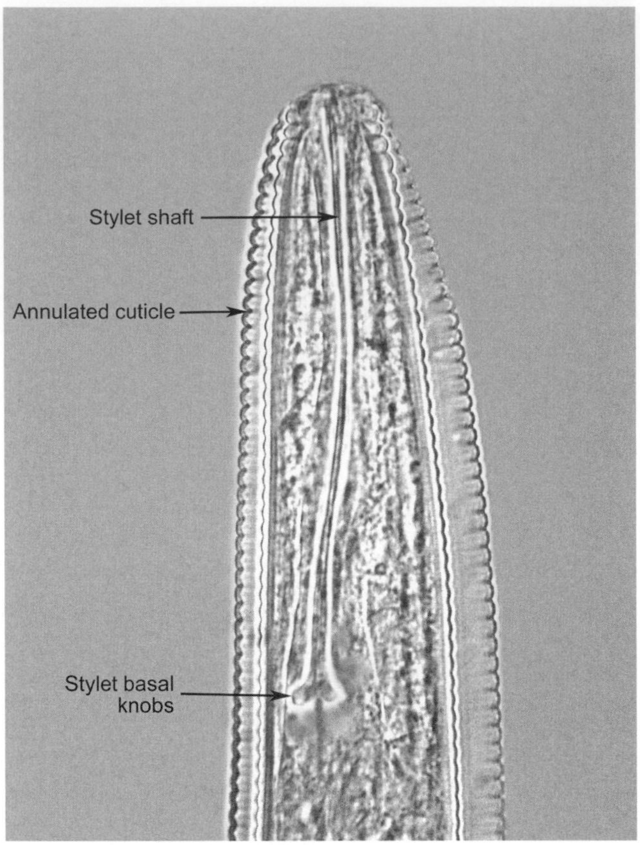

FIGURE 9.2 A protrusible stomatostylet bearing stylet knobs is found in many members of the Tylenchina.

TERMINOLOGY AND MORPHOLOGY

The basic structure of a nematode is a tube-within-a-tube since they lack appendages (Fig. 9.1). The outer tube represents the body wall and the inner one the alimentary tract (Fig. 9.2). The body is transparent to translucent, allowing a clear view of the internal organs (Figs. 9.3 and 9.4). The surface of the outer cuticular layer can be smooth or variously marked with transverse or longitudinal striations or ornamentations in the form of dots and folds (Figs. 9.5 and 9.6). The lateral fields are longitudinal lines extending in the cuticle over the lateral hypodermal cords. The hypodermal cords are protrusions of tissue between the longitudinal muscles. The most important characters are the structure of the stoma, stylet, and pharynx (which vary depending on the nematode's diet) (Figs. 9.2 and 9.7) and the structure of the female and male sexual organs, especially the latter (Figs. 9.6 and 9.8).

Around the mouth opening (top of stoma) are a series of cephalic papillae that comprise the lip region. The stoma is lined with a series of rhabdions that are usually fused but may show signs of separation. The top rhabdions are called the cheilorhabdions and surround the cheilostom. The stoma may contain a mural tooth (a cutting organ situated on the wall of the pharynx), a stomatostylet, which is derived from and situated within the stoma, or an odontostylet, which is a stylet that originates in the pharyngeal wall and has a dorsally oblique aperture. The odontostylet is supported by a ring of stiff extended tissue known as the odontophore. In some groups, the middle section of the stoma is called a stegostom and the lower portion a gymnostom. The pharynx (or esophagus) is composed of an anterior procorpus followed by a metacorpus that often contains a valve (valvated metacarpus) and is connected to the basal bulb by an isthmus. Diminutive teeth (toothlets) may occur in the pharyngeal lumen of some nematodes. The position and structure of the chemoreceptors known as amphids are also important characters since they are usually much more developed in freshwater than terrestrial nematodes. The accompanying figures (Figs. 9.3, 9.4, and 9.5) illustrate diagnostic features in the head region of several groups of freshwater nematodes.

Regarding the gonads, the male reproductive system consists of one or two testes. Sperm is carried out of the testes, down the vas deferens and released into the cloaca, a common terminal chamber of the intestinal and reproductive systems. Just before the rectum may be a prerectum, which is a segment of gut tissue between the intestine and rectum. The spicules are intermittent organs that are inserted into the vulva of the female during mating. At this time, sperm is released into the vagina, which extends from the uterus to the vulva (genital orifice). Many males have extensions of cuticle or bursa around their cloaca to grasp

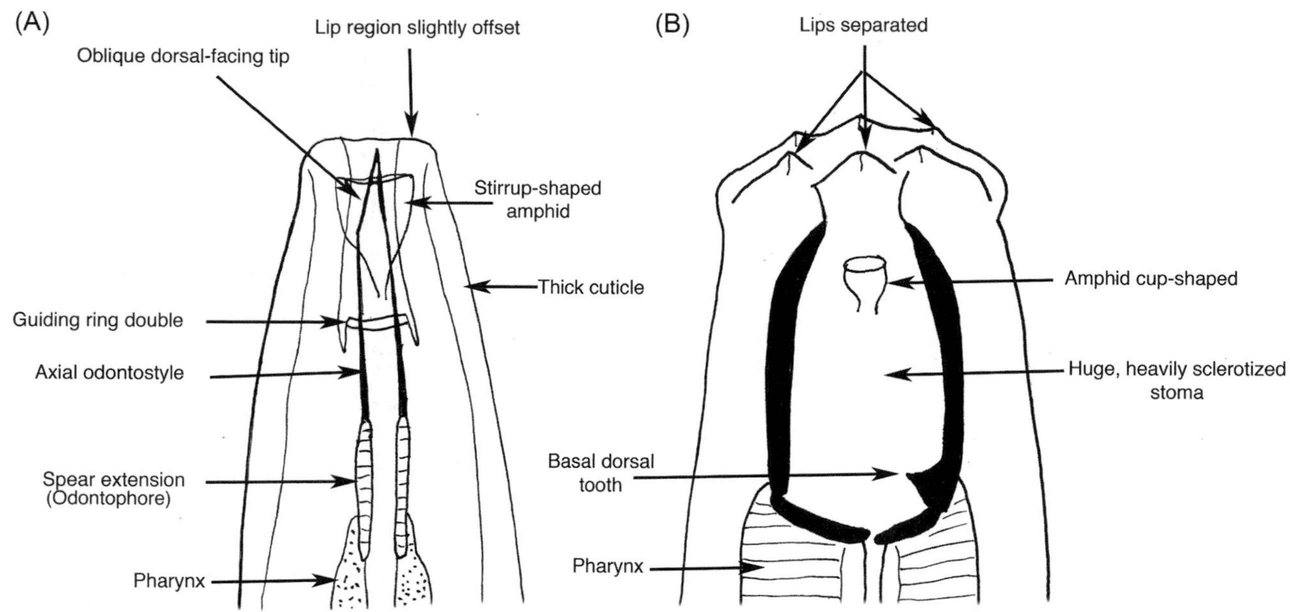

FIGURE 9.3 (A) An axial odontostyle with an oblique tip is characteristic of members of the Dorylaimida. (B) A wide, heavily sclerotized stoma with attached teeth is characteristic of members of the Mononchida.

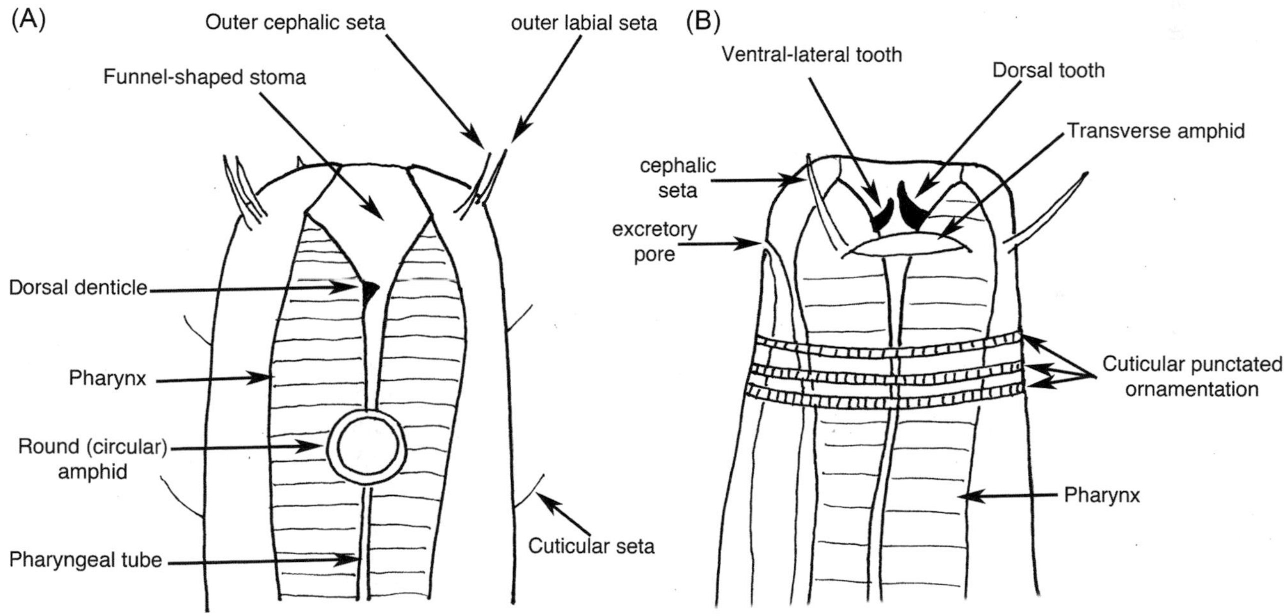

FIGURE 9.4 (A) A funnel-shaped stoma, round amphids and cuticular setae are characters found in the Monhysterida. (B) Transverse amphids and a cuticle with punctated ornamentation are characters of many Chromodorida.

the female. Many males also possess a sclerotized structure know as a gubernaculum, which supports the tips of the spicules as they leave the body during mating. Didelphic refers to two gonads and monodelphic to one. In opisthodelphic females, there is a single ovary that extends posterior toward the tail; while in prodelphic forms, the ovary extends toward the head.

Most freshwater nematodes (those that belonged to the old Aphasmidian group) have three caudal glands located in the tail region (Fig. 9.6). These glands produce a sticky secretion that passes down a tube and exits at the spinneret. This secretion attaches the nematode to various substrates. These caudal glands are not to be confused with the three rectal glands, whose secretions exit into the rectum of "phasmidian" nematodes.

One unusual feature of some freshwater nematodes is paired eyespots (ocelli), small masses of black or colored granules embedded in the pharynx. The function of these

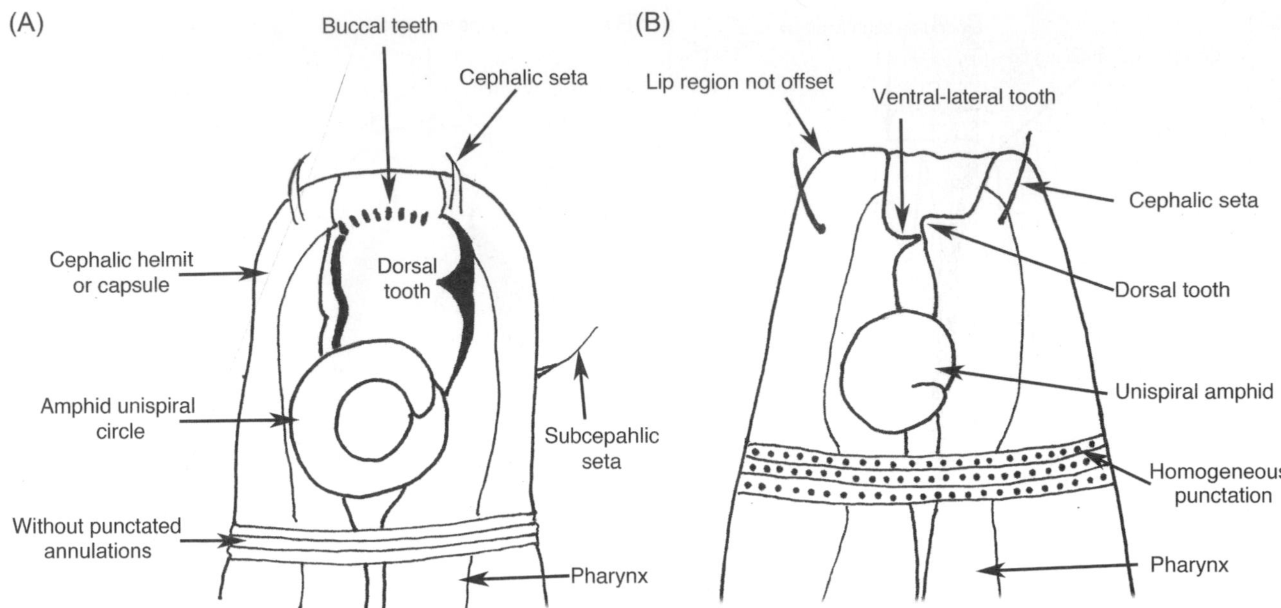

FIGURE 9.5 (A) Unispiral, circular amphids and transverse cuticular annulations lacking punctations are found in members of the Desmodorida. (B) Unispiral amphids and homogeneous cuticular punctation are characters of many Ethmolaimidae.

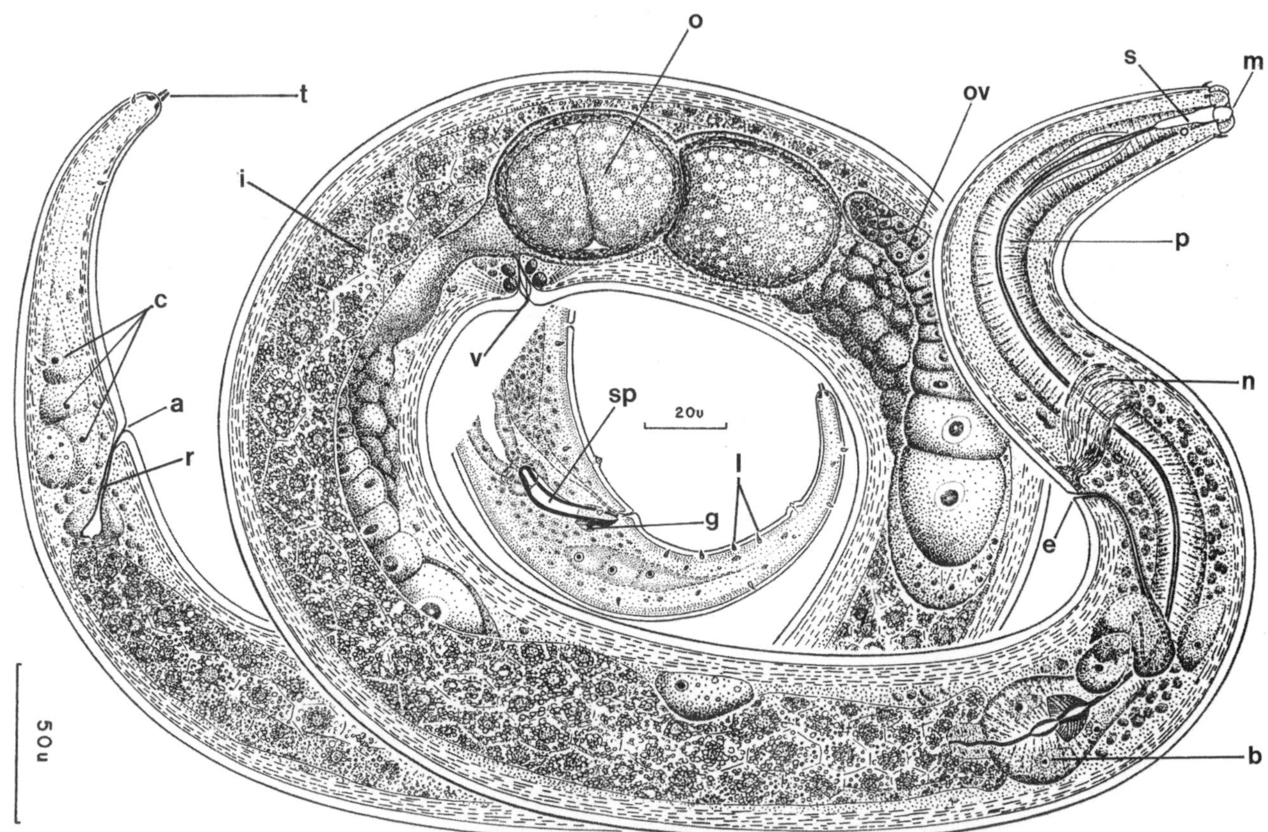

FIGURE 9.6 Adult female of *Plectus* (Plectidae) (center insert shows tail of male). a, anus; b, valvated basal bulb of the pharynx; c, caudal glands; sp, spicule; e, excretory pore; g, gubernaculum; i, intestine; l, genital papillae; m, mouth; n, nerve ring; o, developing ovum; ov, ovary; p, pharynx; r, rectum; s, stoma; t, spinneret; v, vulva. Drawing by the late A. Maggenti; labeling by G. Poinar.

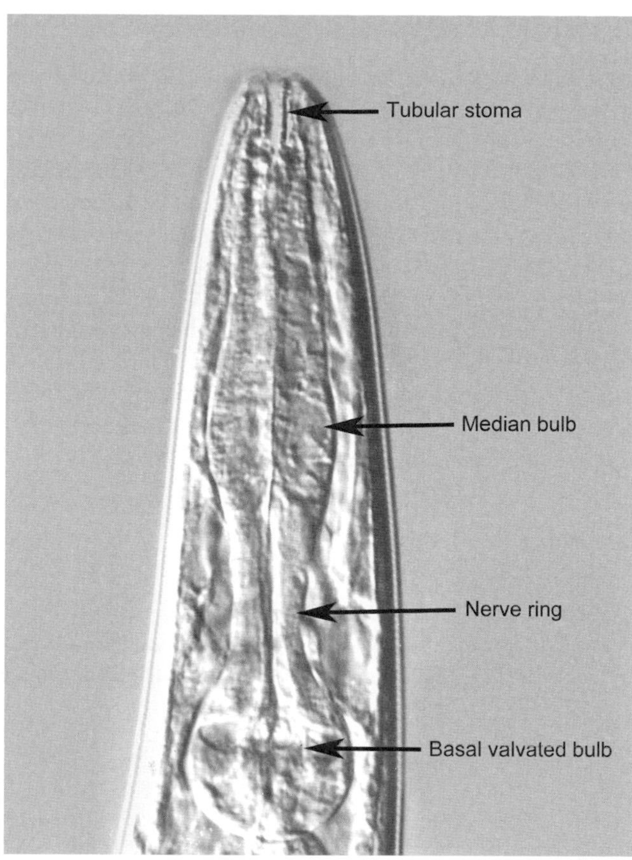

FIGURE 9.7 A tubular stoma and pharynx with a nonvalvated median bulb and a basal valvated bulb are characters found in the Rhabditomorpha.

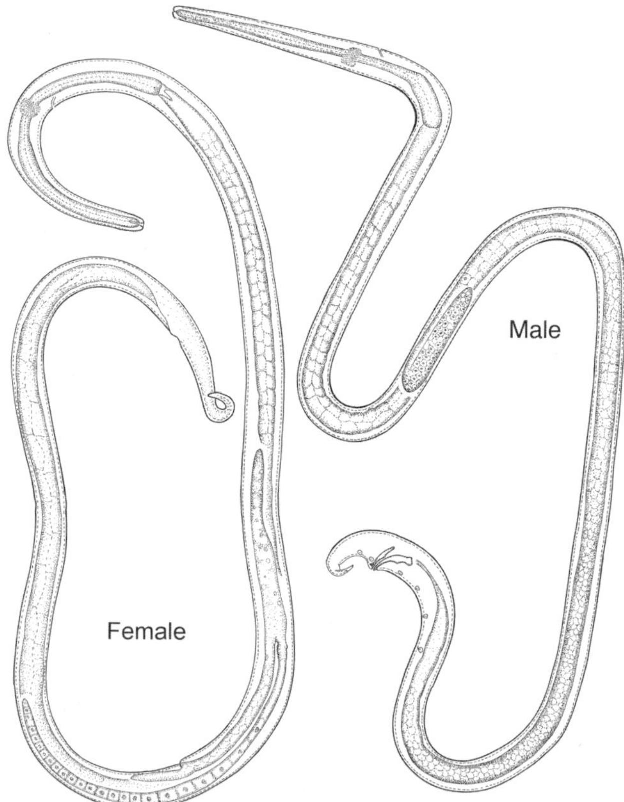

FIGURE 9.8 Slender, slow moving nematodes with an elongate pharynx lacking valves are characters of snail parasites in the genus *Daubaylia*.

eyespots is not clear, although some studies suggest that the pigments are light sensitive. Other body inclusions are crystalloids, small objects that occur within cells, and various types of parasites, especially fungal and protozoan pathogens.

MATERIAL PREPARATION AND PRESERVATION

Nematodes should be killed (relaxed) before being placed in fixative; otherwise, they become distorted and difficult to examine. They are best killed in hot water (60–70 °C) and immediately afterward should be transferred to the fixative. A good fixative for nematodes is TAF (7 ml 40% formalin, 2 ml triethanolamine, and 91 ml distilled water); however, 3–5% formalin or 70% ethanol can also be used if TAF is unavailable.

Nematodes should be left in the fixative for at least 2–3 days before being processed to glycerin for mounting on microscope slides. Nematodes are most easily processed to glycerin by the evaporation method, which requires little handling. Fixed specimens are transferred to a small container containing a solution of 70 ml ethanol (95%), 5 ml glycerol, and 25 ml water. The container is partly covered for the first 3 days to allow the alcohol and water to slowly evaporate, then the cover is removed, and the container is left open for the next 14 days. To remove all the water, the containers (now with mostly glycerin) are placed in a desiccator or oven (35 °C) for another 2 weeks. The nematodes can then be mounted directly on permanent microscope slides. Temporary slides are made by mounting the nematodes directly after fixation in a drop of the fixing solution. They will last several months if the ringing seal is tight. The following steps can be followed to make permanent slides with specimens in glycerin (modified from Poinar, 1983).

1. With a small pointed instrument (dental pulp canal file, small insect pin mounted on a wooden splint, needle), transfer the nematode(s) to a small drop of glycerin in the center of a microscope slide.
2. Push the nematode(s) to the bottom of the drop. Then place small supports (coverslip pieces, wire) with a width equal to or slightly wider than that of the nematode at three equidistant points around the nematode(s). Push these supports also to the bottom of the drop.
3. Place a coverslip over the drop and lower it slowly at a 30° angle so that air bubbles will not become entrapped in the glycerin.
4. Add more glycerin, if needed, by placing a small drop at the edge of the coverslip and allowing it to slowly move under the glass and spread. If there is too much glycerin, remove the excess by blotting it at the edge of the coverslip with a moistened (water) piece of tissue.
5. Carefully seal the edges of the coverslip with a ringing compound, such as nail polish or Turtox slide ringing cement (General Biological Supply House, Chicago, Illinois).

6. Label the slide with information regarding the date, locality, and collector.

Once mounted, the specimens can be examined under a compound microscope. Because identification may require examining fine details, a microscope equipped with oil immersion is beneficial (1000×). The use of dioscopic illumination after Nomarski or differential interference contrast can better reveal minute features of the cuticle, amphids, and stoma. Since certain characters are much clearer on living specimens, it is useful to examine them in water. Their movements can be reduced by removing water from under the coverslip so that the nematode is compressed between the microscope slide and coverslip.

KEYS TO FRESHWATER NEMATA

In the following key, the classification system follows that found in the book "Freshwater Nematodes" by Eyualem-Abebe, Andrássy, and Traunspurger (2006). This work, which includes world genera of free-living freshwater nematodes, can be consulted for detailed descriptions of the genera cited in the following key, as well as a list of species within these genera. Other books that cover identification and morphology of freshwater nematodes are Chitwood & Chitwood (1950), Esser & Buckingham (1987), Ferris et al. (1973), Goodey (1963), Maggenti (1981), and Poinar (1983). A glossary of nematological terms prepared by Caveness (1964) is helpful for clarification of many characters mentioned in the following keys.

Nemata: Orders

1	Not elongate, generally under 1 cm in length; mouth open, actively feeding stages	2
1'	Elongated and thread-like, generally over 1 cm in length; mouth closed, non-feeding in adult stage, invertebrate parasitic as juveniles	Mermithida, one family; **Mermithidae [p. 175]**
2(1)	Head bearing a stylet (Figs. 9.2 and 9.3 A)	3
2'	Head lacking a stylet, however mural teeth and other armature may be present (Figs. 9.3 B, 9.4 A, B, 9.5 A, B, 9.6, 9.7, and 9.8)	5
3(2)	With protrusible stomatostylet containing an aperture: stylet knobs usually present (Fig. 9.2); pharynx composed of a valvated metacorpus followed by a slender isthmus and basal glandular bulb leading into the intestine	4
3'	With protrusible axial odontostyle containing an aperture often with an extension (Fig. 9.3 A) or a non-aperture mural tooth; pharynx bottle-like, composed of narrow anterior portion and expanded posterior portion at junction with intestine; gubernaculum rare; valvated metacorpus and caudal glands absent	**Dorylaimida [p. 176]**
4(3)	Dorsal gland outlet in precorpus; metacorpus moderate in size (less than 3/4 body width)	**Tylenchida [p. 176]**
4'	Dorsal gland outlet in metacorpus; metacorpus large (3/4 of body width or more)	**Aphelenchida [p. 176]**
5(2)	Stoma cylindrical, tubular or funnel-shaped, not a heavily sclerotized oval cavity	6
5'	Stoma usually large and oval in outline, heavily sclerotized and armed with one or more teeth, denticles, etc. (Fig. 9.3 B)	**Mononchida [p. 178]**
6(5)	Cuticle with punctated ornamentation (Figs. 9.4 B and 9.5 B)	7
6'	Cuticle without punctated ornamentation (Fig. 9.5 A)	11
7(6)	Stoma with dorsal and ventrolateral teeth; amphids variable	8
7'	Stomal teeth absent; amphids unispiral or circular	Araeolaimida, Diplopeltidae, one genus; *Cylindrolaimus*
8(7)	One testis	9
8'	Two testes	10
9(8)	Cuticle with punctated ornamentation; amphids a transverse slit (Fig. 9.4 B)	Chromodorida, Chromadoridae, one genus; *Punctodora*
9'	Cuticle without punctated ornamentation (Fig. 9.5 A); amphids circular to unispiral (Fig. 9.4 A)	Desmodorida, Desmodoridae, one genus; *Prodesmodora*
10(8)	Cuticle with punctated ornamentation; amphids unispiral; male precloacal supplements present (Figs. 9.5 B and 9.6)	Ethmolaimida, Ethmolaimidae, one genus; *Ethmolaimus*
10'	Cuticle with punctated ornamentation; amphids multispiral; male precloacal supplements absent	Achromadorida, Achromadoridae, one genus; *Achromadora*
11(6)	Cephalic setae and caudal glands present; bursa usually absent	12
11'	Cephalic setae and caudal glands absent; bursa usually present	**Rhabditida [p. 179]**
12(11)	Stoma tubular or conical; amphids round (Fig. 9.4 A); cuticular setae common	Monhysterida, one family; **Monhysteridae [p. 179]**
12'	Stoma cylindrical, amphids unispiral (Fig. 9.5 A); cuticular setae rare	**Plectida [p. 180]**

Nemata: Mermithida: Mermithidae: Genera

1 Adult cuticle without cross fibers; slender, white, green, pink, brown, or yellow nematodes found in various water sources 2

1' Adult cuticle thick, with cross fibers; robust white nematodes usually found along the edges of bogs, springs, and streams (parasites of wasps, ants, and horseflies) (Fig. 1) .... *Pheromermis*

2(1) With six cephalic papillae 3

2' With four cephalic papillae .... *Pseudomermis*

3(2) With single or fused spicules 4

3' With paired, separate spicules 8

4(3) Spicule medium to long, more than twice anal body width 5

4' Spicule short, less than twice anal body width 6

5(4) Mouth terminal; spicule J-shaped; vulva flap present; [parasites of midges (Chironomidae)] .... *Lanceimermis*

5' Mouth shifted ventrally; spicule curved but not J-shaped; vulva flap absent; [parasites of blackflies (Simuliidae), midges (Chironomidae) and mayflies (Ephemeroptera)] .... *Gastromermis*

6(4) Spicule longer than anal body width; amphids medium to large 7

6' Spicule shorter than anal body width; amphids small; [parasites of mosquitoes Culicidae] .... *Perutilimermis*

7(6) Tail pointed; eight hypodermal cords; [parasites of midges (Chironomidae) and mosquitoes (Culicidae)] .... *Hydromermis*

7 Tail rounded; six hypodermal cords; [parasites of midges (Chironomidae) and blackflies (Simuliidae)] .... *Limnomermis*

8(3) Vagina straight or nearly so, barrel- or pear-shaped 9

8' Vagina S-shaped, U-shaped, or elongate with both ends curved 13

9(8) Spicules shorter than cloacal body width 10

9' Spicules equal to or longer than cloacal body width 11

10(9) Head expanded into a bulbiform shape with very thick cuticle; male lacking dorsolateral genital papillae [parasites of midges (Chironomidae)] .... *Capitomermis*

10' Head not expanded into a bulbiform shape with thick cuticle; male with genital papillae on dorsolateral surface; [parasites of biting midges (Ceratopogonidae)] .... *Heleidomermis*

11(9) Eight hypodermal cords 12

11' Six hypodermal cords; [parasites of blackflies (Simuliidac)] .... *Mesomermis*

12(11) Spicules 2–4 times anal body width; [parasites of mosquitoes (Culicidae)] .... *Romanomermis*

12b. Spicules 1–2 times anal body width; [parasites of mosquitoes (Culicidae) and midges (Chironomidae)] .... *Octomyomermis*

13(8) Six hypodermal cords 14

13' Eight hypodermal cords 16

14(13) Spicule length three or more times cloacal diameter 15

14' Spicule length less than two times anal diameter; [parasites of midges (Chironomidae) and mosquitoes (Culicidae)] .... *Strelkovimermis*

15(14) Spicules ten or more times body width at cloaca; vagina elongate with bends at both ends; postparasitic juveniles with distinct tail appendage; [parasites of diving beetles (Dytiscidae)] .... *Drilomermis*

15' Spicules 3–10 times body width at cloaca; vagina elongate with 3–6 irregular bends; postparasitic juveniles with indistinct or no tail appendage; [parasites of spiders] .... *Aranimermis*

16(13) Vagina S-shaped, ends distinctly curved; spicules range from shorter than anal body diameter to about 10 times tail diameter 17

16' Vagina elongate, slightly curved; spicules shorter than anal body diameter; [parasites of mosquitoes (Culicidae)] .... *Culicimermis*

17(16) Spicule length equal to or shorter than tail diameter; amphids small; cephalic crown well developed; [parasites of mosquitoes (Culicidae)] .... *Empidomermis*

17' Spicule length between one and ten times anal diameter; amphids medium to large, cephalic crown absent or only slightly developed; [parasites of blackflies (Simuliidae)] .... *Isomermis*

Nemata: Tylenchida: Families

1 Head without setae 2

1' Head bearing distinct setae .... Atylenchidae, one species *Atylenchus*

2(1)	Cuticle not strongly annulated; stomatostylet short to medium length; procorpus distinct and narrow before reaching expanded metacorpus	3
2'	Cuticle strongly annulated; stomatostylet very long, with basal knobs; procorpus fused with large, oval metacorpus (Fig. 9.2)	Hemicycliophoridae, one species *Hemicycliophora*
3(2)	Very large distinct stylet; ovaries paired, amphidelphic; tail tapering but not filiform	Pratylenchidae, one genus; *Hirschmanniella*
3'	Style small or medium: ovary single, prodelphic: tail ventrally arcuate of filiform	**Tylenchidae [p. 176]**

Nemata: Tylenchida: Tylenchidae: Genera

1	Stylet small, inconspicuous; tail ventrally arcuate	*Filenchus* Andrássy 1954
1'	Stylet medium, tail straight, filiform	*Tylenchus* Bastian, 1865

Nemata: Aphelenchida: Families

1	Stylet usually with faint or inconspicuous knobs; tail tip usually pointed; males lacking bursa and gubernaculum	**Aphelenchoididae [p. 176]**
1'	Stylet without knobs; tail tip usually rounded; males with bursa and gubernaculum	Aphelenchidae, one genus; *Aphelenchus*

Nemata: Aphelenchida: Aphelenchoididae: Genera

1	Tail elongate, filiform	*Seinura*
1'	Tail short, conical	*Aphelenchoides*

Nemata: Dorylaimida: Families

1	Stoma with a mural tooth	2
1'	Stoma with an axial odontostyle	3
2(1)	Mural tooth with diverging basal projections	Aporcelaimidae, one genus; *Sectonema*
2'	Mural tooth without diverging basal projections	**Nygolaimidae [p. 177]**
3(1)	Lip region continuous, cup-like or cap-like	4
3'	Lip region otherwise	7
4(3)	Lip region cup-like or cap-like	5
4'	Lip region continuous, lips amalgamated	6
5(4)	Lip region cup-like	**Tylencholaimellidae [p. 177]**
5'	Lip region cap-like	**Leptonchidae [p. 177]**
6(4)	Stylet shorter than or longer than lip region	**Aulolaimoididae [p. 177]**
6'	Stylet as long as lip region	**Qudsianematidae** (in part) **[p. 177]**
7(3)	Anterior pharynx not constricted before joining basal expansion (bulb)	8
7'	Anterior pharynx constricted before joining basal expansion	Axonolaimidae, one genus; *Axonchium*
8(7)	Lip region variable	9
8'	Lip region offset, discoid	Aetholaimidae, one genus; *Aetholaimus*
9(8)	Tail filiform or conical	10
9'	Tail short, roundish, clavate	**Belondiridae [p. 177]**
10(9)	Tail long, filiform in both sexes	11
10'	Tail short, conical in both sexes	**Qudsianematidae** (in part) **[p. 177]**
11(10)	Cuticle smooth, finely transversely striated or with superficial cross lines	12
11'	Cuticle otherwise, or with longitudinal ridges	**Dorylaimidae [p. 177]**
12(11)	Cuticle with superficial cross lines	**Aporcelaimidae [p. 178]**
12'	Cuticle smooth or finely transversely striated	**Nordiidae [p. 178]**

Nemata: Dorylaimida: Nygolaimidae: Genera

1	Small nematodes, under 2.5 mm in length	2
1'	Large nematodes, over 3 mm in length	*Feroxides*
2(1)	Lip region offset by constriction	*Nygolaimus*
2'	Lip region not offset, nearly continuous	*Aquatides*

Nemata: Dorylaimida: Tylencholaimellidae: Genera

1	Tail short and rounded	2
1'	Tail long and slender	*Oostenbrinkella*
2(1)	Odontophore with an accessory stiffening piece	*Tylencholaimellus*
2'	Odontophore without an accessory stiffening piece	*Doryllium*

Nemata: Dorylaimida: Leptonchidae: Genera

1	Cheilostom a truncate cone, females didelphic	2
1'	Cheilostom flask or goblet-shaped; females monodelphic	*Proleptonchus*
2(1)	Vulva longitudinal	*Funaria*
2'	Vulva transverse	*Leptonchus*

Nemata: Dorylaimida: Aulolaimoididae: Genera

1	Stylet longer than lip region	*Aulolaimoides*
1'	Stylet shorter than lip region	*Adenolaimus*

Nemata: Dorylaimida: Qudsianematidae: Genera

1	Posterior part of tail not appearing empty	2
1'	Posterior part of tail appearing empty	*Boreolaimus*
2(1)	Vulva transverse; supplements not contiguous	3
2'	Vulva longitudinal; supplements contiguous	*Labronema*
3(2)	Tail conical, longer than anal body width	4
3'	Tail rounded, about as long as body width	*Takamangai*
4(3)	Tail shorter, 1–3 anal body widths	5
4'	Tail longer than 3 body widths	*Epidorylaimus*
5(4)	Most posterior male supplements within range of spicule	6
5'	Most posterior male supplements anterior to spicule	*Eudorylaimus*
6(5)	Body less than 1 mm in length; pharynx expansion at 2/5 its length	*Microdorylaimus*
6'	Body greater than 1 mm in length; pharynx expansion at 1/2 its length	*Allodorylaimus*

Nemata: Dorylaimida: Belondiridae: Genera

1	Odontostylet narrow, less than half diameter of stoma; odontophore with 3 basal flanges	*Dorylaimellus*
1'	Odontostylet robust, greater than half diameter of stoma; odontophore lacking flanges	*Belondira*

Nemata: Dorylaimida: Dorylaimidae: Genera

1	Cuticle lacking longitudinal ridges	2
1'	Cuticle with longitudinal ridges	*Dorylaimus*

2(1)	Prerectum long	3
2'	Prerectum short	*Paradorylaimus*
3(2)	Prerectum beginning at level of supplements	*Mesodorylaimus*
3'	Prerectum beginning before supplements	*Laimydorus*

Nemata: Dorylaimida: Aporcelaimidae: Genera

1	Lips without inner liplets; vulva with smooth cuticle	2
1'	Lips with protruding inner liplets; vulva with wrinkled cuticle	*Epaerolaimus*
2(1)	Large nematodes from 3 to 10 mm in length	*Aporcelaimus*
2'	Smaller nematodes under 3 mm in length	*Aporcelaimellus*

Nemata: Dorylaimida: Nordiidae: Genera

1	Tail short, rounded or conoid	2
1'	Tail long, filiform	*Lenonchium*
2(1)	No sclerotized platelets around stoma	3
2'	Sclerotized platelets around stoma	*Pungentus*
3(2)	Cuticle smooth near vulva	4
3'	Cuticle coarsely striated near vulva	*Rhyssocolpus*
4(3)	Odontophore with basal flanges	*Enchodelus*
4'	Odontophore simple	*Dorydorella*

Nemata: Mononchida: Families

1	Stoma narrow, nearly all embedded in pharynx	2
1'	Stoma large and wide, only 1/4 or less embedded in pharynx (Fig. 9.3 B)	4
2(1)	Stoma moderately sclerotized, lacking a tooth	3
2'	Stoma strongly sclerotized with a large tooth on the ventrolateral side	**Mononchulidae [p. 178]**
3(2)	Stoma a long cylinder, female prodelphic, tail elongate	Cryptonchidae, one genus; *Cryptonchus*
3'	Stoma open anteriorly; female didelphic; tail short, rounded	Bathyodontidae, one genus; *Bathyodontus*
4(1)	Stoma broad, flattened at base, pharyngeal base with 3 tuberculi	**Anatonchidae [p. 178]**
4'	Stoma tapering at base; pharyngeal base lacking tuberculi	**Mononchidae [p. 178]**

Nemata: Mononchida: Mononchulidae: Genera

1	Six rows of transverse denticles in anterior part of stoma; tail elongate, cylindrical	*Mononchulus*
1'	Two to four rows of transverse denticles in anterior part of stoma; tail short, rounded	*Oionchus*

Nemata: Mononchida: Anatonchidae: Genera

1	Stoma with one dorsal tooth and sometimes many denticles on ventral wall	2
1'	Stoma with one dorsal tooth and 2–4 ventral-lateral teeth	3
2(1)	Small dorsal tooth near base of stoma (Fig. 9.3 B).	*Iotonchus*
2'	Dorsal tooth at mid-point or within anterior half of stoma	*Jensenonchus*
3(1)	All three teeth positioned at about same level in stoma	4
3'	Dorsal tooth in anterior half of stoma; ventral-lateral teeth basal	*Promiconchus*
4(3)	All teeth pointed forward	*Miconchus*
4'	All teeth pointed backward (retrorse)	*Anatonchus*

Nemata: Mononchida: Mononchidae: Genera

1 Ventral side of stoma with longitudinal rows of denticles or scattered denticles 2
1' Ventral side of stoma without denticles 3
2(1) Stoma with row of subventral denticles; tail lacking spinneret .... *Prionchulus*
2' Stoma with scattered series of subventral denticles; tail with spinneret .... *Actus*
3(1) Dorsal tooth in anterior part of stoma 4
3' Doral tooth in mid or lower part of stoma (Fig. 9.3 B) .... *Judonchulus*
4(3) Lip region not sharply offset 5
4' Lip region variable 6
5(4) Toothlet opposite dorsal tooth .... *Paramononchus*
5' No toothlet opposite dorsal tooth .... *Mononchus*
6(4) Tail elongate, cylindroid; with distinct spinneret 7
6' Tail conoid, lacking spinneret 8
7(6) Ventrolateral side of stoma with protruding ridge .... *Clarkus*
7' Ventrolateral side of stoma with thin transparent longitudinal ridge .... *Coomansus*
8(6) Teeth arranged differently than below 9
8' Three equal teeth in mid stoma; tail without spinneret .... *Comiconchus*
9(8) Ventral side of stoma with scattered denticles .... *Granonchulus*
9' Ventral side of stoma with transverse rows of denticles .... *Mylonchulus*

Nemata: Rhabditida: Families

1 Pharynx ranges from elongate without a bulb to short to medium with a valve in the median and/or basal portion 2
1' Pharynx with an expanded median bulb (metacorpus) containing a longitudinal valve and a glandular non-valvated basal bulb .... Rhabditolaimidae, one genus *Rhabditolaimus*
2(1) Pharynx short to medium in length, with a valve in the basal bulb (Fig. 9.7); stouter, rapid moving forms 3
2' Pharynx elongate, lacking valves; slender; slow moving forms (Fig. 9.8) (members of this family are parasites of aquatic snails and leeches) .... Daubayliidae, one genus *Daubaylia*
3(2) Stoma with rhabdions separate, female with single ovary, bursa absent .... **Cephalobidae** **[p. 179]**
3' Stoma tubular, with rhabdions fused (Fig. 9.7), female with single or paired ovaries, bursa usually present; rectal glands usually present .... Suborder Rhabditinamorpha

[Most genera in this large suborder are soil forms. While some appear from time to time in aquatic habitats, they can hardly be considered freshwater nematodes, since their life cycle is rarely completed in an aquatic habitat. They are certainly much less "aquatic" than nematode parasites of invertebrates and vertebrates whose eggs and early juvenile stages occur in the aquatic environment. Traditionally, the latter are not considered in works on aquatic nematodes. Even the mermithid parasites, with free-living eggs, first and last stage juveniles and adults in many freshwater habitats, are rarely given equal consideration in treatises on freshwater nematodes.]

Nemata: Rhabditida: Cephalobidae: Genera

1 Lateral fields reach to tail tip; female tail rounded .... *Cephalobus*
1' Lateral fields end at the phasmids; female tail usually pointed .... *Eucephalobus*

Nemata: Monhysterida: Monhysteridae: Genera

1 Small to medium species; stoma wide (Fig. 9.4 A); spinneret short 2
1' Small species (under 0.5 mm); stoma narrow; spinneret long .... *Monhystrella*
2(1) Crystalloids in cell bodies of somatic muscles absent 3
2' Crystalloids in cell bodies of somatic muscles; ocelli often present .... *Monhystera*
3(2) Stoma subdivided into 2 chambers, posterior portion denticulate; epizoic on gills of crustaceans .... *Gammarinema*
3' Stoma not subdivided into 2 chambers; at most only a single denticle in stoma; not epizoic on gills of crustaceans .... *Eumonhystera*

Nemata: Plectida: Families

1 Pharynx with distinct pharyngeal tube; each genital branch with single spermatheca, when latter present; male with at least 1 pair of postcloacal genital papillae (Fig. 9.6) 2

1' Pharynx with indistinct pharyngeal tube; each genital branch with two spermathecae; male without postcloacal genital papillae 3

2(1) Posterior stegostom long; female monoprodelphic .... Chronogastridae, one genus; *Chronogaster*

2' Posterior stegostom short; female didelphic (Fig. 9.6) .... **Plectidae [p. 180]**

3(1) Stoma short, excretory pore opens in cheilostom .... **Aphanolaimidae [p. 180]**

3' Stoma long, tubular; excretory pore opens to exterior through cuticle .... Leptolaimidae, one genus; *Paraplectonema*

Nemata: Plectida: Plectidae: Genera

1 Amphid unispiral (circular in lateral view) (Fig. 9.5 A and B) .... *Plectus*

1' Amphid a transverse slit .... *Anaplectus*

Nemata: Plectida: Aphanolaimidae: Genera

1 Gymnostom broad, cylindrical .... *Paraphanolaimus*

1' Gymnostom short, narrow .... *Aphanolaimus*

REFERENCES

Blair, J. E., K. Ikeo, T. Gojobori & S. B. Hedges. 2002. The Evolutionary position of Nematodes. BMC Evolutionary Biology 2: 7 doi:10.1186/1471-2148-2-7.

Caveness, F. E. 1964. A glossary of Nematological Terms. Moor Plantation Press, Ibadan.

Chitwood, B. G. & M. B. Chitwood.1950. Introduction to Nematology. University Park Press, Baltimore, Maryland.

Esser, R. P. & G. R. Buckingham. 1987. Genera and species of free-living nematodes occupying fresh-water habitats in North America. Pages 477–487 *in*: J. A. Veech, and D. W. Dickson (eds.), Vistas on Nematology. Society of Nematologists, Hyattsville, Maryland.

Eyualem-Abebe, I. Andrássy & W. Traunspurger. 2006. Freshwater Nematodes: Ecology and Taxonomy. CABI publishing, Cambridge.

Ferris, V. R., J. M. Ferris, & J. P. Tjepkema. 1973. Genera of freshwater nematodes (Nematoda) of Eastern North America. U.S. Environmental Protection Agency Identification Manual No. 10. 38 p.

Goodey, J. B. 1963. Soil and Freshwater Nematodes. Methuen, London.

Maggenti, A. 1981. General Nematology. Springer-Verlag, New York.

Poinar, G. O., Jr. 1983. The Natural History of Nematodes. Prentice-Hall. Englewood Cliffs, New Jersey.

Poinar, G. O., Jr. 2011. The Evolutionary History of Nematodes. Brill, Leiden.

Chapter 10

Phylum Nematomorpha

Andreas Schmidt-Rhaesa
Zoological Museum, University Hamburg, Hamburg, Germany

L. Cristina de Villalobos
Museo de Ciencias Naturales, La Plata, Argentina

Fernanda Zanca
Centro de Estudios Parasitológicos y Vectores (CEPAVE), CCT-La Plata, CONICET, FCNyM, Gordiida UNLP, La Plata, Argentina

Ben Hanelt
Department of Biology, University of New Mexico, Albuquerque, NM, USA

Matthew G. Bolek
Department of Zoology, Oklahoma State University, Stillwater, OK, USA

PHYLUM NEMATOMORPHA

Chapter Outline

INTRODUCTION

Twenty-two species of freshwater horsehair worms (Gordiida) are known from the Nearctic. Together with five marine species from the genus *Nectonema*, Gordiida form the taxon Nematomorpha. They are parasites of insects (mostly grasshoppers, crickets, and beetles) and emerge from these hosts at maturity. The usually terrestrial host is driven towards water, where the horsehair worm is released. Reproduction takes place in the water, where the tiny and morphologically distinct larvae infect paratenic hosts. Although relatively little is known about the gordiid life cycle, the intermediate host (e.g., mosquitos, mayflies) generally spans the gap between the aquatic and the terrestrial environment. Recent reviews were published by Hanelt et al. (2005) and Schmidt-Rhaesa (2012). The Nearctic nematomorphs have been reviewed by Schmidt-Rhaesa et al. (2003).

LIMITATIONS

Sexes, but usually not species, can be distinguished with simple optics (e.g., a magnifying glass). In females, the cloacal opening is always terminal on the rounded posterior end, with the exception of *Paragordius varius*, where the posterior end has three lobes, and the cloacal opening is located between these lobes. In males, the cloacal opening is always subterminal and on the ventral side. Most Nearctic species have two tail lobes posterior of the cloacal opening (*Gordius, Gordionus, Parachordodes*, and *Paragordius*). In contrast, representatives of *Chordodes, Neochordodes*, and *Pseudochordodes* have no tail lobes but still possess a ventral cloacal opening. The genus *Gordius* can be recognized by a semicircular fold, the postcloacal crescent.

Gordiids are sometimes confused with large nematodes from the taxon Mermithida. In premature, whitish stages, these taxa may be indistinguishable. Some hints for a proper distinction are a usually darker color in gordiids, pointed (Mermithida) versus rounded (Gordiida) body ends, the presence of tail lobes (some Gordiida), or the presence of a white anterior tip followed by a dark brown ring (Gordiida) (Fig. 10.1). In addition, mermithids are usually transparent under transmitted light, whereas gordiids will appear opaque due to their thick layering of cuticle.

Thorp and Covich's Freshwater Invertebrates. http://dx.doi.org/10.1016/B978-0-12-385028-7.00010-X

FIGURE 10.1 Ventral view on head end of an undetermined gordiid species showing the whitish anterior tip, followed by a darker collar. A dark ventral line extends from the collar along the body.

Proper identification requires higher magnifying techniques. A superficial cuticle sample with attaching musculature removed gives a good impression of the surface structure, especially the pattern of areoles. Nevertheless, scanning electron microscopy (SEM) has become the standard tool for species identification of gordiids, as fine structures can be documented most reliably.

One problem is that the significance of intraspecific morphological variation of characters in gordiids is not well understood. Thus, it is not clear whether minute morphological differences among specimens indicate species differences or are due to intraspecific variation.

Molecular "barcoding," the comparison of fast evolving sequences among specimens, has recently become more common in gordiids (e.g., Begay et al., 2012) and may help identify the numerous cryptic species within this phylum. For example, the common species "*Gordius robustus*" likely represents a large cryptic species complex composed of at least eight distinct genetic lineages (Hanelt et al., 2015) with very few diagnostic characters (e.g., smooth cuticle, absence of bristles). Some further species names are regarded as synonyms (see Schmidt-Rhaesa et al., 2003). For a review on the genus *Gordius*, see Schmidt-Rhaesa (2010).

Not all species of *Gordionus* are well and reliably described (see Schmidt-Rhaesa et al., 2003).

At the parasitic stage, gordiids can usually not be distinguished morphologically. Due to the parasitic lifestyle, adult, free-living specimens are seasonal and rarely found. As gordiid larvae encyst in a variety of paratenic hosts, a search for cysts in squeezed host tissue can provide information on the presence of cysts. At least the larvae of the three most common species of gordiids can be distinguished on the morphology of the encysted larva (see Hanelt & Janovy, 2002; Szmygiel et al., 2014).

TERMINOLOGY AND MORPHOLOGY

Gordiids have comparably few diagnostic characters. In the anterior end, the shape (distinctly tapering or more stout), as well as the coloration, is important. The anterior tip is usually whitish, in several species; this is followed by a dark brown or black ring often called dark collar. A ventral and probably also a dorsal longitudinal line of darker coloration can be present. In the posterior end, tail lobes can be present in females of *Paragordius varius* and in males of *Gordius, Paragordius, Gordionus*, and *Parachordodes* (Fig. 10.2 A). The single cloacal opening in the posterior end leads into the intestinal tract as well as to the gonads. A semicircular fold posterior of the cloacal opening in males is called the postcloacal crescent (Fig. 10.2 C). Other structures in the posterior end of males are cuticular bristles or conical spines. In some species, such structures are clustered in certain regions. Spines are scattered on the inner side of the tail lobes; these are often called the postcloacal spines (Fig. 10.2 A). Bristles can be present on the entire ventral side of the posterior end of the tail lobes or in rows anterolateral of the cloacal opening; these are the precloacal bristlefields (Fig. 10.2 A).

The cuticle is smooth in few species but features regular elevations ("areoles") in others (Fig. 10.3 A–G). Areoles can be present as polygonal structures covering the entire cuticle (Fig. 10.3 A, E–G). They are separated by grooves, in which interareolar structures can be present (Fig. 10.3 G). In *Pseudochordodes*, two different types of areoles are present, with the larger one called a megareole (Fig. 10.3 C, D). In *Chordodes*, elevated areoles ("crown areole") have an apical "crown" of short or long bristles (Fig. 10.3 B).

The larvae are morphologically distinct from adults. They are very small (50–100 μm) and are composed of two body regions, a preseptum and a postseptum (Fig. 10.5 B–E). The preseptum includes rings of hooks and three stylets to penetrate the intestinal wall of the intermediate host (Fig. 10.5 C). A large glandular structure, the pseudointestine, occurs in the postseptum (Fig. 10.5 B–D) and varies structurally among species. Anterior of the pseudointestine, another gland can be present. In the intermediate host, the larvae are coiled inside cysts (Fig. 10.5 A). Larvae can be distinguished by their size, the number of folds within the cyst, and by the structure of the pseudointestine (see Hanelt & Janovy, 2002).

MATERIAL PREPARATION AND PRESERVATION

Nematomorphs are best preserved in 70% ethanol, except for DNA work, for which material should be preserved in 100% ethanol and placed into the freezer. For scanning electron microscopy, follow the usual protocols of dehydration, critical point drying, and sputtering. For light microscopy, cuticular samples can either be removed by a tangential section with a razor blade when the worm is stretched over a finger, or by cutting a complete section from the midbody region. This section (length about 1 mm) should be cut longitudinally and internal tissue scraped off with a scalpel. The cuticle can then be observed on a slide either in ethanol or preferably after transfer into glycerol.

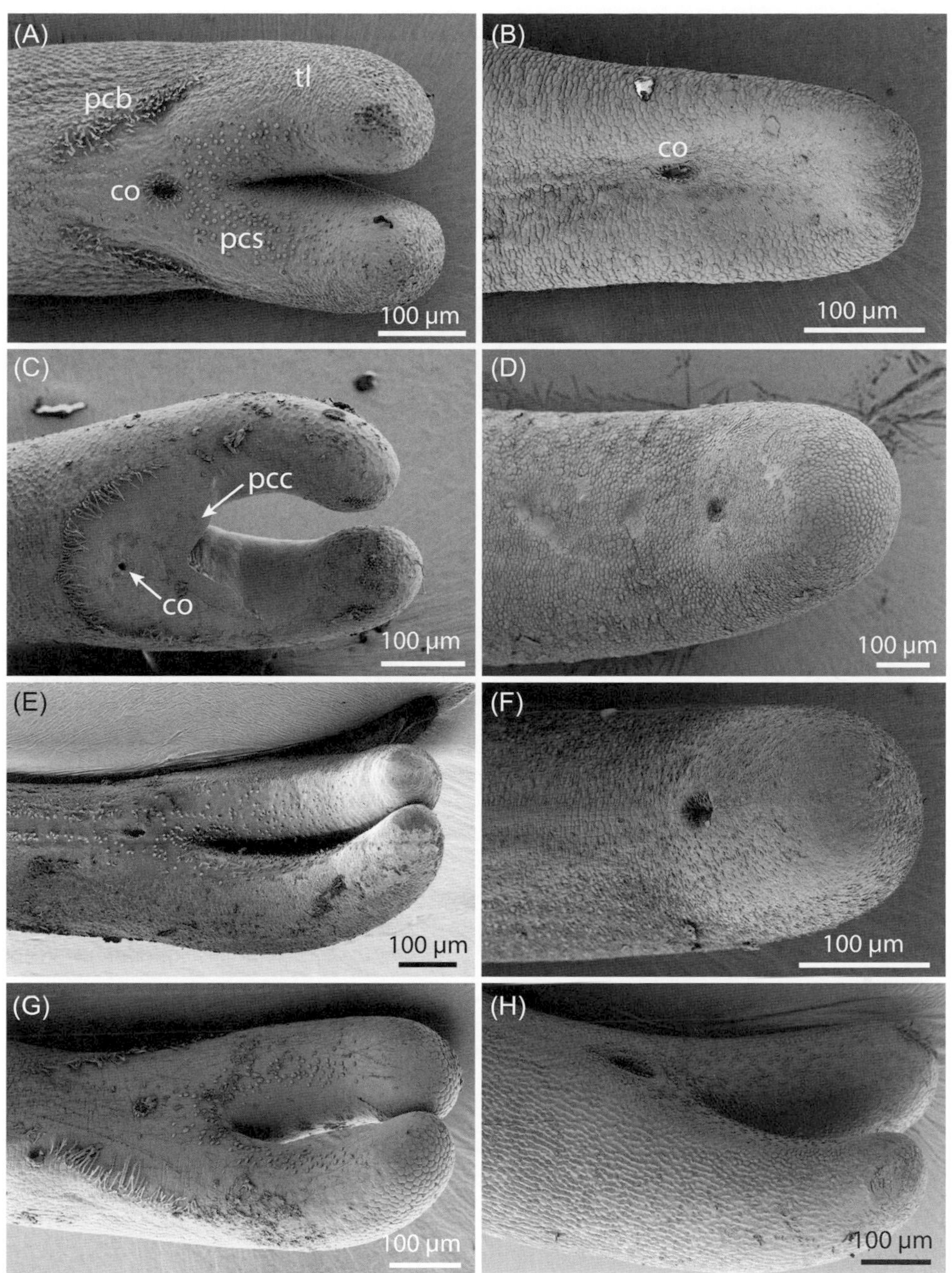

FIGURE 10.2 Ventral side of posterior ends in males. (A) *Gordionus violaceus*. (B) *Neochordodes occidentalis*. (C) *Gordius difficilis*. (D) *Pseudochordodes bulbareolatus*. (E) *Paragordius varius*. (F) *Chordodes morgani*. (G) *Gordionus lineatus*. (H) *Gordionus sinepilosus*. Abbreviations: co = cloacal opening, pcb = precloacal bristles, pcc = postcloacal crescent, pcs = postcloacal spines, tl = tail lobes. All images SEM.

KEYS TO GORDIIDA

Gender

1	Ventral cloacal opening (Fig. 10.2 A–H)	male
1'	Terminal cloacal opening (Fig. 10.4 B)	female

Gordiida Males: Genera

1	Paired tail lobes present (Fig. 10.2 A)	2
1'	Tail lobes absent, posterior end rounded (Fig. 10.2 B, D, F)	4

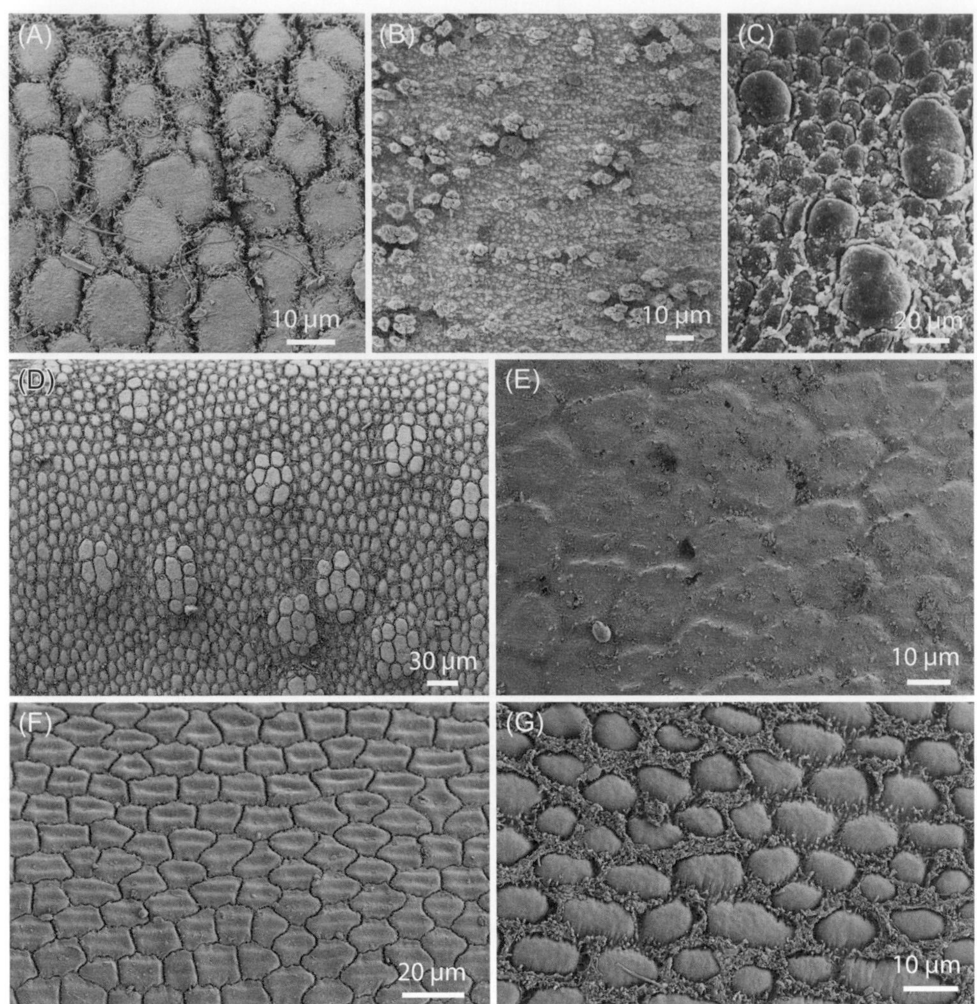

FIGURE 10.3 Areoles on the cuticular surface. (A) *Neochordodes occidentalis*. (B) *Chordodes morgani* with groups of elevated areoles among flatter parts. (C) *Pseudochordodes manteri*, elevated large megareoles among smaller simple areoles. (D) *Pseudochordodes bulbareolatus*, clusters of megareoles bulge out. (E) *Gordius attoni*. (F) *Gordionus lineatus*. No interareolar structures present. (G) *Gordionus violaceus* with interareolar bristles. All images SEM.

2(1) Postcloacal crescent absent (Fig. 10.2 A) 3

2' Postcloacal crescent present (Fig. 10.2 C) .... ***Gordius* [p. 185]**

3(2) Tail lobes longer than two times the width of one lobe (Fig. 10.2 G) .... *Paragordius varius* Leidy, 1851
[Canada, USA, Mexico]

3' Tail lobes shorter than two times the width of one lobe (Fig. 10.2 A, G, H) .... ***Gordionus* or *Parachordodes* [p. 185]**

4(1) More than one type of areoles present (Fig. 10.3 B–D) 5

4' Only one type of areoles present (Fig. 10.3 A) .... ***Neochordodes* [p. 187]**

5(4) Crown areoles present (Fig. 10.3 B) .... *Chordodes morgani* Montgomery, 1898
[USA: Florida, Iowa, Indiana, Kentucky, Maryland, Michigan, Missouri, Nebraska, North Carolina, Pennsylvania, South Carolina, Tennessee, Texas, Wisconsin]

5' Crown areoles absent (Fig. 10.3 C, D) .... ***Pseudochordodes* [p. 187]**

Gordiida Females: Genera

1 Posterior end rounded (Fig. 10.4 B) 2

1' Three tail lobes present (Fig. 10.4 A) .... *Paragordius varius* Leidy, 1851
[Canada, USA, Mexico]

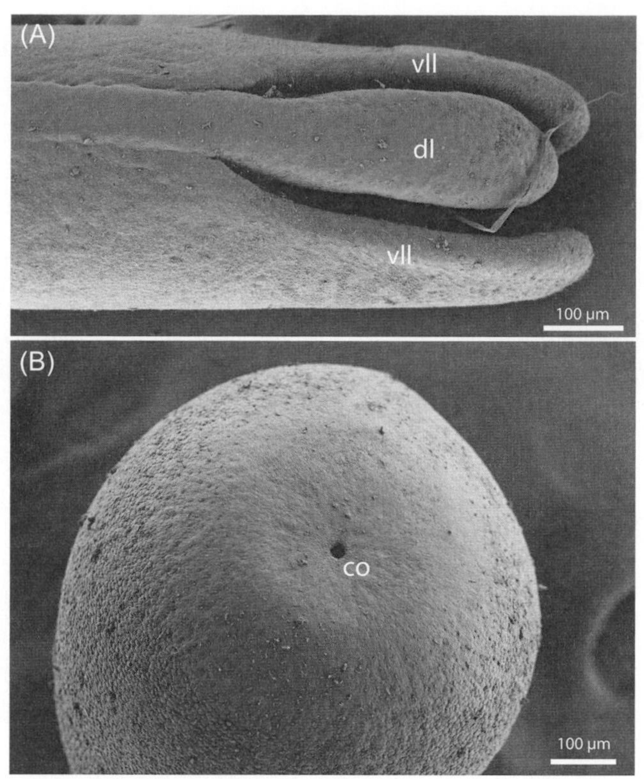

FIGURE 10.4 Posterior end of females. (A) *Paragordius varius* with three lobes, one dorsal (dl) and two larger ventrolateral ones (vll). (B) Round posterior end of a non-Nearctic *Chordodes* female. The cloacal opening (co) is terminal. All images SEM.

2(1) Cuticle with areoles 3

2' Cuticle smooth .... *Gordius robustus* Leidy, 1851

3(2) Crown areoles absent 4

3' Crown areoles present (Fig. 10.2 C) .... *Chordodes morgani* Montgomery, 1898

[USA: Florida, Iowa, Indiana, Kentucky, Maryland, Michigan, Missouri, Nebraska, North Carolina, Pennsylvania, South Carolina, Tennessee, Texas, Wisconsin]

4(1) Only one type of areoles present (Fig. 10.3 A, E–G) .... ***Gordius, Gordionus,*** or ***Neochordodes*** **[p. 187]**

4' More than one type of areoles present (Fig. 10.3 C, D) .... Pseudochordodes or *Parachordodes tegonotus* Poinar, Rykken & LaBonte, 2004

[USA: Oregon]

Gordiida: Male *Gordius:* Species

1 No row, semicircular or parabolic postcloacal crescent of bristles present anterior of cloacal opening 2

1' Semicircular continuous row of bristles anterior of cloacal opening, angled postcloacal crescent (Fig. 10.2 C) .... *Gordius difficilis* Smith, 1994

[USA: Massachussetts, Nebraska, North Carolina, Wisconsin]

2(1) Cuticle smooth .... *Gordius robustus* Leidy, 1851

[Canada: Alberta, probably New Brunswick. USA. Mexico.]

2' Cuticle with polygonal areoles (Fig. 10.3 E) .... *Gordius attoni* Redlich, 1980

[USA: New York, Canada: Saskatchewan]

Gordiida: Male *Gordionus:* Species

1 Precloacal bristle rows present 2

1' Precloacal bristle rows absent 6

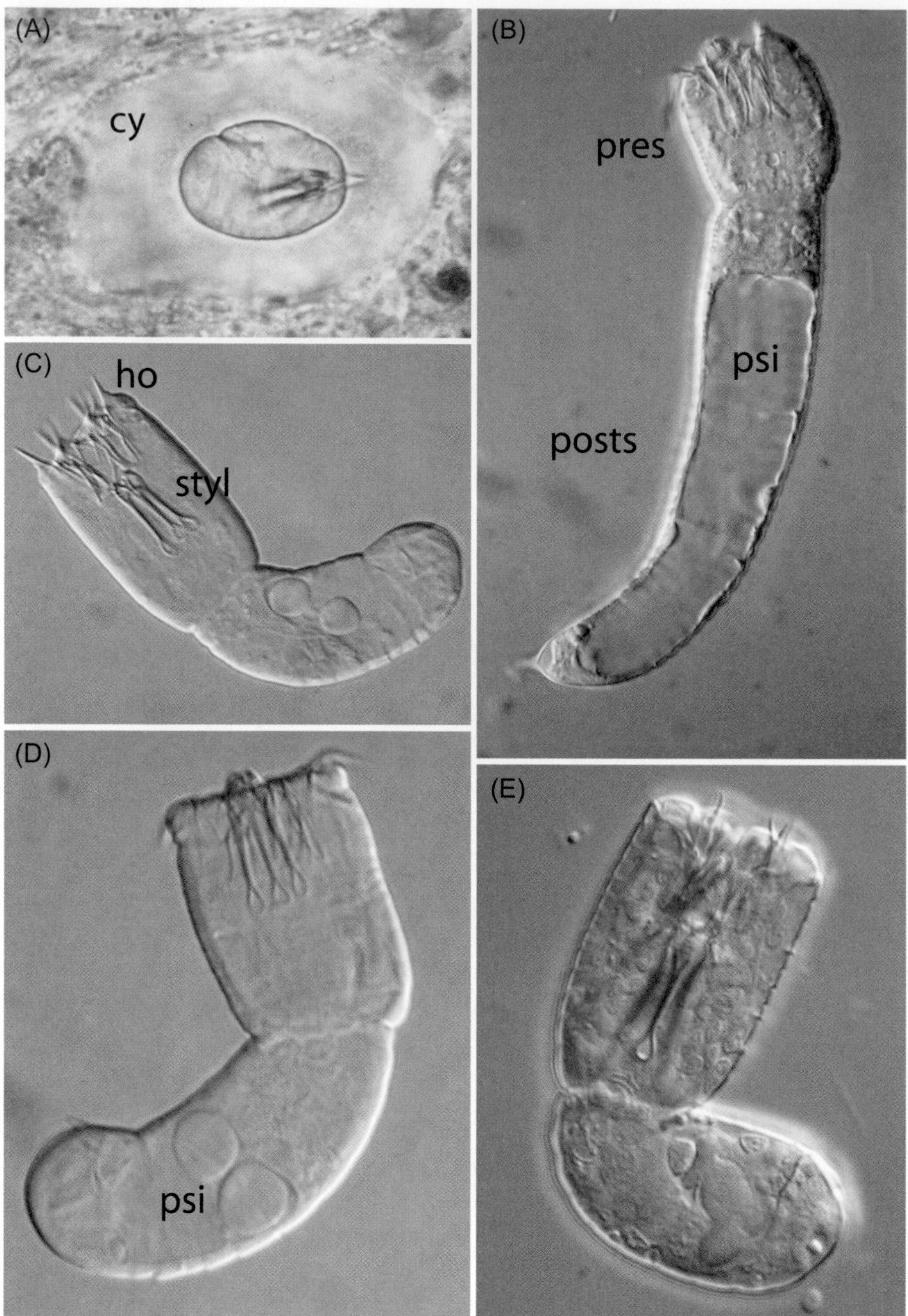

FIGURE 10.5 Larvae. (A) Larva of *Paragordius varius* coiled inside cyst (cy). (B) Larva of *Gordius* sp., entire length about 100 μm. Body is divided into preseptum (pres) and postseptum (posts); latter includes the pseudointestine (psi). (C) Larva of *P. varius* with partly everted hooks (ho) and stylet (styl). (D) Larva of *P. varius* with spherical granules in pseudointestine (psi). (E) Larva of *Neochordodes occidentalis*. Larval length in 3–5 approximately 50 μm.

2(1)	Circumcloacal spines slender and pointed	3
2'	Cloacal opening surrounded by broad, stout spines	*Gordionus lokaaus* Begay, Schmidt-Rhaesa, Bolek & Hanelt, 2012
	[USA: New Mexico]	
3(2)	Areolar surface smooth	4
3'	"Warty" surface on areoles	*Gordionus platycephalus* (Montgomery, 1898)
	[USA: Montana, Wyoming, Canada: Quebec]	
4(3)	Interareolar spaces narrow, with no or few interareolar structures	5
4'	Interareolar spaces broad, with interareolar structures (Fig. 10.3 G)	*Gordionus violaceus* (Baird, 1853)
	[USA: Arizona, California, Massachusetts, Missouri, Nebraska]	

5(4) Interareolar spaces completely surround areoles (Fig. 10.3 E) .... *Gordionus lineatus* (Leidy, 1851)
[USA: Maryland, Massachusetts, Michigan, New York, Pennsylvania, Tennessee]

5' Areoles partly fused with each other .... *Gordionus bilaus* Begay, Schmidt-Rhaesa, Bolek & Hanelt, 2012
[USA: New Mexico]

6(1) No precloacal ridge 7

6' Precloacal ridge present .... *Gordionus alascensis* (Montgomery, 1907)
[USA: Alaska]

7(6) Polygonal areoles, interareolar structures present, no precloacal bristles (Fig. 10.2 H).................. *Gordionus sinepilosus* Schmidt-Rhaesa, Hanelt & Reeves, 2003
[Canada: British Columbia]

7' Elongate areoles, interareolar structures absent .... *Gordionus longareolatus* (Montgomery, 1898)
[USA: California]

Gordiida: Male *Neochordodes:* Species

1 Cuticle areoles equally distributed 2

1' Areoles may be clustered, with broad spaces between .... *Neochordodes nietoi* Caballero, 1936
["Mexico." The cuticular structure of *N. nietoi* needs confirmation by SEM]

2(1) Areoles smooth (Fig. 10.3 A) .... *Neochordodes occidentalis* (Montgomery, 1898)
[USA: California, Montana, Nebraska, Nevada, Texas, Utah, Washington, Wyoming. Mexico: San Francisco, Veracruz]

2' Areoles with small marginal knobs .... *Neochordodes californensis* De Miralles & De Villalobos, 1995
[USA: California]

Gordiida: Male *Pseudochordodes:* Species

1 Both areole types smooth, second type enlarged, darker and bulging 2

1' First type of areoles smooth, second type with "warty" surface .... *Pseudochordodes gordioides* (Montgomery, 1898)
[USA: Arizona, California, Montana, Utah]

2(1) Larger areole clusters (of 3 or more) abundant 3

2' Majority of larger areoles in clusters of two, single ones and few larger clusters can occur (Fig. 10.3 C).................. *Pseudochordodes manteri* Carvalho, 1942
[USA: Nebraska, Mexico: Veracruz]

3(2) Areoles in large clusters 4

3' Many single large areoles and moderate number of larger clusters .... *Pseudochordodes meridionalis* Carvalho & Feio, 1950
[Mexico: Tamanlipas]

4(3) Most clusters contain 4 areoles .... *Pseudochordodes texanus* Schmidt-Rhaesa, Hanelt & Reeves, 2003
[USA: Texas]

4' Most clusters contain more than 4 areoles (Fig. 10.3 D).................. *Pseudochordodes bulbareolatus* Schmidt-Rhaesa & Menzel, 2005
[Mexico: Vcracruz]

Gordiida: Female: Species

Apart from the three species *Gordius robustus*, *Paragordius varius*, and *Chordodes morgani* that can be determined directly with keys above, female diagnosis is quite difficult. Please note that even species from different genera cannot be distinguished with certainty in females. The following key therefore has to be used with great caution.

1 Areoles of two types 2

1' All areoles of one type 7

2(1) The two types of areoles differ in size, both have a smooth surface 3

2' Both areole types with different surfaces: first type of areoles smooth, second type with "warty" surface.................. *Pseudochordodes gordioides* (Montgomery, 1898)
[USA: Arizona, California, Montana, Utah]

3(2) Larger areoles are separated from each 4
3' Larger areoles are partly fused with each other .... *Parachordodes tegonotus* Poinar, Rykken & LaBonte, 2004
[USA: Oregon]
4(3) Larger clusters (of 3 and more areoles) are more abundant 5
4' Majority of larger areoles in clusters of two, single ones and few larger clusters can occur (Fig. 10.3 C) .... *Pseudochordodes manteri* Carvalho, 1942
[USA: Nebraska, Mexico: Veracruz]
5(4) Larger clusters of areoles 6
5' Many single large areoles and moderate number of larger clusters .... *Pseudochordodes meridionalis* Carvalho & Feio, 1950
[Mexico: Tamanlipas]
6(5) Most clusters contain 4 areoles .... *Pseudochordodes texanus* Schmidt-Rhaesa, Hanelt & Reeves, 2003
[USA: Texas]
6' Most clusters contain more than 4 areoles (Fig. 10.3 D) .... *Pseudochordodes bulbareolatus* Schmidt-Rhaesa & Menzel, 2005
[Mexico: Veracruz]
7(1) Areoles cover the entire body, with no or only small grooves between them 8
7' Areoles may be clustered, with broad spaces to other areoles .... *Neochordodes nietoi* Caballero, 1936
["Mexico." The cuticular structure of *N. nietoi* needs confirmation by SEM]
8(7) Areole surface smooth 9
8' Areoles with small marginal knobs .... *Neochordodes californensis* De Miralles & De Villalobos, 1995
[USA: California]
9(8) Areoles separated from each other 10
9' Areoles partly fuse with each other .... *Gordionus bilaus* Begay, Schmidt-Rhaesa, Bolek & Hanelt, 2012
[USA: New Mexico]
10(9) Interareolar structures present .... *Gordionus longareolatus, G. platycephalus, G. sinepilosus* or *G. violaceus*
10' Interareolar structures absent .... *Gordius difficilis, G. attoni, Gordionus alascensis, G. lineatus, G. lokaaus* or *Neochordodes occidantalis*

ACKNOWLEDGMENTS

We thank the National Science Foundation for funding our Nematomorph work (award numbers DEB-0949951 to MGB and DEB- 0950066 to BH and AS-R).

REFERENCES

Begay, A.C., A. Schmidt-Rhaesa, M.G. Bolek & B. Hanelt. 2012. Two new *Gordionus* species (Nematomorpha: Gordiida) from the southern Rocky Mountains (USA). Zootaxa 3406: 30–38.

Hanelt, B. & J. Janovy. 2002. Morphometric analysis of nonadult characters of common species of American gordiids (Nematomorpha: Gordioidea). Journal of Parasitology 88: 557–562.

Hanelt, B., F. Thomas & A. Schmidt-Rhaesa. 2005. Biology of the phylum Nematomorpha. Advances in Parasitology 59: 243–305.

Hanelt, B., A. Schmidt-Rhaesa & M.G. Bolek. 2015. Cryptic spesies of hairworm parasites revealed by molecular data and crowdsourcing of specimen collections. Molecular Phylogenetics and Evolution 82: 211–218.

Schmidt-Rhaesa, A. 2010. Considerations on the genus *Gordius* (Nematomorpha, horsehair worms), with the description of seven new species. Zootaxa 2533: 1–35.

Schmidt-Rhaesa, A. 2012. Nematomorpha. Pages 29–145 *in*: A. Schmidt-Rhaesa (ed.), Handbook of Zoology. Volume Gastrotricha, Cycloneuralia and Gnathifera. De Gryter, Berlin.

Schmidt-Rhaesa, A., B. Hanelt & W. Reeves. 2003. Redescription and cpompiliation of Nearctic freshwater Nematomorpha (Gordiida), with the description of two new species. Proceedings of the Academy of Natural Sciences of Philadelphia 153: 77–117.

Szmygiel, C., A. Schmidt-Rhaesa, B. Hanelt & M.G. Bolek. 2014. Comparative descriptions of non-adult stages of four genera of gordiids (phylum: Nematomorpha). Zootaxa 3768: 101–118.

Chapter 11

Phylum Mollusca

Chapter Outline

Introduction to Mollusca

James H. Thorp

Kansas Biological Survey and Department of Ecology and Evolutionary Biology, University of Kansas, Lawrence, KS, USA

INTRODUCTION

The Mollusca is well represented in freshwaters by two classes of molluscs (or mollusks, depending on your preference): Gastropoda (meaning "stomach foot") and Bivalvia (referring to "two valves" or shells). The freshwater Gastropoda of the Nearctic is divided into two major groups: (1) the pulmonates which evolved from terrestrial snails and still possess a mantle cavity modified as a pulmonary cavity or "lung;" and (2) caenogastropods (formerly called prosobranchs), which have a gill and evolved directly from marine snails. The Nearctic bivalves include the native order Unionoida (primarily the abundant pearly mussels) and one order of both native (seed and mussel clams) and invasive clams and mussels. The latter (Asian clams, zebra mussels, and quagga mussels) have caused major losses of native pearly mussels as well extensive economic damage.

All freshwater molluscs have one or two shells composed of a thin outer layer of proteinaceous perisostracum and a strong inner layer of mostly crystalline calcium carbonate.

Molluscs are present in most Nearctic freshwater habitats other than hypersaline lakes, which are colonized by only a few snail species (*Assiminea* spp.). Their diversity tends to be lower in extremely soft waters, higher in lotic versus lentic habitats, and greatest in the southeastern USA. Some bivalves can survive outside of water (e.g., buried in a river bank until exposed by rising waters), and pulmonate snails regularly climb out of water on emergent vegetation or rocks to gain better access to oxygen or to avoid predators.

LIMITATIONS

Identification of freshwater molluscs is primarily based on shell morphology and secondarily on color. Reliance on shell morphology may pose a serious problem, especially for snails, because the shape of the shell can be modified substantially by water currents (e.g., Britton & McMahon, 2004). Moreover, dried and preserved shells lose much of their identifying colors. Consequently, most taxonomic keys to extant

Thorp and Covich's Freshwater Invertebrates. http://dx.doi.org/10.1016/B978-0-12-385028-7.00011-1

molluscs are based on shell morphology. Identification of specimens beyond a preliminary level within each class should be done by comparing your specimens with shells in museums or in other places where a highly trained specialist can provide identified conspecifics for comparison.

MATERIAL PREPARATION AND PRESERVATION

When retrieving molluscs from any aquatic habitat, try to collect individuals from nearby macro- and microhabitats with different current velocities. We recommend collecting recently dead molluscs (especially the bivalves) if possible. This will prevent killing an individual of one of the many threatened and endangered species. After gently cleaning the shell, allow the labeled shell to dry without any artificial coating. Store the soft tissue from any live specimens collected in labeled jars with the appropriate preservative (see recommendations later in this chapter). The preservatives may differ depending on whether molecular analysis will subsequently be performed.

KEYS TO MOLLUSCA

Mollusca: Classes

1	Single shell present which is coiled (snails) or uncoiled (freshwater "limpets")	**Gastropoda [p. 192]**
1'	Two shells (valves) present and connected by a ligament hinge; mussels and clams	**Bivalvia [p. 211]**

REFERENCE

Britton, D.K. & R.F. McMahon. 2004. Environmentally and genetically induced shell-shape variation in the freshwater pond snail *Physa* (*Physella*) *virgata* (Gould, 1855). American Malacological Bulletin 19: 93–100.

PHYLUM MOLLUSCA

Class Gastropoda

D. Christopher Rogers
Kansas Biological Survey, Kansas University, Lawrence, KS, USA; The Biodiversity Institute, Kansas University, Lawrence, KS, USA

INTRODUCTION

Strong et al. (2008) reported 585 native and invasive species of snails and limpets from the Nearctic, although controversy surrounds the taxonomy. Depending on the author, family, genus, and species definitions and numbers will vary greatly. Freshwater and amphibious gastropods may be found in any temporary or permanent aquatic habitats, except phytotelmata. Most are found on and around aquatic macrophytes or on rocks, but some live in soft muds. Several species are protected by environmental laws, some species are important vectors of wildlife disease, and others are invasive.

LIMITATIONS

These keys are to adult freshwater gastropods. The Nearctic taxa are, for the most part, easily identified to family level. Much revisionary taxonomic work needs to be done in most gastropod groups. The extreme age and tremendous morphological plasticity of gastropods make the definition of most taxonomic categories difficult to quantify. Particularly vexing is the strange confusion in the taxonomic literature, with many taxa accepted or rejected without explanation, and good quantitative taxonomic revisions ignored without explicit justifications. Because of this issue, the keys are taxonomically conservative, often terminating with species groups rather than species.

In Canada (Alberta, British Colombia, and Manitoba) and the USA (Alabama, Georgia, Idaho, New Mexico, Tennessee, and Utah) are several species that are protected under environmental law. Many species are protected at the state and province level in these countries.

TERMINOLOGY AND MORPHOLOGY

Defining characters are presented in Figs. 11.1–11.3[1]. Much of gastropod identification is dependent on shell structures; however, there is considerable overlap of shell characters,

1. Most figures in this chapter came from Brown & Lydeard (2010) or Cummings & Graf (2010). However, some figure legends have been changed to reflect current taxonomy.

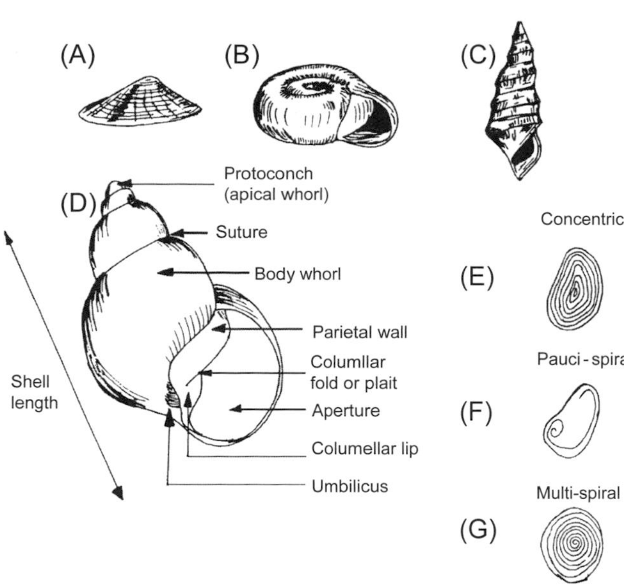

FIGURE 11.1 Basic anatomy of the shell, including shell architecture (conical, A; planospiral, B; spiral C, D), major features of the shell (D), and three types of opercula: (E)=concentric; (F)=paucispiral; and (G)=multispiral.

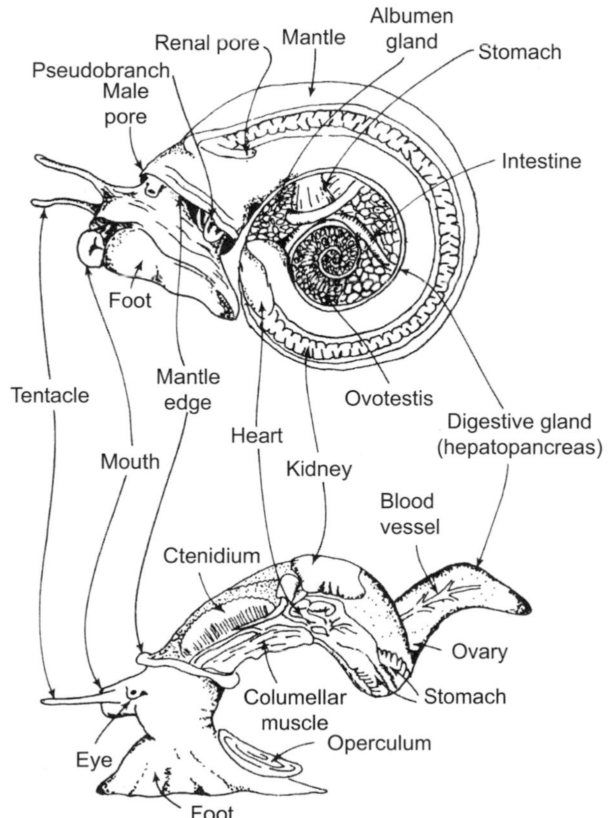

FIGURE 11.2 Basic internal anatomy of a planorbid pulmonate (above, after Burch, 1989), and a pleurocerid prosobranch (below, after Pechenik, 1985).

so they cannot be used exclusively. The basic snail shell is a coiled tube. Each coil of the tube is called a whorl. The last 360 degrees of the coiled shell is called the body whorl and is the last whorl of the tube that does not have the shell coiled completely around it. The dorsolateral surface of the whorl is the shoulder. The shell opening with the animal inside is the aperture. It may contain "teeth," which are flattened lamellae, projecting from the inside surface.

In caenogastropods, this aperture can be sealed shut with a shell "door" called an operculum, which grows on the back of the animal's foot. Operculae may grow spirally, concentrically, or paucispirally, wherein the successive whorls increase in size. If a shell is placed such that the spire is directed upward or away from the observer and the aperture is facing the observer, a shell with the aperture on the left and spiraling towards the right is sinistral in orientation, whereas a shell with the aperture on the right and spiraling to the left is dextral.

Because the shell is basically a tube, lines or grooves (called striae) that grow in the direction of the expanding tube are called spiral, while those that grow across the tube are called transverse. A shell may have a carina or ridgeline that follows the length of the tube, or one or more low lirae that are basically the same as a carina, only much smaller. If the ridges are transverse, they are called costae.

The entire coiled shell, other than the body whorl, is the spire. The line demarcating where one whorl is fused to the previous whorl is called the suture. The end of the spire is the apex. The structural point around which the shell spirals is called the columella. If this opens as a tube, it opens ventrally, and this opening is called the umbilicus. An open umbilicus is termed perforate, whereas a closed umbilicus is called imperforate. An aperture lip that reflects over the umbilicus is called a columellar lip.

Whorls may be rounded in cross section with deep sutures, or whorls may be flattened with shallow sutures. Flattened whorls give the shell a conical shape, while moderately rounded whorls give the shell a subglobose shape, and nearly spherical shells are globose in shape, and of course, a continuum exists between these three shell shapes. In the limpets, the tube is not coiled, but reduced to a conical cap, referred to as patelloid. Ramshorn snails typically coil as a disc, which form is referred to as planospiral.

The animal within the shell is composed of the eversible headfoot and mantle, and the non eversible visceral mass. The primary soft anatomy characters are within the headfoot and the mantle. The head portion contains one or more pairs of sensory tentacles, and generally, the eyes are placed at the bases of the primary pair. Anterior to the primary tentacle pair is the mouth, which may be flanked by a secondary tentacle pair. Inside the mouth is the radula, which is similar to a tongue. The radula is lined with transverse rows of radular teeth, which may each have one or more cusps for rasping periphyton from surfaces.

Most Neartic freshwater gastropods are typically dioecious, with the males having an enlarged right tentacle as a copulatory organ or possess a penis or a specialized verge, or have no copulatory organ at all. Pulmonates, in contrast, are all monoecious. The basic components of the pulmonate reproductive system are shown in Fig. 11.3.

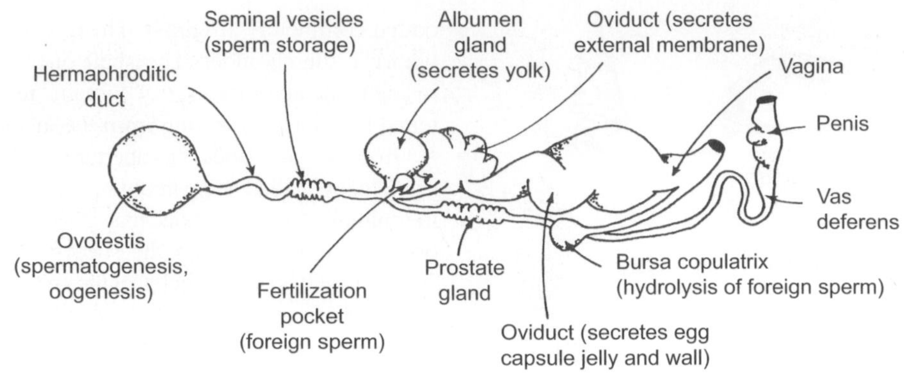

FIGURE 11.3 Anatomy of the reproductive system of the pulmonate snail *Physa* (from Brown & Lydeard, 2010).

Taxonomic diagnosis may require dissection of the genitalia. The penis or associated genitalic structures may have one or more obvious glands, often on projections. In the Physidae, the genitalic structures important for diagnosis are the preputium, the preputial gland and the penal sheath. The preputium is a tube that opens to the outside of the animal for releasing or receiving sperm. The preputium may or may not have a dorsal preputial gland. Posterior to the preputium is the penial sheath, which may be glandular or muscular, or both.

MATERIAL PREPARATION AND PRESERVATION

Whole animals can be preserved in ethyl or isopropyl alcohol. Formalin should never be used, as it will dissolve the shells. Empty shells can be kept dry. A bit of cotton stuffed carefully in the aperture and daubed with glue will hold the operculum in place. Shells that have had the animal removed should be cleaned with alcohol inside and let air dry to reduce odors.

Specimens where the animal needs to be extracted from the shell can be boiled in pure bleach for a few seconds or dipped in concentrated hydrochloric acid for a few seconds. Alternatively, the shell can be gently cracked with forceps and the shell pieces removed. The radula can be viewed by cutting open the top of the head and teasing the structure out with a needle, probe, or forceps.

The genitalia sit inside the headfoot just above and behind the eyes. Using a fine scalpel or a razor, gently cut the head open posteriorly between the eyes. The genitalic penal structures lie on the left side of the animal, behind and at the base of the left tentacle.

KEYS TO GASTROPODA

Gastropoda: Superfamilies

1 Operculum present (Fig. 11.1) 2

1' Operculum absent 7

2(1) Shell rough, or if smooth, not globose; spire prominent or shell patelliform; columella rarely produced as a flat septum; operculum without internal processes 3

2' Shell globose, smooth, like porcelain or a marble; spire reduced, rounded; columella transverse, projecting as a septum (shelf like); operculum calcareous with small digitiform projections along the lower margin (Fig. 11.4 L) .... Neritoidea, one family: Neritidae; one genus: ***Neritina*** **[p. 198]**

3(2) Operculum oval and concentric, paucispiral, or multispiral (Fig. 11.1 E–G) 4

3' Operculum circular and spiral; plumose gill extends outside of shell (Fig. 11.5 A, B) .... Valvatoidea, one family: Valvatidae; one genus: *Valvata*

[Widespread. Genus in need of revision, 8–12 species. Not keyed further.]

4(3) Operculum concentric, corneous (Fig. 11.1 E) 5

4' Operculum paucispiral, or multispiral; operculum calcareous or corneous (Fig. 11.1 F, G) 6

5(4) Aperture broader than spire is long; shell with short spire or planospiral in form; oral tentacles present (Fig. 11.6 A, B) .... Ampullaroidea, one family: **Ampullariidae [p. 198]**

5' Aperature narrower or subequal to length of spire; oral tentacles always absent (Fig. 11.6 C–G) .... Viviparoidea, one family: **Viviparidae [p. 198]**

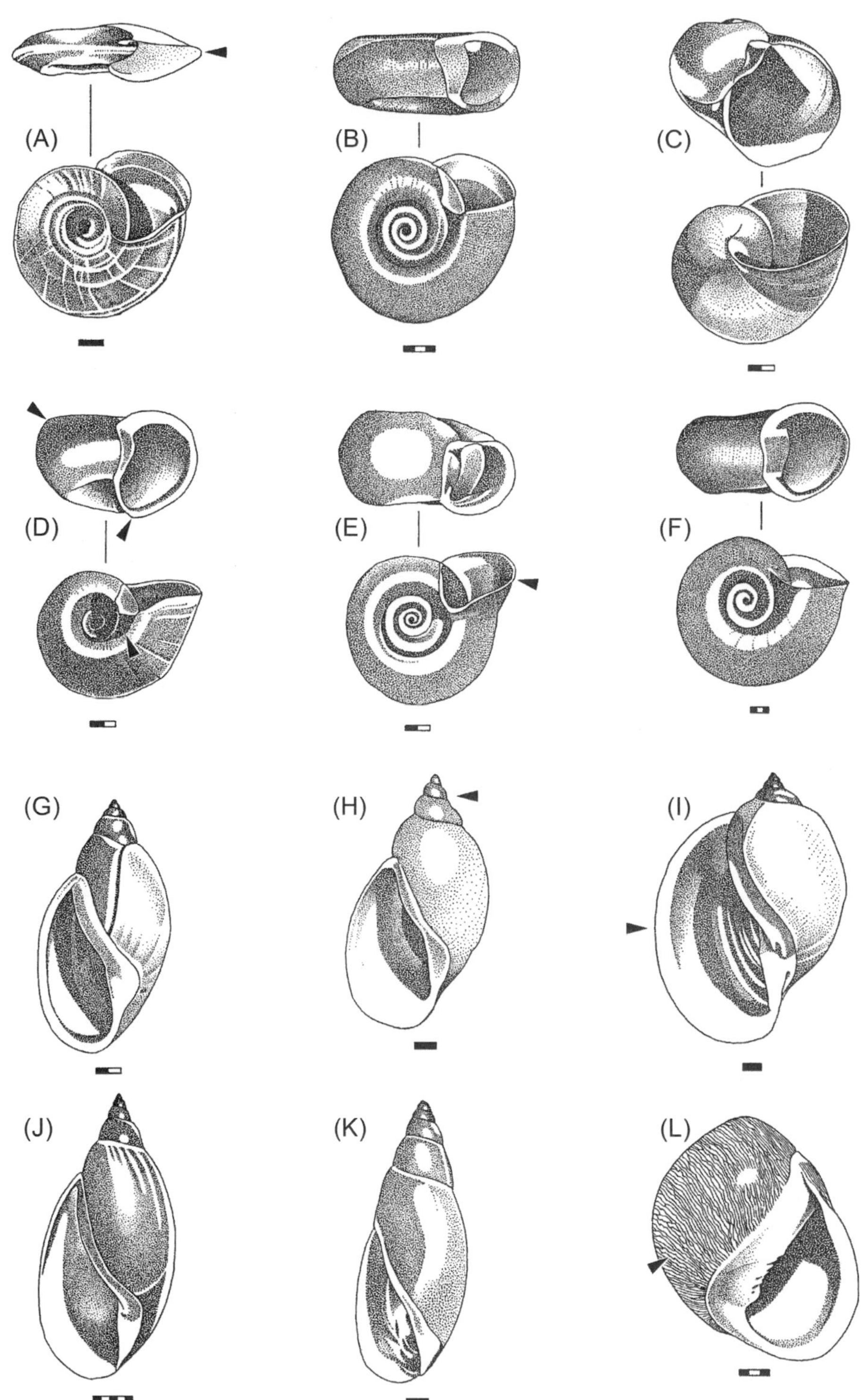

FIGURE 11.4 Representative physids and planorbids and neritinids. (A) *Promenetus exacuous* (note flaired body whorl and carina); (B) *Biomphalaria glabrata*; (C) *Vorticifex effusa*; (D) *Helisoma anceps* (note strong growth lines and carina); (E) *Helisoma companulata* (sometimes called *companulatum*, note flared lip of aperture); (F) *Helisoma trivolvis* (this species reaches 20 mm in diameter and is extremely common); (G) *Physa gyrina*; (H) *Physa integra*; (I) *Physa* sp.; (J) *Physa* sp.; (K) *Aplexa elongata* (note the bullet shape and lustrous black shell); (L) *Neritina reclivata* (note markings on shell and teeth on parietal wall). The scale bars equal 1 mm. *Figures A–I after Burch, 1989.*

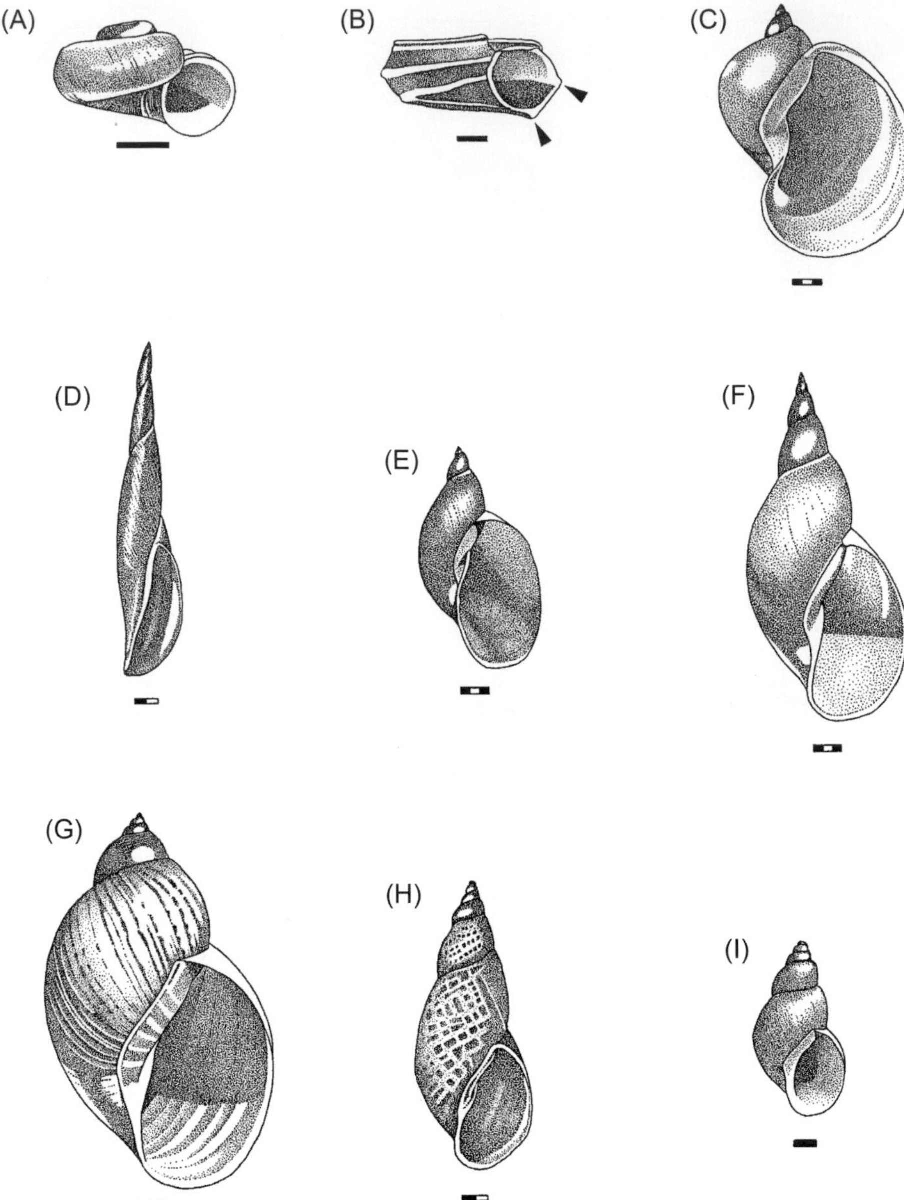

FIGURE 11.5 Representative valvatids and lymnaeids. (A) *Valvata sincera* (note multispiral operculum); (B) *Valvata tricarinata* (note carina); (C) *Radix auricularia* (note expanded body whorl); (D) *Lymnaea haldemani* (note extremely narrow shell); (E) *Lymnaea columella* (note thin, transparent shell and amphibious habit); (F) *Lymnaea stagnalis* (note large and fragile shell) (G) *Lymnaea megasoma* (note thick, large shell); (H) *Lymnaea elodes* (note malleations sometimes present); (I) *Lymnaea humilis* (note small size). Scale bars equal 1 mm. *Figures A–I after Burch, 1989.*

6(4) Males if present with a penis; adult often under 10 mm, but may reach 15 mm; shell rarely with any sculpturing (Fig. 11.7 A–E) .... **Rissooidea [p. 199]**

6' Males if present without a verge; adults typically 15 to 75 mm; shell typically with strong sculpturing (Fig. 11.8 B–I) .... **Cerithoidea [p. 202]**

7(1) Shell not patelliform, or if patelliform, then spire sinistral (apex centered or to right of midline) and blunt, with adult patelliform shell larger than 7 mm (Fig. 11.9 A, B, D–G) 8

7' Shell patelliform with spire dextral (apex to left of midline), acute; adult shell less than 7 mm in length (Fig. 11.9 C) .... Acroloxidea, one family; Acroloxidae: *Acroloxus coloradensis* (Henderson, 1939)

[Canada: Alberta, British Columbia. USA: Idaho, Colorado, Montana]

8(7) Tentacles narrow, filiform 9

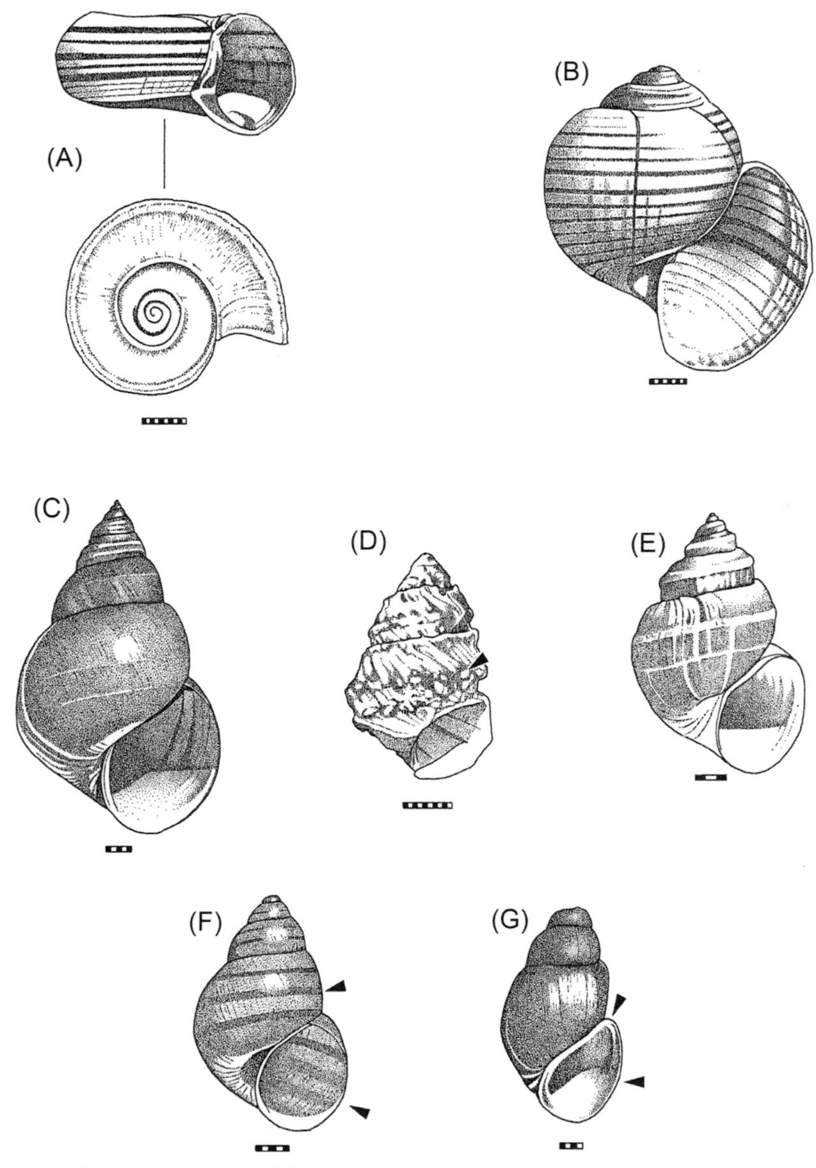

FIGURE 11.6 Representative ampullarids and viviparids. (A) *Marisa cornuarietis* (note large, planospiral shell); (B) *Pomacea paludosa* (large "apple" snail common in Florida), (C) *Bellamya japonica* (introduced but now widespread species in North America); (D) *Tulotoma magnifica* (note tubercules which may be absent in some morphs or species, recently rediscovered in Alabama rivers); (E) *Lioplax subcarinata*; (F) *Viviparus georgianus* (note circular operculum, and bands which may disappear in adults, common throughout eastern U. S.); (G) *Campeloma decisum* (sometimes referred to as *decisa*; note operculum is longer than it is wide and shouldered junction of aperture and body whorl, quite common in rivers and lakes in eastern states). Scale bars equal 1 mm. *Figures A–G after Burch, 1989.*

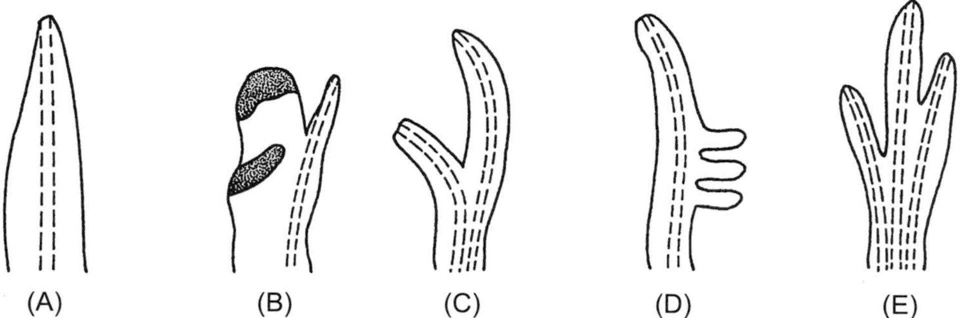

FIGURE 11.7 Structure of the penis (verge) in various hydrobiids. (A) simple penis in Lithoglyphidae; (B) glandular crests of penis in subfamily Nymphophilinae; (C) two-ducted penis in Amnicolidae; (D) penis with accessory lobes in Hydrobiinae; and (E) three ducted penis of hydrobiid subfamily Fontigentinae. *Figures A–E after Burch, 1989.*

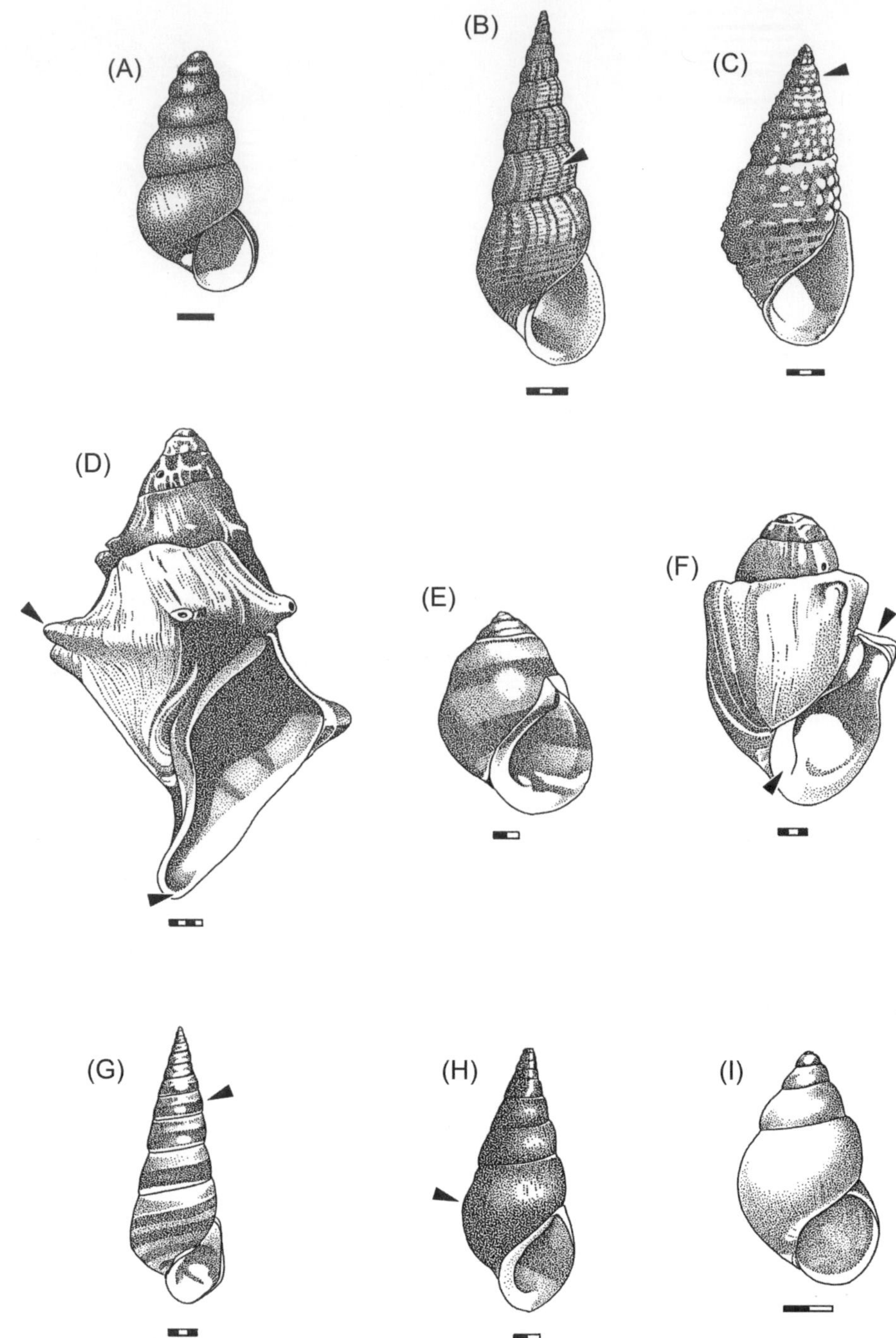

FIGURE 11.8 Representative Potamiopsids, thiarids and pleurocerids. (A) *Pomatiopsis lapidaria* (note paucispiral operculum); (B) *Melanoides tuberculata* (note costae and lirae); (C) *Thiara granifera* (note tubercules and flattened whorls near apex); (D) *Io fluvialis* (length of spines variable); (E) *Leptoxis carinata*; (F) *Lithasia*/*Pleurocera* (note thickened anterior aperture lip); (G) *Lithasia*/*Pleurocera* (note acute angle on anterior aperture); (H) *Lithasia*/*Pleurocera* (note there is tremendous variation in shell sculpture in this genus). (I) *Bithynia tentaculata*. *Figures A–I after Burch, 1989.*

8' Tentacles, broad, flat, triangular; haemoglobin absent; coiled shell always dextral, patelliform shell with apex central or sinestral; never planospiral (Fig. 11.5 C–I) .... Lymnaeoidea, one family: **Lymnaeidae [p. 203]**

9(8) Aperture with three teeth on columella only; shell elongate; haemoglobin absent from blood .... Ellobioidea, one family: Ellobiidae: *Myosotella myosotis* (Draparnaud, 1801)

[Palaearctic. USA: Invasive in California, Oregon]

9' Aperture without teeth, or with teeth set back, inside aperture, never on columella; shell elongate, patelliform, planospiral, fusiform or succiniform; if patelliform, then shell apex central or sinistral; coiled shell dextral or sinestral (Figs. 11.9 D–L and 11.4 A–K); haemoglobin present in blood or not .... **Planorbiodea [p. 205]**

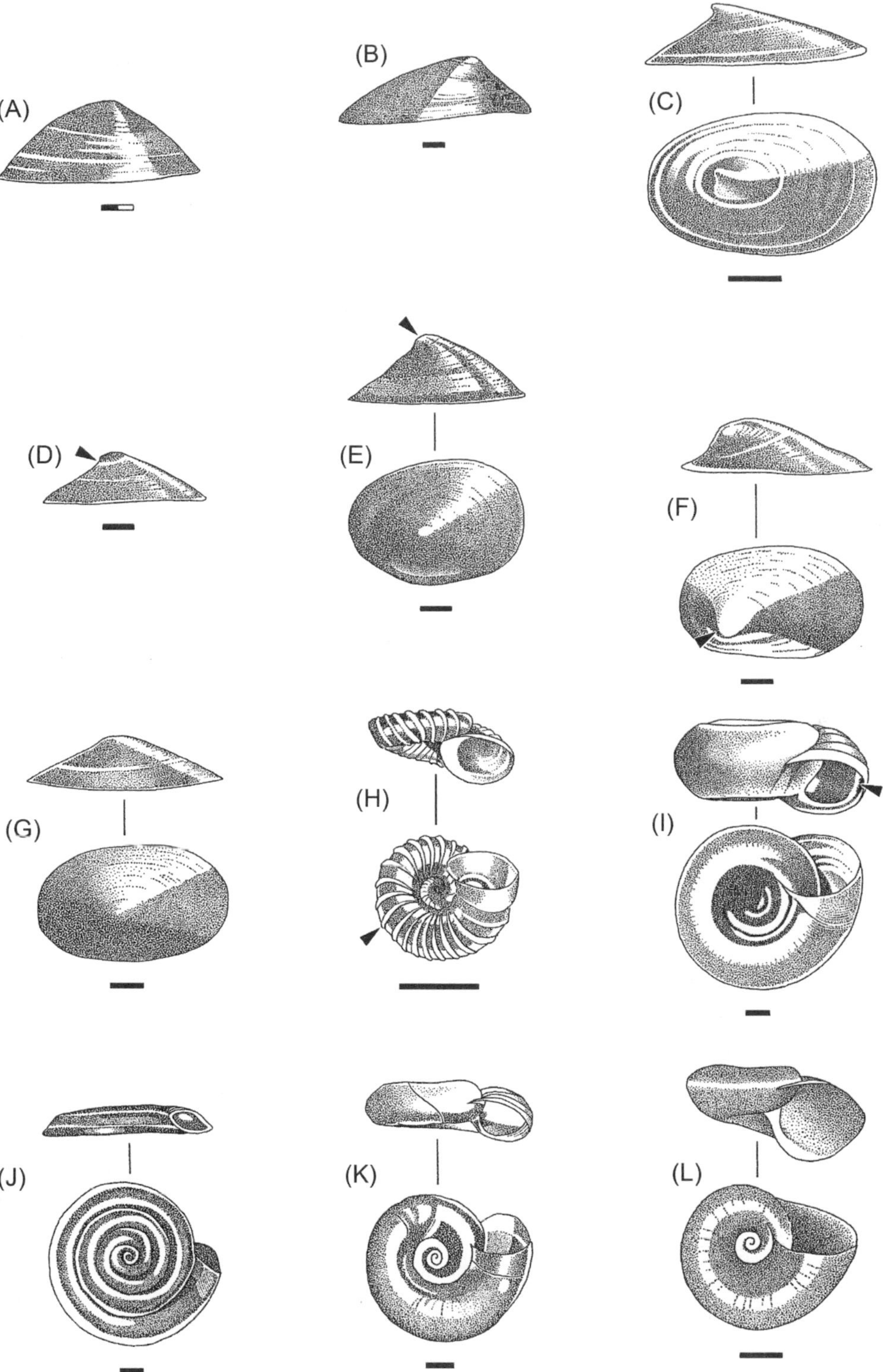

FIGURE 11.9 Representative limpets and planorbids: (A) lymaeid limpet *Lanx patelloides* (note large size, west coast distribution); (B) limpet *Fisherola nutalli*; (C) *Acroloxus*; (D) limpet *Rhodacmea rhodacme* (note notched depression and southeastern distribution); (E) limpet *Ferrissia rivularis* (note elevated shell, may possess posterior "shelf" in shell, wide distribution); (F) limpet *Hebetancylus excentricus* (note depressed apex to the right of midline, colorless tentacles, and southern distribution); (G) *Laevapex fuscus* (note obtuse apex near midline of shell, black pigmented tentacles, and widespread distribution in eastern backwaters and southern, slow flowing streams); (H) *Gyraulus crista* (note costae); (I) *Planorbula armigera* (note teeth in aperture); (J) *Drepanotrema kermatoides*; (K) *Gyraulus deflectus*; (L) *Menetus dilatatus*. *Figures A–L after Burch, 1989.*

Gastropoda: Neritoidea: Neritidae: *Neritina* Species

1 Shell black with scattered white triangular spots .... *Neritina clenchi* Russel, 1940
[USA: Florida. Neotropics]

1' Shell olive green with thin black transverse lines .... *Neritina usnea* (Röding, 1798)
[USA: Alabama, Florida, Louisiana, Mississippi, Texas. Mexico: Gulf coast. Neotropics]

Gastropoda: Ampullaroidea: Ampullariidae: Genera

1 Shell not planospiral (Fig. 11.6 B) .... ***Pomacea* [p. 198]**

1' Shell planospiral (Fig. 11.6 A) .... *Marisa cornuauarietus* (Linnaeus, 1758)
[Invasive, native to Neotropics. USA: California, Colorado, Idaho, Florida, Nevada]

Gastropoda: Ampullariidae: *Pomacea:* Species

1 Sutures and whorls normal, not deeply impressed 2

1' Sutures deeply impressed forming a channel around the spire .... *Pomacea maculata* (Perry, 1810)
[Neotropical. Invasive in USA: Alabama, Arizona, California, Colorado, Idaho, Indiana, Florida, Nevada]

2(1) Shell whorl shoulders angular, flattened .... *Pomacea bridgesii* (Reeve, 1856)
[Neotropical. Invasive in USA: California, Florida]

2' Shell whorl shoulders rounded .... *Pomacea paludosa* (Say, 1829)
[USA: Florida. Neotropical. Invasive in USA: California, Florida, Georgia, Louisiana, Nevada, Oklahoma]

Gastropoda: Viviparoidea: Viviparidae: Genera

1 Operculum concentric 2

1' Operculum concentric with nucleus spiral (Fig. 11.6 E) .... ***Lioplax* [p. 198]**

2(1) Whorls with a distinct shoulder (Fig. 11.6 D, F, G) 3

2' Whorls rounded, shoulder lacking (Fig. 11.6 C) .... ***Bellamya* [p. 199]**
[Oriental and Palaearctic. Invasive in Nearctic. Canada: southern border. USA: Widespread. Mexico(?)]

3(2) Carinae absent or present, but never tuberculate (Fig. 11.6 F, G) 4

3' Carinae broken up into tubercles, may be partially obliterated (Fig. 11.6 D) .... *Tulotoma magnifica* (Conrad, 1834)
[USA: Alabama.]

4(3) Operculum circular, aperture subcircular (Fig. 11.6 F) .... ***Viviparus* [p. 199]**

4' Operculum and aperture oval (Fig. 11.6 G) .... *Campeloma*
[In need of revision, not keyed further. Canada: Great Lakes region. USA: east of the Mississippi to the Atlantic, Great Lakes/St. Lawrence River, to Gulf Coast]

Gastropoda: Viviparoidea: Viviparidae: *Lioplax:* Species

1 Shell form subcylindrical 2

1' Shell form subglobose 3

2(1) Body whorl with shoulder rounded .... *Lioplax cyclostomaformis* (Lea, 1841)
[USA: Alabama, Georgia]

2' Body whorl with shoulder angulate .... *Lioplax talquinensis* Vail, 1979
[USA: Alabama, Florida]

3(1) Shoulder subcarinate or smooth 4

3' Shoulder strongly carinate .... *Lioplax subcarinata* (Say, 1816)
[USA: Atlantic drainages from New York to South Carolina]

4(3) Shell dark green/brown to black, length to 30 mm .... *Lioplax pilsbryi* Walker, 1905
[USA: Alabama, Florida, Georgia]

4' Shell brown to fawn or dark green, length less than 25 mm .... *Lioplax sulculosa* (Menke, 1827)
[USA: Arkansas, Illinois, Indiana, Iowa, Kentucky, Minnesota, Missouri, Ohio, Wisconsin]

Gastropoda: Viviparoidea: Viviparidae: *Bellamya:* Species

1 Whorls malleate; whorls lacking spiral striae; spire obtuse .... *Bellamya chinensis* (Reeve, 1863)
[Canada: southern border regions. USA: (invasive) widespread. Native to China]

1' Whorls not malleate; whorls with spiral striae; spire acute .... *Bellamya japonica* (von Martens, 1861)
[USA: (invasive) widespread. Native to Japan]

Gastropoda: Viviparoidea: Viviparidae: *Viviparus:* Species

1 Whorls rounded, but not globose; broad spiral banding present or absent 2

1' Whorls broadly rounded; broad spiral banding absent .... *Viviparus intertextus* (Say, 1829)
[USA: Alabama, Georgia, Louisiana, Mississippi, South Carolina, Texas]

2(1) Shell yellowish to olive green; typically four color bands (if present) .... *Viviparus georgianus* (Lea, 1834)
[Canada: Quebec. USA: Gulf Coast to Great Lakes and New England]

2' Shell yellowish to greenish brown; three color bands (if present) .... *Viviparus subpurpureus* (Say, 1829)
[USA: Arkansas, Illinois, Iowa, Kentucky, Louisiana, Mississippi, Missouri, Tennessee, Texas]

Gastropoda: Rissooidea: Families

Most families are very difficult to separate, and there are many undescribed taxa. Use caution in identifying these animals. The key below follows the taxonomy of Wilke et al., 2013.

1 Operculum multispiral or paucispiral (Fig. 11.1 F, G) 2

1' Operculum concentric (Fig. 11.1 E) .... Bithyniidae, one species *Bithynia tentaculata* (L., 1758)
[Canada, USA: Great Lakes region, mid-Atlantic states]

2(1) Operculum without a calcareous white smear; separate sexes 3

2' Operculum with a calcareous white smear; almost always females only, males extremely rare .... Tateidae, one species: *Potamopyrgus antipodarum* (Gray, 1853)
[Invasive exotic species. Canada: Ontario. USA: Great Lakes region and scattered localities in the western states, expanding]

3(2) Omniphoric groove absent; tentacles normal 4

3' Omniphoric groove present; tentacles very short to absent .... Assimineidae; one genus: ***Assiminea* [p. 200]**
[Estuarine and saline spring habitats]

4(3) Tentacles with sides gradually converging distally, with or without penal lobes or internal tubular glands (Fig. 11.7 A, C–E); operculum variable 5

4' Tentacles with sides parallel; penis with a single medial or basal lobe, and an internal tubular gland that extends back into the head (Fig. 11.7 C); operculum paucispiral .... **Amnicolidae [p. 200]**

5(4) Radula with marginal teeth bearing larger cusps on inner margin; outer marginal teeth with cusps on both sides 6

5' Radula with marginal teeth with cusps subequal; outer marginal teeth with cusps on inner side (Fig. 11.8 A) .... Pomatiopsidae, one genus: **Pomatiopsis [p. 200]**

6(5) Penis with external glands, often on lobes or crests (Fig. 11.7 B, C) 7

6' Penis lacking external glands (Fig. 11.7 A) .... **Lithoglyptidae [p. 201]**

7(6) Penis with narrow glandular fields, sometimes ridged, or large circular glandular areas sometimes borne on crests .... Hydrobiidae
[Widespread, numerous genera, many in need of revision, and with many undescribed taxa]

7' Penis papillate and/or with several gland types present, often with glands on stalks .... Cochliopidae
[Widespread, numerous genera, many in need of revision and with many undescribed taxa]

Gastropoda: Rissooidea: Assimineidae: *Assiminea:* Species

1 Estuarine species 2
1' Desert spring species 3
2(1) Pacific estuaries .... *Assiminea californica* (Tryon, 1865)
[Canada: British Columbia. USA: California, Oregon, Washington. Mexico: Baja California]
2' Atlantic estuaries .... *Assiminea succinea* (Pfieffer, 1840)
[USA: Massachusetts, south along Altantic and Gulf coasts. Mexico: Gulf coast. Neotropics]
3(1) From Death Valley, California .... *Assiminea infima* Berry, 1947
[USA]
3' From New Mexico and Texas .... *Assiminea pecos* Taylor, 1987
[USA]

Gastropoda: Rissooidea: *Amnicolidae:* Genera

1 Animal lacking eyes and pigment; subterranean 2
1' Animal with eyes and pigment; protoconch with weak spiral lines 3
2(1) Protoconch transversely fimbriate .... *Dasyscias franzi* Thompson & Hershler, 1991
[USA: Florida]
2' Protoconch wrinkled or pustulate, with spiral grooves .... *Antroselates spiralis* Hubricht, 1971
[USA: Indiana, Kentucky]
3(1) Shell 2.9 to 4.8 mm in length; female with one large seminal receptacles 4
3' Shell 1.3 to 3.3 mm in length; female with two small seminal receptacles .... ***Colligyrus*** **[p. 200]**
[USA: northern Great Basin springs]
4(3) Mantle with bands of pigment .... *Amnicola*
[Many undescribed species. Canada: southern regions. USA: widespread]
4' Mantle with pigment pattern diffuse .... *Lyogyrus*
[Many undescribed species. Canada: southern regions. USA: widespread]

Gastropoda: Rissooidea: Amnicolidae: *Colligyrus:* Species

1 Shell globose to depressed conic 2
1' Shell conical .... *Colligyrus greggi* (Pilsbry, 1935)
[USA: Idaho, Utah]
2(1) Aperture sinuate; teleoconch sutures deeply impressed .... *Colligyrus convexus* Hershler, Frest, Liu & Johannes, 2003
[USA: California]
2' Aperture simple; teleoconch sutures normal .... *Colligyrus depressus* Hershler, 1999
[USA: Oregon]

Gastropoda: Rissooidea: Pomatiopsidae: *Pomatiopsis:* Species

Modified from Burch (1982)

1 Distribution west of the continental divide 2
1' Distribution east of the continental divide 4
2(1) Umbilicus open 3
2' Umbilicus imperforate; adult shell 3 mm in length .... *Pomatiopsis binneyi* Tryon, 1863
[USA: California]
3(2) Shell chestnut brown .... *Pomatiopsis californica* Pilsbry, 1899
[USA: California]
3' Shell olive brown .... *Pomatiopsis chacei* Pilsbry, 1937
[USA: California]

4(1) Aperture oval; whorls flattened; shell elongate 5

4' Aperture round; whorls rounded, shell broadly conical .... *Pomatiopsis cincinnatiensis* (Lea, 1840)
[USA: Illinois, Iowa, Michigan, Tennessee, Virginia, Wisconsin]

5(4) Umbilicus broadly open; teleoconch narrow, subacute, spire up to five whorls; length at least twice body whorl width .... *Pomatiopsis lapidaria* (Say, 1817)
[Canada: Ontario, Quebec. USA: east of Continental Divide, except northern Great Plains]

5' Umbilicus narrow; spire up to four whorls, obtuse; length approximately 1.5 times the width of body whorl .... *Pomatiopsis lapidaria* (Say, 1817)

Gastropoda: Rissooidea: *Lithoglyphidae:* Genera

1 Eyes and pigment present 2

1' Eyes and pigment absent; adults minute; occurring in ground water .... **Phreatodrobia [p. 201]**
[USA: Texas]

2(1) Shell not neritiform, columella never produced as a septum 3

2' Shell neritiform; spire reduced, rounded; columella transverse, projecting as a septum (shelf-like) .... *Lepyrium showalteri* (Lea, 1861)
[USA: Alabama]

3(2) Umbilicus closed or very narrow 4

3' Umbilicus open, broad .... *Clappia*
[USA: Alabama, possibly extinct. Two poorly defined species]

4(3) Shell aperture margin thick .... *Fluminicola*
[Large genus in need of revision, many undescribed species. Canada: British Columbia. USA: California, Idaho, Nevada, Oregon, Utah, Washington, Wyoming]

4' Shell aperture thin .... *Gillia altilis* (Lea, 1841)
[USA: Mid-Atlantic States]

Gastropoda: Rissooidea: Lithoglyptidae: *Phreatodrobia:* Species

1 Protoconch normal 2

1' Protoconch with apex uncoiled and protruding distally .... *Phreatodrobia coronae* Hershler & Longley, 1987
[USA: Texas]

2(1) Distal body whorl thickened, appearing whitish and opaque 3

2' Body whorl simple 4

3(2) Shell trochoid to planospiral or low conical; aperture flared, may touch body whorl .... *Phreatodrobia nugax nugax* (Pilsbry & Ferriss, 1906)
[USA: Texas]

3' Shell globose; aperture fused to body whorl, and flared only at fusion point .... *Phreatodrobia nugax inclinata* Hershler & Longley, 1986
[USA: Texas]

4(2) Shell planospiral 5

4' Shell conical 7

5(4) Operculum without striae; aperture not emarginate in dorsal view 6

5' Operculum with short, deep striae across the growth lines; aperture in dorsal view broadly emarginate .... *Phreatodrobia rotunda* Hershler & Longley, 1986
[USA: Texas]

6(5) Aperture simple, circular; operculum circular; body whorl with weak axial growth lines .... *Phreatodrobia micra* (Pilsbry & Ferriss, 1906)
[USA: Texas]

6' Aperture flared apically, transverse to narrowly oval; operculum broader than wide; body whorl with thick, wrinkled colabrial lines .... *Phreatodrobia plana* Hershler & Longley, 1986
[USA: Texas]

7(4) Operculum flared ... 8

7' Operculum simple; teleoconch with numerous short ridges ... *Phreatodrobia conica* Hershler & Longley, 1986

[USA: Texas]

8(7) Operculum as long as wide; teleoconch with collabral costae and spiral lines ... *Phreatodrobia imitata* Hershler & Longley, 1986

[USA: Texas]

8' Operculum broader than wide; teloconch punctate ... *Phreatodrobia punctata* Hershler & Longley, 1986

[USA: Texas]

Gastropoda: Cerithoidea: Families

1 Mantle margin smooth; separate sexes, males common; females deposit eggs; female foot with right side bearing an egg-laying sinus ... 2

1' Mantle margin papillate; parthenogenic, with males extremely rare or absent; females brood eggs and hatchlings in a brood pouch dorso-posteriorly to the head ... **Thiaridae [p. 202]**

2(1) Female with egg laying sinus a deep, "H"-shaped lumen, seminal receptacle absent; alimentary system smooth ... **Pleuroceridae [p. 202]**

[Canada: Ontario, Quebec. USA: east of Continental Divide]

2' Female with egg-laying sinus a shallow, simple, flattened lumen, seminal receptacle present; alimentary system bearing dorsally with deep folds and clefts ... Semisulcospiridae, one genus: ***Juga* [p. 202]**

[Genus in need of revision. Canada: British Columbia. USA: west of Continental Divide]

Gastropoda: Cerithoidea: Thiaridae: Genera

1 Shell whorls flattened; whorl spiral sculpture of transverse tubercle rows (Fig. 11.8 C) ... *Tarebia granifera* (Lamarck, 1816)

[Oriental and Pacific islands. Invasive in USA: California (?), Florida, Idaho, Texas]

1' Shell whorls rounded; whorls bearing spiral groves and fine ridges, and bearing transverse lines that may develop into costae in older animals; groove/ridge intersections with transverse costae may give a tuberculate appearance (Fig. 11.8 B) ... *Melanoides tuberculata* (Müller, 1774)

[Afrotropical, Palaearctic, Oriental. Invasive in Nearctic. USA: southern US and California, Colorado, Idaho, Montana, Nevada, North Carolina, spreading. Mexico: widespread, spreading]

PHYLUM MOLLUSCA

Gastropoda: Cerithoidea: Pleuroceridae: Genera

The key is partially based on Burch (1982), and no attempt was made to resolve the species. Taxonomy in part follows Dillon (2011) and is conservative. The family needs revision, and the genera and many species are difficult to distinguish (Hoznagel & Lydeard, 2000; Lydeard et al., 2002; Ó Foighil et al., 2009; Dillon, 2011, 2014).

1 Aperture with an anterior canal present or absent, if present, never produced anteriorly; aperture rounded or quadrate ... 2

1' Aperature with an anterior canal present and produced anteriorly; aperture fusiform; whorls generally with a single row of large projecting spines (Fig. 11.8 D) ... *Io fluviatilis* (Say, 1825)

[USA: Alabama (extinct), Tennessee, West Virginia]

2(1) Body whorl posterior suture margin without a long slit ... 3

2' Body whorl posterior suture margin with a long slit ... *Gyrotoma*

[Probably extinct due to impoundments. USA: Alabama]

3(2) Radula lateral teeth with median cusps narrow, pointed, triangular or cordate/hastate; shell elongate or narrowly conic to broadly conic, ovate or cylindrical (Fig. 11.8 F–H) ... *Pleurocera*/*Lithasia*

3' Radula lateral teeth with median cusps broadly rounded, blunt; shell globose to subglobose, ovate, or broadly conic (Fig. 11.8 E) ... *Leptoxis*

[USA: Atlantic, Gulf, and lower Mississippi River Drainages]

Gastropoda: Cerithoidea: Semisulcospiridae: *Juga:* Species

This genus needs revision (Strong & Kohler, 2009). There are possible undescribed species. *Juga acutifilosa* and *J. occata* are in the subgenus *Calibasis*, which may represent a separate genus (Strong & Frest, 2007).

1 Adult shell smooth 2
1' Adult shell with sculpture, sometimes fine, sometimes limited to early whorls 4
2(1) Shell whorls with shoulder rounded; shell greenish-brown to brown to black 3
2' Shell whorls with shoulder flattened; shell black; montane headwater streams and rivers .... *Juga nigrina* (Lea, 1856)
[USA: California, Oregon]
3(2) Shell color uniformly brown to black; Great Basin Desert springs .... *Juga laurae* (Goodrich, 1944)
[USA: California, Nevada]
3' Shell light to dark greenish-brown, sometimes with yellow, and often with darker bands of color .... *Juga bulbosa* (Gould, 1847)
[USA: California, Oregon, Washington]
4(1) Shell with spiral ridges on all whorls, costae present or not 5
4' Shell with costae, and bearing spiral ridges on early whorls only .... *Juga hemphilli* (Henderson, 1935)
[USA: Oregon, Washington]
5(4) Shell with costae transverse to the spiral ridges 6
5' Shell without costae 8
6(5) Shell with costae on all whorls 7
6' Shell with costae limited to early whorls .... *Juga silicula* (Gould, 1847)
[USA: Oregon, Washington]
7(6) Shell with spire low and rounded, usually not more than three whorls; spiral ridges large, lamellar, generally projecting beyond and partially obscuring costae .... *Juga occata* (Hinds, 1844)
[USA: California]
7' Shell with spire long, usually composed of four or more whorls; spiral ridges small, flat, never projecting beyond costae; costae never obscured .... *Juga plicifera* (Lea, 1838)
[USA: Oregon, Washington]
8(5) Shell with spiral ridges fine, discontinuous and unevenly spaced on body whorl, and only apparent under low magnification; Great Basin Desert springs .... *Juga interioris* (Goodrich, 1944)
[USA: Nevada]
8' Shell with spiral ridges large, obvious without magnification; mountain springs and rivers .... *Juga acutifilosa* (Stearns, 1890)
[USA: California]

Gastropoda: Lymnaeoidea: Lymnaeidae: Genera

Taxonomy follows Hubendick (1951) and Correa et al. (2010).

1 Shell patelliform (limpets) (Fig. 11.9 A, B) 2
1' Shell spiral (Fig. 11.5 C–I) 3
2(1) Shell apex eccentric (Fig. 11.9 B) .... *Fisherola nuttalli* (Haldeman, 1841)
[USA: Idaho, Oregon, Washington]
2' Shell apex subcentral (Fig. 11.9 A) .... ***Lanx* [p. 203]**
[USA: California, Oregon]
3(1) Adult shell lacking spiral striae; body whorl globose; spire very short (Fig. 11.5 C) .... *Radix auricularia* (Linnaeus, 1758)
[Holarctic, widespread]
3' Adult shell lacking spiral striae; body whorl globose or not (Fig. 11.5 D–I) .... ***Lymnaea* [p. 204]**
[Taxonomy confused, genus needs revision]

Gastropoda: Lymnaeoidea: Lymnaeidae: *Lanx:* Species

1 Shell not depressed, margin not flared, shell height in lateral view one-third or more shell height 2
1' Shell depressed, flattened, margin flared, shell height in lateral view one-third or less shell width .... *Lanx klamathensis* Hannibal, 1912
[USA: Oregon, California (?)]

2(1) Adult shell robust ... 3

2' Adult shell thin, fragile ... *Lanx subrotundata* (Tryon, 1865)

[USA: California, Oregon]

3(2) Shell red to brown ... 4

3' Shell mostly black, often with margin reddish to brown, sometimes with scattered, small green dots ... *Lanx patelloides* (Lea, 1856)

[USA: California]

4(3) Klamath River system ... *Lanx alta* (Tryon, 1865)

[USA: California, Oregon]

4' Snake River system ... *Lanx* sp.

[Typically undescribed species are not considered in this series. However, this undescribed species, the "Banbury Lanx," is federally listed under the U.S. Endangered Species Act.]

Gastropoda: Lymnaeoidea: Lymnaeidae: *Lymnaea:* Species

This genus was revised by Hubendick (1951). Although most recent authors have ignored that work, a recent molecular assessment by Correa et al. (2010) supported Hubendick (1951), at least as far as genus level treatments are concerned, and their work is followed here.

1 Entire shell never attenuated, although spire may be, aperture width generally one-half or more length; body whorl shoulders rounded ... 2

1' Shell strongly attenuated, extremely narrow, aperture width one-third or less length; body whorl shoulders flattened; spire long (Fig. 11.5 D) ... *Lymnaea haldemani* Binney, 1867

[Canada, USA: Great Lakes region]

2(1) Shell not translucent, not thin and fragile; shell periostracum not sculptured with microscopic spiral threads ... 3

2' Shell translucent or nearly transparent, thin and fragile; shell periostracum sculptured with microscopic spiral threads (Fig. 11.5 E) ... *Lymnaea columella* Say, 1817

[Widespread]

3(2) Columella not projecting ... 4

3' Columella with a large projecting fold, nearly a tooth; penis and penal sheath very short and pyriform ... *Lymnaea arctica* Lea, 1864

[Canada, USA: arctic and subarctic]

4(3) Adult shell of variable length, but if greater than 35 mm, the spire longer than aperture ... 5

4' Adult shell length greater than 35 mm, with spire shorter than aperture length; umbilicus imperforate (Fig. 11.5 G) ... *Lymnaea megasoma* Say, 1824

[Canada: Manitoba, Northwest Territories, Nova Scotia, Ontario, Quebec. USA: Great Lakes Region]

5(4) Spire not attenuated ... 6

5' Spire narrow, attenuate; aperture flared; umbilicus generally imperforate (Fig. 11.5 F) ... *Lymnaea stagnalis* Linnaeus, 1758

[Canada: widespread except Arctic. USA: northern states]

6(5) Body whorl usually lacking distinct transverse ridges, but if present then umbilicus barely open; spiral striae present or absent ... 7

6' Body whorl with distinct transverse ridges and spiral striae; umbilicus imperforate ... *Lymnaea palustris* (Müller, 1774)

[Canada: widespread except arctic. USA: northern and western states]

7(6) Umbilicus imperforate to barely open ... 8

7' Umbilicus open ... *Lymnaea bulimoides* Lea, 1841

8(7) Columella with an obvious twist or plait; spiral sculpture very fine to microscopic, but present (Fig. 11.5 I) ... 9

8' Columella generally without a twist or plait; spiral sculpture weak, obscured, or absent (Fig. 11.5 H) ... *Lymnaea catascopium* Say, 1817 species group

[This group contains *L. eleodes, L. emarginata, L. gabbi*, and *L. utahensis*, which cannot be reliably and consistently separated. Canada: widespread. USA: northern and western states]

PHYLUM MOLLUSCA

9(8) Radula with lateral teeth with three large and obvious cusps .... *Lymnaea truncatula* (Müller, 1774)/*Lymnaea humilis* Say, 1822

[Canada: southern provinces. USA: widespread. Mexico: Chihuahua, Coahuila, Nuevo Leon, Tamaulipas]

9' Radula with lateral teeth with two large and obvious cusps .... *Lymnaea cubensis* Pfieffer, 1839

[USA: Florida, Louisiana, Texas. Mexico: widespread. Neotropics]

Gastropoda: Planorbiodea: Families

1 Pseudobranch absent; shell obviously sinistral (coiling to the left), with an acute, posteriorly directed spire; mantle margin generally with digitate or serrate lobes; blood clear, lacking haemoglobin (Fig. 11.4 G–K) .... **Physidae [p. 205]**

1' Pseudobranch present as a cup-like lobe projecting from under the mantle posteriolaterally on the left side; shell patelliform (limpets), discoidal (ram's horn snails), or if with a projecting spire, then appearing dextral (spiraling to the right); blood red with haemoglobin (Figs. 11.4 A–F and 11.9 D–L) .... **Planorbidae [p. 206]**

Gastropoda: Planorbiodea: Physidae: Genera

1 Shell elongated, glossy, shiny, and smooth; spire conical; aperture no more than half the total shell length; mantle never with lobes (Fig. 11.4 K) .... *Aplexa elongata* (Say, 1821)

[Canada: Alberta, British Columbia, Manitoba, Ontario, Saskatchewan. USA: Alaska, Idaho, Montana, North Dakota, South Dakota, Utah, Washington, Wyoming, Great Lakes region, and New England states]

1' Shell variable; aperture length more than half the total shell length; mantle usually with digitate or serrate lobes (Fig. 11.4 G–J) .... ***Physa*** **[p. 205]**

Gastropoda: Planorbiodea: Physidae: *Physa:* Species

Follows Wethington & Lydeard (2007) and Wethington et al. (2009). Shell morphology is highly variable (reviewed in Rogers & Wethington, 2007).

1 Aperture length at most 75% of total shell length, aperture not as wide as widest shell width; animal not adapted as below (Fig. 11.4 G, H, J) 2

1' Aperture length nearly as long as total shell length, aperture width equal to widest shell width; foot very broad, adapted for living on vertical cliffs under a thing water film; animal very black (Fig. 11.4 I) .... *Physa zionis* Pilsbry, 1926

[USA: Utah]

2(1) Penis preputium with a large and obvious preputial gland; mantle variable 3

2' Penis preputium without a large and obvious preputial gland; mantle extending around much of body whorl .... *Physa marmorata* Guilding, 1828

[USA: Texas. Mexico: Gulf Coast and tropics. Neotropical]

3(2) Penal sheath entirely glandular 4

3' Penal sheath either partially glandular distally, or not glandular at all 6

4(2) Mantle not extending externally around most of body whorl 5

4' Mantle extending externally around most of body whorl .... *Physa megalochlamys* Taylor, 1988

[Canada: Saskatchewan. USA: Colorado, Idaho, Oregon, Utah, Wyoming]

5(4) Arctic .... *Physa jennessi* Dall, 1919

[Canada: Arctic. USA: Alaska. Palaearctic]

5' Not found in arctic habitats.. *Physa vernalis* Taylor & Jokinen, 1984

[Canada: Great Lakes Region, Newfoundland. USA: Great Lakes region, New England]

6(3) Penal sheath or not glandular at all; spire acute 7

6' Penal sheath glandular in distal one half to one fourth; spire subacute to rounded 8

7(6) Surface waters (Fig. 11.4 H) .... *Physa acuta* Draparnaud, 1805

[Cosmopolitan]

PHYLUM MOLLUSCA

7' Subterranean .... *Physa spelunca* Turner & Clench, 1974
[USA: Wyoming]

8(6) Penal sheath glandular in distal fourth .... *Physa pomilia* Conrad, 1834 species group
[USA: eastern states and California]

8' Penal sheath glandular in distal half (Fig. 11.4 G, J) .... *Physa gyrina* Say, 1821 species group
[Nearctic]

Gastropoda: Planorbiodea: Planorbidae: Genera

Taxonomy follows Walther et al. (2006), Albrecht et al. (2006), and Dillon & Herman (2009).

1 Shell patelliform (limpets) (Fig. 11.9 D–G) 2

1' Shell spiral (Figs. 11.4 A–F and 11.9 H–L) 5

2(1) Three or more adductor muscles, or adductor muscle scars in shell 3

2' One crescent shaped adductor muscle scar nearly circling the inside of the shell (Fig. 11.9 D) .... ***Rhodacmea* [p. 207]**

3(2) Three primary adductor muscle scars (two anterior, one posterior) with a smaller secondary adductor muscle scar between the anterior two, and one between the right anterior and the posterior primary scars 4

3' Three primary adductor muscle scars (two anterior, one posterior) with no smaller secondary scars (Fig. 11.9 E) .... ***Ferrissia* [p. 207]**

4(3) Shell apex at or just to right of shell midline, and just posterior of center; tentacles with prominent pigment cores (Fig. 11.9 G) .... *Laevapex fuscus* (Adams, 1841)
[Canada, USA: east of the Great Plains]

4' Shell apex far to right of midline, and in posterior fourth of shell; tentacles without a pigmented core (Fig. 11.9 F) .... *Hebetancylus excentricus* (Morelet, 1851)
[USA: Florida, Georgia, Texas. Mexico: Gulf states. Neotropics]

5(1) Shell with body whorl and aperture greatly and abruptly larger than preceding whorl; aperture length greater than one-half the shell width (Fig. 11.4 C) 6

5' Shell with whorls gradually and uniformly increasing in size (Figs. 11.9 H–L and 11.4 A, B, D–F) 8

6(5) Aperture length more than one-half, but less than two-thirds body whorl width 7

6' Aperture length greater than two-thirds body whorl width; adult shell 2 mm in diameter .... *Amphigyra alabamensis* Pilsbry, 1906
[USA: Alabama (possibly extinct)]

7(6) Adult shell 2 mm in diameter .... ***Neoplanorbis* [p. 207]**
[USA: Alabama]

7' Adult shell more than 8 mm in diameter .... ***Vorticifex* [p. 207]**
[USA: California, Nevada, Oregon]

8(5) Aperture without teeth 9

8' Aperture with teeth set back from aperture (aperture may need to be broken to view this character) (Fig. 11.9 I) .... ***Planorbula* [p. 208]**

9(8) Shell aperture width less than one-half shell diameter 10

9' Shell aperture width greater than one-half shell diameter 13

10(9) Adult shell greater than 8 mm in diameter; adult shell with four or five whorls, never with numerous low spiral striae (Fig. 11.4 B, D–F) 11

10' Adult shell less than 8 mm in diameter; adult shell with five or six whorls, if only four or five, then bearing numerous low spiral striae (Fig. 11.9 J) .... ***Drepanotrema* [p. 208]**

11(10) Shell thick, not easily crushed 12

11' Shell thin, fragile (Fig. 11.4 B) .... ***Biomphalaria* [p. 208]**

12(11) Shell shoulder with a carina; spire always present .... *Carinifex*
[In need of revision. USA: California, Nevada, Oregon, Utah, Wyoming]

12' Shell shoulder lacking a carina; spire rarely present, shell generally discoidal (Fig. 11.4 D–F) .... *Helisoma*
[Nearctic: widespread. Large genus in great need of revision. Many "species" cannot be reliably separated from each other]

13(12) Shell whorls without large transverse ridges 14

13' Shell costate, i.e., whorls with large transverse ridges (Fig. 11.9 H) .... ***Gyraulus* (in part) [p. 208]**

14(13) Shell with umbilicus deep and narrow, body whorl on opposite side of shell from aperture slightly wider than umbilicus; shoulder carinate or not 15

14' Shell with umbilicus shallow and wide, body whorl on opposite side of shell from aperture slightly narrower than umbilicus; shoulder not carinate (Fig. 11.9 K) .... ***Gyraulus*** (in part) **[p. 208]**

15(14) Mantle with a large, well defined lobe anterior of the anal pore and pseudobranch; shell with shoulder angular, and bearing a carina between midpoint and shoulder (Fig. 11.9 L) .... *Menetus*

[Genus in need of revision. Canada: widespread, except Arctic. USA: widespread]

15' Mantle without a well defined lobe anterior of the anal pore and pseudobranch; shell with body whorl rounded or carinate medially (Fig. 11.4 A) .... ***Promenetus*** **[p. 208]**

Gastropoda: Planorbiodea: Planorbidae: *Rhodacmea:* Species

Genus needs revision. Modified from Burch (1982).

1 Shell smooth .... *Rhodacmea elatior* (Anthony, 1855)/*Rhodacmea hinkleyi* (Walker, 1908)

[USA: Alabama, Tennessee]

1' Shell ribbed .... *Rhodacmea filosa* (Conrad, 1834)

[USA: Alabama, Tennessee (?)]

Gastropoda: Planorbiodea: Planorbidae: *Ferrissia:* Species

Follows Walther et al. (2010).

1 Shell apex on midline .... *Ferrissia rivularis* (Say, 1817)

[Widespread except in Arctic and subarctic]

1' Shell apex to right of midline .... *Ferrissia fragilis* (Tryon, 1863)

[USA: southern Atlantic drainages]

Gastropoda: Planorbiodea: Planorbidae: *Neoplanorbis:* Species

Modified from Burch (1982). This genus needs to be revised (it may only represent one or two variable species). It is only known from the Coosa River in Alabama and is probably extinct.

1 Umbilicus closed; columella without teeth 2

1' Umbilicus open; columella with teeth 3

2(1) Shell with spiral striae; shoulder with a carina .... *Neoplanorbis tantillus* Pilsbry, 1906

[USA: Alabama]

2' Shell without spiral striae; shoulder rounded .... *Neoplanorbis smithi* Walker, 1908

[USA: Alabama]

3(1) Shoulder with a carina .... *Neoplanorbis carinatus* Walker, 1908

[USA: Alabama]

2' Shoulder rounded .... *Neoplanorbis umbilicatus* Walker, 1908

[USA: Alabama]

Gastropoda: Planorbiodea: Planorbidae: *Vorticifex:* Species

Modified from Burch (1982). This genus needs to be revised.

1 Body whorl with a ventral carina or angle around the umbilicus .... *Vorticifex solida* (Dall, 1870)

[USA: California, Nevada]

1' Body whorl lacking carinae .... *Vorticifex effusa* (Lea, 1856)

[USA: California, Oregon]

Gastropoda: Planorbiodea: Planorbidae: *Planorbula:* Species

1 Shell thin, fragile, up to 20 mm in diameter; low teeth present inside aperture on outer whorl wall .... *Planorbula campestris* (Dawson, 1875)

[Canada: Alberta, British Columbia, Manitoba, Saskatchewan. USA: western states]

1' Shell solid, up to 8 mm in diameter; aperture teeth on all sides .... *Planorbula armigera* (Say, 1821)

[Canada: widespread, except southwest. USA: east of Great Plains]

Gastropoda: Planorbiodea: Planorbidae: *Drepanotrema:* Species

Modified from Burch (1982).

1 Adult shell extremely flattened, bearing four or five whorls, lacking spiral striae 2

1' Adult shell not extremely flattened, bearing five or six whorls, bearing deep, obvious spiral striae .... *Drepanotrema aeruginosum* (Morelet, 1851)

[USA: Arizona, Texas. Mexico: widespread. Neotropics]

2(1) Adult shell shoulder with an obvious and strong keel .... *Drepanotrema kermatoides* (d'Orbigny, 1835)

[USA: Florida, Texas. Mexico: widespread. Neotropics]

2' Adult shell shoulder rounded or angular .... *Drepanotrema cimex* (Moricand, 1839)

[USA: Texas. Mexico: widespread. Neotropical]

Gastropoda: Planorbiodea: Planorbidae: *Biomphalaria:* Species

Modified from Burch (1982).

1 Adult shell (five or more whorls) 15 mm in diameter or larger .... *Biomphalaria glabrata* (Say, 1818)

[Invasive. USA: Florida. Neotropics]

1' Adult shell (five or more whorls) 10 mm in diameter or smaller .... *Biomphalaria havanensis* (Pfieffer, 1839)

[USA: Arizona, Florida, Louisiana, Texas. Mexico: widespread. Neotropics]

Gastropoda: Planorbiodea: Planorbidae: *Gyraulus:* Species

1 Shell with umbilicus deep and narrow, body whorl on opposite side of shell from aperture slightly wider than umbilicus; shoulder carinate or not 2

1' Shell with umbilicus shallow and wide, body whorl on opposite side of shell from aperture slightly narrower than umbilicus; shoulder not carinate .... *Gyraulus crista* (Linnaeus, 1758)

[Holarctic]

2(1) Periostracum glabrous; shoulder rounded to subangular 3

2' Periostracum hirsute on at least the youngest whorls; shoulder flattened obliquely .... *Gyraulus deflectus* (Say, 1824)

3(2) Shell in lateral view, with aperture one-third or more shell width 4

3' Shell in lateral view, with aperture one-fourth or less shell width; shell with spire and umbilicus nearly identical in general appearance .... *Gyraulus circumstriatus* (Tryon, 1866)

[Canada: widespread, except Arctic. USA: northern states, and Rocky Mountains to New Mexico]

4(3) Aperture width and length subequal .... *Gyraulus hornensis* Baker, 1934

[Canada: Manitoba, Northwest Territories, Ontario, Saskatchewan. USA: North Dakota, Wisconsin]

4' Aperture width subequal to half its length .... *Gyraulus parvus* (Say, 1817)

Gastropoda: Planorbiodea: Planorbidae: *Promenetus:* Species

1 Shell body whorl peripherally carinate .... *Promentus exacuous* (Say, 1821)

[Canada: widespread, except Arctic. USA: Alaska, and east of the Continental Divide]

1' Shell body whorl with shoulder rounded, no carina .... *Promenetus umbilicatellus* (Cockerell, 1877)

[Canada: Alberta, British Columbia, Manitoba, Saskatchewan. USA: Alaska, Oregon, Utah, Washington, northern central states]

REFERENCES

Albrecht, C., K. Kuhn & B. Street. 2006. A molecular phylogeny of Planorboidea (Gastropoda, Pulmonata): insights form enhanced taxon sampling. Zoologica Scripta 36: 27–39.

Baker, F. C. 1945. The Molluscan Family Planorbidae. University of Illinois Press, Urbana. 530 pp.

Brown, K.M. & C. Lydeard. 2010. Mollusca: Gastropoda. Chapter 10 Pages 277–307 *in*: Bivalvia. J.H. Thorp & A.P. Covich (eds.), Ecology and Classification of North American Freshwater Invertebrates, Third Edition, Academic Press, Boston, MA.

Burch, J.B. 1982. Freshwater snails (Mollusca: Gastropoda) of North America. Environmental Monitoring and Suppor Laboratory, Office of Research and Development, US Environmental Protection Agency, Cincinnati, OH, USA. 294 pp.

Burch, J. B. 1989. North American Freshwater Snails. Malacological Publications, Hamburg, Michigan, USA.

Correa C. A., J. S. Escobar, P.Durand, F. Renaud, P. David, P. Jarne, J.-P. Pointier & S. Hurtrez-Boussès. 2010. Bridging gaps in the molecular phylogeny of the Lymnaeidae (Gastropoda: Pulmonata), vectors of *Fascioliasis*. BMC Evolutionary Biology 10: 381.

Dillon, R.T., Jr. 2011. Robust shell phenotypes is a local response to stream size in the genus *Pleurocera* (Rafinesque, 1818). Malacologia 53: 265–277.

Dillon, R.T., Jr. 2014. Cryptic phenotypic plasticity in populations of the North American freshwater gastropod, *Pleurocera semicarinata*. Zoological Studies 53: 31–38.

Dillon, R.T., Jr, & J.J. Herman. 2009. Genetics, shell morphology and life history of the freshwater pulmonate limpets *Ferrissia rivularis* and *Ferrissia fragilis*. Journal of Freshwater Ecology 24: 261–271.

Holznagel, W. E. & C. Lydeard. 2000. A molecular phylogeny of North American Pleuroceridae (Gastropoda: Cerithioidea) based on mitochondrial 16s rDNA sequences. Journal of Molluscan Studies 66: 233–257.

Hubendick, B. 1951. Recent Lymnaeidae. Their variation, morphology, taxonomy, nomenclature and distribution. Kungliga Svenska Vetenskapsakademien Handlingar 3: 5–223, plus five plates.

Lydeard, C., W. E. Holznagel, M. Glaubrecht & W. F. Ponder. 2002. Molecular phylogeny of a circum-global, diverse gastropod superfamily (Cerithioidea: Mollusca: Caenogastropoda): Pushing the deepest phylogenetic limits of mitochondrial LSU rDNA sequences. Molecular Phylogenetics and Evolution 22: 399–406.

Meier-BrookK, C. 1983. Taxonomic studies on *Gyraulus* (Gastropoda: Planorbidae). Malacologia 24: 1–113.

Ó Foighil, D., T. Lee, D. C. Campbell & S.A. Clark. 2009. All voucher specimens are not created equal: A cautionary tale involving North American pleurocerid gastropods. Journal of Molluscan Studies 75: 305–306.

Pechenik, J. A. 1985. Biology of the Invertebrates. Prindle, Weber, and Schmidt, Boston, Massachusetts, USA.

Rogers, D.C. & A. Wethington. 2007. *Physa natricina* Taylor, 1988, junior synonym of *Physa acuta* Draparnaud, 1805 (pulmonata: Physidae). Zootaxa 1662: 45–51.

Strong, E., & T.J. Frest. 2007. On the taxonomy and systematics of *Juga* from western North America (Gastropoda: Cerithoidea: Pleuroceridae). The Nautilus 121: 43–65.

Strong, E. & F. Kohler. 2009. Morphological and molecular analysis of '*Melania*' *jacqueti* Dautzenberg and Fischer, 1906: from anonymous orphan to critical basal offshoot of the Semisulcospiridae (Gastropoda: Cerithioidea). Zoologica Scripta 38: 483–502.

Strong, E., O. Gargominy, W. F. Ponder & P. Bouchet. 2008. Global diversity of gastropods (Gastropoda; Mollusca) in freshwater. Hydrobiologia 595: 149–166.

Wilke, T., M. Haase, R. Hershler, H.-P. Liu, B. Misof & W. Ponder. 2013. Pushing short DNA fragments to the limit: phylogenetic relationships of 'hydrobioid' gastropods (Caenogatropoda: Rissooidea). Molecular Phylogenetics and Evolution 66: 715–736.

Walther, A.C., T. Lee, J.B. Burch & D. Ó Foighil. 2006. *E Pluribus Unum*: a phylogentic reassessment of *Laevapex* (Pulmonata: Ancylidae), a North American genus of freshwater limpets. Molecular Phylogenetics and Evolution 40: 501–516.

Walther, A. C., J. B. Burch & D. Ó Foighil. 2010. Molecular phylogenetic revision of the freshwater limpet genus *Ferrissia* (Planorbidae: Ancylinae) in North America yields two species: *Ferrissia* (*Ferrissia*) *rivularis* and *Ferrissia* (*Kincaidilla*) *fragilis*. Malacologia 53: 25–45.

Wethington, A. & C. Lydeard. 2007. A molecular phylogeny of Physidae (Gastropoda: Basommatophora) based on mitochondrial DNA sequences. Journal of Molluscan Studies 73: 241–257.

Wethington, A., J. Wise & R.T. Dillon, Jr. 2009. Genetic and morphological characterization of the Physidae of South Carolina (Gastropoda: Pulmonata: Basommatophora), with description of a new species. The Nautilus 123: 282–292.

Class Bivalvia

James H. Thorp

Kansas Biological Survey and Department of Ecology and Evolutionary Biology, University of Kansas, Lawrence, KS, USA

D. Christopher Rogers

Kansas Biological Survey and The Biodiversity Institute, Kansas University, Lawrence, KS, USA

INTRODUCTION[2]

The Nearctic bivalve molluscs (or mollusks) consist of five families in two orders: Unionoida (Margartiferidae and Unionidae) and Veneroida (Sphaeriidae, Corbiculidae, and Dreissenidae) (Turgeon et al., 1998; Strayer, 1999; Grigorovich et al., 2000).

The vast majority of species in these families are in the Unionidae with its nearly 300 Nearctic species, compared to about 837 species worldwide (Graf & Cummings, 2002).

2. The prekey material in this section is based on various sources, especially Cummings & Graf (2010).

PHYLUM MOLLUSCA

The distribution of Nearctic unionids is shown in Fig. 19.38 of Volume I of this edition. Most occur in rivers of the southeastern USA. These are known as pearly mussels, naiads, or unionids. Another major center of unionid diversity is China.

The Margaritiferidae bear a close resemblance to some unionids, but they lack a posterior mantle fusion. These "pearl mussels" are represented by one genus and five Nearctic species.

The small bivalves of Sphaeriidae (fingernail, pea, pill, or seed clams) are divided into four genera and about 40 species in the Nearctic: *Eupera, Pisidium, Sphaerium*, and *Musculium*, with about 200 species worldwide in 4–5 genera. The sphaeriids are most abundant in the Nearctic within the Rocky Mountains (where unionoid mussel diversity is low) and in the formerly glaciated regions of the continent.

The Dreissenidae contains only a few freshwater forms. In the Nearctic, there is one species native to the Nearctic and two invasive species from the Palaearctic.

LIMITATIONS

Freshwater bivalve molluscs are generally identified by shell characteristics. Unfortunately, shell shape can be variable within a taxon depending on environmental conditions (e.g., see Fig. 19.6 in Vol. I), and the external sculpture can be eroded or otherwise obscured, especially for animals living in turbulent streams or poorly buffered waters of low alkalinity (i.e., with little calcium carbonate for shell formation). Color patterns can also be an element of identification of mussels, but these may not survive drying or preservation. Therefore, the best way to identify bivalves is to consult a specialist and view a variety of shells of the likely taxa and similar ones known from the same locality.

TERMINOLOGY AND MORPHOLOGY

The bivalve shell is composed of two separate halves called valves. The oldest portion of the valve (the juvenile shell) is the umbo or beak, which lies on the dorsal surface of the shell. A diagram defining the anterior, posterior, and right and left sides of the animal is presented in Fig. 11.10. If the soft tissues are still present and attached to the shell, then the incurrent and excurrent apertures exit posteriorly, that is to the right side of the right valve (or left side of the left valve).

Shells in which the umbo is in the middle of each valve are called "equilateral" (equal size right and left valves are termed "equivalve"). The overall outline of a valve is often important. A valve may be rhomboidal, triangular (or trigonal), round, quadrate, oval (or ovoid) or elliptical.

The two valves are connected by a hinge that extends from the beak towards the posterior end of the animal. The hinge is connected by a ligament, which is a flexible structure joining the two valves. This ligament may be internal (inside the hinge) or external (outside the hinge).

Inside the hinge line, on the interior surface of the valve, one or more longitudinal ridges or grooves may be present. These are the lateral teeth. One or more transverse, rounded projections, termed pseudocardinal teeth, may be present just below and sometimes slightly anterior of the umbo. These teeth and the ligament lie on a surface called the umbo cavity.

On the interior surface of each valve are muscle scars. At the anterior end, there may be a large anterior adductor scar and a smaller protractor scar. At the posterior end are the large posterior adductor scar and smaller posterior retractor scar. These scars appear as round to oblong polygons. A long pallial line runs in an arc inside the valve, from the anterior to the posterior muscle scars. The substance making up the inside of the shell is called nacre.

The external shell surface sculpture is important for diagnosis. The external surface of the valves is covered in a thin layer called the periostracum. Most shells have a series of growth lines, which run parallel to and generally mirror the valve margins. The shell may bear transverse or radial ridges, a secondary ridge that runs from the umbo to the posterior margin called a posterior ridge, rounded bumps called pustules, or a flattened flange extending dorsally from the hinge line, called a wing.

Sometimes confusion arises between small juvenile mussels, sphaeriid clams, and small *Corbicula*, but the presence of both posterior and anterior lateral teeth easily identifies all clams. Asian clams (*Corbicula*) grow much larger and thicker than sphaeriids. The external surface of sphaeriids is always smooth or with finely striated ridges, whereas *Corbicula* has evenly spaced, thicker, elevated ridges on the shell surface.

MATERIAL PREPARATION AND PRESERVATION

Detailed methods for collecting, curation, and rearing bivalve molluscs are discussed in Cummings & Graf (2014) in Chapter 19 of Volume I. Sturm et al. (2006) published a guide to collecting and preserving mussels. The following section is a summary of Cummings & Graf (2014).

Focus wherever possible on collecting the shells from recently dead animals to avoid collecting live molluscs, many of which are threatened or endangered in nature. Clean the shell of all specimens with a toothbrush, vegetable brush, or a nylon scrub pad (but not wire brushes or harsh scrub pads) to reveal characters used in identification. Avoid all harsh cleaning agents that may damage the shell, and do not coat the shell with any substances, including paraffin, oil (baby, linseed, etc.), or petroleum jelly. Keep dry shells in cabinets out of direct light in a temperature and humidity controlled environment wherever possible. Catalog number should be written

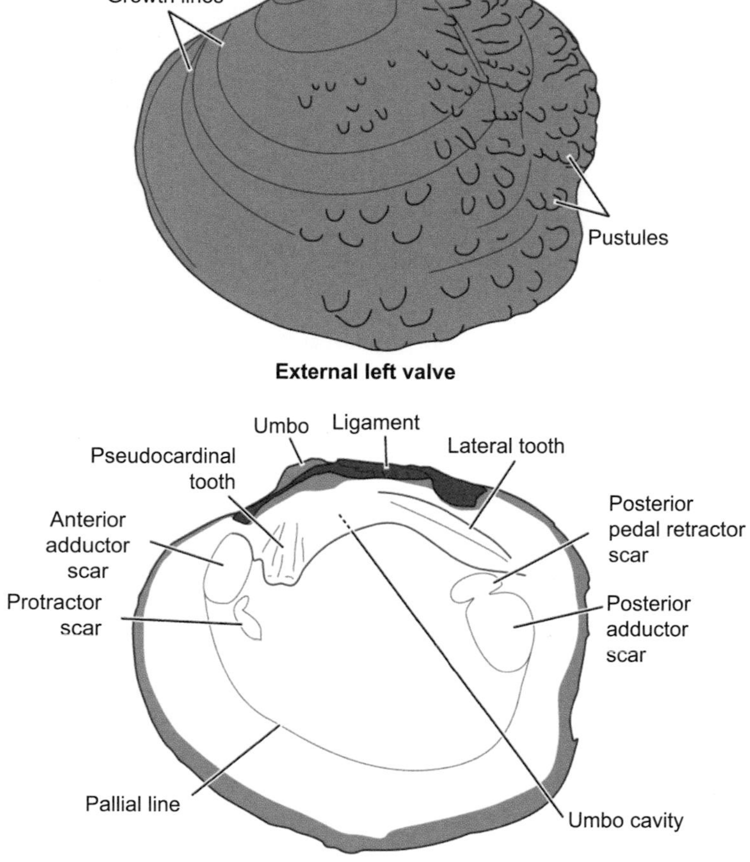

FIGURE 11.10 Gross anatomy of the exterior and interior of a valve (shell) of the freshwater unionid mussel *Theliderma intermedia. From Cummings & Graf, 2010.*

on the inside shell surface with India ink or other permanent marker, such that the internal characters are not obscured; this eliminates the problem of losing tags over the years.

For anatomical studies of live collected specimens, narcotize and relax the specimens before fixing them by using agents such as MS-222, chloroform, menthol crystals, and phenobarbital. Preserve soft tissue (with or without the shells) in 10% buffered formalin (except when planning genetic analyses). If the whole animals are to be preserved, first freeze live animals with the shell slightly open (wedged open slightly with a wooden peg) and then transfer the specimens afterward into formalin. After a few days in formalin, soak the specimens in freshwater with several rinses over a few days, and then transfer the whole animal or soft parts into 70% ethanol for long term storage. For molecular genetic studies, fix and store the animal in 95% ethanol.

KEYS TO BIVALVIA

As described in Cummings & Graf (2010), many state or regional monographs are available, but many are dated, unavailable, or in need of revision. Williams et al. (1993) provides a listing by U.S. states of nearly 200 references on freshwater mussels. Regional monographs on mussels published after 1993 include: Florida (in part) (Williams & Butler, 1994), Vermont (Fichtel & Smith, 1995), Texas (Howells et al., 1996), Maine (Martin, 1997; Nedeau et al., 2000), Minnesota (Graf, 1997), New York (Strayer & Jirka, 1997), Tennessee (Parmalee & Bogan, 1998), Kansas (Bleam et al., 1999), the Apalachicola River and Rio Grande drainages (Johnson, 1999; Brim Box & Williams, 2000), and others (Ohio and Alabama) are in preparation or press. The essential references for tracking freshwater mussel nomenclature through the ages are Simpson (1900, 1914), Frierson (1927), Haas (1969a,b), and Burch (1975) (Cummings & Graf, 2010).

Bivalvia: Orders

1 Anterior and posterior lateral teeth present on inside of valves .... **Veneroida [p. 212]**

1' Anterior pseudocardinal tooth and posterior lateral tooth present on valves .... **Unionoida [p. 213]**

Bivalvia: Veneroida: Families

1 Shell hinge external; hinge teeth present or absent; shell not triangular in cross section 2

1' Shell hinge internal; hinge teeth absent; shell triangular to somewhat oval in cross section; stripes often present on exterior of shell .... **Dreissenidae [p. 212]**

2(1) Lateral teeth serrate; external surface with raised concentric ridges; thick shell yellow to brown, often with a purple stripe on umbo; adults generally >2.5 cm (Fig. 11.11 D) .... Corbiculidae, one genus: *Corbicula*

[Nearctic: invasive. Palaearctic, Oriental]

2' Lateral teeth smooth; external shell smooth or with fine, concentric striae; thin shell; adults generally <2.5 cm .... **Sphaeriidae [p. 212]**

Bivalvia: Veneroida: Dreissenidae: Genera

1 Myopore plate (internal plate at the base of the umbo) lacking a tooth projecting ventrally into the internal shell cavity; umbo pointed .... ***Dreissena* [p. 212]**

1' Myopore plate (internal plate at the base of the umbo) bearing a triangular tooth projecting ventrally into the internal shell cavity; shell umbo rounded .... *Mytilopsis leucophaeata* (Conrad, 1831)

[USA, Mexico: Gulf Coast rivers and estuaries; Invasive in Canada and USA along the Atlantic coast, in the Great Lakes, and in the Mississippi drainage]

Bivalvia: Veneroida: Dreissenidae: *Dreissena:* Species

1 Shell ventral margin flat to concave and flattened; ventrolateral shoulder with a ridge; shell triangular in cross section; variable colors but often with prominent stripes; zebra mussel .... *Dreissena polymorpha* (Pallas, 1771)

1' Shell ventral margin rounded, convex, never flattened; ventrolateral shoulder without a ridge; color pale (especially near hinge) but may have dark concentric rings; quagga mussel .... *Dreissena rostriformis bugensis* (Andrusov, 1897)

Bivalvia: Veneroida: Sphaeriidae: Genera

1 Umbo anterior 2

1' Umbo distinctly posterior; generally small (0.5–10 mm) (Fig. 11.11 C) .... *Pisidium*

[Nearctic]

2(1) Two cardinal teeth in one valve and one in opposing valve; shell not mottled (Fig. 11.11 B) .... *Sphaerium/Musculium*

[Recent studies show extreme overlap in genus and species level characters. Nearctic: widespread]

2 One cardinal tooth per valve; shell interior usually mottled and sometimes visible to exterior (Fig. 11.11 A) .... *Eupera cubensis* (Prime, 1865)

[USA: Atlantic and Gulf Coastal Plain. Neotropical]

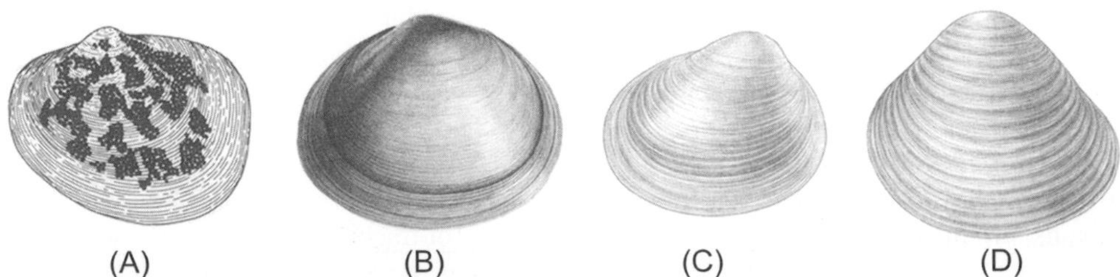

FIGURE 11.11 External view of shells of: (A) *Eupera*, (B) *Sphaerium*, (C) *Pisidium*; and (D) *Corbicula*. *After Burch, 1975.*

Bivalvia: Unionoida: Families

Modified from Burch (1975).

1 Midinterior scars present; mantle not dorsally united to form a separate opening; posteriomedial mantle margins not forming distinct siphons; pseudocardinal and lateral teeth present, at least in juveniles; shell typically thick, dark, and shell elliptical (elongate) with brown or black and relatively thick; umbo displaced anteriorly .... Margaritiferidae, one genus: ***Margaritifera*** **[p. 213]**

1' Midinterior scars absent; mantle united posteriorly to form incurrent and excurrent siphons; pseudocardinal and lateral teeth present or not; shells various; umbo central or displaced anteriorly (Fig. 11.13) .... **Unionidae [p. 213]**

Bivalvia: Unionoida: Margaritiferidae: *Margaritifera:* Species

1 Lateral teeth reduced or absent in adults; shell sculpturing absent 2

1' Lateral teeth always present in the adult; shell posteriodorsal surface at least faintly sculptured and plicate 4

2(1) Pseudocardinal teeth broad; shell thick 3

2' Pseudocardinal teeth reduced, pseudocardinal tooth in right valve subacute; shell thin .... *Margaritifera monodonta* (Say, 1829)
[USA: Upper Mississippi River system]

3(2) East of Continental Divide .... *Margaritifera margaritifera* (Linnaeus, 1758)
[Canada: New Brunswick, Newfoundland and Labrador, Nova Scotia, Prince Edward Island, Quebec. USA: New England]

3' West of Continental Divide .... *Margaritifera falcata* (Gould, 1850)
[Canada: British Columbia. USA: Alaska, California, Idaho, Montana, Nevada, Oregon, Utah, Washington, Wyoming]

4(1) Ventral margin arcuate .... *Margaritifera hembeli* (Conrad, 1838)
[USA: Arkansas, Louisiana]

4' Ventral margin straight or slightly curved .... *Margaritifera marrianae* Johnson, 1983
[USA: Alabama]

Bivalvia: Unionoida: Unionidae: Genera[3]

Kevin S. Cummings
Illinois Natural History Survey, Center for Biodiversity, Champaign, IL, USA

Daniel L. Graf
Biology Department, University of Wisconsin-Stevens Point, Stevens Point, WI, USA

The following keys are divided by subregions within the Nearctic. A map of these subregions is shown in Fig. 11.12.

Pacific Subregion

1 Shell angular, with a strongly developed posterior ridge, reduced lateral teeth .... *Gonidea angulate* (Lea, 1838)
[Canada: British Columbia. USA: California, Idaho, Nevada, Oregon, Washington]

1' Shell not angular, without well-developed posterior ridge, lateral teeth completely absent .... *Anodonta*
[Canada: British Columbia. USA: Alaska, California, Idaho, Nevada, Oregon, Washington]

Atlantic Subregion

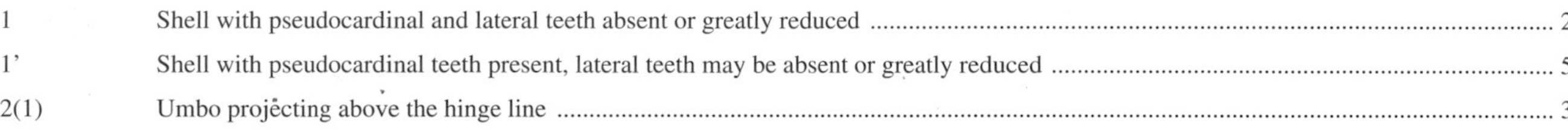

1 Shell with pseudocardinal and lateral teeth absent or greatly reduced 2

1' Shell with pseudocardinal teeth present, lateral teeth may be absent or greatly reduced 5

2(1) Umbo projecting above the hinge line 3

3. This section was edited by Thorp & Rogers from the original text appearing in the third edition (2010).

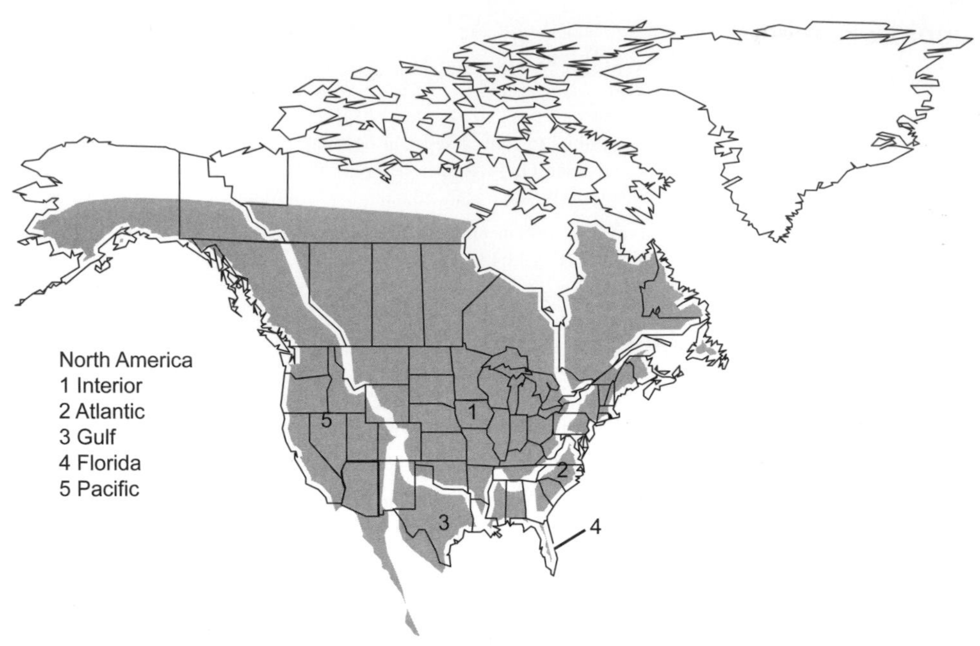

FIGURE 11.12 Map of geographic areas covered in the key (original elements of the map rendered by Jerry Graf).

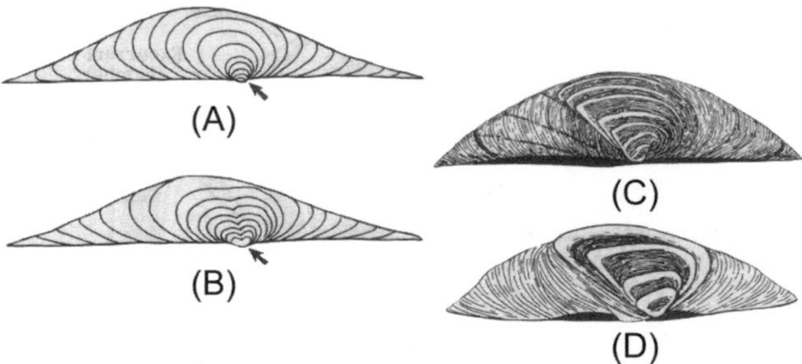

FIGURE 11.13 Beak sculpture of freshwater bivalves: (A) single-looped concentric ridges; (B) double-looped concentric ridges; (C) fine concentric ridges; and (D) coarse concentric ridges. *From Burch, 1975.*

2’ Umbo not projecting above the hinge line .... *Utterbackia*

3(2) Umbo sculpture consists of concentric bars (Fig. 11.13) 5

3’ Umbo sculpture double looped (Fig. 11.13 B), shell thin .... *Pyganodon*

4(3) Nacre usually orange or salmon colored; pseudocardinal and lateral teeth reduced to a thickened ridge; umbo sculpture prominent or pronounced .... *Strophitus*

4’ Nacre bluish or white; pseudocardinal and lateral teeth completely absent; umbo sculpture fine .... *Anodontoides*

5(3) Shell truncated, with well-developed lateral teeth in left valve 6

5’ Right valve with two lateral teeth or lateral teeth absent or reduced and not interlocking .... *Alasmidonta*

6(5) Shell with spines 7

6’ Shell without spines 8

7(6) Shell thick and often large .... *Elliptio*

7’ Shell and spines small .... *Pleurobema collina* Conrad, 1837

8(6) Left valve without extra interdental tooth 9

8’ Left valve hinge line with an additional small interdental or accessory tooth, giving the appearance of three pseudocardinal teeth, shell more or less compressed, rhomboid, periostracum dark green with numerous green rays, umbo sculpture with heavy bars .... *Lasmigona* (in part)

9(8)	Shell shape broadly triangular	10
9'	Shell shape oval, round, or rhomboid	11
10(9)	From the James River	*Lexingtonia*
10'	From an area extending from the Roanoke River south to the headwaters of the Savannah River Basin	*Fusconaia masoni* (Conrad, 1834)
11(10)	Shell shape rhomboid or rectangular	12
11'	Shell shape oval or round	15
12(11)	Shell usually more than twice as long as broad	13
12'	Shell usually less than twice as long as broad; periostracum light in color with numerous green rays; shell relatively small and thin, oval to elongate oval; blade-like pseudocardinal teeth	*Villosa*
13(12)	Nacre color white, shell inflated	14
13'	Nacre color typically some shade or purple, but ranges from salmon to purple	*Elliptio* (in part)
14(13)	Periostracum not rayed and mat or dull, shell moderately thick and slightly rectangular, posterior end angled with two grooves running along the posterior slope	*Uniomerus*
14'	Periostracum not mat and often rayed, particularly in juveniles, posterior end tapered to a sharp point	*Ligumia*
15(13)	Adult shell typically >40 mm in length, with a smooth periostracum	16
15'	Adult shell typically <40 mm in length, with a dark cloth-like or mat periostracum	*Toxolasma*
16(15)	Shell relatively large, oval to elongate oval in shape, periostracum very shiny to mat with rays	*Lampsilis*
16'	Shell thin and oval in shape, periostracum dull yellow, rayless or with very fine rays all over the shell found in or near the tidewater, nacre often salmon colored	*Leptodea ochracea* Say, 1817

Interior Subregion

1	Shell without lateral teeth or all pseudocardinal teeth	3
2'	Shell with both pseudocardinal and lateral teeth present	9
3(2)	Shell with pseudocardinal teeth, but lacking lateral teeth	4
3'	Shell with greatly reduced or completely lacking both pseudocardinal and lateral teeth	5
4 (3)	Shell <25 mm, periostracum often eroded, umbo sculpture consists of heavy bars [Upper Cumberland and Tennessee River basins.]	*Pegias fabula* (Lea, 1838)
4'	Shell relatively large >25 mm, umbo sculpture consists of heavy bars [Widespread in Mississippi River basin; also found in the Tennessee River drainage.]	*Lasmigona*
5(3)	Hinge teeth completely absent	6
5'	Shell with pseudocardinal tooth consisting of a swelling or knob, lateral teeth reduced to a thickened ridge, umbo sculpture heavy "v"-shaped ridges	*Strophitus undulates* Say, 1817
6(5)	Umbo projects above the hinge line	7
6'	Umbo even with or below the hinge line	*Utterbackia imbecillis* (Say, 1829)
7(6)	Umbo projecting only slightly above the hinge line	8
7'	Umbo projects well above the hinge line	*Pyganodon grandis* (Say, 1829)
8(7)	Umbo projecting only slightly above the hinge line, shell elongate, umbo sculpture consists of faint "v"-shaped ridges	*Anodontioides ferussacianus* (Lea, 1834)
8'	Umbo barely above the hinge line, shell rounded in shape, umbo sculpture with double looped ridges	*Utterbackia suborbiculata* Say, 1831
9(2)	Shell with external surface sculpture (plications, pustules, ridges)	10
9'	Shell without surface sculpture	27
10(9)	Shell with prominent external surface pustules, knobs or bumps and/or plications or ridges	11
10'	Shell lacking plicae	19
11(10)	Shell with prominent plications or ridges but lacking pustules, knobs, or bumps	12
11'	Shell with prominent plications or ridges and pustules, knobs or bumps on the anterior part of the shell	17

12(11) Shell with prominent plications or ridges confined to the posterior slope 13

12' Shell with prominent plications or ridges not confined to the posterior slope 14

13(12) Shell elongate, inflated, with bluegreen nacre color .... *Medionidus conradicus* (Lea, 1834)
[Restricted to the Cumberland and Tennessee River basins.]

13' Shell elongate, inflated, nacre not as above, posterior slope and ridge rounded .... *Ptychobranchus subtentum* Say, 1825
[Restricted to the Cumberland and Tennessee River basins.]

14(12) Shell without a distinct posterior ridge; nacre white 15

14' Shell shape rectangular with a distinct posterior ridge; nacre purplish .... *Plectomerus dombeyanus* (Humboldt & Bonpland, 1827)
[Lower Ohio and Mississippi River basins]

15(14) Shell not as below 16

15' Shell thick, small <50 mm, with many low plications all over the posterior surface of the shell .... *Lemiox rimosus* (Rafinesque, 1831)

16(15) Shell laterally compressed to inflated, round to quadrate in shape; heavy pseudocardinal teeth, well-developed lateral teeth; deep umbo cavity, shell surface typically with three prominent, well-developed ridges .... *Amblema plicata* (Say, 1817)
[Widespread in the Mississippi River basin.]

16' Shell inflated and somewhat thin, rounded, approaching oval, pseudocardinal teeth curved, surface with indistinct undulations or ridges .... *Arcidens wheeleri* (Ortmann & Walker, 1912)
[Confined to the Red and Ouachita River basins.]

17(11) Shell not elongate, quadrate or rounded in shape and the pustules less numerous or confined to the anterior part of the shell 18

17' Shell elongate with numerous small pustules over surface .... *Tritogonia verrucosa* (Rafinesque, 1820)

18(17) Shell large and quadrate in shape with the pustules confined to the anterior part of the shell .... *Megalonaias nervosa* (Rafinesque, 1820)

18' Shell more or less rounded in shape with indistinct undulations on the surface; lateral teeth poorly developed .... *Arcidens confragosus* (Say, 1829)

19(10) Shell surface with generally distributed pustules 20

19' Shell with pustules restricted to a single or double row 23

20(19) Shell rounded, compressed to slightly inflated; pustules covering most of the shell; nacre white; umbo cavity deep 21

20' Shell rounded, compressed to slightly inflated, pustules covering most of the shell, nacre purple, umbo cavity deep .... *Cyclonaias tuberculata* (Rafinesque, 1820)

21(20) Shell rounded; pustules generally variable in shape 22

21' Shell rounded, pustules small and covering most of the shell; elevated ridges at the growth lines; nacre white; umbo cavity deep .... *Cyprogenia*

22(21) Shell rounded, compressed to slightly inflated; pustules covering most of the shell; nacre white; umbo cavity deep; a conspicuous green stripe always present on the umbo .... *Amphinaias pustulosa* (Lea, 1831)

22' Shell rounded; pustules covering most of shell; nacre white, pink or salmon; umbo cavity deep; green stripe absent; periostracum reddish chestnut in color .... *Plethobasus cooperianus* (Lea, 1834)

23(19) Pustules forming two distinct rows 24

23' Pustules forming a single row 25

24(23) Shell rounded to quadrate; two rows of pustules, with one row on the posterior ridge, a sulcus or shallow depression may be present between the rows of pustule .... *Quadrula* (in part)

24' Shell rounded to quadrate in shape; two rows of paired pustules, with one row on the posterior ridge, sulcus absent .... *Amphinaias nodulata* (Rafinesque, 1820)

25(23) Valve with a single median row of large pustules 26

25' Shell quadrate to elongate and rectangular in shape, often with chevron markings; posterior ridge with a single row of pustules or small pustules spread across surface .... *Theliderma*

26(25) A single row of three large pustules running down the center of the valve which alternate on each valve .... *Obliquaria reflexa* Rafinesque, 1820

26' A single row of numerous large pustules running down the center of the valve which do not alternate between the valves .... *Dromus*/*Plethobasus*

27(9) Shell with a well-developed dorsal wing projecting above the hinge line, usually found posterior to the umbo but a small wing may also be present anterior to the umbo 28

27' Shell without a prominent dorsal wing 29

28(27) Shell and hinge teeth thin, pseudocardinal teeth compressed; periostracum dull and typically yellow; nacre white with suffusions of pink ..... *Leptodea*

28' Shell and hinge teeth thin, pseudocardinal teeth compressed; periostracum shiny and typically olive, brown, or black; nacre evenly colored light or dark purple ..... *Potamilus*

29(27) Shell shape round 30

29' Shell shape not round 36

30(29) Shell inflated or various shaped 31

30' Shell compressed and thick, usually triangular in shape with a sharp posterior ridge; yellow in color with "v"-shaped rays ..... *Ellipsaria lineolata* (Rafinesque, 1820)

31(30) Umbo nearly central 32

31' Umbos located anteriorly 34

32(31) Shell without raised ridge or hump 33

32' Shell with a raised ridge or hump located near the middle running parallel with the growth lines ..... *Dromus dromas* (Lea, 1834)

[Restricted to the Cumberland and Tennessee River basins.]

33(32) Shell surface smooth, round, without a sulcus ..... *Obovaria*

33' Shell with a broad shallow sulcus; fine green rays; shallow umbo cavity; females with a swollen, broadly expanded posterioventral margin ..... *Epioblasma*

34(33) Umbo cavity relatively shallow 35

34' Umbo cavity deep and compressed; a shallow sulcus present in some species, nacre white ..... *Fusconaia*

35(34) Shell yellowish with green rays near the umbo; thick, well-developed lateral teeth ..... *Lexingtonia dolabelloides* (Lea, 1840)

[Restricted to the Tennessee River basin.]

35' Shell rounded or quadrate; umbo cavity relatively shallow, without a sulcus; nacre white to pink or salmon ..... *Pleurobema*

36(29) Shell shape not oval 37

36' Shell shape oval to oblong 51

37(36) Shell shape rectangular, triangular, or quadrate with sulcus 38

37' Shell shape elongate, two to four times as long as high 41

38(37) Umbo cavity deep 39

38' Umbo cavity shallow 40

39(38) Shell shape rectangular or slightly rounded; umbo with green rays ..... *Fusconaia*

39' Shell shape triangular ..... *Pleurobema*

40(38) Posterior slope angled; periostracum highly variable in color and may be green, brown, or yellow with chevron markings; nacre variable from white to pink ..... *Truncilla*

40' Shell rectangular, thick; nacre white with purple ..... *Elliptio crassidens* (Lamarck, 1819)

41(37) Shell thin 42

41' Shell thick 45

42(41) Shell inflated 43

42' Shell compressed, thin; periostracum yellow with green rays; nacre white with suffusions of pink ..... *Hemistena lata* (Rafinesque, 1820)

43(42) Adult shell >50 mm; periostracum usually with rays 44

43 Adult shell small <50 mm; hinge teeth thin; periostracum light brown and rayless ..... *Simpsonaias ambigua* (Say, 1825)

44(43) Adult shell relatively small <70 mm; hinge teeth typically thin; periostracum rayed ..... *Villosa*

44' Adult shell relatively small <70 mm; hinge teeth thin; periostracum yellow and rayed or rayless ..... *Lampsilis*

45(41) Shell compressed 46

45' Shell inflated 48

46(45) Adult shell small >40 mm 47

46' Adult shell small <40 mm; periostracum rayed ..... *Villosa fabalis* (Lea, 1831)

47(46) Shell thick; hinge teeth stout and curved; nacre white, periostracum yellow with green rays ..... *Ptychobranchus*

47' Shell thick; hinge teeth large, straight or slightly curved; nacre white or purple; periostracum dark brown or black and rayless; umbo sculpture consists of heavy bars .... *Elliptio dilatata* (Rafinesque, 1820)

48(45) Adult shell length <55 mm 49

48' Adult shell length >55 mm 50

49(48) Periostracum olive green with numerous dark green rays .... *Villosa*

[Restricted to the Tennessee River basin.]

49' Periostracum ranges from greenish to dark brown and rayless, nacre white to purple .... *Toxolasma*

50(48) Shell inflated, periostracum rayless, umbo sculpture consists of heavy "v"-shaped ridges .... *Uniomerus tetralasmus* (Say, 1831)

50' Shell inflated, periostracum dark green with faint rays in juveniles and rayless in adults, umbo sculpture consists of numerous wavy w-shaped ridges, nacre white or pink to light purple .... *Ligumia*

51(36) Shell shape oval 52

51' Shell shape oblong 53

53(52) Shell thin to thick; posterior ridge rounded to angular; periostracum usually yellow with green rays, nacre varies from white to pink .... *Lampsilis*

53' Shell elongate oval .... *Villosa*

53(51) Shell oblong, not inflated, thick 54

53' Shell oblong, inflated, thin, rayless; hinge line sinuate .... *Potamilus capax* (Green, 1832)

54(53) Adult shell length <70 mm; shell elliptical, thick; nacre white; hinge teeth stout; periostracum dark yellow to greenish with wavy rays on the posterior end .... *Venustaconcha*

55' Adult shell length >70 mm; shell oval, thick; nacre white; hinge teeth stout; periostracum dark yellow without wavy rays on the posterior end .... *Actinonaias*

Gulf Coast and Florida Subregion

1 Shell with pseudocardinal and lateral teeth reduced or absent 2

1' Shell with both pseudocardinal and lateral teeth present 5

2(1) Shell with pseudocardinal and lateral teeth present but greatly reduced 3

2' Shell completely lacking both pseudocardinal and lateral teeth 4

3(2) Shell with pseudocardinal tooth consisting of a swelling or knob, lateral teeth reduced to a thickened ridge; umbo sculpture heavy "v"-shaped ridges .... *Strophitus*

3' Shell with pseudocardinal tooth thin, reduced and compressed; umbo projecting only slightly above the hinge line; shell elongate, umbo sculpture very fine "v"-shaped ridges .... *Anodontioides radiata* (Conrad, 1834)

4(2) Umbo projects well above the hinge line; shell thin, large, "w"-shaped umbo sculpture .... *Pyganodon*

4' Umbo even with or below the hinge line .... *Utterbackia*

5(1) Shell surface with sculpture (plications, pustules, ridges) 6

5' Shell surface without sculpture 18

6(5) Shell surface with prominent pustules, knobs or bumps and/or plications or ridges 7

6' Shell with only prominent pustules, knobs, or bumps 14

7(6) Shell with prominent plications or ridges but not pustules, knobs or bumps 8

7' Shell anteriorly with prominent plications or ridges and pustules, knobs or bumps 12

8(7) Shell with prominent plications or ridges confined to the posterior slope 9

8' Shell with prominent plications or ridges not confined to the posterior slope 10

9(8) Shell round or oval, compressed; periostracum dark brown or black; posterior wing may be present .... *Lasmigona complanata* (Barnes, 1823)

9' Shell small, elongate, slightly inflated: nacre bluegreen .... *Medionidus*

10(8) Shell round to rectangular without a distinct posterior ridge; nacre white 11

10' Shell thick, rectangular with a distinct posterior ridge; nacre purplish .... *Plectomerus dombeyanus* (Valenciennes, 1827)

11(10) Shell thick, elongate rectangular; posterior slope with placations or wrinkles; nacre white with purple around the distal half .... *Elliptoideus sloatianus* (Lea, 1840)

11'	Shell laterally compressed to inflated; round to quadrate in shape; heavy pseudocardinal teeth, well-developed lateral teeth; deep umbo cavity; shell surface typically with three prominent well-developed ridges .... *Amblema*
12(7)	Shell not elongate, quadrate or rounded in shape; pustules less numerous or confined to the shell anterior 13
12'	Shell elongate with numerous small pustules all over the surface of the shell .... *Tritogonia verrucosa* (Rafinesque, 1820)
13(12)	Shell large and quadrate in shape; pustules confined to the anterior part of the shell .... *Megalonaias*
13'	Shell surface more or less rounded in shape with indistinct corrugations; lateral teeth poorly developed .... *Arcidens confragosus* (Say, 1829)
14(6)	Shell surface with pustules 15
14'	Shell with pustules restricted to a single or double row 16
15(14)	Shell <60 mm; rounded, compressed to slightly inflated; chevron-shaped pustules covering most of the shell; nacre purple; umbo cavity deep .... *Quincuncina*
15'	Shell >60 mm; rounded, slightly inflated; pustules covering most of the shell; nacre white to slightly pink; umbo cavity deep .... *Amphinaias*
16(14)	Pustules forming two distinct rows, especially near the umbo 17
16'	Pustules large, usually three in number, forming a single row down the center of the shell which alternate on each valve .... *Obliquaria reflexa* (Rafinesque, 1820)
17(16)	Shell rounded to quadrate in shape; two rows of pustules, with one row on the posterior ridge .... *Quadrula* (in part)
17'	Shell rounded to quadrate in shape; two rows of pustules, with one row on the posterior ridge, a sulcus or shallow depression may be present between the rows of pustules; nacre purple .... *Amphinaias refulgens* (Lea, 1868)
18(5)	Shell without a prominent dorsal wing 19
18'	Shell with a well-developed dorsal wing projecting above the hinge line (occasionally absent) usually found posterior to the umbo, but a small wing may also be present anterior to the umbo .... *Leptodea fragilis* (Rafinesque, 1820)
19(18)	Shell shape round 20
19'	Shell shape not round 25
20(19)	Shell inflated, not compressed round to oval 21
20'	Shell thick, compressed, round to triangular in shape; sharp posterior ridge; yellow in color with "v"-shaped rays .... *Ellipsaria lineolata* (Rafinesque, 1820)
21(20)	Umbo nearly central 22
21'	Umbos anterior 24
22 (21)	No broad green ray on the umbo 23
22'	Shell with a broad green ray on the umbo; without pustules .... *Amphinaias asperata* (Lea, 1861)
23(22)	Shell surface smooth, round, without a sulcus .... *Obovaria*
23'	Shell with a broad shallow sulcus and fine green rays; umbo cavity shallow; females with a swollen, broadly expanded posterioventral margin .... *Epioblasma*
24(21)	Umbo cavity deep and compressed; a shallow sulcus present in some species; nacre white .... *Fusconaia*
24'	Umbo cavity relatively shallow; shell wedge-shaped to triangular or quadrate, without a Sulcus; nacre white to pink or salmon .... *Pleurobema* (in part)
25(19)	Shell shape not oval 26
25'	Shell shape oval to oblong 38
26(25)	Shell shape rectangular, triangular, or quadrate with sulcus 27
26'	Shell shape elongate, two to four times as long as high 31
27(26)	Umbo cavity deep 28
27'	Umbo cavity shallow or broad and open 29
28(27)	Shell shape rectangular, with green rays on the umbo .... *Fusconaia cerina* (Conrad, 1838)
28'	Shell shape triangular .... *Pleurobema* (in part)
29(27)	Posterior ridge very sharp, posterior slope steep 30
29'	Posterior slope angled; periostracum highly variable in color and may be green, brown, or yellow with chevron markings; nacre also variable from white to pink .... *Truncilla*
30(29)	Shell thin, inflated, abruptly truncate posteriorly .... *Alasmidonta*
30'	Shell rectangular, thin to thick, inflated posterior end; nacre ranges from salmon to purple .... *Elliptio crassidens* (Lamarck, 1819)

31(26)	Shell thin	32
31'	Shell thick	34
32(31)	Shell slightly inflated	33
32'	Shell compressed, thin, elongate to rectangular; periostracum yellowish with green rays; nacre white to light purple	*Popenaias popei* (Lea, 1857)
33(32)	Adult shell relatively small <70 mm; hinge teeth thin; periostracum rayed	*Villosa*
33'	Adult shell relatively large >70 mm; rounded to oval in shape; hinge teeth not thin; periostracum yellow, rayed or rayless	*Lampsilis/Hamiota* spp.
34(31)	Shell compressed	35
34'	Shell slightly inflated	36
35(31)	Shell thick; hinge teeth stout and curved; nacre white: periostracum yellow or brown, with or without green rays	*Ptychobranchus*
35'	Shell thick; hinge teeth large, straight or slightly curved; nacre white or purple; periostracum dark brown or black and rayless; umbo sculpture consists of heavy bars	*Elliptio*
36(34)	Shell rayless	37
36'	Shell elongate, inflated; periostracum dark green with faint rays in juveniles and rayless in adults; umbo sculpture consists of numerous wavy "w"-shaped ridges; nacre white or pink to light purple	*Ligumia recta* (Lamarck, 1819)
37(36)	Shell inflated; posterior end pointed; umbo sculpture consists of heavy "v"-shaped ridges	*Uniomerus*
37'	Adult shell relatively small <50 mm; periostracum ranges from greenish to dark brown	*Toxolasma*
38(25)	Shell shape oval	39
38'	Shell shape oblong	41
39(38)	Pseudocardinal tooth not divided as below	40
39'	Posterior half of the pseudocardinal tooth divided into several parallel vertical ridges; shell thick, inflated; periostracum cloth-like and light to dark brown and rayless	*Glebula rodundata* (Lamarck, 1819)
40(39)	Shell thin to thick; posterior ridge rounded to angular; periostracum usually yellow with green rays; nacre varies from white to pink	*Lampsilis*
40'	Shell shape elongate oval	*Villosa*
41(38)	Shell shape oblong, not inflated or globose	42
41'	Shell shape oblong, inflated to globose; unrayed	*Potamilus*
42(41)	Shell shape elongate oval; relatively thin, compressed; pseudocardinal teeth compressed; periostracum tan to brown, with dark green rays	*Disconaias*
42'	Shell shape oval; relatively thick, moderately inflated; periostracum brown with very faint green rays in juveniles	*Cyrtonaias*

REFERENCES

Bleam, D.E., K.J. Couch & D.A. Distler. 1999. Key to the unionid mussels of Kansas. Transactions of the Kansas Academy of Science 102: 83–91.

Brim Box, J. & J.D. Williams. 2000. Unionid mollusks of the Apalachicola Basin in Alabama, Florida, and Georgia. Bulletin of the Alabama Museum of Natural History 21: 1–143.

Burch, J.B. 1975. Freshwater Unionacean Clams (Mollusca: Pelecypoda) of North America, revised ed. Malacological Publications, Hamburg, MI.

Cummings, K.S. & D.L. Graf. 2010. Mollusca: Bivalvia. Chapter 11. Pages 309–384 *in*: J.H. Thorp & A.P. Covich (eds.), Ecology and Classification of North American Freshwater Invertebrates, Third Edition, Academic Press, Boston, MA.

Fichtel, C. & D.G Smith. 1995. The freshwater mussels of Vermont. Nongame & Natural Heritage Program, Vermont Fish and Wildlife Department Technical Report 185:1–54.

Graf, D.L. 1997. Distribution of unionoid (Bivalvia) faunas in Minnesota, USA. Nautilus 110: 45–54.

Grigorovich, I.A., A.V. Korniushin & H.J. MacIsaac. 2000. Moitessier's pea clam Pisidium moitessierianum (Bivalvia, Sphaeriidae): a cryptpgenic mollusc in the Great Lakes. Hydrobiologia 435: 153–165.

Haas, F. 1969a. Superfamilia Unionacea. Das Tierreich, Lief. 88. Walter de Gruyter and Co., Berlin.

Haas, F. 1969b. Superfamily Unionacea Fleming, 1828. Pages 411–467 *in*: Moore, R.C. (ed.), The Geological Society of America, Inc. and the University of Kansas Treatise on Invertebrate Paleontology, Vol. I, Part N, Mollusca 6: Bivalvia. Lawrence, KS.

Howells, R.G., R.W. Neck & H.D. Murray. 1996. Freshwater Mussels of Texas. Texas Parks and Wildlife Press, Austin, TX.

Johnson, R.I. 1999. Unionidae of the Rio Grande (Rio Bravo del Norte) system of Texas and Mexico. Occasional Papers on Mollusks Museum Comparative Zoology Harvard University 6:1–65.

Martin S.M. 1997. Freshwater mussels (Bivalvia: Unionoida) of Maine. Northeastern Naturalist 4: 1–34.

Nedeau, E.J., M.A. McCullough & B.I. Swartz. 2000. The Freshwater Mussels of Maine. Maine Department of Inland Fisheries and Wildlife, Augusta, ME. 118 p.

Nichols, S.J. & M.G. Black. 1994. Identification of larvae: the zebra mussel (*Dreissena polymorpha*), quagga mussel (*Dreissena rosteriformis bugensis*), and Asian clam (*Corbicula fluminea*). Canadian Journal of Zoology 72: 406–417.

Parmalee, P.W. & A.E. Bogan. 1998. The Freshwater Mussels of Tennessee. University of Tennessee Press, Knoxville, TN, 328 p.

Simpson, C.T. 1900. Synopsis of the Naiades, or pearly fresh-water mussels. Proceedings of the US Natural Museum 22: 501–1044.

Strayer, D.L. 1999. Effects of alien species on freshwater mollusks in North America. Journal of the North American Benthological Society 18: 74–98.

Strayer, D.L. & K.J. Jirka. 1997. The Pearly Mussels of New York State. New York State Museum Memoir, Albany, NY. 140 p.

Sturm, C.F., T.A. Pearce & A. Valdes. 2006. The Mollusks: A Guide to Their Study, Collection, and Preservation. American Malacological Society, Universal Publishers, Boca Raton, FL.

Turgeon, D.D., J.F. Quinn Jr, A.E. Bogan, E.V. Coan, F.G. Hochberg, W.G. Lyons, P.M. Mikkelsen, R.J. Neves, C.F.E. Roper, G. Rosenberg, B. Roth, A. Scheltema, F.G. Thompson, M. Vecchione & J.D. Williams. 1998. Common and Scientific Names of Aquatic Invertebrates from the United States and Canada: Mollusks, second edition. American Fisheries Society Species Publication 26. American Fisheries Society, Bethesda, MD.

Williams, J.D. & R.S. Butler. 1994. Class Bivalvia, Order Unionoida, Freshwater Bivalves. Pages 53-128, 740-742 in: M. Deyrup & R. Franz (eds.), Rare and Endangered Biota of Florida. IV. Invertebrates. University Press of Florida, Gainesville, FL.

Williams, J.D., M.L. Warren Jr, K.S. Cummings, J.L. Harris & R.J. Neves. 1993. Conservation status of the freshwater mussels of the United States and Canada. Fisheries 18: 6–22.

PHYLUM MOLLUSCA

Chapter 12

Phylum Annelida[1]

Chapter Outline

Introduction to the Phylum

James H. Thorp
Kansas Biological Survey and Department of Ecology and Evolutionary Biology, University of Kansas, Lawrence, KS, USA

Lawrence L. Lovell
Research and Collections, Natural History Museum of Los Angeles County, Los Angeles, CA, USA

INTRODUCTION

Annelida is a diverse phylum, which contains the segmented worms. These animals live in marine, freshwater, and semi-terrestrial (damp soil) habitats. Inland waters of the Nearctic are home to many free-living aquatic worms (Oligochaeta), ectoparasitic and predaceous leeches (Hirudinida), and ectosymbiont crayfish worms (Branchiobdellidea). Also present are a few species of free-living bristle worms and fan worms (Polychaeta)—a group almost exclusively confined to marine and estuarine waters—and one species of the globally monogenetic Acanthobdellida, which are exclusively external parasites of salmonid fishes. The Nearctic has a minority of the total global diversity of all these groups except for the

1. Citation of keys in this chapter should be made to the designated author(s) of the specific section rather than to the chapter as a whole.

Thorp and Covich's Freshwater Invertebrates. http://dx.doi.org/10.1016/B978-0-12-385028-7.00012-3

Branchiobdellidea, where two-thirds of the global diversity in this group is present in this zoogeographic region (Gelder & Williams, 2015).

Current systematic studies support the Polychaeta, Aphanoneura, and Clitellata as classes of Annelida, with Oligochaeta, Acanthobdellida, Branchiobdellidea, and Hirudinida as subclasses within the Clitellata (WoRMS, 2012).

LIMITATIONS

The initial steps in separating subphyla and classes of Annelida by external structures are quite easy and should pose few problems if any for most users of these keys. One can usually also separate the groups at the subclass level by habitat: free-living and non-parasitic (a few Polychaeta and the many Oligochaeta), presence externally on crayfish (Branchiobdellidea), parasitic on salmon and trout (Acanthobdellida), or either predators of invertebrates or external parasites on many fish species (Hirudinida).

TERMINOLOGY AND MORPHOLOGY

Knowledge of only five morphological terms is required at this higher level: annuli, chaetae (also spelled chaete), clitellum, parapodia, and peristomium. Annuli are ring-like body segments. Chaetae are bristles which can have different shapes, depending on the taxon. A clitellum is a modification of the epidermis, which develops as an easily visible, glandular girdle partly behind the female pores; it secretes a cocoon in which eggs are laid. Parapodia are paired, unjointed lateral appendages found in polychaete worms, which are often fleshy (especially in marine polychaetes) and used for locomotion, respiration, and other functions. The peristomium is the second body segment of annelids. It may be combined with the anterior segment, or prostomium.

MATERIAL PREPARATION AND PRESERVATION

Annelids can be preserved in 70–80% ethyl alcohol for long periods; however, proper fixation for long-term preservation requires fixing in 5–10% buffered formalin solution for a few days, then transferring them to 70% ethyl alcohol. Annelids can easily fragment during fixation and lose key taxonomic characters. Fragmentation can be reduced by relaxing specimens in a 7% solution of $MgSO_4$ for 30 min before fixation.

Specimen preparation for molecular work requires avoiding exposure of the tissues to formalin. If molecular analysis is planned, a small piece of the specimen (preferably without key taxonomic characters) can be pulled and preserved in 95% ethyl alcohol or frozen at −80 °C. The rest of the specimen (with key taxonomic characters intact) can be fixed in formalin and preserved in ethyl alcohol as a voucher specimen for morphological use.

Due to their small size and taxonomic characters, identification and observation of annelids requires a dissecting microscope with 6–50× magnification. A compound microscope with 10–1000× magnification is required for viewing whole slide mounts of small-bodied individuals or parts of specimens, such as parapodia with chaete. For short-term examinations, use ethyl alcohol as a medium for slide mounts. When extended examination is anticipated, mounting using a 50/50 solution of glycerin and ethyl alcohol is recommended to avoid drying of the specimen. Permanent mounts for teaching or later examination can be made using commercial mounting media such as Permount®.

KEYS TO ANNELIDA

This higher classification is preliminary and a rough amalgam of the classification schemes used elsewhere (Williams et al., 2013; Timm & Martin, 2015; Verdonschot, 2015).

Annelida: Classes

1 Prostomium bearing cilia, papillae, tentacles, or scales; reproductive organs absent or associated with various structures or organs; parapodia well-developed, reduced, or absent; chaetae relatively abundant; clitellum always absent 2

1' Prostomium simple; reproductive organs within a tumid region called a clitellum; parapodia absent; chaetae absent or relatively rare .... **Clitellata [p. 225]**

2(1) Minute worms, 1–2 mm or chains of animals up to 10 mm; hair chaetae in both dorsal and ventral bundles; worms move in a gliding motion using cilia; no eversible pharyngeal pad, prostomium with cilia only; ventral copulatory glands in rare mature specimens; nervous system ladder-like (Fig. 12.1) .... **Aphanoneura [p. 259]**

2' Minute to large worms; prostomium with papillae, tentacles, or scales .... **Polychaeta [p. 261]**

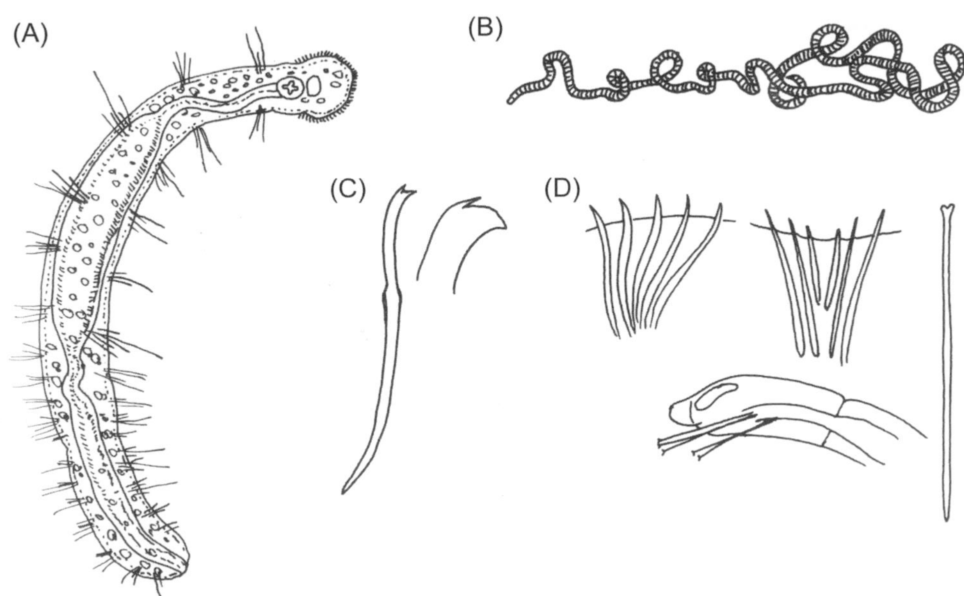

FIGURE 12.1 (A) *Aeolosoma*; note ciliated prostomium, lack of septa, simple pharynx, dorsal and ventral hair chaetae, and colored body wall inclusions (irregular circles); (B) *Haplotaxis*; whole specimens can be even longer in relation to the breadth; (C) Bifid chaetae in Lumbriculidae; and (D) Enchytraeid chaetae. In the latter, these may be straight, curved, and with or without nodulus, and may differ in length within a bundle. All chaetae are usually identical. The anterior end of *Barbidrilus* (below) has ventral bundles of strange rod-like chaetae with bifid tips only in II and III, as shown on the right.

REFERENCES

Gelder, S.R. Monograph of the Branchiobdellida (Annelida: Clitellata) or crayfish worms. in preparation.

Timm, T. & P. Martin. 2015. Clitellata: Oligochaeta. Chapter 21, Pages 529–549 *in*: J.H. Thorp and D. Christopher Rogers (eds.), Thorp and Covich's Freshwater Invertebrates, Volume I: Ecology and General Biology. Academic Press, Elsevier, Waltham, MA, USA.

Verdonschot, P.F.M. 2015. Introduction to Annelida and the class Polychaeta. Chapter 20, pages 509–528 *in*: J.H. Thorp and D. Christopher Rogers (eds.), Thorp and Covich's Freshwater Invertebrates, Volume I: Ecology and General Biology. Academic Press, Elsevier, Waltham, MA, USA.

Williams, B.W., S.R. Gelder, H.C. Proctor & D.W. Coltman. 2013. Molecular phylogeny of North American Branchiobdellida (Annelida: Clitellata). Molecular Phylogenetics and Evolution 66: 30–42.

Class Clitellata

ANNELIDA: CLITELLATA: SUBCLASSES

1 With one or more external suckers 2

1' Without external suckers .... **Oligochaeta [p. 229]**

2(1) Twenty or more body segments 3

2' Eleven body segments .... Branchiobdellidea, one family: **Branchiobdellidae [p. 237]**

3(2) Body divided into 32 post-oral segments; segments subdivided superficially into 3–16 annuli (Figs. 12.2 A and 12.3); anterior sucker present consists of four segments: mouth on ventral surface of anterior sucker (Fig. 12.16 B); jaws present or absent; posterior sucker consists of seven segments; chaetae absent from entire body; median ventral unpaired male and female gonopores (Figs. 12.3 and 12.5) .... **Hirudinida [p. 248]**

3' Body divided into 29 post-oral segments; segments subdivided superficially into four annuli (a1, a2, b5, b6) (Fig. 12.2 B): no anterior sucker (Fig. 12.6 B), mouth on ventral surface of segment III; no jaws; posterior sucker present consisting of four segments: two pairs of chaetae on five consecutive anterior segments; distal ends of chaetae bent to form hooks (Fig. 12.6 B); chaetae absent from remainder of the body; paired ventral male gonopores: single median ventral male gonopore; (Fig. 12.6 A) length 22 mm; ectoparasite of salmonids .... Acanthobdellida, one species: *Acanthobdella peledina* Grube, 1851

[Ectoparasites on salmonid fishes. USA: Alaska]

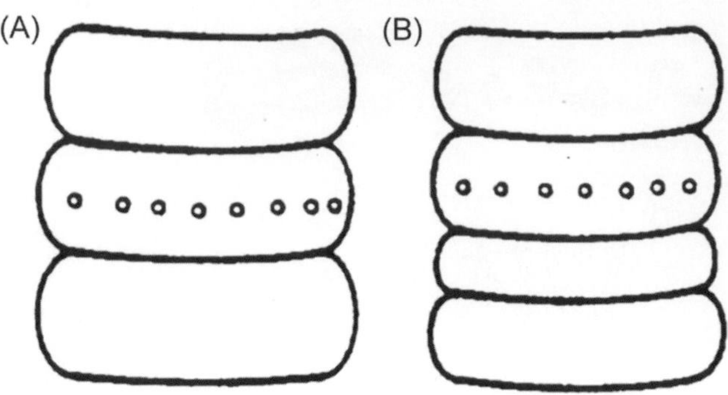

FIGURE 12.2 Annulation showing: (A) 3-annulate condition in leeches; and (B) 4-annulate condition in *Acanthobdella peledina*.

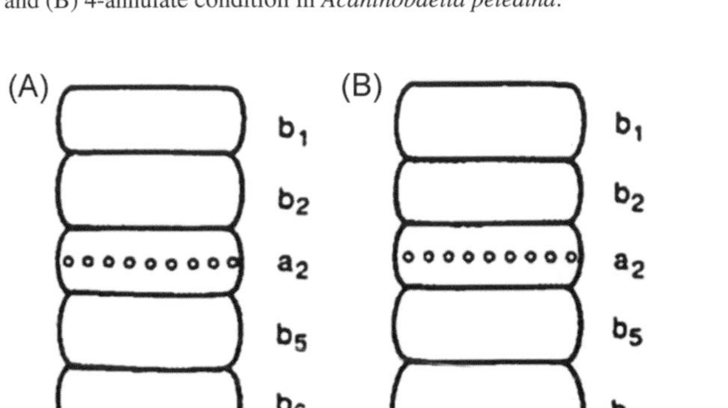

FIGURE 12.3 Annulation of *Erpobdella* with: (A) all annuli (b1, b2, a2, b5, b6) equal in length; and (B) the annuli (b1, b2, a2, b5, c11, c12) of unequal lengths.

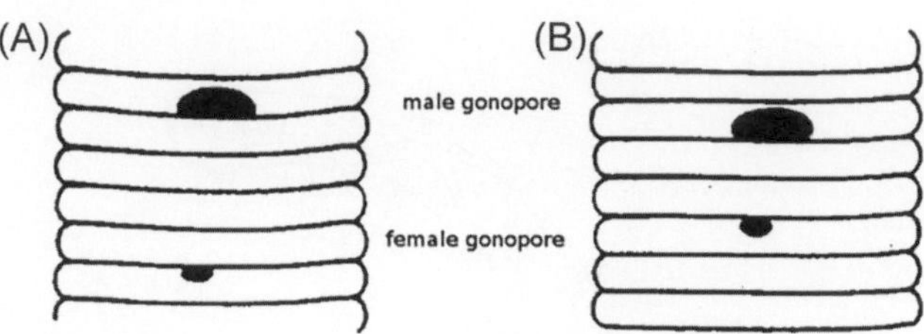

FIGURE 12.5 Ventral view of the anterior male and posterior female gonopores separated by four (A) or two (B) annuli.

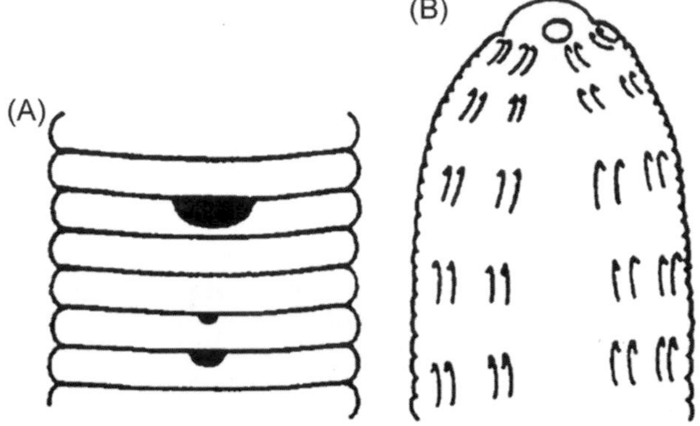

FIGURE 12.6 *Acanthobdella peledina*, ventral view of: (A) the male and female gonopores with a spermatheca; and (B) the anterior view showing the absence of an anterior sucker, the presence of a mouth pore, and five pairs of hooked chaetae.

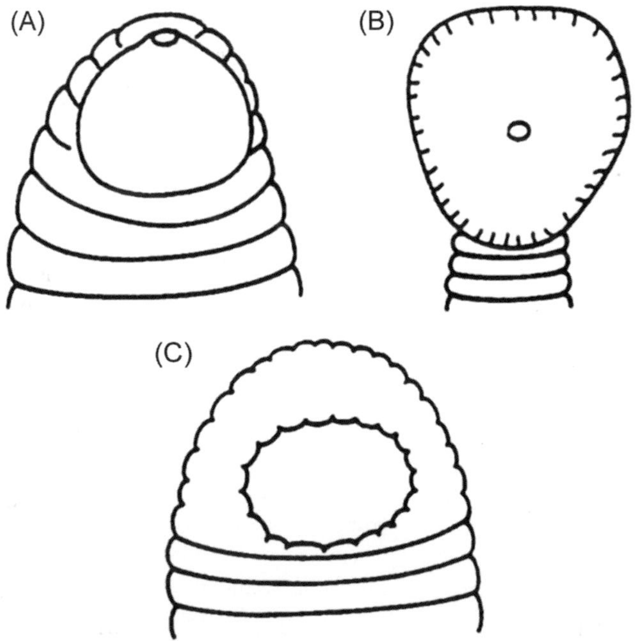

FIGURE 12.4 Ventral view of the anterior sucker showing: (A) the mouth pore on the anterior rim of the sucker (e.g., *Marvinmeyeria, Placobdella*); (B) the center of the sucker (Piscicolidae, *Helobdella*); and (C) the mouth occupying the entire cavity (e.g., Hirudinidae, Erpobdellidae).

Subclass Oligochaeta[2]

Bonnie A. Bain
Department of Biological Sciences, Southern Utah University, Cedar City, UT, USA

Ralph O. Brinkhurst
205 Cameron Court, Hermitage, TN, USA

INTRODUCTION

Freshwater oligochaetes can be found in both standing and flowing water and are often an important component of the aquatic food web. There are currently over 200 oligochaete species reported from North America that inhabit freshwater and estuarine bodies, including those in the lesser known groundwater and organic mud habitats. Most oligochaetes feed by ingesting sediment, but several North American genera prey on other worms or on a variety of small invertebrates. Non-freshwater oligochaetes include terrestrial earthworms and marine naidids.

LIMITATIONS

The taxonomy of North American freshwater oligochaetes is currently being revised, and many of the upcoming changes will not be reflected in this edition due to the recent and (currently) unpublished nature of these changes. As part of this work, a supplementary publication, including web-based resources, is planned that will better reflect the ongoing changes in oligochaete taxonomy. Many of the family, genus, and species definitions are provisional and will be further clarified with each new analysis and inclusion of more taxa.

TERMINOLOGY AND MORPHOLOGY

Oligochaetes are bilaterally symmetrical, segmented coelomates with usually four bundles of chaetae on every segment except the first. In a few exceptional forms, the chaetae are absent in some or even all segments. Each bundle contains several to more than a dozen chaetae. A reproductive structure, the clitellum, is located well behind the gonadal segments. When worms copulate, sperm is passed to the partner and is usually deposited in spermathecae. There may be one or more pairs of these structures located in or near the gonadal segments, commonly in front of the testicular segments or in them. After a worm has copulated and has sperm in its spermathecae, the clitellum secretes a cocoon.

As most worms burrow through soft substrates, the ranges of externally visible anatomical adaptations are few because of limitations on body shape imposed by the habitat. The prostomium lies in front of the mouth. The prostomium may bear a median anterior prolongation (proboscis). In aquatic species, the peristomium is completely separate from the prostomium. It is also presegmental and bears no chaetae. This is important to remember because it is often easier to locate specific segments by counting chaetal bundles (which begin in segment II).

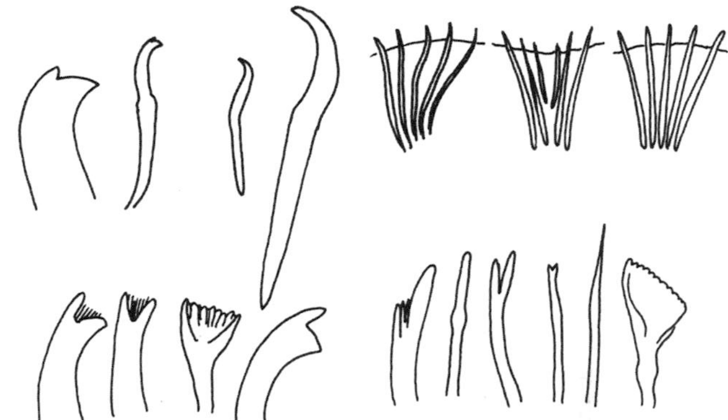

FIGURE 12.7 Representative chaetae. From the left, top row: lumbriculid bifid with reduced upper tooth, head end, and entire; small dorsal and large ventral sickle-shaped chaetae of Haplotaxis; three types of bundles from the Enchytraeidae. Bottom row: two pectinates, a palmate, and a bifid dorsal chaeta, Naididae (formerly Tubificidae); six types of dorsal needles, Naididae.

Chaetae vary in form as well as number. This variation provides very useful key characters, but there is increasing evidence that the fine details of chaetal form, or even the presence or absence of a particular type of chaetae such as hairs, can be affected by quite simple environmental variables such as pH or conductivity (Chapman & Brinkhurst, 1987). The Enchytraeidae have simple pointed chaetae, which may be straight or sigmoid. The chaetae in a bundle may vary in size (Fig. 12.7). Hair chaetae are elongate, simple shafts. They are found only in the dorsal bundles of some species of Naididae and in Crustipellis (Opistocystidae). They are present in both dorsal and ventral bundles in the Aeolosomatidae (Aphanoneura, Aclitellata), this being one of a series of fundamental differences between these minute freshwater worms and true oligochaetes.

2. The section on Oligochaeta was shortened and revised by D.C. Rogers and J.H. Thorp (editors) from material in Chapter 12 in the third edition of Thorp and Covich (2010).

Hair chaetae are termed hispid or serrate when they bear a series of fine lateral "hairs." This condition may vary within a species depending on seasonal or environmental variables, but it is still used as a taxonomic character for many naidids for which the limits of intraspecific variability have yet to be established. In a few naidids, the lateral hairs may be especially prominent and plumose along one side of the hair chaeta. In one or two instances, minute bifid tips have been observed at the ectal end of hair chaetae, indicating their probable origin from the much more ubiquitous bifid chaetae. This bifid form, with the ectal end divided into two teeth, is characteristic of aquatic as opposed to terrestrial oligochaetes, although the functional reason for this is unknown. Bifid chaetae may be the only type present in all bundles, and the ventral bundles usually contain nothing else. Some simple pointed chaetae may be found in a few anterior ventral bundles (e.g., in the naidid genus Spirosperma), or they may accompany the hair chaetae in the dorsal bundles in some Naididae. A common combination in the dorsal bundles is hair chaetae accompanied by pectinate chaetae in at least the preclitellar bundles. Pectinate chaetae are usually of the same general form as the bifids for the particular taxon, but the space between the two teeth is filled with either a comb of thin teeth or, more rarely, a ridged web. In a few naidids, these webbed chaetae may be expanded as palmate chaetae. In general, the dorsal chaetae accompanying the hair chaetae in the majority of naidid species differ markedly from the ventral chaetae. For this reason, they have come to be termed needles, whether they are palmate, pectinate, bifid, or simple pointed. Some authors like to substitute the terms fascicle, capilliform, bifurcate, and crotchet for bundle, hair, bifid, and chaeta, respectively. They also use the terms distal and proximal for upper and lower tooth, the designations used in describing the variations in the length and thickness of the teeth, often a useful characteristic. The normal ventral chaetae may be replaced by specialized genital chaetae, usually beside the spermathecal or penial pores. Genital chaetae do not vary much in form in the Naididae, where there are often two or three blunt penial chaetae in each bundle beside the male pore. These chaetae are quite long proximally, but the distal end beyond the nodulus (the swelling on the chaeta at the point where it emerges from the chaetal sac) is short.

Penial chaetae are usually somewhat long and thicker than the normal ventral chaetae. In the Rhyacodrilinae there are usually several to a bundle. They are usually arranged with the outer ends close together, the inner ends spread out. In Tubificinae, the ventral chaetae in the penial segment are usually lost, but the spermathecal chaetae are often modified, with elaborate distal ends shaped like trowels. In general, the number and diversity of chaetae is reduced toward the posterior end of the worm, and modified teeth on the bifids commonly become "normal." There are a few exceptions. The upper teeth of the chaetae of Amphichaeta get longer posteriorly, and the posterior segments of Telmatodrilus vejdovskyi bear chaetae with almost brush-like tips.

Few other anatomical features are used in aquatic oligochaete taxonomy. External gills are found in a few taxa. The posterior segments of the naidid *Branchiura* bear single dorsal and ventral gill filaments. These are longest in the median segments of the gill-bearing region, very short at each end of the row. *Dero* (Naididae) has gills at the posterior end of the body, usually enclosed in a gill chamber (branchial fossa).

MATERIAL PREPARATION AND PRESERVATION

While many biologists believe identifying oligochaetes is difficult, the advantages of this group are the absence of both larvae and sexual dimorphism (all are hermaphrodites). About 60% of the species can be identified from superficial somatic characters, but the remainder require mature specimens and using mature specimens is preferable for all taxa. We recommend basing identification on slide-mounted specimens. External characteristics are used in many keys to identify the roughly 150 freshwater oligochaete species in the Nearctic, but precise identification may require examining the soft parts of the male reproductive structures.

After initially preserving the worms in alcohol, they should be either immersed in a temporary mounting medium by replacing the alcohol with Amman's lactophenol, or they may be washed with water and then placed in separate drops of lactophenol on slides. It is usually possible to mount live worms under a single coverslip (with two coverslips per slide) unless the worms are very large. These wet mounts should be set aside for one to two days on stacking trays. Amman's medium is quite corrosive, so it should be handled with care and any spillage should be wiped off microscope stages immediately. It consists of 20% phenol, 20% lactic acid, 20% water, and 40% glycerine. It is best stored in dark bottles; however, it is hygroscopic and does not have a good shelf-life. Somewhat more permanent mounts can be made with media such as CMCP and polyvinyl lactophenol. Such slides should usually be sealed with a ring of material (Glyceel, nail polish, etc.) around the edges of the coverslips to prevent the media from drying.

None of these methods produce satisfactory permanent slides, nor are they adequate for working with the Enchytraeidae or the marine Naididae, in which internal anatomy must be seen. For this type of material, permanent whole mounts using stained and dehydrated worms are mounted in Canada Balsam or one of its more recent substitutes. To prevent worms from becoming brittle, they should be cleared in methyl salicylate rather than xylene. This will enable advanced students to try bisecting (sagittally) the separated genital region of each worm or even dissecting out the male ducts when necessary. This should be limited to identification of new taxa to the generic level, when details of the male reproductive system must be established. Serial sections (usually sagittal longitudinal) are also employed in taxonomic work, but not in routine identification of well-known species. It is important to be knowledgeable about the expected locations of the various organs before attempting dissection or interpretation of serial sections.

The procedure for examining a worm on a slide is as follows. Check the prostomium for a proboscis or sense organ. Next, examine the ventral chaetae of the first two or three bundles and determine the form of the dorsal bundles and where they begin (usually in II or between IV and VI). Establish the number and form of these chaetae (bifid, pectinate, hair chaetae, etc.). Determine the relative lengths of the teeth. As these are illustrated by taxonomists with the ectal or outer end of the whole chaeta toward the top edge or corner of the figure, the upper tooth is always drawn lying above the lower tooth. When viewing a whole-mounted specimen, the chaetae often lie flat on the body wall, and so no distinction of position can be made; therefore, always visualize a chaeta in the upright position. Care should be taken to examine several chaetae from an exactly lateral aspect, because slight deviations can produce apparent distortion of the relative lengths of the teeth. Measurements often betray the bias of the eye of the observer and are recommended where an eyepiece micrometer scale is available. The worm should then be searched for genital characters. Carefully check the appropriate segments to see if the ventral chaetae are modified. If the dorsal and ventral chaetae are alike, make sure that you have examined at least three bundles in those segments to ensure that a ventral bundle has been examined. Check the penial segment to see if there is a penis sheath (they should be paired of course) even if it is thin and inconspicuous. Check the posterior chaetae and the body itself for gills or any other special features. Identifications to major groups or to genera can often be made with a little practice. Identifications will require a properly aligned microscope, as chaetae and penis sheaths have refractive indices that do not differ much from the background, and too much light makes them impossible to see. Oil immersion lenses must be used to see pectinations on chaetae in the former tubificids (although with practice they can be identified without), and must also be used to determine the form of the needles (dorsal chaetae) in naidids.

KEYS TO OLIGOCHAETA

Oligocheata: Families

1 All chaetae paired, or multiple in a bundle, unless partially or totally absent; prostomium short, often conical, without a transverse furrow but with or without a proboscis; body form very rarely elongate and slender 2

1' Ventral chaetae large, sickle-shaped, single (i.e., two per segment) dorsal chaetae small, straight, often missing from some or all segments; prostomium very long, furrowed, no proboscis; mouth large, muscular pharynx present; body elongate, slender, resembling a gordian worm (Fig. 12.1 B) .... Haplotaxidae, one genus: *Haplotaxis*

[In need of revision]

2(1) Chaetae all paired from II (Fig. 12.8) or, if more numerous, all simple pointed 3

2' Chaetae more than two per bundle, usually bifid, sometimes with pectinate and hair chaetae dorsally, simple pointed chaetae rare (limited to a few anterior ventral bundles, or single needles with the hairs dorsally) 6

3(2) Chaetae simple pointed 4

3' Chaetae paired, bifid, with small to rudimentary upper teeth (Fig. 12.1 C) .... **Lumbriculidae** [in part] **[p. 231]**

4(3) Thin-bodied worms with clitellum one cell layer thick and in region of gonopores (X–XII or further forward) 5

4' Thick-bodied worms with clitellum several segments behind the gonopores (XII–XIV) .... Megadrili

[Mostly terrestrial but some aquatic species]

5(4) Larger worms (difficult to mount under a coverslip) with sigmoid, nodulate chaetae; proboscis present or absent; spermathecae in or adjacent to the gonadal segments, male pores between VIII and X .... **Lumbriculidae** [in part] **[p. 231]**

5' Smaller worms with chaetae often straight, commonly not nodulate, sometimes missing from some if not all segments (in *Barbidrilus* Loden & Locy, 1980 with uniquely forked chaetae in II and III ventrally, others missing); no proboscis; spermathecae open in V, male pores in XII (Figs. 12.1 D and 12.7 and 8) .... Enchytraeidae

6(2) Posterior end naked, or with gills, or with two lateral processes plus gills, no median process (Fig. 12.9 A); spermathecae in the testicular segment, atria in the ovarian segment, these being in X–XI or further forward (Fig. 12.8) .... **Naididae [p. 232]**

6' Worms (1.7–3.0 mm) with posterior end bearing one median and two lateral processes; genital region with testes in XI, ovaries and atria in XII, spermathecae in XIII (Fig. 12.10 A) .... Opistocystidae, one species: *Crustipellis tribranchiata* (Harman, 1970)

[USA: Florida: Louisiana, Mississippi, North Carolina]

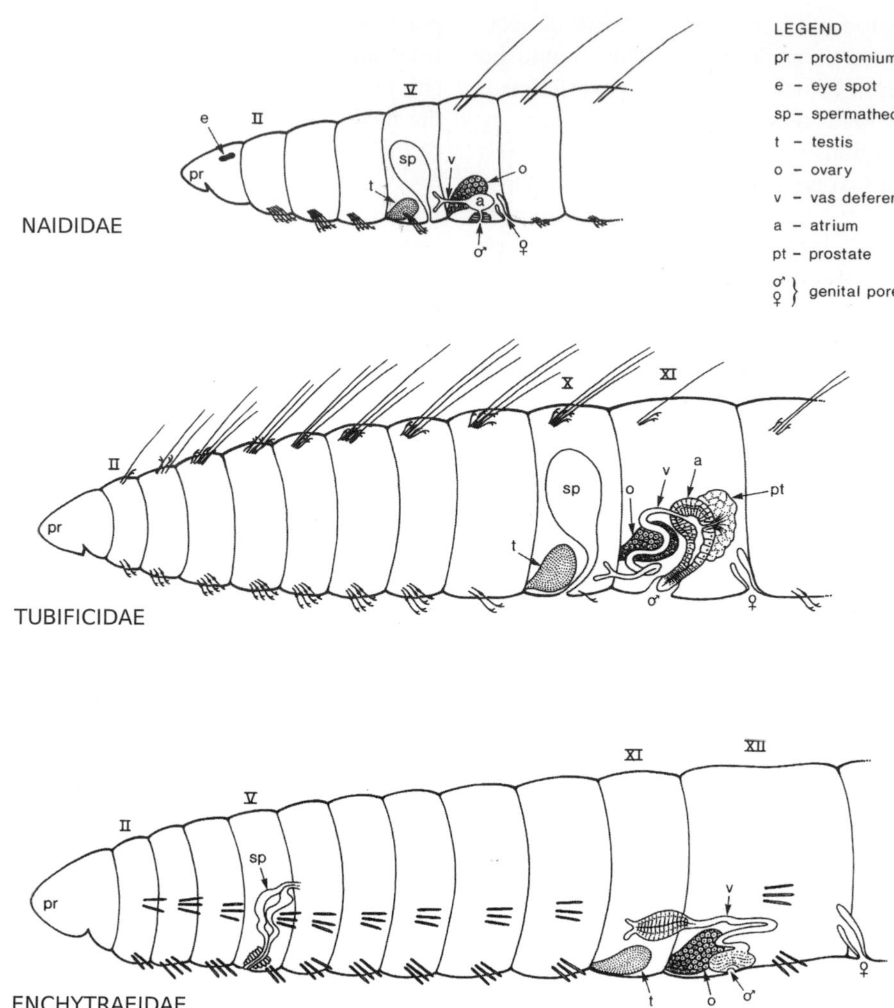

FIGURE 12.8 Reproductive systems and external characteristics of three major families of Oligochaeta. The worms are shown with the anterior to the left, and only the structures of the left side are visible. The dorsal and ventral chaetae are illustrated. In the Naididae, an example with dorsal chaetae beginning in VI is shown. Segments are identified with Roman numerals by convention (septa would be VII/VIII, VIII/IX, etc.).

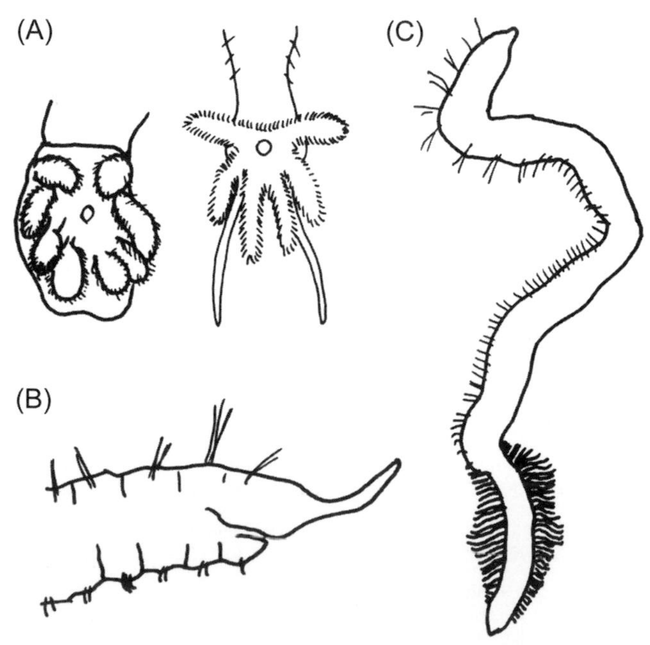

FIGURE 12.9 (A) Perianal gills of *Dero*. *Dero* (*Aulophorus*) with long palps (right) and *Dero* (*Dero*) without palps (left). (B) Proboscis on the prostomium of a naidid, this one with dorsal chaetae from II, but this can be variable. (C) *Branchiura*, showing posterior fan of gill filaments.

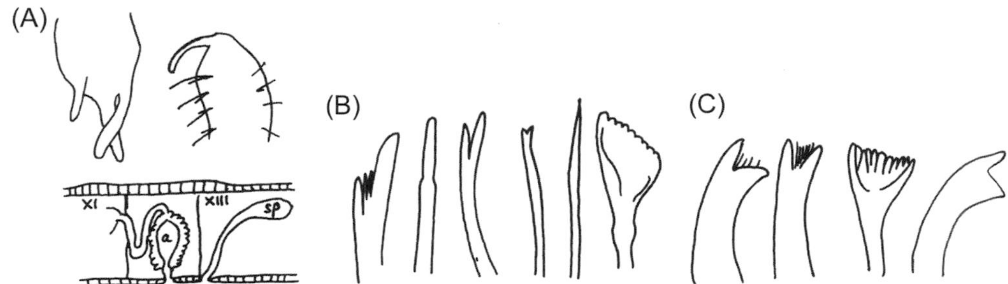

FIGURE 12.10 (A) *Crustipellis* (Opistocystidae). Tail end to left, head end with proboscis to right, genital segments below (a, atrium, sp, spermatheca, both of left side). (B) Various forms of dorsal chaetae, needles, of Naididae. These usually differ from the more normally bifid ventrals. (C) Dorsal chaetae of Naididae (formerly Tubificidae). Pectinates, on the left, differ relatively little from a bifid, shown on the right. Ventral bundles usually consist of bifids, sometimes with minute pectinations in those species with pectinate dorsals. Pectinates are usually accompanied by hairs in the dorsal bundles, rarely are bifid dorsals accompanied by hairs. All chaetae may be bifid.

Oligocheata: Lumbriculidae: Genera

1 Prostomium with a proboscis 2

1' Prostomium without a proboscis 6

2(1) Chaetae bifid, at least in anterior segments 3

2' Chaetae all simple pointed 4

3(2) Proboscis elongate; chaetae bifid in anterior segments .... *Kincaidiana hexatheca* Altman, 1936

[Pacific Northwest]

3' Proboscis short; all chaetae bifid, but with collateral teeth (set side by side) .... *Rhynchelmis* [in part]

4(2) Surface water species with long proboscis 5

4' Cave dwelling species with short proboscis .... *Eremidrilus allegheniensis* (Cook, 1971)

[USA: Tennessee]

5(4) Atria of male reproductive system with spiral muscles .... *Eclipidrilus* [in part]

[USA: southeastern states, Montana]

5' Atria without spiral muscles .... *Rhynchelmis* [in part]

6(1) Chaetae bifid 7

6' Chaetae simple 8

7(6) Elongate (to 100 mm or more) slender worms, the front end often green in life, the rest dark red to black; elaborately branched blood vessels laterally in the body wall of posterior segments; reproduces sexually and asexually (fragmentation with or without encystment); no permanent everted penes .... *Lumbriculus*

[USA: Alaska]

7' Short (25–40 mm), tapering worms, pale to white in color; blood vessels in the posterior lateral body wall with short, unbranched lateral diverticulae; reproduction sexual; permanently everted soft penes on X close together near the midline on mature specimens .... *Stylodrilus heringianus* Claparède, 1862

[Canada: widespread. USA: Great Lakes region. Possibly invasive. Palaearctic]

8(6) Groundwater species 9

8' Surface water species 11

9(8) Male pores paired in X, spermathecal pores paired in either IX and XI or XI and XII 10

9' Four pairs of male pores on X, two pairs of spermathecal pores on IX .... *Spelaedrilus multiporus* Cook, 1975

[USA: Virginia]

10(9) Spermathecal pores paired in IX .... *Stylodrilus* [in part]

[Widespread]

10' Spermathecal pores paired in XI or XI and XII .... *Trichodrilus* [in part]

[Widespread]

11(8) Male pores paired in VIII 12

11' Male pores paired in X 13

12(11) Spermathecal pores in XI, ovaries in X .... *Styloscolex opisthothecus* Sokolskaya, 1969

[USA: Alaska. Palaearctic]

12' Spermathecal pores paired in IX, ovaries in IX .... *Altmanella freidris* (Cook, 1965)

[USA: California]

13(11) Spermathecal pores paired in IX 14

13' Spermathecal pores either single median in VIII–IX or in IX, or paired in IX .... *Eclipidrilus* [in part]

[Canada: Ontario. USA: California, Idaho, northeastern states]

14(13) Arctic .... *Stylodrilus* [in part]

14' Southeastern USA .... *Tenagodrilus musculus* Eckroth & Brinkhurst, 1996

[USA: Alabama]

Oligocheata: Naididae: Genera

1 Length usually 1 cm; hair chaetae usually present in dorsal bundles but absent in some genera; hair chaetae commonly associated with simple pointed or bifid chaetae rather than the rarer palmate or pectinate chaetae, these dorsal chaetae (needles) very often differ from the ventrals in form (Fig. 12.9 B); dorsal chaetae may begin behind II, often in V or VI, sometimes elsewhere; dorsal chaetae often limited to one or two needles, and one or two hairs per bundle; ventral bundles with numerous bifid chaetae, those of II or even II–V often differ in form and thickness from the rest; asexual reproduction by budding forms chains of individuals; mature specimens have spermathecae in IV, V, or VI with the male pores one segment behind them (Fig. 12.8); eyespots may be present; gills may surround the anus (Fig. 12.9 A); prostomium may bear a proboscis (Fig. 12.9 B) 2

1' Length usually 0.1 cm, width usually 0.5–1.0 mm; when hair chaetae are present dorsally, they are usually accompanied by pectinate chaetae that closely resemble the ventral chaetae, apart from having intermediate teeth (Fig. 12.10 C); when hair chaetae are absent dorsally, all bundles usually contain similar bifid chaetae; dorsal chaetae begin in II, normally several per bundle; reproduction normally sexual, rarely by fragmentation; spermathecal pores normally on X, male pores on XI (Fig. 12.8); no eyes or proboscis; no gills around anus, single dorsal and ventral gill filaments on some posterior segments of one species (Fig. 12.9 C) 22

2(1) Dorsal chaetae present (rare in *Ophidonais*) 3

2' No dorsal chaetae present (in some rare specimens the dorsal chaetae are recovered, but these can be recognized because of the absence of ventral chaetae in III–V and the presence of an enlarged pharynx and reduced prostomium associated with a predatory habit) .... *Chaetogaster*

3(2) Dorsal chaetal bundles without hair chaetae 4

3' Dorsal chaetal bundles with hair chaetae 8

4(3) Dorsal chaetae begin in II or III 5

4' Dorsal chaetae begin in V or VI 6

5(4) Dorsal chaetae begin in II; no gap between chaetal bundles of III and IV .... *Homochaeta naidina* Bretscher, 1896

5' Dorsal chaetae begin in III; gap between chaetal bundles of III and IV .... *Amphichaeta*

6(4) Dorsal chaetae begin in VI; freshwater species 7

6' Dorsal chaetae begin in V; estuarine species .... *Paranais*

7(6) Dorsal chaetae stout, solitary, bluntly simple or notched at the outer end .... *Ophidonais serpentina* (Müller, 1773)

7' Dorsal chaetae curved, with bifid tips, two to four per bundle .... *Uncinais uncinata* (Ørsted, 1842)

[See also *Piguetiella*, in which hair chaetae may be missing in most, if not all, dorsal bundles; some *Uncinais* specimens are said to have dorsal chaetae in V]

8(3) With a proboscis on the prostomium (visible as a stump if broken off) 9

8' Without a proboscis on the prostomium 12

9(8) Dorsal chaetae begin in VI (observe with care as ventral chaetae of IV and/or V may be missing) 10

9' Dorsal chaetae begin in II .... *Pristina* [in part]

10(9) No giant hair chaetae 11

10' Dorsal bundles of VI–VIII with 2–16 giant hair chaetae .... *Ripistes parasita* (Schmidt, 1874)

11(10) Ventral chaetae with a characteristic double bend; dorsal bundles with one to three hairs and three to four shorter, setiform needles .... *Stylaria lacustris* (Linnaeus, 1767)

11' Ventral chaetae slightly curved; dorsal bundles with 8–18 hairs and 9–12 setiform needles .... *Arcteonais lomondi* (Martin, 1907)

12(8) Dorsal chaetae begin in II or III 13

12’ Dorsal chaetae begin in IV or further back 15

13(12) Body wall naked 14

13’ Body wall covered with foreign matter adhering to glandular secretions ..... *Stephensoniana*

14(13) Dorsal chaetae begin in II; hair chaetae in all bundles, none especially thin, either none especially elongate “or” elongate hair chaetae in III ..... *Pristina* [in part]

14’ Dorsal chaetae either begin in III, hairs of III elongate “or” begin in II, the hairs being longest in midbody, and often missing from a number of segments, and exceptionally thin when present ..... *Bratislavia*

15(12) No gills; free-living species 16

15’ Posterior end of the body with a branchial fossa, normally with gills; some species symbionts in tree frogs, these develop gills once outside their host ..... *Dero*

16(15) Dorsal chaetae begin in V, VI, or VII (note that ventral chaetae of IV and/or V may be missing) 17

16’ Dorsal chaetae begin in XVIII–XX, each bundle with a single short hair and a robust needle ..... *Haemonais waldvogeli* Bretscher, 1900

17(16) Hair chaetae one-to-three per bundle at most, thin with or without fine lateral hairs 18

17’ Hair chaetae up to nine per bundle, thick, with long lateral hairs ..... *Vejdovskyella*

18(17) No elongate hair chaetae 19

18’ One to three especially long hair chaetae in each dorsal bundle of VI, one to two ordinary hair chaetae in the remainder ..... *Slavina appendiculata* (d’Udekem, 1855)

19(18) Hair chaetae mostly one to two per bundle (two species with up to five), long (180–200 µm or more), present in all bundles from V or VI 20

19’ Hair chaetae very short (84–120 µm), present in only one or two bundles, some with no hair chaetae at all (see Fig. 12.10 B) ..... *Piguetiella*

20(19) Needle chaetae simple, bifid, or faintly pectinate; either ventral chaetae of II differ from the rest OR those of II–V differ markedly in length, width, and form from the rest 21

20’ Characteristic needle chaetae with thick, unequal teeth, sometimes clearly pectinate; ventral chaetae change form and length slightly between V and VI ..... *Allonais*

21(20) Needle chaetae bifid or faintly pectinate; ventral chaetae of II may differ from the rest ..... *Specaria*

21’ Needle chaetae simple pointed, bifid, or pectinate; ventral chaetae of II–V differ strongly in length, thickness, and form from the rest in the majority of species ..... *Nais*

[Species of *Dero* with the posterior end of the body missing may key out here because there will be no gills. Compare specimens to those with gills; the anterior ventral chaetae in *Dero* species have very long upper teeth.]

22(1) No gill filaments 23

22’ Single dorsal and ventral gill filaments on posterior segments ..... *Branchiura sowerbyi* Beddard, 1892

[Note: broken anterior fragments may be mistaken for *Aulodrilus pluriseta*]

23(22) Body wall naked 24

23’ Body wall papillate (with projections usually covered with foreign matter attached by secretions) ..... *Telmatodrilus* [in part]/ *Spirosperma* / *Quistadrilus*

24(23) Spermathecal chaetae replace normal ventral chaetae in the spermathecal segment, very rarely in adjacent segments also; with or without modified penial chaetae and cuticular penis sheaths 25

24’ No modified spermathecal chaetae 29

25(24) Spermathecal chaetae not duplicated (or, if so, on VII–VIII and very rarely scattered from VI to XII); with or without cuticular penis sheaths 26

25’ Thin, hollow tipped spermathecal chaetae in X, similar chaetae in XI together with apparent cuticular penis sheaths ..... *Haber speciosus* (Hrabe, 1931)

26(25) Without this unique set of characters described below combined 27

26’ Spermathecal chaetae broad, spatulate, one each side of IX associated with long tubular glands; ventral chaetae of X similar or absent; several penial chaetae in each ventral bundle of XI with short, knobbed distal ends; male ducts open into large median inversion of the body wall of XI ..... *Rhizodrilus lacteus* F. Smith, 1900

27(24) Spermathecal chaetae longer than the normal ventral chaetae, slender distally, somewhat like a hypodermic needle; true or apparent cuticular penis sheaths present 28

27’ Spermathecal chaetae much broader and longer than the normal ventral chaetae; distal ends hollow, located on X (or in one instance on VII–VIII or even irregularly between VI and XII); no cuticular penis sheaths ..... *Potamothrix*

[Great Lakes region. Possibly invasive from Palaearctic]

28(27) True cuticular penis sheaths present, but not much thicker than the normal cuticular layer of the body wall .... *Isochaetides*

28' Apparent penis sheaths present (probably the cuticular linings of eversible penes), look like crumpled cylinders in whole mounts of most specimens .... *Psammoryctides*

29(24) Penial chaetae present replacing the normal ventral chaetae of XI 30

29' Penial chaetae absent, ventral chaetae of XI usually missing in mature specimens 33

30(29) Penial chaetae strongly modified, with knobbed or hooked tops to short distal ends, elongate proximal ends, arranged fanwise with heads close together (as if they function as claspers) "or" large, single and sickle shaped 31

30' Penial chaetae bifid but with shortened distal ends and elongate proximal ends, somewhat thicker than the normal ventral chaetae; with short penis sheaths on the ends of erectile penes .... *Varichaetadrilus* [in part]

31(30) No median copulatory chamber or sensory pit; sperm loose in spermathecae after copulation 32

31' Male pores open into an eversible chamber in XI that contains so called paratria and penial chaetae when present; sensory pit on the prostomium; sperm attached to the body wall close to the male pore in external spermatophores .... *Bothrioneurum vejdovskyanum* Stolc, 1888

32(31) Penial chaetae in XI, three to six simple chaetae with hooked tips; somatic chaetae three to five bifids from II to XX, simple chaetae from XXV to -XXX posteriad, no dorsal hair chaetae .... *Thalassodrilus hallae* (Cook & Hiltunen, 1975)

32' Penial chaetae with knobbed tips, or single, sickle shaped; no simple somatic chaetae (except in the ventral bundles of one species, when accompanied by bifid chaetae, and the dorsal bundles of that species possess hair chaetae) .... *Rhyacodrilus* [in part]

33(29) Cuticular penis sheaths present in XI 34

33' Cuticular penis sheaths absent 36

34(33) Penis sheaths annular to conical, attached to the surface of the penis .... *Varichaetadrilus* [in part]/*Tubifex*/ *Ilyodrilus*

34' Penis sheaths more or less elongate, cylindrical, with penes free within them; all chaetae bifid .... *Limnodrilus*

35(33) Chaetae bifid, or with upper tooth divided, or pectinate to palmate, all with the teeth in a single plane 36

35' Chaetae simple anteriorly, behind the clitellum the chaetae have brush like tips .... *Telmatodrilus vejdovskyi* Eisen, 1879

[Canada: British Columbia. USA: California, Oregon, Washington]

36(35) Pharynx and mouth not enlarged, prostomium well developed; chaetae of III not modified; with spermathecae 37

36' Pharynx and mouth enlarged, prostomium reduced; chaetae of III broad, curved, with large lower teeth, the rest thinner, straighter, and with teeth more nearly equal; no spermathecae, presumably parthenogenetic .... *Teneridrilus mastix* (Brinkhurst, 1978)

37(36) No simple ventral chaetae 38

37' Ventral chaetal bundles with up to four simple pointed or bifid chaetae with reduced upper teeth .... *Rhyacodrilus* [in part]

38(37) Dorsal chaetal bundles with two to four pectinate to palmate chaetae from II to XIV, from there posteriad all dorsal chaetae bifid; anterior ventral chaetae three to five per bundle, bifid with the upper teeth thinner than but only a little longer than the lower, but beyond XIV the upper teeth become much longer than the lower .... *Arctodrilus wulikensis* Brinkhurst & Kathman, 1983

[USA: Alaska]

38' Chaetae progressively enlarged from III posteriad, becoming very broad with large lower teeth and recurved distal ends, "or" no hair chaetae dorsally and large numbers of bifid chaetae with thin, short upper teeth, becoming palmate beyond VI in one species, OR hair chaetae present but all ventral chaetae again characteristically numerous and with short upper teeth .... *Aulodrilus*

REFERENCE

Chapman, P.M. & R.O. Brinkhurst. 1987. Hair today, gone tomorrow: induced chaetal changes in tubificid oligochaetes. Hydrobiologia 155: 45-55.

Subclass Branchiobdellidea

Stuart R. Gelder

Department of Science and Math, University of Maine at Presque Isle, Presque Isle, ME, USA

INTRODUCTION

Branchiobdellidans (crayfish worms) are small (adults: 0.8–12.0 mm long) annelids that form an ectosymbiotic association on the exposed surface or in branchial chambers of freshwater crustacean hosts, principally astacoidean crayfishes, but also shrimps, crabs, and isopods. Branchiobdellidans have been reported from the Prairie Provinces of Canada to Costa Rica, thus spanning the Nearctic and northern Neotropical regions in North America. The Nearctic Branchiobdellidea is represented by 16 genera and 102 species, or over two-thirds of the order's global taxonomic diversity (Gelder & Williams, 2015).

Information on the distribution of branchiobdellidans in North America is based on published reports, while some unpublished morphological observations are used to make the key less ambiguous. However, it should be remembered that future investigations will extend known species boundaries and a significant number of new species are waiting to be described. Branchiobdellidans form a single family order, but to be consistent with the other clitellate taxa in this chapter, branchiobdellidans have been raised in rank to subclass, Branchiobdellidea. In Chapter 20 of Volume I (p. 511), the group was called Branchiobdellida, which is a correct name if the group is considered an order rather than a subclass.

LIMITATIONS

With few exceptions, identification of crayfish worms requires microscopical examination of either preserved or live mounted specimens. As many species are morphologically similar, determination of several morphological characters is necessary for identification, including body length and shape, jaw structure, anterior nephridial pore number, male and female reproductive structures, and other morphological features. Species identification depends on recognition of various reproductive organ structures located in body segments 5 and 6 in adults; identification of juveniles is inferential at best. There are no legally restricted branchiobdellidans in the Nearctic region. However, collection of their crustacean hosts may be restricted (see Chapter 16), and being acquainted with, and following local regulations, are the collector's responsibility.

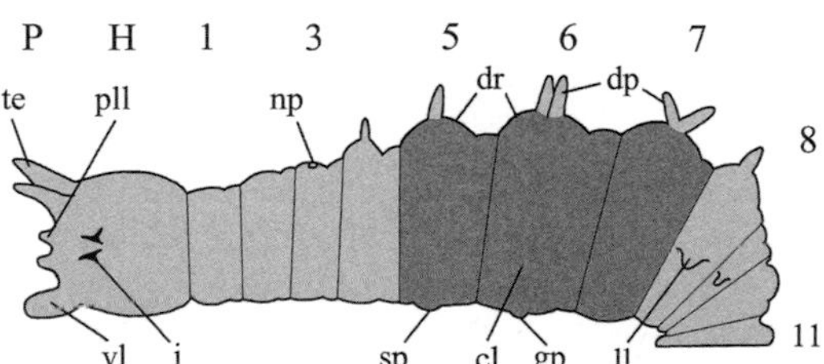

FIGURE 12.11 Lateral view of generalized branchiobdellidan to show most of the external features. P, peristomium; H, head; 1–11, body segment numbers; cl, clitellum; dp, dorsal projections; dr, dorsal ridges; gp, genital pore; j, jaw; pll, peristomial lateral lobe; ll, lateral segmental lobes; np, anterior nephridial pore; sp, spermatheca pore; te, dorsal peristomial lip tentacle; vl, ventral peristomial lip. *Gelder, in preparation; with permission. Copyright © 2015 Dr Stuart R. Gelder. Published by Elsevier Inc. All rights reserved.*

TERMINOLOGY AND MORPHOLOGY

Terms found in branchiobdellidan morphological descriptions differ from those used for other annelids and can lead to confusion. Figure 12.11 illustrates the external body features used in species identification. Branchiobdellidan segment numbering differs from that of both oligochaetes and leeches. Branchiobdellidans have a distinct head, consisting of a peristomium and three segments, followed by 11 numbered body segments. These numbers are given as Arabic numerals to avoid confusion with the Roman numeral-based segment numbering systems in other annelids. Body length and shape are important features as are ornamentation on the peristomium (lobes and tentacles) and segments (two or three annuli, dorsal ridges formed by supernumerary muscles; Fig. 12.15, sm), dorsal projections, and lateral lobes. Fig. 12.12 A–E shows examples of species with a selection of these features. The first identifying character in the dichotomous key is presence of one or two anterior nephridial pores on the dorsal surface of segment 3 (Figs. 12.11 and 12.13). Internal identifying characters include the paired jaws, and the size, shape and number of teeth (Fig. 12.14 A–C). These features are best observed from a dorsal or ventral view when each jaw is slightly displaced. The dental formula consists of the numbers

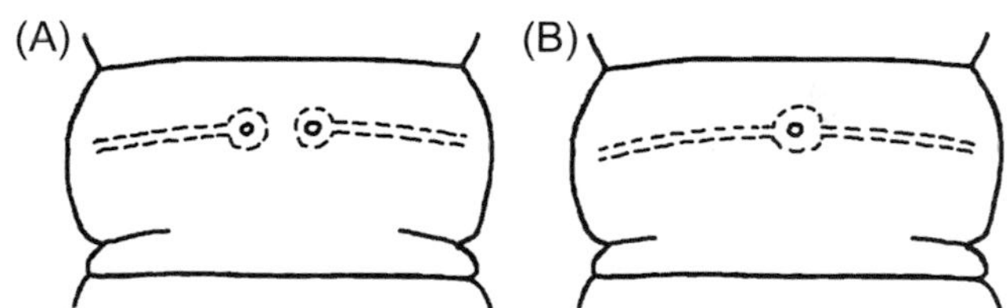

FIGURE 12.13 Anterior nephridial pores (A) or pore (B) on the dorsal surface of segment 3 with underlying collection bulb and duct (dashed lines). *Gelder, in preparation; with permission. Copyright © 2015 Dr Stuart R. Gelder. Published by Elsevier Inc. All rights reserved.*

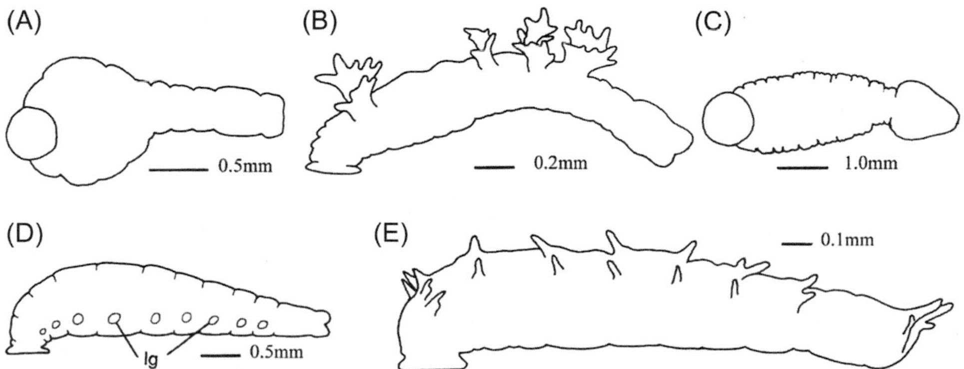

FIGURE 12.12 (A) ventral view of *Xironogiton instabilis* (from Govedich et al., 2010); (B) lateral view of *Pterodrilus alcicornus* (redrawn from Moore, 1895); (C) ventral view of *Triannulata magna* (redrawn from Holt, 1974); (D) lateral view of *Bdellodrilus illuminatus*; lg, lateral glands (from Govedich et al., 2010); (E) lateral view of *Ceratodrilus thysanosomus* (redrawn and modified from Holt, 1960).

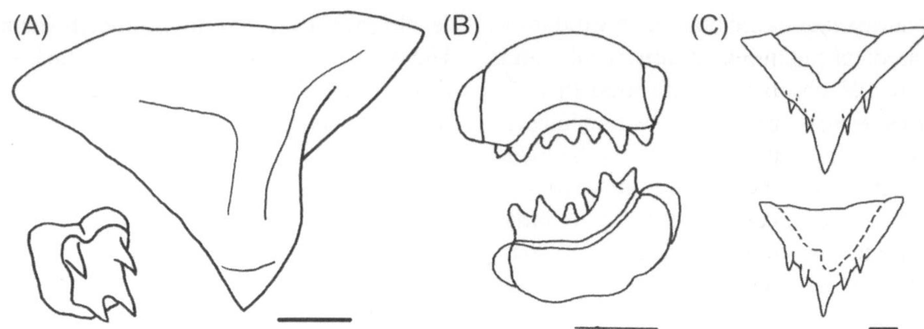

FIGURE 12.14 (A) *Cambarincola heterognathus* jaws (dorsal over ventral, teeth pointing posteriorly), dental formula 1/4, bar = 20 μm; (B) *Ceratodrilus thysanosomus* jaws, dental formula 7/6, bar = 30 μm (Hall, 1914); (C) *Cambarincola fallax* jaws, dental formula 5/5, bar = 20 μm. *"A" and "C" Gelder, in preparation; with permission. Copyright © 2015 Dr. Stuart R. Gelder. Published by Elsevier Inc. All rights reserved.*

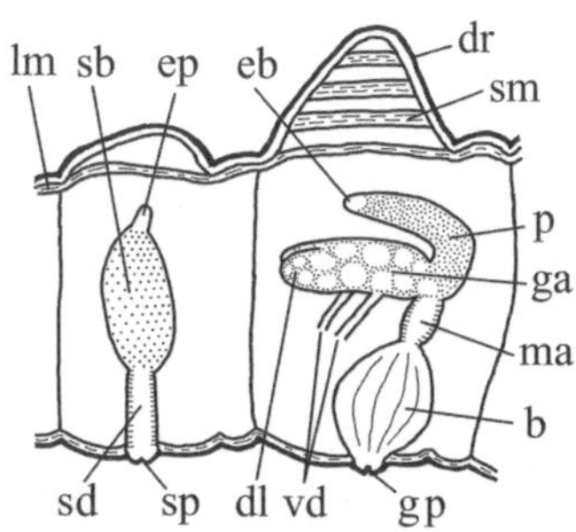

FIGURE 12.15 Lateral view of body segments 5 and 6 showing spermatheca (or female spermatozoa receptacle) and male genitalia: b, bursa; dl, deferent lobes; dr, dorsal ridge; eb, ental bulb; ep, ental process; ga, glandular atrium; gp, genital pore; lm, longitudinal body muscles; ma, muscular atrium; p, prostate gland (shown as "differentiated"); sb, spermatheca bulb; sd, spermatheca duct; sm, supernumerary muscle; sp, spermatheca pore; vd, vasa deferentia. *Gelder, in preparation; with permission. Copyright © 2015 Dr. Stuart R. Gelder. Published by Elsevier Inc. All rights reserved.*

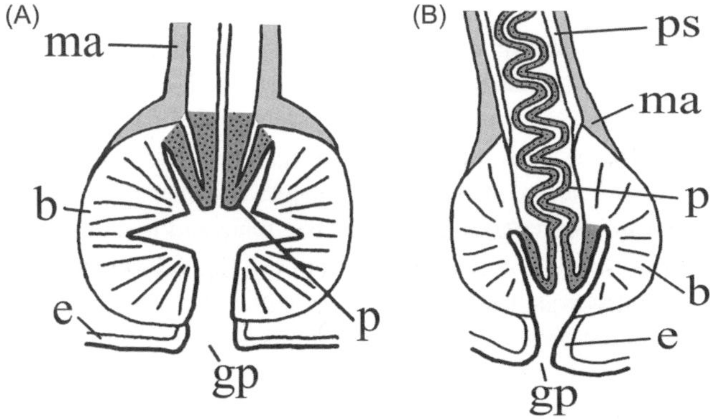

FIGURE 12.16 Schematic longitudinal section through bursa showing a generalized retracted protrusible (A) and eversible (B) penes: b, bursa, e, epidermis; gp, genital pore; ma, muscular atrium (light gray); p, penis (dark gray); ps, penial sheath. *Gelder, in preparation; with permission. Copyright © 2015 Dr. Stuart R. Gelder. Published by Elsevier Inc. All rights reserved.*

of teeth on dorsal and ventral jaws, shown as 1/4, 7/6, and 5/5 in Fig. 12.4 A–C. Dentition in several species is variable.

Figure 12.15 shows all reproductive characters, excluding the penis, used in the key. The relative position of a structure can be referred to as "ental"—nearest the exterior (= proximal)—or "ectal"—farthest from the exterior (= distal). For example, vasa deferentia enter glandular atrium ectally, as shown in Fig. 12.15. The glandular atrium has a free ental end; however, it may be bifurcated to form two deferent lobes. Granule-containing cells line the glandular atrium, and where these also line the prostate gland, the gland is called "undifferentiated." In contrast, when the prostate gland is lined with vacuolar cells, it is referred to as "differentiated." Some species have a prostatic ental bulb, but its size can vary greatly depending on preservation method. A muscular atrium and its ectal bursa connect the glandular atrium with ventral epidermis. A penis is located in the bursa, but in some species it extends entally into the muscular atrium. The penis is characterized by its method of extension: either "protrusible" or "eversible" with the typical appearance of each shown in Fig. 12.16 A, B, respectively. In a few species, penial length approaches total body length, while in *Ellisodrilus*, a penis is absent.

The spermatheca or sperm receptacle (Fig. 12.15) is the only female organ in the key and consists of a pore (sp), duct (sd), bulb (sb) and, when present, an ental process (ep). In some species, the spermatheca is highly variable in size and shape due to the volume of sperm it contains and preservation conditions. The spermatheca is located in segment 5, except in *Ellisodrilus*, where it is absent.

MATERIAL PREPARATION AND PRESERVATION

Although branchiobdellidan identification is frequently accomplished with preserved material, several distinguishing characters are most easily determined using wet-mount preparations of live worms. To prepare a live branchiobdellidan for wet-mounting, place the worm in a drop of water on a microscope slide and gently add a cover-glass. The amount of water on the slide, and thus the mobility of the worm, can be controlled by withdrawing water with a paper tissue or adding drops from a pipette. Too little water, however, will squash and damage the worm. Examine the specimen under a compound light microscope. A specimen must be observed from both dorso-ventral and lateral aspects to see all important characters. Changing the aspect simply requires pushing the cover glass gently so the specimen rolls through a quarter turn. Although a compound microscope with differential interference contrast (DIC) illumination is best for seeing the structures, ordinary bright and dark field illumination will enable all features to be adequately observed. Examination of live branchiobdellidans can greatly reduce the time needed for species identification and leaves the specimen available for subsequent processing (e.g., establishing a culture, for museum curation, histology, electron microscopy, molecular sequencing).

The best general preservative for studying branchiobdellidan morphology is 70% ethanol. It is also ideal for long-term preservation, not carcinogenic, and allows specimens to be mounted for either identification or molecular sequencing. Ten percent formalin is a less desirable, as it is carcinogenic, needs prolonged washing, and can promote formation of insoluble crystals on the surface of the worm that can obscure characters. Occasionally, live branchiobdellidans need to be immobilized or relaxed, and this is most easily achieved by immersing them briefly, 0.5–4.0 min, in carbonated water, e.g., soda water.

To produce permanent museum standard preparations, dehydrate unstained specimens in ethanol solutions, clear in clove oil or oil of wintergreen (methyl salicylate) to remove preservation opacity, and mount them in Canada balsam on microscope slides. The clearing agents also add contrast to a specimen's internal organs. As all permanent mounting methods only allow each worm to be seen from one aspect, multiple mounted specimens are needed to see all the identifying characters. Genitalia in large whole branchiobdellidans (6–10 mm long) may be too thick to be seen on slide preparations. In such cases, the organs should be removed while they are immersed in water, 70% ethanol, or glycerine by micro-dissection. Unpreserved specimens should then be irrigated with preservative and slide mounted, as described above. Instruments for this procedure can be made from razor blade fragments and insect (very small) pins glued to the end of tooth picks.

Alternative preparation methods for examining branchiobdellidans include staining and using clearing mixtures that make semi-permanent microscopical mounts (see microtechnique textbooks or on-line). Specimens in these mountants last from a few days to several months as some of the component chemicals damage soft tissues making the branchiobdellidans questionable or useless for future reference.

KEYS TO BRANCHIOBDELLIDEA

Keys modified from Gelder (In Preparation)

Branchiobdellidea: Branchiobdellidae: Subfamilies

1	Two anterior nephridial pores (Fig. 12.13 A)	2
1'	One anterior nephridial pore (Fig. 12.13 B)	3
2(1)	Vasa deferentia enter glandular atrium ectally (not terminally, Fig. 12.15)	**Branchiobdellinae [p. 237]**
2'	Vasa deferentia enter glandular atrium entally (terminally)	Xironodrilinae, one genus: ***Xironodrilus* [p. 238]**
3(1)	Vasa deferentia enter glandular atrium ectally (not terminally)	**Bdellodrilinae [p. 238]**
3'	Vasa deferentia enter glandular atrium entally (terminally)	**Cambarincolinae [p. 238]**

Branchiobdellidea: Branchiobdellina: Genera

1	Body ventrally flattened, terete-shaped	***Ankyrodrilus* [p. 237]**
1'	Body dorsoventrally flattened	***Xironogiton* [p. 237]**

Branchiobdellidea: Branchiobdellinae: *Ankyrodrilus*: Species

1	Dental formula 3/4 [USA: Tennessee, Virginia, West Virginia]	*Ankyrodrilus koronaeus* Holt, 1965
1'	Dental formula 5/4 or 5/5 [USA: Tennessee, Virginia, West Virginia]	*Ankyrodrilus legaeus* Holt, 1965

Branchiobdellidea: Branchiobdellinae: *Xironogiton*: Species

1	Body shape pyriform	2
1'	Body tennis racket-shaped	3

2(1) Body length greater than 4.5 mm .... *Xironogiton occidentalis* Ellis, 1919
[USA: Oregon, Washington]
2' Body length less than 2.5 mm .... *Xironogiton cassiensis* Holt, 1974
[USA: Idaho, Wyoming]
3(1) Spermatheca less than 1/8 width of segment 5 4
3' Spermatheca about 1/4 width of segment 5 .... *Xironogiton fordi* Holt, 1974
[USA: Idaho, Wyoming]
4(3) Glandular atrium ental tip not touching lateral body wall 5
4' Glandular atrium ental tip touching lateral body wall .... *Xironogiton kittitasi* Holt, 1974
[Canada: British Columbia. USA: Washington]
5(4) Ental tip about midline of body .... *Xironogiton instabilis* (Moore, 1894)
[Canada: New Brunswick. USA: Appalachian Mountains]
5' Ental tip between midline but not touching lateral body wall .... *Xironogiton victoriensis* Gelder & Hall, 1990
[Canada: British Columbia. USA: California, Idaho, Nevada, Oregon, Washington, Wyoming]

Branchiobdellidea: Xironodrilinae: *Xironodrilus*: Species

1 Muscular atrium, present 2
1' Muscular atrium, absent .... *Xironodrilus formosus* Ellis, 1919
[USA: Arkansas, Illinois, Indiana, Kentucky, Michigan, Missouri, New York, Tennessee, West Virginia.]
2(1) Muscular atrium length, less than 0.4× segment width 3
2' Muscular atrium length, greater than 0.7× segment width .... *Xironodrilus bashaviae* Holt & Wiegl, 1979
[North Carolina]
3(2) Dental formula, 3/3 4
3' Dental formula, greater than 4/4 .... *Xironodrilus dentatus* Goodnight, 1940
[USA: Arkansas, Oklahoma, Missouri, Virginia]
4(3) Median tooth, shorter than laterals .... *Xironodrilus pulcherrimus* (Moore, 1894)
[USA: North Carolina, West Virginia]
4' Median tooth, longer than laterals .... *Xironodrilus appalachius* Goodnight, 1943
[USA: Georgia, North Carolina, Tennessee, Virginia]

Branchiobdellidea: *Bdellodrilinae*: Species

1 Epidermis semitransparent (gill dwellers) 2
1' Epidermis opaque (not dwelling in gills) .... *Cronodrilus ogygius* Holt, 1968
[USA: Georgia]
2(1) Lateral segmental glands absent on segments 1–9 .... *Uglukodrilus hemophagus* (Holt, 1977)
[USA: California, Idaho, Oregon, and Washington]
2' Lateral segmental glands present on segments 1–9 .... *Bdellodrilus illuminatus* (Moore, 1894)
[Canada: New Brunswick. USA: Appalachian Mountains. Mexico: Guadalajara?]

Branchiobdellidea: Cambarincolinae: Genera

1 Spermatheca present 2
1' Spermatheca absent .... ***Ellisodrilus* [p. 239]**
2(1) Penis protrusible (Fig. 12.16 A) 3
2' Penis eversible (Fig. 12.16 B) 5
3(2) Prostate gland arises from mid-glandular atrium 4
3' Prostate gland arises adjacent to muscular atrium (Fig. 12.5) .... ***Cambarincola* [p. 239]**

4(3)	Dorsal ridge present across segment 8 [Eastern North America]	***Pterodrilus* [p. 242]**
4'	Dorsal ridge absent across segment 8 [Eastern North America]	***Forbesodrilus* [p. 243]**
5(2)	Dorsal segmental appendages absent	6
5'	Dorsal segmental appendages present	***Ceratodrilus* [p. 243]**
6(5)	Body segments with 2 annuli	7
6'	Body segments with 3–5 annuli [USA: Oregon, Washington]	*Triannulata magna* Goodnight, 1940
7(6)	Bursa large, length equal segment diameter	8
7'	Bursa smaller, length 0.7× or less than segment diameter	9
8(7)	Penis filling about half ental bursa	***Oedipodrilus* [p. 243]**
8'	Penis small [USA: California]	*Magmatodrilus obscurus* Holt, 1967
9(7)	Penis attached by fibrous muscles to bursa wall	***Sathodrilus* [p. 243]**
9'	Penis coiled very long, in long ental bursal sheath [USA: Tennessee]	*Tettodrilus friaufi* Holt, 1968

Branchiobdellidea: Cambarincolinae: *Ellisodrilus*: Species

1	Prostate gland thick and not tapering	2
1'	Prostate gland narrow and tapering entally [USA: Indiana, Michigan]	*Ellisodrilus durbini* (Ellis, 1919)
2(1)	Bursa coma-shaped [USA: Kentucky, Tennessee]	*Ellisodrilus clitellatus* Holt, 1960
2'	Bursa short pyriform [USA: Tennessee]	*Ellisodrilus carronamus* Holt, 1988

Branchiobdellidea: Cambarincolinae: *Cambarincola*: Species

1	Prostate gland cells undifferentiated	2
1'	Prostate gland cells differentiated	24
2(1)	Ental bulb absent	3
2'	Ental bulb present	12
3(2)	Glandular atrium length and prostate gland about equal	4
3'	Glandular atrium length and prostate gland not equal	8
4(3)	Deferent lobes absent	5
4'	Deferent lobes present	6
5(4)	Glandular atrium tubular [Mexico: San Luis Potosí]	*Cambarincola speocirolanae* Holt, 1984
5'	Glandular atrium slim ovoid [USA: Kentucky, Tennessee]	*Cambarincola alienus* Holt, 1963
6(4)	Dorsal peristomial lobes present	7
6'	Dorsal peristomial lobes absent [USA: Kansas, Texas]	*Cambarincola ouachita* Hoffman, 1963
7(6)	Head normal size (length equal or less than segments 1–3) [USA: Virginia]	*Cambarincola branchiophilus* Holt, 1954
7'	Head large (length equal to segments 1–4) [USA: Arkansas]	*Cambarincola restans* Hoffman, 1963

8(3) Prostate gland longer than glandular atrium 9

8' Prostate gland about half length glandular atrium 10

9(8) Dorsal jaw triangular .... *Cambarincola mesochoreus* Hoffman, 1963

[Canada: Ontario. USA: Alabama, Arkansas, California, Connecticut, Indiana, Iowa, Kansas, Kentucky, Louisiana, Maine, Massachusetts, Michigan, Minnesota, Mississippi, Missouri, Nebraska, New Hampshire, New Mexico, Oklahoma, South Dakota, Texas, Vermont, Wisconsin]

9' Dorsal jaw reniform .... *Cambarincola pamelae* Holt, 1984

[USA: Arkansas, California, Louisiana, Maryland]

10(8) Glandular atrium short, not reflexed 11

10' Glandular atrium long, slim, reflexed .... *Cambarincola ellisi* Holt, 1973

[Mexico: Nuevo Leon]

11(10) Deferent lobes present .... *Cambarincola shoshone* Hoffman, 1963

[USA: Idaho]

11' Deferent lobes absent .... *Cambarincola carcinophilus* Holt, 1973

[Mexico: Veracruz]

12(2) Glandular atrium tubular 13

12' Glandular atrium ovoid .... *Cambarincola demissus* Hoffman, 1963

[USA: Virginia, West Virginia]

13(12) Glandular atrium reflexed, double length of prostate gland 14

13' Glandular atrium about equal with prostate gland 16

14(13) Spermatheca length greater than segment diameter 15

14' Spermatheca length less than segment diameter .... *Cambarincola susanae* Holt, 1973

[Mexico: Campeche, Nuevo Leon, Puebla, San Luis Potosí, Veracruz]

15(14) Jaws similar size .... *Cambarincola goodnighti* Holt, 1973

[USA: Florida]

15' Jaws dissimilar size .... *Cambarincola barbarae* Holt, 1981

[USA: California, Louisiana]

16(13) Dental formula 1/2 17

16' Dental formula greater than 1/2 19

17(16) Glandular atrium shorter than prostate gland 18

17' Glandular atrium equal to prostate gland .... *Cambarincola leoni* Holt, 1973

[USA: Florida]

18(17) Glandular atrium shorter than muscular atrium .... *Cambarincola micradenus* Holt, 1973

[Mexico: Puebla]

18' Glandular atrium about equal to muscular atrium .... *Cambarincola hoffmani* Holt, 1973

[Mexico: Puebla]

19(16) Dental formula 5/4 20

19' Dental formula 5/5 .... *Cambarincola leptadenus* Holt, 1973

[USA: Tennessee]

20(19) Glandular atrium same thickness as prostate gland 21

20' Glandular atrium thicker than prostate gland 23

21(20) Ventral jaw shape not triangular 22

21' Ventral jaw shape triangular .... *Cambarincola jamapaensis* Holt, 1973

[Mexico: Puebla, Veracruz]

22(21) Muscular atrium about half length of glandular atrium .... *Cambarincola olmecus* Holt, 1973

[Mexico: San Luis Potosí, Veracruz]

22' Muscular atrium about a quarter length of glandular atrium .... *Cambarincola manni* Holt, 1973

[USA: Florida]

23(20) Jaw shape triangular .... *Cambarincola illinoisensis* Holt, 1982

[USA: Illinois]

23’ Jaw shape not triangular .... *Cambarincola marthae* Holt, 1973
[USA: Tennessee]

24(1) Ental bulb absent 25

24’ Ental bulb present 29

25(24) Deferent lobes present 26

25’ Deferent lobes absent 28

26(25) Four peristomial tentacles present 27

26’ Four peristomial lobes present .... *Cambarincola okadai* Yamaguchi, 1933
[Canada: British Columbia. USA: California, Idaho, Oregon, Washington]

27(26) Head large, length about 25% of body .... *Cambarincola macrocephalus* Goodnight, 1943
[USA: Idaho, Wyoming]

27’ Head normal, length about 15% of body .... *Cambarincola holti* Hoffman, 1963
[USA: Kentucky, Tennessee]

28(25) Dorsal jaw multiple teeth .... *Cambarincola serratus* Holt, 1981
[USA: Idaho]

28’ Dorsal jaw one median, large tooth .... *Cambarincola gracilis* Robinson, 1954
[Canada: British Columbia. USA: California, Oregon, Washington]

29(24) Prostate gland length half the length of glandular atrium 30

29’ Prostate gland length subequal or longer than glandular atrium 35

30(29) Glandular atrium reflexed 31

30’ Glandular atrium curved 33

31(30) Deferent lobes absent 32

31’ Deferent lobes present .... *Cambarincola virginicus* Hoffman, 1963
[USA: Virginia]

32(31) Male organs almost filling segment 6 .... *Cambarincola vitreus* Ellis, 1919
[Canada: Alberta, Manitoba, Ontario, Saskatchewan. USA: Arkansas, Colorado, Illinois, Kansas, Michigan, Minnesota, Montana, Oklahoma, North & South Dakota, Wisconsin]

32’ Male organs filling about half segment 6 .... *Cambarincola osceola* Hoffman, 1963
[USA: Delaware, Florida, Georgia, Massachusetts, North Carolina, South Carolina, Virginia]

33(30) Dental formula 5/5 34

33’ Dental formula 1/4 .... *Cambarincola heterognathus* Hoffman, 1963
[USA: Florida, Georgia, North Carolina, Tennessee, Virginia, West Virginia]

34(33) Male organs almost filling segment 6 .... *Cambarincola bobbi* Holt, 1988
[USA: Tennessee, Virginia]

34’ Male organs filling about half of segment 6 .... *Cambarincola dubius* Holt, 1973
[USA: Indiana]

35(29) Prostate gland length shorter than glandular atrium 36

35’ Prostate gland length longer than glandular atrium .... *Cambarincola ingens* Hoffman, 1963
[USA: North Carolina, Tennessee, Virginia, West Virginia]

36(35) Dental formula 5/5 37

36’ Dental formula 5/4 39

37(36) Peristomial lobes present 38

37’ Peristomial tentacles present .... *Cambarincola fallax* Hoffman, 1963
[Canada: British Columbia, New Brunswick, Ontario, Quebec. USA: California, Georgia, Kentucky, Maine, Massachusetts, Michigan, New York, North Carolina, Ohio, South Carolina, Tennessee, Vermont, Virginia, West Virginia]

38(37) Glandular atrium long, thin .... *Cambarincola sheltensis* Holt, 1973
[USA: Alabama]

38’ Glandular atrium short, thick .... *Cambarincola holostomus* Hoffman, 1963
[USA: Tennessee, Virginia]

39(36) Spermatheca ental process present 40

39' Spermatheca ental process absent 42

40(39) Peristomial lobes present 41

40' Peristomial lobes absent ..... *Cambarincola macrodontus* Ellis, 1912

[USA: Colorado, New Mexico. A wider distribution in the central drainage has been question by Hoffman (1963). Mexican records are also questionable.]

41(40) Male organs almost filling segment 6 ..... *Cambarincola meyeri* Goodnight, 1942

[USA: Kentucky]

41' Male organs filling about half segment 6 ..... *Cambarincola philadelphicus* (Leidy, 1851)

[Canada: Ontario, Quebec. USA: Connecticut, Kentucky, Maine, Massachusetts, Maryland, Minnesota, New York, North Carolina, Pennsylvania, South Carolina, Tennessee, Vermont, Virginia, West Virginia. Mexican records are questionable.]

42(39) Dorsal jaw larger than ventral 43

42' Jaws about same size ..... *Cambarincola floridanus* Goodnight, 1941

[USA: Florida, Georgia]

43(42) Spermatheca length equal segment diameter ..... *Cambarincola chirocephalus* Ellis, 1919

[Canada: Manitoba, Ontario, Saskatchewan. USA: Alabama, Arkansas, Florida, Illinois, Indiana, Iowa, Kansas, Kentucky, Michigan, Mississippi, Missouri, North Dakota, New York, Oklahoma, Pennsylvania, Tennessee, Virginia]

43' Spermatheca length half segment diameter ..... *Cambarincola toltecus* Holt, 1973

[Mexico: Veracruz]

Branchiobdellidea: Cambarincolinae: *Pterodrilus*: Species

1 Dorsal projections on any of segments 2–5 2

1' Dorsal projections absent from segments 2–5 5

2(1) Two digitiform projections on a segment 3

2' Explanate projections on segment 3–5, and 8 4

3(2) Projections on each segment 2–7 ..... *Pterodrilus distichus* Moore, 1895

[Canada: Ontario. USA: Illinois, Indiana, Kentucky, Michigan, New York, Ohio]

3' Projections on segments 4 and 5 ..... *Pterodrilus robinae* Williams & Gelder, 2011

[USA: Tennessee]

4(2) Dorsal projections on segment 2 ..... *Pterodrilus simondsi* Holt, 1968

[USA: Georgia, North Carolina]

4' Dorsal projections absent from segment 2 ..... *Pterodrilus alcicornus* Moore, 1895

[USA: North Carolina, Tennessee, Virginia, West Virginia]

5(1) Dorsal digitiform projections on prominent ridge of segment 8 6

5' Dorsal projections absent from prominent ridge of segment 8 8

6(5) Five digitiform projections present 7

6' Four digitiform projections present ..... *Pterodrilus mexicanus* Ellis, 1919

[USA: Arkansas, Missouri, Oklahoma. Mexico: Veracruz]

7(6) Ridges present on segments 2–7 ..... *Pterodrilus cedrus* Holt, 1968

[USA: Tennessee]

7' Ridges absent on segments 2–7 ..... *Pterodrilus hobbsi* Holt, 1968

[USA: Alabama, Kentucky, North Carolina, Tennessee, Virginia]

8(5) Prominent ridges absent from segments 2–7 9

8' Prominent ridges on segments 2–7 ..... *Pterodrilus annulatus* Gelder, 1996

[USA: Arkansas, Georgia, Tennessee]

9(8) Prostate gland differentiated ..... *Pterodrilus missouriensis* Holt, 1968

[USA: Maine, Missouri]

9' Prostate gland undifferentiated ..... *Pterodrilus choritonamus* Holt, 1968

[USA: Tennessee]

Branchiobdellidea: Cambarincolinae: *Forbesodrilus*: Species

1 Dental formula 5/4 .... *Forbesodrilus nanagnathus* (Holt, 1973)
[Mexico: Veracruz. Nicaragua]

1' Dental formula 7/7 .... *Forbesodrilus acudentatus* (Holt, 1973)
[Mexico: Tamaulipas]

Branchiobdellidea: Cambarincolinae: *Ceratodrilus*: Species

1 Peristomial tentacles and dorsal projections short .... *Ceratodrilus thysanosomus* Hall, 1914
[USA: Idaho, Oregon, Utah, Wyoming]

1' Peristomial tentacles and dorsal projections long .... *Ceratodrilus ophiorhysis* Holt, 1960
[USA: Idaho, Oregon, Wyoming]

Branchiobdellidea: Cambarincolinae: *Oedipodrilus*: Species

1 Dental formula 5/4 2

1' Dental formula 2/1 .... *Oedipodrilus anisognathus* Holt, 1988
[USA: Tennessee]

2 Prostate gland present 3

2' Prostate bulb absent .... *Oedipodrilus cuetzalanae* Holt, 1984
[Mexico: Puebla]

3(2) Length of bursa 0.9× segment diameter .... *Oedipodrilus oedipus* Holt, 1967
[USA: Tennessee]

3' Length of bursa 1.4× segment diameter .... *Oedipodrilus macbaini* (Holt, 1955)
[USA: Illinois, Indiana, Kentucky, Pennsylvania, Tennessee]

Branchiobdellidea: Cambarincolinae: *Sathodrilus*: Species

1 Prostate absent 2

1' Prostate present 6

2(1) Glandular atrium length equal or longer than muscular atrium plus bursa 3

2' Glandular atrium length shorter than bursa 4

3(2) Dental formula 1/1 .... *Sathodrilus shastae* Holt, 1981
[USA: California]

3' Dental formula 5/4 .... *Sathodrilus hortoni* Holt, 1973
[USA: Florida]

4(2) Dental formula 5/4 5

4' Dental formula 3/4 .... *Sathodrilus okaloosae* Holt, 1973
[USA: Florida]

5(4) Dorsal ridge segment 8 absent .... *Sathodrilus. veracruzicus* (Holt, 1968)
[Mexico: Puebla, Veracruz]

5' Dorsal ridge segment 8 present .... *Sathodrilus nigrofluvius* Holt, 1989
[USA: Missouri]

6(1) Prostate protuberance present 7

6' Prostate gland present 10

7(6) Glandular atrium length longer than bursa 8

7' Glandular atrium length shorter than bursa 9

8(7) Dorsal ridge segment 8 absent .... *Sathodrilus megadenus* Holt, 1968
[USA: Georgia]

8' Dorsal ridge segment 8 present .... *Sathodrilus norbyi* Holt, 1977
[USA: Idaho, Washington]

9(7) Dorsal jaw triangular .... *Sathodrilus villalobosi* Holt, 1968
[Mexico: Hidalgo, Puebla]

9' Dorsal jaw rectangular .... *Sathodrilus carolinensis* Holt, 1968
[USA: South Carolina]

10(6) Prostate gland ental bulb present 11

10' Prostate gland ental bulb absent 14

11(10) Dorsal jaw 5 teeth 12

11' Dorsal jaw 2 teeth .... *Sathodrilus wardinus* Holt, 1968
[USA: Washington]

12(11) Dental formula 5/4 13

12' Dental formula 5/2 .... *Sathodrilus chehalisae* Holt, 1981
[USA: Washington]

13(12) Deferent lobes absent .... *Sathodrilus prostates* Holt, 1973
[Mexico: Puebla, Veracruz]

13' Deferent lobes present .... *Sathodrilus attenuatus* Holt, 1981
[USA: Idaho, Oregon, Washington, Wyoming]

14(10) Glandular atrium length longer than bursa 15

14' Glandular atrium length equal or shorter than bursa .... *Sathodrilus dorfus* Holt, 1977
[USA: Oregon]

15(14) Glandular atrium and prostate gland similar thickness 16

15' Glandular atrium much thicker than prostate gland .... *Sathodrilus elevatus* (Goodnight, 1940)
[Canada: Ontario. USA: Arkansas, Iowa, Illinois, Indiana, Michigan, Minnesota, Ohio, South Dakota, Wisconsin]

16(15) Peristomial lobes absent 17

16' Peristomial lobes present .... *Sathodrilus lobatus* Holt, 1977
[USA: Oregon, Washington]

17(16) Dorsal ridge segment 8 absent .... *Sathodrilus rivigeae* Holt, 1988
[USA: Arkansas]

17' Dorsal ridge segment 8 present .... *Sathodrilus inversus* (Ellis, 1919)
[USA: Idaho, Oregon, Washington]

REFERENCES

Gelder, S.R. Monograph of the Branchiobdellida (Annelida: Clitellata) or crayfish worms. In preparation.

Hall, M.C. 1914. Description of a new genus and species of the discodrilid worms. Proceedings of the United States National Museum. 48(2071): 187–193.

Holt, P.C. 1960. The genus *Ceratodrilus* Hall (Branchiobdellidae, Oligochaeta) with the description of a new species. Virginia Journal of Science 11, new series (2):53–73.

Holt, P.C. 1974. An emendation of the genus *Triannulata* with the assignment of *Triannulata montana* to *Cambarincola* Ellis 1912 (Clitellata: Branchiobdellida). Proceedings of the Biological Society of Washington 87:57–72.

Subclass Hirudinida

William E. Moser
Department of Invertebrate Zoology, Smithsonian Museum of Natural History, Smithsonian Institution, Washington, DC, USA

Fredric R. Govedich
Department of Biological Sciences, Southern Utah University, Cedar City, UT, USA

Alejandro Oceguera-Figueroa
Laboratorio de Helmintología, Instituto de Biología, Universidad Nacional Autónoma de México, Ciudad Universitaria, Coyoacán, DF, Mexico

Dennis J. Richardson
School of Biological Sciences, Quinnipiac University, Hamden, CT, USA

Anna J. Phillips
Department of Invertebrate Zoology, National Museum of Natural History, Smithsonian Institution, Washington, D.C., USA

INTRODUCTION

Leeches are easily recognized by their segmented bodies and attachment suckers. There are ~100 described freshwater leech species in North America and three terrestrial to semi-terrestrial species. Leeches are an important component of aquatic ecosystems, functioning as predators or ectoparasites. They are indicators of aquatic chemistry and biodiversity, and the presence of specific leech species are connected with the basic aquatic conditions and the occurrence of certain animals. Leeches inhabit lakes, ponds, swamps, rivers, streams, and moist soil. They are most abundant along the shallow vegetated shoreline. The full distribution of most leech species is unknown. The paucity of knowledge is due to the lack of regional collecting investigations and the difficulty of identifying some species. Misidentifications have also muddied species concepts and geographic range. In addition, recent molecular and morphological studies have revealed numerous errors in the classification of leeches and leech species concepts, cryptic speciation, and unexpected relationships. It is therefore important that specimen vouchers are deposited in a natural history museum for future analysis. These keys include updated taxonomy and, where possible, use external characteristics to separate taxa.

TERMINOLOGY AND MORPHOLOGY

Knowledge of the following terms is needed to use the Hirudinida keys. Accessory eyes: Two pair of variable concentrations of dark pigment situated between 2 and 5 annuli behind the single functional pair of eyes (giving the impression of 3 pairs of eyes) in the cephalic (head) region of *Placobdella hollensis*. These are also called accessory ocelli, accessory eyespots or supplementary eyes. Agnathus: Complete absence of jaw-like prominences in the pharynx cavity, as in members of the family Erpobdellidae. Annuli (annulus - singular): External body rings or superficial transverse furrows subdividing each somite. There are essentially 3 primary annuli, labeled a1, a2, a3. Each can be further subdivided into secondary annuli (b1, b2, to b6) and further subdivided into tertiary annuli (c1, c2, to c12) and rarely subdivided into quaternary annuli (d1, d2 to d24). Anterior Sucker: The attachment or suction device found at the anterior end of the leech and contains the mouth or mouth pore. It is also called oral sucker or cephalic sucker. Atrial Cornua: Horn-like prolongations of the atrium where the sperm ducts (ejaculatory ducts) attach to the atrium; also called horn of atrium. Atrium: Male reproductive organ consisting of a thin-walled bursa (may be eversible), a thick-walled muscular and glandular medium chamber and a pair of atrial cornua. The atrium opens externally through the male gonopore and in some species, modified into an eversible penis. Buccal Cavity: The mouth chamber which contains muscular ridges that may or may not have jaws attached. Leeches with a proboscis (order Rhynchobdellida) do not possess a buccal cavity. Caeca (caecum): Diverticula (pouches) of the crop (stomach) or intestine, which increases the surface area of the digestive system. The opposite is acaecate (without caeca). Caudal Ocelli: Eyespots (ocelli) on the caudal sucker of certain Piscicolids. These may be either crescent-shaped (crescentiform) or dot-shaped (punctiform), and are also called oculiform spots. Caudal Sucker: The attachment or suction device found at the posterior end of the leech. These may be called posterior sucker or subanal sucker. Cephalic Region: The region of the head. Chromatophore: A cutaneous pigment cell or group of pigment cells which provide coloration in leeches and can be altered in shape and size (under the control of the nervous system). There are 3 basic types of leech chromatophores: brown cutaneous, cream-colored cutaneous, and green hypodermal. Crop: A major section of the digestive system adapted for the storage of blood or invertebrate body fluids. These typically have one or more pairs of caeca, but are acaecate in most Erpobdellid leeches. These are also called the stomach. Clitellum: A swollen glandular region or saddle of epidermal tissue in the area of the gonopores that secretes material to form cocoons. The area that contains the clitellum is called the clitellar region. Cocoon: Chitinous, membranous, or spongy structure that is formed by clitellar secretions and contains fertilized eggs. Cocoons are either deposited on a submerged object, moist area or brooded by the adult. Copulatory Depressions: A ventral glandular area around the gonopores and copulatory pits of *Philobdella* spp. Copulatory Gland Pores: External openings of the copulatory glands, located on the anterior mid-body ventral surface and a few annuli posterior to the female gonopore in *Macrobdella* spp. The number and arrangement of copulatory gland pores are a diagnostic character of Macrobdella species and exist in a linear or transverse pattern of either 4, 6, 8, or 24. Copulatory Glands: Glands whose contents exit through the copulatory gland pores, with an uncertain function (may secrete a glandular adhesive—to stick leeches together during copulation). Copulatory Pits: Thick depressions and prominences of the glandular area around the gonopores of *Philobdella* spp. Crop

Caeca: Paired diverticula (pouches) of the crop. The last pair (postcaeca) typically extends to XIX/XX in blood-sucking leeches and acaecate in most Erpobdellid leeches. These are also called gastric caeca. Denticles: Small dentiform processes on the jaws of Hirudinid and Haemadipsid leeches. Denticles typically are in one row (monostichodont) (such as *Hirudo* spp. or *Macrobdella* spp.) or two rows (distichodont) (such as some species of *Philobdella* or *Haemopis*). Digitate Processes: Approximately 30 or 60 retractable digitform papillae along the inner margin of the caudal sucker in the genus *Actinobdella*. These are generally retracted in preserved specimens. Discoid Head: Anterior somites and oral sucker expanded, circular, and demarcated from the body with a nuchal (neck) constriction; disciform, as in *Placobdella montifera* and *P. nuchalis*; distichodont: see denticles. Diverticula (diverticulum): Blind pouches or sacs that extend out from the crop or intestine (see also caeca, crop caeca). Eyes: Photoreceptors of the dorsal cephalic (head) region that are formed from a number of light sensitive cells backed by a pigment cup. Eyes are used as higher level identification characters, as they vary in number and arrangement. They are also called ocelli or eyespots. Ejaculatory Duct: Paired ducts that run from the epididymis (or sperm sacs in some species) to the male atrium. The arrangement and location of ejaculatory ducts are important species identification characters, especially in some erpobdellids. These are also called preatrial loops. Epididymis: Large paired, sperm-storage ducts (sometimes coiled) connecting to ejaculatory ducts or sperm sacs (in some Gnathobdellid species). These are also called seminal vesicles. Furrow: Narrow groove between two annuli (body rings). Ganglia (ganglion): Enlarged sections of the ventral nerve cord made up of concentrations of neural cells. Leeches have 34 ganglia (6 coalesced ganglia in the cephalic region designated as the brain, nerve ring, pharyngeal ganglionic mass, supraoesophageal ganglionic mass, or anterior ganglionic mass), 21 free body ganglia, and 7 coalesced ganglia in the caudal region (called posterior ganglionic mass or suboesophageal mass). Ganglia are given in Roman numerals (e.g., the first ganglia after anterior ganglionic mass is VII). Gnathous: The presence of jaws in the pharynx cavity, as in members of the Macrobdellidae. The opposite is agnathus. Gonopores: External openings of the male and female reproductive systems, located on the anterior mid-body ventral surface. The male pore is larger and anterior to the less conspicuous female pore; however, in a few species, the male and female gonopore is fused in a single pore. Integument: The outer layers of protective covering of leeches comprised of an epidermis blanketed by an elastic cuticle (with epicuticular projections). Internal Ridges: See pharynx folds. Intestine: Section of the digestive system where digestion and absorption of ingested blood or invertebrate body tissue occurs. Intestinal Caeca: Paired diverticula (pouches) of the intestine. There are typically 4 pairs of intestinal caeca in Glossiphoniids, caecate in Piscicolids, and acaecate in Arhynchobdellids. Jaws: Cutting apparati (that resemble half circular saw blades) with numerous denticles that typically occur in groups of three in the mouth cavity of most Hirudinid and Haemadipsid leeches. Jaws are arranged with one median dorsal and the other two ventrolateral, and create a tripartite incision. Lateral Ocelli: Eyespots found on the lateral margin of the urosome in some piscicolids, such as *Cystobranchus meyeri*. Metamere: see somite. Metameric: The division of the body into a series of similar or identical repeating units (e.g., segments or somites). Metameric dots, spots, patches, or prominences are metamerically (segmentally) repeating external dorsal pigment patterns or structures. Monostichodont: see denticles. Mouth: A medium to large opening on the entire ventral surface of the oral sucker of Arhynchobdellids with rounded lips around the edges. Mouth pore: A small opening on the ventral surface of the oral sucker of Rhynchobdellids where a proboscis is protruded and retracted. It may be located either in the center or the anterior rim of the oral sucker and is also called a proboscis pore. Mycetomes: Structures (generally paired) in blood-feeding Rhynchobdellids that are located in the oesophageal region and harbor symbiotic microflora, which aide in blood digestion. Neural Annulus: The annulus that contains the ganglion of each somite internally and the metameric sensory organs or sensillae externally. It is the middle annulus in 3 or 5 annulate somites. These are also called sensory annulus. Nephridia (nephridium): Metameric excretory organs, opening to the outside via small pores of the body called nephridiopores. Nuchal plate: see scute. Ocelli (ocellus): Photoreceptor cells found on the cephalic, caudal, or lateral margins of the body of certain leeches. Oculiform spots: see caudal ocelli. Oesophagus: Narrow duct where ingested blood or invertebrate body fluids pass from the pharynx or proboscis to the crop. It may contain mycetomes or oesophageal organ in some blood-feeding species. Oral Sucker: The attachment or suction device found at the anterior end of the leech and contains the mouth or mouth pore. It is also called anterior sucker or cephalic sucker. Oviduct: Paired ducts that run from the ovisacs to the female gonopore. Ovisacs: Specialized coelomic sacs that produce and store eggs. Papillae: Protrusible sensory organs that are typically on the dorsal surface, either scattered or metamerically arranged. Pedicel: A narrow stalk of annuli supporting the caudal sucker in some leeches (e.g., *Placobdella pediculata*). It is also called a peduncle. Penis: Protrusible organ for transfer of sperm in Hirudinid and Haemadipsid Gnathobdellids. Not present in Rhynchobdellids and Erpobdellids, which transfer sperm via spermatophores (the male atrium is protrusible like a penis in some Theromyzon species). Pharynx: Anterior muscular section of the digestive system, following the mouth and anterior to the oesophagus. Pharynx folds: Internal muscular ridges of the pharynx of

Haemopis species. These are also called internal ridges or pharynx pods. Pigment: A dorsal or ventral cutaneous structure of color found in chromatophores (typical colors: brown, black, red, green, blue). Postcaeca: The last pair of caeca in the crop. These are also called posterior crop caeca. Posterior Sucker: The attachment or suction device found at the posterior end of the leech. These are also called the caudal sucker. Proboscis: A modified muscular, tubular protrusible pharynx with a blunt tip and triradiate lumen. This is used for feeding by Rhynchobdellid leeches (families Glossiphoniidae and Piscicolidae) via insertion in the outer integument of an animal. Pulsatile Vesicles: Hemispherical extensions of the coelomic system along the lateral body margins of some Piscicolids (e.g., *Cystobranchus* spp. and *Piscicola* spp.). They rhythmically pulsate and function as respiratory organs. Rays: A pigmented pattern on the caudal sucker of some Piscicolids. Salivary Glands: Numerous spherical salivary cells (consisting of a cell soma and elongated ductule) that are either scattered throughout the anterior portion of the body (diffuse) or aggregated in discrete masses (compact), and aide in the ingestion of blood or invertebrate body fluids by secreting salivary chemicals. Scute: Dark brown chitinous structure on the dorsal nuchal (neck) region of *Helobdella stagnalis*. These are also called nuchal scute or dorsal plaque. Segment: see somite. Sensillae: Metameric sensory structures located on the integument of the neural annulus, which provide chemo- and/or mechanoreception. Also called segmental receptors. Somite: A serially repeated body segment that corresponds with one ganglion in the central nervous system. All leeches have 34 somites, which correspond to 34 ganglia in the nervous system. Somite numbers are given in Roman numerals. They are also called segments or metameres. Spermatophore: A packet of spermatozoa enclosed in a capsule that is used in Rhychobdellids and Erpobdellids to transport sperm to another adult by hypodermic implantation of the ventral or dorsal body surface (typically the clitellar region in some species). Sperm Sac: Bulbous, muscular paired sperm-storage structure in Hirudinid leeches. Size and arrangement with epididymis is an important species identification character in *Haemopis*. Stomach: see crop. Teeth: see denticles. Testisacs: Specialized coelomic sacs that produce and store spermatozoa. These may occur in pairs or grape-like clusters. Trachelosome: Narrow, cylindrical neck region (comprised of 6 somites) in some Piscicolids. Tubercles: Large papillae. Urosome: Thick, flat mid-body region (comprised of 12 somites) in some Piscicolids. Vagina: Enlargement of the female gonopore where the penis is introduced in Hirudinid copulation. Vas Deferens: Ducts that run anteriorly to form the large epididymis (seminal vesicle). Vas Efferens: Short ducts that connect the testisacs to the vas deferens each side of the body. Velum: A transverse flap of tissue separating the buccal cavity from the mouth (oral-opening cavity) in Arhynchobbdellids.

MATERIAL PREPARATION AND PRESERVATION

Leeches can often be studied when they are alive, and this can provide information on reproduction, feeding, etc.; however, preserved specimens, dissections, permanently stained slides, and serial sections may be required to identify some species. Colors and patterns should be noted before leeches are preserved because the chromatophores can be dissolved or altered during the preservation process. Slides that have been prepared without stains may need to be cleared to determine eye number and placement.

Preservation of leech specimens can be difficult, as they will contract when placed into fixatives. When you are preparing to preserve leeches, it is best to first narcotize the leeches, and then once relaxed, the leeches can be placed into fixatives. Narcotizing and relaxing leeches can be problematic, but the following alternative methods have been used successfully:

1. Add drops of 95% ethanol slowly to the water containing the leech, gradually increasing the concentration for about 30 min until movement ceases. When the leech is limp and no longer responds to touch, pass it between the fingers to straighten it and to remove the excess mucus.
2. Add carbonated water or bubble in CO_2 until movement of the leech stops. Straighten the leech out on a slide and slowly add warm 70% ethanol and a few drops of glacial acetic acid until covered.
3. Add a drop or two of 6% nembutal until movements stop, and then straighten the leech on a slide.

Once relaxed, leeches should be fixed in 5 to 10% buffered formalin for at least 24 hr depending on the size of the individuals. For future molecular analyses, a few specimens should be placed directly in 95% ethanol. Leeches should be kept in 70–75% ethanol for long-term preservation storage (colors will be bleached out over time). Larger specimens should be injected with formalin to preserve the internal organs. For whole mount slide preparations, leeches should be flattened. This can be done by carefully placing the specimens between two glass slides and adding additional weights if necessary. Staining with Semichon's acetocarmine, clearing in methyl salicylate, and mounting in Canada Balsam or Damar Balsam have shown good results. Fleming's or Bouin's fixatives can be used for histological preparations with specimens being stained in Mayer's paracarmine, borax carmine, or Hams' hematoxylin for 12–94 hr and then destained in 1% HCl–70% ethanol and neutralized in a 1% NH_4OH–70% ethanol solution. Once-stained specimens can be counterstained in fast green or eosin and then dehydrated in progressively higher concentrations of ethanol, cleared in methyl salicylate, and mounted in a neutral pH mounting medium (Govedich et al., 2010).

KEYS TO HIRUDINIDA

Hirudinida: Orders

1 Mouth a small pore on ventral surface of anterior sucker through which a muscular pharyngeal proboscis can be protruded (Fig. 12.19 A, B); no jaws or teeth .... **Rhynchobdellida [p. 248]**

1' Mouth large, occupying the entire cavity of anterior sucker (Fig. 12.14 C); no protrusible proboscis; jaws with teeth either present or absent (Fig. 12.17) .... **Arhynchobdellida [p. 254]**

Hirudinida: Rhynchobdellida: Families

1 Body flattened dorsoventrally and much wider than head (Fig. 12.18 A) (except *Placobdella montifera* and *Placobdella nuchalis*); body not cylindrical (except for *Helobdella elongata*, which is subcylindrical); not differentiated into two body regions: anterior (oral) sucker ventral, more or less fused to body and narrower than body; body never divided into anterior trachelosome and posterior urosome; eggs in membranous cocoons and young brooded on ventral surface of parent; 1–4 pairs of eyes; no oculiform eyespots on posterior sucker; segments 3-annulate (a1, a2, a3) (Fig. 12.2 A) (except *Placobdella biannulata* which is 2-annulate) .... **Glossiphoniidae [p. 248]**

1' Body cylindrical and usually long and narrow; body sometimes divided into a narrow anterior trachelosome and a wider posterior urosome (*Myzobdella*) (Figs. 12.18 and 12.27) anterior (oral) sucker expanded and distinct from body: 0–2 pairs of eyes; pulsatile vesicles along the lateral margins present (*Piscicola* and *Cystobranchus*) or absent; seven or more annuli per segment (except *Myzobdella reducta* which is 3-annulate); oculiform eyespots sometimes present on posterior (caudal) sucker (Figs. 12.27 B and 12.29 B–D) no brooding of cocoons or young; parasitic primarily on fishes (except for *Cystobranchus virginicus*, which apparently feeds on fish eggs) .... **Piscicolidae [p. 253]**

Hirudinida: Rhynchobdellida: Glossiphoniidae: Genera

1 Posterior sucker without a marginal circle of glands or retractile papillae (digitate processes) 2

1' Posterior sucker conspicuous with a marginal circle of 30–60 glands and retractile papillae (digitate processes), their positions being indicated dorsally by faint radiating ridges (Fig. 12.21) .... ***Actinobdella* [p. 250]**

2(1) Zero, one, or two pairs of eyes; a series of paired accessory eyes sometimes present along body (Fig. 12.22 C) 3

2' Three or four pairs of eyes 7

3(2) Eyes close together (separated by less the diameter of one eye, touching, or coalesced); proboscis pore (mouth) apical or subapical on rim of anterior sucker (Fig. 12.19 A) or just below rim of anterior sucker 4

3' Eyes distinctly separated or separated by at least the diameter of one eye; proboscis pore (mouth) within anterior sucker and clearly not on rim (Fig. 12.19 B) 6

4(3) Two pairs of eyespots; eyespots separated, close together or confluent, but not fused into a single spot 5

4' One pair of eyespots fused into fused forming a single spot; bacteriomes (mycetomes) two pairs of spheroidal masses connected to the esophagus through thin ducts .... ***Haementeria* [p. 250]**

[Mexico]

5(4) Two distinct pair of eyes, the first always on segment III, the second on IV; dorsum with longitudinal brown lines; seven pairs of crop caeca; papillae absent; length 20 mm .... *Batracobdelloides moogi* Nesemann & Csányi, 1995

[Canada: Nova Scotia]

5' Two pair of eyes (one pair large on segment III and one pair small on segment II) (Fig. 12.22 A) [*Placobdella hollensis* has several pairs of accessory eyes on the neck (Fig. 12.22 C)]; eyes close together or confluent (Fig. 12.22 B) (except *Placobdella montifera* and *Placobdella nuchalis* which have eyes well separated); male and female gonopores separated by two annuli (Fig. 12.20 B); body usually papillated; seven pairs of crop caeca; bifurcated ovisacs .... ***Placobdella* [p. 250]**

6(3) Male and female gonopores united in common bursal pore; one pair of eyes well separated; body smooth without papillae; pigmented with black chromatophores and dark, thin paramedial lines; six pairs of crop caeca; length 22 mm .... *Marvinmeyeria lucida* (Moore 1954)

6' One pair of eyes which are well separated (Fig. 12.25); gonopores separated by one annulus .... ***Helobdella* [p. 252]**

7(2) Three pairs of eyes (Fig. 12.26 B, C) (coalesced eyes sometimes occur but the lobed nature indicates the original condition); body firm 8

7' Four pairs of eyes on paramedian lines of segments II–V (Fig. 12.26 A); body very soft .... ***Theromyzon* [p. 253]**

8(7) Eyes equidistant in two paramedian rows (Fig. 12.26 B) .... ***Glossiphonia* [p. 253]**

8' First pair of eyes closer together than succeeding two pairs, i.e., eyes arranged in triangular pattern (Fig. 12.26 C); no papillae; male and female ducts open into a common gonopore; little pigmentation; generally amber colored; length 10 mm .... *Alboglossiphonia heteroclita* (Linnaeus, 1761)

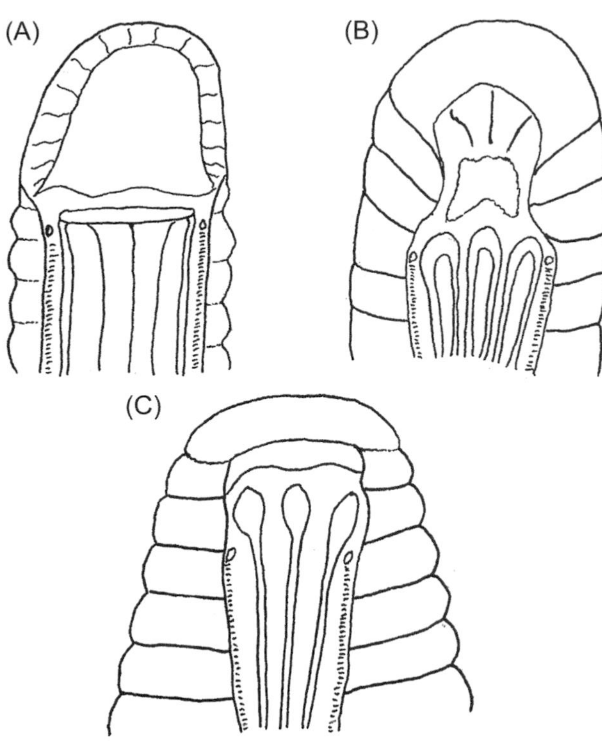

FIGURE 12.17 Ventral view of the dissection of the mouth and buccal cavity of: (A) *Haemopis grandis*; (B) *H. marmorata*; and (C) *Macrobdella decora* showing the velum, and the relative size of the jaws in *Macrobdella decora* and *H. marmorata* and the absence of jaws in *H. grandis*.

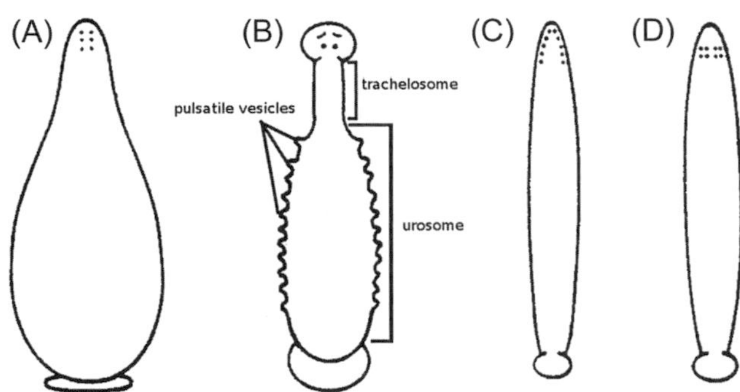

FIGURE 12.18 Dorsal view of the general body shape of a member of each of the families: (A) Glossiphoniidae; (B) Piscicolidae; (C) Macrobdellidae; and (D) Erpobdellidae.

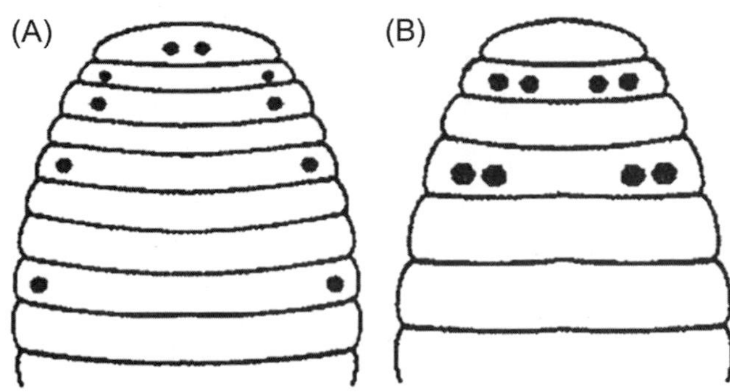

FIGURE 12.19 Arrangement of eyes in: (A) Haemopidae and Macrobdellidae; and (B) Erpobdellidae.

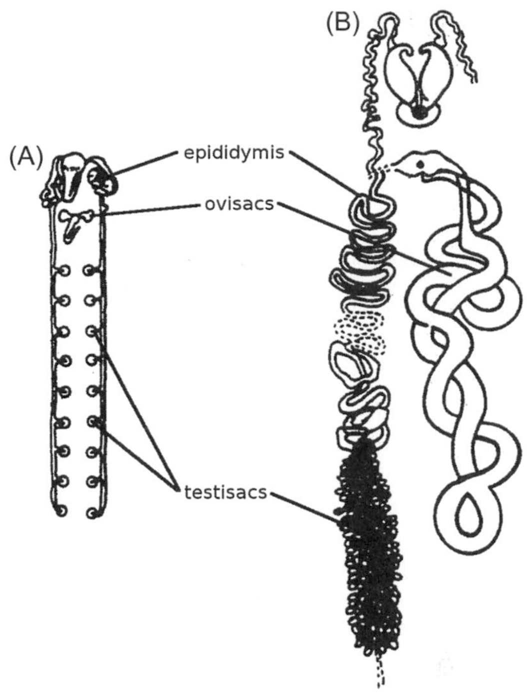

FIGURE 12.20 Reproductive systems of: (A) *Hirudo medicinalis* (Hirudinidae) and (B) *Erpobdella obscura* (Erpobdellidae).

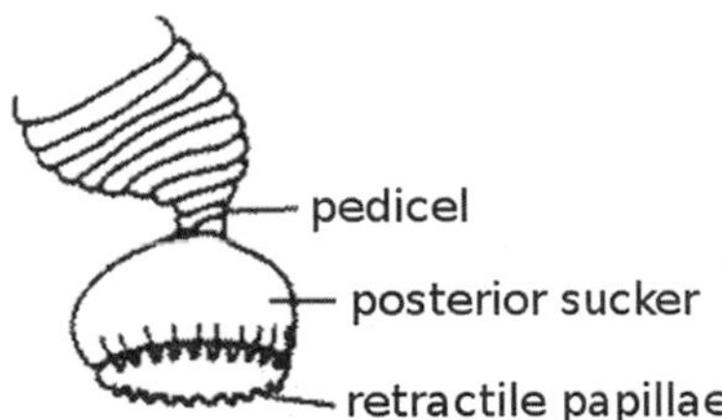

FIGURE 12.21 Lateral view of the posterior sucker of *Actinobdella inequiannulata* showing the pedicel and the retractile papillae around the margin.

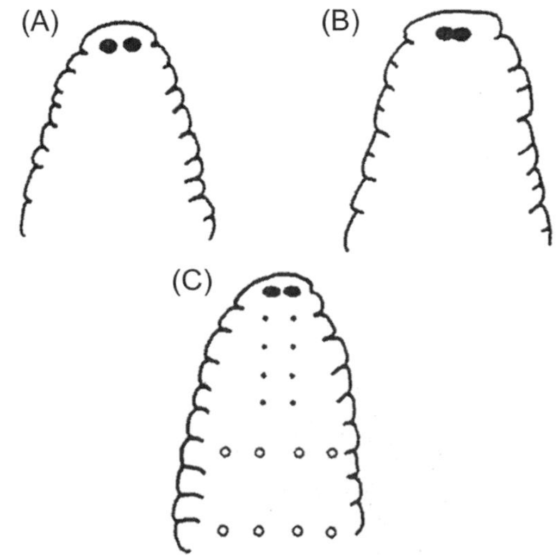

FIGURE 12.22 Dorsal view of the head of *Placobdella* with a single pair of eyes: (A) close together, (B) lobed, and (C) *Placobdella hollensis* with one pair of eyes followed by pairs of accessory eyes.

Hirudinida: Rhynchobdellida: Glossiphoniidae: *Actinobdella*: Species

1 Posterior sucker on short, distinct pedicel with 29–31 digitate processes on rim (Fig. 12.21); somites 3- or 6-annulate; dorsal papillae absent or in 1–5 longitudinal rows (some individuals with middorsal ridge); length 22 mm .... *Actinobdella inequiannulata* Moore 1901

1' Posterior sucker on short distinct pedicel with about 60 digitate processes on rim; somites 6-annulate with b3 and b5 the largest and most conspicuous; length 11 mm; dorsal papillae in five longitudinal rows .... *Actinobdella annectens* Moore 1906

[Canada: Ontario]

Hirudinida: Rhynchobdellida: Glossiphoniidae: *Haementeria*: Species

1 Dorsal surface 5-annulate (a1 and a3 subdivided) (Blood feeding) .... *Haementeria officinalis* de Fillipi, 1849

[Mexico]

1' Dorsal surface 3-annulate (a1 and a3 not subdivided) (Blood-feeding, found feeding on *Rhinella marina*, cane toad) .... *Haementeria lopezi* Oceguera-Figueroa, 2006

[Mexico]

Hirudinida: Rhynchobdellida: Glossiphoniidae: *Placobdella*: Species

1 One small pair of eyes on segment II and one pair of larger sometimes coalesced eyes; no supplementary eyes 2

1' One pair of small eyes on segment II and a larger sometimes coalesced pair of eyes on segment III followed by up to five pairs of accessory eyes (Fig. 12.26 C); three rows of dorsal papillae and three pairs of pre-anal papillae; length 30–55 mm .... *Placobdella hollensis* (Whitman, 1892)

2(1) Anus close to the posterior sucker; no pedicel 3

2' Anus between segments XXIII and XXIV with the 16 postanal annuli forming a slender stalk (pedicel) which bears the posterior sucker; no papillae; length 35 mm; typically found on its host *Aplodinotus grunniens* (freshwater drum) .... *Placobdella pediculata* Hemingway, 1908

3(2) Margins of posterior sucker denticulate (Fig. 12.23); head expanded and discoid and set off from body by a narrow neck (Fig. 12.24) 4

3' Margins of posterior sucker not denticulate; head not distinctly expanded or otherwise set off from body by a narrow neck, but more or less continuous with body 5

4(3) Dorsum with three prominent tuberculate keels or ridges (Fig. 12.24); length 16 mm .... *Placobdella montifera* Moore, 1906

4' Dorsum smooth; no keels or ridges; length 25 mm .... *Placobdella nuchalis* Sawyer & Shelley, 1976

5(3) Dorsum smooth or with numerous small sensillae 6

5' Dorsum moderately to roughly papillated 7

6(5) Dorsum brownish yellow and extremely variable pigment from simple (medial pigmented yellow line) to elaborate (medial and paramedial marbled patches), ventral surface with 8 to 12 blue, brown or green stripes; length 60 mm .... *Placobdella parasitica* (Say, 1824)

6' Dorsum olive green; 2-annulate; posterior sucker large; length 7 mm .... *Placobdella biannulata* (Moore, 1900)

7(5) Dorsum with conspicuous white genital and anal patches, one or more medial white patches and white band in neck region 8

7' Dorsum without white patches and with or without white band in neck region 12

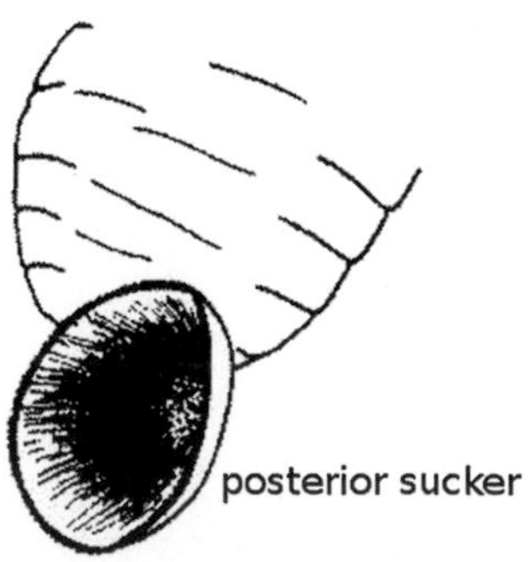

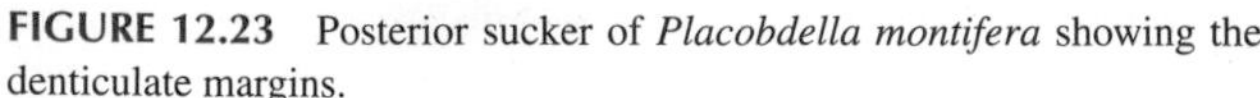
FIGURE 12.23 Posterior sucker of *Placobdella montifera* showing the denticulate margins.

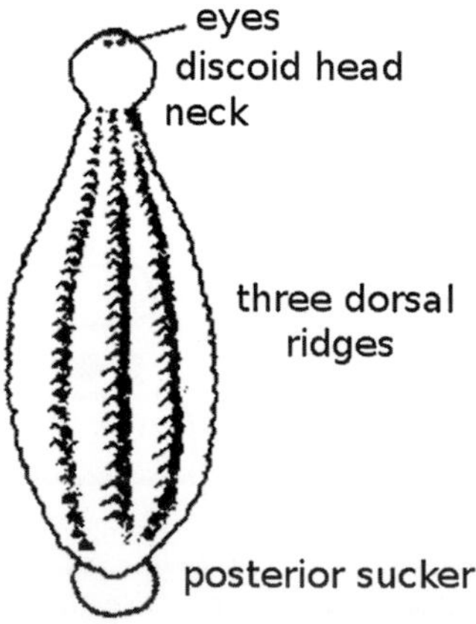

FIGURE 12.24 *Placobdella montifera* showing the expanded discoid head and dorsum with three prominent tuberculate ridges (keels).

8(7)	Doral papillae present	9
8'	Lanceolate body and smooth dorsum (without papillae); unpigmented nuchal band and anal patch with three unpigmented patched in between; length 10–20 mm	*Placobdella nuchalis* Sawyer & Shelley, 1976
9(8)	Dorsal papillae in 2–3 rows	10
9'	Dorsal papillae in 5–6 rows	11
10(9)	Dorsum reddish brown with/without dorsal medial pigment line and unpigmented nuchal band; unpigmented genital bar and anal patch with/without 2–3 unpigmented patches in between; 3 rows of papillae (paralateral papillae white tipped to genital bar and black tipped to nuchal band; medial papillae are black tipped); five pair of pre-anal papillae; length 16–26 mm [USA: New England]	*Placobdella ornata* (Verrill, 1872)
10'	Dorsum rust to reddish brown with 2 lateral rows of unpigmented papillae, two unpigmented nuchal bands and unpigmented genital bar and anal patch; four pair of pre-anal papillae; length 13–17 mm	*Placobdella cryptobranchii* (Johnson & Klemm, 1977)
11	Ovate and dorsoventrally flattened body with 5 rows of white tipped papillae surrounded by yellowish dots on dorsum; unpigmented nuchal band; unpigmented genital bar and anal patch with unpigmented patched in between; length 10 mm	*Placobdella michiganensis* (Sawyer, 1972)
11'	Dorsum chocolate to russet brown with 6 rows of papillae; many thin unpigmented vertical lines; unpigmented nuchal band, small genital unpigmented bar and unpigmented genital patch with few scattered unpigmented patches in between; no pre-anal papillae; length 10 mm.	*Placobdella appalachiensis* Moser & Hopkins, 2014
12(7)	Body with ventral stripes	13
12'	Without ventral stripes	17
13(12)	Dorsum with 7 longitudinal rows of small white tipped papillae	14
13'	Medial and paramedial series of tubercles on all midbody annuli; tubercles bear a full cap of five to six sensory papillae; two pair of prominent paramedial pre-anal papillae followed by two rows of three papillae [USA: New York, New England]	*Placobdella ali* Hughes & Siddall, 2007

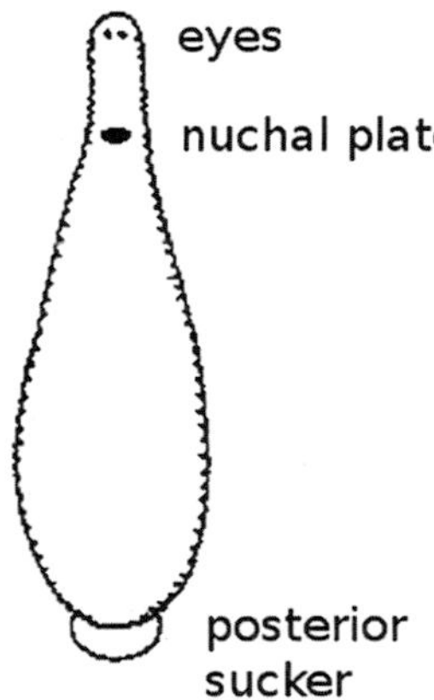

FIGURE 12.25 *Helobdella stagnalis* (or *H. bowermani, H. californica, H. modesta*) showing the chitinous scute (nuchal plate) on the dorsal surface and the single pair of eyes.

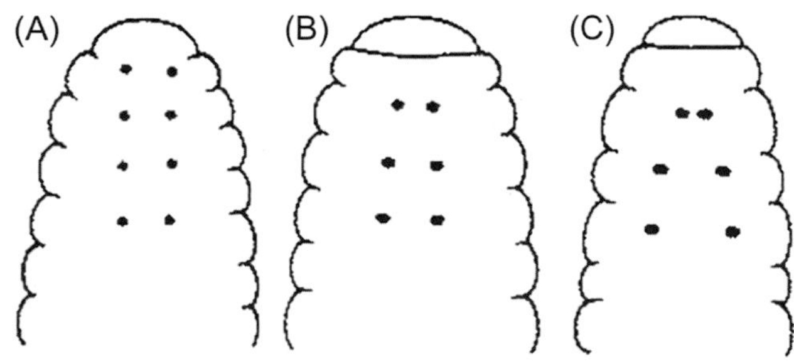

FIGURE 12.26 Dorsal view of: (A) *Theromyzon trizonare*; (B) *Glossiphonia elegans*; and (C) *Alboglossiphonia heteroclita* showing the arrangement of the eyes.

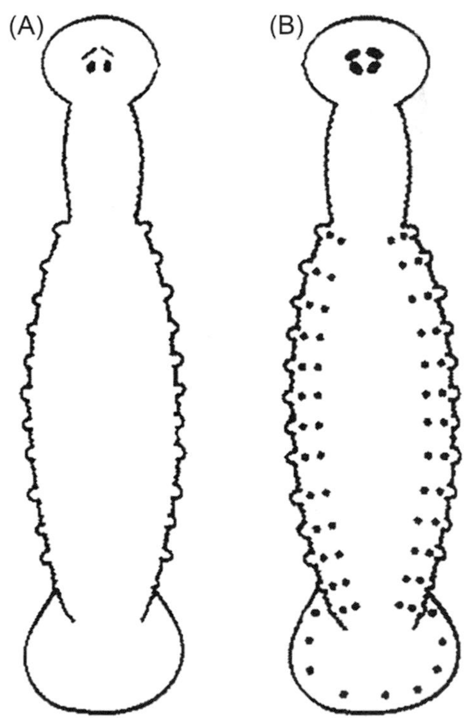

FIGURE 12.27 (A) *Cystobranchus verrilli* and (B) *Cystobranchus meyeri* showing the pulsatile vesicles on the urosome and the presence on *C. meyeri* of oculiform spots on the posterior sucker and the paired lateral oculiform spots on the urosome.

14(13) Dorsum smooth with few papillae, or numerous small papillae making a rough surface 15

14' Dorsum with 7 longitudinal rows of small white tipped papillae; interrupted dorsomedial line; alternating lateral rows of light and dark pigmentation; three pair of pre-anal papillae; length 45 mm .... *Placobdella papillifera* (Verrill, 1872)

15(14) Dorsum with few or numerous small papillae arranged in rows 16

15' Dorsum smooth, few papillae or tubercles .... *Placobdella mexicana* (Moore, 1898)

16(15) Dorsum with few or numerous small papillae arranged in five distinct longitudinal rows; length 50 mm .... *Placobdella multilineata* Moore, 1953

16' Dorsum with three inconspicuous longitudinal rows of papillae and small papillae irregularly dispersed in the space not occupied by rows of papillae .... *Placobdella lamothei* Oceguera-Figueroa & Siddall, 2008

17(12) Dorsum with dorsal medial line, typically interrupted by papillae 18

17' Body near transparent green with 6 rows of dorsal papillae, no nuchal band or pigment patches .... *Placobdella sophieae* Oceguera-Figueroa, Kvist, Watson, Sankar, Overstreet & Siddall, 2010

[USA: Oregon, Washington]

18(17) Dorsal papillae in 6–7 rows 19

18' Dorsal papillae in 5 rows 20

19(18) Dorsum brownish-green, variegated with orange with thin dark dorsal medial line and 6 to 7 rows of white tipped papillae; nuchal band; length 13–25 mm .... *Placobdella picta* (Verrill, 1872)

19' Dorsum brownish with seven rows of papillae (papillar rows formed by a2 papillae); interrupted dorsal medial pigment line; one pair of salivary glands; length 21 mm .... *Placobdella kwetlumye* Oceguera-Figueroa, Kvist, Watson, Sankar, Overstreet & Siddall, 2010

[USA: Washington]

20(18) Dorsum brownish with yellowish/cream and heavily papillated with 5 rows of papillae; interrupted dorsal medial pigment line; two pair of prominent paramedial pre-anal papillae followed by two rows of three papillae; length 30–50 mm .... *Placobdella rugosa* (Verrill, 1874)

20' Five longitudinal rows of papillae; up to 19 papillae on all mid-body annuli; thin medial stripe interrupted only at a2; ventral surface unpigmented or lightly pigmented without well-defined stripes .... *Placobdella burresonae* Siddall & Bowerman, 2006

[USA: Oregon]

Hirudinida: Rhynchobdellida: Glossiphoniidae: *Helobdella*: Species

1 Segment VIII dorsal surface with a brown, horny, chitinous scute (nuchal plate) (Fig. 12.25) 2

1' Without a nuchal plate 6

2(1) Dorsum with middorsal row of papillae or with stripes 3

2' No longitudinal stripes or papillae on dorsum 5

3(2) Body lacking lateral projections 4

3' Body with lateral projections on a2 and a3, particularly conspicuous at the posterior half; testisacs 6 pairs; postcaeca (diverticula) absent .... *Helobdella atli* Oceguera-Figueroa & León-Règagnon, 2005

[Mexico]

4(3) Dorsum with mid dorsal row of papillae (a1 small papillus, a2-a3 large papillae) and scattered black chromatophores; length 5–10 mm .... *Helobdella bowermani* Moser, Fend, Richardson, Hammond, Lazo-Wasem, Govedich, & Gullo, 2013

[USA: Oregon]

4' Pigment dorsum with pair of longitudinal stripes; diverticulated crop caeca; posterior sucker pigmented on dorsum; length 18 mm .... *Helobdella californica* Kutschera, 1988

5(2) Testisacs 4 or 5 pairs .... *Helobdella octatestisaca* Lai & Chang, 2009

[Mexico]

5' Testisacs 6 pairs .... *Helobdella* cf. *stagnalis* (Linnaeus, 1758)

[Nearctic contain a complex of undescribed cryptic species]

6(1) Dorsal surface smooth 7

6' Dorsal surface with 3–7 longitudinal series of papillae, or with scattered papillae 10

7(6) Body pigmented, with or without longitudinal or transverse bands; body flat with posterior wider than tapering anterior; six pairs of crop caeca 8

7' Body unpigmented and translucent; body rounded and subcylindrical; lateral margins almost parallel; posterior sucker small and terminal; one pair of crop caeca; length 25 mm .... *Helobdella elongata* (Castle, 1900)

8(7) Dorsum without transverse pigmentation, bearing longitudinal stripes 9

8' Dorsum with transverse brown interrupted stripes alternating with irregular white banks; length 10 mm .... *Helobdella transversa* Sawyer, 1972

9(8) Dorsum without transverse pigmentation; dorsum with six prominent longitudinal white stripes alternating with brown stripes or uniformly brown with unpigmented patches; length 14 mm .... *Helobdella fusca* (Castle, 1900)

9' Dorsum with approximately 12 longitudinal brown stripes and nearly smooth with few papillae .... *Helobdella lineata* (Verrill, 1874)

10(6) Dorsal surface with three or fewer incomplete series of small papillae or with scattered papillae; length 25–30 mm 11

10' Dorsal surface with 3–9 longitudinal rows of papillae irrespective of size and pigmentation; length 14 mm .... *Helobdella papillata* Moore, 1952

11(10) Ratio of body width to body length 0.26 ± 0.02 SD 12

11' Ratio of body width to body length 0.37 ± 0.01 SD 13

12(11) Dorsal surface with longitudinal stripes many of which are interrupted by circular zones of unpigmented skin. Five pairs of crop caeca which have numerous secondary diverticula with a bumpy outline; with cephalic transverse banding .... *Helobdella triserialis* (Blanchard, 1849)

[Neotropical. Possible invasive in California]

12' Dorsal surface with complex pattern of longitudinal stripes and metameric papillae, longitudinal stripes many of which are interrupted by circular zones of unpigmented skin. Five pairs of simplg crop caeca .... *Helobdella socimulcensis* (Caballero, 1931)

[Mexico]

13(11) Dorsal surface with five narrow longitudinal stripes, which extend unbroken along the body; white pigment spots may occur on every annulus and are concentrated in the regions between the stripes; brown chromatophores are grouped on the dorsal surface into a reliable pattern of narrow longitudinal stripes which extend unbroken throughout most of the body's length; five pairs of crop caeca which are bilobed and smooth in outline .... *Helobdella robusta* Shankland, Bissen, & Weisblat, 1992

13' Dorsum with longitudinal lines and irregularly arranged black and white papillae, white pigment spots on just the central annulus; length 13–17 mm .... *Helobdella austinensis* Kutschera, Langguth, Weisblat, & Shankland, 2013

Hirudinida: Rhynchobdellida: Glossiphoniidae: *Theromyzon*: Species

1 Gonopores separated by three or four annuli 2

1' Gonopores separated by two annuli 3

2(1) Gonopores separated by three annuli, cocoons attached to ventral body wall, two female pores .... *Theromyzon trizonare* (Davies & Wilkialis, 1992)

2' Gonopores separated by four annuli, cocoons attached directly to substrate, single female pore .... *Theromyzon tessulatum* (Muller, 1776)

3(1) One female pore 4

3' Two female pores .... *Theromyzon maculosum* (Rathke, 1862)

4(3) Female atrium cylindrical .... *Theromyzon bifarium* Oosthuizen & Davies, 1993

4' Female atrium spherical .... *Theromyzon rude* (Baird, 1869)

Hirudinida: Rhynchobdellida: Glossiphoniidae: *Glossiphonia*: Species

1 Dorsum with papillae on annulus a2 in six longitudinal rows; pair of paramedial stripes on dorsum and ventrum; seven pairs of crop caeca; gonopores separated by two annuli (Fig. 12.20 B); length 25 mm .... *Glossiphonia elegans* (Verrill, 1872)

1' Dorsum with large, distinct papillae on annuli a2 and a3; dorsum with numerous, irregularly shaped whitish spots; seven pairs of crop caeca; length 25 mm .... *Glossiphonia verrucata* (Müller, 1844)

[Canada: British Columbia. USA: Alaska]

Hirudinida: Rhynchobdellida: Piscicolidae: Genera

1 Posterior sucker flattened, as wide or wider than the widest part of body (Fig. 12.18 B); Pulsatile vesicles on lateral margins of neural annuli of urosome (Fig. 12.27 A, B); zero or two pairs of eyes 2

1' Posterior sucker concave, weakly developed, and narrower than widest part of the body; body may be divided into small trachelosome and larger urosome (Fig. 12.28 B); no pulsatile vesicles (Fig. 12.28); zero or one pair of eyes .... ***Myzobdella* [p. 254]**

2(1) Body divided into anterior trachelosome and posterior urosome; 11 pairs of small pulsatile vesicles not very apparent in preserved specimens; each pulsatile vesicle covers two annuli; body cylindrical or sometimes slightly flattened; two pairs of eyes; both anterior and posterior suckers wider than body; oculiform spots present on posterior sucker (Fig. 12.29 C–D) (except *Piscicola punctata*) (Fig. 12.29 A): 14-annulate (except *Piscicola punctata* which is 3-annulate) .... ***Piscicola* [p. 254]**

2' Body divided into distinct anterior trachelosome and posterior urosome with distinct shoulders at junction; 11 pairs of large and distinct pulsatile vesicles easily seen in preserved and live specimens (Figs. 12.18 B and 12.27); each pulsatile vesicle covers four annuli; well developed anterior and posterior suckers; no papillae; zero or two pairs of eyes (Figs. 12.27 and 12.29 B); oculiform spots on posterior sucker present or absent (Fig. 12.37); 7-annulate; length 80 mm .... ***Cystobranchus*** **[p. 254]**

Hirudinida: Rhynchobdellida: Piscicolidae: *Myzobdella*: Species

1 Uniformly brown, lacking stripes; twelve or fourteen annuli per segment length ≥9 mm .... *Myzobdella lugubris* Leidy, 1851

[Common and widely distributed; may represent a species complex]

1' Longitudinal stripes and 3 annuli per segment, length 6–8 mm .... *Myzobdella reducta* Meyer, 1940

[Common and widely distributed throughout eastern North America]

Hirudinida: Rhynchobdellida: Piscicolidae: *Piscicola*: Species

1 Posterior sucker with 8–14 oculiform spots (Fig. 12.29 B–D) 2

1' Posterior sucker without oculiform spots (Fig. 12.29 A); 3-annulate (Fig. 12.2 A); two pairs of crescent-shaped eyes: gonopores separated by three or four annuli; length 16 mm .... *Piscicola punctata* (Verrill, 1872)

[Widely distributed; most common in northern U.S. and Canada]

2(1) With 10–12 (usually 10) punctiform oculiform spots on posterior sucker; dark rays absent from posterior sucker (Fig. 12.29 C): gonopores separated by two annuli (Fig. 12.20 A); anterior pair of eyes like heavy dashes twice as long as wide; length 24 mm .... *Piscicola milneri* (Verrill, 1874)

2' With 12–14 punctiform oculiform spots on posterior sucker, separated by an equal number of dark pigmented rays (Fig. 12.29 D); gonopores separated by three annuli; anterior pair of eyes like fine dashes five times as long as wide; length 30 mm .... *Piscicola geometra* (Linnaeus, 1758)

Hirudinida: Rhynchobdellida: Piscicolidae: *Cystobranchus*: Species

1 With oculiform spots on posterior sucker 2

1' Without oculiform spots on posterior sucker 4

2(1) Oculiform spots on the posterior sucker not crescent-shaped 3

2' Eight to ten oculiform spots on the posterior sucker crescent-shaped (Fig. 12.29 B); gonopores separated by two annuli; 14-annulate .... *Cystobranchus salmositicus* (Meyer, 1946)

[USA: western states]

3(2) Eight oculiform spots on posterior sucker; two rows of 12 lateral oculiform spots on each side of body (Fig. 12.27 B); two pairs of eyes; length 7 mm .... *Cystobranchus meyeri* Hayunga & Grey, 1976

[Canada: Quebec. USA: Maryland, New York]

3' Ten oculiform spots on posterior sucker; no lateral ocelli; two pairs of eyes; length 15 mm .... *Cystobranchus virginicus* Hoffman, 1964

[Canada: Quebec. USA: Virginia, West Virginia. Associated with fish nests]

4(1) Eyes present 5

4' Eyes absent; caudal sucker orbicular; length 30 mm .... *Cystobranchus mammillatus* (Malm, 1863)

[Canada: Northwest Territories. Palaearctic]

5(4) Two pairs of eyes; the first pair forming conspicuous dashes at 45° to the longitudinal axis; the second part of eyes ovoid (Fig. 12.27 A); gonopores separated by two annuli (Fig. 12.20 B); Caudal sucker orbicular; length 30 mm .... *Cystobranchus verrilli* Meyer, 1940

[Canada: midwestern. USA: Arkansas, New York, Virginia, West Virginia, midwestern states]

5' Two pairs of eyes on the oral sucker; black stellate chromatophores scattered throughout the body, except on unpigmented longitudinal medial band; 13 pair of testisacs; caudal sucker shallowly to broadly obdeltoid (Fig. 12.30) .... length 5.5 to 12 mm .... *Cystobranchus klemmi* (Williams & Burreson, 2005)

[USA: Arkansas, Illinois, Missouri, Oklahoma]

Hirudinida: Arhynchobdellida: Families

1 Five pairs of eyes arranged in an ocular arch; body elongate (Fig. 12.18 C); jaws with denticles (teeth) either present or absent; nine or ten pairs of testisacs arranged metamerically (Fig. 12.20 A); pharynx short 2

1' Zero, three, or four pairs of eyes in separate labial and buccal groups (Fig. 12.19 B); body elongate (Fig. 12.18 D); no jaws (agnathus) 5

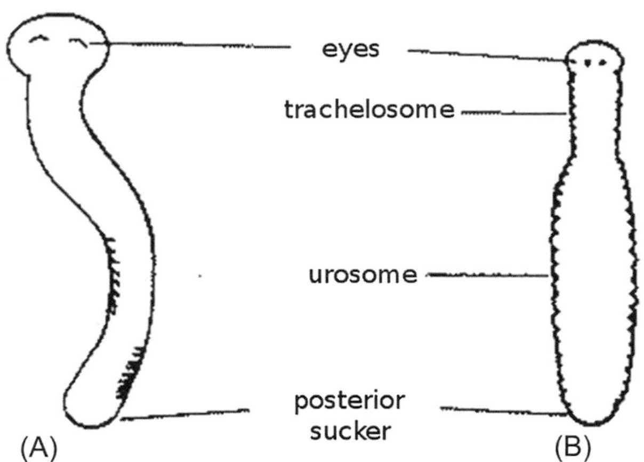

FIGURE 12.28 (A) *Myzobdella lugubris* and (B) *Myzobdella reducta* showing the trachelosome and urosome without pulsatile vesicles, the relative sizes of the eyes, and the weakly developed posterior sucker.

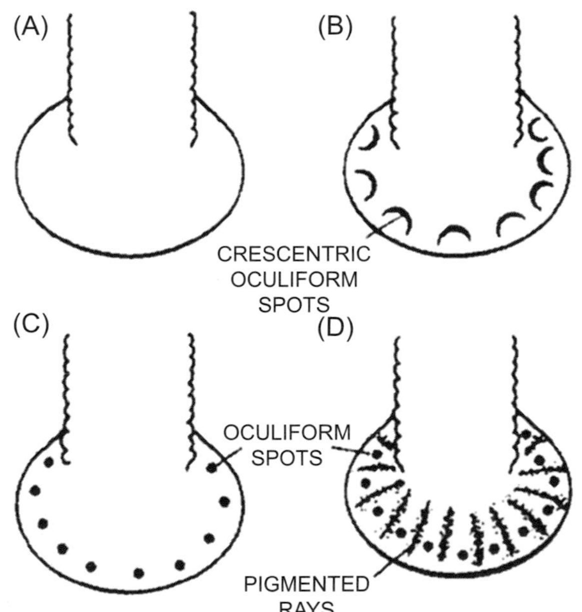

FIGURE 12.29 Posterior suckers of: (A) *Piscicola punctata* without oculiform spots; (B) *Cystobranchus salmositica* with 8–10 crescentiform oculiform spots; (C) *P. milnera* with 10–12 punctiform oculiform spots on posterior sucker; and (D) *P. geometra* with 12–14 oculiform spots separated by pigmented rays.

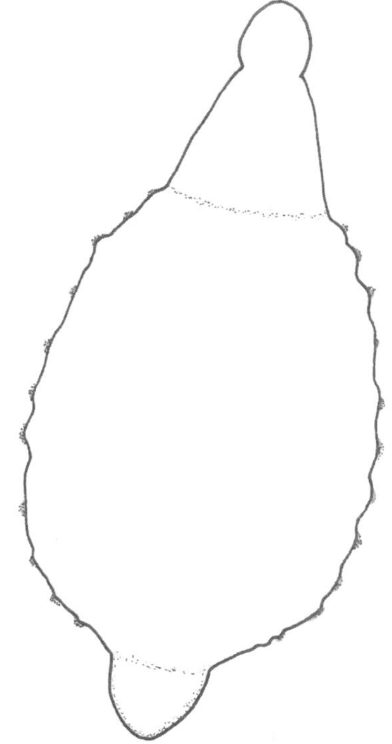

FIGURE 12.30 Body shape of *Cystobranchus klemmi. Drawn from Richardson et al., 2013.*

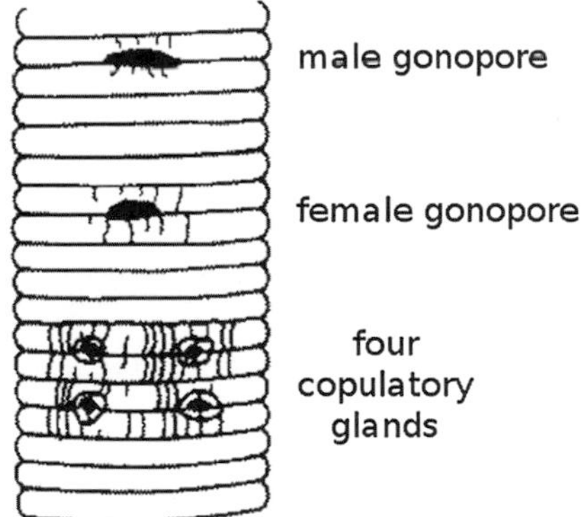

FIGURE 12.31 Ventral view of the male and female gonopores of *Macrobdella decora.*

2(1) Ventral paired nephridia including 17th pair, five pairs of eyespots in an ocular arch with 1–2 annuli between fourth and fifth pairs 3

2' Ventral paired nephridia, but lacking the 17th pair, instead have medioventral common pore at the base of the caudal sucker; five pairs of eyespots arranged in "haemadipsine" ocular arch (*sensu* Blanchard, 1917) with the fourth and fifth pair of eyespots separated by two annuli; 8–12 annuli per somite .... Xerobdellidae, one species: *Diestecostoma mexicana* (Baird, 1869)

3(2) One or two crop caeca per somite 4

3' Crop acecate except for postcaeca; five pairs of eyespots arranged in an arch on segments II–VI with the third and fourth pairs of eyespots separated by one annulus (Fig. 12.19 A),; gonopores separated by 5–7 annuli; jaws absent or present (distichodont, 9–25 pairs of teeth per jaw) (Figs. 12.17 A, B and 12.33 B); dorsum uniform color with or without black longitudinal middorsal stripe; uniformly dark olive green with faint longitudinal stripes along midline and numerous small irregular scattered black flecks; or without dorsal stripes but with few moderately to extensively blotched, spotted, or mottled with olive, yellow, dark gray or black .... Haemopidae, one genus: ***Haemopis* [p. 256]**

4(3) Ventral surface with glandular area and external copulatory gland depressions or pores, in some species arranged in rows, either surrounding and directly posterior to female gonopore or ten or eleven annuli posterior to the male gonopore (Figs. 12.31–12.32); lacking penis

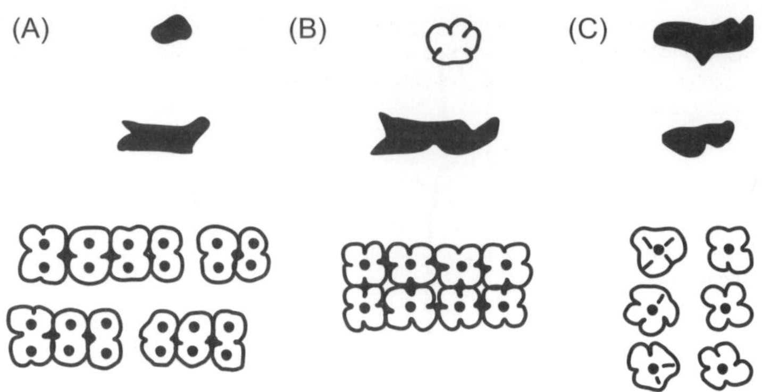

FIGURE 12.32 Diagrammatic representation of the relative slopes and positions of the male and female gonopores and copulatory glands of: (A) *Macrobdella sestertia*; (B) *M. ditetra*; and (C) *M. diplotertia*.

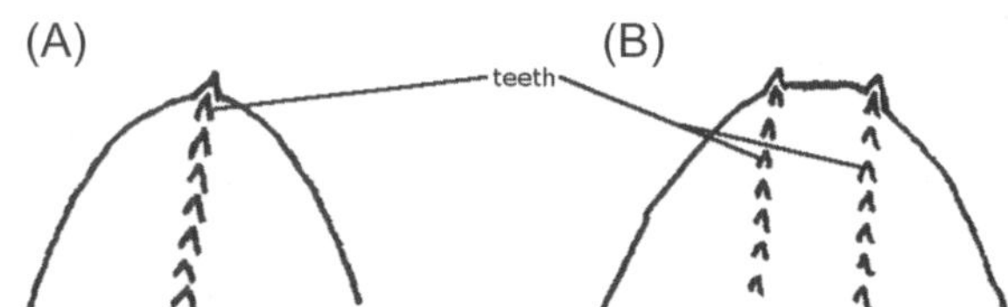

FIGURE 12.33 Surface view of the jaws of Hirudiniformes showing the teeth arranged in: (A) one (monostichodont) or (B) two (distichodont) rows.

sheath; gonopores separated by 2–5½ annuli (Figs. 12.31–12.32); monostichodont except for single partially distichodont species, 20–65 teeth per jaw (Fig. 12.33 A); crop with 2 pairs of caeca per somite, equal in size; vagina fusiform, vaginal duct short; length 40–150 mm .... **Macrobdellidae [p. 257]**

4' Ventral surface without glandular area or external copulatory glands, reduced number of teeth per jaw (usually less than 12, but always less than 35); caudal sucker wider than maximum body width, blood feeding with preference for feeding from mucous membranes of mammals .... Praobdellidae, one species: *Limnobdella mexicana* Blanchard, 1893

[Species complex in need of revision]

5(1) Ventral surface with a pair of accessory pits (gastropores?), one anterior and one posterior of the gonopores, respectively (predaceous) .... Salifidae, one species: *Barbronia weberi* Blanchard, 1897

[USA, Mexico: Invasive. Palaearctic]

5' Ventral surface without accessory pits (gastropores?) .... **Erpobdellidae [p. 258]**

Hirudinida: Arhynchobdellida: Haemopidae: *Haemopis*: Species

1 Jaws absent (Fig. 12.17 A) 2

1' Jaws present, 9–25 pairs of teeth per jaw (distichodont) 3

2(1) Lower surface of velum smooth; 5 annuli between gonopores; gonopores in the furrows between annuli; pharynx with 12 internal ridges; epididymis massive and extending well beyond posterior end of penis sheath; color variable, usually shades of green or gray, sometimes with black blotches; no dorsal stripes; length 150–300 mm .... *Haemopis grandis* (Verrill, 1874)

2' Lower surface of velum closely and finely papillated; 5–5½ annuli between gonopores; gonopores in middle of annuli; pharynx with 15 internal ridges; epididymis not extending beyond sperm sac; dorsum uniform gray with few or no black blotches, with red or yellow orange bands along lateral margins; length 140–200 mm .... *Haemopis plumbeus* Moore, 1912

3(1) Gonopores separated by 6½–7 annuli; female gonopore large, conical and nipple-like in adults, flattened in immatures 4

3' Gonopores separated by 5–5½ annuli 5

4(3) Approximately 16 white tipped papillae on a2 of midbody somites (absent in median field), distributed bilaterally and wrapping around periphery of dorsum and venter; jaws with approximately 10 pairs of teeth each; pharynx with 15 internal ridges; gonopores separated by 7 annuli; male reproductive system notably large with long tubing, muscular penis sheath terminating in bulbous prostate; epididymis relatively large, more than twice the size of the sperm sac; female reproductive system consisting of relatively uncoiled tubing; lacking a distinct vagina; caudal sucker narrower that width of body at maximum; dorsum medium to dark brown with variable black longitudinal mid dorsal stripe, moderate to extensive black mottling; length 120–300 mm, body firm and muscular; amphibious to terrestrial .... *Haemopis ottorum* Wirchansky & Shain, 2010

4' Body smooth, without pronounced papillae; jaws with 15 pairs of teeth each; gonopores separated by 6½–7 annuli; dorsum uniformly dark olive-green with faint dark longitudinal middorsal stripe and numerous small irregularly scattered black flecks that continue onto dorsal side of caudal sucker. Venter lighter olive green without black flecks, occasionally with yellow lateral margins; length 200 mm, firm and muscular; amphibious to terrestrial .... *Haemopis septagon* Sawyer & Shelley, 1976

5(3) Dorsum with black longitudinal middorsal stripe 6

5' Dorsum with irregular, black blotches, without longitudinal stripes 7

6(5) Jaws with 9–14 pairs of teeth each; caudal sucker as wide as body width; dorsum brown green to olive green with black longitudinal mid-dorsal stripe, scattered black and yellow orange blotches (usually more black than yellow orange), and sometimes paired dark longitudinal lateral stripes; margins conspicuously mottled with yellow orange forming broken longitudinal stripes; venter uniform, dark gray, occasional yellow orange blotches .... *Haemopis kingi* Mathers, 1954

6' Jaws with 20–25 pairs of teeth each; caudal sucker width narrower than maximum body width; dorsum uniformly black or dark gray with black longitudinal middorsal stripe; length 250 mm, body firm and muscular; amphibious to terrestrial .... *Haemopis terrestris* Forbes, 1890

7(5) Caudal sucker small ½ or less than of the width of the body 8

7' Jaws with 10–12 pairs of teeth each; caudal sucker discoid, approximately ¾ as wide as body width, broadly attached by a very short pedicel at XXVII; gonopores separated by 5 annuli; epididymis large, extend well beyond posterior end of sperm sac; dorsum olive green, irregular moderate to extensive black mottling and few scattered yellow blotches, yellow marginal stripes; venter uniform gray with few indistinct black or yellow blotches. Length 50–85 mm, body soft and limp .... *Haemopis lateromaculata* Mathers, 1963

8(7) Jaws with 12–13 short blunt teeth; caudal sucker less than ½ of the width of the body; gonopores separated by 5 annuli; body background color is almost black with pale yellowish-brown patches with a pale venter. Length up to 180 mm in life .... *Haemopis caballeroi* (Richardson 1971)

8' Jaws with 12–16 pairs of teeth each; caudal sucker approximately half as wide as body width; dorsum with variable color, usually olive green, yellow gray with irregular moderate to extensive black mottling dorsally and ventrally, or uniform dark gray with few irregular black blotches; amphibious to terrestrial .... *Haemopis marmorata* (Say, 1824)

Hirudinida: Arhynchobdellida: Macrobdellidae: Genera

1 External copulatory gland pores on ventral surface arranged in rows, ten or eleven annuli posterior to the male gonopore (Figs. 12.31–12.32); gonopores separated by 2–5 ½ annuli (Figs. 12.31 and 12.32); monostichodont, 46–65 teeth per jaw (Fig. 12.33 A); crop with 2 pairs of caeca per somite, equal in size; vagina fusiform, vaginal duct short; length 50–50 mm .... ***Macrobdella* [p. 257]**

1' External copulatory gland depressions surrounding and posterior to female gonopore, male gonopore formed by a deep copulatory pit; monostichodont or partially distichodont, 20–48 teeth per jaw (Fig. 12.33 B); gonopores separated by 3–4 annuli; crop with two pairs of caeca per somite, anterior caeca smaller than posterior; fused male bursa; small vaginal caecum, no common oviduct; dorsum with yellow or brown longitudinal middorsal stripe; length 40–85 mm .... ***Philobdella* [p. 257]**

Hirudinida: Arhynchobdellida: Macrobdellidae: *Macrobdella*: Species

1 Two or 2½ annuli between gonopores; 24 or 8 copulatory gland pores on ventral surface (Figs. 12.31–12.32) 2

1' Four and a half to 5½ annuli between gonopores; 4 or 6 copulatory gland pores on ventral surface (Figs. 12.31–12.32) 3

2(1) Twentyfour copulatory gland pores (4 rows with 6 gland pores each, in 2 groups) on raised pads on ventral surface (Fig. 12.32); 2–2½ annuli between gonopores; 39–46 teeth per jaw; male gonopore on annulus, female gonopore in furrow between gonopores; length 74–116 mm .... *Macrobdella sestertia* Whitman, 1886

2' Eight copulatory gland pores (2 rows of 4) on ventral surface (Fig. 12.32 B); 2 annuli between gonopores; 46–55 teeth per jaw; dorsum dark olive green without red or orange spots, brown or gray, usually with lateral irregular black spots; venter yellow or rusty orange with some or without black blotches, lateral margins bright yellow or orange bordered by black stripes; length 110–150 mm .... *Macrobdella ditetra* Moore, 1953

3(1) Four copulatory gland pores (2 rows of 2) on ventral surface (Fig. 12.31); 5–5½ annuli between gonopores. 50–65 teeth per jaw; dorsum green with approximately 20 red or orange dots down midline, black spots laterally; venter red or orange with some black spots; length 110–150 mm .... *Macrobdella decora* (Say, 1824)

3' Six copulatory gland pores (3 transverse rows of 2 each) on ventral surface (Fig. 12.32 C); 4½–5 annuli between gonopores; approximately 57 teeth per jaw; dorsum olive green with median row of red or orange spots, row of approximately 19 black spots in dorsolateral fields; venter lighter yellow or gray, with few irregular black blotches. Length 110–150 mm .... *Macrobdella diplotertia* Meyer, 1975

Hirudinida: Arhynchobdellida: Macrobdellidae: *Philobdella*: Species

1 Monostichodont, 20–26 teeth per jaw; dorsum with dark brown longitudinal mid dorsal stripe (if present) and two faint reddish-brown lateral bands; lateral margins with irregular black stripes, sometime broken but no discrete spots .... *Philobdella floridana* (Verrill, 1874)

1' Partially distichodont towards proximal end of jaw, 35–48 teeth per jaw; dorsum with light yellow longitudinal middorsal stripe and brownish black irregular flecks or spots in dorsolateral field; venter light yellow with some irregular black flecks or spots near lateral margins .... *Philobdella gracilis* Moore, 1901

Hirudinida: Arhynchobdellida: Erpobdellidae: Genera

1 Somites 5-annulate (b1, b2, a2, b5, b6) (Fig. 12.3 A) with all annuli equal in length; three pairs of eyes; gonopores separated by two annuli (Fig. 12.20 B); length 100 mm 2

1' Somites 6- or 7-annulate (Fig. 12.3 B); annuli of unequal length units will be either subdivided or longer than the others; in any group of six consecutive annuli, at least one annulus narrower or wider than the others; three or four pairs of eyes .... **Erpobdella** [in part] **[p. 258]**

2(1) Eyes all similar in size .... **Erpobdella** [in part] **[p. 258]**

2' Eyes differ in size with the second and third pairs smaller than the anterior pair; each annulus raised on dorsum with 10–18 small white-tipped papillae; anus located at base of posterior sucker; one or two pairs of crop caeca gonopores in furrows separate by two annuli .... **Motobdella [p. 259]**

Hirudinida: Arhynchobdellida: Erpobdellidae: *Erpobdella*: Species

1 Somites 5-annulate (b1, b2, a2, b5, b6) (Fig. 12.3 A) with all annuli equal in length; three pairs of eyes; gonopores separated by two to five annuli (Fig. 12.20 B); length 100 mm 2

1' Somites 6- or 7-annulate (Fig. 12.3 B); annuli of unequal length units will be either subdivided or longer than the others; in any group of six consecutive annuli, at least one annulus narrower or wider than the others; three or four pairs of eyes 5

2(1) Annuli not raised on dorsum and without papillae; anus located three or four segments anterior to posterior sucker; mouth small; gonopores in furrows separated by two to three and a half annuli 3

2' Five annuli between gonopores; and preatrial loops of the male paired ducts extend to XI .... *Erpobdella lahontana* Hovingh & Klemm, 2000

[USA: western]

3(2) Gonopores in furrows separated by 2 annuli; preatrial loops of the male paired ducts extend to XI 4

3' Three or three and a half annuli between gonopores; mouth circular, small; no preatrial loops of the male paired ducts .... *Erpobdella ochoterenai* Caballero, 1932

[Mexico: Central Plateau]

4(3) Preatrial loops of the male paired ducts simple, not coiled .... *Erpobdella punctata* (Leidy, 1870)

4' Atrial cornua spirally coiled like the horn of a ram .... *Erpobdella mexicana* (Dugès, 1872)

[Mexico: Central Plateau]

5(1) Four pairs of eyes, two labial and two buccal, of similar size 6

5' Zero or three pairs of eyes 8

6(5) Three or more annuli between gonopores; atrial cornua simply curved (Fig. 12.29 B) 7

6' Two annuli between gonopores (Fig. 12.20 A); anus one segment anterior to poster sucker; atrial cornua spirally coiled like the horn of a ram (Fig. 12.34 A); length 100 mm .... *Erpobdella obscura* (Verrill, 1872)

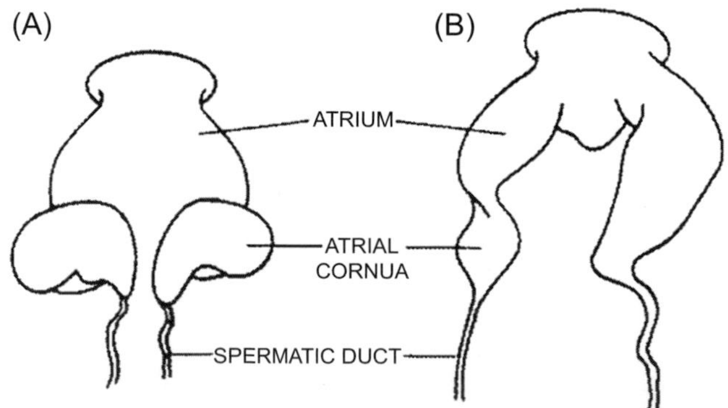

FIGURE 12.34 (A) Spirally curved atrial cornua of *Erpobdella obscura* and (B) the simply curved atrial cornua of most *Erpobdella*.

7(6) Three and a half to four annuli between gonopores; body heavily blotched with a median stripe; anus large, opening on a conical tubercle; length 60 mm .... *Erpobdella dubia* (Moore & Meyer, 1951)

7' Three to three and a half annuli between gonopores; nearly pigmentless or with a few dark spots; anus small, not on tubercle; length 30 mm .... *Erpobdella parva* (Moore, 1912)

8(5) Dorsum lacking scattered black pigment 9

8' Dorsum with scattered black pigment; two annuli between gonopores (Fig. 12.5 B); length 55 mm .... *Erpobdella melanostoma* (Sawyer & Shelley, 1976)

9(8) Two to two and a half annuli between gonopores 10

9' Three to four and a half annuli between gonopores 12

10(9) No papillae around male gonopore; three (rarely four) pairs of eyes 11

10' Male gonopore surrounded by circle of papillae; zero or three pairs of eyes; two annuli between gonopores (Fig. 12.20 A); length 15 mm .... *Erpobdella anoculata* (Moore, 1898)

11(10) Two annuli between gonopores (Fig. 12.20 A); three (rarely four) pairs of eyes; length 50 mm .... *Erpobdella fervida* (Verrill, 1871)

11' Either 2 or 2½ annuli between gonopores; three pairs of eyes; length 30 mm .... *Erpobdella bucera* (Moore, 1949)

12(9) Three pairs of eyes; gonopores separated by three annuli; length 50 mm .... *Erpobdella microstoma* (Moore, 1901)

12' Gonopores separated by 4–4½ annuli; three pairs of eyes; length 40 mm .... *Erpobdella tetragon* (Sawyer & Shelley, 1976)

Hirudinida: Arhynchobdellida: Erpobdellidae: *Motobdella*: Species

1 14 to 18 small, white tipped papillae in a ring about each annulus; mouth very large (4.0 mm); two pairs of crop caeca nephridiopores not visible, internal surface of posterior sucker pigmented .... *Motobdella montezuma* (Davies, Singhal, & Blinn, 1985)

1' 10 to 14 small, white tipped papillae on dorsal surface only of each annulus; mouth 1.1 mm in diameter; nephridiopores visible as small white spots; internal surface of posterior sucker not pigmented; one or two pars of crop caeca .... *Motobdella sedonensis* Govedich, Blinn, Keim, & Davies, 1998

REFERENCES

Govedich, F.R., W.E. Moser, S. Gelder, B.A. Bain, R. Brinkhurst & R.W. Davies. 2010. Freshwater Annelida. Pages 385–436 *in*: Ecology and Classification of North American Freshwater Invertebrates. 3rd Edition. J.H. Thorp & A.P. Covich (eds.). Elsevier, Boston.

Richardson, D.J., R. Tumlison, J.W. Allen Jr, W.E. Moser, C.T. McAllister, S.E. Trauth & H.W. Robison. 2013. New host records for the fish leech *Cystobranchus klemmi* (Hirudinida: Piscicolidae) on cyprinid fishes from Arkansas and Oklahoma. Journal of the Arkansas Academy of Science 67:211–213.

Class Aphanoneura

Lawrence L. Lovell

Research and Collections, Natural History Museum of Los Angeles County, Los Angeles, CA, USA

INTRODUCTION

The Aeolosomatidae (Fig. 12.1 A) and Potamodrillidae have genetically been shown to be more closely allied with polychaetes, not the clitellates. Van der Land (1971) presented a thorough review of the aeolosomatids.

TERMINOLOGY AND MORPHOLOGY

Terminology follows that found in the glossary of character terms in Fauchald (1977) and Rouse & Pleijel (2001), with a preference for chaete and chaetiger over setae and setiger. Fauchald (1977) provided companion figures to illustrate the glossary.

KEYS TO APHANONEURA

Aphanoneura: Families

1 Body with 8 or fewer segments .... Potamodrillidae, one genus: *Potamodrilus* sp.

1' Body with 13–20 segments .... Aeolosomatidae, one genus: ***Aeolosoma*** **[p. 260]**

Aphanoneura: Aeolosomatidae: *Aeolosoma*: Species

1	Chaetal segments with both long, flexible hair and short, hard sigmoid chaete	2
1'	Chaetal segments with only hair chaete, sigmoid chaete completely absent	5
2(1)	Sigmoid chaete smooth or with a single row of small distal teeth	3
2'	Sigmoid chaete with thin pointed tip with several rows of subdistal teeth present on the concave side	*Aeolosoma travancorense* Aiyer, 1926
3(2)	Body large, width 120–200 μm	4
3'	Body small, width 50–100 μm sigmoid chaete smooth with bent	*Aeolosoma beddardi* Michaelsen, 1900
4(3)	Sigmoid chaete present in all segments with the exception of segment 2	*Aeolosoma leidyi* Cragin, 1887
4'	Sigmoid chaete occur only in posterior segments, after segment 5	*Aeolosoma tenebrarum* Vejdovský, 1882
5(1)	Hair chaete <160 μm, body 40–175 μm wide	6
5'	Some hair chaete >175 μm, body 150–300 μm wide	*Aeolosoma headleyi* Beddard, 1888
6(5)	Epidermal glands present	7
6'	Epidermal glands absent	*Aeolosoma niveum* Leydig, 1865
7(6)	Epidermal glands colored orangish-red	*Aeolosoma hemprichi* Ehrenberg, 1831
7'	Epidermal glands colored greenish-yellow	*Aeolosoma variegatum* Vejdovský, 1884

REFERENCES

Fauchald, K. 1977. The polychaete worms. Definitions and keys to the orders, families, and genera. Natural History Museum of Los Angeles County. Science Series 28: 1–188.

Rouse & Pleijel, 2001. Polychaetes. Oxford University Press. 354 p.

Van der Land, J. 1971. Family Aeolosomatidae. Pages 665–706 *in*: R.O. Brinkhurst & B.G.M. Jamieson (eds.), Aquatic Oligochaeta of the World. Oliver and Boyd, Edinburgh, Scotland.

Class Polychaeta

Lawrence L. Lovell

Research and Collections, Natural History Museum of Los Angeles County, Los Angeles, CA, USA

INTRODUCTION

Polychaetes are primarily marine, with an estimated 10,000+ species, but some freshwater and freshwater tolerant species exist. Two excellent papers have addressed non-marine polychaete species. Glasby & Timm (2008) reviewed the global diversity of freshwater polychaetes, indicating that 168 species live in freshwater. Glasby et al. (2009) presented a detailed, updated list of the non-marine polychaetes of the world. It contains ecological, distributional, and habitat information on 197 species worldwide.

There are currently 15 freshwater and 21 freshwater tolerant polychaete species reported for the Nearctic region. Some of these species have very restricted distributions and are found only in the Nearctic, whereas others are very broadly distributed. New species will undoubtedly be discovered. The currently accepted classification scheme for the Polychaeta includes the subclasses Scolecida (containing several families having no ordinal assignment) and Palpata (containing orders Aciculata and Canalipalpata), and several *incertae sedes* families (including Potamodrilidae) (Fauchald & Rouse, 1997; Rouse & Pleijel, 2001). Ten families of polychaete are represented in the Nearctic.

LIMITATIONS

Terrestrial, commensal, and parasitic species are not included in the keys given below for Polychaeta.

TERMINOLOGY AND MORPHOLOGY

Terminology follows that found in the glossary of character terms in Fauchald (1977) and Rouse & Pleijel (2001), with a preference for chaete and chaetiger over setae and setiger. Fauchald (1977) provided companion figures to illustrate the glossary. The section on polychaete anatomy in Rouse & Pleijel (2001) provided illustrations of various anterior regions, parapodial configurations, and chaetal types for representative families.

Polychaeta morphology consists of three body regions: the head end composed of the prostomium and peristomium, the body composed of serially repeated segments, and the posterior composed of the pygidium (Fig. 12.35). The head end may possess palps, antennae, nuchal organs, tentacles, cirri, or eyes associated with the prostomium and/or peristomium. Branchiae are associated with the serial body segments occurring in pairs. They may arise directly from the body wall or be associated with the parapodial lobes. Parapodial lobes are located laterally on the body segments, but their orientation can be dorsally or ventrally shifted. They can be greatly reduced or well developed. Parapodial lobes occur in several arrangements: uniramous, subbiramous, and biramous. One or more internal acicula support the notopodial and neuropodial lobes. Each lobe may have associated branchiae and cirri. There are chaete of many types. They may be similar for the entire length of the body (capillary chaete), they may change types along the body region (capillary chaete changing to hooks), or there may be several chaetal types distributed in the notopodial and neuropodia rami.

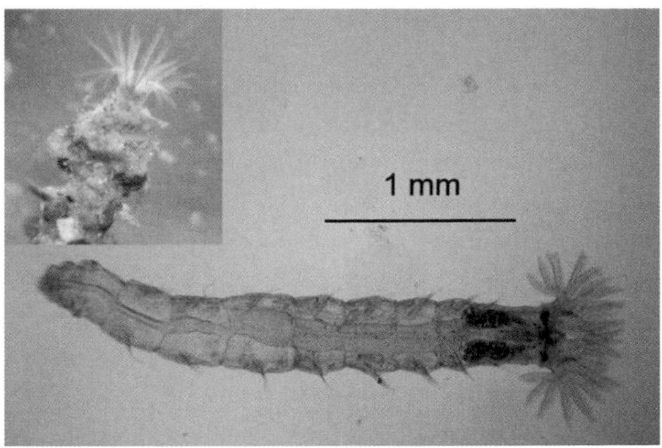

FIGURE 12.35 The freshwater polychaete *Manayunkia speciosa*. Shown here is an adult worm outside of its tube. The head with its tentacles is prominent on the right. Early egg development in this female is shown as paired dark masses to the left of the tentacles. The inset photo shows an adult worm inside its tube while in a feeding posture. Photographs are courtesy of Sarah and David Malakauskas and appeared in Malakauskas et al. (2013).

KEYS TO POLYCHAETA

Polychaeta: Families

1	Body with numerous segments; larger macrobenthic forms	2
1'	Body with 8 segments, antennae absent, palps present, all chaete simple capillaries	Nereillidae, one species: *Troglochaetus beranecki* Delachaux, 1921
2(1)	Body not divided into regions, segments similar; prostomium not reduced	3
2'	Body with distinct thoracic and abdominal regions, with multiple dorsal branchial pairs or branchial crown around the mouth; prostomium reduced	7
3(2)	Pharynx well-developed, muscular, with or without jaws; head end with antennae and/or palps; parapodial lobes well developed with simple and compound	4
3'	Prostomium reduced, antenna lacking; palps, if present occur as single dorsal pair; parapodial lobes reduced, with capillary chaete or hooks	5
4(3)	Two antennae, 2 pair of tentacular cirri, adults large >5 mm	**Nereididae [p. 261]**
4'	Three antennae, 4 pair of tentacular cirri, adults small <5 mm	Hesionidae, one species: *Hesionides riegerorum* Westheide, 1979
5(3)	Head end with paired palps or grooved tentacles on anterior chaetigers	6
5'	Head end without palps or grooved feeding tentacles	**Capitellidae [p. 262]**
6(5)	Anterior end with conical or with frontal horns and paired palps; cirriform or pinnate branchiae on one or more segments; biramous parapodia well developed, with capillary chaete and dentate hooks	**Spionidae [p. 262]**
6'	Anterior end lacking appendages; palps present as a pair just before or on chaetiger 1 or as two multiple clusters on chaetiger 6; slender filiform branchiae begin on chaetiger 1; biramous parapodia reduced with capillary chaete only or mixed with acicular spines	**Cirratulidae [p. 262]**
7(2)	Anterior with branchial crown composed of multiple radiole pairs	8
7'	Anterior without branchial crown, with multiple dorsal branchial pairs on anterior segments	**Ampharetidae [p. 262]**
8(7)	With stalked operculum, thoracic collar, with calcareous tube	**Serpulidae [p. 263]**
8'	Without stalked operculum, thoracic collar absent, mud or sand tubes	**Sabellidae [p. 263]**

Polychaeta: Nereididae: Species

1	Notopodia strongly reduced or absent, notopodial lobes absent, neuropodia supported by noto- and neuroacicula	2
1'	Parapodia biramous, notopodial lobes present, both noto- and neuropodia with acicula	5

2(1) Antennae short, subconical; eyes present; dorsal cirri anteriorly with cylindrical cirrophores, posteriorly cirrophores become elongate and leaf-like; notochaete usually present 3

2' Antennae longer than prostomium; eyes absent; dorsal cirri lack cirrophores; notochaeta absent 4

3(2) Anterior heterogomph falcigers replaced by heterogomph falcigers in posterior parapodia .... *Namalycastis borealis* Glasby, 1999

3' Heterogomph falcigers present in all parapodia .... *Namalycastis intermedia* Glasby, 1999

4(2) Eyes present .... *Namanereis littorealis* complex

4' Eyes absent .... *Namanereis cavernicola* (Solis-Weiss & Espinosa, 1991)

5(1) Pharynx with either papillae or paragnaths 6

5' Dorsal cirrophores elongated with distal styles; pharynx without papillae or paragnaths .... *Steninonereis martini* Wesenberg-Lund, 1958

6(5) Pharynx without papillae, with paragnaths 7

6' Pharynx with papillae, without paragnaths .... *Laenonereis culveri* (Webster, 1879)

7(6) Proboscis with large paragnaths, posterior notopodial lobe enlarged .... *Alitta succinea* (Leukart, 1847)

7' Proboscis with small paragnaths, posterior notopodial lobe not enlarged .... *Hediste limnicola* (Johnson, 1903)

Polychaeta: Capitellidae: Species

1 Prostomium small, pointed; parapodia biramous, chaetigers 1–4 or 5 with capillaries only, hooks thereafter 2

1' Prostomium conical, bluntly rounded; parapodia biramous, chaetigers 1–4 with capillaries only, mixed capillaries and hooks in following segments; genital spines present in notopodial of chaetigers 8–9 of males and some females .... *Capitella capitata* complex

2(1) First 4 chaetigers with capillary chaete only .... *Mediomastus californiensis* Hartman, 1944

2' First 5 chaetigers with capillary chaete only .... *Heteromastus filiformis* (Claparède, 1964)

Polychaeta: Spionidae: Species

1 Fifth chaetiger modified with specialized chaete 2

1' Fifth chaetiger not modified, chaeta similar to others 3

2(1) Spines of chaetiger 5 simple, falcate, without bristles distally .... *Dipolydora socialis* (Schmarda, 1861)

2' Spines of chaetiger 5 large, falcate, curved tip with bushy top .... *Dipolydora caulleryi* (Mesnil, 1897)

3(1) With many pairs of branchiae beginning on chaetiger 1 4

3' With a single pair of branchiae posterior to the palps, second segment with dorsal membrane .... *Streblospio gynobranchiata* Rice & Levin, 1998

4(3) Neurochaetal hooks begin no more than 15 segments after notochaetal hooks begin 5

4' Neurochaetal hooks begin more than 20 segments after notochaetal hooks begin .... *Marenzelleria bastropi* Bick, 2005

5(4) Nuchal organs not extending past midsegment of chaetiger 2 6

5' Nuchal organs extend to the end of chaetiger 3 .... *Marenzelleria neglecta* Sikorski & Bick, 2004

6(5) Branchiae present on 30–40 segments .... *Marenzelleria arctica* (Chamberlin, 1920)

6' Branchiae present on 60–120 segments .... *Marenzelleria viridis* (Verrill, 1873)

Polychaeta: Cirratulidae: Species

1 Palps present as a pair just before or on chaetiger 1; biramous parapodia with capillary chaete only .... *Monticellina* sp of Karlan in Glasby et al., 2009

1' Palps present as two multiple clusters on chaetiger 6; biramous parapodia with capillary chaete and mixed acicular spines appearing in median and posterior chaetigers .... *Cirriformia moorei* Blake, 1996

Polychaeta: Ampharetidae: Species

1 Dorsal fold present behind branchiae .... *Melinna maculata* Webster, 1879

1' Dorsal fold absent .... *Hypania florida* (Hartman, 1951)

Polychaeta: Serpulidae: Species

1 Operculum with spines .... *Ficopomatus enigmaticus* (Fauvel, 1922)

1' Operculum smooth .... *Ficopomatus miamiensis* (Treadwell, 1914)

Polychaeta: Sabellidae: Species

1 Body with 8 thoracic and 3 abdominal segments, radioles undivided (Fig. 12.35) .... *Manayunkia speciosa* Leidy, 1858

1' Body with 8 thoracic and 8 abdominal segments, radioles pinnately divided .... *Chone* sp. of Holmquist 1973, 1975

REFERENCES

Fauchald, K. 1977. The polychaete worms. Definitions and keys to the orders, families, and genera. Natural History Museum of Los Angeles County. Science Series 28: 1–188.

Fauchald, K. & G.W. Rouse. 1997. Polychaete systematics: past and present. Zoological Scripta 26: 71–138.

Glasby, C.J. & T. Timm. 2008. Global diversity of polychaetes (Polychaeta; Annelida) in freshwater. Hydrobiologia 595: 107–115.

Glasby, C.J., T. Timm, A.I. Muir & J. Gil. 2009. Catalogue of non-marine Polychaeta (Annelida) of the World. Zootaxa 2070: 1–52.

Malakauskas, D.M., S.J. Willson, M.A. Wilzbach & N.A. Som. 2013. Flow variation and substrate type affect dislodgement of the freshwater polychaete, *Manayunkia speciosa*. Freshwater Science 32: 862–873.

Rouse & Pleijel, 2001. Polychaetes. Oxford University Press. 354 p.

Van der Land, J. 1971. Family Aeolosomatidae. Pages 665–706. *in*: Brinkhurst, R.O. and B.G.M. Jamieson (eds.), Aquatic Oligochaeta of the World. Oliver and Boyd, Edinburgh, Scotland.

WoRMS. 2012. Annelida. Accessed through: World Register of Marine Species on 10/18/2012 at: http://www.marinespecies.org/aphia.php?p=taxdetails&id=882.

Chapter 13

Phylum Ectoprocta (Bryozoa)

Timothy S. Wood
Department of Biological Sciences, Wright State University, Dayton, OH, USA

Chapter Outline

INTRODUCTION

The Phylum Ectoprocta, or Bryozoa, is primarily a marine group with over 8000 recognized species and a rich fossil record with many thousands more. The 100 or so bryozoan species known from freshwater would seem to be overshadowed by such a robust family tree. In fact, however, freshwater bryozoans are an important part of the benthic community in lakes, ponds, and rivers worldwide. They filter suspended particles from the water, and their fecal pellets provide nourishment for a wide variety of scavengers. Bryozoans are also significant biofouling organisms that often interfere with the function of irrigation, water treatment, and industrial cooling systems (Wood, 2005a).

The phylum Entoprocta is sometimes grouped with the phylum Ectoprocta (Bryozoa) because of certain similarities, but because these are superficial, this group is discussed in the separate Chapter 14.

LIMITATIONS

In this chapter, I recognize 29 species of freshwater bryozoans in the Nearctic region. Many are well known and easily identified, while others are distinguished only by minute and sometimes tentative features. Most of the difficult taxonomic issues occur in the family Plumatellidae, which comprises about half the Nearctic species. Over the past several decades, much progress has been made in the taxonomy of the entire group, but many questions remain unresolved.

Freshwater bryozoans are unequally classified among two major groups: (1) the exclusively freshwater class Phylactolaemata; and (2) the mostly marine order Ctenostomata within the class Gymnolaemata. In the Nearctic region, phylactolaemates are said to be represented by at least 32 species and the ctenostomes by 4 species (Massard & Geimer, 2008). In years to come, these numbers will surely be revised. My own current estimate is 25 phylatolaemate species and 4 ctenosomes.

The soft tissues of bryozoans offer few reliable features for species identification. Hard parts, when they are available, tend to be much more consistent and verifiable. In fact, this general principle holds true for most invertebrate animals. It is surely not by coincidence that animal groups lacking hard anatomical features have relatively few known species. The likely reason is that many species simply cannot be distinguished by morphology alone, and the tools of molecular genetics may not be sufficiently refined to detect cryptic species. Among freshwater bryozoans, the ctenostomes, with no hard anatomical parts, are often very difficult to identify with certainty. It is not surprising that the species number in this group is relatively low. The traditional taxonomist is forced to rely on such variable features as tentacle counts and the layout of the gut.

By contrast, phylactolaemate bryozoans have taxonomically useful hard parts in the encapsulated products of asexual reproduction called statoblasts. These provide a wealth of characters, such as relative shapes and dimensions, as well as tiny tubercles and net-like lines that are believed to be species specific. In most cases, the statoblast alone is sufficient to identify a species. In fact, when statoblasts are absent, many phylactolaemates cannot be positively identified at the species level.

However, even statoblasts have their limitations. The search for ever more elusive taxonomic features has inevitably led to important features being detected only through scanning electron microscopy (SEM). This becomes a

Thorp and Covich's Freshwater Invertebrates. http://dx.doi.org/10.1016/B978-0-12-385028-7.00013-5

challenge to those investigators without access to such an instrument. In the following key, I have tried to avoid using characters that rely on SEM examination, but in a few cases, it simply was not possible.

The reported species distributions naturally depend on the accuracy of species identification. Our understanding of taxonomically significant features has evolved over many years. Even now the value of floatoblast nodules, for example, is not universally accepted. Partly for this reason, the species distributions stated in the key below should be regarded as tentative.

As explained below, it must be remembered that different preservation techniques will alter certain key morphological characters.

TERMINOLOGY AND MORPHOLOGY

The bryozoan colony is composed of numerous zooids, all physically and physiologically connected (Figs. 13.1 and 13.2). The feeding organ is a cluster of ciliated tentacles called a lophophore, which captures particles suspended in the water. To accommodate many tentacles the dorsal side of the lophophore is deflected inward to make a horseshoe shape (Fig. 13.3 A). In species with few tentacles the lophophore is circular in outline (Fig. 13.3 B).

The outermost part of a zooid is a non-living layer called an ectocyst, composed of either chitin or a slimy mucopolysaccharide. The ectocyst ranges from transparent to opaque, sometimes within the same species, depending on age, growth rate, and other factors. The inner part of the zooid is the polypide, which includes the lophophore and the entire digestive tract. The polypide can be shifted towards the zooid tip to extend the lophophore. When the zooid is alarmed, the entire polypide is quickly retracted and the lophophore becomes fully protected (Fig. 13.1).

Statoblasts are essentially tiny packages of yolky material and germinal cells enclosed within a bilayered case. The case is the "hard part" mentioned earlier, and it can be essential for species identification. There are several types of statoblasts: floatoblasts, sessoblasts, and piptoblasts.

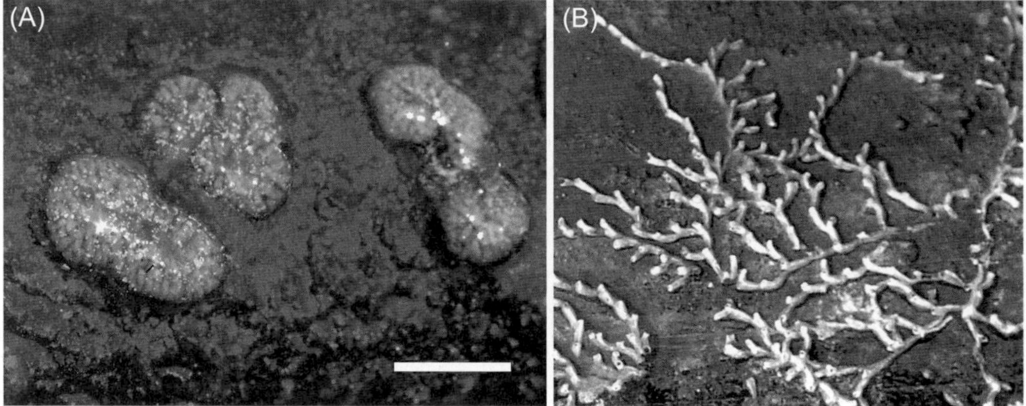

FIGURE 13.1 Two basic morphologies of bryozoan colonies. (A) Globular colonies of *Asajirella gelatinosa*; (B) tubular colonies of *Plumatella javanica*. Scale bar = 1 cm.

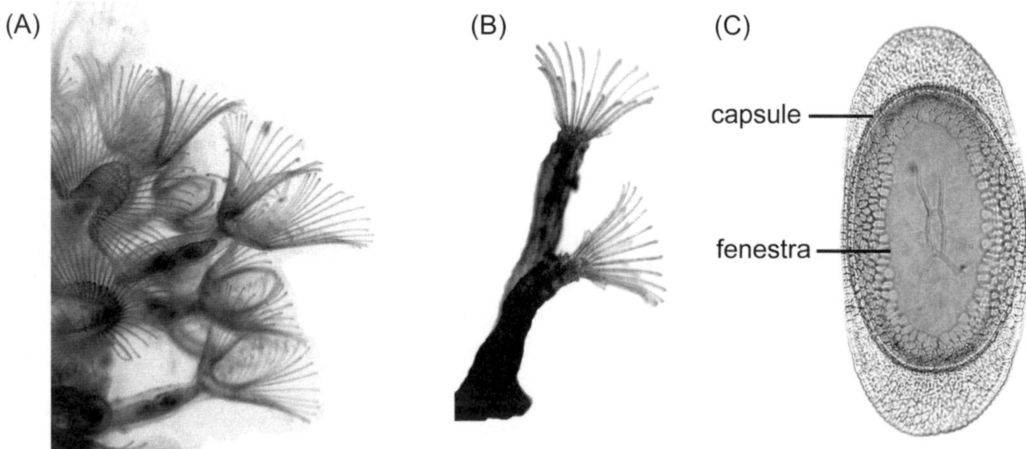

FIGURE 13.2 Aspects of bryozoan colonies. (A) Horseshoe-shaped lophophores typical of most phylactolaemate species; (B) circular lophophores of fredericellid species; (C) uninflated floatoblast of *Plumatella fruticosa* distinguishing the capsule from the fenestra. Scale bar = 0.5 mm.

Floatoblasts are freely released and have a peripheral ring of gas-filled chambers for buoyancy (Figs. 13.4 A and 13.5). Sessoblasts are firmly attached to the substrate (Figs. 13.4 B and 13.10). Zooids in the family Fredericellidae produce bean-like piptoblasts that generally remain with the substrate but are not firmly attached (Fig. 13.9).

All statoblasts are formed like two halves of a clam shell, which split apart at the time of germination. Each half, or valve, is normally composed of two layers: an inner capsule and an outer periblast (Fig. 13.4). The capsule has no taxonomic value, but the periblast bears a wide variety of morphological features. In floatoblasts, the periblast includes the ring of gas-filled chambers. In the center of this ring is an area called the fenestra. The relative dimensions of the periblast and its fenestra can be useful taxonomic features (Fig. 13.12). By convention, the valve with the smallest fenestra is considered dorsal, while the opposite side is ventral (Figs. 13.5 and 13.12 A).

Several terms are useful in describing statoblasts. The line where the valves join together is called a suture (Figs. 13.6 and 13.7). Tubercles are pimple-like protuberances that may occur on various regions of the statoblast surface (Fig. 13.11 A). The reticulum is a net-like pattern of raised lines (Fig. 13.5). Pitting is the appearance of small depressions seen in certain fredericellid statoblasts (Fig. 13.9 A). Nodules are very small, rash-like dots visible only with scanning electron microscopy (Fig. 13.11 A).

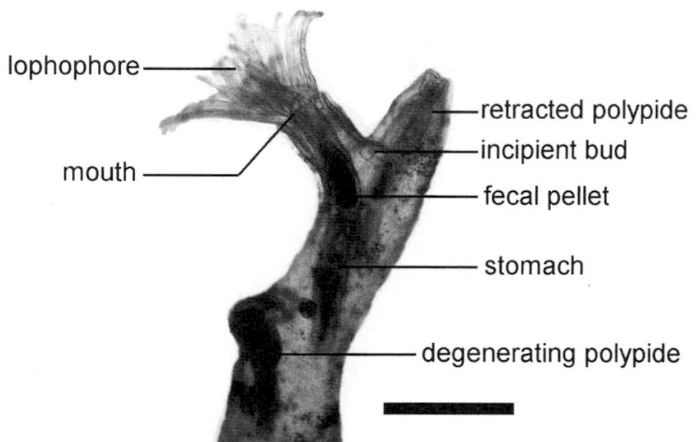

FIGURE 13.3 Zooids of a phylactolaemate bryozoan. Scale bar = 0.5 mm.

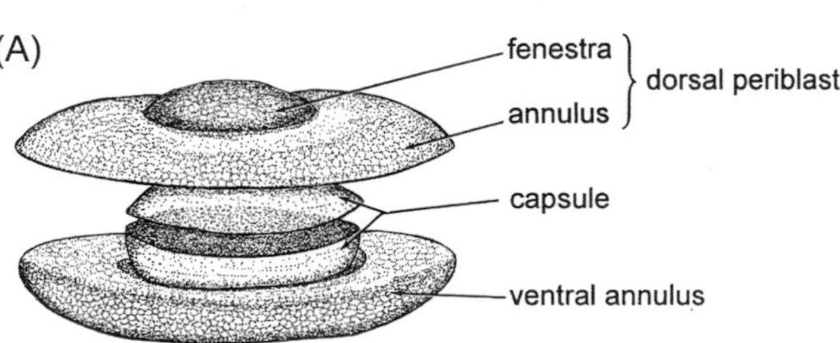

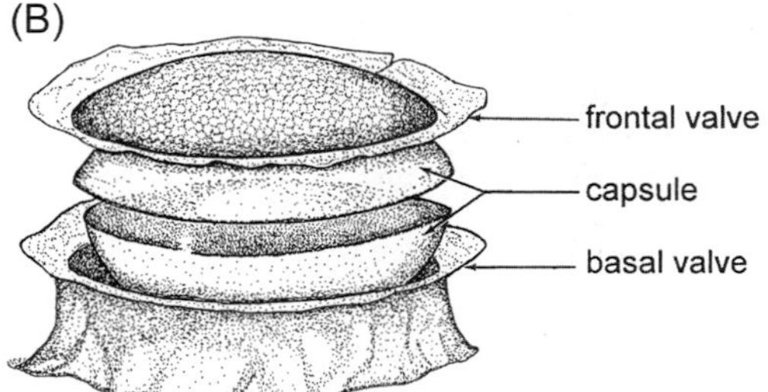

FIGURE 13.4 Sclerotized components of generalized phylactolaemate statoblasts in exploded view. (A) Floatoblast; (B) Sessoblast.

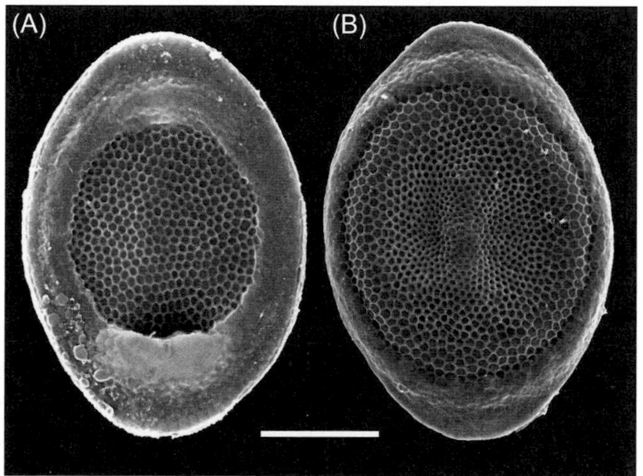

FIGURE 13.5 Floatoblast valves of *Plumatella vaihiriae*. (A) Dorsal valve, (B) ventral valve. Scale bar = 100 µm.

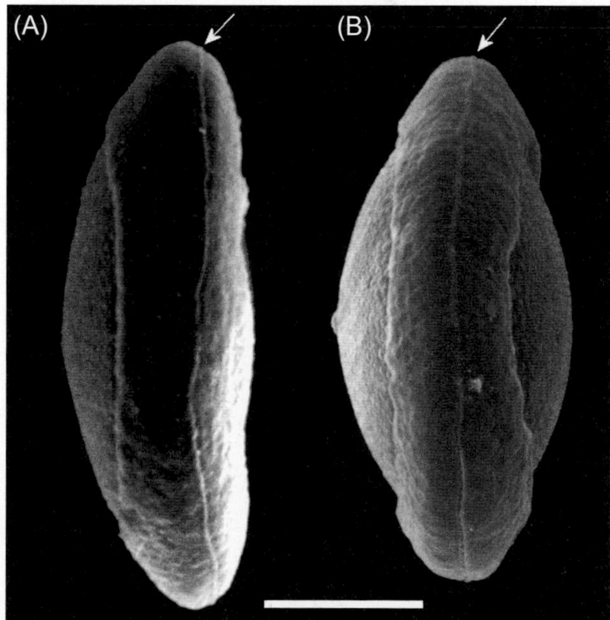

FIGURE 13.6 Floatoblast lateral symmetry. Arrows indicate the suture between valves. (A) Typical asymmetry of *Plumatella fungosa*, (B) typical symmetry of *Plumatella nitens*. Scale bar = 100 µm.

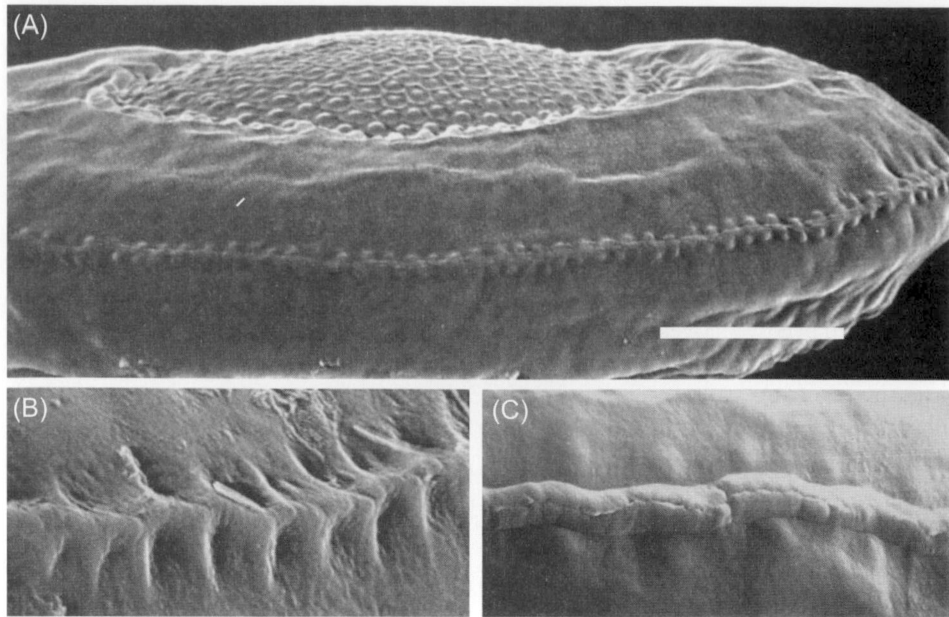

FIGURE 13.7 Floatoblast sutures. (A) *Plumatella repens*, scale bar = 50 μm; (B) *Plumatella fungosa*; (C) *Plumatella rugosa*.

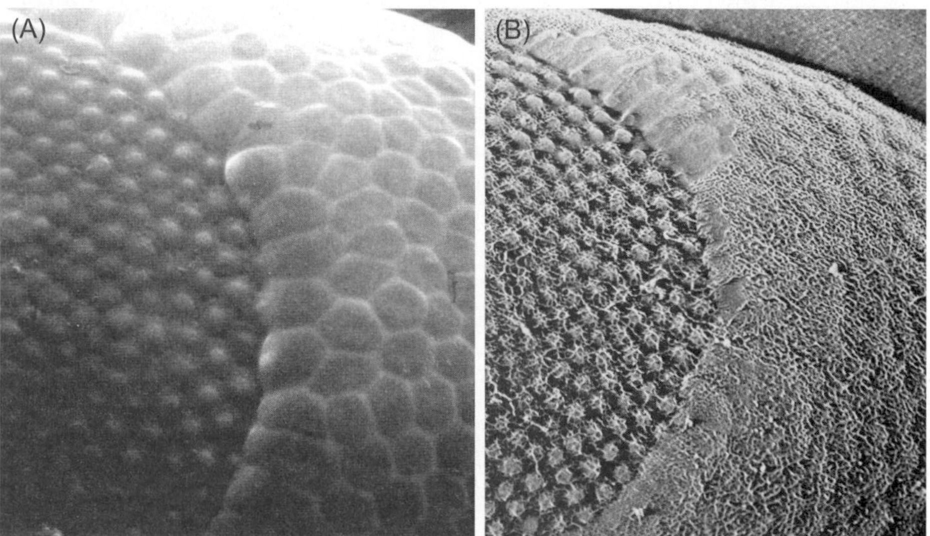

FIGURE 13.8 Floatoblast surface features. (A) *Plumatella emarginata* showing typical "paved" annulus; (B) *Plumatella mukai* showing typical wrinkled features. Scale bar = 50 μm.

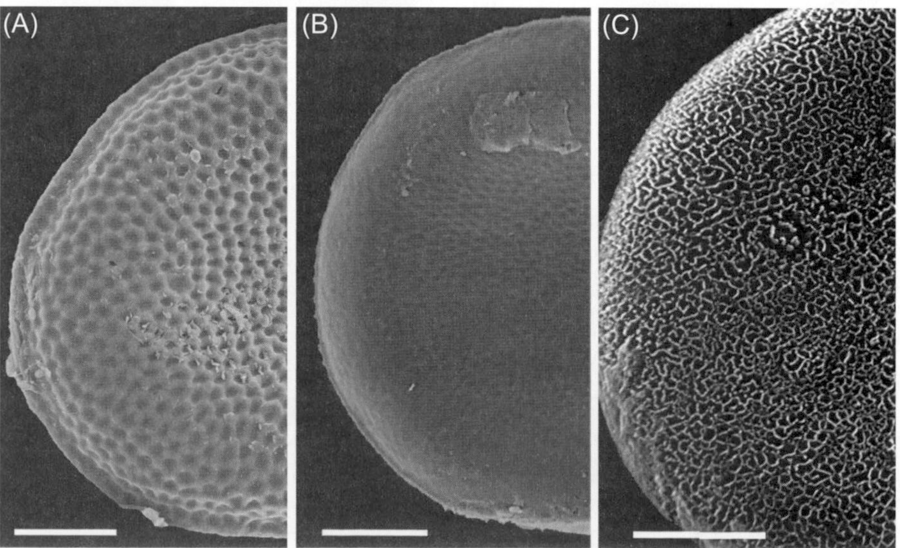

FIGURE 13.9 Three fredericellid species. (A) *Fredericella indica*, (B) *Fredericella toriumii*, (C) *Fredericella browni*. Scale bar = 50 μm. *Modified from Hirose & Mawatari, 2011, used by permission from The Japanese Society of Systematic Zoology.*

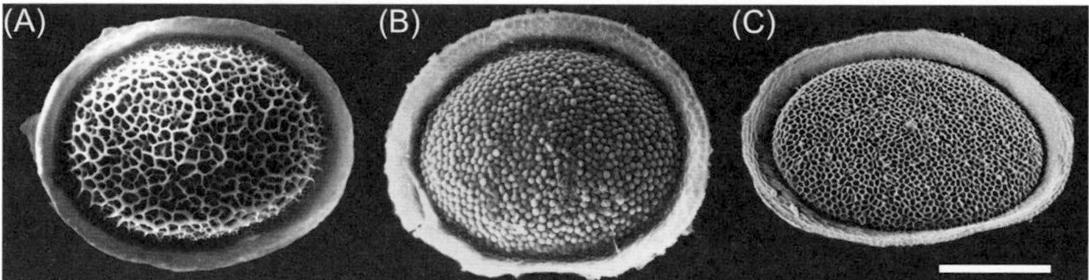

FIGURE 13.10 Sessoblast surface features. (A) Irregular reticulation of *Plumatella reticulata*, (B) tubercles common in many species, in this instance *Plumatella repens*, (C) regular reticulation of *Plumatella vaihiriae*, which sometimes thickens to form pores. Scale bar = 100 μm.

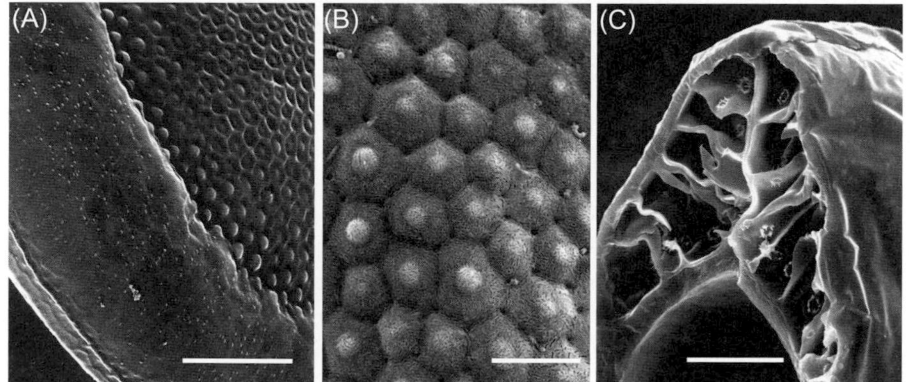

FIGURE 13.11 Small features on statoblasts. (A) Partial view of floatoblast in *Plumatella semilirepens* showing nodules on the annulus and both tubercles and reticulation on the fenestra; scale bar = 25 μm. (B) Hypertubercles on floatoblast fenestra, characteristic of *Rumarcanella* species, in this instance *R. siamensis*; scale bar = 10 μm. (C) A section of the floatoblast annulus showing chamber pores; scale bar = 20 μm. *Photo by M. Taticchi.*

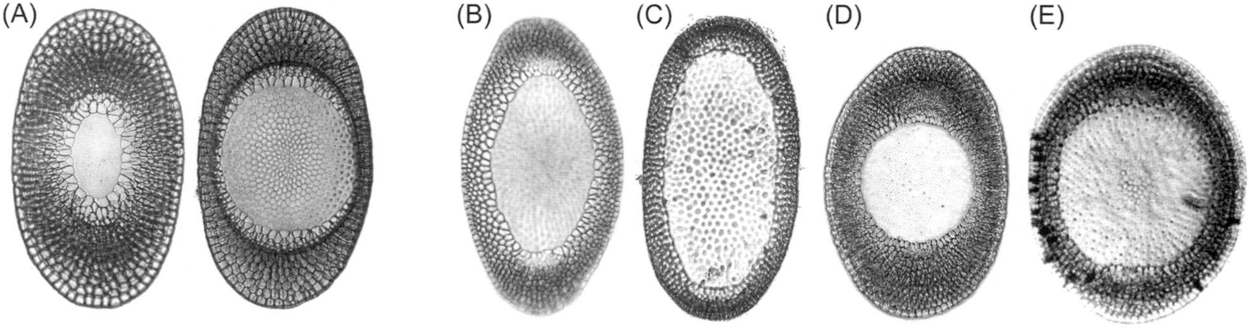

FIGURE 13.12 Floatoblast valves. (A) *Plumatella emarginata*, dorsal valve on the left, ventral valve at right. (B) *P. casmiana*, capsuled floatoblast, dorsal valve. (C) *P. casmiana*, leptoblast, dorsal valve. (D) *P. fungosa* dorsal valve; (E) *P. nitens* dorsal valve. Scale bar = 200 μm.

In addition, there are pores within the floatoblast annulus chambers that may have taxonomic significance (Fig. 13.11 C) (see Rubini et al., 2011). A more recently used structure is a hypertubercle, defined as a small tubercle atop a larger one (Fig. 13.11 B), a feature that defines the genus *Rumarcanella* (Hirose & Mawatari, 2011). Examining a floatoblast in side view (suture toward the viewer) can help determine whether or not the valves are laterally symmetrical (Fig. 13.6).

Statoblasts are never formed by freshwater ctenostome bryozoans. Instead, the dormant phase of the life cycle is represented by modified parts called hibernaculae. Currently, the variation in hibernaculum morphology is too poorly known for these to be assigned any taxonomic value.

MATERIAL PREPARATION AND PRESERVATION

Standard procedure is to fix and preserve bryozoans in either 5% formalin or 70% ethyl alcohol. For molecular genetics, 100% ethyl alcohol is the preferred preservative. For histological work, the best fixative is buffered glutaraldehyde, with the buffer concentration at 0.01 M, not 0.1 M as for many other invertebrates.

Many investigators like to preserve specimens with the lophophores extended, and for this to happen, the material must be narcotized. Menthol (crystals or molded wafers) is very effective and convenient, although it is not recommended for fine histological work. Also, menthol has a low melting point and tends to melt under hot field conditions (Wood, 2005b). The solid menthol is floated on the water surface in a small, closed container with the specimen. Exposure time varies with temperature and species but seldom exceeds one hour. The other alternative is to administer a chloral hydrate solution, drop by drop, and mix it over several hours. This is much less damaging to sensitive tissues than menthol, but chloral hydrate is difficult to obtain in countries that list it as a controlled substance.

For close examination of statoblasts, the valves must often be separated and cleaned in a strong solution of potassium (or sodium) hydroxide. If you are in a hurry, heat the statoblasts in the solution in a spoon over a flame for no more than a minute. Then transfer the statoblasts to freshwater and tease apart the valves. The capsule valves may separate from the periblast or not, depending on the species.

When examining statoblast valves with light microscopy, keep in mind that you can detect structures only on the fenestra, not on the annulus. Even then, you will not see faint reticulation or nodules. Tubercles and reticulation can be easily confused because under the light microscope they appear very similar. Tubercles often act as tiny lenses; and as you focus up and down, you can reduce them to little points of light. With most reticulation, you can focus up and down, and the pattern will just go in and out of focus without forming points of light. However, there are times when you really cannot tell the difference between tubercles and reticulation without scanning electron microscopy. It is a common mistake to confuse the statoblast capsule with the fenestra. This can lead to errors in species identification. In fresh material, the amber-colored capsule is generally hidden by gas chambers of the annulus. But when the gas is lost in preserved specimens, the capsule is fully visible and the outlines of the fenestra are less distinct (Fig. 13.3 C).

Statoblasts can be prepared for scanning electron microscopy by simply drying and sputtering on aluminum pins. Normally the ventral valve will collapse, which does not affect species identification. However, if you are finicky about such things, you can always use critical point drying to retain the original shape.

Identifying bryozoan species often requires taking measurements of the statoblast using an optimal micrometer, and this brings up a sensitive subject. The literature is filled with incorrect measurements because statoblast valves being measured had been either dried or flattened under a coverslip when the data were taken. Simply separating the valves also can greatly alter their shape. Even under normal conditions, statoblast size can vary considerably within a species and often within a single colony. There are several things to be learned from this:

- whenever possible, take measurements from fresh material in water;
- measure numerous statoblasts and run the statistics;
- never rely on dimensions from separated valves or from statoblasts prepared for SEM;
- approach published statoblast measurements with some caution.

KEYS TO FRESHWATER ECTOPROCT BRYOZOANS

Ectoprocta: Classes

1 Extended lophophore U-shaped in outline (Fig. 13.3 A) with the exception of family Fredericellidae; lophophore with more than 20 tentacles; orifice is circular when lophophore is withdrawn; internal septa infrequent; statoblasts may be present .... **Phylactolaemata [p. 270]**

1' Extended lophophore always circular in outline with fewer than 20 tentacles (Fig. 13.13); orifice appears quadrangular when lophophore is withdrawn; statoblasts never present; individual zooids separated by internal septa, although this is not always evident .... Gymnolaemata, one order; **Ctenostomata [p. 274]**

Ectoprocta: Phylactolaemata: Families

1 Colony composed of a linear, branching series of zooids (Fig. 13.1 B); branches sometimes adhering to each other; statoblasts (if present) with smooth margins; tentacles fewer than 65 2

1' Colony globular and either lobed or entire in outline (Fig. 13.1 A); statoblasts with peripheral spines, hooks, or pointed extensions; tentacles more than 65 4

2(1) Lophophore U-shaped in outline, tentacles more than 22; statoblasts either cemented to the substrate or released freely into the water, often buoyant 3

2' Lophophore circular in outline, tentacles fewer than 25; statoblasts reniform and retained within the colony, neither cemented to the substrate nor released freely into the water, never buoyant (Note: Statoblasts may be infrequent and difficult to find. If you cannot see through the ectocyst and strong substage light does not help, you may need to dissect the colony. Also, check the surrounding substrate for brown to orange, reniform statoblasts left behind by a disintegrated part of the colony) .... **Fredericellidae [p. 271]**

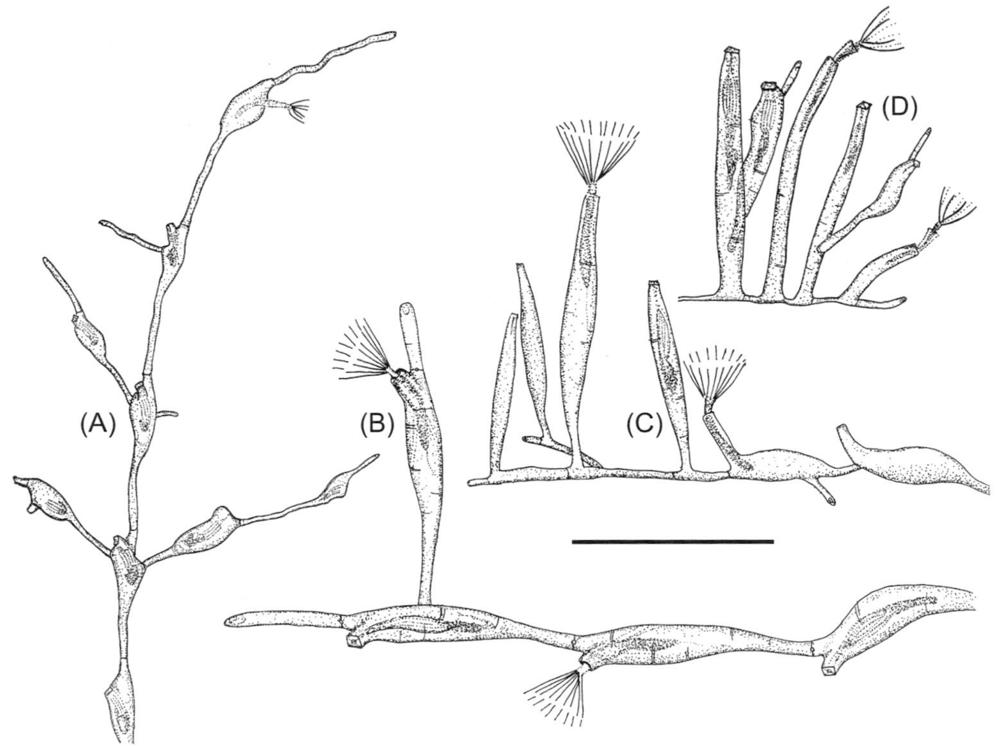

FIGURE 13.13 Ctenostome bryozoan colonies. (A) *Sineportella forbesi*; (B) *Paludicella articulata*; (C) *Pottsiella erecta*; (D) *Victorella pavida*. Scale bar = 1 mm.

3(2) Colony composed of delicate, nearly invisible, transparent tubes extending from a soft, gelatinous matrix; polypide and lophophore relatively small; floatoblast circular and covered with net-like reticulation and lacking tubercles; sessoblast with needle-like spikes in the fenestra .... Stephanellidae, one species; *Stephanella hina* Oka, 1908

[USA: Massachusetts, Oregon, Virginia, Washington]

3' Zooids and statoblasts not as above .... **Plumatellidae [p. 272]**

4(1) Mouth region without red pigmentation, lophophore lacking pair of white spots 5

4' Mouth region with red pigmentation; prominent pair of white spots at end of each arm of lophophore; statoblasts round with hooked spines radiating from outer margin of annulus. Colony gelatinous and slimy, ranging from a flat sheet to football-sized mass .... Pectinatellidae, one species; *Pectinatella magnifica* (Leidy, 1851)

[Canada: southern third; USA. Palaearctic]

5(4) Colony distinctly linear, often exceeding 2 cm; statoblasts round with wiry, hooked spines radiating beyond periphery from margin of fenestrae of both valves; occurring mainly in oligotrophic waters .... Cristatellidae, one species; *Cristatella mucedo* Cuvier, 1798

[Canada: southern tier; USA: northern border states, Colorado, New England]

5' Colony not distinctly linear; statoblasts oblong, not round .... **Lophopodidae [p. 274]**

Ectoprocta: Phylactolaemata: Fredericellidae: Genera

1 Piptoblast oval to elongate, seldom more than one per zooid, valves lacking a wrinkled mantle 2

1' Piptoblast broadly oval to round, often more than one per zooid; valves covered by a minutely wrinkled mantle (Fig. 13.9 C), which is easily dissolved in hot, concentrated KOH solution to reveal smooth sclerotized surface; basal valve with prominent attachment ring .... *Fredericella browni* (Rogick, 1945)

[USA: Illinois, Ohio, and southern states. Mexico: Chihuahua]

2(1) Piptoblast surface appears dull when dry and viewed with reflected light; strongly and uniformly pitted even after cleaning with KOH (Fig. 13.9 A) .... *Fredericella indica* (Annandale, 1909)

[Canada. USA. Mexico: Sinaloa]

2' Cleaned frontal valve appears shiny when dry and viewed with reflected light, but showing a trace of weak pitting when viewed with substage illumination (Fig. 13.9 B) In most cases to see this feature, it is best to separate and clean the valves in hot KOH or NaOH, then place them in clean freshwater and examine the frontal valve with substage illumination .... *Fredericella toriumii* Hirose & Mawatari, 2011

[USA: Oregon]

Ectoprocta: Phylactolaemata: Plumatellidae: Genera

1 Colony wall thick, soft, and transparent; colony entirely adherent to substratum; sparsely branched; zooids forming low mounds when polypides are retracted; floatoblast width greater than 300 μm, length over 450 μm; floatoblasts dark and seldom buoyant upon release from colony, sessile statoblasts never present .... *Hyalinella punctata* (Hancock, 1850)

[Canada: southern third; USA. Two species are described in the genus *Hyalinella*: *H. punctata* and *H. lendenfeldi*. The colonies are almost identical in structure: sparsely branching, fully recumbent, with a soft, swollen-looking ectocyst, a large, free statoblasts, and no sessile statoblasts at all]

1' Floatoblast and colony not as above, sessile statoblasts may be present .... ***Plumatella* [p. 272]**

Ectoprocta: Phylactolaemata: Plumatellidae: *Plumatella:* Species

1 Floatoblasts circular or nearly so, the annulus on each side uniform in width; uncommon throughout the Nearctic region (not to be confused with *Stephanella*—see Phylactolaemata Families, Couplet 3) 2

1' Floatoblasts oval or oblong; fenestrae of two valves distinctly different in size; abundant and widespread 3

2(1) Floatoblast fenestra tuberculated, annulus adorned with minute nodules, sessoblast circular or nearly so; thick walls of large colonies fusing into a firm, transparent matrix .... *Plumatella orbisperma* (Kellicott, 1882)

[Canada: Ontario. USA: Alaska, Michigan]

2' Floatoblast fully reticulated and without tubercles; sessoblast long oval to rectangular. Known only from small ponds in forested sites of New England .... *Plumatella recluse* Smith, 1992

[USA: Massachusetts, New Hampshire]

3(1) Floatoblast dorsal annulus width at the poles greater than length of dorsal fenestra (Fig. 13.12 A) 4

3' Floatoblast dorsal annulus width at the poles less than length of dorsal fenestra (Fig. 13.12 B–E) 9

4(3) Colony and sessoblast not as below 5

4' Colony with long, slender, branches often free of the substratum; sessoblast length more than twice the width; often found in cool, oligotrophic waters .... *Plumatella fruticosa* Allman, 1844

[Canada: New Brunswick, Quebec. USA: Michigan, New England]

5(4) Floatoblast laterally asymmetrical (Fig. 13.6 A) dorsal valve nearly flat, ventral valve deeply convex 6

5' Floatoblast valves almost equally convex (Fig. 13.6 B); sessoblast surface roughened by distinctive network of raised lines (Fig. 13.10 A); internal septa slightly oblique .... *Plumatella reticulata* Wood, 1988

[Canada. USA. Mexico]

6(5) Floatoblast dorsal annulus extending evenly from suture to fenestra without bulging around the inner capsule; sessoblast surface densely tuberculate 7

6' Floatoblast dorsal annulus bulging around inner capsule to form a distinct shoulder (Fig. 13.5); Entire floatoblast surface covered with raised reticulum (Fig. 13.5). Sessoblast frontal valve with deep pits, but lacking tubercles (Fig. 13.10 C) (Note: Several species will show a "capsule bulge" when the floatoblast is dry. Only this species has such a bulge when wet. The feature is best seen on the floatoblast dorsal side). Uncommon but often locally abundant, especially in warm, eutrophic waters .... *Plumatella vaihiriae* Hastings, 1929

[USA]

7(6) Colony and floatoblasts not as below 8

7' Colony forming tight masses of fused tubules without free branches, colonies capable of exceeding 1 cm thick; tubercles on floatoblast annulus seen only by SEM (floatoblast shown in Figs. 13.6 A and 13.14 D) .... *Plumatella fungosa* (Pallas, 1768)

[Canada. USA]

8(7) Floatoblast surface smooth except for fenestra tubercles (sometimes absent on the dorsal fenestra) (Fig. 13.8 A) .... *Plumatella emarginata* Allman, 1844

[Canada. USA]

8' Floatoblast surface, including tubercles, appear wrinkled, a feature conspicuous with SEM and otherwise visible by reflected light (Fig. 13.8 B) (Note: the wrinkled surface feature is best seen with scanning electron microscopy. However, it can sometimes be detected with a compound microscope using reflected (not transmitted) light) .... *Plumatella mukaii* Wood, 2001

[USA: California, Oregon]

9(3) Floatoblast dorsal fenestra less than 1.5 times the width (Fig. 13.12 D, E) 10

9' Floatoblast fenestra length more than 1.5 times the width (Fig. 13.12 B, C); tentacles fewer than 28 .... *Plumatella casmiana* Oka, 1907

[Canada: Ontario, Quebec. USA]

10(9) Floatoblast dorsal annulus extending evenly from suture to fenestra without bulging around the inner capsule; sessoblast surface densely tuberculate 11

10' Floatoblast dorsal annulus bulging around inner capsule to form a distinct shoulder (Fig. 13.5) Entire floatoblast surface covered with raised reticulum (Fig. 13.5). Sessoblast with deep pits, but lacking tubercles (Fig. 13.10 C). (Note: Several species will show a "capsule bulge" when the floatoblast is dry. Only this species has such a bulge when wet. The feature is best seen on the floatoblast dorsal side). Uncommon but often locally abundant, especially in warm, eutrophic waters .... *Plumatella vaihiriae* (Hastings, 1929)

[USA]

11(10) Floatoblast ventral annulus distinctly wider at the poles than along the sides 12

11' Floatoblast ventral annulus width uniformly narrow all around, or nearly so .... *Plumatella nitens* Wood 1996

[Canada: Ontario. USA: Illinois. Indiana, Massachusetts, Michigan, Minnesota, New York, Ohio, Wisconsin]

12(13) Floatoblast laterally symmetrical, or nearly so (Fig. 13.6 B); Note: the following couplets (13–16) require statoblast examination by scanning electron microscopy. If this point is reached and a scanning electron microscope is not available, tentative species identity may be inferred by the geographic reference, or any of the species following may be referred as belonging to the "*Plumatella repens* group" 13

12' Floatoblasts laterally asymmetrical (Fig. 13.6 A); colony forming tight masses of fused tubules without free branches, colonies capable of exceeding 1 cm thick. (SEM of floatoblast shown in Figs. 13.12 B and 13.14 D) .... *Plumatella fungosa* (Pallas, 1768)

[Canada. USA]

13(14) Floatoblast annulus cell outlines not easily distinguished by SEM (Fig. 13.14 B, E) 14

13' Floatoblast annulus cells outlines easily distinguished by SEM (Fig. 13.14 A, C) 15

14(13) Floatoblast length to width ratio greater than 1.5; floatoblast annulus densely populated with nodules, fenestra with prominent, well-spaced tubercles, reticulation absent (Fig. 13.14 B, F)................................ *Plumatella bushnelli* Wood, 2001

[USA: North Carolina south to Florida]

14' Floatoblast length to width ratio less than 1.5; nodules usually present on floatoblast annulus; floatoblast fenestra with both tubercles and reticulation, tubercles becoming more prominent at the fenestra margins (Fig. 13.14 E) .... *Plumatella repens* (L., 1758)

[Canada: Ontario, Quebec; USA]

15(13) Nodules present on floatoblast annulus and/or fenestra 16

15' Nodules absent from any part of the floatoblast; floatoblast annulus cell walls depressed to form lines of reticulation where adjacent cells meet (Fig. 13.14 C); sessoblast frontal valve strongly reticulated .... *Plumatella rugosa* Wood et al., 1998.

[Canada. USA]

16(15) Floatoblast annulus "paved"; (Fig. 13.8 A) fenestra with both tubercles and reticulation (Fig. 13.14 G) .... *Plumatella similirepens* Wood, 2001

[USA: Illinois]

16' Floatoblast surface covered with reticulation, tubercles, and nodules (Fig. 13.14 A) .... *Plumatella nodulosa* Wood, 2001

[USA: Illinois, New York, Ohio]

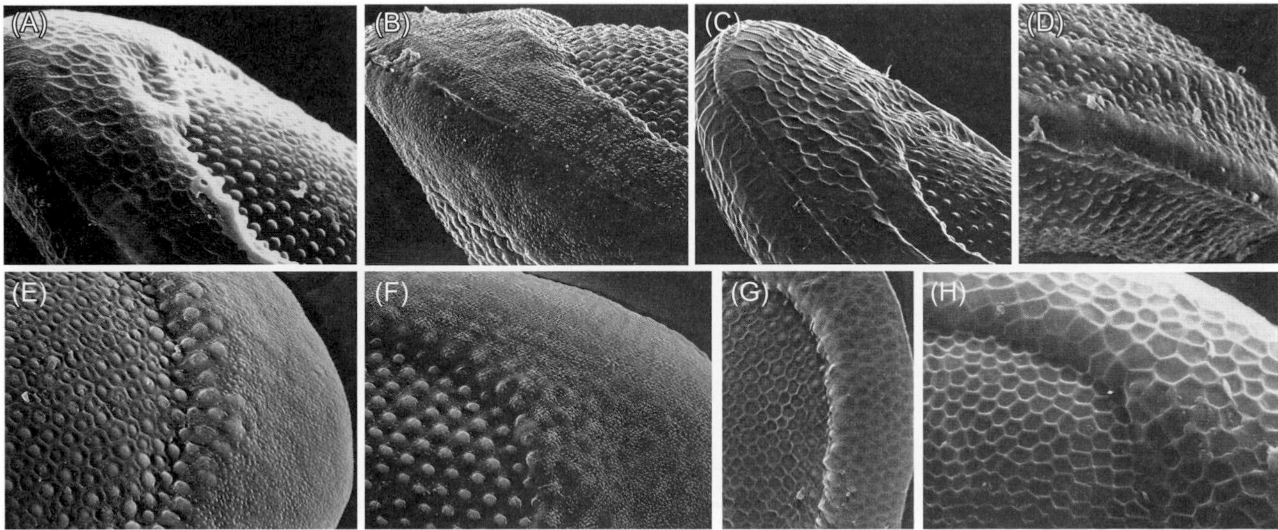

FIGURE 13.14 Floatoblasts of the "repens" group of species showing features accessible by SEM. (A) *Plumatella nodulosa*; (B) *Plumatella bushnelli*; (C) *Plumatella rugosa*; (D) *Plumatella fungosa*; (E) *Plumatella repens*; (F) *Plumatella bushnelli*; (G) *Plumatella similirepens*; (H) *Plumatella vaihiriae*.

Ectoprocta: Phylactolaemata: *Lophopodidae:* Genera and Species

1 Colony compact, seldom larger than pea-sized, often divided by clefts which eventually deepen and pinch off to form daughter colonies. Statoblasts with a series of small hooks localized along the polar margins. Uncommon, but can be locally abundant .... *Lophopodella carteri* (Hyatt, 1866)

[Canada: Ontario, Quebec. USA: Illinois, Kentucky, New Jersey, Ohio, Pennsylvania, Virginia]

1' Colony irregularly shaped, only loosely attached to substratum; statoblasts tapering to a single point at each end; very uncommon ...... *Lophopus crystallinus* (Pallas, 1768)

[Canada: Quebec. USA: Illinois]

Ectoprocta: Gymnolaemata: Ctenostomata: Families

1 Tentacles numbering more than 8 2

1' Tentacles numbering exactly 8 .... **Victorellidae [p. 274]**

2(1) Zooids branching at nearly right angles to form a diffuse, rambling colony with many free branches (Fig. 13.13 B) .... Paludicellidae, one species; *Paludicella articulata* (Ehrenberg, 1831)

[Canada. USA]

2' Zooids arising from narrow, stolon-like parts of the colony; zooids ranging from bulbous and recumbant to spindle-shaped and erect (Fig. 13.13 C) .... Pottsiellidae, one species; *Pottsiella erecta* (Potts, 1884)

[USA: eastern states]

Ectoprocta: Gymnolaemata: Ctenostomata: Victorellidae: Species

1 Erect portion of zooid 0.6–1.6 mm long and only slightly contractile (Fig. 13.13 D); hibernaculae with short, marginal, forked projections; occurring in both fresh and brackish water .... *Victorella pavida* (Kent, 1870)

[USA: Chesapeake Bay region]

1' Erect portion of zooid never more than 0.3 mm long and highly contractile (Fig. 13.13 A); hibernaculae with smooth outer margins. Known only from two sites in Illinois .... *Sineportella forbesi* Wood & Marsh, 1996

[USA: Illinois]

REFERENCES

Hirose, M. & S.F. Mawatari. 2011. Freshwater Bryozoa of Lake Biwa, Japan. Species Diversity 16: 1–37.

Massard, J. A. & G. Geimer, 2008. Global diversity of bryozoans (Bryozoa or Ectoprocta) in freshwater: an update. Bulletin de la Société des Naturalistes luxembourgeois 109: 139–148.

Rubini, A., G. Pieroni, A. Elia, L. Zippilli, F. Paolocci, & M. Taticchi. 2011. Novel morphological and genetic tools to discriminate species among the family Plumatellidae (Phylactolaemata, Bryozoa). Hydrobiologia 664: 81–93.

Vinogradov, A. V. 2004. Taxonomical structure of Bryozoans Phylactoalemata.. Vestnik Zoologii 38: 3–14.

Wood. T.S. 2005a. Study methods for freshwater bryozoans. Denisia 16: 103–110.

Wood, T.S. 2005b. The pipeline menace of freshwater bryozoans. Denisia 16: 203–208.

Chapter 14

Phylum Entoprocta

Timothy S. Wood
Department of Biological Sciences, Wright State University, Dayton, OH, USA

Chapter Outline

INTRODUCTION

Entoprocta is a small phylum of about 150 described species (Nielsen, 2011) known almost entirely from marine habitats, including several estuarine species. Only two species have been described from freshwater: *Loxosomatoides sirindhornae* Wood, 2005 in Thailand, and *Urnatella gracilis* Leidy, 1847, first reported from southeastern United States and now distributed worldwide (Oda, 1982). At least one additional freshwater species is known but not yet described.

In general, entoprocts are small, sessile, filter feeding animals. The freshwater species grow as colonies of interconnected zooids. They are most commonly found in flowing or lightly turbulent water. A dense population of *Urnatella* can cover all available hard substrates in a felt-like layer (Emschermann, 1987).

For many years, entoprocts were included in the phylum Bryozoa along with a larger group called Ectoprocta (see Chapter 13 in this volume and Chapter 16 in Volume I of this series). Similarities between the groups include sessile colonies (although many marine species are solitary), a U-shaped gut, and the use of ciliated tentacles to gather food. Potentially more significant are the apparent similarities in larval development (see Nielsen, 2011). However, despite much debate, the relationship has never been conclusive, and the weight of new evidence now places the entoprocts in a wholly independent phylum (Waeschenbach et al., 2012).

The discussion below focuses on *Urnatella gracilis*, the best-known and most widely distributed freshwater entoproct, and the only species occurring in the Nearctic.

TERMINOLOGY AND MORPHOLOGY

Urnatella gracilis grows as a stalk of articulating segments capped by a bulbous head bearing ciliated tentacles. The stalk is often branched, with each segmented branch terminating in its own head, also called an atrium or calyx (Fig. 14.1). The stalk is almost entirely muscular except for a branching protonephridial system. The head includes all the organs and tissues required for the normal life functions of feeding, digestion, excretion, sexual reproduction, sensory perception, and nervous coordination. Tentacles bear several rows of cilia of varying sizes. The feeding current flows in a direction opposite from ectoproct bryozoans, drawn from the near the base of the head and discharged away from the mouth. Particles trapped by cilia are thrown into a food groove, along which they travel down the tentacle to the mouth opening.

Reproduction appears to be mostly asexual. Under certain optimal conditions, the branching colonies of *Urnatella* develop special buds of 1–3 zooids that detach from the parent and drift away. These free buds may also creep slowly along the substrate by alternating contractions of longitudinal and transverse muscles (Emschermann, 1987). By this means, the local population can grow very rapidly.

Sexual reproduction in *Urnatella* also has been described, but it appears to be a rare occurrence (Emschermann, 1965). Both ova and sperm have been observed in the same colony. It is believed that embryos are brooded one at a time and released as free swimming larvae.

When environmental conditions become unfavorable, the head drops off and each of the stalk segments becomes a dormant hibernaculum with powers of regeneration.

Thorp and Covich's Freshwater Invertebrates. http://dx.doi.org/10.1016/B978-0-12-385028-7.00014-7

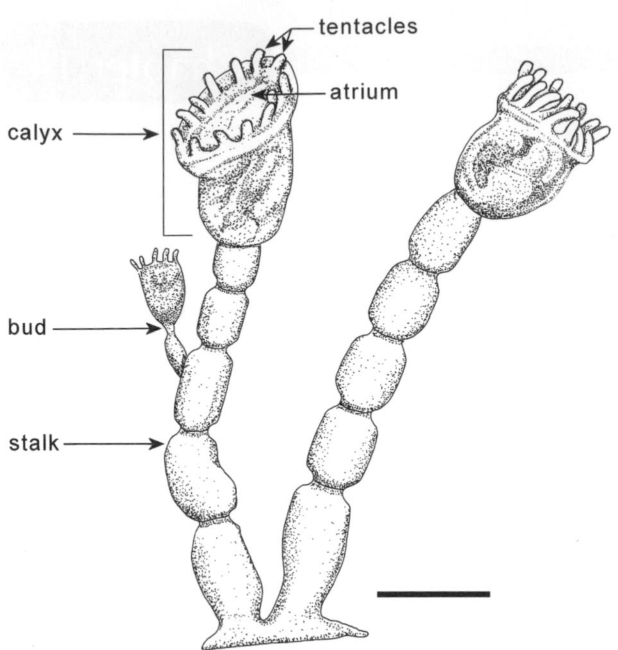

FIGURE 14.1 Small entoproct colony (*Urnatella gracilis*) showing external structures. Scale bar = 0.5 mm.

Overwintering occurs in this form. Resistance of the hibernaculum to desiccation and other environmental extremes has not been tested.

MATERIAL PREPARATION AND PRESERVATION

In the laboratory, dormant stalks of *Urnatella* colonies can remain viable in water for at least 3 years at 0–4 °C. After germination at 15–20 °C, they should be kept in circulating water and fed twice weekly from a culture of *Chlamydomonas* (Emschermann, 1987).

Entoproct colonies may be fixed and preserved in either 5% formalin or 70% ethyl alcohol. Attempts to narcotize with menthol are seldom very successful. For histological work, the best fixative is 2% glutaraldehyde solution in 0.01 M sodium-cacodylate buffer at a pH of 7.4 for 1 hour at around 23 °C (Schwaha et al., 2010). Molecular genetics requires preservation in 100% ethyl alcohol.

REFERENCES

Emscherman, P. 1965. Über die sexuelle Fortpflanzung und die Larve von *Urnatella gracilis* Leidy (Kamptozoa). Zeitschrift für Morphologie und Ökol der Tierre 55: 100–114.

Emschermann, P. 1987. Creeping propagation stolons – an effective propagation system of the freshwater enroproct *Urnatella gracilis* Leidy (Barentsiidae). Archiv fur Hydrobiologie 108(3): 439–448.

Nielsen, C. 2012. Animal evolution: interrelationships of the living phyla. Oxford University Press, Oxford. 464 pp.

Oda, S. 1982. *Urnatella gracilis*, a freshwater kamptozoan occurring in Japan. Annotationes Zoologicae Japonenses 55(3): 151–166.

Schwaha, T., T.S. Wood, & A. Wanninger. 2010. Trapped in freshwater: the internal anatomy of the entoproct *Loxosomatoides sirindhornae*. Frontiers in Zoology 7(1): 7.

Waeschenbach, A., P.D. Taylor & D.T.J. Littlewood. 2012. A molecular phylogeny of bryozoans. Molecular Phylogenetics and Evolution 62(2): 718–735.

Chapter 15

Phylum Tardigrada

Diane R. Nelson
Department of Biological Sciences, East Tennessee State University, Johnson City, TN, USA

Roberto Guidetti, Lorena Rebecchi
Department of Life Sciences, University of Modena and Reggio Emilia, Modena, Italy

Chapter Outline

INTRODUCTION

Tardigrades are xerophilous, hygrophilous, and hydrophilous micrometazoans categorized into two main classes (Eutardigrada and Heterotardigrada) with over 1200 species described from terrestrial, freshwater, and marine habitats. Commonly called "water bears," they must be surrounded by a film of water in order to be active. Cryptobiosis allows tardigrades to survive periods of desiccation in terrestrial habitats, but few freshwater and marine species are known to have this adaptation. Encystment occurs in some freshwater, moss-dwelling, and soil tardigrades, possibly in response to stressful environmental conditions (e.g., oxygen depletion, pH alteration, and temperature variation). Some typically terrestrial species may inhabit freshwater environments, but most are probably "accidentals" washed into water during precipitation events (rainfall, flooding), although several are limnoterrestrial and colonize both terrestrial and freshwater habitats. The fewest number of species are obligate limnic species; however, the freshwater habitat has been investigated less frequently so it is difficult to estimate the total number of species. In the literature, the number of Nearctic species in freshwater habitats is over 70, with about 30 species found exclusively or primarily in aquatic environments, although we expect the actual number to be much higher. These species are primarily eutardigrades that belong to 2 orders and 7 families (Nelson et al., 2015, Table 17.2). We expect the global zoogeographic distribution of limnoterrestrial tardigrades, as previously reported in the literature, to undergo further analysis and revision as species groups and cryptic species are investigated more thoroughly (McInnes, 1994; McInnes & Pugh, 2007).

Hydrophilous tardigrades are benthic organisms that inhabit permanent freshwater habitats, crawling on aquatic plants or in the interstitial spaces of sediments in lakes, rivers, streams, groundwater, and temporary or permanent ponds. They also live in subterranean caves, the hyporheic zone, activated sludge of sewage treatment plants, and in cryoconite holes in glaciers.

LIMITATIONS

Identification of tardigrades is based primarily on variations in the morphology of the claws, buccal-pharyngeal apparatus, cuticle, and egg shell. These characters are included in the key and shown in Figs. 15.1–15.14. We have included a key to genera since many species are difficult to identify even by tardigrade taxonomists. Cryptic species are present, and many species groups need taxonomic revision. Although adult specimens are recommended for use in identification, juvenile eutardigrades are similar to adults. Sexual dimorphism in characters is known only for a few species. Revisions of genera, families and superfamilies in recent years are based on a combination of molecular and morphological evidence (Sands et al., 2008; Guidetti et al., 2009; Jørgensen et al., 2010; Marley et al., 2011; Vicente et al., 2013).

Thorp and Covich's Freshwater Invertebrates. http://dx.doi.org/10.1016/B978-0-12-385028-7.00015-9

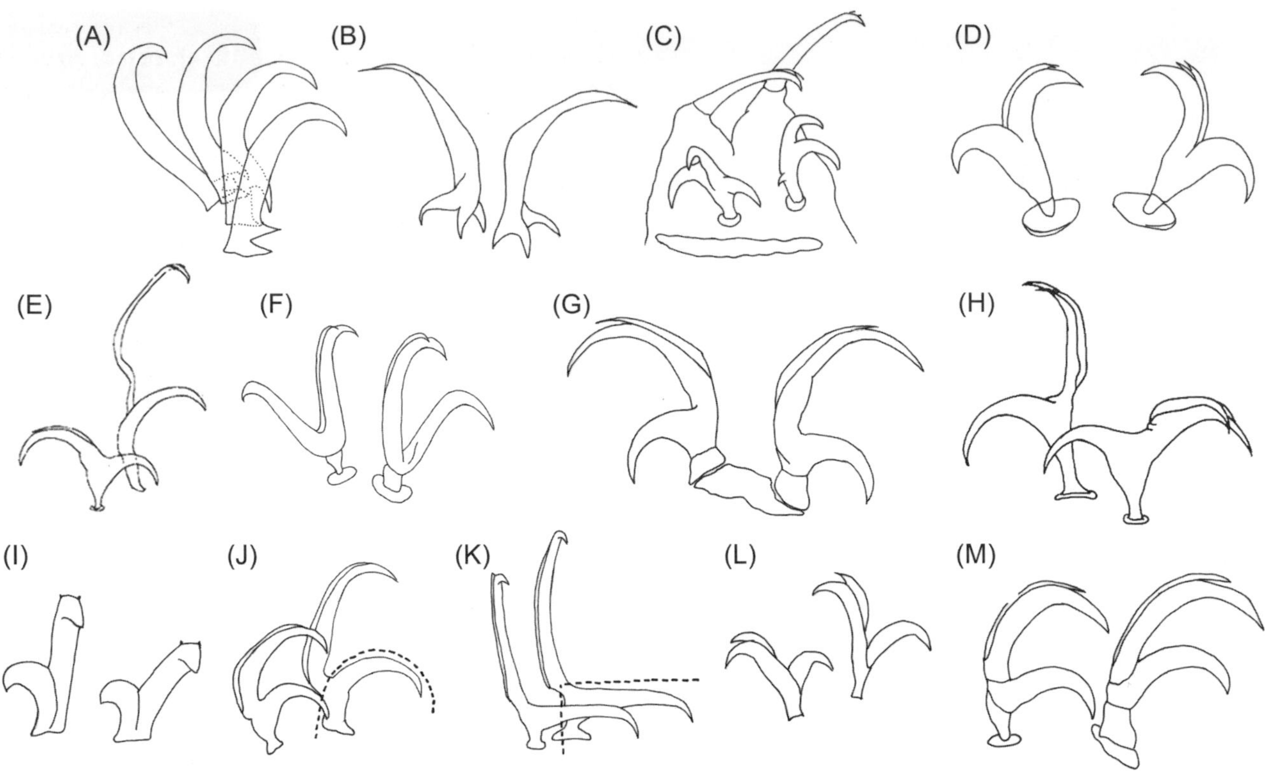

FIGURE 15.1 Diagrammatic representation of claw morphologies in tardigrades. (A and B) Heterotardigrada: (A) *Echiniscus*; (B) *Carphania*. (C–M) Eutardigrada: (C) *Milnesium*; (D) *Macrobiotus*; (E) *Ramajendas*; (F) *Murrayon*; (G) *Dactylobiotus*; (H) *Thulinius*; (I) *Calohypsibius*; (J) *Hypsibius*; (K) *Isohypsibius*; (L) *Microhypsibius*; (M) *Bertolanius*. *A–D, F–K, M, redrawn from Bertolani, 1982; E, redrawn from Pilato & Binda, 1990; L, redrawn from Pilato, 1998.*

TERMINOLOGY AND MORPHOLOGY

Tardigrades range in size from 50 µm in juveniles to over 1.5 mm. They have four pairs of legs, usually with claws and a buccal-pharyngeal apparatus designed for piercing-sucking. The body is covered by a cuticle, which molts periodically.

In echiniscid heterotardigrades (occasionally found in freshwater), each leg terminates in four unbranched claws in adults but only two claws per leg in the first instar. The freshwater genus *Carphania* has two claws on legs I-III and only one claw on reduced leg IV; all claws have two accessory points (as in eutardigrades) and four basal spurs (Fig. 15.1B). Accessory points, which are thin cuticular structures that extend their distal portion free of the claw, are present only in *Carphania*.

The limnoterrestrial eutardigrades usually possess two double-claws on each leg, arranged as external and internal claws on the first three pairs of legs and as anterior and posterior claws on the hind legs. The claws on each leg may be similar or different in shape and size (Figs. 15.1 C–M and 15.3). In true limnic species of eutardigrades, the claws are usually long or very long. Each double-claw consists of a longer primary branch with two accessory points and a secondary branch without accessory points (Guidetti & Bertolani, 2005; Pilato & Binda, 2010).

In the eutardigrade order Apochela (family Milnesiidae), there are no true double-claws, since the two claw branches are separate from each other. The primary branch is long and thin and usually has accessory points, while the secondary branch is short, stout, and bears two or three hooks or spurs; the two claws of each leg are symmetrical (Fig. 15.1C). In the eutardigrade order Parachela, the primary and secondary branches are connected and usually arise from a common basal tract. The secondary branch has no hooks; however, in the limnic species *Isohypsibius deflexus* Mihelčič, 1960, the two branches of the external claws are not connected (Guidetti & Bertolani, 2005; Pilato & Binda, 2010). In many cases, the base of the claw is surrounded by a cuticular lunule (Fig. 15.4). Cuticular bars, with taxonomic value at the species level, can be present on the legs in various positions.

Within the Parachela, the families Macrobiotidae and Murrayidae have symmetrical claws with a lateral-medial sequence of claw branches as follows: secondary branch of the external claw and its primary branch, primary branch of the internal claw and its secondary branch ("2-1-1-2 claws") (Fig. 15.1 D, F, G). Macrobiotidae have Y-shaped

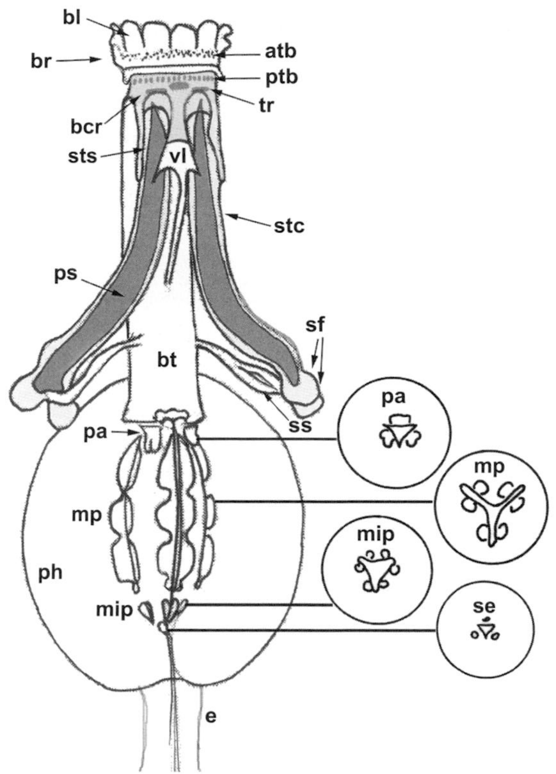

FIGURE 15.2 Diagrammatic representation of a tardigrade buccal-pharyngeal apparatus. Note: For this and subsequent figures in Chapter 15, the following abbreviations apply: *ap*=accessory points; *atb*=anterior band of teeth; *b*=basal portion; *bc*=buccal cirri; *bcr*=buccal crown; *bl*=peribuccal lamellae; *bp*=buccal papillae; *br*=buccal ring; *bt*=buccal tube; *c*=claws; *cA*=cirrus A; *cl*=clava; *co*=common tract; *ct*= cuticular plate; *e*=esophagus; *f*=filament; *l*=lunule; *lp*=lateral papillae; *mb*=main branch; *mip*=microplacoid; *mp*=macroplacoid; *op*=oral papillae; *pa*=pharyngeal apophyses; *ph*=pharynx; *pl*=placoids; *ptb*=posterior band of teeth; *ps*=piercing stylets; *s*=stylet; *sb*=secondary branch; *se*=septulum; *sf*=stylet furca; *sn*=spine; *ss*=stylet supports; *st*=stalk; *stc*=stylet coat; *sts*=stylet sheath; *tb*=teeth band; *tr*=transverse ridges; *vl*=ventral lamina. *Redrawn from Guidetti et al., 2012.*

double-claws (Figs. 15.1 D and 15.3 C), with the claw branches fused over a basal tract of variable length. Murrayidae have double-claws with the two branches diverging immediately after the basal tract (V-shaped, Fig. 15.1 F, and L-shaped double-claws, Figs. 15.1 G and 15.3 D). In the limnic *Dactylobiotus* (Figs. 15.1 G and 15.3 D), and *Macroversum* (Murrayidae), a cuticular link between the bases of the claw pair is present. The basal tract may consist of three parts: a proximal stalk connecting the basal tract to the lunule, a distal portion (which may be delimited by a transverse septum), and a common tract after which the two claw branches diverge (Fig. 15.4) (Guidetti & Bertolani, 2005; Pilato & Binda, 2010).

All other families are characterized by asymmetrical claws ("2-1-2-1" claw sequence: secondary branch of the external claw and its primary branch, secondary branch of the internal claw and its primary branch; Fig. 15.1 E, H–M). In Isohypsibiidae, the secondary branch is inserted perpendicularly to the common basal tract of the claw; the external and internal claws are often similar in size and shape (*Isohypsibius*-type claw, Figs. 15.1 H, K and 15.3 B). In Hypsibiidae, the secondary branch of the external double-claw forms a common arc with the basal tract; the external and internal claws are different in shape and size (*Hypsibius*-type claw, Fig. 15.1 E, J) (Guidetti & Bertolani, 2005; Pilato & Binda, 2010).

The buccal-pharyngeal apparatus is taxonomically important, especially in the eutardigrades. Basically, it consists of a mouth, buccal ring, buccal tube (totally or partially rigid) with protuberances for muscle attachments (e.g., buccal crown, dorsal and ventral apophyses), stylet mechanism, and a muscular sucking pharynx (usually containing bars or placoids) (Figs. 15.2, 15.5, 15.6, 15.7, and 15.8). The presence and the number of cuticular structures around the mouth (peribuccal structures; Fig. 15.5) are significant at the generic level (Guidetti & Bertolani, 2005; Pilato & Binda, 2010; Guidetti et al., 2012, 2013).

The interior of the mouth may contain complex cuticular structures (buccal armature) consisting of anterior and posterior bands of small cuticular teeth (mucrones) followed by dorsal and ventral transverse ridges (teeth), which are species-specific characteristics (Figs. 15.2 and 15.5). Most eutardigrades have a completely rigid buccal tube, which is supported by a ventral lamina (strengthening bar) (Figs. 15.2, 15.6 E–G, O, 15.7 A, and 15.8 A, B) in macrobiotids, murrayids, one genus of isohypsibiids (*Doryphoribius*), some calohypsibiids, and in *Apodibius*. The murrayids *Dactylobiotus* and *Murrayon* have a deep hollow in the ventral lamina that in lateral view forms an evident hook (Fig. 15.8 A) (Guidetti & Bertolani, 2005; Pilato & Binda, 2010; Guidetti et al., 2012, 2013).

The wall of the anterior portion of the buccal tube is ventrolaterally surrounded by the buccal crown, which is composed of cuticular crests and laminae on which the stylet protractor muscles are attached (Figs. 15.2, 15.7, and 15.8) (Guidetti et al., 2012, 2013). In lateral view, the buccal crown shows apophyses that are taxonomically significant and vary in size and shape (e.g., crests as in *Isohypsibius*, *Thulinius* and *Ramajendas*, hooks as in *Hypsibius*, lobes as in *Borealibius*, and others) (Fig. 15.8).

In some genera in different eutardigrade families (Milnesiidae, Hypsibiidae, Ramazzottiidae, Macrobiotidae, Eohypsibiidae) and in some echiniscid heterotardigrades, a rigid buccal tube between the buccal ring and the stylet supports is followed by a flexible, annulated pharyngeal tube from just below the stylet supports to the pharynx (Figs. 15.6 F, G, I, K; and 15.7 C).

The buccal tube enters the tripartite pharynx, often terminating with three buccal tube apophyses functioning as valves. Lining the pharyngeal lumen are three cuticular bars in heterotardigrades (Fig. 15.6 A, B), including

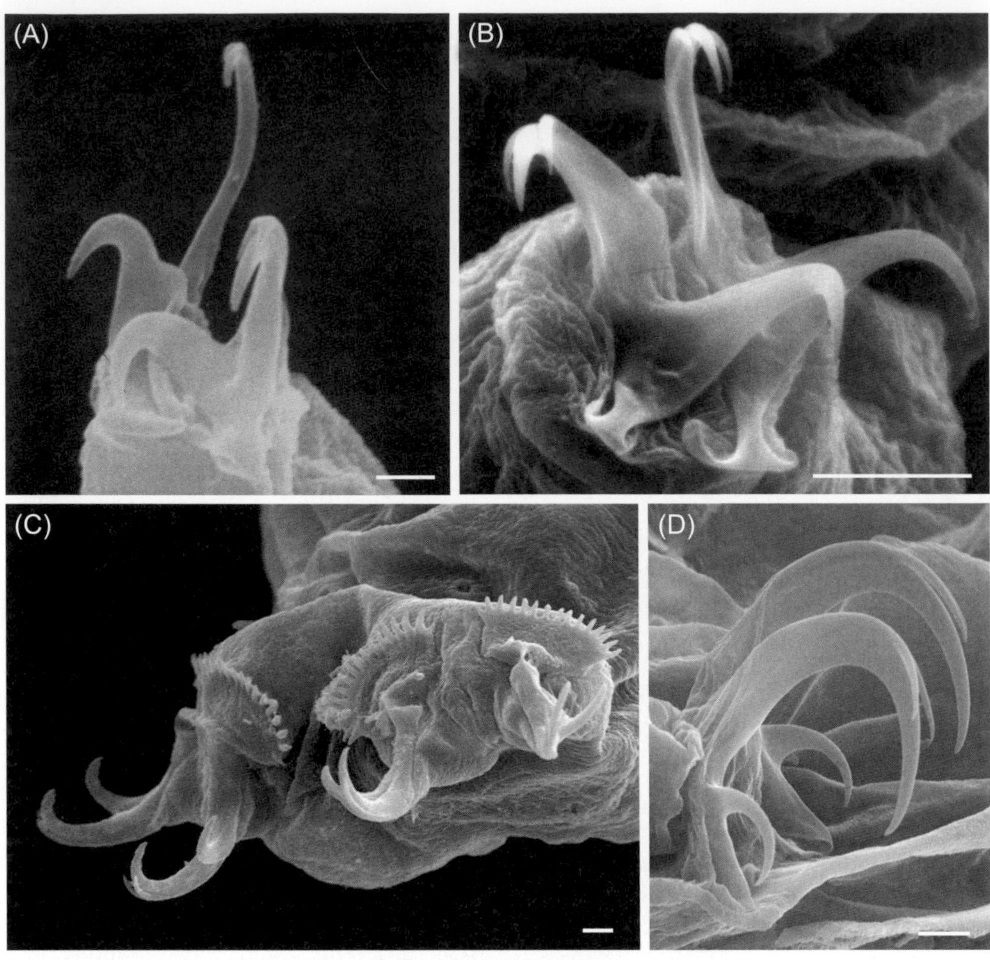

FIGURE 15.3 Scanning electron micrographs of eutardigrade claws: (A) *Ramazzottius oberhaeuseri*; (B) *Thulinius* sp.; (C) *Richtersius coronifer*; (D) *Dactylobiotus* sp. Scale bars = 5 μm.

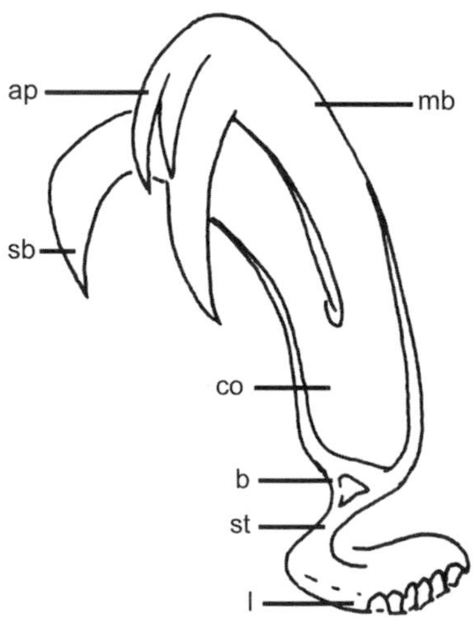

FIGURE 15.4 Diagrammatic representation of tardigrade claw structure. *Redrawn from Nelson & McInnes, 2002.*

the limnic *Carphania fluviatilis* Binda, 1978 (Fig. 15.6 C) or three double rows of cuticular thickenings called placoids (Fig. 15.2), which are taxonomically significant in the Parachela (Eutardigrada). The placoids alternate in position with the three buccal tube apophyses. The larger, anterior placoids called macroplacoids are in two or three transverse rows. Smaller, more posterior placoids called microplacoids may be present and only in a single transverse row. In the same plane as each apophysis, posterior to the placoids, a single median thickening (septulum; three septula in a transverse row) may be present in some species. Microplacoids and/or septula can be present or absent and are taxonomically significant at the species level. In Apochela (e.g., *Milnesium*) and a few Parachela (e.g., *Itaquascon, Parascon, Astatumen*), the cuticular thickenings within the pharynx are reduced or absent (Fig. 15.6 L) (Guidetti & Bertolani, 2005; Pilato & Binda, 2010; Guidetti et al., 2012, 2013).

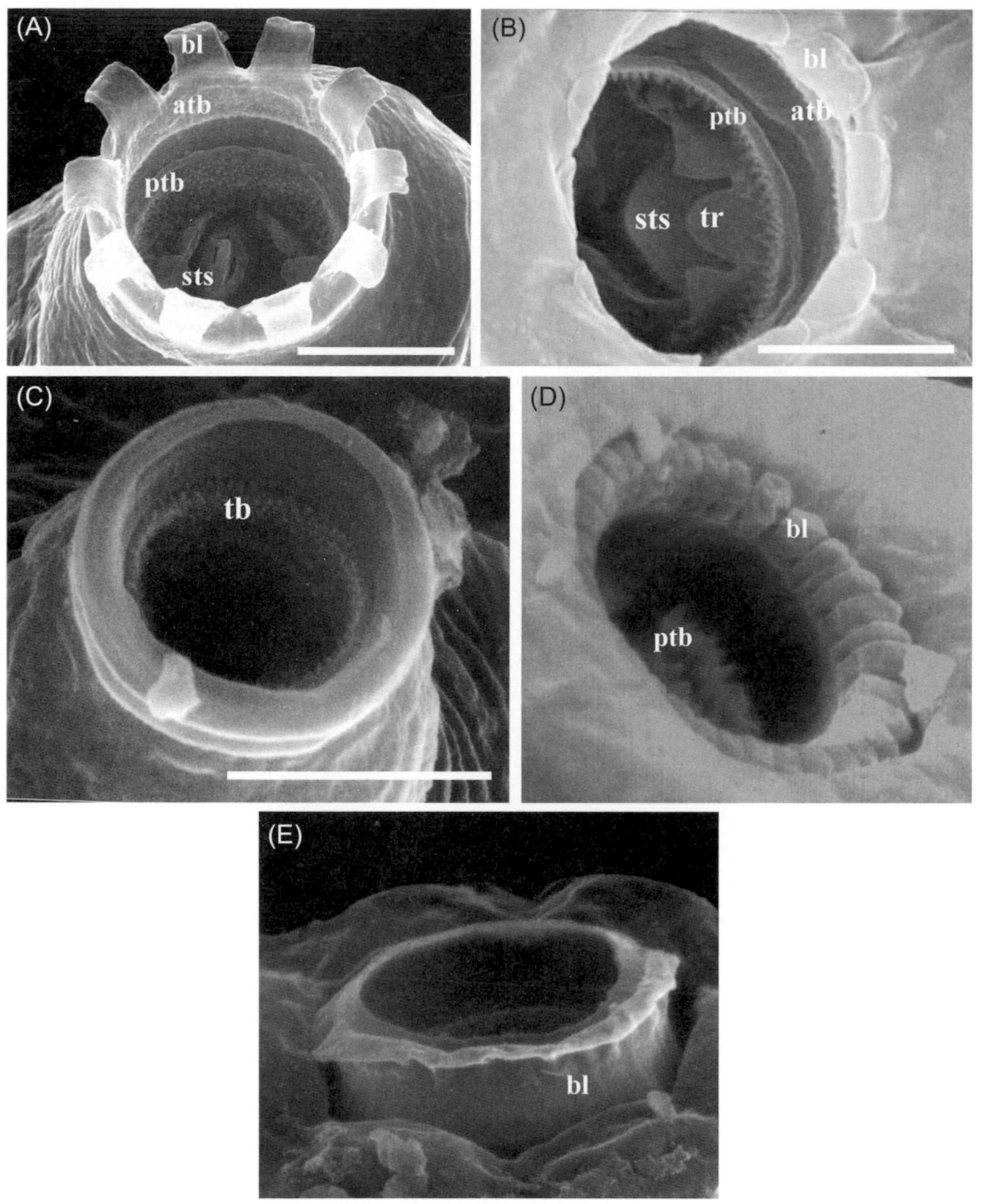

FIGURE 15.5 Scanning electron micrographs of eutardigrade mouth openings: (A) *Dactylobiotus selenicus*; (B) *Macrobiotus* sp.; (C) *Isohypsibius lunulatus*; (D) *Pseudobiotus* sp.; (E) *Thulinius* sp. Scale bars = 5 μm.

The cuticle and its processes are taxonomically important for identifying genera and species and often are different in the two extant classes of Tardigrada, Heterotardigrada, and Eutardigrada (Nelson et al., 2015, Table 17.1). The classes are distinguished by the presence (in Heterotardigrada) or absence (in Eutardigrada) of a pair of cuticular cephalic cirri (called cirri A) (Figs. 15.9, 15.10 A, and 15.11 A). Heterotardigrada includes mainly the marine and armored terrestrial tardigrades, while its sister group, Eutardigrada, primarily encompasses unarmored freshwater and other terrestrial species.

Eggs are often essential for the identification of freshwater and terrestrial species (Fig. 15.12), especially in some eutardigrade genera in which the animals are morphologically very similar. Ornamented eggs are usually deposited freely and singly or in small groups. The ornamentation may include pores, reticulations, and processes with a wide variety of shapes, which have taxonomic significance. The laid eggs are often spherical or oval, with a diameter usually 50–100 μm but sometimes up to 235 μm, including the processes of the shell. Other eutardigrades and limnoterrestrial heterotardigrades (except *Carphania*, whose eggs

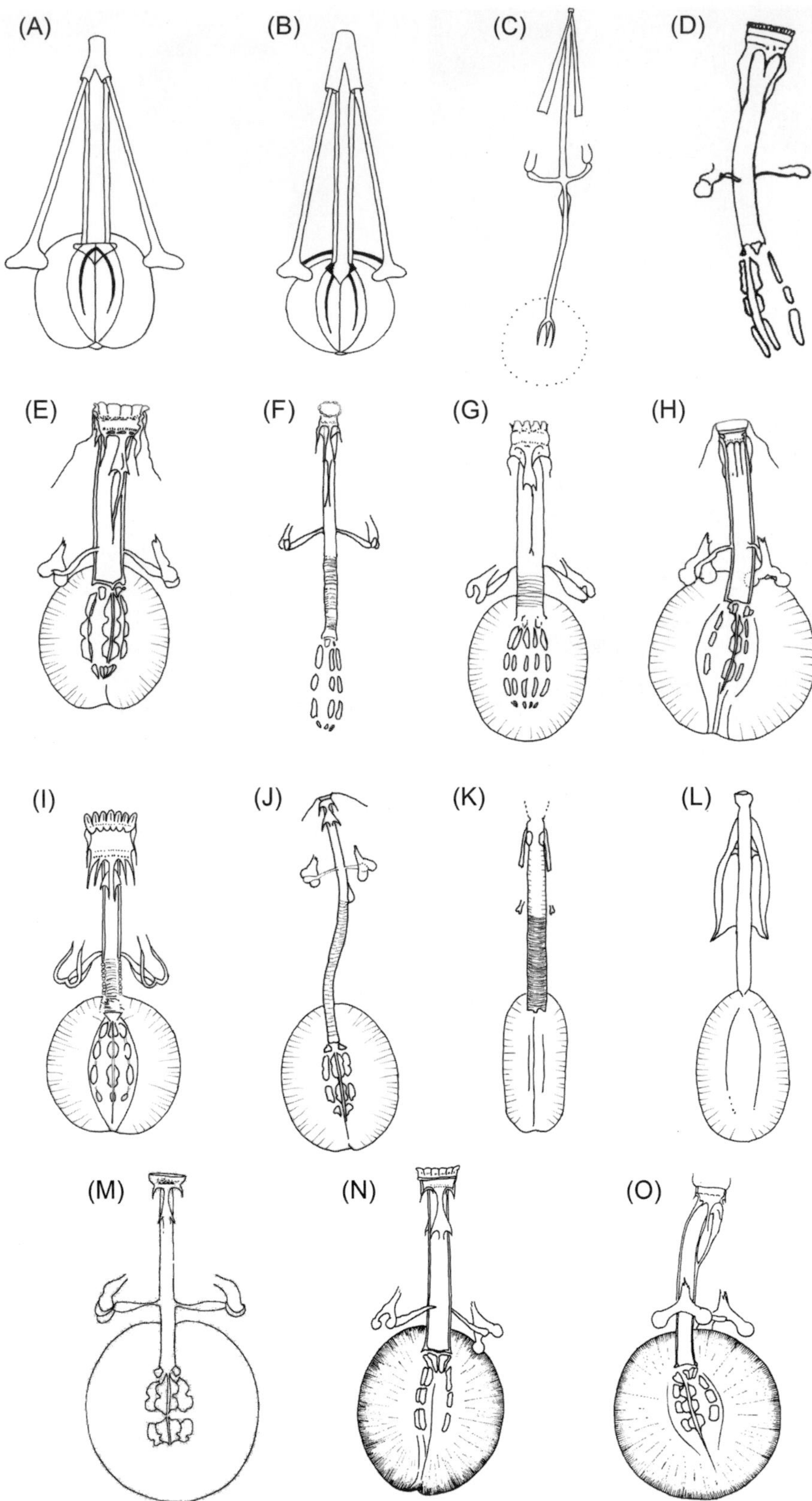

FIGURE 15.6 Diagrammatic representation of buccal-pharyngeal apparatuses: (A–C) Heterotardigrada: (A) *Echiniscus*; (B) *Pseudechiniscus*; (C) *Carphania*. (D–L) Eutardigrada: (D) *Pseudobiotus*; (E) *Macrobiotus*; (F) *Biserovus*; (G) *Insuetifurca*; (H) *Isohypsibius*; (I) *Eohypsibius*; (J) *Diphascon*; (K) *Astatumen*; (L) *Parascon*; (M) *Ramajendas*; (N) *Thulinius*; (O) *Doryphorybius*. *A, B, redrawn from Kristensen, 1987; C, redrawn from Binda & Kristensen, 1986; D, E, H, J, K, N, O, redrawn from Bertolani, 1982; F, G, redrawn from Guidetti & Pilato, 2003; I, redrawn from Bertolani & Kristensen, 1987; L, redrawn from Pilato & Binda, 1987; M, redrawn from Pilato & Binda, 1990.*

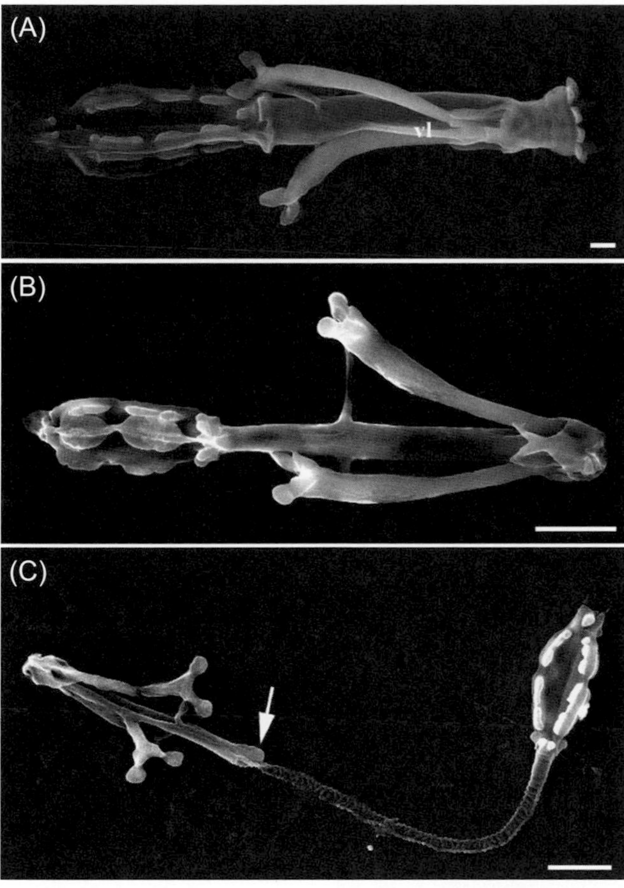

FIGURE 15.7 Scanning electron micrographs of eutardigrade buccal-pharyngeal apparatuses: (A) *Paramacrobiotus richtersi* with a ventral lamina on the buccal tube; (B) *Hypsibius dujardini* and (C) *Diphascon* sp. without a ventral lamina on the buccal tube. Arrow = drop-like thickening. Scale bars = 5 μm.

are unknown and *Oreella*, which has ornamented eggs laid freely) lay smooth eggs within the exuvium (Fig. 15.13).

MATERIAL PREPARATION AND PRESERVATION

If possible, mount individual specimens collected in the field (see techniques in Nelson et al., 2015) on microscope slides immediately and observe with phase contrast and/or differential interference contrast (DIC) microscopy. Hoyer's mounting medium (distilled water, 50 ml; gum arabic, 30 g; chloral hydrate, 150 g; glycerol 20 ml) produces cleared, distended specimens, but seal the coverslip (e.g., with epoxy paint) to ensure permanence. Reduce the amount of chloral hydrate to decrease the extent of clearing of specimens, and add potassium iodide (KI, 1 g) to stain specimens slightly to increase the resolution of morphological structures. Since chloral hydrate is a controlled substance in many countries, other recommended media include Faure's medium, glycerin, polyvinyl alcohol, and polyvinyl lactophenol.

If not mounted on slides after extraction, preserve tardigrades in small amounts of water, with Carnoy's fluid, boiling water, or boiling alcohol (85% or higher) for tardigrade fixation, but use alcohol to preserve specimens to be mounted later. Fixation shrinks the tardigrades, making them more difficult to identify, and specimens lose color in alcohol, and some characters (e.g., fine cuticular structures) are less visible. For molecular analysis (DNA extraction), preserve tardigrades in a very small amount of distilled water and freeze at −20 °C, or preserve in ethanol.

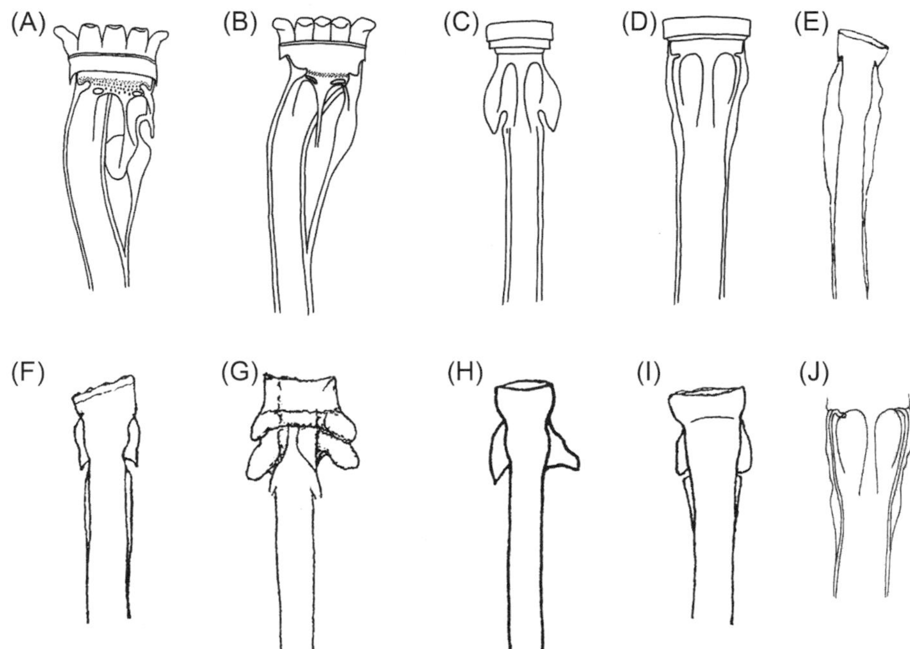

FIGURE 15.8 Diagrammatic representation of apophyses for the insertion of the stylet muscles on the wall of the buccal tube: (A) *Dactylobiotus*; (B) *Macrobiotus*; (C) *Hypsibius* and *Diphascon*; (D) *Isohypsibius*; (E) *Ramajendas*; (F) *Acutuncus*; (G) *Borealibius*; (H) *Bindius*; (I) *Mixibius*; (J) *Thulinius*. *A–C, redrawn from Bertolani, 1982; D, redrawn from Nelson et al., 1999; E, redrawn from Pilato & Binda, 1990; F–I, redrawn from Pilato & Binda, 2010; J, redrawn from Bertolani, 1982.*

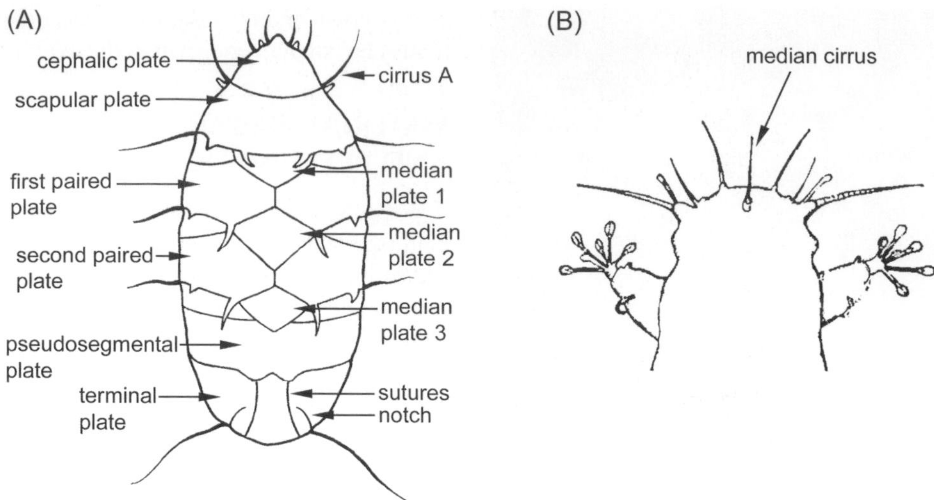

FIGURE 15.9 (A) Diagrammatic representation of dorsal cuticular plates of an echiniscid; (B) Diagrammatic representation of a head of an arthrotardigrade. *A, B, redrawn from Ramazzotti & Maucci, 1983.*

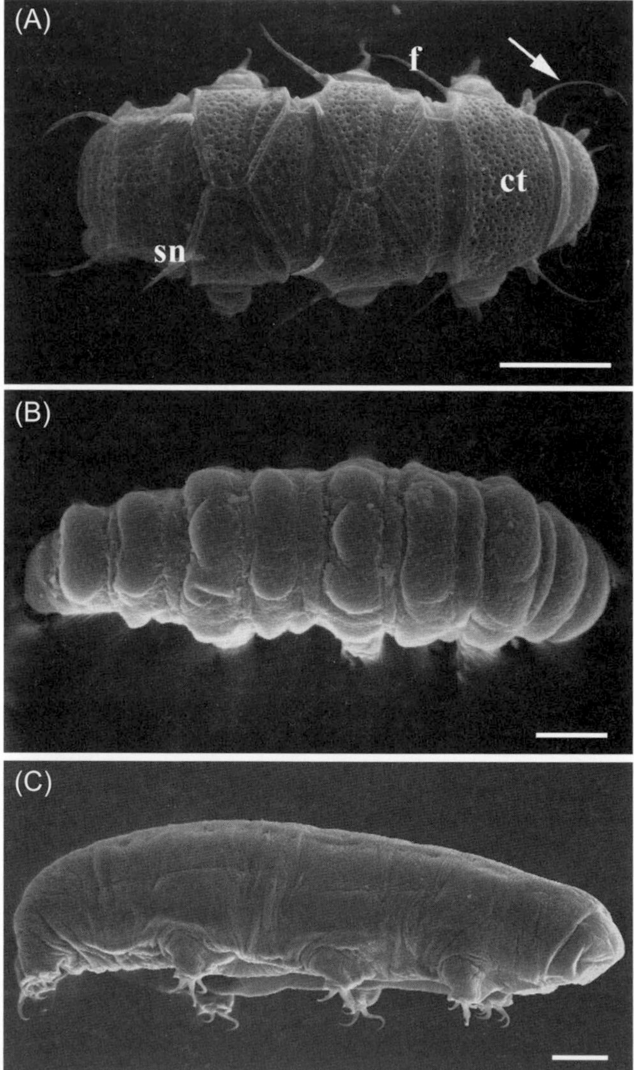

FIGURE 15.10 Scanning electron micrographs of tardigrades: (A) dorsal view of the heterotardigrade *Echiniscus duboisi* (arrow = cirrus A); (B) dorsal view of the eutardigrade *Isohypsibius lunulatus*; (C) lateral view of the limnic eutardigrade *Borealibius zetlandicus*. Scale bars = 25 μm.

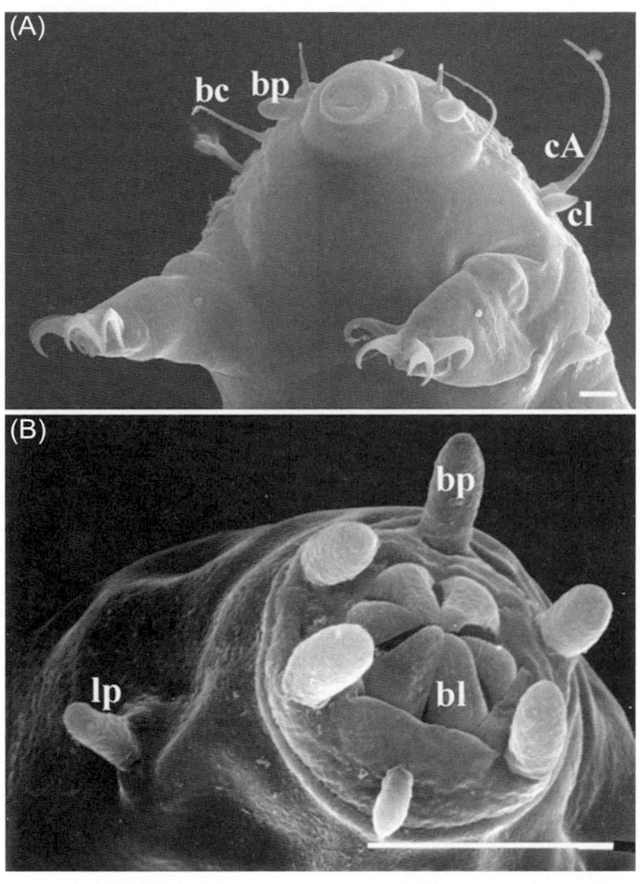

FIGURE 15.11 Scanning electron micrographs of tardigrade heads: (A) the echiniscid heterotardigrade *Antechiniscus parvisentus*; (B) the milnesiid eutardigrade *Milnesium* cf. *tardigradum*. *B, from Bertolani et al., 2009.*

KEYS TO FRESHWATER TARDIGRADA

In the following keys we use "obligate" for the genera with species that typically live in freshwater habitats, "limnoterrestrial" for the genera with species that may be found in both freshwater and terrestrial habitats, and "accidental" for the genera with species that typically do not live in freshwater habitats but may be rarely found or are accidently washed or blown into the aquatic habitat.

Tardigrada: Classes

1 Lateral cirri A present (Figs. 15.9, 15.10 A, and 15.11 A) .... **Heterotardigrada [p. 285]**

1' Lateral cirri A absent .... **Eutardigrada [p. 287]**

Tardigrada: Heterotardigrada: Orders

1 Median cirrus usually present (Fig. 15.9 B); digits or claws directly inserted on the leg .... Arthrotardigrada, one species: *Styraconyx hallasi* [accidental, marine but also found in Greenland springs]

1' Median cirrus absent; each claw on a papilla inserted on the leg .... **Echiniscoidea [p. 285]**

Tardigrada: Heterotardigrada: Echiniscoidea: Families

1 Four claws per leg in adults, cirrus A short to very long; with dorsal-lateral plates (Figs. 15.1 A, 15.9 A, and 15.10 A) .... **Echiniscidae [p. 286]**

1' Two claws on legs I-III, one claw on leg IV in adults; cuticle without dorsal plates; cirrus A very short; obligate .... Carphaniidae, one genus: *Carphania*

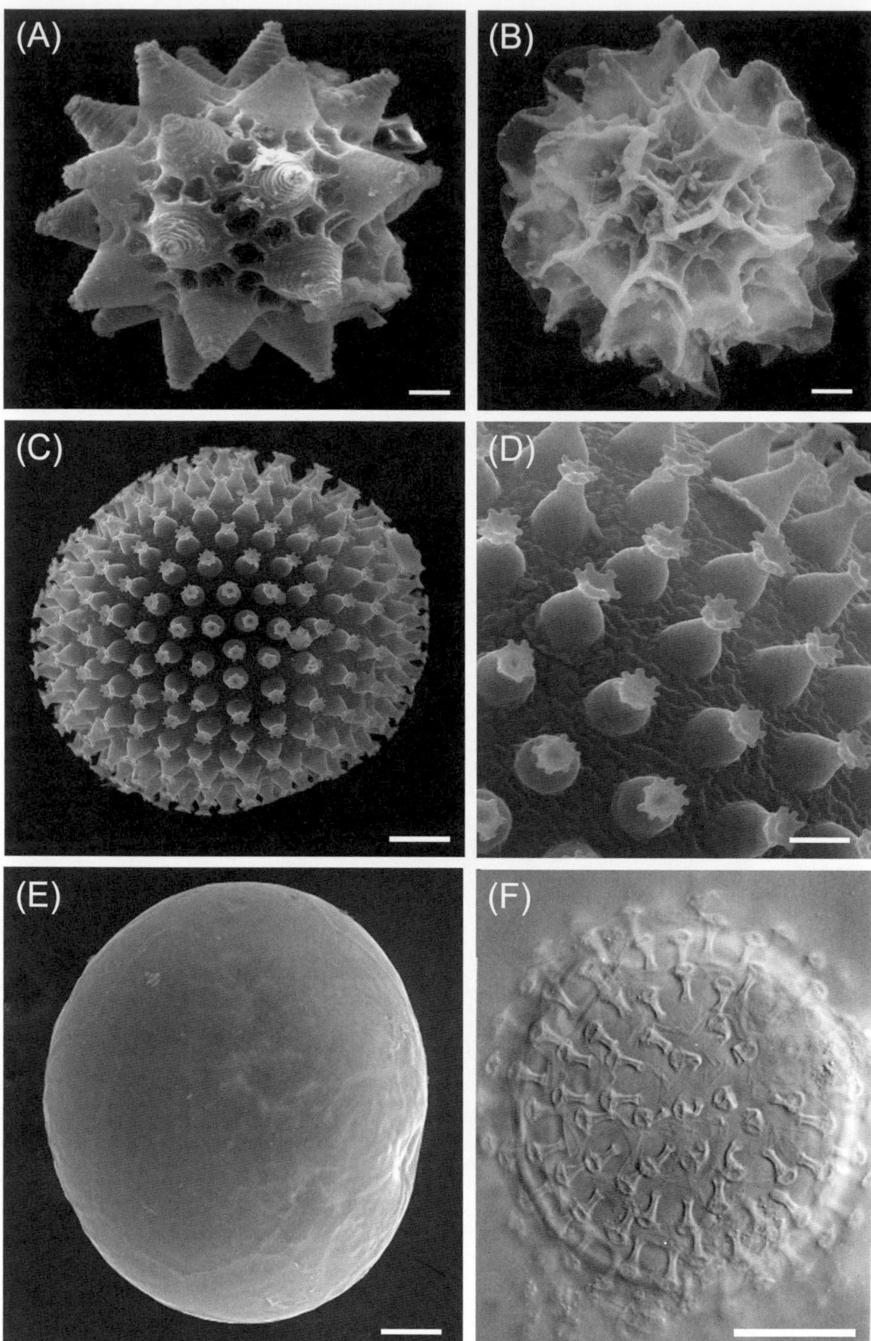

FIGURE 15.12 (A–D) Scanning electron micrographs of ornamented tardigrade eggs: (A) *Paramacrobiotus richtersi*; (B) *Bertolanius weglarskae*; (C) *Macrobiotus* gr. *hufelandi*; (D) magnification of C; (E) *Milnesium* cf. *tardigradum* smooth egg extracted from exuvium (F) light micrograph (DIC) of the ornamented egg of *Murrayon hastatus*. Scale bars: A–D, F = 25 μm, E = 8 μm. *F, from Bertolani & Rebecchi, 1999.*

Tardigrada: Heterotardigrada: Echiniscoidea: Echiniscidae: Genera

1 Pseudosegmental plate absent; median plates not subdivided; black or red eyespots (Fig. 15.14 A) 2

1' Pseudosegmental plate present between second paired plate and terminal plate (Fig. 15.9 A); black eyespots; accidental .... *Pseudechiniscus*

2 Terminal plate notched (Fig. 15.9 A); red eyespots; accidental .... *Echiniscus*

2' Terminal plate divided by sutures (Fig. 15.9 A); black eyespots (Fig. 15.14 A); accidental .... *Hypechiniscus*

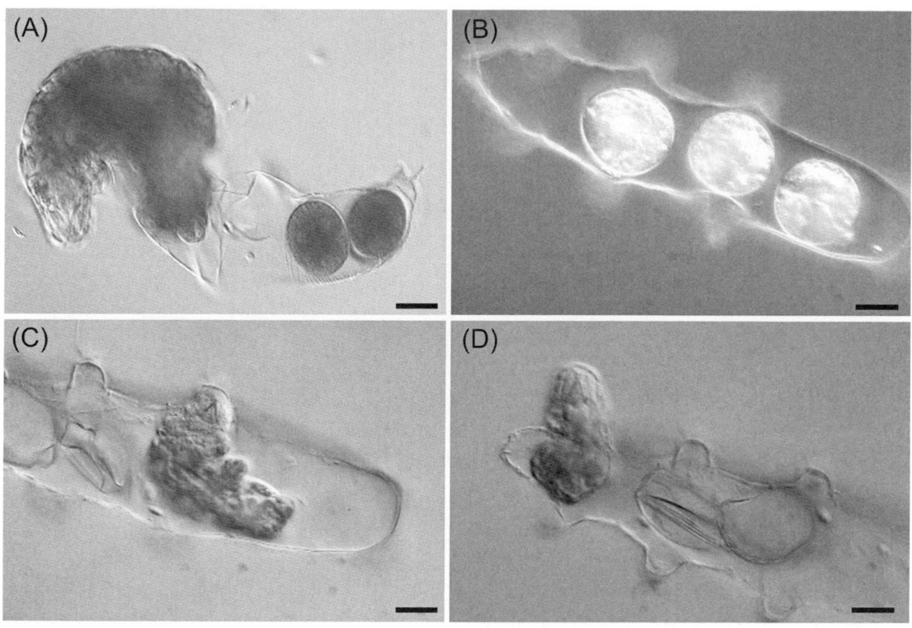

FIGURE 15.13 *In vivo* micrographs of *Diphascon* sp. (DIC). (A) Animal releasing the eggs in the exuvium; (B) exuvium with three embryonate eggs; (C) newborn (juvenile) trying to find the exit from the exuvium; (D) newborn (juvenile) crawling out of exuvium. Scale bars = 50 μm.

Tardigrada: Eutardigrada: Orders

1 Head with cephalic papillae, including two lateral papillae (Fig. 15.11 B); claws with the secondary branch not connected to the primary branch (Fig. 15.1 C) .... Apochela, one genus: *Milnesium* [limnoterrestrial]

1' Head without cephalic papillae; usually two double claws per leg (Fig. 15.1 D–M) .... **Parachela [p. 287]**

Tardigrada: Eutardigrada: Parachela: Superfamilies

1 Two double claws on each leg "asymmetrical," with sequence 2-1-2-1 (Figs. 15.1 E, H–M and 15.3 A, B); ventral lamina absent; smooth eggs laid within the exuvium (Fig. 15.13) or ornamented eggs laid freely (Fig. 15.12) 2

1' Two double claws on each leg similar in size and shape and "symmetrical," with sequence 2-1-1-2 (Figs. 15.1 D, F, G and 15.3 C, D); ventral lamina present (Figs. 15.2, 15.7 A, and 15.8 A, B); ornamented eggs laid freely (Fig. 15.12) .... **Macrobiotoidea [p. 287]**

2(1) Internal and external claws of similar size and shape (Figs. 15.1 H, K–M and 15.3 B) 3

2' Internal and external claws of clearly dissimilar size and shape (Figs. 15.1 E, J and 15.3 A); most with external claws of *Hypsibius*-type, with secondary branch forming a common arc with the basal portion; most with smooth eggs laid within the exuvium .... **Hypsibioidea [p. 289]**

3(2) Each double claw with three distinct but connected parts (*Eohypsibius*-type claw; Fig. 15.1 M); the internal claw can rotate and simulate a symmetrical arrangement with the external claw on the same leg; 14 peribuccal lamellae; eggs laid freely and ornamented .... Eohypsibioidea, one family: **Eohypsibiidae [p. 289]**

3' Claws with primary and secondary branch continuous with the basal portion; secondary branch of each claw forming a right angle (or wider) with claw base (Fig. 15.1 H, K); peribuccal lamellae present or absent; eggs laid within the exuvium .... Isohypsibioidea, one family: **Isohypsibiidae [p. 289]**

Tardigrada: Eutardigrada: Parachela: Macrobiotoidea: Families

1 L or V-shaped claws, i.e., with the two claw branches diverging immediately after the basal portion (Fig. 15.1 F, G); evident hook on the ventral lamina (Fig. 15.8 A) .... **Murrayidae [p. 288]**

1' Y-shaped claws, i.e., with claw branches fused over a tract of variable length (common tract) (Figs. 15.1 D, 15.3 C, and 15.4); no evident hook on the ventral lamina .... **Macrobiotidae [p. 288]**

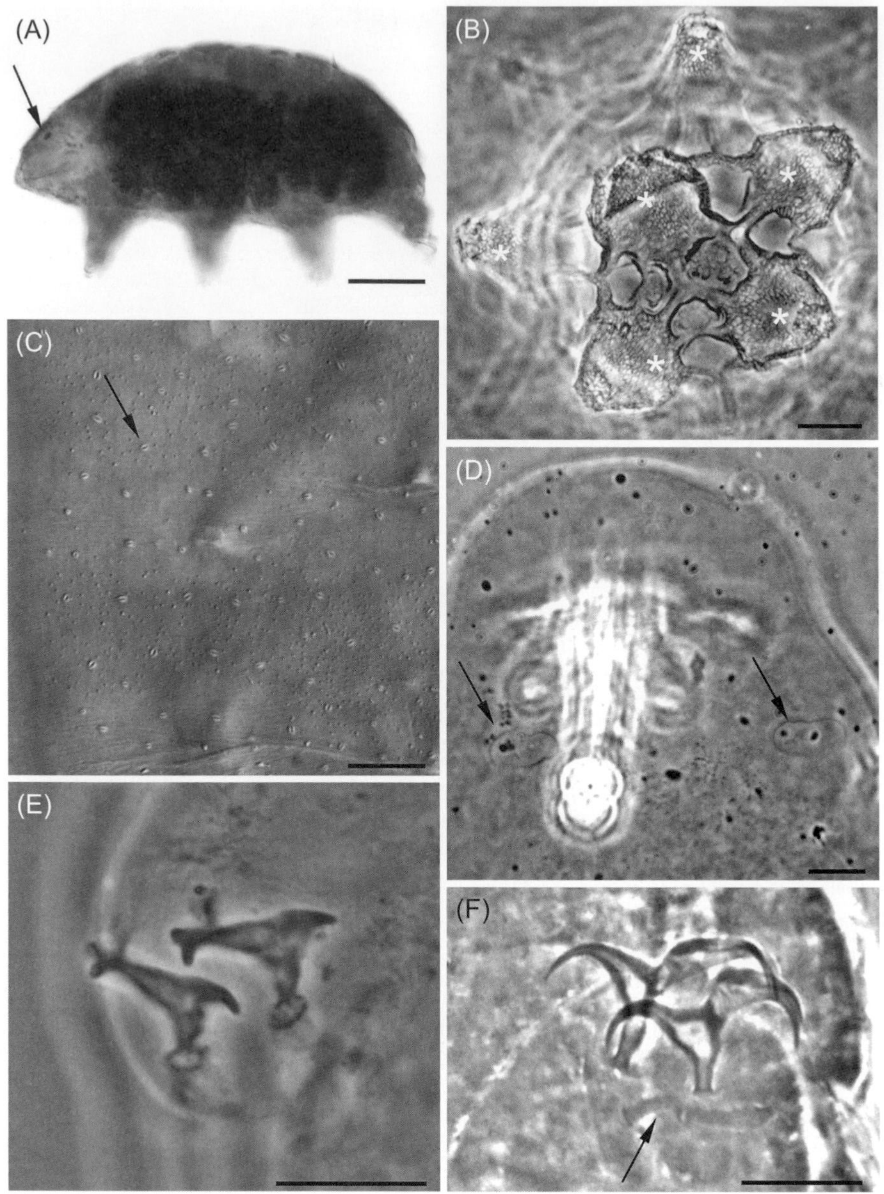

FIGURE 15.14 (A) *Echiniscus* sp. *in vivo* with an eye spot (arrow) (DIC). (B) Egg processes (asterisks) with reticulated walls of a *Paramacrobiotus* species (PhC). (C) Dorsal cuticle with pores of a *Macrobiotus* species (arrow) (DIC). (D) Elliptical dorsolateral sense organs on the head of *Ramazzottius* sp. (PhC). (E) Hind claws of *Eremobiotus* sp. (PhC). (F) Cuticular bars under the claw bases (arrow) of *Thulinius* sp. (PhC). Scale bars: A = 100 μm, B–F = 10 μm.

Tardigrada: Eutardigrada: Parachela: Macrobiotoidea: Murrayidae: Genera

1 Very well-developed L-shaped claws, with cuticular thickenings connecting the base of the claws on each hind leg (Figs. 15.1 G and 15.3 D); with or without lunules 2

1' V-shaped claws, with lunules (Fig. 15.1 F); limnoterrestrial .... *Murrayon*

2(1) With lunules around the claw bases; obligate .... *Macroversum*

2' Without lunules; obligate .... *Dactylobiotus*

Tardigrada: Eutardigrada: Parachela: Macrobiotoidea: Macrobiotidae: Genera

1 Cuticle without pores, three rod-shaped macroplacoids; microplacoid, if present, distant from the third macroplacoid more than its length (Fig. 15.7 A); eggs always with large reticulated processes (Fig. 15.14 B); limnoterrestrial........ *Paramacrobiotus*

1' Cuticle with (Fig. 15.14 C) or without pores; two or three macroplacoids; microplacoid, if present, close to the third macroplacoid; eggs with processes of various types; limnoterrestrial .... *Macrobiotus*

Tardigrada: Eutardigrada: Parachela: Hypsibioidea: Families

1 Medium-sized or large claws 2

1' Very small claws, with a narrow common portion continuous with the primary branch, on which the secondary branch is rigidly connected (Fig. 15.1 L); buccal tube narrow, three macroplacoids; limnoterrestrial........ Microhypsibiidae, one genus: *Microhypsibius*

2(1) External claws of *Hypsibius*-type (with secondary branch forming a common arc with the basal portion; Fig. 15.1 J) or *Isohypsibius*-type (with secondary branch forming a right angle with the basal portion; Fig. 15.1 H, K); most with smooth eggs laid within the exuvium .... **Hypsibiidae [p. 289]**

2' Claws of *Ramazzottius*-type, with primary branch of external claw very long and narrow, with very thin flexible connection to upper basal portion; basal portion of the external claw long and straight (Fig. 15.3 A); buccal tube rigid; elliptical dorsolateral sense organs on head (Fig. 15.14 D); ornamented eggs laid freely; accidental .... Ramazzottiidae, one genus: *Ramazzottius*

Tardigrada: Eutardigrada: Parachela: Hypsibioidea: Hypsibiidae: Genera

1 Buccal tube rigid, without posterior part of spiral composition; placoids in the pharyngeal bulb 2

1' Buccal tube with posterior part of spiral composition (Figs. 15.6 J, K and 15.7 C); placoids in the pharyngeal bulb present, thin or absent 4

2(1) Internal claws not of *Isohypsibius*-type 3

2' External claws of *Hypsibius*-type, but internal claws similar to *Isohypsibius*-type; main branch of the external claw particularly long and not connected to the basal portion (Fig. 15.1 E); apophyses for the insertion of the muscles of the stylets on the anterior part of the buccal tube crest-shaped (Figs. 15.6 M and 15.8 E); limnoterrestrial .... *Ramajendas*

3(2) Apophyses for the insertion of the muscles of the stylets on the anterior part of the buccal tube hook-shaped (Fig. 15.8 C); limnoterrestrial .... *Hypsibius*

3' Apophyses for the insertion of the muscles of the stylets on the anterior part of the buccal tube almost cylindrical in shape with a large rounded or bilobed projection (Fig. 15.8 G); limnoterrestrial .... *Borealibius*

4(1) More than one row of macroplacoids present in the pharyngeal bulb; annulation of the pharyngeal tube clearly visible at maximum magnification 5

4' Macroplacoids absent, or present as a single thin and long strand; very fine annulation of the pharyngeal tube 7

5(4) Pharyngeal tube very thin and longer that the buccal (rigid) tube 6

5' Pharyngeal tube shorter or as long as the buccal (rigid) tube, furcae of the stylets with posterolateral processes spoon-like; limnoterrestrial .... *Platicrista*

6(5) Apophyses for the insertion of the muscles of the stylets on the anterior part of the buccal tube in the shape of "semilunar hooks" (Fig. 15.8 C); limnoterrestrial .... *Diphascon*

6' Asymmetrical apophyses for the insertion of the muscles of the stylets, the dorsal is triangular with rectilinear dorsal margin and posterior apex very distant from buccal tube, the ventral with arched margin and posterior apex nearer to buccal tube (Fig. 15.8 H); limnoterrestrial .... *Bindius*

7(4) Stylet supports very thin, but present; limnoterrestrial .... *Itaquascon*

7' Stylet supports completely absent; only the most anterior part of the buccal tube is not annulated (Fig. 15.6 K); limnoterrestrial .... *Astatumen*

Tardigrada: Eutardigrada: Parachela: Eohypsibioidea: Eohypsibiidae: Genera

1 Posterior part of buccal tube flexible, of very fine spiral composition (Fig. 15.6 I); limnoterrestrial .... *Eohypsibius*

1' Buccal tube rigid, without spiral composition 2

2(1) Large trumpet-shaped mouth; crest-shaped buccal tube apophyses for insertion of muscles of the stylets with a single pair of caudal hooks; limnoterrestrial .... *Bertolanius*

2' Tubular mouth; 2 to 6 hook-shaped appendages for insertion of the stylet muscles .... *Austeruseus*

Tardigrada: Eutardigrada: Parachela: Eohypsibioidea: Isohypsibiidae: Genera

1 Secondary branch of each claw forming right angle with base: *Isohypsibius*-type claws (Fig. 15.1 H, K) 2

1' Claws with very wide angle (about 180°) between the primary and the secondary branch at least on the internal claw on leg IV (Fig. 15.14 E); limnoterrestrial .... *Eremobiotus*

2(1) Ventral lamina absent 3

2' Ventral lamina present (Fig. 15.6 O); limnoterrestrial .... *Doryphoribius*

3(2)	Apophyses for the insertion of the muscles of the stylets on the anterior part of the buccal tube crest-shaped and symmetrical (Fig. 15.8 H, L) 4
3'	Apophyses for the insertion of the muscles of the stylets on the anterior part of the buccal tube hook-shaped and asymmetrical (Fig. 15.8 I); obligate .... *Mixibius*
4(3)	Peribuccal lamellae around the mouth opening present (caution, may be difficult to see) (Fig. 15.5 A, B, D, E); very long claws (Fig. 15.1 H) 5
4'	Peribuccal lamellae around the mouth opening absent (Fig. 15.5 C); limnoterrestrial .... *Isohypsibius*
5(4)	About 30 peribuccal lamellae (Fig. 15.5 D); placoid row in a straight line; obligate .... *Pseudobiotus*
5'	Twelve peribuccal lamellae present, often fused (Fig. 15.5 E) and then difficult to recognize with light microscopy; placoid row in the shape of a Grecian urn (Fig. 15.6 N); one or two cuticular bars under (but not connected to) the claw bases on legs I-III (Fig. 15.14 F); limnoterrestrial .... *Thulinius*

REFERENCES

Bertolani, R. 1982. Tardigradi. Guide per il Riconoscimento delle Specie Animali delle Acque Interne Italiane. Consiglio Nazionale Delle Ricerche, Verona, Italy. 104 pp. [In Italian.]

Bertolani, R. & R. Kristensen. 1987. New records of *Eohypsibius nadjae* Kristensen, 1982, and revision of the taxonomic position of two genera of Eutardigrada (Tardigrada). Pages 359–372 *in*: R. Bertolani (ed.), Biology of Tardigrades. Selected Symposia and Monographs, Unione Zoologica Italiana, 1. Mucchi, Modena, Italy.

Bertolani, R. & L. Rebecchi. 1999. Tardigrada. Pages 703–717 *in*: E. Knobil and J.D. Neill (eds.), Encyclopedia of Reproduction, Vol. 4. Academic Press, San Diego, California, USA.

Bertolani, R., T. Altiero & D.R. Nelson. 2009. Tardigrada (Water Bears). Pages 443–455 *in*: G.E. Likens (ed.), Encyclopedia of Inland Waters, Vol. 2. Elsevier, Oxford.

Binda, M.G. & R. Kristensen. 1986. Notes on the genus *Oreella* (Oreellidae) and the systematic position of *Carphania fluviatilis* Binda, 1978 (Carphanidae fam. nov., Heterotardigrada). Animalia 13:9–20 (in Italian).

Guidetti, R., T. Altiero, T. Marchioro, L. Sarzi Amadè, A.M. Avdonina & L. Rebecchi. 2012. Form and function of the feeding apparatus in Eutardigrada (Tardigrada). Zoomorphology 131: 127–148.

Guidetti, R. &R. Bertolani. 2005. Tardigrade taxonomy: an updated check list of the taxa and a list of characters used in their identification. Zootaxa 845: 1–46.

Guidetti, R., R. Bertolani & L. Rebecchi. 2013. Comparative analysis of the tardigrade feeding apparatus: adaptive convergence and evolutionary pattern of the piercing stylet system. Journal of Limnology 72(s1): 24–35.

Guidetti, R. & G. Pilato. 2003. Revision of the genus *Pseudodiphascon* (Tardigrada, Macrobiotidae), with the erection of three new genera. Journal of Natural History 37: 1679–1690.

Guidetti, R., R.O. Schill, R. Bertolani, T. Dandekar & M. Wolf. 2009. New molecular data for tardigrade phylogeny, with the erection of *Paramacrobiotus* gen. n. Journal of Zoological Systematics and Evolutionary Research 47:315–321.

Kinchin, I.M. 1994. The Biology of Tardigrades. Portland Press, London, UK. 186 pp.

Kristensen, R. 1987. Generic revision of the Echiniscidae (Heterotardigrada), with a discussion of the origin of the family. Pages 261–335 *in*: R. Bertolani (ed.). Biology of Tardigrades. Selected Symposia and Monographs, Unione Zoologica Italiana, 1. Mucchi, Modena, Italy.

Jørgensen, A., S. Faurby, J.G. Hansen, N. Møbjerg & R.M. Kristensen. 2010. Molecular phylogeny of Arthrotardigrada (Tardigrada). Molecular Phylogenetics and Evolution 54: 1006–1015.

Marley, N.J., S.J. McInnes & C.J. Sands. 2011. Phylum Tardigrada: a re-evaluation of the Parachela. Zootaxa 2819: 51–64.

McInnes, S.J. 1994. Zoogeographic distribution of terrestrial/freshwater tardigrades from current literature. Journal of Natural History 28: 257–352.

McInnes, S.J. & P. Pugh. 2007. An attempt to revisit the global biogeography of limnoterrestrial Tardigrada. Journal of Limnology 66 (Suppl. 1): 90–96.

Nelson, D.R., N.J. Marley & R. Bertolani. 1999. Re-description of the genus *Pseudobiotus* (Eutardigrada, Hypsibiidae) and of the new type species *Pseudobiotus kathmanae* sp. n. Zoologischer Anzeiger 238: 311–317.

Nelson, D.R. & S.J. McInnes. 2002. Chapter 7. Tardigrada. Pages 177–215 *in*: S.D. Rundle, A.L. Robertson, and J.M. Schmid-Araya (eds.), Freshwater Meiofauna: Biology and Ecology. Backhuys Publishers, Leiden, The Netherlands.

Nelson, D.R., R. Guidetti, & L. Rebecchi. 2015. Chapter 17. Tardigrada *in*: J.H. Thorp and D.C. Rogers (eds.), Thorp and Covich's Freshwater Invertebrates, Vol. I, Academic Press, Elsevier, Boston, MA.

Pilato, G. 1998. Microhypsibiidae, new family of eutardigrades, and description of the new genus *Fractonotus*. Spixiana 21: 129–134.

Pilato, G. & M.G. Binda. 1987. *Parascon schusteri* n. gen. n. sp. (Eutardigrada, Hypsibiidae, Itaquasconinae). Animalia 14: 91–97 (in Italian with English summary).

Pilato, G. & M.G. Binda. 1990. Tardigradi dell'Antartide. I. *Ramajendas*, nuovo genere di Eutardigrado. Nuova posizione sistematica di *Hypsibius remaudi* Ramazzotti, 1972 e descrizione di *Ramajendas frigidus* n. sp. Animalia, 17: 61–71.

Pilato, G. & M.G. Binda. 2010. Definition of families, subfamilies, genera and subgenera of the Eutardigrada, and keys to their identification. Zootaxa 2404: 1–54.

Ramazzotti, G. & W. Maucci. 1983. Il Phylum Tardigrada. III edizione riveduta e aggiornata. Memorie dell'Istituto Italiano di Idrobiologia 41: 1–1012. [An English translation may be obtained from Dr. Diane Nelson, East Tennessee State University, Johnson City, Tennessee, USA].

Sands, C.J., S.J. McInnes, N.J. Marley, W.P. Goodall-Copestake, P. Convey & K. Linse. 2008. Phylum Tardigrada: an ''individual'' approach. Cladistics 24: 1–11.

Vicente, F., P. Fontoura, M. Cesari, L. Rebecchi, R. Guidetti, A. Serrano & R. Bertolani. 2013. Integrative taxonomy allows the identification of synonymous species and the erection of a new genus of Echiniscidae (Tardigrada, Heterotardigrada). Zootaxa 3613: 557–572.

Chapter 16

Phylum Arthropoda

Chapter Outline

Introduction to the Phylum

D. Christopher Rogers

Kansas Biological Survey, University of Kansas, Lawrence, KS, USA; The Biodiversity Institute, University of Kansas, Lawrence, KS, USA

INTRODUCTION

The Arthropoda constitute the largest and most morphologically diverse group of animals. Specific relationships between the taxa of this clade are only partially understood, as are the relationships between the Arthropoda and other possibly related phyla. One hypothesis unites the Arthropoda and the Annelida as the "Articulata" (reviewed in Scholtz, 2002), uniting the two groups based upon segmentation. More recently, the Arthropoda and the Nemata have been united in a larger group as the "Ecdysozoa," being all animals that molt (Aguinaldo et al., 1997).

Aquatic Arthropods include the subphyla Chelicerata (mites and spiders), Myriapoda (aquatic millipedes), and Crustacea (crustaceans, apterygotans, and insects).

TERMINOLOGY AND MORPHOLOGY

Arthropods all bear segmented bodies with segmented limbs. The Chelicerata body is divided into a cephalothorax, which contains the head, limbs, and thoracic organs, and an abdomen. The segmentation may be obscured due to body segments fusing as in the mites. Chelicerates lack antennae, although the oral palpi may be elongated and serve a similar sensory function.

The Myriapoda have elongated bodies with serially homologous body segments, each bearing two pairs of limbs. In reality, each segment only has one limb pair, but the body "segments" are actually each two segments fused into one. Only one pair of antennae is present. Freshwater Myriapoda are absent from the Nearctic, and thus not discussed further in this chapter.

The Crustacea have the body divided into three divisions: head, thorax, and abdomen. The thorax typically may have three to seven pairs of ambulatory limbs; however, up to 70 pairs can be present in the Branchiopoda. The abdomen may have limbs as well, called pleopods or "swimmerettes," which can be used to propel the animal. Crustacea *sensu stricto* generally has two pairs of antennae, and the Hexapoda have one pair.

Thorp and Covich's Freshwater Invertebrates. http://dx.doi.org/10.1016/B978-0-12-385028-7.00016-0

KEYS TO ARTHROPODA SUBPHYLA

1 Animal body variously separated into two divisions, cephalothorax and abdomen in spiders and gnathosoma and idiosoma in mites; antennae absent; four pairs of legs (except larval mites which have three); never with a bivalved carapace that can close entirely around the animal .... Chelicerata, one class: **Arachnida [p. 292]**

1' Animal body divided into three or more divisions, if fused, then body completely enclosed in a bivalved carapace; one or two pairs of antennae present; insects, springtails, branchiopods, copepods, crabs, etc. .... **Crustacea [p. 413]**

REFERENCES

Aguinaldo, A. M., J. M. Turbeville, L. S. Linfoot, M. C. Rivera, J. R. Garey, R. A. Raffe & J. A. Lake. 1997. Evidence for a clad of nematodes, arthropods and other molting animals. Nature 387: 489–493.

Scholtz, G. 2002. The Articulata Hypothesis – or what is a segment? Organisms, Diversity & Evolution 2:197–215.

Subphylum Chelicerata

CLASS ARACHNIDA

D. Christopher Rogers

Kansas Biological Survey, University of Kansas, Lawrence, KS, USA; The Biodiversity Institute, University of Kansas, Lawrence, KS, USA

INTRODUCTION

Arachnids are a large group of mainly terrestrial organisms classified in 15 extant orders (see Proctor et al., 2015). Representatives of four of these orders, Araneae (true spiders) and three orders of the subclass Acari (mites)—namely Mesostigmata, Sarcoptiformes, and Trombidiformes—are the only chelicerates habitually found in and around freshwater habitats. Among these, members of only a few groups of Sarcoptiformes and Trombidiformes are truly aquatic.

Phylogenetic relationships of arachnid orders are not clearly understood, and the classification is undergoing substantial revision based on interpretation of morphological, behavioral, and genetic data. Here we follow the classification outlined by Proctor et al. (2015) in Volume 1 of this series.

KEYS TO ARACHNIDA SUBCLASSES

1 Body conspicuously divided into cephalothorax and abdomen, spinnerets present, spiders .... **Araneae [p. 292]**

1' Body less conspicuously divided into gnathosoma (capitulum) and idiosoma, spinnerets absent, mites .... **Acari [p. 294]**

Subclass Araneae

D. Christopher Rogers

Kansas Biological Survey, Kansas University, Lawrence, KS, USA; The Biodiversity Institute, Kansas University, Lawrence, KS, USA

Arachnida: Aranae: Families

1 Eight eyes; upper eye row of four eyes with lateral most eyes set posterior to middle eyes, such that it appears that there are three rows of eyes .... Lycosidae

1' Eight eyes; upper row of four eyes with lateral most eyes posteriolateral to middle eyes .... Pisauridae, one genus ***Dolomedes* [p. 292]**

Arachnida: Aranae: Pisauridae: *Dolomedes*: Species

Modified from Carico (1973).

1 Carapace submarginal bands light, distinct, continuous, with entire margins 2

1' Carapace submarginal bands absent or, if present with margins not entire 7

2(1) Carapace with longitudinal light vittae, continuous on abdomen 3

2' Carapace longitudinal light vittae not continuing on abdomen 5

3(2) Abdominal longitudinal light vittae with short medial branches; abdomen lacking dorsal circular spots 4

3' Abdominal longitudinal light vittae without medial branches; abdomen medial dark band with or without dorsal circular spots .... *Dolomedes striatus* Giebel, 1869

[Canada: Alberta: New Brunswick, Nova Scotia, Ontario, Quebec. USA: Great Lakes region through New England, south to Virginia]

PHYLUM ARTHROPODA

4(3) Male carapace length/cymbium (pedipalp tarsal segment that is hollow and contains the palpal organs) length generally greater than 2.2 .... *Dolomedes vittatus* Walckenaer, 1837

[Canada: Ontario. USA: east of Mississippi River, Arkansas, Louisiana (?), Oklahoma,Texas]

4' Male carapace length /cymbium length generally less than 2.2 .... *Dolomedes holti* Carico, 1973

[Mexico: Nuevo Leon]

5(2) Abdomen dorsum without distinct white spots; abdominal dorsum bearing a transverse "W"-shaped, dark vittae 6

5' Abdomen dorsum with distinct circular white spots, often surrounded with dark markings; abdominal dorsum lacking a transverse "W"-shaped vittae .... *Dolomedes triton* (Walckenaer, 1837)

6(5) Males with carapace length/cymbium length generally greater than 1.5 .... *Dolomedes scriptus* Henz, 1845

[Canada: Manitoba, Ontario, Quebec. USA east of continental divide]

6' Males with carapace length/cymbium length generally less than 1.5 .... *Dolomedes gertschi* Carico, 1973

[USA: Arizona, western New Mexico]

7(1) Carapace submarginal region lacking light markings; abdomen dorsum with distinct white spots 8

7' Carapace submarginal region with light markings, often rhomboidal in shape, connected into an emarginate vitta; abdomen dorsum lacking distinct white spots 9

8(7) Females: epigynum width/width between spiracles greater than 1.9 .... *Dolomedes vittatus* Walckenaer, 1837

[Canada: Ontario. USA: east of Mississippi River, Arkansas, Louisiana (?), Oklahoma, Texas]

8' Females: epigynum width/width between spiracles less than 1.9 .... *Dolomedes holti* Carico, 1973

[Mexico: Nuevo Leon]

9(7) Specimen from east of continental divide 10

9' Specimen from west of continental divide .... *Dolomedes gertschi* Carico, 1973

[USA: Arizona, western New Mexico]

10(9) Carapace with cephalic region indistinct and on same plane as thoracic region 11

10' Carapace with cephalic region distinct, prominently elevated above thoracic region .... *Dolomedes albineus* Henz, 1845

[USA: Texas, southeastern states]

11(10) Abdomen dorsum with transverse bands discontinuous medially 12

11' Abdomen dorsum with transverse bands continuous medially .... *Dolomedes scriptus* Henz, 1845

[Canada: Manitoba, Ontario, Quebec. USA east of continental divide]

12(11) Male pedipalp, median apophysis narrowed apical region bent acutely 180 degrees .... *Dolomedes tenebreosus* Hentz, 1843

[Canada: Manitoba, Ontario, Quebec. USA: Great Plains and eastern states, except peninsular Florida]

12' Male pedipalp, median apophysis narrowed apical region broadly arcing ~160 degrees .... *Dolomedes okefinokensis* Bishop, 1924

[USA: Georgia, Florida]

REFERENCE

Carico, J. E. 1973. The Nearctic spiders of the genus *Dolomedes* (Aranae: Pisauridae). Bulletin of the Museum of Comparative Zoology, Harvard University 144: 435–488.

Subclass Acari

Ian M. Smith

Canadian National Collection of Insects, Arachnids and Nematodes, Agriculture and Agri-Food Canada, Ottawa, ON, Canada

INTRODUCTION

The following treatment provides keys to the groups of mites (subclass Acari, superorder Acariformes) that comprise the vast majority of truly aquatic arachnids inhabiting freshwater in North America. Three of these groups, namely the halacarid mites (supercohort Eupodides, superfamily Halacaroidea, family Halacaridae[1]), the true water mites (supercohort Anystides,

1. The huge size of this chapter on Arthropoda and the many authors of separate keys and figure formats necessitated a different way of expressing figure numbers. Consequently, figures of taxa that generally occur with other figures on a single plate are referred to by their chapter, plate (Pl.), and/or figure number or letter. For example, Fig. 16.25.110 refers to Chapter 16, Plate 25, and Figure 110. The editors apologize for the resulting complexity.

4' Opisthonotal gland opens without conspicuous protuberance; with 3 pairs of adanal, 2 pairs of anal setae; peranal segment absent (anal segment paraproctal); 2 pairs of coxisternum II setae (Fig. 16.1.3, 4) .... Gehypochthoniidae, one genus: *Gehypochthonius*

5(3) Body not cylindrical, often globose or dorsally flattened; notogaster distinct from ventral region throughout its length; adanal plates distinct; genital and anal plates adjacent or only narrowly separated; prodorsum without posteriorly directed tubercles 6

5' Body cylindrical, somewhat elongated; notogaster completely fused to ventral region in posterior half, such that adanal plates are distinguishable only by differences in cuticular pattern; soft cuticle delimiting notogastral margin in anterior half curves ventrad toward genital plates, such that in ventral aspect a pair of crescentiform scissures are directed toward wide space between genital and anal plates (Fig. 16.2.11; arrow); prodorsum posteriorly with paired tooth or ridge-like tubercles projecting over dorsosejugal groove (Fig. 16.2.10; arrow) .... Nanhermanniidae

6(5) Epimere II with 1 (usually) or 0 pair of setae; with 4–24 pairs of genital setae, but all near medial edge; aggenital setae present or absent; preanal plate present or absent; rostrum without medial incision; integument various, rarely foveolate...7

6' Epimere II with 3 or more pairs of setae (Fig. 16.2.14); with 9 pairs of genital setae, one posterior seta far lateral of others, away from medial edge of plate (arrow) ; aggenital setae absent; preanal plate distinct (Fig. 16.2.14; pr); rostrum with short medial incision (Fig. 16.2.12; arrow); integument distinctly foveolate (Fig. 16.2.13) .... Nothridae, one genus: *Nothrus*

7(6) Genital plate without carina that delimits marginal seta-bearing region; anal plates reduced, much narrower than adanal plate, sometimes inconspicuous; aggenital setae absent...8

7' Genital plates with medial margin bearing all genital setae in narrow region delimited from rest of plate by distinct carina (Fig. 16.2.16); anal plate wider or only slightly narrower than adanal plate, clearly not greatly reduced; aggenital setae present (Fig. 16.2.16; ag) found in peatlands and other aquatic and semiaquatic habitats .... Crotoniidae, one species: *Platynothrus peltifer* (C. L. Koch, 1839)

8(7) Bothridial seta and bothridium lost without trace (Figs. 16.2.17, 18); subcapitulum anarthric, without labiogenal articulation; cuticle of notogaster not imbricate; cerotegument present in form of large, thin plates that appear waxy, without excrescences, strongly birefringent in polarized light .... **Malaconothridae [p.304]**

8' Bothridial seta present, usually with well developed bothridium (in some populations of *Trhypochthoniellus* these are vestigial or absent); subcapitulum with labiogenal articulation (Fig. 16.3.21); cuticle of notogaster at least partly imbricate, cerotegument, if present, of different form, not birefringent .... **Trhypochthoniidae [p. 305]**

9(1) Pteromorphs absent; notogaster without octotaxic system of porose areas or saccules; discidium and circumpedal carina usually absent...........10

9' Pteromorphs (Figs. 16.5.43, 50; ptm) present; notogaster with (Figs. 16.5.48, 49) or without octotaxic system of porose areas or saccules; discidium (di) and circumpedal carina (cc) usually present (Figs. 16.6.55, 59) 12

10(9) Lamella present (Fig. 16.4.32; La). Genital plates close to and larger than anal plates; epimeral setae *4a* and *4b* closely adjacent (Fig. 16.4.33; arrow); pedotectum I expressed as narrow dorsal lamina and wider ventral lamina (Fig. 16.4.34; pd) .... Tegeocranellidae, one genus: *Tegeocranellus*

10' Costula present (Fig. 16.4.35; co). Genital plates distant from and smaller than anal plates; epimeral setae *4a* and *4b* not closely adjacent; pedotectum I as single lamina (as in Fig. 16.4.40; pd) ....Hydrozetidae, one genus: *Hydrozetes*

11(9) Seta *d* not most proximal seta on femora; bothridial seta well developed 12

11' Seta *d* inserted on proximal fifth of femora I-III, proximal to other femoral setae (Fig. 16.4.41); bothridial seta often weakly developed, reduced, or absent (Fig. 16.4.38) .... Limnozetidae, one genus: *Limnozetes*

12(11) Prodorsum with genal notch (Fig. 16.6.54; gn); pteromorphs with or without hinge; if with hinge, notogaster with posterior tectum (Fig. 16.5.47) 13

12' Prodorsum without genal notch; pteromorphs with hinge (Fig. 16.5.43; arrow); notogaster without posterior tectum (Fig. 16.5.46) .... Haplozetidae, one species; *Rostrozetes ovulum* (Berlese, 1908)

[Canada: Nova Scotia, Ontario, Quebec. USA. Mexico. Neotropics]

13(12) Notogaster with posterior tectum (Fig. 16.5.47); subcapitular mentum with or without tectum; epimere IV with at most 3 pairs of setae; 10 pairs normal notogastral setae or setae reduced to alveoli 14

13' Notogaster without posterior tectum; subcapitular mentum with tectum (Fig. 16.5.52; arrow); epimere IV neotrichous (4 or more pairs of setae); 10 pairs of minute notogastral setae or setae reduced to alveoli (sockets); length: 1000–1200 μm (Fig. 16.5.48) .... Euzetidae, one genus: *Euzetes*

14(13) Legs I and IV often with different number of claws; if with same number of claws, either porose area A1 positioned medially, anterior to seta *lp* and medial to Aa and A2 (Fig. 16.6.56) or tutorium brush-like anteriorly (Fig. 16.6.54); subcapitular mentum without tectum (Fig. 16.6.60); length: 250–750 μm .... **Zetomimidae [p. 305]**

14' Legs I and IV with same number of claws; porose area A1 (or saccule S1) positioned laterally, in antero-posterior line with Aa and A2 (Fig. 16.5.49, I); subcapitular mentum with or without tectum; length: 400–650 μm .... **Punctoribatidae [p. 305]**

Arachnida: Acari: Oribatida: Malaconothridae: Genera

1 Legs tridactylous; leg tarsi elongated, gradually narrowed distally...*Trimalaconothrus*

1' Legs monodactylous; leg tarsi short, broad, not noticeably tapering.. *Malaconothrus*

Arachnida: Acari: Oribatida: Trhypochthoniidae: Genera

1 Rostral setae well separated, mutual distance usually equal to or greater than that of lamellar setae (Figs. 16.3.23, 26, 30); rostrum not excised laterally to form an anterior mucro; bothridial seta and bothridium well formed or (rarely) highly regressed, represented by vestiges or absent (some *Trhypochthoniellus*). Epimere III with 3 pairs of setae; legs tridactylous 2

1' Rostral setae (*ro*) closely adjacent, inserted on anterior mucro formed by laterally excised rostrum (Fig. 16.3.20); bothridial seta (*bo*) small, setiform; its alveolus narrow, simple, not a cup-like bothridium, and located on small, disciform elevation; epimere III with 2 pairs of setae; legs monodactylous (Fig. 16.3.22) .... *Mucronothrus*

2(1) Subcapitulum stenarthric (Figs. 16.3.28, 31); three pairs of adanal setae; body slightly flattened. Anal plates well formed, only slightly narrower than adanal plates (Fig. 16.3.27); epimere IV with 2 or 3 pairs of setae 3

2' Subcapitulum diarthric (Fig. 16.3.25); two pairs of adanal setae (Fig. 16.3.24); body strongly flattened. Anal plates unusually narrow, strap-like; epimere IV with 2 pairs of setae .... *Trhypochthoniellus*

3(2) One pair of anal setae, 6–18 pairs of genital setae, three pairs of epimere IV setae; subcapitular setae *m* vestigial (Fig. 16.3.31; arrow) .... *Trhypochthonius*

3' Two pairs of anal setae, 6 pairs of genital setae (Fig. 16.4.39), two pairs of epimere IV setae; two pairs of subcapitular setae *m* closely adjacent (Fig. 16.4.40; arrow), semiaquatic, typically in *Sphagnum* moss .... *Mainothrus*

[Canada. USA. Temperate]

Arachnida: Acari: Oribatida: Zetomimidae: Genera

1 Tutorium pointed anteriorly (Fig. 16.6.57), never brush-like. Porose area A1 positioned medially, anterior to seta *lp* and medial to Aa and A2 (Fig. 16.6.56), or porose areas absent 2

1' Tutorium brush-like anteriorly (Figs. 16.6.53, 54; tu). Porose area A1 positioned laterally, posterior to seta *lp* and in longitudinal line with Aa and A2 (Fig. 16.6.53) pond and lake edges, marshes .... *Naiazetes reevesi* Behan-Pelletier, 1996

[Canada: Quebec. USA: Alabama, Florida, New York]

2(1) Notogaster with octotaxic system porose areas absent; notogastral setae at most length of their alveolus (Figs. 16.6.58, 59) .... *Heterozetes*

2' Notogaster with octotaxic system porose areas present; notogastral setae medium length (Fig. 16.6.56) .... *Zetomimus*

Arachnida: Acari: Oribatida: Punctoribatidae: Genera

1 Interlamellar setae densely barbed, bifurcate distally (Fig. 16.5.49; *in*); mentum of subcapitulum without tectum (as in Fig. 16.4.35) .... *Pelopsis*

[Widespread]

1' Interlamellar seta not bifurcate distally; mentum of subcapitulum with tectum (Fig. 16.5.52; arrow); anterior notogastral tectum with large medial process (Fig. 16.5.50; arrow) .... *Punctoribates*

[Widespread]

REFERENCES

Behan-Pelletier, V.M. & B. Eamer. 2007. Aquatic Oribatida: adaptations, constraints, distribution and ecology. Pages 71–82 *in*: Morales-Malacara, J.B., Behan-Pelletier, V., Ueckermann, E., Pérez, T.M., Estrada-Venegas & E.G., Badil, M. (eds.), Acarology XI: Proceedings of the International Congress. Instituto de Biología and Facultad de Ciencias, Universidad Nacional Autónoma de México; Sociedad Latinoamericana de Acarología, México.

Grandjean, F. 1934. La notation des poils gastronotiques et des poils dorsaux du propodosoma chez les Oribates (Acariens). Bulletin de la Société Zoologique de France 59: 12–44.

Norton, R.A. & V.M. Behan-Pelletier. 2009. Chapter 15, Oribatida. Pages 421–564 *in*: G.W. Krantz, Walter, D.E. (ed.), A Manual of Acarology 3rd Edition. Texas Tech. University Press, Lubbock.

Norton, R.A., G. Alberti, G. Weigmann & S. Woas. 1997. Porose integumental organs of oribatid mites (Acari, Oribatida). 1. Overview of types and distribution. Zoologica, Stuttgart 146: 1–31.

Pugh, P.J.A., P.E. King & M.R. Fordy. 1990. Respiration in *Fortuynia maculata* Luxton (Fortuyniidae, Cryptostigmata, Acarina) with particular reference to the role of van der Hammen's organ. Journal of Natural History 24: 1529–1547.

Schatz, H. & V.M. Behan-Pelletier. 2008. Global diversity of oribatids (Oribatida; Acari, Arachnida). In: Balian, E.V., C. Lévêque, H. Segers, K. Martens (editors). Freshwater Animal Diversity Assessment. Hydrobiologia 595: 323–328.

Solhøy, I.W. & T. Solhøy. 2000. The fossil oribatid fauna (Acari: Oribatida) in late-glacial and early-Holocene sediments in Kråkenes Lake, western Norway. Journal of Paleolimnology 23: 35–47.

Trombidiformes: Prostigmata

Ian M. Smith
Canadian National Collection of Insects, Arachnids and Nematodes, Agriculture and Agri-Food Canada, Ottawa, ON, Canada

David R. Cook
Paradise Valley, AZ, USA

Ilse Bartsch
Forschungsinstitut Senckenberg, Hamburg, Germany

KEY TO PROSTIGMATA COHORTS: LARVAE

1 Idiosoma bearing a pair of epimeral pores and lacking urstigmata (Fig. 16.8.7) .... supercohort Eupodides, superfamily Halacaroidea, family **Halacaridae [p. 308]**

1' Idiosoma bearing paired urstigmata between first two pairs of coxal plates and lacking epimeral pores (Figs. 16.11.4, 13.10, 16, 15.26, 16.41, 17.45, 20.72, 24.100, 32.165, 35.192) .... supercohort Anystides, cohort Parasitengonina, subcohorts **Hydrachnidiae & Stygothrombiae [p. 312]**

Prostigmata Cohorts: Adults

1 Idiosoma variously shaped but never vermiform and lacking glandularia (Figs. 16.7.1, 2) .... supercohort Eupodides, superfamily Halacaroidea, family **Halacaridae [p. 308]**

1' Idiosoma variously shaped and bearing paired series of well-defined glandularia (Figs. 16.37.200, 38.207, 41.240, 42.244, 63.484) or vermiform and bearing paired series of stomatoid lyrifissures (Fig. 16.38.204) .... supercohort Anystides, cohort Parasitengonina, subcohorts **Hydrachnidiae & Stygothrombiae [p. 312]**

Family Halacaridae

Ilse Bartsch
Forschungsinstitut Senckenberg, Hamburg, Germany

INTRODUCTION

The Halacaridae (superfamily Halacaroidea) comprises a large group of aquatic acariform mites ranging from 150–2000 μm in length. The phylogenetic position of Halacaridae among Acariformes is not well understood, but they are evidently not closely related to Hydrachnidiae or any other group of mites with aquatic representatives. Halacarids are primarily marine, and more than 1000 species have been described from salt water habitats worldwide. Another 50–60 species occur exclusively, mainly, or at least regularly, in freshwater (Bartsch, 2009). Formerly, genera adapted for life in freshwater were classified in the subfamily Limnohalacarinae Viets, 1927 (Viets, 1927). In 1933, Viets proposed the family Porohalacaridae for halacaroid mites with external genital acetabula to include all freshwater species. This family proved to be a very artificial taxon including several paraphyletic genera whose members were closely related to marine species with internal genital acetabula classified in the Halacaridae (Newell, 1947; Bartsch, 1989). Currently freshwater species and genera are assigned to four essentially freshwater subfamilies of Halacaridae, namely Astacopsiphaginae, Limnohalacarinae, Porolohmannellinae, and Ropohalacarinae, and three mainly marine subfamilies, namely Copidognathinae, Halacarinae, and Lohmannellinae (Viets, 1933; Bartsch, 1989, 2006).

The first species of freshwater halacarid reported from North America was the cavernicolous species *Hamohalacarus subterraneus* Walter, 1931. Newell's keys to North American freshwater halacarids (1947, 1959) included members of six genera. The present work includes ten species representing eight genera that are found exclusively in freshwater (Bartsch, 2011b).

The known species diversity of freshwater halacarid mites is substantially lower in the Nearctic, with only ten described species, than in the Palearctic, with more than 30 described species. All but one of the genera found in the Nearctic also occur in the Palearctic, and in many cases they are represented by the same species in both regions. Many species of freshwater halacarids have broad global distributions spanning several continents (Bartsch, 2007b, 2009).

TERMINOLOGY AND MORPHOLOGY

Adult halacarids exhibit the basic acarine body plan with an idiosoma bearing four pairs of legs and a gnathosoma bearing the palps and chelicerae (Figs. 16.7.1, 4). The idiosoma is oblong and slightly to strongly flattened dorsoventrally. Freshwater halacarids have a body length between 150 and 600 μm, with the notable exception of *Astacopsiphagus parasiticus* Viets, 1931, a poorly known species found in the gill chambers of an Australian decapod crustacean that may reach a length of 2000 μm (Viets, 1931).

The idiosomal dorsum typically bears smooth, reticulate or foveate plates including unpaired anterior and posterior plates and a pair of ocular plates that may be reduced or absent (Fig. 16.10.1). The dorsal integument usually bears serial arrangements of six pairs of idiosomatic setae and five pairs of gland pores, although the numbers may be reduced in some species and supernumerary setae may be present in others. The dorsal as well as ventral plates may be divided, reduced, or enlarged and fused to a dorsal or ventral shield.

The idiosomal venter bears an anterior epimeral plate formed by the fusion of epimera I and II, a pair of posterior epimeral plates formed by fusion of epimera III and IV, and a genital or genitoanal plate formed by fusion of the genital and anal plates. Epimeral pores may be present or absent. The genital plate bearing the genital opening (Fig. 16.7.4) is sexually dimorphic, with females typically having a larger genital opening and a lower number of setae around the genital opening than males. Internal to the genital opening females have an ovipositor and males have an elaborate spermatopositor, both of which are visible through the genital plate.

The gnathosoma consists of the gnathosomal base, rostrum, palps, and chelicerae. There are two pairs of maxillary setae, with one pair inserted on the gnathosomal base and the other on the rostrum, or with both pairs inserted on the rostrum. The four segmented palps are inserted either laterally or dorsally on the gnathosomal base. The first and third palpal segments are relatively short compared to the longer second segment (Fig. 16.7.3). The chelicerae are elongate and typically bear a terminal claw, but in some genera are styliform.

The first and second pairs of legs are directed anteriorly and the third and fourth pairs are directed posteriorly. The legs have six segments, namely the trochanter, basifemur, telofemur, genu, tibia, and tarsus (Figs. 16.7.1, 4). The tarsi bear paired claws terminally. The leg segments bear arrays of setae of various lengths, thicknesses and form (smooth, plumose or pectinate).

Halacarid mites are usually found in small numbers and apparently play relatively minor roles in freshwater communities compared to Hydrachnidiae and Oribatida. Halacarids differ from other freshwater mites by the following combination of character states: idiosomal plates present but relatively weakly sclerotized, epimeral plates present; first and second legs directed anteriorly and third and fourth legs directed posteriorly; idiosomal dorsum bearing no more than five pairs of gland pores and seven pairs of setae; palps four segmented; and legs six segmented.

Most species of freshwater halacarids have a life history with one larval and two nymphal stages in addition to the adult. Larvae are smaller than adults and have three pairs of five-segmented legs. They also differ from adults in the shape and size of the dorsal and ventral idiosomal plates (Figs. 16.8.5, 7; cf. Figs. 16.12.6, 8), the lack of a genital plate, the invariable presence of paired epimeral pores, incomplete setation of the ventral idiosomal plates and legs, and often in the relative lengths of the leg segments (Figs. 16.8.11, 12). The shape and setation of the larval gnathosoma is similar to that of adults (Figs. 16.8.9, 10) (Bartsch, 2007a, 2011a).

Larvae often can be associated with adults of the same species using the shape and setation of the tarsi and their claws, the shape of the gnathosoma, and the arrangement of setae and gland pores on the dorsum.

Protonymphs are characterized by four pairs of legs, with the first three pairs six-segmented and the fourth pair five-segmented. They exhibit epimeral pores in species that retain these structures as adults, but have them reduced in size or absent in other species. Protonymphs have a relatively small genital plate bearing only one pair of genital acetabula and exhibit incomplete chaetotaxy of the idiosoma and legs.

Deutonymphs have four pairs of six-segmented legs. Their dorsal and ventral idiosomal plates are slightly smaller than those of adults and are always separate from one another. The numbers of setae on the idiosoma and leg segments are often similar to those in the adults.

MATERIAL PREPARATION AND PRESERVATION

Halacarid mites should be preserved and stored in 70% ethanol. Fixatives such as Formalin make the mites difficult to clear.

Reliable identification of halacarid mites requires microscopic examination. Specimens can be examined in a drop of glycerine, but for species level identification must be cleared and slide-mounted. Halacarids are typically cleared in pepsin or lactic acid and gentle warming (40–50 °C) may facilitate the process. The gnathosoma is then removed using a sharp needle and the body contents are squeezed out with help of gentle pressure using a blunt needle.

Cleared specimens can be mounted for study and permanent storage in glycerine jelly or Hyrax. After initial hardening of the mounting medium, the coverslip should be ringed using a standard microscopical sealing agent. Berlese's medium, modified Hoyer's fluid, polyvinyl lactophenol, or glycerine can be used for temporary mounts. Specimens cleared in lactic acid must be rinsed to prevent formation of crystals in the mounting medium. Staining with chlorazol black often facilitates the interpretation of morphological structures in over-cleared specimens.

Mounting specimens between two cover slips is recommended to facilitate examination of both dorsal and ventral surfaces. These preparations can be stored in metal slide frames or attached to a glass slide using a drop of glycerine.

KEYS TO HALACARIDAE

Arachnida: Acari: Halacaridae: Genera and Species

1 Gnathosoma relatively wide, only 0.9–1.8 times longer than wide and little more than one-third as long as idiosoma (Figs.16.7.1, 4); integument pale in colour and with brown or green spots indicating gut contents and eye pigment visible through integument 2

1' Gnathosoma relatively slender, 2.5 times longer than wide and about half as long as idiosoma (Figs. 16.8.8, 10); integument violet or pink in colour .... *Porolohmannella violacea* (Kramer, 1879)

[Canada: Alberta, British Columbia, Manitoba, Newfoundland, Northwest Territories, Ontario, Quebec. USA: New Hampshire, New York, Rhode Island]

2(1) Palps inserted laterally on gnathosoma and in ventral view with most of first segment visible (Fig. 16.9.13); genital acetabula borne on genital sclerites in females (Fig. 16.9.16) and on genital plate immediately posterior to genital opening in known males 3

2' Palps inserted dorsally on gnathosoma and in ventral view with first segment almost entirely invisible (Figs. 16.7.1, 4); genital acetabula borne on genital plate, posterior, lateral, or anterior to genital opening in females and males (Figs. 16.7.4, 9.24) 5

3(2) Idiosoma anterior margin arched or truncate (Figs. 16.7.1, 9.14); first leg with tarsus lateral fossary membrane not enlarged and with tarsus bearing one ventral seta and pair of parambulacral setae (Fig. 16.9.15) 4

3' Idiosoma anterior margin with a frontal spine (Fig. 16.9.16); first leg with tarsus lateral fossary membrane enlarged to about twice height of dorsomedial membrane (Fig. 16.9.17, arrow) and with tarsus bearing one ventromedial spiniform seta, two ventral setae, and pair of parambulacral setae .... *Lobohalacarus weberi* (Romijn & Viets, 1924)

[Canada: British Columbia, New Brunswick, Newfoundland, Ontario, Quebec. USA: Alabama, Arizona, California, Colorado, Georgia, Illinois, New Hampshire, New Mexico, New York, North Carolina, Rhode Island, Tennessee, Virginia]

4(3) Ocular plates with conspicuous cornea and black eye spots (Fig. 16.9.14) .... *Porohalacarus alpinus* (Thor, 1910)

[Canada: Ontario, Quebec. USA: New Hampshire, Rhode Island]

4' Ocular plates reduced to delicate sclerites (Fig. 16.9.19) lacking eye spots .... *Ropohalacarus uniscutatus* (Bartsch, 1982)

[USA: New York, Rhode Island]

5(2) First leg with telofemur and tibia cylindrical and less than twice as long as genu (Figs. 16.9.21, 23) 6

5' First leg with telofemur and tibia clavate and more than three times as long as very short genu (Fig. 16.9.20) .... *Parasoldanellonyx parviscutatus* (Walter, 1917)

[USA: Rhode Island]

6(5) First leg with tarsal claws bearing either thick tines or slender tines and a lamellar ventral process (Figs. 16.7.2, 9.25) 7

6' First leg with tarsal claws very slender and with tines very delicate or absent (Fig. 16.9.18) .... *Hamohalacarus subterraneus* Walter, 1931

[USA: Indiana]

7(6) First leg with tarsal claws pectinate and bearing thick tines (Fig. 16.7.2); anal sclerites at least half as large as genital sclerites (Fig. 16.7.4); genital acetabula borne in posterolateral regions of genital plate .... ***Soldanellonyx* [p. 308]**

7' First leg with tarsal claws bearing slender tines and a lamellar ventral process (Fig. 16.9.25); anal sclerites minute (Fig. 16.9.24, arrow), much smaller than genital sclerites; genital acetabula borne along lateral margins of genital plate .... *Limnohalacarus cultellatus* Viets, 1940

[USA: Georgia, Wisconsin]

Arachnida: Acari: Halacaridae: *Soldanellonyx*: Species

1 First leg with telofemur 1.7–1.8 times longer than genu and tibia with four ventral spines (Fig. 16.9.23) 2

PHYLUM ARTHROPODA

1' First leg with telofemur of leg I about 1.3 times longer than genu and tibia with two ventral setae (Fig. 16.9.21) .... *Soldanellonyx monardi* Walter, 1919

[Canada: British Columbia, New Brunswick, Newfoundland, Ontario, Quebec. USA: Alabama, Arizona, California, Georgia, Indiana, Missouri, New Hampshire, New York, North Carolina, Oregon, Pennsylvania, Rhode Island, Tennessee, Texas, Virginia]

2(1) First leg ventral spines with blunt or slightly spinose tips; palp with third segment bearing spine near or just proximal to mid-length .... *Soldanellonyx chappuisi* Walter, 1917

[Canada: Nova Scotia, Ontario, Quebec. USA: Colorado, Indiana, North Carolina, Oregon]

2' First leg ventral spines with tapering tips; palp with third segment bearing spine well proximal to mid-length (Fig. 16.9.22) .... *Soldanellonyx visurgis* Viets, 1959

[USA: Arizona, Georgia, New York, Rhode Island]

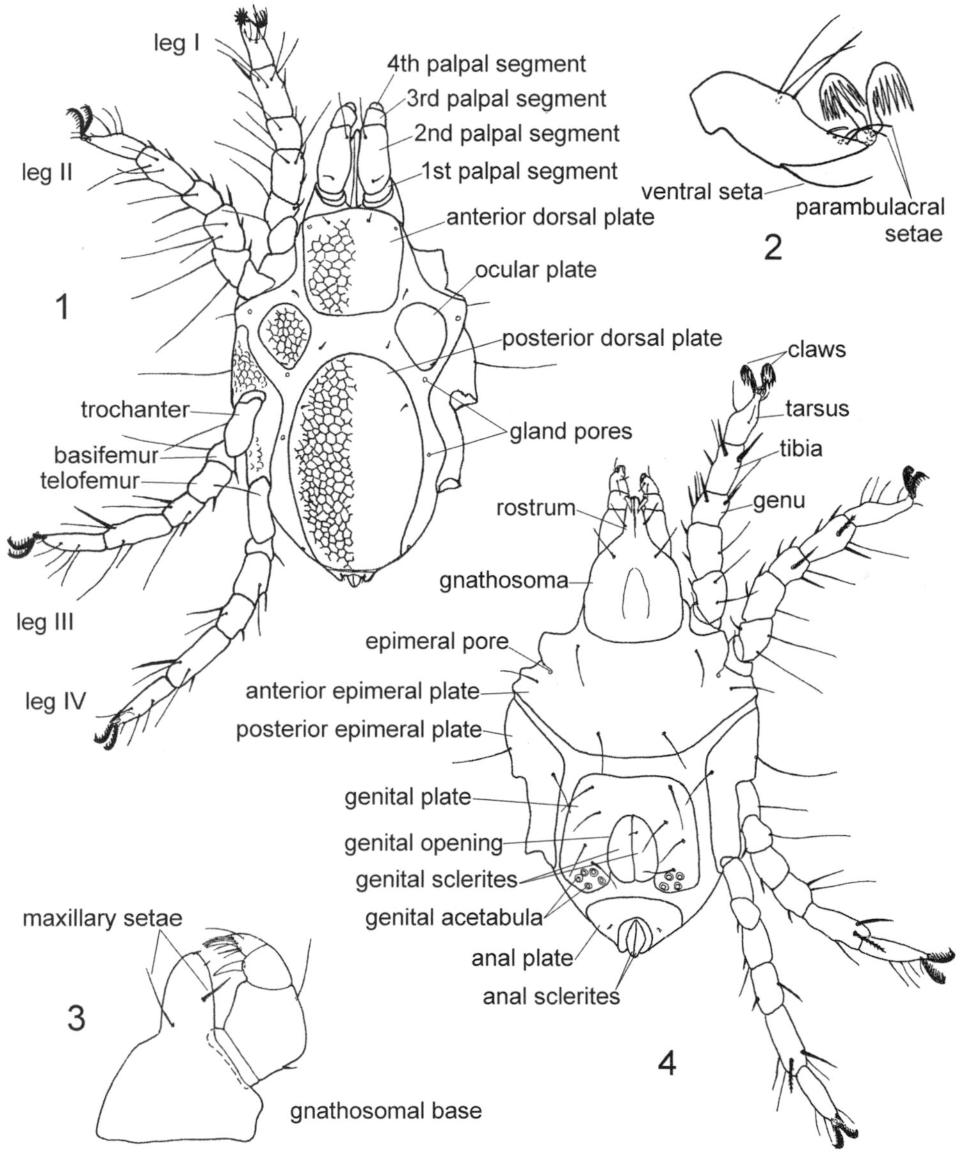

PLATE 16.07 **Figures 1–4:** *Soldanellonyx monardi* Walter, female adult. **Fig. 1** dorsum of gnathosoma and idiosoma. **Fig. 2** tarsus of first leg, lateral view. **Fig. 3** gnathosoma, lateral view. **Fig. 4** venter of gnathosoma and idiosoma.

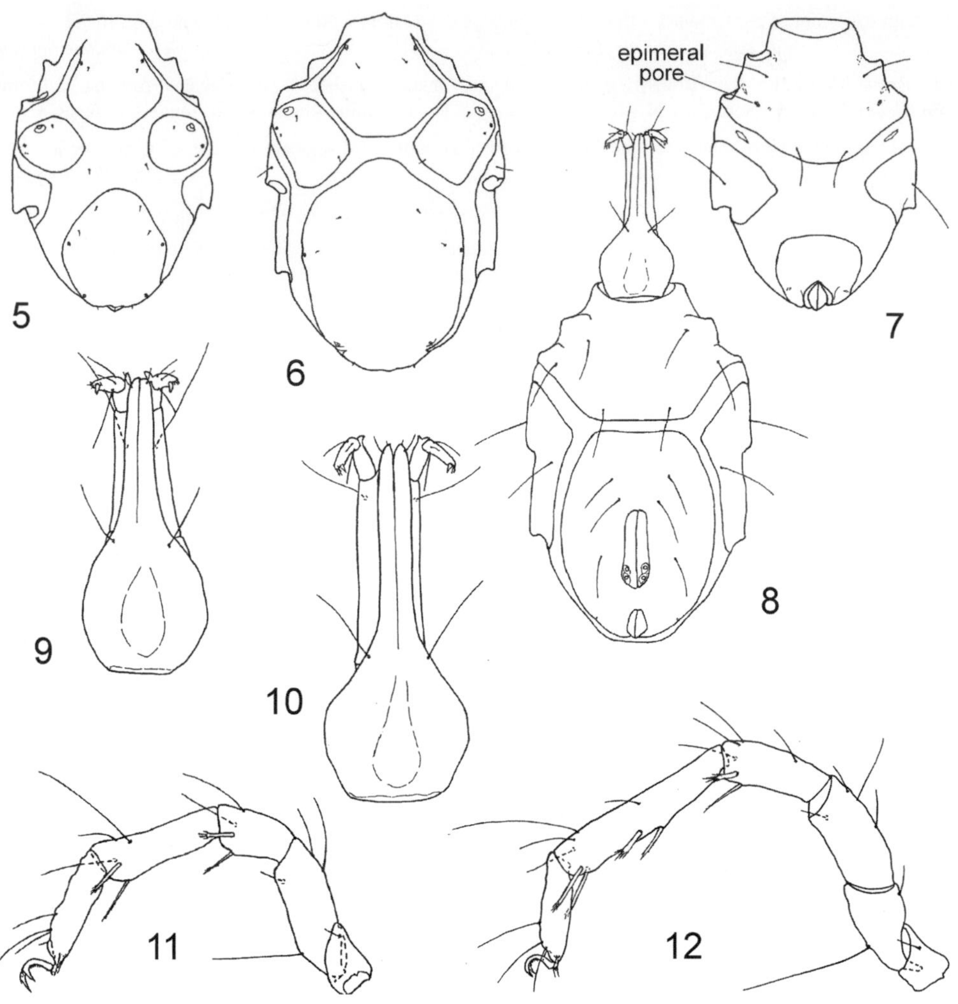

PLATE 16.08 Figures 5–12: *Porolohmannella violacea* (Kramer). **Fig. 5** larva, dorsum of idiosoma. **Fig. 6** female adult, dorsum of idiosoma. **Fig. 7** larva, venter of idiosoma. **Fig. 8** female adult, venter of gnathosoma and idiosoma. **Fig. 9** larva, venter of gnathosoma. **Fig. 10** female adult, venter of gnathosoma. **Fig. 11** larva, first leg, medial view. **Fig. 12** female adult, first leg, medial view.

REFERENCES

Bartsch, I. 1982. Halacariden (Acari) im Süßwasser von Rhode Island, USA, mit einer Diskussion über Verbreitung und Abstammung der Halacaridae. Gewässer und Abwässer 68/69: 41–58.

Bartsch, I. 1989. Süsswasserbewohnende Halacariden und ihre Einordnung in das System der Halacaroidea (Acari). Acarologia 30: 217–239.

Bartsch, I. 2006: 5. Acari: Halacaroidea. Pages 113–157 *in*: Gerecke, R. (ed.), Chelicerata: Araneae, Acari I. Süßwasserfauna Mitteleuropas, 7/2-1. Spektrum Elsevier, Heidelberg.

Bartsch, I. 2007a. The freshwater mite *Porolohmannella violacea* (Kramer, 1879) (Acari: Halacaridae), description of juveniles and females and notes on development and distribution. Bonner zoologische Beiträge 55: 47–59.

Bartsch, I. 2007b. Freshwater Halacaridae (Acari) from New Zealand rivers and lakes, with notes on character variability. Mitteilungen aus dem Hamburgischen Zoologischen Museum und Institut 104: 73–87.

Bartsch, I. 2009. Checklist of marine and freshwater halacarid mite genera and species (Halacaridae: Acari) with notes on synonyms, habitats, distribution and descriptions of the taxa. Zootaxa 1998: 1–170.

Bartsch, I. 2011a. The freshwater halacarid mite *Soldanellonyx chappuisi* Walter, 1917 (Acari: Halacaridae), character development from larva to adult and comparison with other halacarids. Entomologische Mitteilungen aus dem Zoologischen Museum Hamburg 15(184): 223–235.

Bartsch, I. 2011b. North American freshwater Halacaridae (Acari): literature survey and new records. International Journal of Acarology 37: 490–510.

Kramer, P. 1879. Ueber die Milbengattungen *Leptognathus* Hodge, *Raphignathus* Dug., *Caligonus* Koch und die neue Gattung *Cryptognathus*. Archiv für Naturgeschichte 45: 142–157.

Newell, I. M. 1947. A systematic and ecological study of the Halacaridae of eastern North America. Bulletin of the Bingham Oceanographic Collection 10: 1–232.

Newell, I. M. 1959. Chapter 42: Acari. Pages 1080–1116 in: H.B. Ward, H.B & G.C. Whipple (eds.), Fresh-water Biology. Second edition. John Wiley and Sons, New York.

Romijn, G., K.H. Viets. 1924. Neue Milben. Archiv für Naturgeschichte 90: 215–225.

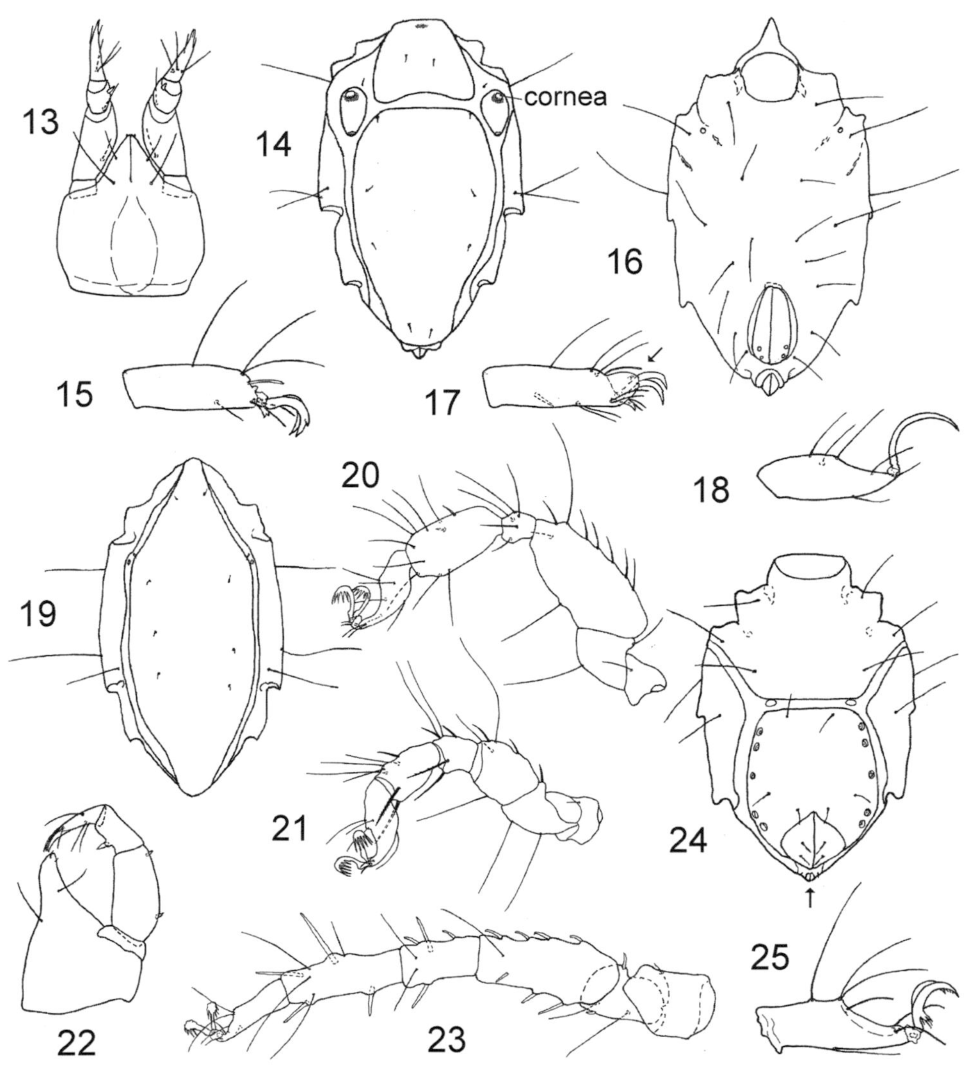

PLATE 16.09 **Figures 13–15:** *Porohalacarus alpinus* (Thor), female adult. **Fig. 13** venter of gnathosoma. **Fig. 14** dorsum of idiosoma. **Fig. 15** tarsus of first leg, lateral view. **Figures 16–17:** *Lobohalacarus weberi* (Romijn & Viets), female adult. **Fig. 16** venter of idiosoma. **Fig. 17** tarsus of first leg, lateral view. **Fig. 18:** *Hamohalacarus subterraneus* Walter, female adult, tarsus of first leg (from Walter, 1931). **Fig. 19** *Ropohalacarus uniscutatus* (Bartsch), female adult, dorsum of idiosoma. **Fig. 20** *Parasoldanellonyx parviscutatus* (Walter), female adult, first leg, medial view. **Fig. 21** *Soldanellonyx monardi* Walter, female adult, first leg, medial view. **Fig. 22** *Soldanellonyx visurgis* Viets, female adult, gnathosoma, lateral view. **Fig. 23** *Soldanellonyx chappuisi* Walter, female adult, first leg, dorsal view. **Figures 24–25:** *Limnohalacarus cultellatus* Viets, female adult. **Fig. 24** venter of idiosoma. **Fig. 25** tarsus of first leg, lateral view.

Thor, S. 1910. Die erste norwegische Süßwasserform der Halacariden. Zoologischer Anzeiger 36: 348–351.

Viets, K. 1927. Die Halacaridae der Nordsee. Zeitschrift für Wissenschaftliche Zoologie 130: 83–173.

Viets, K. 1931. Über eine an Krebskiemen parasitierende Halacaride aus Australien. Zoologischer Anzeiger 96: 115–120.

Viets, K. 1933. Vierte Mitteilung über Wassermilben aus unterirdischen Gewässern (Hydrachnellae et Halacaridae, Acari). Zoologischer Anzeiger 102: 277–288.

Viets, K. 1940. Zwei neue Porohalacaridae (Acari) aus Südamerika. Zoologischer Anzeiger 130: 191–201.

Viets, K. 1959. Die aus dem Einzugsgebiet der Weser bekannten oberirdisch und unterirdisch lebenden Wassermilben. Veröffentlichungen des Instituts für Meeresforschung in Bremerhaven 6: 303–513.

Walter, C. 1917. Schweizerische Süßwasserformen der Halacariden. Revue Suisse de Zoologie 25: 411–423.

Walter, C. 1919. Schweizerische Süßwasserformen der Halacariden. Revue Suisse de Zoologie 27: 235–242.

Walter, C. 1931. Arachnides halacariens. Biospeologica LVI. Campagne spéologique de C. Bolivar et R. Jeannel dans l'Amérique du Nord (1928). Archives de Zoologie Expérimentale et Générale 71: 375–381.

Parasitengonina: Hydrachnidiae and Stygothrombiae

Ian M. Smith
Canadian National Collection of Insects, Arachnids and Nematodes, Agriculture and Agri-Food Canada, Ottawa, ON, Canada

David R. Cook
Paradise Valley, AZ, USA

INTRODUCTION

In the Nearctic, water mites (Hydrachnidiae) are among the most abundant and diverse benthic arthropods in many freshwater habitats. One square meter areas of substratum from littoral weed beds in eutrophic lakes may contain as many as 2000 deutonymphs and adults representing up to 75 species in 25 or more genera. Comparable samples from an equivalent area of substratum in rocky riffles of streams often yield over 5000 individuals of more than 50 species in over 30 genera (including both benthic and hyporheic forms). Water mites have coevolved with some of the dominant insect groups in freshwater ecosystems, especially nematocerous Diptera, and typically interact intensively with these insects at all stages of their life histories.

The parasitic larvae and predaceous deutonymphs and adults of water mites have direct and almost certainly significant effects on the size and structure of insect populations in many habitats. Unfortunately, their impact has rarely been measured accurately because of the routine neglect of mites in ecological studies of freshwater communities. Due to their small size and often cryptic habits, mites are usually absent or seriously underrepresented in samples that are collected using standard techniques for capturing insects and crustaceans. Most entomologists are unfamiliar with mites and tend either to disregard them as too poorly known ecologically and difficult taxonomically or to lump them together in a meaningless way when conducting community studies. It is safe to assume that failure to include realistic assessment of the roles played by water mites routinely results in seriously flawed analyses of the structure and dynamics of freshwater communities.

Hydrachnidiae, along with the aquatic Stygothrombidioidea and terrestrial Calyptostomatoidea, Trombidioidea, and Erythraeoidea, belong to a remarkably diverse natural group of actinedid acariform mites, the Parasitengonina (Walter et al., 2009). The complex life history (Fig. 16.10.1) with profound metamorphosis that characterizes this group is unique among Acari. Typically, after emerging from the egg membrane, the hexapod larva seeks out an appropriate host and becomes an ectoparasite, which is passively transported while feeding on host fluids (Smith & Oliver, 1986; Proctor et al., 2015). When fully engorged the larva transforms to the quiescent protonymph (or nymphochrysalis). Radical structural reorganization occurs during this stage, giving rise to the active deutonymph which resembles the adult in being octopod and typically predaceous but is sexually immature and exhibits incomplete sclerotization and chaetotaxy. The deutonymph feeds and grows in size before entering another quiescent stage, the tritonymph (or imagochrysalis). After completion of metamorphosis during this stage, the mature adult emerges.

Larval water mites can be distinguished morphologically from those of other parasitengones by having two rather than one setae on the genu of the pedipalp. Deutonymphal and adult water mites differ from all other Acari in having a series of paired glandularia on the idiosoma.

The following key to adults of North American genera includes a number of taxa that have recently been elevated in rank from subgenus to genus (Smith et al., 2015). Readers are encouraged to consult the extensive list of references cited by Smith et al. (2010) for other recently published information on North American water mites.

LIMITATIONS

The following keys for larvae and adults are updated versions of the generic keys previously published by Smith et al. (2010). They incorporate a number of new or previously unreported genera and employ many additional characters and illustrations. We focus on the generic level because most of the 150 water mite genera that occur in the Nearctic appear to be monophyletic and can be diagnosed using characters that are reasonably easy to use. It is worth noting that members of nearly 35 new or unreported genera have been discovered in the Nearctic since the publication of Cook's global review in 1974, an average of almost one each year.

The keys are artificial in that genera do not always key out with others currently placed in the same family. In fact, after considerable thought, we concluded that it would not be helpful to provide a separate key to families for the reasons outlined below. However, we do include the currently accepted family and subfamily affiliations of all genera at appropriate couplets in the keys to provide access to the higher classification for those with a need for that information.

The keys are intended for use by non-specialists and employ characters that should be obvious on slide-mounted specimens to careful observers. Stygothrombidiid mites are included in the same keys as the true water mites because they are frequently collected with water mites and closely resemble them in morphology, behavior, and ecology.

The larval key is included to encourage and enable introductory study of the parasitic associations of larval

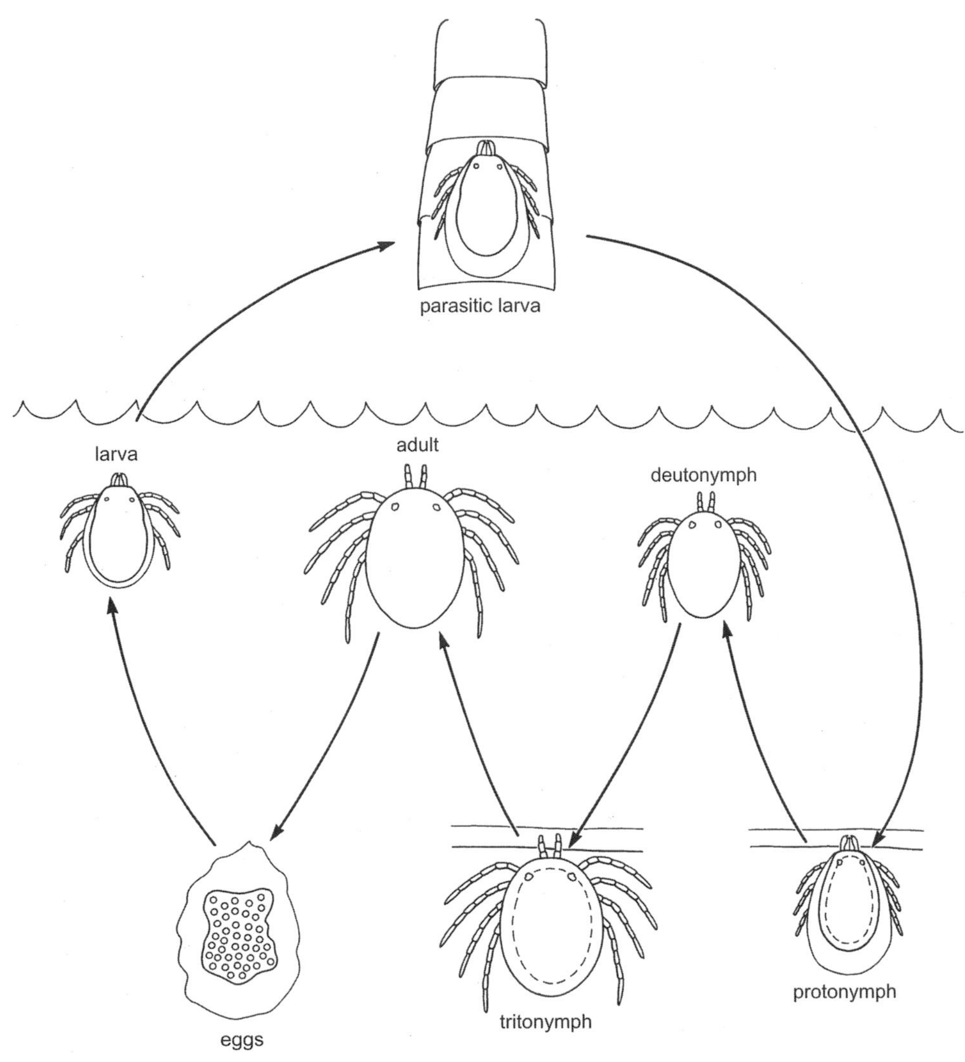

PLATE 16.10 **Fig. 1** Diagrammatic illustration of a generalized water mite history. *Redrawn and modified from Smith, 1976.*

water mites with aquatic insects, an essential aspect of water mite ecology and evolution and a major focus of the arachnid chapter of Volume 1 of this edition (Proctor et al., 2015). It is based on previously published descriptions of larvae of several hundred species in a wide range of genera that have been associated with conspecific adults by rearing. It covers a large number of taxa that parasitize aquatic insects but does not include all genera that appear in the adult key as larvae of many genera remain unknown. Consequently, the key for larvae will be subject to considerable expansion, revision, and refinement as larvae of species representing additional taxa are reared and described.

It is worth mentioning here that the larval stage is known to be suppressed to varying degrees in several water mite taxa (Proctor et al., 2015). Females of some genera with unknown larvae, such as *Kongsbergia* (Radwell & Smith, 2012), produce relatively large eggs that are thought to hatch as deutonymphs. Efforts to rear larvae from females of several rheophilic and interstitial genera have been repeatedly unsuccessful, possibly indicating that the larval stage is bypassed in a number of them.

Users of the larval key should proceed with caution, bearing in mind that it is intended as an introduction to the study of water mite larvae.[2] It has proven to be particularly useful at sites where the taxonomic composition of the fauna is well-known based on adults. Larval identifications should ultimately be confirmed by associating specimens with conspecific adults either by rearing or genetic sequence data. This key is a starting point for investigating water mite life histories and their parasitic associations with aquatic insects.

The family level classification of water mites is currently undergoing critical reassessment and is in need of substantial revision. In his major review and revision of the generic classification of water mites, Cook (1974) noted that several of the traditionally accepted families

2. The editors wish to reinforce the authors' cautionary statements on the use of larval keys. Knowledge of larvae in a number of genera is based on only one or a few species, and larvae of many genera remain unknown.

that were originally based on knowledge of the Palearctic fauna were not monophyletic and that numerous genera could not be assigned to established families without considerable difficulty. At that time, he expressed the hope that improved knowledge of the morphologically conservative larval instar would ultimately permit better insight into phylogeny as the foundation for a more natural family level classification.

There has been promising progress in larval taxonomy since then that has helped in testing and strengthening phylogenetic hypotheses at the family level (Prasad & Cook, 1972; Smith, 1976, 1978, 1982, 1984; Simmons & Smith, 1984). However, during the same period of time knowledge of global water mite diversity and regional endemism at the species and generic levels has greatly increased (Cook 1980, 1983, 1986, 1988, 1991, 1992; Viets, 1987; Smith et al., 2015). Efforts to accommodate this ever expanding taxonomic diversity within the existing family structure have resulted in an increasingly unworkable classification.

Some progress has been made in updating the family classification of water mites (Smith, 1972, 1992b; Cook et al., 2000) mainly by segregating small apparently holophyletic groups of genera from large polyphyletic families. However, recent efforts to produce technically correct larval and adult keys for families (see Walter et al., 2009) necessitated the use of several complex and cumbersome couplets that present daunting challenges for non-specialists. A number of currently accepted families remain polyphyletic or paraphyletic, and some have become too broadly defined to be very useful for diagnostic purposes.

Further revision of the classification will only resolve these problems and achieve stability if it is comprehensive and based on substantially improved understanding of phylogeny. An initiative to construct a robust phylogeny for water mites and their relatives by employing integrated analysis of morphological and behavioral characters along with molecular and genetic data for representative species of a broad range of taxa from all areas of the world is currently underway. The results of this study should ultimately permit construction of a more natural family level classification. In the meantime, the easiest and most reliable way to identify specimens is to focus on the generic level.

We do not attempt to provide keys to the species level in this treatment. We conservatively estimate that more than 2000 species of water mites and stygothrombidiids are currently represented in collections from the Nearctic. More than half of these species, including many that are common and widespread, remain undescribed and unnamed.

Undescribed species abound in collections from most parts of the Nearctic and numerous species names reported in the literature are based on misidentifications, many of which resulted from uncritical and incorrect application of names from the European literature to Nearctic species. Recent taxonomic revisions of some relatively small genera (Smith, 1989, 1991a,b, 1992a,b, 2003; Smith & Cook, 1994, 2000, 2006; Radwell & Smith, 2012) showed that 50–80% of the included species were new to science. The most diverse genera, such as *Sperchon*, *Torrenticola*, and *Aturus* in lotic habitats and *Piona* and *Arrenurus* in lentic habitats, each contain 100 or more Nearctic species, probably more than 400 species in the case of *Arrenurus*.

Additional taxa await discovery throughout the Nearctic in all habitats, especially springs and interstitial situations. The least explored areas, such as the southeastern coastal and piedmont regions, California, northern Mexico, and the boreal and arctic regions of northern Canada and Alaska, undoubtedly will yield additional unexpected taxa when more thoroughly studied.

Efforts to produce regional lists and keys for water mite species in all areas of the Nearctic will necessarily be highly provisional—and potentially misleading—until comprehensive systematic revisions have been completed for many more genera. Although extensive, modern, and well curated water mite collections are now available, taxonomic progress continues to be limited by a continuing shortage of trained experts in positions to carry out the needed long term species level research on the Nearctic fauna.

TERMINOLOGY AND MORPHOLOGY

Water mites exhibit the characteristic acarine body plan comprising the gnathosoma, or mouth region, and the idiosoma, or body proper. The standard morphological terminology proposed by Grandjean (see Krantz et al., 2009) has been widely adopted by most authors dealing with prostigmatic mites, but until recently little effort was made to apply this terminology to water mites. The extreme plasticity of the adult exoskeleton during water mite evolution has obscured many homologies with other Prostigmata, and water mite workers have developed a highly useful set of peculiar morphological terms for external structures of this instar. Early taxonomic studies of larval water mites generated an additional set of special morphological terms for the exoskeleton of this strongly heteromorphic instar, and we used a modified version of this terminology in an earlier edition. However, it has become evident that the relatively plesiotypical body plan of larvae permits much of the standard idiosomal terminology applied to other mites to be used with confidence and we employ it here.

Larvae

The short gnathosoma bears the stocky pedipalps, which have five free segments (trochanter, femur, genu, tibia, tarsus) that flex ventrally. The tarsus of the pedipalp is relatively long and cylindrical in some early derivative clades but is typically reduced to a dome- or button-shaped pad in most genera. A highly modified thick, curved seta is present dorsally at the end of the tibia representing the homologue of the tibial claw, which characterizes the pedipalp of most terrestrial Anystoidea, Parasitengona and related groups. The paired chelicerae, each consisting of a cylindrical basal segment and a movable terminal claw, lie between the pedipalps.

The idiosoma is plesiotypically mainly unsclerotized in basal taxa but is extensively sclerotized in many families. The dorsum bears a medial eye (reduced or absent in some taxa), two pairs of lens-like lateral eyes, four pairs of lyrifissures and the following complement of setae: four pairs on the propodosoma, two pairs of verticils, the internals (*vi*) or anteromedials (*AM*) and externals (*ve*) or anterolaterals (*AL*), and two pairs of scapulars, the internals (*si*) or sensillae (*SS*) and externals (*se*) or posterolaterals (*PL*); and eight pairs on the hysterosoma, three pairs in the *c*-row (*c1, c2* and *c3*), two pairs in the *d*-row (*d1* and *d2*), two pairs in the *e*-row, (*e1* and *e2*), and one pair from the *f*-row (*f1*). Ventrally the integument bears three pairs of coxal plates, paired urstigmata (*Ur*) laterally between the first and second coxal plates, the excretory pore (*EP*), one pair of lyrifissures, and the following complement of setae: three to six pairs on the coxal plates, two in the *1*-row (*1a* and *1b*), zero, one or two in the *2*-row (*2b* usually present, *2a* and *2c* usually absent), and one or two in the *3*-row (*3a* always present, *3b* usually absent), six pairs on the hysterosoma, one pair of preanals (*pa*), two pairs associated with the excretory pore (*ps1* and *ps2*), one pair from the *f*-row (*f1*), and two pairs in the *h*-row (*h1* and *h2*). The size, shape, and position of these dorsal and ventral structures, and the degree of fusion of the sclerites associated with them, provide useful taxonomic characters. The legs are inserted laterally on the coxal plates, and in the plesiotypical condition have six movable segments, namely trochanter (*Tr*), basifemur (*BFe*), telofemur (*TFe*), genu (*Ge*), tibia (*Ti*), and tarsus (*Ta*), that articulate to permit ventral flexion. The segments have characteristic complements of setae and solenidia. The tarsi bear paired claws and claw-like empodia terminally.

We follow the convention for numbering the setae on the leg segments of larvae that was established by Prasad and Cook (1972) and modified slightly by Smith (1976) and Walter et al. (2009). The setae are numbered starting proximally on the posterolateral surface, proceeding distally along the dorsal half of the segment to the end, then returning proximally along the ventral half of the segment. Exceptions are made in the cases of solenidia (setiform chemoreceptors), which have their own standard notation as follows: "σ" on the genu, "φ" on the tibia, and "ω" on the tarsus. Eupathids are always numbered second, that is as "2" on the tarsi. Where it has been necessary to use leg chaetotaxy in the key, the number of solenidia present on a segment is indicated in brackets (e.g., "+2φs") immediately following the information on the number of setae, so that all setiform structures can be accounted for. We usually have omitted the empodia and claws from the illustrations of the legs of larvae so that the number and positions of distal setae on the tarsi can be seen clearly. The names of the segments of the appendages are consistently abbreviated as follows: trochanter, Tr; femur, Fe (basifemur, BFe; telofemur, TFe); genu, Ge; tibia, Ti; and tarsus, Ta.

Deutonymphs and Adults

The gnathosoma consists of the gnathosomal base, or capitulum, and associated appendages, namely the chelicerae and pedipalps. The capitulum is plesiotypically a simple, short channel, derived from extensions of the pedipalpal coxae, leading to the oesophagus. A protrusible tube of integument connecting the capitulum to the idiosoma has developed independently in several distantly related genera. The paired pedipalps, inserted on the capitulum, have both tactile and raptorial functions. In the plesiomorphic condition, the pedipalps have five movable segments, namely trochanter, femur, genu, tibia, and tarsus, that are essentially cylindrical and articulate to allow ventral flexion. The tibia bears a thick, blade-like, dorsal seta distally in many ancient genera. As in larvae, this seta is the homologue of the tibial claw of terrestrial relatives, and it often makes the pedipalps appear chelate. In derivative groups, other setae along with various denticles and tubercles may be elaborated to enhance the raptorial function of the pedipalps. Segmentation of the pedipalps is reduced by fusion in a few genera. A modification that has developed independently in various groups of Arrenuroidea is the so-called uncate condition. In these genera, the tibia is expanded and produced ventrally to oppose the tarsus, permitting the mites to grasp and hold slender appendages of prey organisms securely. The pedipalps are highly modified in some interstitial genera, presumably to facilitate prey capture in confined hyporheic spaces. The paired chelicerae lie in longitudinal grooves between the pedipalps on the dorsal surface of the capitulum. Plesiotypically they consist of a cylindrical basal segment bearing a movable terminal claw. This cheliceral structure is designed for tearing the integument of prey organisms, and is retained in nearly all derivative groups. Hydrachnidae are unique in having unsegmented and stilletoform chelicerae, an obvious adaptation for piercing the insect eggs upon which they feed. The chelicerae are separate in all groups except Limnocharidae and Eylaidae where they are fused medially.

The idiosoma, or body proper, is plesiotypically round or ovoid in outline, slightly flattened dorsoventrally, and mostly unsclerotized. The dorsum bears an unpaired medial eye, paired lateral eyes that are usually enclosed in capsules, paired preocular and postocular setae, and longitudinal series of paired glandularia (six dorsoglandularia, five lateroglandularia), muscle attachment sites (five dorsocentralia, four dorsolateralia), and lyrifissures (five). The venter bears the paired coxal plates (fused into anterior and posterior groups on each side), the genital field (comprising the gonopore, three pairs of acetabula, and paired genital valves), five pairs of ventroglandularia (including coxoglandularia I between the anterior and posterior coxal groups and coxoglandularia II behind the posterior groups) and the excretory pore. As in larvae, these idiosomal structures provide a wealth of useful taxonomic characters.

The legs are inserted laterally on the coxae, and plesiotypically articulate on a vertical major axis and have six movable segments that articulate to permit ventral flexion. The segments are essentially cylindrical and have variable complements of setae. Though chaetotaxy of the legs provides a variety of taxonomic characters in deutonymphs and adults, the expression and position of individual setae are highly variable within taxa and even on opposite sides of the same specimen. Consequently, the rigorous analysis of chaetotactic patterns that proves so useful in the case of larvae is not practicable for later instars. The leg tarsi plesiotypically bear paired claws terminally. Walking forms have relatively short, stocky leg segments and setae, whereas swimmers have longer segments bearing fringes of slender swimming setae.

Adaptation to habitats such as seepage areas and springs or substrata in streams required a change in locomotor habits from walking or swimming. Mites adapted for living in moss mats and wet litter habitats developed hydrophilic integument, which draws a film of water over the dorsal surface, creating sufficient downward force to press the body to the substratum and prevent efficient walking. Groups that invaded streams evolved a wedge-shaped body designed for negotiating confined spaces to avoid exposure to water turbulence and strong currents. Locomotion in both of these types of habitats necessitated evolution of a crawling gait. This change involved a shift in orientation of the major axis of the legs, shortening and thickening of leg segments and setae, enlargement of tarsal claws, and development of stronger, more massive muscles to control leg movements. Expansion of coxal plates and sclerites associated with glandularia, setal bases, and the genital field, along with sclerotization of the dorso- and laterocentralia, occurred to provide rigid exoskeletal support for these muscles. Fusion of these sclerotized areas led to development of complete dorsal and ventral shields in adults of many taxa.

Certain groups are adapted for exploiting interstitial habitats in subterranean waters and the hyporheic zone of rheocrenes and streams. These mites tended to lose eyes and integumental pigmentation, and required further streamlining of the body to facilitate locomotion in interstitial spaces. Consequently, soft-bodied forms adopted a vermiform shape, while sclerotized mites became extremely compressed laterally or dorsoventrally. The crawling mode of locomotion became secondarily modified in several groups of well sclerotized interstitial genera to permit rapid and agile running in hyporheic and groundwater habitats.

Members of different clades have developed superficially similar sclerite arrangements in adapting to lotic or interstitial habitats while others have undergone homoplastic reduction or loss of sclerites in secondarily invading lentic habitats. Modern communities are highly heterogeneous phylogenetically and morphologically, with each monophyletic component exhibiting unique exoskeletal adaptations for living in its particular habitat.

MATERIAL PREPARATION AND PRESERVATION

For morphological study, deutonymphal and adult mites should be preserved in modified Koenike's solution (or GAW), consisting of 5 parts glycerin, 4 parts water, and 1 part glacial acetic acid, by volume so that they retain their shape and can be efficiently cleared in 10% KOH, dissected, and slide-mounted in glycerin jelly. Specimens preserved in alcohol or other hygroscopic agents become distorted and brittle, making subsequent preparation difficult. Glycerin jelly is the recommended mounting medium because it permits manipulation and orientation of dissected body sclerites and appendages prior to hardening.

Reared larvae are often slide-mounted directly, but can be preserved in 70% ethanol for subsequent mounting without being seriously damaged. Parasitic larvae should be preserved with the host in 70% ethanol, although larvae removed from hosts that have been pinned and dried often can be slide-mounted successfully. Before larvae are removed from hosts, the attachment sites should be noted and recorded. Larvae should be mounted whole in Hoyer's medium which also acts as an efficient clearing agent for these small specimens.

Specimens required for molecular or genetic analysis should be preserved in 95% ethanol and stored in a freezer as soon as possible.

KEYS TO PARASITENGONINA

Keys to Larvae

1 Dorsal plate bearing three pairs of setae (external verticils, scapulars) and an unpaired anteromedial seta (probably representing internal verticils) (Fig. 16.13.9, arrow); urstigmata (Ur) stalked (Figs. 16.13.9, 10); pedipalp tibia with terminal claw-like seta four-pronged (Fig. 16.13.11, arrow), legs with five movable segments (Fig. 16.13.12); leg tarsi bearing paired pectinate claws and a simple empodium (Fig. 16.13.12) .... Stygothrombiae .... Stygothrombidioidea .... Stygothrombidiidae .... *Stygothrombium*

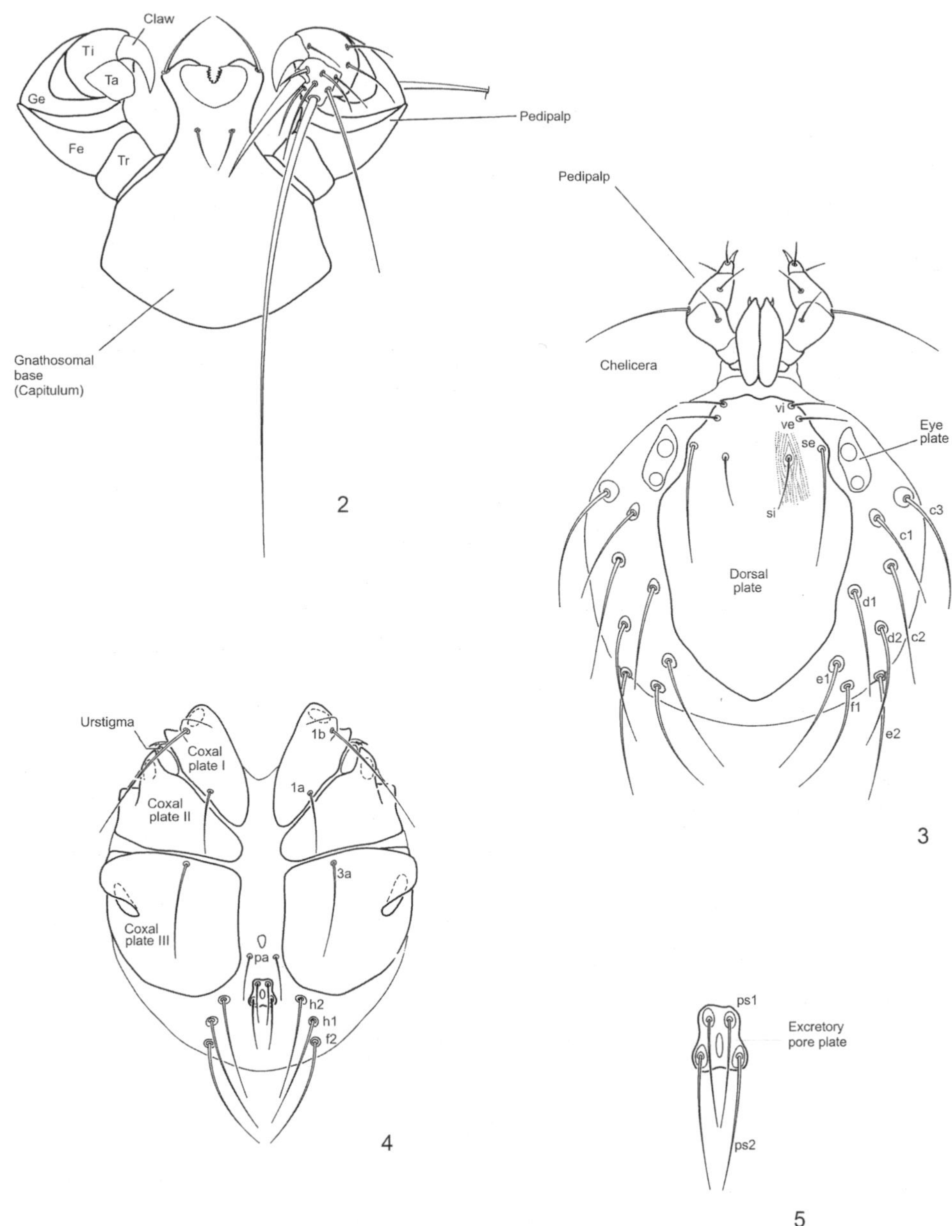

PLATE 16.11 *Sperchonopsis ecphyma* Prasad and Cook (Sperchontidae) larva. **Fig. 2** venter of gnathosoma. **Fig. 3** dorsum of idiosoma and gnathosoma. **Fig. 4** venter of idiosoma. **Fig. 5** excretory pore plate. See text for terminology and abbreviations of setae.

1'	Dorsal plate bearing at least four pairs of setae (verticils and scapulars), and in some cases one or more additional pairs of hysterosomal setae, but no unpaired setae (Figs. 16.16.38, 19.65, 20.70); urstigmata sessile (Figs. 16.16.41, 19.66); pedipalp tibia with terminal claw-like seta undivided (Fig.16.19.67, arrow), or deeply bisected (Figs.16.16.40, 20.71, arrows) but never four-pronged; legs with five or six movable segments; leg tarsi bearing paired simple claws (Fig. 16.33.175) or lacking claws (Fig. 16.16.39), but always bearing a simple empodium........ Hydrachnidiae........	2
2(1)	Legs with six movable segments, with basifemur (BFe) and telofemur (TFe) separated (Figs. 16.13.13, 14.21, 15.31, 18.57)	3
2'	Legs with five movable segments, with basifemur and telofemur fused (Figs. 16.12.6–8, 16.37, 19.64)	16
3(2)	Gnathosoma with elaborate camerostome enclosing chelicerae and with pedipalps inserted ventrally (Figs. 16.11.4, 14.17, 18); dorsal plate present and bearing only two pairs of setae (verticils—ve and vi) near anterior edge (Figs. 16.13.14 and 16.14.17); third coxal plates located posteriorly on idiosoma with insertions of third legs located at posterolateral edges (posterior to level of excretory pore) and third legs directed posteriorly (Figs. 16.13.16, 14.19) .... Hydrovolzioidea	4
3'	Gnathosoma lacking elaborate camerostome; dorsal plate absent (Figs. 16.17.50, 18.53), or present and bearing more than two pairs of setae with verticils (ve and vi) anteriorly and at least internal scapulars (si) in posterior half near edge (Figs. 16.17.46, 52, 18.59); third coxal plates located near mid-length of idiosoma with insertions of third legs located anterolaterally (anterior to level of excretory pore) and third legs directed laterally (Fig. 16.18.62)	5

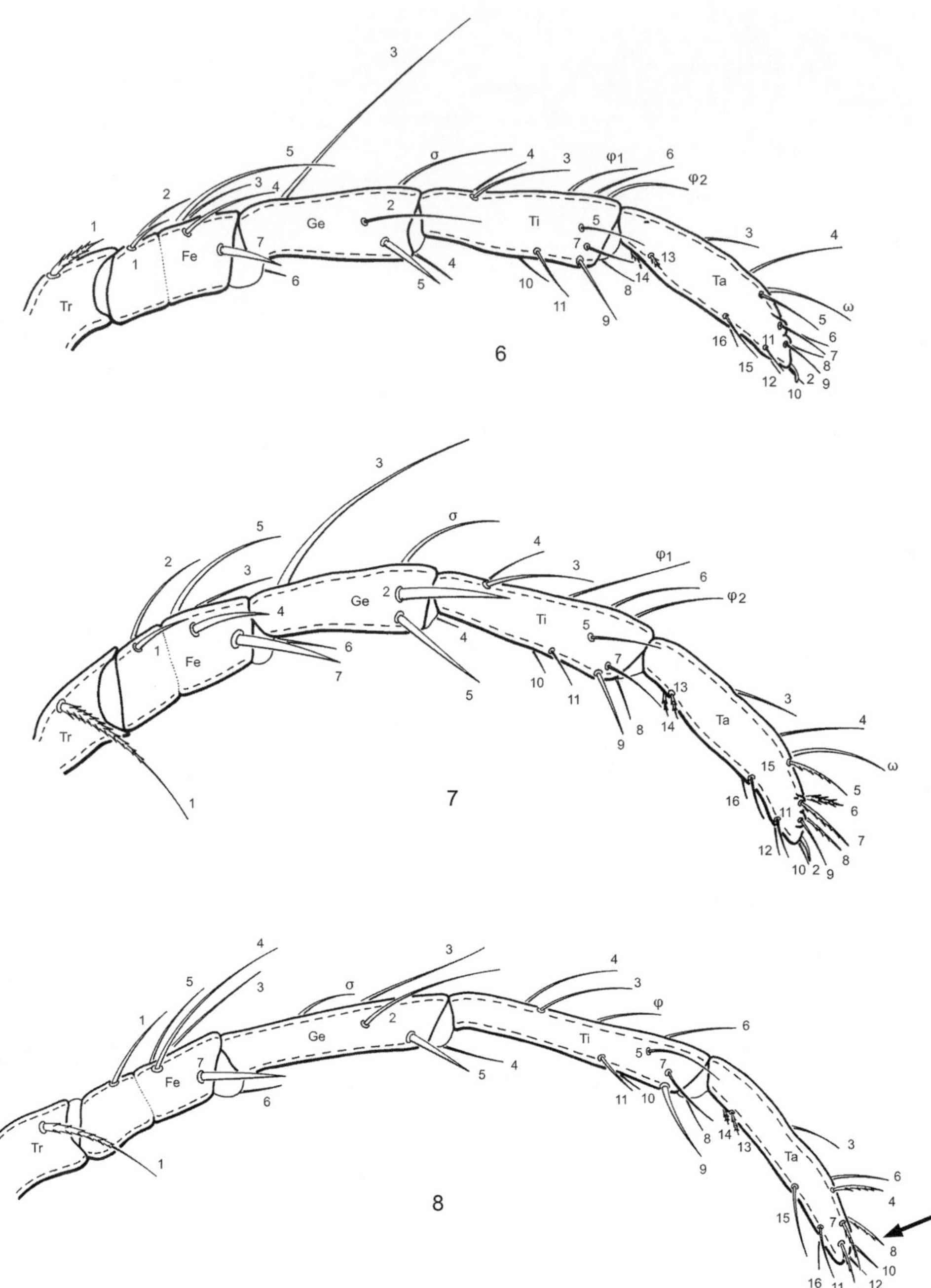

PLATE 16.12 Figures 6–8: *Sperchonopsis ecphyma* Prasad and Cook (Sperchontidae) larva antero-lateral views of legs. **Fig. 6** first leg. **Fig. 7** second leg. **Fig. 8** third leg. Fe, femur; Ge, genu; Ta, tarsus; Ti, tibia; Tr, trochanter.

4(3) Dorsal plate relatively small (Fig. 16.13.14); venter with rows of numerous small urstigmata borne between second and third coxal plates (Fig. 16.13.16, arrow); first and second coxal plates separated from one another on each side and from opposite members medially (Fig. 16.13.16); pedipalps massive, with tibia (Ti) bearing four long, thick, curved setae (Fig. 16.13.15); setae c3 on idiosomal dorsum often foliate (Fig. 16.13.14) .... Hydrovolziidae .... *Hydrovolzia*

4' Dorsal plate relatively large (Fig. 16.14.17); venter with urstigmata either inconspicuous or absent (Fig. 16.14.19); first and second coxal plates fused together to form single anterior plate (Fig. 16.14.19); pedipalps moderate in size, with tibia (Ti) bearing two slightly thickened, straight setae (Fig. 16.14.18) .... Acherontacaridae .... *Acherontacarus*

5(3) Dorsal plate absent (Figs. 16.17.50, 18.53), or small, covering less than one-third length of idiosoma and usually with internal scapular setae (si) at posterior edge (Figs. 16.17.46, 52, 18.59); lateral eyes borne on separate platelets (Figs. 16.17.46, 50, 18.53, 59); leg tarsi with unmodified claws and empodium (emp) (Fig. 16.18.60) .... Hydryphantoidea 6

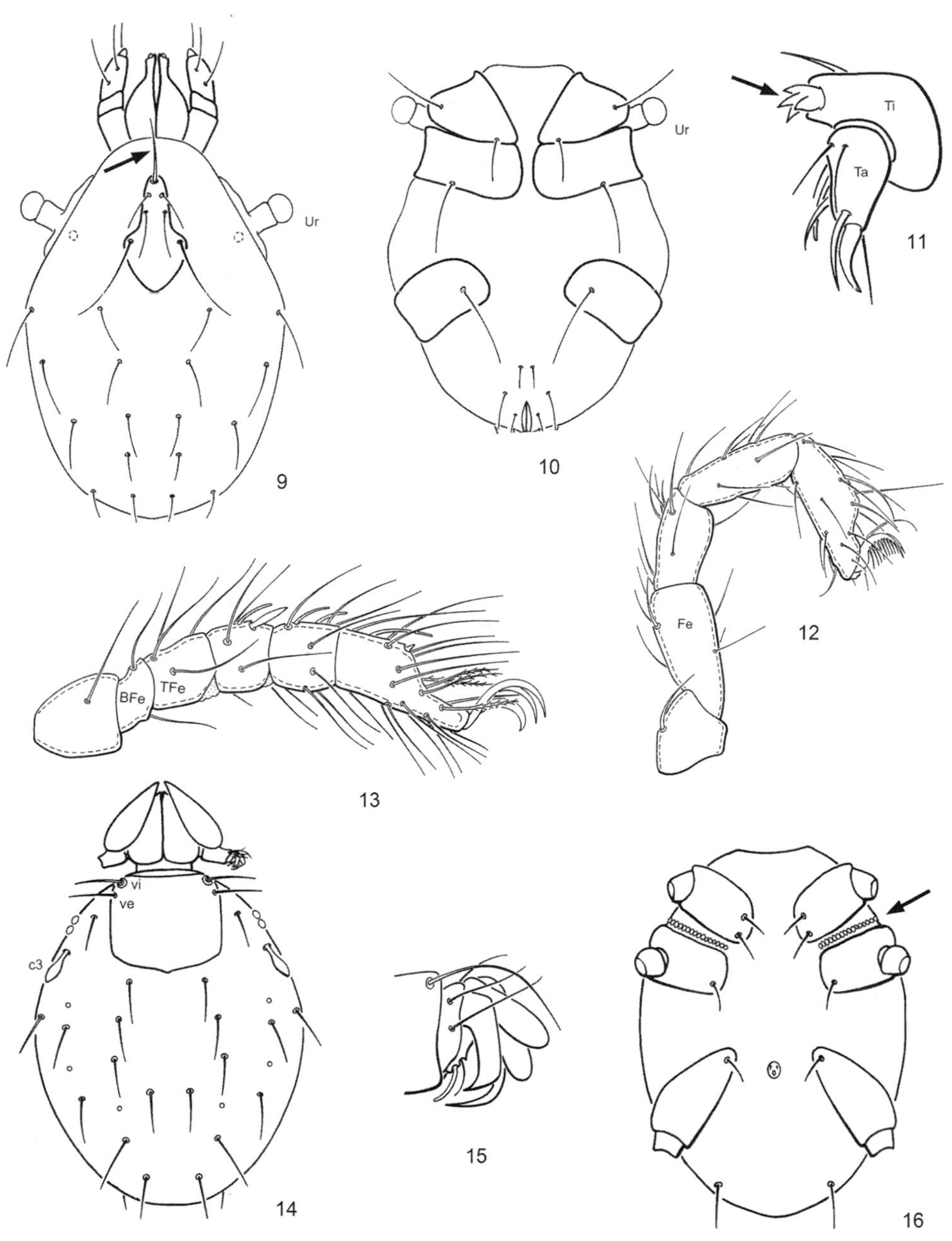

PLATE 16.13 Figures 9–12: *Stygothrombium* sp. (Stygothrombidiidae) larva. **Fig. 9** dorsum of idiosoma and gnathosoma. **Fig. 10** venter of idiosoma. **Fig. 11** tibia and tarsus of pedipalp. **Fig. 12** first leg. **Figures 13–16:** *Hydrovolzia gerhardi* Mitchell (Hydrovolziidae) larva. **Fig. 13** first leg anterolateral view. **Fig. 14** dorsum of idiosoma and gnathosoma. **Fig. 15** distal segments of pedipalp. **Fig. 16** venter of idiosoma. *Redrawn and modified from Wainstein, 1980.*

5' Dorsal plate large, covering well over one-third length of idiosoma and with internal scapular setae (si) near mid-length (Figs. 16.14.22, 15.25, 29, 34, 16.35); lateral eyes borne on single eye plates (Figs. 16.14.22, 15.25, 29); leg tarsi with claws modified or reduced (Figs. 16.14.23, 15.30, 33) .... Eylaoidea 12

6(5) Dorsal plate roughly quadrangular, bearing four pairs of setae (ve, vi, se, si) (Figs. 16.17.52, 18.59), or three pairs with external scapulars (se) borne on separate platelets near posterolateral angles of plate 7

6' Dorsal plate triangular, fragmented, or absent, bearing fewer than three pairs of setae (Figs. 16.17.46, 50, 18.53) 10

7(6) Claw (movable digit) of chelicera over one-half length of basal segment (Fig. 16.17.42); excretory pore setae (ps1 and ps2) and their alveoli absent (Fig. 16.17.43) .... Thermacaridae .... *Thermacarus*

7' Claw (movable digit) of chelicera less than one-third length of basal segment (Fig. 16.17.47); excretory pore setae (ps1 and ps2) or at least their alveoli, present (Figs. 16.17.51, 18.55, 62) .... Hydryphantidae (in part) 8

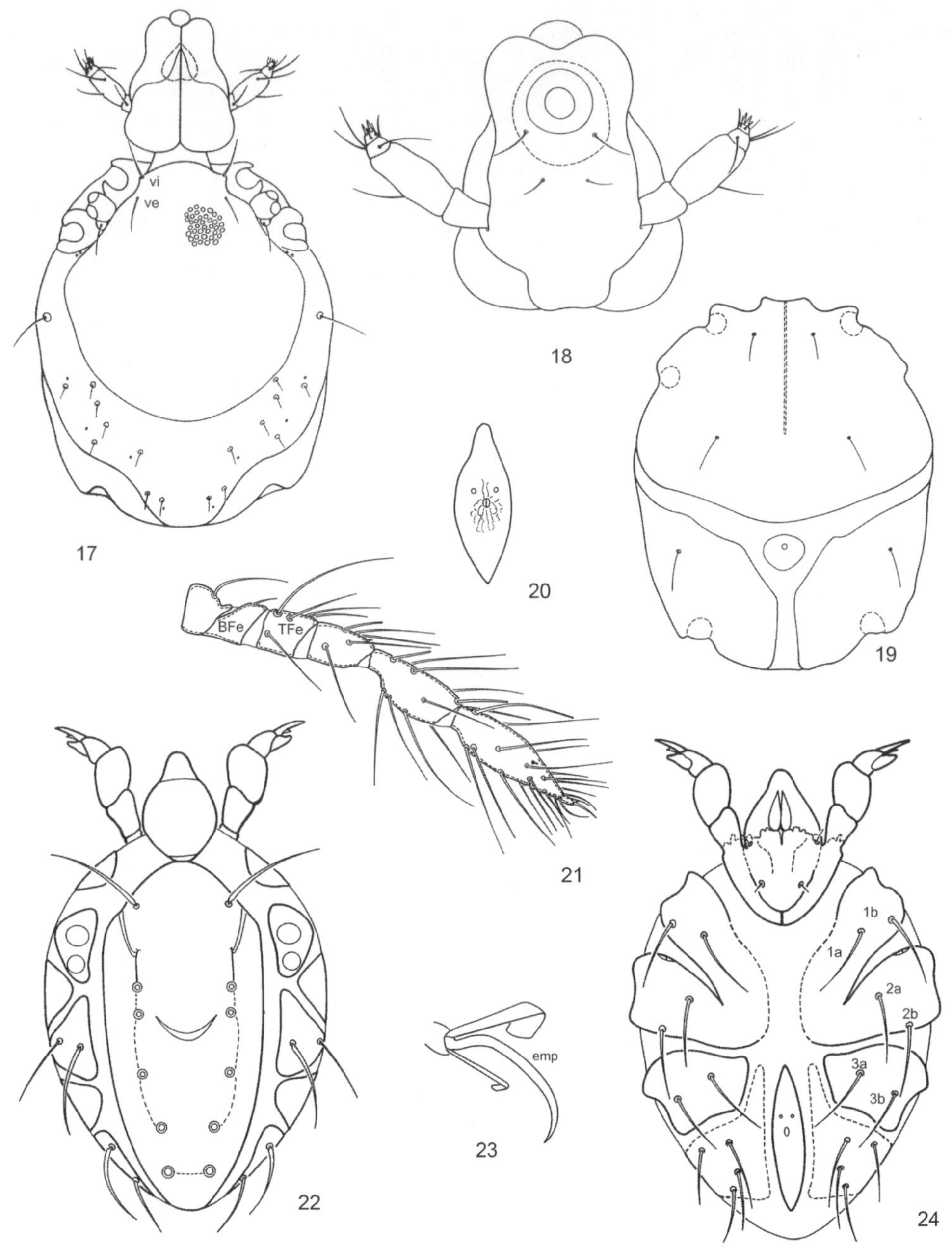

PLATE 16.14 **Figures 17–19:** *Acherontacarus* sp. (Acherontacaridae) larva. **Fig. 17** dorsum of idiosoma and gnathosoma. **Fig. 18** venter of gnathosoma. **Fig. 19** venter of idiosoma. **Figures 20–24:** *Eylais major* Lanciani (Eylaidae) larva. **Fig. 20** excretory pore plate. **Fig. 21** first leg. **Fig. 22** dorsum of idiosoma and gnathosoma. **Fig. 23** claws and empodium of first leg. **Fig. 24** venter of idiosoma and gnathosoma.

8(7) Second coxal plates lacking setae (2b absent) (Fig. 16.17.45); urstigmata relatively large (Fig. 16.17.45, arrow); pedipalp tibia with terminal claw-like seta deeply bifurcate (Fig. 16.17.44, arrow) and tarsus with only one thick seta; excretory pore plate absent but setae ps1 and ps2 present (Fig. 16.17.45) .... Wandesiinae .... *Wandesia*

8' Second coxal plates bearing setae 2b (Fig. 16.18.62); urstigmata relatively small (Fig. 16.18.62, arrow); pedipalp tibia with terminal claw-like seta only slightly bifurcate or undivided, and tarsus with two thick setae (Fig. 16.18.61, arrows); excretory pore plate present; setae ps1 and ps2 present or absent, but their alveoli always present (Fig. 16.18.62) 9

9(8) Medial eye present (Fig. 16.18.59, arrow) .... Euthyadinae .... *Panisus, Panisopsis, Todothyas, Thyasides, Euthyas, Zschokkea*

9' Medial eye reduced (Fig. 16.17.52) .... Tartarothyadinae .... *Tartarothyas*

10(6) Dorsal plate triangular, bearing scapular setae (se and si), with verticil setae (ve and vi) borne on small platelets flanking apex of plate (Fig. 16.17.46); excretory pore setae present, with ps1 borne on plate (Fig. 16.17.48) and ps2 borne on small separate platelets; basal segment of chelicerae massive and longitudinally striate .... Hydryphantidae (in part) .... Hydryphantinae .... *Hydryphantes*

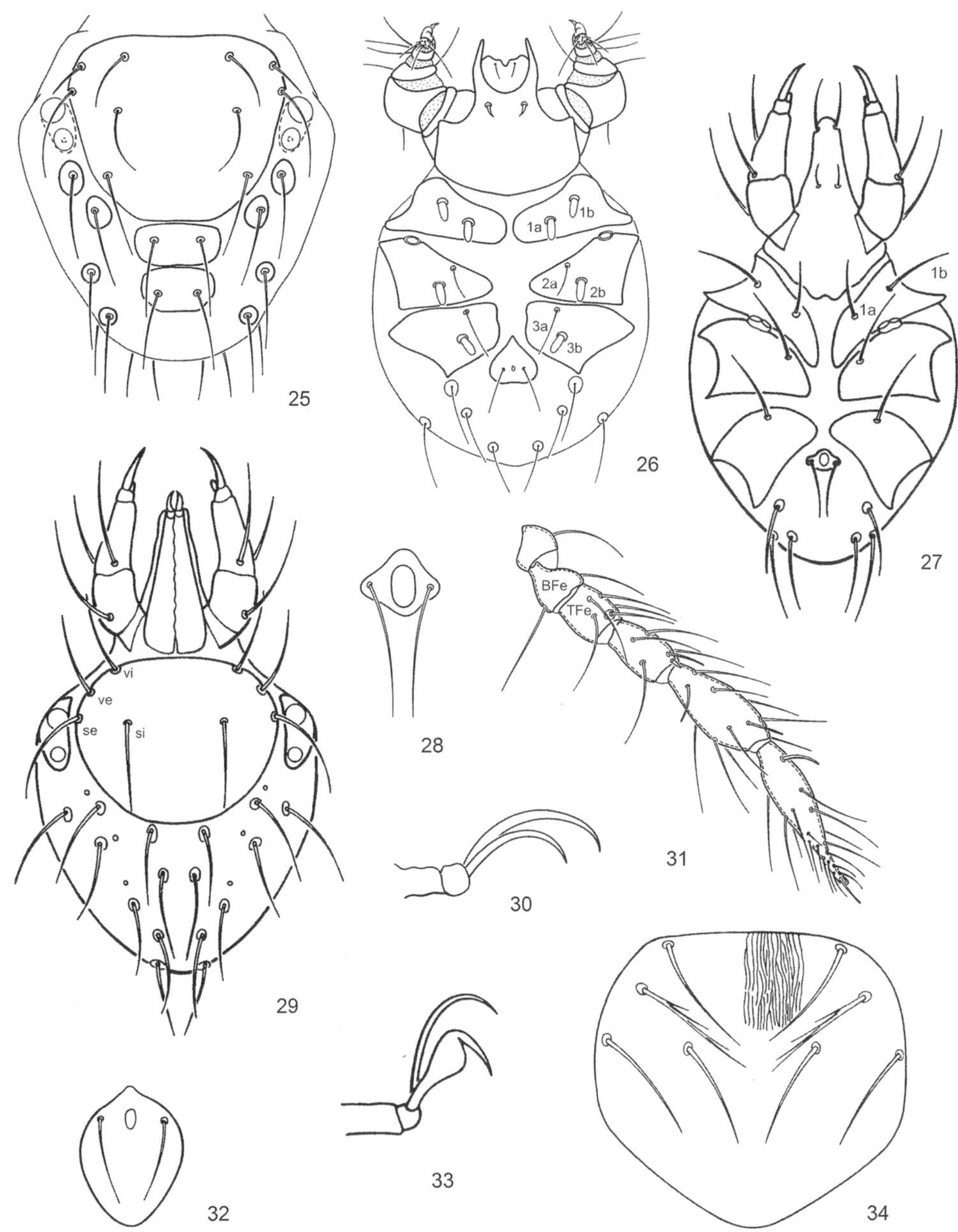

PLATE 16.15 **Figures 25–26:** *Piersigia* sp. (Piersigiidae) larva. **Fig. 25** dorsum of idiosoma. **Fig. 26** venter of idiosoma and gnathosoma. **Figures 27–31:** *Limnochares americana* Lundblad (Limnocharinae) larva. **Fig. 27** venter of idiosoma and gnathosoma. **Fig. 28** excretory pore plate. **Fig. 29** dorsum of idiosoma and gnathosoma. **Fig. 30** claw and empodium of first leg. **Fig. 31** first leg. **Figures 32–34:** *Neolimnochares johnstoni* Smith & Cook (Limnocharinae) larva. **Fig. 32** excretory pore plate. **Fig. 33** claw and empodium of first leg. **Fig. 34** dorsal plate.

10' Dorsal plate fragmented or absent, with verticil and scapular setae borne on paired platelets (Figs. 16.17.50, 18.53); excretory pore setae reduced to vestiges or absent but their alveoli present (Figs. 16.17.51, 18.55); basal segment of chelicerae relatively slender and smooth (Fig. 16.18.59) 11

11(10) Medial eye present (Fig. 16.17.50, arrow); excretory pore plate well sclerotized and nearly triangular (Fig. 16.17.51); pedipalp tarsus with all setae slender; solenidia on leg tarsi slender (Fig. 16.17.49, arrow) .... Hydryphantidae (in part) .... Protziinae .... *Protzia*

11' Medial eye absent (Fig. 16.18.53); excretory pore plate weakly sclerotized and oblong (Fig. 16.18.55); pedipalp tarsus with two thickened, blade-like setae (Fig. 16.18.56, arrows); solenidia on tarsi of first and second legs very thick (Fig. 16.18.54, arrow) .... Hydrodromidae .... *Hydrodroma*

12(5) First, second and third coxal plates on each side bearing two, two, and two setae respectively (Figs. 16.14.24, 15.26); tarsi of legs bearing paired claws and claw-like empodium (emp) (Fig. 16.14.23) 13

12' First, second and third coxal plates on each side bearing two or one, one, and one setae respectively (Figs. 16.15.27, 16.36); tarsi of legs bearing two dissimilar claw-like structures terminally (Fig. 16.15.30, 33) .... Limnocharidae 14

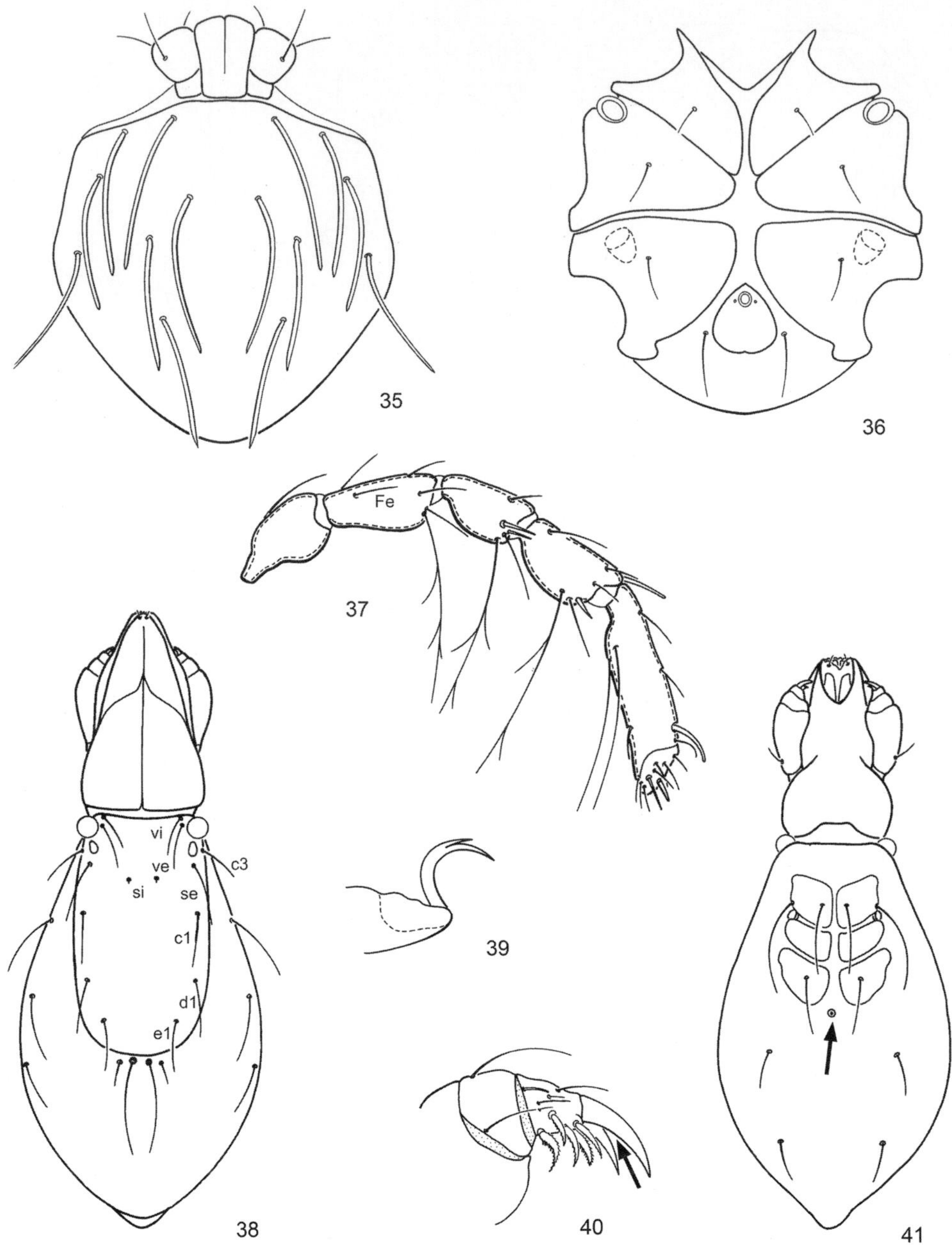

PLATE 16.16 **Figures 35–36:** *Rhyncholimnochares kittatinniana* Habeeb (Rhyncholimnocharinae) larva. **Fig. 35** dorsum of idiosoma and gnathosoma. **Fig. 36** venter of idiosoma. **Figures 37–41:** *Hydrachna magniscutata* Marshall (Hydrachnidae) larva. **Fig. 37** first leg. **Fig. 38** dorsum of idiosoma and gnathosoma. **Fig. 39** empodium of first leg. **Fig. 40** distal segment of pedipalp. **Fig. 41** venter of idiosoma and gnathosoma.

13(12) Dorsum of idiosoma nearly covered (in unengorged larvae) by single, elongate dorsal plate bearing seven pairs of setae (or their alveoli) (Fig. 16.14.22); coxal plates with setae all simple (Fig. 16.14.24); excretory pore plate elongate (Figs. 16.14.20, 24) .... Eylaidae..*Eylais*

13' Dorsum of idiosoma with three separate plates bearing five, one, and one pairs of setae respectively (Fig. 16.15.25); coxal plates with setae 1a, 1b, 2b and 3b blunt and conical in shape (Fig. 16.15.26); excretory pore plate obcordate (Fig. 16.15.26) .... Piersigiidae..*Piersigia*

14(12) Dorsal plate covering no more than anterior half of idiosomal dorsum, bearing four pairs of setae (verticils and scapulars) (Figs. 16.15.29, 34); first coxal plates bearing two pairs of setae, 1a and 1b (Fig. 16.15.27); excretory pore plate diamond-shaped (Fig. 16.15.28) or nearly oval (Fig. 16.15.32)...Limnocharinae...15

14' Dorsal plate covering entire idiosomal dorsum in unengorged larvae, bearing seven pairs of setae, including verticils, scapulars, and three pairs of hysterosomal setae (Fig. 16.16.35); first coxal plates bearing one pair of setae (Fig. 16.16.36); excretory pore plate pyriform (Fig. 16.16.36) ....Rhyncholimnocharinae...*Rhyncholimnochares*

15(14) Dorsal plate (Fig. 16.15.29) finely punctate; excretory pore plate relatively small, little larger than excretory pore (Fig. 16.15.28); leg tarsi bearing 2 similar claw-like structures with empodium and single ambulacral claw both slender (Fig. 16.15.30) ....*Limnochares*

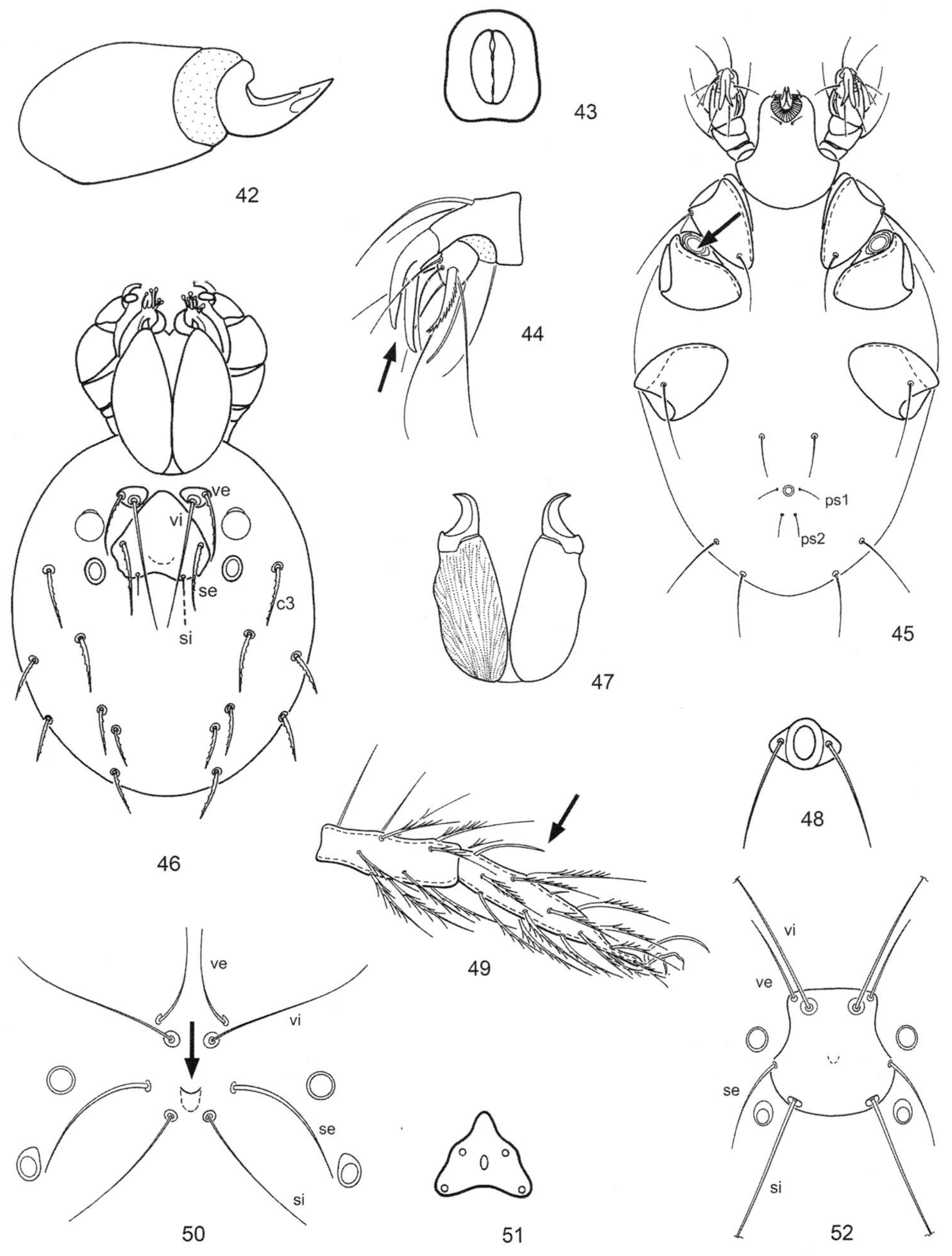

PLATE 16.17 Figures 42–43: *Thermacarus nevadensis* Marshall (Thermacaridae) larva. **Fig. 42** chelicera. **Fig. 43** excretory pore plate. **Figures 44–45:** *Wandesia* sp. (Wandesiinae) larva. **Fig. 44** tibia and tarsus of pedipalp. **Fig. 45** venter of idiosoma and gnathosoma. **Figures 46–48:** *Hydryphantes ruber* (de Geer) (Hydryphantinae) larva. **Fig. 46** dorsum of idiosoma and gnathosoma. **Fig. 47** chelicerae. **Fig. 48** excretory pore plate. **Figures 49–51:** *Protzia* sp. (Protziinae) larva. **Fig. 49** tibia and tarsus of first l29.eg. **Fig. 50** prodorsal region of idiosoma. **Fig. 51** excretory pore plate. **Figure 52:** *Tartarothyas martini* Smith and Cook (Tartarothyadinae) larva dorsal plate.

15’ Dorsal plate (Fig. 16.15.34) strongly striate; excretory pore plate relatively large, much larger than excretory pore (Fig. 16.15.32); leg tarsi bearing 2 dissimilar claw-like structures, with empodium wider than single ambulacral claw (Fig. 16.15.33) .... *Neolimnochares*

16(2) Dorsal plate bearing eight pairs of setae, including verticils, scapulars, c3, and three additional pairs of hysterosomal setae (Fig. 16.16.38); leg tarsi lacking paired claws (Fig. 16.16.39); excretory pore plate tiny, and bearing neither setae nor their alveoli (Fig. 16.16.41, arrow) .... Hydrachnoidea...Hydrachnidae ....*Hydrachna*

16’ Dorsal plate bearing four pairs of setae, including only verticils and scapulars (Fig. 16.19.65) or, in some cases, up to six pairs of setae including verticils, scapulars and one or two pairs of hysterosomal setae (Fig. 16.20.70); leg tarsi bearing paired claws (Figs. 16.29.143, 33.175); excretory pore plate small to large, always bearing setae ps1 and ps2 or at least their alveoli (Figs. 16.19.63, 21.80, 24.104, 25.109, 113–114, 33.172, 176, 34.185, 36.193) 17

17(16) First, second, and third coxal plates on each side all separate (Figs. 16.19.66, 69, 20.72, 33.172, 34.179–180, 35.186), or all fused (Fig. 16.34.185) *and* plates of two sides fused medially *and* dorsal plate round and bearing setae c1 laterally (Fig. 16.34.184) 18

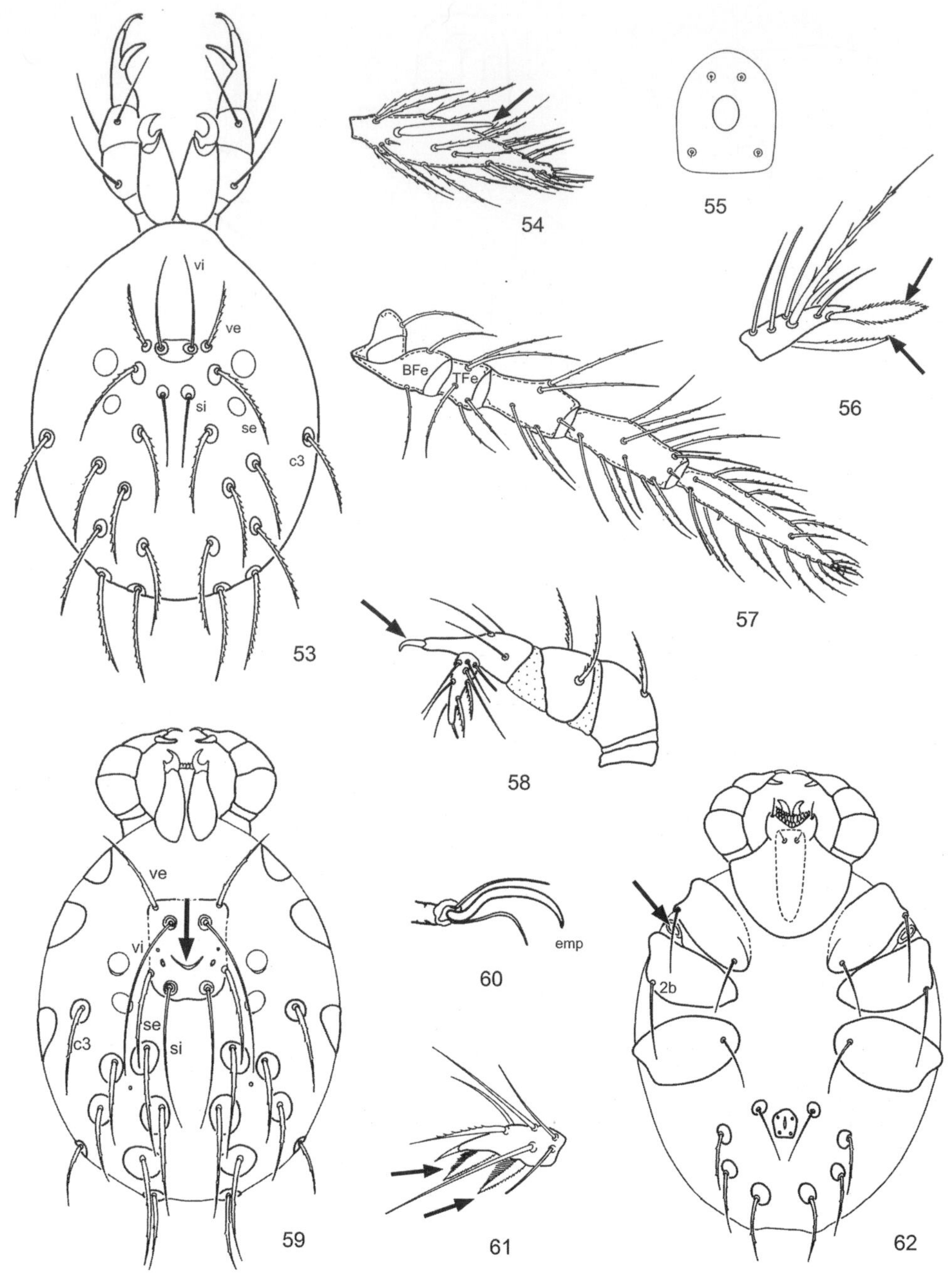

PLATE 16.18 **Figures 53–56:** *Hydrodroma despiciens* (Müller) (Hydrodromidae) larva. **Fig. 53** dorsum of idiosoma and gnathosoma. **Fig. 54** tarsus of first leg. **Fig. 55** excretory pore plate. **Fig. 56** tarsus of pedipalp. **Figures 57–62:** *Todothyas stolli* (Koenike) (Euthyadinae) larva. **Fig. 57** first leg. **Fig. 58** pedipalp. **Fig. 59** dorsum of idiosoma and gnathosoma. **Fig. 60** claws and empodium. **Fig. 61** tarsus of pedipalp. **Fig. 62** venter of idiosoma and gnathosoma.

17' First and second coxal plates on each side separate and second and third coxal plates on each side fused at least medially (Figs. 16.21.83, 24.106, 29.145, 31.157), or all plates on each side fused (Figs. 16.23.92, 24.100) *and* plates of two sides separate medially *and* dorsal plate elliptical and not bearing setae c1 (Figs. 16.23.91, 24.101, 29.144, 30.150, 153, 32.168) 34

18(17) Third legs with tarsi bearing 12 or more setae, including Ta8 (Fig. 16.12.8, arrow); dorsal plate usually elongate and elliptical and bearing only four pairs of setae (verticils and scapulars) (Figs. 16.11.3, 19.65, 68); when dorsal plate round and bearing six pairs of setae, including c1 laterally and e1 posterolaterally (Fig. 16.20.70), then third coxal plates bearing setae pa and h2 in addition to 3a (Fig. 16.20.72) and all setae on pedipalp tarsus short and blade-like (Fig. 16.20.71) ....Lebertioidea (in part) 19

18' Third legs with tarsi usually bearing 11 or fewer setae, lacking at least Ta8 (Figs. 16.33.174, 34.181, 36.196, 198); dorsal plate usually nearly round (Figs. 16.34.178, 184, 35.188–189, 191), rarely elongate and elliptical (Fig. 16.33.171); when tarsi of third legs bearing 12 setae then dorsal plate round and bearing five pairs of setae, including c1 laterally (Fig. 16.36.195) *and* third coxal plates bearing only setae 3a (Fig. 16.36.193) *and* at least one seta on pedipalp tarsus long and whip-like (similar to Fig. 16.33.177) .... Arrenuroidea......................................23

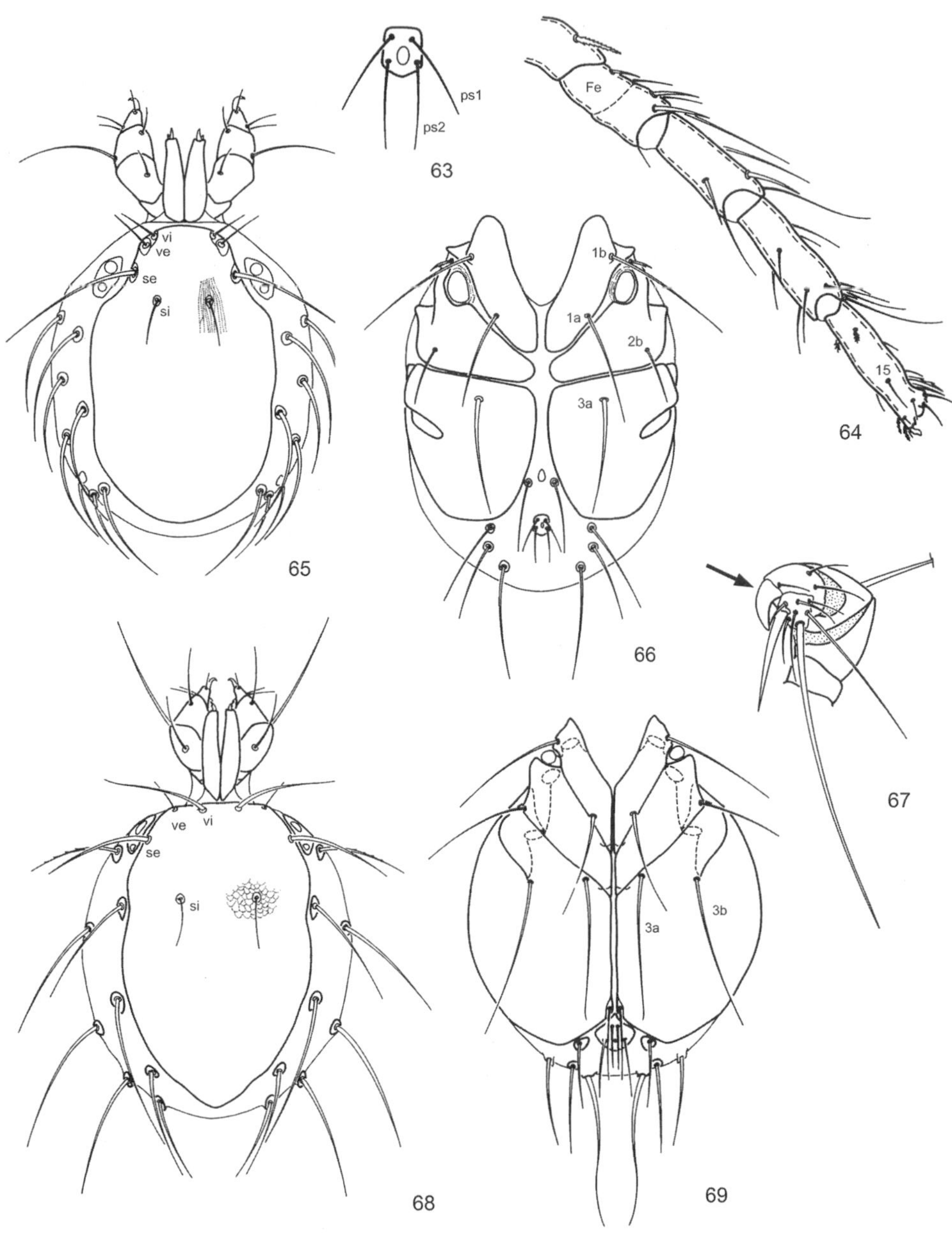

PLATE 16.19 Figures 63–67: *Sperchon glandulosus* Koenike (Sperchontidae) larva. **Fig. 63** excretory pore plate. **Fig. 64** first leg. **Fig. 65** dorsum of idiosoma and gnathosoma. **Fig. 66** venter of idiosoma. **Fig. 67** pedipalp. **Figure 68-69:** *Teutonia lunata* Marshall (Teutoniidae) larva. **Fig. 68** dorsum of idiosoma and gnathosoma. **Fig. 69** venter of idiosoma.

19(18) Pedipalp tarsus with no setae as long as pedipalp (Fig. 16.20.71); tarsi of first, second and third legs bearing 13 or 14 (+ω), 13 or 14 (+ω), and 12 setae respectively, lacking Ta15 .... Anisitsiellidae (in part) 20

19' Pedipalp tarsus with at least one seta as long as pedipalp (Figs. 16.11.2, 19.67); tarsi of first, second, and third legs bearing 15 (+ω), 15 (+ω), and 13 or 14 setae respectively, including Ta15 (Figs. 16.12.6–8, 19.64) 21

20(19) Dorsal plate bearing six pairs of setae, including verticils, scapulars and both c1 and e1 on lateral edge, and with setae si long, thick, and located near mid-length (Fig. 16.20.70); third coxal plates bearing 3 pairs of setae, including 3a and both pa and h2 near posterior edges (Fig. 16.20.72) .... *Bandakiopsis*

20' Dorsal plate bearing four pairs of setae (verticils and scapulars), with setae si short, slender, and located in anterior third of plate (Fig. 16.20.73); third coxal plates bearing only setae 3a (Fig. 16.20.74) .... *Utaxatax*

21(19) Dorsal plate longitudinally striated (Figs. 16.11.3, 19.65); third coxal plates bearing only setae 3a (Figs. 16.11.4, 19.66); excretory pore plate small and usually quadrangular (Figs. 16.11.5, 19.66) .... Sperchontidae 22

21' Dorsal plate reticulate (Fig. 16.19.68); third coxal plates bearing two pairs of setae, 3a and supernumerary setae 3b (Fig. 16.19.69); excretory pore plate relatively large and diamond-shaped (Fig. 16.19.69) .... Teutoniidae .... *Teutonia*

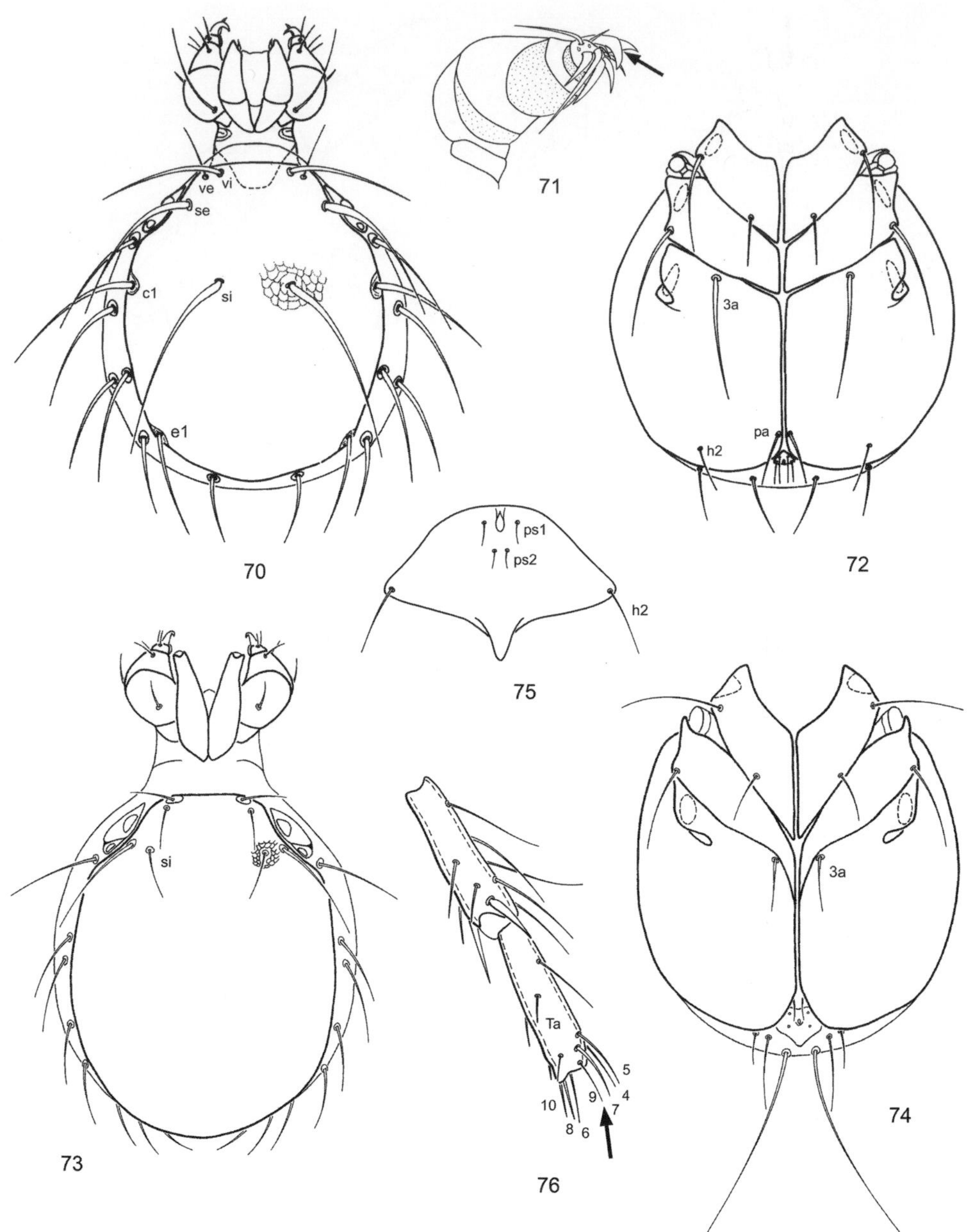

PLATE 16.20 Figures 70–72: *Bandakiopsis fonticola* Smith (Anisitsiellidae) larva. **Fig. 70** dorsum of idiosoma and gnathosoma. **Fig. 71** pedipalp. **Fig. 72** venter of idiosoma. **Figures 73–74:** *Utaxatax newelli* Habeeb (Anisitsiellidae) larva. **Fig. 73** dorsum of idiosoma and gnathosoma. **Fig. 74** venter of idiosoma. **Figures 75–76:** *Bandakia phreatica* Cook (Anisitsiellidae) larva. **Fig. 75** excretory pore plate. **Fig. 76** tibia and tarsus of third leg.

22(21) Second coxal plates bearing one pair of setae, 2b, posterolaterally (Fig. 16.19.66) .... *Sperchon*

22' Second coxal plates lacking setae (Fig. 16.11.4) .... *Sperchonopsis*

23(18) Coxal plates on each side all fused and coxal plates of two sides fused medially, with setae 1a, 2b and 3a reduced to vestiges (Fig. 16.34.185); pedipalp tarsus with no setae as long as pedipalp (Fig. 16.34.182); dorsal plate with setae si reduced to vestiges (Fig. 16.34.184)........ Acalyptonotidae........ *Acalyptonotus*

23' Coxal plates on each side all separate and coxal plates of two sides separate, with setae 1a, 2b, and 3a not reduced (Figs. 16.33.172, 35.186, 190, 192, 36.193); pedipalp tarsus with at least one seta as long as pedipalp (Fig. 16.33.177) 24

24(23) Idiosoma moderately flattened dorsoventrally; gnathosoma projecting beyond anterior edge of dorsal plate, entirely exposed in dorsal view (Figs. 16.33.171, 36.195); excretory pore plate ariously-shaped (Figs. 16.33.172, 36.193, 197) but usually not attenuate anteriorly; pedipalp tarsus with long, thick, distal seta straight or only slightly bowed basally; leg genua with setae Ge5 borne on tubercles that usually are prominent (Fig. 16.33.173, arrow) 25

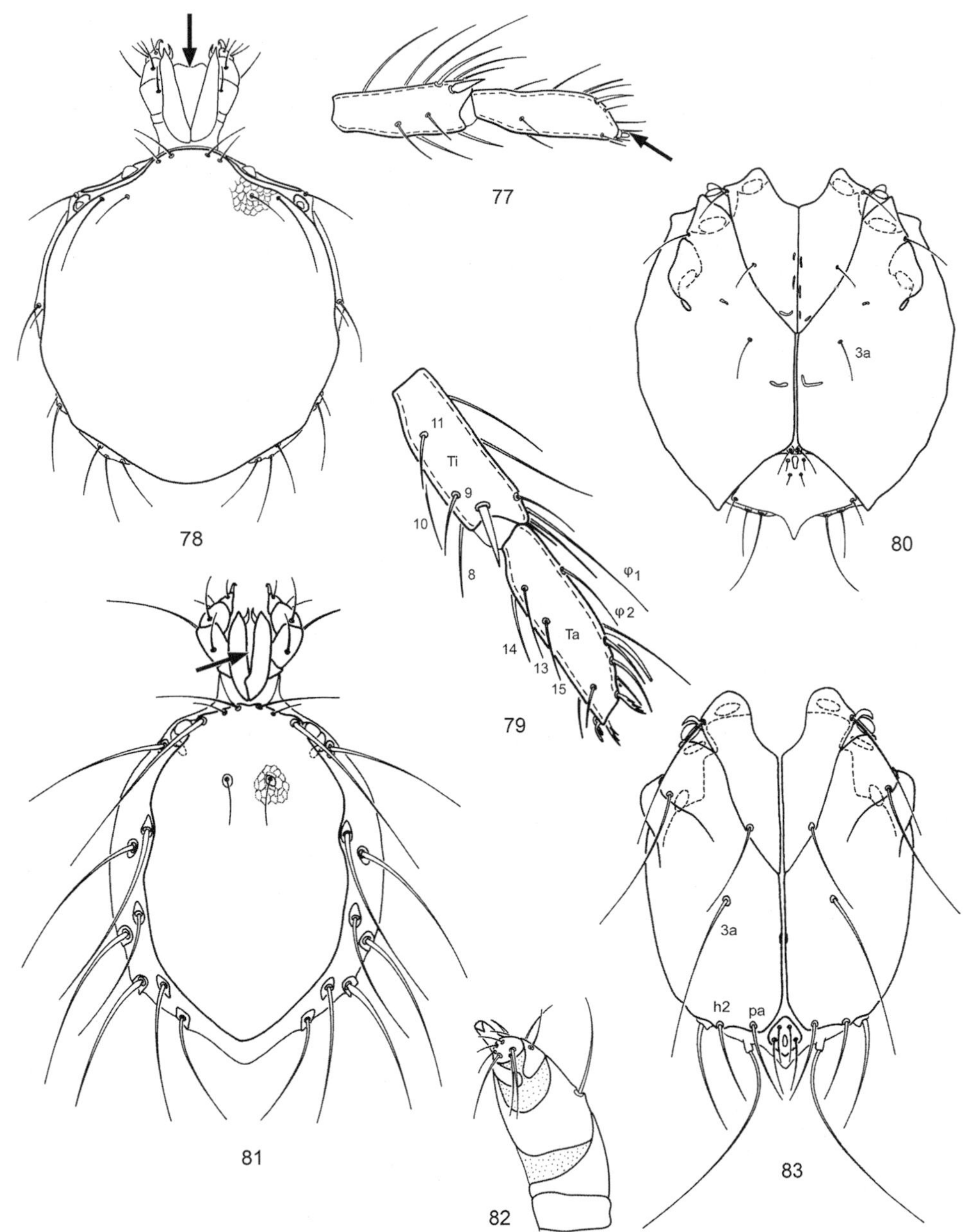

PLATE 16.21 Figures 77–78: *Bandakia phreatica* Cook (Anisitsiellidae) larva. **Fig. 77** tibia and tarsus of first leg. **Fig. 78** dorsum of idiosoma and gnathosoma. **Fig. 79:** *Lebertia* sp. (Lebertiidae) larva. tibia and tarsus of first leg. **Fig. 80:** *Bandakia phreatica* Cook (Anisitsiellidae) larva. venter of idiosoma. **Fig. 81:** *Lebertia* sp. (Lebertiidae) larva. dorsum of idiosoma and gnathosoma. **Fig. 82:** *Bandakia phreatica* Cook (Anisitsiellidae) larva. pedipalp. **Fig. 83:** *Lebertia* sp. (Lebertiidae) larva. venter of idiosoma.

24' Idiosoma extremely flattened dorsoventrally; gnathosoma recessed beneath anterior edge of dorsal plate, partially or entirely concealed in dorsal view (in unmounted specimens) (Fig. 16.35.188); excretory pore plate triangular with anterior apex attenuate (Figs. 16.33.176, 35.192); pedipalp tarsus with long, thick, distal seta strongly bowed, and usually lobed, basally (Figs. 16.33.177, 34.183, arrows); leg genua with setae Ge5 not borne on tubercles 29

25(24) Dorsal plate bearing four pairs of setae (verticils and scapulars), with setae c1 on lateral membranous integument (Fig. 16.33.171); setae c2 and d2 thick and long relative to other hysterosomal setae (Fig. 16.33.171); excretory pore setae ps1 and ps2 absent, represented by their alveoli on excretory pore plate (Fig. 16.33.172); tibia of first leg bearing eight setae (+2φ), including Ti11; leg tarsi lacking setae Ta14 (Figs. 16.33.173, 175) .... Momoniidae........ 26

25' Dorsal plate bearing five pairs of setae, including the verticils, scapulars and setae c1 laterally (Fig. 16.36.195); setae c2 and d2 similar to other hysterosomal setae (Fig. 16.36.195); excretory pore setae ps1 and ps2 present on plate (Figs. 16.36.193, 197), tibia of first leg bearing seven setae (+2φ), lacking Ti11; leg tarsi bearing setae Ta14 (Figs. 16.36.196, 198; arrows) 27

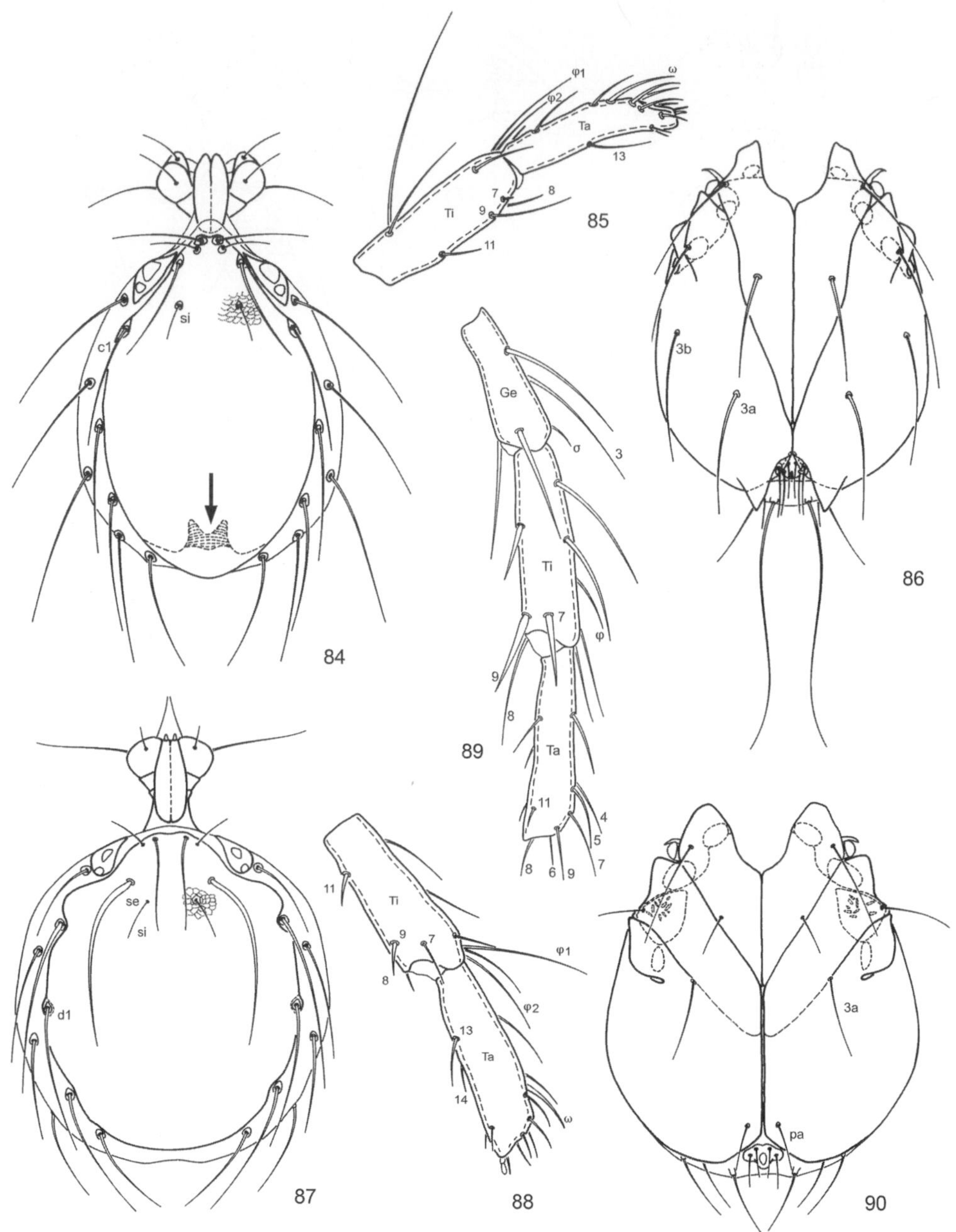

PLATE 16.22 Figures 84–86: *Oxus elongatus* Marshall (Oxidae) larva. **Fig. 84** dorsum of idiosoma and gnathosoma. **Fig. 85** tibia and tarsus of first leg. **Fig. 86** venter of idiosoma. **Figures 87–90:** *Testudacarus americanus* Marshall (Testudacarinae) larva. **Fig. 87** dorsum of idiosoma and gnathosoma. **Fig. 88** tibia and tarsus of first leg. **Fig. 89** distal segments of third leg. **Fig. 90** venter of idiosoma.

26(25) Legs with setae Tr1, Ge5, Ti9 and Ti11 much longer than respective segments; setae IITi9, IITi11, IIIGe5, IIITi9 and IIITi11 plumose (Fig. 16.33.173) .... Momoniinae.. *Momonia*

26' Legs with setae Tr1, Ge5, Ti9 and Ti11 shorter than respective segments; setae IITi9, IITi11, IIIGe5, IIITi9 and IIITi11 blade-like (Fig. 16.33.175) ....Stygomomoniinae .... *Stygomomonia*

27(25) Third leg with tarsus bearing nine setae, lacking Ta12 (Fig. 16.36.198); second and third legs with tarsal setae Ta4 and Ta6 long (usually as long as respective segments) (Fig. 16.36.198); first and second coxal plates with conspicuous denticulate projections on posterior edges (Figs. 16.36.197, 199, arrows); third coxal plates with transverse muscle attachment scars (TMAS); excretory pore plate subtriangular (Figs. 16.36.197, 199) .... Krendowskiidae..28

27' Third leg with tarsus bearing at least 11 setae, including Ta12 (Fig. 16.36.196); second and third legs with tarsal setae Ta4 and Ta6 usually short (conspicuously shorter than respective segments) (Fig. 16.36.196); first and second coxal plates lacking denticulate projections on posterior edges; third coxal plates lacking transverse muscle attachment scars (TMAS); excretory pore plate variously shaped, but rarely triangular (Fig. 16.36.193) .... Arrenuridae... *Arrenurus*

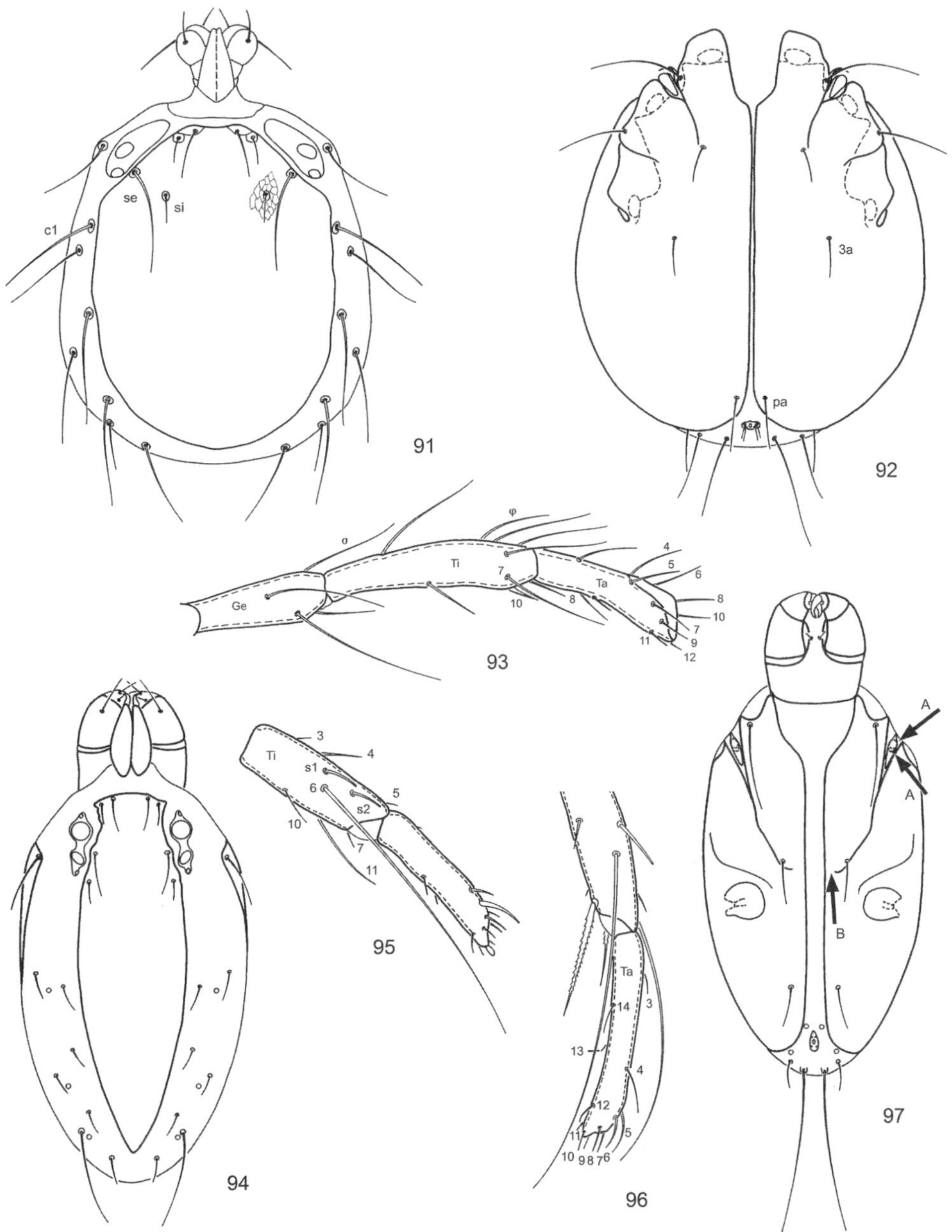

PLATE 16.23 Figures 91–93: *Torrenticola* sp. (Torrenticolidae) larva. **Fig. 91** dorsum of idiosoma and gnathosoma. **Fig. 92** venter of idiosoma. **Fig. 93** distal segments of third leg. **Figures 94–97:** *Limnesia marshalliana* Lundblad (Limnesiinae) larva. **Fig. 94** dorsum of idiosoma and gnathosoma. **Fig. 95** tibia and tarsus of second leg. **Fig. 96** tibia and tarsus of third leg. **Fig. 97** venter of idiosoma and gnathosoma.

28(27) First and second coxal plates each with row of denticulate projections along entire posterior edge (Fig. 16.36.197, arrows); TMAS weakly developed and short (Fig. 16.36.197); tubercles bearing setae f2 large, prominent, and medially attenuate (Fig. 16.36.197) .... *Krendowskia*

28' First and second coxal plates each bearing only few, denticulate projections in medial third of posterior edge (Fig. 16.36.199); TMAS strongly developed and long (Fig. 16.36.199); tubercles bearing setae f2 not unusually prominent (Fig. 16.36.199) .... *Geayia*

29 (24) Dorsal plate bearing five pairs of setae, including verticils, scapulars and setae c1 laterally (Fig. 16.35.188) .... Mideopsidae .... *Mideopsis*

29' Dorsal plate bearing four pairs of setae (verticils and scapulars), with setae c1 on lateral membranous integument (Figs. 16.34.178, 35.189) 30

30(29) Second and third legs with tibia bearing nine setae (+2φ and 1φ, respectively), including both Ti10 and Ti11 (Figs. 16.33.174, 34.181) 31

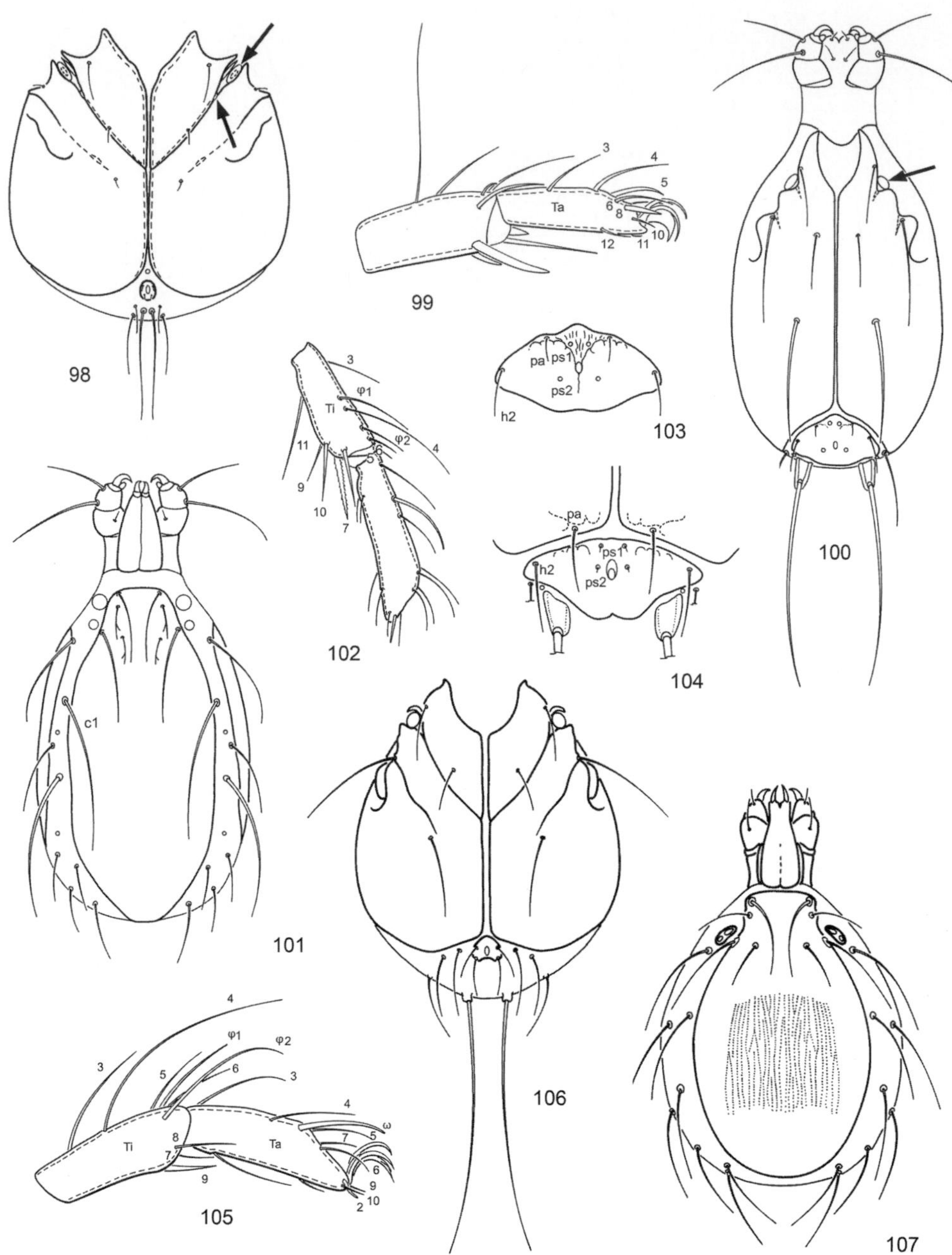

PLATE 16.24 **Figures 98–99:** *Tyrrellia* sp. (Tyrrelliinae) larva. **Fig. 98** venter of idiosoma. **Fig. 99** tibia and tarsus of third leg. **Figures 100–101:** *Atractides grouti* Habeeb (Hygrobatidae) larva. **Fig. 100** venter of idiosoma and gnathosoma. **Fig. 101** dorsum of idiosoma and gnathosoma. **Figure 102:** *Hygrobates* sp. (Hygrobatidae) larva; tibia and tarsus of second leg. **Fig. 103:** *Atractides grouti* Habeeb (Hygrobatidae) larva; excretory pore plate. **Fig. 104:** *Hygrobates neocalliger* Habeeb (Hygrobatidae) larva; excretory pore Plate 16.region. **Figures 105–107:** *Feltria* sp. (Feltriidae) larva. **Fig. 105** tibia and tarsus of first leg. **Fig. 106** venter of idiosoma. **Fig. 107** dorsum of idiosoma and gnathosoma.

30'	Second and third legs with tibia bearing fewer than nine setae (+2φ and +1φ respectively), lacking either Ti10 or Ti11 or both (Fig. 16.35.187, 36.194) 32
31(30)	Pedipalp tarsus with long, thick, distal seta bowed, but not deeply lobed or fringed basally, and with most medial seta moderately thick but not fringed (Fig. 16.34.183); second and third legs with setae Ge5, Ti9 and Ti11 much longer than respective segments and plumose, and with setae Ta4 and Ta6 much longer than respective segments (Fig. 16.34.181) .... Mideidae .... *Midea*
31'	Pedipalp tarsus with long, thick, distal seta bowed, deeply lobed and fringed basally, and most medial seta thick and fringed (Fig. 16.33.177); second and third legs with setae Ge5, Ti9 and Ti11 shorter than respective segments and simple, and with setae Ta4 and Ta6 shorter than respective segments (Fig. 16.33.174) .... Nudomideopsidae .... *Nudomideopsis, Paramideopsis*
32(30)	Second leg with tibia bearing seven setae (+2φ), lacking Ti10 and Ti11 (Fig. 16.36.194); third legs with tibia bearing eight setae (+1φ), lacking Ti10 .... Laversiidae........ *Laversia*

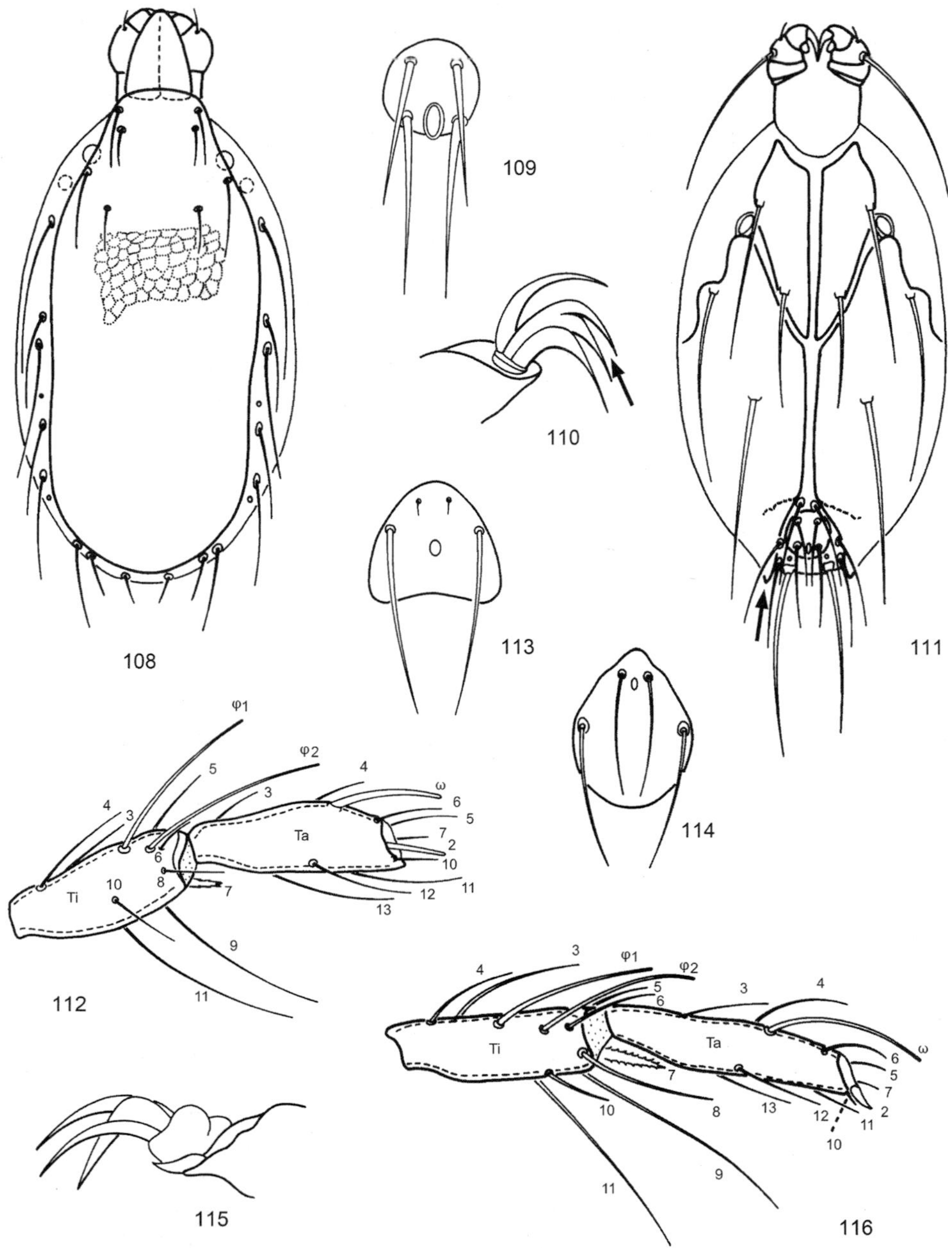

PLATE 16.25 **Figures 108–112:** *Unionicola* sp. (Unionicolinae) larva. **Fig. 108** dorsum of idiosoma and gnathosoma. **Fig. 109** excretory pore plate. **Fig. 110** claws and empodium of first leg. **Fig. 111** venter of idiosoma and gnathosoma. **Fig. 112** tibia and tarsus of first leg. **Fig. 113:** *Koenikea marshallae* Viets (Pionatacinae) larva excretory pore plate. **Figures 114–116:** *Neumania punctata* Marshall (Pionatacinae) larva. **Fig. 114** excretory pore plate. **Fig. 115** claws and empodium of first leg. **Fig. 116** tibia and tarsus of first leg.

32' Second and third legs with tibia bearing eight setae (+2φ and +1φ respectively), including Ti10 but lacking Ti11 (Fig. 16.35.187) 33

33(32) Dorsal plate with setae si located well in anterior half of plate and setae c1 located in lateral integument near mid-length of plate, posterior to level of setae si (Fig. 16.35.191) .... Neoacaridae .... *Neoacarus, Volsellacarus*

33' Dorsal plate with setae si located near mid-length of plate and setae c1 located in lateral integument near setae c3, anterior to level of setae si (Fig. 16.35.189) .... Athienemanniidae............ Athienemanniinae .... *Chelomideopsis, Platyhydracarus*

34(17) Third coxal plates bearing setae 3a and at least one other pair of setae (pa or 3b) and excretory pore plate bearing only setae ps1 and ps2 (Figs. 16.21.83, 22.86, 90, 23.92) .... Lebertioidea (in part) 35

34' Third coxal plates usually bearing only one pair of setae, 3a (Figs. 16.21.80, 23.97, 24.100, 29.145, 30.152), when third coxal plates also bearing setae pa then excretory pore plate bearing setae h2 in addition to ps1 and ps2 (Figs. 16.24.104, 31.159–163, 32.166) 38

35(34) Third coxal plates truncate posteriorly, bearing three pairs of setae, including 3a and both pa and h2 on posterior edges (Fig. 16.21.83); all legs with tibia bearing nine setae (+2φ, +2φ, and +1φ respectively), including both Ti9 and Ti10 (Fig. 16.21.79); first and second legs with tarsus bearing 14 setae (+ω), including Ta14 and Ta15 (Fig.. 16.21.79); cheliceral bases separate (Fig. 16.21.81, arrow) .... Lebertiidae .... *Lebertia, Estelloxus*

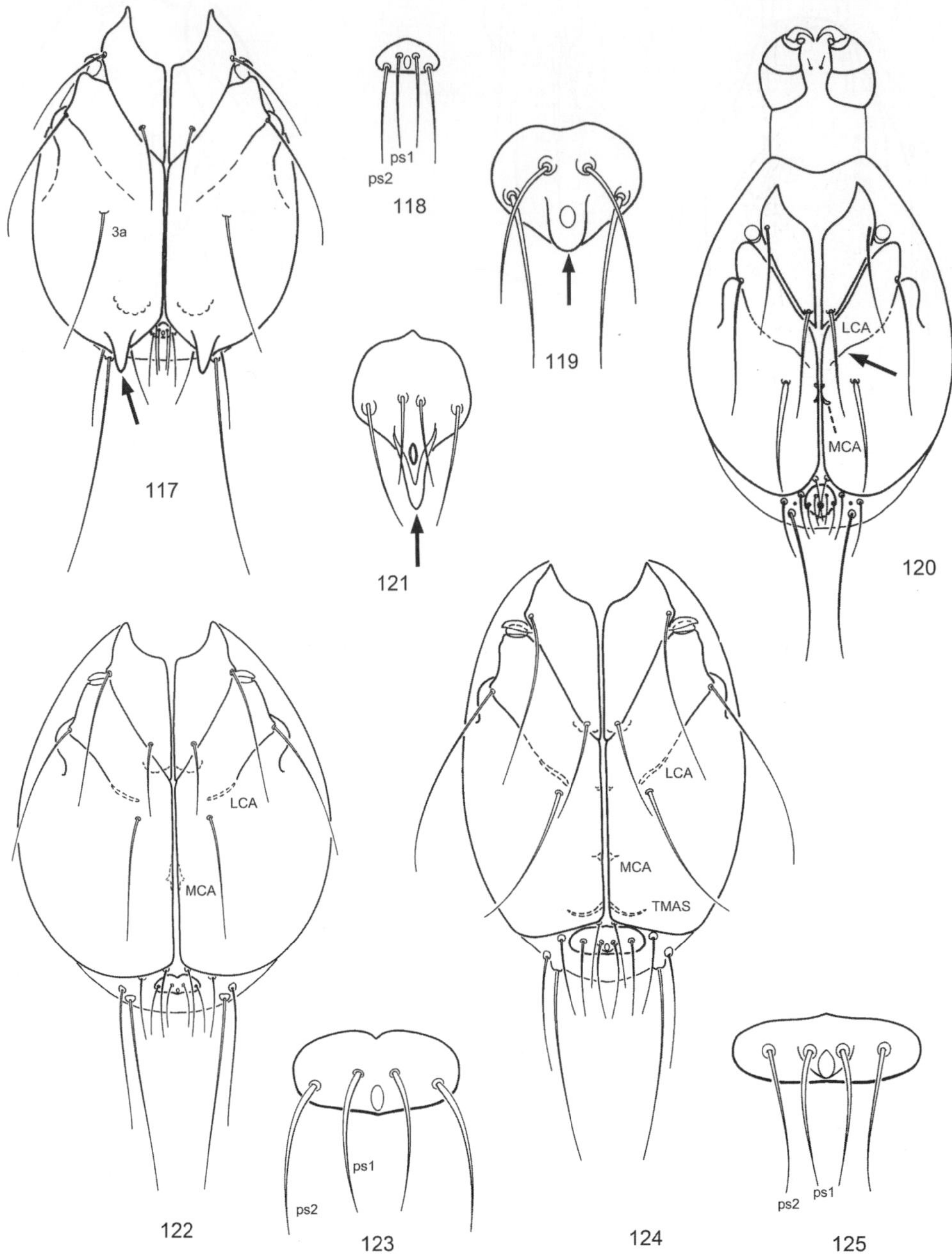

PLATE 16.26 **Figures 117–118:** *Wettina ontario* Smith (Wettinidae) larva. **Fig. 117** venter of idiosoma. **Fig. 118** excretory pore plate. **Figures 119–120:** *Hydrochoreutes minor* Cook (Hydrochoreutinae) larva. **Fig. 119** excretory pore plate. **Fig. 120** venter of idiosoma and gnathosoma. **Fig. 121** *Hydrochoreutes microporus* Cook (Hydrochoreutinae) larva; excretory pore plate. **Figures 122–123:** *Huitfeldtia rectipes* Thor (Huitfeldtiinae) larva. **Fig. 122** venter of idiosoma. **Fig. 123** excretory pore plate. **Figures 124–125:** *Neotiphys pionoidellus* (Habeeb) (Tiphyinae) larva. **Fig. 124** venter of idiosoma. **Fig. 125** excretory pore plate.

35' Third coxal plates rounded or pointed posteriorly, bearing two pairs of setae, including setae 3a and either 3b laterally or pa posteromedially (Figs. 16.22.86, 90, 23.92); all legs with tibia bearing eight setae (+2φ, +2φ, and +1φ, respectively), lacking Ti9 or Ti10 (Figs. 16.22.85, 88, 89); first and second legs with tarsus bearing 13 setae (+ω), lacking Ta15 (Fig. 16.22.88), or 12 setae (+ω1s), lacking both Ta15 and Ta14 (Fig. 16.22.85); cheliceral bases fused (Figs. 16.22.84, 87, 23.91) 36

36(35) Dorsal plate bearing five pairs of setae including verticils, scapulars, and setae c1 laterally; integument beneath posterior edge of dorsal plate intricately folded (Fig. 16.22.84, arrow); first coxal plates elongate, extending posteriorly nearly to level of excretory pore plate; third coxal plates bearing setae 3b laterally (Fig. 16.22.86); first and second legs with tarsus bearing 12 setae (+ω), lacking Ta14 (Fig. 16.22.85) .... Oxidae .... *Oxus*, *Frontipoda*

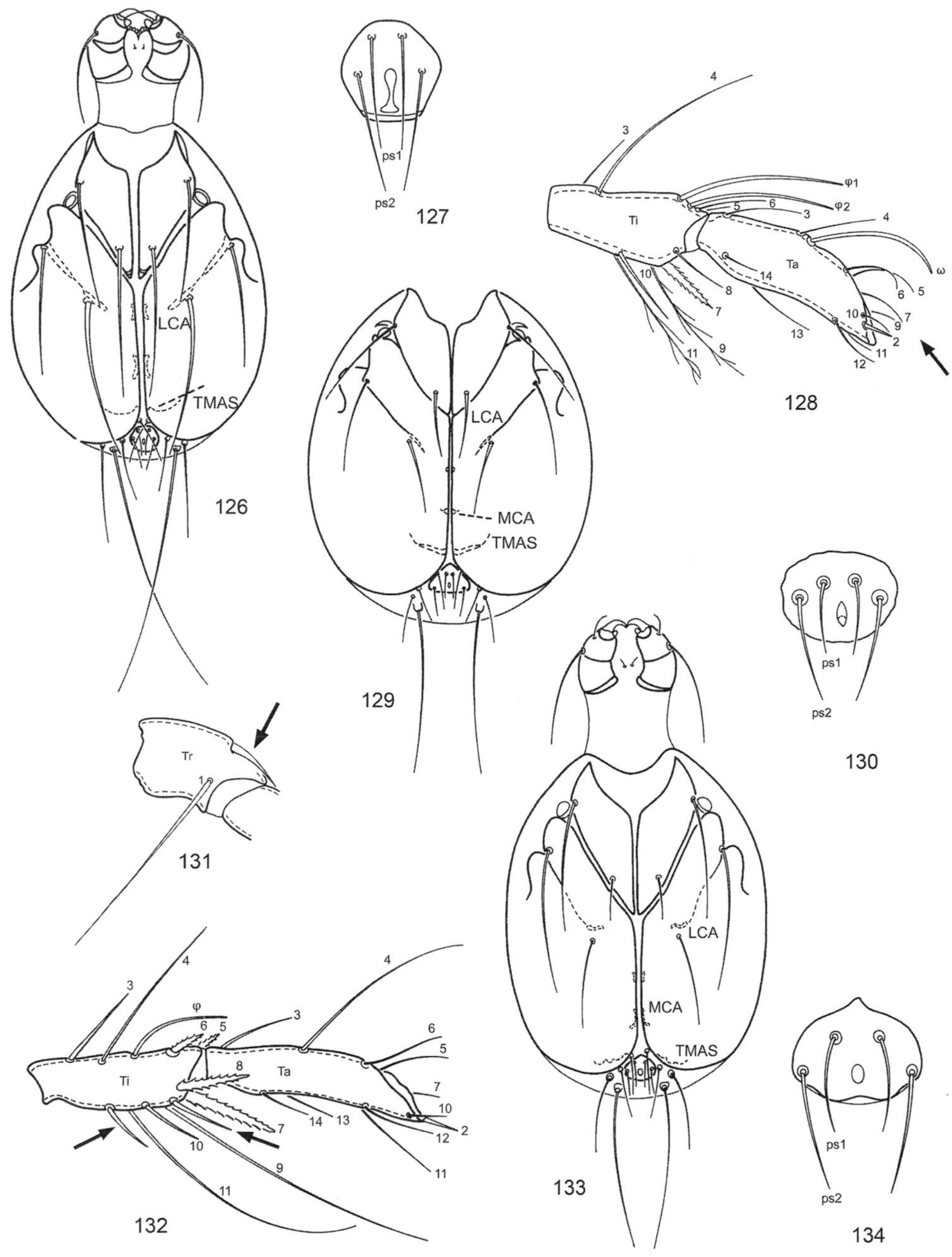

PLATE 16.27 **Figures 126–127:** *Tiphys americanus* (Marshall) (Tiphyinae) larva. **Fig. 126** venter of idiosoma. **Fig. 127** excretory pore plate. **Figures 128–130:** *Tiphys ornatus* (Koch) (Tiphyinae) larva. **Fig. 128** tibia and tarsus of first leg. **Fig. 130** excretory pore plate. **Figures 129–131:** *Pseudofeltria multipora* Cook (Foreliinae) larva. **Fig. 129** venter of idiosoma. **Fig. 131** trochanter of third leg. **Figures 132–134:** *Forelia ovalis* Marshall (Foreliinae) larva. **Fig. 132** tibia and tarsus of third leg. **Fig. 133** venter of idiosoma. **Fig. 134** excretory pore plate.

36' Dorsal plate bearing four pairs of setae (verticils and scapulars), or five pairs including verticils, scapulars and setae d1 laterally; integument beneath posterior edge of dorsal plate not intricately folded (Figs. 16.22.87, 23.91); first coxal plates not elongate, extending posteriorly only to level of insertion of third legs (Fig. 16.22.90), or fused with second coxal plates (Fig. 16.23.92); third coxal plates bearing setae pa posteromedially (Figs. 16.22.90, 23.92); first and second legs with tarsus bearing 13 setae (+ω), including Ta14 (Fig. 16.22.88) .... Torrenticolidae 37

37(36) Dorsal plate bearing five pairs of setae including verticils, scapulars, and setae d1 laterally, and with setae se long and thick (Fig. 16.22.87); first coxal plates clearly delineated by complete suture lines, and second and third coxal plates on each side fused with suture line incomplete medially (Fig. 16.22.90); excretory pore plate with convex projection posteromedially (Fig. 16.22.90); all legs with tibia lacking setae Ti10; third leg with genu bearing four setae (+σ) including Ge3, and tarsus bearing 11 setae, lacking Ta10 (Fig. 16.22.89) .... Testudacarinae .... *Testudacarus*

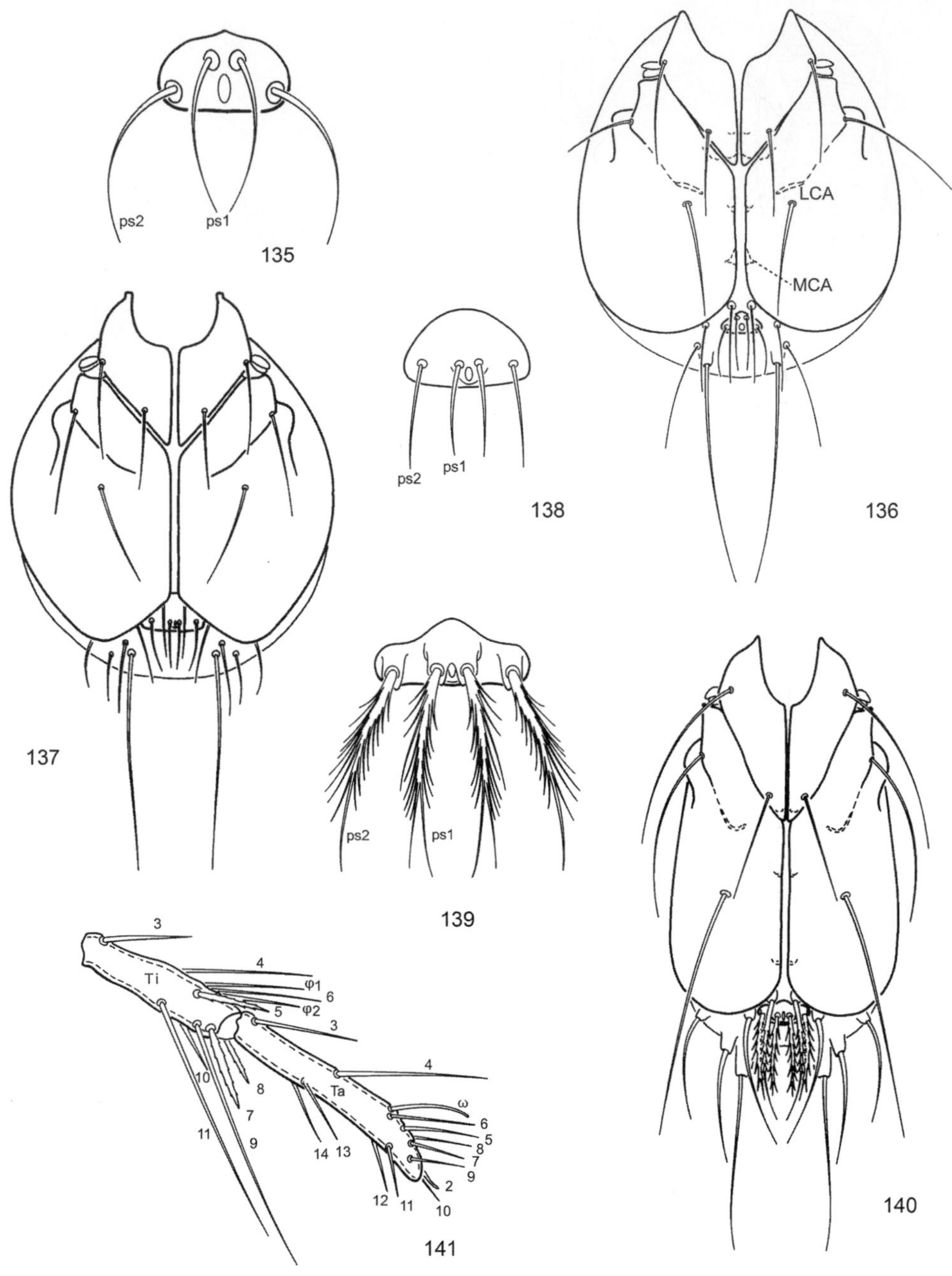

PLATE 16.28 **Figures 135–136:** *Forelia onondaga* Habeeb (Foreliinae) larva. **Fig. 135** excretory pore plate**. Fig. 136** venter of idiosoma. **Figures 137–138:** *Najadicola ingens* Koenike (Najadicolinae) larva. **Fig. 137** venter of idiosoma. **Fig. 138** excretory pore plate. **Figures 139–140:** *Nautarachna muskoka* Smith (Pioninae) larva. **Fig. 139** excretory pore plate. **Fig. 140** venter of idiosoma. **Fig. 141:** *Najadicola ingens* Koenike (Najadicolinae) larva; tibia and tarsus of second leg.

37' Dorsal plate bearing only four pairs of setae (verticils and scapulars), and with setae se relatively short and slender (Fig. 16.23.91); coxal plates on each side all fused, with suture lines obliterated except laterally (Fig. 16.23.92); excretory pore plate without projection posteromedially (Fig. 16.23.92); all legs with tibia lacking setae Ti9; third leg with genu bearing three setae (+σ), lacking Ge3, and tarsus bearing 12 setae, including Ta10 (Fig. 16.23.93) .... Torrenticolinae.. *Torrenticola*

38(34) Cheliceral bases separate (Fig. 16.21.78, arrow); third leg with tarsus bearing 12 setae, including Ta9 (Fig. 16.20.76, arrow); excretory pore plate very large (equal in width to coxal plates of one side), bearing setae ps1 and ps2 anteromedially and occasionally also setae h2 at posterolateral angles (Figs. 16.20.75, 21.80); first coxal plates separate from posterior coxal groups, and third coxal plates bearing only setae 3a (Fig. 16.21.80) ....Lebertioidea (in part).. Anisitsiellidae (in part) .... Anisitsiellinae (in part).. *Bandakia*

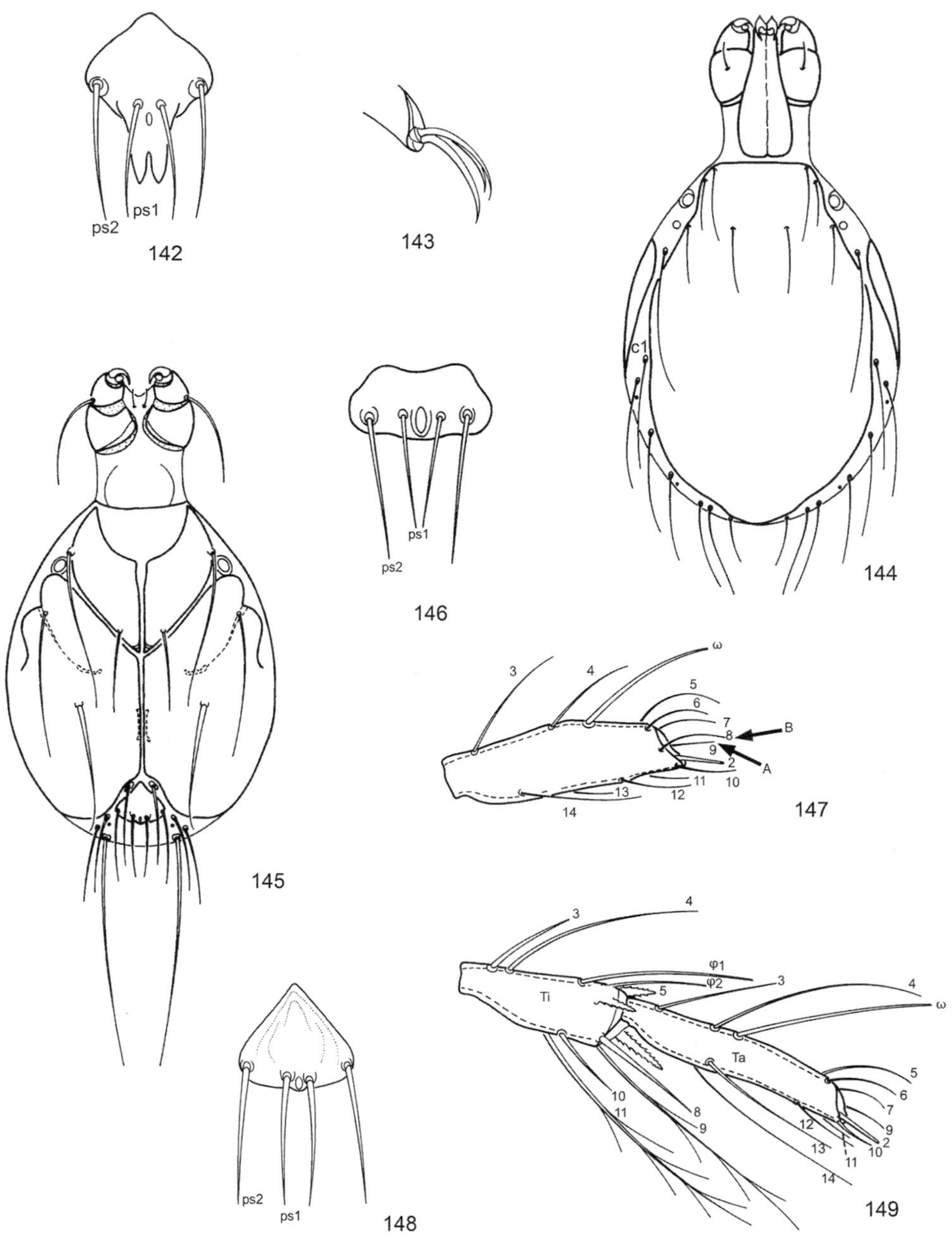

PLATE 16.29 **Figures 142–143:** *Piona carnea* (Koch) (Pioninae) larva. **Fig. 142** excretory pore plate. **Fig. 143** claws and empodium of third leg. **Figures 144–145:** *Piona interrupta* Marshall (Pioninae) larva. **Fig. 144** dorsum of idiosoma and gnathosoma. **Fig. 145** venter of idiosoma and gnathosoma. **Fig. 146:** *Piona constricta* (Wolcott) (Pioninae) larva; excretory pore plate. **Figure 147:** *Piona mitchelli* Cook (Pioninae) larva; tarsus of first leg. **Fig. 148:** *Piona interrupta* Marshall (Pioninae) larva; excretory pore plate. **Fig. 149:** *Piona mitchelli* Cook (Pioninae) larva; tibia and tarsus of second leg.

38' Cheliceral bases fused (Figs. 16.24.101, 24.107, 25.108, 29.144, 30.150); third leg with tarsus usually bearing 11 or fewer setae, lacking Ta9 (Fig. 16.27.132) (exception Limnesiidae); excretory pore plate small to very large, when equal in width to coxal plates of one side (measured from midline to insertion of third leg) (Figs. 16.24.100, 103, 104, 30.152, 32.164) then either excretory pore located near posterior edge of plate (Fig. 16.30.152, arrow), or first coxal plates fused to posterior coxal groups (Fig. 16.24.100), or third coxal plates bearing setae pa in addition to setae 3a (Fig. 16.32.164) ....Hygrobatoidea...39

39(38) Two pairs of urstigmata borne distally between first and second coxal plates (Fig. 16.23.97, arrows A; Fig. 16.24.98, arrows); first and second legs with tibia bearing seven or eight setae (+2φ), including only three ventral setae, lacking either Ti8 and Ti9 or Ti10 and Ti11 (Figs. 16.23.95, 24.99) ....Limnesiidae 40

39' One pair of urstigmata borne distally between first and second coxal plates (Figs. 16.24.100, arrow; 24.106, 25.111, 27.133, 32.164); first and second legs with tibia usually bearing nine or more setae (+2φ), including five ventral setae (Ti7, Ti8, Ti9, Ti10, and Ti11) (Figs. 16.24.102, 28.141, 29.149) (exception Feltriidae) 41

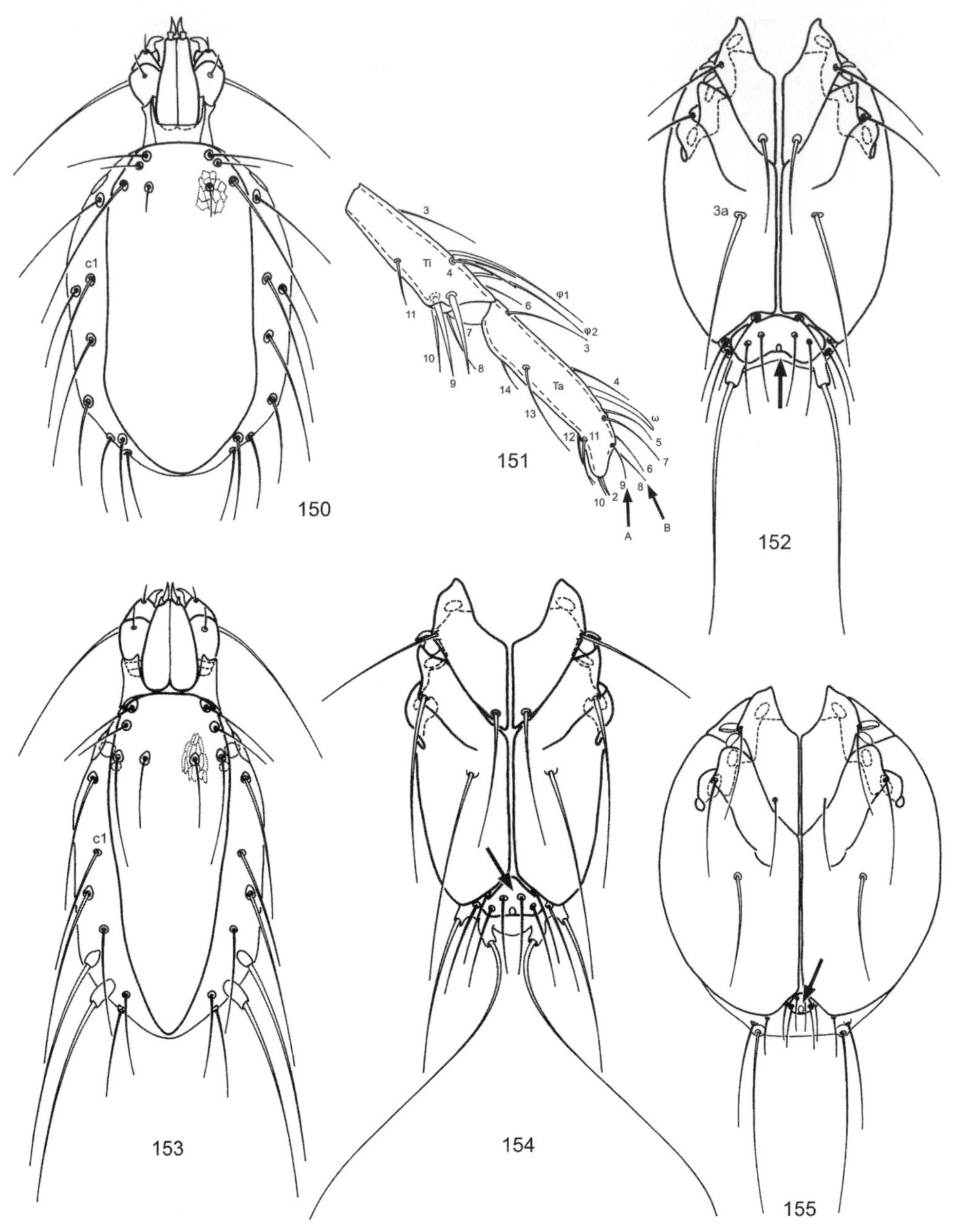

PLATE 16.30 Figures 150–152: *Ljania bipapillata* Thor (Axonopsinae) larva. **Fig. 150** dorsum of idiosoma and gnathosoma. **Fig. 151** tibia and tarsus of first leg. **Fig. 152** venter of idiosoma. **Figures 153–154:** *Albia neogaea* Habeeb (Albiinae) larva. **Fig. 153** dorsum of idiosoma and gnathosoma. **Fig. 154** venter of idiosoma. **Figure 155:** *Neobrachypoda ekmani* (Walter) (Axonopsinae) larva; venter of idiosoma.

40(39) First and second coxal plates on each side completely separate (Fig. 16.24.98); third leg with tarsus bearing nine setae, lacking Ta9, Ta13 and Ta14 (Fig. 16.24.99); dorsal plate extending laterally well beyond level of eye plates (contrast with Fig. 16.23.94) .... Tyrrelliinae .... *Tyrrellia*

40' First and second coxal plates on each side fused, with suture lines obliterated medially (Fig. 16.23.97, arrow B); third leg with tarsus bearing 12 setae, including Ta9, Ta13 and Ta14 (Fig. 16.23.96); dorsal plate narrow, usually confined to region between eye plates (Fig. 16.23.94) .... Limnesiinae .... *Limnesia*

41(39) Coxal plates on each side all fused (Fig. 16.24.100) .... Hygrobatidae 42

41' First coxal plates separate from posterior coxal group on each side (Figs. 16.24.106, 25.111, 29.145, 30.152, 32.169) 43

42(41) Excretory pore plate bearing three pairs of setae, including ps1 and ps2 medially and h2 near anterolateral angles, with setae pa on posterior edges of third coxal plates (Fig. 16.24.104) .... *Hygrobates*

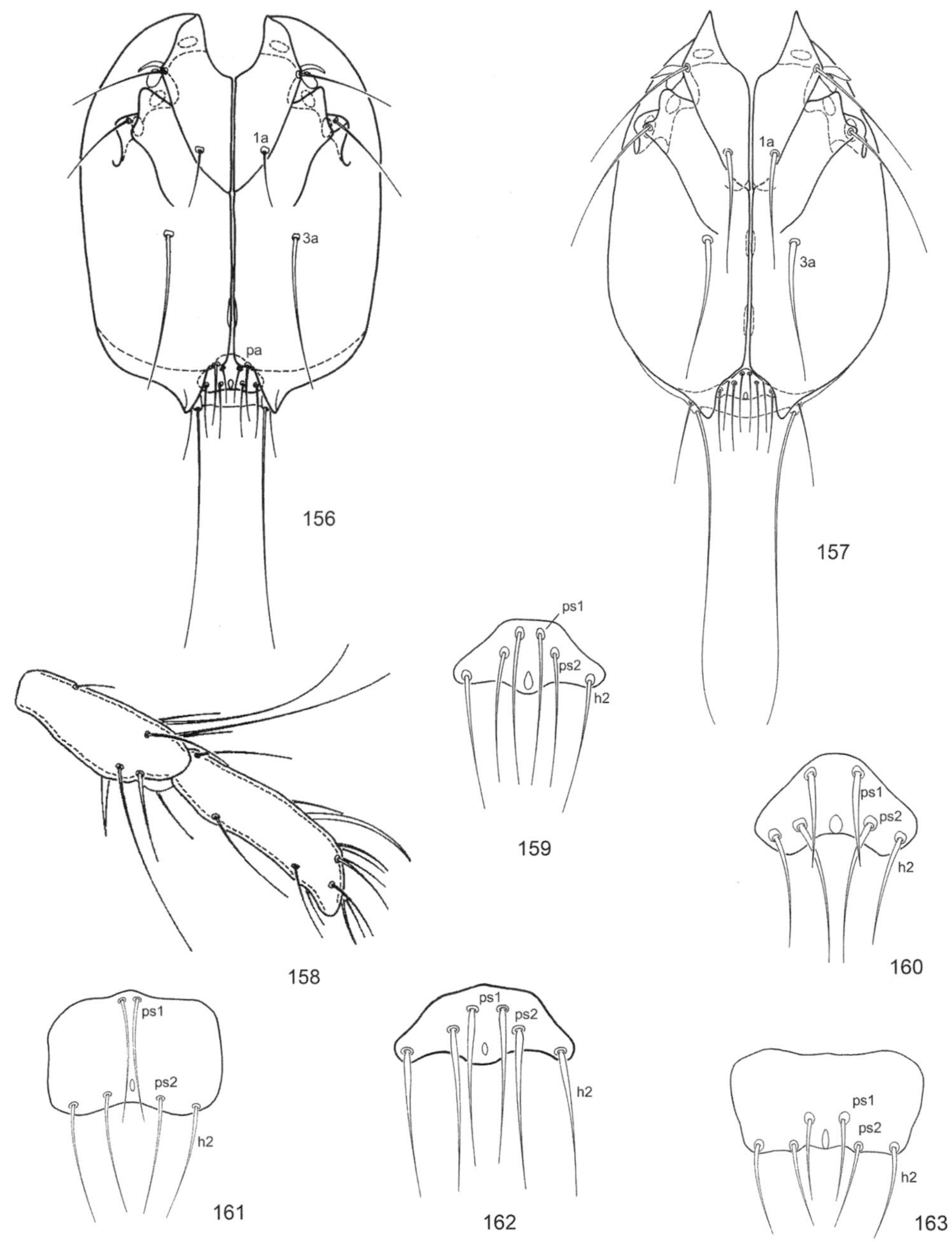

PLATE 16.31 **Fig. 156:** *Woolastookia setosipes* Habeeb (Axonopsinae) larva; venter of idiosoma. **Figures 157–159:** *Estellacarus unguitarsus* (Habeeb) (Axonopsinae) larva. **Fig. 157** venter of idiosoma. **Fig. 158** tibia and tarsus of first leg. **Fig. 159** excretory pore plate. **Fig. 160:** *Woolastookia setosipes* Habeeb (Axonopsinae) larva; excretory pore plate. **Fig. 161:** *Woolastookia pilositarsa* (Habeeb) (Axonopsinae) larva excretory pore plate. **Fig. 162:** *Brachypoda cornipes* Habeeb (Axonopsinae) larva excretory pore plate. **Fig. 163:** *Ochybrachypoda setosicauda* (Habeeb) (Axonopsinae) larva excretory pore plate.

42' Excretory pore plate bearing four pairs of setae, including ps1 and ps2 medially, pa near anterior edge, and h2 near lateral angles (Figs. 16.24.100, 103) .... *Atractides*

43(41) First and second legs with tarsus bearing 10 setae (+ω), lacking Ta14, and either Ta12 (Fig. 16.24.105) or Ta9 (Figs. 16.25.112, 116).......... 44

43' First and second legs with tarsus bearing 11 or more setae (+ω), including Ta9 (Figs. 16.27.128, 32.170, arrows; Figs. 16.29.147, 30.151, arrows A) 47

44(43) Dorsal plate, coxal plates, and leg sclerites conspicuously longitudinally striated (as in Fig. 16.24.107); third coxal plates rounded posteriorly (Fig. 16.24.106); first leg with tibia bearing seven setae (+2φ), lacking Ti10 and Ti11 (Fig. 16.24.105) ....
Feltriidae .... *Feltria*

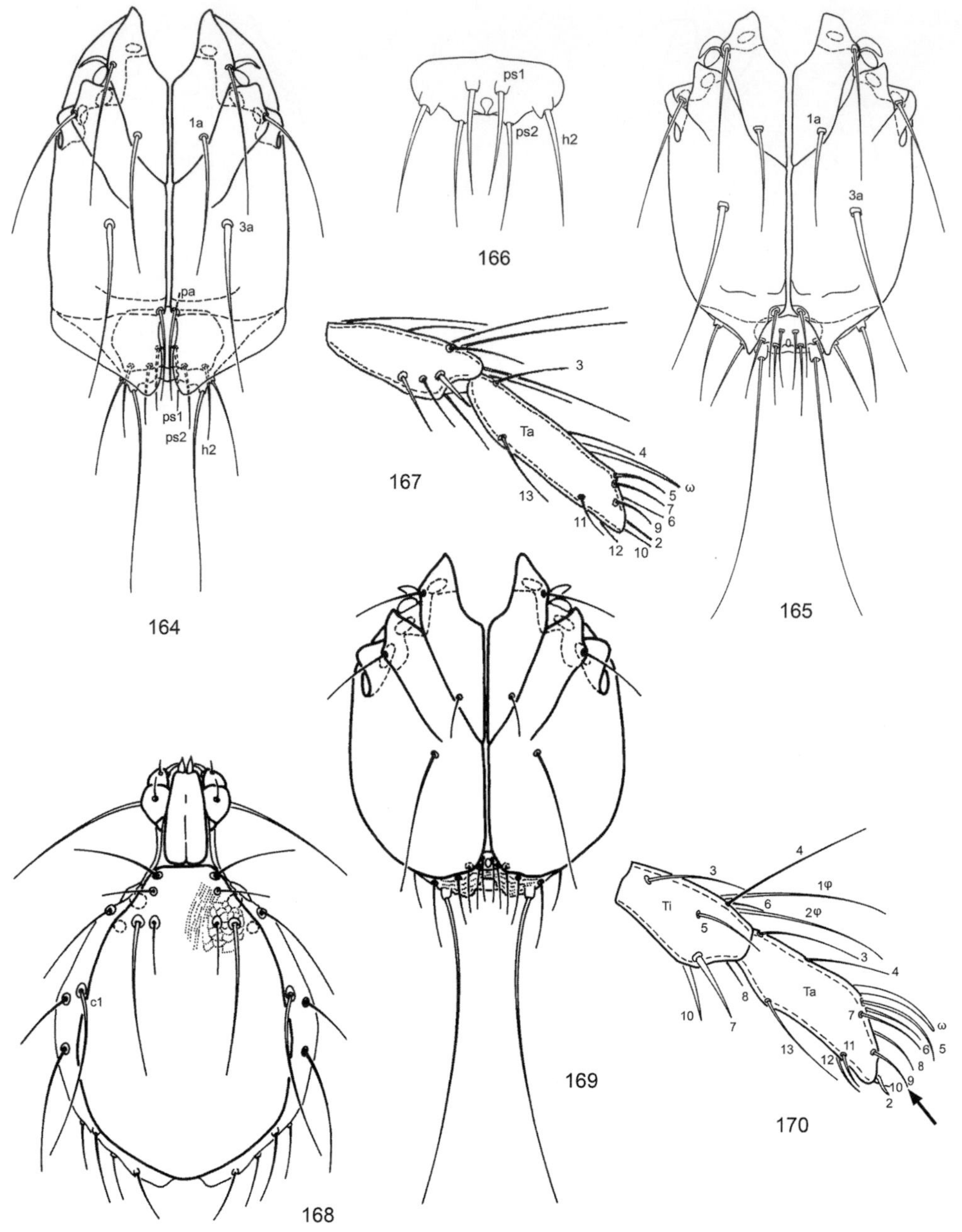

PLATE 16.32 **Fig. 164:** *Ocybrachypoda setosicauda* (Habeeb) (Axonopsinae) larva; venter of idiosoma. **Figures 165–167:** *Brachypodopsis setoniensis* (Habeeb) (Axonopsinae) larva. **Fig. 165** venter of idiosoma. **Fig. 166** excretory pore plate. **Fig. 167** tibia and tarsus of first leg. **Figures 168–170:** *Aturus* sp. (Aturinae) larva. **Fig. 168** dorsum of idiosoma and gnathosoma. **Fig. 169** venter of idiosoma. **Fig. 170** tibia and tarsus of first leg.

44'	Dorsal plate and coxal plates reticulate (Fig. 16.25.108); third coxal plates bearing pointed projections posteriorly (Fig. 16.25.111, arrow); first leg with tibia bearing 9 setae (+2φ), including Ti10 and Ti11 (Figs. 16.25.112, 116) .... Unionicolidae 45	
45(44)	Tarsi of legs with empodium undivided (Fig. 16.25.115) ....Pionatacinae 46	
45'	Tarsi of legs with empodium bifurcate (Fig. 16.25.110, arrow) ....Unionicolinae....................................*Unionicola*	
46(45)	Excretory pore plate usually oval in shape with excretory pore borne near anterior edge (Fig. 16.25.114) ....*Neumania*	
46'	Excretory pore plate obcordate with excretory pore borne near middle of plate (Fig. 16.25.113) .... *Koenikea*	
47(43)	All legs with tibia bearing seven setae (+2φ, +2φ, and +1φ, respectively), lacking Ti9 and Ti11 (Fig. 16.32.170) .... Aturidae (in part) .... Aturinae ....*Aturus*	
47'	All legs with tibia bearing eight or more setae (+2φ, +2φ, and +1φ, respectively), including Ti9 and Ti11 (Fig. 16.30.151) 48	

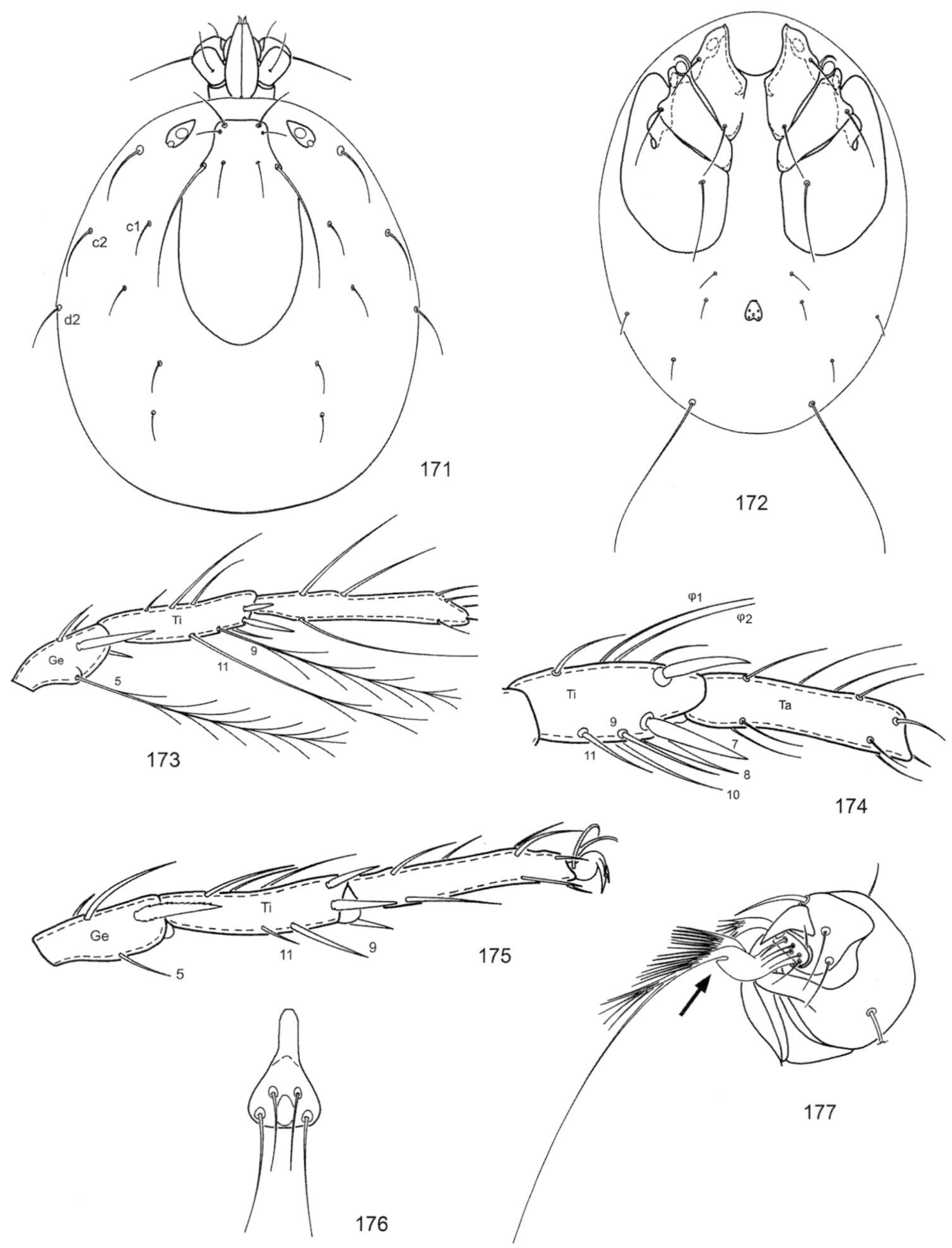

PLATE 16.33 **Figures 171–172:** *Stygomomonia mitchelli* Smith (Stygomomoniinae) larva. **Fig. 171** dorsum of idiosoma and gnathosoma. **Fig. 172** venter of idiosoma. **Fig. 173:** *Momonia campylotibia* Smith (Momoniinae) larva; distal segments of third leg. **Fig. 174:** *Paramideopsis susanae* Smith (Nudomideopsidae) larva; tibia and tarsus of third leg. **Fig. 175:** *Stygomomonia mitchelli* Smith (Stygomomoniinae) larva; distal segments of third leg. **Figure 176:** *Nudomideopsis magnacetabula* (Smith) (Nudomideopsidae) larva; excretory pore plate. **Figure 177:** *Paramideopsis susanae* Smith (Nudomideopsidae) larva; pedipalp.

48(47)	Third coxal plates with pointed or lobed projections posteriorly, bearing two pairs of setae, 3a anteromedially and pa posteromedially (Figs. 16.31.156, 32.164); excretory pore plate large, bearing three pairs of setae, ps1 and ps2, and h2 posterolaterally (Figs. 16.31.159–163, 32.166) ....Aturidae (in part)..Axonopsinae (in part)...49
48’	Third coxal plates without, or with small, projections posteriorly, bearing only setae 3a (Figs. 16.26.117, 29.145, 30.152–154); excretory pore plate small or large, bearing only setae ps1 and ps2 (Figs. 16.26.118, 123, 28.135, 138, 29.142, 146) 53
49(48)	First and second legs with tarsus bearing 12 setae, including setae Ta8 (Fig. 16.31.158); excretory pore plate with setae ps1 at or near anterior edge (Figs. 16.31.159, 160) 50
49’	First and second legs with tarsus bearing 11 setae, lacking setae Ta8 (Fig. 16.32.167); excretory pore plate with setae ps1 near anterior edge to posterior to mid-length (Figs. 16.31.162–163, 32.166) 51

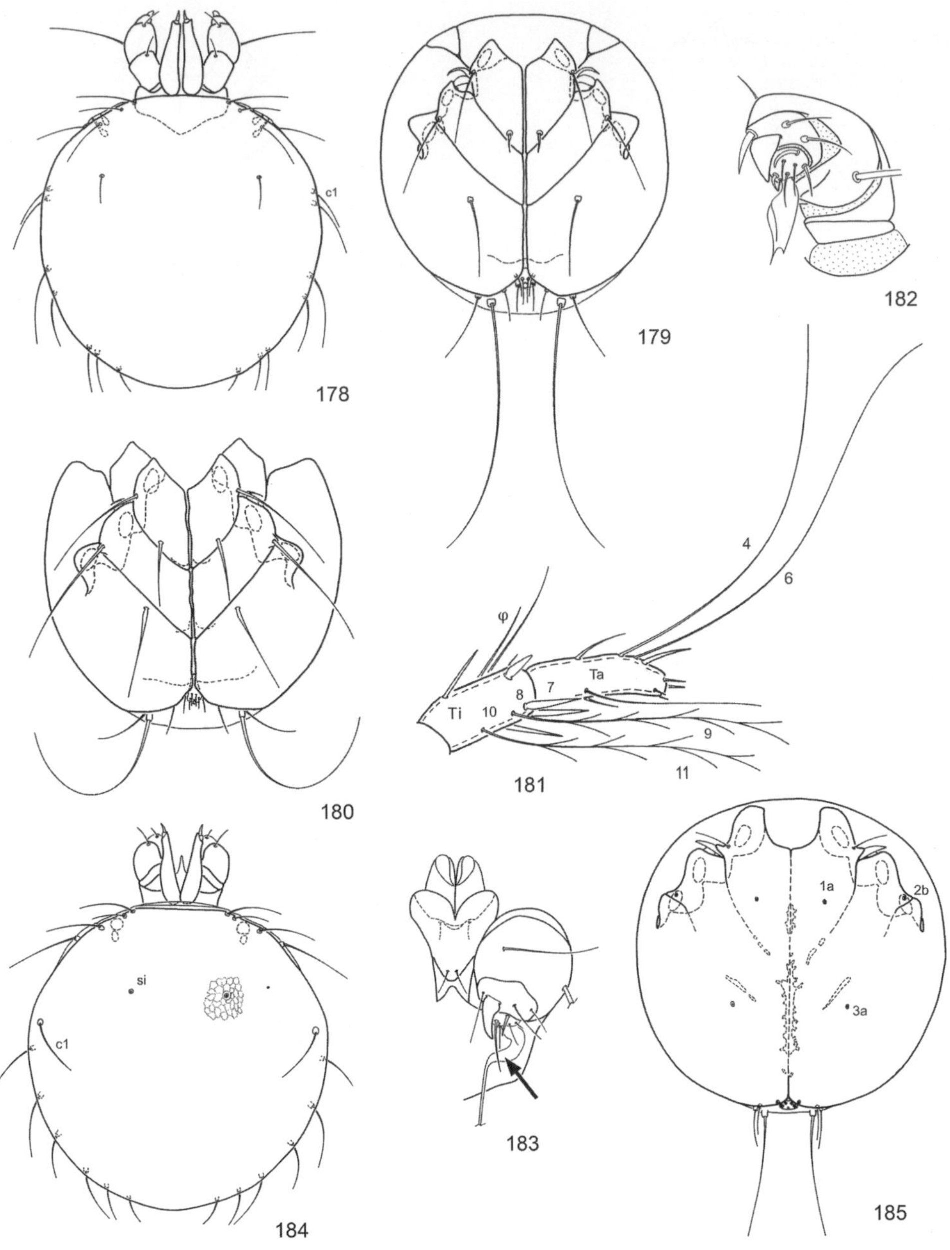

PLATE 16.34 **Figures 178–179:** *Nudomideopsis magnacetabula* (Smith) (Nudomideopsidae) larva. **Fig. 178** dorsum of idiosoma and gnathosoma. **Fig. 179** venter of idiosoma. **Figures 180–181:** *Midea expansa* Marshall (Mideidae) larva. **Fig. 180** venter of idiosoma. **Fig. 181** tibia and tarsus of third leg. **Fig. 182:** *Acalyptonotus neoviolaceus* Smith (Acalyptonotidae) larva; pedipalp. **Fig. 183:** *Midea expansa* Marshall (Mideidae) larva; venter of gnathosoma. **Figures 184–185:** *Acalyptonotus neoviolaceus* Smith (Acalyptonotidae) larva. **Fig. 184** dorsum of idiosoma and gnathosoma. **Fig. 185** venter of idiosoma.

50(49) Excretory pore plate subtriangular (Fig. 16.31.159) with acutely rounded posterolateral angles, with setae ps2 anterior to mid-length between setae ps1 and h2; dorsal plate not reticulate; first coxal plates weakly fused medially with second plates, and with setae 1a long, extending posteriorly beyond level of setae 3a (Fig. 16.31.157) .... *Estellacarus*

50' Excretory pore plate obcordate (Fig. 16.31.160) or nearly quadrangular (Fig. 16.31.161), with broadly rounded posterolateral angles, and with setae ps2 posterior to mid-length, not between setae ps1 and h2; dorsal plate reticulate; first coxal plates separate from second plates, and with setae 1a relatively short, not extending posteriorly beyond level of setae 3a (Fig. 16.31.156) .... *Woolastookia*

51(49) Suture line between second and third coxal plates short, not extending medially to region of base of setae 3a (Fig. 16.32.165); excretory pore plate moderately large and broadly elliptical or subquadrangular (Fig. 16.32.166) .... *Brachypodopsis*

51' Suture line between second and third coxal plates long, extending medially to near base of setae 3a (Fig. 16.32.164); excretory pore plate either moderately large and subtriangular (Fig. 16.31.162) or very large and subquadrangular (Fig. 16.31.163) 52

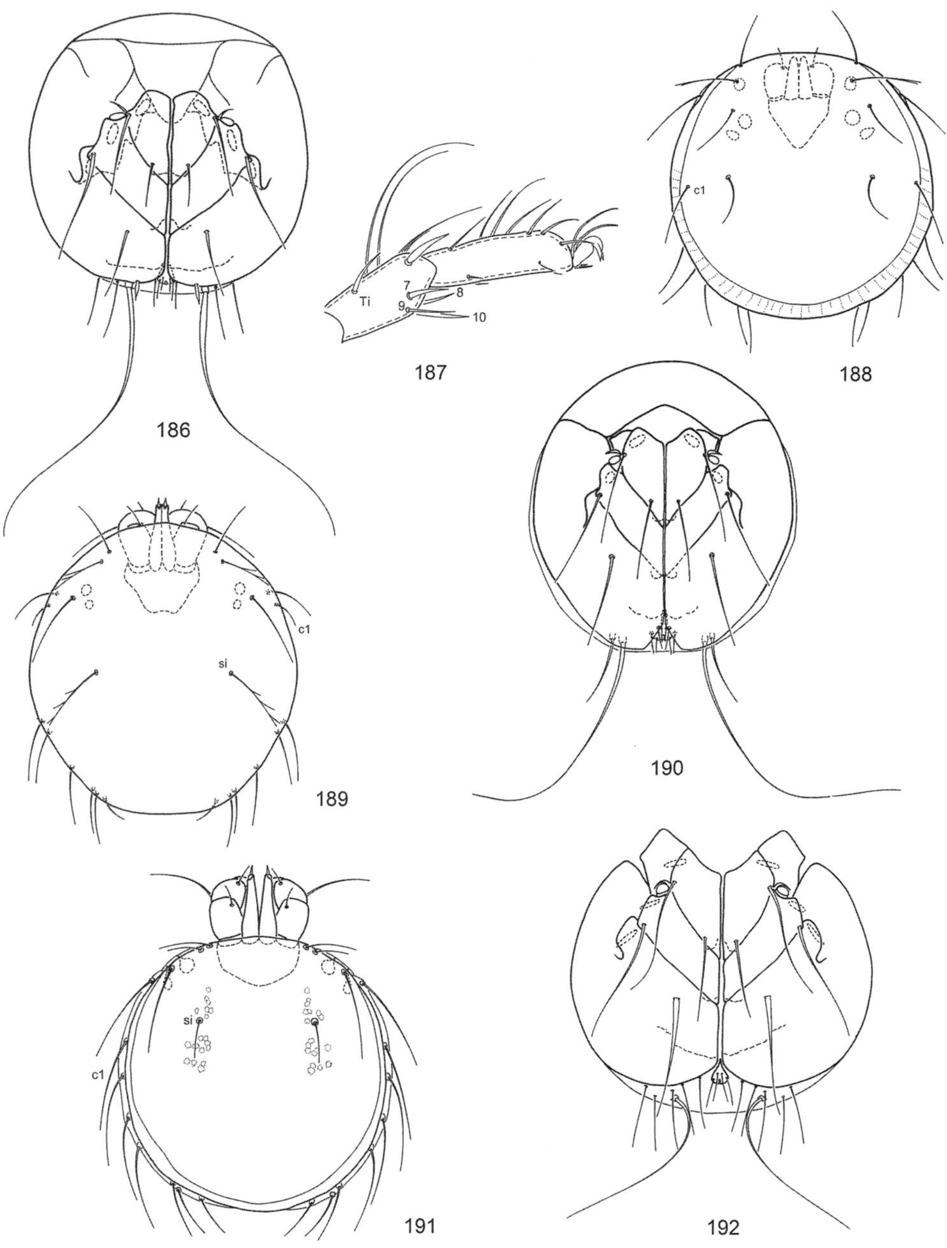

PLATE 16.35 **Fig. 186:** *Mideopsis borealis* Habeeb (Mideopsinae) larva; venter of idiosoma. **Figure 187:** *Platyhydracarus juliani* Smith (Athienemanniinae) larva; tibia and tarsus of second leg. **Fig. 188:** *Mideopsis borealis* Habeeb (Mideopsinae) larva; dorsum of idiosoma and gnathosoma. **Figures 189–190:** *Platyhydracarus juliani* Smith (Athienemanniinae) larva. **Fig. 189** dorsum of idiosoma and gnathosoma. **Fig. 190** venter of idiosoma. **Fig. 191:** *Neoacarus occidentalis* Cook (Neoacaridae) larva; dorsum of idiosoma and gnathosoma. **Fig. 192:** *Laversia berulophila* Cook (Larversiidae) larva; venter of idiosoma.

52(51) Excretory pore plate moderately large and subtriangular (Fig. 16.31.162) .... *Brachypoda*

52' Excretory pore plate very large and subquadrangular (Fig. 16.31.163) .... *Ocybrachypoda*

53(48) Third coxal plates with small projections posteriorly (Fig. 16.26.117, arrow); excretory pore plate small, little larger than combined area of excretory pore and bases of setae ps1 and ps2 (Fig. 16.26.118) .... Wettinidae .... *Wettina*

53' Third coxal plate without projections posteriorly (Figs.16.26.124, 27.133, 28.137, 140, 29.145, 30.152, 154); excretory pore plate considerably larger than combined area of excretory pore and bases of setae ps1 and ps2 (Figs. 16.26.123, 125, 27.129, 28.138, 29.142, 30.152) 54

54(53) First leg with tarsus bearing 13 setae (+ω), including Ta8 (Figs. 16.29.147, 30.151, arrows B); excretory pore plate large, and usually triangular or obcordate (Figs. 16.28.138, 139, 29.148, 30.152) 55

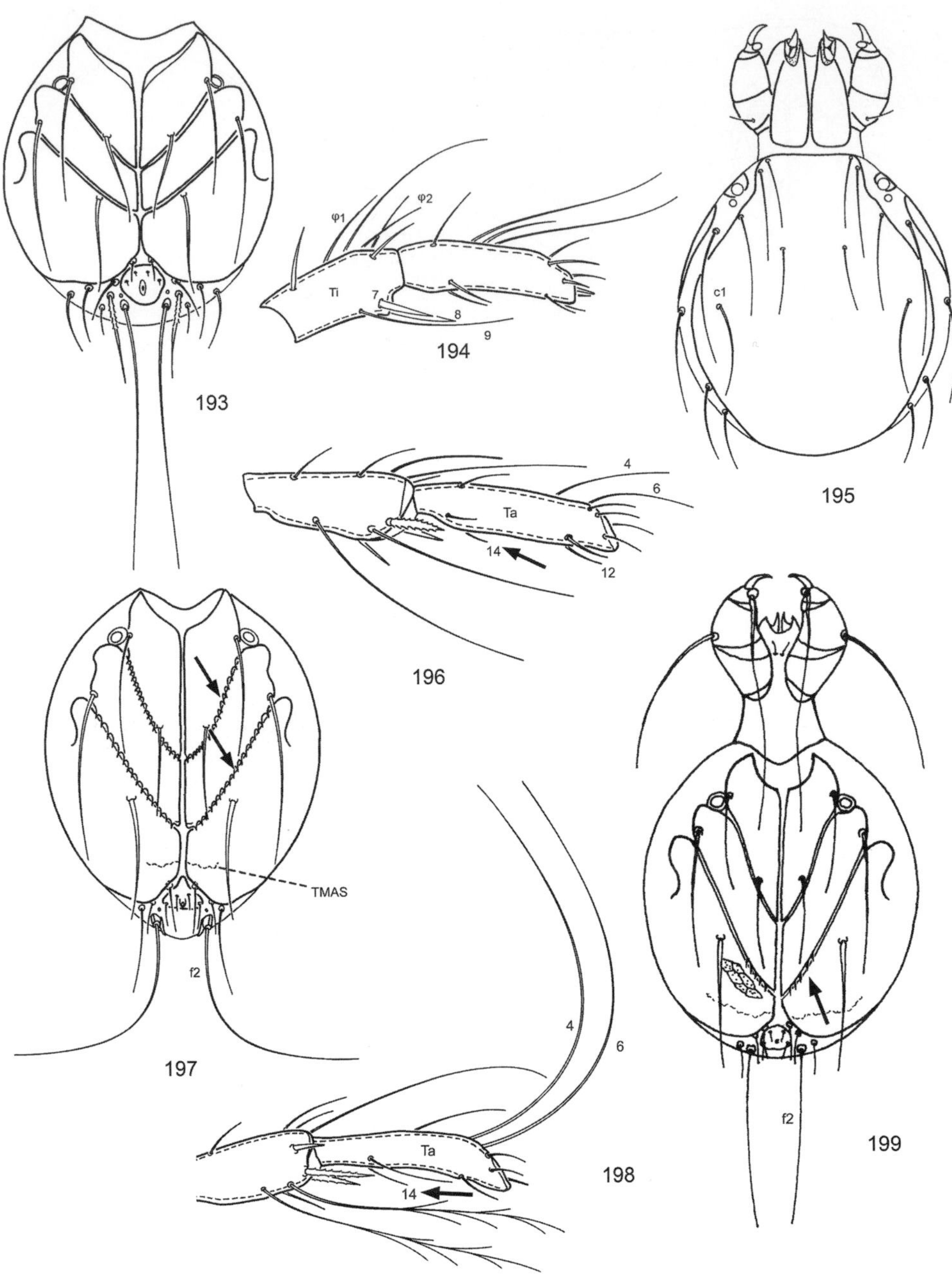

PLATE 16.36 **Fig. 193:** *Arrenurus planus* Marshall (Arrenuridae) larva; venter of idiosoma. **Figure 194** *Laversia berulophila* Cook (Laversiidae) larva; tibia and tarsus of second leg. **Figures 195–196:** *Arrenurus planus* Marshall (Arrenuridae) larva. **Fig. 195** dorsum of idiosoma and gnathosoma. **Fig. 196** tibia and tarsus of third leg. **Figures 197–198:** *Krendowskia similis* Viets (Krendowskiidae) larva. **Fig. 197** venter of idiosoma. **Fig. 198** tibia and tarsus of third leg. **Fig. 199:** *Geayia ovata* (Wolcott) (Krendowskiidae) larva; venter of idiosoma and gnathosoma.

54' First leg with tarsus bearing 12 setae (+ω), lacking Ta8 (Fig. 16.27.128); excretory pore plate small to large, and variously shaped (Figs. 16.26.119, 121, 123, 125, 27.134, 28.135) 59

55(54) Excretory pore plate with setae ps1 close together adjacent to excretory pore in posterior half of plate (Figs. 16.28.138, 139, 29.142, 146, 148) .... Pionidae (in part) 56

55' Excretory pore plate with setae ps1 widely separated and removed from excretory pore in anterior half of plate (Fig. 16.30.154, arrow) .... Aturidae (in part) 58

56(55) Second leg with tarsus bearing 13 setae (+ω), including Ta8 (Fig. 16.28.141); all legs with tibial setae Ti9 and Ti11 simple (Fig. 16.28.141) .... Najadicolinae .... *Najadicola*

56' Second leg with tarsus usually bearing 12 setae (+ω), lacking Ta8 (Fig. 16.29.149), when bearing 13 setae then all legs with tibial setae Ti9 and Ti11 long and plumose (as in Fig. 16.29.149) .... Pioninae 57

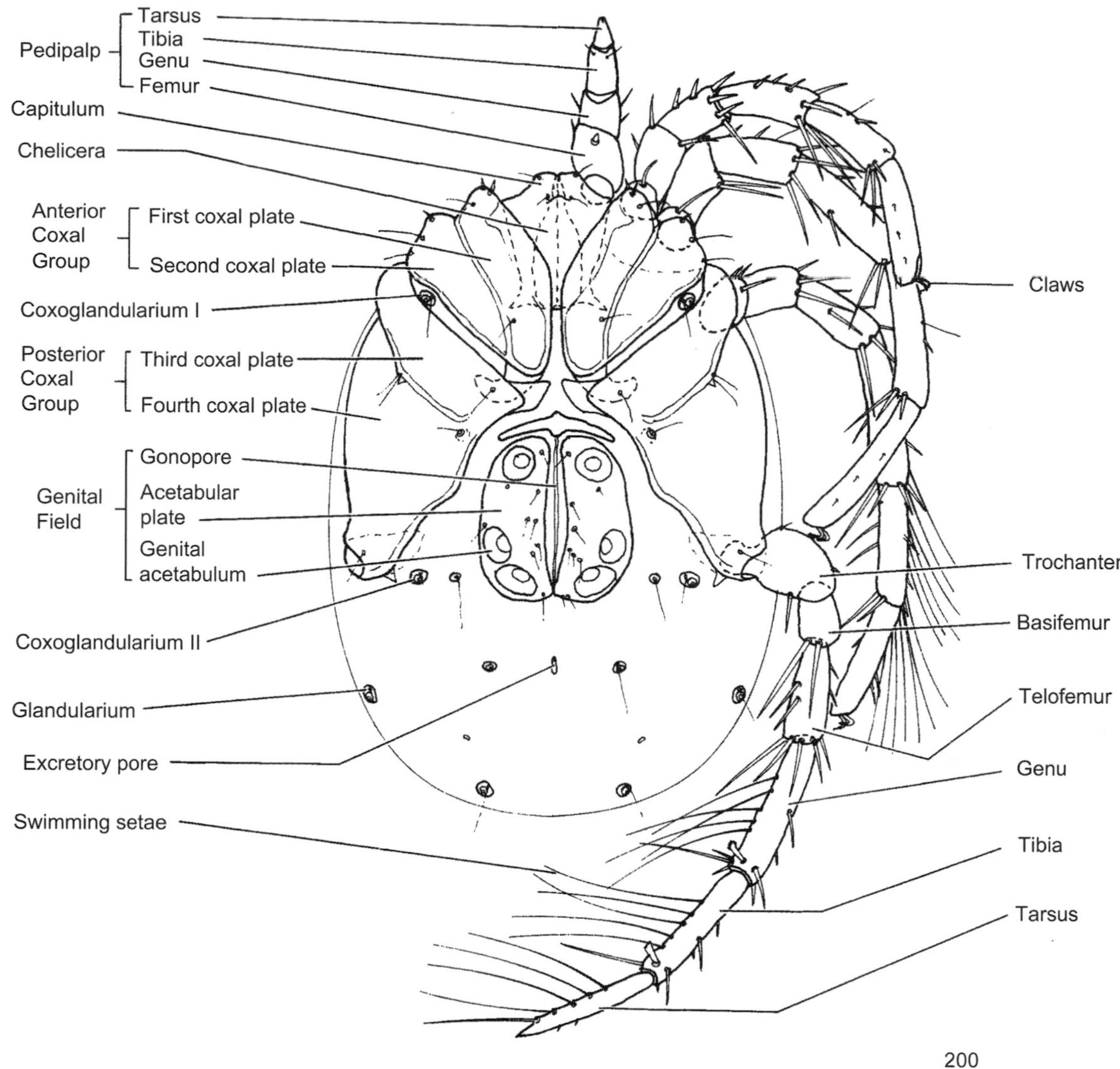

PLATE 16.37 **Figure 200:** *Limnesia* sp. (Limnesiidae) adult female; venter of idiosoma.

57(56) Second leg with femoral seta Fe5 shorter than segment; excretory pore plate usually much less than twice as wide as long, when nearly twice as wide as long then plate is concave or transverse anteromedially (Fig. 16.29.146) .... *Piona*

57' Second leg with femoral seta Fe5 longer than segment; excretory pore plate twice as wide as long and convex anteromedially (Fig. 16.28.139) .... *Nautarachna*

58(55) Excretory pore plate broadly obcordate (Fig. 16.30.152) .... Axonopsinae (in part) .... *Ljania*

58' Excretory pore plate triangular (Fig. 16.30.154) .... Albiinae .... *Albia*

59(54) Excretory pore plate small and subtriangular, with setae ps2 borne near posterolateral angles of plate (Fig. 16.30.155); suture lines between second and third coxal plates parallel to anterior edge of second plate, coxal plates without distinct lateral coxal apodemes and third coxal plates lacking medial coxal apodemes and transverse muscle attachment scars (Fig. 16.30.155) .... Aturidae (in part) .... Axonopsinae (in part) .... *Neobrachypoda*

59' Excretory pore plate variously shaped, when triangular with setae ps2 borne near posterolateral angles of plate then suture lines between second and third coxal plates terminating in distinct lateral coxal apodemes (LCA) that are usually nearly transverse and third coxal plates bearing at least medial coxal apodemes (MCA) (Figs. 16.26.120, 124, 27.126, 129, 133, 28.136) .... Pionidae (in part) 60

60(59) Suture line between second and third coxal plates on each side extending medially beyond lateral coxal apodeme nearly to midline (Fig. 16.26.120, arrow); excretory pore borne on elevated projection which extends to or beyond posterior edge of plate (Figs. 16.26.119, 121) .... Hydrochoreutinae .... *Hydrochoreutes*

60' Suture line between second and third coxal plates on each side extending medially only as far as lateral coxal apodeme (Figs. 16.26.122, 124, 27.126, 129, 133, 28.136); excretory pore usually not borne on elevated projection, when on small projection this does not extend to posterior edge of plate 61

61(60) Third leg with trochanter bearing two setae, including supernumerary dorsal seta (Fig. 16.27.131, arrow), *or* second and third legs with tibia bearing 11 or 12 setae (+2φ and +1φ, respectively), including two supernumerary posteroventral setae (Fig. 16.27.132, arrows), *or* third coxal plates lacking transverse muscle attachment scars and excretory pore plate with setae ps2 near posterior edge of plate (Fig. 16.28.136) .... Foreliinae 62

61' Third leg with trochanter bearing only one seta; second and third legs with tibia bearing only nine setae (+2φ and +1φ, respectively) (as in Figs. 16.28.141, 29.149); third coxal plates usually with transverse muscle attachment scars (Figs. 16.26.124, 27.126, 133), when coxal plates lacking these scars (as in Fig. 16.26.122) then excretory pore plate with setae ps2 removed from posterior edge of plate (Fig. 16.26.123) 63

62(61) Third leg with trochanter bearing two setae, including supernumerary dorsal seta (Fig. 16.27.131, arrow) .... *Pseudofeltria*

62' Third leg with trochanter bearing one seta .... *Forelia, Madawaska*

63(61) Excretory pore plate transversely elliptical or semicircular (Fig. 16.26.123) .... Huitfeldtiinae .... *Huitfeldtia*

63' Excretory pore plate broadly elliptical, subcircular, or subtriangular .... Tiphyinae 64

64(63) Excretory pore plate three times as wide as long and broadly elliptical or dumbbell-shaped (Figs. 16.26.124, 125) .... *Neotiphys*

64' Excretory pore plate less than twice as wide as long and subcircular (Figs. 16.27.127, 130) to subtriangular .... *Pionopsis, Tiphys, Acercopsis*

Keys to Parasitengonina: Adults

1 Legs with tarsus bearing paired claws and claw-like empodium (Fig. 16.38.201); pedipalp with tarsus bearing a single long, rod-shaped seta terminally (Fig. 16.38.202, arrow); idiosoma vermiform and integument soft (Fig. 16.38.204); idiosomal dorsum bearing a spindle-shaped anterior crista that does not bear lateral eyes (Fig. 16.38.203) .... Stygothrombiae .... Stygothrombidioidea .... Stygothrombidiidae .... *Stygothrombium*

1' Legs with tarsus usually bearing paired claws but always lacking empodium (Figs. 16.42.245, 43.258, 44.268, 59.439, 61.465); claws simple (Fig. 16.42.245) or variously modified (Figs. 16.42.248, 61.461, 66.532, 68.546); pedipalp with tarsus bearing various setae but never with a single long, terminal seta as above; idiosoma variously shaped and sclerotized but never bearing a spindle-shaped anterior crista that does not bear lateral eyes .... Hydrachnidiae 2

2(1) Genital field bearing movable genital flaps (Figs. 16.38.206, 210) but genital acetabula borne on coxal plates (Figs. 16.38.210, 39.212; arrow) .... Hydrovolzioidea 3

2' Genital field bearing or lacking genital flaps, when genital flaps are present acetabula are borne on flaps (Fig. 16.37.200), beneath flaps (Fig. 16.50.341) or medial to flaps in onopore (Fig. 16.49.326) but never on coxal plates 4

3(2) Dorsal shield comprising a large central plate surrounded by a smaller platelet anteriorly and numerous pairs of small similar platelets laterally and posteriorly (Fig. 16.38.209); glandularia represented only by setae with gland portion absent; ventral shield comprising an array of closely fitting plates (Fig. 16.38.210); pedipalp with tibia subequal in length to genu and with tarsus relatively short and bearing a group of similar long, blunt, rod-shaped setae distally (Fig. 16.38.211, arrow) .... Acherontacaridae .... *Acherontacarus*

3' Dorsal shield comprising a large central plate surrounded by a smaller platelet anteriorly and several pairs of small dissimilar platelets laterally and posteriorly (Fig. 16.38.205); glandularia with separate gland and seta bearing platelets (Fig. 16.38.207); ventral shield comprising an array of plates separated from one another by soft integument (Fig. 16.38.206); pedipalp with tibia longer than genu and with tarsus relatively long and bearing an array of dissimilar slender and spine-like setae distally (Fig. 16.38.208) .... Hydrovolziidae .... *Hydrovolzia*

4(2) Idiosoma nearly spherical (Fig. 16.39.213); pedipalp with tibia much shorter than genu and bearing a dorsodistal spine-like seta that extends well beyond insertion of tarsus (Fig. 16.39.214); chelicera one-segmented (Fig. 16.39.215, arrow) .... Hydrachnoidea .... Hydrachnidae .... *Hydrachna*

4' Idiosoma variously shaped, but rarely spherical; pedipalp with tibia usually longer than genu and when shorter than genu then lacking a dorsodistal projecting, spine-like seta; chelicera two-segmented (Fig. 16.59.438) 5

5(4) Idiosomal dorsum with lateral eye capsules present and usually borne on a medial prodorsal platelet (Figs. 16.39.217, 40.222, 225, 226), rarely on moderately large, rugose, paired platelets associated with a complex of six prodorsal sclerites (Fig. 16.39.218) .... Eylaoidea 6

5' Idiosomal dorsum with lateral eye capsules present or absent and when present not borne on a medial prodorsal platelet (Figs. 16.41.234, 42.244, 44.263) 11

6(5) Lateral eyes borne on lateral platelets separated by a medial triangular platelet (Fig. 16.39.218, arrows); idiosoma bearing twenty-six pairs of glandularia on large rugose platelets (Figs. 16.39.218, 219); gnathosoma with capitulum wider than long and pedipalp with only a single movable segment that flexes medially (Fig. 16.39.219) .... Apheviderulicidae .... *Apheviderulix* (deutonymph)

6' Lateral eyes borne on a medial prodorsal platelet (Figs. 16.39.217, 40.222, 225, 226); idiosoma bearing fewer than twenty-six pairs of glandularia on small sclerites; gnathosoma with capitulum longer than wide and pedipalp with 3 to 5 movable segments that flex ventrally (Figs. 16.40.228–230) 7

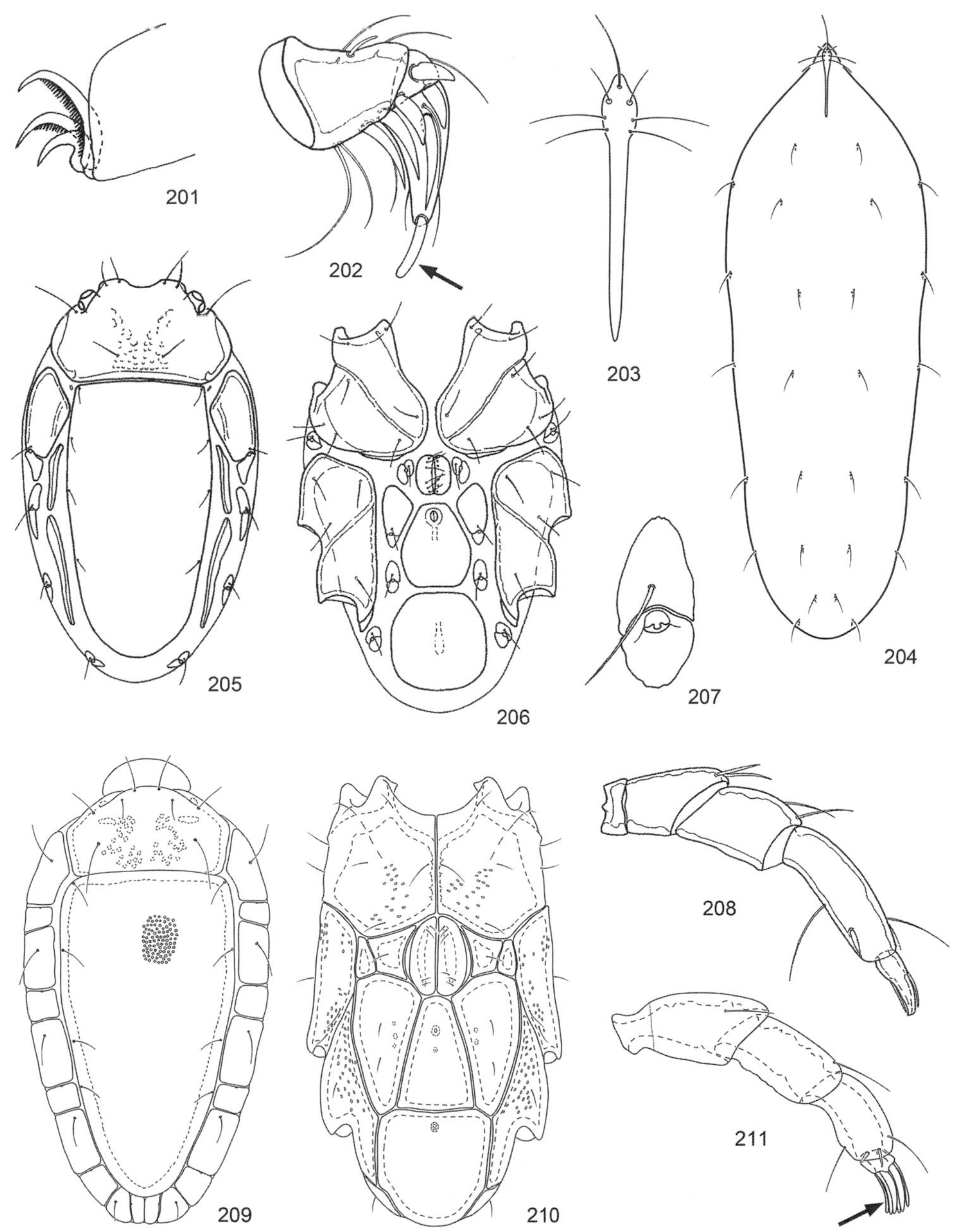

PLATE 16.38 **Figures 201–202:** *Stygothrombium* sp. (Stygothrombidiidae) adult female. **Fig. 201** claws and empodium of first leg. **Fig. 202** pedipalp. **Figures 203–204:** *Stygothrombium estellae* (Habeeb) (Stygothrombidiidae) adult female. **Fig. 203** prodorsal sclerite ("crista metopica"). **Fig. 204** dorsum of idiosoma. **Figures 205–208:** *Hydrovolzia marshallae* Cook (Hydrovolziidae) adults. **Fig. 205** female dorsum of idiosoma. **Fig. 206** male venter of idiosoma (acetabula not shown). **Fig. 207** male glandularium. **Fig. 208** female pedipalp. **Figures 209–211:** *Acherontacarus smokeyensis* Smith (Acherontacaridae) adult female. **Fig. 209** dorsum of idiosoma. **Fig. 210** venter of idiosoma. **Fig. 211** pedipalp.

7(6)	Lateral eyes borne on a medial prodorsdal platelet that is constricted medially and bears one pair of setae (Fig. 16.39.217); idiosoma with integument soft and lacking dorsal platelets (Fig. 16.39.216) .... Eylaidae .... *Eylais*
7'	Lateral eyes borne on a medial prodorsal platelet that is not constricted medially and bears four pairs of setae (Figs. 16.40.222, 225, 226); idiosoma with integument soft and lacking dorsal platelets (Fig. 16.40.223) or bearing an array of dorsal platelets (Figs. 16.40.224, 227) 8
8(7)	Prodorsal plate approximately as wide as long (Fig. 16.40.222); genital acetabula on distinct acetabular plates (Fig. 16.40.221); idiosomal dorsum bearing a characteristic array of platelets (Fig. 16.40.220) .... Piersigiidae .... *Piersigia*
8'	Prodorsal plate much longer than wide (Figs. 16.40.225, 226); genital acetabula scattered on ventral integument (Fig. 16.40.231); gnathosoma slightly or greatly protrusible (Fig. 16.40.223); idiosoma with integument soft and either lacking dorsal platelets (Fig. 16.40.223) or bearing an array of dorsal platelets (Figs. 16.40.224, 227) .... Limnocharidae 9
9(8)	Gnathosoma greatly protrusible (Fig. 16.40.223); pedipalp three-segmented, with tarsus rod-shaped and inserted dorsally on tibia (Fig. 16.40.229, arrow) .... Rhyncholimnocharinae .... *Rhyncholimnochares*

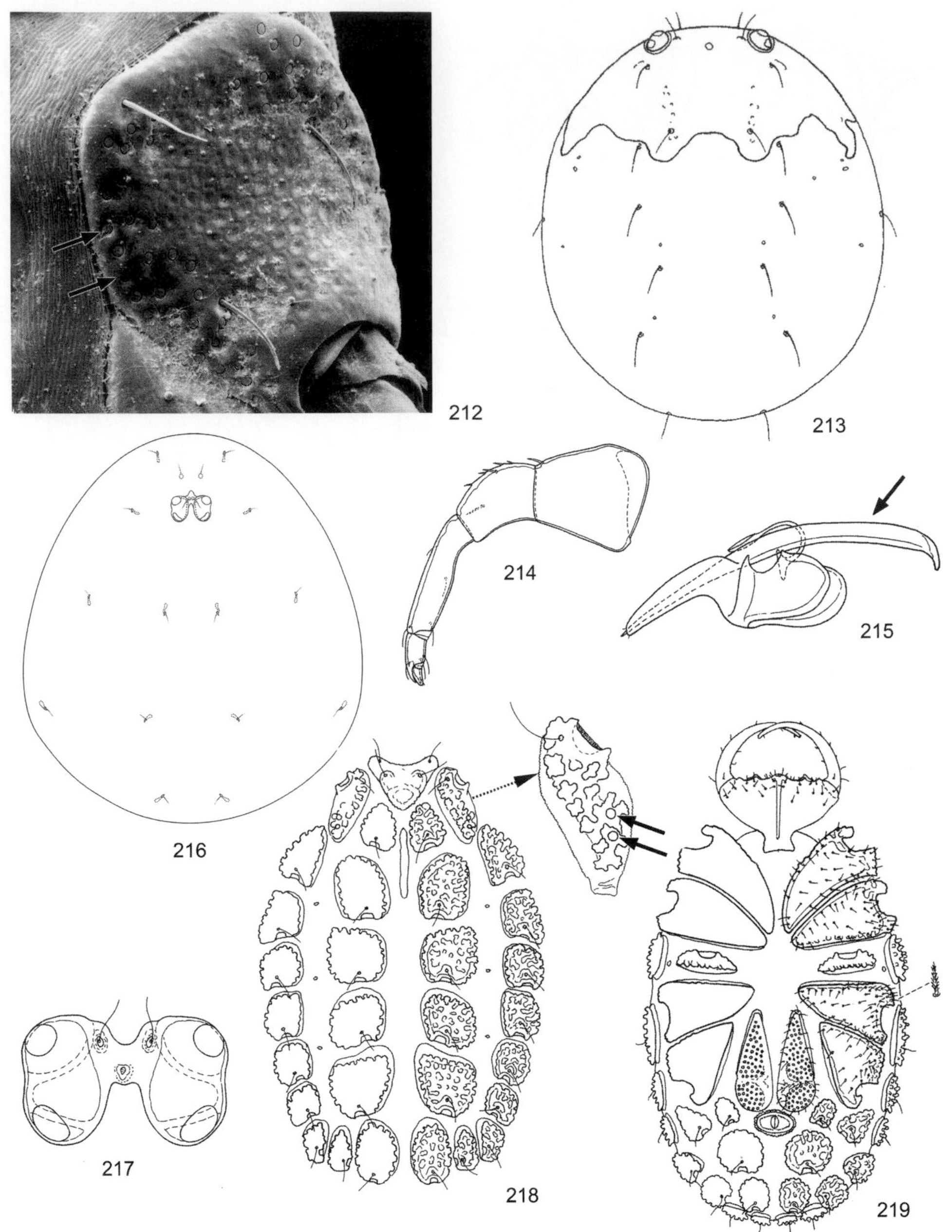

PLATE 16.39 **Fig. 212:** *Hydrovolzia marshallae* Cook (Hydrovolziidae) adult female third coxal Plate 16.showing acetabula. **Figures 213–215:** *Hydrachna* sp. (Hydrachnidae) adult. **Fig. 213** dorsum of idiosoma. **Fig. 214** pedipalp. **Fig. 215** gnathosomal base (capitulum) and chelicera. **Figures 216–217:** *Eylais* sp. (Eylaidae) adult female. **Fig. 216** dorsum of idiosoma. **Fig. 217** ocular plate. **Figures 218–219:** *Apheviderulix santana* Gerecke Smith and Cook (Apheviderulicidae) deutonymph. **Fig. 218** dorsum of idiosoma showing detail of ocular plate. **Fig. 219** venter of gnathosoma and idiosoma.

9'	Gnathosoma slightly protrusible; pedipalp four- or five-segmented, with tarsus inserted at distal end of tibia (Figs. 16.40.228, 230) .... Limnocharinae 10
10(9)	Pedipalp four-segmented (Fig. 16.40.228); idiosomal dorsum with an array of dorsal platelets similar to that in Fig. 16.40.224 .... *Neolimnochares*
10'	Pedipalp five-segmented (Fig. 16.40.230); idiosomal dorsum lacking platelets or with an array of platelets similar to that in Fig. 16.40.227 .... *Limnochares*
11(5)	Idiosoma bearing a characteristic array of reticulate platelets (Figs. 16.41.232, 233) ....Hydryphantoidea (in part)..........................Rhynchohydracaridae ....Clathrosperchontinae...................... *Clathrosperchon*
11'	Idiosoma various but never with an array of reticulate platelets as shown above 12

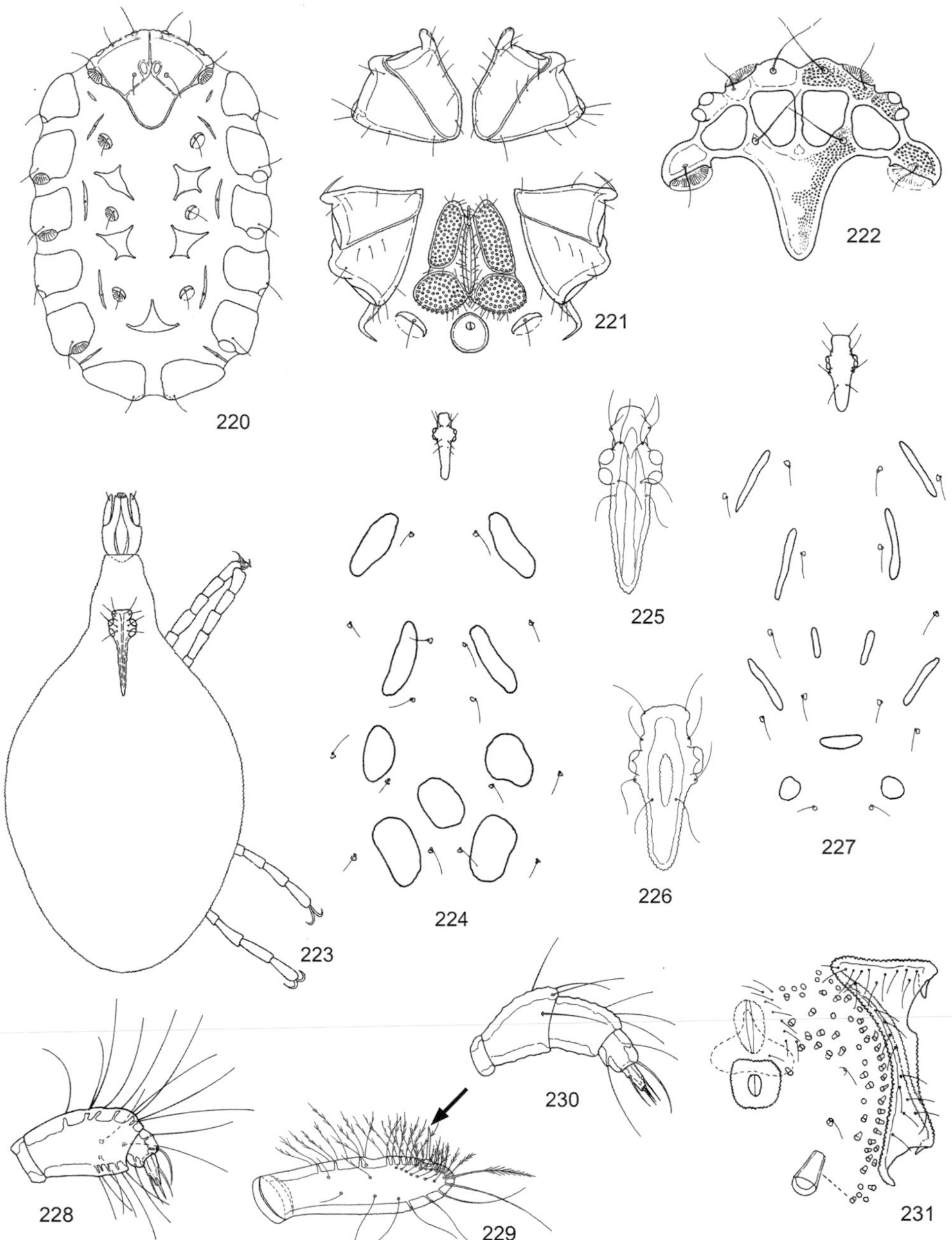

PLATE 16.40 **Figures 220–21:** *Piersigia crusta* Mitchell (Piersigiidae) adult female. **Fig. 220** dorsum of idiosoma. **Fig. 221** venter of idiosoma. **Fig. 222:** *Piersigia limnophila* Protz (Piersigiidae) adult female prodorsal plate. **Fig. 223:** *Rhyncholimnochares kittatinniana* Habeeb (Rhyncholimnocharinae) deutonymph dorsum of idiosoma. **Figures 224, 225, 228:** *Neolimnochares johnstoni* Smith and Cook (Limnocharinae) adult female. **Fig. 224** dorsum of idiosoma. **Fig. 225** prodorsal plate. **Fig. 228** pedipalp. **Figures 226, 227, 230:** *Limnochares anomala* Habeeb (Limnocharinae) adult female. **Fig. 226** prodorsal plate. **Fig. 227** dorsum of idiosoma. **Fig. 230** pedipalp. **Fig. 229** *Rhyncholimnochares* sp. (Rhyncholimnocharinae) adult pedipalp. **Fig. 231** *Limnochares appalachiana* Smith and Cook (Limnocharinae) adult female genital field region.

12(11) Pedipalp usually appearing chelate with tibia bearing a long distodorsal projection (Fig. 16.41.236, arrow) or a long, thick, spine-like dorsodistal seta extending well beyond insertion of tarsus (Figs. 16.42.251, 43.254, 45.279, 47.298, 48.313; arrows); when pedipalp not appearing chelate (as in some *Trichothyas*-like genera) (Fig. 16.45.283) then capitulum lacking an anchoral process (similar to Fig. 16.69.556, arrow) and genital acetabula borne either in gonopore beneath movable genital flaps (Fig. 16.46.287) or attached to but not on flaps (Fig. 16.46.290) .... Hydryphantoidea (in part) 13

12' Pedipalp usually not appearing chelate, but when appearing chelate with tibia bearing a long dorsodistal seta that extends beyond insertion of tarsus (as in some species of pionid subfamily Tiphyinae) (Fig. 16.65.518) then capitulum with well-developed anchoral process (Fig. 16.64.496, arrow) and genital acetabula borne on acetabular plates flanking gonopore (Fig. 16.64.496) 39

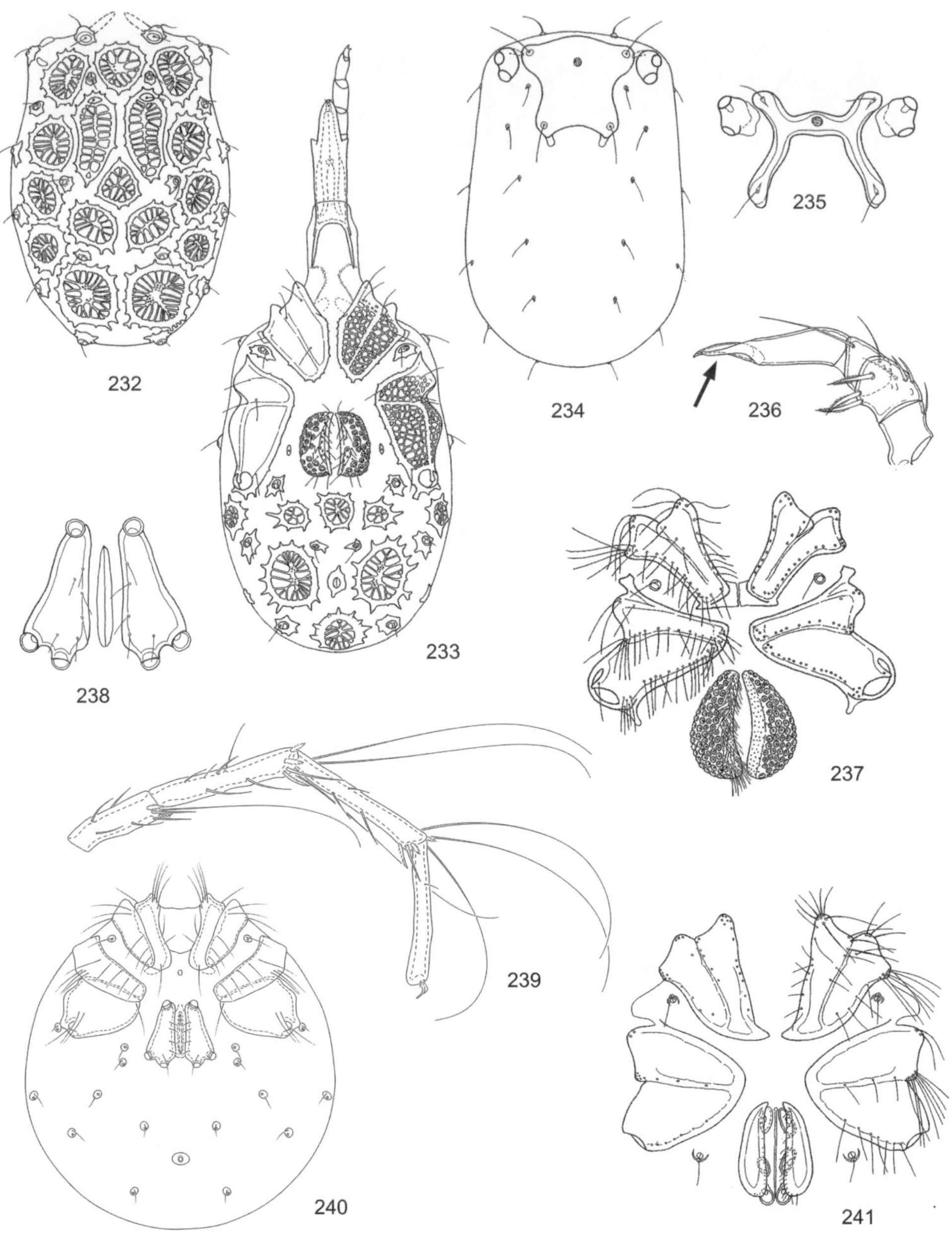

PLATE 16.41 **Figures 232–233:** *Clathrosperchon ornatus* Cook (Clathrosperchontinae) adult male. **Fig. 232** dorsum of idiosoma. **Fig. 233** venter of gnathosoma and idiosoma. **Fig. 234:** *Hydryphantes ruber* (de Geer) (Hydryphantinae) adult female dorsum of idiosoma. **Fig. 235:** *Hydryphantes* sp. (Hydryphantinae) adult male prodorsal region. **Figures 236–237:** *Hydrodroma* sp. (Hydrodromidae) adult female. **Fig. 236** pedipalp. **Fig. 237** venter of idiosoma. **Figures 238–240:** *Chimerathyas cooki* Mitchell (Chimerathyadinae) adult female. **Fig. 238** genital field. **Fig. 239** distal segments of fourth leg. **Fig. 240** venter of idiosoma. **Fig. 241:** *Pseudohydryphantes* sp. (Pseudohydryphantinae) adult male venter of idiosoma.

13(12) Pedipalp with a relatively long spine-like dorsodistal extension of tibia (Fig. 16.41.236, arrow); idiosomal venter with coxal plates arranged as in Fig. 16.41.237 .... Hydrodromidae .... *Hydrodroma*

13' Pedipalp with a relatively short spine-like dorsal seta on tibia (Figs. 16.42.251, 43.254, 259, 45.279, 47.298, 48.313; arrows); idiosomal venter various .... Hydryphantidae 14

14(13) Some or all legs bearing swimming setae on distal segments (Fig. 16.41.239) 15

14' All legs lacking swimming setae (Figs. 16.44.268, 47.300, 48.316) 17

15(14) Idiosomal dorsum with a medial eye platelet between lateral eyes bearing a pigmented medial eye and two pairs of setae (Figs. 16.41.234, 235) .... Hydryphantinae .... *Hydryphantes*

15' Idiosomal dorsum lacking a medial eye platelet between lateral eyes 16

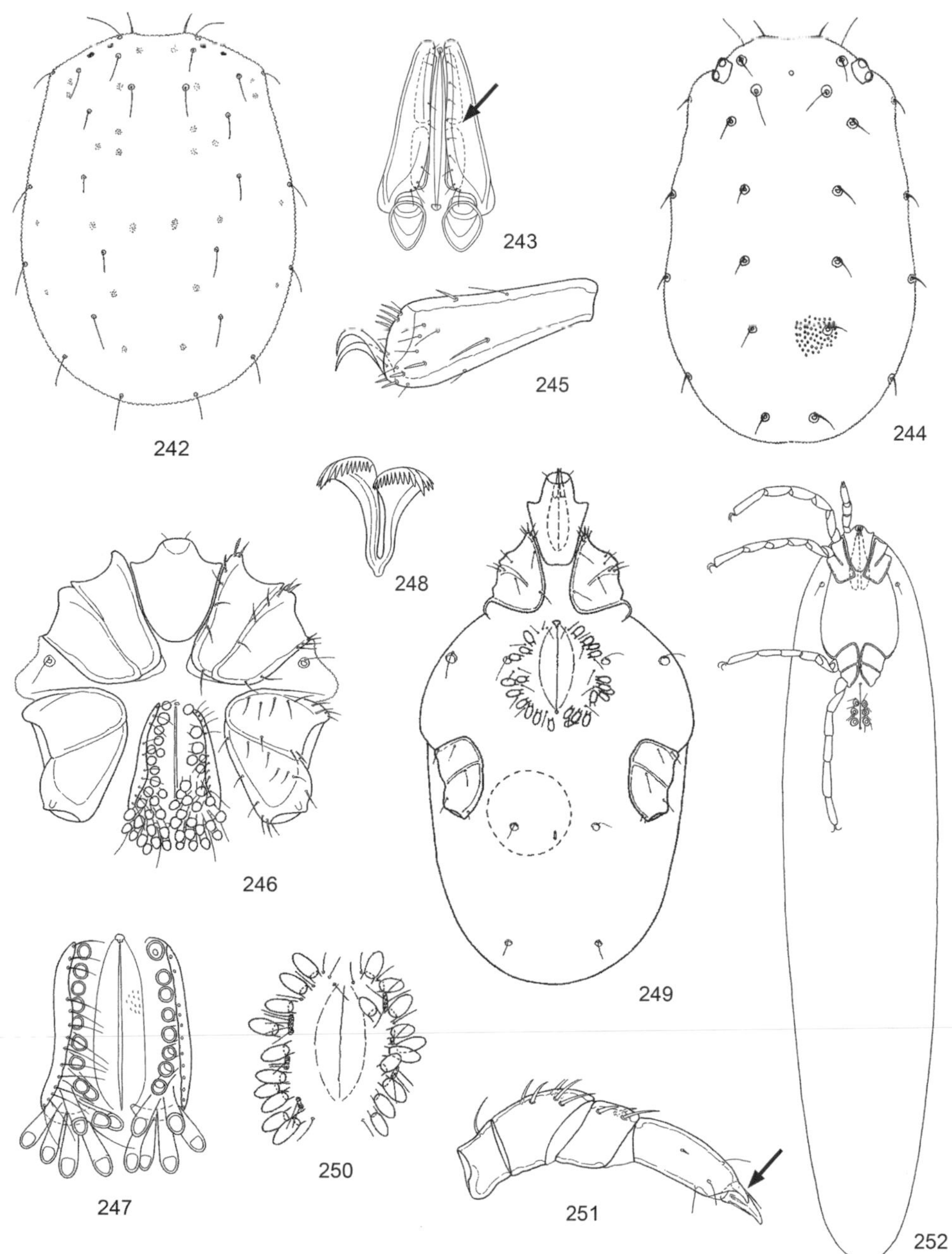

PLATE 16.42 **Fig. 242:** *Tartarothyas* sp. (Tartarothyadinae) adult female dorsum of idiosoma. **Fig. 243:** *Tartarothyas occidentalis* Smith and Cook (Tartarothyadinae) adult male genital field. **Figures 244–247:** *Partnunia steinmanni* Walter (Protziinae) adult female. **Fig. 244** dorsum of idiosoma. **Fig. 245** tarsus of first leg. **Fig. 246** venter of idiosoma. **Fig. 247** genital field. **Figures 248–250:** *Protzia* sp. (Protziinae) adults. **Fig. 248** female claws of first leg. **Fig. 249** female venter of idiosoma. **Fig. 250** male genital field. **Figures 251–252:** *Wandesia* sp. (Wandesiinae) adult female. **Fig. 251** pedipalp. **Fig. 252** venter of idiosoma.

16(15) Genital field with first and second pairs of acetabula borne in soft integument and seperated from genital flaps (Fig. 16.41.241) .... Pseudohydryphantinae .... *Pseudohydryphantes*

16' Genital field with all three pairs of acetabula borne on genital flaps (Figs. 16.41.238, 240)............ Chimerathyadinae .... *Chimerathyas*

17(14) Third and fourth coxal plates with medial margins extensive and close together with genital field located posterior to them (Fig. 16.43.255); genital flaps very large and noticeably narrower anteriorly than posteriorly (Fig. 16.43.255); idiosomal dorsum with a large dorsal shield (Fig. 16.43.253); lateral eyes borne in capsules .... Cowichaniinae .... *Cowichania*

17' Posterior coxal groups usually well separated from each other with genital field extending between them (Figs. 16.46.288, 47.301), when coxal groups close together then lateral eyes not in capsules and idiosoma soft and greatly elongated (Fig. 16.42.252) 18

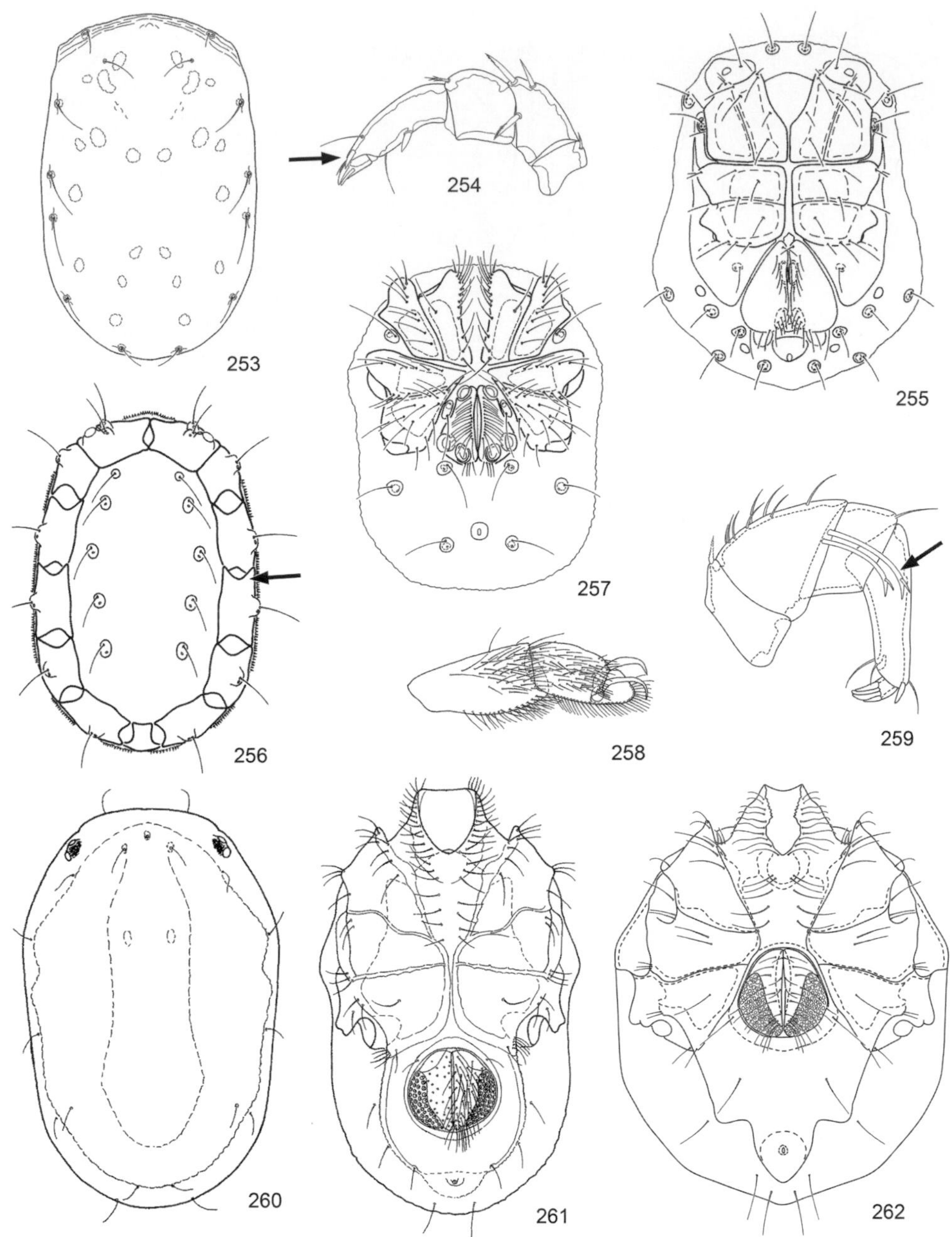

PLATE 16.43 Figures 253–255: *Cowichania interstitialis* Smith (Cowichaniinae) adults. **Fig. 253** female dorsal shield. **Fig. 254** male pedipalp. **Fig. 255** female ventral shield. **Figures 256–257:** *Cyclothyas siskiyouensis* Smith adult female. **Fig. 256** dorsum of idiosoma. **Fig. 257** venter of idiosoma. **Figures 258–262:** *Thermacarus nevadensis* Marshall adults. **Fig. 258** male tibia and tarsus of fourth leg. **Fig. 259** female pedipalp. **Fig. 260** male dorsal shield. **Fig. 261** male ventral shield. **Fig. 262** female ventral shield.

18(17) Lateral eyes present and with lenses borne in capsules that are usually raised well above surface of integument and located dorsally (Fig. 16.42.244), but rarely flattened and located at anterior edges of idiosomal dorsum (Figs. 16.48.308, arrow; 311) 19

18' Lateral eyes present or absent, when present then with lenses lying below integument and not borne in eye capsules (Fig. 16.42.242) 38

19(18) Idiosomal dorsum lacking dorsalia (Figs. 16.42.244, 44.263, 265) 20

19' Idiosomal dorsum bearing dorsalia of various sizes and arrangements (Figs. 16.43.256, 44.272, 45.278) 23

20(19) Genital field bearing three pairs of acetabula (Figs. 16.44.264, 266) .... Euthyadinae (in part) 21

20' Genital field bearing numerous stalked acetabula (Figs. 16.42.247, 250) .... Protziinae 22

21(20) Idiosomal dorsum bearing a medial eye (Fig. 16.44.265); genital field with acetabula all borne on genital flaps (Figs. 16.44.266, 267) .... *Zschokkea*

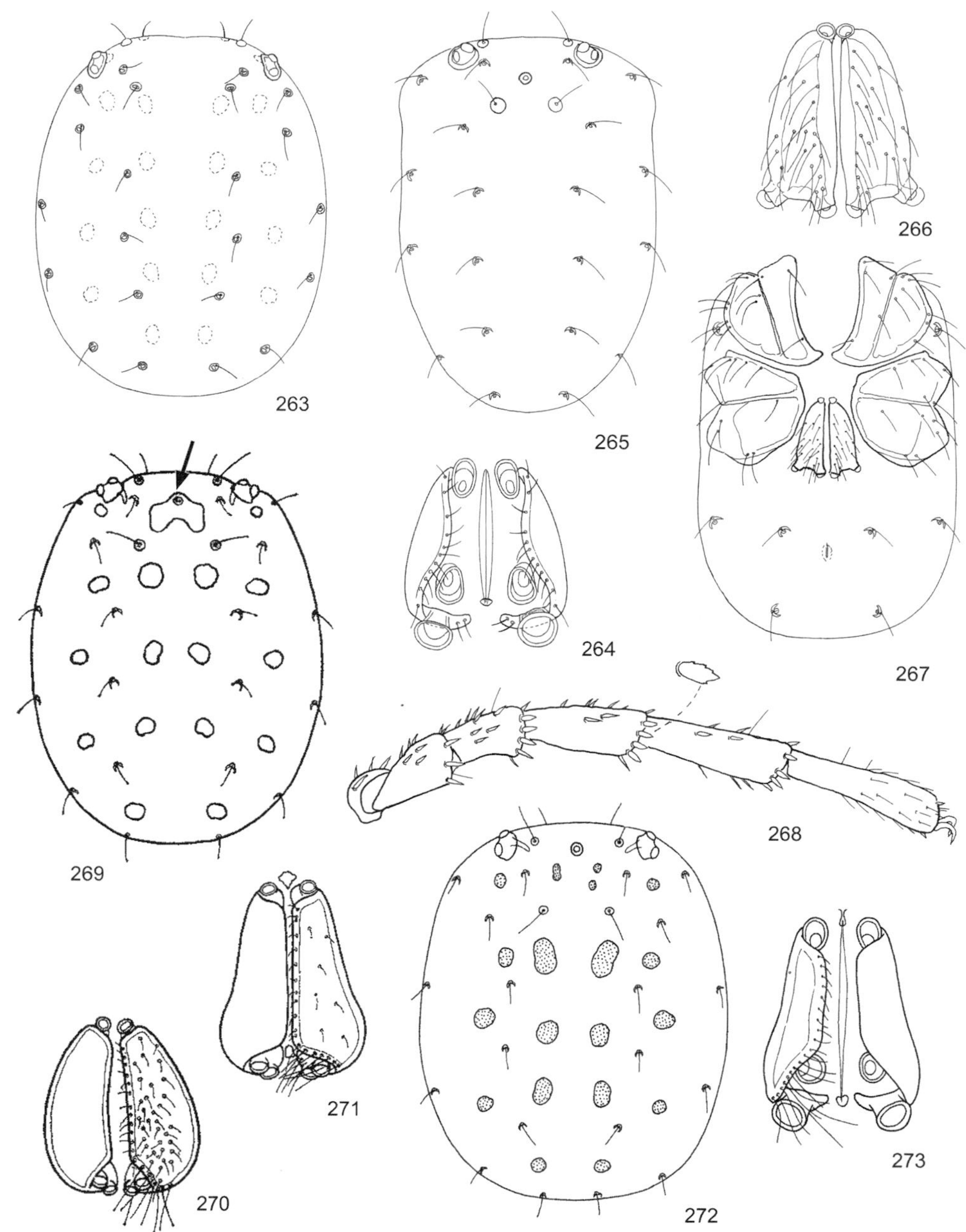

PLATE 16.44 **Figures 263–264:** *Albertathyas montanus* Smith and Cook (Euthyadinae) adult female. **Fig. 263** dorsum of idiosoma. **Fig. 264** genital field. **Figures 265–268:** *Zschokkea ontarioensis* Smith and Cook (Euthyadinae) adults. **Fig. 265** male dorsum of idiosoma. **Fig. 266** male genital field. **Fig. 267** male venter of idiosoma. **Fig. 268** female first leg. **Figures 269–271** *Thyasides sphagnorum* Habeeb (Euthyadinae) adults. **Fig. 269** male dorsum of idiosoma. **Fig. 270** male genital field. **Fig. 271** female genital field. **Fig. 272:** *Todothyas stolli* (Koenike) male dorsum of idiosoma. **Fig. 273:** *Todothyas rivalis* (Koenike) female genital field.

21' Idiosomal dorsum lacking a medial eye (Fig. 16.44.263); genital field with second pair of acetabula separated from genital flaps (Fig. 16.44.264) .... *Albertathyas*

22(20) Legs with tarsal claws expanded distally and bearing numerous terminal clawlets (Fig. 16.42.248); coxal plates relatively small and covering only a small area of idiosomal venter anterior and posterolateral to genital field (Fig. 16.42.249); genital field lacking genital flaps (Fig. 16.42.250) .... *Protzia*

22' Legs with tarsal claws simple and unmodified (Fig. 16.42.245); coxal plates relatively large and covering most of idiosomal venter anterior and lateral to genital field (Fig. 16.42.246); genital field bearing slender sclerotized flaps flanking acetabula (Fig. 16.42.247) .... *Partnunia*

23(19) Idiosomal dorsum with dorsalia forming a peripheral ring (Fig. 16.43.256, arrow); genital field with acetabula large and borne on genital flaps (Fig. 16.43.257) .... Cyclothyadinae .... *Cyclothyas*

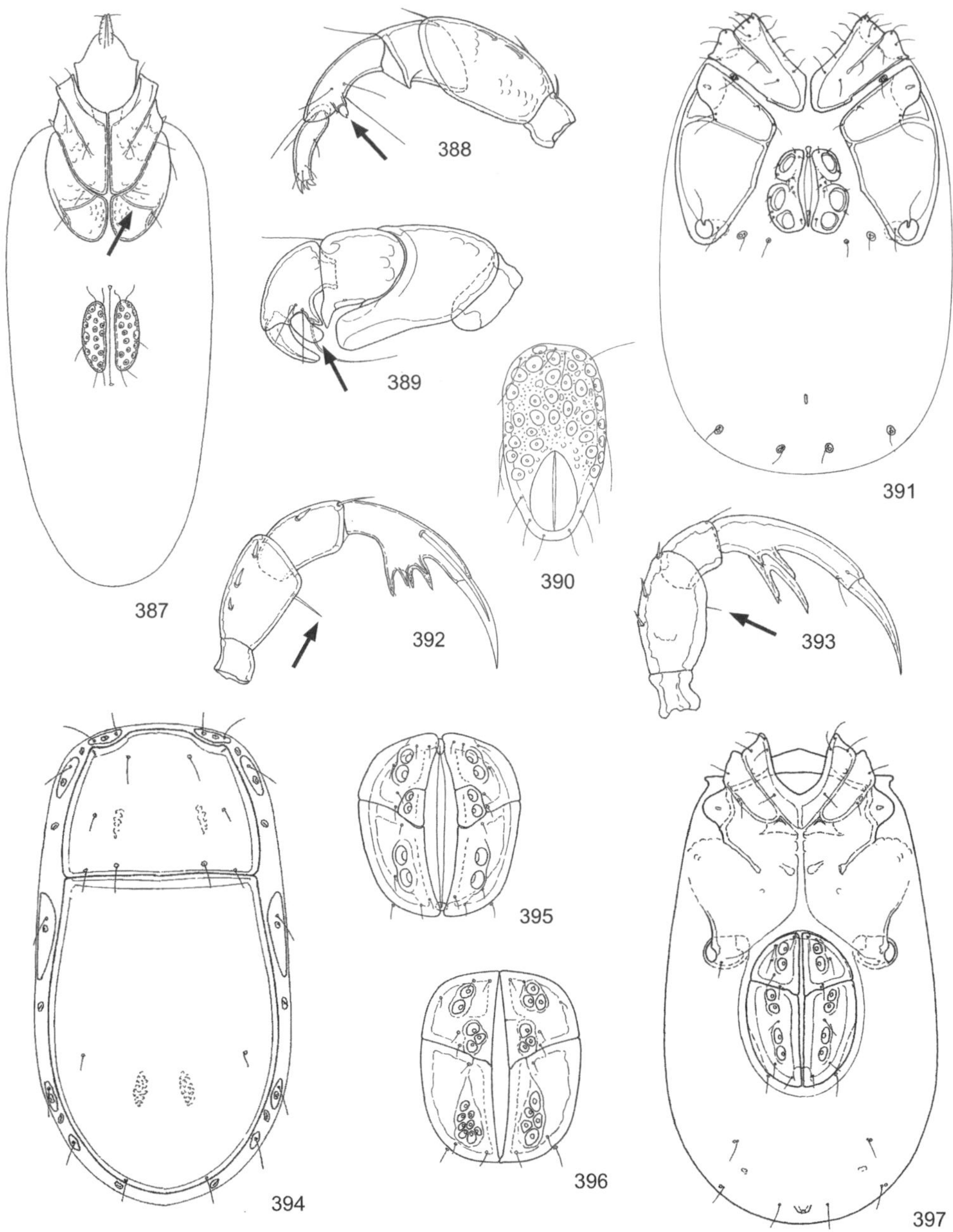

PLATE 16.55 **Figures 387–388:** *Omartacurus elongatus* Cook (Omartacaridae) adult female. **Fig. 387** venter of idiosoma. **Fig. 388** pedipalp. **Figures 389–390:** *Omartacarus brevipalpis* Cook (Omartacaridae) adult male. **Fig. 389** pedipalp. **Fig. 390** genital field. **Fig. 391:** *Kawamuracarus cramerae* Smith and Cook (Kawamuracarinae) adult male venter of idiosoma. **Fig. 392:** *Kawamuracarus gilaensis* Smith and Cook (Kawamuracarinae) adult female pedipalp. **Fig. 393:** *Neomamersa cramerae* Smith and Cook (Neomamersinae) adult male pedipalp. **Fig. 394:** *Neomamersa chihuahua* Smith and Cook (Neomamersinae) adult female dorsum of idiosoma. **Figures 395 and 397:** *Neomamersa californica* Smith and Cook (Neomamersinae) adults. **Fig. 395** male genital field. **Fig. 397:** female ventral shield. **Fig. 396:** *Neomamersa lundbladi paucipora* Cook (Neomamersinae) adult male genital field.

67(64) Genital field bearing acetabula on immovable acetabular plates that are joined anteriorly and posteriorly to completely surround gonopore (Figs. 16.57.415, 419) .... MALES 68

67' Genital field bearing acetabula on movable genital flaps (Figs. 16.37.200, 57.418, 420) .... FEMALES .... *Limnesia*, *Centrolimnesia*

68(67) Idiosomal venter with coxoglandularia II closely flanking genital field (Fig. 16.57.419, arrow) .... *Centrolimnesia*

68' Idiosomal venter with coxoglandularia II widely separated from genital field (Fig. 16.57.415, arrow) .... *Limnesia*

69(63) Idiosoma bearing dorsal and ventral shields, with dorsal shield divided into anterior and posterior platelets (Fig. 16.56.406) and ventral shield as in Fig. 16.56.407; pedipalp with femur bearing a ventral seta inserted directly on segment (Fig. 16.56.408, arrow); leg segments relatively long and slender .... Protolimnesiinae .... *Protolimnesia*

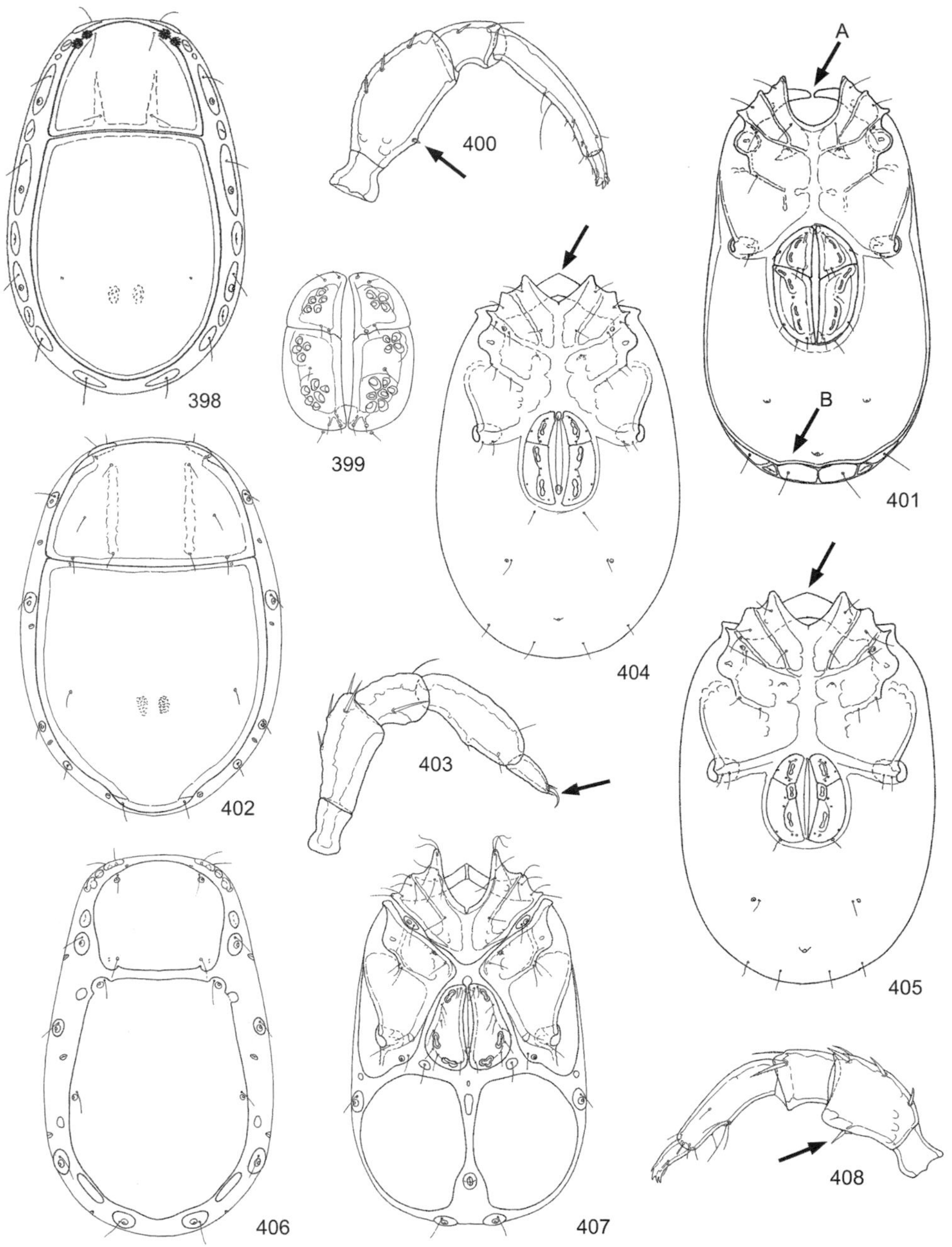

PLATE 16.56 **Fig. 398:** *Meramecia oculatus* (Cook) (Neomamersinae) adult female dorsum of idiosoma. **Figures 399–400:** *Meramecia occidentalis* Smith and Cook (Neomamersinae) adult female. **Fig. 399** genital field. **Fig. 400** pedipalp. **Fig. 401:** *Meramecia multipora* Smith and Cook (Neomamersinae) adult male ventral shield. **Figures 402–405:** *Arizonacarus chiricahuensis* Smith and Cook (Neomamersinae) adults. **Fig. 402** female dorsum of idiosoma. **Fig. 403** female pedipalp. **Fig. 404** female ventral shield. **Fig. 405** male ventral shield. **Figures 406–408:** *Protolimnesia ventriplacophora* Smith and Cook (Protolimnesiinae) adults. **Fig. 406** male dorsum of idiosoma. **Fig. 407** male venter of idiosoma. **Fig. 408** female pedipalp.

69' Idiosomal dorsum with integument exhibiting a range of sclerotization from partly covered by enlarged platelets showing various degrees of fusion (Fig. 16.57.409) to covered by an entire dorsal shield (Fig. 16.57.413); pedipalp with femur bearing a peg-like seta inserted on a prominent ventral tubercle (Fig. 16.57.410, arrow); leg segments relatively short and stout (Fig. 16.57.414) .... Tyrrelliinae 70

70(69) Third coxal plates bearing a pair of accessory glandularia (Fig. 16.57.417, arrow); genital field with acetabula small and numerous (Figs. 16.57.412, 417) .... *Neotyrrellia*

70' Third coxal plates lacking a pair of accessory glandularia (Fig. 16.57.411); genital field usually with three pairs of large acetabula (Fig. 16.57.411) but occasionally with as many as seven pairs .... *Tyrrellia*

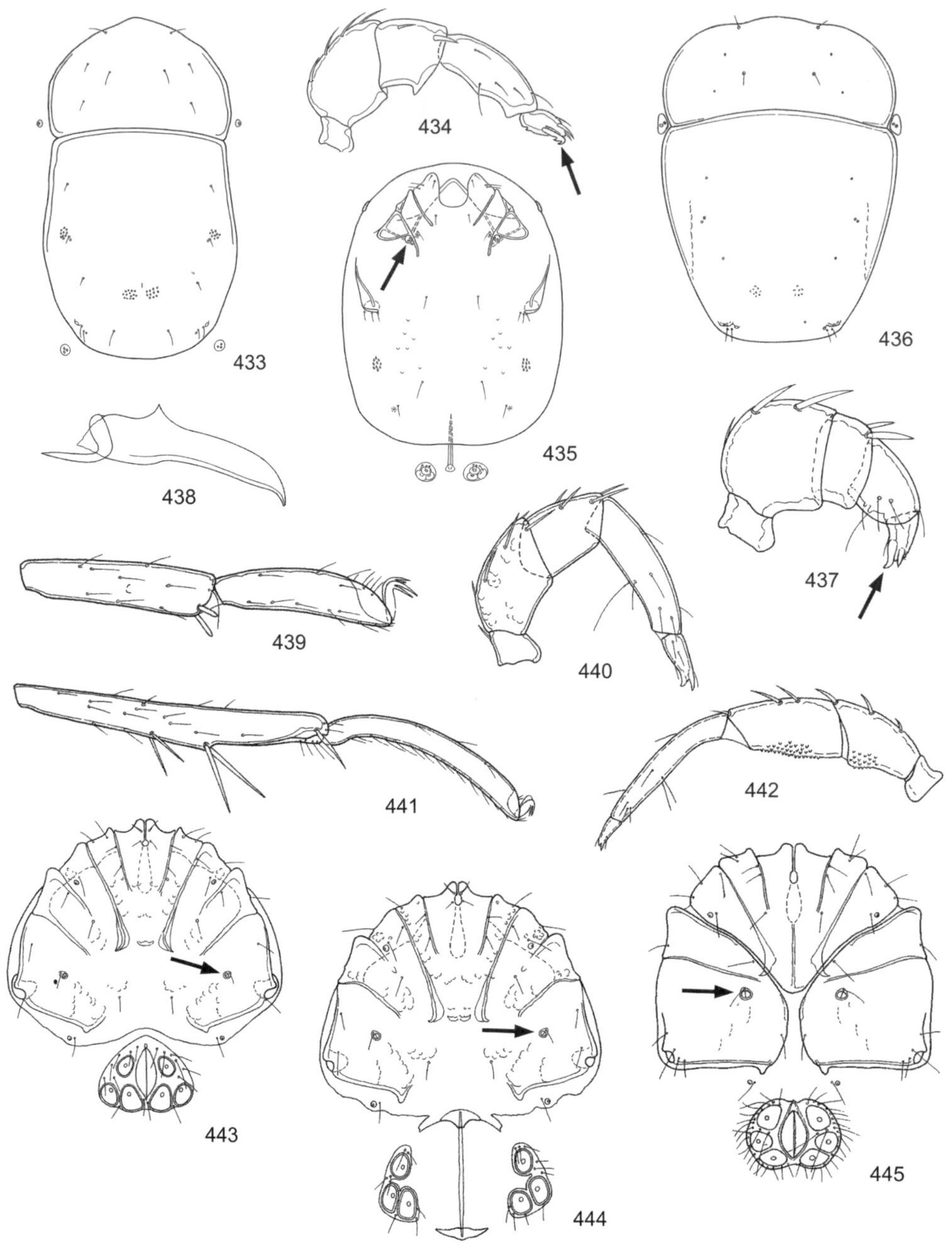

PLATE 16.59 **Figures 433–435:** *Diamphidaxona cavecreekensis* Smith and Cook (Hygrobatidae) adult female. **Fig. 433** dorsum of idiosoma. **Fig. 434** pedipalp. **Fig. 435** venter of idiosoma. **Fig. 436:** *Diamphidaxona imamurai* Cook (Hygrobatidae) adult male dorsum of idiosoma. **Fig. 437:** *Diamphidaxona neomexicana* Smith and Cook (Hygrobatidae) adult female pedipalp. **Fig. 438:** *Hygrobates* sp. (Hygrobatidae) adult chelicera. **Figures 439–440 and 443–444:** *Hygrobates inermis* Cook (Hygrobatidae) adults. **Fig. 439** female distal segments of first leg. **Fig. 440** female pedipalp. **Fig. 443** male venter of idiosoma. **Fig. 444** female venter of idiosoma. **Figures 441–442 and 445:** *Mesobates forcipatus* Thor (Hygrobatidae) adult male. **Fig. 441** distal segments of first leg. **Fig. 442** pedipalp. **Fig. 445** venter of idiosoma.

74(73) Idiosomal venter with suture lines between third and fourth coxal plates complete and ending at medial margins of fourth coxal plates (Figs. 16.67.533, 542); medial margins of fourth coxal plates reduced to, or nearly to, medial angles (Figs. 16.67.533, 542; arrows); fourth leg with tarsus bearing sickle-shaped claws (Figs. 16.67.537, 543); genital field with acetabular plates bearing more than three, and usually numerous, acetabula (Figs. 16.67.533, 542) 75

74' Idiosomal venter with suture lines between third and fourth coxal plates incomplete (Figs. 16.66.524, 529); medial margins of fourth coxal plates extensive (Fig. 16.66.524) or reduced to medial angles (Fig. 16.66.529); fourth leg with tarsus bearing modified claws (Figs. 16.66.525, 531, 532); genital field with acetabular plates usually bearing three pairs of acetabula (Fig. 16.66.524) but occasionally bearing up to seven pairs (Fig. 16.66.529) 76

75(74) Idiosoma with dorsal and ventral shields; third leg with tarsus similar to those of first and second legs, with claw socket large and dorsal in position and with claws unmodified (Figs. 16.67.535, 538); ventral shield as in Fig. 16.67.533; pedipalp as in Fig. 16.67.534; leg segments lacking swimming setae .... *Pseudofeltria*

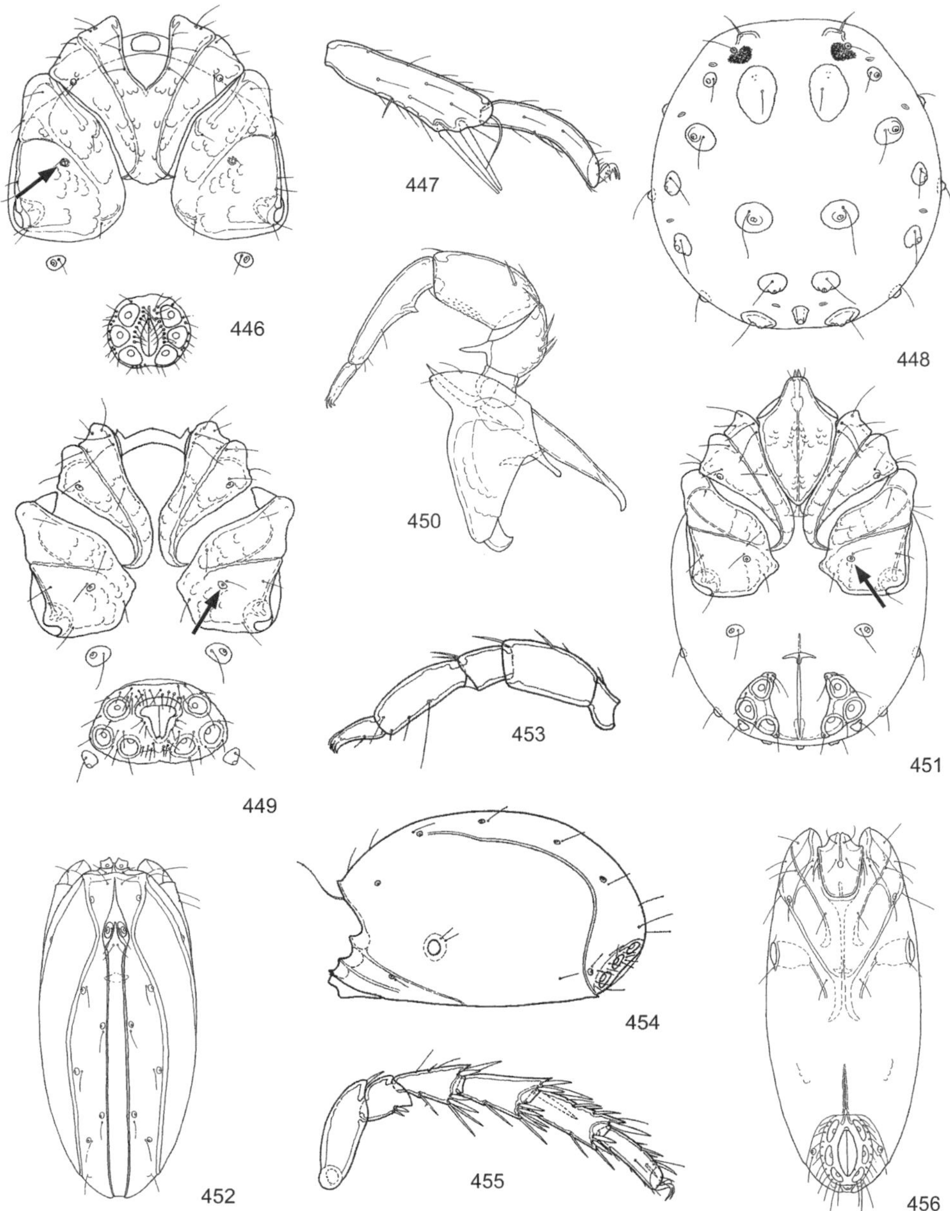

PLATE 16.60 **Figures 446–447:** *Atractides* spp. (Hygrobatidae) adult males. **Fig. 446** venter of idiosoma. **Fig. 447** distal segments of first leg. **Figures 448–451:** *Corticacarus delicatus* Habeeb (Hygrobatidae) adults. **Fig. 448** female dorsum of idiosoma. **Fig. 449** male venter of idiosoma. **Fig. 450** male gnathosoma and pedipalp. **Fig. 451** female venter of idiosoma. **Figures 452 and 456:** *Frontipodopsis* sp. (Frontipodopsidae) adult male. **Fig. 452** dorsum of idiosoma. **Fig. 456** venter of idiosoma. **Figures 453–455:** *Frontipodopsis nearctica* Cook (Frontipodopsidae) adult male. **Fig. 453** pedipalp. **Fig. 454** lateral view of idiosoma. **Fig. 455** fourth leg.

75' Idiosoma with or without dorsal and ventral shields; third leg with tarsus not similar to those of first and second legs, with claw socket small and terminal in position and with claws modified (Fig. 16.67.540 cf. 67.541); idiosomal venter as in Fig. 16.67.542; pedipalp as in Fig. 16.67.539; distal segments of third and fourth legs bearing swimming setae .... *Forelia*

76(74) Genital field with acetabular plates bearing three pairs of acetabula (Fig. 16.66.524); medial margins of fourth coxal plates extensive (Fig. 16.66.524); fourth leg with genu lacking a projection that extends well beyond insertion of tibia, tibia bearing peg-like setae both proximal to and distal to deepest part of curved portion, and genu and tibia both bearing swimming setae (Fig. 16.66.525); pedipalp as in Fig. 16.66.527 .... *Pionacercus*

76' Genital field with acetabular plates bearing five to seven pairs of acetabula; medial margins of fourth coxal plates relatively short (Fig. 16.66.529); fourth leg with genu bearing a large projection extending well beyond insertion of tibia, tibia bearing peg-like setae only proximal to deepest part of curved portion, and only tibia bearing swimming setae (Figs. 16.66.531, 532); pedipalp as in Fig. 16.66.528 .... *Madawaska*

77(73) Third leg with tarsus bearing modified claws as in Figs. 16.68.546, 69.561, 564; fourth leg with genu notched and bearing short, peg-like setae on both sides of, and occasionally in, notch (Figs. 16.68.552, 554, 69.563) .... Hygrobatidae (in part).......... Pionidae (in part) .... Pioninae (in part) .... MALES 78

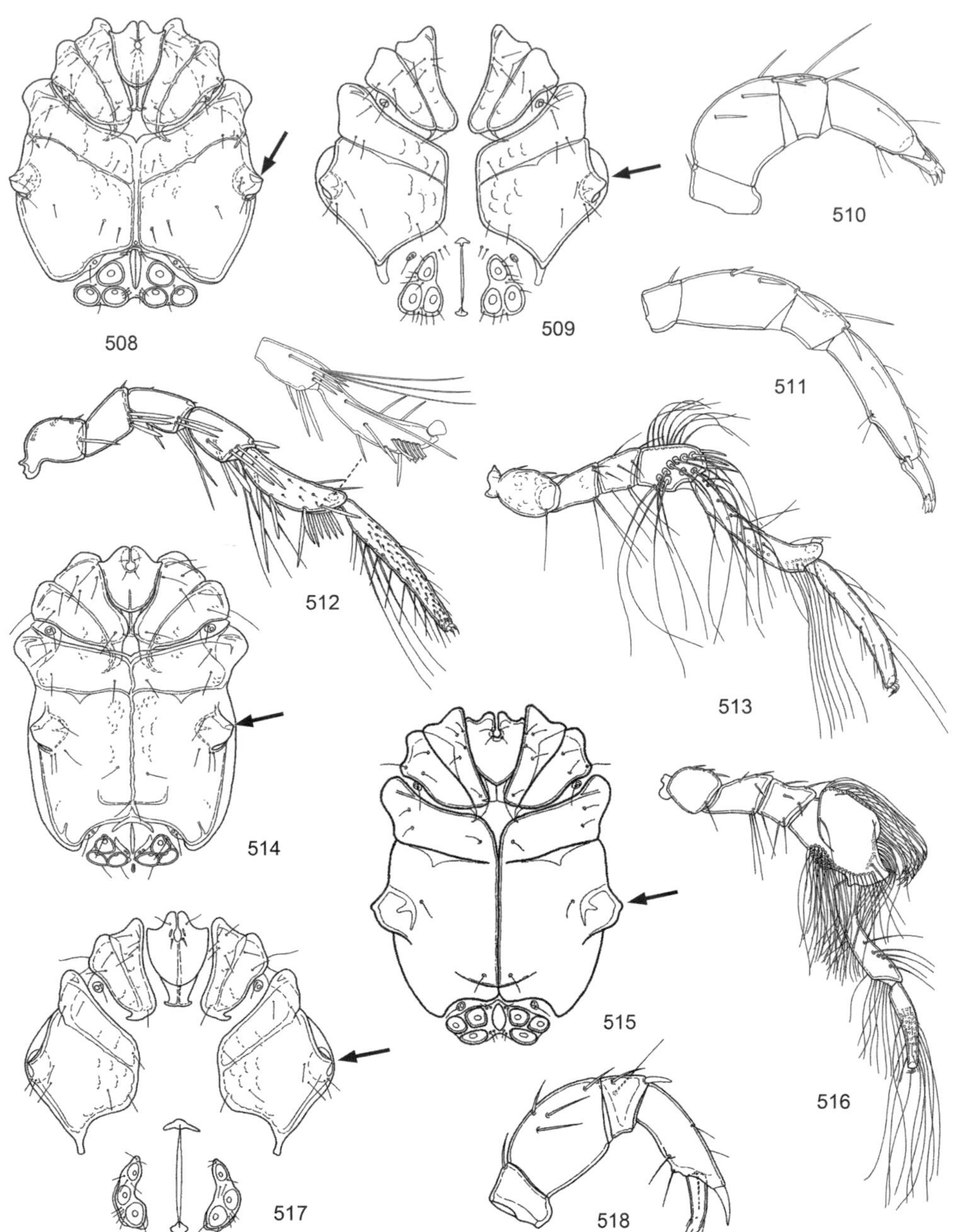

PLATE 16.65 **Figures 508–509, 511–512:** *Pionopsis paludis* Habeeb (Tiphyinae) adults. **Fig. 508** male venter of idiosoma. **Fig. 509** female venter of idiosoma. **Fig. 511** male pedipalp. **Fig. 512** male fourth leg showing both anterior and posterior views of genu and tibia. **Fig. 510:** *Tiphys diversus* (Marshall) male pedipalp. **Figures 513–514:** *Tiphys gracilipes* Cook (Tiphyinae) adult male. **Fig. 513** fourth leg. **Fig. 514** venter of idiosoma. **Figures 515 & 517:** *Tiphys americanus* (Marshall) (Tiphyinae) adults. **Fig. 515** male venter of idiosoma. **Fig. 517** female venter of idiosoma. **Figures 516 & 518:** *Tiphys weaveri* Cook (Tiphyinae) adults. **Fig. 516** male fourth leg. **Fig. 518** female pedipalp.

85(82) Dorsal shield transversely divided into anterior and posterior plates (Figs. 16.59.433, 436); ventral shield with suture lines between third and fourth coxal plates looped forward around a pair of glandularia (Fig. 16.59.435, arrow); pedipalp usually as in Fig. 16.59.434, rarely with segments relatively short and stocky (Fig. 16.59.437); pedipalp tarsus bearing a thick, usually serrated seta distoventrally (Figs. 16.59.434, 437; arrows) .... Hygrobatoidea (in part) .... Hygrobatidae (in part) .... *Diamphidaxona*

85' Idiosoma with dorsal shield usually entire, never transversely divided as above; venter with suture lines between third and fourth coxal plates very rarely looped forward around a pair of glandularia; pedipalp various 86

86(85) Outer edges of anterior coxal plates forming a smooth arc that does not extend beyond anterior edge of idiosoma (Figs. 16.82.704, 706; arrows); genital field with three pairs of acetabula borne in gonopore in males (Fig. 16.82.704) and on edge of ventral shield flanking gonopore in females (Fig. 16.82.706); pedipalp slightly modified (Fig. 16.82.708) to highly modified (Fig. 16.82.702), with femur expanded and bearing a row of long, thick setae ventrally that may be blade-like or long and serrate, genu very short, tibia slightly to greatly expanded and flattened, and bearing a thick seta near distal edge (Figs. 16.82.702, 708; arrows) that may be peg-like, bifurcate or serrate, and tarsus relatively long, slender and curved; dorsal shield bearing four pairs of glandularia (Fig. 16.82.703) .... Arrenuroidea (in part) .... Chappuisididae (in part) .... Uchidastygacarinae .... *Uchidastygacarus*

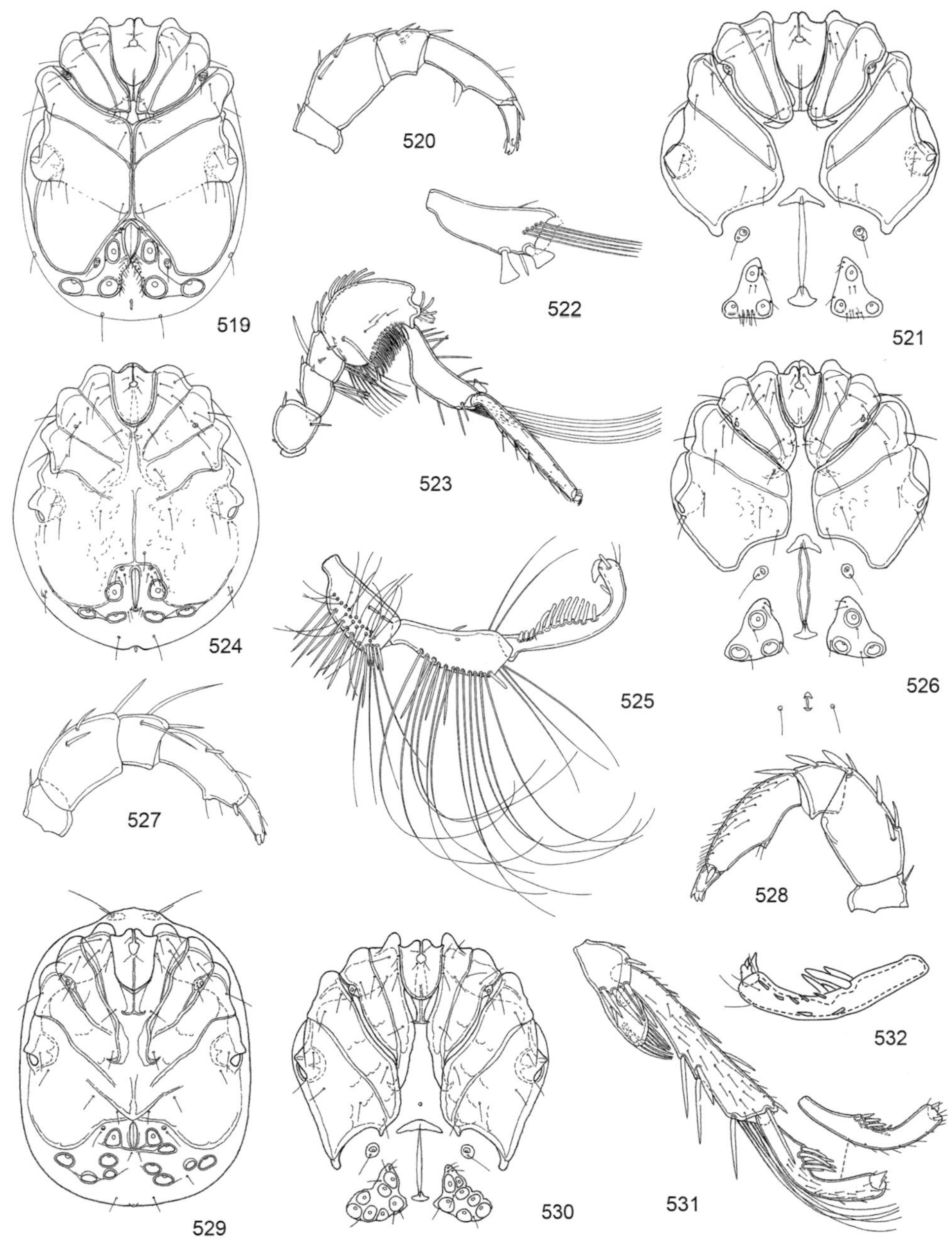

PLATE 16.66 **Figures 519–523**: *Acercopsis vernalis* Habeeb (Tiphyinae) adults. **Fig. 519** male venter of idiosoma. **Fig. 520** female pedipalp. **Fig. 521** female venter of idiosoma. **Fig. 522** male distal segments of second leg. **Fig. 523** male fourth leg. **Figures 524–527:** *Pionacercus* sp. (Foreliinae) adults. **Fig. 524** male venter of idiosoma. **Fig. 525** male distal segments of fourth leg. **Fig. 526** female venter of idiosoma. **Fig. 527** female pedipalp. **Figures 528–532** *Madawaska borealis* Habeeb (Foreliinae) adults. **Fig. 528** male pedipalp. **Fig. 529** male venter of idiosoma. **Fig. 530** female venter of idiosoma. **Fig. 531** male distal segments of fourth leg. **Fig. 532** male tarsus of fourth leg.

86' Outer edges of anterior coxal plates not forming a smooth arc; pedipalp not modified as described and illustrated above 87

87(86) Idiosomal venter with a pair of glandularia (coxoglandularia 2) located near midline at junction of third and fourth coxal plates (Figs. 16.82.699, 700; arrows); genital field with three or four pairs of acetabula borne freely in a row near midline in gonopore in males (Fig. 16.82.699) and in an arc near lateral edges of gonopore, either freely or on small acetabular plates, in females (Fig. 16.82.700); pedipalp (Fig. 16.82.701) with femur bearing two long, blade-like setae ventrally and tibia expanded, flattened, and with a prominent rounded flange ventrally; dorsal shield bearing four pairs of glandularia (Fig. 16.82.698) .... Arrenuroidea (in part) .... Chappuisididae (in part) .... Chappuisidinae .... *Chappuisides*

87' Idiosomal venter with coxoglandularia 2 not in position described and illustrated above; pedipalp not modified as described and illustrated above 88

88(87) Idiosomal venter with suture lines between third and fourth coxal plates oblique, extending posteromedially to genital field region and well separated from each other medially (Figs. 16.83.718–720, 723); genital field with three pairs of genital acetabula borne freely in gonopore in males (Figs. 16.83.718, 720) and with five to nine pairs of acetabula arranged in a single row on slender acetabular plates flanking

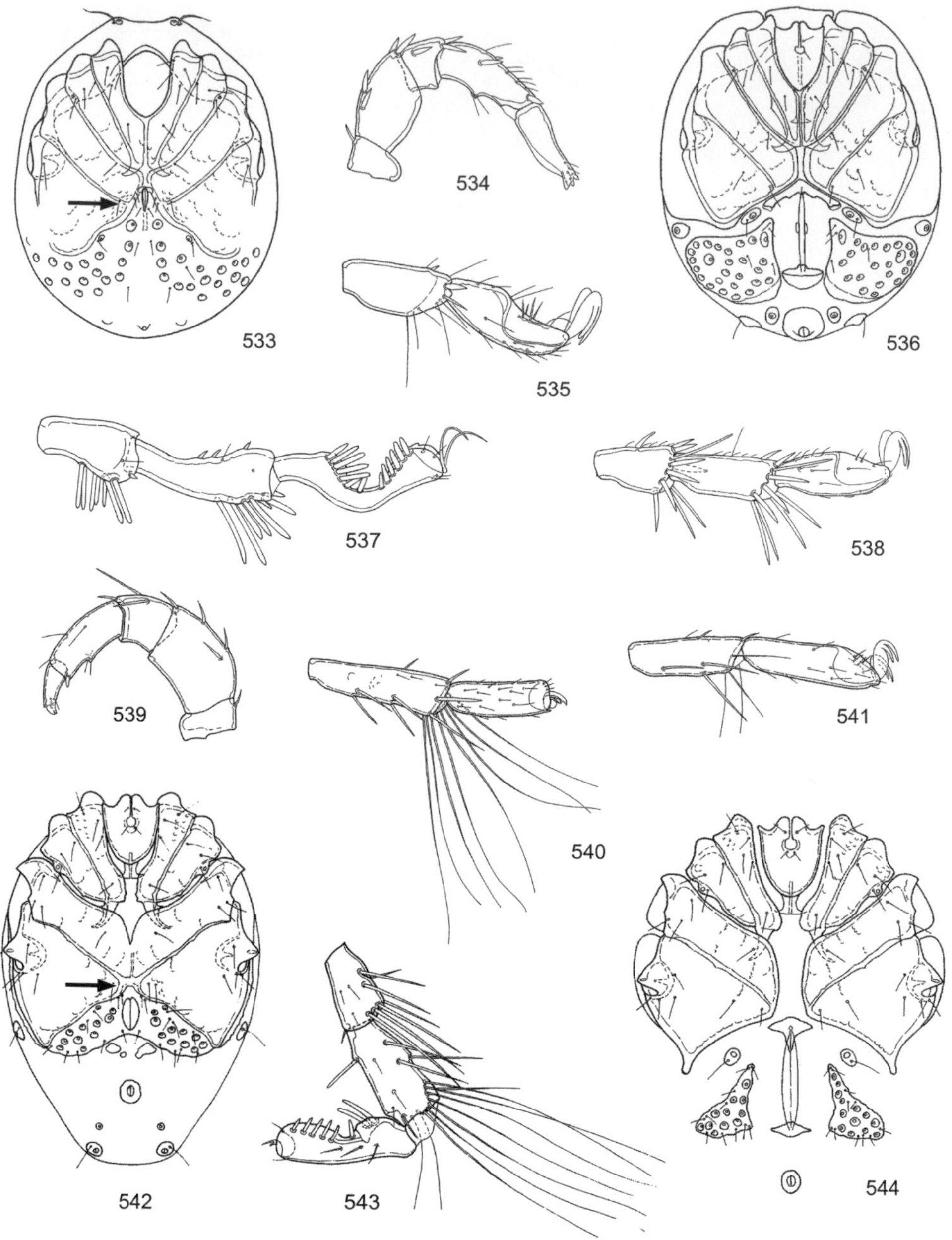

PLATE 16.67 Figures 533–538: *Pseudofeltria laversi* Cook (Foreliinae) adults. **Fig. 533** male venter of idiosoma. **Fig. 534** male pedipalp. **Fig. 535** female distal segments of first leg. **Fig. 536** female venter of idiosoma. **Fig. 537** male distal segments of fourth leg. **Fig. 538** male distal segments of third leg. **Figures 539–544:** *Forelia floridensis* Cook (Foreliinae) adults. **Fig. 539** female pedipalp. **Fig. 540** male distal segments of third leg. **Fig. 541** female distal segments of first leg. **Fig. 542** male venter of idiosoma. **Fig. 543** male distal segments of fourth leg. **Fig. 544** female venter of idiosoma.

gonopore in females (Figs. 16.83.719, 723); dorsal shield bearing three pairs of glandularia (Fig. 16.83.717) .... Arrenuroidea (in part) .... Neoacaridae 89

88' Idiosomal venter not with above combination of characters 90

89(88) Pedipalp with tibia bearing a more or less prominent distoventral projection that opposes tarsus (Fig. 16.83.716); third leg with tibia moderately to greatly expanded and flattened and bearing rows of both slender and modified setae in males (Fig. 16.83.715) .... *Neoacarus*

89' Pedipalp with tibia and tarsus highly modified to form a chelate appendage (Fig. 16.83.722); third leg with tibia slightly expanded and bearing rows of slender setae in males (Fig. 16.83.721) .... *Volsellacarus*

90(88) Idiosomal venter with coxoglandularia 1 shifted far forward on second coxal plates to a position near suture lines between first and second coxal plates (Figs. 16.80.686, 81.687, 690, 692, 695; arrows); genital field with acetabula borne in gonopore (Figs. 16.81.687, 688, 690, 692, 695, 696), on slender plates at lateral edges of gonopore (Fig. 16.80.686), on edge of ventral shield flanking gonopore (Figs. 16.81.691, 697), or on highly modified plates protruding ventrally from genital field (Fig. 16.80.684) 91

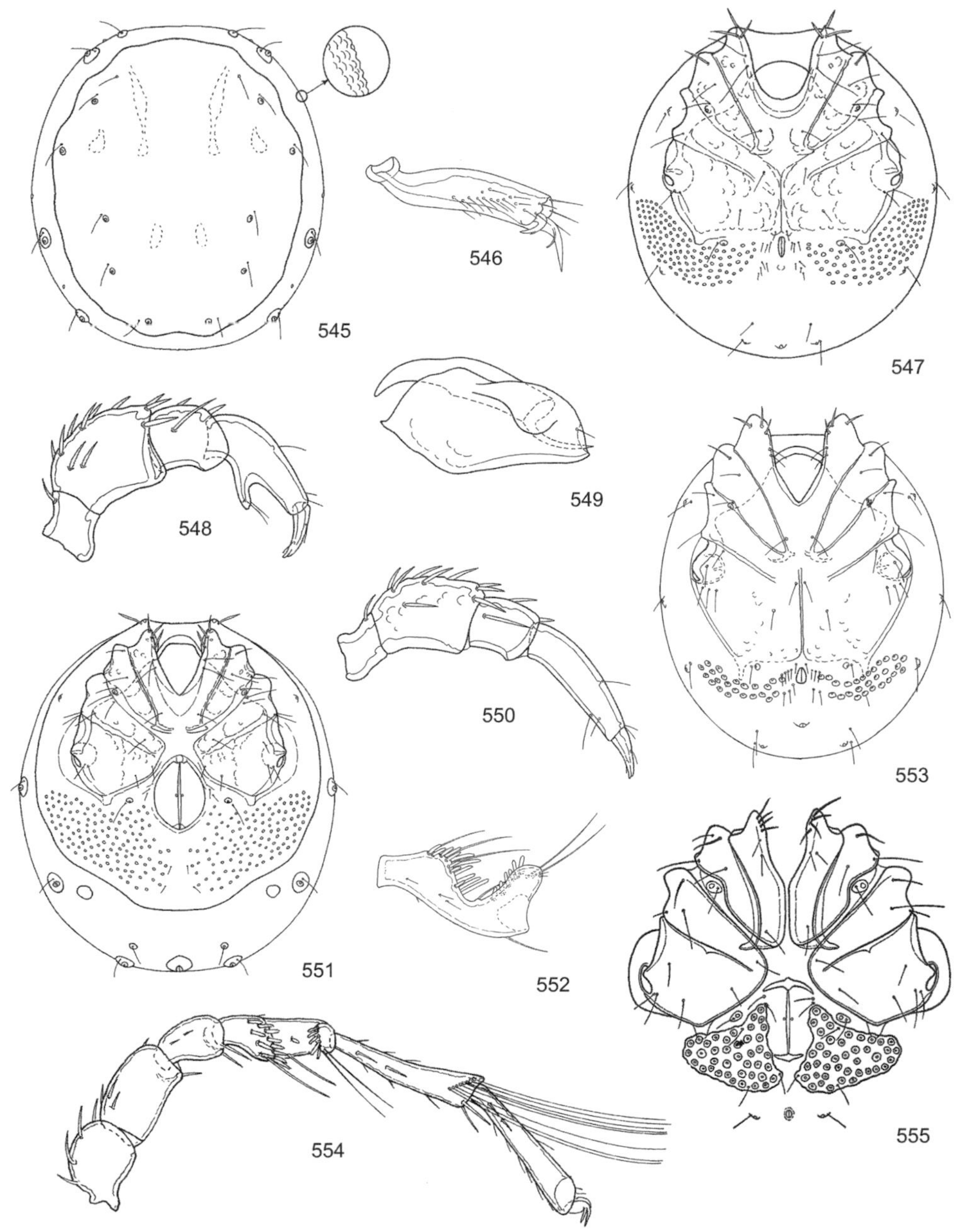

PLATE 16.68 Figures 545–551: *Nautarachna queticoensis* Smith (Pioninae) adults. **Fig. 545** male dorsum of idiosoma. **Fig. 546** male tarsus of third leg. **Fig. 547** male venter of idiosoma. **Fig. 548** male pedipalp. **Fig. 549** female gnathosomal base (capitulum). **Fig. 550** female pedipalp. **Fig. 551** female venter of idiosoma. **Figures 552–554:** *Nautarachna janae* (Habeeb) (Pioninae) adults. **Fig. 552** male genu of fourth leg. **Fig. 553** male venter of idiosoma. **Fig. 554** male fourth leg. **Fig. 555:** *Nautarachna pioniformis* Cook (Pioninae) female venter of idiosoma.

90' Idiosomal venter with coxoglandularia 1 usually located between second and third coxal plates (Fig. 16.89.784, arrow D), when shifted slightly onto posterior edges of second coxal plates (Fig. 16.84.727, arrow) then genital acetabula very numerous and located on wing-like acetabular plates flanking gonopore (Figs. 16.84.724, 727) 95

91(90) Genital field extending well forward and widely separating fourth coxal plates (Fig. 16.80.686), highly modified in males (Fig. 16.80.684) and with numerous acetabula borne on slender plates flanking gonopore in females (Fig. 16.80.686); third leg with tarsus highly modified in males (Fig. 16.80.683); pedipalp as in Fig. 16.80.685 .... Arrenuroidea (in part) .... Mideidae .... *Midea*

91' Genital field posterior in position and not separating fourth coxal plates (Figs. 16.81.687, 692, 695, 697), with three pairs to many acetabula borne freely in gonopore in males (Figs. 16.81.687, 690, 692, 695) and either freely in gonopore (Figs. 16.81.688, 696) or on edges of ventral shield flanking gonopore (Figs. 16.81.691, 697) in females; dorsal shield bearing three pairs of glandularia (Fig. 16.81.689); pedipalp with tibia bearing two slender setae proximoventrally (Figs. 16.81.693, 694) .... Arrenuroidea (in part) .... Nudomideopsidae 92

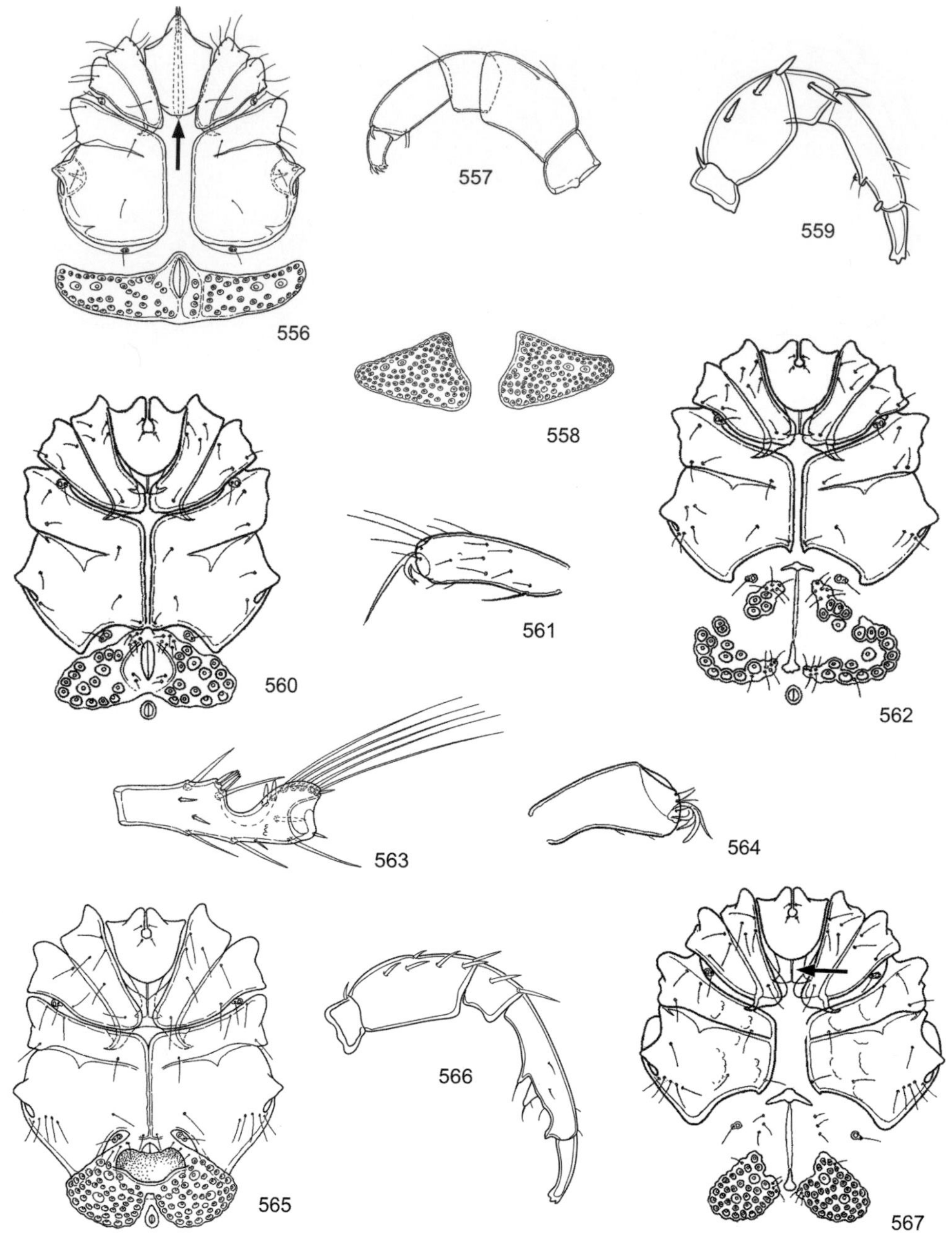

PLATE 16.69 Figures 556–558: *Najadicola ingens* Koenike (Najadicolinae) adults. **Fig. 556** male venter of idiosoma. **Fig. 557** female pedipalp. **Fig. 558** female acetabular plates. **Figures 559–562:** *Piona inconstans* (Wolcott) (Pioninae) adults. **Fig. 559** male pedipalp. **Fig. 560** male venter of idiosoma. **Fig. 561** male tarsus of third leg. **Fig. 562** female venter of idiosoma. **Fig. 563:** *Piona lapointei* Smith and Cook (Pioninae) adult male genu of fourth leg. **Figures 564–567:** *Piona interrupta* Marshall (Pioninae) adults. **Fig. 564** male tarsus of third leg. **Fig. 565** male venter of idiosoma. **Fig. 566** female pedipalp. **Fig. 567** female venter of idiosoma.

92(91)	Genital field bearing three pairs of acetabula	93
92'	Genital field bearing more than three pairs of acetabula	94
93(92)	Idiosoma round or oval in outline; ventral shield with coxal plates separated medially by a distinct suture line (Figs. 16.81.687, 688); genital field with acetabula borne freely and close together in narrow gonopore in males (Fig. 16.81.687) and at edges of wide gonopore in females (Fig. 16.81.688)	*Nudomideopsis*
93'	Idiosoma elliptical in outline; ventral shield with coxal plates fused medially with suture lines obliterated (Figs. 16.81.690, 691); genital field with acetabula borne freely and close together in narrow gonopore in males (Fig. 16.81.690) and on edges of ventral shield flanking wide gonopore in females (Fig. 16.81.691)	*Allomideopsis*

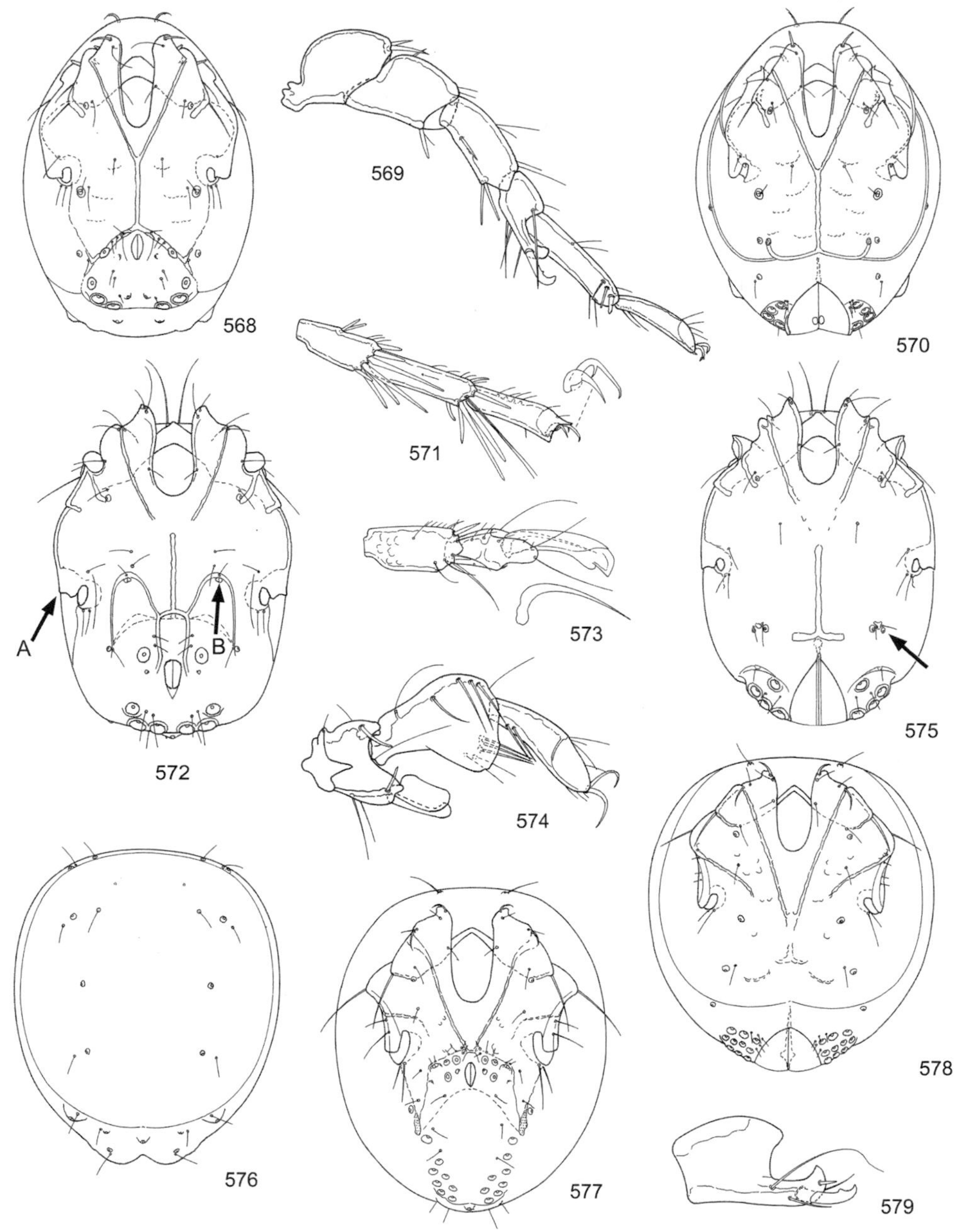

PLATE 16.70 Figures 568–571: *Axonopsella bakeri* Smith and Cook (Axonopsinae) adults. **Fig. 568** male ventral shield. **Fig. 569** male fourth leg. **Fig. 570** female ventral shield. **Fig. 571** male distal segments of second leg. **Figures 572 & 575:** *Stygalbiella yavapai* Smith and Cook (Axonopsinae) adults. **Fig. 572** male ventral shield. **Fig. 575** female ventral shield. **Fig. 573:** *Stygalbiella arizonica* Cook (Axonopsinae) adult male distal segments of second leg showing detail of unmodified claw. **Fig. 574:** *Stygalbiella affinis* Cook (Axonopsinae) adult male distal segments of fourth leg. **Figures 576–579:** *Submiraxona gilana* (Habeeb) (Axonopsinae) adults. **Fig. 576** female dorsal shield. **Fig. 577** male ventral shield. **Fig. 578** female ventral shield. **Fig. 579** male genu of fourth leg.

94(92) Suture lines between third and fourth coxal plates ending medially at midline (Figs. 16.81.692, 696); genital field with four to six pairs of acetabula borne freely and in single rows that are close together in narrow gonopore in males (Fig. 16.81.692) and at edges of wide gonopore in females (Fig. 16.81.696) .... *Paramideopsis*

94' Suture lines between third and fourth coxal plates ending medially at posterior edges of posterior coxal groups, widely separated from midline (Figs. 16.81.695, 697); genital field with numerous acetabula borne in several rows that are close together in narrow gonopore in males (Fig. 16.81.695) and on narrow plates that are fused with edges of ventral shield flanking wide gonopore in females (Fig. 16.81.697) .... *Neomideopsis*

95(90) Pedipalp distinctly uncate, with distal edge of tibia much longer than proximal edge of dorsally inserted tarsus (Figs. 16.86.748, arrow A; also 91.798, 803, 92.811, 812, 93.822, 823, 94.830, 834, 95.840, 841, 844; appearing as in Figs. 16.86.750, 87.756, 764 in genera with tibia rotated 90° relative to genu) 96

95' Pedipalp not distinctly uncate, with distal edge of tibia only slightly longer than proximal edge of tarsus (Figs. 16.84.725, 729, 85.736, 742); tibia with or without a ventral projection near mid-length of segment as in Fig. 16.85.742 102

96(95) Fourth coxal plates bearing a pair of glandularia (Figs. 16.91.799, 800, 804, 805; arrows A); genital field extending between fourth coxal plates and widely separating them, bearing three to six pairs of acetabula borne freely in gonopore in both sexes and only slightly sexually dimorphic (Figs. 16.91.799, 800, 804, 805); dorsal shield as in Figs. 16.91.797, 802 .... Arrenuroidea (in part) ....Krendowskiidae 97

96' Fourth coxal plates lacking glandularia; genital field extending at most only slightly between fourth coxal plates and not separating them medially, bearing numerous acetabula and relatively strongly sexually dimorphic 98

97(96) Gnathosoma elongate and protrusible (Fig. 16.91.806); sclerotized bridge forming dorsal side of camerostome relatively wide (Figs. 16.91.804, 805; arrows B); dorsal shield relatively small (Fig. 16.91.802); pedipalp with trochanter short and with dorsal edge subequal in length to, or slightly longer than, ventral edge, femur with two long medial setae located near mid-length of segment, and tarsus with distoventral projection relatively short (Figs. 16.91.803, 806); ventral shield as in Figs. 16.91.804, 805 .... *Geayia*

97' Gnathosoma relatively short and not protrusible (Fig. 16.91.801); sclerotized bridge forming dorsal side of camerostome relatively narrow (Figs. 16.91.799, 800; arrows B); dorsal shield relatively large (Fig. 16.91.797); pedipalp with trochanter relatively long and with dorsal edge noticeably longer than ventral edge, femur with two long medial setae located proximally on segment, and tarsus with distoventral projection relatively long (Figs. 16.91.798, 801); ventral shield as in Figs. 16.91.799, 800 ....*Krendowskia*

98(96) Gnathosomal base (capitulum) bearing a pair of long setae distoventrally (Fig. 16.86.750); pedipalp with tibia rotated up to 90 degrees relative to genu; genital field with acetabula borne freely in gonopore (Figs. 16.86.745, 751–753, 87.762, 763) or both freely in gonopore and on acetabular plates at edges of relatively narrow gonopore (Fig. 16.87.757) in males, and on acetabular plates at edges of relatively wide gonopore in females (Figs. 16.86.747, 754, 87.759); acetabular plates not fused with ventral shield; dorsal shield as in Figs. 16.86.746, 749, 87.755, 761; leg segments lacking swimming setae (Fig. 16.87.760) .... Arrenuroidea (in part).... Athienemanniidae 99

98' Gnathosoma lacking a pair of long setae distoventrally; pedipalp with tibia not rotated relative to genu; genital field with acetabula borne on acetabular plates that are incorporated into ventral shield and extend laterally from edges of gonopore, never borne freely in gonopore (Figs. 16.92.808, 809, 813, 815, 816, 93.819, 820, 824, 827, 94.829, 838, 95.843, 846); idiosoma highly modified to form a more or less complex and prominent cauda in males (Figs. 16.92.807, 810, 814, 817, 93.818, 821, 824, 825, 94.828, 831–833, 835–837, 95.839, 842, 845, 847, 848), and relatively unmodified in females (Figs. 16.92.809, 813, 816, 93.820, 826, 827, 94.829, 838, 95.843, 846); distal segments of at least fourth legs usually bearing swimming setae .... Arrenuroidea (in part) .... Arrenuridae ....Arrenurinae .... *Arrenurus*

99(98) Pedipalp with genu bearing a pronounced ventral projection (Fig. 16.86.748; arrow B); genital field with four to seven anterior pairs of acetabula borne freely in single rows followed by paired posterior groups of five to seven acetabula borne on small plates in narrow gonopore in males (Figs. 16.86.745, 752), and with eight to ten pairs of acetabula borne in an irregular row on narrow acetabular plates at lateral edges of large gonopore in females (Fig. 16.86.747) .... Stygameracarinae.... *Stygameracarus*

99' Pedipalp with genu not bearing pronounced ventral projection (Figs. 16.86.750, 87.756, 764); genital field with numerous acetabula borne in one to several rows (Figs. 16.86.751, 753, 754, 87.757–759, 762, 763) .... Athienemanniinae....100

100(99) Idiosoma elliptical in outline (Figs. 16.87.761, 762); first coxal plates extending anteriorly well beyond anterior edge of ventral shield (Fig. 16.87.762); genital field with eleven or twelve acetabula borne freely in one or two rows in gonopore in males (Fig. 16.87.762, 763); female unknown ....*Chelohydracarus*

100' Idiosoma oval or nearly round in outline (Figs. 16.86.749, 751, 754, 87.755, 758, 759); first coxal plates not extending beyond anterior end of ventral shield (Figs. 16.86.751, 754, 87.758, 759); genital field with numerous acetabula borne in several rows freely in gonopore (Figs. 16.86.751, 753) or in two or three rows freely in gonopore and several rows on acetabular plates at edges of gonopore (Figs. 16.87.757, 758) in males, and on acetabular plates at lateral edges of gonopore in females (Figs. 16.86.754, 87.759) 101

101(100) Genital field bearing all acetabula freely in gonopore in males (Fig. 16.86.753), and in two to five rows on acetabular plates at edges of gonopore in females (Fig. 16.86.754) .... *Chelomideopsis*

101' Genital field bearing acetabula in two or three rows freely in gonopore and several rows on acetabular plates at lateral edges of gonopore in males (Fig. 16.87.757), and in five to eight rows on acetabular plates at lateral edges of gonopore in females (Fig. 16.87.759) .... *Platyhydracarus*

102(95) Pedipalp with femur and tibia bearing distinctive ventral projections (Fig. 16.85.742); fourth coxal plates bearing large projections associated with insertions of fourth legs that extend laterally nearly at right angles to long axis of body (Figs. 16.85.740, 744); suture lines between third and fourth coxal plates distinct but none of coxal suture lines extending to midline (Figs. 16.85.740, 744); idiosoma highly modified in males (Figs. 16.85.739, 740); genital field located at posterior edge of ventral shield (Figs. 16.85.740, 744), bearing two or three pairs of acetabula freely in gonopore and twelve to fifteen pairs of acetabula on acetabular plates that are fused with ventral shield flanking gonopore in males (Figs. 16.85.740, 743), and bearing around twenty acetabula borne on acetabular plates near lateral edges of gonopore and not fused with ventral shield in females (Fig. 16.85.744); fourth leg highly modified in males with tarsus bearing a large flat flange laterally (Fig. 16.85.741) ....Arrenuroidea (in part) .... Amoenacaridae ....*Amoenacarus*

102' Not with above combination of characters 103

103(102) Genital field located near middle of ventral shield and bearing numerous acetabula on distinct acetabular plates flanking gonopore in both sexes (Figs. 16.84.724, 727) and a few additional acetabula freely in gonopore in males (Figs. 16.84.724, 726); ventral shield with coxoglandularia 1 shifted slightly anteriorly onto second coxal plates (Figs. 16.84.724, 727; arrows); pedipalp as in Fig. 16.84.725 .... Arrenuroidea (in part) ....Laversiidae ....*Laversia*

103' Not with above combination of characters 104

104(103) Ventral shield with first three pairs of coxal plates located far forward and with capitular bay shallow and wide (Figs. 16.84.730, 731); coxoglandularia 1 located on third coxal plates (Figs. 16.84.730, 731; arrows); pedipalp stocky with femur bearing a crenellate ventral projection (Fig. 16.84.729, arrow); genital field bearing several pairs of acetabula on ventral shield flanking gonopore in both sexes (Figs. 16.84.730, 731) and an additional one or two pairs of acetabula freely in gonopore in males (Fig. 16.84.730); dorsal shield as in Fig. 16.84.728 ....Arrenuroidea (in part) .... Arenohydracaridae ....*Arenohydracarus*

104' Not with above combination of characters 105

105(104) Ventral shield with medial edges of posterior coxal group forming a rounded angle and with suture lines between second and third coxal plates complete but those between first and second and third and fourth coxal plates incomplete (Figs. 16.84.733, 734); genital field with three pairs of acetabula borne freely in gonopore in males (Fig. 16.84.733) and on small plates at posterolateral edges of large gonopore in females (Fig. 16.59.433); idiosomal dorsum with a large dorsal plate bearing only postocular setae and surrounded by ring of dorsoglandularia (Fig. 16.84.732); pedipalp similar to Fig. 16.85.736 .... Arrenuroidea (in part)... Acalyptonotidae (in part) .... *Paenecalyptonotus*

105' Not with above combination of characters 106

106(105) Dorsal shield bearing four pairs of glandularia including one pair at anterior edge of shield (Figs. 16.82.707, 83.711; arrows); genital field bearing three pairs of acetabula (Figs. 16.82.709, 710, 83.712, 714); pedipalp with femur bearing several thick setae distodorsally (Figs. 16.82.705, 83.713) ....Arrenuroidea (in part) .... Chappuisididae (in part) .... Morimotacarinae 107

106' Not with above combination of characters 108

107(106) Pedipalp four-segmented with tibia and tarsus fused (Fig. 16.83.713); genital field with acetabula borne freely in gonopore in males (Fig. 16.83.712) and on acetabular plates fused with ventral shield flanking gonopore in females (Fig. 16.83.714) .... *Yachatsia*

107' Pedipalp five-segmented (Fig. 16.82.705); genital field with acetabula borne freely in gonopore in both sexes (Figs. 16.82.709, 710) ....*Morimotacarus*

108(106) Dorsal shield comprising a large central plate completely surrounded by a ring of small paired platelets (Figs. 16.61.463, 468) .... Hygrobatoidea (in part) ....Lethaxonidae 109

108' Dorsal shield not as described and illustrated above 110

109(108) Genital field with three pairs of acetabula arranged in triangles borne on ventral shield near posterior edge of fourth coxal plates and posterolateral to gonopore (Figs. 16.61.464, 466); fourth coxal plates with a pair of glandularia located immediately posterior to insertions of fourth legs (Figs. 16.61.464, 466; arrows); pedipalp unmodified in females and in males of one species (Fig. 16.61.469), highly modified in males of other species (Fig. 16.61.467) ....*Lethaxona*

109' Genital field bearing four to six pairs of acetabula arranged in single rows on ventral shield along suture lines indicating posterior edges of fourth coxal plates (Fig. 16.61.470); fourth coxal plates with pair of glandularia located immediately posterior to insertions of fourth legs reduced to setae with gland portion absent (Fig. 16.61.470, arrow); pedipalp unmodified in both sexes ....*Lethaxonella*

110(108) Ventral shield with suture lines between first and second coxal plates and those between third and fourth coxal plates distinct, and all extending to midline at same level as insertions of fourth legs (Figs. 16.75.633, 634); with posterior suture lines of fourth coxal plates located near posterior end of ventral shield (Figs. 16.75.633, 634); genital field bearing numerous acetabula on acetabular plates flanking gonopore (Figs. 16.75.633, 634); dorsal shield as in Fig. 16.75.632; pedipalp as in Fig. 16.75.631 ....Hygrobatoidea (in part) Aturidae (in part)................................Albiinae.. ....*Albia*

110' Ventral shield not as above..111

111(110) Dorsal shield bearing six pairs of glandularia of which three pairs are close together in an arc or triangle on each side (Fig. 16.63.484, arrow); projections associated with openings for insertion of fourth legs long, wide and extending at right angles to long axis of idiosoma (Figs. 16.63.485, 486); ventral shield as in Figs. 16.63.485, 486; gnathosomal base without (Fig. 16.63.489) or with (Fig. 16.63.487) a long rostrum; pedipalp with segments usually slender (Figs. 16.63.487, 489) but occasionally stocky (Fig. 16.63.488) .... Hygrobatoidea (in part).......................Unionicolidae (in part)................................ Pionatacinae (in part).....................................*Koenikea*

111' Dorsal shield not as above 112

112(111) Genital field bearing three or four pairs of genital acetabula...113

112' Genital field bearing more than four pairs of acetabula...131

113(112) Genital field with three pairs of acetabula borne freely in gonopore that is completely surrounded by ventral shield (Fig. 16.89.780) .... Arrenuroidea (in part)............................ Mideopsidae (in part).................................Mideopsinae..114

113' Genital field bearing three or four pairs of acetabula on acetabular plates flanking gonopore (Figs. 16.72.597, 604) and often fused with ventral shield (Figs. 16.72.596, 603) .... Hygrobatoidea (in part) ....Aturidae (in part) ....Axonopsinae 117

114(113) Idiosomal venter with coxoglandularia 2 at same level as, or slightly anterior to level of, first pair of genital acetabula (Figs. 16.88.770, 772; arrows); fourth coxal plates lacking projections covering insertions of fourth legs (Figs. 16.88.770, 772); dorsal shield with one or more pairs of prominent ridges that converge posteriorly (Fig. 16.88.769); pedipalp with tibia bearing a prominent ventral projection (Fig. 16.88.771); distal segments of legs lacking swimming setae .... *Xystonotus*

114' Idiosomal venter with coxoglandularia 2 (in some cases lacking gland portion) located anterior to genital field (Figs. 16.88.766, 767, 773, 774, 89.778, 781, 784; arrows A); fourth coxal plates bearing projections at least partially covering insertions of fourth legs (Figs. 16.88.766, 767, 773, 774, 89.778, 781, 784; arrows B); dorsal shield smooth or with ridges variously developed; pedipalp with tibia bearing or lacking a ventral projection; distal segments of at least third and fourth legs bearing swimming setae (Fig. 16.88.775) 115

115(114) Fourth coxal plates bearing relatively large projections nearly covering insertions of fourth legs (Figs. 16.88.766, 767; arrows B); dorsal shield smooth and lacking ridges (Fig. 16.88.765); pedipalp with tibia lacking ventral projections and bearing two small setae proximally on ventral surface (Fig. 16.88.768) .... *Mideopsides*

115' Fourth coxal plates bearing relatively small projections only partially covering insertions of fourth legs (Figs. 16.88.773, 774, 89.778, 781, 784, arrows B); dorsal shield smooth (Fig. 16.89.777) or with ridges variously developed (Fig. 16.89.779); pedipalp with tibia bearing two ventral setae near mid-length that are usually borne on a ventral bulge or projection (Figs. 16.88.776, 89.782, 783, 785, 786) 116

116(115) Ventral shield with two most posterior pairs of glandularia complete, comprising both glands and associated setae (Figs. 16.88.773, 774; arrows C); genital field with only two or three pairs of setae flanking gonopore (Figs. 16.88.773, 774); dorsal shield usually with posterior two pairs of glandularia grouped together near posterior edge of shield in males; pedipalp with tibia bearing a long, blunt ventral projection (Fig. 16.88.776); fourth leg modified in males (Fig. 16.88.775) .... *Neoxystonotus*

116' Ventral shield with one or both of two most posterior two pairs of glandularia incomplete, lacking gland component (Figs. 16.89.778, 781, 784; arrows C); genital field with numerous setae flanking gonopore (Figs. 16.89.778, 781, 784); dorsal shield with posterior two pairs of glandularia not grouped together near posterior edge of shield in males; pedipalp with tibia bulging ventrally (Figs. 16.89.785, 786) or bearing a relatively short ventral projection (Figs. 16.89.782, 783); fourth leg not modified in males .... *Mideopsis*

117(113) Ventral shield with posteriorly directed projections associated with insertions of fourth legs (Fig. 16.70.572, arrow A) 118

117' Ventral shield lacking posteriorly directed projections associated with insertions of fourth legs 119

118(117) Ventral shield with first coxal plates projecting beyond anterior end of shield, and suture lines between first and second coxal plates not extending to midline (Figs. 16.70.572, 575); idiosoma and legs strongly sexually dimorphic with males having distal segments of second leg modified and with one tarsal claw much longer than tarsus (Fig. 16.70.573), and fourth leg highly modified (Fig. 16.70.574) .... *Stygalbiella*

118' Ventral shield with first coxal plates not extending beyond anterior end of shield and suture lines between first and second coxal plates extending to midline (Figs. 16.70.568, 570); idiosoma and legs moderately sexually dimorphic with males having distal segments of second leg only slightly modified and claws much shorter than tarsus (Fig. 16.70.571), and fourth leg less highly modified (Fig. 16.70.569) .... *Axonopsella*

119(117) Ventral shield with posterior edges of fourth coxal plates clearly indicated by suture lines that are indented posteromedially to accommodate a pair of glandularia (Figs. 16.71.587, 588; arrows); pedipalp as in Fig. 16.71.589 .... *Ljania*

119' Ventral shield with posterior edges of fourth coxal plates only faintly indicated by suture lines or obliterated by fusion with ventral shield, never indented to accommodate a pair of glandularia as described and illustrated above 120

120(119) Ventral shield with distinct ridges extending from lateral ends of suture lines between third and fourth coxal plates to edges of shield (Fig. 16.74.620, arrow) 121

120' Ventral shield lacking ridges as described and illustrated above 130

121(120) Ventral shield with ridges extending from lateral ends of suture lines between third and fourth coxal plates directed anteriorly and extending nearly to anterior edge of shield (Fig. 16.73.612, arrow); fourth leg with segments flattened and with telofemur relatively small (Fig. 16.73.614); dorsal shield as in Fig. 16.73.610; genital field as in Figs. 16.73.611, 612; pedipalp as in Fig. 16.73.613 .... *Erebaxonopsis*

121' Ventral shield with ridges extending from lateral ends of suture lines between third and fourth coxal plates directed anterolaterally and extending to lateral edges of shield (Fig. 16.74.620, arrow); fourth leg with segments not flattened and with telofemur not reduced in size 122

122(121) Pedipalp with femur bearing a spine-like ventral projection that is usually large and pointed (Figs. 16.71.591, 72.594, 600) but rarely short and rounded 123

122' Pedipalp with femur lacking a distinct spine-like ventral projection (Figs. 16.73.608, 74.616, 619, 621) 126

123(122) Genital field bearing four pairs of acetabula (Figs. 16.71.590, 592); legs not highly modified in males; pedipalp as in Fig. 16.71.591 .... *Neobrachypoda*

123' Genital field bearing three pairs of acetabula (Figs. 16.72.596, 597, 599, 603, 604); fourth legs often highly modified in males (Figs. 16.72.595, 598, 601) 124

124(123) Fourth coxal plates bearing short ridges extending posteriorly from region of insertion of fourth legs (Figs. 16.72.603, 604, arrows); fourth leg with tarsus greatly expanded distally in males (Fig. 16.72.601); dorsal shield as in Fig. 16.72.602 .... *Estellacarus*

124' Fourth coxal plates lacking ridges extending posteriorly from region of insertion of fourth legs (Figs. 16.72.596, 597, 599); fourth leg with distal segments highly modified in males (Figs. 16.72.595, 598) 125

125(124) Ventral shield bearing a pronounced ridge extending anterolaterally from genital field in males (Fig. 16.72.596, arrow A) and with insertions of fourth legs bearing condyles in both sexes (Fig. 16.72.596, arrow B); genital field relatively compact with no acetabula extending laterally beyond level of insertions of fourth legs, and lacking a patch of long setae immediately lateral to acetabula in males (Fig. 16.72.596); fourth leg with genu bearing a distal projection extending well beyond level of insertion of tibia in males (Fig. 16.72.595); dorsal shield bearing two enlarged glandularia in males (Fig. 16.72.593, arrow) .... *Brachypoda*

125' Ventral shield lacking a ridge extending anterolaterally from genital field in males (Fig. 16.72.599) and with insertions of fourth legs lacking condyles in both sexes (Fig. 16.72.599); genital field relatively wide with acetabula extending laterally beyond level of insertions of fourth legs, and bearing a patch of long setae immediately lateral to acetabula in males (Fig. 16.72.599); fourth leg with genu lacking a distal projection extending well beyond level of insertion of tibia in males (Fig. 16.72.598); dorsal shield lacking enlarged glandularia in males .... *Ocybrachypoda*

126(122) Excretory pore borne on a small separate platelet at posterior end of dorsal shield (Figs. 16.73.605, 606; arrows); first coxal plates not extending to anterior edge of ventral shield (Figs. 16.73.606, 607); posterior region of ventral shield highly modified in males (Figs. 16.73.606 cf. 73.607); fourth leg with distal segments modified in males (Fig. 16.73.609); pedipalp as in Fig. 16.73.608 .... *Woolastookia*

126' Excretory pore borne posteriorly on dorsal shield (Figs. 16.74.615, 618, 622, 75.626); first coxal plates as described above or extending to or beyond anterior edge of ventral shield (Fig. 16.74.620); posterior region of ventral shield only slightly modified in males (Figs. 16.74.620, 624, 75.629, 630 cf. 74.617, 625, 75.628) 127

127(126) Idiosomal dorsum with dorsal furrow bearing two pairs of glandularia on small platelets (Fig. 16.74.615, arrow); dorsal shield bearing three or four pairs of glandularia with one pair flanking excretory pore (Fig. 16.74.615); ventral shield lacking glandularia in region between genital field and insertions of fourth legs (Fig. 16.57.417); pedipalp with tibia relatively long and slender, lacking ventral projection, and bearing two slender setae distoventrally (Fig. 16.74.616) .... *Axonopsis*

127' Idiosomal dorsum with dorsal furrow lacking glandularia (Figs. 16.74.618, 622, 75.626); dorsal shield bearing three to six pairs of glandularia but without glandularia flanking excretory pore (Figs. 16.74.618, 622, 75.626); ventral shield bearing one or two pairs of glandularia in region between genital field and insertions of fourth legs (Figs. 16.74.620, 625, 75.628); pedipalp with tibia relatively short and stocky and bearing one or two setae proximal to mid-length on a ventral swelling or projection (Figs. 16.74.619, 621, 75.627) 128

128(127) Gnathosomal base (capitulum) with apodemata (anchoral process) extremely elongate, extending posteriorly to level of suture lines indicating posterior edges of third coxal plates (Fig. 16.74.623); pedipalp with tarsus as long as tibia and tapered distally to a slender tip (Fig. 16.74.621); dorsal shield as in Fig. 16.74.622; ventral shield as in Fig. 16.74.625 .... *Vicinaxonopsis*

128' Gnathosomal base (capitulum) with apodemata (anchoral process) relatively short, not extending posteriorly beyond region of first coxal plates; pedipalp with tarsus much shorter than tibia and tapered distally to a blunt tip bearing several thick setae (Figs. 16.74.619, 75.627) 129

129(128) Dorsal shield bearing three pairs of glandularia (Fig. 16.75.626); ventral shield bearing one pair of glandularia in region between genital field and insertions of fourth legs (Fig. 16.75.628); genital field of males, and to a lesser extent females, borne on a short cauda (Figs. 16.75.629, 630 cf. 75.628) .... *Paraxonopsis*

129' Dorsal shield bearing four to six pairs of glandularia (Fig. 16.74.618); ventral shield bearing two pairs of glandularia in region between genital field and insertions of fourth legs (Fig. 16.74.620); genital field not borne on a cauda (Figs. 16.74.617, 620) .... *Brachypodopsis*

130(120) Anterior coxal plates extending beyond anterior edge of ventral shield (Figs. 16.71.581, 582); genital field borne on a separate platelet posterior to ventral shield in males (Fig. 16.71.581); dorsal shield as in Fig. 16.71.580; pedipalp as in Fig. 16.71.586 .... *Albaxona*

130' Anterior coxal plates not extending beyond anterior edge of ventral shield (Figs. 16.71.584, 585); genital field incorporated into ventral shield in males (Fig. 16.71.584); dorsal shield as in Fig. 16.71.583 .... *Javalbia*

131(112) First coxal plates not extending beyond anterior edge of ventral shield (Figs. 16.70.577, 578); ventral shield bearing large posteriorly directed projections associated with insertions of fourth legs (Figs. 16.70.577, 578); genital field bearing around ten pairs of acetabula and strongly sexually dimorphic, with a small gonopore located near midlength of body between posterior coxal plates and acetabula distributed throughout posteromedial region of ventral shield in males (Fig. 16.70.577), and a large gonopore located posteriorly and acetabula borne on ventral shield flanking gonopore in females (Fig. 16.70.578); fourth leg with genu bearing a thick, curved, serrated seta distoventrally in males (Fig. 16.70.579); dorsal shield as in Fig. 16.70.576 in females, similar in males but with excretory pore borne on ventral shield .... Hygrobatoidea (in part) Aturidae (in part) .... Axonopsinae (in part) .... *Submiraxona*

131' First coxal plates extending slightly or well beyond anterior edge of ventral shield (Figs. 16.76.636, 640, 645, 77.648, 651, 652, 78.658, 659, 664, 665); ventral shield bearing or lacking posteriorly directed projections associated with insertions of fourth legs; genital field slightly to strongly sexually dimorphic, but never as described and illustrated above; fourth leg unmodified or variously modified in males but never with genu modified as described and illustrated above; dorsal shield various .... Hygrobatoidea (in part) .... Aturidae (in part) .... Aturinae 132

132(131) Dorsal shield with edges finely crenellated, bearing only postocularia and one pair of glandularia (Figs. 16.77.646, 647); dorsal furrow bearing pair of glandularia on small platelets near anterolateral angles of dorsal shield (Figs. 16.77.646, 647); ventral shield with edges finely crenellated, capitular bay very deep and narrow (Fig. 16.77.648); pedipalp with all segments long and slender and with femur bearing a slender projection proximoventrally (Fig. 16.77.649) .... *Bharatalbia*

132' Dorsal shield bearing three or more pairs of glandularia (Figs. 16.76.635, 639, 641, 642, 77.650, 78.656, 661); ventral shield and pedipalp not as described and illustrated above 133

133(132) Fourth coxal plates bearing moderately large projections associated with insertions of fourth legs (Figs. 16.77.651, 652, 78.658, 659; arrows) 134

133' Fourth coxal plates lacking projections associated with insertions of fourth legs (Figs. 16.76.636, 640, 645, 78.664, 665) 135

134(133) Dorsal shield bearing seven or more pairs of glandularia (Fig. 16.77.650); ventral shield bearing large projections associated with insertions of fourth legs that extend to lateral edges of shield (Figs. 16.77.651, 652; arrows); genital field with acetabula borne in one or two rows along posterolateral edges of ventral shield (Figs. 16.77.651, 652); fourth leg with tibia highly modified and bearing one or more long, thick, blade-like setae ventrally in males (Fig. 16.77.653); pedipalp often highly modified in males (Fig. 16.77.655 cf. 77.654) ....*Kongsbergia*

134' Dorsal shield bearing six pairs of glandularia (Fig. 16.78.656); ventral shield bearing relatively small projections associated with insertions of fourth legs that do not extend to lateral edges of ventral shield (Figs. 16.78.658, 659, arrows); genital field with acetabula borne in several rows flanking gonopore (Figs. 16.78.658, 659); fourth leg with tibia relatively little modified and lacking long, thick, blade-like setae in males (Figs. 16.78.658, 660); pedipalp as in Fig. 16.78.657 and not sexually dimorphic ....*Neoaturus*

135(133) Dorsal shield bearing four pairs of glandularia (Fig. 16.78.661); dorsal furrow bearing four pairs of glandularia on small platelets (Fig. 16.78.661); dorsal setae variously modified in males (Fig. 16.78.661); ventral shield with a posteromedial cleft in males (Fig. 16.78.664); genital field with acetabula borne in a single row along posterolateral edges of ventral shield (Figs. 16.78.664, 665); pedipalp with tibia not bulging ventrally and with tarsus bearing relatively slender and straight setae terminally (Fig. 16.78.662); fourth leg with genu and tibia variously modified and bearing several long, thick setae in males (Figs. 16.78.663, 664) ....*Aturus*

135' Dorsal shield bearing three pairs of glandularia (Figs. 16.76.635, 639, 641, 642); dorsal furrow bearing several pairs of small platelets of which two pairs bear glandularia (Figs. 16.76.635, 639, 641, 642); genital field with ten to thirteen pairs of acetabula borne on acetabular plates flanking gonopore and variously incorporated into posterodorsal urogenital plate or ventral shield in males (Figs. 16.76.640, 641, 645), and separate or variously fused with ventral shield in females (Fig. 16.76.636); excretory pore borne on urogenital plate when present, or incorporated into dorsal or ventral shield in males (Figs. 16.76.635, 639, 641, 645; arrows); pedipalp with tibia bulging ventrally and bearing two ventral setae and with tarsus bearing one or two thick curved setae terminally (Figs. 16.76.638, 644); fourth leg slightly to moderately modified and with at least tibia bearing several long, thick, blade-like setae in males (Figs. 16.76.637, 643) 136

136(135) Ventral shield bearing a pronounced ridge originating anterior to insertions of fourth legs and extending laterally at right angles to long axis of idiosoma to edges of shield (Fig. 16.76.645); surfaces of dorsal and ventral shields rugose; dorsal shield bearing two pairs of sub-parallel longitudinal ridges (Fig. 16.76.642); pedipalp with tibia bearing two slender setae distoventrally and with tarsus relatively slender and tapered distally (Fig. 16.76.644); fourth leg with distal segments straight and only slightly modified in males (Fig. 16.76.643) ....*Ameribrachypoda*

136' Ventral shield bearing a ridge that may be pronounced or weakly expressed originating anterior to insertions of fourth legs and extending anterolaterally to edges of shield (Figs. 16.76.636, 640); surfaces of dorsal and ventral shields smooth; dorsal shield bearing one pair of idges outlining medial depression of shield in males (Figs. 16.76.635, 639, 641); pedipalp with tibia bearing two slender setae ventrally near mid-length and tarsus relatively stout and only slightly tapered distally (Fig. 16.76.638); fourth leg with distal segments variously curved and expanded and bearing modified setae in males (Fig. 16.76.637) ....*Phreatobrachypoda*

137(79) Fourth coxal plates bearing a pair of glandularia near middle or anterior edges of plates (Figs. 16.59.443–445, 60.446, 449, 451; arrows) ....Hygrobatoidea (in part)............................... Hygrobatidae 138

137' Fourth coxal plates lacking glandularia near middle or anterior edges of plates, although rarely bearing a pair of glandularia near posterolateral angles of plates (some species of *Unionicola*) (Fig. 16.62.476, arrow B) 141

138(137) First coxal plates separated medially (Figs. 16.60.449, 451); idiosomal dorsum bearing an array of enlarged paired dorsalia and glandularia platelets (Fig. 16.60.448); pedipalp femur bearing a ventral projection (Fig. 16.60.450) .... *Corticacarus*

138' First coxal plates fused medially (Figs. 16.59.443–445, 60.446); dorsum various; pedipalp femur bearing or lacking (Fig. 16.59.440) a ventral projection 139

139(138) Gnathosomal base (capitulum) broadly fused with first coxal plates (Figs. 16.59.443–445) 140

139' Gnathosomal base (capitulum) separated from first coxal plates or fused with them by an extremely narrow strip of sclerotization; first leg with tibia bearing two long, thick setae and a slender down-turned seta distoventrally and tarsus straight to noticeably curved (Fig. 16.60.447) ....*Atractides*

140(139) First leg with tibia bearing two short, thick setae distoventrally and with tarsus not noticeably curved (Fig. 16.59.439); fourth coxal plates bearing glandularia near middle of plates (Figs. 16.59.443, 444) or near suture lines between third and fourth coxal plates; genital field usually with three pairs of acetabula (Figs. 16.59.443, 444) but occasionally with numerous acetabula; pedipalp femur bearing or lacking (Fig. 16.59.440) a ventral projection ....*Hygrobates*

140' First legs with tibia bearing two short, thick setae ventrally near mid-length of segment and with tarsus noticeably curved (Fig. 16.59.441); fourth coxal plates bearing glandularia anteromedially on plates (Fig. 16.59.445); pedipalp femur lacking a ventral projection (Fig. 16.59.442) .... *Mesobates*

141(137) First leg with tarsal claw socket large and occupying more than one half of dorsal surface of tarsus (Fig. 16.61.461); genital field bearing three or four pairs of acetabula (Figs. 16.61.458–460); idiosomal dorsum as in Fig. 16.61.457; pedipalp as in Fig. 16.61.462 .... Hygrobatoidea (in part).......................Wettinidae .... *Wettina*

141' First leg with tarsal claw socket occupying much less than one half of dorsal surface of segment 142

142(141) Idiosoma bearing extensive ventral shield (Figs. 16.85.737, 738) but with dorsum covered by soft integument bearing numerous small dorsalia (Fig. 16.85.735); genital field with three pairs of acetabula or around 10 acetabula per side, borne freely in small gonopore in males (Fig. 16.85.737) and on acetabular plates near posterolateral edges of wide gonopore in females (Fig. 16.85.738); pedipalp with distal edge of tibia about twice as long as proximal edge of tarsus and with tarsus inserted distodorsally on tibia (Fig. 16.85.736) .... Arrenuroidea (in part) ....Acalyptonotidae (in part) ....*Acalyptonotus*

142' Not with above combination of characters 143

143(142) First coxal groups bearing large apodemes extending posteriorly to or near mid-length of fourth coxal plates (Figs. 16.62.473, 474; arrows); first leg bearing numerous long, thick setae (Fig. 16.62.475); idiosomal integument usually soft except for coxal plates, rarely with extensive secondary sclerotization dorsally (Fig. 16.62.471) and/or ventrally (Figs. 16.62.472, 473), but never forming entire dorsal *and* ventral shields; genital field with numerous acetabula borne on round or irregularly shaped acetabular plates flanking gonopore (Figs. 16.62.472, 473) ....Hygrobatoidea (in part)..................Unionicolidae (in part) .... Pionatacinae (in part) ....*Neumania*

143' First coxal groups bearing small apodemes extending posteriorly only slightly beyond anterior margin of third coxal plates (Fig. 16.62.476) 144

144(143) Idiosomal venter bearing a pair of glandularia on or closely flanking posteromedial corners of fourth coxal plates (Fig. 16.62.476, arrow B); suture lines between third and fourth coxal plates short and curved anteriorly (Fig. 16.62.476, arrow A); genital field usually with five or six pairs of acetabula, rarely with numerous acetabula; acetabular plates fused around gonopore in males (Fig. 16.62.481), and usually divided into two pairs of small platelets flanking gonopore (Fig. 16.62.482), rarely fused into a single pair of plates flanking gonopore (Fig. 16.62.483), in females; acetabular plates bearing specialized setae for piercing host tissue and oviposition in females (Figs. 16.62.476, 482); pedipalp highly variable with segments usually relatively long and slender in free-living species (Fig. 16.62.477) and relatively short and stout in parasitic species (Fig. 16.62.479); legs variable with segments usually long and slender and setae long in free-living species (Fig. 16.62.478), and segments relatively short and stout and setae short in parasitic species (Fig. 16.62.480); free-living, commensal in sponges or parasitic in mussels and snails ....Hygrobatoidea (in part) .... Unionicolidae (in part)..................Unionicolinae.................. ....*Unionicola*

144' Idiosomal venter lacking a pair of glandularia on or closely flanking posteromedial corners of fourth coxal plates .... Hygrobatoidea (in part)..................Pionidae (in part)..................145

145(144) Gnathosomal base lacking an anchoral process (Fig. 16.69.556, arrow); fourth coxal plates rounded posteromedially (Fig. 16.69.556); genital field bearing numerous acetabula, with two pairs noticeably larger than others, on large tongue-shaped plates extending laterally from gonopore (Figs. 16.69.556, 558); pedipalp as in Fig. 16.69.557; parasitic in mussels .... Najadicolinae ....*Najadicola*

145' Not with above combination of characters 146

146(145) Pedipalp with all segments, but especially tibia, very long and slender (Fig. 16.63.493); genital field bearing three pairs of acetabula and highly sexually dimorphic (Figs. 16.63.490, 492), bearing a complex sclerotized petiole in males (Figs. 16.63.490, 491, 494); third leg with genu highly modified in males (Fig. 16.63.495) ....Hydrochoreutinae ....*Hydrochoreutes*

146' Pedipalp with segments much shorter than described and illustrated above; genital field with three pairs to many acetabula and lacking a petiole in males; third leg with genu not modified as described and illustrated above 147

147(146) Gnathosomal base (capitulum) lacking an anchoral process (Fig. 16.68.549 and similar to Fig. 16.69.556, arrow); pedipalp with tibia lacking a thick seta distally (Figs. 16.68.548, 550); idiosoma with (Fig. 16.68.551) or without (Fig. 16.68.549) dorsal and ventral shields but with ventral shield never incorporating excretory pore ....Pioninae (in part).................. *Nautarachna* .... FEMALES

147' Gnathosomal base (capitulum) bearing an anchoral process (Fig. 16.64.496, arrow A); pedipalp with tibia bearing a thick seta distally (Figs. 16.64.500, 501, 507, 65.510, 511, 518, 66.520, 527, 528, 67.534, 539, 69.559, 566); idiosomal sclerotization various 148

148(147) Second and third coxal plates each bearing one to three noticeably thickened setae (Figs. 16.64.496, arrows B; Fig. 16.64.499, arrows) ....Huitfeldtiinae 149

148' Coxal plates bearing only slender setae..................150

149(148) Genital field bearing eight to ten pairs of acetabula (Figs. 16.64.496, 497); pedipalp with genu bearing a long slender seta (Fig. 16.64.501, arrow); leg segments lacking spatulate setae in males (similar to Fig. 16.64.506) .... *Huitfeldtia*

149' Genital field bearing three pairs of acetabula (Figs. 16.64.498, 499); pedipalp with genu lacking a long slender seta (Fig. 16.64.500); legs with femur, genu, and tibia bearing spatulate setae (Fig. 16.64.502) and second, third, and fourth legs slightly modified in males .... *Gereckea*

150(148) Genital field usually bearing numerous acetabula, when bearing as few as five pairs then fourth coxal plates with medial edges reduced to medial angles (Fig. 16.66.530) 151

150' Genital field usually bearing three pairs of acetabula, when bearing up to six pairs then fourth coxal plates with medial edges relatively long (as in Figs. 16.65.514, 515, 517) .... Pionidae (in part)..................Tiphyinae (in part)..................154

151(150) Medial edges of fourth coxal plates long (Figs. 16.69.562, 567); genital field with acetabula usually borne on acetabular plates (Figs. 16.69.562, 567), rarely freely in ventral integument ....Pionidae (in part).................. Pioninae (in part)..................*Piona*.................. .... FEMALES

151' Medial edges of fourth coxal plates reduced to medial angles (Figs. 16.66.530, 536, 67.544)..................Pionidae (in part) ....Forellinae (in part) ....FEMALES..................152

152(151) First, second, and third coxal plates usually close together (Fig. 16.67.536) or fused medially, when slightly separated then pedipalp with femur bearing a prominent ventral projection ....*Pseudofeltria*

152' First, second, and third coxal plates usually well separated (Figs. 16.66.530, 67.544), when only slightly separated then pedipalp with femur lacking a prominent ventral projection (Figs. 16.66.528, 67.539) 153

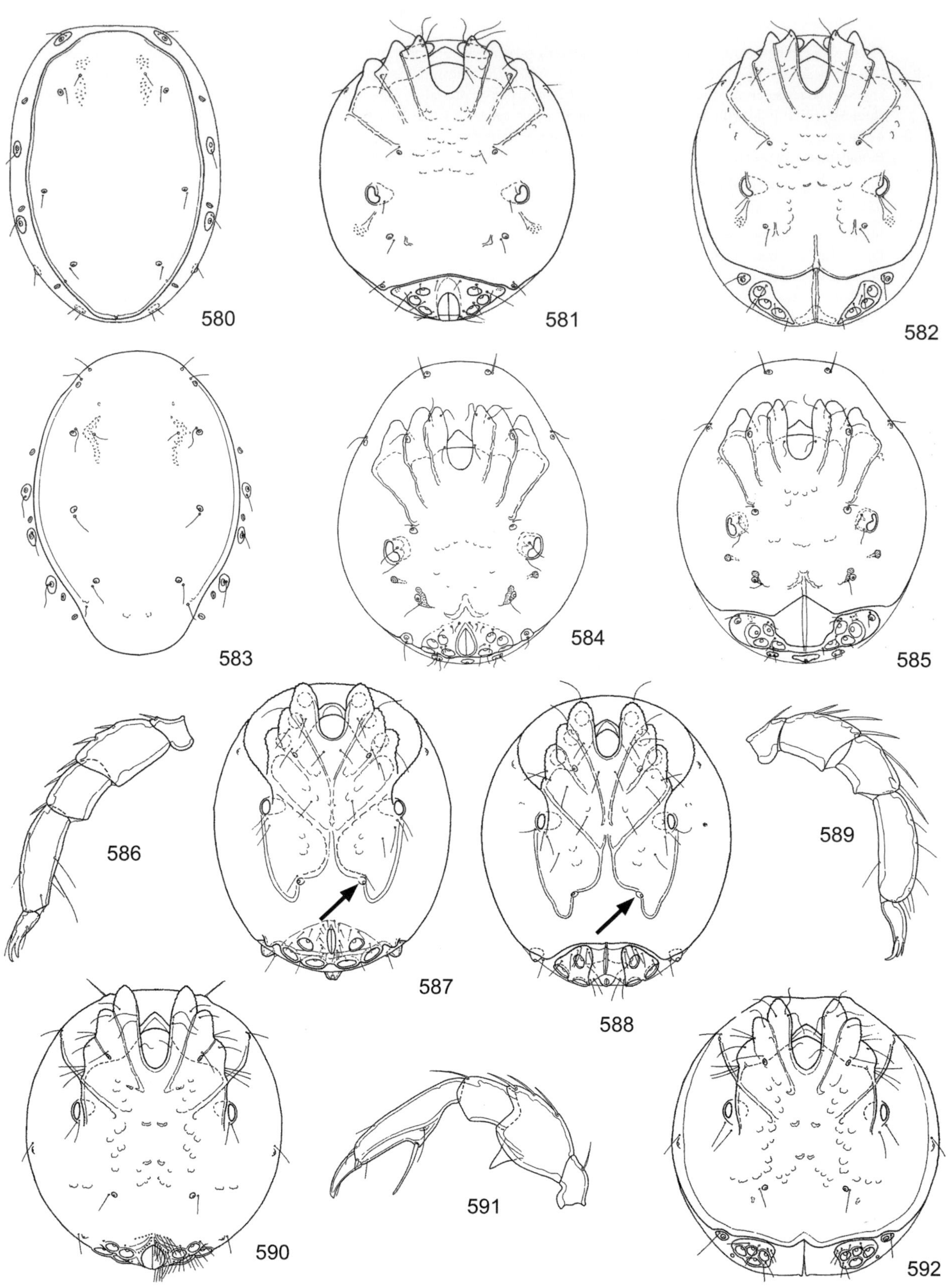

PLATE 16.71 Figures 580–582 & 586: *Albaxona nearctica* Cook (Axonopsinae) adults. **Fig. 580** male dorsum of idiosoma. **Fig. 581** male venter of idiosoma. **Fig. 582** female venter of idiosoma. **Fig. 586** male pedipalp. **Figures 583–585:** *Javalbia* sp. (Axonopsinae) adults. **Fig. 583** female dorsum of idiosoma. **Fig. 584** male venter of idiosoma**. Fig. 585** female venter of idiosoma. **Figures 587–589:** *Ljania michiganensis* Cook (Axonopsinae) adults. **Fig. 587** male venter of idiosoma. **Fig. 588** female venter of idiosoma. **Fig. 589** male pedipalp. **Figures 590–592:** *Neobrachypoda eckmani* (Walter) (Axonopsinae) adults. **Fig. 590** male ventral shield. **Fig. 591** male pedipalp. **Fig. 592** female venter of idiosoma.

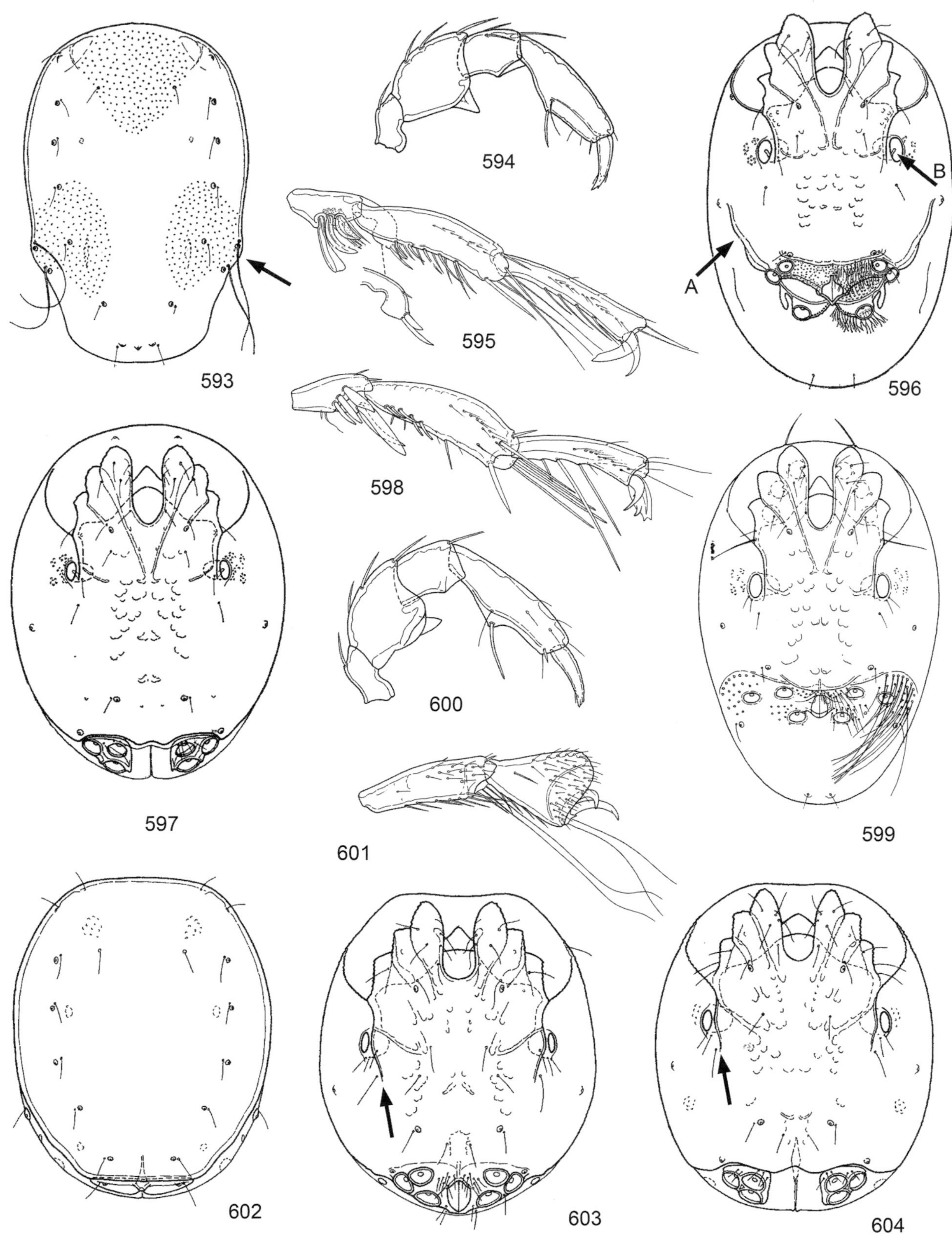

PLATE 16.72 **Figures 593–597:** *Brachypoda cornipes* Habeeb (Axonopsinae) adults. **Fig. 593** male dorsal shield. **Fig. 594** female pedipalp. **Fig. 595** male distal segments of fourth leg showing detail of distal projection on genu. **Fig. 596** male ventral shield. **Fig. 597** female venter of idiosoma. **Figures 598–599:** *Ocybrachypoda laversi* (Cook) (Axonopsinae) male adult. **Fig. 598** distal segments of fourth leg. **Fig. 599** ventral shield. **Figures 600–604:** *Estellacarus unguitarsus* Habeeb (Axonopsinae) adults. **Fig. 600** male pedipalp. **Fig. 601** male distal segments of fourth leg. **Fig. 602** female dorsum of idiosoma. **Fig. 603** male ventral shield. **Fig. 604** female venter of idiosoma.

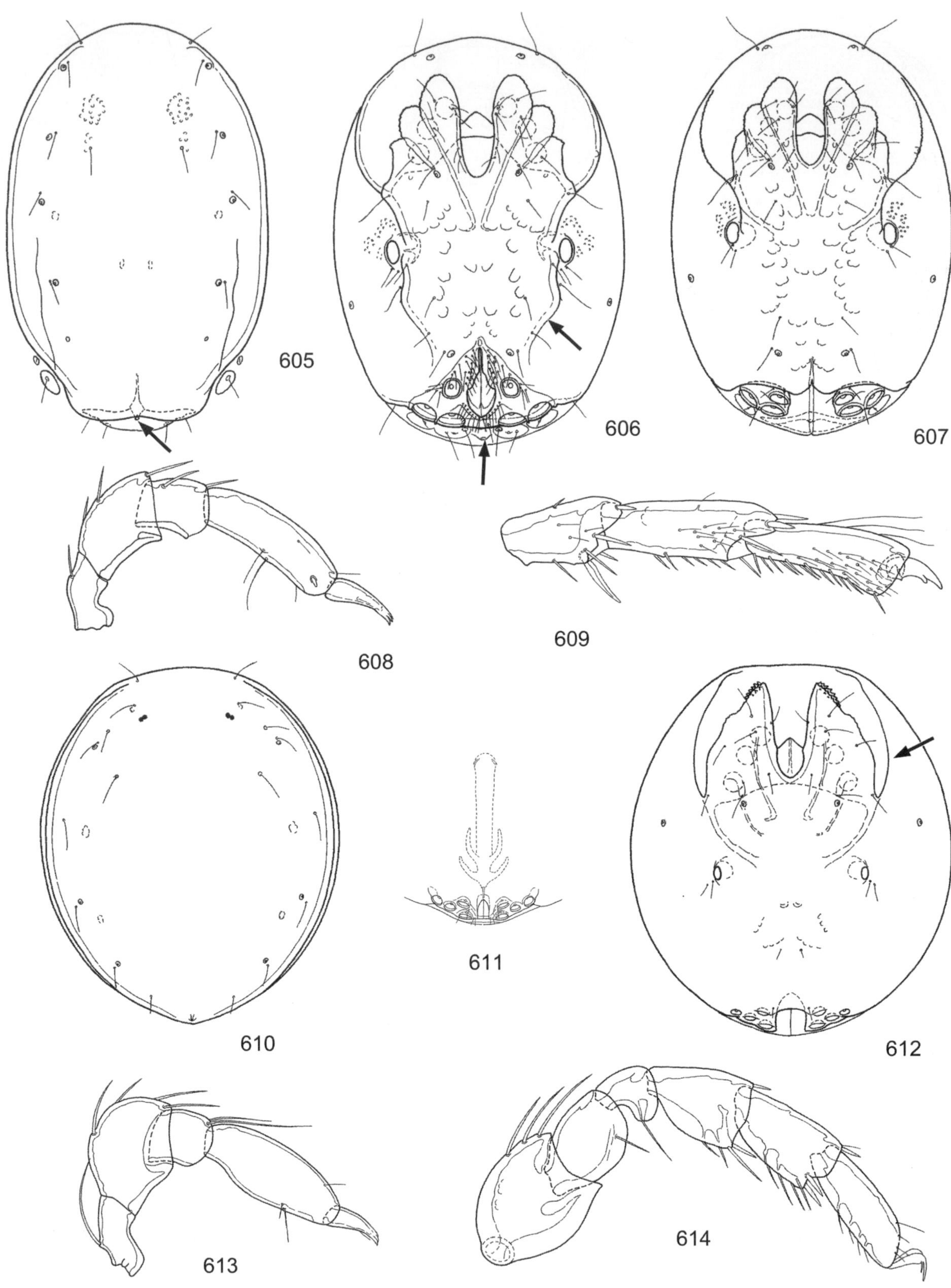

PLATE 16.73 Figures 605–609: *Woolastookia pilositarsa* (Habeeb) (Axonopsinae) adults. **Fig. 605** female dorsum of idiosoma**. Fig. 606** male ventral shield. **Fig. 607** female venter of idiosoma. **Fig. 608** female pedipalp. **Fig. 609** male distal segments of fourth leg. **Figures 610–614:** *Erebaxonopsis nearctica* Cook (Axonopsinae) adults. **Fig. 610** male dorsal shield. **Fig. 611** male genital field region. **Fig. 612** female ventral shield. **Fig. 613** female pedipalp. **Fig. 614** male fourth leg.

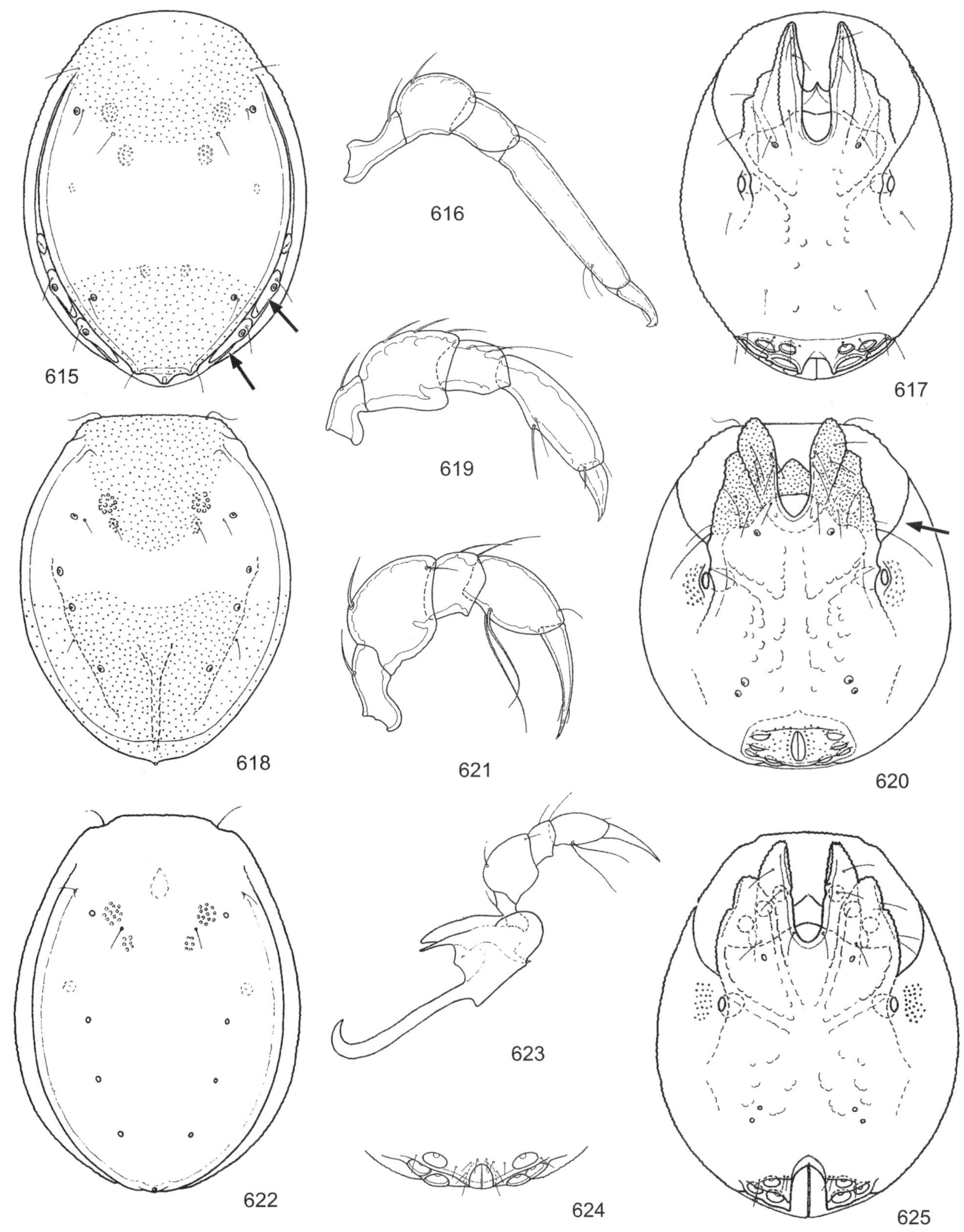

PLATE 16.74 **Figures 615–617:** *Axonopsis sabulonis* Cook (Axonopsinae) adult female. **Fig. 615** dorsum of idiosoma. **Fig. 616** pedipalp. **Fig. 617** ventral shield. **Figures 618–620:** *Brachypodopsis beltista* (Cook) (Axonopsinae) adult male. **Fig. 618** dorsal shield. **Fig. 619** pedipalp. **Fig. 620** ventral shield. **Figures 621–625:** *Vicinaxonopsis californiensis* (Cook) (Axonopsinae) adults. **Fig. 621** male pedipalp. **Fig. 622** female dorsal shield. **Fig. 623** female gnathosomal base (capitulum) and pedipalp. **Fig. 624** male genital field. **Fig. 625** female ventral shield.

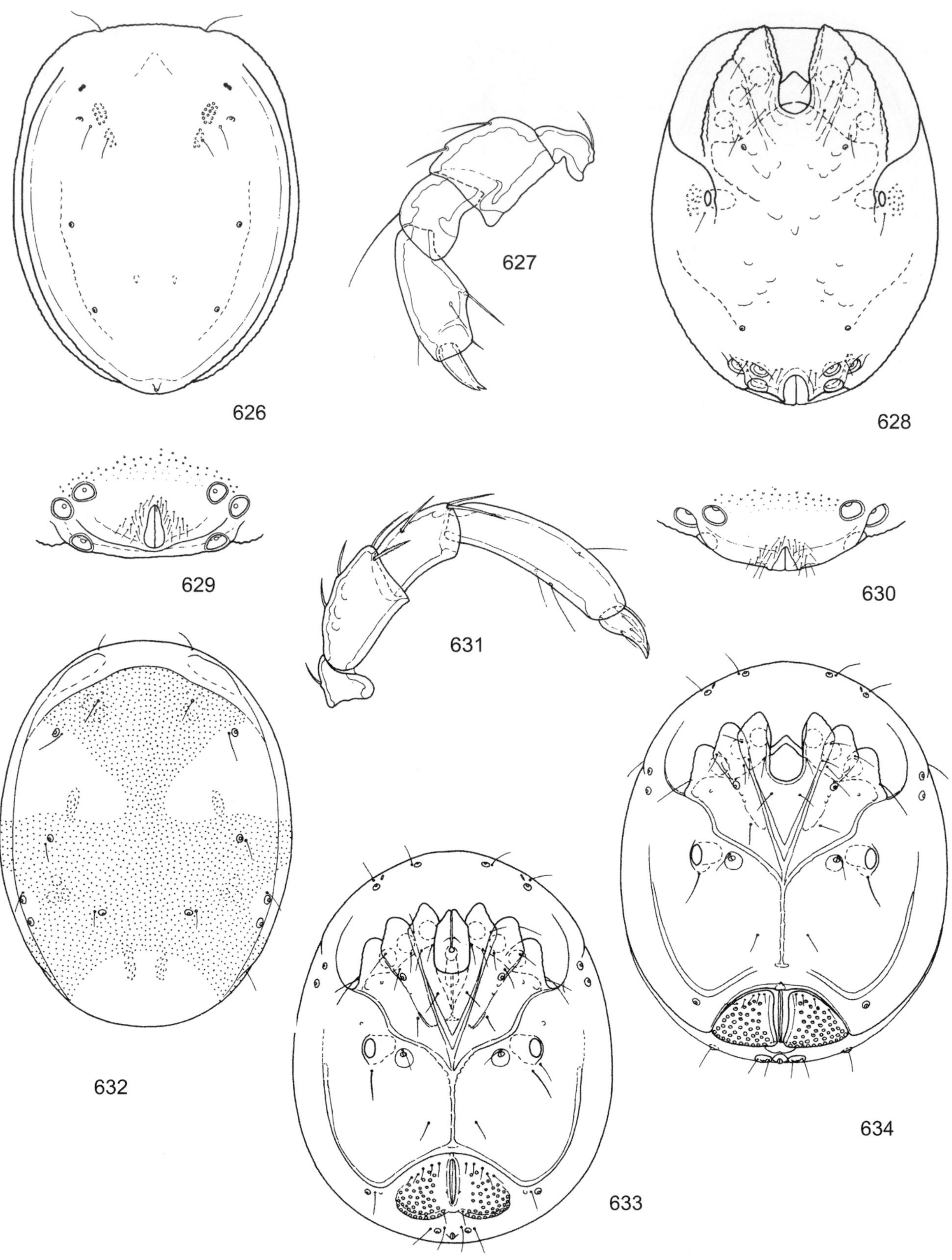

PLATE 16.75 **Figures 626–630:** *Paraxonopsis pumila* (Cook) (Axonopsinae) adults. **Fig. 626** female dorsal shield. **Fig. 627** male pedipalp. **Fig. 628** female ventral shield. **Fig. 629** male ventral view of genital field. **Fig. 630** male posteroventral view of genital field. **Figures 631–634:** *Albia iantha* Cook (Albiinae) adults. **Fig. 631** male pedipalp. **Fig. 632** male dorsal shield. **Fig. 633** male ventral shield. **Fig. 634** female venter of idiosoma.

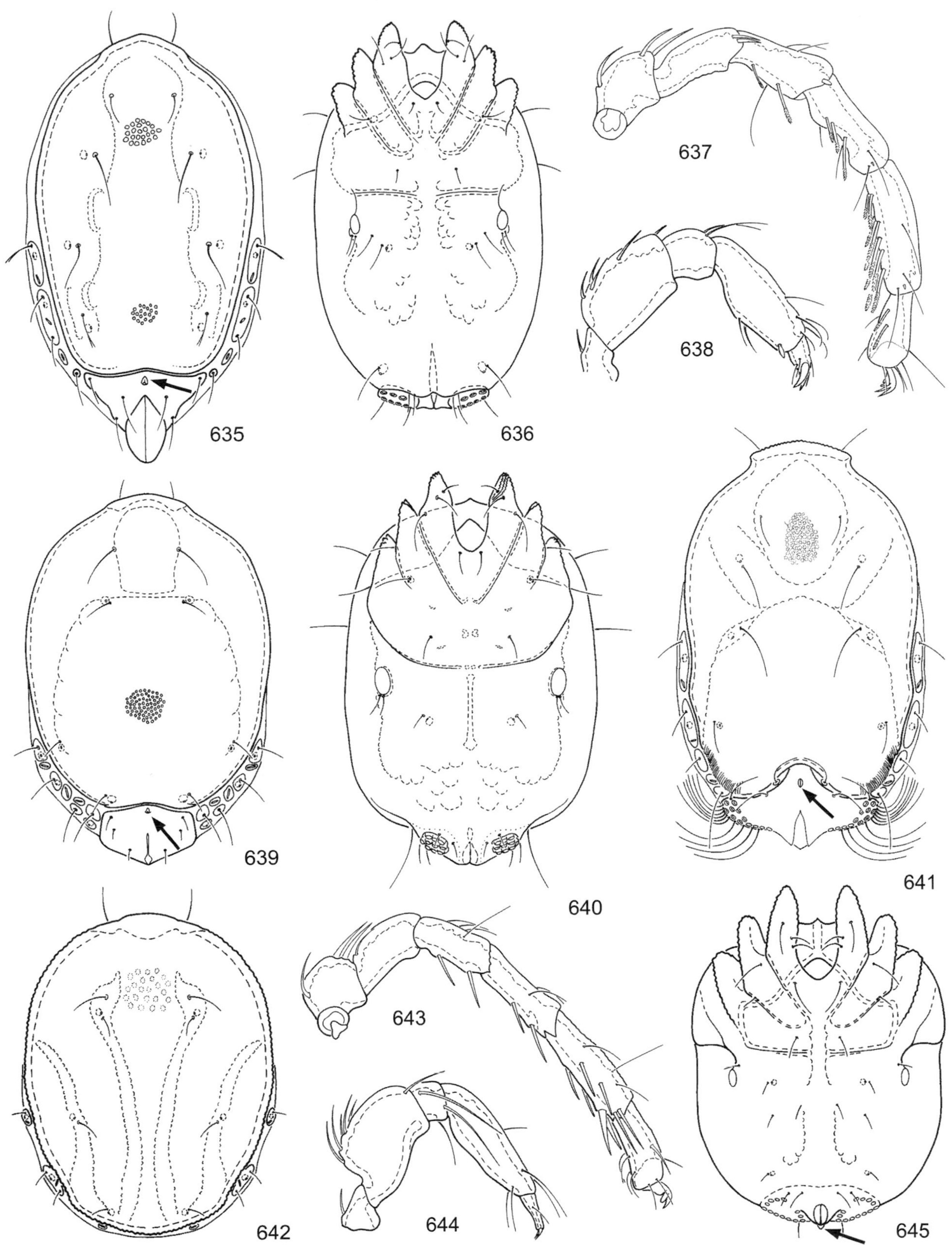

PLATE 16.76 **Figures 635–637:** *Phreatobrachypoda multipora* Cook (Aturinae) adults. **Fig. 635** male dorsum of idiosoma. **Fig. 636** female venter of idiosoma. **Fig. 637** male fourth leg. **Figures 638–640:** *Phreatobrachypoda oregonensis* Smith (Aturinae) adult male. **Fig. 638** pedipalp. **Fig. 639** dorsum of idiosoma. **Fig. 640** ventral shield. **Fig. 641:** *Phreatobrachypoda gledhilli* Smith (Aturinae) adult male dorsum of idiosoma. **Figures 642–645:** *Ameribrachypoda robusta* (Cook) (Aturinae) adult male. **Fig. 642** dorsum of idiosoma. **Fig. 643** fourth leg. **Fig. 644** pedipalp. **Fig. 645** ventral shield.

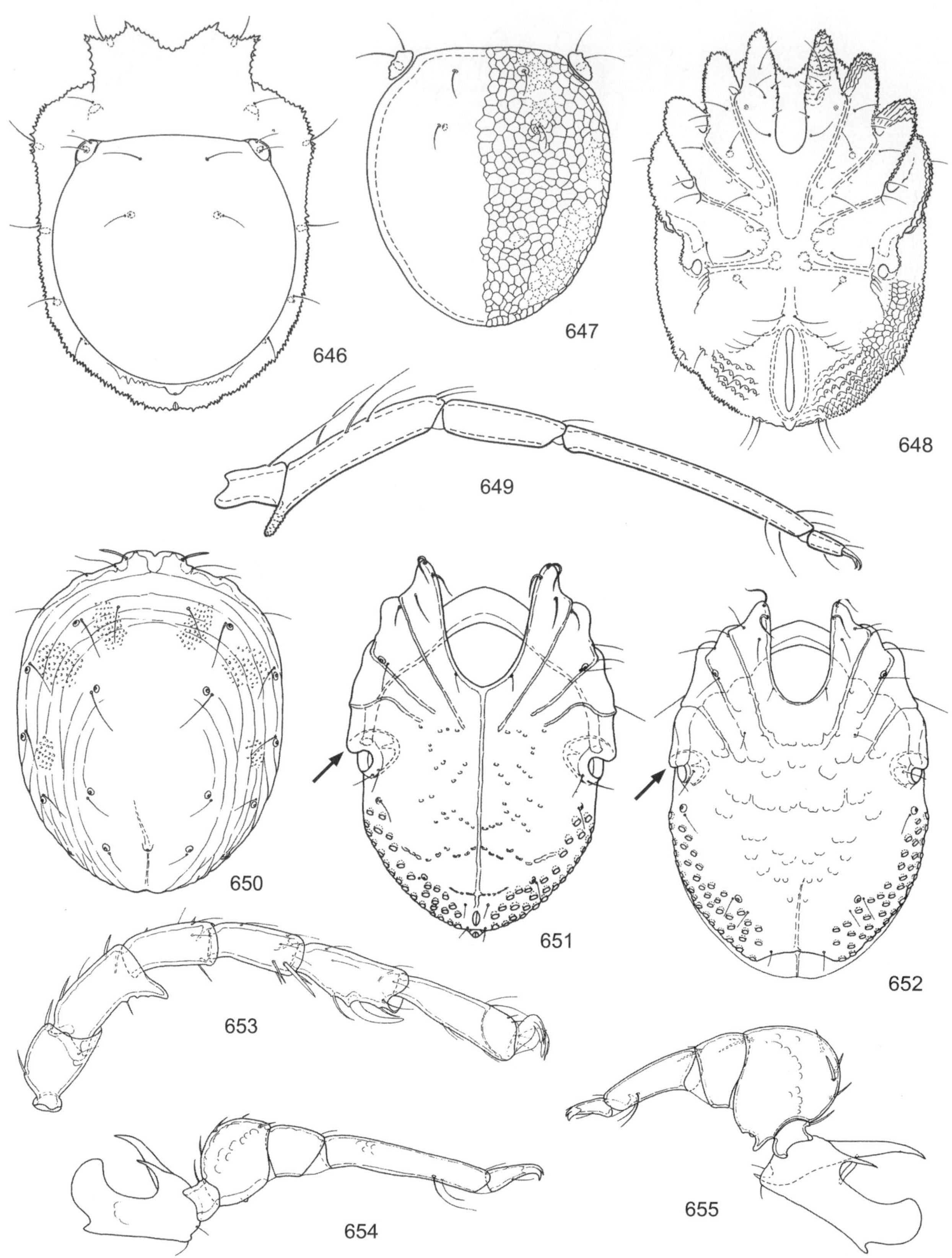

PLATE 16.77 **Figures 646–649:** *Bharatalbia cooki* Smith (Aturinae) adults. **Fig. 646** female dorsum of idiosoma. **Fig. 647** male dorsal shield. **Fig. 648** male ventral shield. **Fig. 649** male pedipalp. **Figures 650–655:** *Kongsbergia globipalpis* Lundblad (Aturinae) adults. **Fig. 650** female dorsal shield. **Fig. 651** male ventral shield. **Fig. 652** female ventral shield. **Fig. 653** male fourth leg. **Fig. 654** female gnathosomal base (capitulum) and pedipalp. **Fig. 655** male gnathosomal base (capitulum) and pedipalp.

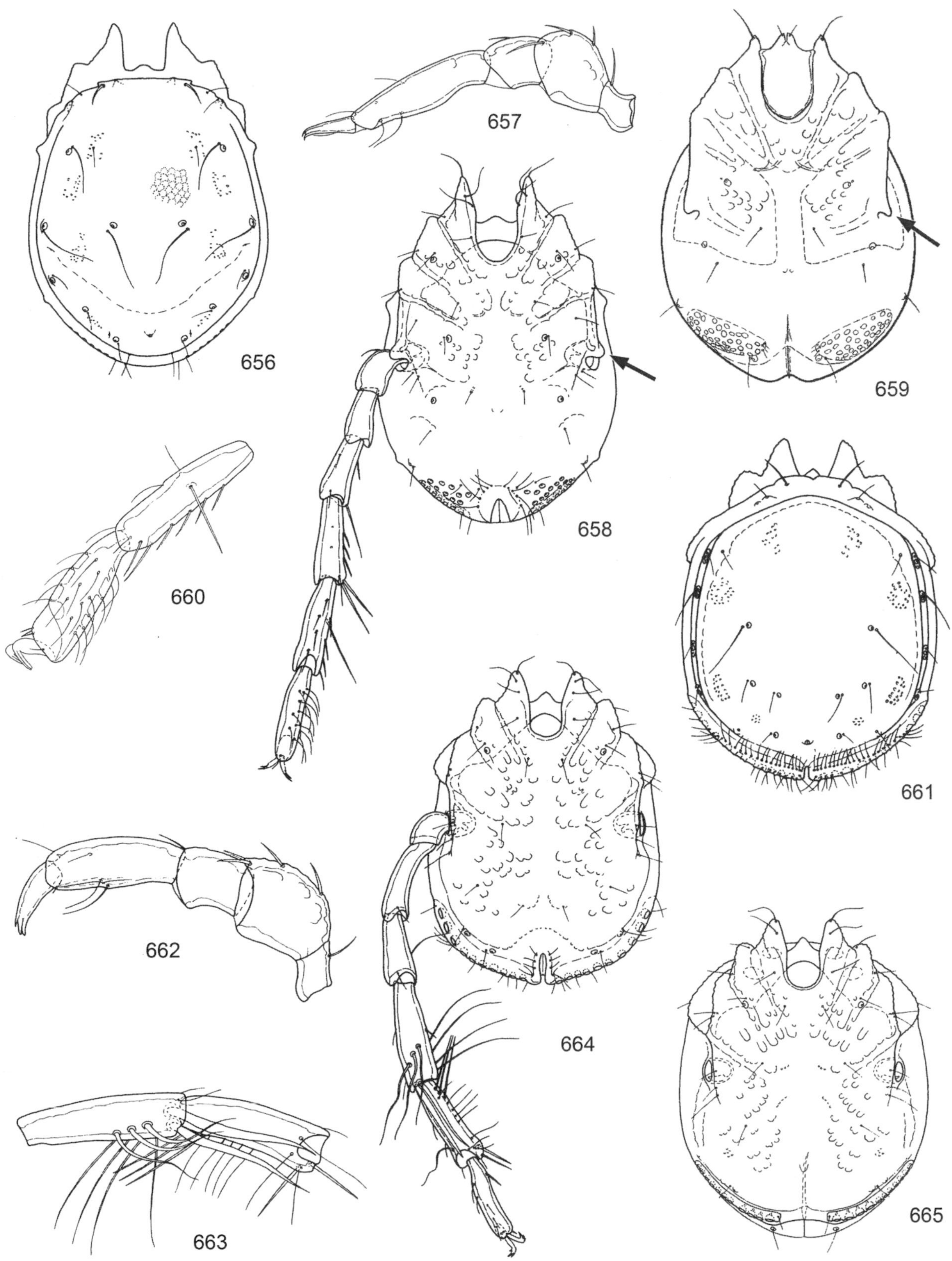

PLATE 16.78 **Figures 656–660:** *Neoaturus kurtvietsi* Cook (Aturinae) adults. **Fig. 656** male dorsum of idiosoma. **Fig. 657** male pedipalp. **Fig. 658** male venter of idiosoma and fourth leg. **Fig. 659** female ventral shield. **Fig. 660** male distal segments of fourth leg. **Figures 661–665:** *Aturus mexicanus* Cook (Aturinae) adults. **Fig. 661** male dorsal shield. **Fig. 662** male pedipalp. **Fig. 663** male genu and tibia of fourth leg. **Fig. 664** male venter of idiosoma and fourth leg. **Fig. 665** female ventral shield.

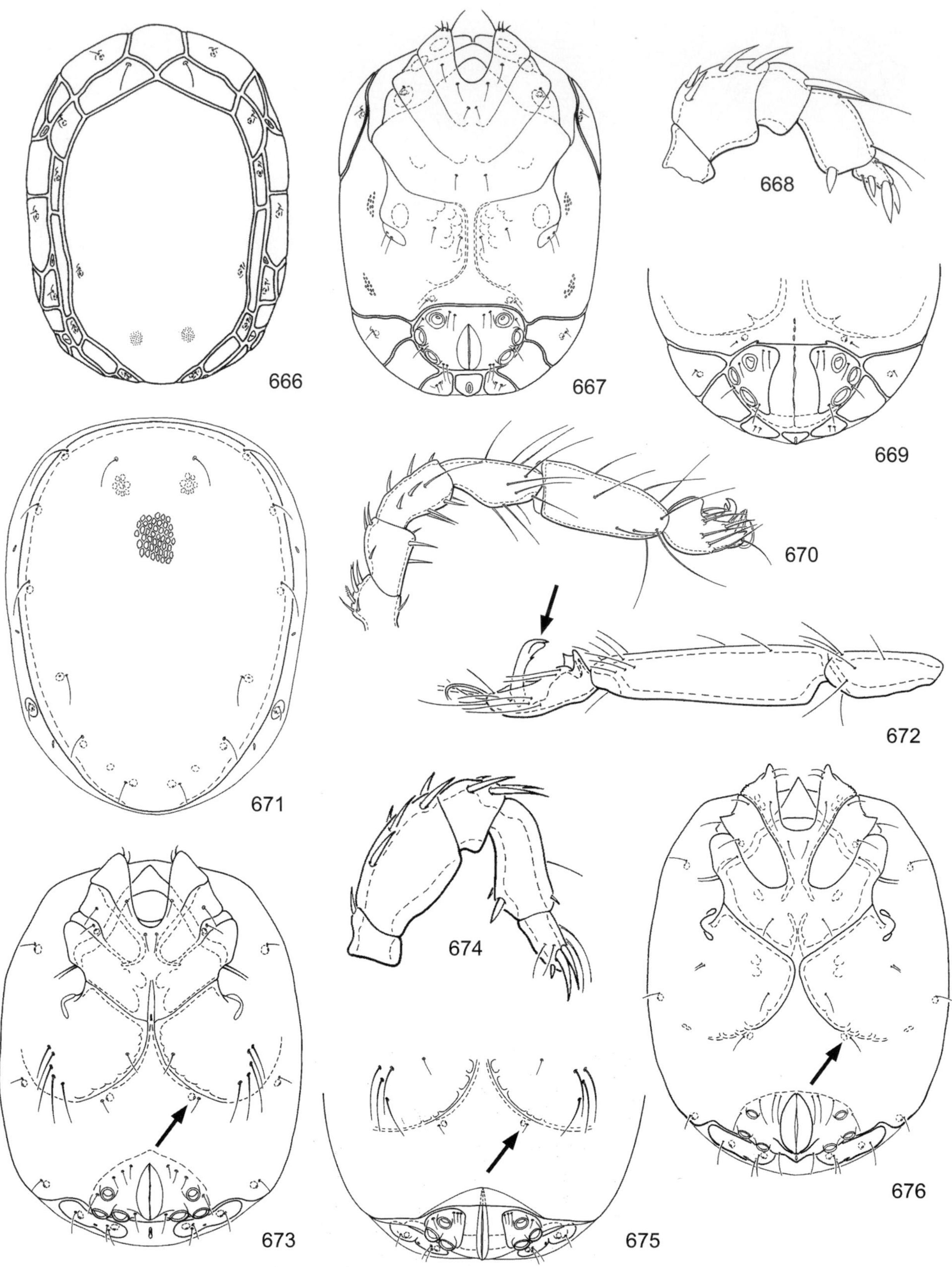

PLATE 16.79 **Figures 666–670:** *Cyclomomonia andrewi* Smith (Cyclomomoniinae) adults. **Fig. 666** male dorsal shield. **Fig. 667** male ventral shield. **Fig. 668** male pedipalp. **Fig. 669** female genital field region. **Fig. 670** male first leg. **Fig. 671:** *Stygomomonia californiensis* Smith (Stygomomoniinae) adult male dorsum of idiosoma. **Fig. 672:** *Stygomomonia riparia* Habeeb (Stygomomoniinae) adult male distal segments of first leg. **Figures 673 & 675:** *Stygomomonia mitchelli* Smith (Stygomomoniinae) adults. **Fig. 673** male venter of idiosoma. **Fig. 675** female venter of idiosoma. **Fig. 674:** *Stygomomonia neomexicana* Cook (Stygomomoniinae) adult male pedipalp. **Fig. 676:** *Stygomomonia mendocinoensis* Smith (Stygomomoniinae) adult male venter of idiosoma.

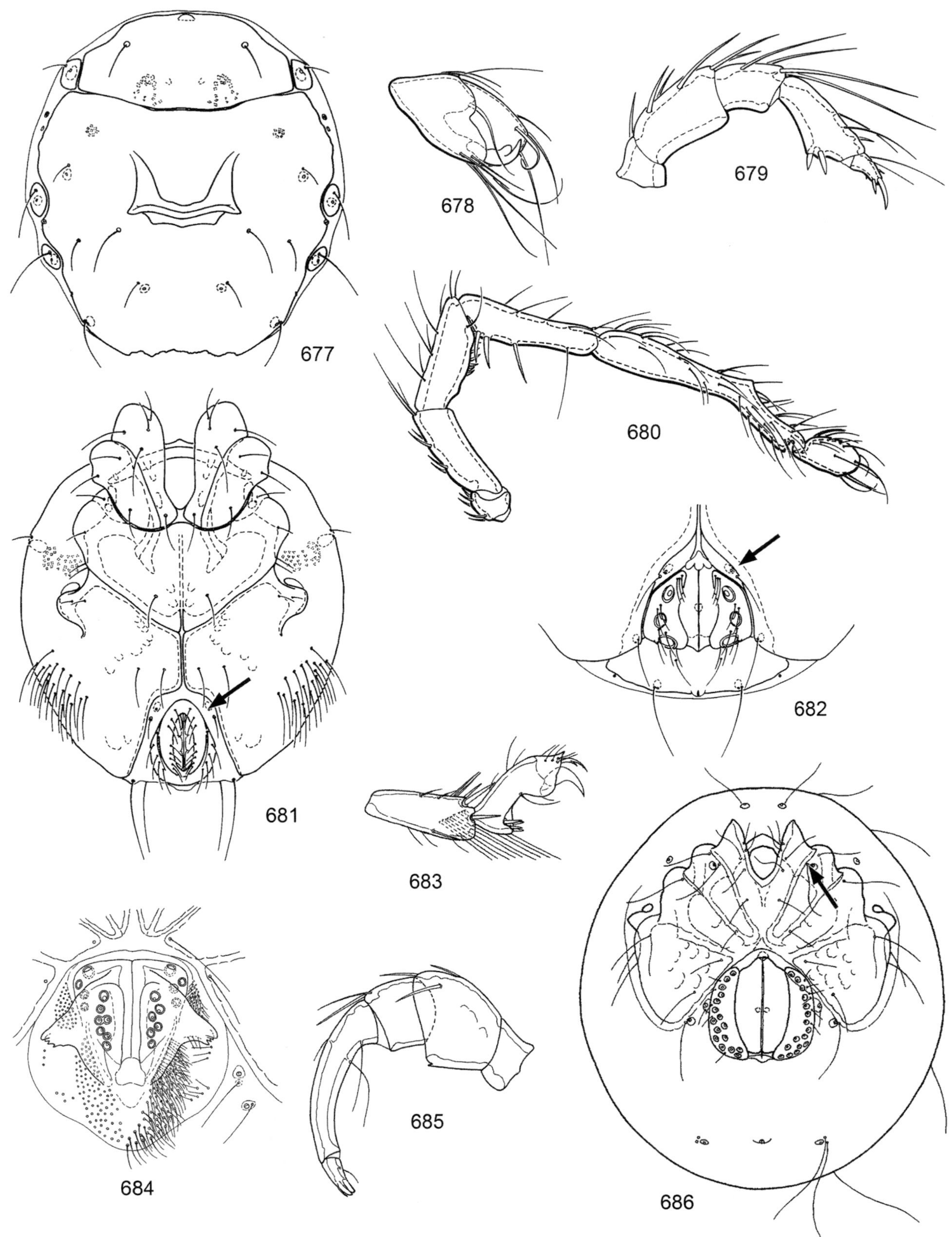

PLATE 16.80 **Figures 677, 679, 681–682:** *Momonia projecta* Cook (Momoniinae) adults. **Fig. 677** male dorsum of idiosoma. **Fig. 679** male pedipalp. **Fig. 681** male venter of idiosoma. **Fig. 682** female genital field region. **Figures 678 & 680** *Momonia campylotibia* Smith (Momoniinae) adult male. **Fig. 678** tarsus of first leg. **Fig. 680** first leg. **Figures 683–684:** *Midea* sp. (Mideidae) adult male. **Fig. 683** distal segments of third leg. **Fig. 684** genital field. **Figures 685–686:** *Midea expansa* Marshall (Mideidae) adult female. **Fig. 685** pedipalp. **Fig. 686** ventral shield.

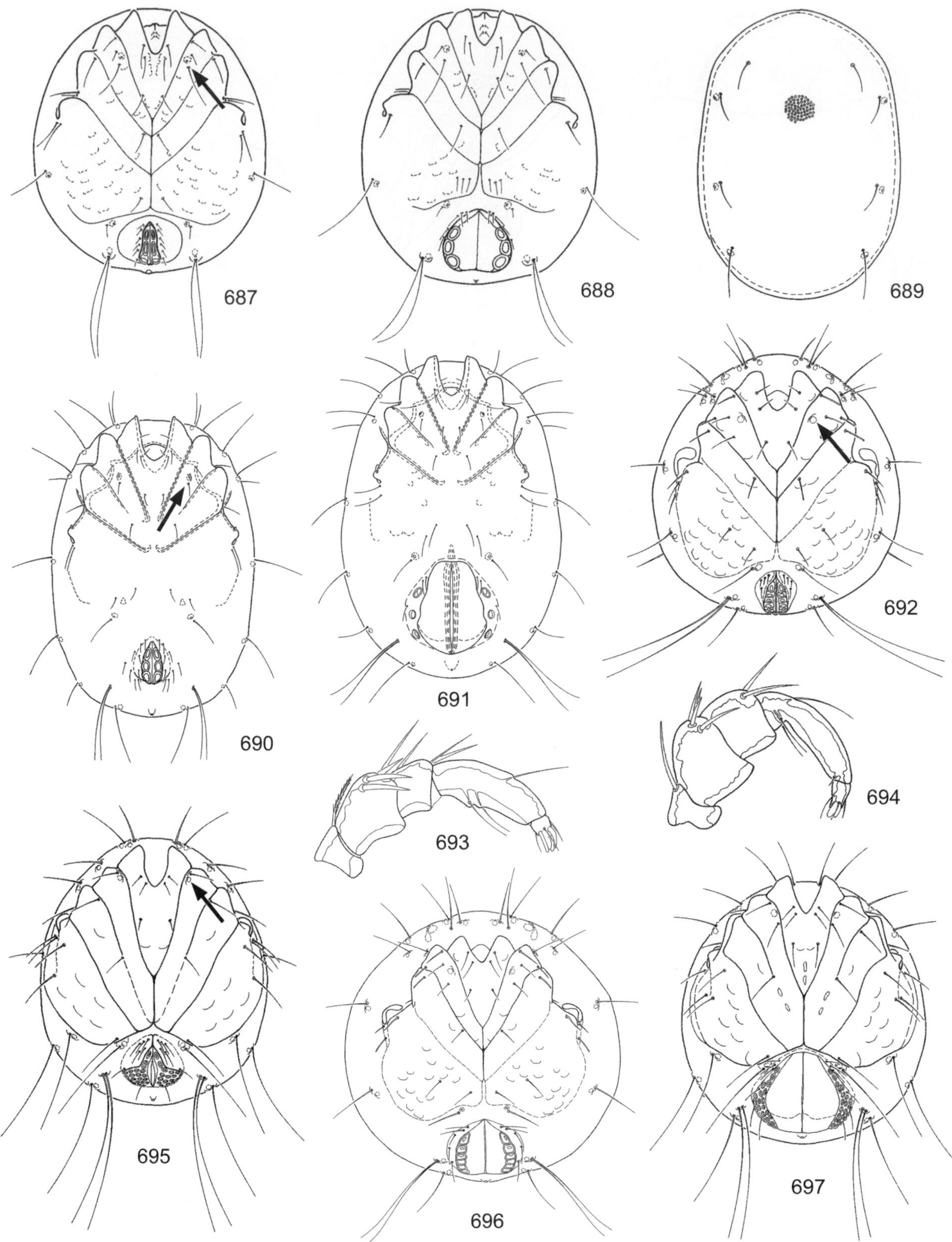

PLATE 16.81 **Figures 687–688:** *Nudomideopsis magnacetabula* (Smith) (Nudomideopsidae) adults. **Fig. 687** male ventral shield. **Fig. 688** female ventral shield. **Figures 689–691:** *Allomideopsis wichitaensis* (Smith) (Nudomideopsidae) adults. **Fig. 689** male dorsal shield. **Fig. 690** male ventral shield. **Fig. 691** female ventral shield. **Figures 692–693 & 696:** *Paramideopsis susanae* Smith (Nudomideopsidae) adults. **Fig. 692** male ventral shield. **Fig. 693** male pedipalp. **Fig. 696** female ventral shield. **Figures 694–695 & 697:** *Neomideopsis siuslawensis* Smith (Nudomideopsidae) adults. **Fig. 694** male pedipalp. **Fig. 695** male ventral shield. **Fig. 697** female ventral shield.

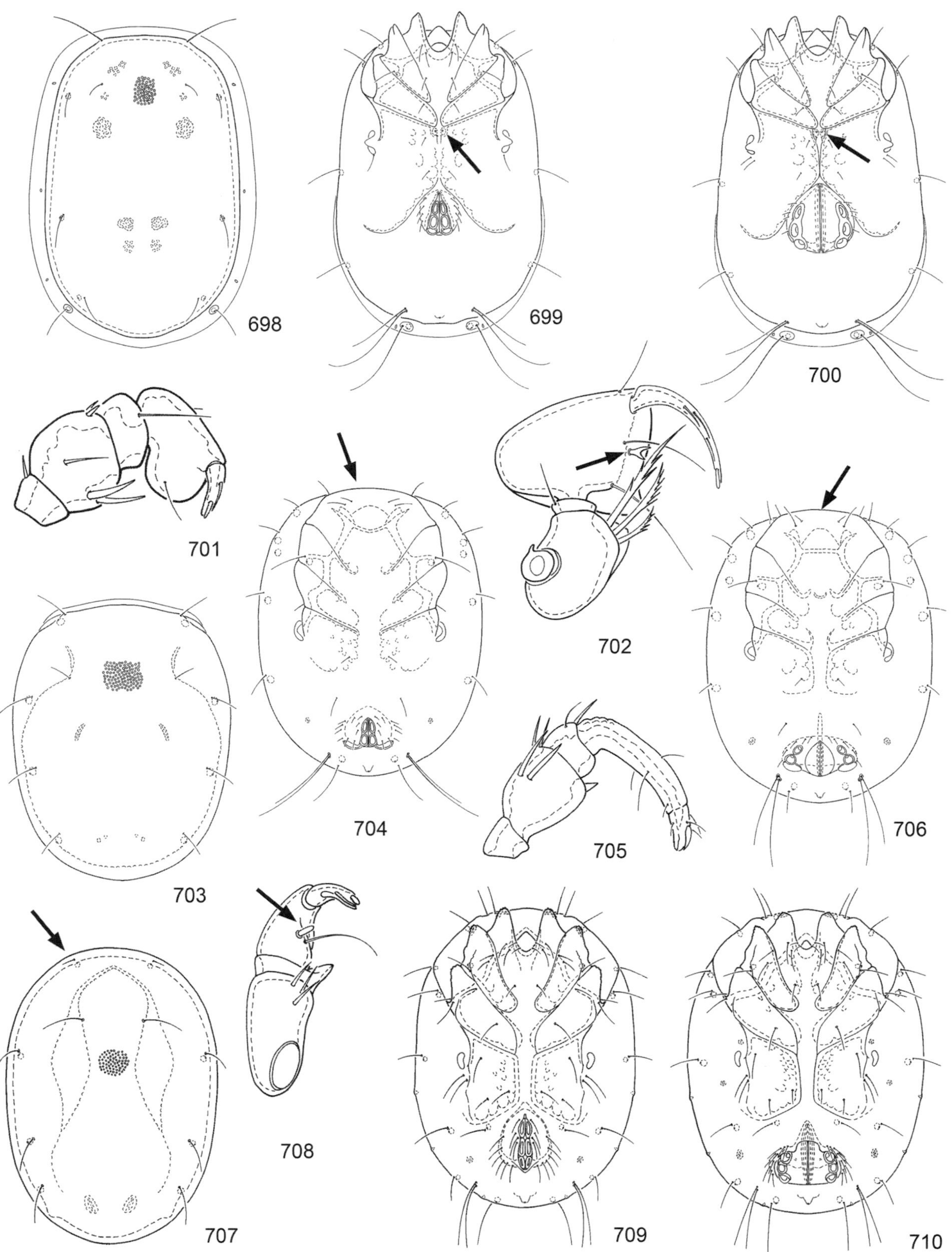

PLATE 16.82 **Figures 698–701:** *Chappuisides cooki* Smith (Chappuisidinae) adults. **Fig. 698** male dorsum of idiosoma. **Fig. 699** male venter of idiosoma. **Fig. 700** female venter of idiosoma. **Fig. 701** male pedipalp. **Fig. 702:** *Uchidastygacarus imamurai* Cook (Uchidastygacarinae) adult male pedipalp. **Figures 703–704 & 706:** *Uchidastygacarus acadiensis* Smith (Uchidastygacarinae) adults. **Fig. 703** male dorsal shield. **Fig. 704** male ventral shield. **Fig. 706** female ventral shield. **Figures 705, 707 & 709–710:** *Morimotacarus nearcticus* Smith (Morimotacarinae) adults. **Fig. 705** male pedipalp. **Fig. 707** male dorsal shield. **Fig. 709** male ventral shield. **Fig. 710** female ventral shield. **Fig. 708** *Uchidastygacarus ovalis* Cook (Uchidastygacarinae) adult male pedipalp.

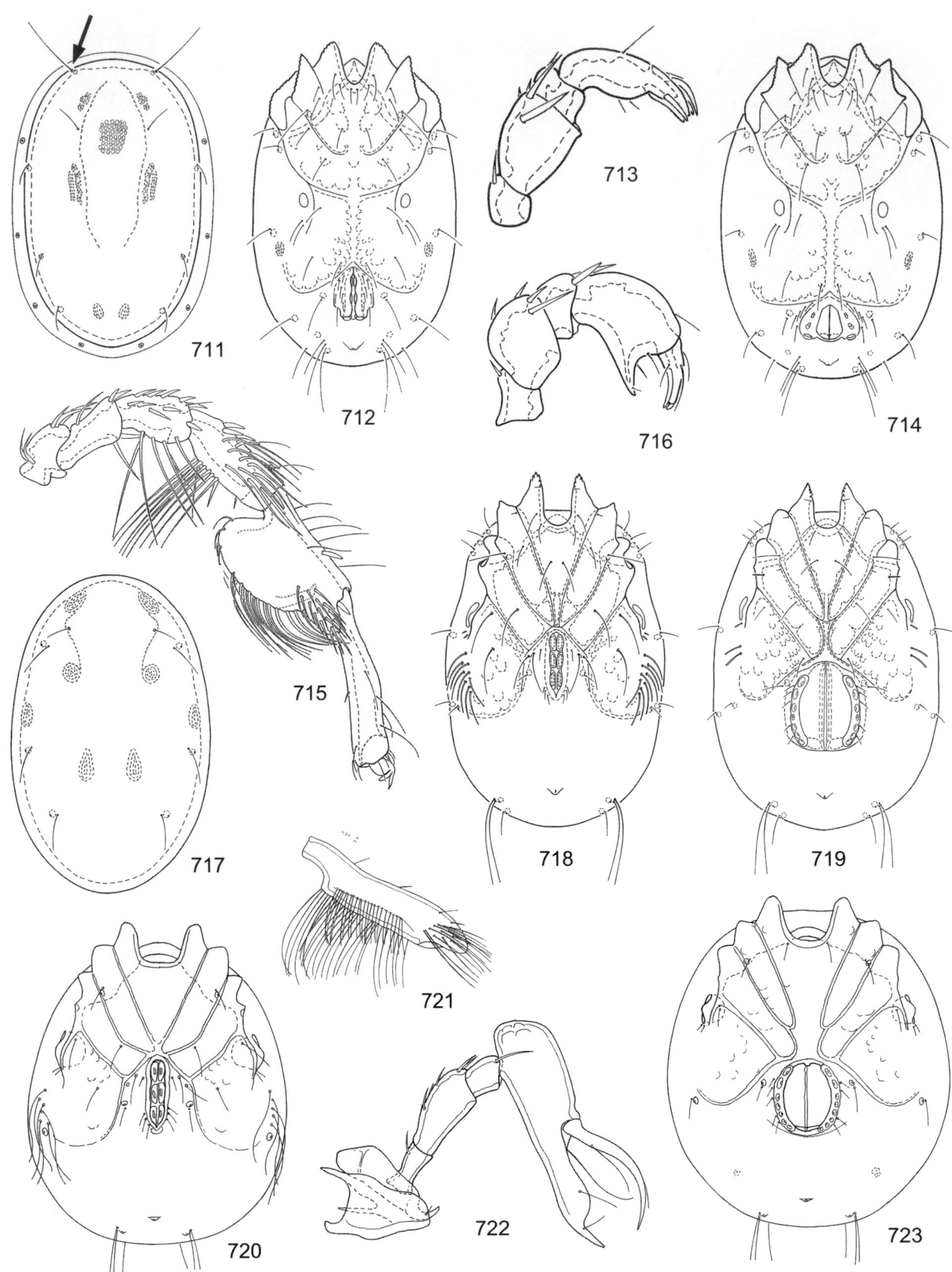

PLATE 16.83 **Figures 711–714:** *Yachatsia mideopsoides* Cook (Morimotacarinae) adults. **Fig. 711** male dorsum of idiosoma. **Fig. 712** male ventral shield. **Fig. 713** male pedipalp. **Fig. 714** female ventral shield. **Figures 715–719:** *Neoacarus californicus* Smith (Neoacaridae) adults. **Fig. 715** male third leg. **Fig. 716** male pedipalp. **Fig. 717** male dorsal shield. **Fig. 718** male ventral shield. **Fig. 719** female ventral shield. **Figures 720–723:** *Volsellacarus ovalis* Cook (Neoacaridae) adults. **Fig. 720** male ventral shield. **Fig. 721** male tibia of third leg. **Fig. 722** male pedipalp. **Fig. 723** female ventral shield.

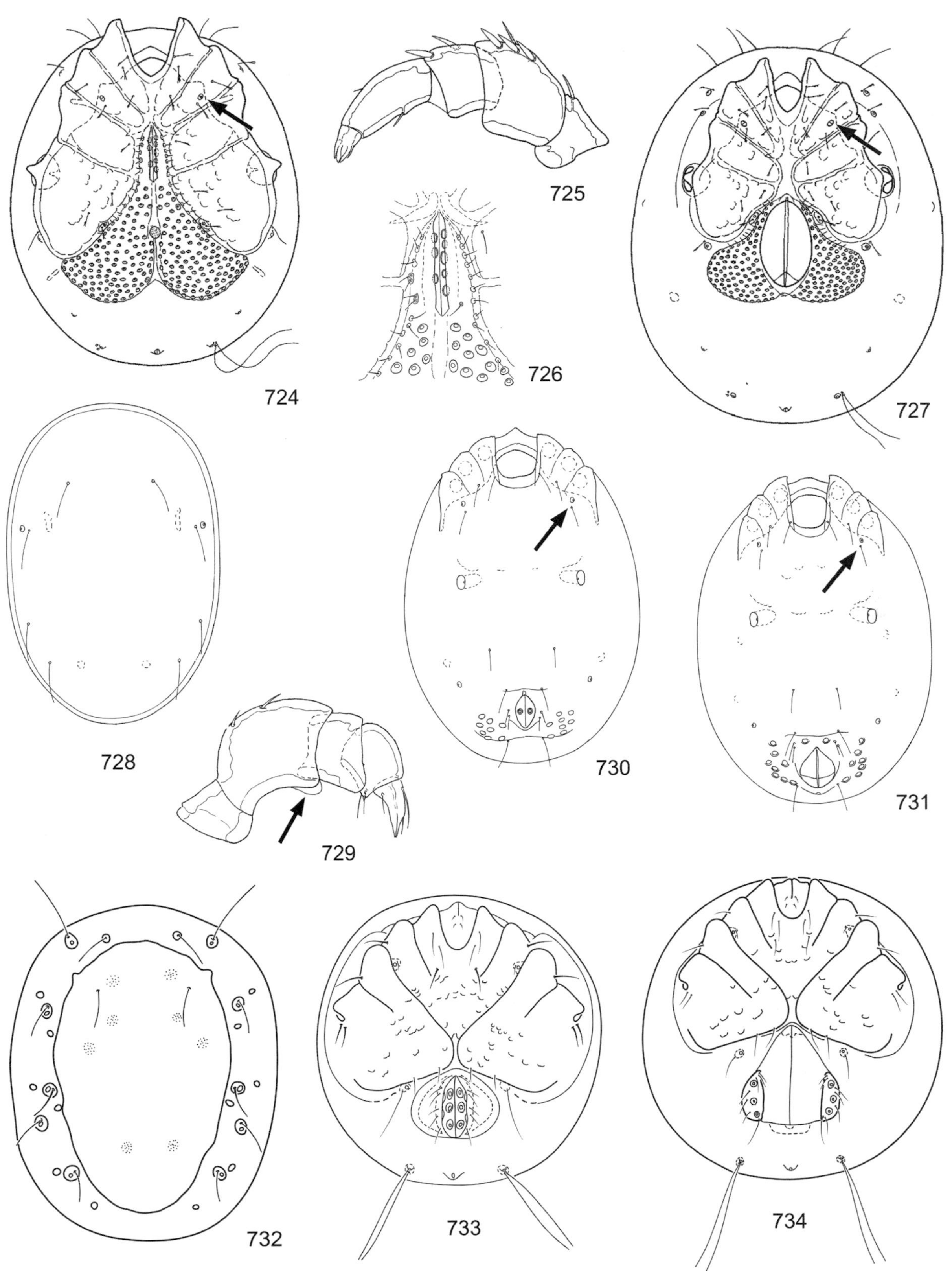

PLATE 16.84 Figures 724–727: *Laversia berulophila* Cook (Laversiidae) adults. **Fig. 724** male ventral shield. **Fig. 725** female pedipalp. **Fig. 726** male detail of genital field region. **Fig. 727** female ventral shield. **Figures 728–731:** *Arenohydracarus eremitus* Cook (Arenohydracaridae) adults. **Fig. 728** male dorsal shield. **Fig. 729** male pedipalp. **Fig. 730** male ventral shield. **Fig. 731** female ventral shield. **Figures 732–734:** *Paenecalyptonotus fontinalis* Smith (Acalyptonotidae) adults. **Fig. 732** male dorsum of idiosoma. **Fig. 733** male venter of idiosoma. **Fig. 734** female venter of idiosoma.

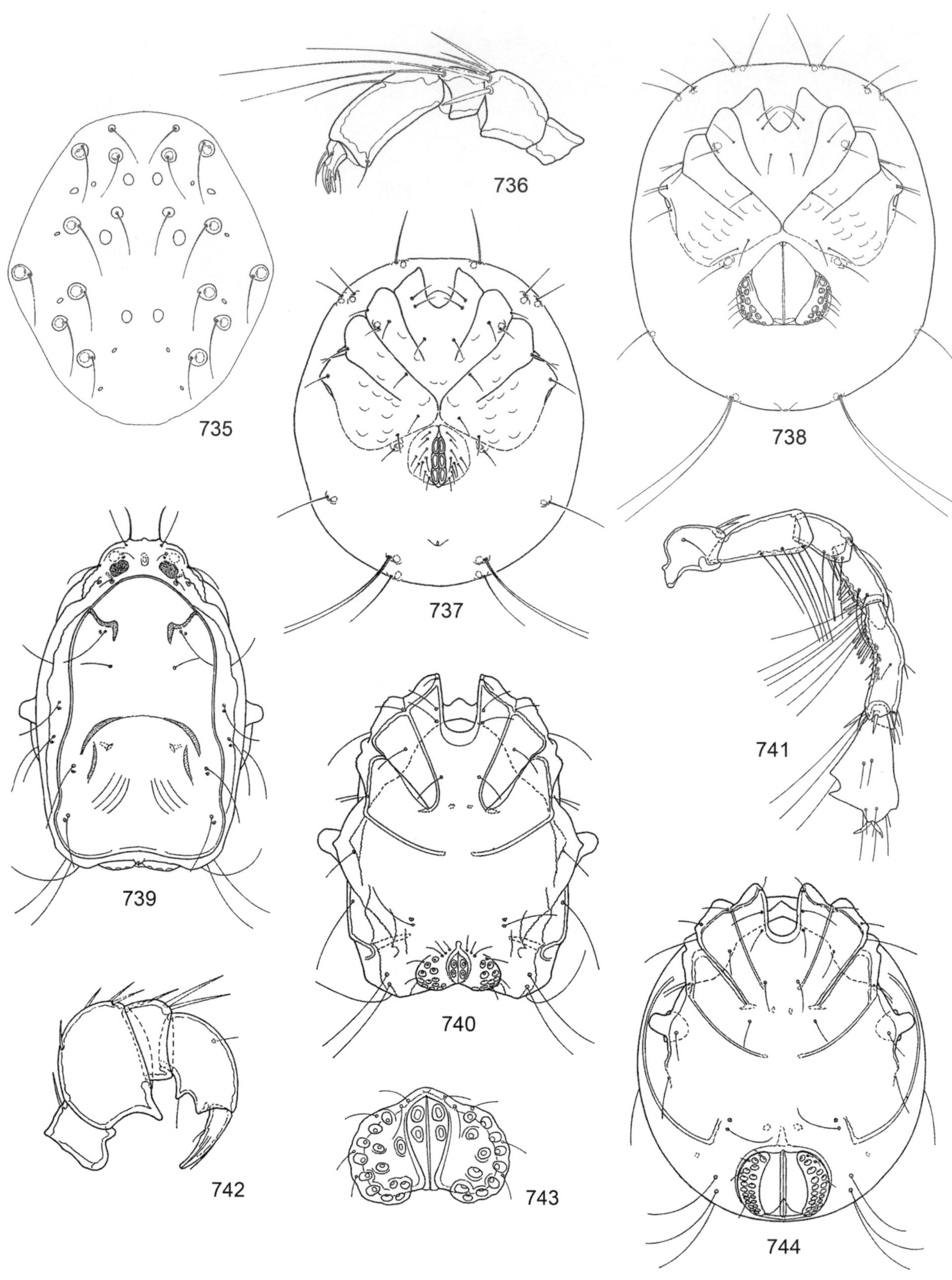

PLATE 16.85 Figures 735–737: *Acalyptonotus pacificus* Smith (Acalyptonotidae) adult male. **Fig. 735** dorsum of idiosoma. **Fig. 736** pedipalp. **Fig. 737** ventral shield. **Fig. 738:** *Acalyptonotus neoviolaceus* Smith (Acalyptonotidae) adult female ventral shield. **Figures 739–744:** *Amoenacarus dixiensis* Smith and Cook (Amoenacaridae) adults. **Fig. 739** male dorsum of idiosoma. **Fig. 740** male ventral shield. **Fig. 741** male fourth leg. **Fig. 742** male pedipalp. **Fig. 743** male genital field. **Fig. 744** female ventral shield.

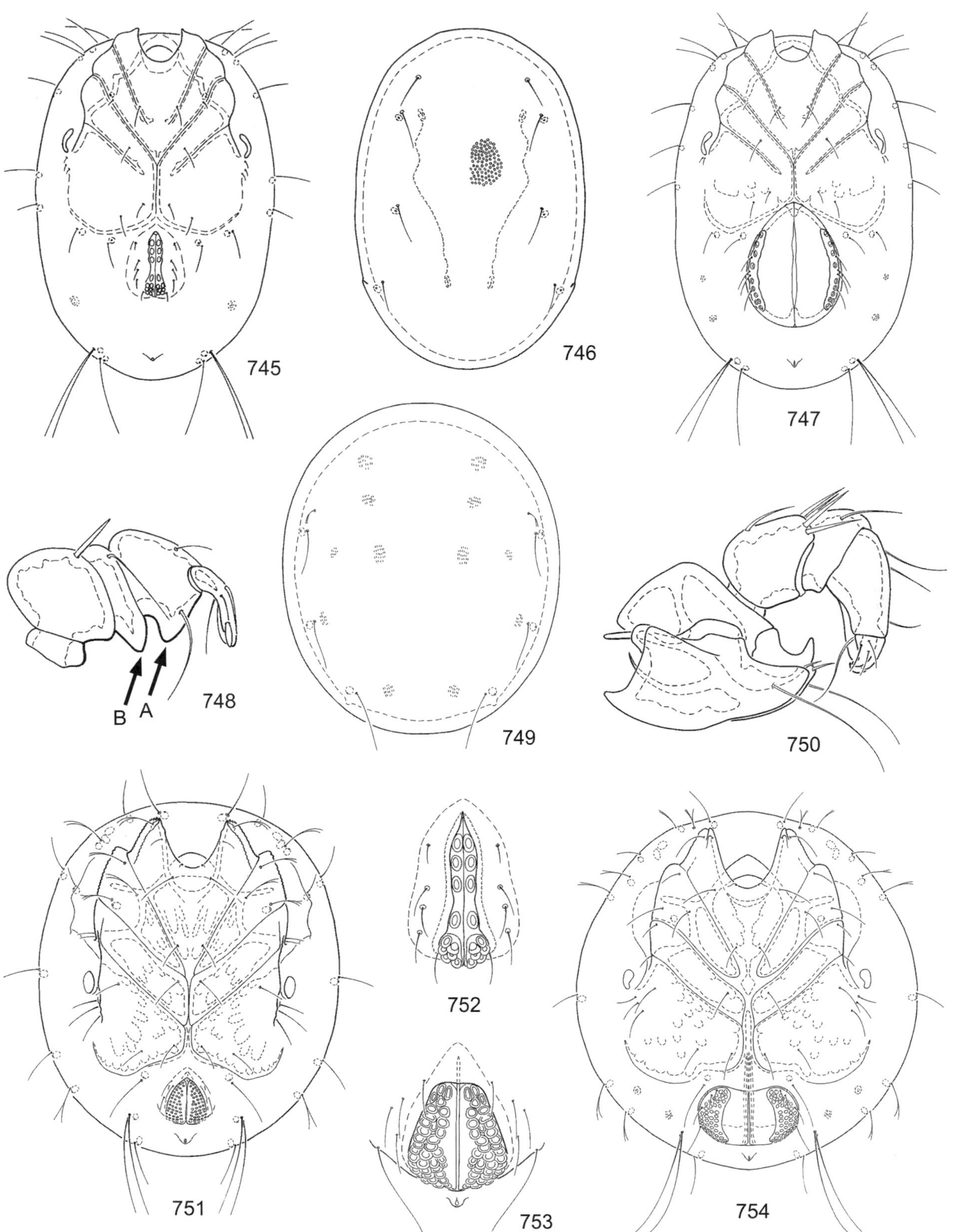

PLATE 16.86 **Figures 745–748 & 752**: *Stygameracarus cooki* Smith (Stygameracarinae) adults. **Fig. 745** male ventral shield. **Fig. 746** male dorsal shield. **Fig. 747** female ventral shield. **Fig. 748** male pedipalp. **Fig. 752** male genital field. **Figures 749 & 753–754:** *Chelomideopsis besselingi* (Cook) (Athienemanniinae) adults. **Fig. 749** male dorsal shield. **Fig. 753** male genital field. **Fig. 754** female ventral shield. **Figures 750–751:** *Chelomideopsis siskiyouensis* Smith (Athienemanniinae) adult male. **Fig. 750** gnathosomal base (capitulum) and pedipalp. **Fig. 751** ventral shield.

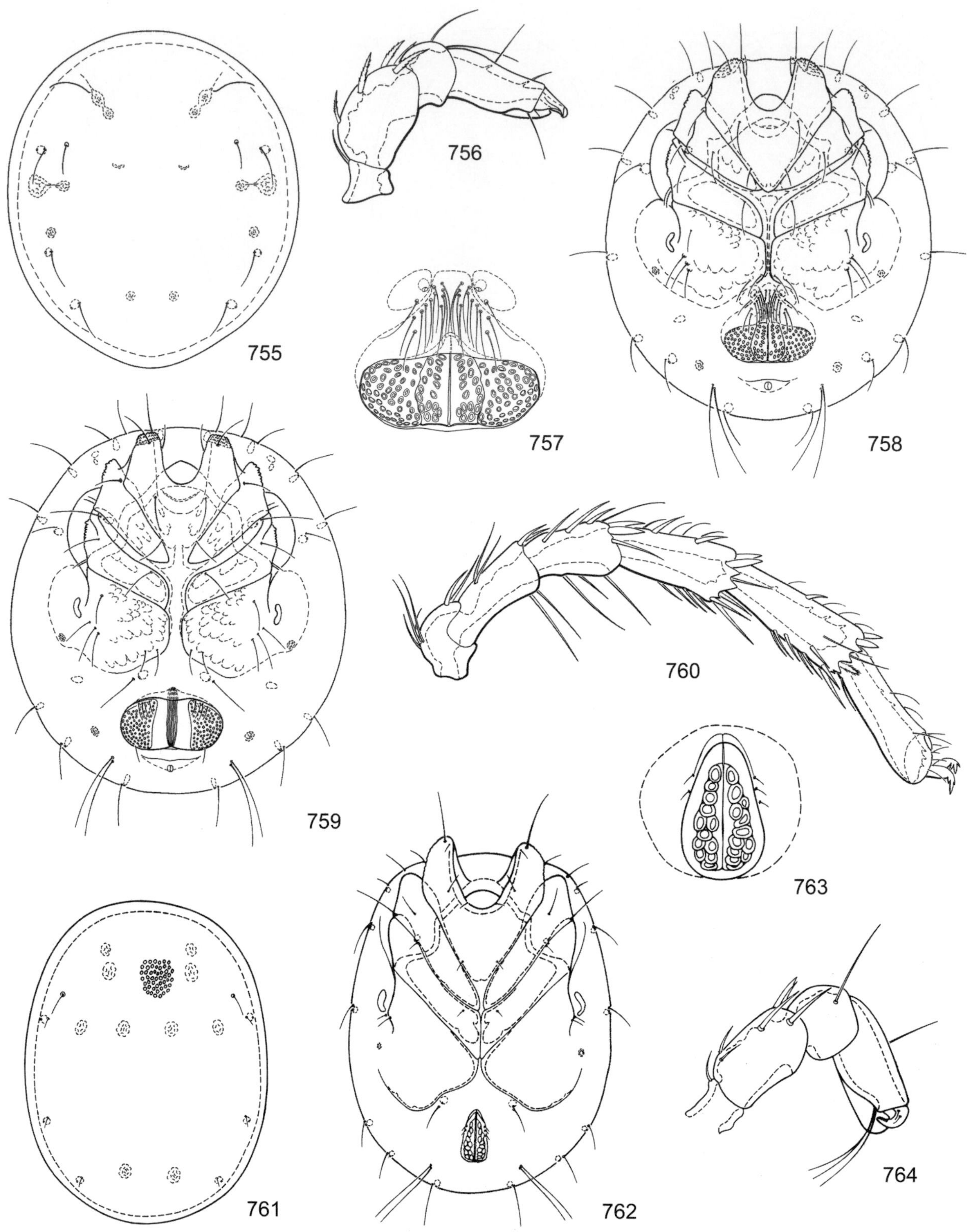

PLATE 16.87 **Figures 755–760:** *Platyhydracarus juliani* Smith (Athienemanniinae) adults. **Fig. 755** male dorsal shield. **Fig. 756** male pedipalp. **Fig. 757** male genital field. **Fig. 758** male ventral shield. **Fig. 759** female ventral shield. **Fig. 760** male first leg. **Figures 761–764:** *Chelohydracarus navarrensis* Smith (Athienemanniinae) adult male. **Fig. 761** dorsal shield. **Fig. 762** ventral shield. **Fig. 763** genital field. **Fig. 764** pedipalp.

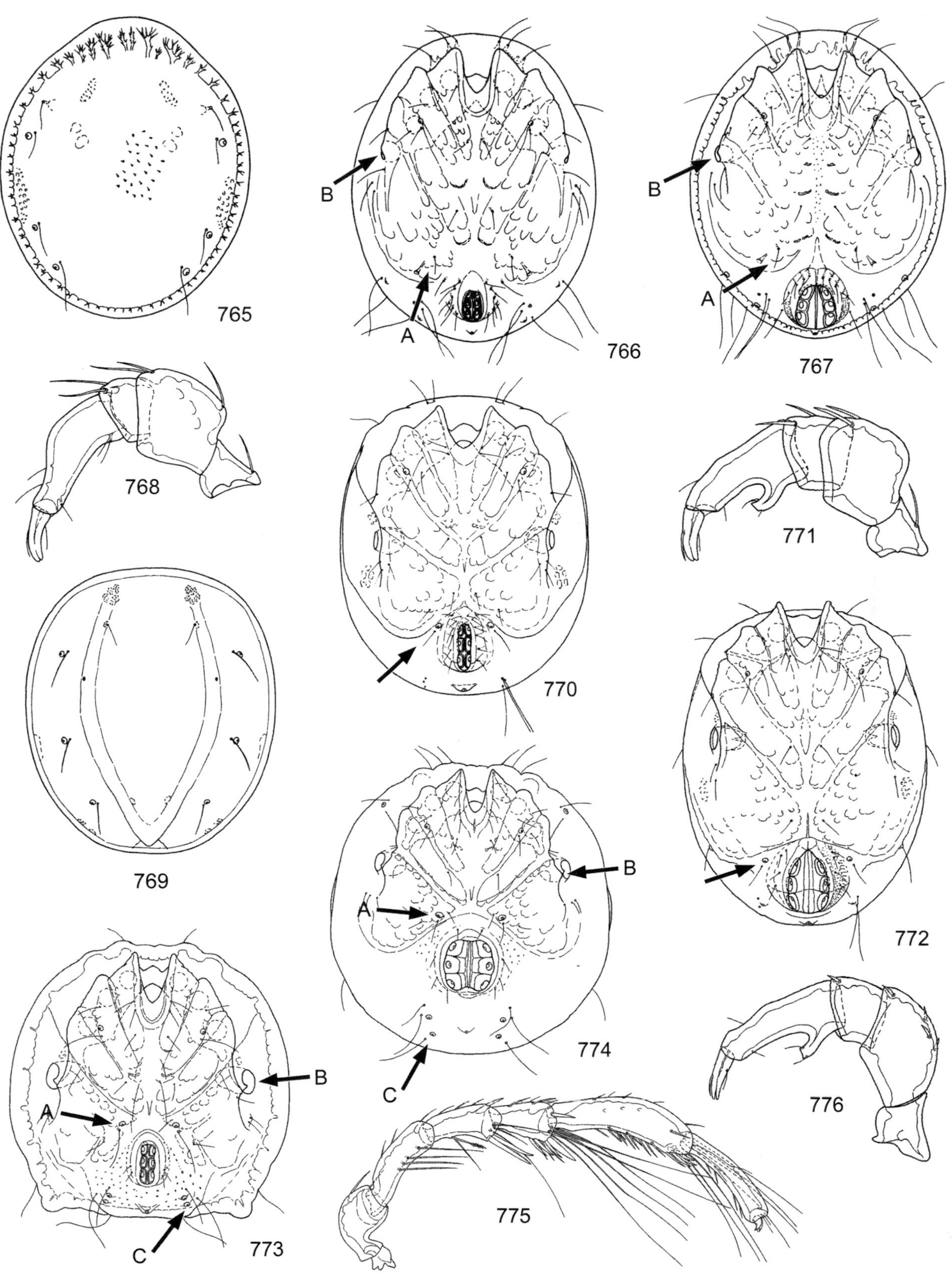

PLATE 16.88 **Figures 765–768:** *Mideopsides beta* Cook (Mideopsidae) adults. **Fig. 765** male dorsal shield. **Fig. 766** male ventral shield. **Fig. 767** female ventral shield. **Fig. 768** female pedipalp. **Fig. 769–770:** *Xystonotus mexicana* (Cook) (Mideopsidae) adults. **Fig. 769** female dorsal shield. **Fig. 770** male ventral shield. **Figures 771–772:** *Xystonotus interstitialis* Cook (Mideopsidae) adult female. **Fig. 771** pedipalp. **Fig. 772** ventral shield. **Figures 773–776:** *Neoxystonotus reelfootensis* (Hoff) (Mideopsidae) adults. **Fig. 773** male ventral shield. **Fig. 774** female ventral shield. **Fig. 775** male fourth leg. **Fig. 776** female pedipalp.

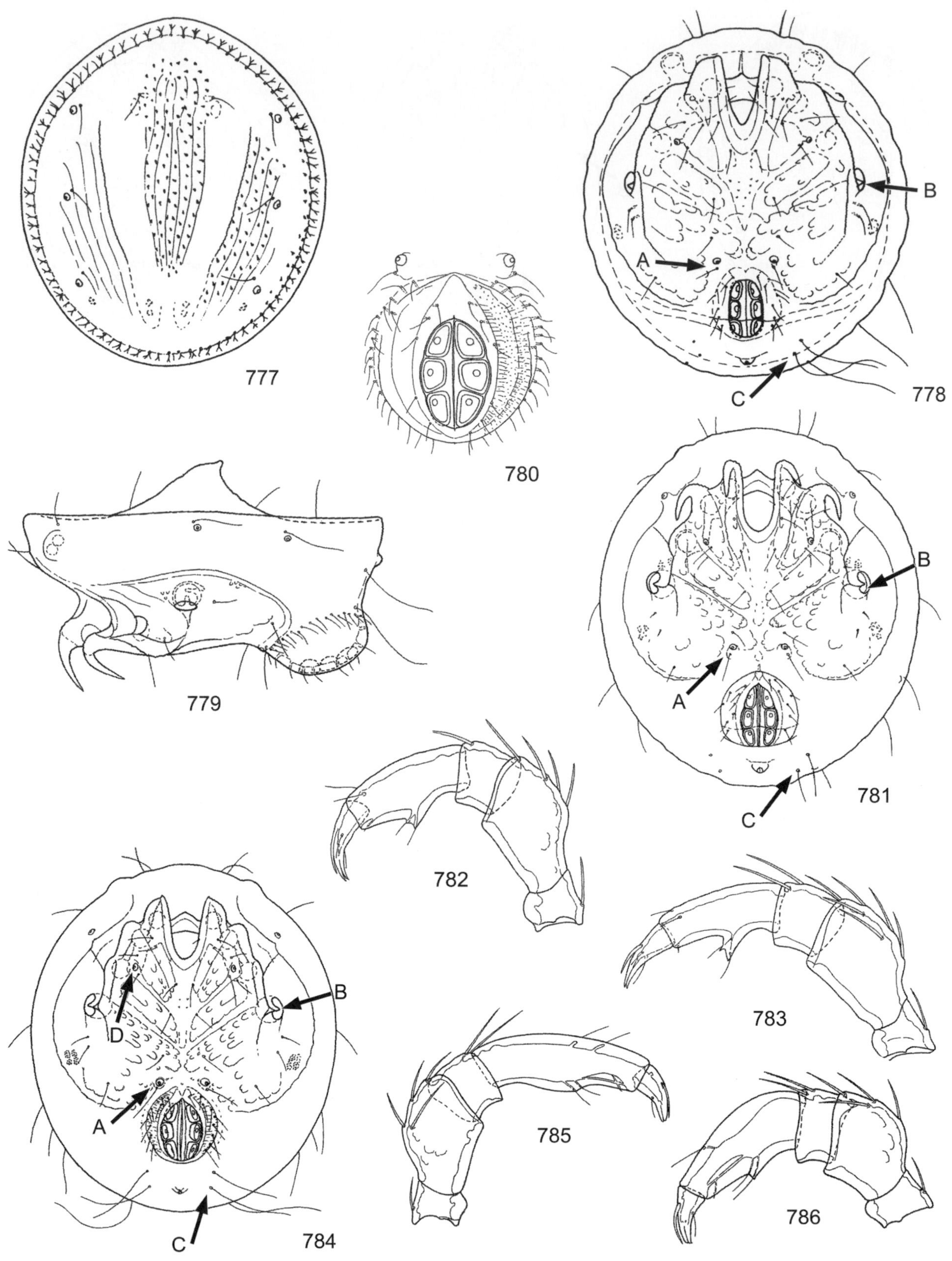

PLATE 16.89 **Figures 777–778:** *Mideopsis cartesa* Cook (Mideopsinae) adult female. **Fig. 777** dorsal shield. **Fig. 778** ventral shield. **Figures 779–780:** *Mideopsis gladiator* Habeeb (Mideopsinae) adults. **Fig. 779** male lateral view of idiosoma. **Fig. 780** male genital field. **Fig. 781** female ventral shield. **Fig. 782** female pedipalp. **Fig. 783:** *Mideopsis jacunda* Cook (Mideopsinae) adult male pedipalp. **Figures 784–785:** *Mideopsis marshallae* Cook (Mideopsinae) adult female. **Fig. 784** ventral shield. **Fig. 785** pedipalp. **Fig. 786:** *Mideopsis lamellipalpis* Lundblad (Mideopsinae) adult male pedipalp.

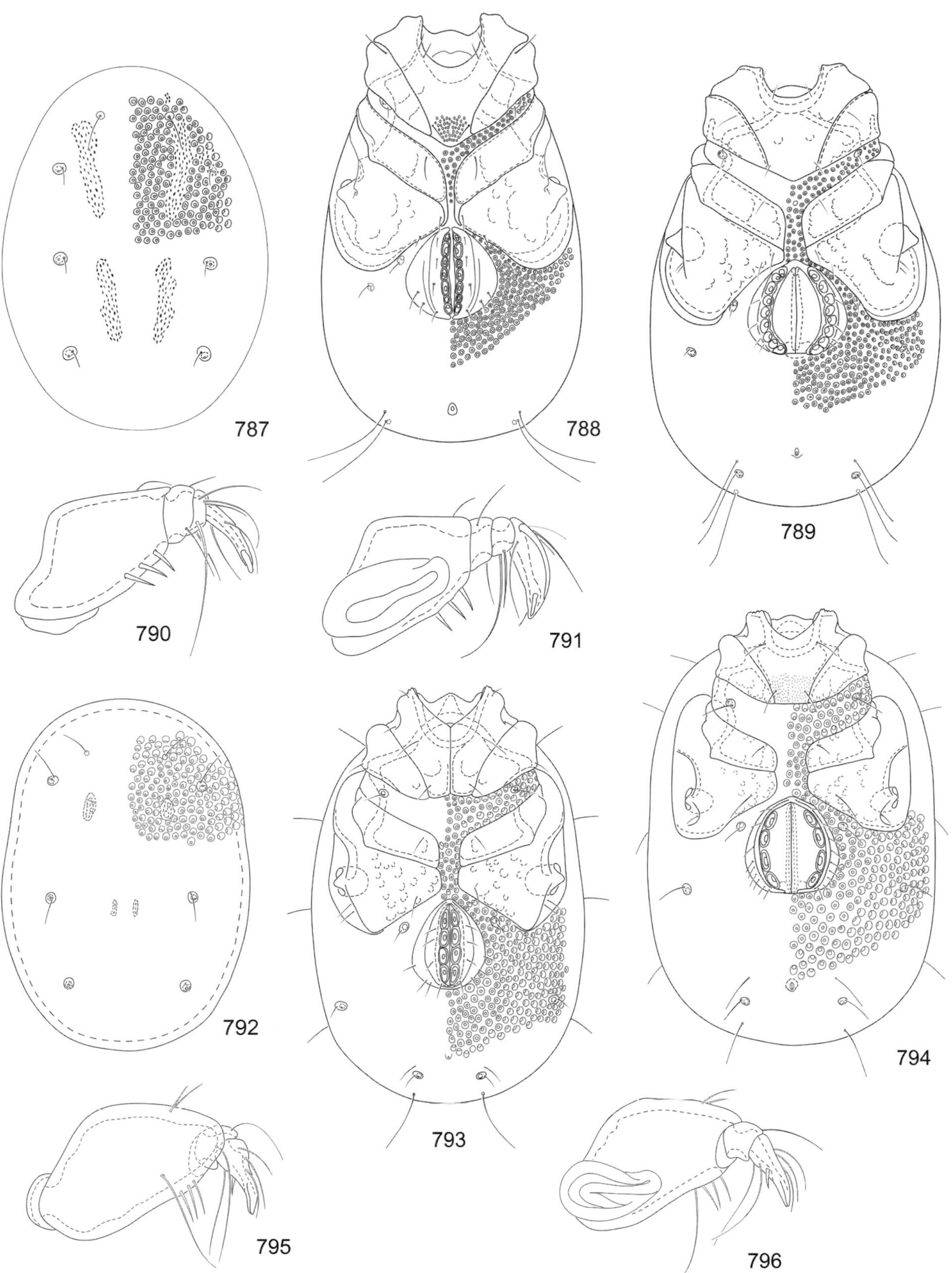

PLATE 16.90 **Figures 787–791:** *Bogatia appalachiana* Smith and MacKenzie (Bogatiinae) adults. **Fig. 787** male dorsal shield. **Fig. 788** male ventral shield. **Fig. 789** female ventral shield. **Fig. 790** male pedipalp dorsal view. **Fig. 791** male pedipalp ventral view. **Figures 792–796:** *Horreolanus orphanus* Mitchell (Horreolaninae) adults. **Fig. 792** male dorsal shield. **Fig. 793** male ventral shield. **Fig. 794** female ventral shield. **Fig. 795** male pedipalp dorsal view. **Fig. 796** male pedipalp ventral view.

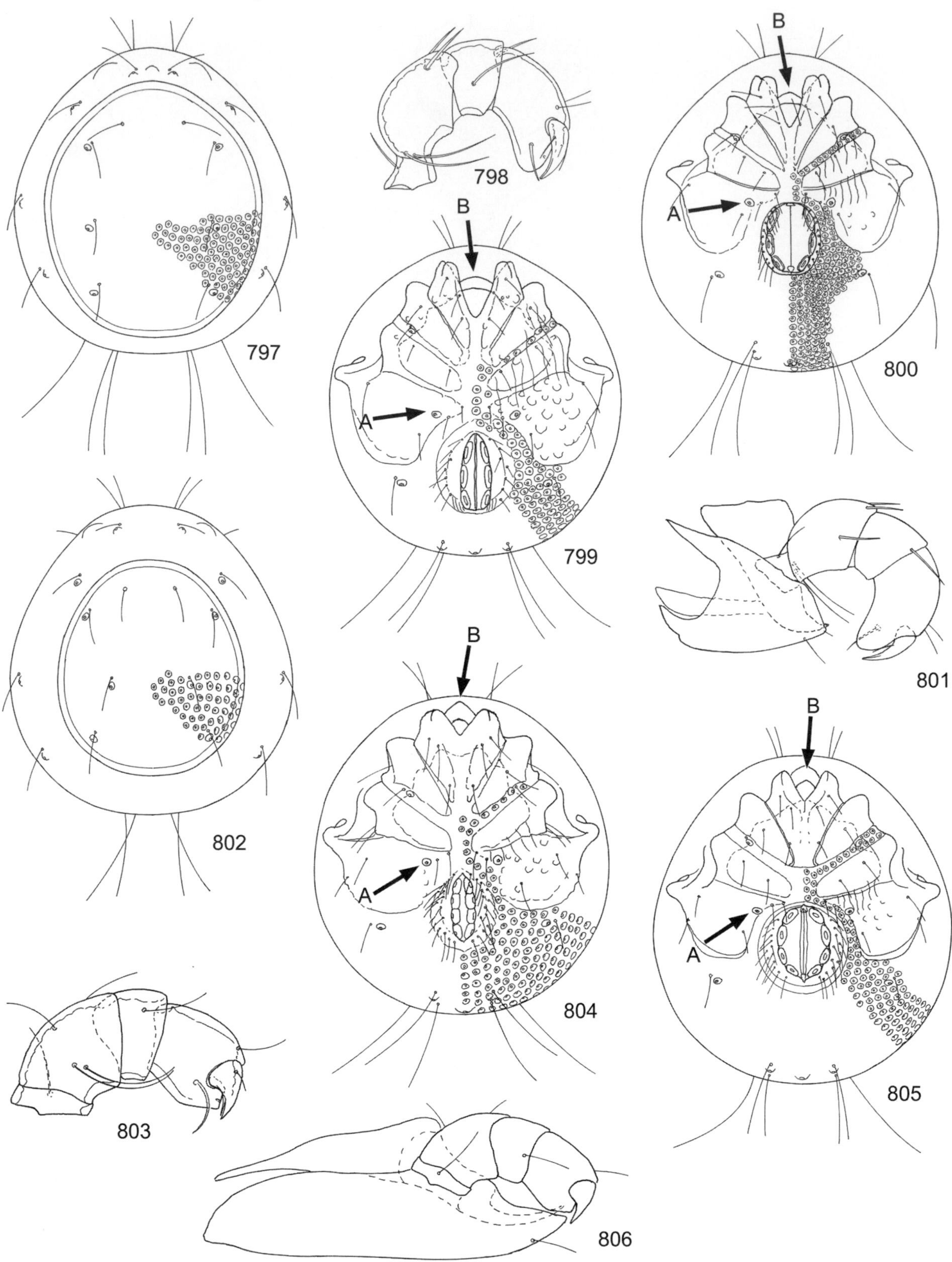

PLATE 16.91 Figures 797–801: *Krendowskia similis* Viets (Krendowskiidae) adults. **Fig. 797** male dorsum of idiosoma. **Fig. 798** male pedipalp. **Fig. 799** male ventral shield. **Fig. 800** female ventral shield. **Fig. 801** male gnathosomal base (capitulum) and pedipalp. **Figures 802–806:** *Geayia ovata* Wolcott (Krendowskiidae) adults. **Fig. 802** male dorsal shield. **Fig. 803** male pedipalp. **Fig. 804** male ventral shield. **Fig. 805** female ventral shield. **Fig. 806** male gnathosomal base (capitulum) and pedipalp.

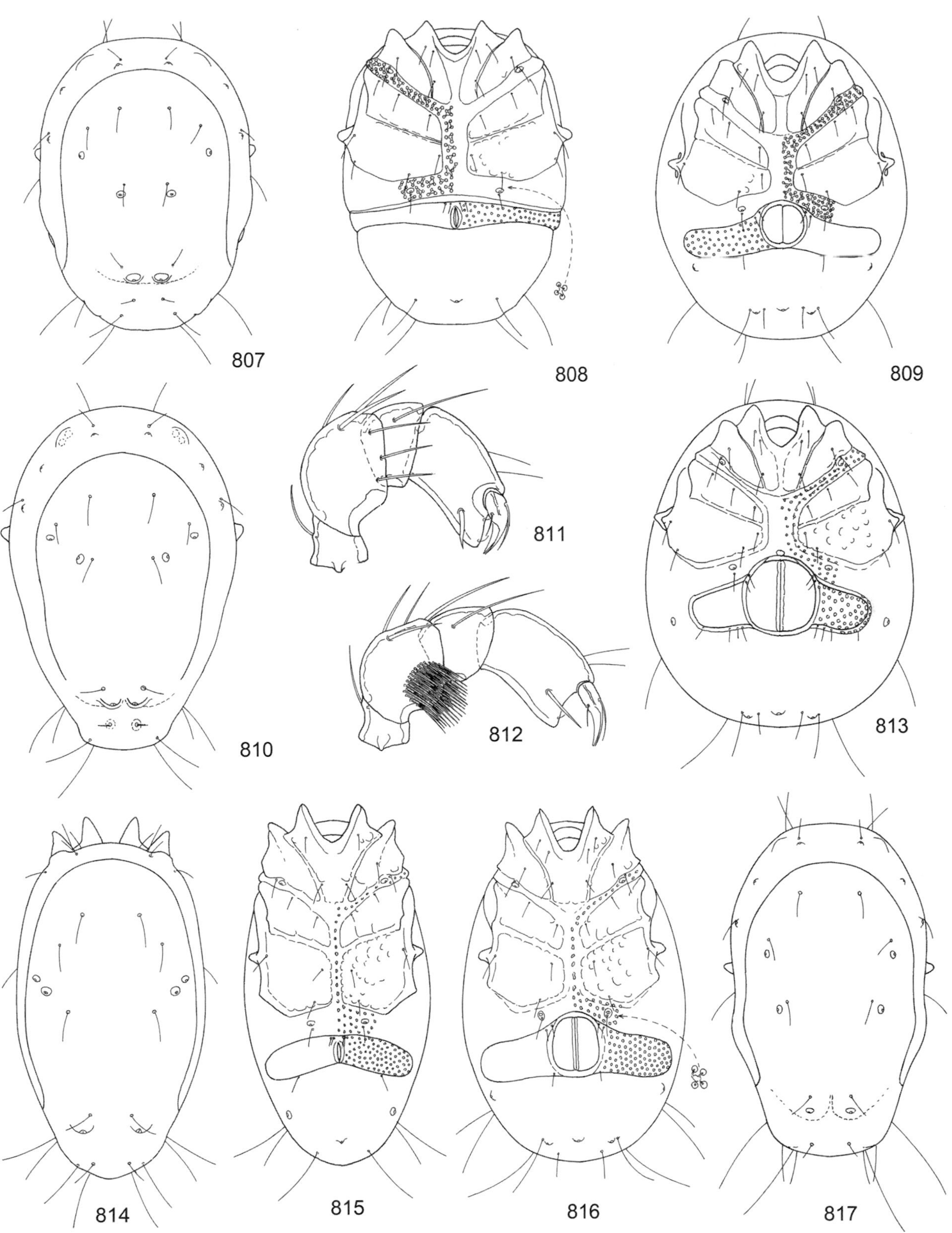

PLATE 16.92 Figures 807–809 & 811: *Arrenurus kenki* Marshall (Arrenuridae) adults. **Fig. 807** male dorsum of idiosoma. **Fig. 808** male venter of idiosoma. **Fig. 809** female venter of idiosoma. **Fig. 811** male pedipalp. **Figures 810 & 813:** *Arrenurus rufopyriformis* Habeeb (Arrenuridae) adults. **Fig. 810** male dorsum of idiosoma. **Fig. 813** female ventral shield. **Figures 812 & 817:** *Arrenurus ringwoodi* Mullen (Arrenuridae) adult male. **Fig. 812** pedipalp. **Fig. 817** dorsum of idiosoma. **Figures 814–816:** *Arrenurus loticus* Mullen (Arrenuridae) adults. **Fig. 814** male dorsum of idiosoma. **Fig. 815** male venter of idiosoma. **Fig. 816** female ventral shield.

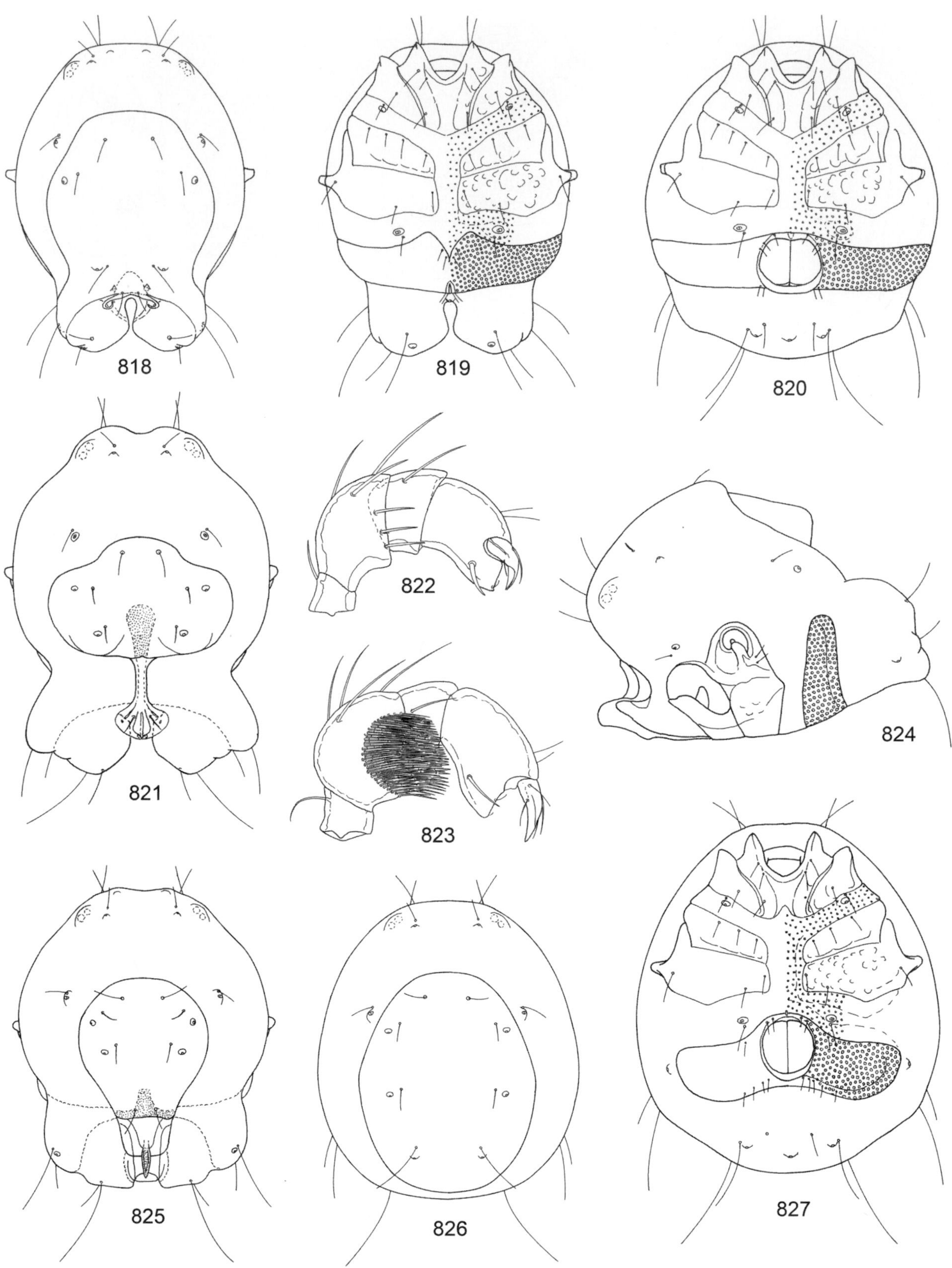

PLATE 16.93 Figures 818–820 & 822: *Arrenurus acutus* Marshall (Arrenuridae) adults. **Fig. 818** male dorsum of idiosoma. **Fig. 819** male venter of idiosoma. **Fig. 820** female venter of idiosoma. **Fig. 822** male pedipalp. **Figures 821, 823–824 & 827:** *Arrenurus hiatocaudautus* Cook (Arrenuridae) adults. **Fig. 821** male dorsum of idiosoma. **Fig. 823** male pedipalp. **Fig. 824** male lateral view of idiosoma. **Fig. 827** female ventral shield. **Figures 825–826:** *Arrenurus cheboyganensis* Cook (Arrenuridae) adults. **Fig. 825** male dorsum of idiosoma. **Fig. 826** female dorsum of idiosoma.

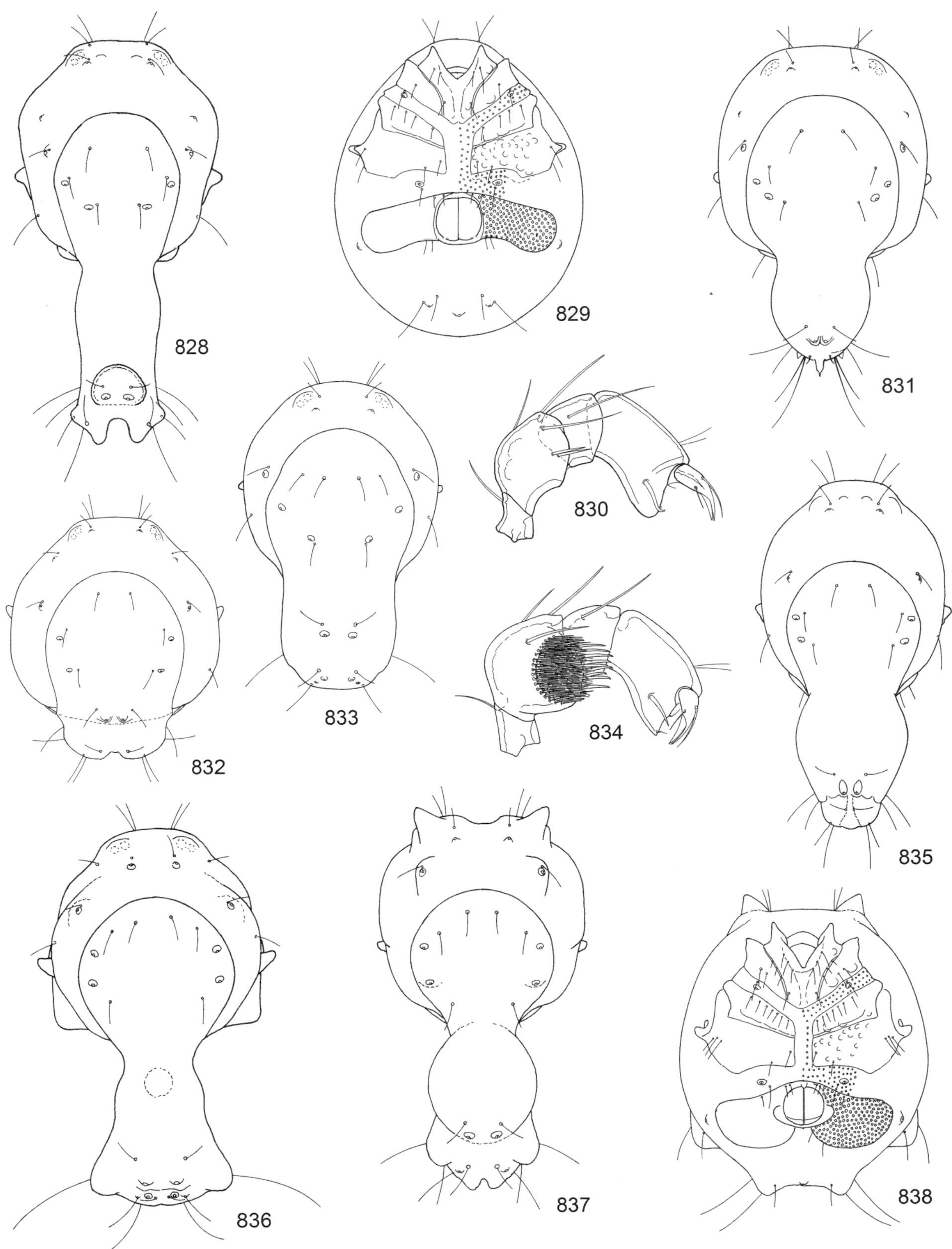

PLATE 16.94 **Figures 828–830:** *Arrenurus cardiacus* Marshall (Arrenuridae) adults. **Fig. 828** male dorsum of idiosoma. **Fig. 829** female ventral shield. **Fig. 830** male pedipalp. **Fig. 831:** *Arrenurus neomamillanus* Cook (Arrenuridae) adult male dorsum of idiosoma. **Fig. 832:** *Arrenurus rotundus* Marshall (Arrenuridae) adult male dorsum of idiosoma. **Fig. 833:** *Arrenurus birgei* Marshall (Arrenuridae) adult male dorsum of idiosoma. **Figures 834 & 837–838:** *Arrenurus megalurus* Marshall (Arrenuridae) adults. **Fig. 834** male pedipalp. **Fig. 837** male dorsum of idiosoma. **Fig. 838** female venter of idiosoma. **Fig. 835:** *Arrenurus pseudoconicus* Piersig (Arrenuridae) adult male dorsum of idiosoma. **Fig. 836:** *Arrenurus expansus* Marshall (Arrenuridae) adult male dorsum of idiosoma.

Smith, I.M. 1978. Descriptions and observations on host associations of some larval Arrenuroidea (Prostigmata: Parasitengona), with comments on phylogeny in the superfamily. Canadian Entomologist 110: 957–1001.

Smith, I.M. 1982. Larvae of water mites of the genera of the superfamily Lebertioidea (Prostigmata: Parasitengona) in North America with comments on phylogeny and higher classification of the superfamily. Canadian Entomologist 114: 901–990.

Smith, I.M. 1984. Larvae of water mites of some genera of Aturidae (Prostigmata: Hygrobatoidea) in North America with comments on phylogeny and classification of the family. Canadian Entomologist 116: 307–374.

Smith, I.M. 1989. North American water mites of the family Momoniidae Viets (Acari: Arrenuroidea). III. Revision of species of *Stygomomonia* Szalay, 1943, subgenus *Allomomonia* Cook, 1968. Canadian Entomologist 121: 989–1025.

Smith, I.M. 1991a. North American water mites of the genera *Phreatobrachypoda* Cook and *Bharatalbia* Cook (Acari: Hygrobatoidea: Aturinae). Canadian Entomologist 123: 465–499.

Smith, I.M. 1991b. North American water mites of the family Momoniidae Viets (Acari: Arrenuroidea). IV. Revision of species of *Stygomomonia* (sensu stricto) Szalay, 1943. Canadian Entomologist 123: 501–558.

Smith, I.M. 1992a. North American species of the genus *Chelomideopsis* Romijn (Acari: Arrenuroidea: Athienemanniidae). Canadian Entomologist 124: 451–490.

Smith, I.M. 1992b. North American water mites of the family Chappuisididae Motas and Tanasachi (Acari: Arrenuroidea). Canadian Entomologist 124: 637–723.

Smith, I.M. 2003. North American species of *Neoacarus* Halbert, 1944 (Acari: Hydrachnida: Arrenuruoidea: Neoacaridae). pp. 257–302 *in*: I.M. Smith (ed.). An Acarological Tribute to David R. Cook (From Yankee Springs to Wheeny Creek). Indira Publishing House. xiv + 331 pp.

Smith, I.M. & D.R. Cook. 1994. North American species of Neomamersinae Lundblad (Acari: Hydrachnida: Limnesiidae). Canadian Entomologist 126: 1131–1184.

Smith, I.M. & D.R. Cook. 1999. An assessment of global distribution patterns in water mites (Acari: Hydrachnida), pages 109–124 *in*: G.R. Needham, R.D. Mitchell, D.J. Horn and W.C. Welbourn (eds), Acarology IX, Volume 2, Symposia. Ohio Biological Survey, Columbus.

Smith, I.M. & D.R. Cook. 2006. North American species of *Diamphidaxona* (Acari: Hydrachnida: Hygrobatidae). Zootaxa 1279: 1–44.

Smith, I.M. & D.R. Cook. 2000. North American species of *Kawamuracarus* Uchida, 1937 (Acari: Hydrachnida: Limnesiidae: Kawamuracarinae). International Journal of Acarology 26: 63–71.

Smith, I.M., Cook, D.R., Smith, B.P. (2010) Water mites and other arachnids. Chapter 15, pages 485–586 *in*: Thorp, J. & Covich, A. (eds.), Ecology and classification of North American freshwater invertebrates. 3rd Edition. Academic Press. 1021 pp.

Smith, I.M. & D.R. Oliver. 1986. Review of parasitic associations of larval water mites (Acari: Parasitengona: Hydrachnida) with insect hosts. Canadian Entomologist 118: 407–472.

Smith, I.M., D.R. Cook, R. Gerecke. 2015. Revision of the status of some genus-level water mite taxa in the families Pionidae Thor, 1900, Aturidae Thor, 1900, and Nudomideopsidae Smith, 1990 (Acari: Hydrachnidiae). Zootaxa 3919: 111–156.

Smith, I.M., D.R. Cook, B.P. Smith. 2010. Water mites (Hydrachnidiae) and other arachnids, Chapter 15, Pages 485–586 *in*: J.H. Thorp & A.P. Covich. (eds.), Ecology and classification of North American Freshwater Invertebrates, Third Edition. Academic Press, London, Burlington, San Diego.

Walter, D.E., E.E. Lindquist, I.M. Smith, D.R. Cook, G.W. Krantz. 2009. Order Trombidiformes, Chapter 13. In: Krantz, G.W. and D.E. Walter (editors). A Manual of Acarology, Third Edition. Texas Tech University Press, Lubbock: 233–420.

ACKNOWLEDGMENTS

We thank Michelle MacKenzie for her invaluable assistance during preparation of the plates of figures and for proofreading the keys.

Subphylum Crustacea

D. Christopher Rogers

Kansas Biological Survey, University of Kansas, Lawrence, KS, USA; The Biodiversity Institute, University of Kansas, Lawrence, KS, USA

INTRODUCTION

Within the Arthropoda, recent molecular, morphological, and paleontological studies have suggested many classification hypotheses that upend traditional views. The preponderance of evidence demonstrates that the Hexapoda (Apterygota + Insecta) represent a clad nested within the Crustacea (Reiger et al., 2005). This Crustacea + Hexapoda clade has been referred to "Tetraconata" or "Pancrustacea" (Reiger et al., 2005; Oakley et al., 2013), and this nomenclature was followed in "Volume I: Ecology and General Biology" of *Thorp and Covich's Freshwater Invertebrates* (4th edition). However, in retrospect, we will henceforth treat the Crustacea + Hexapoda clade under the name "Crustacea." We believe that creating a new nomen for a previously named clade just because some other clade is found to nest within it will only create confusion. Instead we prefer to apply the principles of the International Code of Zoological Nomenclature (which do not officially apply to higher level taxonomy) and treat the Hexapoda as a Class within the Crustacea, as would be done for lower taxonomic levels.

As relationships within the Crustacea are still unclear, we revert to a more traditional organization (Ahyong et al., 2011) for all categories encompassed within this clade.

TERMINOLOGY AND MORPHOLOGY

Hexapods and crustaceans are fundamentally separated by the position of the genitalia. In the crustaceans the genitalia are distoventrally positioned on the thorax. Malacostracan males typically have the first pleopod pair (abdominal limbs) modified to transfer sperm from the thoracic gonopods to the female. The Hexapoda have the genitalia terminally positioned at the end of the abdomen, often withdrawn internally.

PHYLUM ARTHROPODA

KEYS TO CRUSTACEA: CLASSES

1	Genitalia positioned at posterior end of thorax; limbs present, more than three pairs; two pairs of antennae; free-living or parasitic; wings never present	2
1'	Genitalia positioned at abdominal terminus; limbs present or absent, if present, not more than three pairs (Fig. 16.96.1–2); pseudolegs may be present in some larval dipterans (Fig. 16.116.163) and lepidopterans (Fig. 16.111.116); one pair of antennae (Fig. 16.96.1–2); wings present or absent	**Hexapoda [p. 415]**
2(1)	Thoracopods segmented, never lamellar; carapace bivalved or not	3
2'	Thoracopods lamellar, not segmented (Pl. 16.118); carapace bivalved or not (Pl. 16.119); **OR** if thoracopods are segmented, then bivalved carapace greatly reduced (Pl. 16.119 J)	**Branchiopoda [p. 439]**
3(2)	Carapace not bivalved	4
3'	Carapace bivalved, enclosing entire animal (Pls. 16.135–155)	**Ostracoda [p. 486]**
4(3)	Naupliar eye present, or if absent, then animal entirely enclosed in a fused exoskeleton, with a single distal opening (barnacles; (Pl. 16.157–163); telson and uropods always absent (Pl. 16.164)	**Maxillopoda [p. 514]**
4'	Naupliar eye absent in adults, telson or pleotelson present, uropods present or absent (Pls. 16.181, 182.03–04, 183.01–16, 185.01–06, 186.03–05, 189.01–04, 191.01)	**Malacostraca [p. 575]**

REFERENCES

Ahyong, S.T., J.K. Lowry, M. Alonso, R.N. Bamber, G.A. Boxshall, P. Castro, S. Gerken, G. S. Karaman, J. W. Goy, D. S. Jones, K. Meland, D.C. Rogers & J. Svavarsson. 2011. Subphylum Crustacea Brünnich, 1772. Pages 1–237 *in*: Zhang, Z.-Q. (Ed.), Animal biodiversity: An outline of higher-level classification and survey of taxonomic richness. Zootaxa 3148.

Oakley, T. H., J. M. Wolfe, A. R. Lindgren & A. K. Zaharoff. 2013. Phylotranscriptomics to bring the understudied into the fold: monophyletic ostracoda, fossil placement, and pancrustacean phylogeny. Molecular Biology & Evolution 30: 215–233

Reiger, J. C., J. W. Schultz & R. E. Kambic. 2005. Pancrustacean phylogeny: hexapods are terrestrial crustaceans and maxillopods are not monophyletic. Proceedings of the Royal Society B, 272: 395–401.

Class Hexapoda

R. Edward DeWalt
Illinois Natural History Survey, Champaign, IL, USA

Vincent H. Resh
Department of Environmental Science, Policy, and Management, University of California, Berkeley, CA, USA

INTRODUCTION

This chapter includes orders and families in which one or more life stages are truly aquatic and adapted for survival under or on the water surface. Not included are families that live on or burrow into emergent aquatic vegetation, internal parasites of aquatic animals, and riparian families that are closely associated with water but do not inhabit it. Insects that are in the first two of these categories are covered elsewhere (Merritt et al., 2008). Entognatha, which was previously included in the Insecta, is currently considered a separate subclass within Hexapoda.

Ten insect orders contain aquatic or semiaquatic species. Five (Ephemeroptera, Odonata, Plecoptera, Trichoptera, and Megaloptera) are orders in which almost all species have aquatic larvae. The remaining five orders (Coleoptera, Diptera, Heteroptera, Lepidoptera, Neuroptera) are primarily terrestrial. However, these orders contain species or entire families that have one or more life stages adapted for living in or beside aquatic environments.

Three aquatic orders (Ephemeroptera, Odonata, and Plecoptera) have a hemimetabolous life cycle, which includes three developmental stages: egg, larva (nymph), and adult. The term larva is used here instead of nymph or naiad to simply the text. The larva nearly always lives in the water, while the adult is nearly always terrestrial. The other two aquatic orders (Trichoptera and Megaloptera) have a holometabolous life cycle that includes four developmental stages: egg, larva, pupa, and adult.

Four of the five partially aquatic orders (Coleoptera, Diptera, Lepidoptera, and Neuroptera) also have holometabolous life cycles. Heteroptera is hemimetabolous.

LIMITATIONS

The family level keys emphasize aquatic larvae and adults with aquatic stages. Keys to eggs, Heteroptera larvae, pupae, and terrestrial stages of aquatic species are not included. The most recent keys available to identify genera of aquatic insects in North America appear in Merritt et al. (2008). Regional keys to genera or species of aquatic insects include those in Usinger (1956), Brigham et al. (1982), Peckarsky et al. (1990), Clifford (1991), and Hilsenhoff et al. (1995). Unfortunately, reliable species level keys do not exist for many families and genera. References to recent revisions, regional species checklists, and descriptions of new species may be found in the individual order bibliographies in Merritt et al. (2008) and in bibliographies published annually by the Society for Freshwater Science (formerly North American Benthological Society) (www.freshwater-science.org). It is becoming increasingly easy to access digital copies of taxonomic works.

TERMINOLOGY AND MORPHOLOGY

Morphological terms used for identification of aquatic insects vary somewhat among orders, but several terms are used in common. A plecopteran larva (Pls. 16.96.1 & 16.97.2) is used to illustrate these terms. Insects have three body regions: head, thorax, and abdomen. Dorsally, the head (Pl. 16.96.1) has a pair of antennae, a pair of compound eyes, and up to three ocelli or simple eyes. On the head are several mouthparts (Pl. 16.97.2), including a labrum (or upper lip), a pair of mandibles, a pair of maxillae, and a labium (lower lip). The mandibles usually are heavily sclerotized, tooth-like structures. The maxillae and labium are usually multiarticulate and frequently have a pair of elongate palps.

The thorax is immediately posterior to the head and consists of three large segments (Pl. 16.96.1): prothorax (anterior), mesothorax, and metathorax (posterior). The dorsal sclerite of each segment is referred to as a notum, the ventral sclerite as a sternum, and the lateral sclerites as pleura (singular pleuron). Each thoracic segment has a pair of legs, with each leg being divided into five parts (Pls. 16.96.1 & 16.97.2): a coxa (plural coxae), trochanter, femur (femora), tibia (tibiae), and tarsus (tarsi). The femur and tibia are usually elongate and the tarsus has from one to five articles, depending on the order. Each tarsus has one or two tarsal claws. The prefixes pro-, meso-, and meta- refer to the respective thoracic segments. Thus, a pronotum would be on the prothorax, and a mesofemur would be on the mesothorax. In hemimetabolous insect larvae, wingpads (developing wings) are usually present dorsally on the meso- and metathorax.

The abdomen shares the same sclerite arrangement as the thorax, but may lack pleura in some groups. It may have as many as ten visible segments, which are numbered consecutively and posteriorly from the thorax. Terminal abdominal appendages vary widely or may be absent; cerci (singular cercus) are present in several other orders. In some orders, gills of various types may be present on the abdomen, but they may also be located on the thorax, or even on the head.

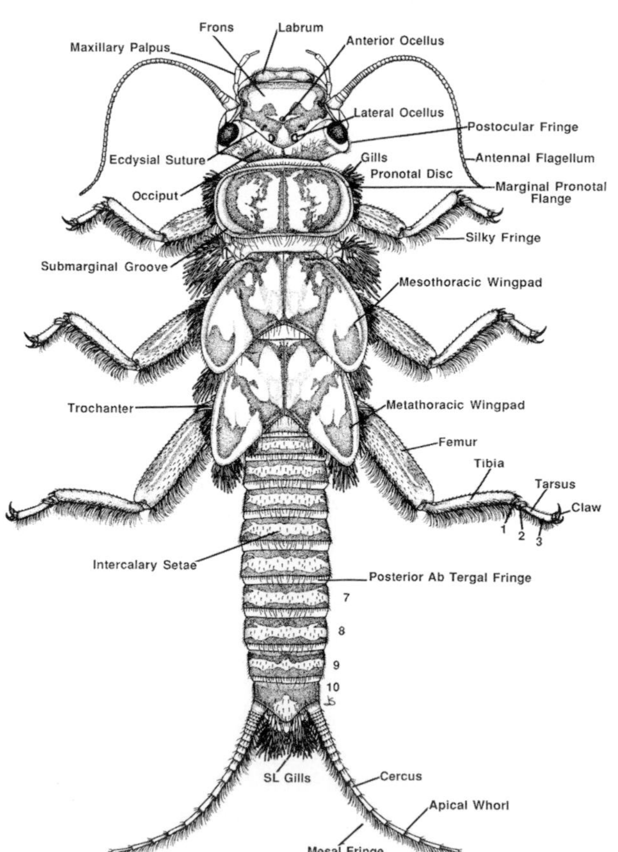

PLATE 16.96 Fig. 1 Plecoptera larva: dorsal view.

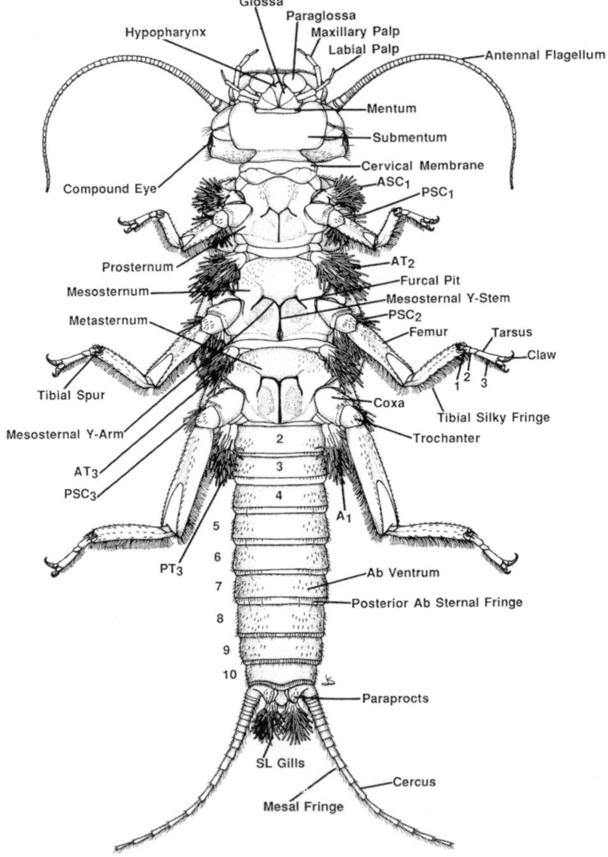

PLATE 16.97 Fig. 2 Plecoptera larva: ventral view.

MATERIAL PREPARATION AND PRESERVATION

Preservative type depends on the intended use of the specimens. If DNA is to be extracted, specimens are best put into 95% EtOH (ethanol) or frozen. High concentrations of EtOH leave specimens brittle and difficult to manipulate without damage. A concentration of 70–80% EtOH is preferable for teaching, reference collections, or life history study measurements. A less expensive alternative is the use of 5–10% buffered formalin or Kahle's fluid. These are fixatives that create crossbridges in proteins, resulting in supple specimens that also retain internal structures better than those in EtOH. However, fixatives that contain formalin require more careful storage and disposal, and are not desirable for permanent storage. Other fixatives and preservatives for aquatic specimens are reported elsewhere (Smith, 2001).

Permanent storage of aquatic larvae and most adults should be in at least 70–80% EtOH. The type of vial and stopper or cap used is extremely important in preventing desiccation of specimens. Many museums use 3 or 4 dram, glass patent lip vials with neoprene, or rubber stoppers. Neoprene stoppers tend to swell in alcohol and some bleed color as well. The right vial and stopper combination will be viable for many decades without preservative loss. Screw cap vials with flat paper or plastic liners and shell vials that are stoppered with cork or plastic snap caps should be avoided because of gradual evaporation of the alcohol. With much use, the plastic threads of snap caps are damaged and evaporation of alcohol begins. Screw cap vials with beveled polyethylene inserts are acceptable alternatives to patent lip vials. Three dram screw cap vials are not wide enough to pass standard forceps to the bottom.

Adult aquatic Coleoptera and Heteroptera are normally mounted on pins, but may be preserved in EtOH. Several specimens of a given species from a given site/date collection effort can be kept in a single vial to reduce storage space and curation time. Specialists collect adult Odonata into labeled paper triangles and then soak them in acetone. Acetone helps to preserve color and extracts fats that leave specimens looking greasy and dull. The specimen should then be dried with wings folded above the thorax and stored in clear, glycine envelopes. A card of acid free paper with location and determination information should be inserted into the envelope and the envelopes stored as index cards in a pest proof container.

A dissecting microscope is used to examine most specimens. Those specimens stored in liquid should be placed in a watch glass, petri dish, or similar container and completely covered with EtOH. The use of a layer of very small glass beads or silica sand in the watch glass permits insects to be imbedded for viewing from any angle. For some fragile insects, especially mayflies, it may be better to crimp small fishing weights (called "split shots") around minuten pins as anchors to keep specimens in place. Whole specimens are generally not slide mounted, with the exception of some Diptera larvae, but the mouthparts and other sections can be teased apart for viewing on slides, in glycerin, water, or alcohol.

KEYS TO HEXAPODA

Crustacea: Hexapoda: Subclasses

1 Mouthparts internal; abdomen forked distally, and bent ventrally to meet the collophore (tubular projection) descending from the first abdominal segment; wings absent (e.g., Fig. 16.98.3–7) .... Entognatha, one order: **Collembola [p. 415]**

1' Mouthparts external; abdomen various, collophore absent; wings present or not (e.g., Figs. 16.96.1, 97.2) .... **Insecta [p. 417]**

Crustacea: Hexapoda: Entognatha: Collembola: Families

Although most Nearctic springtail species are not aquatic, seven families have one or more species that live in, on, or near aquatic habitats (Christiansen & Snider, 2008). Since all Nearctic families of springtails could be encountered while sampling, they have been included in the key. This key is modified from that of Christiansen & Snider (2008). The complexity of characters required to separate Collembola families necessitates the use "in part" after several family names—some families key out in multiple locations in the key.

1 Body elongate; thorax and abdomen distinctly segmented (Fig. 16.98.3) 2

1' Body somewhat globular or oval (Fig. 16.98.4) 11

2(1) First thoracic segment visible dorsally, with dorsal setae (Fig. 16.98.3) 3

2' First thoracic segment often not visible dorsally, without dorsal setae 5

3(2) Dens absent or <2.5× as long as manubrium; mouthparts directed forward; furcula apex not distinctly convergent 4

3' Dens >3× long as manubrium (Fig. 16.98.6); mouthparts directed downward; furcula apex distinctly convergent (Fig. 16.98.7) .... Poduridae, one species: *Podura aquatica* Linnaeus, 1758

4(3) Pseudocelli present; eyes always absent (Fig. 16.99.8) .... Onychiuridae

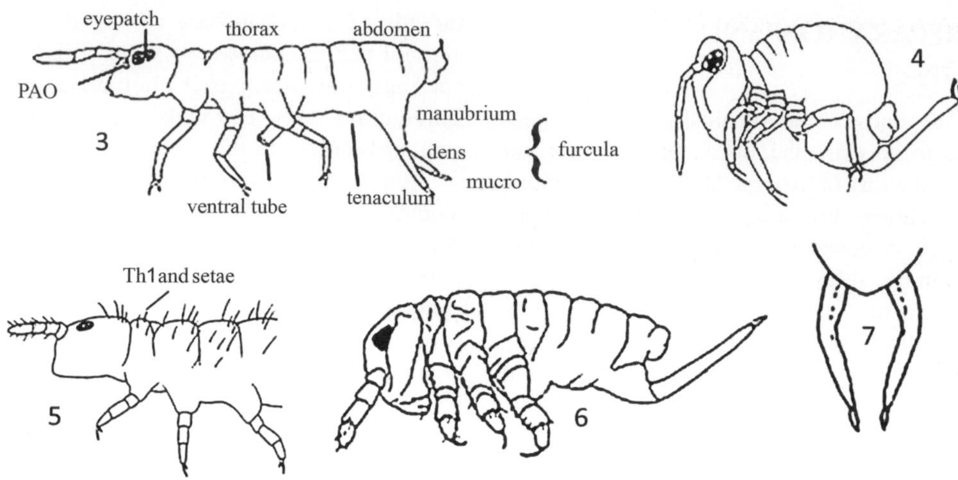

PLATE 16.98 **Fig. 3** Generalized Collembola, lateral (modified from Christiansen & Snider, 2008). **Fig. 4** Sminthuridae body form, lateral. Modified from Christiansen and Snider (2008). **Fig. 16.5** Collembolan with thoracic segment one (Th1) visible and setae, lateral (modified from Christiansen & Snider, 2008). **Fig. 6** *Podura aquatica* Linnaeus (Poduridae) with dens 3× longer than mucro, lateral (modified from Christiansen & Snider, 2008). **Fig. 7** *Podura aquatica* Linnaeus (Poduridae) with dens convergent, dorsal view. Modified from Christiansen and Snider, 2008.

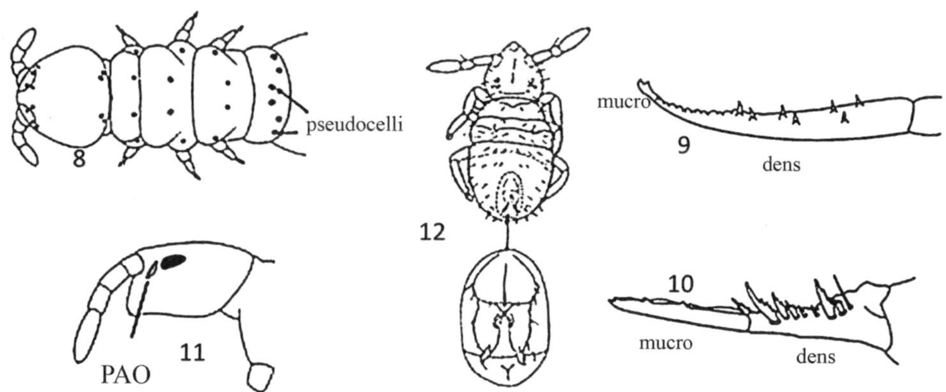

PLATE 16.99 **Fig. 8** Onychiuridae with pseudocelli, dorsal (modified from Christiansen & Snider, 2008). **Fig. 9** Isotomidae dens and mucro, ventrolateral (modified from Christiansen & Snider, 2008). **Fig. 10** Entomobryidae dense and mucro, ventolateral (modified from Christiansen & Snider, 2008). **Fig. 11** Isotomidae head with postantennal organ (PAO) (modified from Christiansen & Snider, 2008). **Fig. 12** Mackenziellidae (*Mackenziella*) ventral. Inset showing short dens with three setae (modified from Christiansen & Snider, 2008).

4'	Pseudocelli absent; eyes present or rarely absent	Hypogastruridae
5(2)	Dens with spines (Figs. 16.99.9, 99.10)	6
5'	Dens without spines	7
6(5)	Mucro short and bidentate (Fig. 16.99.9)	Isotomidae (in part)
6'	Mucro elongate with 3 or more teeth (Fig. 16.99.10)	Entomobryidae (in part)
7(5)	Body without scales	8
7'	Body with scales	Entomobryidae (in part)
8(7)	Postantennal organ absent	9
8'	Postantennal organ present or rarely absent (Fig. 16.99.11)	Isotomidae (in part)
9(8)	Abdominal segments 4–6 separate	10
9'	Abdominal segments 4 and 5 fused, segment 6 reduced	Actaletidae (*Spinactaletes*)
10(9)	Antennae longer than head; eyes present	11
10'	Antennae shorter than head; eyes absent	Neelidae
11(10)	Body elongate oval; furcula short, 20% of body length, dens with 3 setae (Fig. 16.99.12)	Mackenziellidae (*Mackenziella*)
11'	Body globular; furcula long, near half body length, dens with numerous setae (Fig. 16.98.4)	Sminthuridae

Crustacea: Hexapoda: Insecta: Orders

1 Mummy-like, often encased in cocoon (Lepidoptera, some Trichoptera), terrestrial cell (Megaloptera), or puparium (Diptera, most Trichoptera) but may also be free (some Diptera); often with developing wings and legs appressed to or free from body .... pupae (not keyed)

1' Not mummy-like, not encased in cocoon, cell, or puparium; legs and wings, if present, fully developed .... larvae or adults 2

2(1) Thorax with jointed legs (e.g., Figs. 16.97.2, 109.98, 112.121) 3

2' Mobile larvae with prolegs, pseudopods, or creeping welts on one or more segments (Figs. 16.115.151.3, 16.116.157, 158, 160–164, 16.117.165, 168–179) .... **Diptera [p. 417]**

3(1) With large functional wings, may be rigid and chitinized or thickened in the proximal half (16.110.100, 113.129) 4

3' Wingless (16.115.149) or with developing wings (Fig. 16.96.1) (wingpads) 6

4(3) Mesothoracic wings rigid and chitinized (Fig. 16.113.129) or thickened in basal half (Fig. 16.110.100) 5

4' All wings completely membranous, with numerous veins .... terrestrial adults (not keyed)

5(4) Chewing mouthparts; mesothoracic wings rigid, chitinized (Figs. 16.113.129) .... **Coleoptera** (in part)

5' Sucking mouthparts formed into a broad or narrow tube (Figs. 16.110.101, 103); mesothoracic wings hardened in basal half (Figs. 16.110.100, 110.106–109) .... **Heteroptera** (in part)

6(3) With 2 or 3 long, multisegmented terminal appendages (Figs. 16.103.35, 105.56) 7

6' Terminal appendages absent, or each appendage composed of only 1 or 2 segments (Figs. 16.104.54, 109.98) 8

7(6) Abdomen with plumose, forked, or lamellar gills (Figs. 16.100.14, 101.21–25, 102.29, 103.34–39); usually with 2 cerci and a median caudal filament, the latter occasionally reduce to few segments or absent (Fig. 16.103.35); tarsi with 1 claw (Figs. 16.100.15–19, 102.30–32, 103.33) .... **Ephemeroptera [p. 430]**

7' Gills, if present, never lamellar, but digitiform, singly or in clusters at the base of the head, thorax, on abdominal segments 1–2 or 1–3, and/or about the anus (Figs. 16.105.55, 56, 59, 60, 62, 106.65); two cerci, no median filament; tarsi with 2 claws............ **Plecoptera [p. 432]**

8(6) Without extensile labium; abdomen not terminating in lamellae or points 9

8' Labium an extensible grasping organ (Figs. 16.104.42–44); abdomen terminating in 3 lamellae (Fig. 16.104.41) or 5 usually short, triangular points (Fig. 16.104.54) .... **Odonata [p. 432]**

9(8) Mouthparts sucking, formed into a broad or narrow tube (Figs. 16.110.101, 103) or a pair of long stylets (Fig. 16.111.115) 10

9' Mouthparts not sucking, not formed into a tube or pair of stylets 11

10(9) Parasitic on sponges; mouthparts a pair of long stylets (Fig. 16.111.115); all tarsi with 1 claw............... Neuroptera, one family: Sisyridae

10' Free-living; mouthparts a broad or narrow tube (Figs. 16.110.101, 103); mesotarsi with 2 claws (Figs. 16.110.100, 106–109) .... **Heteroptera** (in part)

11(9) Abdomen without ventral prolegs on abdominal segments 3–6 12

11' Ventral prolegs on abdominal segments 3–6, each with a ring of fine hooks (Fig. 16.111.116) .... Lepidoptera

12(11) Body not caterpillar-like; not housed in cases; antennae elongate, with 3 or more segments 13

12' Body caterpillar-like (Fig. 16.107.80); sometimes housed in organic or inorganic cases (Figs. 16.107.72, 74, 79); antennae relatively short, inconspicuous, one-segmented (Figs. 16.107.76, 108.91, 93) .... **Trichoptera [p. 433]**

13(12) With long lateral filaments (Figs. 16.109.98, 99, 114.131) 14

13' Without long lateral filaments (Figs. 16.114.132–142, 144, 115.147, 149) .... **Coleoptera** (in part)

14(13) Each tarsus with a single claw (Figs. 16.114.135–144) *or* tarsi two clawed and abdomen terminating in 2 slender filaments (Fig. 16.114.134) *or* tarsi two clawed with a single median proleg terminating in 4 hooks (Fig. 16.114.131) .... **Coleoptera** (in part)

14' Each tarsus with 2 claws; abdomen terminating in a single slender filament (Fig. 16.109.98) or in 2 prolegs, each with 2 hooks (Fig. 16.109.99) .... **Megaloptera [p. 436]**

Crustacea: Hexapoda: Insecta: Diptera: Families

1 Larvae with more than 7 apparent segments; without 6 ventral suckers 2

1' First segment comprised of head, thorax and first abdominal segment, larvae appear 7-segmented; first 6 segments each with a prominent ventral sucker (Fig. 16.116.153) .... Nematocera, Blephariceridae

2(1) Sclerotized head capsule absent (Figs. 16.116.157, 158, 160, 161, 163, 164), or posteriorly incomplete (Figs. 16.116.154, 155), and retracted at least partially into thorax, one exception with full head capsule (Fig. 16.116.156); mandibles parallel, without secondary apical teeth, and moving in a vertical plane or in a partially sclerotized retracted head may be opposed and moving in a horizontal plane 3

2' Head capsule completely sclerotized and fully visible (Figs. 16.117.165–177); mandibles moving in a horizontal plane, and usually with two or more apical teeth; body never flattened and with a posterior spiracular chamber margined with long, soft hairs Nematocera 14

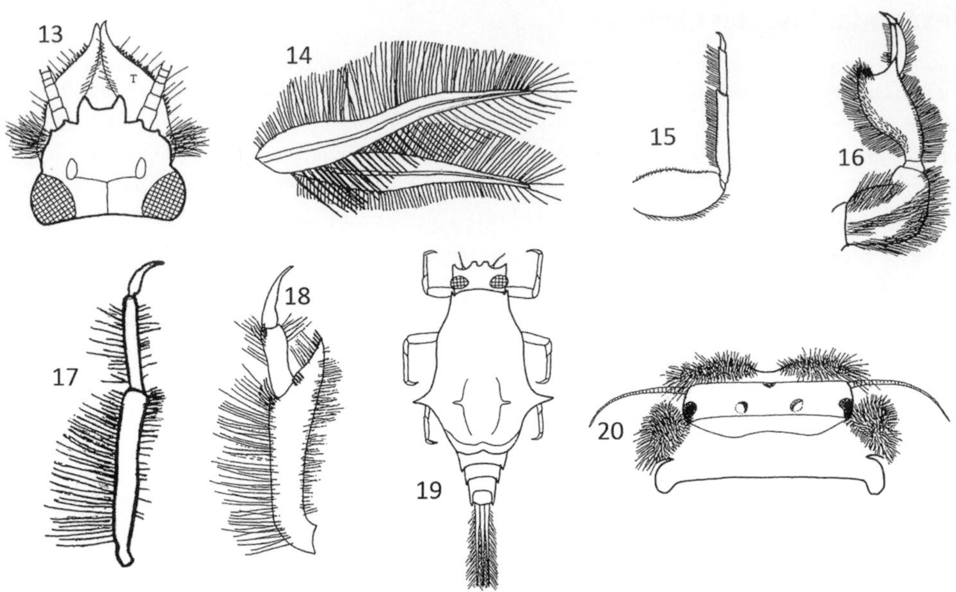

PLATE 16.100 **Fig. 13** Palingeniidae; head of *Pentagenia vittigera* (dorsal view) showing mandibular tusk (T). **Fig. 14** Ephemeridae; gills of *Hexagenia*. **Fig. 15** Potamanthidae; prothoracic leg of *Anthopotamus*. **Fig. 16** Ephemeridae; prothoracic leg of *Hexagenia*. **Fig. 17** Polymitarcyidae; metatibia and tarsus of *Ephoron*. **Fig. 18** Ephemeridae; metatibia and tarsus of *Ephemera*. **Fig. 19** Baetiscidae; *Baetisca* (dorsal view). **Fig. 20** Behningiidae; head and pronotum of *Dolania americana*.

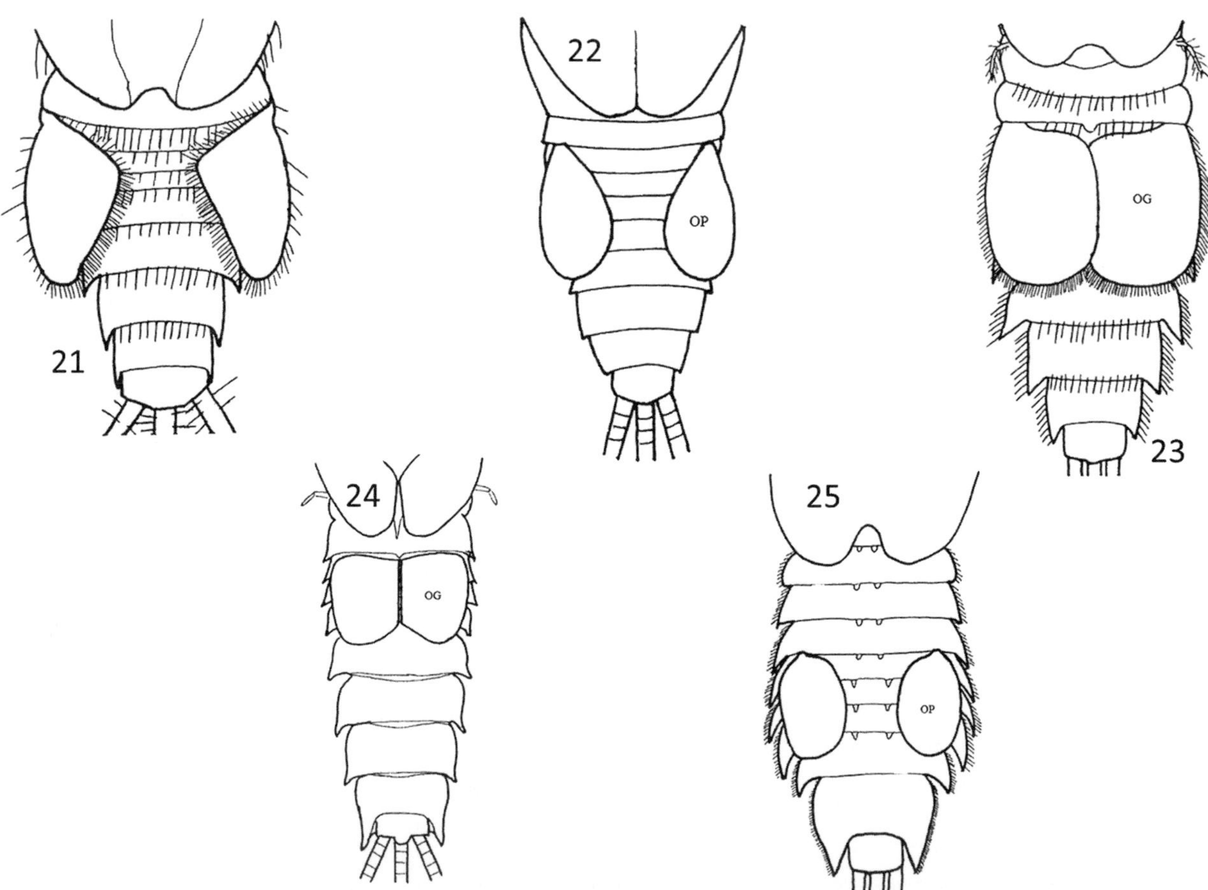

PLATE 16.101 **Fig. 21** Leptohyphidae; abdomen of *Tricorythodes* (dorsal view) showing operculate gill (OG). **Fig. 22** Leptohyphidae; abdomen of *Leptohyphes* (dorsal view) showing operculate gill (OG). **Fig. 23** Caenidae; abdomen of *Caenis* (dorsal view) showing operculate gill (OG). **Fig. 24** Neoephemeridae; abdomen of *Neoephemera* (dorsal view) showing operculate gill (OG). **Fig. 25** Ephemerellidae; abdomen of *Eurylophella* (dorsal view) showing operculate gill (OG).

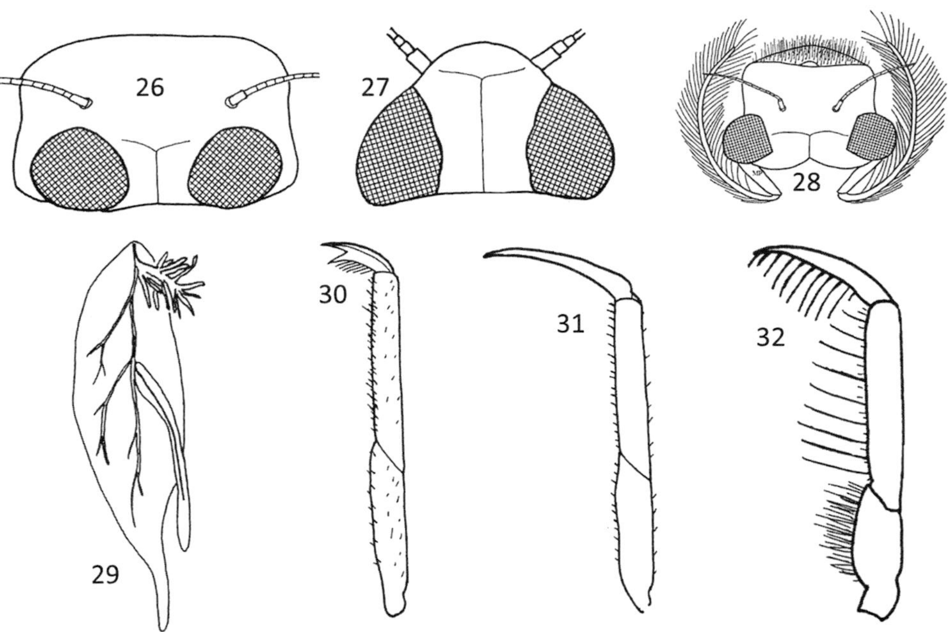

PLATE 16.102 **Fig. 26** Heptageniidae; head (dorsal view). **Fig. 27** Siphlonuridae; head of *Siphlonurus* (dorsal view). **Fig. 28** Arthropleidae; head of *Arthroplea* (dorsal view) showing maxillary palp (MP). **Fig. 29** Pseudironidae; gill lamella of *Pseudiron centralis* (ventral view). **Fig. 30** Metretopodidae; protarsus and tibia of *Siphloplecton*. **Fig. 31** Metretopodidae; metatarsus and tibia of *Siphloplecton*. **Fig. 32** Ametropodidae; protarsus and tibia of *Ametropus*.

3(2) Mandibles parallel, moving vertically, and without secondary apical teeth; larvae narrowed anteriorly with head capsule lacking or poorly developed and incompletely sclerotized or body terminating in a long respiratory tube (Fig. 16.116.157) or a posterior spiracular chamber margined with long, soft setae .... Brachycera 4

3' Mandibles opposed, moving in a horizontal plane, and with two or more apical teeth; larvae truncate anteriorly with head capsule retracted into thorax and incomplete posteriorly (Figs. 16.116.154–155) .... Nematocera: Tipulidae

4(3) Head mostly retracted into thorax and elongate, or indistinguishable; body nearly circular in cross section; without a posterior spiracular chamber margined with long, soft setae; integument lacking calcium carbonate crystals 5

4' Head mostly visible, truncate in shape (Fig. 16.116.156); body somewhat flattened; posterior spiracular chamber margined with long, soft setae; integument covered with shiny calcium carbonate crystals .... Stratiomyidae

5(4) Larvae without a long respiratory tube, if a short tube is present, it is divided apically 6

5' Larvae with a partially retractile caudal respiratory tube at least one-half as long as body (Fig. 16.116.157) .... Syrphidae

6(5) Body not terminating in a short tube or a pair of spines 7

6' Body terminating in a short tube that is divided apically (Fig. 16.116.158), or in a pair of spines .... Ephydridae

7(6) Caudal spiracular disk without palmate setae, if surrounded by lobes, body not wrinkled 8

7' Caudal spiracular disk with palmate setae and surrounded by 8–10 lobes, some of which may be very short (Fig. 16.116.159); body wrinkled .... Sciomyzidae

8(7) Abdominal segments lacking girdles of welts or prolegs, or present on abdominal segments 2–7, possible multiple pair per segment; integument without striations or extensive pubescence 9

8' First 7 abdominal segments girdled by 3 or 4 pairs of fleshy creeping welts (spine-bearing swellings) (Fig. 16.116.160); integument with longitudinal striations or with a short pubescence covering the body .... Tabanidae

9(8) Abdominal segments with slender lateral and dorsolateral filaments or short dorsal crocheted knobs on segments 6–7; abdominal segments with ventral prolegs bearing crotchets (Figs. 16.116.161, 163, 164, 117.179) 10

9' Larva cylindrical, with smooth shiny integument and with segmentation beadlike; lacking prolegs (Fig. 16.117.178) .... Pelecorynchidae

10(9) Abdomen lacking slender filaments; prolegs or creeping welts present on abdominal segments 2–7 (Fig. 16.116.161) 11

10' Abdomen with slender lateral and dorsolateral filaments and pairs of ventral prolegs with crotchets on abdominal segments 1–7; segment 8 with two longer filaments, fringed filaments (Fig. 16.116.162) .... Athericidae

11(10) Body without dorsal crotcheted knobs; anal division with short setae; abdominal segments 2–7 a single pair of short prolegs or creeping welts present 12

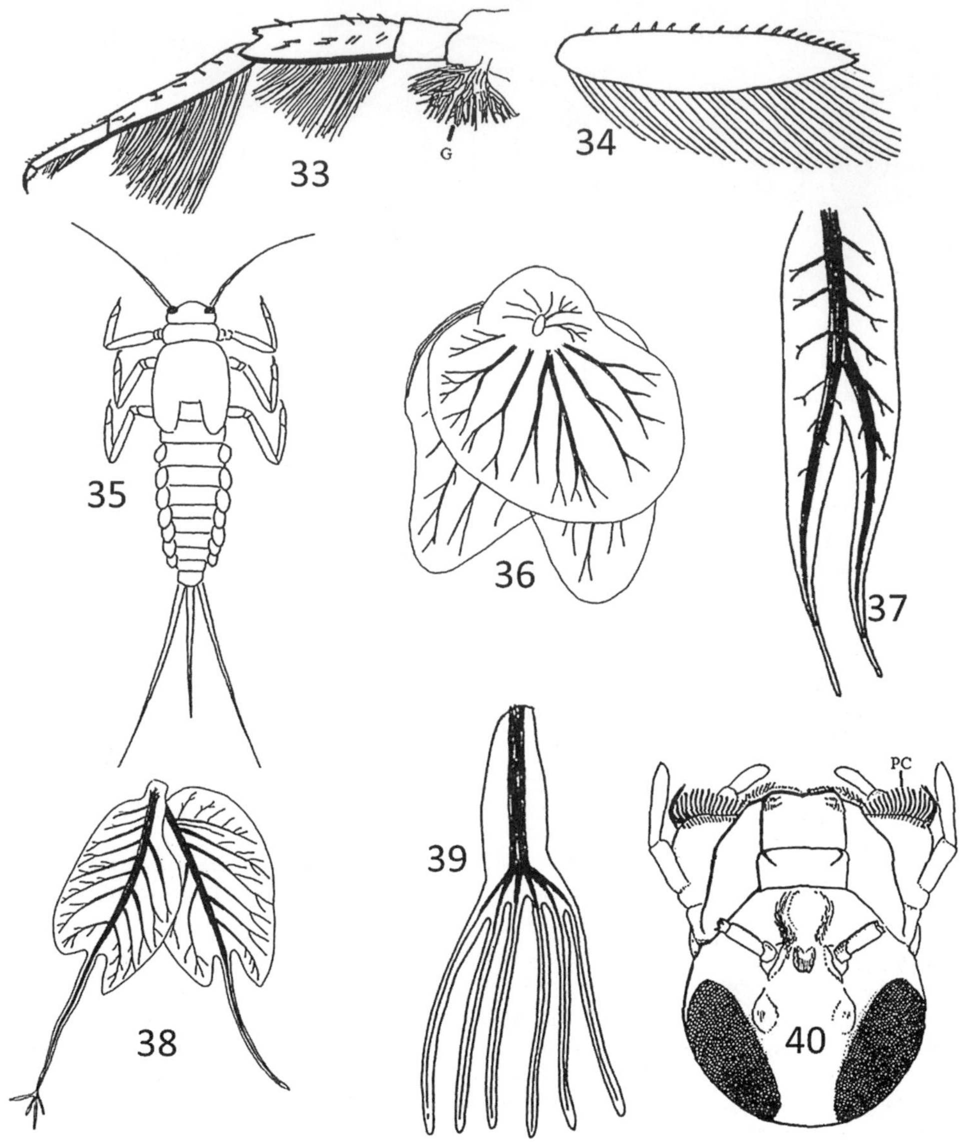

PLATE 16.103 **Fig. 33** Isonychiidae; prothoracic leg of *Isonychia* showing gill (G). **Fig. 34** Oligoneuriidae; gill on abdominal segment 3 of *Homoeoneuria*. **Fig. 35** Baetidae; *Baetis* (dorsal view). **Fig. 36** Siphlonuridae; gills on abdominal segment 2 of *Siphlonurus*. **Fig. 37** Leptophlebiidae; gill on abdominal segment 3 of *Paraleptophlebia*. **Fig. 38** Leptophlebiidae; gills on abdominal segment 3 of *Leptophlebia*. **Fig. 39** Leptophlebiidae; gill on abdominal segment 3 of *Habrophlebia*. **Fig. 40** Ameletidae; *Ameletus* pectinate comb (PC) setae on maxilla. *Modified from Burks, 1953.*

11'	Abdominal segments 6–7 with short dorsal crotcheted knobs; anal division bearing long apical setae; abdominal segments 2–7 with two pair of ventral, long slender prolegs present (Fig. 16.117.179) .... Oreoleptidae, one species: *Oreoleptis torrenticola* Zloty, Sinclair & Pritchard, 2005
12(11)	Body not terminating in a spiracular pit with 4 lobes 13
12'	Body terminating in a spiracular pit surrounded by 4 pointed lobes (Fig. 16.116.161) .... Dolichopodidae
13(12)	Some external head structure visible, with palps and antennae usually present; abdomen usually with ventral prolegs and elongate, paired, terminal appendages (Fig. 16.116.163) or a bulbous segment .... Empididae
13'	No visible external head structure; abdomen often with ventral prolegs and usually short, paired, terminal appendages (Fig. 16.116.164) .... Muscidae
14(2)	Prolegs absent (Figs. 16.117.165, 117.168, 169) 15
14'	Prolegs present on thorax and/or abdomen (Figs. 16.117.170–178) 19
15(14)	Thoracic segments fused and distinctly thicker than abdomen (Fig. 16.117.165) 16
15'	Thoracic segments not fused, equal to abdomen in diameter (Figs. 16.117.168, 169) 18

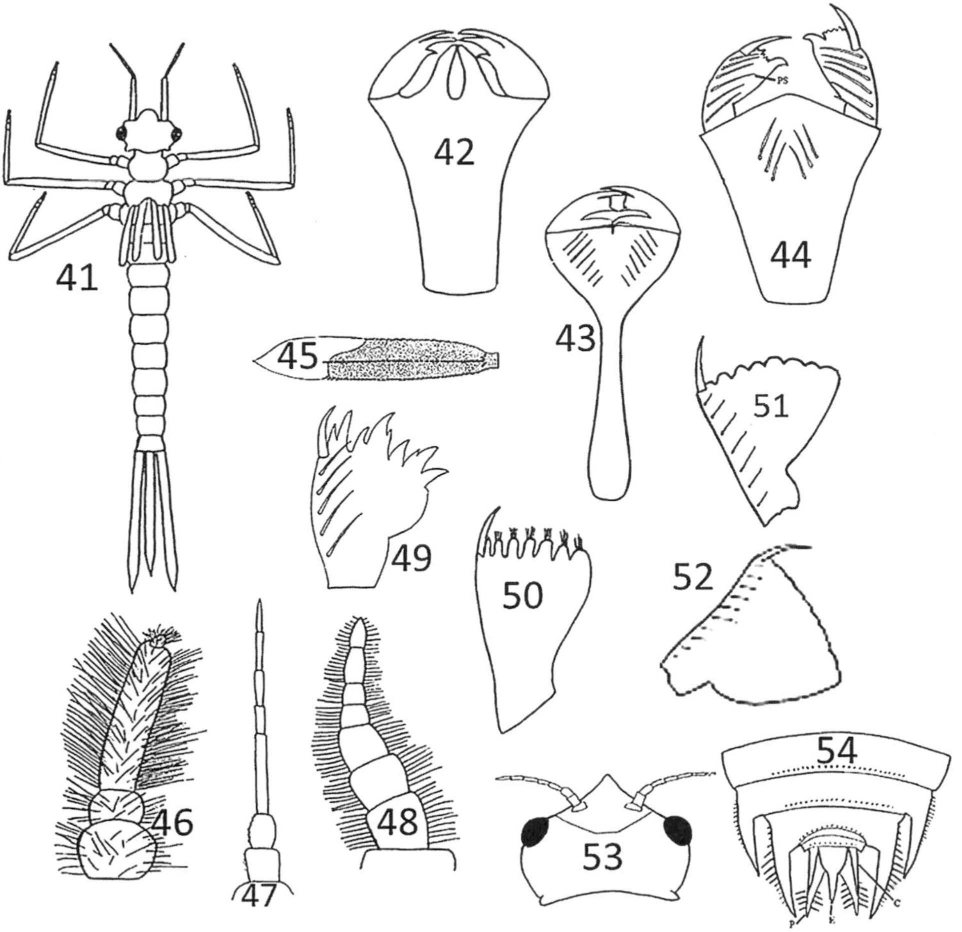

PLATE 16.104 **Fig. 41** Calopterygidae; *Calopteryx* (dorsal view). **Fig. 42** Calopterygidae; prementum of *Hetaerina* (dorsal view). **Fig. 43** Lestidae; prementum of *Lestes* (dorsal view). **Fig. 44** Coenagrionidae; prementum of *Enallagma* (dorsal view) showing dorsal setae (DS), palpal lobe (PL), and palpal setae (PS). **Fig. 45** Protoneuridae; caudal lamella of *Protoneura* (lateral view). **Fig. 46** Gomphidae; antenna of *Gomphus*. **Fig. 47** Aeshnidae; antenna of *Anax*. **Fig. 48** Petaluridae; right antenna of *Tachopteryx* (dorsal view). **Fig. 49** Cordulegastridae; palpal lobe of *Cordulegaster* (dorsal view). **Fig. 50** Corduliidae; palpal lobe of *Neurocordulia* (dorsal view). **Fig. 51** Corduliidae; palpal lobe of *Cordulia* (dorsal view). **Fig. 52** Libellulidae; palpal lobe of *Tramea* (dorsal view). **Fig. 53** Macromiidae; head of *Macromia* (dorsal view). **Fig. 54** Libellulidae; abdominal segments 7–10 of *Pantala* (dorsal view) showing epiproct (E), cerci (C), and paraprocts (P).

16(15)	Antennae prehensile, with long, strong apical spines (Fig. 16.117.166)	17
16’	Antennae not prehensile, lacking long apical spines (Fig. 16.117.165)	Culicidae
17(16)	Antennae bases close together; a row of spinose setae on dorsolateral margin of head	Corethrellidae
17’	Antennae at anterolateral margins of head (Fig. 16.117.166); head without spinose setae dorsolaterally	Chaoboridae
18(15)	Thoracic and abdominal segments each distinctly divided into 2 or 3 annuli, with sclerotized dorsal plates on some annuli (Fig. 16.117.168)	Psychodidae
18’	Abdomen without secondary annulations (Fig. 16.117.169)	Ceratopogonidae (in part)
19(14)	Intermediate body segments with prolegs (Figs. 16.117.170–173)	20
19’	Anterior and/or posterior ends of body with prolegs only (Figs. 16.117.174–176)	23
20(19)	Abdomen with seven or eight pairs of distinct prolegs (Figs. 16.117.170, 171)	21
20’	Abdomen with two or three pairs of weak prolegs (Figs. 16.117.172, 173)	22
21(20)	Eight pairs of slender, ventrally projecting prolegs (Fig. 16.117.170) [Canada: NB, QC. USA: ME]	Nymphomyiidae
21’	Seven pairs of stout, ventrolaterally projecting prolegs (Fig. 16.117.171) [Western Mountains]	Deuterophlebiidae

22(20)	Abdominal segment 1 and usually also segment 2 with paired ventral prolegs; posterior end of body with 2 pairs of fringed processes (Fig. 16.117.172)	Dixidae
22'	Abdominal segments 1–3 with paired ventral prolegs; body terminating in a long respiratory tube (Fig. 16.117.173)	Ptychopteridae
23(19)	Posterior prolegs usually present; posterior of abdomen not swollen and terminating in a ring of hooks	24
23'	Prothorax with a single proleg present only; posterior of abdomen swollen and terminating in a ring of numerous small hooks (Fig. 16.117.174)	Simuliidae
24(23)	Only posterior prolegs present (Fig. 16.117.175)	25
24'	Anterior and usually posterior prolegs present (Figs. 16.117.176, 177)	26
25(24)	Last two abdominal segments and prolegs with long filamentous processes (Fig. 16.117.175)	Tanyderidae
25'	Last two abdominal segments without filamentous processes	Ceratopogonidae (in part)
26(24)	Body at most covered with setae	27
26'	Body covered with long, strong spines	Ceratopogonidae (in part)
27(26)	At least one pair of prolegs separated distally (Fig. 16.117.176); stalked spiracles lacking	Chironomidae
27'	Prolegs unpaired (Fig. 16.117.177); short, stalked spiracles dorsolaterally on the prothorax	Thaumaleidae

Crustacea: Hexapoda: Insecta: Coleoptera: Families

Adults in the following families are not keyed since they generally inhabit moist areas, but not within water: Carabidae, Limnichidae, Scarabaeidae, Scirtidae, Staphilinidae, and Ptiliidae. Hydrophilidae are treated in a broad sense, the Hydrochidae and Helophoridae of others being considered subfamilies. Larvae in the following families are not keyed since they generally inhabit moist areas, but not within water: Carabidae and Heteroceridae.

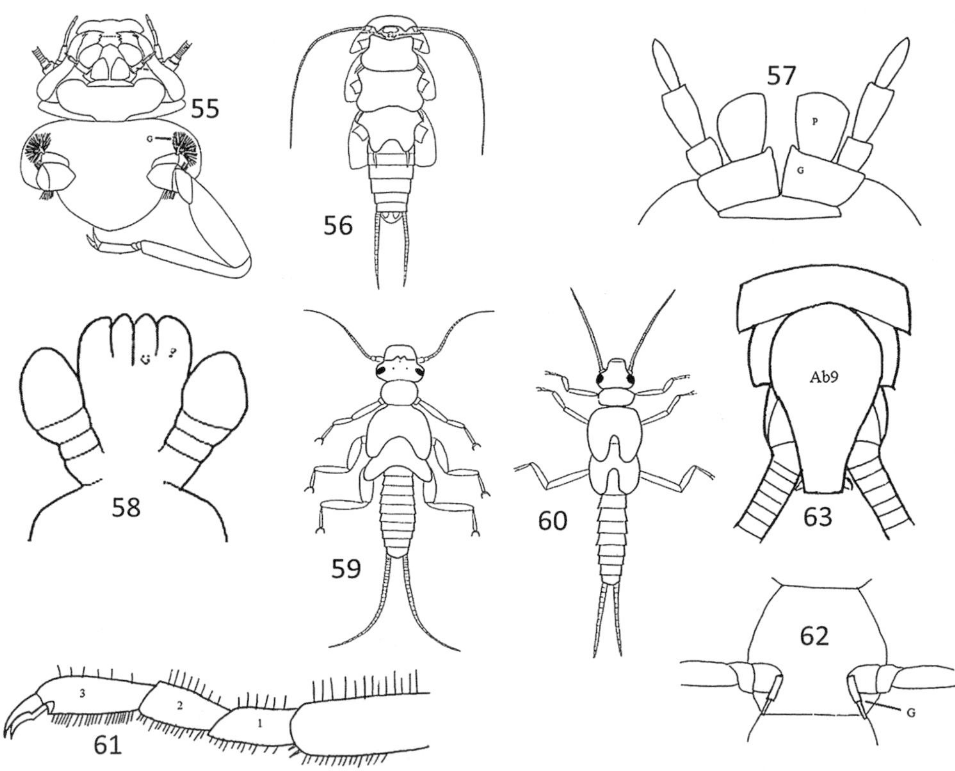

PLATE 16.105 **Fig. 55** *Acroneuria* head and prosternum with filamentous tuft of gill (G) (ventral). **Fig. 56** Peltoperlidae; *Peltoperla* (ventral view) with legs removed beyond coxae. **Fig. 57** Chloroperlidae; labium of *Alloperla* (ventral view) showing glossae (G) and paraglossae (P). **Fig. 58** Nemouridae; labium (ventral view) showing glossae (G) and paraglossae (P). **Fig. 59** Perlodidae; *Isoperla* (dorsal view). **Fig. 60** Chloroperlidae; *Haploperla* (dorsal view). **Fig. 61** Taeniopterygidae; tarsal segments (1–3) of *Taeniopteryx* (lateral view). **Fig. 62** Taeniopterygidae; mesosternum of *Taeniopteryx* showing telescopic gills (G) at base of coxae. **Fig. 63** Taeniopterygidae; ninth abdominal sternum (9) of *Strophopteryx* (ventral view).

1	Adults	2
1'	Larvae	16
2(1)	Compound eyes not divided; meso- and metathoracic legs not flattened, with tarsi not folded fanlike	2
2'	Compound eyes divided into dorsal and ventral segments; meso- and metathoracic legs short, extremely flattened, with tarsi folding fan-like	Gyrinidae
3(2)	Elytra covering all or nearly all of abdomen; size variable, but generally larger; varied habitats	4
3'	Elytra short, exposing 2–4 abdominal segments (Fig. 16.113.130); extremely small, <1.5 mm long; all tarsi 3 segmented	Hydroscaphidae, one genus: *Hydroscapha*
	[Mexico. USA: AZ, CA, ID, NV]	
4(3)	Head not formed into a beak or snout; antennae not geniculate	5
4'	Head formed as a snout or beak anteriorly; antennae elbowed (Fig. 16.112.117)	Curculionidae
5(4)	Metacoxae expanded into large coxal plates that cover all or part of abdominal sternites 1–3 (Fig. 16.112.118)	6
5'	Metacoxae not expanded into large plates	7
6(5)	Hind coxal plates covering abdominal sternites 1–2, but exposing those segments laterally; bases of hind femora exposed; small (<1.5 mm length), oval, highly convex beetles; inhabits stream margins	Sphaeriusidae, one genus: *Sphaerius*
6'	Hind coxal plates covering abdominal sternites 1–3 completely (Fig. 16.112.118); bases of hind femora hidden; larger beetles; fully aquatic	Haliplidae
7(6)	Prosternum with a postcoxal process extending posteriorly to mesocoxae (Fig. 16.112.119); first visible abdominal sternum completely divided by metacoxal process	8
7'	Prosternum with postcoxal process absent or short; first visible abdominal sternum extending for its entire breadth behind metacoxae (Fig. 16.112.120)	11
8(7)	Pronotum posterior margin subequal in width to base of elytra; metatarsi flattened and fringed with long setae, or beetles <3 mm long	9
8'	Pronotum posterior margin much narrower than base of elytra (Fig. 16.112.121); metatarsi rounded, not fringed with long setae; large, >10 mm long	Amphizoidae, one genus; *Amphizoa*
	[Canada: western. USA: western]	

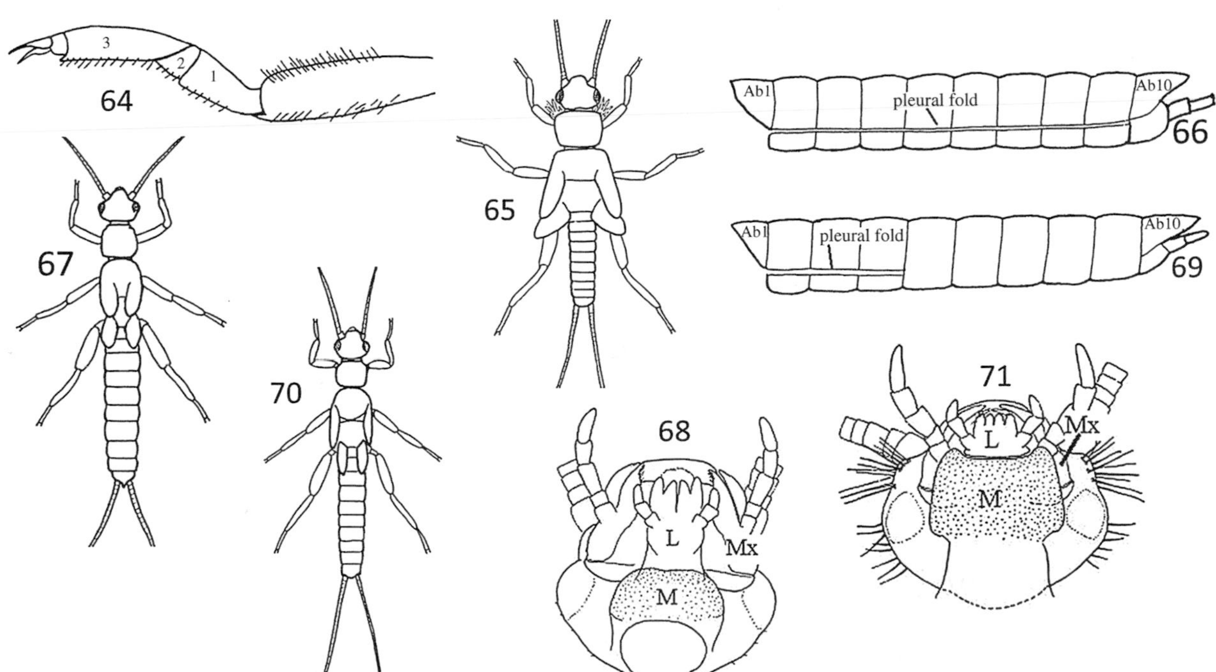

PLATE 16.106 **Fig. 64** Nemouridae; tarsal segments (lateral view). **Fig. 16.65** Nemouridae; *Amphinemura* (dorsal view). **Fig. 16.66** Capniidae; abdomen of *Allocapnia* (lateral view) showing pleural fold. **Fig. 16.67** Capniidae: *Paracapnia* (dorsal view). **Fig. 16.68** Capniidae; *Capnia* ventral view of mentum (M), labium (L), and maxilla (Mx) modified from Zwick, 2006. **Fig. 16.69** Leuctridae; abdomen of *Leuctra* (lateral view) showing pleural fold. **Fig. 16.70** Leuctridae; *Leuctra* (dorsal view). **Fig. 16.71** Leuctridae; *Leuctra* (ventral view) mentum (M), labium (L), and maxilla (Mx) modified from Zwick, 2006.

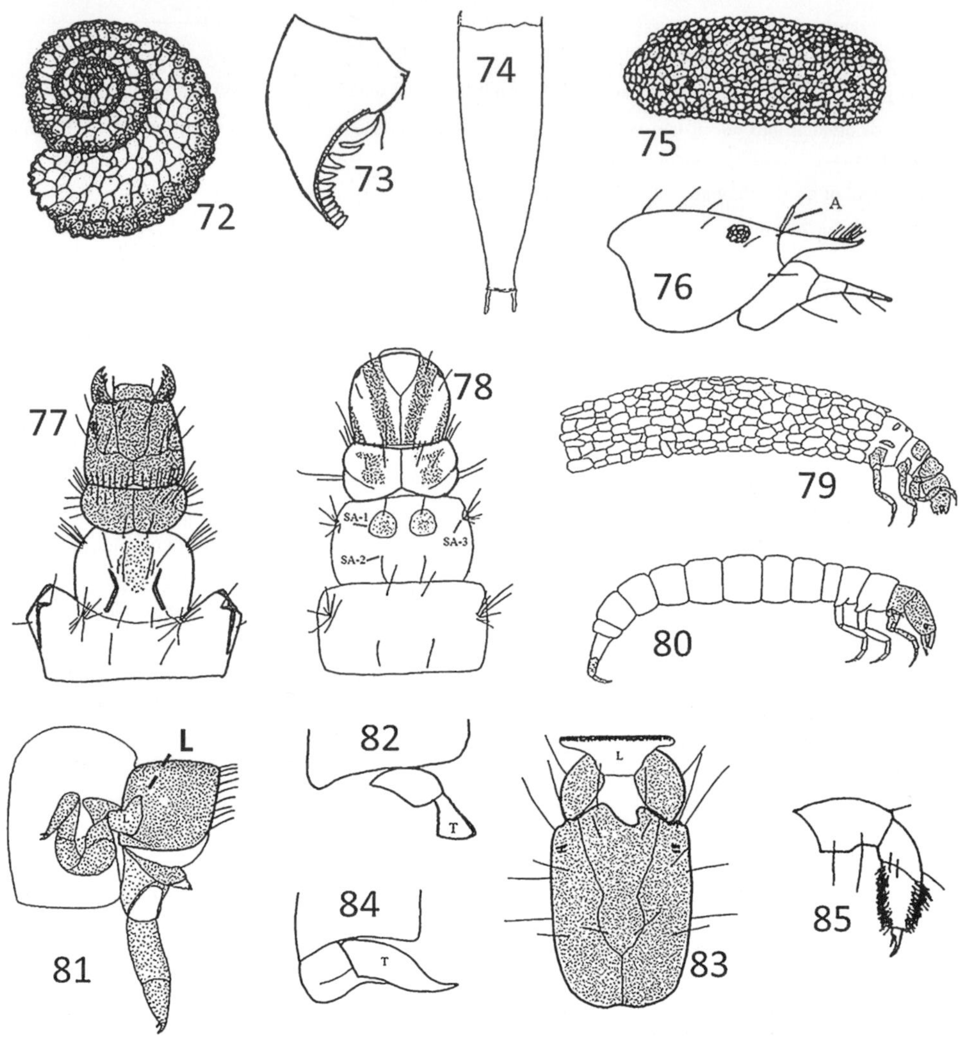

PLATE 16.107 **Fig.72** Helicopsychidae; case of *Helicopsyche* (dorsal view). **Fig. 73** Helicopsychidae; claw on anal proleg of *Helicopsyche*. **Fig. 74** Hydroptilidae; case of *Oxyethira* (lateral view). **Fig. 75** Hydroptilidae; case of *Hydroptila* (lateral view). **Fig. 76** Leptoceridae; head of *Oecetis* (lateral view) showing antenna (A). **Fig. 77** Leptoceridae; head and thoracic terga of *Ceraclea*. **Fig. 78** Phryganeidae; head and thoracic terga of *Oligostomis* showing location of setal areas (SA). **Fig. 79** Limnephilidae; *Hesperophylax* (lateral view). **Fig. 80** Polycentropodidae; *Polycentropus* (lateral view). **Fig. 81** Xiphocentronidae; pro- and mesothorax of *Xiphocentron* (lateral view) showing projecting lobe (L). **Fig. 82** Psychomyiidae; pronotum (P) and protrochantin (T) of *Psychomyia*. **Fig. 83** Philopotamidae; head of *Chimarra* (dorsal view) showing T-shaped labrum (L). **Fig. 84** Polycentropodidae; pronotum (P) and protrochantin (T) of *Polycentropus*. **Fig. 85** Dipseudopsidae; tarsus of *Phylocentropus*.

9(8) Prosternum with postcoxal process and metasternum in same plane anteriorly (Fig. 16.112.122); pro- and mesotarsi distinctly 5-segmented, segment 4 as long as 3 10

9' Prosternum greatly depressed and not in same plane as its postcoxal process and metasternum (Fig. 16.112.123); pro- and mesotarsi appear to be 4-segmented, 4th hidden in bilobed 3rd segment .... Dytiscidae (in part)

10(9) Prosternal process pointed or nearly so (Fig. 16.112.119); no curved spur or hooked apex on protibiae; >4 mm long .... Dytiscidae (in part)

10' Prosternal process truncate or rounded posteriorly (Fig. 16.113.124); protibiae with large setae, curved spur, or hooked apex (Fig. 16.113.125) or beetles <3 mm long .... Noteridae

[Canada: Southeast. USA: Eastern, Southwest. Mexico.]

11(7) Antennae short, clavate, with antennomere 4 or 6 modified to form a cupule (Fig. 16.113.124); maxillary palpi usually longer than antennae 12

11' Antennae pectinate or filiform (Figs. 16.113.128, 129), usually longer than maxillary palps 14

12(11) Antennae with 3 antennomeres past cupule (Fig. 16.113.126); 1.5–40.0 mm long 13

12' Antennae with 5 antennomeres past cupule (be careful not to confuse the long maxillary palps of some *Hydraena* with the antennae); <2.5 mm long .... Hydraenidae

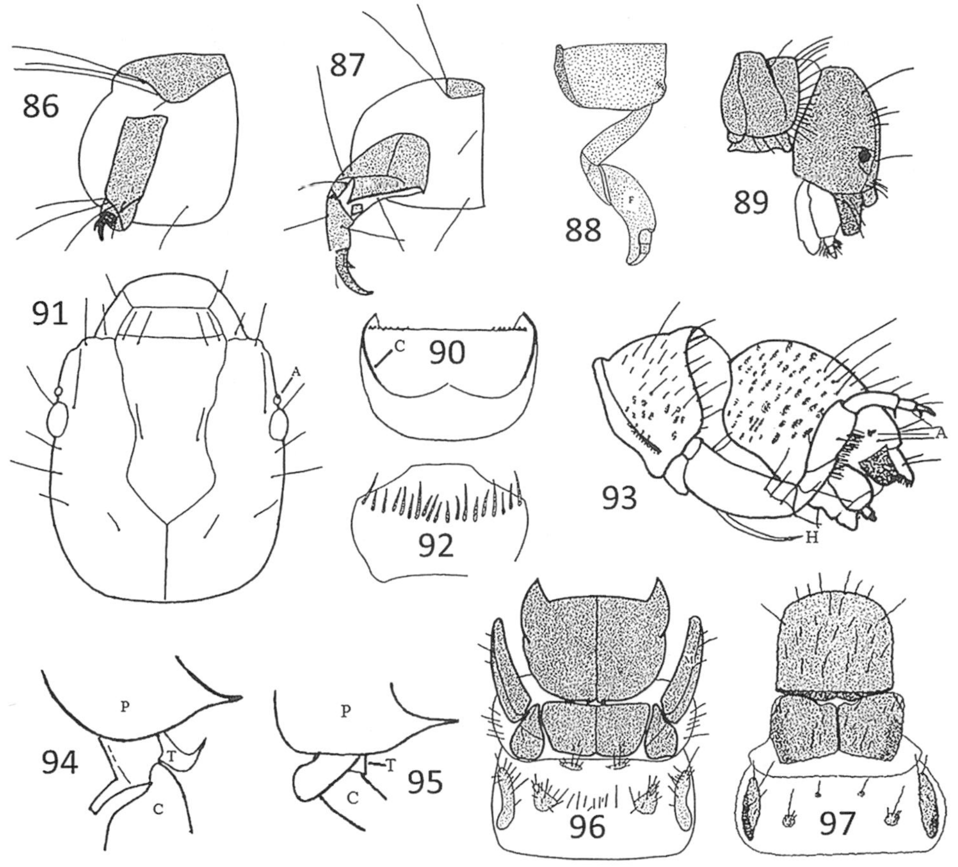

PLATE 16.108 **Fig. 86** Glossosomatidae; last abdominal segment of *Glossosoma* (lateral view) showing anal proleg. **Fig. 87** Rhyacophilidae; last abdominal segment of *Rhyacophila* (lateral view) showing anal proleg. **Fig. 88** Hydrobiosidae; pronotum and prothoracic leg with chelate femur (F) of *Atopsyche* (posterolateral view). **Fig. 89** Brachycentridae; pronotum and head of *Brachycentrus* (lateral view). **Fig. 90** Beraeidae; pronotum of *Beraea* (dorsal view) showing carina (C). **Fig. 91** Lepidostomatidae; head of *Lepidostoma* (dorsal view) showing antenna (A). **Fig. 92** Calamoceratidae; labrum of *Heteroplectron* (dorsal view). **Fig. 93** Limnephilidae; head and prothorax of *Platycentropus radiatus* (lateral view) showing antenna (A) and prosternal horn (H). **Fig. 94** Sericostomatidae; pronotum (P), protrochantin (T), and coxa (C) of *Agarodes*. **Fig. 95** Odontoceridae; pronotum (P), protrochantin (T), and coxa (C) of *Psilotreta*. **Fig. 96** Goeridae; thoracic terga of *Goera* showing mesepisternum (M). **Fig. 97** Thremmatidae; thoracic terga of *Neophylax*.

13(12)	Four protarsi; metacoxae widely separated, intercoxal process broadly truncate (Fig. 16.113.127); body often covered with sand grains .... Georissidae, one genus: *Georissus*
13'	Five protarsi; metacoxae nearly contiguous; intercoxal process variable .... Hydrophilidae
14(11)	Antennae slender, filiform; (Fig. 16.113.129); <4.5 mm long 15
14'	Antennae short with pectinate club (Fig. 16.113.128); >5.0 mm long .... Dryopidae
15(14)	Five tarsomeres (Fig. 16.113.129); antennae subequal to head in length .... Elmidae
15'	Four tarsomeres, the bilobed 3rd tarsomere hiding the miniature 4th; antennae longer than head and pronotum combined .... Chrysomelidae
16(1)	Tarsi with 2 claws 17
16'	Tarsi with 1 claw 20
17(16)	Last abdominal segment lacking hooks; if lateral abdominal filaments are present, there are only 6 pairs 18
17'	Abdomen with 4 conspicuous hooks on last segment; abdominal segments with at least 8 pairs of lateral filaments (Fig. 16.114.131) .... Gyrinidae
18(17)	Abdominal and thoracic terga not flattened and expanded laterally 19
18'	Abdominal and thoracic terga flattened and expanded laterally (Fig. 16.114.132) .... Amphizoidae, one genus; *Amphizoa*
19(18)	Urogomphi shorter than last abdominal segment; legs short, stout, adapted for digging (Fig. 16.114.133); mandibles short, adapted for chewing .... Noteridae
19'	Urogomphi usually longer than last abdominal segment (Fig. 16.114.134); if shorter, legs are elongate with setal fringe for swimming; mandibles elongate, pointed, for piercing .... Dytiscidae

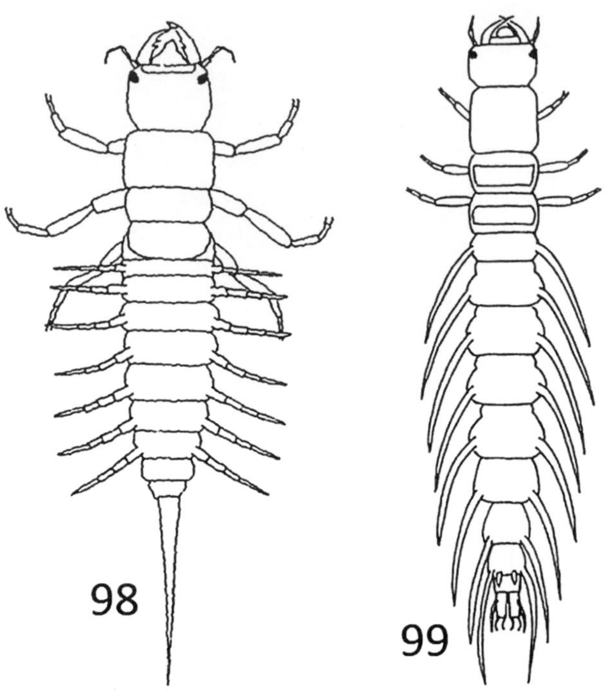

PLATE 16.109 **Fig. 98** Sialidae; *Sialis* (dorsal view). **Fig. 99** Corydalidae; *Nigronia* (dorsal view).

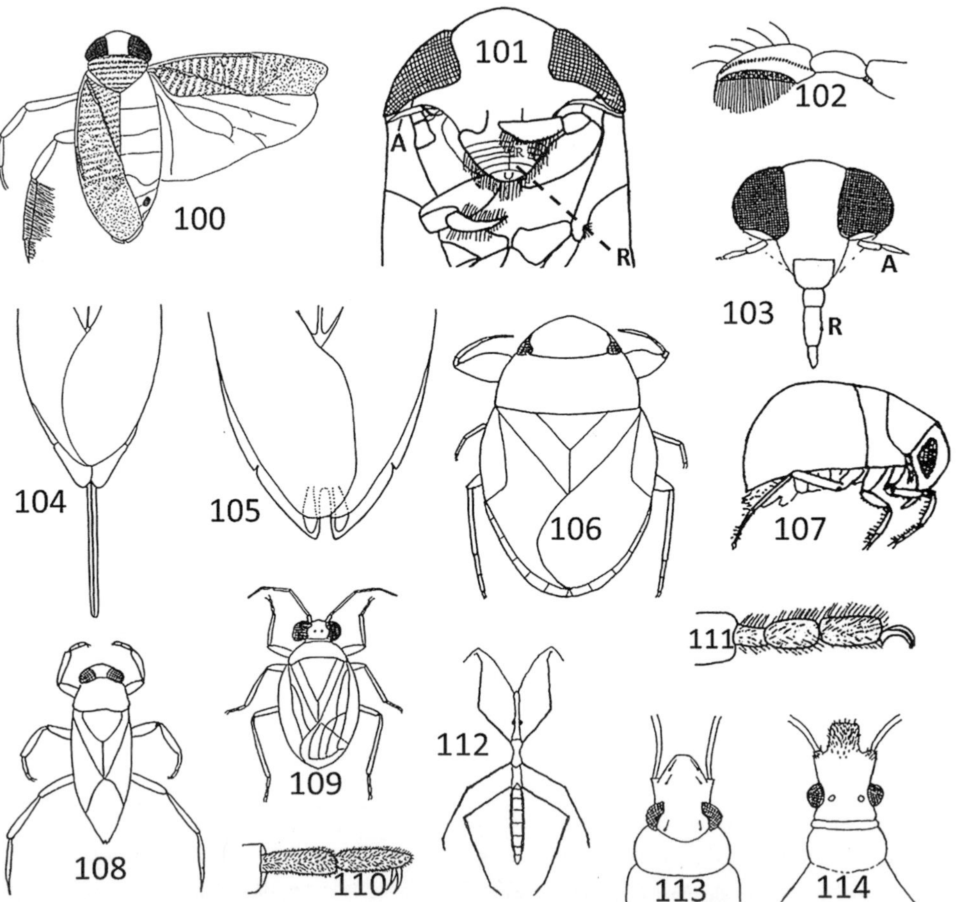

PLATE 16.110 **Fig. 100** Corixidae; *Sigara* (dorsal view), with right wings extended laterally. **Fig. 101** Corixidae; head and prothorax of *Sigara* (ventral view) showing rostrum (R) and antenna (A). **Fig. 102** Corixidae; pala of male *Sigara*. **Fig. 103** Notonectidae; head of *Notonecta* (ventral view) showing rostrum (R) and antenna (A). **Fig. 104** Nepidae; apex of abdomen of *Nepa* (dorsal view). **Fig. 105** Belostomatidae; apex of abdomen of *Belostoma* (dorsal view). **Fig. 106** Naucoridae; *Pelocoris* (dorsal view). **Fig. 107** Pleidae; *Neoplea* (lateral view). **Fig. 108** Notonectidae; *Notonecta* (dorsal view). **Fig. 109** Saldidae; *Salda* (dorsal view) showing membrane (M). **Fig. 110** Gerridae: protarsus of *Gerris*. **Fig. 111** Mesoveliidae; protarsus of *Mesovelia*. **Fig. 112** Hydrometridae; *Hydrometra* (dorsal view). **Fig. 113** Mesoveliidae; head of *Mesovelia* (dorsal view). **Fig. 114** Macroveliidae; head of *Macrovelia* (dorsal view).

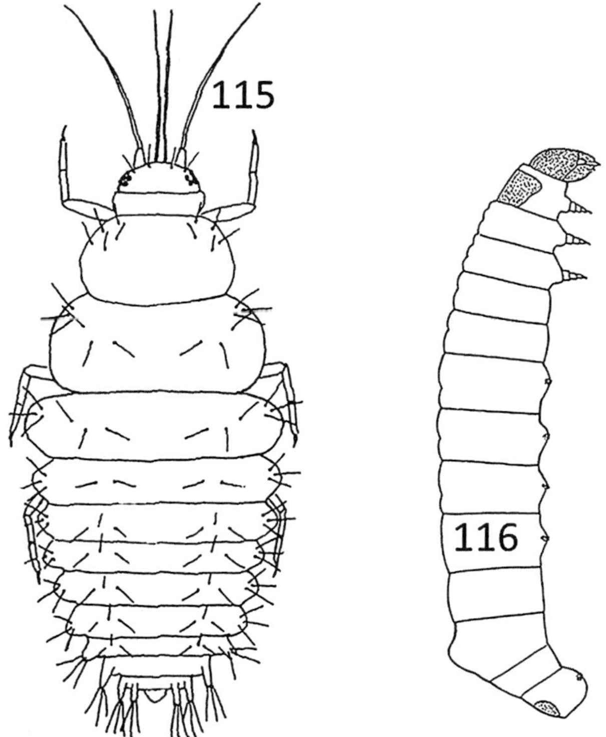

PLATE 16.111 **Fig. 115** Neuroptera Larva (Sisyridae); *Climacia* (dorsal view). **Fig. 116** Lepidoptera Larva (Nymphalidae); *Nymphula* (lateral view).

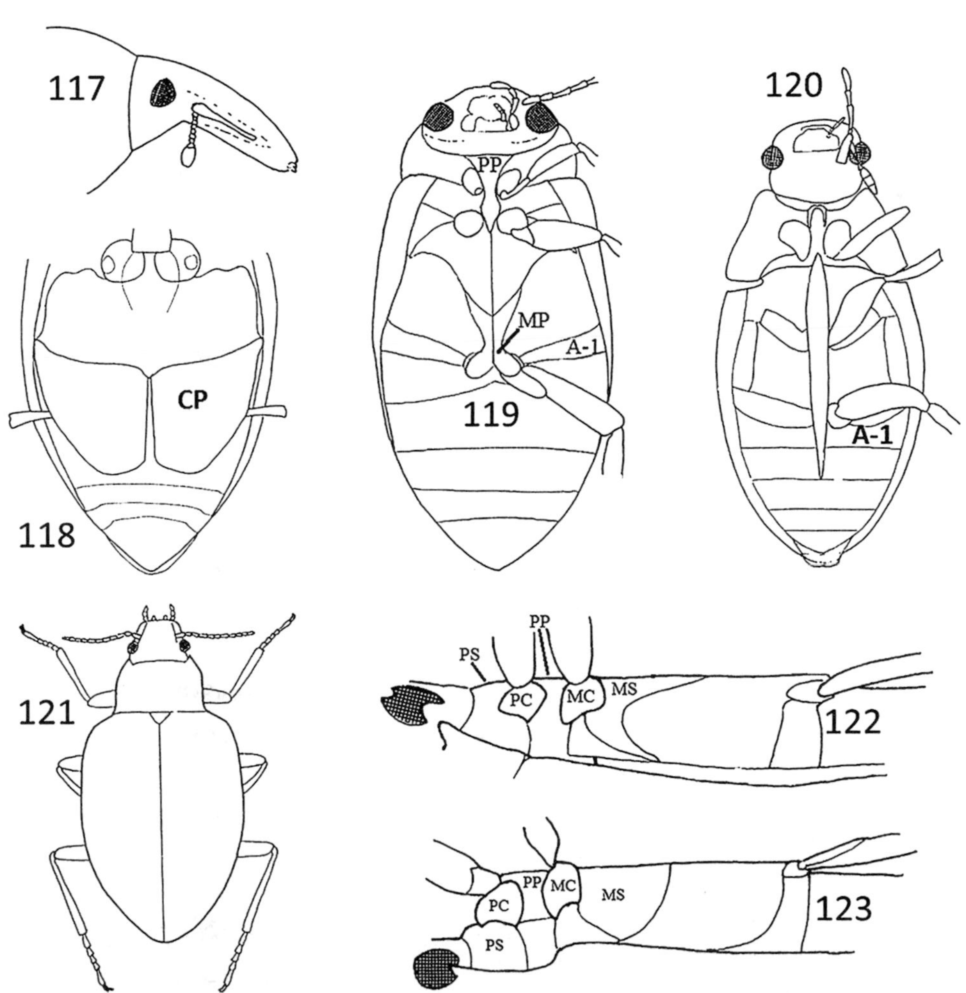

PLATE 16.112 **Fig. 117** Curculionidae; head (lateral view). **Fig. 118** Haliplidae; metathorax and abdomen of *Haliplus* (ventral view) showing metacoxal plate (CP). **Fig. 119** Dytiscidae; *Agabus* (ventral view) showing prosternal process (PP), first abdominal sternum (A-1), and metacoxal process (MP). **Fig. 120** Hydrophilidae; *Tropisternus* (ventral view) showing first abdominal sternum (A-1). **Fig. 121** Amphizoidae; *Amphizoa* (dorsal view). **Fig. 122** Dytiscidae; *Agabus* (lateral view) showing prosternum (PS), prosternal process (PP), metasternum (MS), procoxa (PC), and mesocoxa (MC). **Fig. 123** Dytiscidae; *Hydroporus* (lateral view) showing prosternum (PS), prosternal process (PP), metasternum (MS), procoxa (PC), and mesocoxa (MC).

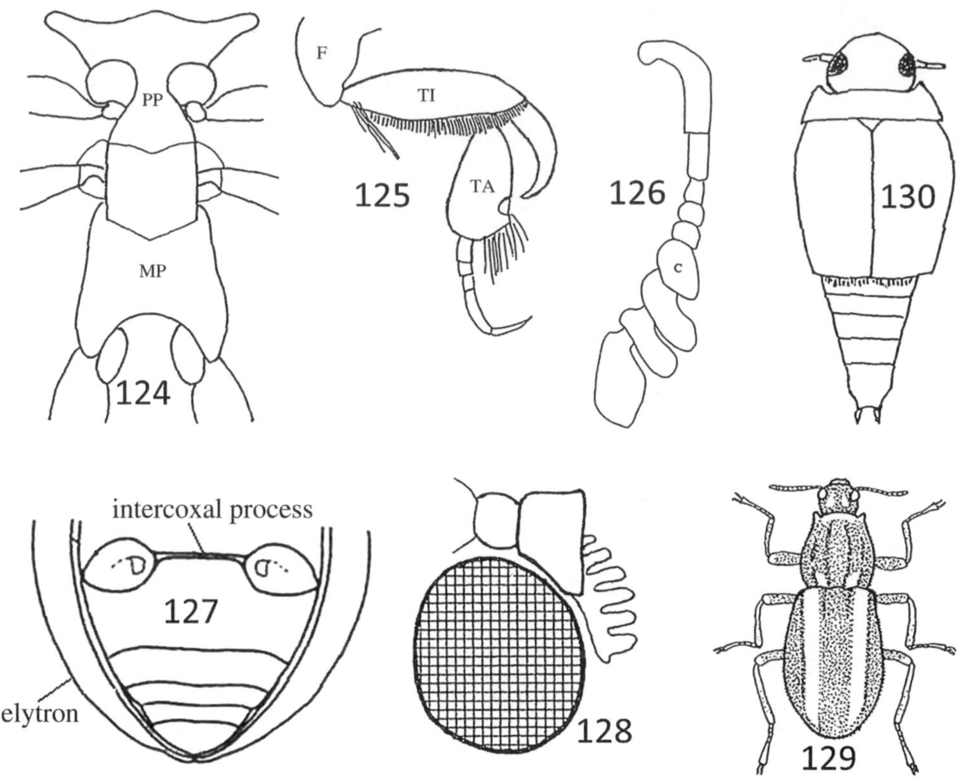

PLATE 16.113 **Fig. 124** Noteridae; thorax of *Hydrocanthus* (ventral view) showing prosternal process (PP) and metasternal plate (MP). **Fig. 125** Noteridae; prothoracic leg of *Hydrocanthus* showing profemur (F), tibia (TI), and tarsus (TA). **Fig. 126** Hydrophilidae; antenna of *Tropisternus* showing cupule (C). **Fig. 127** Georissidae; abdominal sternites of *Georissus* sp. (ventral view). Modified from White & Roughley, 2008. **Fig. 128** Dryopidae; right antenna and eye of *Helichus* (dorsal view). **Fig. 129** Elmidae; *Stenelmis* (dorsal view). **Fig. 130** Hydroscaphidae; *Hydroscapha* (dorsal view).

20(16)	Legs apparently of four articles; abdomen not terminating in long filaments	21
20'	Legs distinctly of five articles	30
21(20)	Mandibles not readily visible in dorsal view	22
21'	Mandibles large, readily visible in dorsal view (Fig. 16.114.137)	24
22(21)	Antennae much shorter than head and thorax combined	23
22'	Antennae long, filiform, as long as head and thorax combined (Fig. 16.114.138)	Scirtidae
23(22)	Abdomen with 10 segments; urogomphi short, papilliform (Fig. 16.115.151); legs short (Fig. 16.115.152)	Georissidae, one genus: *Georissus*
23'	Abdomen 8 segmented, or rarely if 10 segmented, the urophomphi long, with 2–3 articles; legs long	Hydrophilidae
24(21)	Body elongate and round or triangular in cross section; head exposed, except mostly concealed in Lampyridae	25
24'	Body oval and extremely flat; head completely concealed from dorsal view (Figs. 16.114.139, 140)	Psephenidae
25(24)	Thoracic and abdominal segments not covered by flat, plate-like sclerites; head visible from above	26
25'	Each thoracic and abdominal segment covered by a flat, plate-like sclerite dorsally; prothoracic plate mostly or completely concealing head from above (Fig. 16.114.141)	Lampyridae
26(25)	Body elongate and sclerotized; no spines on last abdominal tergum (Figs. 16.114.143, 144)	27
26'	All terga rounded and pale; larvae with 2 spines on last abdominal tergum (Fig. 16.114.142)	Chrysomelidae
27(26)	Last abdominal segment lacking an operculum	28
27'	Last abdominal segment having an operculum	29
28(27)	Abdominal sterna 1–7 with distinct gill tufts [USA: California]	Eulichadidae

PHYLUM ARTHROPODA

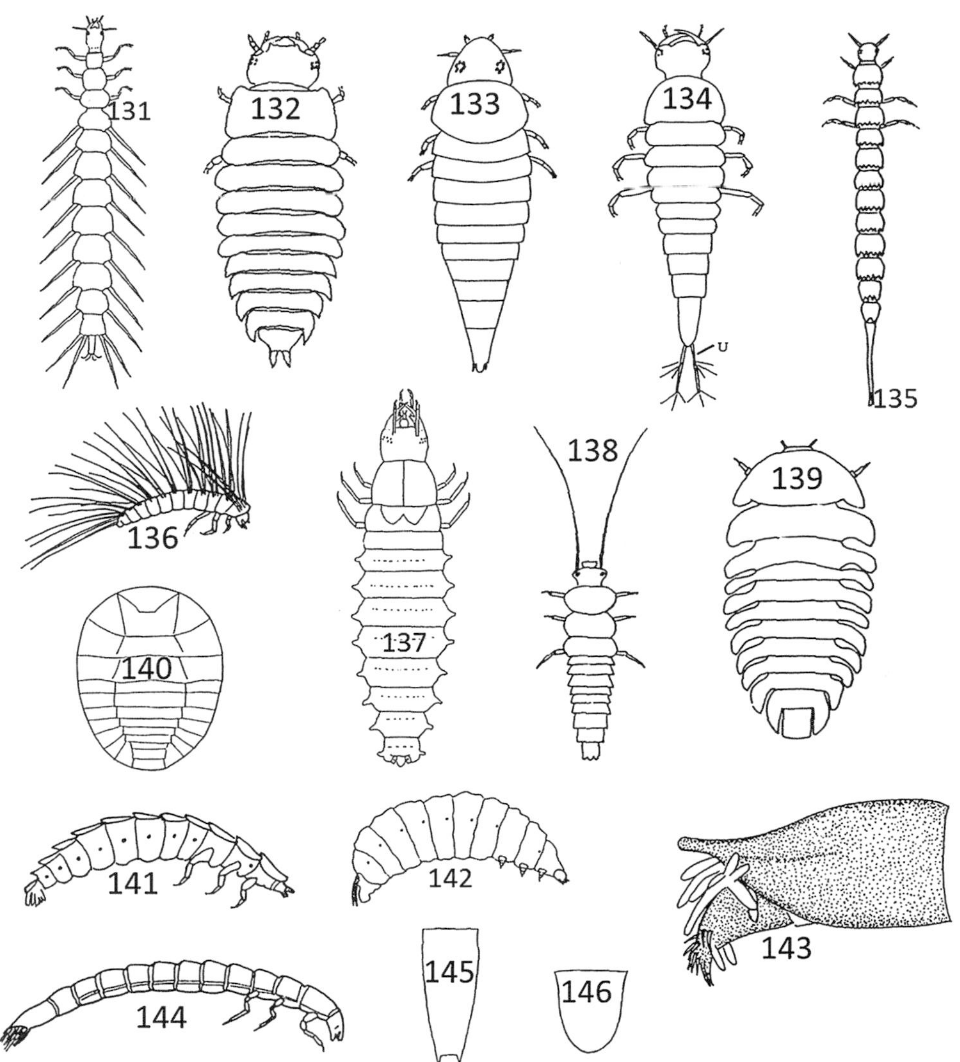

PLATE 16.114 **Fig. 131** Gyrinidae; *Dineutus* (dorsal view). **Fig. 132** Amphizoidae; *Amphizoa* (dorsal view). **Fig. 133** Noteridae; *Hydrocanthus* (dorsal view). **Fig. 134** Dytiscidae; *Agabus* (dorsal view) showing urogomphi (U). **Fig. 135** Haliplidae; *Haliplus* (dorsal view). **Fig. 136** Haliplidae; *Peltodytes* (lateral view). **Fig. 137** Hydrophilidae; *Tropisternus* (dorsal view). **Fig. 138** Scirtidae; *Cyphon* (dorsal view). **Fig. 139** Psephenidae; *Ectopria* (dorsal view). **Fig. 140** Psephenidae; *Psephenus* (dorsal view). **Fig. 141** Lampyridae (lateral view). **Fig. 142** Chrysomelidae; *Donacia* (lateral view). **Fig. 143** Ptilodactylidae; last abdominal segment of *Anchytarsus* (lateral view). **Fig. 144** Elmidae; *Stenelmis* (lateral view). **Fig. 145** Elmidae; last abdominal tergum of *Stenelmis*. **Fig. 146** Lutrochidae; last abdominal tergum of *Lutrochus*.

28' Abdominal sterna 1–7 without gill tufts (Fig. 16.114.143) .... Ptilodactylidae

[Canada. USA. Mexico]

29(27) Last abdominal tergum bifid or notched apically (Fig. 16.114.145) .... Elmidae

29' Last abdominal tergum rounded apically (Fig. 16.114.146) .... Lutrochidae

[USA. Mexico]

30(20) Abdomen not terminating in 1 or 2 long filaments 31

30' Abdomen terminating in 1 or 2 long filaments (Figs. 16.114.135, 136) .... Haliplidae

31(30) Abdominal tergites 1–8 with lateral digitiform articulated lobes (Fig. 16.115.147); antennomere 2 two or three times longer than broad, with a thin lateral appendage (Fig. 16.115.148) .... Sphaeriusidae, one genus; *Sphaerius*

31' Abdominal tergites 1 and 8 only with lateral digitiform articulated lobes (Fig. 16.115.149); antenna short, composed of 2 antennomeres, 1 being short and broad, 2 much longer with distal sensory appendage adjacent to antenna (Fig. 16.115.150) .... Hydroscaphidae, one genus; *Hydroscapha*

[USA: Arizona, California, Idaho, Nevada. Mexico]

Crustacea: Hexapoda: Insecta: Heteroptera: Families

1 Antennae shorter than head, inserted beneath eyes and (except Ochteridae) not visible from above (Figs. 16.110.101, 110.103, 107) ... suborder Nepomorpha ... 2

1' Antennae longer than head, inserted in front of eyes and visible from above (Figs. 16.110.109, 112–114) ... 9

2(1) Rostrum cylindrical or conical, distinctly of 3 or 4 articles (Fig. 16.110.103) ... 3

2' Rostrum broad, blunt, and triangular, not distinctly segmented (Fig. 16.110.101); each front tarsus of one article, fringed with setae (Fig. 16.110.102) ... Corixidae

3(2) Apical respiratory appendages absent, or if present, short and flat (Fig. 16.110.105) ... 4

3' Abdominal apex with a long, slender, tubular respiratory appendage (Fig. 16.110.104) ... Nepidae

4(3) Meso- and metathoracic legs fringed with swimming setae; ocelli absent; aquatic ... 5

4' Meso- and metathoracic legs without swimming setae; ocelli usually present; riparian ... 8

5(4) Dorsoventrally flattened, ovate insects; profemora broad, raptorial (Fig. 16.110.106) ... 6

5' Elongate or hemispherical insects, not flattened dorsoventrally (Figs. 16.110.107, 108); profemora slender, similar to other legs ... 7

6(5) Length >18 mm; short, flat, strap-like apical respiratory areas present (Fig. 16.110.105); eyes protrude from head margin ... Belostomatidae

6' Length <16 mm; apical respiratory appendages absent; eyes not protruding (Fig. 16.110.106) ... Naucoridae

7(5) Body hemispherical (Fig. 16.110.107); length <3 mm ... Pleidae

7' Body elongate (Fig. 16.110.108); length >5 mm ... Notonectidae

8(4) Profemora broad, raptorial; antennae concealed from above ... Gelastocoridae

8' Profemora slender, similar to other legs; antennae visible from above ... Ochteridae

9(1) Membrane of hemelytra without veins or with dissimilar sized cells; metacoxae small, conical; semiaquatic or riparian ... suborder Gerromorpha ... 10

9' Membrane of hemelytra with 4 or 5 equal-sized cells (Fig. 16.110.109); metacoxae large, transverse; riparian ... Leptopodomorpha, Saldidae

10(9) Claws of at least protarsi inserted before apex (Fig. 16.110.110) ... 11

10' Claws of all tarsi inserted at apex (Fig. 16.110.111) ... 12

11(10) Metafemora long, greatly surpassing apex of abdomen ... Gerridae

11' Metafemora short, not, or only slightly surpassing apex of abdomen ... Veliidae

12(10) Head short and stout, eyes near posterior margin ... 13

12' Head as long as entire thorax, slender, with eyes set about halfway to base (Fig. 16.110.112) ... Hydrometridae

13(12) Head not grooved ventrally; three tarsomeres; >2.5 mm long ... 14

13' Head grooved ventrally to receive rostrum; two tarsomeres; <2.5 mm long ... Hebridae

14(13) Eye medial margins converge anteriorly (Fig. 16.110.113); femora with 1 or more dorsal black spines distally ... Mesoveliidae

14' Eye medial margins rounded (Fig. 16.110.114); femora without black spines ... Macroveliidae

[Western North America]

Crustacea: Hexapoda: Ephemeroptera: Families

1 Mandibles with large, forward projecting tusks (Fig. 16.100.13); abdominal segments 2–7 with gills forked, margins fringed (Fig. 16.100.14) and projecting laterally or dorsally over abdomen ... 2

1' Mandibles usually lacking tusks (exceptions are a few Leptophlebiidae in western North America); fringed gills absent or projecting ventrolaterally ... 6

2(1) Gills projecting laterally; protibiae slender, subcylindrical (Fig. 16.100.15) ... 3

2' Gills held dorsally; protibiae fossorial (Fig. 16.100.16) ... 4

3(2) Mandibular tusks >2× length of head with many long setae ... Euthyplociidae

[USA: Southwest. Mexico]

3' Mandibular tusks without many long setae ... Potamanthidae

[USA: except Southwest]

4(2) Apex of metatibiae rounded (Fig. 16.100.17) ... 5

4' Apex of metatibiae forming an acute point ventrally (Fig. 16.100.18); mandibular tusks curved upward or outward apically ... Ephemeridae

5(4) Mandibular tusk cross section somewhat rounded, length curved inward and downward apically, crenulations on dorsal and lateral surfaces ... Polymitarcyidae

5' Mandibular tusks flattened, curved inward, then laterally, crenulation restricted to lateral edges of tusk .... Palingeniidae

6(1) Mesonotum not modified into a carapace, gills exposed 7

6' Mesonotum modified into a carapace-like structure that covers the gills on abdominal segments 1–6 (Fig. 16.100.19) .... Baetiscidae

7(6) Head and pronotum without pads of long, dark setae; gills not fringed and projecting ventrolaterally 8

7' Head and pronotum with pads of long, dark setae on each side (Fig. 16.100.18); fringed gills projecting ventrolaterally .... Behningiidae

[USA: Southeast, Wisconsin]

8(7) Gills on abdominal segment 2 operculate or semioperculate, covering or partially covering gills on succeeding segments (Figs. 16.101.21, 22, 101.23, 24) 9

8' Gills on abdominal segment 2 similar to those on succeeding segments (Fig. 16.103.35) or absent (Fig. 16.101.25); operculate gills may be present, but on segments 3 or 4 11

9(8) Operculate gills quadrate and meeting along medial edge (Figs. 16.101.23, 24); gills on segments 3–6 with fringed margins 10

9' Operculate gills oval or subtriangular and well-separated from each other mesally (Figs. 16.101.21, 22); gills on segments 3–6 without fringed margins .... Leptohyphidae

10(9) Operculate gills fused medially (Fig. 16.101.24) .... Neoephemeridae

[USA: Eastern]

10' Operculate gills not fused medially, but may overlap (Fig. 16.101.23) .... Caenidae

11(8) Abdominal segments 1 or 2 to 7 with gills (Fig. 16.103.35) 12

11' Gills on abdominal segments 1–3 often missing, those on segment 3, when present, or on segment 4, are operculate, covering succeeding gills (Fig. 16.101.25); head, thorax, and/or abdomen often with paired, median tubercles .... Ephemerellidae

12(11) Head flattened dorsoventrally; eyes and antennae dorsal (Fig. 16.102.26); gills a single lamella, some genera with gills bearing a ventral fibrilliform tuft 13

12' Head and body not dorsoventrally flattened; eyes, and usually also antennae, along lateral margin of head (Fig. 16.102.27) 15

13(12) Maxillary palp distal palpomere short, usually not visible in dorsal view 14

13' Maxillary palp distal palpomeres setose and extending back over thorax and visible in dorsal view (Fig. 16.102.28) .... Arthropleidae, one species: *Arthroplea bipunctata* (McDunnough, 1924)

[Canada: Far North, Northeast, Northwest. USA: Midwest, Northeast, Southeast]

14(13) Gill lamellae with a ventral digitiform projection (Fig. 16.102.29); tarsal claws very long .... Pseudironidae, one species: *Pseudiron centralis* McDunnough, 1931

[Canada. USA]

14' Gill lamellae without such a projection; claws not exceptionally long .... Heptageniidae

15(12) Protarsal claws much shorter than tibiae and those on other tarsi (Figs. 16.102.30, 31); meso- and metatarsal claws long and slender, about as long as tibiae (Fig. 16.102.31) 16

15' All tarsal claws similar in structure and length 17

16(15) Protarsal claws bifid (Fig. 16.102.30); procoxae without a spinose pad .... Metretopodidae

[Canada. USA: except Southwest]

16' Protarsal claws simple, with long, slender denticles (Fig. 16.102.31); procoxae with a spinose pad .... Ametropodidae (*Ametropus*)

[Canada: Northwest. USA: Midwest, Northeast, Southeast]

17(15) Prothoracic femora and tibiae with dense, double row of long setae along inner surface (Fig. 16.103.33) 18

17' Prothoracic legs without dense row of setae along inner surface 19

18(17) Abdominal segment 1 gills dorsolateral; gills lamellate, with a basal fibrilliform tuft; gill tuft at base of procoxae (Fig. 16.103.33) .... Isonychiidae

18' Abdominal segment 1 gills ventral; gills on abdominal segments 2–7 small, either lanceolate with a posterior fringe (Fig. 16.103.34), or rounded and plate-like; no gill tuft at base of procoxae .... Oligoneuriidae

19(18) Gills single (Fig. 16.103.35) or double lamellae (Fig. 16.103.36) 20

19' Gills forked (Fig. 16.103.37), or bilamellate and terminating in a point (Fig. 16.103.38), or terminating in filaments (Fig. 16.103.39) .... Leptophlebiidae

20(19) Tibiae and tarsi not bowed; claws not as below 21

20' Tibiae and tarsi bowed; claws very long and slender; metatarsal claws about as long as tarsi .... Acanthametropodidae

[Canada: Northwest. USA: except Southwest]

21(18) Abdominal segments 8 and 9 produced posterolaterally into distinct, flattened spines (similar to Fig. 16.101.25); if spines are weak, antennae more than twice width of head 22

21'	Abdominal segments 8 and 9 without such spines (Fig. 16.103.35), if weak spines are present, antennal length less than twice width of head .... Baetidae
22(21)	Maxillae with crown of pectinate spines; gills single, with sclerotized band on ventral margin and little or no tracheation (Fig. 16.103.40) .... Ameletidae
22'	Maxillae without a crown of pectinate spines; gills single or double with well developed tracheation (Fig. 16.103.36) .... Siphlonuridae

Crustacea: Hexapoda: Plecoptera: Families

1	Thoracic segments with ventral or lateral conspicuous, finely branched gills (Fig. 16.105.55) 2
1'	Gills absent, confined ventrally to base of head (Fig. 16.106.65) are unbranched or fork at leg bases or rarely on abdomen (Figs. 16.105.56, 105.62) 3
2(1)	Abdominal sterna 1–2 or 1–3 with finely branched gills .... Pteronarcyidae
2'	Anterior abdominal sterna lack gills, but anal gills may be present .... Perlidae
3(1)	Thoracic sterna not produced into overlapping plates; coxal gills, if present, telescopic (Fig. 16.105.62); body form usually elongate, not roach-like 4
3'	Thoracic sterna produced into plates that overlap succeeding segment (Fig. 16.105.56); single, double, or forked gills at least behind meso- and metacoxae; form roach-like .... Peltoperlidae
	[Canada: East, West. USA: East, West (absent from mid-continent)]
4(3)	Glossae much shorter than paraglossae (Fig. 16.105.57) 5
4'	Glossae nearly as long as paraglossae (Fig. 16.105.58) 6
5(4)	Cerci almost as long or longer than abdomen; metathoracic wingpads with inner and outer margins strongly diverging from body axis (Fig. 16.105.59); head and thorax usually with a pigmented pattern .... Perlodidae
5'	Cerci distinctly shorter than abdomen; metathoracic wingpads with inner and outer margins nearly parallel to axis of body (Fig. 16.105.60); head and thorax usually not patterned .... Chloroperlidae
6(4)	Tarsomere two much shorter than first (Fig. 16.106.64); wingpads parallel or divergent 7
6'	Tarsomere two about as long as or longer than first (Fig. 16.105.61); wingpads strongly divergent (as in Fig. 16.106.65) .... Taeniopterygidae
7(6)	Gracile larvae; abdomen long, metathoracic legs not reaching tip of abdomen; metathoracic wingpads nearly parallel to axis of body (Figs. 16.106.67, 70) 8
7'	Robust larvae; abdomen short, extended metathoracic legs reaching abdominal tip; metathoracic wingpads divergent from axis of body (Fig. 16.106.65) .... Nemouridae
8(7)	Mentum small, not covering base of maxillae (Fig. 16.106.68); abdominal segments 1–9 divided by a membranous pleural fold (Fig. 16.106.66); metathoracic wingpads, if present, about same distance apart as mesothoracic wingpads (Fig. 16.106.67) .... Capniidae
8'	Mentum larger, covering bases of maxillae (Fig. 16.106.71); abdominal segments 1–4 or 1–7 divided by pleural fold (Fig. 16.106.69); metathoracic wingpads usually contained within the tips of the mesothoracic ones (Fig. 16.106.70) .... Leuctridae

Crustacea: Hexapoda: Odonata: Families

1	Abdomen terminating in 3 caudal gills (Fig. 16.104.41) .... suborder Zygoptera 2
1'	Abdomen terminating in 5 points (Fig. 16.104.54) .... suborder Anisoptera 6
2(1)	Antennomere one shorter than others combined; prementum with at most a very small median cleft (Figs. 16.104.43, 44) 3
2'	Antennomere one as long as, or longer than, remaining antennomeres combined (Fig. 16.104.41); prementum with deep, median cleft (Fig. 16.104.42) .... Calopterygidae
3(2)	Prementum with proximal half not greatly narrowed (Fig. 16.104.44); prementum in repose extends only to procoxae 4
3'	Prementum with proximal half greatly narrowed and elongate (Fig. 16.104.43); prementum in repose extends back to or past mesocoxae .... Lestidae
4(3)	Prementum wider distally than basally and ligula without median cleft 5
4'	Prementum wider basally than distally and ligula with distinct median cleft .... Platysticidae
	[USA: Arizona. Mexico]
5(4)	Gills divided into a thick basal half and thin, lighter colored distal half (Fig. 16.104.45); palpal lobes of prementum with one dorsal seta .... Protoneuridae
	[USA: Texas. Mexico]
5'	Gills not distinctly divided as above; palpal lobes of prementum usually with 2 or more dorsal setae (Fig. 16.104.44) .... Coenagrionidae

6(1) Prementum flat or nearly so, without dorsal setae; palpal lobes also flat and usually without stout setae 7

6' Prementum rounded, spoon-shaped, and usually with dorsal setae; palpal lobes also rounded, always with stout setae, and covering face to base of antennae 9

7(6) Antennae with 6 or 7 antennomeres (Figs. 16.104.47, 48); pro- and mesothoracic legs with 3 tarsomeres 8

7' Antennae with 4 antennomeres, often with antennomere three longer than other antennomeres (Fig. 16.104.46); pro- and mesothoracic leg pairs with 2 tarsomeres ....Gomphidae

8(7) Antennal segments slender, not setose (Fig. 16.104.47) ....Aeshnidae

8 Antennal segments short, thick, and setose (Fig. 16.104.48) ....Petaluridae

9(6) Palpal lobes with distal margins having small, even crenulations (Figs. 16.104.50–52) or nearly straight (Fig. 16.104.52) 10

9' Palpal lobe with distal margins having large, irregular teeth (Fig. 16.104.49) .... Cordulegastridae

10(9) Head without a prominent frontal process (except in *Neurocordulia molesta*); legs shorter, apex of metafemora usually not reaching apex of abdominal segment 8; metasternum without a median tubercle..11

10' Head with a prominent, almost erect, thick frontal process between bases of antennae (Fig. 16.104.53); legs very long, apex of each metafemur reaching to or beyond apex of abdominal segment 8; metasternum with broad, median tubercle .... Macromiidae

11(10) Palpal lobe crenulations large, at least 1/4 as long as wide (Figs. 16.104.50, 51) 12

11' Palpal lobe crenulations very shallow, 1/6 to <1/10 as long as wide (Fig. 16.104.52) or separated by minute notches bearing setae .... Libellulidae (in part)

12(11) Abdominal segment 8 with lateral spines as long or longer than mid-dorsal length of segment 9 (Fig. 16.104.54) .... Libellulidae (in part)

12' Abdominal segment 8 with lateral spines shorter than mid-dorsal length of segment 9 ....Corduliidae

Crustacea: Hexapoda: Trichoptera: Families

1 Case not spiral or case absent; anal claws not forming a comb, apex hook-shaped..2

1' Larval case spiral, resembling a snail shell made of sand grains (Fig. 16.107.72); anal claws with short teeth forming a comb, apex hook-shaped (Fig. 16.107.73) ....Helicopsychidae

2(1) Each thoracic segment covered with a single dorsal plate, which may have a mesal or transverse suture line 3

2' Metanotum mostly membranous, having only scattered hairs or small plates, or with 2 or more sclerites 5

3(2) Abdomen without branched gills; often with a portable case..4

3' Abdomen with rows of branched gills ventrally; no portable case.. Hydropsychidae

4(3) Mature larvae <5 mm long; anal prolegs short and with a stout claw; abdominal tergum 9 sclerotized; fifth instar larvae with abdomen much enlarged (prior instars free-living); in a barrel or purse-like case (Figs. 16.107.74, 75) .... Hydroptilidae

4' Mature larvae >5 mm long; anal prolegs elongate with a projecting claw; abdominal tergum 9 entirely membranous; abdomen not enlarged; larvae in sand retreats..Ecnomidae

[USA: Southwest. Mexico]

5(2) Antennae relatively short, ≤3× long as wide, often inconspicuous and arising at various points between eye and mandible base; mesonotum never with a pair of sclerotized, narrow, curved or angled bars; with or without case 6

5' Antennae relatively long, at least 6× long as wide, and arising near base of mandibles (Fig. 16.107.76); and/or mesonotum membranous, except for a pair of sclerotized, narrow, curved or angled bars (Fig. 16.107.77); larvae in a case .... Leptoceridae

6(5) Meso- and metanotum entirely membranous, or with only weak sclerites at SA-1 of mesonotum (Fig. 16.107.78); pronotum never with an anterolateral projection; larvae with or without a case 7

6' Meso- and metanotum with some conspicuous sclerotized plates (Figs. 16.108.96, 97); pronotum sometimes with an anterolateral projection; larvae in a case..15

7(6) Abdominal tergum 9 entirely membranous; no portable case (Fig. 16.107.80) 8

7' Abdominal tergum 9 bearing a sclerotized dorsal plate; with or without portable case..12

8(7) Tibiae and tarsi distinct on all legs; mesopleura without a projecting lobe..9

8' Tibiae and tarsi fused to form a single segment on all legs; mesopleura with a forward projecting lobe (Fig. 16.107.81) ....Xiphocentronidae

[USA: Southwest. Mexico: Northwest]

9(8) Protrochantin pointed or poorly developed..10

9' Protrochantin broad, hatchet-shaped (Fig. 16.107.82) .... Psychomyiidae

10(9) Protrochantin pointed anteriorly (Fig. 16.107.84); head usually with dark markings or dark or light muscle scars; labrum sclerotized and widest near base 11

10' Protrochantin poorly developed; head without markings; labrum membranous and T-shaped (Fig. 16.107.83) .... Philopotamidae

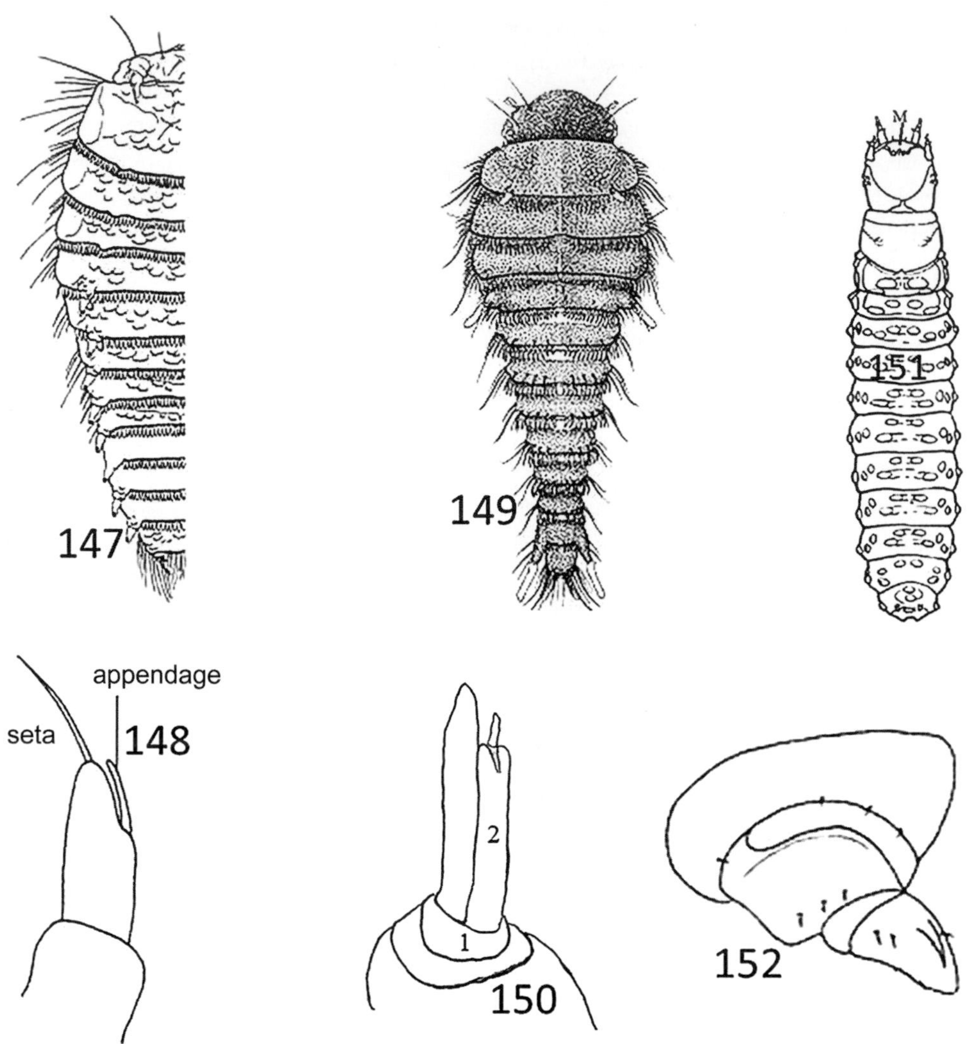

PLATE 16.115 **Fig 147** Sphaeriusidae; larva of *Sphaerius* (dorsal) (modified from Beutel & Leschen, 2011). **Fig. 148** Sphaeriusidae; antenna of larva of *Sphaerius* (dorsal) (modified from Beutel & Leschen, 2011). **Fig. 149** Hydroscaphidae; *Hydroscapha* larva, dorsal (modified from White & Roughley, 2008). **Fig. 150** Hydroscaphidae; *Hydroscapha* larva, antenna (modified from Beutel & Haas, 1998). **Fig. 151** Georissidae; *Georissus* larva (dorsal). Modified from White & Roughley, 2008. **Fig. 152** Georissidae; *Georissus* larva leg (dorsal). Modified from White & Roughley, 2008.

11(10)	Tarsi broad and densely pilose with a small, short claw (Fig. 16.107.85); mandibles short and triangular, each with a large, thick mesal brush	Dipseudopsidae
	[Canada: Eastern. USA: Eastern]	
11'	Tarsi with little or few hairs and with a large claw (Fig. 16.107.80); mandibles elongate	Polycentropodidae
12(7)	Prosternal horn absent; SA-3 on meso- and metanotum without a sclerite and usually with a single seta; case, if present, of sand and pebbles	13
12'	Prosternal horn present (as in Fig. 16.108.93); SA-3 on meso- and metanotum with a small sclerite and a cluster of setae (Fig. 16.107.78); case of vegetation, spirally wound or a series of rings	Phryganeidae
13(12)	Anal claws as long as elongate sclerite on anal legs (Fig. 16.108.87); larvae without a case	14
13'	Anal claws much shorter than elongate sclerite on anal legs (Fig. 16.108.86); larvae with turtle-like case of coarse sand with ventral strap of sand	Glossosomatidae
14(13)	Profemora with ventral projection to form chelate legs (Fig. 16.108.88)	Hydrobiosidae
	[USA: Southwest. Mexico]	
14	Profemora not chelate	Rhyacophilidae
15(6)	Claws of metathoracic legs as long as those on mesothoracic legs; case never of sand with lateral flanges	16

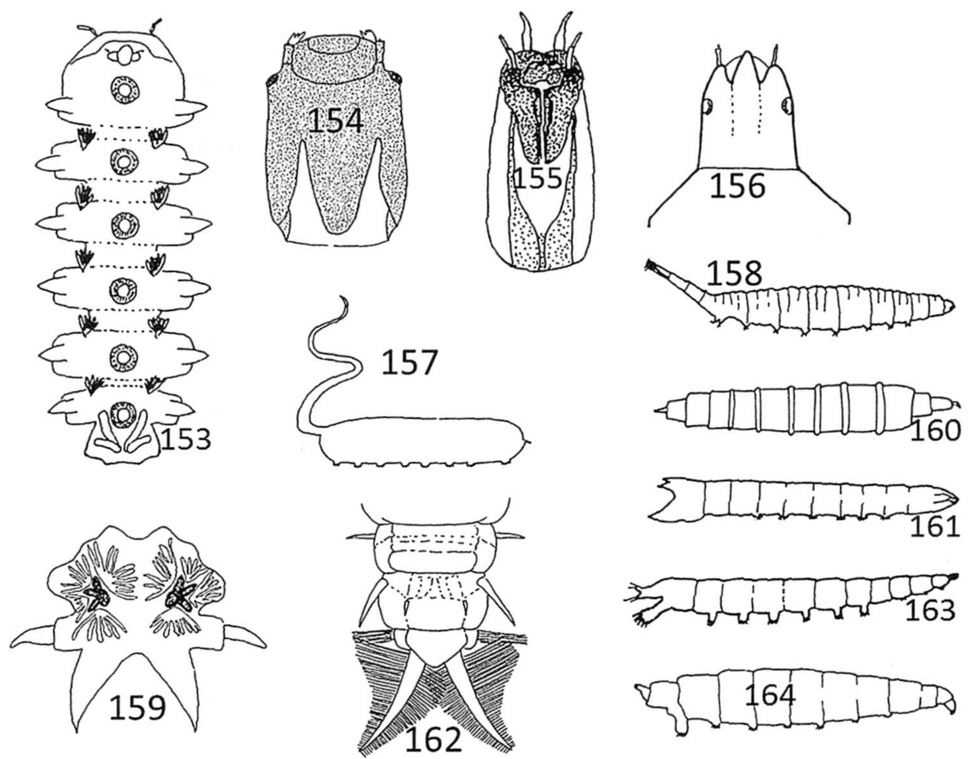

PLATE 16.116 **Fig. 153** Blephariceridae; *Blepharicera* (ventral view). **Fig. 154** Tipulidae; head of *Limonia* (dorsal view). **Fig. 155** Tipulidae; head of *Hexatoma* (dorsal view). **Fig. 156** Stratiomyidae; head of *Odontomyia* (dorsal view). **Fig. 157** Syrphidae; *Eristalis* (lateral view). **Fig. 158** Ephydridae (lateral view). **Fig. 159** Sciomyzidae; spiracular disc of *Sepedon*. **Fig. 160** Tabanidae; *Chrysops* (lateral view). **Fig. 161** Dolicopodidae (lateral view). **Fig. 162** Athericidae; terminal segments of *Atherix* (dorsal view). **Fig. 163** Empididae (lateral view). **Fig. 164** Muscidae; *Limnophora* (lateral view).

15' Claws of metathoracic legs a short stub or a slender filament, different from that of other legs; case of sand with lateral flanges and a hood .... Molannidae

16(15) Pronotum with at most a shallow furrow; lateral humps and usually dorsal hump present on abdominal segment 1 17

16' Pronotum divided by a sharp furrow across middle, area in front of furrow depressed (Fig. 16.108.89); no dorsal or lateral humps on abdominal segment 1 .... Brachycentridae

17(16) Pronotum without a distinct carina terminating in a rounded anterolateral lobe 18

17' Pronotum divided obliquely by a sharp carina that terminates as a projecting, rounded anterolateral lobe (Fig. 16.108.90) .... Beraeidae

[Canada: Eastern. USA: Eastern]

18(17) Antennae midway between eyes and base of mandibles or near base of mandibles; dorsal hump present on abdominal segment 1 19

18' Antennae located extremely close to eyes (Fig. 16.108.91); dorsal hump absent from abdominal segment 1 .... Lepidostomatidae

19(18) Anterolateral angles of pronotum if produced, not divergent; labrum without a dorsal row of long setae; case of various materials 20

19' Anterolateral angles of pronotum produced and divergent; labrum with transverse dorsal row of about 18 long setae (Fig. 16.108.92); case of vegetation .... Calamoceratidae

20(19) Prosternal horn absent; antennae close to base of mandibles 21

20' Prosternal horn present (Fig. 16.108.93); antennae about midway between base of mandibles and eye (Fig. 16.108.93) 22

21(20) Protrochantin large, hook-shaped (Fig. 16.108.94); anal prolegs each with about 30 long setae .... Sericostomatidae

21' Protrochantin small, not hook-shaped (Fig. 16.108.95); anal prolegs each with about 5 long setae .... Odontoceridae

22(20) Mesepisternum not enlarged anteriorly 23

22' Mesepisternum formed anteriorly into a sharp, elongate process (Fig. 16.108.96) or a rounded spiny prominence .... Goeridae

23(22) Mesonotum without an anterior notch; SA-1 of metanotum with a sclerotized plate and/or more than 2 setae 24

23' Mesonotum with an anteromesal notch (Fig. 16.108.97); SA-1 of metanotum unsclerotized and with only 1 or 2 setae .... Thremmatidae

24(23) Head, pronotum, and mesepisternum without prominent surface sculpturing 25

24' Head, pronotum, and mesepisternum with prominent surface sculpturing .... Rossianidae

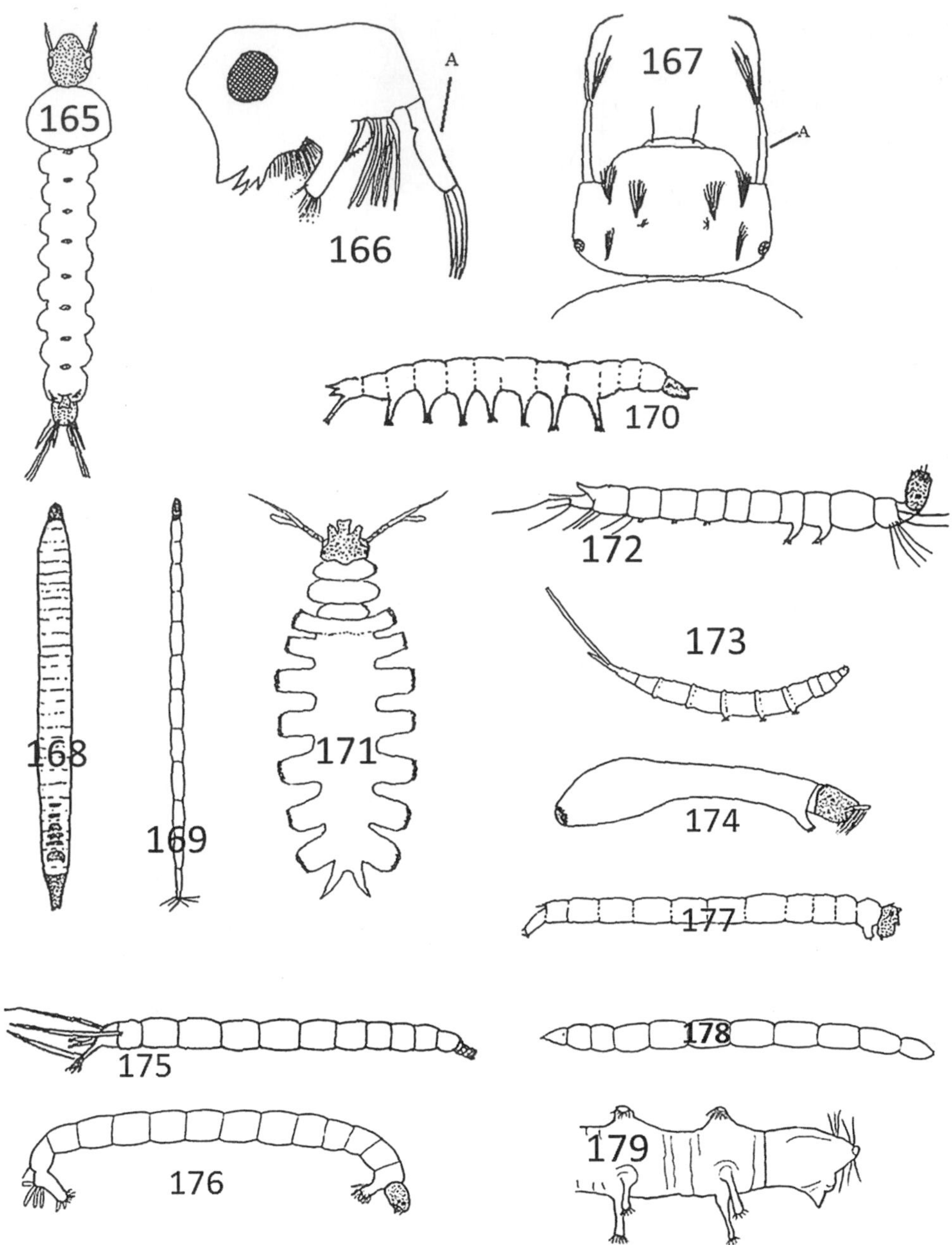

PLATE 16.117 **Fig. 165** *Anopheles* (dorsal view). **Fig. 166** Chaoboridae; head of *Chaoborus* (lateral view) showing antenna (A). **Fig. 167** Culicidae; head of *Coquillettidia* (dorsal view) showing antenna (A). **Fig. 168** Psychodidae; *Psychoda* (dorsal view). **Fig. 169** Ceratopogonidae; *Palpomyia* (dorsal view). **Fig. 170** Nymphomyiidae; *Palaeodipteron* (lateral view). **Fig. 171** Deuterophlebiidae; *Deuterophlebia* (dorsal view). **Fig. 172** Dixidae; *Dixella* (lateral view). **Fig. 173** Ptychopteridae; *Ptychoptera* (lateral view). **Fig. 174** Simuliidae; *Simulium* (lateral view). **Fig. 175** Tanyderidae; *Protoplasa* (lateral view). **Fig. 176** Chironomidae; *Chironomus* (lateral view). **Fig. 177** Thaumaleidae; *Thaumalea* (lateral view). **Fig. 178** Pelecorynchidae; *Glutops* (lateral view) (redrawn from Agriculture Canada, Research Branch, 1981). **Fig. 179** Oreoleptidae; *Oreoleptis torrenticola* (lateral view), abdominal segments 6–8 (redrawn from Zloty et al., 2005).

25(24) Mandibles with uniform scraper blades or if toothed, SA-1 of metanotum with 25+ setae on membrane between sclerites, SA-1 sclerites lacking in some genera; case cornucopia-shaped and mostly of sand grains; larvae <12 mm long .... Apataniidae

25' Mandibles almost always toothed; setae usually absent from between metanotal SA-1 sclerites, or if present are <25; case of vegetation or mineral materials, larvae often >12 mm long and with a transverse depression of the pronotum in most species.................. Limnephilidae

Crustacea: Hexapoda: Megaloptera: Families

1 Abdominal segments 1–7 with lateral filaments, last segment with a long, terminal filament (Fig. 16.109.98) .... Sialidae, one genus: *Sialis*

1' Abdominal segments 1–8 with lateral filaments, last segment without a median filament; but with a pair of prolegs, each with a pair of claws (Fig. 16.109.99) .... Corydalidae

ACKNOWLEDGMENTS

The author of this chapter in editions one and two, and coauthor of edition three (DeWalt et al., 2010), Dr. William L. Hilsenhoff, passed away in June of 2011. Bill was a passionate student of aquatic insects and shared his enthusiasm with dozens of students, many of which are still working in aquatic science today. He was one of the first to empirically develop tolerance values for aquatic insects and with his students made great contributions to the study of aquatic beetles, aquatic and semiaquatic Heteroptera, Plecoptera, Ephemeroptera, and Trichoptera in Wisconsin and the Midwest. He has left behind a tremendous legacy in his research collection of aquatic insects at the University of Wisconsin at Madison. His contributions will stand the test of time.

REFERENCES

Agriculture Canada, Research Branch. 1981. Manual of Nearctic Diptera Vol. 1. Monograph No. 27. Canadian Government Publishing Centre, Supply and Services Canada, Hull, Que., Canada, vi + 674 pp.

Beutel, R.G. & A. Haas. 1998. Larval head morphology of *Hydroscapha natans* (Coleoptera, Myxophaga) with reference to miniaturization and the systematic position of Hydroscaphidae. Zoomorphology 118:103–116.

Beutel, R.G. (Ed.) & R. Leschen (ed.). 2011. Teilband/Part 38 Volume 1: Morphology and Systematics (Archostemata, Adephaga, Myxophaga, Polyphaga partim). Berlin, Boston: De Gruyter. Retrieved 31 Oct. 2014, from http://www.degruyter.com/view/product/14421

Brigham, A. R., W.U. Brigham & A. Gnilka (eds.). 1982. The aquatic insects and oligochaetes of North and South Carolina. Midwest Aquatic Enterprises, Mahomet, IL, xi + 837 pp.

Burks, B.D. 1953. The mayflies, or Ephemeroptera, of Illinois. Illinois Natural History Survey Bulletin 26:1–216.

Christiansen, K.A. & R.J. Snider. 2008. Aquatic Collembola, pp. 165–179, *in*: R.W. Merritt, K.W. Cummins, and M.B. Berg (eds.). An Introduction to the Aquatic Insects of North America. Fourth Edition. Kendall/Hunt, Dubuque, Iowa. 1000 pp.

Clifford, H. F. 1991. Aquatic Invertebrates of Alberta. University of Alberta Press, Edmonton, 538 pp.

DeWalt, R.E., V.H. Resh & W.L. Hilsenhoff. 2010. Diversity and Classification of Insects and Collembola. pp. 587–657, *in*: J.H. Thorp & A.P. Covich (eds.). Ecology and Classification of North American Freshwater Invertebrates. Third Edition. Elsevier. 1021 pp.

Peckarsky, B. L., Fraissinet, P., Penton, M. A., Conklin, D. J., Jr. 1990. Freshwater Macroinvertebrates of Northeastern North America. Cornell University Press, Ithaca, NY, xi + 422 pp.

Hilsenhoff, W. L. 1995. Aquatic insects of Wisconsin. Keys to Wisconsin genera and notes on biology, habitat, distribution and species. Publication 3 of the Natural History Museums Council, University of Wisconsin-Madison, 79 pp.

Merritt, R.W., K.W. Cummins, & M.B. Berg (eds.). 2008. An Introduction to the Aquatic Insects of North America. Fourth Edition. Kendall/Hunt, Dubuque, Iowa. 1000 pp.

White, D.S. & R.E. Roughley. 2008. Aquatic Coleoptera. Pages 571–671 *in*: Merritt, R.W., K.W. Cummins, & M.B. Berg (eds.), An Introduction to the Aquatic Insects of North America. Fourth Edition. Kendall/Hunt, Dubuque, Iowa. 1000 pp.

Usinger, R. L., Ed. 1956. Aquatic insects of California with keys to North American genera and California species. University of California Press, Berkeley, ix + 508 pp.

Zloty, J., B.J. Sinclair & G. Pritchard. 2005. Discovered in our back yard: a new genus and species of a new family from the Rocky Mountains of North America (Diptera, Tabanomorpha). Systematic Entomology 30:248–266.

Zwick, P. 2006. New family characters of larval Plecoptera, with an analysis of the Chloroperlidae: Paraperlinae, Aquatic Insects 28:13–22.

Class Branchiopoda

D. Christopher Rogers

Kansas Biological Survey, University of Kansas, Lawrence, KS, USA; The Biodiversity Institute, University of Kansas, Lawrence, KS, USA

Brenda J. Hann

Department of Biological Sciences, University of Manitoba, Winnipeg, MB, Canada

INTRODUCTION

Approximately 400 species of branchiopod crustaceans are found in Nearctic temporary and permanent aquatic habitats, from small seasonal wetlands, to lakes, ponds, and large rivers. Most are planktonic or semibenthic and are often among the most abundant invertebrates in those habitats. Branchiopods are often used as indicators of aquatic ecosystem health, which ultimately reflects on land use practices and management (Rogers, 2009). Several species are protected by environmental laws and some species are economically important.

LIMITATIONS

The Nearctic anostracans are well known and for the most part are readily identified to species level. The anostracan keys presented here are based primarily on characters of the males, but females can be identified in some taxa. One group of *Branchinecta* species must be separated by female characters. There are 72 Nearctic species (Brendonck et al., 2008); however, a new species is discovered and described approximately every three years.

Nearctic notostracans are represented by two genera: *Lepidurus* and *Triops*. *Lepidurus* is well known for North

America (Rogers, 2001); however, *Triops* needs revision and is not reliably identifiable beyond genus. Both males and female laevicaudatan clam shrimp and the Cyclestherida are easily identified. The spinicaudatan clam shrimp are more difficult: the genus *Cyzicus* is in need of revision, and the species of *Eulimnadia* require specimens with mature eggs for identification.

In the USA, California and Oregon support several species of anostracan and one species of notostracan that are federally listed under the Endangered Species Act. Some species of clam shrimp are protected at the state level in some parts of New England. Permits must be acquired prior to collecting any legally protected taxa.

Cladocera keys are primarily based upon adult females, which are readily identified to genus. Species level identifications typically require use of a compound microscope, and many genera need to be revised or have incompletely known species. Nearctic cladocerans in the Daphniidae (e.g., *Daphnia, Simocephalus, Ceriodaphnia*) are particularly variable morphologically; many *Daphnia* species are known to hybridize, further complicating identification beyond genus or species group. Several families (e.g., Ilyocryptidae, Chydoridae) are prevalent in wetlands and other littoral habitats, which can be difficult to sample. Hence, these taxa also need revision.

TERMINOLOGY AND MORPHOLOGY

In the Nearctic, branchiopods range in size from less than a millimetre to 17 cm. Adult branchiopods retain the reduced first antennae of the larvae. In notostracans and laevicaudatans the second antennae are reduced. Diplostracan second antennae are typically branched, with an anterior endopod and a posterior exopod, and are used for propulsion. Anostracan second antennae are robust, and in males may have prehensile or cheliform antennal appendages (sometimes fused medially into a central cephalic appendage), posteriolateral basal projections called apophyses, and/or anteriomedial spiny patches called pulvilli (singular: pulvillus) (Pls. 16.118 B, C; 120 E–G, J; 121 A, J, K; 122 A–C). In *Streptocephalus*, the antennal appendages are cheliform, divided into two rami, referred to as a thumb and finger (Pl. 16.125 F, H, I). The thumb is typically longer than the finger and may have a posterior spur. The finger may have one or more teeth on the anterior margin.

The head is proportioned to the body in anostracans, notostracans, and spinicaudatans, but is massive in the laevicaudatans and may be larger than the rest of the animal. A rostrum is absent in anostracans and notostracans.

The nauplii have a single median eye, which is reduced in most adult large branchiopods, but in the Notostraca becomes a large tubercle on the nuchal organ, behind the eyes. The adult compound eyes are fused in laevicaudatans and diplostracans, but are separate and borne on stalks in anostracans. The nuchal organ lies on the dorsal surface of the head; although in the spinicaudatan family Limnadiidae, it may be borne on a short stalk. A naupliar stage is absent in Cladocera, except *Leptodora*, which has a metanauplius stage. The adult compound eyes are fused and an ocellus may be present. Some cave dwelling cladocerans lack eyes entirely.

Nearctic anostracans have eleven pairs of thoracopods (Pl. 16.119 A) except for *Polyartemiella*, which has 17 (Pl. 16.120 K).

Cladoceran sexes are morphologically similar, with small differences in appendages and body shape. Because most cladocerans reproduce asexually at least part of the time, most individuals will be females.

MATERIAL PREPARATION AND PRESERVATION

All branchiopods can be preserved in ethyl alcohol long term; for purposes of identification all can be manipulated in alcohol. Dissection is seldom necessary, but it is a good idea to have several animals available for identification. This is especially true for the cladocerans because these small, delicate animals are easily crushed. Cladocerans smaller than about 1 mm should be slide mounted for high magnification viewing; larger species can often be identified without being mounted, using a dissecting microscope with a magnification of ≤50 diameters. The best cladoceran mounting medium is one which is more or less permanent and soluble in both water and alcohol solutions so specimens can be added directly to the medium without tedious dehydration. If you have access to chloral hydrate (currently a controlled substance), a good medium is Hoyer's. Divide the Hoyer's into two batches. Let one sit in a warm place until it is as thick as honey, and keep the second batch in a closed jar so it remains thin. You can always add more water to thin either batch. Keep at least some of the thick and thin Hoyer's in screw top eye dropper bottles.

To make a slide, use the eye dropper to put a small streak of dilute Hoyer's (about a quarter of a drop) toward one end of the slide. Arrange the specimens in this streak. It is wise to place four or so of what you think are the same kind of animal on a slide to allow for comparison and viewing of difficult characters. Delicate specimens can be protected from compression by putting a few pieces of broken cover slips around the specimens. Let the Hoyer's streak dry on a slide warmer or in a warm place (below 100 °C). When the streak is dry, add a drop or two of the thick Hoyer's and gently lower a cover glass over the Hoyer's. Put the slide back into a warm place to dry again. Label the slide as to the date and location of the collection. It is important to use thickened Hoyer's in the last step; otherwise, the Hoyer's will shrink as it dries and produce large bubbles in the final preparation. However, if you use thickened Hoyer's, small bubbles produced will disappear as the medium dries. These slides are semi permanent. If the climate is humid, the Hoyer's will thin and cover glasses will slip off the slide. If the climate is arid, the Hoyer's will desiccate enough so that the cover glasses pop off the slide. These problems may be avoided by storing the slides horizontally in a climate that has moderate humidity, and by ringing the slide with clear nail polish.

An alternate mounting method, especially useful for bosminids, chydorids, and macrothricids, is polyvinyl lactophenol stained with lignin pink. Temporary slides can be made using glycerol or glycerine jelly. Best of all for museum specimens is possibly Canada Balsam, although this requires the specimens be dehydrated through an alcohol series.

KEYS TO BRANCHIOPODA

Crustacea: Branchiopoda: Orders

1 Compound eyes sessile; carapace present or absent (Pl. 16.119 C, E–J) 2

1' Compound eyes on stalks separated; carapace absent (Pl. 16.119 A) .... **Anostraca [p. 439]**

2(1) Carapace bivalved, folded or reduced; eyes not projecting dorsally through carapace (Pl. 16.119 D–J) 3

2' Carapace broad, never folded or bivalved; eyes projected dorsally through carapace (Pl. 16.119 C) .... **Notostraca [p.453]**

3(2) Carapace with a true, interlocking dorsal hinge; carapace subglobular, smooth, without growth lines, containing entire animal including head (Pl. 16.119 D, E) .... **Laevicaudata [p. 454]**

3' Carapace folded, no true hinge present laterally flattened OR carapace reduced or absent (Pl. 16.119 F–J) .... **Diplostraca [p. 456]**

Crustacea: Branchiopoda: Anostraca: Families

1 Praeepipodites not cleft or divided (Pl. 16.120 B) 9

1' Praeepipodites (Pl. 16.120 A) obviously cleft, or entirely divided into two .... **Chirocephalidae [p. 439]**

2(1) Second antennal distal antennomere not broadly triangular; brood pouch without large pair of spines (Pl. 16.120 D–G) 3

2' Second antennal distal antennomere broadly triangular, apex acute (Pl. 16.120 C); brood pouch with a ventral pair of large, recurved spines, halophiles, widespread .... Artemiidae, one genus: ***Artemia* [p. 446]**

3(2) Gonopod rigid basal portions parallel, directed ventrally, with the bases closely united; male antennal and or frontal appendages present or not; female second antennae lamellar 4

3' Gonopods directed ventrolaterally, widely separated at the base; male frontal and antennal appendages always absent, although rigid projections may be present; female second antennae subcylindrical in cross section; widespread (Pl. 16.120 D) .... Branchinectidae, one genus: ***Branchinecta* [p. 446]**

4(3) Male second antennal proximal antennomere with or without appendages, if present, then appendages cylindrical or lamellar, but never cheliform; second antennae and appendages not laterally compressed .... **Thamnocephalidae [p. 450]**

4' Male second antennal proximal antennomere terminating in a large, cheliform, multiramal appendage directed ventrally (Pl. 16.120 E–G); second antennae and appendages laterally compressed; widespread .... Streptocephalidae, one genus: ***Streptocephalus* [p. 452]**

Crustacea: Branchiopoda: Anostraca: Chirocephalidae: Genera

1 Eleven pairs of thoracopods; abdomen at least as long as thorax 2

1' Seventeen pairs of thoracopods; abdomen reduced, much shorter than thorax; male second antennal antennomeres fused and branched (Pls. 16.120 K, 121 G, H) .... ***Polyartemiella* [p. 444]**

2(1) Male second antenna with a lamellar or triangular appendage; brood pouch variable (Pls. 16.120 I, J; 121 A–D, J, K; 122 A–C) 3

2' Male second antenna without a lamellar or triangular appendage; brood pouch transverse, about twice as broad as long (Pls. 16.120 H, 121 E) .... ***Artemiopsis* [p. 445]**

3(2) Males: second antennae with two antennomeres; gonopods present (Pl. 16.119 A, lower) 4

3' Females: second antennae of one antennomere; brood pouch present (Pl. 16.119 A, upper) 6

4(3) Second antennal appendage 0.5 to 2.5 times as long as the proximal antennomere (Pls. 16.120 J, 121 A, 122 A–C) 5

4' Second antennal appendage less than 0.5 as long as the proximal antennomere (Pls. 16.120 I, 121 J, K) .... ***Linderiella* [p. 445]**

5(4) Second antennal distal antennomere straight, curved, or bent no more than 65° in apical fourth, widespread, often with a proximal branch (Pl. 16.122 A–C) .... ***Eubranchipus* [p. 445]**

5' Second antennal distal antennomere bent medially 90° in proximal third (Pl. 16.121 A–C) .... *Dexteria floridana* (Dexter, 1953) [USA: Florida]

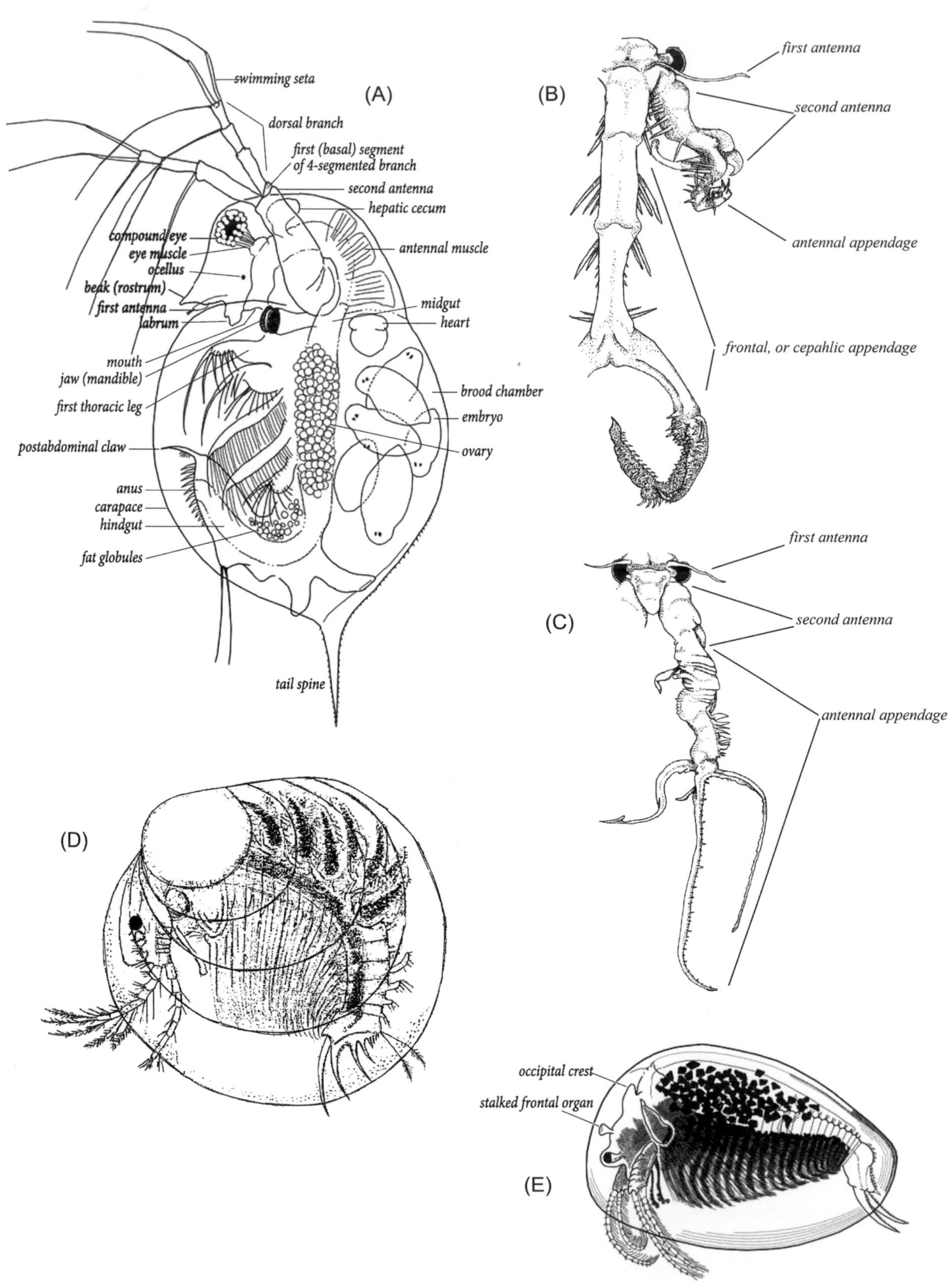

PLATE 16.118 (A) *Daphnia*, female, left lateral view. (B) *Branchinella ondonaguae* (from Africa) male head, left anterior view. (C) *Streptocephalus moorei*, male head, left anterior view. (D) *Cyclestheria* sp., left lateral view. (E) *Limnadia lenticularis*, left lateral view.

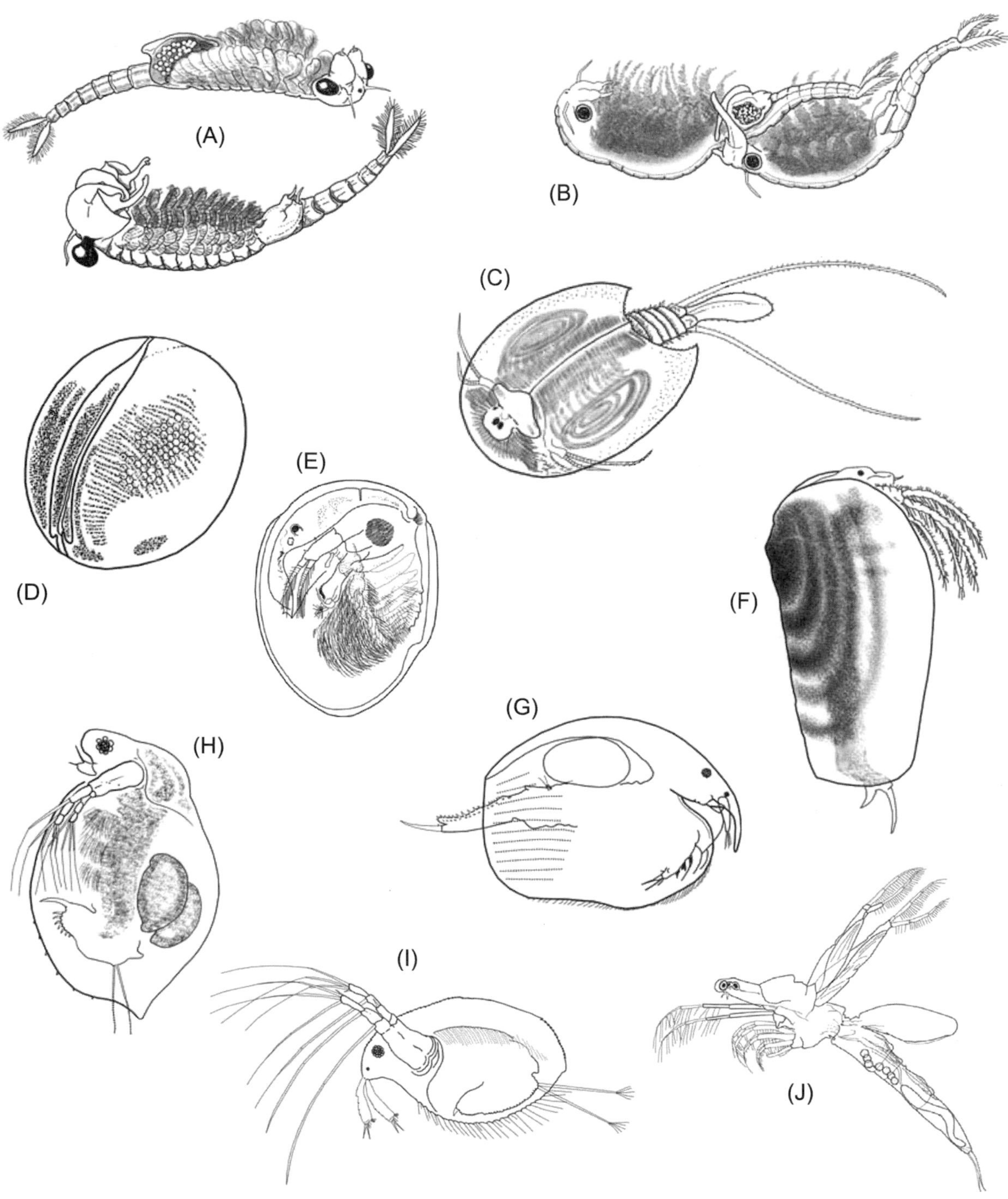

PLATE 16.119 (A) *Eubranchipus* sp., female above, male below. (B) *Linderiella occidentalis*, male amplexing female. (C) *Lepidurus*, dorsal view. (D) *Lynceus* sp., anteriolateral view. (E) *Lynceus* sp., lateral view, left carapace valve removed. (F) *Leptestheria compleximanus*, right lateral view. (G) *Kurzia media*, right lateral view. (H) *Ceriodaphnia* sp., left lateral view. (I) *Macrothrix* sp., left lateral view. (J) *Leptodora kindti*, left lateral view.

6(3)	Second antenna with a small, digitiform, medial appendage (Pl. 16.121 B)	7
6'	Second antenna without an appendage, widespread	***Eubranchipus* [p. 445]**
7(8)	Brood pouch subspherical (Pl. 16.121 D)	***Linderiella* [p. 445]**
7'	Brood pouch broadly oval, dorsoventrally compressed (Pl. 16.121 C) [USA: Florida]	*Dexteria floridana* (Dexter, 1953)

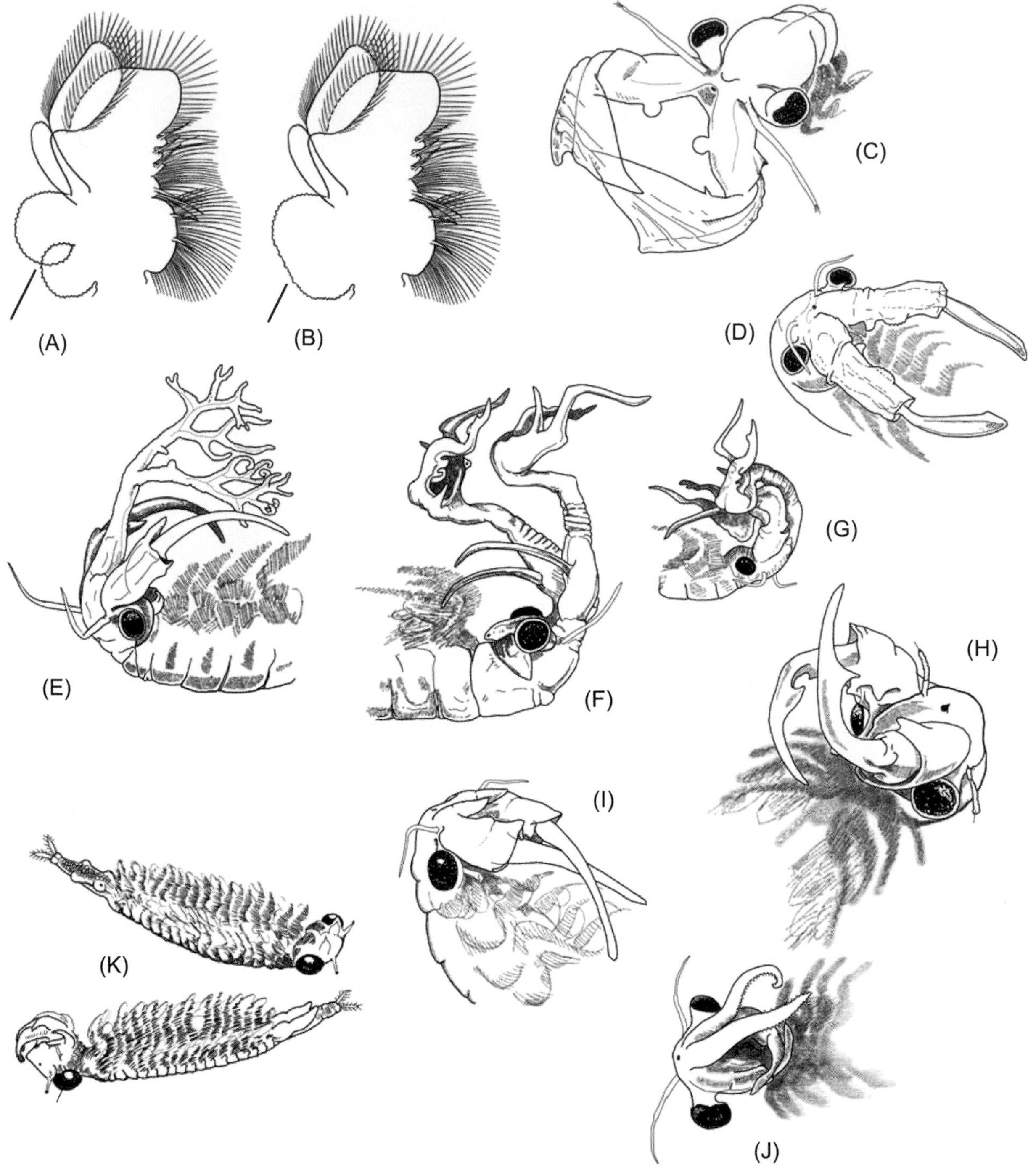

PLATE 16.120 (A) *Eubranchipus* sp., fifth limb (praeepipodite indicated). (B) *Branchinecta* sp., fifth limb (praeepipodite indicated). (C) *Artemia* sp., male, head, anterior view. (D) *Branchinecta coloradensis*, male head, anterioventral view. (E) *Thamnocephalus platyurus*, male, head, anteriolateral view. (F) *Streptocephalus* sp., male, head, anteriolateral view. (G) *Streptocephalus* sp., male, head, anteriolateral view. (H) *Artemiopsis steffanssoni*, male head, anterioventral view. (I) *Linderiella occidentalis*, male, head, anterioventrolateral view. (J) *Eubranchipus* sp., male head, anterioventral view. (K) *Polyartemiella hazeni*, female above, male below.

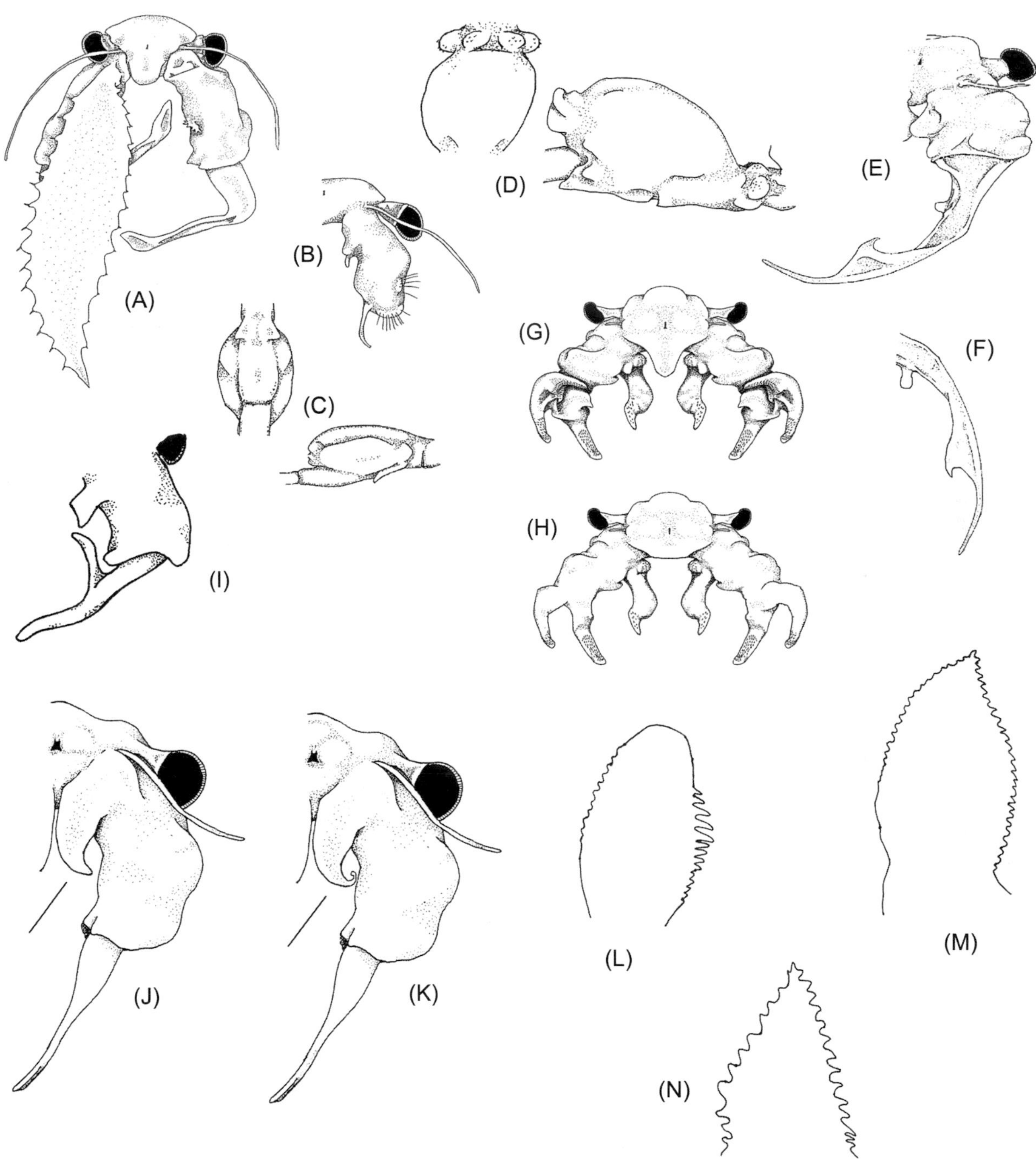

PLATE 16.121 *Dexteria floridana*: (A) male head, anterior view; (B) female head, left anterior view; (C) brood pouch, dorsal view (above), lateral view (below). (D) *Linderiella occidentalis*, brood pouch, ventral view (above), lateral view (below). (E) *Artemiopsis steffanssoni*, male head, left anterior view. (F) *Artemiopsis bungei*, male, second antenna distal antennomere, anterior view. (G) *Polyartemiella hazeni* male head, anterior view. (H) *Polyartemiella judayi* male head, anterior view. (I) *Eubranchipus hesperius*, male, second antenna, posterior view. (J) *Linderiella occidentalis*, male head, left anterior view (antennal appendage indicated). (K) *Linderiella santarosae*, male head, left anterior view (antennal appendage indicated). (L) *Eubranchipus neglectus*, male antennal appendage. (M) *Eubranchipus vernalis*, male antennal appendage. (N) *Eubranchipus oregonus*, male antennal appendage.

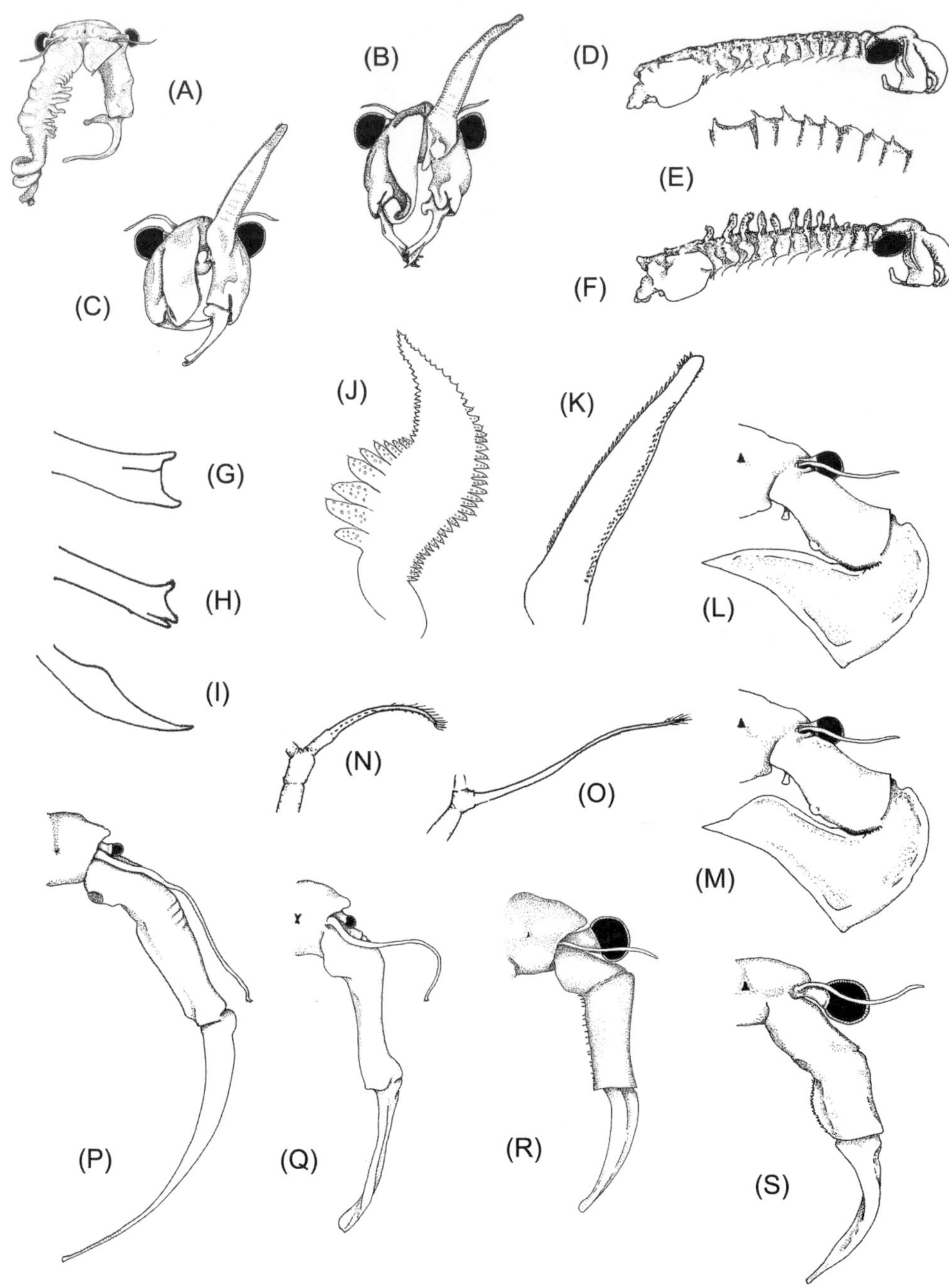

PLATE 16.122 (A) *Eubranchipus holmani* male head, anterior view. (B) *Eubranchipus intricatus*, male head, anterior view, left antennal appendage extended. (C) *Eubranchipus bundyi*, male head, anterior view, left antennal appendage extended. (D) *Eubranchipus holmani*, thorax, lateral view. (E) *Eubranchipus moorei*, thorax, lateral view. (F) *Eubranchipus stegosus*, thorax, lateral view. (G) *Eubranchipus serratus* male, second antenna, apex. (H) *Eubranchipus bundyi* male, second antenna, apex. (I) *Eubranchipus ornatus* male, second antenna, apex. (J) *Eubranchipus serratus*, male antennal appendage. (K) *Eubranchipus bundyi*, male antennal appendage. (L) *Artemia franciscana* male, head, anteriolateral view. (M) *Artemia* sp. male, head, anteriolateral view. (N) *Branchinecta gigas* cercopod. (O) *Branchinecta raptor* cercopod. (P) *Branchinecta raptor* male, head, anteriolateral view. (Q) *Branchinecta gigas* male, head, anteriolateral view. (R) *Branchinecta paludosa* male, head, anteriolateral view. (S) *Branchinecta serrata* male, head, anteriolateral view.

Crustacea: Branchiopoda: Anostraca: Chirocephalidae: *Polyartemiella*: Species

1 Male head without an anterior projection; antennal appendage (anterior branch of second antenna) biramous (Pl. 16.121 H); female with dorsal projection on genital segments at least one fourth the length of brood pouch .... *Polyartemiella judayi* Daday, 1910

[USA: Alaska: Nunivak Island, Pribilof Islands]

1' Male head with an anterior projection; antennal appendage triramous (Pl. 16.121 G); female with dorsal projection on genital segments lacking or less than 16 the length of brood pouch .... *Polyartemiella hazeni* (Murdoch, 1884)

[Canada: Alberta, Northwest Territories, Yukon. USA: Alaska. Palaearctic]

Crustacea: Branchiopoda: Anostraca: Chirocephalidae: *Artemiopsis*: Species

1 Male second antennal distal antennomere with two medial projections, distal most projection with apex directed distally (Pl. 16.121 F) .... *Artemiopsis bungei* Sars, 1897

[USA: Alaska. Palaearctic]

1' Male second antennal distal antennomere with two medial projections, distal most projection with apex recurved, directed proximally (Pl. 16.121 E) .... *Artemiopsis steffansoni* Johansen, 1921

[Canada: Alberta, Newfoundland, Northwest Territories, Quebec. USA: Alaska. Greenland]

Crustacea: Branchiopoda: Anostraca: Chirocephalidae: *Linderiella*: Species

1 Male second antenna with antennal appendages directed ventromedially; apices of distal antennomere on second antennae of male acute; female with amplexial groove bearing evenly rounded lateral tubercles (Pl. 16.121 D, J) .... *Linderiella occidentalis* (Dodds, 1923)

[USA: California]

1' Male second antenna with antennal appendages extended and curling dorsolaterally; male second antennal distal antennomere apices slightly inflated; female amplexial groove with lateral tubercles with the posterior distal surface at 90° angles to the abdomen (Pl. 16. 121 K) .... *Linderiella santarosae* Thiery & Fugate, 1994

[USA: California]

Crustacea: Branchiopoda: Anostraca: Chirocephalidae: Eubranchipus: Species

1 Male antennal appendage shorter than second antennal proximal antennomere 2

1' Male antennal appendage longer than second antennal proximal antennomere 4

2(1) Male second antennal distal antennomere with a basomedial projection (Pl. 16.121 I); antennal appendage apically acute or subacute 3

2' Male second antennal distal antennomere with a posteriomedial basal projection; antennal appendage apically rounded (Pl. 16.121 L) .... *Eubranchipus neglectus* Garman, 1926

[Canada: Ontario. USA: Alabama, Arkansas, Indiana, Michigan, Ohio, Pennsylvania]

3(2) Male antennal appendages triangular, serrate to apices; male second antennal distal antennomere curved medially. Male second antennal distal antennomere basal protrusion directed ventrally (Pl. 16.121 N) .... *Eubranchipus oregonus* Creaser, 1930

[Canada: British Columbia, Vancouver Island. USA: California, Oklahoma, Oregon, Washington]

3' Male antennal appendages lamelliform, not triangular, apices acute; male second antennal distal antennomere with apex directed laterally; male second antennal distal antennomere basal protrusion directed medially (Pl. 16.121 M) .... *Eubranchipus vernalis* (Verrill, 1869)

[Canada: Ontario. USA: Connecticut, Massachusetts, New Jersey, New York, Pennsylvania, Rhode Island, South Carolina, Tennessee, West Virginia]

4(1) Male second antennal appendages subcylindrical distally and ringed with spines (Pl. 16.122 A) 5

4' Male second antennal appendage entirely lamellar (Pls. 16.121 L–N, 122 B, C, J, K) 7

5(4) Male with five or more thoracic segments each bearing one or more dorsal projections (Fig. 16.122 E, F) 6

5' Male thorax with dorsum relatively smooth, lacking projections (Pl. 16.122 D) .... *Eubranchipus holmani* (Ryder, 1979)

[USA: Minnesota, Ohio to New England, south to Alabama and Georgia]

6(5) Male posterior thoracic segments and first abdominal segments, with paired conical projections (Pl. 16.122 A, E) .... *Eubranchipus moorei* Brtek, 1967

[USA: Alabama, Arkansas, Georgia, Louisiana, South Carolina]

6' Male thoracic segments each with a dorsal, transverse, subquadrate lamella bearing denticulate margins (Pl. 16.122 F) .... *Eubranchipus stegosus* Rogers, Jensen & Floyd, 2004

[USA: Georgia]

7(4) Male second antennal distal antennomere bifid or trifid apically (Pl. 16.122 G, H) 8

7' Male second antennal distal antennomere aciculate, apically acute (Pl. 16.122 I) .... *Eubranchipus ornatus* Holmes, 1910

[Canada: Alberta, Manitoba. USA: Minnesota, North Dakota, Nebraska, Wisconsin]

8(7) Male antennal appendage lateral margin without papillae, or with very small, mound-like papillae (Pl. 16.122 B, C, K) 9

8' Male antennal appendage lateral margin with long, digitiform papillae, each papilla at least three times as long as broad (Pl. 16.122 J) 10

9(8) Male labrum not produced anteriorly (Pl. 16.122 B); female 11th thoracic segment with cylindrical dorsolateral processes .... *Eubranchipus intricatus* Hartland-Rowe, 1967

[Canada: Alberta, Manitoba, Saskatchewan. USA: Massachutesettes, Minnesota, Montana, Wyoming]

9' Male labrum produced anteriorly as a large knob (Pl. 16.122 C), with second antennae heavily chitinized at points of contact with labrum; female 10th thoracic segment with broad, flat, dorsolateral processes .... *Eubranchipus bundyi* Forbes, 1876

[Generally distributed across Canada and the northern three-fourths of the USA]

10(9) Abdominal segments projecting distolaterally; telson broadly explanate, roughly triangular in outline in dorsal or ventral view .... *Eubranchipus serratus* Forbes, 1876

[USA: Arkansas, Illinois, Indiana, Kansas, Maryland, Michigan, Minnesota, Missouri, Nebraska, Oklahoma, Tennessee, Wisconsin, Virginia]

10' Abdominal segments subcylindrical, lacking lateral projections; telson not broadly explanate .... *Eubranchipus hesperius* Rogers, 2014

[Canada: British Colombia. USA: Arizona, California, Idaho, Montana, Nevada, Oregon, Washington]

Crustacea: Branchiopoda: Anostraca: Artemiidae: *Artemia*: Species

1 Mature eggs smooth 2

1' Mature eggs with an undulating surface, with prominent "button-like" structures .... *Artemia monica* Verill, 1869

[USA: California: Mono Lake]

2(1) Male second antennae with distal antennomere broadly triangular (Pl. 16.122 L), female first antenna length 2–3 times as long as length of eye plus peduncle .... *Artemia franciscana* Kellogg, 1906

[Canada: Alberta, British Colombia, Saskatchewan. USA: Arizona, California, Connecticut, Montana, Nevada, New Mexico, North Dakota, Oregon, Texas, Utah, Washington. Mexico: Baja California Sur, Coahuila, Oaxaca. Neotropical. Invasive species in Hawaii, Palaearctic, Australia]

2' Male second antennae with distal antennomere subrectangular (Pl. 16.122 M), female first antenna length 3.5–4.5 times as long as length of eye plus peduncle .... *Artemia "parthenogenetica"*

[USA: Utah (Invasive in Great Salt Lake). Native to Palaearctic, Australia (?). A parthenogenic form of one of the Eurasian species, probably *Artemia salina*]

Crustacea: Branchiopoda: Anostraca: Branchinectidae: *Branchinecta*: Species

1 Cercopods glaborous, or setose along the apical third of the medial margin; eyes reduced; endopodites not sexually dimorphic; usually greater than 5 cm in length (Pl. 16.122 N–Q) 2

1' Cercopods with plumose setae along entire lateral and medial margins; eyes; female second antennae shorter than male second antennae; endopodites sexually dimorphic; usually less than 3 cm in length normal (Pls. 16.122 R, S; 123 A–Q; 124 A–D, K, M, N) 3

2(1) Male second antennal distal antennomere longer than proximal antennomere by a third; pulvilli large and obvious; second antennal proximal antennomere with medial surface bearing a longitudinal, ridge-like pulvinus; female with second antennae one third the length of the second antennae (Pl. 16.122 O, P) .... *Branchinecta raptor* Rogers et al., 2006

[USA: Idaho]

2' Male second antennal distal antennomere subequal or shorter than proximal antennomere; pulvilli obscure or absent; second antennal proximal antennomere without a pulvinus; female with second antennomere longer than first antennae (Pl. 16.122 N, Q) .... *Branchinecta gigas* Lynch, 1937

[Canada: Alberta, Saskatchewan. USA: California, Montana, Nevada, North Dakota, Oregon, Utah, Washington. Mexico: Baja (Norte)]

3(1) Male second antennal apices subacute or acute (Pl. 16.122 Q, R) 4

3' Male second antennal apices truncate, broadly rounded, bilobed, bent medially or laterally (Pls. 16.123 D–Q, 124 A–R) 8

4(3) Female second antennae with a prominent medial, subapical spine; male second antennal distal antennomere 1/3 the length of the proximal antennomere; male second antennal proximal antennomere distomedial surface with or without spines, but the spines are never in one discreet longitudinal row (Pl. 16.123 A, B) 5

4' Females without a prominent medial subapical spine on the second antennae; male second antennal distal antennomere 2/3 to subequal the length of the proximal antennomere; male second antennal proximal antennomere with a longitudinal row of or spines on medial surface (Pls. 16.122 R, S, 123 C) 6

5(4) Male second antenna distal antennomere flattened laterally; female without cephalic projections (Pl. 16.123 A, B) .... *Branchinecta hiberna* Rogers & Fugate, 2001

[USA: California, Idaho, Nevada, Oregon]

5' Male second antenna distal antennomere flattened anteroposteriorly; female with large dorsolateral corneous cephalic projections (Pl. 16.123 F) .... *Branchinecta cornigera* Lynch, 1958

[USA: Oregon, Washington]

6(4) Male second antennal proximal antennomere without an anteriomedial, blunt conical process; female with dorsum smooth or not (Pl. 16.122 R, S) 7

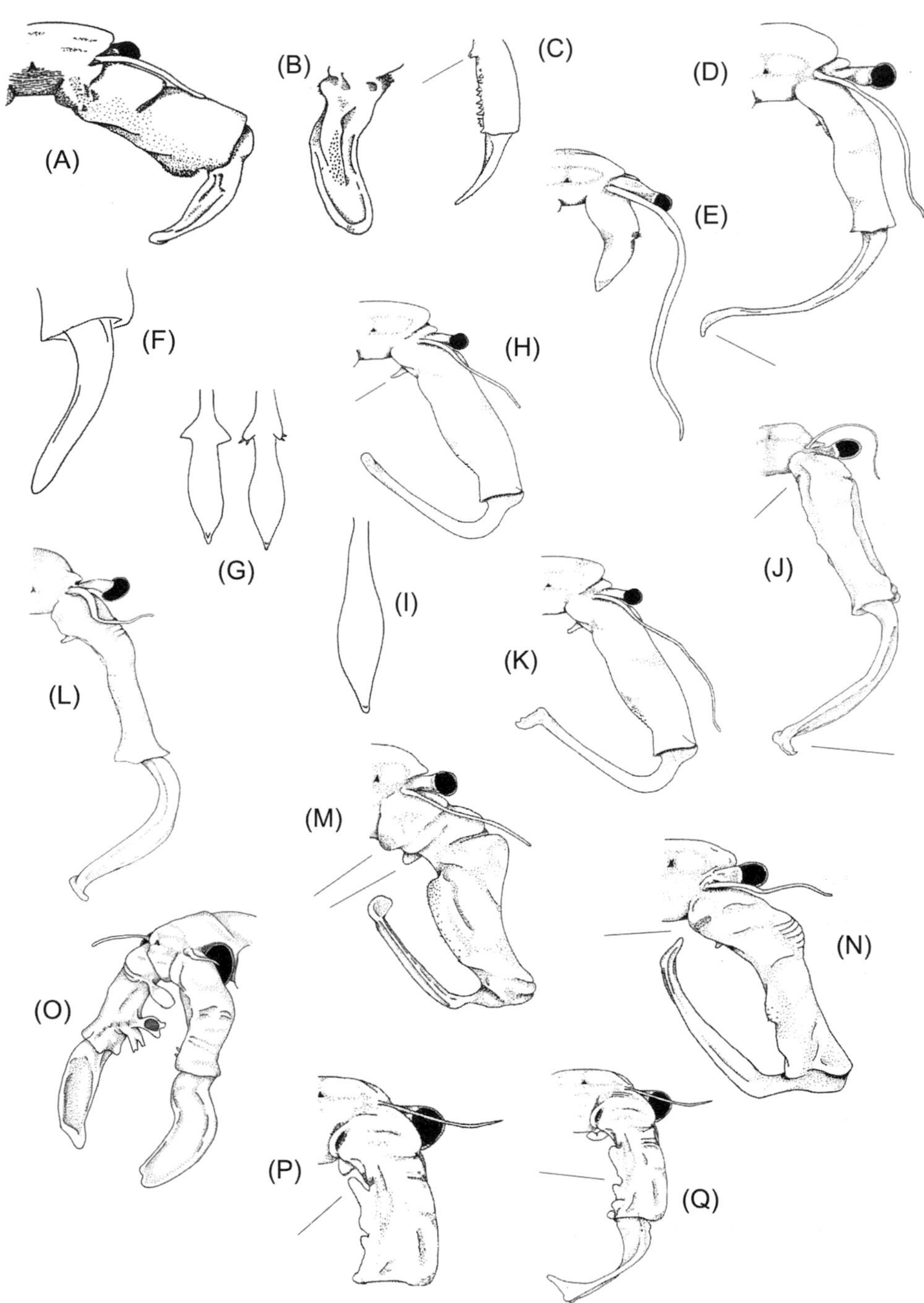

PLATE 16.123 (A) *Branchinecta hiberna* male, head, anteriolateral view. (B) *Branchinecta hiberna* male, second antenna, distal antennomere, medial view. (C) *Branchinecta kaibabensis* male, second antenna, anterior view (conical process with pulvillus indicated). (D) *Branchinecta readingi* male, head, anteriolateral view (second antenna distal antennomere apex indicated). (E) *Branchinecta mackini* female, head, anteriolateral view. (F) *Branchinecta cornigera* male, second antenna, distal antennomere, medial view. (G) *Branchinecta campestris* two forms of the brood pouch. (H) *Branchinecta mackini* male, head, anteriolateral view (apophysis indicated). (I) *Branchinecta mackini* female, brood pouch. (J) *Branchinecta potassa* male, head, anteriolateral view (pulvillus and second antenna distal antennomere apex indicated). (K) *Branchinecta campestris* male, head, anteriolateral view. (L) *Branchinecta lateralis* male, head, anteriolateral view. (M) *Branchinecta coloradensis* male, head, anteriolateral view (pulvillus and apophysis indicated). (N) *Branchinecta lynchi* male, head, anteriolateral view (pulvillus indicated). (O) *Branchinecta belki* male, head, anteriolateral view. (P) *Branchinecta packardi* male, head, anteriolateral view (medial process indicated). (Q) *Branchinecta mexicana* male, head, anteriolateral view (cleft medial process indicated).

6' Male second antennal proximal antennomere with a basal, anteriomedial, blunt conical process with an apical pulvillus; female thoracic dorsum smooth (Pl. 16.123 C) .... *Branchinecta kaibabensis* Belk & Fugate, 2000

[USA: Arizona]

7(6) Male second antenna distal antennomere evenly curved medially; female with dorsum smooth (Pl. 16.122 R) .... *Branchinecta paludosa* (Müller, 1788)

[Circumpolar. Canada: Baffin Island, Canadian Archipelago, Labrador, Manitoba, Nova Scotia, Northwest Territories, Quebec, Yukon. USA: Alaska, Colorado, Idaho, Montana, Wyoming. Greenland. Palaearctic]

7' Male second antenna distal antennomere with medial surface twisted anteriorly; female with a row of dorsolateral, subconical thoracic projections on segments III to VII (Pl. 16.122 S) .... *Branchinecta serrata* Rogers, 2006

[USA: Wyoming]

8(3) Male second antennae without a pulvillus, proximal antennomere with a few scattered spines; apophyses present, brood pouch fusiform (Pl. 16.123 D, G–I, K, L) 9

8' Male second antennae with a pulvillus (may be very difficult to see) (Pls. 16.123 J, M–O, 124 A–D, K, M, N); male second antennae armed or unarmed, apophyses present or absent, brood pouch variable 12

9(8) Female brood pouch fusiform, with lateral outpocketings; female first antennae subequal in length to second antennae (Pl. 16.123 G) 10

9' Female brood pouch fusiform, smooth; female first antennae at least 50% longer than second antennae (Pl. 16.123 E, I) 11

10(9) Male second antenna with distal antennomere apex bent laterally 90° (Pl. 16.123 L) .... *Branchinecta lateralis* Rogers, 2006

[Canada: Alberta, Saskatchewan. USA: Montana, Texas, Wyoming]

10' Male second antenna with distal antennomere apex truncate to laterally bent 60° (Fig. 16.123 K) .... *Branchinecta campestris* Lynch, 1960

[Canada: British Columbia. USA: California, Oregon, Utah, Washington]

11(9) Male first antennae less than 75% the length of second antenna proximal antennomere; distal segment apex truncate (Pl. 16.123 H) .... *Branchinecta mackini* Dexter, 1956

[USA: California, Idaho, Nevada, Oregon, Utah, Washington. Mexico: Baja California]

11' Male first antennae subequal in length to second antenna proximal antennomere; male second antenna distal segment apex bent laterally (Pl. 16.123 D) .... *Branchinecta readingi* Belk, 2001

[Canada: Alberta, Saskatchewan. USA: Montana, Nebraska, North Dakota, Wyoming]

12(8) Male second antennae apex directed medially or distally (Pls. 16.123 M–Q, 124 A–D, K, M, N) 13

12' Male second antennae apex bent 60° laterally; apophyses absent; females with first antennae longer than second antennae (Pl. 16.123 J) .... *Branchinecta potassa* Belk, 1979

[USA: Nebraska]

13(12) Male with apophyses (may be very small or absent in specimens not fully mature) (Fig. 16.123 M–Q) 14

13' Male without apophyses (Pl. 16.124 A–D, K, M, N) 18

14(13) Male second antennal proximal antennomere with a large medial process (Pl. 16.123 O–Q) 15

14' Male second antennal proximal antennomere with a medial convexity, bearing small spines (Pl. 16.123 M, N) 17

15(14) Male second antennae proximal antennomere with a large medial process dorsally directed, process may be cleft (Pl. 16.123 P, Q) 16

15' Male second antennae proximal antennomere with a large medial process branched, and medially directed; male second antennae proximal antennomere with a distal row of spines, anterior to the large medial process (Pl. 16.123 O) .... *Branchinecta belki* Maeda-Martínez, Obergón-Barboza & Dumont, 1992

[Mexico: Coahulia]

16(15) Male second antennae proximal antennomere with medial process not cleft, but may bear a peg-like appendix, second antennal distal antennomere with apex bent medially; females may have enlarged maxillary glands; brood pouch extending to fourth or fifth abdominal segment (Pl. 16.123 P) .... *Branchinecta packardi* Pearse, 1912

[Canada: Alberta, Saskatchewan. USA: Arkansas, Arizona, Colorado, Kansas, Montana, Nebraska, New Mexico, North Dakota, Oklahoma, Texas, Utah, Wyoming. Mexico: Coahuila, Durango, San Luis Potosi, Sonora, Zacatecas]

16' Male second antennae proximal antennomere with medial process cleft, giving a lobed appearance, second antennal distal antennomere apically truncate; female brood pouch extending to abdominal segment six (Pl. 16.123 Q) .... *Branchinecta mexicana* Maeda-Martínez, Obergón-Barboza & Dumont, 1993

[Mexico: Puebla, Tlaxcala]

17(14) Male second antennal proximal antennomere with a large, dorsally directed medial spiny convexity, extending at least 40% the length of the antennal antennomere; apophyses sub-cylindrical, and may be reduced; brood pouch sub-cylindrical to fusiform (Pl. 16.123 M) .... *Branchinecta coloradensis* Packard, 1874

[Canada: Alberta, British Columbia, Saskatchewan. USA: Arizona, California, Colorado, Idaho, Montana, Washington, Nevada, New Mexico, Oregon, Utah, Wyoming]

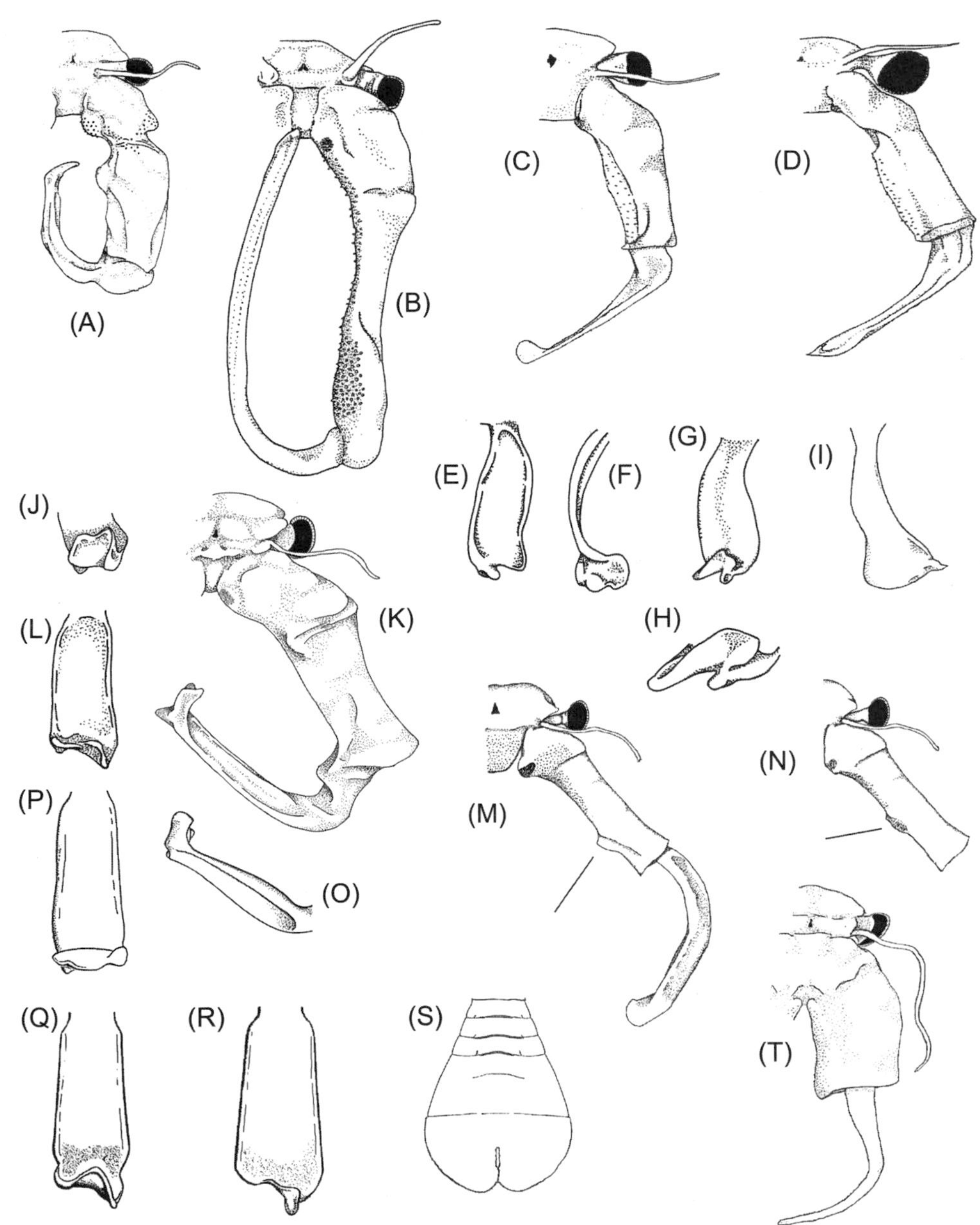

PLATE 16.124 (A) *Branchinecta constricta* male, head, anteriolateral view. (B) *Branchinecta longiantenna* male, head, anteriolateral view. (C) *Branchinecta oriena* male, head, anteriolateral view. (D) *Branchinecta dissimilis* male, head, anteriolateral view. (E–H) *Branchinecta lutulenta*, male, left second antenna distal antennomere in: (E) lateral; (F) anterior; (G) medial, and; (H) distal views. (I) *Branchinecta dissimilis*, male, left second antenna distal antennomere, medial view. (J) *Branchinecta mesovallensis*, male, left second antenna distal antennomere, distal view. (K) *Branchinecta mesovallensis* male, head, anteriolateral view. (L) *Branchinecta mesovallensis*, male, left second antenna distal antennomere, medial view. (M) *Branchinecta mediospinosa* male, head, anteriolateral view. (N) *Branchinecta oterosanvicentei* male, head, anteriolateral view. (O) *Branchinecta conservatio*, male, left second antenna distal antennomere, anterior view. (P) *Branchinecta conservatio*, male, left second antenna distal antennomere, medial view. (Q) *Branchinecta lindahli*, male, left second antenna distal antennomere, medial view. (R) *Branchinecta sandiegonensis*, male, left second antenna distal antennomere, medial view. (S) *Thamnocephalus platyurus*, abdomen and cercopods. (T) *Phallocryptus sublettei* male, head, anteriolateral view.

17' Male second antennal proximal antennomere with a small medial spiny convexity on the distal third of the segment; apophyses flattened, transverse; brood pouch pyriform (Pl. 16.123 N) .... *Branchinecta lynchi* Eng, Belk & Eriksen, 1990

[USA: California, Oregon]

18(13) Male second antennae proximal antennomere without a constriction 19

18' Male second antennae proximal antennomere bearing a constriction just distal to the pulvillus; male second antennae proximal antennomere bearing a lateral spiny tubercle between the constriction and the head; male second antennae distal antennomere curved medially (Pl. 16.124 A) .... *Branchinecta constricta* Rogers, 2006

[USA: Idaho, Wyoming]

19(18) Male second antennae proximal antennomere with medial surface covered in spines; male second antennal distal antennomere with apex bent medially or not (Pl. 16.124 B–D) 20

19' Male second antennae proximal antennomere with medial surface bearing a few scattered or no spines; male second antennal distal antennomere with apex bent medially (Pl. 16.124 K, M, N) 22

20(19) Male second antennae not extending to genitalia; male second antenna proximal antennomere with distal two-thirds to three-fourths of medial surface spinose; brood pouch fusiform, elongate (Pl. 16.124 C, D) 21

20' Male second antennae elongate, reaching genitalia, proximal antennomere distomedial surface with "wart like" mounds; brood pouch short, extending to abdominal segment two or three (Pl. 16.124 B) .... *Branchinecta longiantenna* Eng, Belk & Eriksen, 1990

[USA: California]

21(20) Male second antennal distal antennomere narrow, margins roughly parallel, apex truncate, bent medially and rotated anteriorly (Pl. 16.124 C) .... *Branchinecta oriena* Belk & Rogers, 2002

[USA: California, Nevada, Oregon (?)]

21' Male second antennal distal segment broadly explanate, with a subapical, anteriorly directed digitiform process (Pl. 16.124 D, I) .... *Branchinecta dissimilis* Lynch, 1972

[USA: California, Oregon]

22(19) Male second antennal distal antennomere with apex bilobed or trilobed (Pl. 16.124 E–G, J, L, P) 23

22' Male second antennal distal antennomere with apex not bilobed (Pl. 16.124 M, N, Q, R) 25

23(22) Male second antennal distal antennomere with apex bilobed (Pl. 16.124 J–L, O, P) 24

23' Male second antennal distal antennomere with apex trilobed (Pl. 16.124 E–H) .... *Branchinecta lutulenta* Rogers & Hill, 2013

[USA: Washington]

24(23) Male second antennal distal antennomere bilobed apex with anterior lobe broader than posterior lobe; brood pouch pyriform (Pl. 16.124 J–L) .... *Branchinecta mesovallensis* Belk & Fugate, 1999

[USA: California]

24' Male second antennal distal antennomere bilobed apex with anterior lobe narrower than posterior lobe (Pl. 16.124 O, P) .... *Branchinecta conservatio* Eng, Belk & Eriksen, 1990

[USA: California]

25(22) Male second antennal proximal antennomere without a medial pulvinus (Pl. 16.124 M) 26

25' Male second antenna proximal antennomere with a medial pulvinus situated halfway down the segment (Pl. 16.124 N) .... *Branchinecta oterosanvicentei* Obregon-Barbóza, Maeda-Martínez, García-Velazco & Dumont, 2001

[Mexico: Coahuila]

26(25) Male second antennal proximal antennomere without a medial projection 27

26' Male second antennal proximal antennomere with a medial spiniform tubercle, apically directed (Pl. 16.124 M) .... *Branchinecta mediospinosa* Rogers, 2011

[USA: Kansas]

27(26) Male second antennal distal antennomere in lateral view widening towards apex; female with paired dorsolateral conical thoracic processes (Pl. 16.124 R) .... *Branchinecta sandiegonensis* Fugate, 1993

[USA: California. Mexico: Baja California (Norte)]

27' Male second antennal distal antennomere in lateral view with a slight subapical constriction; female with single dorsolateral conical thoracic processes (Pl. 16.124 O) .... *Branchinecta lindahli* Packard, 1883

[Canada: Alberta. USA: western USA, from Great Plains to Pacific. Mexico: Baja California (Norte), Baja California Sur]

Crustacea: Branchiopoda: Anostraca: *Thamnocephalidae*: Genera

1 Cercopods free, subcylindrical 2

1' Cercopods broadly transverse and fused with abdomen into a broad, flat 'paddle'(Pl. 16.124 S) .... ***Thamnocephalus* [p. 452]**

2(1) Male without an antenna-like appendage; gonopods with medial surface of rigid proximal portion without a dense patch of denticles (Pl. 16.124 T) .... *Phallocryptus sublettei* (Sissom, 1976)

[USA: New Mexico, Texas]

2' Male with an 'antennae-like' appendage between eyestalk and antennae; gonopods with medial surface of rigid proximal portion without a dense patch of denticles (Pl. 16.125 C–E) .... ***Dendrocephalus* [p. 452]**

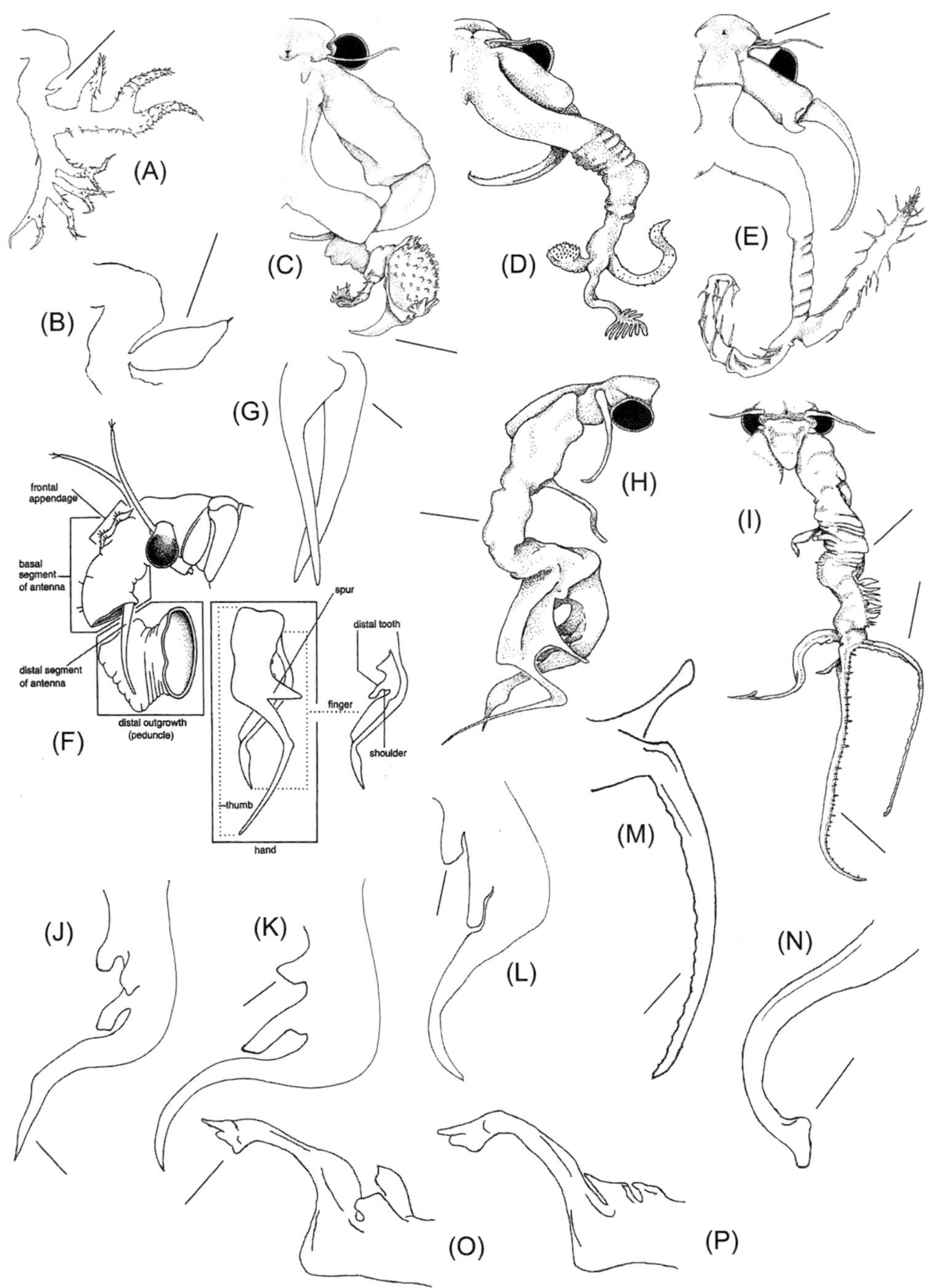

PLATE 16.125 (A) *Thamnocephalus platyurus* male, frontal appendage, right half (branch M1 indicated). (B) *Thamnocephalus mexicanus* male, frontal appendage, basal portion (branch M1 indicated). (C) *Dendrocephalus lithacus* male, head, anteriolateral view. (D) *Dendrocephalus alachua* male, head, anteriolateral view. (E) *Dendrocephalus acacioidea* male, head, anteriolateral view. (F) *Streptocephalus coloradensis*, exploded lateral view. (G) *Streptocephalus kargesi*, distal portion of antennal appendage, lateral view. (H) *Streptocephalus dorothae*, male, head, left lateral view. (I) *Streptocephalus moorei* male, head, anteriolateral view. (J) *Streptocephalus similis*, male, 'finger.' (K) *Streptocephalus sealii*, male, 'finger.' (L) *Streptocephalus woottoni*, male, 'finger.' (M) *Streptocephalus mattoxi*, male, 'thumb.' (N) *Streptocephalus guzmani*, male, 'finger.' (O) *Streptocephalus potosinensis*, male, 'finger.' (P) *Streptocephalus texanus*, male, 'finger.'

Crustacea: Branchiopoda: Anostraca: Thamnocephalidae: *Thamnocephalus*: Species

1 Male cephalic appendage branch 1MD broadly lamellar, foliate (Pl. 16.125 B) .... *Thamnocephalus mexicanus* Linder, 1941

[USA: Arizona, New Mexico, Texas. Mexico: Baja California Sur, Coahuila, Durango, Nayarit, Nuevo León, Sinaloa, Sonora]

1' Male cephalic appendage branch 1MD subcylindrical, slightly bent, with an obvious apical spine (Pl. 16.125 A) .... *Thamnocephalus platyurus* Packard, 1877

[USA: Arizona, Arkansas, California, Colorado, Kansas, Montana, Nebraska, New Mexico, Nevada, Oklahoma, Texas, Wyoming. Mexico: Aguascalientes, Baja California (Norte), Coahuila, Chihuahua, Durango, Guanajuato, Nuevo León, San Luis Potosi, Tamaulipas, Zacatecas]

Crustacea: Branchiopoda: Anostraca: Thamnocephalidae: *Dendrocephalus*: Species

1 Male frontal appendage with branches lacking a terminal chitinized hook (Pl. 16.125 D, E) 2

1' Male frontal appendage with largest branch bearing a terminal chitinized hook (Pl. 16.125 C) .... *Dendrocephalus lithacus* (Creaser, 1940)

[USA: Georgia (possibly extinct)]

2(1) Male frontal appendage with two main branches, each terminating in three sub-branches (Pl. 16.125 D) .... *Dendrocephalus alachua* (Dexter, 1953)

[USA: Florida]

2' Male frontal appendage with two main branches, each terminating in two sub-branches (Pl. 16.125 E) .... *Dendrocephalus acacioidea* (Belk & Sissom, 1992)

[USA: Texas]

Crustacea: Branchiopoda: Anostraca: Streptocephalidae: *Streptocephalus*: Species

Please refer to Pl. 16.125 F for descriptive terms.

1 Male second antennal appendage with peduncle subequal to or shorter than second antennal proximal antennomere 2

1' Male second antennal appendage with peduncle 1.5 times longer than second antennal proximal antennomere 6

2(1) Peduncle subequal in length to second antennal proximal antennomere; thumb with spur present (Pl. 16.125 H, I) 3

2' Peduncle shorter than second antennal proximal antennomere; 'thumb' without a spur (Fig. 16.125 G) .... *Streptocephalus kargesi* Spicer, 1985

[Mexico: Veracruz]

3(2) Finger with two teeth; genitalia with linguiform outgrowths; cephalic appendage bilobed (Pls. 16.125 K, L, O, P; 126 B–E) 4

3' Finger with three teeth; genitalia without linguiform outgrowths; cephalic appendage truncate; mature specimens with cercopods bearing stout marginal spines on the distal half (Pl. 16.125 J) .... *Streptocephalus similis* Baird, 1852

[USA: Texas. Mexico: Nuevo Leon, San Luis Potosi, Tamaulipus. Antigua, Belize, Hispaniola, Jamaica, Puerto Rico, Santa Domingo]

4(3) Finger with proximal tooth triangular, distal tooth digitate; mature male specimens with cercopods bearing stout marginal spines on the distal half (Pl. 16.125 K) 5

4' Finger with proximal tooth short and spatulate, distal tooth semirectangular; mature specimens with cercopod margins entirely setose (Pl. 16.124 L) .... *Streptocephalus woottoni* Eng, Belk & Eriksen, 1990

[USA: California. Mexico: Baja California (Norte)]

5(1) Thumb margin smooth between spur and main ramus; fully mature, large males with frontal appendage cleft .... *Streptocephalus sealii* Ryder, 1879

[USA: Atlantic and Gulf Coast states, Coastal Plain. Mexico: Gulf Coast slope]

5' Thumb with small protrusion between spur and main ramus; fully mature, large males with frontal appendage entire .... *Streptocephalus coloradensis* Dodds, 1916

[Canada: Alberta, British Colombia, Manitoba. USA: Alaska, western and central states, from Kentucky, westward to California, south to Texas, north to Washington, Montana, North Dakota, Minnesota]

6(1) Thumb with anterior margin crenate or spinose (Pl. 16.125 I, M) 7

6' Thumb with anterior margin inerm (Pl. 16.125 H) 9

7(6) Anterior margin of thumb spinose (Pl. 16.125 I) 8

7' Anterior margin of thumb crenate; medial surface of peduncle with a series of ~14 conical protuberances; surface of 'hand' with an anteromedial spinose structure (Pl. 16.125 M) .... *Streptocephalus mattoxi* Maeda-Martínez, Belk, Obregon-Barbóza & Dumont, 1995

[USA: Texas]

8(7) Thumb bearing a basal dorsolateral curved process; spur short, acuminate; peduncle with lateral vermiform processes (Pl. 16.125 I) .... *Streptocephalus moorei* Belk, 1973

[USA: New Mexico. Mexico: Chihuahua]

8' Thumb without a basal process; 'thumb' with two digitiform projections; spur truncate .... *Streptocephalus antilliensis* Mattox, 1950

[USA: Puerto Rico]

9(6) Finger apex with a lateral lobe, lamella or one or more spines (Pls. 16.125 N–P, 126 A–C)........ 10

9' Finger without any subtending structures (Pl. 16.126 D, E) 15

10(9) Finger apex with a lateral lobe or lamella (Pl. 16.124 O, P) 11

10' Finger apex with one or more subtending spines (Pl. 16.126 A–C) 13

11(10) Finger apex with a lateral lamella, lying along the apex (Pl. 16.125 O, P) 12

11' Finger apex with a lateral lobe, pediform (Pl. 16.125 N); peduncle without conical projections; proximal projection on 'finger' rounded .... *Streptocephalus guzmani* Maeda-Martínez, Belk, Obregon-Barbóza & Dumont,1995

[Mexico: Coahuila]

12(11) Finger proximal tooth larger than distal tooth (Pl. 16.125 O) .... *Streptocephalus potosinensis* Maeda-Martínez, Belk, Obregon-Barbóza & Dumont, 1995

[Mexico: San Luis Potosi]

12' Finger proximal tooth smaller than distal tooth (Pl. 16.125 P) .... *Streptocephalus texanus* Packard, 1871

[USA: Arkansas, Arizona, southern California, Colorado, Kansas, Missouri, Montana, New Mexico, Nebraska, Oklahoma, Texas, Utah and Wyoming. Mexico: Baja California Sur, Chihuahua, Cohuila, Durnago, Nuevo León, Oaxaca, San Luis Potosí and Tamualipas. Caribbean French Département d'outre-mer Guadeloupe: Désirade. Antigua and Barbuda: Island of Barbuda]

13(10) Finger apex with a single, large, subtending spine (Pl. 16.126 B, C) 14

13' 'Finger' apex with a posterior row of subtending spines, the largest spine being furthest from the apex; 'finger' strongly recurved; eight conical projections on medial surface of peduncle; 'thumb' long and acuminate (Pl. 16.125 A) .... *Streptocephalus linderi* Moore, 1966

[USA: Texas. Mexico: Coahuila, Nuevo León, Tamaulipas]

14(13) Proximal tooth truncate; distal tooth with base 1.5 times the height; apex of finger with lateral subtending spur; anterior margin of finger evenly curved (Pl. 16.126 B) .... *Streptocephalus mackini* Moore, 1966

[USA: Arizona, New Mexico, Texas. Mexico: Aguascalientes, Chihuahua, Coahuila, Distrito Federal, Durango, Estado de Mexico, Guanajuato, Hidalgo, Jalisco, Morelos, Nuevo León, Puebla, Querétaro, San Luis Potosí, Sinaloa, Sonora, Tlaxcala, and Zacatecas]

14' Proximal tooth subtriangular; distal tooth with base as wide as projection height; apex of 'finger' with posteriolateral subtending spur; anterior margin of 'finger' flattened (Pl. 16.126 C) .... *Streptocephalus henridumontis* Maeda-Martínez, Obregon-Barbóza, Prieto-Salazar & García-Velazco, 2005

[USA: Arizona, New Mexico. Mexico: Baja California Norte, Sinaloa, Sonora]

15(9) Proximal tooth truncate, distal tooth lying against finger (Pls. 16.125 H, 126 D) .... *Streptocephalus dorothae* Mackin, 1942

[USA: Arizona, California, New Mexico, Oklahoma, Texas, Utah, Wyoming. Mexico: Baja California Sur]

15' Proximal tooth arcuate, distal tooth erect, acute (Pl. 16.126 E) .... *Streptocephalus thomasbowmani* Maeda-Martínez, Obregon-Barbóza, Prieto-Salazar & García-Velazco, 2005

[USA: Arizona, New Mexico.]

Crustacea: Branchiopoda: Notostraca: Genera

One family: Triopsidae Kielhack, 1910

1 Telson without a caudal lamina projecting between cercopods (Pl. 16.126 F) .... *Triops*

[Taxonomy confused, genus needs revision. Canada: Alberta, Saskatchewan. USA: arid and semiarid west, Illinois, Missouri. Mexico: arid and semiarid regions.]

1' Telson with a caudal lamina projecting between cercopods (Pl. 16.119 C) .... ***Lepidurus* [p. 454]**

Crustacea: Branchiopoda: Notostraca: *Lepidurus*: Species

From Rogers, 2001.

1 More than 50 pairs of legs; 30–36 body rings 2

1' Less than 50 pairs of legs; 24–29 body rings 3

2(1) Nuchal organ behind a line drawn between the posterior apices of eyes; caudal lamina truncate; living animals tend to be shining yellow or greyish silver to light green (Pl. 16.126 H) .... *Lepidurus lemmoni* Holmes, 1894

[Canada: Alberta. USA: Arizona, California, Idaho, Nevada, Oregon, Washington. Mexico: Baja California (Norte)]

2' Dorsal organ intersected by a line drawn between the posterior apices of eyes (Pl. 16.126 G); caudal lamina subquadrate .... *Lepidurus bilobatus* Packard, 1883

[USA: Colorado, Idaho, Montana, Nevada, Oregon, Wyoming]

3(1) Endites of second thoracic appendage long, unequal in length, extending beyond the margin of the carapace (endites are fragile and may be broken off) (Pl. 16.119 C); sulcus with large spines 1 to 1.3 times as long as broad; 30 to 39 pairs of legs 4

3' Endites of second thoracic appendage short, subequal, not quite extending beyond the margin of the carapace; sulcus with acute spines, 1.5 to 2.5 times as long as broad (Pl. 16.126 I); strong carinal spine; 40 to 45 pairs of legs .... *Lepidurus arcticus* (Pallas, 1793)

[Canada: Northwest Territories, Yukon. USA: Alaska. Greenland. Iceland, Norway, Russia]

4(3) Sulcus margined with large triangulate spines separate by two to three times their width, with numerous, variously shaped small spines in between the larger spines (Pl. 16.126 K); caudal lamina equal to or less than 0.3 times the mid-dorsal length of carapace 5

4' Sulcus margined with large rounded spines separate by twice their width, with an occasional single small spine between (Pl. 16.126 J); caudal lamina 0.3 to 0.7 times the mid-dorsal length of carapace .... *Lepidurus couesii* Packard, 1875

[Canada: Alberta, Manitoba, Saskatchewan. USA: Idaho, Minnesota, Montana, North Dakota, Utah]

5(4) 12S rDNA with adenine bases at positions 11, 93, 148, 160, 168, 229, and 263; thymine at position 161 and 305; guanine at position 209 .... *Lepidurus packardi* Simon, 1886

[USA: Great Central Valley of California]

5' 12S rDNA with guanine bases at positions 11, 93, 148, 160, 168, 229, and 263; cytosine at position 161 and 305; adenine at position 209 .... *Lepidurus cryptus* Rogers, 2001

[USA: Intermontane California, Idaho, Oregon, Washington]

Crustacea: Branchiopoda: Laevicaudata: Genera

One family: Lynceidae Baird, 1845 (Pl. 16.119 D, E). Modified from Martin & Belk, 1988.

1 Male second thoracopod unmodified, similar to subsequent thoracopods .... ***Lynceus* [p. 454]**

1' Male second thoracopod obviously modified, but unequally (Pl. 16.126 V) .... ***Paralimnetis* [p. 456]**

Crustacea: Branchiopoda: Laevicaudata: *Lynceus*: Species

Modified from Martin & Belk, 1988 and Rogers et al. in prep.

1 Males: first thoracopods modified into clapers to amplex the female (Pl. 16.126 L) 2

1' Females: first thoracopods not modified; eggs often present 5

2(1) Right and left claspers subequal in size 3

2' Right and left claspers obviously dimorphic .... *Lynceus gracilicornis* (Packard, 1871)

[USA: Alabama, Florida, Georgia, South Carolina, Texas. Mexico: Puebla, Tlaxcala (?)]

3(2) Rostrum with medial carina straight, not bifurcated (Pl. 16.126 P–R) 4

3' Rostrum with medial carina bifurcated distally, making rostrum appear truncated in lateral view (Pl. 16.126 T) .... *Lynceus brevifrons* (Packard, 1877)

[USA: Arizona, Colorado, Kansas, New Mexico, Oklahoma, Texas. Mexico: Durango, Chihuahua, Guanajuato, San Luis Potosí]

4(3) Clasper with small spiniform process extending distally from fixed finger at base of movable finger and with small, smooth bump on outer border of movable finger (Pl. 16.126 M, R); last abdominal appendage of male with obvious stout upturned hooked process .... *Lynceus mucronatus* (Packard, 1875)

[Canada: Alberta, British Columbia. USA: Montana]

PHYLUM ARTHROPODA

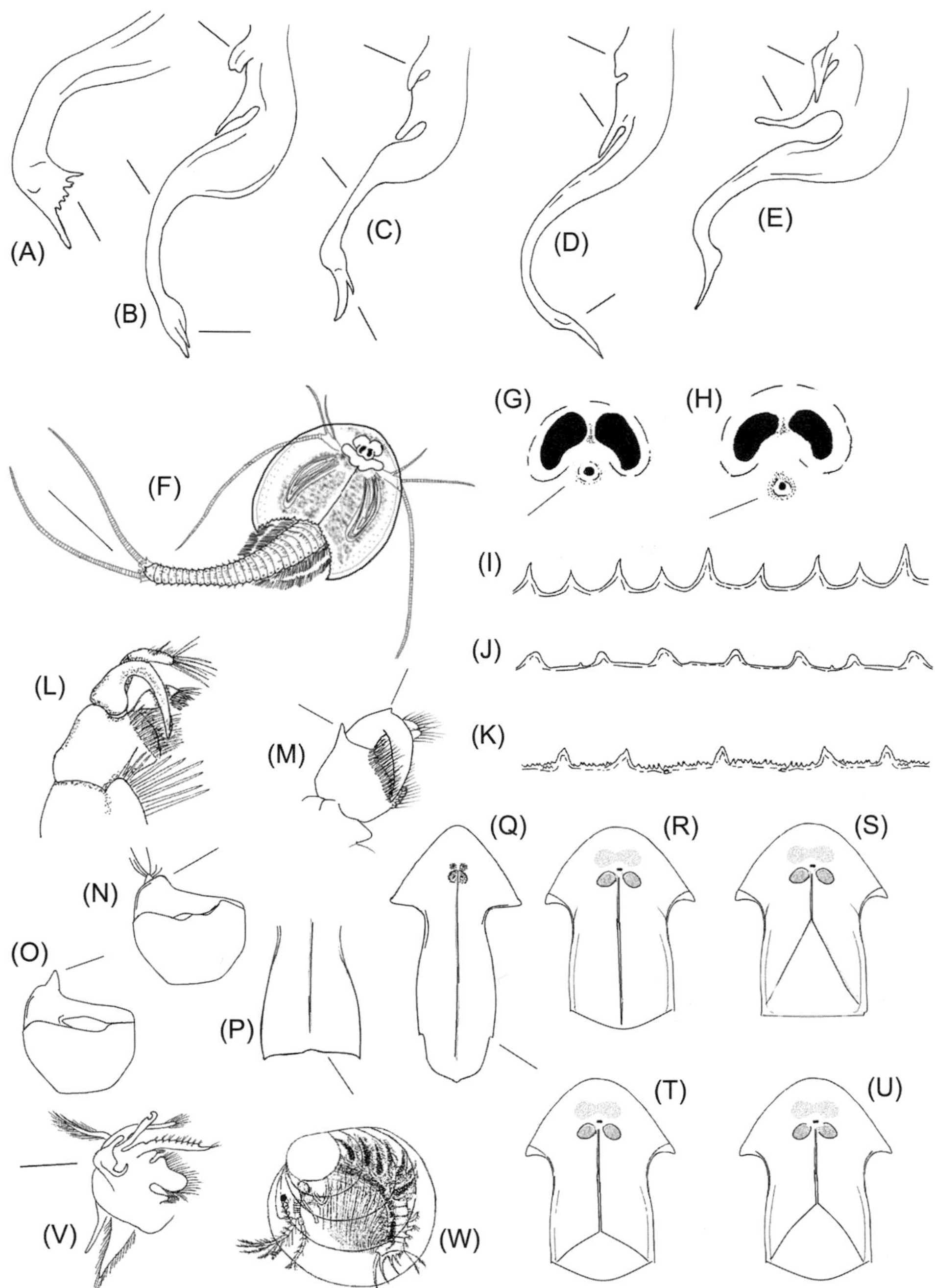

PLATE 16.126 (A) *Streptocephalus linderi*, male, 'finger.' (B) *Streptocephalus mackini*, male, 'finger.' (C) *Streptocephalus henridumontis*, male, 'finger.' (D) *Streptocephalus dorothae*, male, 'finger.' (E) *Streptocephalus thomasbowmani*, male, 'finger.' (F) *Triops* sp., dorsal view. (G) *Lepidurus bilobatus*, ocular tubercle (nuchal organ indicated). (H) *Lepidurus lemmoni*, ocular tubercle (nuchal organ indicated). (I) *Lepidurus arcticus*, carapace sulcus marginal spines. (J) *Lepidurus couesii*, carapace sulcus marginal spines. (K) *Lepidurus cryptus*, carapace sulcus marginal spines. (L) *Lynceus brachyurus*, male thoracopod I. (M) *Lynceus mucronatus*, male thoracopod I. (N) *Paralimnetis mapimi*, male thoracopod I. (O) *Paralimnetis texana*, male thoracopod I. (P) *Lynceus brachyurus*, male rostrum, anterior view. (Q) *Lynceus mucronatus*, male head, anterior view. (R) *Lynceus mucronatus*, female head, anterior view. (S) *Paralimnetis texana*, male head, anterior view. (T) *Lynceus brevifrons*, male head, anterior view. (U) *Paralimnetis mapimi*, male head, anterior view. (V) *Paralimnetis texana*, male thoracopod II. (W) *Cyclestheria* sp.

4' Clasper without spiniform process extending distally from immovable finger; outer border of movable finger smoothly curving, lacking bump (Pl. 16.126 L); last abdominal appendage of male similar to preceding ones, without upturned hooked process .... *Lynceus brachyurus* Müller, 1776

[Canada: Alberta, Ontario, Quebec, Yukon. USA: Alaska, Arizona, California, Colorado, Indiana, Illinois, Massachusetts, Michigan, Montana, New Hampshire, New Mexico, New York, Ohio, Rhode Island, Washington, Wyoming. Palaearctic]

5(1) Rostral margin with an anteriorly directed lateral spine, apex acute (Pl. 16.126 Q) 6

5' Rostral margin without lateral spines .... *Lynceus gracilicornis* (Packard, 1971)

[USA: Alabama, Florida, Georgia, South Carolina, Texas. Mexico: Puebla, Tlaxcala (?)]

6(5) Acumen margin straight or concave 7

6' Acumen margin convex (Pl. 16.126 Q) .... *Lynceus mucronatus* (Packard, 1975)

[Canada: Alberta, British Columbia. USA: Montana]

7(6) Acumen margin concave .... *Lynceus brachyurus* Müller, 1776

[Canada: Alberta, Ontario, Quebec, Yukon. USA: Alaska, Arizona, California, Colorado, Indiana, Illinois, Massachusetts, Michigan, Montana, New Hampshire, New Mexico, New York, Ohio, Rhode Island, Washington, Wyoming. Palaearctic]

7' Acumen margin straight .... *Lynceus brevifrons* (Packard, 1877)

[USA: Arizona, Colorado, Kansas, New Mexico, Oklahoma, Texas. Mexico: Durango, Chihuahua, Guanajuato, San Luis Potosí]

Crustacea: Branchiopoda: Laevicaudata: *Paralimnetis*: Species

1 Males: first thoracopods modified into clapers to amplex the female 2

1' Females: first thoracopods not modified; eggs often present 3

2(1) Movable finger of major clasper with smooth, small outward protrusion at about level of distal palp (endite 5) (Pl. 16.126 N) .... *Paralimnetis mapimi* Maeda-Martínez, 1987

[Mexico: Durango, Chihuahua, Sinaloa, Sonora, Tamaulipas]

2' Movable finger of major clasper with large, sharp outward protrusion at about level of distal palp (endite 5) (Pl. 16.126 O) .... *Paralimnetis texana* Martin & Belk, 1988

[USA: Texas]

3(1) Distal branches of rostral carina shorter than unbranched portion of carina; apex of rostrum slightly truncate (Pl. 16.126 U) .... *Paralimnetis mapimi* Maeda-Martinez, 1987

[Mexico: Durango, Chihuahua, Sinaloa, Sonora, Tamaulipas]

3' Distal branches of rostral carina as long as unbranched portion of carina; apex of rostrum slightly emarginate (Pl. 16.126 S) .... *Paralimnetis texana* Martin & Belk, 1988

[USA: Texas]

Crustacea: Branchiopoda: Diplostraca: Suborders

1 Carapace encompassing head, growth lines present, although may be faint (Pls. 16.118 D, E; 119 F; 132 A–D, J, K) 2

1' Carapace, if present, not encompassing head, growth lines absent (Pls. 16.119 G–J, 127 A–I, 128 A–M, 129 A–L, 130 A–N, 131 A–O) .... **Cladocera [p. 456]**

2(1) First antenna smooth, without lobes along anterior surface (Pls. 16.118 D, 126 W) .... Cyclestherida, one genus: *Cyclestheria* sp.

[USA: Florida (?), Texas. Mexico: Tamaulipas]

2' First antenna with a longitudinal series of lobes on anterior surface (Pls. 16.118 E, 119 F, 132 A–D, J, K) .... **Spinicaudata [p. 476]**

Crustacea: Branchiopoda: Diplostraca: Cladocera: Infraorders

Note: based upon females.

1 Abdomen long or short, not distinctly segmented; male thoracopod I with clasper; metanauplius absent (Pls. 16.119 G–I, 127 A–I, 128 A–M, 129 A–L, 130 A–N, 131 A–O) 2

1' Abdomen elongate, clearly segmented; male thoracopod I without clasper; metanauplius present (Pl. 16.119 J) .... Haplopoda one family: Leptodoridae, one species: *Leptodora kindti* (Focke, 1844)

[Canada. USA. Nearctic, Palaearctic]

2(1) Trunk and 4–6 pairs of thoracopods enclosed in a carapace that opens ventrally 3

2' Trunk and 4 pairs of cylindrical thoracopods not enclosed by a carapace (Pl. 16.127 A–D) .... **Onchyopoda [p. 457]**

3(2) 6 pairs of foliaceous limbs; antennae with more than 10 natatory setae on both rami (except *Holopedium*, which has only one ramus) (Pl. 16.127 E, G, H) .... **Ctenopoda [p. 457]**

3' 4–6 pairs of differentiated limbs; antennae with fewer than 10 natatory setae on both rami .... **Anomopoda [p. 460]**

Crustacea: Branchiopoda: Diplostraca: Cladocera: Onchyopoda: Families

1 Adult female total length greater than 2 mm, abdominal process longer than the body (Fig. 16.127 A–C) .... **Cercopagidae [p. 457]**

1' Adult female total length 1–2 mm, abdominal process slender and about half as long as the rest of the body (Pl. 16.127 D) .... Polyphemidae, one species: *Polyphemus pediculus* (Linnaeus, 1761)

[Canada. USA. Nearctic, Palaearctic]

Crustacea: Branchiopoda: Diplostraca: Cladocera: Onchyopoda: Cercopagidae: Genera

1 Abdominal processes (the long filament with spines, which extends posterior to the abdomen, starting at the anus) more than 4 times as long as the combined head, thorax, and abdomen; brood chamber almost twice as long as wide, and ending in a point (Pl. 16.127 A, B) .... *Cercopagis pengoi* (Ostroumov, 1891)

[Invasive in the Laurentian Great Lakes. Palaearctic]

1' Abdominal process about 2 times as long as the head and thorax combined; brood chamber round (Pl. 16.127 C) .... *Bythotrephes longimanus* Leydig, 1860

[Invasive in the Laurentian Great Lakes. Palaearctic]

Crustacea: Branchiopoda: Diplostraca: Cladocera: Ctenopoda: Families

1 Second antenna (the swimming antenna) uniramous (Pl. 16.127 E) .... Holopedidae, one genus: ***Holopedium* [p. 457]**

1' Second antenna with two rami .... **Sididae [p. 457]**

Branchiopoda: Diplostraca: Cladocera: Ctenopoda: Holopedidae: *Holopedium*: Species

1 Carapace ventral margin with a row of fine spines; postabdomen with more than 10 anal teeth; transparent gelatinous envelope (Pl. 16.127 E) .... *Holopedium gibberum* Zaddach, 1855

[Nearctic]

1' Carapace ventral margin lacking spines; postabdomen with 6–9 anal teeth; gelatinous envelope absent .... *Holopedium amazonicum* Stingelin, 1904

[USA: Louisiana, North Carolina]

Branchiopoda: Diplostraca: Cladocera: Ctenopoda: Holopedidae: Sididae: Genera

1 Second antennae rami each with 3 antennomeres (proximal antennomere of lower ramus may be very small), no process; carapace valves lacking long marginal setae and not forming a ventral inflexion directed medially 2

1' Second antenna dorsal ramus with 2 antennomeres, large process; ventral ramus with 3 antennomeres; carapace valves with long marginal setae and forming a ventral inflexion directed medially; postabdominal claw with two basal spines (Pl. 16.127 F) .... ***Latona* [p. 458]**

2(1) Postabdomen without dorsal marginal teeth; clusters or patches of submarginal spines (fascicles) present 3

2' Postabdomen with 20 marginal teeth (denticles); no clusters of submarginal spines (fascicles) (Pl. 16.127 G) .... *Sida crystallina* (O.F. Müller, 1776)

[Canada. USA. Nearctic, Palaearctic]

3(2) Female first antennae long, almost reaching or exceeding length of basal antennomere of second antenna 4

3' Female first antennae short, less than half length of basal antennomere of second antenna (Pl. 16.128 A) .... ***Diaphanosoma* [p. 458]**

4(3) Postabdominal claw with 3 long, thin basal teeth .... *Sarsilatona serricauda* (Sars, 1901)

[USA: Florida, Louisiana, Texas]

4' Postabdominal claw with 2 long, thin basal teeth, and a short basal tooth less than half as long as the width of the claw base (Pl. 16.127 H, I) .... *Pseudosida bidentata* Herrick, 1884

[Canada. USA: Alabama, Georgia, Louisiana, South Carolina, Texas]

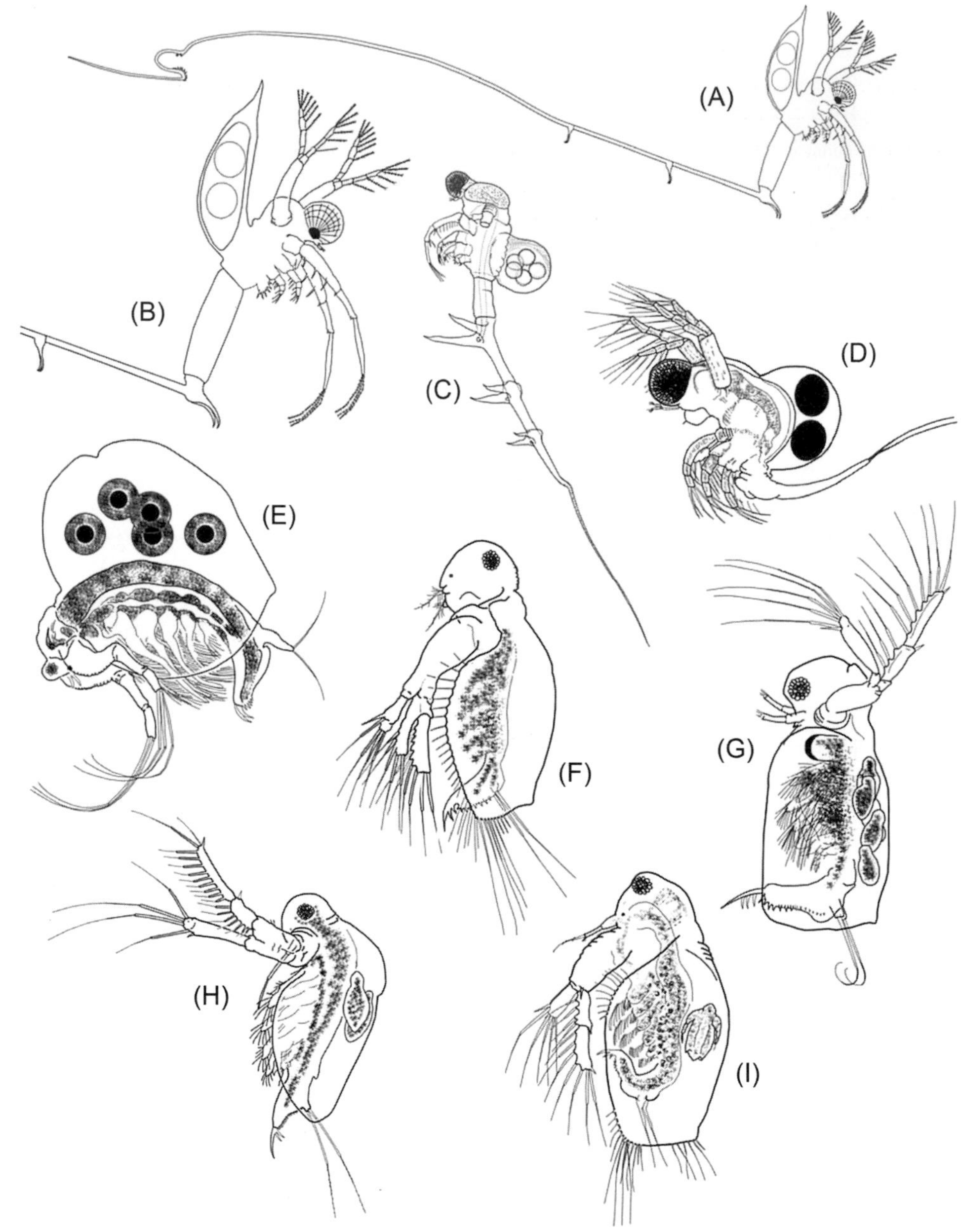

PLATE 16.127 (A and B) *Cercopagis pengoi*. (C) *Bythotrephes longimanus*. (D) *Polyphemus pediculus*. (E) *Holopedium gibberum*. (F) *Latona setifera*. (G) *Sida crystallina*. (H and I) *Pseudosida bidentata*.

Crustacea: Branchiopoda: Cladocera: Ctenopoda: Holopedidae: Sididae: *Latona*: Species

1 Female first antennae long and bent at right angles (Pl. 16.127 F) .... *Latona setifera* (O.F. Müller, 1776)
[Canada. USA. Nearctic, Palaearctic]

1' Female first antennae short, straight .... *Latona parviremis* Birge, 1910
[Canada: Nova Scotia, Newfoundland. USA: New York, Wisconsin, Michigan, Maine]

Branchiopoda: Cladocera: Ctenopoda: Holopedidae: Sididae: *Diaphanosoma*: Species

1 Ventral carapace margins forming a narrow inflexion 2

1' Ventral carapace margins forming a wide, often flap-shaped inflexion 4

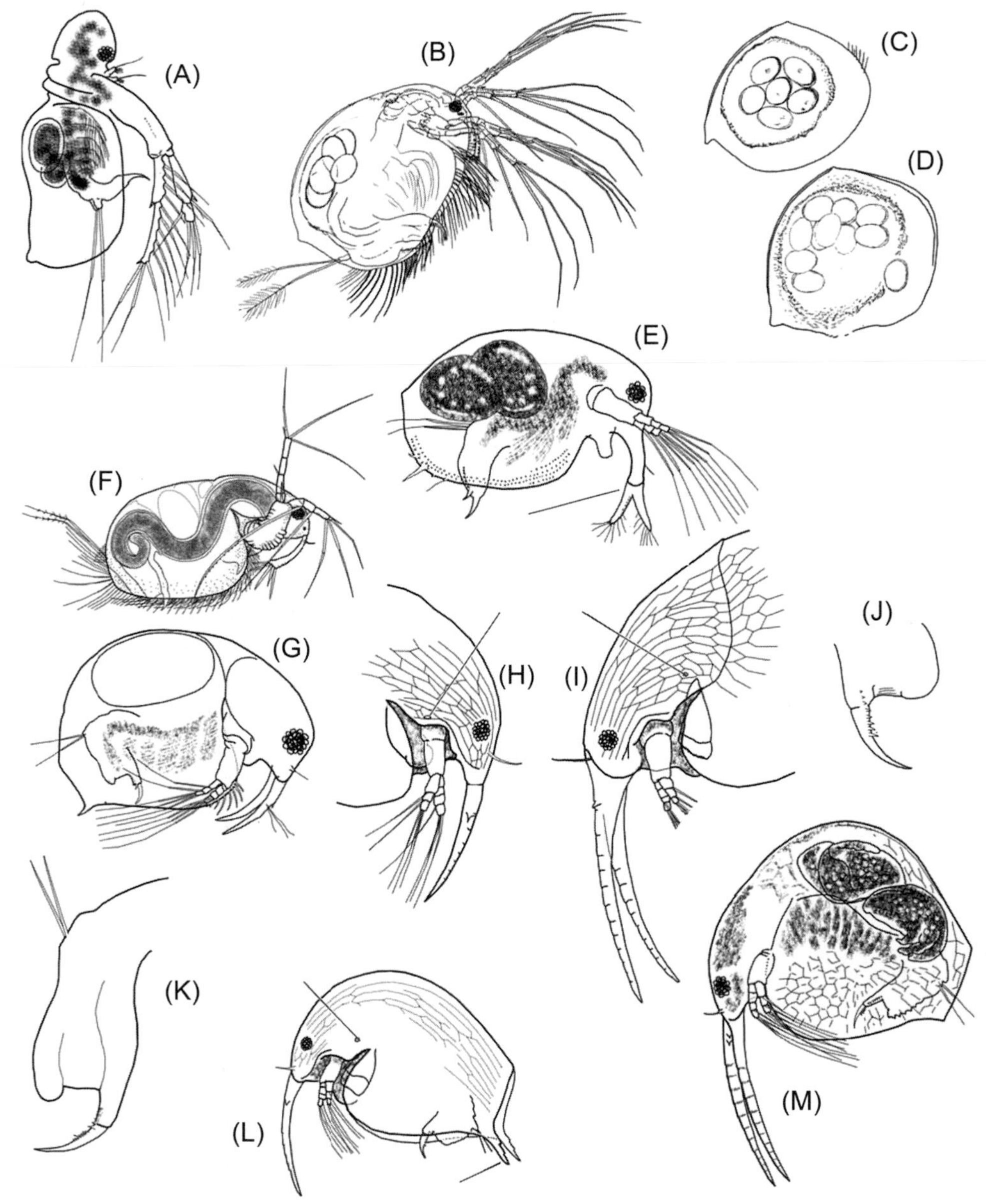

PLATE 16.128 (A) *Diaphanosoma birgei*. (B,C, & D) *Dumontia oregonensis*. (E) *Bosminopsis deitersi*. (F) *Acantholeberis curvirostris*. (G and H) *Bosmina longirostris*. (I and J) *Bosmina coregoni*. (K) *Bosmina longirostris*. (L) *Bosmina hagmanni*. (M) *Bosmina coregoni*.

2(1)	Posterioventral carapace margin with variable sized denticles	6
2'	Posterioventral carapace margin with uniform denticles	3
3(2)	Head roundish-rectangular	*Diaphanosoma fluviatile* Hansen, 1899
	[southern USA]	
3'	Head protruding	*Diaphanosoma oligosetum* Korovchinsky, 2005
	[USA: Louisiana]	
4(1)	Distal antennomere of dorsal antennal ramus with 7 or 8 setae; postabdomen with no anal teeth	5
4'	Distal antennomere of dorsal antennal ramus with 8 setae; postabdomen with anal teeth	*Diaphanosoma freyi* Korovchinsky, 2002
	[USA: Louisiana, Missouri]	
5(4)	Posterioventral margin of valves with denticles	*Diaphanosoma brevireme* Sars, 1901
	[USA, Mexico]	

5' Posterioventral margin of valves with denticles, distal antennomere with 7 or 8 setae .... *Diaphanosoma dorotheae* Korovchinsky, 2005
[USA: North Carolina]

6(2) Head protruding dorsally .... *Diaphanosoma birgei* Kořinek, 1981
[Canada. USA: eastern states. Nearctic]

6' Head rounded-rectangular, not protruding .... *Diaphanosoma brachyurum* (Lievin, 1848)
[Canada. USA. Nearctic, Palaearctic]

Crustacea: Branchiopoda: Diplostraca: Cladocera: Anomopoda: Families

1 First antennae not fused with the rostrum 2

1' First antennae fused with the rostrum to form two long and pointed tusk-like structures (Pl. 16.128 E, G–M) .... **Bosminidae [p. 460]**

2(3) First antennae of one antennomere, which may be several times longer than wide, to so short it is reduced to a mound with a cluster of setae.. 3

2' First antennae with two antennomeres (a nearly square basal antennomere) (Pl. 16.129 A) .... Ilyocryptidae, one genus: ***Ilyocryptus* [p. 462]**

3(2) Second antenna ventral ramus with 3 antennomeres, dorsal ramus with 4 antennomeres (look for a small basal antennomere, as in Pl. 16.129 B-J) 4

3' Second antenna rami with 3 antennomeres 5

4(3) Thoracic legs 3 and 4 with more than 40 setae total, mostly a group of closely-set setae forming a filter-like comb (Pl. 16.129 B–J) .... **Daphniidae [p. 464]**

4' Thoracic legs 3 and 4 with fewer than 30 setae total, filter combs not present (Fig.16.128 B–D) .... Dumontidae, one species: *Dumontia oregonensis* Santos Flores & Dodson, 2003
[USA: Oregon, California]

5(3) First antenna originating just ventral to the anterior tip of the head, rostrum lacking (Pl. 16.128 F) 6

5' First antenna originating on head ventral margin, and more or less covered by the rostrum (Pl. 16.130 A) 7

6(5) Anus in terminal position on postabdomen 8

6' Anus in medial position on postabdomen.. **Ophryoxidae [p. 475]**

7(5) Postabdomen strongly flattened, broad, with a single row of 80–150 marginal denticles, resulting in a saw-like appearance; anus distal (Pl. 16.130 A).. Eurycercidae, one genus: ***Eurycercus* [p. 468]**

7' Postabdomen thicker, not so broad, and usually with two rows of marginal denticles, mostly fewer than 20 on each side; anus proximal (Pls. 16.130 A–N, 131 L).. **Chydoridae [p. 469]**

8(6) Six pairs of trunk limbs (P6 rudimentary); P3-P5 with large exopodites (Fig.16.128 F) .... Acantholeberidae, one species *Acantholeberis curvirostris* (O. F. Müller, 1776)

8' Five pairs of trunk limbs; all exopodites small .... **Macrothricidae [p. 475]**

Crustacea: Branchiopoda: Diplostraca: Cladocera: Anomopoda: Bosminidae: Genera

1 Lateral head pores; accessory claw at base of postabdominal claw absent; first antennae separately attached to the head, straight or curved toward the body (Pl. 16.128 G–M) .... ***Bosmina* [p. 460]**

1' No lateral head pores; accessory claw at base of postabdominal claw; first antennae fused together in the basal half, separated and curved outward away from the head and body in the distal half (resembling a mermaid's tail; Pl. 16.128 E) ....*Bosminopsis deitersi* Richard, 1895
[USA: southwest. Neotropical]

Branchiopoda: Diplostraca: Cladocera: Anomopoda: Bosminidae: *Bosmina*: Species

1 Lateral head pore nearer mandible's proximal tip than to base of second antenna (Pl. 16.128 I, L); mucro (the posterio-ventral spine on carapace) with a row of teeth 2

1' Lateral head pore at edge of head shield, above base of second antennae (Pl. 16.128 G, H, K); mucro lacking teeth (sometimes called microspinules) .... *Bosmina* (*Bosmina*) *longirostris* (O.F. Müller, 1776)
[This species group also includes *B. liederi* and *B. freyi*, best treated as a sibling species complex. Canada. USA. Nearctic]

2(1) Mucro always present 3

2' Mucro absent, or if present with a row of teeth along ventral margin (Pl. 16.128 I, J, M) .... *Bosmina (Eubosmina) coregoni* (Baird, 1850)
[Canada: Manitoba, Laurentian Great Lakes. *B. longispina* and *B. maritima* are perhaps a sibling species group, with morphologies similar to *coregoni*.]

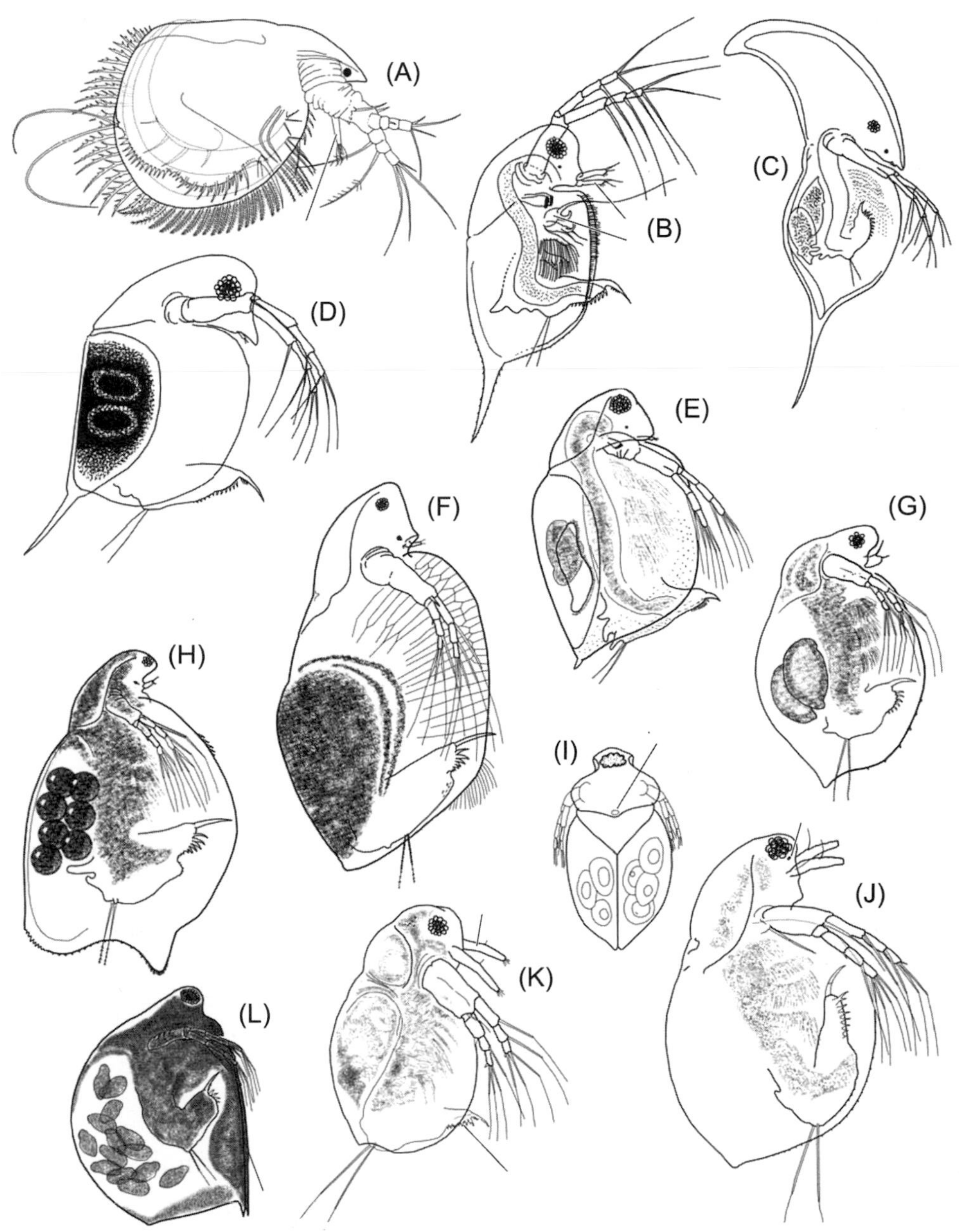

PLATE 16.129 (A) *Ilyocryptus* sp. (B) *Daphnia* sp. (C) *Daphnia retrocurva*. (D) *Daphnia pulicaria*. (E) *Daphniopsis ephemeralis*. (F) *Simocephalus* sp. (G) *Ceriodaphnia* sp. (H) *Simocephalus* sp. (I) *Megafenestra nasuta*. (J) *Moinodaphnia macleayi*. (K) *Moina* sp. (L) *Scapholeberis* sp.

3(2) Mucro and with a row of teeth along the dorsal margin (at least in juvenile specimens, Pl. 16.128 L) .... *Bosmina (Liederobosmina)* Brtek, 1997 4

3' Mucro long with truncated tip and 1–2 teeth on the ventral margin; few strong denticles on inner surface of mucro .... *Bosmina (Lunobosmina) oriens* (DeMelo & Hebert, 1994)

[Canada: Ontario. USA]

4(3) First antenna up to 1.5 times as long as the mucro .... *Bosmina (Liederobosmina) tubicen* (Brehm, 1953)

[Canada: Ontario, Nova Scotia. USA: Arkansas, Florida]

4' First antenna more than twice as long as the mucro (Pl. 16.128 L) .... *Bosmina (Liederobosmina) hagmanni* (Stingelin, 1904)

[USA: South Carolina]

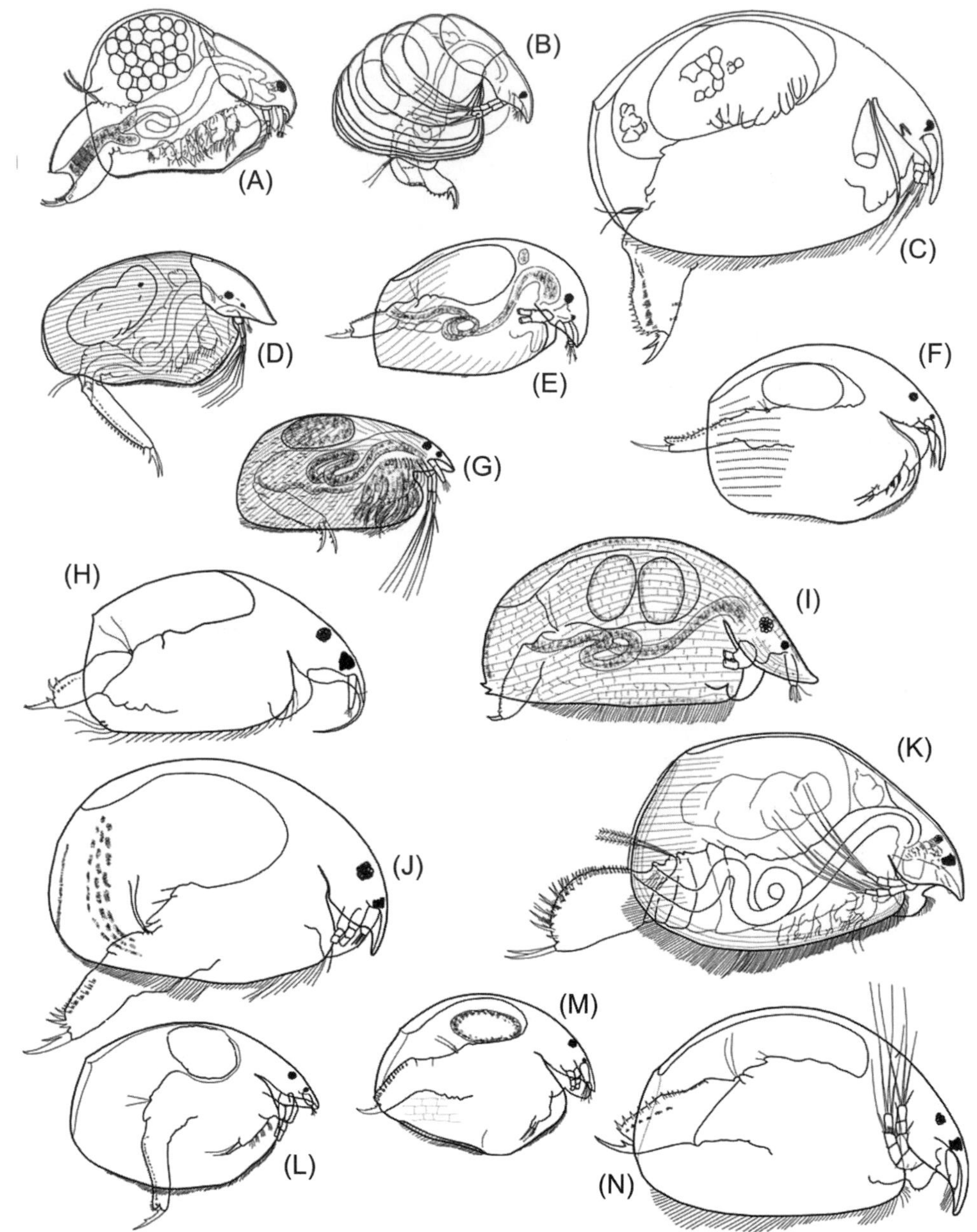

PLATE 16.130 (A) *Eurycercus* sp. (B) *Monospilus dispar*. (C) *Bryospilus repens*. (D) *Camptocercus* sp. (E) *Acroperus harpae*. (F) *Kurzia media*. (G) *Alonopsis americana*. (H) *Rhynchotalona* sp. (I) *Graptoleberis testudinaria*. (J) *Oxyurella brevicaudis*. (K) *Leydigia* sp. (L) *Euryalona orientalis*. (M) *Notoalona freyi*. (N) *Coronatella circumfimbriata*.

Branchiopoda: Diplostraca: Cladocera: Anomopoda: Ilyocryptidae: *Ilyocryptus*: Species

Kotov & Štifter (2006) provide keys for all species; however, morphological variation is not yet understood; minute differences between species may be due to phenotypic plasticity.

1 Moulting incomplete (carapace with growth lines) 2

1' Moulting complete; dorsal keel on carapace well developed .... *Ilyocryptus agilis* Kurz, 1878

[USA: Maryland, Virginia]

2(1) Postabdominal claw proximoventral setules long (maximum length greater than or equal to claw diameter) 3

2' Postabdominal claw proximoventral setules short (maximum length less than claw diameter); postabdomen distal postanal margin with 3–5 long lateral setae (4–5 times length of preanal spines), proximal most located very far from anus .... *Ilyocryptus spinifer* Herrick, 1882

[Canada. USA]

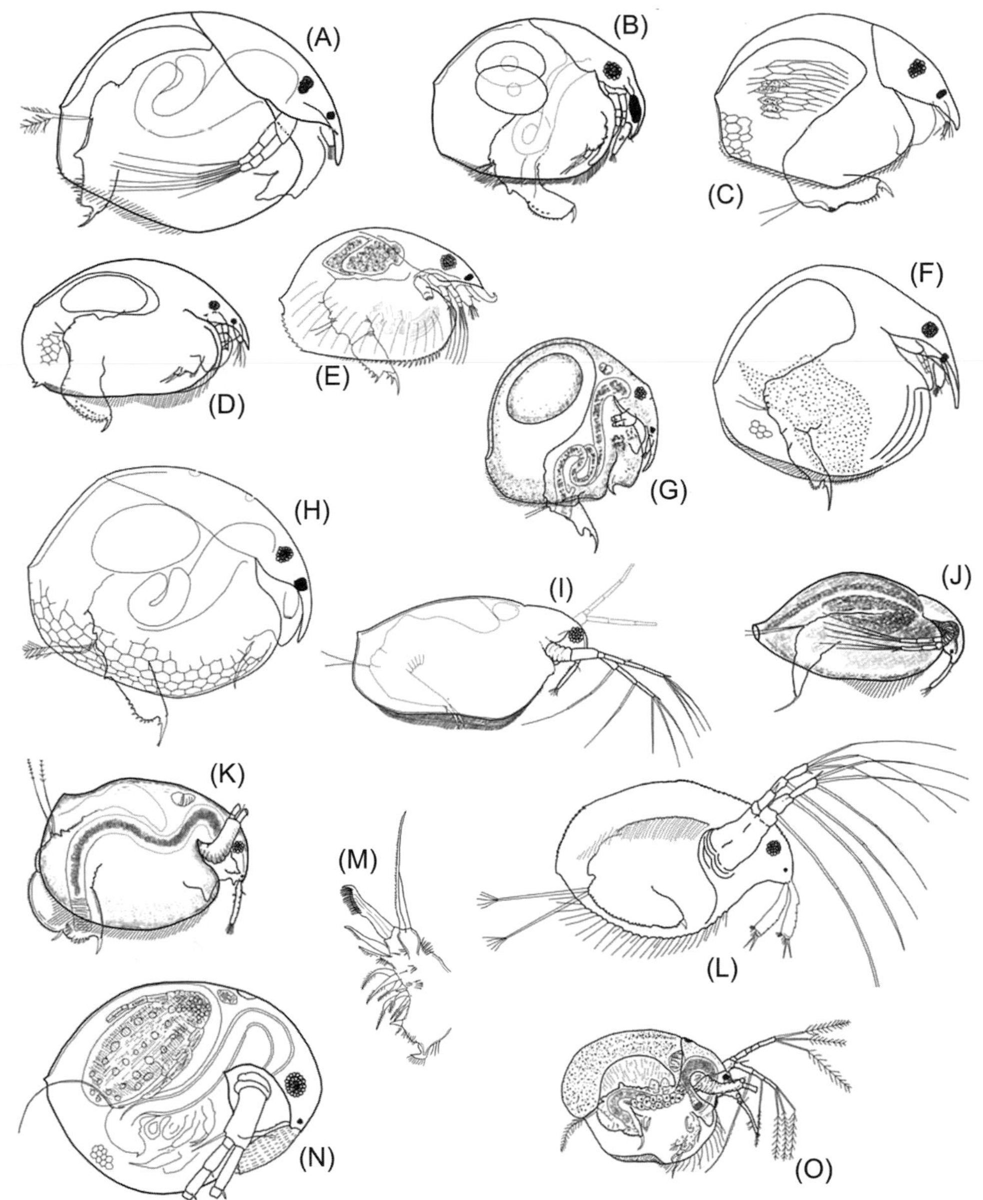

PLATE 16.131 (A) *Ephemeroporus* sp. (B) *Dadaya macrops*. (C) *Alonella* sp. (D) *Dunhevedia* sp. (E) *Pleuroxus* sp. (F) *Pseudochydorus globosus*. (G) *Anchistropus emarginatus*. (H) *Chydorus* sp. (I) *Ophryoxus gracilis*. (J) *Parophryoxus tubulatus*. (K) *Grimaldina brazzai*. (L and M) *Macrothrix* sp. (N) *Guernella raphaellis*. (O) *Bunops serricaudata*.

3(2) Preanal teeth approximately equal size 4

3' Preanal teeth markedly increasing in size proximally .... *Ilyocryptus bernerae* Kotov, Elias-Gutierrez & Williams, 2002
[USA: South Carolina]

4(3) Preanal teeth without close spinules or setules; preanal teeth normally double, or mix of single and double 5

4' Preanal teeth with adjacent spinules or setules; preanal teeth single 7

5(4) Lateral swimming setae with short setules only 6

5' Lateral swimming setae unilaterally armed with strong setules; 8–9 preanal teeth, increasing in length proximally .... *Ilyocryptus gouldeni* Williams, 1978
[Canada: British Columbia. USA: Texas, Alabama, Virginia, Louisiana, Missouri, Maryland, New York. Mexico]

6(5) Antennal exopod, second antennomere, with spine short; preanal teeth numerous (9–12) .... *Ilyocryptus cuneatus* Štifter, 1988
[Canada: British Columbia. USA: Missouri]

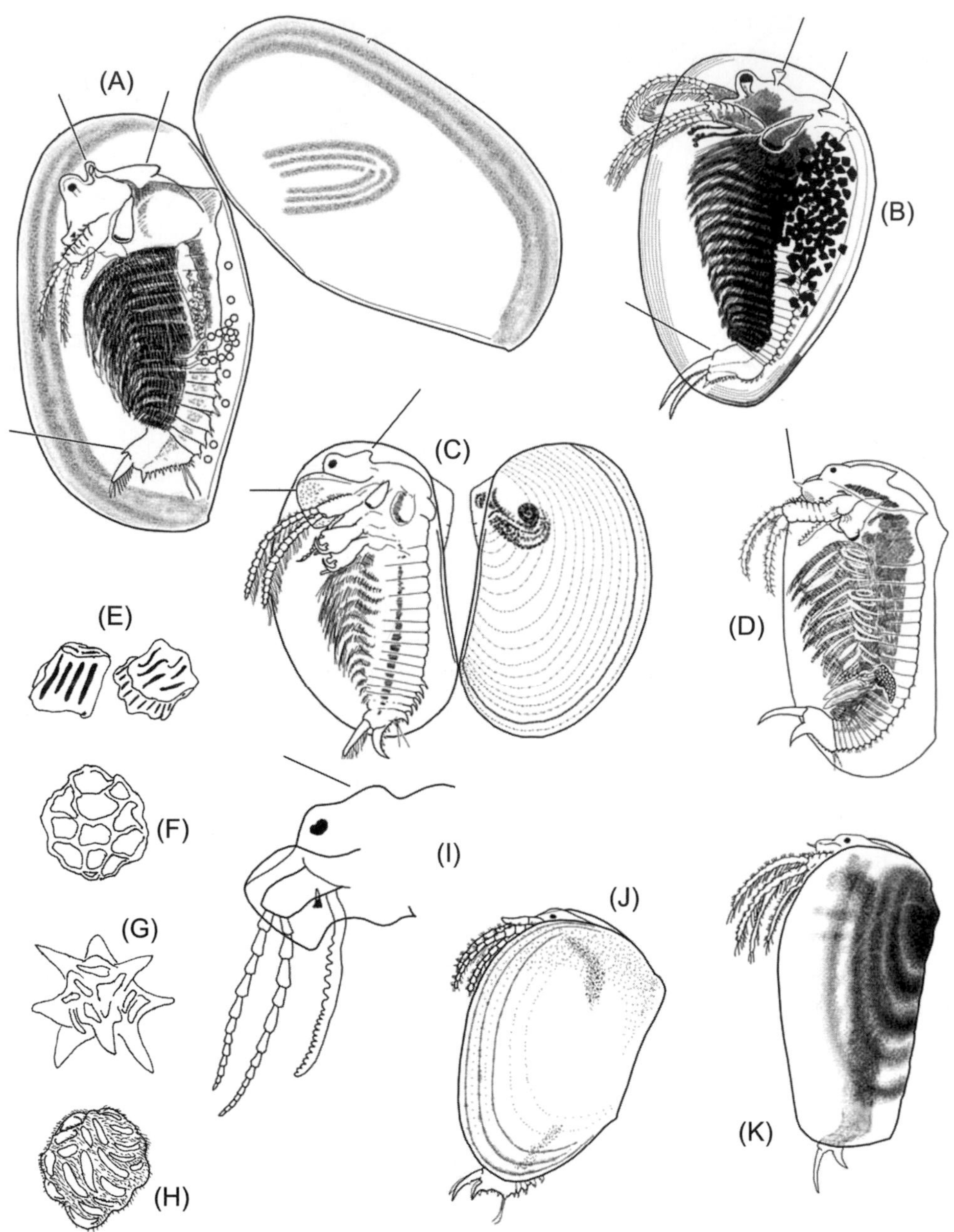

PLATE 16.132 (A) *Eulimnadia* sp., left lateral view, with left carapace valve opened. (B) *Limnadia lenticularis*. (C) *Cyzicus* sp. (D) *Leptestheria compleximanus*, left lateral view, with left carapace valve removed. (E) *Eulimandia texana*, eggs. (F) *Eulimnadia diversa*, eggs. (G) *Eulimandia astraova*, eggs. (H) *Eulimandia agassizii*, eggs. (I) *Eocyzicus digueti*, head, left lateral view. (J) *Eocyzicus digueti*, left lateral view. (K) *Leptestheria compleximanus*, left lateral view.

6' Antennal exopod, second antennomere, with spine long; 5–8 preanal teeth .... *Ilyocryptus silvaeducensis* Kotov & Štifter, 2005
[Canada: Newfoundland]

7(5) Preanal teeth 12–14; postabdominal claw distal and medioventral margins without denticles .... *Ilyocryptus sordidus* (Liévin, 1848)
[Canada? USA? Palaearctic]

7' Preanal teeth numerous, 17–19; postabdominal claw ventral margin with 1–2 denticles distally .... *Ilyocryptus spinosus* Štifter, 1988
[Canada: British Columbia. USA: Washington]

Crustacea: Branchiopoda: Diplostraca: Cladocera: Anomopoda: Daphniidae: Genera

1 First antenna less than half as long as width of head; postabdomen teeth always single 2

1' First antenna about as long as width of head; postabdomen posterior angle (near claw) usually bearing a tooth with two points 7

2(1) Ventral margin of carapace rounded and not pigmented........................ 3

2' Ventral margin of carapace straight and usually black 6

3(2) Adult tail spine, if present, less than three times as long as broad 4

3' Adult with a tail spine, at least three times as long as broad and pointed (Pls. 16.118 A, 129 B) .... ***Daphnia* [p. 465]**

4(3) Second antenna, four antennomere ramus, with fourth (distal) antennomere shorter than the third 5

4' Second antenna, four antennomere ramus, with fourth (distal) antennomere longer than the third (Pl. 16.129 E) .... *Daphniopsis ephemeralis* Schwartz & Hebert, 1987

[Canada: Ontario]

5(4) Second antenna, four antennomere ramus, with second antennomere with an apical spine, about 1/4 as long as second antennomere (Pl. 16.129 F, H) .... ***Simocephalus* [p. 466]**

5' Second antenna, four antennomere ramus without an apical spine (Pl. 16.129 G) .... ***Ceriodaphnia* [p. 467]**

6(2) Head shield dorsally with a median oval plate about 20 μm in diameter, marked by slightly raised edges (Pl. 16.129 I) .... *Megafenestra nasuta* Dumont & Pensaert, 1983

[Canada: USA]

6' Headshield dorsum without such a plate (Pl. 16.129 L) .... ***Scapholeberis* [p. 468]**

7(1) Second antenna, four antennomere ramus with four terminal setae; ocellus absent (Fig. 16.129 K) .... ***Moina* [p. 468]**

7' Second antenna, four antennomere ramus with three setae and a terminal spine shorter than its antennomere; ocellus present (Pl. 16.129 J) .... *Moinodaphnia macleayi* (King, 1853)

[USA: southwest]

Branchiopoda: Diplostraca: Cladocera: Anomopoda: Daphniidae: *Daphnia*: Species

Taxa within *Daphnia* (*Daphnia*) could be partitioned (according to Colbourne & Hebert, 1996) into three species groups: (1) *Daphnia pulex* complex (chromosome # 2n = 24; including *middendorffiana, pulex, pulicaria, latispina, minnehaha, arenata, catawba, melanica, tenebrosa, villosa*), (2) *Daphnia obtusa* complex (including *obtusa, neo-obtusa, pileata, prolata, cheraphila*), and (3) *Daphnia longispina* complex (chromosome # 2n = 20; including *dubia, mendotae, laevis, longiremis, dentifera, thorata, "umbra," parvula, retrocurva, ambigua*).

1 Head shield extends as acute V shape along dorsal ridge, pointing toward posterior; fornices rounded 2

1' Carapace extends anteriorly along dorsal ridge, separating head shield into right and left lobes; fornices pointed extensions of head shield *Daphnia* (*Ctenodaphnia*) 8

2(1) Postabdominal claw with fine teeth decreasing regularly in size from the base to tip, teeth in middle pecten less than 1.5 times as long as those of proximal pecten 3

2' Postabdominal claw teeth in three groups (pectens), with middle teeth of middle pectin at least 1.5 times as long as teeth in proximal pectens .... *Daphnia* (*Daphnia*) 7

3(2) Tips of antennal swimming setae do not reach to base of tail spine; swimming seta arising from base of second antennomere of ramus with 3 antennomeres is as long as other swimming setae and reaches past end of ramus; head variable in shape and size; ocellus present 4

3' Tips of swimming setae of antennae longer than carapace, extending posteriorly past base of tail spine; swimming seta arising from base of second antennomere of ramus with 3 antennomeres is shorter than other swimming setae and does not reach end of ramus; head never higher than wide; ocellus absent .... *Daphnia (Hyalodaphnia) longiremis* (*longispina* group) Sars, 1862

[Canada. USA: Great Lakes region. Holarctic]

4(2) Head always wider than long, never ending in a point 5

4' Dorsally-curving helmet longer than wide and ending in a point .... *Daphnia (Daphnia) retrocurva* Forbes, 1882

[Canada. USA: northern states. Holarctic]

5(4) Body size at first reproduction from 0.5 to 2 mm, normally >1.25 mm; rostrum acute, with an angle of around 40 degrees; anterior and posterior margins of rostrum sometimes nearly parallel proximal to tip 6

5' Body size at first reproduction about 1 mm; rostrum blunt, with an angle of about 60 degrees, and wedge-shaped, never giving the impression that the anterior and posterior margins are nearly parallel proximal to tip .... *Daphnia (Daphnia) parvula* Fordyce, 1901

[Canada: Alberta, Saskatchewan, Manitoba, Ontario, Nova Scotia. USA]

6(5) Setae attached sub-marginally inside carapace ventral margin; setae are curved and hyaline, but extend past the margin .... *Daphnia (Daphnia) obtusa* (*obtusa* group) Kurz, 1874

[Canada. USA: temperate regions]

6' Sub-marginal setae inside of carapace rim are equal in length, not extending beyond carapace margin .... *Daphnia (Daphnia) pulex* (*pulex* group) Leydig, 1860

[Canada. USA: temperate regions]

7(2) Small form (reproducing at ~1 mm in length); head always shorter than wide but during the warm season a point near the midline may be present on adults .... *Daphnia (Daphnia) ambigua* Scourfield, 1947

[Canada: British Columbia, Manitoba, Ontario, New Brunswick, Nova Scotia. USA]

7' Adults begin reproducing at about 1.25 mm to 2.5 mm; head variable in shape, but never with a point sitting on a head that is shorter than wide; head shape is very variable from a low curve to a helmet longer than wide; helmet may be rounded or may come to a point, either near the main axis of the body, or even recurved dorsally .... *Daphnia (Daphnia) mendotae* Birge, 1918

[Canada. USA: temperate regions]

8(1) Head without a helmet (low and rounded) or with a rounded cone-shaped helmet, but not with a sharp point 9

8' Head ending in a sharp point; helmet often several times as long as wide .... *Daphnia lumholtzi* Sars, 1885

[Invasive: Laurentian Great Lakes to southern states. Palaearctic]

9(8) Posterior margin of rostrum with teeth (appearing almost like a "moustache") 10

9' Posterior margin of rostrum without teeth 11

10(9) Anteriodorsal extension of the carapace does not increase in width toward the anterior end; rostrum with 9–12 teeth on each side .... *Daphnia barbata* Weltner, 1898

[Mexico: invasive: Lake Cuitzeo, Marcelo; non-indigenous]

10' Anteriodorsal extension of the carapace doubles in width toward the anterior end; rostrum with 12–15 teeth on each side .... *Daphnia brooksi* Dodson, 1985

[USA: Utah?]

11(9) Dorsal margin of postabdomen not sinuate in lateral view 12

11' Dorsal margin of postabdomen deeply sinuate in lateral view, anal teeth small or absent .... *Daphnia magna* Straus, 1820

[Canada: widespread; USA: northern states. Possible invasive. Holarctic]

12(11) Rostrum short with antennular aesthetascs exceeding tip 13

12' Rostrum longer, antennular aesthetascs reaching but not exceeding tip; tail spine long (greater than half carapace length); ventral and dorsal margin of carapace with stout spinules .... *Daphnia similis* Claus, 1876

[Canada: British Columbia, Alberta, Saskatchewan. USA: California, Nebraska]

13(12) Rostrum blunt, rounded, tip of antennule reaches end of rostrum .... *Daphnia salina* Hebert & Finston, 1993

[Canada: Saskatchewan. USA: western states, Montana, Utah]

13' Rostrum sharply pointed, antennules do not reach end of rostrum .... *Daphnia exilis* Herrick, 1895

[USA: California, Nebraska, New Mexico, Colorado, Oklahoma, Texas]

Branchiopoda: Diplostraca: Cladocera: Anomopoda: Daphniidae: *Simocephalus*: Species

1 Postabdominal claw with a basal pecten either twice or one-half as long as the more distal teeth 2

1' Postabdominal claw with tooth size decreasing evenly from the base to the tip 5

2(1) Postabdominal claw with a basal pecten teeth about twice as long as the more distal teeth; head rounded or pointed, but without teeth at the apical corner 3

2' Postabdominal claw with a basal pecten teeth about half as long as the more distal teeth; head has a right angled anterior corner, often with spines (teeth) near the angle .... *Simocephalus serrulatus* (Koch, 1841)

[Canada. USA. Nearctic]

3(2) Adults with a low rounded mound along the posterior margin of the carapace 4

3' Adults with a pointed triangular tail-like ending of the carapace .... *Simocephalus daphnoides* Herrick, 1883

[USA: Alabama; south into South America]

4(3) Head bulges anterior to eye, evenly rounded (obtuse angle) .... *Simocephalus exspinosus* (DeGeer, 1778)

[Nearctic]

4' Head with nipple-like anterior extension (acute angle, but with a rounded tip) .... *Simocephalus rostratus* Herrick, 1884

[Nearctic]

5(1) Posterior end of the carapace with a tail-like rounded mound 6

5' Posterior end of the carapace is an even curve, no tail-like mound .... *Simocephalus mixtus* Sars, 1903

[Nearctic]

6(5) Ocellus round .... *Simocephalus punctatus* Orlova-Bienkowskaja, 1998

[USA: western California]

6' Ocellus several times longer than wide, a rhomboidal mass with a tail several times as long as the main mass .... *Simocephalus mirabilis* Orlova-Bienkowskaja, 1998

[USA: southern states. Mexico. Neotropical]

Branchiopoda: Diplostraca: Cladocera: Anomopoda: Daphniidae: *Ceriodaphnia*: Species

1 Cervical fenestra absent 2

1' Cervical fenestra present 6

2(1) Along the ventral-posterior margin of the carapace, submarginal microspinules all the same size, without occasional longer microspinules (looking like exclamation points along the line of microspinules) 3

2' Ventroposterior submarginal microspinules with occasional longer microspinules "punctuating" the row of smaller spinules 4

3(2) Carapace either not reticulated, or if reticulated, then the reticulation line nodes have no projecting setae or spines projecting; head never with a forward-projecting horn (observation at different seasons in the year are necessary to confirm absence); fornices may have rather stout straight or hooked pointed spines pointing forward, back, or laterally .... *Ceriodaphnia quadrangula* (O.F. Müller, 1785)

[Canada; USA]

3' Carapace often reticulated, with reticulation line nodes bearing microsetae or spines (usually less than half as long as the width of the reticulation cells); head sometimes with a projection resembling a forward-projecting sharp horn; fornices often with sharp teeth .... *Ceriodaphnia cornuta* Sars, 1885

[USA: southern states]

4(2) Carapace ventral margin with 7 to 9 long marginal setae (difficult to see, but always present); postabdominal claw with teeth decreasing in size evenly 5

4' Carapace ventral margin with 15 long marginal setae near the margin center (about as long as the postabdominal claw); postabdominal claw with teeth in first and second pectens about 1.5 times longer than in the third and distal pectin .... *Ceriodaphnia laticaudata* P.E. Müller, 1867

[USA: Virginia, Texas. Mexico: Nuevo Leon]

5(4) Proximal anal denticles alternating with with long, fine setaform accessory spines; postabdomen dorsal margin smooth (not serrate) proximal to anal denticles; posterioventral carapace with about 40 submarginal microspinules between each of the larger "punctuating" microspinules; fornices smoothly rounded or sometimes with small, laterally pointed teeth ....*Ceriodaphnia pulchella* Sars, 1862

[Canada. USA]

5' Proximal anal denticles lacking long, fine hair-like accessory spines; postabdomen dorsal margin, proximal to anal denticles serrrated; posterioventral carapace submarginal spines with about 20 of the smaller spines between each of the larger "punctuating" microspinules; fornices often extended into sharp points or hooks .... *Cerriodaphnia megops* Sars, 1862

[Canada. USA: northern states]

6(1) Carapace posterioventral margin with spines, submarginal microspinules with the larger punctuating spines less than twice as thick as the smaller spines; postabdominal claw second pecten teeth always about as thick as teeth in the first and third pectens 7

6' Carapace posterioventral margin without spines, some or all larger anterior submarginal punctuating spines are more than twice as thick as the smaller spines; teeth in second pecten of postabdominal claw vary (even within populations and seasonally) from about as thick as teeth in the first and third pectens, to several times thicker and longer 8

7(6) Carapace strongly reticulated, with spines projecting from the nodes of the reticulation lines (projections are usually less than half as long as the width of the reticulation cells); anal denticles nearly straight .... *Ceriodaphnia acanthina* Ross, 1897

[Canada: western provinces to Manitoba; USA: Pacific Northwest to California]

7' Carapace reticulation, if visible, lacking projecting spines; anal denticles curved .... *Ceriodaphnia lacustris* Birge, 1893

[Canada: Alberta eastwards. USA]

8(6) Postabdominal claw, middle pecten varies from about 22 fine spinules slightly longer and heavier than those of the 3rd pecten to 8–14 heavy, prominent, ovate teeth more than twice as long as those of the 3rd pecten; first antenna long and thick with anterior sensory setae arising from distinct peduncle near midpoint; anal denticles nearly straight .... *Ceriodaphnia dubia* Richard, 1894

[Canada. USA]

8' Middle pecten with 2–8 prominent, straight-edged teeth (similar to large-toothed morph of *C. dubia*); first antenna with anterior sense hair nearly terminal; anal denticles curved .... *Ceriodaphnia reticulata* (Jurine, 1820)

[Canada. USA]

Branchiopoda: Diplostraca: Cladocera: Anomopoda: Daphniidae: *Scapholeberis*: Species

1 Carapace distal margin with a conspicuous hyaline membrane; submarginal denticulated membrane 2

1' Denticulated membrane along distal margin of carapace 3

2(1) Carapace and head reticulated but smooth; rostrum triangularly produced in front; depression between eyes and sides of rostrum; second groove along the head behind the implant of antennules .... *Scapholeberis rammneri* Dumont & Pensaert, 1983

[Canada: Alberta]

2' Head large; antennule small, not protruding beyond rostrum; front of rostrum rectilinear; postabdominal pectens composed of numerous spinules of equal size .... *Scapholeberis mucronata* (O.F. Müller, 1776)

[Canada. USA]

3(1) Carapace and head reticulated but smooth; rostrum trilobite with wide angular middle lobe; denticulated membrane marginal .... *Scapholeberis armata* Herrick, 1882

[Canada; USA; Mexico: Chihuahua]

3' Denticulate membrane at posterior margin of carapace markedly thick, with conspicuous underlying hyaline membrane .... *Scapholeberis duranguensis* Quiroz-Vazquez & Elias-Gutierrez, 2009

[Mexico: Durango]

Branchiopoda: Diplostraca: Cladocera: Anomopoda: Daphniidae: *Moina*: Species

1 Postabdomen dorsal margin tooth row with distal tooth bident 2

1' Postabdomen dorsal margin tooth row with distal tooth simple .... *Moina hutchinsoni* Brehm, 1937

[USA: western states. Mexico]

2(1) First thoracic leg last and penultimate articles with marginal plumose setules; ephippium with one or two eggs 3

2' First thoracic leg last and penultimate articles with marginal spines; ephippium with two eggs .... *Moina macrocopa* (Straus, 1820)

[Canada. USA: widespread]

3(2) Head and carapace with long setae 4

3' Head and carapace glabrous 5

4(3) Never more than 1.2 mm long, 1 egg per ephippium .... *Moina affinis* Birge, 1893

[USA: midwest to southeast. Mexico]

4' Some adults longer than 1.2 mm; 2 eggs per ephippium .... *Moina wierzejskii* Richard, 1895

[USA: southwest. Mexico]

5(3) Length less than 1.2 mm; head with supraocular depression; 1 egg per ephippium 6

5' Adults longer than 1.2 mm; head large, with no supraocular depression; 2 eggs per ephippium .... *Moina brachycephala* Goulden, 1968

[USA: California]

6(5) Ocellus absent .... *Moina micrura* Kurz, 1875

[USA: widespread]

6' Ocellus present .... *Moina dumonti* Kotov, Elias-Gutierrez & Granados-Ramirez, 2005

[Mexico: southeast]

Branchiopoda: Diplostraca: Cladocera: Anomopoda: Eurycercidae: *Eurycercus*: Species

1 Carapace broadly rounded, no dorsal keel except in ephippial females; head shield rounded behind 2

1' Carapace with prominent and sharp median dorsal keel; median pore projecting from the head and directed posteriorly with a prominent indentation immediately behind; head shield pointed behind .... subgenus *Eurycercus* .... *Eurycercus microdontus* Frey, 1978

[USA: southeastern]

2(1) Head pore not projecting; intestine with double loop .... subgenus *Teretifrons* 3

[Undescribed *Teretifrons* species exist in eastern and western Arctic regions]

2' Head pore projecting like a bubble; intestine with a single loop .... subgenus *Bullatifrons* 4

3(2) Postabdominal teeth pointed; first antennae shorter than labrum .... *Eurycercus glacialis* Lilljeborg, 1887

[Canada: Nunavut]

3' Postabdominal teeth rounded, deeply pigmented, more than 100; first antennae longer than labrum............ *Eurycercus nigracanthus* Hann, 1990

[Canada: Newfoundland]

4(2') Bubble-like projecting head pore on prominent transverse fold...................................... *Eurycercus beringi* Bekker, Kotov & Taylor, 2012

[USA: Alaska]

4' Bubble-like projecting head pore on flat surface of head shield.. *Eurycercus longirostris* Hann, 1982

[Canada, USA, Mexico]

Crustacea: Branchiopoda: Diplostraca: Cladocera: Anomopoda: Chydoridae: Genera

1 Mandible proximal tip articulates where headshield and carapace touch; midline with two or three major headpores, which usually are connected by a double sclerotized ridge, resembling a channel; carapace free posterior margin not greatly less than maximum height of animal; postabdominal claw with one (sometimes absent) basal spine; postabdomen nearly always with lateral fascicles ....subfamily Aloninae... 2

1' Mandible proximal tip articulates in a special pocket on the headshield, some distance in a dorsal direction from the point of contact between the headshield and carapace; two major headpores on midline, completely separated from one another and with no sclerotized ridge connecting them; minor pores most commonly near midline between major pores; free posterior margin of shell usually considerably less than maximum height of animal; typically two basal spines on postabdominal claw; articulated lateral fascicles on postabdomen lacking .... subfamily Chydorinae 18

2(1) Ocellus present; compound eye absent 3

2' Compound eye and ocellus both present 4

3(2) Carapaces from previous instars firmly nested dorsally; headshield transversely truncate posteriorly; single median headpore (Pl. 16.130 B) .... *Monospilus dispar* Sars, 1862

[Nearctic]

3' Carapaces not nested; two separated median pores without sclerotized connecting ridge; minor pores located far laterally; very short antennae (Pl. 16.130 C) .... *Bryospilus repens* Frey, 1980

[USA: Puerto Rico. Neotropical]

4(2) Body strongly compressed laterally, at times almost wafer thin; postabdominal claw with a secondary spine midway along concave margin, and usually with a comb of fine setules decreasing in length proximally between it and the basal spine 5

4' Body not so strongly compressed; postabdominal claw having only a basal spine of variable length (except in *Euryalona*); sometimes none at all 8

5(4) Carapace strongly keeled, and head also usually strongly keeled 6

5' Carapace weakly keeled, and head without a keel 7

6(5) Postabdomen narrow, elongate, tapered distally, with many (generally about 15 or more) marginal denticles (Pl. 16.130 D) .... *Camptocercus*

[Except for *C. oklahomensis*, the species in N.A. are completely unknown and undescribed, although certainly more than five species are present. *C. rectirostris, C. lilljeborgi*, and *C. macrurus*, which were described from Europe, do not occur in North America.]

6' Postabdomen shorter and broader, parallel sided, without any distinct marginal denticles (Pl. 16.130 E) .... *Acroperus harpae* (Baird, 1834)

[Nearctic]

7(5) Postabdomen rather slender, elongate, somewhat tapered distally; body high; carapace widely open behind (Pl. 16.130 F); dorsal margin of postabdomen extends distal to the base of the claw and basal claw spine .... *Kurzia media* Birge, 1879

[Canada. USA]

7' Postabdomen broader, elongate, parallel sided; carapace with parallel striae sloping downward posteriorly, and with short longitudinal scratch marks in between (Pl. 16.130 G) .... *Alonopsis americana* Kubersky, 1977

[Canada: maritime provinces. USA: New England states]

8(4) Rostrum rounded, barely or only slightly exceeding first antennae in length 9

8' Rostrum long, attenuate, recurved, greatly exceeding first antennae in length (Pl. 16.130 H); postabdomen with 2–4 stout marginal denticles distally and with an almost continuous row of long, setae laterally .... ***Rhynchotalona* [p. 471]**

9(8) Rostrum narrowly rounded; carapace may be weakly reticulated, but not head 10

9' Rostrum very broad in dorsal view, semicircular, wider than body; carapace and head strongly reticulated; carapace ventral margin with a dense setal fringe, and generally with two posterioventral, large, sharp, triangular spines; postabdomen tapered distally, marginal denticles small, lateral fascicles minute (Pl. 16.130 I)... *Graptoleberis testudinaria* (Fischer, 1851)

[Nearctic]

10(9) Postabdomen elongate and rather narrow; with marginal denticles well-developed 11

10' Postabdomen much less elongate (except in *Leydigia*); marginal denticles usually well-developed, but in some species greatly reduced 12

11(10) Postabdomen dorsal margin straight or slightly convex; marginal denticles very short proximally, increasing in length distally to three very long curved denticles; four median headpores (middle one is divided into two), usually completely separated from one another, or with middle pores joined by a sclerotized ridge (Pl. 16.130 J); postabdominal claw basal spine long, slender, attached some distance from base of claw; head and shell generally have a yellowish, hyaline appearance ....*Oxyurella brevicaudis* Michael & Frey, 1983

[Canada: Ontario, east. USA: eastern states. Mexico]

11' Postabdomen dorsal margin distinctly concave; marginal denticles increase in length distal, but never reach large size as in *Oxyurella*; one median headpore, sometimes completely absent (Pl. 16.130 L); postabdominal claw proximal half with a comb of setae on concave margin, increasing in length distally; trunklimb I with a stout spine on distomedial lobe, with coarse, rounded tubercles on concave edge .... *Euryalona orientalis* (Daday, 1898)

[USA: southeastern states. Mexico]

12(10) Postabdomen smaller, and with shorter setae in lateral fascicles; headpores variable 13

12' Postabdomen large, broad, flattened, almost semicircular; armed laterally with long, spiniform setae in groups, those of distal groups projecting far beyond postabdomen margin; postabdomen margin with short, fine spinules; three median headpores close together, and with minor pores close-in laterally (Pl. 16.130 K); carapace free posterior margin very long; ocellus as large as compound eye or larger .... **Leydigia** **[p. 471]**

13(12) Carapace usually not sculptured at all, or if so, then rather weakly; postabdomen usually with well-developed marginal denticles and lateral fascicles; two or three median headpores, most commonly united by a sclerotized ridge, although sometimes completely separated (Pl. 16. 30 N) 14

13' Carapace strongly sculptured, with longitudinal striations in posterior part and about five vertical striations anteriorly, roughly paralleling anterior margin; dorsal edge of postabdomen minutely serrate; about 14 fascicles laterally; no median headpores, but instead there are two comma-shaped thickenings, which may contain the minor pores (Pl. 16.130 M) .... *Notoalona freyi* Rajapaksa & Fernando, 1987

14(13) Carapace valves with posteroventral corners without denticles, or bearing 1–3 denticles, separated by the width of the denticle base or less 15

14' Carapace valves with posteroventral corners bearing 3–5 denticles, distance between them greater than twice the width of denticle base .... *Karualona penuelasi* Dumont & Silva-Briano, 2000

[Mexico]

15(14) Rostrum very blunt, truncated or with a subapical depression in lateral view; three primary head pores 16

15' Rostrum acute in lateral view; two or three primary head pores 17

16(15) Lateral head pores tiny, point-like; antenna I with terminal aesthetasc, not more than 1.5 times longer than terminal setae ....**Leberis** **[p. 471]**

16' Two lateral headpores, flower-shaped; antenna I with terminal aesthetasc, twice as long as terminal setae .... *Nicsmirnovius fitzpatricki* (Chen, 1970)

17(15) Six limb pairs; postabdomen with merged marginal denticles (Pl. 16.130 N) .... **Alona** **[p. 471]**

17' Five limb pairs; postabdomen with separated marginal denticles .... *Coronatella circumfimbriata* (Megard, 1967)

[Nearctic]

18(1) Headshield not elongated posteriorly; if two headpores are present, postpore distance between posterior headpore and headshield posterior margin usually considerably less than interpore distance 19

18' Headshield elongated posteriorly, extending well beyond the heart to the middle of the dorsum; postpore distance usually greater than interpore distance, sometimes by several fold 22

19(18) Zero or one median headpore 20

19' Two median headpores, well separated and not connected by a sclerotized ridge 21

20(19) One median headpore only in first instar; none in later instars (Pl. 16.131 A); postabdomen with long proximal and distal marginal denticles, and with shorter ones in between; labral keel large, with 1–4 teeth on anterior margin .... **Ephemeroporus** **[p. 472]**

20' One median headpore in all instars (Pl. 16.131 B) .... *Dadaya macrops* (Daday, 1898)

[Mexico]

21(19) Postabdomen of more typical chydorid structure, with pre- and postanal angles and a well developed series of postanal marginal denticles (Pl. 16.131 C) ....**Alonella** **[p. 473]**

21' Postabdomen consisting of a much expanded postanal portion, having a series of short spines along dorsal margin and many small clusters of spinules laterally (Pl. 16.131 D) ....**Dunhevedia** **[p. 473]**

22(18) Body elongate, somewhat flattened; carapace posterioventral angle with one or more spines at the posterior-ventral angle of shell (Pl. 16.131 E) 23

22' Body globular, nearly round; carapace posterioventral angle seldom with any spines 24

23(22) Measure the length from the base of the postabdominal claw to the base of the natatory setae. Divide this length by the width of the postabdomen, measured at the distal edge of the anus. Postabdominal length/width ratio 3.7 to 5.2; on the postabdomen dorsal margin, the two distal most spines (near the claw) about twice as long as the more proximal spines .... **Picripleuroxus** **[p. 474]**

23' Postabdominal length/width ratio 2.3 to 3.3; postabdomen dorsal margin with two distal most teeth (near the claw) no longer than more proximal teeth .... ***Pleuroxus* [p. 474]**

23(22) Carapace ventral margin without a hook; carapace marginal setae arise submarginally posteriorly, forming a distinct duplicature 24

23' Carapace ventral margin anteriorly with hook-like projection containing a groove, in which a stout seta with spinules on trunk limb I operates; postabdomen distal angle with a cluster of long, slender setae (Pl. 16.131 G) .... *Anchistropus emarginatus* Sars, 1862

[Canada. USA]

24(23) Postabdomen shorter, with fewer than 15 marginal denticles, and laterally with crescentic clusters of short spinules, which do not resemble fascicles; carapace is clear to light yellow (Pl. 16.131 H) 25

24' Postabdomen elongate, dorsal and ventral sides parallel; postanal portion with ~15 slender marginal denticles; lateral surface with a row of fascicle-like setal clusters; carapace heavily sclerotized, golden in colour (Pl. 16.131 F) .... *Pseudochydorus globosus* (Baird, 1843)

[Canada. USA]

25(24) Carapace lacks a posterioventral spine; carapace ventral setae submarginal, in an arc inside the carapace margin; ephippium with one or two eggs .... ***Chydorus* [p. 473]**

24' Carapace with posterioventral angle with a spine; carapace ventral setae marginal; first antennae with two submarginal aesthetascs (blunt sensory setae); ephippium with two eggs; carapace often dimpled .... *Paralona pigra* (Sars, 1862)

[Canada. USA]

Branchiopoda: Diplostraca: Cladocera: Anomopoda: Chydoridae: *Rhynchotalona*: Species

1 Rostrum ~1.5 times the length of antenna I; posteriomost valve setae short .... *Rhynchotalona weiri* Sinev & Kotov, 2014

[Canada: Manitoba eastward. USA: northeast]

1' Rostrum ~3 times the length of antenna I; posteriomost valve setae long .... *Rhynchotalona longiseta* Sinev & Kotov, 2014

[Canada: Manitoba eastward. USA: northeast]

Branchiopoda: Diplostraca: Cladocera: Anomopoda: Chydoridae: *Leydigia*: Species

Kotov (2009) has revised the genus *Leydigia*; however, morphological features distinguishing *Leydigia leydigi* from *Leydigia louisi* are minute. Characteristics mentioned in Kotov (2003) are inconsistent with those identified in descriptions contained therein and with those used in the latter paper.

1 Postabdominal claw basal spine large (as long as the diameter of the claw at its base) 2

1' Postabdominal claw basal spine absent .... *Leydigia acanthocercoides* (Schödler, 1862)

2(1) Postabdominal lateral setae short, longest less than half length of postabdominal claw .... *Leydigia leydigi* (Schödler, 1862)

[Canada. USA. Holarctic]

2' Postabdominal lateral setae long, longest greater than half length of postabdominal claw .... *Leydigia louisi mexicana* Kotov, Elias-Gutierrez & Nieto, 2003

[Mexico: central plateau. Neotropical. Afrotropical]

Branchiopoda: Diplostraca: Cladocera: Anomopoda: Chydoridae: *Leberis*: Species

1 Carapace posterior margin with internal spinules short, not forming groups .... *Leberis davidi* (Richard, 1895)

[USA: Puerto Rico. Mexico]

1' Posterior margin of carapace with internal spinules forming groups .... *Leberis chihuahuaensis* Sinev & Silva-Briano, 2012

[Mexico: Durango]

Branchiopoda: Diplostraca: Cladocera: Anomopoda: Chydoridae: *Alona*: Species

Much progress has been achieved in the last decade in resolving the polyphyletic genus *Alona* (e.g., Van Damme & Dumont, 2008), with delineation of new genera and species groups, including *Leberis, Coronatella, Karualona*, all of which occur in North America and Mexico.

1 Three median head pores 2

1' Two median head pores 12

2 Carapace posterioventral angle rounded, without spines 3

2' Carapace posterioventral angle with one or more teeth .... *Alona monacantha* Sars, 1901

[Some individuals in a population may lack posterioventral angle teeth. USA: Wisconsin to New England, south to Florida.]

3(2) Labrum posteriorly lacking a row of marginal setae 4

3' Labrum with posterior straight margin with a row of setae; labrum bearing a protrusion (rounded or rectangular), and a straight part posterior to protrusion .... *Alona barbulata* Megard, 1967

[Canada. USA: Central Plains. Mexico.]

4(3) Postabdomen dorsal margin, distal to anal opening, with 12 or fewer spines (or clusters of denticles) 5

4' Postabdomen dorsal margin, distal to anal opening, with about 15–18 broad and separate spines (often with microserrations along the proximal face); lateral head pores are larger than each of the median head pores .... *Alona quadrangularis* (O.F. Müller, 1776)

[Canada: British Columbia, Ontario. USA: Washington, Puerto Rico]

5(4) Lateral head pores larger than the individual medial head pores 6

5' Lateral head pores smaller than the individual head pores; lateral head pores may be indiscernible or absent 8

6(5) Lateral head pores more or less round; carapace valves with long lines 7

6' Lateral head pores elongated (perpendicular to the midline), about as long as the line of three median head pores; lateral head pores without yellow chitinous thickenings; carapace valves indistinctly reticulate, sometimes with tubercles (parts of the headshield and carapace melanized to a golden colour); postabdomen with ~11 patches of lateral fascicles (bundles of microspinules) with the middle spinules longest .... *Alona bicolor* Frey, 1965

[Canada. USA. Eastern North America.]

7(6) Lateral head pores with a yellow chitinous thickening, lateral pore diameter almost half as long as the line of three median pores; postabdomen with ~5 lateral fascicle patches, with middle setules longest .... *Alona costata* Sars, 1862

[Canada. USA.]

7' Lateral head pores with a straight side about as long as one of the median pores and without a chitinous thickening .... *Alona rustica* Scott, 1895

[Canada: Ontario. USA: New York, Minnesota]

8(5) Carapace ventral marginal setae plumose, especially near the posterioventral angle 9

8' Carapace ventral marginal setae simple .... *Alona poppei* Richard, 1897

[USA: Puerto Rico, Texas, Wisconsin]

9(8) Postabdomen lateral fascicles made up of distinct microspinules, the fascicle bases not sclerotized 10

9' Postabdomen lateral fascicles indistinct, although fascicles bases may be sclerotized .... *Alona guttata* Sars, 1862

[Canada. USA.]

10(9) Head shield posterior margin with notches, and some reticulation 11

10' Head shield posterior margin rounded, without notches or a point, lacking reticulation; median head pores are not completely connected with the sclerotized ridge .... *Alona setulosa* Megard, 1967

[Canada. USA.]

11(10) Head shield posterior margin (viewed from above) rounded, with notches and reticulations; postanal corner prominent .... *Alona lapidicola* Chengalath & Hann, 1981

[Canada: Alberta]

11' Head shield posterior margin (viewed from above) pointed and reticulated; postabdomen without postanal corner .... *Alona borealis* Chengalath & Hann, 1981

[Canada: Saskatchewan]

12(1) Postabdomen postanal portion much longer than anal portion; postabdomen dorsal margin with ~13 spines; carapace posterioventral angle punctuated with submarginal setae .... *Alona affinis* (Leydig, 1860)

[Canada: Ontario. USA: Oregon, Puerto Rico, Wisconsin. Mexico]

12' Postabdomen postanal portion short, expanded; postabdomen dorsal margin with fewer than 12 postabdominal teeth .... *Alona intermedia* Sars, 1862

[Nearctic]

Crustacea: Branchiopoda: Cladocera: Anomopoda: Chydoridae: *Ephemeroporus*: Species

1 Carapace posterioventral angle with spines 2

1' Carapace posterioventral angle without spines; labrum with one tooth .... *Ephemeroporus acanthodes* Frey, 1982

[USA: Louisiana]

2(2) Carapace posterioventral angle with one spine .... *Ephemeroporus hybridus* (Daday, 1905)
[USA: Florida]

2' Carapace posterioventral angle with 3–5 long spines .... *Ephemeroporus archboldi* Frey, 1982
[USA: Florida]

Branchiopoda: Diplostraca: Cladocera: Anomopoda: Chydoridae: *Alonella*: Species

The distinctness of *Disparalona* was questioned by Michael & Frey (1984) on the basis of increased knowledge of trunk limb morphology. Similarities among members of *Alonella* appear to outweigh differences among putative members of *Disparalona* but merit further detailed analysis.

1 Carapace with polygonal, elongated reticulations 2

1' Carapace subglobular; valve striation strongly expressed, curving from anterioventral margin to posteriodorsal region; carapace posterioventral corner with one small tooth .... *Alonella nana* (Baird, 1843)
[Canada. USA.]

2(1) Rostrum and first antennae subequal 3

2' Rostrum much longer than first antennae 5

3(2) Carapace posterioventral angle with 1–3 indentations 4

3' Carapace posterioventral angle with 2–6 small spines; postabdomen with 6–8 small sharp postanal spines .... *Alonella pulchella* Herrick, 1884
[Canada. USA.]

4(3) Carapace polygons with internal striae .... *Alonella excisa* (S. Fischer, 1854)
[Canada. USA.]

4' Carapace polygons without internal striae .... *Alonella exigua* (Lilljeborg, 1853)
[Canada. USA.]

5(2) Postabdomen much longer than wide, many postanal spines 6

5' Postabdomen short .... *Alonella dadayi* Birge, 1910
[Canada. USA.]

6(5) Rostrum very long, slender, recurved .... *Alonella acutirostris* (Birge, 1879)
[Canada. USA.]

6' Rostrum not recurved; postabdomen with 9–11 postanal spines .... *Alonella leei* Chien, 1970
[Canada. USA.]

Branchiopoda: Diplostraca: Cladocera: Anomopoda: Chydoridae: *Dunhevedia*: Species

1 Labral plate with a smooth rounded bump, no spines .... *Dunhevedia crassa* King, 1853
[Canada. USA]

1' Labral plate with pointed spines .... *Dunhevedia americana* Rajapaksa & Fernando, 1987
[USA: southern states]

Branchiopoda: Diplostraca: Cladocera: Anomopoda: Chydoridae: *Chydorus*: Species

1 Carapace and headshield covered with honeycomb-like polygons 2

1' Carapace and headshield without honeycomb-like polygons, carapace without ornamentation, or covered with a polygon mesh pattern of lines 4

2(1) Carapace polygons elongated laterally into horns (wings) near the base of the headshield 3

2' Carapace without lateral horn-like projections .... *Chydorus faviformis* Birge, 1893
[Canada: Manitoba, Ontario, Nova Scotia, Newfoundland. USA]

3(2) Carapace polygons elongated laterally into pointed horns (wings) in dorsal view .... *Chydorus bicornutus* Doolittle, 1909
[Canada: Ontario, New Brunswick, Nova Scotia, Newfoundland. USA: Maine]

3' Carapace polygons elongated along the head-carapace junction, to form a collar ....*Chydorus bicollaris* Frey, 1982

[Canada. USA: North Carolina]

4(2) Second antennal exopod, basal antennomere lacking a seta 5

4' Second antennal exopod, basal antennomere with a seta; carapace golden, sclerotized 6

5(4) Rostrum apex sharply or weakly emarginate; postabdominal claw subapical flagellum prominent 7

5' Rostrum apex not emarginate; postabdominal claw subapical flagellum indistinct or absent 8

6(4) Labrum short, apically blunt; headpores closely spaced; ephippium with one egg .... *Chydorus canadensis* Chengalath & Hann, 1981

[Canada: Ontario]

6' Labrum long; headpores widely spaced; ephippium with two eggs ....*Chydorus ovalis* Kurz, 1875

[Canada. USA]

7(5) Rostrum apex sharply emarginate; few polygons on carapace .... *Chydorus brevilabris* Frey, 1980

[Canada. USA]

7' Rostrum apex weakly emarginate; entire carapace strongly reticulated .... *Chydorus linguilabris* Frey, 1982

[Canada: Nova Scotia, Newfoundland. USA: North Carolina, Florida]

8(5) Postabdominal claw with subapical flagellum weakly developed 9

8' Postabdominal claw without a subapical flagellum; golden color; postabdomen short, broad, with prominent preanal angle ....*Chydorus gibbus* Sars, 1890

[Canada. USA]

9(8) Ephippium with one egg .... *Chydorus sphaericus* (O.F. Müller, 1776)

[Canada. USA]

9' Ephippium with two eggs .... *Chydorus biovatus* Frey, 1985

[Canada: Alberta. USA: Montana, Washington, Alaska]

Crustacea: Branchiopoda: Cladocera: Anomopoda: Chydoridae: *Picripleuroxus*: Species

1 Postabdomen dorsal margin with more than 20 postanal denticles 2

1' Postabdomen dorsal margin with fewer than 15 postanal denticles 3

2(1) Carapace smooth, without longitudinal lines.. *Picripleuroxus denticulatus* (Birge, 1879)

[Canada. USA]

2' Carapace with fine striations.. *Picripleuroxus striatus* (Schödler, 1862)

[Canada: Newfoundland, Nova Scotia, west to British Columbia, NWT. USA]

3(1) Carapace with tiny scratches; carapace posterioventral corner with a right angle spine .... *Picripleuroxus straminius* (Birge, 1879)

[Canada: eastern provinces. USA: eastern states]

3' Carapace without tiny scratches all over; carapace with a rounded spine anteriad of posterioventral corner .... *Picripleuroxus chiangi* (Frey, 1988)

[Canada: British Columbia to Nova Scotia. USA: Minnesota]

Branchiopoda: Diplostraca: Cladocera: Anomopoda: Chydoridae: *Pleuroxus*: Species

1 Rostrum curved toward the ventral and posterior 2

1' Rostrum curved anteriorly, hook-shaped, directed dorsally; carapace with 0–8 denticles, starting at the posterioventral angle, extending along the posterior margin .... *Pleuroxus procurvus* Birge, 1879

[Canada. USA]

2(1) Postabdomen dorsal margin with denticles increasing in size regularly toward apex; carapace posterioventral corner denticles directed ventroposteriorly .... *Pleuroxus trigonellus* (O.F. Müller, 1776)

[Canada: Newfoundland, Quebec, Ontario, Manitoba, Saskatchewan, Alberta. USA]

2' Postabdomen dorsal margin with denticles increasing in size regularly toward apex, but are grouped into fascicles, except for the last 4 or so, which are separate; carapace posterioventral corner denticles directed posteriorly .... *Pleuroxus aduncus* (Jurine, 1820)

[Canada: Manitoba, Saskatchewan, Alberta. USA]

Crustacea: Branchiopoda: Diplostraca: Cladocera: Anomopoda: Ophryoxidae: Genera

1 Antennal setae of females 0-0-0-3/1-1-3 (Pl. 16.131 I) .... *Ophryoxus gracilis* Sars, 1862

[Canada. USA]

1' Antennal setae of females 0-0-0-3/0-0-3 (Pl. 16.131 J) .... *Parophryoxus tubulatus* Doolittle, 1909

[USA: New England]

Crustacea: Branchiopoda: Diplostraca: Cladocera: Anomopoda: Macrothricidae: Genera

1 Second antenna ramus with four antennomeres with 4 setae (0-0-1-3) 2

1' Second antenna ramus with four antennomeres with 2, 3, or 5 setae 4

2(1) Intestine not looped (convoluted) 3

2' Intestine looped (convoluted); postabdomen bilobed, preanal lobe with serrations .... *Streblocerus serricaudatus* (Fischer, 1849)

[Canada. USA]

3(2) Postabdomen dorsal margin bilaterally compressed; postanal margin with widely splayed spines distally; proximal side of anal opening with a single large spine (Fig. 16.131 K) .... *Grimaldina brazzai* Richard, 1892

[USA: Florida]

3' Postabdomen dorsal margin without a long spine; antennal seta formula (0-0-1-3/1-1-3) (Pl. 16.131 L, M) .... ***Macrothrix*** **[p. 475]**

4(1) Second antenna ramus with four antennomeres with 5 (0-1-1-3) setae 5

4' Second antenna ramus with four antennomeres with 2 (0-1-0-1) or 3 (0-0-0-3) setae 6

5(4) First antennae short, thick, with several transverse rows of setules; small body size (to 0.4 mm) (Pl. 16.131 M) .... *Guernella raphaellis* Richard, 1892

[USA: Florida]

5' First antennae long, slender .... *Lathonura rectirostris* (O.F. Müller, 1785)

[Canada. USA]

6(4) Second antenna ramus with four antennomeres with 3 setae 7

6' Second antenna ramus with four antennomeres with 2 setae .... *Wlassicsia kinistinensis* Birge, 1910

[Canada: Manitoba]

7(6) Carapace with dorsal keel; Crapace dorsal margin smooth, no spine; postabdomen preanal margin serrated (Pl. 16.131 O) .... *Bunops serricaudata* (Daday, 1884)

[Canada. USA]

7' Carapace with moderate dorsal keel; dorsal carapace margin with a spine behind cervical depression; postabdomen preanal margin serrated .... *Drepanothrix dentata* (Eurén, 1861)

[Canada. USA]

Crustacea: Branchiopoda: Cladocera: Anomopoda: Macrothricidae: *Macrothrix*: Species

1 First antenna post margin apex with a spinule cluster in an indentation just below terminal sensory setae; up to two transverse rows of spinules near apex; first antennae about 1.5 times wider at apex as at base 2

1' First antenna with either stout spines (at least 1/4 as long as first antenna width) or at least two transverse rows of microspinules; postabdomen dorsal margin spines longest toward natatory setae or toward anus 3

2(1) Carapace with a strong dorsal keel; carapace dorsal margin serrated; valves reticulated and tumid laterally; postabdomen dorsal margin spines longest toward middle (not including larger teeth near anus); seta natatoria with naked proximal article .... *Macrothrix laticornis* (Fischer, 1851)

[Canada. USA]

2' Carapace with a weak dorsal keel; carapace dorsal margin with fine serrations; no reticulations; seta natatoria with proximal article unilaterally setulated .... *Macrothrix sierrafriatensis* Silva-Briano, Dieu & Dumont, 1999

3(1) First antenna with setaform, transverse rows of spinules present (typically not in clusters and thicker than spinules); first antenna with 2–8 rows of ca. 5 spinules running across anterior surface 4

3' First antenna with stout spines near apex just below sensory setae, and along anterior margin; first antenna transverse spinule rows present or absent 5

4(3) Carapace dorsal keel with sharp, sclerotized spine; carapace dorsal margin posterior to spine finely serrate; basal antennomere with 2–7 sclerotized spines .... *Macrothrix mexicanus* Ciros-Perez, Silva-Briano & Elias-Gutierrez, 1996

[USA: Oregon, California. Mexico: central highlands]

4' Carapace without dorsal keel, dorsal margin finely serrated; basal antennomere without spines; carapace never with a dorsal spine; first antenna with ~2 transverse spinule rows near tip .... *Macrothrix spinosa* King, 1853

[Mexico: Sonora, Durango, Coahuila, Nueva Leon]

5(3) First antenna with 4 or more transverse spinule rows 6

5' First antenna with spinules but lacking transverse rows .... *Macrothrix paulensis* Sars, 1900

[USA: Florida. Mexico]

6(5) First antenna with 7–8 transverse rows of spinules; carapace dorsal margin smooth; postabdomen with "heel" at base of natatory setae .... *Macrothrix smirnovi* Ciros-Perez & Elias-Gutierrez, 1997

[USA: Oregon, California. Mexico: central highlands]

6' First antenna with ~4 transverse rows of spinules; carapace dorsal margin not serrated, no heel .... *Macrothrix agsensis* Dumont, Silva-Briano & Babu, 2002

[Mexico: Chihuahua, Durango, central highlands]

Crustacea: Branchiopoda: Diplostraca: Spinicaudata: Families

1 Frontal organ not produced on a peduncle (Pl. 16.132 C, D, I) 2

1' Frontal organ pedunculate (Pls. 16.119 E, 132 A, B) .... **Limnadiidae [p. 476]**

2(1) Rostrum lacking an apical spine (Pl. 16.132 C, I) .... **Cyzicidae [p. 477]**

2' Rostrum with a single apical spine (may be broken off, look for scar) (Pl. 16.132 D, K) .... Leptestheriidae, one species *Leptestheria compleximanus* (Packard, 1880)

[USA: Great Plains, southwest deserts. Mexico: northern deserts]

Crustacea: Branchiopoda: Diplostraca: Spinicaudata: Limnadiidae: Genera

1 Telson with a ventral spiniform projection just anteriad of cercopod base (Pl. 16.132 A) .... ***Eulimnadia* [p. 476]**

1' Telson without a ventral spiniform projection (Pl. 16.132 B) .... *Limnadia lenticularis* (Linneaus, 1761)

[USA: New England south to Florida, east of the Appalachian Mountains]

Crustacea: Branchiopoda: Diplostraca: Spinicaudata: *Eulimnadia*: Species

The *Eulimnadia* species are determined based on egg morphology (Belk, 1989; Rabet, 2010).

1 Egg subspherical (Pl. 16.132 F–H) 2

1' Egg cylindrical (Pl. 16.132 E) 7

2(1) Egg without elongated projections (Pl. 16.132 F, H) 3

2' Egg with elongated projections (Pl. 16.132 G) .... *Eulimnadia astraova* Belk, 1989

[USA: Louisiana]

3(2) Eggs with round depressions (Pl. 16.132 F, G) 4

3' Egg with slits (Pl. 16.132 H) 5

4(3) Eggs with short pilosity .... *Eulimandia antlei* Mackin, 1940

[USA: Arizona, New Mexico, Oklahoma, Texas]

4' Eggs smooth .... *Eulimnadia diversa* Mattox, 1937

[USA: Colorado, New Mexico, Oklahoma, Texas. Mexico: "Northern Mexico"]

5(3) Eggs glabrous 6

5' Eggs with short pilosity (Pl. 16.132 H) .... *Eulimnadia agassizii* (Packard, 1871)

[USA: New England]

6(5) Slits rectilinear, width uniform to truncated ends .... *Eulimnadia graniticola* Rogers, Weeks & Hoeh, 2010

[USA: Georgia, Florida]

6' Slits narrow, with ends widened, broader than width of slit; slit ends rounded .... *Eulimnadia follisimilis* Pereira & García, 2001

[USA: Colorado, New Mexico. Neotropical]

7(1) Cylinders with lateral edges of round ends produced, shaped like a spool .... *Eulimnadia cylindrova* Belk, 1989

[USA: Arizona, New Mexico. Mexico: Baja California Sur, Chiapas, Chihuahua, Coahuila, Jalisco, San Luis Potosí, Sinaloa, Sonora, Tamaulipas]

7' Cylinders with edges of rounded ends smooth, not produced, shaped like a barrel .... *Eulimnadia texana* (Packard, 1871)

[USA: arid west, Great Plains, Florida, Louisiana. Mexico: Baja California (Norte), Baja California Sur, Chihuahua, Coahuila, Durango, Jalisco, Morelos, Nuevo León, Oaxaca, San Luis Potosí, Tamaulipas. Neotropical]

Crustacea: Branchiopoda: Diplostraca: Spinicaudata: Cyzicidae: Genera

1 Occipital condyle a low, broadly rounded mound, no notch present between head posteriodorsal surface and thorax (Pl. 16.132 I, J) .... *Eocyzicus digueti* (Richard, 1895)

[USA: Arizona, California, Colorado, Kansas, Nevada, New Mexico, Oklahoma, Texas. Mexico: Baja California Sur, Chihuahua, Coahulia, Durango, Estado de Mexico, Puebla, Sonora]

1' Occipital condyle acute, creating a deep notch between head posteriodorsal surface and thorax (Pl. 16.132 C) .... *Cyzicus*

[Taxonomy confused, genus in need of revision. Canada: Alberta, British Columbia, Manitoba, Saskatchewan. USA: widely distributed west of the Mississippi, and Great Lakes region. Mexico: Aguascalientes, Baja California (Norte), Chihuahua, Coahuila, Distrito Federal, Durango, Estado de Mexico, Hidalgo, San Luis Potosí, Sonora, Puebla, Zacatecas.]

REFERENCES

Belk, D. 1989. Identification of species in the conchostracan genus *Eulimnadia* by egg shell morphology. Journal of Crustacean Biology 9: 115–125.

Brendonck, L., D. C. Rogers, J. Olesen, S. Weeks & R. Hoeh. 2008. Global diversity of large branchiopods (Crustacea: Branchiopoda) in fresh water. Hydrobiologia 595:167 – 176.

Colbourne, J.K. & P.D.N. Hebert. 1996. The systematics of North American *Daphnia* (Crustacea: Anomopoda): a molecular phylogenetic approach. Phil. Trans. R. Soc. Lond. B 351: 349–360.

Kotov, A.A. 2003. Separation of *Leydigia louisi* Jenkin, 1934 from *L. leydigi* (Schoedler, 1863) (Chydoridae, Anomopoda, Cladocera). Hydrobiologia 490: 147–168.

Kotov, A.A., M. Elias-Gutierrez & M. G. Nieto. 2003. *Leydigia louisi louisi* Jenkin, 1934 in the Neotropics, *L. louisi Mexicana* n. subsp. In the Central Mexican highlands. Hydrobiologia 510: 239–255.

Kotov, A.A. 2009. A revision of *Leydigia* Kurz, 1875 (Anomopoda, Cladocera, Branchiopoda), and subgeneric differentiation within the genus. Zootaxa 2082: 1–84.

Kotov, A.A. & P. Štifter. 2006. Cladocera: Family Ilyocryptidae (Branchiopoda: Cladocera: Anomopoda). Guides to the Identification of the Microinvertebrates of the Continental Waters of the World. H.J. Dumont, editor.

Martin, J. W. & D. Belk. 1988. A review of the clam shrimp family Lynceidae Stebbing, 1902 (Branchiopoda: Conchostraca), in the Americas. Journal of Crustacean Biology 8(3):451–482.

Michael, R.G. & D.G. Frey. 1984. Separation of *Disparalona leei* (Chien, 1970) in North America from *D. rostrata* (Koch, 1841) in Europe (Cladocera, Chydoridae). Hydrobiologia 114: 81–108.

Rabet, N. 2010. Revision of the egg morphology of *Eulimnadia* (Crustacea, Branchiopoda, Spinicaudata). Zoosystema 32:373–391.

Rogers, D.C. 2001. Revision of the Nearctic *Lepidurus* (Notostraca). Journal of Crustacean Biology 21: 991 – 1006.

Rogers, D. C. 2009. Branchiopoda (Anostraca, Notostraca, Laevicaudata, Spinicaudata, Cyclestherida). *In*: Likens, G. F. (editor) Encyclopedia of Inland Waters 2: 242–249.

Van Damme, K. & H.J. Dumont. 2008. Further division of *Alona* Baird, 1843: separation and position of *Coronatella* Dybowski & Grochowski and *Ovalona* gen. n. (Crustacea: Cladocera). Zootaxa 1960: 1–44.

Class Ostracoda

Alison J. Smith

Department of Geology, Kent State University, Kent, OH, USA

David J. Horne

School of Geography, Queen Mary University of London, London, United Kingdom

INTRODUCTION

This section of Chapter 16 focuses on the Order Podocopida, which contains all the extant non-marine ostracode taxa, comprising members of the superfamilies Cypridoidea, Darwinuloidea, and Cytheroidea. Ostracodes are perhaps best known for their distinctive bivalved carapace, thinly or heavily calcified, which is molted and replaced eight times as the animal grows to maturity. Ostracode adult sizes generally range from 0.5 mm to 2–3 mm in length, with a few non-marine taxa reaching 8 mm.

Non-marine ostracodes are found throughout the Nearctic in most aquatic systems ranging from lakes, streams and springs to wetlands, bromeliads, and oxygenated aquifers. Ostracode distributions include endemic and regional taxa, as well as Holarctic and cosmopolitan taxa.

Much remains to be investigated in North America and Greenland regarding ostracode distributions and biogeography. Most collections have centered on the North American midcontinent within the U.S. and Canada. The North American Combined Ostracode Database "NACODE" (Curry et al., 2012) is a rich source of information on species biogeography (Pl. 16.133), but regions remain to be explored in a systematic way. For example, only limited surveys have been made of the fauna in the Arctic, in eastern Canada and the United States, and in large areas of the southern USA. Even so, we know that the Nearctic contains at least 48 genera of non-marine ostracodes (Martens & Savatenalinton, 2011), that some regional species have extensive fossil records that extend back into the Pliocene (Swain, 1999), and that other species are widely distributed throughout the Northern Hemisphere. Hoff (1944) speculated on the origin of the Nearctic ostracode fauna, noting the overwhelming presence of a Northern Hemisphere Holarctic (i.e., Palearctic and Nearctic) cohort of species, and smaller cohorts originating in the Caribbean and in regional endemism. In general, this assessment is still valid, although a larger cohort of regionally endemic species is now apparent. A taxonomic list of the Nearctic genera and common species indicated in these keys is presented in Table 16.1, which also notes co-occurrences in the Palearctic bioregion.

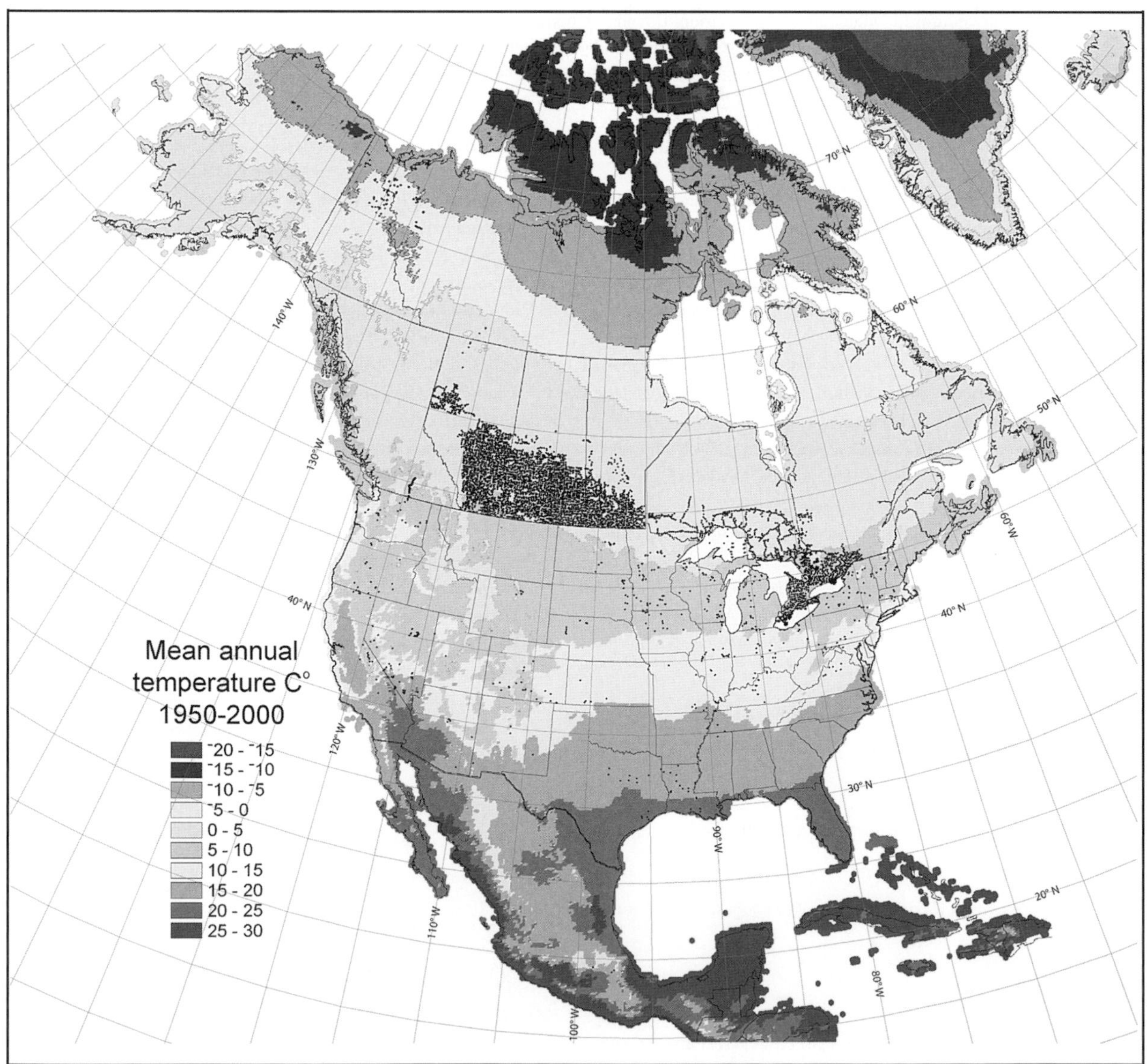

PLATE 16.133 Distribution of samples represented in the The North American Combined Ostracode Database "NACODE," including the Delorme Ostracode Autecological Database housed in the Canadian Museum of Nature, and the North American Non-marine Ostracode Database, housed at Kent State University. *Modified from Figure 6.1 a in Curry et al., 2012.*

TABLE 16.1 Common Non-marine Species of Ostracoda in the Nearctic Ecoregion[a]
Class Ostracoda Latreille, 1806
Subclass Podocopa G.W. Müller, 1894
Order Podocopida Sars, 1866
Suborder Cypridocopina Baird, 1845
Superfamily Cypridoidea Baird, 1856
Family Cyprididae Baird, 1845
Subfamily Cyprettinae Hartmann, 1971
Cypretta Vavra, 1895
Cypretta brevisaepta brevisaepta Furtos, 1934[b]
Subfamily Cypricercinae McKenzie, 1971
Bradleystrandesia Broodbakker, 1983
Bradleystrandesia reticulata (Zaddach, 1854)[b,c]
Spirocypris Sharpe, 1903
Spirocypris horridus (Sars, 1926)
Spirocypris tuberculata Sharpe, 1908[b]
Strandesia Stuhlmann, 1888
Strandesia canadensis (Sars, 1926)[b,c]
Subfamily Cypridinae Baird, 1845
Chlamydotheca Saussure, 1858
Chlamydotheca arcuata (Sars, 1901)[b,c,e]
Cypris O.F. Müller, 1776
Cypris pubera O.F. Müller, 1776[b,c]
Subfamily Cypridopsinae Kaufmann, 1900
Cavernocypris Hartmann, 1964
Cavernocypris wardi Marmonier, Meisch & Danielopol, 1989[b]
Cypridopsis Brady, 1867
Cypridopsis vidua (O.F. Müller, 1776)[b,c]
Cypridopsis okeechobei Furtos, 1936[b]
Potamocypris Brady, 1870
Potamocypris pallida Alm, 1914[b,c]
Potamocypris smaragdina (Vávra, 1891)[b,c]
Potamocypris unicaudata Schäfer, 1943)[b,c]
Sarscypridopsis McKenzie, 1977
Sarscypridopsis aculeata (Costa, 1847)[b,c]
Subfamily Cyprinotinae Bronshtein, 1947
Cyprinotus Brady, 1886
Cyprinotus newmexicoensis Ferguson, 1967
Heterocypris Claus, 1892

Continued

TABLE 16.1 Common Non-marine Species of Ostracoda in the Nearctic Ecoregion[a]—cont'd

Heterocypris incongruens (Ramdohr, 1808)[b,c]
Heterocypris glaucus (Furtos, 1933)[b]
Subfamily Dolerocypridinae Triebel, 1961
Dolerocypris Kaufmann, 1900
Dolerocypris fasciata (O.F. Müller, 1776)[b,c]
Dolerocypris sinensis Sars, 1903[c]
Subfamily Eucypridinae Bronshtein, 1947
Cypriconcha Sars, 1926
Cypriconcha ingens Delorme, 1967[b]
Eucypris Vavra, 1891
Eucypris crassa (O.F. Müller, 1785)[c]
Eucypris virens (Jurine, 1820)[c]
Candocyprinotus Delorme, 1970 (may be junior synonym of *Eucypris*)
Candocyprinotus ovatus Delorme, 1970[b]
Prionocypris Brady & Norman, 1896
Prionocypris longiforma Dobbin, 1941
Tonnacypris Diebel & Pietrzeniuk, 1975
Tonnacypris glacialis glacialis (Sars, 1890)[b,c,d]
Subfamily Herpetocypris Brady & Norman, 1889
Herpetocypris Brady & Norman, 1889
Herpetocypris brevicaudata Kaufmann, 1900[c]
Herpetocypris reptans (Baird, 1835)[b,c]
Stenocypris Sars, 1889
Stenocypris major major (Baird, 1859).[b,c,e]
Subfamily Isocypridinae Hartmann & Puri, 1974
Isocypris G.W. Müller, 1908
Isocypris beauchampi (Paris, 1920)[b,c]
Subfamily Pelocypridinae Triebel, 1962
Pelocypris Klie, 1939
Pelocypris tuberculatum (Ferguson, 1967)[b,e]
Subfamily Scottiinae Bronshtein, 1947
Scottia Brady & Norman, 1889
Scottia pseudobrowniana Kempf, 1971[b,c]
Family Candonidae Kaufmann, 1900
Subfamily Candoninae Kaufmann, 1900
Candona Baird, 1845
Candona acuta Hoff, 1942[b]
Candona acutula Delorme, 1967[b]

TABLE 16.1 Common Non-marine Species of Ostracoda in the Nearctic Ecoregion[a]—cont'd

Candona candida (O.F. Müller, 1776)[b,c]
Candona crogmaniana Turner, 1894[b]
Candona decora Furtos, 1933[b]
Candona distincta Furtos, 1933[b]
Candona elliptica Furtos, 1933[b]
Candona ikpikpukensis Swain, 1963[b,d]
Candona neglecta Sars, 1897[b,c]
Candona ohioensis Furtos, 1933[b]
Candona subtriangulata Benson & MacDonald, 1963[b]
Fabaeformiscandona Krstic, 1972 (included in key as *Candona s.l.*)
Fabaeformiscandona acuminata (Fischer, 1854)[b,c]
Fabaeformiscandona caudata (Kaufman, 1900)[b,c]
Fabaeformiscandona patzcuaro (Tressler,1954)[b]
Fabaeformiscandona protzi (Hartwig, 1898)[b,c,d]
Fabaeformiscandona rawsoni (Tressler, 1957)[b,c]
Nannocandona Ekman, 1914
Nannocandona sp. unnamed
Paracandona Hartwig, 1899
Paracandona euplectella (Robertson, 1889)[b,c]
Pseudocandona Kaufmann, 1900 (included in key as *Candona s.l.*)
Pseudocandona albicans (Brady, 1864)[b,c]
Pseudocandona stagnalis (Sars, 1890)[b,c]
Schellencandona Meisch, 1996
Schellencandona cf. triquetra (Klie, 1936)[b]
Candonopsis (Candonopsis) Vávra, 1891[e]
Subfamily Cyclocypridinae Kaufmann, 1900
Cyclocypris Brady & Norman, 1889
Cyclocypris ampla Furtos, 1933[b]
Cyclocypris ovum (Jurine, 1820)[b,c]
Cyclocypris serena (Koch, 1838)[b,c]
Cypria Zenker, 1854
Cypria ophtalmica (Jurine, 1820)[b,c]
Physocypria Vavra, 1897
Physocypria globula Furtos, 1933[b]
Physocypria pustulosa (Sharpe, 1897)[b]
Family Ilyocyprididae Kaufmann, 1900
Subfamily Ilyocypridinae Kaufmann, 1900
Ilyocypris Brady & Norman, 1889

Continued

TABLE 16.1 Common Non-marine Species of Ostracoda in the Nearctic Ecoregion[a]—cont'd

Ilyocypris bradyi Sars, 1890[b,c]
Ilyocypris gibba (Ramdohr, 1808)[b,c]
Ilyocypris salebrosa Stepanaitys, 1960[b,c]
Family Notodromadidae Kaufmann, 1900
Subfamily Cyproidinae Hartmann, 1963
Cyprois Zenker, 1854
Cyprois marginata (Straus, 1821)[b,c]
Subfamily Notodromadinae Kaufmann, 1900
Notodromas Lilljeborg, 1853
Notodromas monacha (Müller, 1776)[b,c]
Superfamily Darwinuloidea Brady & Robertson, 1885
Family Darwinulidae Brady & Robertson, 1885
Alicenula Rossetti & Martens, 1998
Darwinula Brady & Robertson, 1885
Darwinula stevensoni (Brady & Robertson, 1890)[b,c]
Microdarwinula Danielopol, 1968
Microdarwinula zimmeri Danielopol, 1968[b,c]
Penthesilenula Rossetti & Martens, 1998
Penthesilenula sp. Rossetti & Martens, 1998[b]
Superfamily Cytheroidea
Family Cytheridae Baird, 1850
Perissocytheridea Stephenson, 1938
Perissocytheridea bicelliforma Swain, 1955[b]
Family Cytherideidae
Subfamily Cytherideinae
Cyprideis Jones, 1857
Cyprideis beaconensis (Leroy) 1943[b]
Cyprideis salebrosa (Van den Bold, 1963)[b]
Cytherissa Sars, 1925
Cytherissa lacustris (Sars, 1863)[b,c]
Family Limnocytheridae Klie, 1938
Subfamily Limnocytherinae Klie, 1938
Limnocythere Brady, 1868
Limnocythere bradburyi Forester, 1985[b]
Limnocythere ceriotuberosa Delorme, 1967[b]
Limnocythere friabilis (Benson and MacDonald, 1963)[b]
Limnocythere herricki Staplin, 1963[b]
Limnocythere inopinata (Baird, 1843)

TABLE 16.1 Common Non-marine Species of Ostracoda in the Nearctic Ecoregion[a]—cont'd
Limnocythere itasca Cole, 1949[b]
Limnocythere sappaensis Staplin, 1963 [b] (= syngamic form of *L. inopinata*[b,c])
Limnocythere staplini Gutentag & Benson, 1962[b]
Limnocythere verrucosa Hoff, 1942[b]
Subfamily Timiriaseviinae Mandelstam, 1960
Cytheridella Daday, 1905
Cytheridella ilosvayi Daday, 1905[b]
Elpidium O.F. Müller, 1880
Elpidium maricaoensis (Tressler, 1941)[e]
Family Loxoconchidae Sars, 1925
Cytheromorpha Hirschmann, 1909
Cytheromorpha fuscata (Brady, 1869)[b,c]

[a]Hierarchical taxonomic list follows that of Martens & Savatenalinton (2011).
[b]Listed in tables and illustrated.
[c]Also occurs in Palearctic Ecoregion.
[d]Restricted to Arctic.
[e]Restricted to southern U.S., Caribbean and northern Mexico.

Distribution maps, hydrochemical ranges, and additional images are available as well on NANODe, the North American Non-Marine Ostracode Database (www.kent.edu/nanode). Another extremely valuable resource is the Delorme Ostracode Autecological Database and a comprehensive matching collection of specimens, both housed at the Canadian Museum of Nature's Research and Collections Facility in Gatineau, Quebec, comprising species records with associated environmental and climatic data for more than 6000 sites (http://nature.ca).

LIMITATIONS

The keys presented here conclude at the level of genus and are based, as far as possible, on morphological characteristics of the carapace and appendages that can be determined without the need for dissection. Features carefully observed in the shell are sufficient to identify many common taxa to the species level, but confirmation can be sought by examining major attributes of the soft parts as well. As a valuable source of paleoenvironmental and paleoclimate information, fossil ostracodes recovered from Quaternary aged material are often studied, and soft parts are generally not present. In such cases, the shell alone is available for identification. Keys to species level are inadvisable because many taxa are in need of revision and it is believed that many more are yet to be described; however, illustrations of a selection of common species are provided and their environmental references are tabulated. The keys deal with free-living ostracodes; the commensal Entocytheridae are not represented here, but a key can be found in Smith & Delorme (2010). A detailed review of non-marine ostracodes, their ecology and biology and important references can be found in Chapter 30 of Volume 1 (Smith et al., 2015).

TERMINOLOGY AND MORPHOLOGY

These keys take identification to genus level and are supplemented by photographic and SEM images of commonly encountered species. Images were taken in transmitted light on a binocular dissecting microscope in order that shell features and overall carapace appearance will appear as similar as possible to the features under observation by an aquatic biologist examining a recent sample. Scanning Electron Microscope (SEM) images are also included to provide for the taxonomist more detail than can be easily observed in transmitted light. Both kinds of images are important ways to view specimens, because they highlight different features. Arrows in images indicate the anterior direction unless otherwise noted. Some species are strongly sexually dimorphic, and in such cases the gender is indicated on the images. Seventy species with regional, continental, or hemispherical distribution, and including examples from a range of aquatic habitats, are illustrated here.

All of these keys use adult characters. Juvenile specimens typically are more tapered posteriorly in carapace shape, lack well-developed calcified inner lamellae, and (depending on the growth stage) have fewer pairs of appendages. The following paragraph explains some of the main features that are used in the keys and is not intended to be a comprehensive account of ostracode morphology (for which the reader is referred to Smith et al., 2015).

The ostracode body is completely encased within a bivalved carapace. Each valve of the carapace is divided into two parts, the calcified outer lamella and inner lamella. The outer lamella is the major part of the carapace, whereas the calcified part of the inner lamella bordering the posterior, ventral and anterior margin is a mineralized part of the inner layer of the duplicature, strengthening the free marginal region of the carapace. It forms a part of the free margin and projects toward the center of the carapace. Inward from the free margin, the calcified inner and outer lamellae may be separated in some regions, forming anterior and posterior vestibula. The inner edge of the calcified inner lamella is referred to as the inner margin. Ridge structures (selvage and lists) on the surface of the calcified inner lamellae are important in species level taxonomy. Pustules, teeth, or crenulations may appear on the outer margin of the duplicature.

It is relatively easy to determine the ostracode superfamilies by identifying the central Adductor Muscle Scar (AMS) pattern on the carapace and/or the general arrangement of the appendages and, with careful observation of other morphological features of the carapace and some appendages, to identify common taxa. The AMS often has additional frontal and mandibular scars in close proximity (Pl. 16.134); the actual scars are on the inner surface of the valve, but the pattern is often visible externally as "lucid spots." Examine Plate 16.134 as you use this key.

The carapace may contain additional structures on the outer lamella, such as lateral depressions called sulci. Raised areas variously termed alae, knobs, bosses, papillae, or pustules, depending on their shape, size, and orientation, may also be found on the carapace surface. The surface of the carapace may also be pitted, punctate, wrinkled, or reticulate. Normal pores appear as holes in the outer surface of the shell and are the exit points for sensilla (sensory "hairs") carried by pore canals; sometimes they emerge from pore conuli, which are cones or wart-like protuberances (the latter often given the German term "porenwarzen"). Where such pores pass through the fused zone, where the calcified inner and outer lamellae are in contact, they form radial or marginal pore canals. In the Cypridoidea, traces of the ovaries and testes can often be seen laterally in the posterior of the carapace. These ancillary structures of the carapace are very often used for generic and specific identification.

The carapace dorsal hinge type may be a useful diagnostic feature. An adont hinge lacks terminal teeth and simply has a bar in one valve fitting into a groove in the other. A lophodont hinge has smooth terminal teeth separated by a median bar in one valve, with corresponding sockets and groove in the other. In a merodont (or anti-merodont) hinge the terminal teeth are crenulated or subdivided into several teeth.

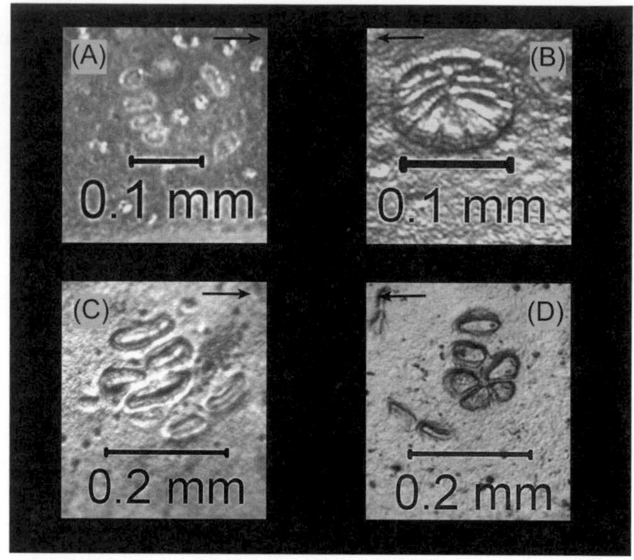

PLATE 16.134 Central muscle scar (AMS, frontal and mandibular scars) patterns of podocopid ostracodes (arrows point toward anterior of carapace): (A) Cytheroidea; note the vertical stack of four to five scars; (B) Darwinuloidea; note the circular to ovate rosette of approximately 10–12 segments; (C) Cypridoidea, family Cyprididae; note elongate scars, openly arranged; and (D) Cypridoidea, family Candonidae; note the more tightly arranged "pawprint". *From Fig. 19.3 in Smith and Delorme, 2010.*

Appendage/limb terminology is summarised in Table 16.2. The terminal pincer on the cypridoidean L7 (the cleaning limb) is a complex structure formed by modification of small terminal setae and is apparently capable of grasping small particles.

The Triebel Loop is a triangular to oval-shaped loop visible in the uropodal attachment of members of the Cyprícercinae.

MATERIAL PRESERVATION AND PREPARATION

The collector has a purpose in collecting that governs the preparation and preservation of aquatic invertebrate material. In the case of ostracode collection, the focus may be on taxonomic identifications for a study of biodiversity or biogeography. The collector may intend these samples to become available for genetic analysis as well. A brief description of common practice is included here; however, the reader is directed to the detailed overview of preparation and preservation methods presented in Volume 1 of this series (Smith et al., 2015).

When conducting field collections, a decision must be made as to whether the ostracodes are to be used for genetic analysis or not. If genetic analysis is planned, then a split of

TABLE 16.2 Common Terminology Used in the Key to Ostracodes in This Chapter

Most Common Terminology	Alternate Terminology in Use
1. A1 – Antennule	Antennula, or First Antenna
2. A2 – Antenna	Second Antenna
3. Md – Mandible	Mandibula
4. Mx1 – Maxillule	Maxillula, or First Maxilla
5. L5 - Fifth Limb	First Thoracic Leg (T1), Maxilla, Second Maxilla, Maxilliped, or Walking Leg
6. L6 - Sixth Limb	Second Thoracic Leg (T2), or Walking Leg
7. L7 - Seventh Limb	Third Thoracic Leg (T3), Walking Leg, or Cleaning Leg
8. Uropod (Uropodal ramus)	Furca, Furcal Rami, or Caudal Rami

the sample from which ostracodes will be selected must be preserved in pure 95–99% ethanol. A lower concentration of ethanol will not be acceptable for samples undergoing genetic analysis. Samples not intended for genetic analysis can be treated initially with 10–30% ethanol, or alternatively a buffered 7–10% formalin solution; the initial use of a lower concentration of ethanol (10–30%) will cause death with the valves gaping open (making later dissection easier), following which long-term preservation in 70% ethanol is recommended.

The sample can also be water washed through a stack of standard 20 mm diameter brass sieves of mesh sizes 850, 150, and 63 μm (the larger opening collects twigs, leaf fragments, and other unwanted debris, and the smallest opening is the size boundary between sand and silt fractions, and can collect juvenile forms as well as other small invertebrates). Most of the ostracodes will be caught in the 150 μm sieve. These can be rinsed into a small Whirlpak® bag, frozen and then freeze-dried, or simply air-dried, and finally placed in vials or micropaleontological slides. Ostracodes can be picked out under a low power stereoscopic binocular microscope, either by pipette from wet samples in a petri dish, or with a wetted fine artist's paintbrush from dry sieved residues scattered on a tray. If placed on micropaleontological slides for long term storage, specimens that are likely to be used for geochemical work (e.g., trace element or isotopic analyses of the shells) should be stored loose and not glued on to the slides; otherwise they are best glued down with a water-soluble glue, so that specimens that have been sorted (e.g., according to taxon, males and females, adults and juveniles) stay in place and can easily be reexamined and counted as necessary; they can be detached and reoriented or moved by adding a drop of water with a fine paintbrush and waiting for a minute or two while the glue softens.

Here we focus on the main preparation techniques needed for taxonomic identification; for more comprehensive and detailed coverage the reader is referred to the descriptions of methods provided by Athersuch et al. (1989), Danielopol et al. (2002), Griffiths & Holmes (2000), and Meisch (2000). A stereoscopic binocular microscope (magnification up to 50×) with a combination of incident and transmitted light should be used for studying whole specimens, whether alive or preserved in ethanol or dry on micropalaeontological slides. Specimens are best manipulated with a moistened fine artist's paintbrush; fragile shells may easily be broken if forceps or tweezers are used. If specimens can be examined alive in a petri dish, their colour and behaviour (e.g., swimming, crawling, or burrowing) may be helpful in identification. A microscope with higher magnifications up to at least 100× (and, ideally, phase contrast) is generally necessary for examining the morphology of dissected appendages. General morphology of whole carapaces or single valves (dry or wet) can be examined in incident light, but features such as adductor muscle scars and marginal pore canals are best observed in transmitted light and can be seen more clearly if the specimen is immersed in glycerine rather than water or ethanol. Staining with a water soluble dye (e.g., food colouring) may help to reveal fine details of surface ornamentation and hinge structure.

Anatomical dissection can be carried out under a stereoscopic binocular microscope (at 20x to 50x depending on the size of the specimen) in combined incident and transmitted light. It is best done with specimens preserved in ethanol, but good results can often be obtained with dry specimens if they are first moistened gradually with distilled water or ethanol, allowing air bubbles to escape (however, air trapped inside limbs may be impossible to disperse); for older dried material (e.g., in museum collections), which may be very brittle, soaking in a 10% solution of trisodium orthophosphate for 24 hr is recommended. Dissection is performed in a drop of mounting medium on a glass microscope slide with a pair of fine needles, with a glass cover slip added after completion. Tungsten needles have the advantage that they can be sharpened by immersion in hot sodium nitrite, but fine entomological pins (no. 000) mounted in pin chucks (or simply fastened to the ends of rods with plasticene or a similar material) are very effective and blunted or bent ones can be replaced easily, even during a dissection. Glycerine is an inexpensive and safe mounting medium. Other media such as Hydro-Matrix may be used, but PolyVinyl Lactophenol (PVL), once popular, is no longer recommended due to its carcinogenic properties. The specimen should be immersed in glycerine on the glass slide and the valves opened with the needles to extract the soft body and appendages; the valves should be removed from the glycerine, washed in a drop of distilled water and stored dry on a labelled micropalaeontological slide. The needles can then be used to tease apart the appendages and arrange them on the glass slide. Any air bubbles should be removed with a needle, before dropping a glass cover slip on top of the completed dissection. Partial dissections or undissected "squash" mounts of the whole body and limbs are useful and easier to perform, but significant morphological features may be obscured. The finished dissection slide is best left for a day or two before sealing round the edges of the cover slip with nail varnish (if this is not done, the glycerine will eventually evaporate and large air pockets will invade the dissection). Dissection is a difficult skill to acquire and is best practiced initially on large specimens that are in plentiful supply. For more detailed explanations of dissection techniques the reader is referred to the excellent, illustrated guide by Namiotko et al. (2011).

KEYS TO OSTRACODA

Ostracoda: Podocopida: Superfamilies

1 AMS pattern not a circular to ovate rosette 2

1' AMS pattern a distinct circular or ovate rosette; carapace with calcified inner lamella very narrow; body with two pairs of walking legs (L6 & L7) .... Darwinuloidea, one family: **Darwinulidae [p. 486]**

2(1) AMS pattern is a cluster arranged like a "pawprint," with scars tightly clustered, or more openly arranged; carapace with moderately wide calcified inner lamella; body with one pair of walking legs (L6); L7 is an inverted cleaning leg .... **Cypridoidea [p. 487]**

2' AMS pattern is a vertically stacked arrangement of 4 scars (some occasionally subdivided in two); carapace with moderately wide calcified inner lamella; body with three pairs of walking legs (L5, L6, & L7) .... **Cytheroidea [p. 500]**

Ostracoda: Podocopida: Darwinuloidea: Darwinulidae: Genera

Darwinulids are benthic, infaunal, or interstitial; common throughout North America in most aquatic settings, often very abundant. At present writing, six extant darwinulid genera are described, of which three are common in North America (*Darwinula, Microdarwinula*, and *Penthesilenula*) and one, *Alicenula*, is rare, reported from Florida (Keyser, 1975; D.G. Smith, 2001) and California (Mark Angelos, personal communication). The representatives of *Penthesilenula* and *Alicenula* in North America have not been formally described. The genera *Vestalenula* and *Isabenula* have not yet been identified in North America. Images of North American Darwinulidae are presented in Plate 16.133. Table 16.3 lists common Nearctic species with habitat information and plate number.

Diagnostic Characters of the Darwinulidae

Carapace small (0.4–0.8 mm length), smooth, white, thinly calcified, calcified inner lamellae very narrow, vestibules absent, distinctive rosette-shaped AMS pattern. Normal pores simple. Median eye present. A1 with six articulated podomeres. A2 endopodite with three articulated podomeres, without swimming setae, exopodite a reduced podomere with two long and one very short setae. Mandibular palp with a row of eight rake-like setae and a small branchial plate with up to eight rays. Maxillula with a large branchial plate with four reflexed, forward-pointing rays. L5 with strong masticatory structure, a leg-like palp of three endopodite podomeres, and a small respiratory plate; L6 and L7 both walking legs similar in structure and direction; uropodal rami reduced to setae or lacking; ovaries do not originate within duplicature, female carries eggs or early-stage juveniles in posterior brood space of carapace, usually visible through shell. Males exceptionally rare, unknown in North American taxa.

TABLE 16.3 Common Nearctic Species in the Superfamily Darwinuloidea, as Illustrated in This Chapter

Family/Subfamily	Species	Notes on Habitat	Figures
Darwinulidae	*Darwinula stevensoni* (Brady & Robertson, 1870)	Common in most non-marine aquatic settings. Cosmopolitan.	Plate 16.135 A, B, F
	Microdarwinula zimmeri Danielopol, 1968	Interstitial in wetland sediments, springs, streams, shallow groundwater. Common in these settings.	Plate 16.135 C, D
	Penthesilenula sp.	Interstitial in wetland sediments, springs, streams, shallow groundwater. Common in these settings.	Plate 16.135 E, G, H

Keys to Darwininulidae: Genera

1 Carapace moderately elongate, subcylindrical or wedge-shaped, tapering anteriorly, with straight or weakly arched dorsal margin; AMS in front of midpoint 2

1' Carapace short, subovate with distinctly arched dorsal margin, equally rounded anteriorly and posteriorly in lateral view, AMS approximately central; left valve interior with one posteroventral and one anteroventral tooth .... *Microdarwinula*

2(1) Left valve overlaps right valve; internal marginal teeth present 3

2' Right valve overlaps left valve; carapace elongate, rounded wedge shape in dorsal and lateral view; valves without internal marginal teeth .... *Darwinula*

3(2) Carapace wedge-shaped in dorsal and lateral views; left valve with anteroventral and posteroventral internal marginal teeth or one posterior internal marginal tooth .... *Penthesilenula*

3' Carapace subrectangular or wedge-shaped in lateral view, wedge-shaped in dorsal view; left valve with anteroventral and posterior internal marginal teeth .... *Alicenula*

Ostracoda: Podocopida: Cypridoid Families

The superfamily Cypridoidea contains the majority of living non-marine species, many of which are active nekto-benthic swimmers with well-developed long swimming setae on the first and second antennae (A1 and A2); in crawlers and burrowers these setae are reduced or absent and these animals living infaunally, interstitially, or as part of the epibenthos, such as the members of the Candoninae. Four families are represented in North America: the Candonidae, Ilyocyprididae, Notodromadidae, and Cyprididae. Table 16.4 lists common Nearctic species with habitat information and plate number.

Diagnostic Characters of the Cypridoidea

Surface of the shell usually smooth or finely ornamented, dorsal margin with weak or simple hinge teeth. Size and shape of carapace variable, with strong sexual dimorphism in some genera. Calcified inner lamella usually well-developed. Basic AMS pattern a "pawprint" with three scars in an arcuate row in front of two scars in a vertical row. Normal pores simple. Lateral ocelli of median eye sometimes with corresponding eyespots developed in the valves. A1 with six to eight podomeres. A2 endopodite with three or four podomeres, exopodite reduced to a small scale-like protuberance bearing at most three setae. Mandibular palp with a small branchial plate with up to six rays. Maxillula with a large branchial plate with up to three reflexed, forward-pointing rays. L5 usually with a small branchial plate, sometimes reduced to one or two setae; endopodite one- or two-segmented, forming a small palp in the female but enlarged to form a clasper in the male. The L6 is a walking leg with an endopodite of three or four podomeres and a strong terminal claw. The L7 is bent dorsally and is a cleaning leg, usually with three terminal setae one of which may be modified as a pincer. The uropod (furcal ramus) is typically well developed with terminal claws but may be reduced to a whip-like structure. The gonads are located within the duplicature of the valves. In the male, a portion of the vas deferens is modified to form an ejaculatory duct, the Zenker's organ, separate from the copulatory appendage. Females never with a brood chamber.

Keys to the Superfamily Cypridoidea: Families

1 Each valve subrectangular with two dorsomedian sulci and a long, fairly straight dorsal hinge line, more than three-quarters of the length 2

1' Dorsomedian sulci absent; valves with short straight (less than two-thirds the length of the carapace) or arched hinge lines 3

2(1) Carapace >1 mm long; valve surface pitted, with small marginal spines or denticles; female L5 endopodite unsegmented; L7 with terminal pincer .... **Cyprididae** (In part)

2' Carapace typically <1 mm long; valve surface pitted, with small marginal spines or denticles, and sometimes ornamented with tubercles or spines; female L5 endopodite with two podomeres, L7 with simple terminal setae .... Ilyocyprididae, one genus: *Ilyocypris*

[Note: Ilyocyprids may be confused with limnocytherines (Superfamily Cytheroidea), which also have long, straight hinges and dorsomedian sulci; they are easily separated by looking at the AMS, and furthermore many limnocytherids have relatively broad flattened areas on the anterior and posteroventral margins of the valves, never seen in ilyocyprids.]

3(1) Carapace ovoid, elongate ovoid or reniform, ventral margin in lateral view convex or sinuous, rarely straight; A2 swimming setae present, reduced, or absent 5

3' Carapace ovoid with a straight ventral margin in lateral view; valve margins rimmed with a flat flange, widest anteriorly; in dorsal view, compressed or tumid; A2 with well developed swimming setae .... **Notodromadidae [p. 495]**

4(3) AMS a loosely arranged "pawprint" of rounded, often elongate scars (see Pl. 16.134); carapace typically with maximum height close to or well in front of midpoint; L7 with a terminal pincer (the position of the frontal and mandibular scars which are always anterior to the AMS (Pl. 16.134) is helpful in determining orientation. To determine the presence or absence of an L7 pincer requires at least partial dissection). .... **Cyprididae** (In part)

4' AMS a tight "pawprint" cluster of rounded wedge-shaped scars (Pl. 16.134); carapace typically with maximum height close to or well behind midpoint; L7 without terminal pincer .... **Candonidae [p. 495]**

TABLE 16.4 Common Nearctic Species in the Superfamily Cypridoidea, as Illustrated in This Chapter

Family/Subfamily	Species	Notes on Habitat	Figures
Ilyocyprididae, Ilyocypridinae	*Ilyocypris bradyi* Sars, 1890	Common in shallow ponds, streams, springs, lake margins. A benthic species. Cosmopolitan.	Plate 16.140 A, B
Ilyocyprididae, Ilyocypridinae	*Ilyocypris gibba* (Ramdohr, 1808)	Common in shallow ponds, streams, springs, lake margins. A nektobenthic species. Cosmopolitan	Plate 16.140 C, D
Ilyocyprididae, Ilyocypridinae	*Ilyocypris salebrosa* Stepanaitys, 1960	Uncommon. Oxbows and shallow, weedy littoral environments. Important in Pleistocene paleolimnologic records.	Plate 16.140 E, F
Notodromadidae, Cyproidinae	*Cyprois marginata* (Straus, 1821)	Common in vernal pools, weedy lake margins and freshwater prairie potholes.	Plate 16.136 A, B
Notodromadidae, Notodromadinae	*Notodromas monacha* (O.F. Müller, 1776)	Uncommon. Vernal pools, weedy lake margins, and freshwater prairie potholes.	Plate 16.136 C, D
Cyprididae, Cypridinae	*Cypris pubera* O.F. Müller, 1776	Common. Weedy lake margins, freshwater prairie potholes.	Plate 16.138 C, D
Cyprididae, Cypridinae	*Chlamydotheca arcuata* (Sars, 1901)	Common. Streams, springs, canals, rice-fields, typically in southern half of North America, as far north as Ohio.	Plate 16.138 A, B
Cyprididae, Pelocypridinae	*Pelocypris tuberculatum* Ferguson, 1967	Common in seasonally filled playas of southwestern U.S. and vernal ponds of southern California.	Plate 16.140 G, H
Cyprididae, Cypricercinae	*Strandesia canadensis* (Sars, 1926)	In permanent freshwater springs and streams in western U.S. and western Canada.	Plate 16.136 E, F
Cyprididae, Cypricercinae	*Bradleystrandesia reticulata* (Zaddach, 1894)	Common in springs, ponds, ditches, littoral zones.	Plate 16.136 G, H
Cyprididae, Cypricercinae	*Spirocypris tuberculatus* Sharpe, 1909	Reported from Canada, in ponds and littoral zone.	Plate 16.137 A, B
Cyprididae, Eucypridinae	*Cypriconcha ingens* Delorme, 1967	Prairie potholes and ponds of the Great Plains, often found in large populations.	Plate 16.138 E, F
Cyprididae, Eucypridinae	*Tonnacypris glacialis glacialis* (Sars, 1890)	High latitude, circumpolar distribution in wetlands and ponds.	Plate 16.146 G, H
Cyprididae, Eucypridinae	*Candocyprinotus ovatus* Delorme, 1970 (possibly a *Eucypris* species)	Mixed-woods zone as well as the southern fringe of the boreal forest, predominantly in intermittent streams.	Plate 16.142 A, B
Cyprididae, Herpetocypridinae	*Herpetocypris reptans* (Baird, 1835)	Prefers permanent, vegetation-rich ponds. Parthenogenetic and cosmopolitan.	Plate 16.139 C, D
Cyprididae, Herpetocypridinae	*Stenocypris major major* (Baird, 1859)	Reported from rice fields in Louisiana and cenotes of the Yucatan. Cosmopolitan.	Plate 16.139 G
Cyprididae, Isocypridinae	*Isocypris beauchampi* (Paris, 1920)	Weedy littoral zones of lakes, ponds and also in rice fields. Holarctic distribution.	Plate 16.137 C, D
Cyprididae, Dolerocypridinae	*Dolerocypris fasciata* (O.F. Müller, 1776)	Associated with weedy ponds and littoral zones in mid-latitudes. Parthenogen. Holarctic distribution.	Plate 16.139 A, B
Cyprididae, Scottinae	*Scottia pseudobrowniana* Kempf, 1971	Common in hydrated fen wetland soils and charophyte mounds, weedy ponds.	Plate 16.139 E, F
Cyprididae, Cyprettinae	*Cypretta brevisaepta brevisaepta* Furtos, 1934	Found in southern U.S. and northern Mexico in lakes and ponds.	Plate 16.148 G, H

TABLE 16.4 Common Nearctic Species in the Superfamily Cypridoidea, as Illustrated in This Chapter—cont'd

Family/Subfamily	Species	Notes on Habitat	Figures
Cyprididae, Cypridopsinae	*Cypridopsis vidua* (O.F. Müller, 1776)	Most common ostracode in the Nearctic. Most freshwater habitats, including wastewater treatment plants. Cosmopolitan.	Plate 16.148 A, B
Cyprididae, Cypridopsinae	*Cypridopsis okeechobei* Furtos, 1936	Common in most freshwater habitats in the U.S., rarely reported in Canada.	Plate 16.148 C, D
Cyprididae, Cypridopsinae	*Sarscypridopsis aculeata* (Costa, 1847)	Saline prairie potholes and littoral zone of saline lakes.	Plate 16.148 E, F
Cyprididae, Cypridopsinae	*Potamocypris pallida* Alm, 1914	Common in springs and seeps, including those along streams and lake edge.	Plate 16.149 A, B
Cyprididae, Cypridopsinae	*Potamocypris smaragdina* (Vávra, 1891)	Common in littoral zones of lakes throughout the U.S. and Canada. Cosmopolitan.	Plate 16.149 C, D
Cyprididae, Cypridopsinae	*Potamocypris unicaudata* Schäfer, 1943	Common in littoral zones of lakes and ponds throughout the U.S. and Canada. Cosmopolitan.	Plate 16.149 E, F
Cyprididae, Cypridopsinae	*Cavernocypris wardi* Marmonier, Meisch & Danielopol, 1989	Hyporheos, springs, seeps, and pumped from aquifers.	Plate 16.149 G, H
Cyprididae, Cyprinotinae	*Heterocypris incongruens* (Ramdohr, 1808)	Very common in puddles, small temporary water bodies. Parthenogenetic, cosmopolitan.	Plate 16.137 E, F
Cyprididae, Cyprinotinae	*Heterocypris glaucus* (Furtos, 1933)	Common in lakes and ponds throughout the Great Plains.	Plate 16.137 G, H
Candonidae, Candoninae	*Candona acuta* Hoff, 1942	Shallow littoral with subaquatic vegetation.	Plate 16.142 G, H
Candonidae, Candoninae	*Candona acuminata* (Fischer, 1854) (assignable to *Fabaeformiscandona*)	Littoral of lakes in forested region of western N.America. Although uncommon, very visible in samples due to large size.	Plate 16.145 G, H
Candonidae, Candoninae	*Candona acutula* Delorme, 1967	Shallow littoral with subaquatic vegetation. Today lives north of U.S./Canada border, extended south of this border in Pleistocene and early Holocene.	Plate 16.142 C, D
Candonidae, Candoninae	*Candona albicans* Brady, 1864 (assignable to *Pseudocandona*)	Common in springs, vernal pools, littoral zones of lakes in forested regions.	Plate 16.141 E, F
Candonidae, Candoninae	*Candona candida* (O.F. Müller, 1776)	Springs and seeps in a range of aquatic settings. Usually parthenogenetic, but males occasionally occur, reported from Canada but no males reported from the U.S. Cosmopolitan.	Plate 16.143 A, B
Candonidae, Candoninae	*Candona caudata* Kaufman, 1900 (assignable to *Fabaeformiscandona*)	Most freshwater habitats. In lakes, to considerable depth. Collected throughout North America.	Plate 16.145 E, F
Candonidae, Candoninae	*Candona crogmaniana* Turner, 1894	Great Lakes and eastern U.S. lakes, vernal ponds, permanent spring-fed ponds. In lakes to considerable depth.	Plate 16.143 C, D
Candonidae, Candoninae	*Candona decora* Furtos, 1933	Most common in vernal ponds of forested regions of North America, but also occurs in wetlands in southwestern U.S.	Plate 16.144 A, B
Candonidae, Candoninae	*Candona distincta* Furtos, 1933	Most common in vernal ponds and permanent lakes of forested regions, mid to high latitudes in North America, but also occurs in wetlands.	Plate 16.144 C, D

Continued

TABLE 16.4 Common Nearctic Species in the Superfamily Cypridoidea, as Illustrated in This Chapter—cont'd

Family/Subfamily	Species	Notes on Habitat	Figures
Candonidae, Candoninae	*Candona elliptica* Furtos, 1933	Most common in lakes of forested regions of midlatitude N. America, but can also be found in streams, wetlands, and springs.	Plate 16.144 E, F
Candonidae, Candoninae	*Candona ikpikpukensis* Swain, 1963	Lakes and ponds of Alaska and northern Canada-Arctic regions. First described as Pleistocene fossil.	Plate 16.144 G, H
Candonidae, Candoninae	*Candona kingsleii* (Brady & Robertson, 1870) (assignable to *Candonopsis*)	One of four *Candonopsis* species reported from the Neotropics, often collected from bromeliads.	
Candonidae, Candoninae	*Candona neglecta* Sars, 1887	Headwater streams and springs of forested regions of Great Lakes area.	Plate 16.143 E, F
Candonidae, Candoninae	*Candona ohioensis* Furtos, 1933.	Freshwater lakes in forested Great Lakes region, Northeastern U.S. and southern Canada.	Plate 16.142 E, F
Candonidae, Candoninae	*Candona patzcuaro* Tressler, 1954 (assignable to *Fabaeformiscandona*)	Temporary ponds and wetlands of southwestern U.S. and Mexico. Note similarity to *Candona rawsoni*, which has a wider geographic range.	Plate 16.145 A, B
Candonidae, Candoninae	*Candona protzi* Hartwig, 1898 (assignable to *Fabaeformiscandona*)	Occurrence in ponds, streams, and littoral zones of lakes primarily in Saskatchewan and northern Canada. Not known from the U.S.	Plate 16.146 E, F
Candonidae, Candoninae	*Candona punctata* Furtos, 1933	Springs, ponds, and littoral zones of lakes in the forested Great Lakes region.	Plate 16.146 A, B
Candonidae, Candoninae	*Candona rawsoni* Tressler, 1957 (assignable to *Fabaeformiscandona*)	Common in wetlands, temporary water bodies, and littoral zones of lakes, wide tolerance for salinity, cosmopolitan.	Plate 16.145 C, D
Candonidae, Candoninae	*Candona rectangulata* Alm, 1914	Arctic distribution, wetlands, ponds, thermokarst pools. Holarctic.	Plate 16.143 G, H
Candonidae, Candoninae	*Candona stagnalis* Sars, 1890 (assignable to *Pseudocandona*)	Common in wetlands, springs, slow moving streams, littoral zones of lakes throughout North America.	Plate 16.141 G, H
Candonidae, Candoninae	*Candona subtriangulata* Benson & MacDonald, 1963	The most common candonid in the Great Lakes, occurring at great depths in Lakes Superior, Huron, and Michigan. Benthos of large lakes in forested regions.	Plate 16.146 C, D
Candonidae, Candoninae	*Schellencandona.cf. triquetra* (Klie, 1936)	Aquifer taxon. Collected by pumping or in rheocrene springs. This form is recovered from wells and deep springs in West Virginia.	Plate 16.141 C, D
Candonidae, Candoninae	*Paracandona euplectella* (Robertson, 1889)	Common in hyporheos of streams, and also in springs, seeps, and shallow groundwater. Often occurs in high abundance in these areas.	Plate 16.141 A, B
Candonidae, Cyclocypridinae	*Cyclocypris ampla* Furtos, 1933	Common in Canada and northern United States to just south of Great Lakes. Littoral zone of lakes, ponds.	Plate 16.147 E, G
Candonidae, Cyclocypridinae	*Cyclocypris ovum* (Jurine, 1820)	Common in Canada and mid to northern U.S., not reported from southern U.S. states. Littoral zone of lakes, ponds.	Plate 16.147 A, B

TABLE 16.4 Common Nearctic Species in the Superfamily Cypridoidea, as Illustrated in This Chapter—cont'd

Family/Subfamily	Species	Notes on Habitat	Figures
Candonidae, Cyclocypridinae	*Cyclocypris serena* (Koch, 1838)	Wetlands and ponds in western North America.	Plate 16.147 C, D, F
Candonidae, Cyclocypridinae	*Physocypria globula* Furtos, 1933	Cosmopolitan, all shallow aquatic habitats, prefers subaquatic vegetation.	Plate 16.150 A, B
Candonidae, Cyclocypridinae	*Physocypria pustulosa* (Sharpe, 1897)	Cosmopolitan, all shallow aquatic habitats, prefers subaquatic vegetation.	Plate 16.150 C, D, E
Candonidae, Cyclocypridinae	*Cypria ophtalmica* (Jurine, 1820)	Cosmopolitan, all shallow aquatic habitats, prefers subaquatic vegetation.	Plate 16.150 F, G

Ostracoda: Podocopida: Cypridoidea: Cyprididae: Genera

Care must be taken not to confuse marginal septa (internal walls between the inner and outer lamellae) with marginal pore canals (internal tubes which open externally on the outer margin and bear sensilla); both features must be viewed in transmitted light.

The subfamily Cypricercinae has recently been revised (Savatenalinton & Martens, 2009) and most Nearctic species previously described as *Cypricercus* have been reclassified into other genera, notably *Brandleystrandesia*, *Strandesia*, or *Spirocypris*. The genus *Cypricercus* (*sensu stricto*) appears to be confined to the southern hemisphere. Nearctic species in the literature identified as *Cypricercus* (e.g., *Cypricercus cheboyganensis, C. burlingtonensis*) will need to be reexamined to assess their taxonomic position within the subfamily.

1	Dorsomedian sulci absent; valves with short straight (less than two-thirds the length of the carapace) or arched hinge lines	2
1'	Each valve subrectangular with two dorsomedian sulci and a long, fairly straight dorsal hinge line, more than three-quarters of the length	*Pelocypris*
	[Note: As the only Nearctic member of the Cyprididae with dorsomedian sulci, *Pelocypris* has sometimes been mistakenly assigned to the Ilyocyprididae, but the L5 and L7 morphology clearly show its true affinities.]	
2(1)	Calcified inner lamellae without inwardly displaced selvage, and inner list inconspicuous or absent	3
2'	Calcified inner lamella in one or both valves ridged with a prominent inwardly displaced selvage and/or conspicuous inner list	13
3(2)	Carapace elongate or ovate, calcified inner lamella broad anteriorly but narrow or absent posteriorly, with conspicuous radial septa	4
3'	Calcified inner lamellae without radial septa	5
4(3)	Carapace ovate in lateral view, compressed in dorsal/ventral view, calcified inner lamella broad anteriorly with short radial septa, absent posteriorly	*Isocypris*
4'	Carapace elongate, calcified inner lamella broad anteriorly with long radial septa, narrow posteriorly	*Stenocypris*
5(3')	Carapace subtrapezoidal, subclavate, subtriangular or subovate in lateral view; without Triebel Loop (to ascertain the presence of a Triebel Loop requires at least that one valve be removed from the animal, but a full dissection may not be necessary)	6
5'	Carapace ovate or elongate in lateral view; uropodal attachment with a distinctive loop structure (the Triebel Loop) (to ascertain the presence of a Triebel Loop requires at least that one valve be removed from the animal, but a full dissection may not be necessary)	11
6(5)	Valve margins smooth	7
6'	One or both valve margins denticulate	9
7(6)	Carapace medium to large (<2.5 mm long), subclavate to subtriangular or subovate	8
7'	Carapace very large (>2.5 mm long), subtrapezoidal, dorsal and ventral margins subparallel	*Cypriconcha*
8(7)	Carapace subclavate, maximum height well in front of midpoint; calcified inner lamella with a small blunt anteroventral internal tooth or peg in left valve (best seen in SEM, but often visible in transmitted light)	*Tonnacypris*
8'	Carapace subtriangular to subovate, maximum height close to midpoint; internal peg absent; anterior external surface of valve often with small rounded pore conuli ("porenwarzen")	*Eucypris*/*Candocyprinotus*
9(6)	Right valve with marginal rounded pustules/denticles anteroventrally and posteroventrally (sometimes weakly developed or absent); carapace subovate in dorsal/ventral view, often with a beak-like protrusion of the anterior end)	10
9'	Both valve margins with small, pointed denticles anteroventrally and posteroventrally, sometimes weakly developed or absent anteriorly in right valve; carapace moderately compressed in dorsal/ventral view, without protrusion of the anterior end	*Prionocypris*
10(9)	Right valve with a dorsal marginal hump	*Cyprinotus*

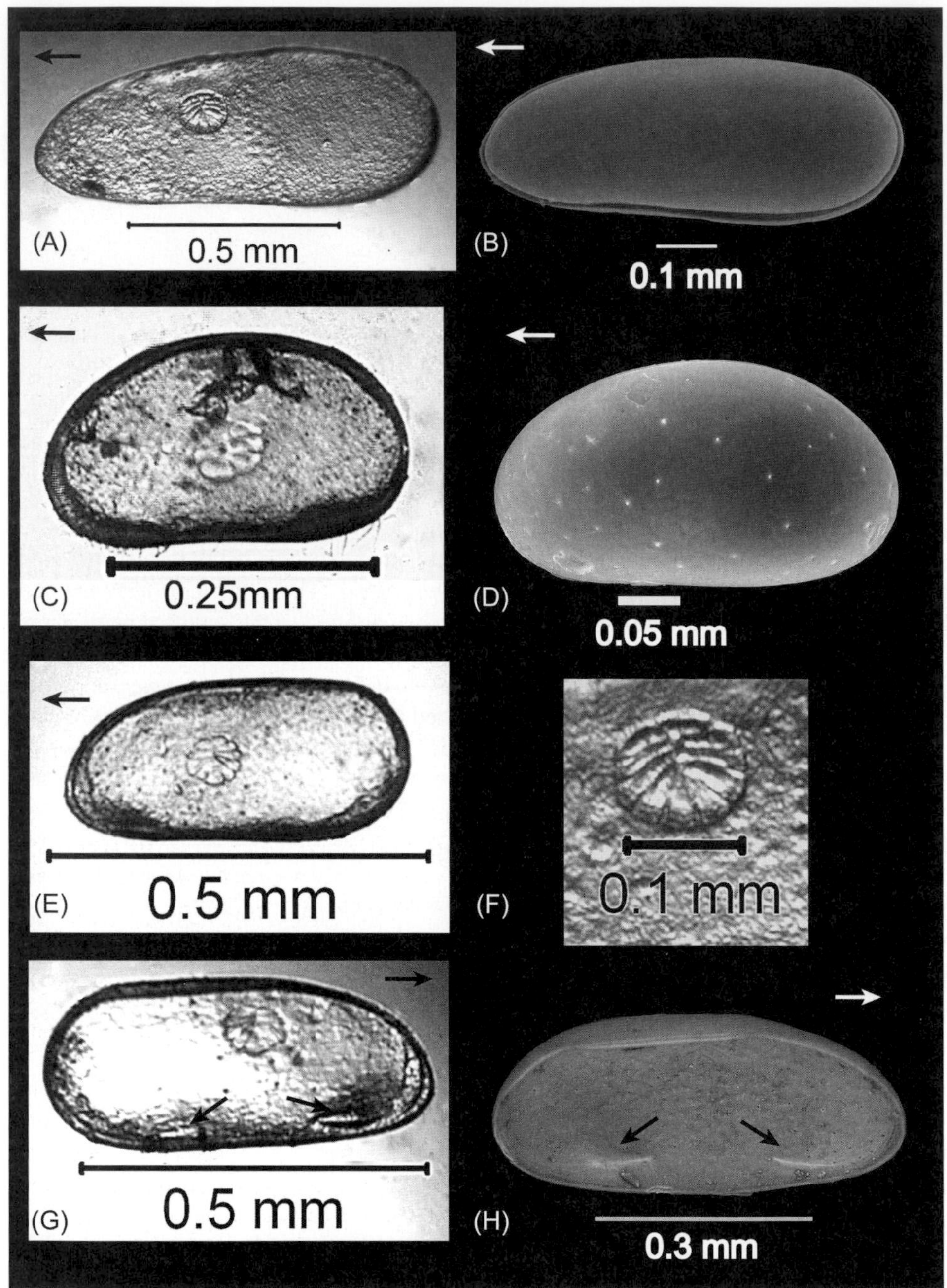

PLATE 16.135 Species of Darwinulidae. (A, B, F) *Darwinula stevensoni* (Brady & Robertson, 1870) showing left valve (A, B) and central muscle scar rosette (F); specimen from littoral benthos, East Twin Lake, Ohio. (C, D): left valve of *Microdarwinula zimmeri* Danielopol, 1968, from hydrated fen wetland sediments, Mantua Bog, Ohio. (E, G, H) *Penthesilenula* sp. Rossetti & Martens, 1998: left valve (E, F) and interior view of left valve (G), showing antero and postero ventral internal teeth (arrows), from hydrated fen wetland sediments, Herrick Fen, Ohio.

10'	Right valve without marginal hump (sometimes present on left valve)	*Heterocypris*
11(9)	Carapace ovate in lateral and dorsal/ventral views; left valve with or without a submarginal internal groove	12
11'	Carapace elongate subtriangular to subovate in lateral view; left valve with submarginal internal groove around free margin	*Strandesia*
12(11)	Carapace surface smooth to faintly reticulate	*Bradleystrandesia*
12'	Carapace surface densely tuberculate	*Spirocypris*
13(2)	Carapace large (>1.5 mm long)	14
13'	Carapace small to medium (<1.0 mm long)	17
14(13)	Carapace subtriangular to subovate; calcified inner lamella in both valves with a prominent inwardly displaced selvage, the anterior margin extending beyond it like a lip	15

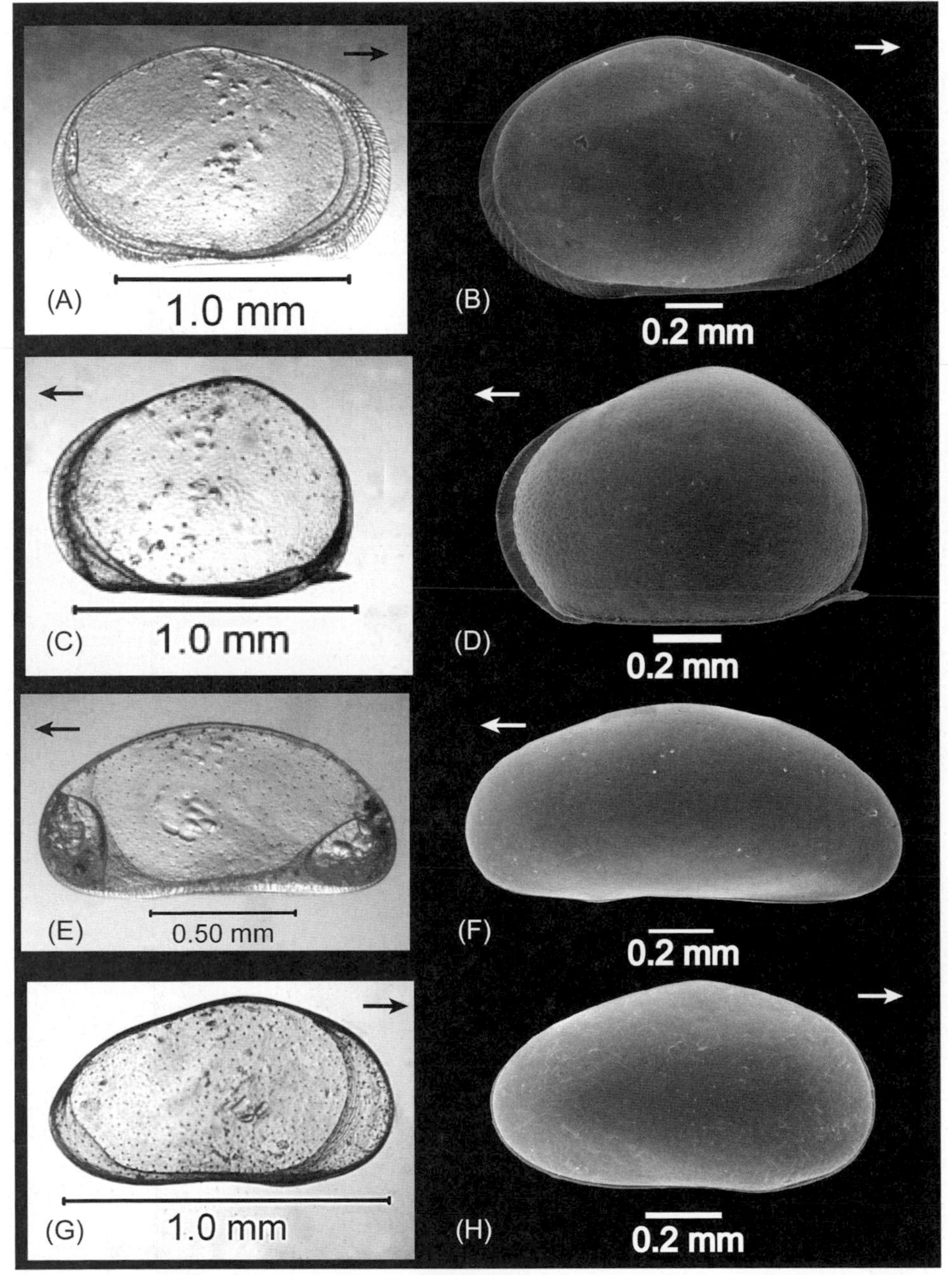

PLATE 16.136 Species of *Cyprois, Notodromas, Strandesia,* and *Bradleystrandesia.* (A, B) Right valve of *Cyprois marginata* (Straus, 1821) from littoral benthos of Elkwater Lake, Alberta. (C, D) Left valve of female *Notodromas monacha* (O.F. Müller 1776). Specimen from Cottonwood Lake wetland P8, North Dakota. (E, F) Left valve of *Strandesia canadensis* (Sars, 1926) from Denton Spring, Colorado. (G, H) Right valve of *Bradleystrandesia reticulata* (Zaddach, 1894) from littoral benthos, Elkwater Lake, Alberta.

14'	Carapace markedly elongate	16
15(18)	Valves with small anterior and posterior marginal denticles, one of which in the right valve is developed into a more conspicuous postero-ventral spine	*Cypris*
15'	Valves without marginal denticles	*Chlamydotheca*
16(14)	Carapace elongate subrectangular, dorsal margin weakly convex to almost straight; left valve with conspicuous inner list	*Herpetocypris*
16'	Carapace markedly elongate, somewhat spindle-shaped, dorsal margin broadly convex; left valve with inwardly displaced selvage	*Dolerocypris*
17(13)	Calcified inner lamella without septa	18
17'	Calcified inner lamella with radial septa anteriorly in both valves; carapace tumid, height greater than half the length	*Cypretta*
18(17)	Carapace subtriangular to subreniform in lateral view with maximum height at or in front of midpoint; L6 with a single long terminal claw, uropodal rami greatly reduced (flagelliform) or absent	19

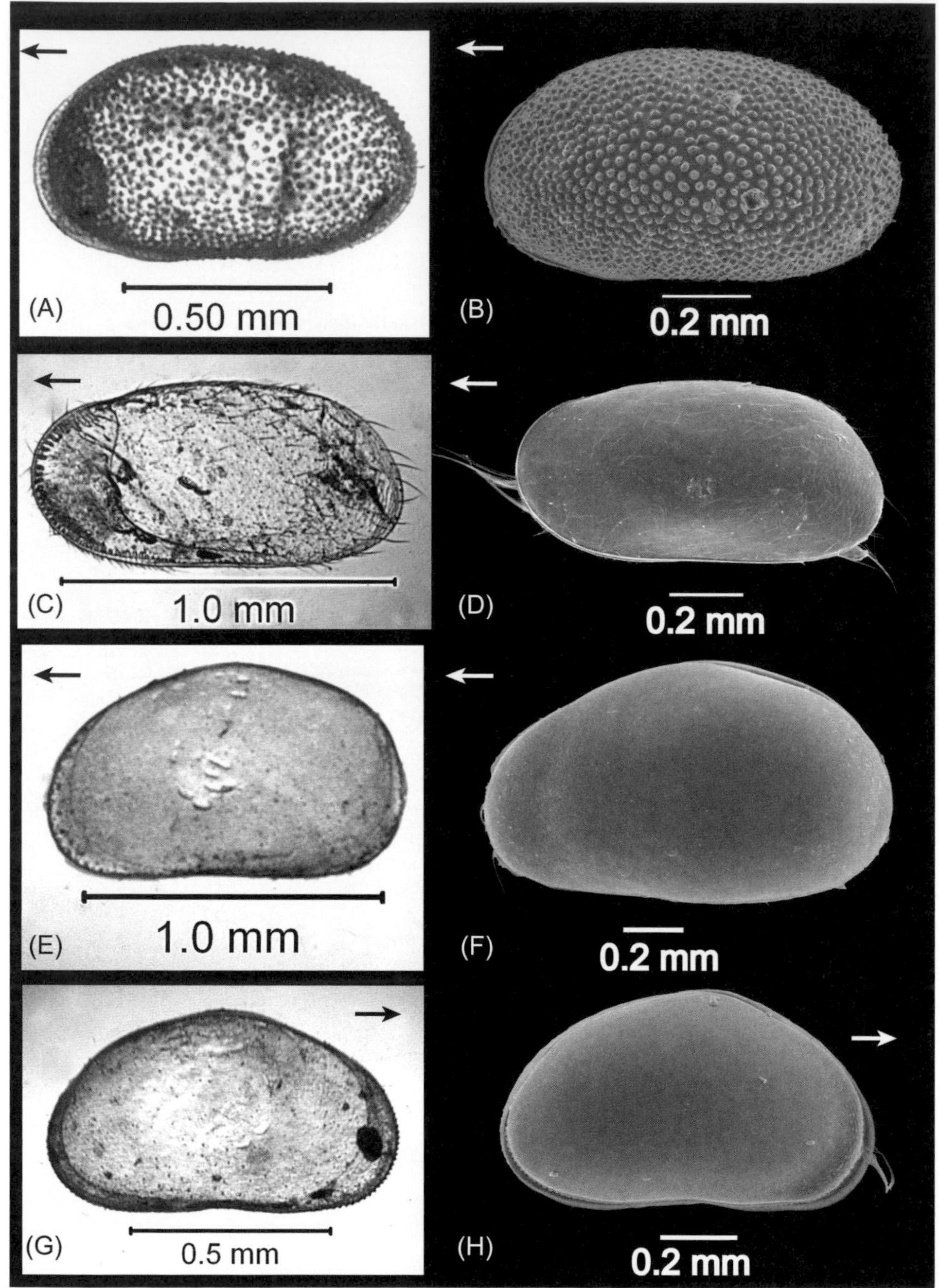

PLATE 16.137 Species of *Spirocypris, Isocypris*, and *Heterocypris.* (A, B) Left valve of *Spirocypris tuberculatus* Sharpe, 1909, from unnamed intermittent stream, Manitoba. LDD collection # 1482. (C, D) Left valve of *Isocypris beauchampi* (Paris, 1920) from littoral, High Point Lake, Pennsylvania. (E, F) Left valve of *Heterocypris incongruens* (Ramdohr, 1808) from Iris Spring, NV (E) and unnamed pond, Alberta: (F) LDD collection # 1790. (G, H) Right valve of *Heterocypris glaucus* (Furtos, 1933) from littoral, Alkaline Lake, North Dakota.

18'	Carapace subovate in lateral view with maximum height slightly behind midpoint; L6 with five podomeres and two long terminal claws, uropodal rami well-developed (L6 terminal claws and well-developed uropodal rami can often be seen protruding from between the valves of whole animals, without recourse to dissection)	*Scottia*
19(18)	Calcified inner lamella more or less equally broad anteriorly and posteriorly	20
19'	Calcified inner lamella broad anteriorly, narrow posteriorly	21
20(19)	Left valve overlaps right valve ventrally, calcified inner lamellae without inwardly displaced selvage	*Cavernocypris*
20'	Right valve overlaps left valve dorsally, left valve calcified inner lamella with inwardly displaced selvage anteriorly and posteriorly	*Potamocypris*

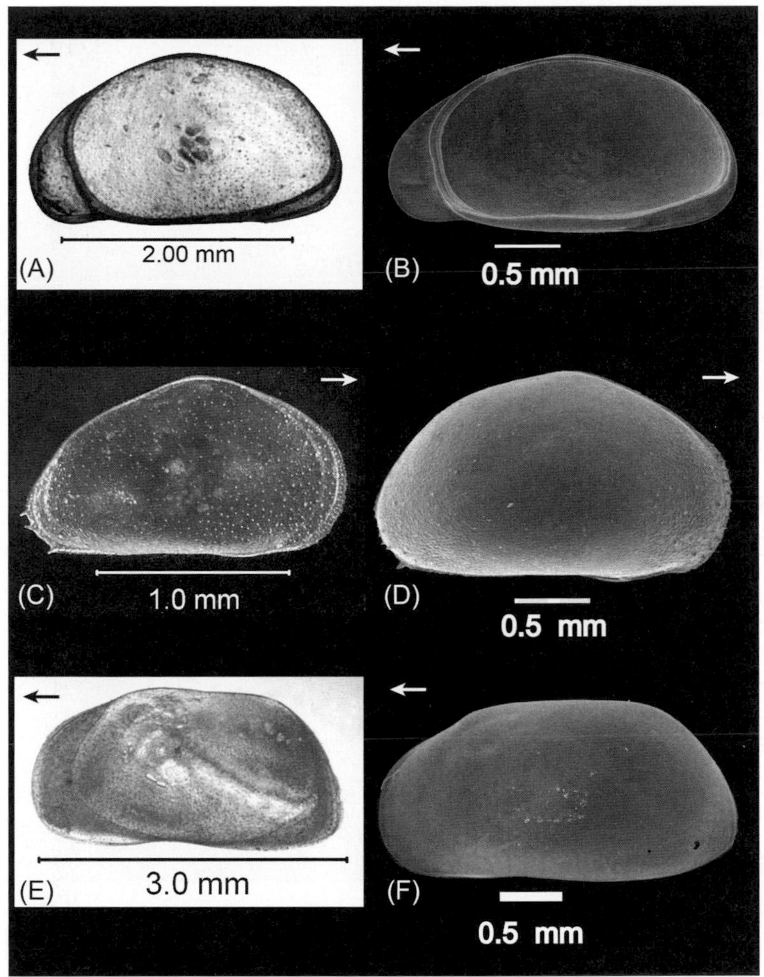

PLATE 16.138 Species of *Chlamydotheca, Cypris*, and *Cypriconcha*. Right valve (A) and right valve interior (B) of *Chlamydotheca arcuata* (Sars, 1901) from Ash Spring, Nevada. (C, D) Right valve of *Cypris pubera* O.F. Müller, 1776, from littoral, Elkwater Lake, Alberta. (E, F) Left valve of *Cypriconcha ingens* Delorme, 1967, from littoral, Lake Ardock, North Dakota.

21(19) Left valve calcified inner lamella with well-developed oblique double folded inner list posteroventrally; both valves without inner list anteriorly .... *Cypridopsis*

21' Both valves with conspicuous anterior inner list running well inside the outer margin; left valve posteroventral inner list simple, inconspicuous .... *Sarscypridopsis*

Ostracoda: Podocopida: Cypridoidea: Notodromadidae: Genera

1 Valves with prominent striated marginal flanges; carapace compressed in dorsal/ventral view; cleaning leg (L7) with a terminal pincer .... *Cyprois*

1' Valve flanges not striated; carapace inflated in dorsal/ventral view, with a flattened ventral area rimmed with ridges; cleaning leg (L7) without a terminal pincer; ommatidia distinctly separated .... *Notodromas*

Ostracoda: Podocopida: Cypridoidea: Candonidae: Genera

Here we use the genus *Candona sensu lato* because although other genera are widely used (particularly *Fabaeformiscandona* and *Pseudocandona*) their diagnostic characters and the species that should be assigned to them are subjects of debate in the literature (see, for example, Meisch, 2000 and Karanovic, 2012) and it remains unclear where many Nearctic species should be correctly assigned. We also include *Candonopsis* in this category of *Candona s.l.* because it is distinguished from *Candona* primarily on the basis of the absence of a posterior seta on the uropod, and thus not distinguishable using any aspects of shell morphology.

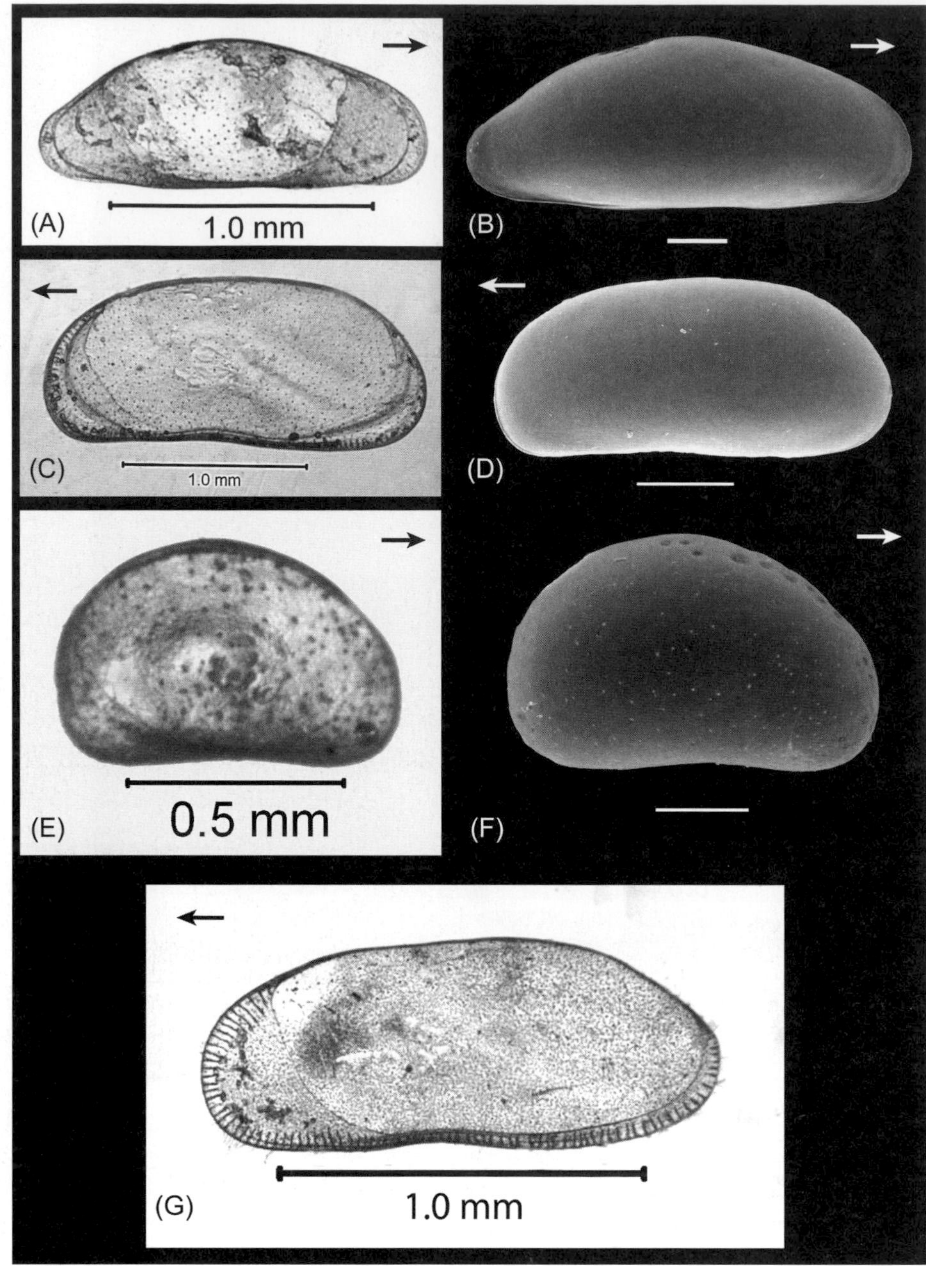

PLATE 16.139 Species of *Dolerocypris, Herpetocypris, Scottia*, and *Stenocypris*. (A, B) Right valve of *Dolerocypris fasciata* (O.F. Müller, 1776). Specimen from littoral zone, Big Twin Lake, Wisconsin. (C, D) Left valve of *Herpetocypris reptans* (Baird, 1835). Specimen from Ocean Point Marsh, Washington. (E, F) Right valve of *Scottia pseudobrowniana* Kempf, 1971, from hydrated fen wetland soil, Herrick Fen, Ohio. (G) Left valve of *Stenocypris major major* (Baird, 1859) from cultivated rice field, Oak Ridge, Louisiana.

1	Carapace length more than 1.5 times the height; subtrapezoidal, subovate, or subreniform in lateral view, often elongate; live specimens typically white	2
1'	Carapace short subovate in lateral view, length less than 1.5 times the height; live specimens often pigmented brown or pink	5
2(1')	Whole carapace surface conspicuously ornamented with net-like reticulation	3
	[At least two Nearctic species of *Candona* s.l. have distinctive reticulate or pitted ornament but should not key out at couplet 3: *C. ikpikpukensis* has a distinctive reticulate ornament that fades out towards the margins, while *C. albicans* (assigned to *Pseudocandona* by Delorme, 1970) has fine pitting that fades out towards the central area.]	
2'	Carapace surface smooth or partially ornamented (typically only faint)	4
3(2)	Carapace dorsal margin highly arched	*Schellencandona*

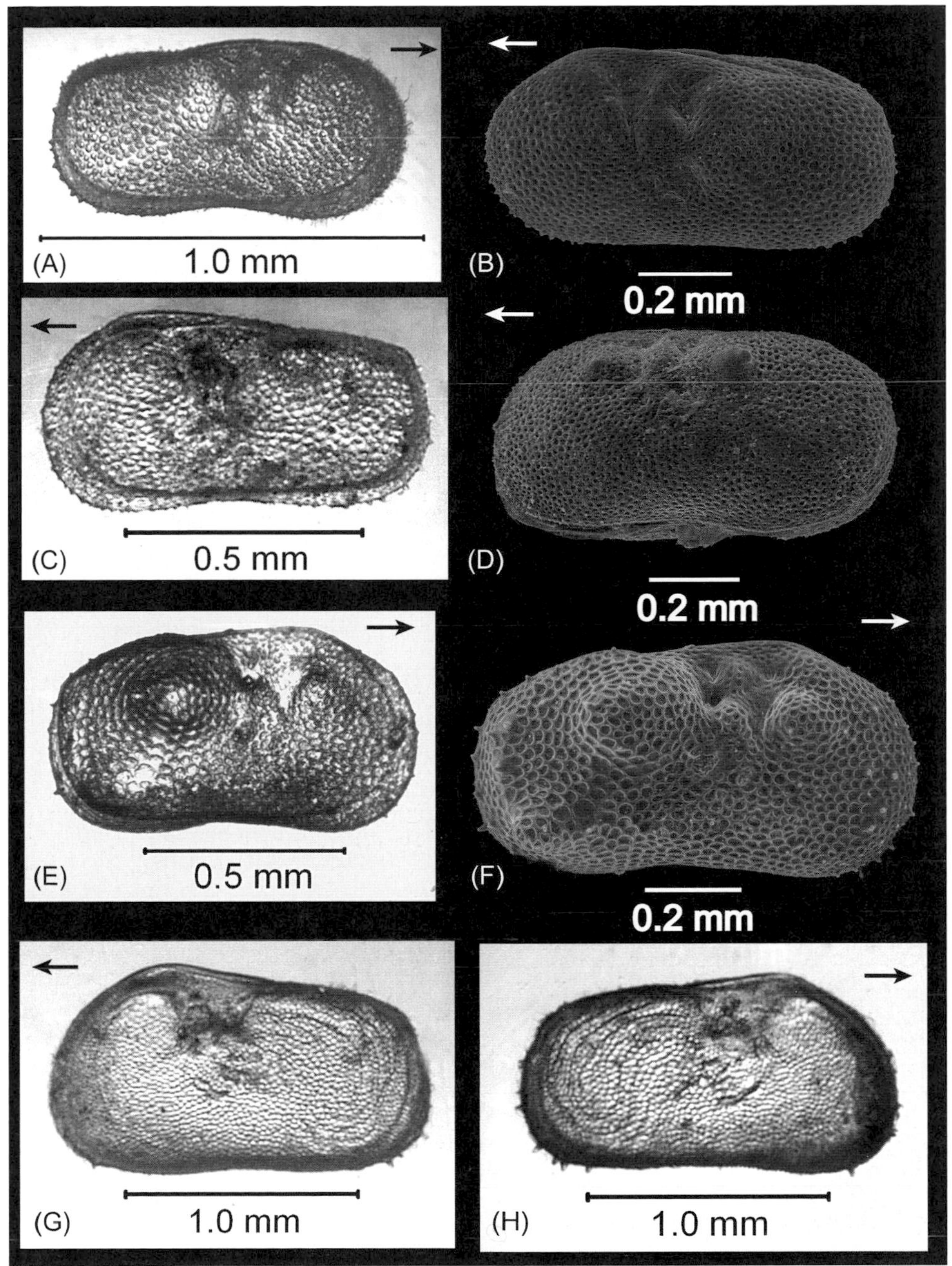

PLATE 16.140 Species of Ilyocypridinae and Pelocypridinae. Right (A) and left (B) valves of *Ilyocypris bradyi* Sars, 1890, from littoral benthos, Bullock Pen Lake, Kentucky. (C, D) left valve of *Ilyocypris gibba* (Ramdohr, 1808) from benthos, Elkwater Lake, Alberta. (E, F) Right valve of *Ilyocypris salebrosa* Stepanaitys, 1960, littoral benthos of Bandwell Reservoir, Texas. Left (G) and right (H) valves of male *Pelocypris tuberculatum* (Ferguson, 1967) from unnamed vernal pool, southern California.

3'	Carapace dorsal margin straight and parallel to ventral margin	*Paracandona*
4(2')	Carapace small (<0.5 mm long), subtrapezoidal in lateral view with short, slightly concave dorsal margin in left valve	*Nannocandona*
4'	Carapace small to large (up to 1.3 mm long), dorsal margin straight or convex in both valves	*Candona sensu lato*
5(1)	Carapace compressed in dorsal/ventral view	6
5'	Carapace tumid in dorsal/ventral view; live specimens brown	*Cyclocypris*
6(5)	Right valve with marginal pustules or denticles anteriorly and/or ventrally	*Physocypria*
6'	Valve margins smooth	*Cypria*

PLATE 16.162 *Amphibalanus inexpectatus*: (A) tergum exterior, (B) tergum interior, (C) scutum exterior, (D) scutum interior, (E) whole shell exterior side view.

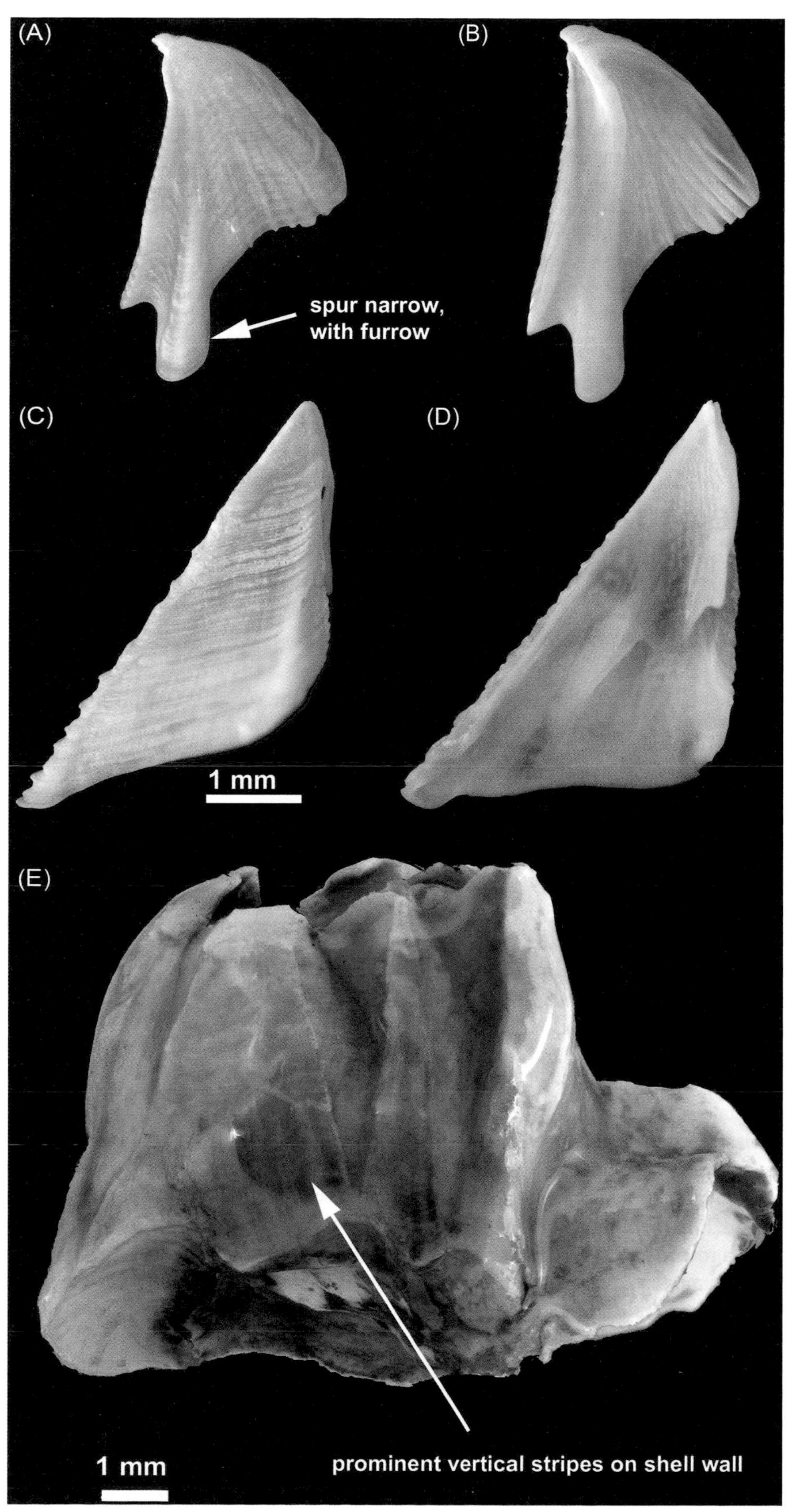

PLATE 16.163 *Fistulobalanus pallidus*: (A) tergum exterior, (B) tergum interior, (C) scutum exterior, (D) scutum interior, (E) whole shell exterior side view.

9(7) A1 ♀, seta of antennomere 1 reaching end of segment 3 or 4; right P5 ♂, medial surface of basis with 1 or 2 small, unornamented, rounded or subrectangular cuticular outgrowths 10

9' A1 ♀, seta of antennomere 1 not reaching beyond middle of segment 2; right P5 ♂, medial surface of basis with a spinulose, denticulate, or serrate protrusion 13

10(9) Genital segment ♀ with large lateral protrusions; right P5 ♂, basis with 1 or 2 rounded processes at midpoint of inner surface, or no process present; and right exp2 with irregularly expanded inner margin 11

10' Genital segment ♀ without lateral protrusions; right P5 ♂, basis with 1 quadrate cuticular process at midpoint of inner surface, and right exp2 with regularly convex inner margin ..... *Hesperodiaptomus schefferi* Wilson, 1953

[USA: Alaska, northern Rocky Mountain states]

11(10) Metasomal wings of ♀ expanded, appearing rounded in dorsal view; right P5 ♂, basis with 1 or 2 rounded to subquadrate processes on inner surface 12

11' Metasomal wings of ♀ not expanded, appearing acute and posteriorly directed in dorsal view; right P5 ♂, basis with a very low process on inner surface ..... *Hesperodiaptomus victoriaensis* (Reed, 1958)

[Canada: Victoria Island, Northwest Territories]

12(11) Genital segment ♀, left protrusion directed dorsally; right P5 ♂, basis about as long as wide and with 1 rounded process at midpoint of inner surface ..... *Hesperodiaptomus californiensis* Scanlin & Reid, 1996

[USA: California]

12' Genital segment ♀, left protrusion directed laterally; right P5 ♂, basis about 1.5 times longer than wide and with 2 subquadrate protrusions on inner surface ..... *Hesperodiaptomus kiseri* (Kincaid, 1953)

[Canada: Saskatchewan. USA: Washington]

13(9) P5 ♀, medial spine of exopod reaching past midlength of claw; right P5 ♂, apical claw sharply bent at about mid-length 14

13' P5 ♀, medial spine of exopod not reaching midlength of claw; right P5 ♂, apical claw gently curved and without distinct angle; right P5 ♂, basis with wide inner flange edges with a row of spinules, and tiny spinous point on caudal face near distal margin of flange ..... *Hesperodiaptomus arcticus* (Marsh, 1920)

[Canada. USA: (Alaska, Pacific coast states, northern Rocky Mountains). Palaearctic]

14(13) Right P5 ♂, basis medial surface smoothly convex, armed only with a medial spinulose protrusion ..... *Hesperodiaptomus breweri* Wilson, 1958

[Canada: Saskatchewan. USA: Nebraska]

14' Right P5 ♂, basis medial surface with a wide, variably shaped spinulose protrusion, caudal surface with longitudinal ridge and a prominent sclerotized spine or denticle ..... *Hesperodiaptomus eiseni* (Lilljeborg, 1889)

[Canada: west coast provinces to Labrador. USA: Alaska, Aleutian Islands, western states. Palaearctic. NOTE: The females of *H. breweri* and *H. eiseni* are very similar.]

15(3) A1 ♀♂ segment 2 with 3 setae (and aesthetasc) 16

15' A1 ♀♂ segment 2 with 4 setae (and aesthetasc) ..... *Hesperodiaptomus caducus* Light, 1938

[Canada: British Columbia. USA: Pacific coast states]

16(15) A1 ♀ and left A1 ♂, segment 3 seta not reaching segment 6; caudal ramus ♀ with no setae dorsally; right P5 ♂, basis caudal surface with no ornament 17

16' A1 ♀, segment 3 seta reaching segment 10; left A1 ♂, segment 3 seta reaching segment 8; caudal ramus ♀ with dorsal setae; right P5 ♂, basis caudal surface with lobe ..... *Hesperodiaptomus hirsutus* Wilson, 1953

[USA: California]

17(16) A1 ♀ and left A1 ♂, segment 19 with 1 seta; genital segment ♀ with lateral protrusions; P5 ♂, apical claw only slightly curved inwards ..... *Hesperodiaptomus shoshone* (Forbes, 1893)

[Canada: Alberta, British Columbia, Ontario, Saskatchewan. USA: Alaska, western states]

17' A1 ♀ and left A1 ♂, segment 19 with 2 setae; genital segment ♀ without lateral protrusions; P5 ♂, apical claw sharply curved inwards at middle ..... *Hesperodiaptomus novemdecimus* Wilson, 1953

[Canada: Rocky Mountains, Saskatchewan. USA: California, Montana, Washington]

Crustacea: Maxillipoda: Copepoda: Calanoida: Diaptomidae: *Leptodiaptomus*: Species

This key is to males only.

1 Right P5 ♂, basipod with or without a process on medial surface, if a process is present it does not exceed half width of basipod, and endopod arises from mediodistal corner; P5 ♀, posterior sensillum of basipod short and bluntly conical 2

1' Right P5 ♂, basipod with large mammiform medial lobe, from which endopod arises medially; P5 ♀, posterior sensillum of basipod spiniform and at least as long as basipod .... *Leptodiaptomus trybomi* (Lilljeborg, 1889)

[USA: Oregon, Multnomah Falls. The assignment to genus *Leptodiaptomus* is tentative]

2(1) Right A1 ♂, segment 19 (antepenultimate segment) without a process (may have a hyaline membrane) 3

2' Right A1 ♂, segment 19 (antepenultimate segment) with a long distal process 5

3(2) Left P5 ♂ exp2, distal and medial processes about the same length, distal process not more than 1/4 length of lateral margin of segment 4

3 Left P5 ♂ exp2, distal process longer than medial process, about 1/2 length of lateral margin of segment .... *Leptodiaptomus tyrrelli* (Poppe, 1888)

[Canada. USA: Alaska west of Rocky Mountains. Palaearctic]

4(3) Right P5 ♂ exp1 with 2 large lobate distal hyaline processes, one on the medial margin and the other on the distal posterior face; left P5 ♂ basipod with tiny lobate process on medial surface .... *Leptodiaptomus coloradensis* (Marsh, 1911)

[USA: Colorado, Utah]

4' Right P5 ♂ exp1 with 2 tiny crescentic distal hyaline processes; left P5 ♂ basipod with large 3-lobed hyaline process .... *Leptodiaptomus pribilofensis* (Juday & Muttkowski, 1915)

[Canada: Manitoba, Northwest Territories. USA: Alaska, Wisconsin]

5(2) Right A1 ♂ segment 15 with a small spinous process, or none 6

5' Right A1 ♂ segment 15 with a prominent spinous process .... *Leptodiaptomus mexicanus* (Marsh, 1929)

[Mexico]

6(5) Right A1 ♂ process of antennomere 19 reaching to middle of antennomere 20 or beyond, its tip swollen or pointed but not outcurved or hook-like 7

6' Right A1 ♂ process of antennomere 19 usually not reaching beyond middle of antennomere 20, usually shorter, its tip outcurved or hook-like 13

7(6) Right A1 ♂ process of antennomere 19 usually with apex swollen, blunt, or rounded; right P5 ♂ coxa (basal segment 1), medial protrusion, if present, small 8

7' Right A1 ♂ process of antennomere 19 usually tapering to aciculate point; right P5 ♂ coxa (basal segment 1) with greatly enlarged medial protrusion that extends posteriorly between the right and left legs .... *Leptodiaptomus spinicornis* Light, 1938

[USA: California, Nevada, Washington]

8(7) Right P5 ♂ basipod, medioproximal surface without process 9

8 Right P5 ♂ basipod, medioproximal surface with process or protrusion 11

9(8) Right P5 ♂ exp1 with large medial process; exp 2 lateral spine distally directed, and inserted at about distal 2/3 of segment 10

9' Right P5 ♂ exp1 with no distal process; exp2 with lateral spine laterally directed, and inserted just proximal to midlength of segment .... *Leptodiaptomus minutus* (Lilljeborg, 1889)

[Canada. USA: east of the Rocky Mountains. Palaearctic]

10(9) Right P5 ♂ exp1 with coniform process on mediodistal corner; left P5 ♂ basis with medioproximal corner not expanded .... *Leptodiaptomus sicilis* (Forbes,1882)

[Canada. USA: Alaska, northern states south to California, Arizona, Missouri, Virginia]

10' Right P5 ♂ exp1 with entire medial margin expanded in subquadrate process; left P5 ♂ basis with medial proximal corner greatly expanded .... *Leptodiaptomus dodsoni* Elías-Gutiérrez, Suárez-Morales and Romano-Márquez, 1999

[Mexico]

11(8) Right P5 ♂ exp2, proximal part of segment more or less widened but not forming distinct angle; lateral spine equal to or shorter than greatest width of the segment 12

11' Right P5 ♂ exp2, proximal part of segment widened and forming distinct angle at point of insertion of lateral spine; lateral spine stout, 1.5–2.5 times greatest width of the segment .... *Leptodiaptomus ashlandi* (Marsh, 1893)

[Canada. USA: northern states]

12(11) Right P5 ♂ exp1, mediodistal margin with small rounded hyaline process, and proximal part of exp2 distinctly wider than distal part; right A1 ♂, process on segment 23 reaching to end of segment 25 .... *Leptodiaptomus insularis* Kincaid, 1956

[USA: Alaska]

12' Right P5 ♂ exp1, mediodistal margin with long pointed hyaline process, and exp 2 not broadened proximally; right A1 ♂, process on segment 23 reaching just past end of antennomere 24 .... *Leptodiaptomus judayi* (Marsh, 1907)

[USA: California, Colorado]

13(6) Right P5 ♂ basipod with no protrusion or process on proximomedial corner 14

13' Right P5 ♂ basipod with conspicuous upwardly projecting protrusion or process on proximomedial corner 16

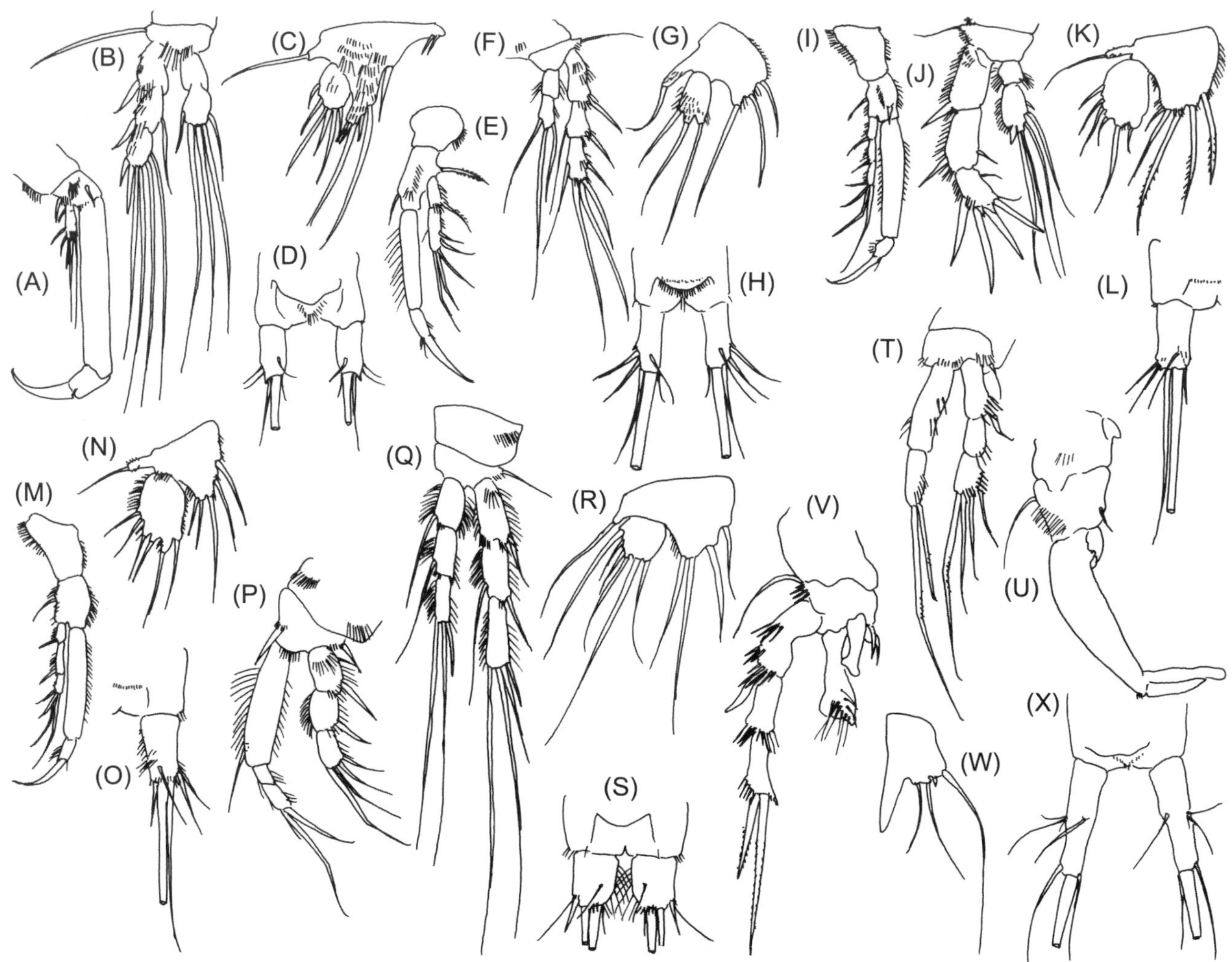

PLATE 16.175 Representative structures of harpacticoid copepods: (A) *Microlaophonte trisetosa* leg 1 female; (B) *M. trisetosa* leg 4 female; (C) *M. trisetosa* leg 5 female; (D) *M. trisetosa* caudal rami female; (E) *Onychocamptus mohammed* leg 1; (F) *O. mohammed* leg 4 female; (G) *O. mohammed* leg 5 female; (H) *O. mohammed* caudal rami female; (I) *Heterolaophonte stroemii* leg 1; (J) *H. stroemii* leg 3 male; (K) *H. stroemii* leg 5 female; (L) *H. stroemii* caudal ramus female; (M) *Pseudonychocamptus proximus* leg 1; (N) *P. proximus* leg 5 female; (O) *P. proximus* caudal ramus female; (P) *Schizopera knabeni* leg 1; (Q) *S. knabeni* leg 4 female; (R) *S. knabeni* leg 5 female; (S) *S. knabeni* caudal rami female; (T) *Parastenocaris brevipes* leg 1 female; (U) *P. brevipes* leg 3 male; (V) *P. brevipes* leg 4 male; (W) *P. brevipes* leg 5 female; (X) *P. brevipes* caudal rami female. [Not to scale.] *Figures redrawn: I–L from Sars, 1907, and P–S from Lang, 1965.*

Crustacea: Maxillipoda: Copepoda: Harpacticoida: Parastenocarididae: *Parastenocaris*: Species

1 Caudal ramus with group of 3 lateral setae inserted at about distal 2/3 of ramus 2

1' Caudal ramus with group of 3 lateral setae inserted in proximal half or at mid-length of ramus 4

2(1) P1 basipod medial surface without spine; P4 ♂ with 1 curved sclerotized spine above complexly branched hyaline endopod 3

2' P1 basipod inner surface with spine (slender and tapering in ♀, club-shaped in ♂); P4 ♂ with 3 curved spines above slender simple hyaline endopod .... *Parastenocaris trichelata* Reid, 1995

[USA: District of Columbia, Virginia]

3(2) P3 ♂ exp1 medial margin with small rounded process on proximal 1/4 and larger process at mid-length .... *Parastenocaris palmerae* Reid, 1992

[USA: Virginia]

3' P3 ♂ exp1 medial margin with only small rounded process on proximal 1/4 .... *Parastenocaris delamarei* Chappuis, 1958

[Great Lakes. USA: New York]

4(1) Caudal ramus with group of 3 lateral setae inserted at midlength of ramus; P5 medial margin bare; caudal ramus ♂ medial margin bare 5

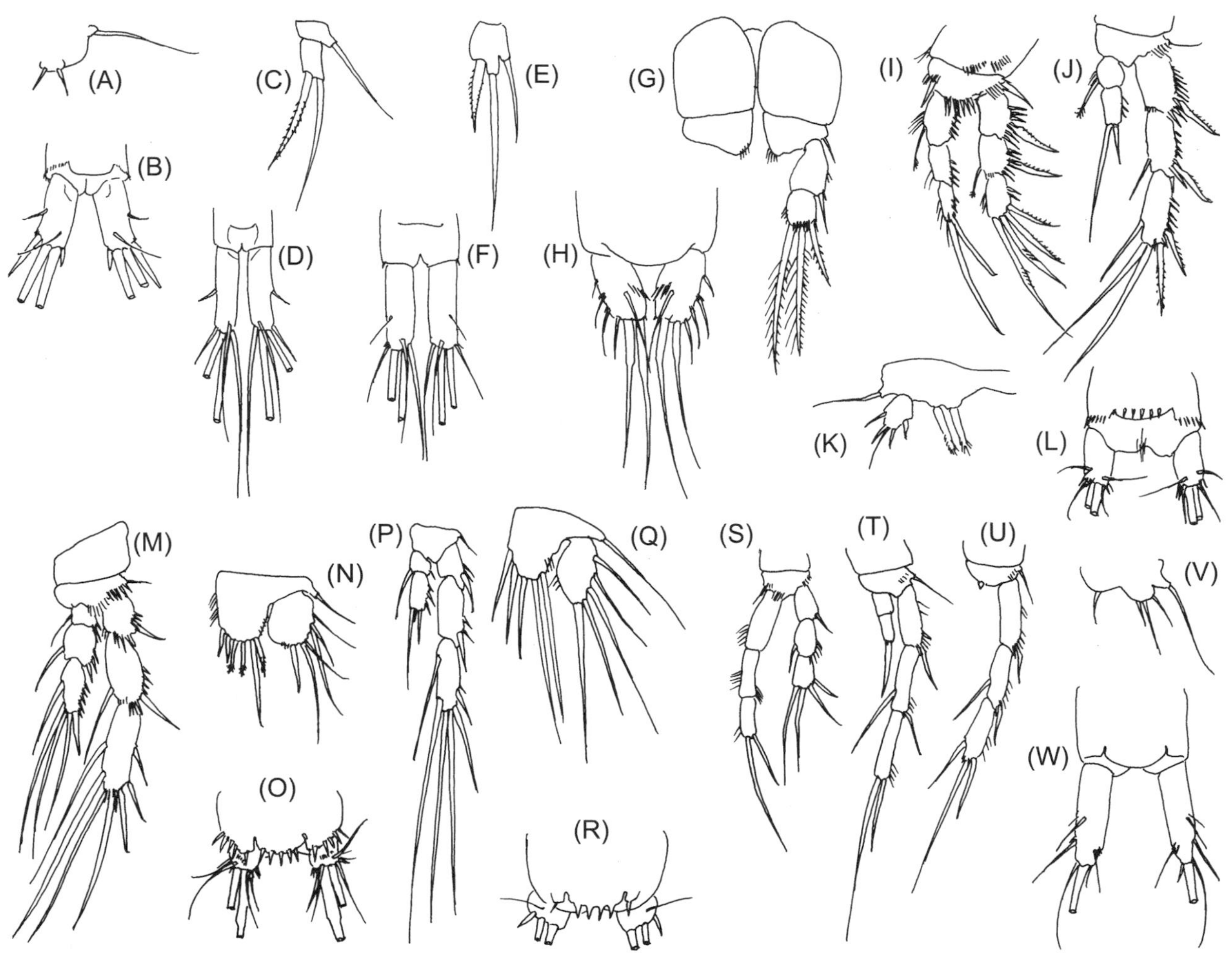

PLATE 16.176 Representative structures of cyclopoid, gelyelloid, and harpacticoid copepods: (A) *Stolonicyclops heggiensis* fifth leg of female; (B) *S. heggiensis* caudal rami; (C) *Thermocyclops tenuis* fifth leg; (D) *T. tenuis* caudal rami; (E) *Tropocyclops prasinus mexicanus* fifth leg; (F) *T. p. mexicanus* caudal rami; (G) *Scaeogelyella caroliniana* leg 1; (H) *S. caroliniana* caudal rami; (I) *Nitocrellopsis texana* leg 1; (J) *N. texana* leg 4 female; (K) *N. texana* leg 5 female; (L) *N. texana* caudal ramus female; (M) *Nitocra spinipes* leg 4 female; (N) *N. spinipes* leg 5 female; (O) *N. spinipes* caudal ramus female; (P) *Novanitocrella aestuarina* leg 4 female; (Q) *N. aestuarina* leg 5 female; (R) *N. aestuarina* caudal rami female; (S) *Psammonitocrella boultoni* leg 1 female; (T) *P. boultoni* leg 3 female; (U) *P. boultoni* leg 4 female; (V) *P. boultoni* leg 5 female; (W) *P. boultoni* caudal rami female. [Not to scale.] *Figures redrawn: I–L from Fiers & Iliffe, 2000; P–R from Coull & Bell, 1979, and S–W from Rouch, 1992.*

4' Caudal ramus with group of 3 lateral setae inserted at proximal 2/3 of ramus; P5 medial margin with row of spines; caudal ramus ♂ with row of spinules along medial proximal margin .... *Parastenocaris lacustris* Chappuis, 1958

[Great Lakes]

5(4) P5 about as long as wide, subquadrate, with 3 setae and 1 tiny spine all on distal margin; P4 ♂ endopod-complex including 2 variously shaped, heavily sclerotized claws, short stocking-shaped hyaline process, and longer hyaline process with subdistal group of spines .... *Parastenocaris brevipes* Kessler, 1913

[USA: Massachusetts, Michigan, New Hampshire, Virginia, Wisconsin. Palaearctic]

5' P5 about twice as long as wide, with apical spiniform projection and 3 or 4 setae on lateral margin; P4 ♂ endopod-complex a proximal sclerotized lamella with 2 projections, and distal clavate hyaline lobe bearing few spines .... *Parastenocaris texana* Whitman, 1984

[USA: Texas]

Crustacea: Maxillipoda: Copepoda: Harpacticoida: Ameiridae: Genera

1 P2 endopods with 3 segments 2

1' P2 endopods with 2 segments 4

2(1) P1 exopod middle segment with a spine on lateral margin 3

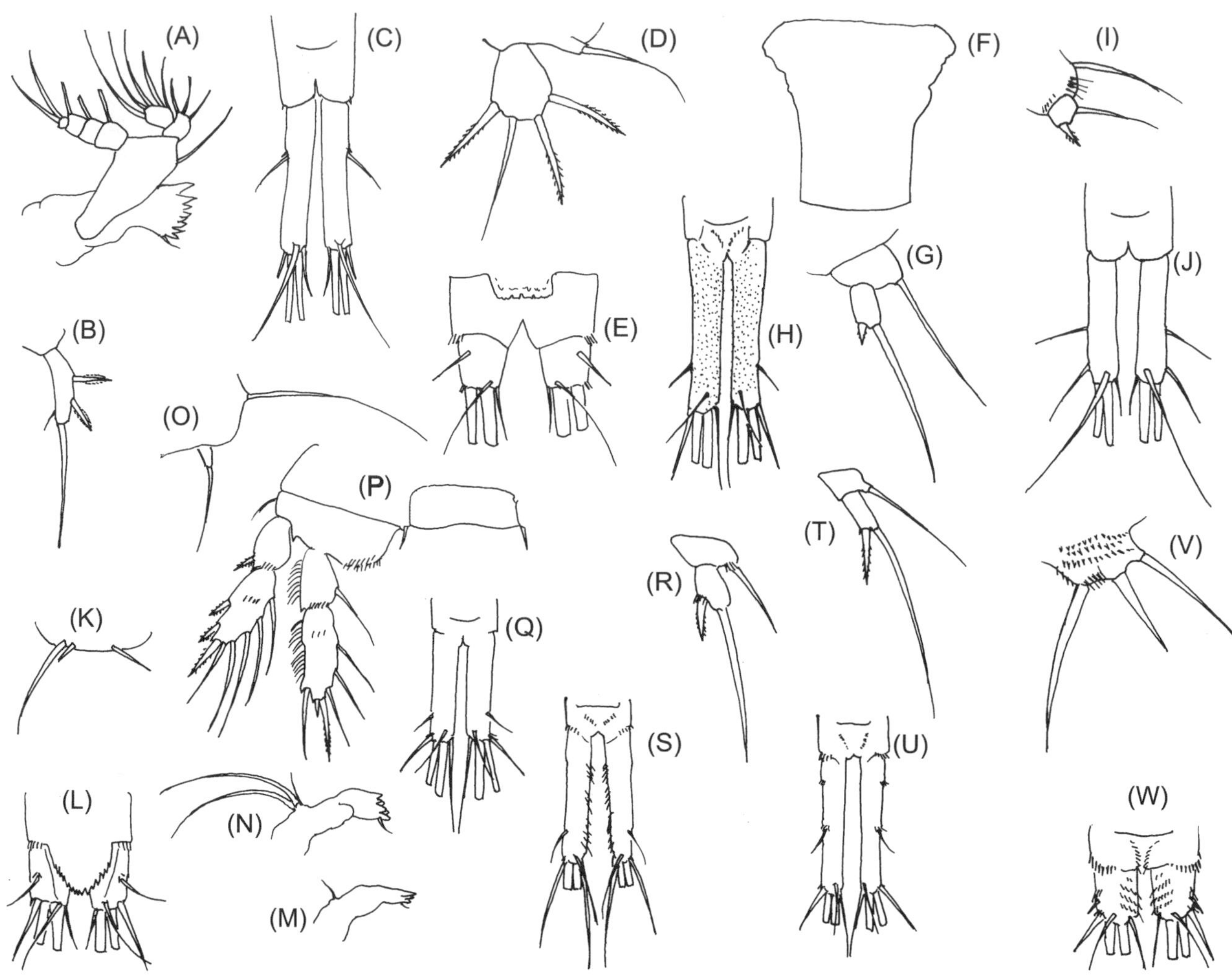

PLATE 16.179 Representative structures of cyclopoid copepods: (A) *Limnoithona tetraspina* mandible; (B) *L. tetraspina* fifth leg; (C) *L. tetraspina* caudal rami; (D) *Halicyclops laminifer* fifth leg female; (E) *H. laminifer* caudal rami; (F) *Acanthocyclops vernalis* female genital double-somite; (G) *A. vernalis* leg 5; (H) *A. vernalis* caudal rami; (I) *Apocyclops panamensis* fifth leg; (J) *A. panamensis* caudal rami; (K) *Bryocyclops muscicola* fifth leg; (L) *B. muscicola* caudal rami; (M) *B. muscicola* mandible; (N) generalized cyclopid mandible with the palp bearing three setae; (O) *Cryptocyclops bicolor* leg 5; (P) *C. bicolor* leg 4 and coupler; (Q) *C. bicolor* caudal rami; (R) *Cyclops strenuus* leg 5; (S) *C. strenuus* caudal rami; (T) *Diacyclops thomasi* leg 5; (U) *D. thomasi* caudal rami; (V) *Ectocyclops phaleratus* leg 5; (W) *E. phaleratus* caudal rami. [Not to scale.] Caudal rami in dorsal view. *Figures redrawn: A–C from Abiahy et al., 2007, and O–Q from Monchenko, 1974.*

5(4) P5 ♀ exopod about 1.5 times longer than wide, and outermost seta on baseoendopod nearly as long as next outermost seta .... *Nitokra lacustris* (Shmankevich, 1875) s.l.

[USA: Florida Keys, Louisiana, New Jersey, South Carolina, Texas. Mexico. Neotropics]

5' P5 ♀ exopod about 2.5 times longer than wide, and outermost seta on baseoendopod less than half length of next outermost seta (about as long as 3 innermost setae) .... *Nitokra lacustris sinoi* Marcus & Por, 1961

[USA: Florida. Mexico: eastern. Neotropical]

6(4) P1 enp1 shorter than exopod segments 1 and 2 combined, and reaching at most to distal end of exp2; ♂ P5 baseoendopod with 4 setae; anal operculum with 7 large teeth .... *Nitokra evergladensis* Bruno & Reid, 2002

[USA: Florida Everglades]

6' P1 enp1 longer than exopod segments 1 and 2 combined, and reaching nearly to mid-length of exp3; ♂ P5 baseoendopod with 3 setae; anal operculum with smooth margin .... *Nitokra bdellurae* (Liddell, 1912)

[USA: southeastern estuaries. Among gill lamellae of *Limulus*]

7(3) P2-4 enp3 with 4,5,5 setae respectively; P5 ♀ exopod with 5 or 6 setae, baseoendopod with 5 setae; P5 ♂ exopod with 6 setae, baseoendopod with 3–5 setae .... *Nitokra spinipes* Boeck, 1865

[Canada: Hudson Bay. USA: Alaska, Atlantic coast to Florida Keys. Mexico: Yucatán]

7' P2-4 enp3 with 3,4,4 setae respectively; P5 ♀ ♂ baseoendopod both with 2 setae .... *Nitokra bisetosa* Mielke, 1993

[USA: Florida Everglades. Neotropics]

Crustacea: Maxillipoda: Copepoda: Harpacticoida: Ameiridae: *Stygonitocrella*: Species

1 P5 ♀ baseoendopod medial lobe absent; caudal ramus ♀ about as long as wide .... *Stygonitocrella* (*Eustygonitocrella*) *mexicana* Suárez-Morales & Iliffe, 2005

[Mexico: Tamaulipas]

1' P5 ♀ baseoendopod medial lobe present, bearing 2 setae; caudal ramus ♀♂ 6 times longer than wide (Pl. 16.178 A–D) .... *Stygonitocrella* (*Fiersiella*) *sequoyahi* Reid, Hunt, & Stanley, 2003

[USA: Arkansas, Oklahoma]

Crustacea: Maxillipoda: Copepoda: Harpacticoida: Tachidiidae: Genera

1 P1 exp3 with 6 setae and spines; ♂ P2 enp2 without inner apophysis 2

1' P1 exp3 with 5 setae and spines; ♂ P2 enp2 with inner apophysis .... ***Tachidius*** **[p. 553]**

2(1) P1-4 endopod 1 small, without medial seta; ♀ with genital double-somite .... *Microarthridion littorale* (Poppe, 1881)

[Coastal and Great Lakes. Palaearctic]

2' P1-4 endopod 1 normal sized, with medial seta; ♀ with genital and first abdominal somites separate .... *Geeopsis incisipes* (Klie, 1913)

[USA: Alaska. Palaearctic]

Crustacea: Maxillipoda: Copepoda: Harpacticoida: Tachidiidae: *Tachidius*: Species

1 A2 exopod with 3 or more setae; anal operculum ♀ with few small spinules, or smooth .... *Tachidius spitzbergensis* Oloffson, 1917

[Canada: Atlantic coast. USA: Alaska Atlantic coast south to New York. Palaearctic]

1' A2 exopod with 2 setae; anal operculum ♀ with 12 or more spinules (Pl. 16.172 M–O) .... *Tachidius discipes* Giesbrecht, 1881

[Canada: Northwest Territories, Ellesmere Island. USA: Alaska]

Crustacea: Maxillipoda: Copepoda: Harpacticoida: Huntemannidae: Genera

1 P1 endopod prehensile (proximalmost segment long; Pl. 16.175 P) *or* not prehensile, with 2 or more apical setae, which may be stout (Pl. 16.175 T) .... ***Huntemannia*** **[p. 553]**

1' P1 endopod prehensile, much longer than exopod, ending in 1 stout claw; if an apical seta is also present, it is tiny (Pl. 16.175 A) .... *Nannopus palustris* Brady, 1880

[Estuaries. Holarctic]

Crustacea: Maxillipoda: Copepoda: Harpacticoida: Huntemannidae: *Huntemannia*: Species

1 P4 exp2 ♂ with 6 setae; P5 ♀ exopod about as long as wide .... *Huntemannia jadensis* Poppe, 1885

[Canada: British Columbia. USA: Alaska, Washington. Palaearctic]

1' P4 exp2 ♂ with 7 setae; P5 ♀ exopod about twice as long as wide (Pl. 16.175 Q–T) .... *Huntemannia lacustris* Wilson, 1958

[USA: Utah]

Crustacea: Maxillipoda: Copepoda: Harpacticoida: Cletodidae: Genera

1 A1 ♀ 5 segmented; ♀ P1-3 endopods 2 segmented, P4 endopod 1 segmented, distal or only segments all with 2 setae .... *Kollerua breviarticulatum* (Shen & Tai, 1964)

1' A1 ♀ 4-segmented; ♀ P1-4 endopods 2-segmented, distal segments with 2,2,3,3 setae .... *Limnocletodes behningi* Borutzky, 1926

Crustacea: Maxillipoda: Copepoda: Harpacticoida: Darcythompsoniidae: *Leptocaris*: Species

1 P1 endopod 1 segmented 2

1' P1 endopod 2 segmented 3

2(1) P1 exp3 with 3 setae and spines, P3 and 4 exp3 with 4 setae and spines .... *Leptocaris mangalis* Por, 1983

[USA: Florida, South Carolina. Neotropics]

Crustacea: Maxillipoda: Copepoda: Harpacticoida: Canthocamptidae: *Bryocamptus*: Species

1 P1 exp2 with inner (medial) seta 2

1' P1 exp2 without inner seta 3

2(1) P1 endopod 2 segmented 8

2' P1 endopod 3 segmented 15

3(1) P2-4 exopods with 3 lateral spines 4

3' P2-4 exopods with 2 lateral spines .... *Bryocamptus (Arcticocamptus) subarcticus* (Willey, 1925)

[Canada: Québec. USA: Alaska]

4(3) P3 ♀ enp2 with 3 or 4, and P4 enp2 with 4 setae; caudal ramus ♀ with or without distal, inner group of spines 5

4' P3 and 4 ♀, enp2 each with 5 setae; caudal ramus ♀♂ with distal, inner group of spines .... *Bryocamptus* (*Limocamptus*) *hiemalis*-complex 6

5(4) Caudal ramus ♀ (and presumably that of ♂) lacking inner group of spines; P3 ♀ apical endopod segment with 3 or 4 setae; P4 ♀ apical endopod segment with 4 setae .... *Bryocamptus* (*Limocamptus*) *morrisoni* (Chappuis, 1929)

[USA: Indiana]

5' Caudal ramus ♀♂ with inner group of spines; ♀ P3 and 4, enp2 with 3 and 4 setae respectively (middle apical seta of P4 enp2 is tiny, hard to see); ♀ P2-4 enp1 with 0,1,0 setae; ♂ P4 enp2 with 4 setae, of which inner apical seta is tiny .... *Bryocamptus* (*Limocamptus*) *elegans* (Chappuis, 1929)

[USA: District of Columbia, Kentucky]

6(4) P4, middle apical seta of enp2 shorter than outer (lateral) spine 7

6' P4, middle apical seta of enp2 longer than outer spine (Pl. 16.178 Q, R) .... *Bryocamptus* (*Limocamptus*) *hiemalis* (Pearse, 1905)

[Canada: British Columbia, Ontario. USA: Colorado, District of Columbia, Nebraska, Minnesota, Montana, North Carolina, Tennessee, Utah. Palaearctic.]

7(6) A1 ♀ 8 segmented .... *Bryocamptus* (*Limocamptus*) *nivalis* (Willey, 1925)

[Canada: Québec. USA: Alaska, District of Columbia, Indiana, Michigan, New York, North Carolina, West Virginia, Great Lakes. Palaearctic]

7' A1 ♀ 7 segmented .... *Bryocamptus* (*Limocamptus*) *douwei* (Willey, 1925)

[Canada: Québec]

8(2) P2 and P3 exp3 with 3 outer marginal spines 9

8' P2 and P3 exp3 with 2 outer marginal spines 10

9(8) ♀ urosome segment 4 with spine row consisting of large lateral spines and small ventral spines, with larger ventromedial spines; P3 ♂ endopod, longer apical seta unmodified (long, slender, tapering) .... *Bryocamptus* (*Rheocamptus*) *zschokkei* (Schmeil, 1893) s.str.

[Canada: British Columbia, Québec. USA: Alaska, Colorado, District of Columbia, Iowa, Maryland, Minnesota, New Hampshire, New York, North Carolina, West Virginia, Great Lakes. Cosmopolitan, except Australia]

9' ♀ urosome segment 4 with spine row consisting of large spines laterally and only small spines ventrally; P3 ♂ endopod, longer apical seta modified (broad and flattened, tapering sharply to slender curved tip) .... *Bryocamptus* (*Rheocamptus*) *alleganiensis* Coker, 1934

[USA: District of Columbia, New York, North Carolina, Virginia, West Virginia, ?Alaska]

10(8) P2 and 3 exp2 with short inner seta (rarely reaching beyond middle of exp3) 11

10' P2 and 3 exp2 with well-developed inner seta (reaching at least near to end of exopod); ♀ P2-4 enp2 with 4,5,5 setae respectively; anal operculum finely denticulate .... *Bryocamptus* (*Rheocamptus*) *pygmaeus* (Sars, 1862)

[USA: New York. Cosmopolitan, except Australia]

11(10) P3 exp3 with total of 6 spines and setae; caudal ramus ♀, outer apical caudal seta overlying middle seta 12

11 P3 exp3 with total of 5 spines and setae; caudal ramus ♀, apical caudal setae not overlying one another 13

12(11) P5 ♀ baseoendopod, outermost seta reaching well past distal end of segment; ♀ anal operculum with about 18 tiny teeth; P5 ♂ baseoendopod, inner seta less than 1.5 times longer than outer seta .... *Bryocamptus* (*Arcticocamptus*) *cuspidatus* (Schmeil, 1893)

[Canada: Québec. USA: New York. Palaearctic]

12' P5 ♀ baseoendopod with outermost seta short, not reaching distal end of segment; ♀ anal operculum with 11–16 (usually 12) teeth; P5 ♂ baseoendopod, inner seta more than 2 times length of outer seta .... *Bryocamptus* (*Arcticocamptus*) *cuspidatus intermedius* Flössner, 1988

[Greenland: Disko Island]

13(11) P5 ♀, seta 4 of baseoendopod similar in length and slenderness to seta 5; P3 ♂ apical endopod setae much shorter than endopod 3 segment 14

13' P5 ♀, seta 4 of baseoendopod much shorter and stouter than setae 3 and 5; P3 ♂ apical endopod setae about as long as endopod 3 segment (Pl. 16.178 S–U) .... *Bryocamptus* (*Arcticocamptus*) *tikchikensis* Wilson, 1958

[USA: Alaska. Greenland.]

14(13) Caudal ramus with crest of spines at medial to dorsal midlength; P4 enp2 with 5 and 4 setae respectively .... *Bryocamptus* (*Arcticocamptus*) *arcticus* (Lilljeborg, 1902)

[Canada: Newfoundland. USA: Alaska. Palaearctic]

14' Caudal ramus without crest of spines; P4 enp2 with 5 setae, of which innermost seta is tiny .... *Bryocamptus* (*Arcticocamptus*) *subarcticus* (Willey, 1925)

[Canada: Québec. USA: Alaska]

15(2) P1 enp1 without inner seta (may have 1–2 long setae) 16

15' P1 enp1 with inner seta 17

16(15) Anal operculum without spines .... *Bryocamptus* (*Bryocamptus*) *newyorkensis* (Chappuis, 1927)

[USA: Florida, Louisiana, Minnesota, New York, Tennessee, Lake Huron. Palaearctic]

16' Anal operculum with spines; P2-4 ♀, endopods 2 segmented .... *Bryocamptus* (*Bryocamptus*) *hiatus* (Willey, 1925)

[Canada: Québec. USA: Alaska, North Carolina, Virginia. Ishida (1992) described a variation from Alaska, with 6 setae on P5 ♀ baseoendopod (2 short inner corner setae); eastern populations have 5 setae (1 short inner corner seta)]

17(15) P5 ♀, baseoendopod with 6 setae; P3 ♂ endopod with normal apical setae; ♀ P2-3 endopods 3 segmented (segmentation is sometimes partial, or more evident on the P3 endopod); P2 ♂ endopod 3 segmented .... *Bryocamptus* (*Bryocamptus*) *minutus*-complex 18

17' P5 ♀, baseoendopod with 5 setae, of which seta 1 is very reduced (and sometimes absent); P3 endopod with one of its apical setae modified (broad, flattened, bifid); ♀ P2-3 endopods 3 segmented; P2 ♂ endopod 2 segmented .... *Bryocamptus* (*Bryocamptus*) *umiatensis* Wilson, 1958

[USA: Alaska. Palaearctic]

18(17) Caudal ramus ♀ with the usual 3 apical setae present, the outer seta slender and normally setiform; apical caudal setae inserted side by side, or the outer seta more or less underlying the longer middle seta 19

18 Caudal ramus ♀ with 2 apically placed caudal setae (the outer seta lacking, the long middle seta usually greatly enlarged, the inner seta reduced as usual), plus accessory piliform seta inserted in a hollow on the ventral surface of the ramus 22

19(18) Caudal ramus ♀ outer distal corner with conical or piliform process 20

19' Caudal ramus ♀ outer distal corner with no process present 21

20(19) Caudal ramus ♀ with an apical, outwardly or somewhat dorsolaterally placed, more or less spiniform process at base of caudal setae, this process not reaching joint of middle apical caudal seta .... *Bryocamptus* (*Bryocamptus*) *hutchinsoni* Kiefer, 1929

[Canada: Northwest Territories, Québec, Saskatchewan. USA: Connecticut, District of Columbia, Florida, Minnesota, Virginia. Palaearctic]

20' Caudal ramus ♀ outer distal corner with long blunt piliform process, reaching past joint of large middle apical caudal seta .... *Bryocamptus* (*Bryocamptus*) *pilosus* Flössner, 1989

[U.S.A. (Montana). Lakes. Length ♀ 0.60–0.66 mm, ♂ 0.53–0.58 mm.]

21(19) Caudal ramus ♀ with caudal setae inserted apically .... *Bryocamptus* (*Bryocamptus*) *washingtonensis* Wilson, 1958

[USA: California, Oregon, Washington]

21' Caudal ramus ♀ with outer caudal seta inserted toward ventral surface (may or may not completely underlie the larger middle seta) .... *Bryocamptus* (*Bryocamptus*) *minutus* (Claus, 1863)

[Canada: possibly Québec. USA: New York, Great Lakes, possibly District of Columbia. Palaearctic]

22(18) Caudal ramus ♀ outer distal corner produced as a spinous point 23

22' Caudal ramus ♀ outer distal corner not produced .... *Bryocamptus* (*Bryocamptus*) *minusculus* (Willey, 1925)

[Canada: Québec. USA: New York. Inadequately described]

23(22) Anal operculum with simple triangular teeth .... *Bryocamptus* (*Bryocamptus*) *vejdovskyi* (Mrázek, 1893)

[Canada: Northwest Territories. USA: Alaska, Minnesota, New York, West Virginia. Palaearctic]

23' Anal operculum with 10 teeth, middle 6 teeth bifid .... *Bryocamptus* (*Bryocamptus*) *vejdovskyi* forma *minutiformis* Kiefer, 1934

[USA: Connecticut, Michigan]

Crustacea: Maxillipoda: Copepoda: Harpacticoida: Canthocamptidae: *Pesceus*: Species

1 P2-4 ♀ and P 2 and 4 ♂ without endopods (Pl. 16.173 I–M) .... *Pesceus reductus* (Wilson, 1956)
[USA: Alaska. Palaearctic]

1' P2-4 ♀ and P 2 and 4 ♂ with 2 segmented endopods .... *Pesceus reggiae* (Wilson,1958)
[USA: Alaska]

Crustacea: Maxillipoda: Copepoda: Harpacticoida: Canthocamptidae: *Moraria*: Species

1 Middle urosomites with spines only on ventral and lateral portions of posterior margins 2

1' Middle urosomites with spines completely encircling posterior margins .... *Moraria hudsoni* Reid & Lesko, 2003
[USA: Michigan, Lakes Huron and Michigan]

2(1) Anal operculum produced posteriorly well past end of anal somite, by half length of operculum 3

2' Anal operculum not extending past end of anal somite 4

3(2) Anal operculum rounded distally; inner surface of caudal ramus with row of spines in ♀, no spines in ♂ *Moraria laurentica* Willey, 1927
[Canada: Québec. USA: New Jersey]

3' Anal operculum pointed distally; inner surface of caudal ramus ♀♂ with row of spines .... *Moraria virginiana* Carter, 1944
[USA: Maryland, North Carolina, Tennessee, Virginia]

4(2) P2-4 enp2 ♀ with 3,3,3 or 3,4,4 setae, or variations of these numbers; P2 enp2 ♂ with 2 apical setae (as far as known) 5

4' P2-4 enp2 ♀ with 4,5,4 setae; P2 enp2 ♂ with 1 inner and 2 apical setae .... *Moraria duthiei* (Scott &Scott, 1896)
[Canada: Alberta, British Columbia, Northwest Territories, Saskatchewan, Yukon. Greenland. USA: Alaska, Lakes Huron and Michigan. Palaearctic]

5(4) Posterior margins of body segments not serrate 6

5b. Posterior margins of body segments strongly serrate .... *Moraria cristata* Chappuis, 1929
[USA: Great Lakes, District of Columbia, Indiana, Maryland, Minnesota, North Carolina, Ohio, Tennessee, Virginia]

6(5) Caudal ramus with row of spines on inner margin (as far as known); P5 exopod ♀ innermost seta like other setae 7

6' Caudal ramus with inner margin smooth; P5 exopod ♀ innermost seta modified (spiniform) .... *Moraria affinis* Chappuis, 1927
[USA: Alaska, New York, Lake Huron, Virginia]

7(6) Anal operculum semicircular, with smooth margin .... *Moraria mrazeki* Scott, 1902
[Canada: Alberta, British Columbia, Northwest Territories, Saskatchewan, Yukon. Greenland. USA: Alaska, Colorado, Minnesota, North Carolina, Virginia, Great Lakes. Palaearctic]

7' Anal operculum quadrate, with toothed margin .... *Moraria arctica* Flössner, 1989
[Canada: Northwest Territories. Greenland: Disko Island]

Crustacea: Maxillipoda: Copepoda: Harpacticoida: Canthocamptidae: *Gulcamptus*: Species

1 P2-4 ♀ enp2 with 3,3,3 setae respectively 2

1' P2-4 ♀ enp2 with 2,4,3 setae respectively .... *Gulcamptus laurentiacus* (Flössner, 1992)
[Canada: Yukon, Northwest Territories]

2(1) P4 endopod ♀ M 2 segmented; anal operculum with 3–8 (usually 5 or 6) teeth .... *Gulcamptus alaskaensis* Ishida, 1996
[USA: Alaska]

2' P4 endopod ♀ 1 segmented (♂ unknown); anal operculum ♀ with 3 large teeth .... *Gulcamptus huronensis* Reid, 1996
[USA: Alaska, Michigan—Lake Huron]

Crustacea: Maxillipoda: Copepoda: Harpacticoida: Canthocamptidae: *Mesochra*: Species

1 A1 ♀ 6-segmented 2

1' A1 ♀ 7-segmented 4

2(1) P2-4 exp3 with 6,7,7 setae and spines 3

2' P2-4 exp3 with 5,6,6 setae and spines .... *Mesochra wolskii* Jakubisiak, 1933
[USA: Louisiana, Texas. Mexico: Yucatán. Palaearctic, Neotropics, Oriental]

3(2) P5 ♀ baseoendopod with 6, exopod with 5 setae; P5 ♂ baseoendopod with 3, exopod with 5 setae .... *Mesochra mexicana* Wilson, 1971

[USA: Louisiana, South Carolina, Texas]

3' P5 ♀ baseoendopod with 5, exopod with 4–5 setae; P5 ♂ baseoendopod with 2, exopod with 5–6 setae .... *Mesochra pygmaea* (Claus, 1863)

[North American Pacific, Arctic, Atlantic coasts]

4(1) P1 endopod 3-segmented; P5 ♀ baseoendopod with 5–6, exopod with 5–6 setae; P5 ♂ baseoendopod with 3, exopod with 4 setae 5

4' P1 endopod 2-segmented; P5 ♀ baseoendopod with 6, exopod with 5 setae; P5 ♂ baseoendopod with 3, exopod with 5 setae .... *Mesochra lilljeborgi* Boeck, 1865

[USA: Massachusetts, Florida, Washington. Palaearctic]

5(4) P1 endopod segment 1 reaching well past end of exopod .... *Mesochra alaskana* Wilson, 1958

[USA: Alaska, Aleutian Islands, Washington, Lakes Erie and Ontario). Palaearctic]

5' P1 endopod segment 1 not reaching end of exopod .... *Mesochra rapiens* (Schmeil, 1894)

[Canada: British Columbia, Northwest Territories. USA: Alaska, Pacific coast, Utah—Bear Lake. Palaearctic]

Crustacea: Maxillipoda: Copepoda: Harpacticoida: Canthocamptidae: *Canthocamptus*: Species

1 Anal somite with spinous processes on lateral part of posterior margin 2

1' Anal somite with only spinules along its posterior margin 3

2(1) Caudal ramus with spinules on inner margin, apical caudal setae jointed at their bases .... *Canthocamptus oregonensis* Wilson, 1956

[Western North America]

2' Caudal ramus with smooth inner margin, apical caudal seta not jointed at their bases .... *Canthocamptus staphylinus* (Jurine, 1820)

[Invasive in the Great Lakes. Palaearctic]

3(1) P2 endopod with 2 setae on distal part of medial margin of terminal segment 4

3' P2 endopod with 1 seta on distal part of inner margin of terminal segment 5

4(3) Caudal ramus ♀♂, medial margin with spinules .... *Canthocamptus staphylinoides* Pearse, 1905

[Canada: Ontario. USA: Alaska, central and northeastern states]

4' Caudal ramus ♀, inner margin with spinules; caudal ramus ♂, with or without spinules .... *Canthocamptus sinuus* Coker, 1934

[USA: Colorado, Connecticut, Maryland, New Jersey, North Carolina, Virginia]

5(3) Caudal ramus outer apical seta stout, spiniform, usually with setules; P5 ♂, where known, exopod with 6 setae 6

5' Caudal ramus outer apical seta extremely slender, usually smooth; P5 ♂ exopod with 5 setae .... *Canthocamptus robertcokeri* Wilson, 1958

[USA: California, Ohio, Missouri, Kansas, North Carolina, Louisiana, Ohio, Utah, Virginia, Great Lakes]

6(5) P5 ♀ baseoendopod, mid-portion of distal margin hardly produced, with all but innermost seta on the same level; P4 ♀ endopod middle apical seta shorter or only slightly longer than outer apical seta .... *Canthocamptus assimilis* Kiefer, 1931

[Canada: western regions. USA: Alaska, Arizona, California, Connecticut, Louisiana, Missouri, Nebraska, New York, New Hampshire, Ohio, Virginia, Great Lakes]

6' P5 ♀ baseoendopod, mid-portion of distal margin bearing setae 2 to 4 produced, setae 5 and 6 inserted next to each other at same level; P4 ♀ endopod middle apical seta 2–4 times longer than outer apical seta .... *Canthocamptus vagus* Coker & Morgan, 1940

[Canada: British Columbia, Newfoundland. USA: Colorado, District of Columbia, North Carolina, Washington, Virginia, Great Lakes]

Crustacea: Maxillipoda: Copepoda: Harpacticoida: Canthocamptidae: *Elaphoidella*: Species

The key is mainly for females, because the males of several species cannot be distinguished or are undescribed. *E. tenuicaudis* (Herrick) may be a valid species, but is inadequately described and is not included in the key.

1 Caudal ramus ♀♂ variously shaped and in some species with dorsal keel, but without prominent dorsal hook; P5 ♀ exp with 3–5 setae, if 5 setae are present then exopod segment is much longer than baseoendopod; P4 enp ♂ 2-segmented, with 3 setae on distal segment 2

1' Caudal ramus ♀♂ slightly longer than wide, with prominent dorsal hook; P5 ♀ exp with 5 setae, exopod segment shorter than baseoendopod; P4 enp ♂ 1-segmented, with 2 apical setae .... *Elaphoidella bidens* (Schmeil, 1894)

[USA: Minnesota to Louisiana and east from New York to Florida. Mexico: Coahuila, Nuevo León, San Luis Potosí. Considered cosmopolitan, but is probably introduced in the Americas]

2(1) P5 exp ♀ longer than inner margin of baseoendopod, with 5 setae; P2 enp2 ♂ with 4 setae, as far as known 3

2' P5 exp ♀ shorter than inner margin of baseoendopod, with 3 or 4 setae; P2 enp2 ♂ with 3 setae, as far as known 6

3(2) P2 enp2 ♀ with 6 setae 4

3b. P2 enp2 ♀ with 5 setae .... *Elaphoidella californica* Wilson, 1975

[USA: California]

4(3) Caudal ramus ♀ 2–3 times longer than wide 5

4b. Caudal ramus ♀ about as long as wide, extending slightly past end of anal somite .... *Elaphoidella wilsonae* Hunt, 1979

[USA: Colorado, New Mexico]

5(4) Caudal ramus ♀ with 3 apical processes, 2 of these digitiform, in addition to 3 apical setae .... *Elaphoidella kodiakensis* Wilson, 1975

[USA: Alaska]

5' Caudal ramus ♀ without apical processes, only 3 apical setae .... *Elaphoidella reedi* Wilson, 1975

[Canada: Saskatchewan]

6(2) P2 enp2 ♀ with 3 or 4 setae 7

6' P2 enp2 ♀ with 5 setae .... *Elaphoidella marjoryae* Bruno & Reid, 2000

[USA: Florida]

7(6) P2 enp2 ♀ with 3 setae 8

7' P2 enp2 ♀ with 4 setae 9

8(7) P3 enp2 ♀ with 5 setae .... *Elaphoidella shawangunkensis* Strayer, 1989

[USA: New York]

8' P3 enp2 ♀ with 3 setae .... *Elaphoidella amabilis* Ishida, 1993

[USA: Maryland]

9(7) Caudal ramus ♀ without inner process; P3 enp2 ♀ with 5 setae 10

9' Caudal ramus ♀ with inner process; P3 enp2 ♀ with 4 setae .... *Elaphoidella carterae* Reid, 1993

[USA: Virginia]

10(9) Caudal ramus ♀ 2–3 times longer than wide, proximally expanded and bottle-shaped, with pronounced longitudinal dorsal keel .... *Elaphoidella subgracilis* (Willey, 1934)

[Canada: Saskatchewan to Québec. USA: Minnesota to New York]

10' Caudal ramus ♀ 2.1 times longer than wide, smoothly tapering, with slight dorsal keel .... *Elaphoidella fluviusherbae* Bruno & Reid, 2000

[USA: Florida]

Crustacea: Maxillipoda: Copepoda: Harpacticoida: Canthocamptidae: *Attheyella*: Species

1 P5 ♀ baseoendopod much wider than exopod; P5 ♂ baseoendopod not produced 2

1' P5 ♀ both exopod and baseoendopod elongate, about equally wide, baseoendopod reaching nearly to end of exopod; P5 ♂ baseoendopod produced into narrow, elongate expansion (Pl. 16.178 E, F) .... subgenus *Attheyella* .... *Attheyella* (*A.*) *idahoensis* (Marsh, 1903)

[Canada: British Columbia. USA: Alaska, Idaho, Montana. Palaearctic.]

2(1) P5 ♀ baseoendopod with 3 to 5 setae; caudal rami ♀♂ similar, and body segments coarsely serrate 3

2' P5 ♀ baseoendopod with 6 setae; caudal rami ♀♂ different from each other, and body segments weakly serrate or smooth 5

3(2) All or most setae on P1-5 slender 4

3' Setae on P1-5 short, stout, spiniform (Pl. 16.178 G, H) .... *Attheyella* (*Ryloviella*) *spinipes* Reid, 1987

[USA: District of Columbia, Maryland, Virginia.]

4(3) P5 ♀ baseoendopod with 3 or 4 setae; P5 exopod ♀♂ about 2 times longer than wide; caudal ramus with 2 or more longitudinal rows of spinules .... *Attheyella* (*Ryloviella*) *carolinensis* Chappuis, 1932

[USA: District of Columbia, Georgia, Kentucky, North Carolina, South Carolina, Tennessee, Virginia, West Virginia]

4' P5 ♀ baseoendopod with 5 (rarely 4) setae; P5 ♀♂ exopod about 1.5 times longer than wide; caudal ramus with 2 or 3 oblique medial setal rows .... *Attheyella* (*Ryloviella*) *pilosa* Chappuis, 1929

[USA: Georgia, Illinois, Indiana, Kentucky, South Carolina, Tennessee, Virginia. Mexico]

5(2) Caudal ramus ♀♂ inner margin smoothly tapering or concave, without a process 6

5' Caudal ramus ♀ with prominent, acute, haired inner process; caudal ramus ♂ with smaller, smooth inner process .... *Attheyella (Neomrazekiella) obatogamensis* (Willey, 1925)

[Canada: British Columbia, Québec. USA: District of Columbia, New Hampshire, New York, North Carolina]

6(5) P5 ♀ baseoendopod produced to middle of exopod segment or beyond; P5 ♂ exopod seta 3 glabrous, more slender than other setae; P2-3 ♀ endopods usually 3 segmented 7

6' P5 ♀ baseoendopod hardly at all produced; P5 ♂ exopod seta 3 usually similar to other setae; P2-3 ♀ endopods usually 2 segmented 8

7(8) Caudal ramus ♀, distal half of outer margin strongly constricted, and outer apical seta outbent at base; P4 exp3 ♂ outer distal and apical spines strongly curved (Pl. 16.178 I–M) .... *Attheyella (Neomrazekiella) nordenskioldii* (Lilljeborg, 1902)

[Canada. USA. Palaeacrtic]

7' Caudal ramus ♀, outer margin evenly rounded, and base of outer apical seta straight; P4 exp3 ♂ outer distal and apical spines straight .... *Attheyella (Neomrazekiella) illinoisensis* (Forbes, 1882)

[Canada: Québec. USA: Alaska, central and eastern states south to Georgia. Palaearctic]

8(6) Caudal ramus, lateral setae inserted next to each other 9

8' Caudal ramus, insertions of lateral setae well separated 10

9(8) P5 ♀ baseoendopod with 6 normal setae; caudal ramus ♀, outer distal corner with rounded sclerotized flange overlying bases of apical setae .... *Attheyella (Neomrazekiella) dogieli* (Rylov, 1923)

[Canada: Yukon. USA: Alaska, Washington. Palaearctic]

9' P5 ♀ baseoendopod with 6 slender spiniform setae, all of them completely fused with baseoendopod; caudal ramus ♀, outer distal corner with only a few spinules .... *Attheyella (Neomrazekiella) ussuriensis* Rylov, 1933

[Canada: Northwest Territories. Palaearctic]

10(9) Caudal ramus about as long as anal somite, smoothly tapering, dorsal surface with prominent subquadrate or crescentic sclerotization distal to dorsal seta .... *Attheyella (Neomrazekiella) dentata* (Poggenpol, 1874)

[Canada: Saskatchewan. USA: Alaska. Palaearctic]

10' Caudal ramus about 1/2 length of anal somite, outer distal margin constricted, dorsal surface with no special structure .... *Attheyella (Neomrazekiella) americana* (Herrick, 1884)

[Canada: central and eastern. USA: central and eastern]

Crustacea: Maxillipoda: Copepoda: Harpacticoida: Miraciidae: *Schizopera*: Species

1 P2-4 enp3 ♀ with 4,4,3 setae respectively 2

1' P2-4 enp3 ♀ with 3,3,3 setae respectively .... *Schizopera borutzkyi* Monchenko, 1967

[Invasive: Great Lakes. Palaearctic]

2(1) P4 exopod ♀ and P3 and 4 exopods ♂, segment 2 without inner seta .... *Schizopera tobae cubana* Petkovski, 1973

[USA: Florida Keys. Mexico: Quintana Roo, Yucatán. Neotropics]

2' P4 exopod ♀ and P3 and 4 exopods ♂, segment 2 with inner seta .... *Schizopera knabeni* Lang, 1965

[USA: Atlantic, Gulf of Mexico, and Pacific coasts]

Crustacea: Maxillipoda: Copepoda: Poecilostomatoida: Ergasilidae

Most of the species in North America belong to the large genus *Ergasilus*. In most species, adult females are ectoparasites of fish, and males and copepodids are planktonic. The exception is *E. chautauquaensis* Fellows, 1887, in which the female is also free-living; this species occurs widely in North America. The Asian *Neoergasilus japonicus* (Harada, 1930) has been reported from many fish hosts in Alabama and the Great Lakes. See the keys by Roberts (1970), and subsequent works by Johnson (1971), Johnson & Rogers (1972), and Hudson and Lesko (2003).

Crustacea: Maxillipoda: Copepoda: Cyclopoida: Families

1 Mandibular palp large, complex, with many plumose setae (Pl. 16.179 A); cephalic shield projecting laterally in 2 flaps .... Oithonidae, one genus: ***Limnoithona* [p. 562]**

1' Mandibular palp consisting of tiny segment bearing 1–3 setae (Pl. 16.179 M, N), or mandibular palp entirely absent; cephalic shield not projecting laterally .... **Cyclopidae [p. 562]**

Crustacea: Maxillipoda: Copepoda: Cyclopoida: Cyclopidae: Cyclopinae: *Thermocyclops*: Species

1 Caudal ramus, outermost apical seta with several spinules at its base 2

1' Caudal ramus, outermost apical seta with no spinules at its base 3

2(1) Caudal ramus, lateral seta with no spinules at its base; P1 basipod inner expansion without spine; tips of 2 longest caudal setae strongly curved........ *Thermocyclops parvus* Reid, 1989

[USA: Florida]

2' Caudal ramus, lateral seta with several spinules at its base; P1 basipod inner expansion with spine; tips of 2 longest caudal setae straight........ *Thermocyclops inversus* (Kiefer, 1936)

[USA: Louisiana. Mexico. Neotropics]

3(1) Caudal ramus 3.5 times longer than wide; tips of longest caudal setae straight; P4 coupler with 2 small rounded expansions, without ornament .... *Thermocyclops tenuis* (Marsh, 1909)

[USA: Arizona, Kentucky, Louisiana, Mississippi, New Mexico. Mexico. Neotropics]

3' Caudal ramus 2.0–2.6 times longer than wide; tips of longest caudal setae strongly curved; P4 coupler with 2 rows of hairs on caudal surface, and several spines on large paired expansions .... *Thermocyclops crassus* (Fischer, 1853)

[Introduced. USA: Lake Champlain, New Hampshire, Vermont. Mexico. Palaearctic, Australia]

Crustacea: Maxillipoda: Copepoda: Cyclopoida: Cyclopidae: Cyclopinae: *Cyclops*: Species

A speciose genus with no recent revision. A record of the Palaearctic *C. insignis* Claus from New York is questionable, and therefore this species is not included in the key.

1 P1-4 spine formula 3,4,3,3; A1 ♀ with 17 antennomeres; surface of maxillular palp with spinules 2

1' P1-4 spine formula 2,3,3,3; A1 ♀ with usually 16 (rarely 17) antennomeres; surface of maxillular palp bare .... *Cyclops kolensis alaskaensis* Lindberg, 1956

[USA: Alaska]

2(1) A1 ♀ longer than cephalothorax; outermost terminal caudal seta inserted distinctly proximal to tip of caudal ramus 3

2' A1 ♀ shorter than cephalothorax; outermost terminal caudal seta inserted nearly at tip of caudal ramus .... *Cyclops canadensis* Einsle, 1988

[Canada: Northwest Territories, Saskatchewan, Yukon. USA: Alaska]

3(2) Pediger 5 ♀ widely pointed in dorsal view, nearly as wide as or even wider than pediger 4; lateral seta of caudal ramus inserted at about distal 63% of ramus 4

3' Pedigers 4 and 5 ♀ not widely pointed in dorsal view; lateral seta of caudal ramus inserted at about distal 73–87% of ramus .... *Cyclops strenuus* Fischer, 1851

[Canada. USA: Alaska, Great Lakes. Palaearctic]

4(3) Caudal ramus about 6–7 times longer than wide; dorsal caudal seta equal to or longer than caudal ramus; A1 ♀ segment 12, aesthetasc relatively long, reaching past mid-length of segment 14 .... *Cyclops columbianus* Lindberg, 1956

[Canada: British Columbia. USA: Alaska (Kodiak Island)]

4 Caudal ramus about 4 times longer than wide; dorsal caudal seta about 80% length of caudal ramus; A1 ♀ segment 12, aesthetasc relatively short, not reaching mid-length of segment 14 .... *Cyclops scutifer* Sars, 1863

[Canada. USA: Alaska, California, New York. Palaearctic]

Crustacea: Maxillipoda: Copepoda: Cyclopoida: Cyclopidae: Cyclopinae: *Diacyclops*: Species

1 At least some rami of swimming legs 2 segmented 2

1' All rami of swimming legs 3 segmented 7

2(1) Both rami of P1 and endopod of P2, 2 segmented; P2 exp and both rami of P3 and P4, 3 segmented 3

2' ♀, all rami of P1-4, 2 segmented; ♂, both rami of P1-3 and P4 enp 2-segmented, P4 exp fully 3 segmented .... *Diacyclops dimorphus* Reid & Strayer, 1994

[USA: Florida]

3(2) A1 ♀ with 11 antennomeres 4

3' A1 ♀ with 16 antennomeres .... *Diacyclops languidus* (G. O. Sars, 1863) s.l.

[Canada: Québec. USA: Alaska, Massachusetts, New York]

4(3) P4 enp3 with 3 setae and 2 apical spines 5

4'	P4 enp3 with 3 setae and 1 apical spine	*Diacyclops albus* Reid, 1992
	[Canada: Ontario. USA: Great Lakes region, Michigan, New York, Virginia]	
5(4)	Caudal ramus lateral seta inserted at distal 2/3 to 3/4 of ramus	6
5b.	Caudal ramus lateral seta inserted at about mid-length of ramus	*Diacyclops nanus* (G. O. Sars, 1863)
	[Canada: Manitoba, Newfoundland, Ontario. USA: Alaska, Great Lakes, Michigan, New Hampshire, North Carolina, Ohio. Mexico]	
6(5)	P4 enp3 about as long as wide	*Diacyclops hypnicola* (Gurney, 1927)
	[Canada: Northwest Territories. Palaearctic]	
6'	P4 enp3, 1.4 times longer than wide	*Diacyclops languidoides* (Lilljeborg, 1901) s.l.
	[Canada: North West Territories, Québec, Saskatchewan. USA: Alaska, Colorado, Michigan, Montana]	
7(1)	P4 enp3 with 1 lateral spine and 2 apical spines, *or* 1 lateral seta, and 1 apical spine and 1 apical seta	8
7'	P4 enp3 with 1 lateral seta and 2 apical spines	15
8(7)	P4 enp3 with 1 lateral seta, and 1 apical spine and 1 apical seta (*jeanneli*-group)	9
8'	P4 enp3 with 1 lateral spine and 2 apical spines (*nearcticus*-group)	10
9(8)	A1 ♀ with 17 antennomeres	*Diacyclops jeanneli* (Chappuis, 1929)
	[USA: Indiana. Springs, drip pools in caves]	
9b.	A1 ♀ with 11 antennomeres	*Diacyclops jeanneli putei* (Yeatman, 1943)
	[USA: North Carolina. From a well]	
10(8)	A1 ♀ with 16 or more antennomeres	11
10'	A1 ♀ 12-segmented	*Diacyclops conversus* Reid, 2004
	[USA: Indiana]	
11(10)	A1 ♀ with 17 antennomeres	12
11'	A1 ♀ with 16 antennomeres	*Diacyclops alabamensis* Reid, 1992
	[USA: Alabama]	
12(11)	P1 enp2 with 1 seta on inner margin	13
12'	P1 enp2 with 2 setae on inner margin	*Diacyclops sororum* Reid, 1992
	[USA: Florida, Indiana, Kentucky, Massachusetts, Texas, Virginia]	
13(12)	P2 enp2 with 1 seta on inner margin	14
13'	P2 enp2 with 2 setae on inner margin	*Diacyclops nearcticus* (Kiefer, 1934)
	[USA: eastern states]	
14(13)	A2 antennomere 1 with total of 2 setae	*Diacyclops chrisae* Reid, 1992
	[Canada: Ontario. USA: Indiana, Maryland]	
14'	A2 antennomere 1 with only 1 seta	*Diacyclops harryi* Reid, 1992
	[USA: District of Columbia, Maryland, New York, North Carolina, Ohio, Virginia]	
15(7)	A1 ♀ with 11, 12, or 14 antennomeres	16
15'	A1 ♀ with 16 or 17 antennomeres	18
16(15)	A1 ♀ with 11 or 12 antennomeres	17
16'	A1 ♀ with 14 antennomeres	*Diacyclops bicuspidatus* cf. *lubbocki* (Brady, 1868)
	[Canada: British Columbia, Ontario, Québec. USA: Ohio, Wisconsin, ?Kentucky, ?New York. Palaearctic]	
17(16)	Caudal ramus 4–5 times longer than wide	*Diacyclops crassicaudis* (G. O. Sars, 1863) s.str.
	[Canada: North West Territories, Saskatchewan. USA: Alaska, Michigan, Mississippi, Ohio, Virginia. Palaearctic]	
17'	Caudal ramus 3.1–3.6 times longer than broad	*Diacyclops crassicaudis* var. *brachycercus* (Kiefer, 1927)
	[Canada. USA]	
18(15)	P4 coupler without conspicuous acute protrusions on distal margin (coupler may have surface spines and/or low, rounded marginal protrusions)	19
18'	P4 coupler with 2 conspicuous acute spiniform protrusions on distal margin	*Diacyclops bernardi* (Petkovski, 1986)
	[USA: Florida, Louisiana. Mexico. Neotropics]	
19(18)	Caudal ramus with rows or groups of small but conspicuous spines or stiff hairs on medial surface	20

abdominal segment appendages are reduced and at most present on the first abdominal segment. The uropods are styliform and have a medial row of spines. Eyes and statocysts are absent.

MATERIAL PREPARATION AND PRESERVATION

Bathynellaceans typically inhabit the hyporheic zone of rivers or lakes, although occasionally occur in caves or in wells. The Karaman-Chappuis method is usually used for collecting specimens, by digging a hole or driving a core to the groundwater level. The water (either through ladling or pumping) is filtered through bolting-silk plankton net (mesh size 50 µm). From cave water or wells, specimens may be collected using a hand net.

These crustaceans are usually vulnerable to abrupt temperature changes. However, they can live for 2–3 days, when kept at a temperature similar to that of their habitat. Bathynellaceans usually have a long and tender body, which may easily deform by a high concentration fixative. We recommend fixing them by stages for morphological work. After fixation, specimens may be transferred into glycerol for preparation and preservation or, ethylene glycol (for DNA analysis).

KEYS TO BATHYNELLACEA

Crustacea: Malacostraca: Bathynellacea: Families

1 Body slender, more or less throughout; pleotelson with 1 or 2 pairs of weaker lateral setae near the furcal rami bases; furcal rami with 3 or more than 5 spines along the medial and terminal margins; first antennae with 6 or 7 antennomeres; second antenna 3, 5, or 6 antennomeres, bent posteriorly, without exopod; labrum rim dentate or with setules; mandiblular palp usually with 1 palpomere tipped with a long seta; first maxilla distal endite with distolateral margin bearing a group of plain setae; thoracopod I coxa without setae; thoracopods I-VII exopods with 2 podomeres, each with dorsal and ventral setae; female thoracopod VIII without epipod; male thoracopod VIII with exopod reduced; pleopod 1 usually setaform or of one podomere, or absent; uropodal sympod elongate, with the spine row arranged longitudinally; uropodal exopod often longer than endopod, sometimes with medial margin lacking setae .... **Parabathynellidae [p. 578]**

1' Body slender with pleon being a little more bulky than thorax; pleotelson with a pair of dorsal stout setae; furcal rami always with 5 terminal spines; first antenna with 7 antennomeres; second antenna with 7antennomeres, directed anteriorly, with a small exopod of one podomere; labrum rim smooth; mandibular palp prehensile, with 2 palpomeres, and tipped with 2 curved, stout claws; first maxilla distal endite with distolateral margin bearing a group of plumose setae; thoracopod I coxa usually with a long plumose seta; thoracopods I-VII exopods of one podomere, with dorsal, ventral and terminal setae; female thoracopod VIII usually with epipod; male thoracopod VIII with distinct exopod; pleopod 1 of 2 podomeres; uropodal sympod rather short, with spine row oblique at its end; uropodal exopod never longer than endopod, always with medial setae .... **Bathynellidae [p. 581]**

Crustacea: Malacostraca: Bathynellacea: Parabathynellidae: Genera

1 Thorax with seven pair of walking legs (thoracopods I-VII) 2

1' Thorax with six pair of walking legs (thoracopods I-VI) (Pls. 16.181.3 A, 7 B, 8 B) .... ***Hexabathynella* [p. 578]**

2(2) Mandible with 5 or more spines in a row; palp inserting at the base of pars incisivus; second maxilla with 3 or 4 maxillomeres 3

2' Mandible with 4 small teeth along cutting edge; palp inserting on level of pars incisivus; second maxilla with two maxillomeres, and long distal prehensile claw (Pls. 16.181.1 B, 5 B, 6 B, 7 A) .... ***Califobathynella* [p. 579]**

3(2) Second antenna with 3 antennomeres 4

3' Second antenna with 7 antennomeres .... *Montanabathynella salish* Camacho, Stanford & Newell, 2009

4(3) First antenna with 6 antennomeres (Pls. 16.181.2 B, 3 C, 4 B, 7 C, 8 D, 9 A, B, 16.182.1 C) .... ***Texanobathynella* [p. 579]**

4' First antenna with 7 antennomeres (Pls. 16.181.2 A, 4 A, 5 A, 6 A, 7 D, 8 C) .... *Californibathynella californica* Camacho & Serban, 1998

[USA: California]

Crustacea: Malacostraca: Bathynellacea: Parabathynellidae: *Hexabathynella*: Species

1 Uropodal exopod with 2 terminal setae 2

1' Uropodal exopod with 3 terminal setae .... *Hexabathynella virginiae* Cho & Schminke, 2006

[USA: Virginia]

2(1) Thoracopod I-VI exopods with 2 podomeres 3

PHYLUM ARTHROPODA

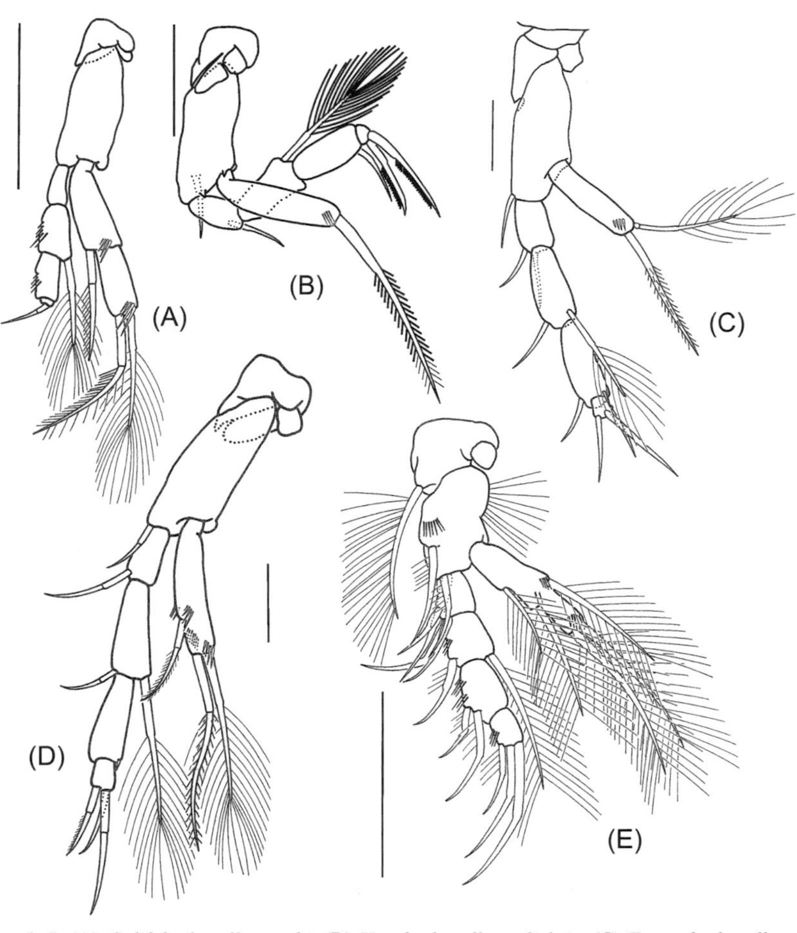

PLATE 16.181.07 Thoracopods I: (A) *Califobathynella noodti*; (B) *Hexabathynella muliebris*; (C) *Texanobathynella sachi*; (D) *Californibathynella californica*; (E) *Bathynella fraterna*. Scale bars = 0.025 mm.

2' Thoracopod I-VI exopods with 1 podomeres (Pl. 16.181.8 B) .... *Hexabathynella otayana* Cho, 2001
[USA: California]

3(2) Uropodal sympod with 2 spines; female thoracopod VIII triangular; body to 1.36 mm .... *Hexabathynella hessleri* Cho, 2001
[USA: California]

3' Uropodal sympod with 3 spines, female thoracopod VIII crumpled; body to 0.76 mm (Pls. 16.181.3 A, 4 C, 7 B) .... *Hexabathynella muliebris* Cho, 2001
[USA: California]

Crustacea: Malacostraca: Bathynellacea: Parabathynellidae: *Califobathynella*: Species

1 Labrum with large medial teeth flanked by 2 smaller teeth, no spinules further laterally; throcaopod I, endopod fourth podopomere with 2 setae (Pls. 16.181.1 B, 3 B, 5 B, 6 B, 7 A, 8 A) .... *Califobathynella noodti* Cho, 1997
[USA: California]

1' Labrum with medial group of 5 teeth not flanked by smaller teeth but with spinules further laterally; thoracopod I, endopod fourth podopomere with 1 seta .... *Califobathynella teucherti* Cho, 1997
[USA: California]

Crustacea: Malacostraca: Bathynellacea: Parabathynellidae: *Texanobathynella*: Species

1 Thoracopods I-VII exopods with podomere formula: 2-3-3-3-3-3-2; abdominal somite I pleopods without rests .... *Texanobathynella bowmani* Delamare Deboutteville et al., 1975
[USA: Texas]

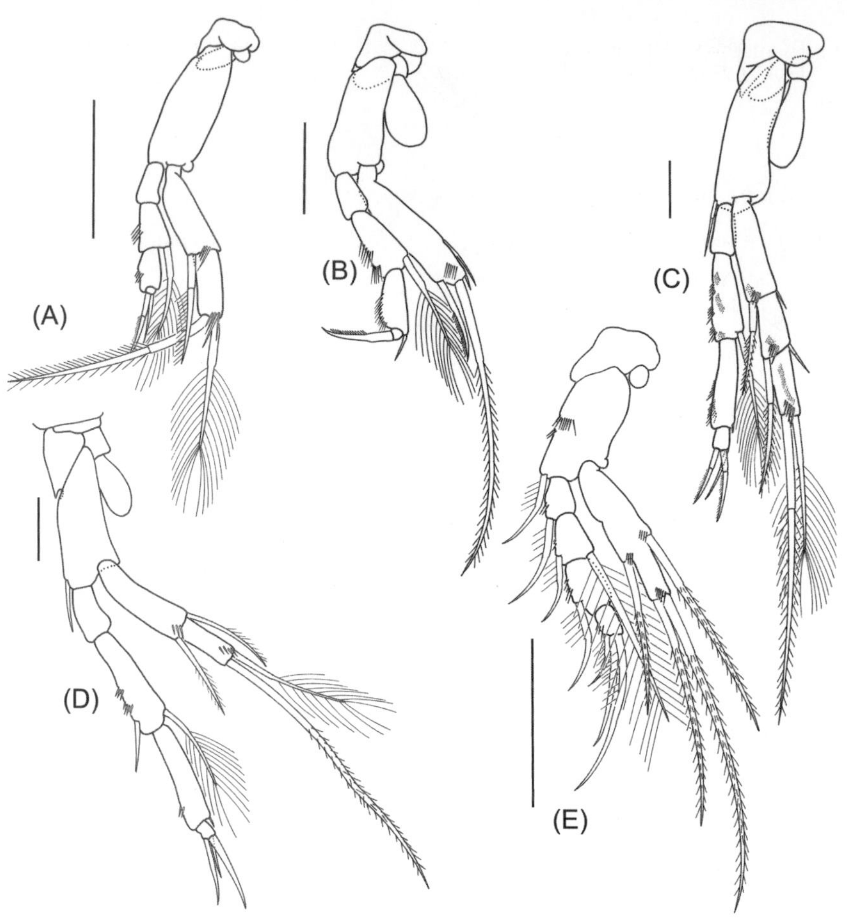

PLATE 16.181.08 Thoracopods II: (A) *Califobathynella noodti*; (B) *Hexabathynella otayana*; (C) *Californibathynella californica*; (D) *Texanobathynella sachi*; (E) *Bathynella fraterna*. Scale bars = 0.025 mm.

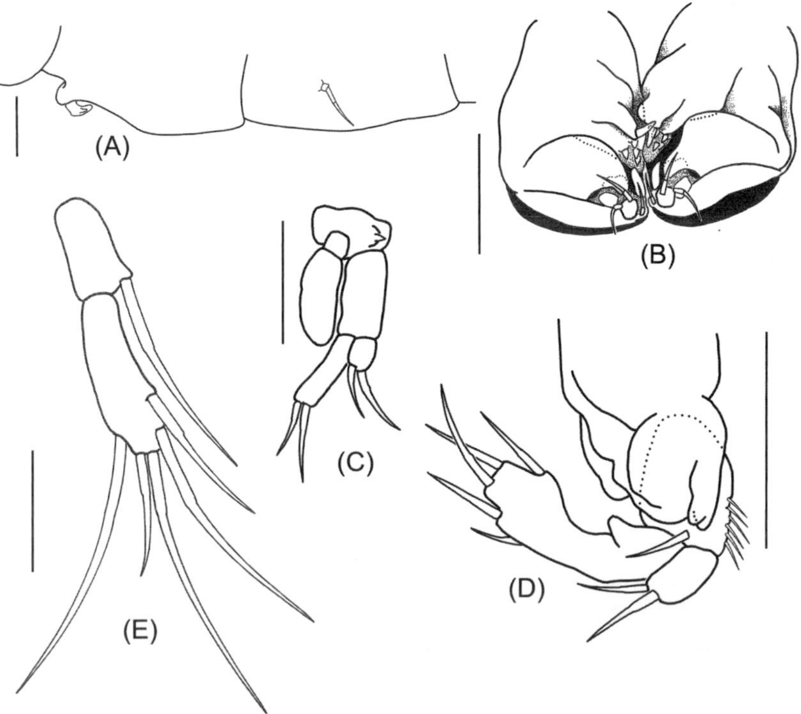

PLATE 16.181.09 Thoracopods VIII and pleopods: (A) female thoracopod VIII of *Texanobathynella sachi*; (B) male thoracopod VII of *Texanobathynella sachi*; (C) female thoracopod VIII of *Bathynella fraterna*; (D) male thoracopod VIII of *Bathynella fraterna*; (E) pleopod of *Bathynella fraterna*. Scale bars = 0.025 mm.

1' Thoracopods I-VII exopods with podomere formula: 1-3-3-2-2-2-2; abdominal somite I pleopods with 2 setae as rests (Pls. 16.181.2 B, 3 C, 4 B, 7 C, 8 D, 9 A, B, 16.182.1 C) .... *Texanobathynella sachi* Cho, 1996

[USA: Texas]

Crustacea: Malacostraca: Bathynellacea: Bathynellidae: Genera

1 Enp of male Th. VI of 4 segments (Pls. 16.181.1 C, 2 C, 3 D, 4 E, 5 C, 6 C, 7 E, 8 E, 9 C–E) .... *Bathynella*

1' Male thoracopod VI endopod of 3 podomeres; first podomere broader than usual; second podomere broad and dilated with strong curved seta at lateral margin .... *Pacificabathynella*

[USA: California]

REFERENCES

Camacho, A.I., R.L. Newell & B. Reid. 2009. New records of Bathynellacea (Syncarida, Bathynellidae) in North America: three new species of the genus *Pacificabathynella* from Montana, USA. Journal of Natural History 43: 1805–1834.

Camacho, A.I., J.A. Stanford & R.L. Newell. 2009. The first record of Syncarida in Montana, USA: a new genus and species of Parabathynellidae (Crustacea, Bathynellacea) in North America 43: 309–321.

Cho, J.-L. 1996. A new species of the genus *Texanobathynella* from California (Crustacea, Malacostraca, Bathynellacea). Korean Journal of Systematic Zoology 12: 389–395.

Cho, J.-L. 1997. Two new species of a new genus of Leptobathynellinae (Crustacea, Bathynellacea) from California, USA. Korean Journal of Biological Sciences 1: 265–270.

Cho, J.-L. 2001. Phylogeny and zoogeography of three new species of the genus *Hexabathynella* (Crustacea, Malacostraca, Bathynellacea) from North America. Zoologica Scripta. 30:145–157.

Cho, J.-L. & H.K. Schminke. 2006. A phylogenetic review of the genus *Hexabathynella* Schminke, 1972 (Crustacea, Malacostraca, Bathynellacea): with a description of four new species. Zoological Journal of the Linnean Society 147: 71–96.

Delamare Deboutteville, C., N. Coineau & E. Serban. 1975. Découverte de la famille des Parabathynellidae (Bathynellacea) en Amérique du Nord: *Texanobathynella bowmani* n. g. n. sp. Comptes-rendus Hebdomadaires des Séances de l'Académie des Sciences de Paris (Sér. D) 280: 2223–2226.

Pennak, R.W. & J.V. Ward. 1985. Bathynellacea (Crustacea: Syncarida) in the United States, and a new species from the phreatic zone of a Colorado Mountain Stream. Transactions of the American Microscopical Society 104: 209–215.

Schminke, H. K. & W. Noodt. 1988. Groundwater Crustacea of the order Bathynellacea (Malacostraca) from North America. Journal of Crustacean Biology 18: 290–299.

Order Amphipoda

D. Christopher Rogers

Kansas Biological Survey, University of Kansas, Lawrence, KS, USA; The Biodiversity Institute, University of Kansas, Lawrence, KS, USA

INTRODUCTION

The Amphipoda (scuds, sideswimmers) are common crustaceans with numerous freshwater, marine, and terrestrial species. About 300 species occur in the Nearctic subterranean and surface waters. However, the true diversity of Nearctic amphipods is undoubtedly much higher because the cave fauna is poorly known.

LIMITATIONS

Many characters needed to determine amphipods to family, genus, or species level are easily broken, broken off, (such as antennae, limbs, uropods, and telson) or worn away (dorsal spines, limb setae). These structures are often shed as a predator avoidance response or just due to the fragility of the animal. Often only one gender can be used to identify a genus or species. An additional limitation is that several genera, including the most common (*Hyallela, Gammarus*), as well as the largest genera (*Crangonyx, Stygobromus*), are in desperate need of revision and cannot be identified to species level primarily due to the large number of undescribed species.

TERMINOLOGY AND MORPHOLOGY

The basic morphological terms are presented in (Pl. 16.182.2). The head bears two pairs of antennae (antenna I upper most, and antenna II lower), with maxillipeds below the antennae. The antennae are divided into antennomeres: the proximal three are the peduncle, the remaining ones are the flagellum. The first antennae may have an accessory flagellum of one to five flagellomeres lying on the dorsomedial surface of the flagellum, coming from the distal end of any of the antenna I first five flagellomeres. Eyes may be present or absent.

PHYLUM ARTHROPODA

The first two pereopods are generally modified as gnathopods, which are for grasping. The proximal most article is the coxa. The last three limb pairs are the uropods. Uropods typically have a basal peduncle with two rami: a lateral exopod and a medial endopod. One or both may be reduced or absent. On the dorsal surface of the last somite is the telson, which generally projects away from the body.

Female amphipods are identified by the presence of oostegites (plates) on the ventral surface between the limbs for carrying eggs and young. These plates should not be confused with coxal and/or sternal gills, which are typically associated with a limb, and are attached by a short stalk. These gills are usually independent of each other and partially joined as in the oostegites.

MATERIAL PREPARATION AND PRESERVATION

Amphipods are best preserved and examined in ethyl alcohol. Occasionally specimens (Hadziidae, Artesiidae) are small enough that they must be slide-mounted. For some genera, the mouthparts must be removed and mounted on slides.

KEYS TO AMPHIPODA

Crustacea: Malacostraca: Amphipoda: Families

1	Body laterally flattened (Pl. 16.182.2)	2
1'	Body subcylindrical (Pl. 16.182.3), or pereopod II bearing long setal fringes longer than the pereopod segments	**Corophiidae [p. 583]**
2(1)	First antenna lacking an accessory flagellum, not subterranean	3
2'	First antenna bearing an accessory flagellum, or if accessory flagellum is lacking, then specimens collected from a subterranean habitat	4
3(2)	Pereopod I simple, pereopod II a gnathopod or subchelate (Pl. 16.182.4)	**Talitridae [p. 584]**
3'	Pereopod I and II gnathopods (Pl. 16.182.2)	Hyalellidae, one genus *Hyalella*
	[Numerous species, with confused taxonomy. Canada. USA. Mexico]	
4(2)	Gnathopods large (first two leg pairs)	5
4'	Gnathopods simple, reduced, similar to other limbs (Pl. 16.182.5)	**Pontoporeiidae [p. 584]**
5(4)	First antennae shorter than body	6
5'	First antennae as long as or longer than body; hypogean	**Hadziidae [p. 585]**
6(5)	First antenna accessory flagellum consisting of one distinct flagellomere and one reduced terminal flagellomere, or just one conical flagellomere; eyes present or absent	7
6'	First antenna accessory flagellum consisting of two to seven distinct flagellomeres; eyes present	12
7(6)	First antenna accessory flagellum consisting of one distinct flagellomere and one reduced terminal flagellomere	8
7'	First antenna accessory flagellum consisting of just one conical flagellomere	Pontogeniidae, one genus ***Paramoera* [p. 586]**
8(7)	Telson cleft, partially cleft, or medially excavate (Pl. 16.182.2)	9
8'	Telson entire, subtruncate to rounded, hypogean	Sebidae, one genus ***Seborgia* [p. 587]**
9(8)	Telson at least partially cleft (Pl. 16.182.2 part 13)	10
9'	Telson medially excavate, with one large plumose seta on each apex of the telson; hypogean	Bogidiellidae, one species *Parabogidiella americana* Holsinger, 1980
	[USA: Texas]	
10(9)	Uropod III endopod and/or exopod absent, if present, then shorter than peduncle; gnathopods palmar margin with spines apically bifurcated	11
10'	Uropod III endopod subequal in length to exopod, both longer than peduncle hypogean	Artesiidae, one genus *Artesia*
	[USA: Texas]	
11(10)	Uropod III exopod elongate, of two articles	Allocrangonyctidae, one genus: ***Allocrangonyx* [p. 587]**
11'	Uropod III exopod present or absent, if present only of one article	**Crangonyctidae [p. 587]**
12(6)	Coxal gills with two or more lamellae	12
12'	Coxal gills with a single lamella	**Gammaridae [p. 588]**

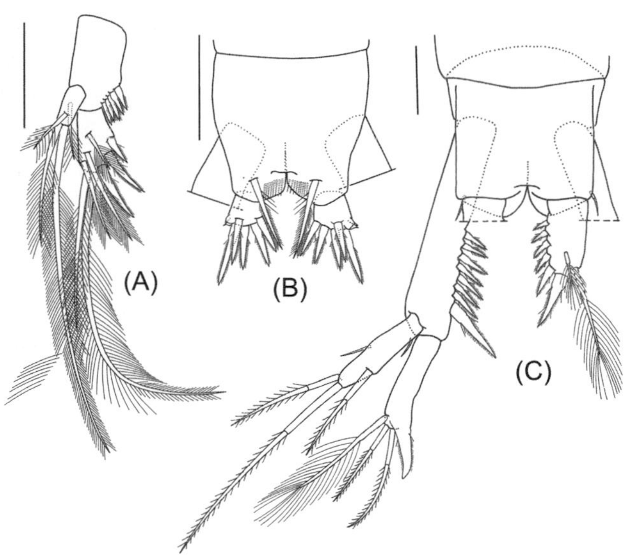

PLATE 16.182.01 Uropods, pleotelsons, and furcas: (A) uropod of *Bathynella fraterna*; (B) pleotelson and furca of *Bathynella fraterna*; (C) uropod, pleotelson, and furca of *Texanobathynella sachi.* Scale bars = 0.05 mm.

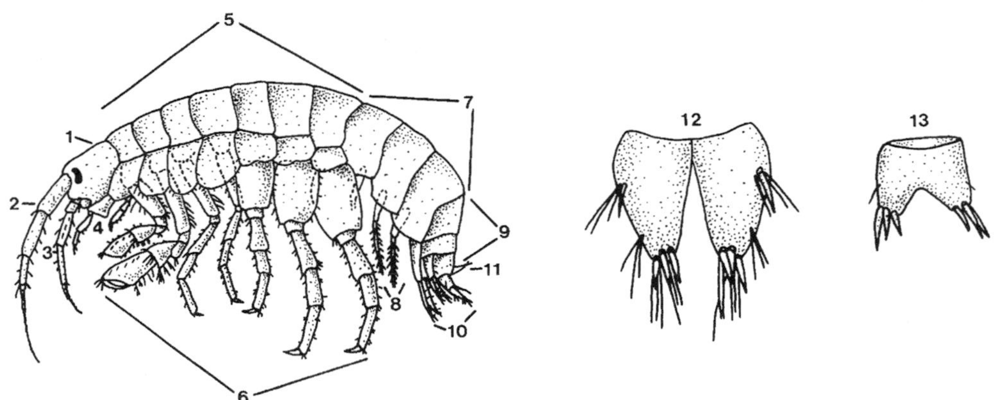

PLATE 16.182.02 Schematic drawings of the external side view of generalized freshwater gammarid amphipods: (1) head, (2) antenna 1, (3) antenna 2, (4) mouth parts, (5) pereonites 1 to 7, (6) pereopods 1 to 7, (7) pleonites 1 to 3, (8) pleopods 1 to 3, (9) uronites 1 to 3, (10) uropods 1 to 3, (11) telson (redrawn from Holsinger, 1972), (12–13) top view of telsons for *Gammarus* and *Crangonyx*, respectively.

13(11) Uropod III endopod reduced, smaller than exopod .... **Anisogammaridae [p. 588]**

13' Uropod III endopod and exopod subequal, large dorsal spines present, sternal gills present .... Gammaracanthidae, one genus *Gammaracanthus*

[Two described and two possibly undescribed species; Arctic]

Crustacea: Malacostraca: Amphipoda: Corophiidae: Genera

1 Urosome segments entirely fused 2

1' Urosome segments free (Pl. 16.182.3) .... ***Americorophium* [p. 584]**

2(1) Pereopod II with article five articulating from the base of an elongated article four, both articles with elongated setal fringes for filter feeding .... *Paracorophium* sp.

[Probably introduced, undescribed species in tidally influenced freshwater. USA: California]

2' Pereopod II not modified for filtering .... *Apocorpohium lacustre* (Vanhöffen, 1911)

[USA: Atlantic coast and Gulf States. Introduced: Mississippi River system to Illinois and Indiana.]

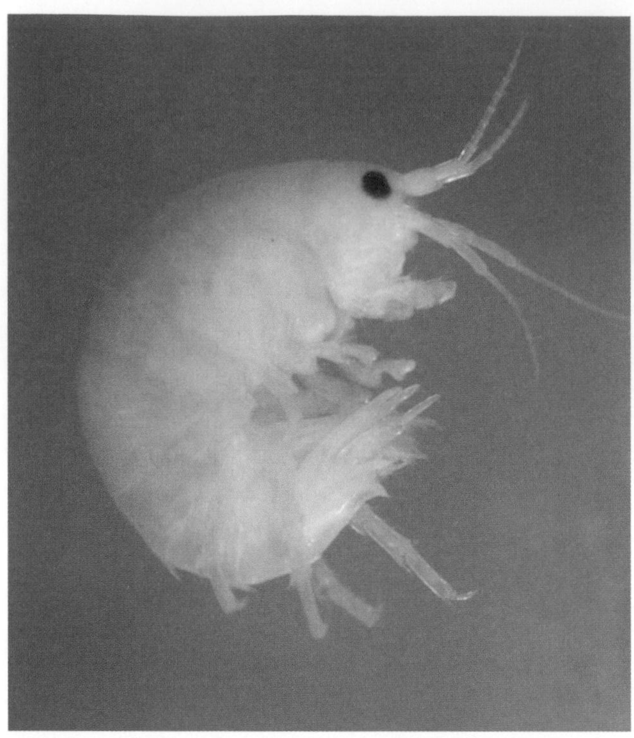

PLATE 16.182.03 The widespread amphipod genus *Hyalella*. *Photograph courtesy of John Pfieffer at EcoAnalysts, Inc.*

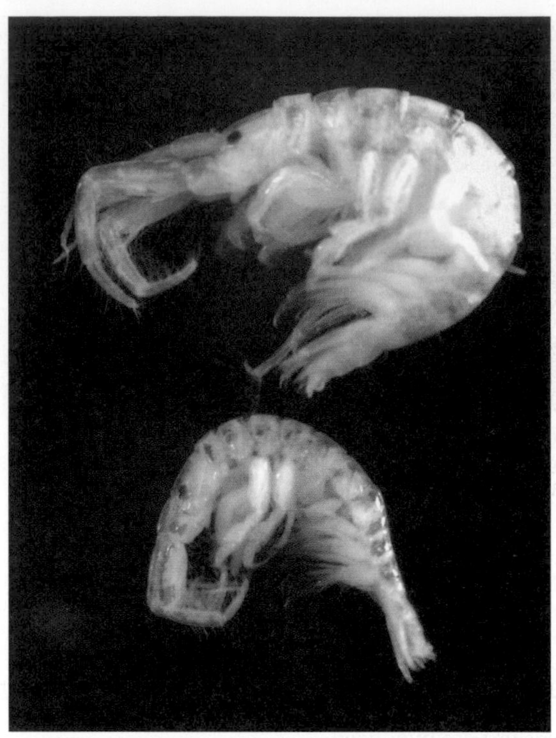

PLATE 16.182.04 Photograph of the amphipod *Americorophium*. *Photograph courtesy of John Pfieffer at EcoAnalysts, Inc.*

Crustacea: Malacostraca: Amphipoda: Corophiidae: *Americorophium*: Species

1 Rostrum projecting narrowly at middle, apically rounded or subacute 2

1' Rostrum not projecting, broadly rounded or bidentate .... *Americorophium spinicorne* (Stimpson, 1857)
[USA: California, Oregon, Washington. Introduced to Idaho]

2(1) Male first antenna proximal antennomere broadly explanate, lacking a spiniform projection .... *Americorophium salmonis* (Stimpson, 1857)
[USA: California, Oregon]

2' Male first antenna proximal antennomere not explanate, sides subparallel, and with a ventral spiniform projection .... *Americorophium stimpsoni* (Shoemaker, 1941)
[USA: California, Oregon]

Crustacea: Malacostraca: Amphipoda: Talitridae: Genera

1 Pereopod II a gnathopod, terrestrial or amphibious 2

1' Pereopod II subchelate; freshwater, brackish water, and semiterrestrial .... *Chelorchestia forceps* Smith & Heard, 2001
[USA: Florida]

2(1) Pereopod I with podomere six shorter than segment five, pleopod I and II subequal in length; terrestrial or amphibious .... *Talitroides* sp.
[At least two invasive species. USA: Arizona, California, Gulf Coast north to the Carolinas. IndoPacific]

2' Pereopod I with podomere six and five subequal, pleopod II smaller than pleopod I; terrestrial or amphibious .... *Arctitalitrus sylvaticus* (Haswell, 1880)
[USA: Invasive in California, Florida, Texas. Australia.]

Crustacea: Malacostraca: Amphipoda: Pontoporeiidae: Genera

1 Telson cleft 50%, eyes black .... *Monoporeia affinis* (Lindstrom, 1855)
[Canada: southern. USA: northern]

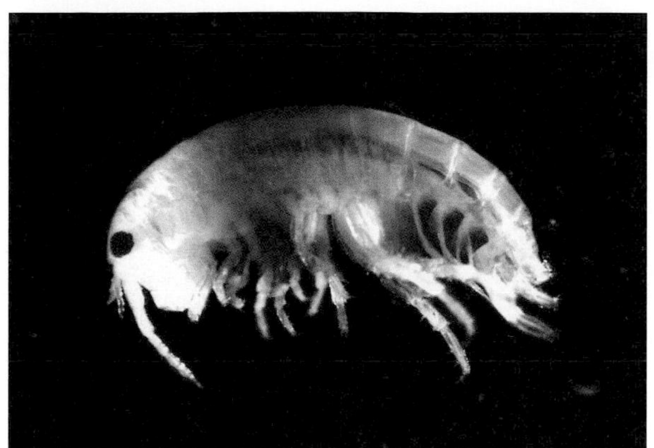

PLATE 16.182.05 Photograph of lawn shrimps or land hoppers in the genera *Talitroides. Photograph courtesy of Matthew A. Hill and EcoAnalysts, Inc.*

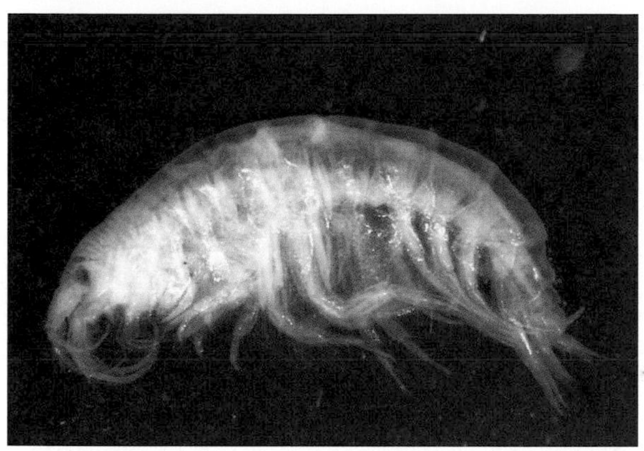

PLATE 16.182.06 Photograph of a planktonic scud in the genus *Monoporeia. Photograph courtesy of Matthew A. Hill and EcoAnalysts, Inc.*

1' Telson cleft to base, eyes red in life .... *Diporeia* sp.
[Six described and undescribed species. Canada: southern. USA: northern]

Crustacea: Malacostraca: Amphipoda: Hadziidae: Genera

All species are hypogean.

1 Second antennal peduncle distal antennomeres without spines, pereopod 5 and 6 with segment 6 posterior margin lacking long setae 2

1' Second antennal peduncle distal antennomeres with spines, pereopod 5 and 6 with segment 6 posterior margin bearing long setae .... *Allotexiweckelia hirsuta* Holsinger, 1980
[USA: Texas]

2(1) Uropod III endopod and exopod subparallel sided 3

2' Uropod III endopod and exopod broad, ovoid .... *Tamaweckelia apalpa* Holsinger, 2005
[Mexico: Tamaulipas]

3(2) Coxa five lobiform 4

3' Coxa five normal 5

4(3) Coxal plates broadly explanate; first antenna lacking an accessory flagellum .... *Holsingerius samacos* (Holsinger, 1980)
[USA: Texas]

4' Coxal plates normal; first antenna with accessory flagellum of one small antennomere or absent .... ***Mexiweckelia* [p. 585]**

5(3) Uropod III with plumose setae and spines; first antenna with accessory flagellum of one small antennomere or absent 6

5' Uropod III with spines, no plumose setae; first antenna with accessory flagellum of one small antennomere .... ***Paramexiweckelia* [p. 586]**

6(5) Coxal plates normal; coxal gills pedunculate; first antenna with accessory flagellum of one small antennomere or absent 7

6' Coxal plates broadly explanate; coxal gills ellipsoidal; first antenna with accessory flagellum absent .... ***Paraholsingerius* [p. 586]**

7(6) Pereopods VI and VII approximately half as long as body .... *Texiweckeliopsis insolita* (Holsinger, 1980)
[USA: Texas]

7' Pereopods VI and VII subequal in length to body .... *Texiweckelia texensis* (Holsinger, 1980)
[USA: Texas]

Crustacea: Malacostraca: Amphipoda: Hadziidae: *Mexiweckelia*: Species

1 Telson deeply cleft 2

1' Telson apically emarginate .... *Mexiweckelia hardeni* Holsinger, 1992
[USA: Texas]

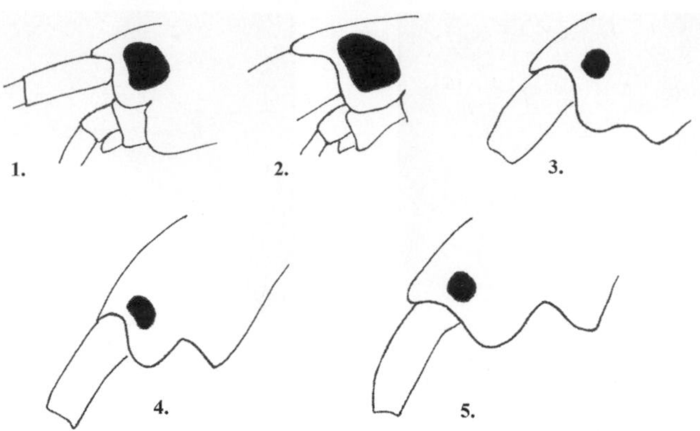

PLATE 16.182.07 Amphipoda, *Paramoera*: head, left lateral view: (1) *P. columbiana*, (2) *P. bousfieldi*, (3) *P. crassicauda*, (4) *P. bucki*, (5) *P. carlottensis*.

2(1) Uropod III 15–20% body length in male, 20–25% body length in female; uropod III endopod with apical spines and lateral setae .... *Mexiweckelia mitchelli* Holsinger, 1971

[Mexico: Durango]

2' Uropod III 15 body length; uropod III endopod with apical and lateral spines .... *Mexiweckelia colei* Holsinger & Minckley, 1971

[Mexico: Coahuila]

Crustacea: Malacostraca: Amphipoda: Hadziidae: *Paramexiweckelia*: Species

1 Telson cleft at least 80% .... *Paramexiweckelia particeps* Holsinger, 1971

[Mexico: Coahuila]

1' Telson cleft no more than 60% .... *Paramexiweckelia ruffoi* Holsinger, 1996

[USA: Texas]

Crustacea: Malacostraca: Amphipoda: Hadziidae: *Paraholsingerius*: Species

1 Uropods I and II with long plumose setae on exopod, endopod with spines only .... *Paraholsingerius smaragdinus* (Holsinger, 1992)

[USA: Texas]

1' Uropods I and II with long plumose setae on both rami .... *Paraholsingerius mexicanus* Sawicki & Holsinger, 2005

[Mexico: Coahulia]

Crustacea: Malacostraca: Amphipoda: Pontogeniidae: *Paramoera*: Species

Pacific Coast of Canada and USA, in tidally influenced freshwater and freshwater pools that receive some salt spray. Modified from Staude, 1986.

1 Head anterior margin below eye, with postantennal lobe separated from lateral lobe by a distinct cleft Pl. 16.182.07 (1 & 2) 2

1' Head anterior margin below eye, without a cleft between the postantennal and lateral lobes Pl.16.182.07 (3–5) 3

2(1) Head with cleft between eye and second antenna extending behind eye Pl. 16.182.07 (2) .... *Paramoera bousfieldi* Staude, 1986

[Canada: British Columbia. USA: Alaska, California, Oregon, Washington]

2' Head with cleft between eye and second antenna extending to just below eye anterior margin Pl. 16.182.07 (1) .... *Paramoera columbiana* Bousfield, 1958

[Canada: British Columbia. USA: California, Oregon, Washington]

3(1) Head anterior margin with emargination between postantennal and lateral lobes angular, deep Pl. 16.182.07 (4–5) 4

3' Head anterior margin with emargination between postantennal and lateral lobes rounded, shallow Pl. 16.182.07 (3) .... *Paramoera crassicauda* Staude, 1986

[USA: Alaska]

4(3) Pereopod VII with a coxal gill; eye separated from postantennal lobe distal margin by width of eye Pl. 16.182.07 (5) .... *Paramoera carlottensis* Bousfield, 1958
[Canada: British Columbia. USA: Alaska]

4' Pereopod VII lacking a gill; eye separated from postantennal lobe distal margin by at least twice width of eye Pl. 16.182.07 (4) .... *Paramoera bucki* Staude, 1986
[Canada: British Columbia. USA: Alaska, Washington]

Crustacea: Malacostraca: Amphipoda: Seborgidae: *Seborgia*: Species

Hypogean species.

1 Mandible molar with a setule .... *Seborgia relicta* Holsinger, 1980
[USA: Texas]

1' Mandible molar lacking a setule .... *Seborgia hershleri* Holsinger, 1992
[USA: Texas]

Crustacea: Malacostraca: Amphipoda: Allocrangonyctidae: *Allocrangonyx*: Species

1 Telson with six or seven unequal spines on each lobe .... *Allocrangonyx pellucidus* (Mackin, 1935)
[USA: Oklahoma]

1' Telson with four unequal spines on each lobe .... *Allocrangonyx hubrichti* Holsinger, 1971
[USA: Missouri]

Crustacea: Malacostraca: Amphipoda: Crangonyctidae: Genera

Modified from Bousefield & Holsinger, 1989.

1 Pereopods VI and VII coxal plates deeper than corresponding body segments; pleonal plates I and II produced and/or acuminate; eyes present or absent 2

1' Pereopods VI and VII coxal plates subequal to corresponding body segments; pleonal plates I and II rounded; eyes absent 3

2(1) Uropod III biramous, exopod longer than peduncle, endopod reduced, scale-like .... *Crangonyx*
[Numerous described and undescribed species, hypogean or epigean. Holarctic]

2' Uropod III uniramous, exopod only, exopod shorter than peduncle .... ***Synurella*** **[p. 587]**
[Four species, eastern USA and Alaska; hypogean.]

3(1) Uropod III uniramous, endopod absent, exopod present or absent 4

3' Uropod III biramous; endopod reduced, scaliform .... ***Bactrurus*** **[p. 588]**

4(3) Uropod III exopod obviously shorter than peduncle or absent .... *Stygobromus*
[Numerous described and undescribed species, hypogean or epigean. Canada. USA. Mexico. Holarctic]

4' Uropod III exopod equal in length to peduncle and bearing two or three spines .... *Stygonyx courtneyi* Bousefield & Holsinger, 1989
[USA: Oregon]

Crustacea: Malacostraca: Amphipoda: Crangonyctidae: *Synurella*: Species

1 Urosomites fused 2

1' Urosomites free .... *Synurella chamberlaini* (Ellis, 1941)
[USA: Atlantic coastal states from Maine to South Carolina]

2(1) Gnathopods with palmar margins distinctly concave .... *Synurella dentata* Hubricht, 1943
[USA: Indiana, Kentucky, Ohio, Tennessee]

2' Gnathopods with palmar margins straight to slightly concave .... *Synurella bifurca* (Hay, 1882)
[USA: Alabama, Arkansas, Louisana, Mississippi, Texas]

Crustacea: Malacostraca: Amphipoda: Crangonyctidae: *Bactrurus*: Species

Modified from Koenemann & Holsinger, 2001. Females are separated from males by the presence of oostegites.

1 Male telson at least twice as long as uropod III; female telson as long as broad, lateral margins convergent 2

1' Male telson subequal or shorter than uropod III; female telson variable, lateral margins parallel or subconvergent 3

2(1) Pereopod VII with gill reduced; pereonites 2 and 3 with medial sternal process .... *Bactrurus pseudomucronatus* Koenemann & Holsinger, 2001

[USA: Arkansas, Missouri]

2' Pereopod VII without gill; pereonites 2 and 3 with medial sternal process absent .... *Bactrurus mucronatus* (Forbes, 1876)

[USA: Illinois, Indiana, Michigan, Ohio]

3(1) Telson as wide as long as or wider than long 4

3' Telson longer than wide 5

4(3) Uropod III exopod longer than or subequal in length to peduncle .... *Bactrurus speleopolis* Holsinger et al., 2006

[USA: Arkansas]

4' Uropod III exopod less than 75% the length of peduncle .... *Bactrurus hubrichti* Shoemaker, 1945

[USA: Kansas, Missouri, Oklahoma]

5(3) Male uropod I peduncle with a serrate distal process 6

5' Male uropod I peduncle lacking a serrate distal process .... *Bactrurus brachycaudus* Hubricht & Mackin, 1940

[USA: Illinois, Missouri]

6(5) Pereopod VII coxal gill less than half the size of pereopod VI coxal gill; gnathopod II carpus distomedial margin with six or more plumose setae 7

6' Pereopod VII coxal gill subequal to, or slightly smaller than pereopod VI coxal gill; gnathopod II carpus distomedial margin with ca. three plumose setae .... *Bactrurus wilsoni* Koenemann & Holsinger, 2001

[USA: Alabama]

7(6) Gnathopod I propodus palm margin angled approximately 120°; uropod III peduncle with one or two distal spines, exopod <80% the length of peduncle, and bearing two to five subapical spines on both margins; maxilla II inner plate with oblique row of seven plumose setae .... *Bactrurus angulus* Koenemann & Holsinger, 2001

[USA: Tennessee, Virginia]

7' Gnathopod I propodus palm margin angled 135–140°; uropod III peduncle with three distal spines, exopod >80% the length of peduncle, and bearing four to eight subapical spines on both margins; maxilla II inner plate with oblique row of ten plumose setae .... *Bactrurus cellulanus* Koenemann & Holsinger, 2001

[USA: Indiana]

Crustacea: Malacostraca: Amphipoda: Gammaridae: Genera

1 Uropod III with endopod 0.20 times the length or less of exopod .... *Echinogammarus ischnus* (Stebbing, 1899)

[Invasive species in the Great Lakes. Palaearctic]

1' Uropod III with endopod 0.40 times the length or more of exopod .... *Gammarus*

[Numerous decribed and undescribed species. Holarctic]

Crustacea: Malacostraca: Amphipoda: Anisogammaridae: Genera

1 Urosomite II without a large medial spine 2

1' Urosomite II with a large medial spine .... *Anisogammarus pugettensis* (Dana, 1853)

[Pacific Coast estuaries and tidally influenced freshwater. Canada. USA]

2(1) Pleon with paired posteriomedial and/or submedial spines 3

2' Pleon segments without posteriomedial or submedial spines, although some setae may be present .... *Eogammarus confervicolus* (Stimpson, 1856)

[Canada: British Columbia. USA: California, Oregon, Washington. Mexico: Baja California]

3(2) Urosomites with distomedial margins bearing spines and/or setae in transverse rows .... ***Ramellogammarus* [p. 589]**

3' Urosomites with medial and submedial oblique rows of spines .... *Locustogammarus locustoides* (Brandt 1851)

[Canada: British Columbia. USA: Alaska]

Crustacea: Malacostraca: Amphipoda: Anisogammaridae: *Ramellogammarus*: Species

Modified from Bousfield & Morino, 1992.

1 Antenna I peduncle antennomere II posterior margin with four or more setal clusters; urosomite III with dorsolateral and medial paired spines or spine clusters 2

1' Antenna I peduncle antennomere II posterior margin with three or less setal clusters; urosome III with lateral paired spines 3

2(1) Pleonites I-III with strong spines; pleonite III posteriorly with eight or more dorsal spines .... *Ramellogammarus oregonensis* (Shoemaker, 1944)

[USA: Oregon, Washington]

2' Pleonites II and III with spines weak or absent; pleonite III posteriorly with six or fewer spines .... *Ramellogammarus setosus* Bousfield & Morino, 1992

[USA: Washington]

3(1) Antenna I peduncle antennomere II posterior margin with zero to two setal clusters; peduncle antennomere III with or without a single seta; urosomite I with medial and lateral pairs of spines 4

3' Antenna I peduncle antennomere II posterior margin with three setal clusters; peduncle antennomere III with one or two seta clusters; urosomite I with medial pairs of spines only 6

4(3) Pleonites II and/or III with medial spines and setae; pleonite III posterior margin with a single setal row; urosomite I and usually II with medial spines paired 5

4' Pleonites I-III dorsally setose; pleonite III posterior margin usually with two or more setal rows; urosomite I and II with medial spines single .... *Ramellogammarus columbianus* Bousfield & Morino, 1992

[Canada: British Columbia. USA: California, Oregon, Washington]

5(4) Pleonite III posterior margin with two spines; urosomite I and II with medial paired spine clusters bearing one to three spines per cluster .... *Ramellogammarus vancouverensis* Bousfield, 1979

[Canada: British Columbia]

5' Pleonites II and III posterior margin with two to four spines; urosomite I with medial paired spine clusters bearing two spines per cluster .... *Ramellogammarus littoralis* Bousfield & Morino, 1992

[USA: Oregon]

6(3) Antenna I peduncle antennomere I with a posteriodistal spine 7

6' Antenna I peduncle antennomere I lacking a posteriodistal spine 8

7(6) Pleonite III posterior margin with a single lateral seta; pereopod VII basis posterior margin with ~13 setae; pleonite III posteriodorsal margin with a single row of weak setae .... *Ramellogammarus californicus* Bousfield & Morino, 1992

[USA: California]

7' Pleonite III posterior margin with a two or more lateral setae; pereopod VII basis posterior margin with ~18 setae; pleonite III posteriodorsal margin with a two or more rows of setae .... *Ramellogammarus campestris* Bousfield & Morino, 1992

[USA: Oregon]

8(6) Pleonite III posterior margin with a single lateral seta; male uropod I endopod lateral margin inerm, glaborous .... *Ramellogammarus ramellus* (Weckel, 1907)

[USA: California]

8' Pleonite III posterior margin with four or more lateral seta; male uropod I endopod lateral margin bearing spines .... *Ramellogammarus similimanus* (Bousfield, 1961)

[USA: Oregon]

REFERENCES

Bousfield, E.L. & J.R. Holsinger, 1989. A new Crangonyctid amphipod from hypogean fresh waters of Oregon. Canadian Journal of Zoology, 67: 963–968.

Bousfield & Morino, 1992. The amphipod genus *Paramoera* Miers (Gammaridea: Eusiroidea: Pontogeniidae) in the eastern North Pacific. Amphipacifica, 1: 61–102.

Holsinger, J.R. 1972. The freshwater amphipod crustaceans (Gammaridae) of North America. Biota of Freshwater Ecosystems, Identification Manual 5. U.S. Environmental Protection Agency.

Koenemann, S. & J.R. Holsinger, 2001. Systematics of the North American subterranean amphipod genus *Bactrurus* (Crangonyctidae). Beaufortia, 51: 1–56.

Staude, C.P. 1986. Systematics and behavioral ecology of the amphipod genus *Paramoera* Miers (Gammaridea: Eusiroidea: Pontogeniidae) in the eastern North Pacific. Ph.D. Dissertation, University of Washington, 325 pp.

Order Tanaidacea

Tom Hansknecht
Barry Vittor and Associates, Inc., Mobile, Alabama, USA

INTRODUCTION

The euryhaline, freshwater tolerant Tanaidacea of the Nearctic consists of species that usually inhabit salt marshes, mangrove areas, bays and tidally influenced rivers. Seasonal variations in freshwater input may lead to oligohaline or freshwater conditions that are the focus for this series. These arthropods are free-living and domicolous, with the Nearctic freshwater species being tube builders. Young are brooded in the tube and are nonplanktonic, emerging as mancas.

LIMITATIONS

Tanaids in the genera *Leptochelia* and *Hargeria* are strongly dimorphic, with males possessing well-developed, diagnostic chelae. Females, however, cannot be reliably separated.

TERMINOLOGY AND MORPHOLOGY

Tanaids are easily recognized and separated from other peracarids by paired sessile compound eyes, body dorsoventrally depressed, consisting of a cephalothorax, pereon, and pleon, with first pair of walking appendages chelate. These true chela make them look like little lobsters. The chela are followed by six pairs of legs in adults. There are no gills. The single pair of uropods are terminal, as in asellote isopods. Tubiculous tanaidaceans like some amphipods have spinning glands in the pereopods and build fine tubes of detritus, silt, and/or sand, sometimes incorporating plant fragments. The estuarine species, *Hargeria rapax*, on the Gulf Coast of northwest Florida has been found to dominate arthropod communities with populations exceeding 40,000 m^2 (Heard, unpublished). This brackish water species is also common in temperate and tropical waters of the northwestern Atlantic Ocean (Heard et al., 2003).

The suborders are easily distinguished by the presence (Apseudomorpha) or absence (Tanaidomorpha) of an antennular accessory flagellum. Only a single apseudomorph, *Halmyrapseudes bahamaensis*, has been recorded from this bioregion (Sieg et. al., 1982).

Some Nearctic tanaidomorphs have wide global distributions with *Sinelobus stanfordi* found nearly worldwide with records in Europe, Asia, Japan, Pacific Ocean, Indian Ocean, Galapagos, and the subantarctic (Gardiner, 1975; Sieg, 1980b, 1983). *Teleotanais gerlachii* was reported from Brazil (type locality) with additional records from Central America (El Salvador), West Africa (Nigeria), and Florida (Sieg & Heard, 1983).

MATERIAL PREPARATION AND PRESERVATION

Tanaidaceans are typically preserved in 10% formalin and stored in 70% ethanol. Material is usually collected in the field by hand cores, or various types of grabs. Sediments containing tanaidaceans can be preserved *in situ*, or tanaids can be extracted by repeated water washings and elutriation. The latter is usually done with a bucket by swirling and decanting of the invertebrates in the samples. The collected material, preserved in situ, is washed with gentle pressure in a graded series of sieves and usually retained by a 500 μm sieve. This sorted material can be saved in ethanol for future study. Tanaidaceans are separated from the other invertebrates under a dissection microscope and saved in separate, labeled vials. Collection locality data is very important for taxonomic studies and should be recorded with permanent ink or pencil.

Dojiri & Sieg (1997) provided a detailed account of external morphology. Dissection is usually done in ethanol with watchmaker forceps or chemically sharpened tungsten needles (Larsen & Hansknecht, 2004). Temporary mounts can be prepared with glycerin alcohol on slides with coverslips.

KEYS TO TANAIDACEA

Crustacea: Malacostraca: Tanaidacea: Suborders

1 First antenna without accessory flagellum (Pl. 16.183.2); thoracopod I not larger than other thoracopods (Pl. 16.183.3) .... **Tanaidomorpha [p. 591]**

1' First antenna with accessory flagellum (Pl. 16.183.04); thoracopod I stronger and larger than following thoracopods (Pl. 16.183.05) .... Apseudomorpha .... Parapsudidiae, one species: *Halmyrapseudes bahamaensis* Bacescu & Gutu, 1974 [USA: Florida]

Crustacea: Malacostraca: Tanaidacea: Tanaidomorpha: Families

1 Uropod biramus (exopod present) (Figs. 16.183.6, 12); abdomen with 5 pleonites; pleonites 1 and 2 lacking transverse row of plumose setae 2

1' Uropod uniramus (no exopod present) (Pl. 16.183.8); abdomen with 4 pleonites; pleonites 1 and 2 each with well-developed transverse row of long plumose setae (Pls. 16.183.1, 8–9) .... Tanaidae, one species: *Sinelobus stanfordi* (Richardson, 1901)

[Pacific, Atlantic and Gulf of Mexico coastal freshwater tidal habitats]

2 Uropod with endopod of 5 articles and exopod of 1 article (Pl. 16.183.6); body not pigmented (Pl. 16.183.7) .... **Leptocheliidae [p. 591]**

2' Uropod with endopod and exopod of 2 articles (Pl. 16.183.12); body strongly pigmented (Pl. 16.183.13) .... Nototanaidae, one species: *Teleotanais gerlachi* Lang, 1956

[Gulf of Mexico freshwater tidal habitats]

Crustacea: Malacostraca: Tanaidacea: Tanaidomorpha: *Leptocheliidae*: Species

1 Male with cheliped longer than body (Pl. 16.183.14); male pleotelson with posterior post anal spatulate process (Pl. 16.183.15) .... *Hargeria rapax* (Harger, 1879)

[Canada: New Brunswick, Nova Scotia. USA: Atlantic and Gulf coasts. Mexico: Gulf Coast]

1' Male with cheliped shorter than body (Pl. 16.183.16); male pleotelson without spatulate process (Pl. 16.183.11) .... *Leptochelia dubia* (Kroyer, 1842)

[Pacific, Atlantic and Gulf of Mexico coastal freshwater tidal habitats]

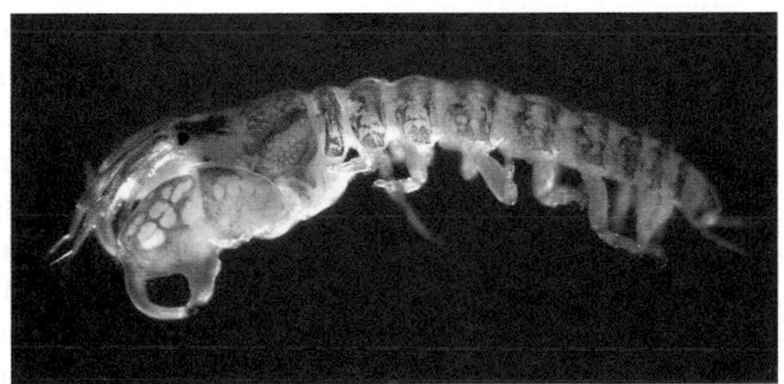

PLATE 16.183.01 *Sinelobius. Photograph courtesy of John Pfieffer at EcoAnalysts, Inc.*

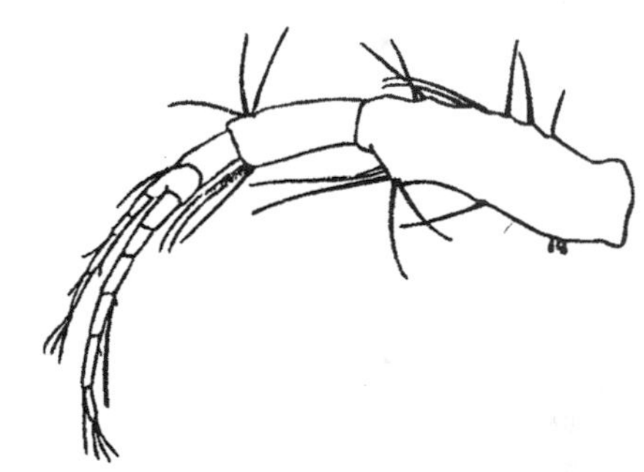

PLATE 16.183.04 Generalized apseudomorph antennule; left, dorsal view. *From Heard et al., 2003.*

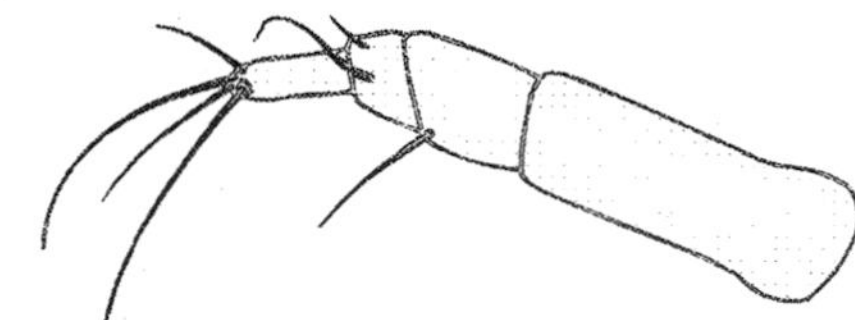

PLATE 16.183.02 Antennule of Tanaidomorpha. *From Heard et al., 2003.*

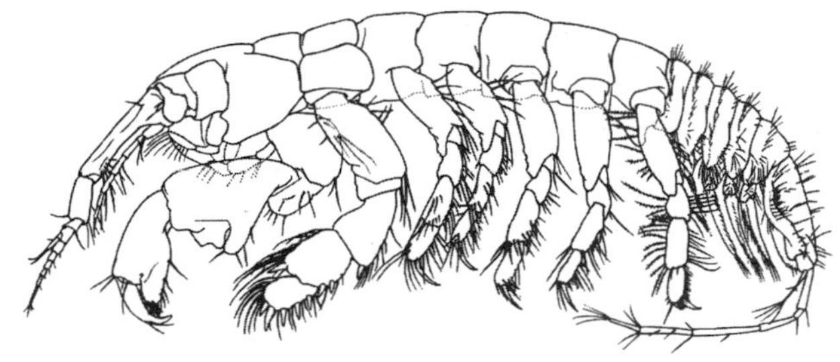

PLATE 16.183.05 *Halmyrapseudes bahamensis*, male; wet mount, lateral view. *From Heard et al., 2003.*

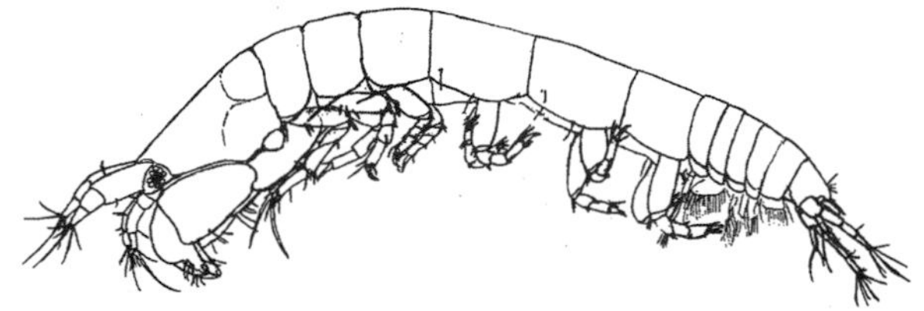

PLATE 16.183.03 Leptochelidae, female; wet mount, lateral view. *From Heard et al., 2003.*

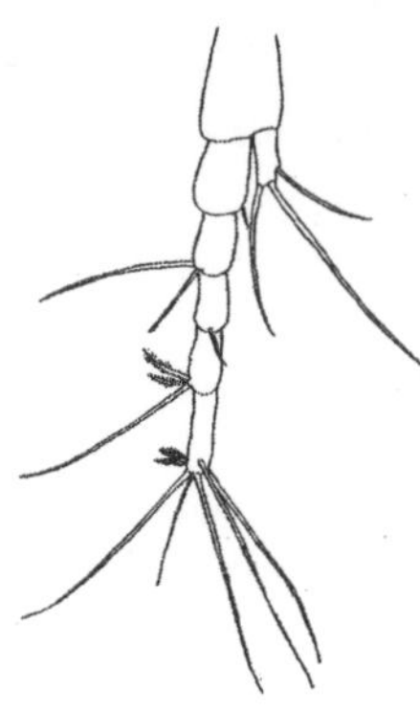

PLATE 16.183.06 Leptochelidae, uropod; right side view. *From Heard et al., 2003.*

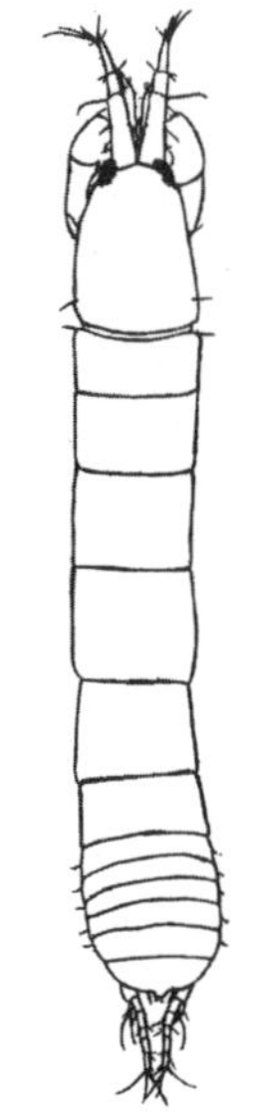

PLATE 16.183.07 Leptochelidae; wet mount, dorsal view. *From Heard et al., 2003.*

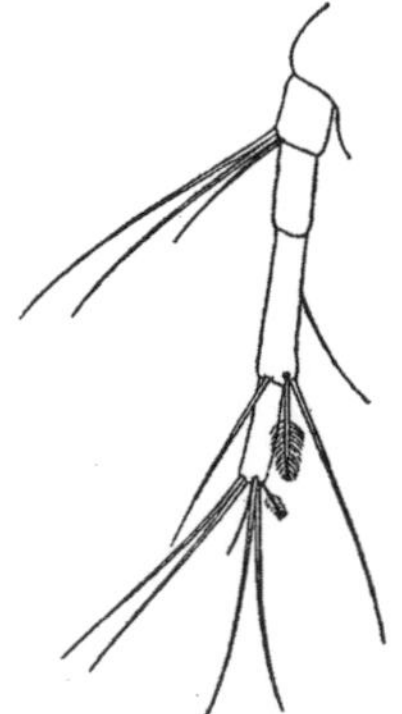

PLATE 16.183.08 *Sinelobus stanfordi*, uropod; left side view. *From Heard et al., 2003.*

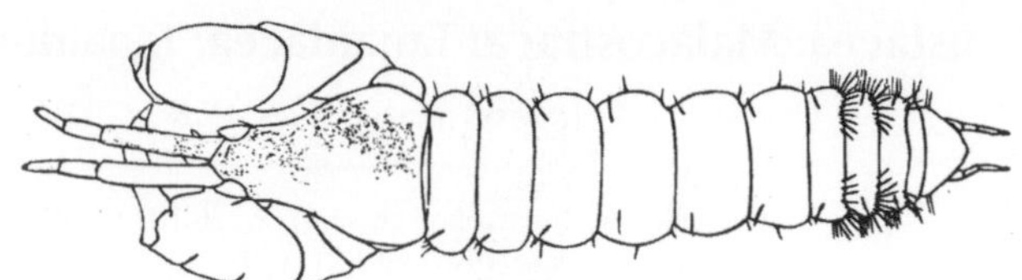

PLATE 16.183.09 *Sinelobus stanfordi*, male; wet mount, dorsal view. *From Heard et al., 2003.*

PLATE 16.183.10 *Leptochelia dubia*, uropod; right side view. *From Heard et al., 2003.*

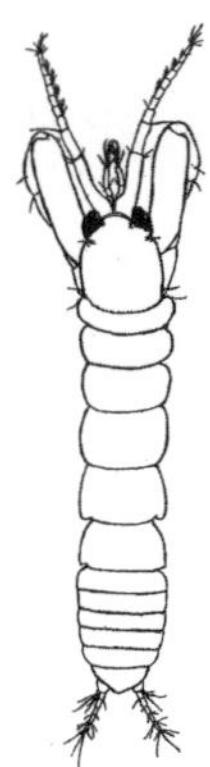

PLATE 16.183.11 *Leptochelia dubia* male; wet mount, dorsal view. *From Heard et al., 2003.*

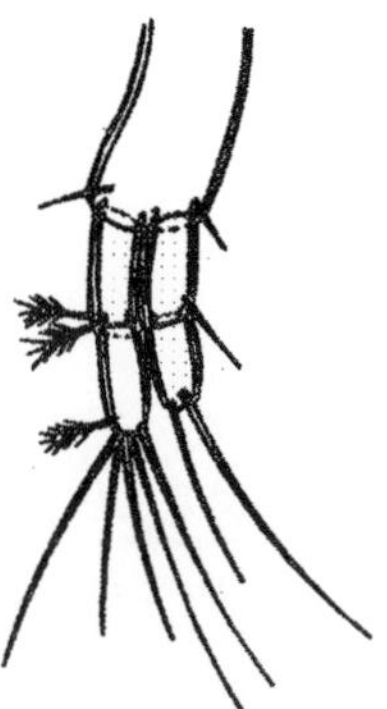

PLATE 16.183.12 *Teleotanais gerlachi*, uropod. *From Heard et al., 2003.*

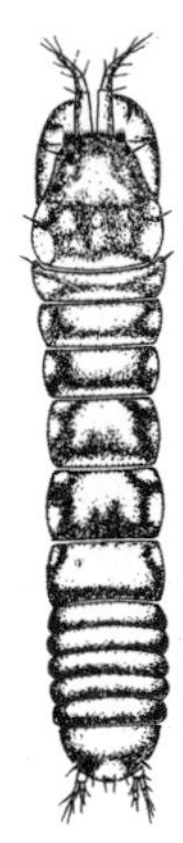

PLATE 16.183.13 *Teleotanais gerlachi*, female; wet mount, dorsal view. *From Heard et al., 2003.*

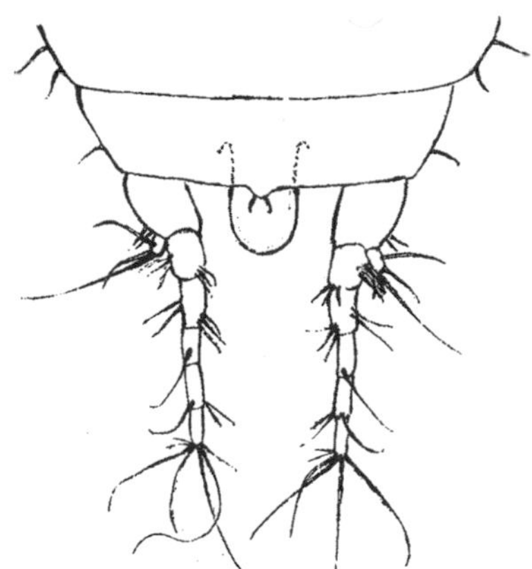

PLATE 16.183.15 *Hargeria rapax*, pleotelson and uropods. *From Heard et al., 2003.*

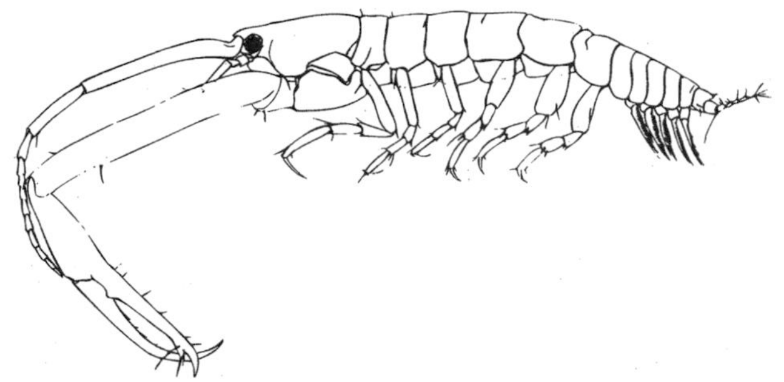

PLATE 16.183.14 *Hargeria rapax*, male; wet mount, lateral view. *From Heard et al., 2003.*

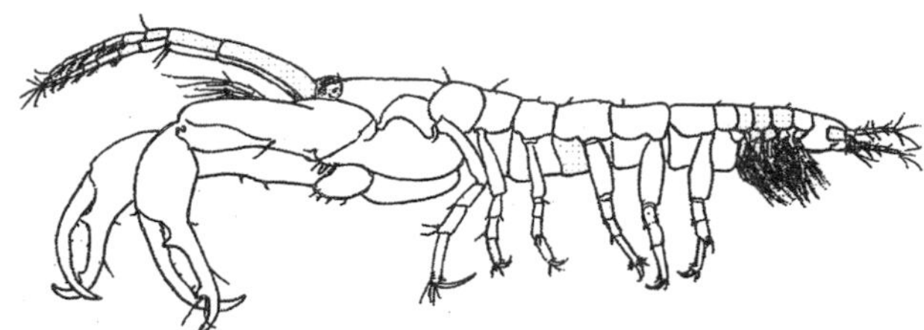

PLATE 16.183.16 *Leptochelia dubia*, male; wet mount, lateral view. *From Heard et al., 2003.*

REFERENCES

Dojiri, M. & J. Sieg. 1997. Chapter 3. The Tanaidacea. pp 181–268 in J.A. Blake and Paul H. Scott, eds. Taxonomic Atlas of the Benthic Fauna of the Santa Maria Basin and Western Santa Barbara Channel. Vol. II. The Crustacea Part 2. The Isopoda, Cumacea and Tanaidacea. Santa Barbara Museum of Natural History, Santa Barbara, CA. I-V. 1–278.

Gardiner, L.F. 1975. A fresh and brackish water tanaidacean, *Tanais stanfordi* Richardson, 1901, from a hypersaline lake in the Galapagos Archipelago, with a report on West Indian species. Crustaceana 29: 127–140.

Heard, R., Hansknecht, T. & K. Larsen. 2003. An illustrated identification guide to Florida Tanaidacea (Crustacea: Peracarida) occurring in depths of less than 200 m. Florida Dept. of Environmental Protection, DEP Contract no. WM828, Tallahassee, Florida. pp. I-V, 1–163.

Larsen, K. & T. Hansknecht. 2004. A new genus and species of freshwater tanaidacea, *Pseudohalmyrapseudes aquadulcis* (Apseudomorpha: Parapseudidae), from Northern Territory, Australia. Journal of Crustacean Biology 24: 567–575.

Sieg, J. 1981. The Tanaidae (Crustacea; Tanaidacea) of California, with a key to the world genera. Proceedings of the Biological Society of Washington 94: 315–343.

Sieg, J. & R. W. Heard. 1983. Tanaidacea of the Gulf of Mexico. III. On the occurrence of *Teleotanais gerlachi* Lang, 1956 (Nototanaidae) in the Eastern Gulf. Gulf Research Reports 7: 267–271.

Sieg, J., Heard, R. W & J. T. Ogle. 1982. Tanaidacea (Crustacea: Peracarida) of the Gulf of Mexico. II. The occurrence of *Halmyrapseudes bahamaensis* Băcescu and Guțu, 1974 (Apseudidae) in the Eastern Gulf with redescription and ecological notes. Gulf Research Reports 7: 105–113.

Order Isopoda

Julian J. Lewis
Lewis & Associates LLC, Borden, IN, USA

D. Christopher Rogers
Kansas Biological Survey, University of Kansas, Lawrence, KS, USA; The Biodiversity Institute, University of Kansas, Lawrence, KS, USA

INTRODUCTION

Over 100 species of isopod crustaceans have been described from the Nearctic, although establishing a precise number would be impossible given the need for revision of several genera. Many undescribed species are known and undoubtedly more remain to be discovered given the propensity of these crustaceans to inhabit groundwater habitats. Most workers typically encounter one of the dozen or so common species of the Asellidae that occur in epigean habitats. These species are inhabitants of streams of all sizes, temporary pools, ponds, lakes, and rivers where they are typically found on the substrate, under stones and other debris, or at times crawling about vegetation. More than half of the described species (from several families) are obligatory inhabitants of groundwater, where they are found in caves or other habitats where subterranean waters flow through unconsolidated deposits (e.g., gravel, sand, soil). One group of isopods inhabits desert hot springs, while another is comprised of parasites. Due to their rarity, several species have state or federal protection and others are likely to be listed in the future. Some species have been transported far from their natural ranges and are becoming invasive species.

LIMITATIONS

The Nearctic isopods can be sorted into families and frequently genera with little or no dissection and using only low power magnification. Some of the groups are monotypic and are readily identified. Range and habitat information are very helpful in identifying isopods. Endemicity is pronounced, and a surprising number of taxa can be identified based on collection location and ecological data. In the interest of completeness, a few species that are primarily allied to marine or brackish environments are included in the following keys, as they are sometimes known to occur in the upstream reaches of estuaries or adjacent freshwater habitats. Likewise, the Oniscoidea are included to encompass some terrestrial taxa that are amphibious or can tolerate immersion.

The predominant isopods in epigean habitats are asellids of the genera *Caecidotea* and *Lirceus*. Adult males are necessary for identification of asellids to species. Most of the epigean *Caecidotea* were characterized sufficiently by Williams (1970) for identification, although one or more of these species may be conspecific. New species are described occasionally, and others await description in museum collections. Over 60 species of *Caecidotea* have been described from caves and other subterranean habitats (Henry et al., 1986), and too many undescribed species remain to make identification practical with a key. The subterranean faunas of a few states have been treated comprehensively (Illinois: Lewis & Bowman, 1981; Texas: Lewis & Bowman, 1996; Virginia: Lewis, 2009; Maryland: Lewis & Bowman, 2010; Lewis et al., 2011). Otherwise, the information concerning most subterranean *Caecidotea* species is scattered in the literature among many accounts of one or a few species.

The most commonly encountered isopods after *Caecidotea* belong to the genus *Lirceus*. The monograph of the genus by Hubricht & Mackin (1949) relied on questionable characteristics and few if any of the epigean taxa are reliably identifiable. Two rare subterranean species endemic to southwestern Virginia are the only adequately described species of *Lirceus* (Holsinger & Bowman, 1973; Estes & Holsinger, 1976).

Six other genera of asellids are known only west of the Mississippi River. The seven species of *Lirceolus* are extremely difficult to identify from their morphology, but a few exceptions can be determined from their mostly disjunct ranges in northern Mexico, Texas, and the Ozarks (Lewis & Bowman, 1996; Lewis, 2001). The other asellid genera of the western USA are monotypic or contain only two species.

TERMINOLOGY AND MORPHOLOGY

Isopods range in size from around 2 mm in some of the *Lirceolus* species to as much as 9 cm in *Saduria entomon*, with 5–20 mm being typical of the common asellid species usually encountered. Epigean species are frequently mottled brown in appearance, with their unpigmented subterranean counterparts a whitish color. Likewise, eyes are present in the surface species, but vary from being reduced to completely absent in subterranean species. The body is comprised of three main sections. The head contains the two pairs of antennae and the mouthparts, called (in sequence) the mandibles (sometimes with palps), two pairs of maxillae, and the maxillipeds. The thorax is comprised of 7 pereonites, each with a pair of legs, i.e., pereopods. The first pair of pereopods may be modified for grasping with the penultimate podomere called the propodus enlarged and opposed by the claw-like final podomere, the dactylus. The abdomen is comprised of 5 segments known as pleonites, with appendages called pleopods. The pleonites in some groups are fused into a pleotelson. In the predominant asellids the second pleopod of the male is modified as a sperm reservoir and transfer mechanism. The tip of the male second pleopod endopodite possesses a sperm transfer tube, called the cannula, which has a unique construction in each species, frequently surrounded by accessory processes. The identification of most asellid species relies on the examination of this structure. The terminal appendages are the uropods and vary in their form and place of attachment to the pleon.

MATERIAL PREPARATION AND PRESERVATION

Isopods should be placed in 70% ethyl alcohol for storage, but 80–95% may be necessary initially if many specimens are collected, especially if any detritus is present in the sample.

Place the specimens in 70% ethanol and view under low power magnification for initial examination. At that point, specimens can usually be identified to the family level and often placed in a genus. Species identification frequently requires dissection of at least the male second pleopod, which can be accomplished by manipulating the specimen with fine (e.g., watch-makers) forceps and fine probes constructed of insect minuten pins. By repeatedly piercing the exoskeleton in the sutures between the ventral pleonites, the tiny second pleopod can be teased away from the specimen. The simplest procedure is to transfer the pleopod to a drop of glycerin on a microscope slide. It is usually a good idea to look at the positioning of the pleopod in the glycerin to assure than nothing needs to be adjusted before adding a coverslip. The pleopod is now ready for examination, which depending on the size of the specimen, may require 400–1000× for evaluation of the endopod apical structures. After examination, the pleopods should be removed from the slide and placed in a microvial within the vial that accompanies the dissected specimen.

KEYS TO ISOPODA

Crustacea: Malacostraca: Isopoda: Suborders

1 Uropods terminal (Pl. 16.185.1, 6 A) 2

1' Uropods lateral, ventral or vestigial 4

2 Body length greater than 2 mm, adults mostly greater than 5 mm 3

2' Body length less than 1 mm .... Microcerberidea, Microcerberidae, one genus and species: .... *Microcerberus carolinensis* Wägaele et al., 1995

[USA: South Carolina; at least one other undescribed species from Alabama]

3 Antenna 1 well formed (Pl. 16.185.1, 6 D) .... **Asellota [p. 597]**

3' Antenna 1 vestigial .... **Oniscoidea [p. 604]**

4 Free-living, morphologically unspecialized 5

4' Parasitic, appendages greatly reduced, modified for existence in branchial chambers of freshwater shrimp (Pl. 16.185.2, 6 B, C) .... Cymothoida, one family: Bopyridae, one genus: *Probopyrus*

[Primarily a marine group, three species recognized from USA Atlantic and Gulf of Mexico coastal states in fresh or brackish waters, but systematic status of these species questionable]

5 Uropods lateral (Pl. 16.185.1) 6

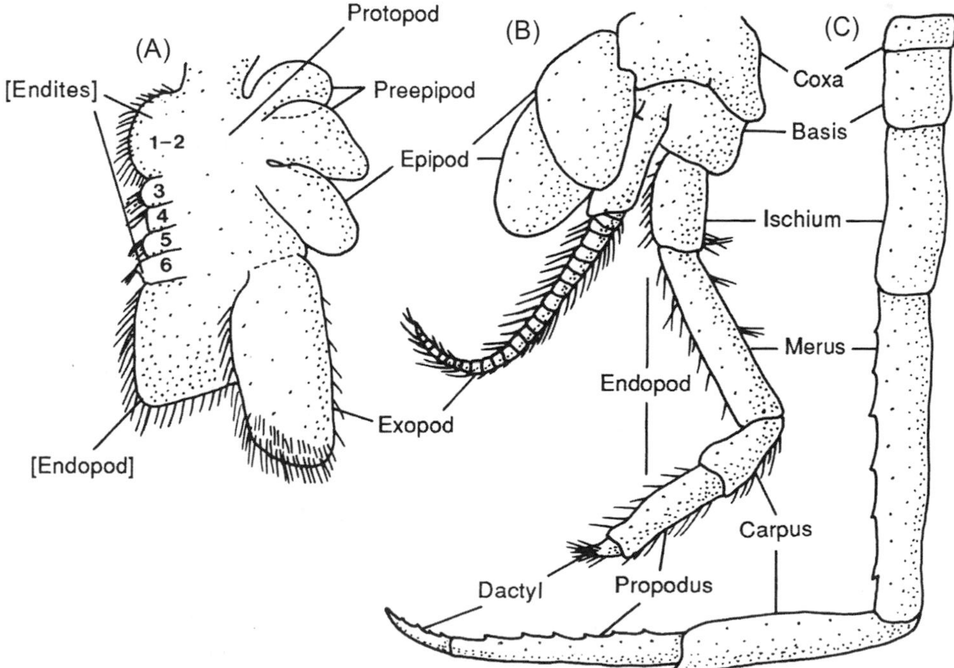

PLATE 16.184.01 Representative crustacean appendages: (A) a phyllopod appendage of Anostraca; (B) biramous appendage of Anaspidacea (superorder Syncarida), (C) uniramous stenopod appendage of a Decapoda. *Redrawn from Mclaughlin, 1982.*

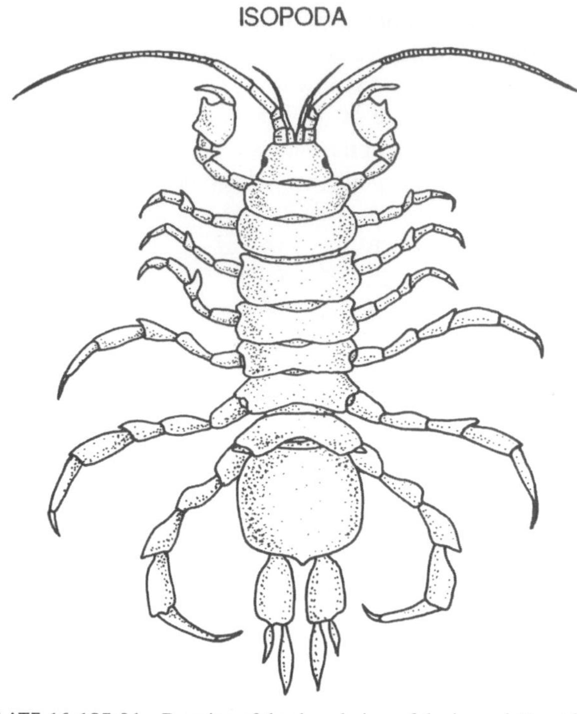

PLATE 16.185.01 Drawing of the dorsal view of the isopod *Caecidotea* (drawn without setation).

PLATE 16.185.02 Parasitic isopod *Probopyrus* sp. (bulge under the carapace of *Macrobrachium ohione*). *Photograph courtesy of Matthew A. Hill & EcoAnalysts, Inc.*

PLATE 16.185.03 Valviferan isopod (*Chirdotea almyra*). *Photograph courtesy of Matthew A. Hill & EcoAnalysts, Inc.*

PLATE 16.185.04 Western aquatic pill bug (*Gnorimosphaeroma* sp.,). *Photograph courtesy of John Pfieffer & EcoAnalysts, Inc.*

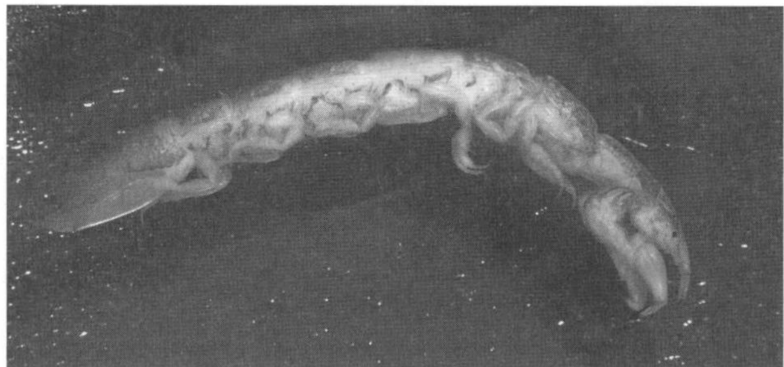

PLATE 16.185.05 The isopod *Cyathura*. *Photograph courtesy of John Pfieffer & EcoAnalysts, Inc.*

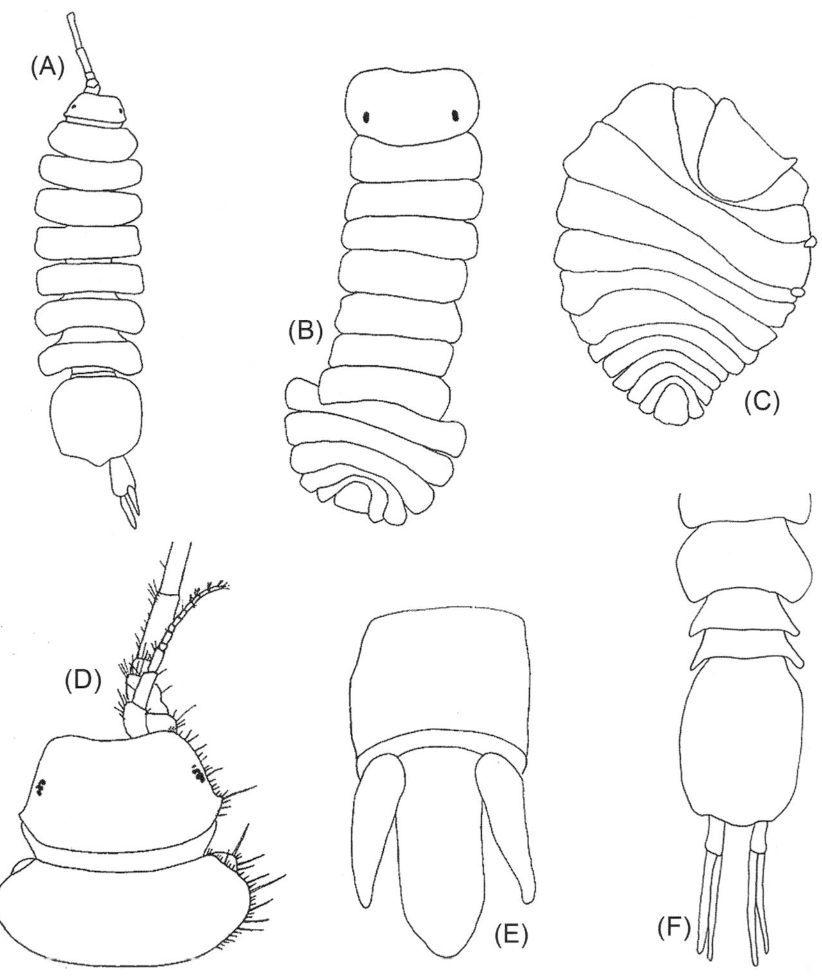

PLATE 16.185.06 Isopoda: (A) suborder Asellota, habitus; uropods terminal, as in *Caecidotea* spp.; (B) *Probopyrus* male; (C) *Probopyrus* female; (D) antenna in Asellota; (E) *Cyathura polita*; (F) *Etlastenasellus* sp.

5'	Uropods ventral (Pl. 16.185.3) .... [brackish habitats along the Atlantic coast]	Valvifera, Chaetiliidae, one species: *Chiridotea almyra* Bowman, 1955
6	Body globose, length typically about twice that of width (Pl. 16.185.4)	**Flabellifera [p. 602]**
6'	Body vermiform, length more than 4× width (Pl. 16.185.5, 6 E) .... [Primarily a marine group, this species occurs in brackist habitats along Atlantic coast]	Anthuridea, Anthuridae, one species: *Cyathura polita* (Stimpson, 1855)

Crustacea: Malacostraca: Isopoda: Asellota: Families

1	Abdomen with pleonites 1–2 thin, not pronounced in dorsal view	**Asellidae [p. 597]**
1'	Abdomen with pleonites 1–2 pronounced in dorsal view (Pl. 16.185.6 F)	**Stenasellidae [p. 601]**

Crustacea: Malacostraca: Isopoda: Asellota: Asellidae: Genera

1	Pigmented, eyes present, epigean (Pl. 16.185.7 A).......	2
1'	Unpigmented, eyes vestigial or absent, subterranean	4
2(1)	Pleopod 3 with transverse suture (Pl. 16.185.7 B)	3
2'	Pleopod 3 with oblique suture (Pl. 16.185.7 C)	*Lirceus* (in part)
3(2)	Pleopod 2 endopod with pronounced basal spur (Pl. 16.185.7 D)	***Asellus* [p. 599]**

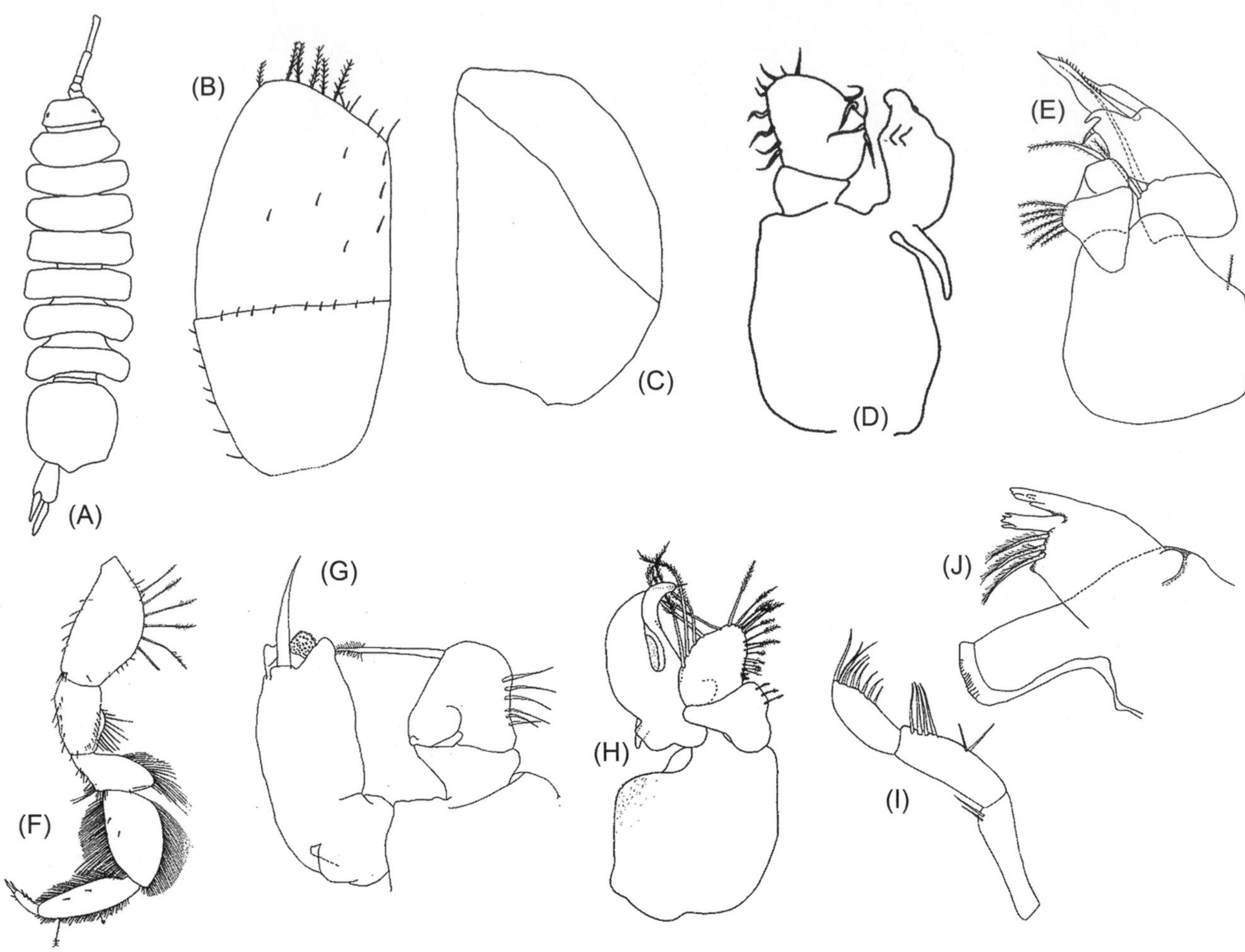

PLATE 16.185.07 (A) Eyes in *Caecidotea* sp.; (B) *Caecidotea* pleopod 3 showing transverse suture; (C) oblique suture on pleopod 3 in *Lirceus* sp.; (D) *Asellus* pleopod #2 with prominent basal spur; (E) *Salmasellus howarthi* showing pleopod 2 with stylet; (F) pleopod modified for swimming in *Remasellus parvus*; (G) endopod of pleopod 2 lacking setae in *Bowmanasellus sequoiaei*; (H) endopod of pleopod 2 with stylet thin at base and lacking prominent labial spur in *Columbasellus acheron*; (I) mandibular palp in *Calasellus longus* with 3 articles; (J) *Oregonasellus elliotti* in which mandibular palp is absent.

3'	Pleopod 2 endopod with basal spur absent or minimal	*Caecidotea* (in part)
	[Widespread. Many described and undescribed species. Genus in need of revision]	
4(1)	Pleopod 2 lacking stylet, species east of Rocky Mountains	5
4'	Pleopod 2 with stylet, west of Rocky Mountains (Pl. 16.185.7 E)	8
5(4)	Pleopod 3 exopod with oblique suture (Pl. 16.185.7 C)	6
5'	Pleopod 3 exopod with transverse suture (Pl. 16.185.7 B)	7
6(5)	Head with dorsal carina	*Lirceus* (in part)
	[East of Continental Divide, introduced in northwest]	
6	Head without dorsal carina	***Lirceolus* [p. 599]**
	[USA: Texas, Missouri. Mexico: Coahuila]	
7(5)	Pereopods densely setose, modified as swimming appendages (Pl. 185.7 F)	*Remasellus parvus* (Steeves, 1964)
	[USA: Florida]	
7'	Pereopods not densely setose, unable to swim in water column	*Caecidotea* (in part)
	[60+ described species, many undescribed species]	
8(4)	Pleopod 2 endopod with stylet thick at base (Pl. 16.185.7 E)	9
8'	Pleopod 2 endopod with stylet thin at base (Pl. 16.185.7 H)	10

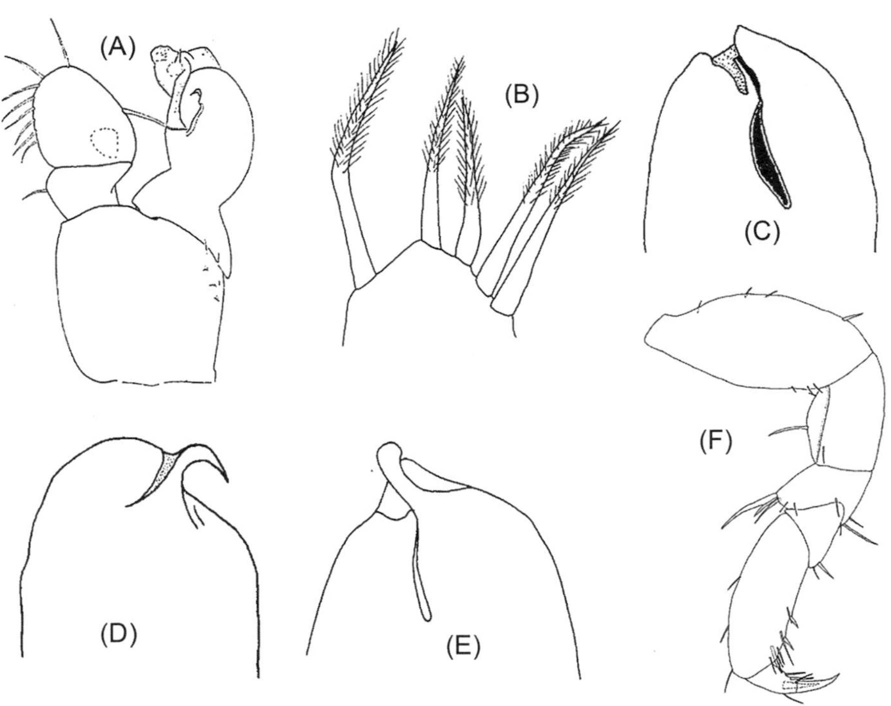

PLATE 16.185.08 Isopoda family Asellidae: (A) pleopod 2 of male *Asellus alaskensis*; (B) plumose setae on medial lobe of maxilla of *Lirceolus bisetus*; (C) pleopod 2 tip in *Lirceolus nidulus*; (D) pleopod 2 tip in *Lirceolus pilus*; (E) pleopod 2 tip in *Lirceolus bisetus*; (F) pereopod 1 in *Salmasellus howarthi*.

9(8) Pleopod 2 endopod with prominent setae (Pl. 16.185.7 E) .... ***Salmasellus* [p. 601]**
[Canda: Alberta; USA: Montana, Washington]

9' Pleopod 2 endopod without setae (Pl. 16.185.7 G) .... *Bowmanasellus sequoiaei* (Bowman, 1975)
[USA: California]

10(8) Pleopod 2 endopod with prominent labial spur adjacent to endopodial groove 11

10' Pleopod 2 endopod without prominent labial spur (Pl. 16.185.7 H) .... *Columbasellus acheron* Lewis et al., 2001
[USA: Washington]

11(10) Mandibular palp with 3 articles (Pl. 16.185.7 I) .... ***Calasellus* [p. 601]**
[USA: California]

11' Mandibular palp vestigial or absent (Pl. 16.185.7 J) .... *Oregonasellus elliotti* Lewis, 2008
[USA: Oregon, Malheur Cave]

Crustacea: Malacostraca: Isopoda: Asellota: Asellidae: *Asellus*: Species

1 Male pleopod 2 exopod distal segment with 1–2 setae along mesial margin (Pl. 16.185.8 A) .... *Asellus alaskensis* Bowman & Holmquist, 1975
[USA: Alaska]

1' Male pleopod 2 exopod distal segment with 4 setae along mesial margin .... *Asellus hilgendorfi* Bovallius, 1886
[Invasive. USA: California. Palaearctic]

Crustacea: Malacostraca: Isopoda: Asellota: Asellidae: *Lirceolus*: Species

1 Maxilla 1, medial lobe with 4–5 plumose setae (Pl. 16.185.8 C) 2

1' Maxilla 1, medial lobe with 8 plumose setae .... *Lirceolus smithii* (Ulrich, 1902)
[USA: Texas]

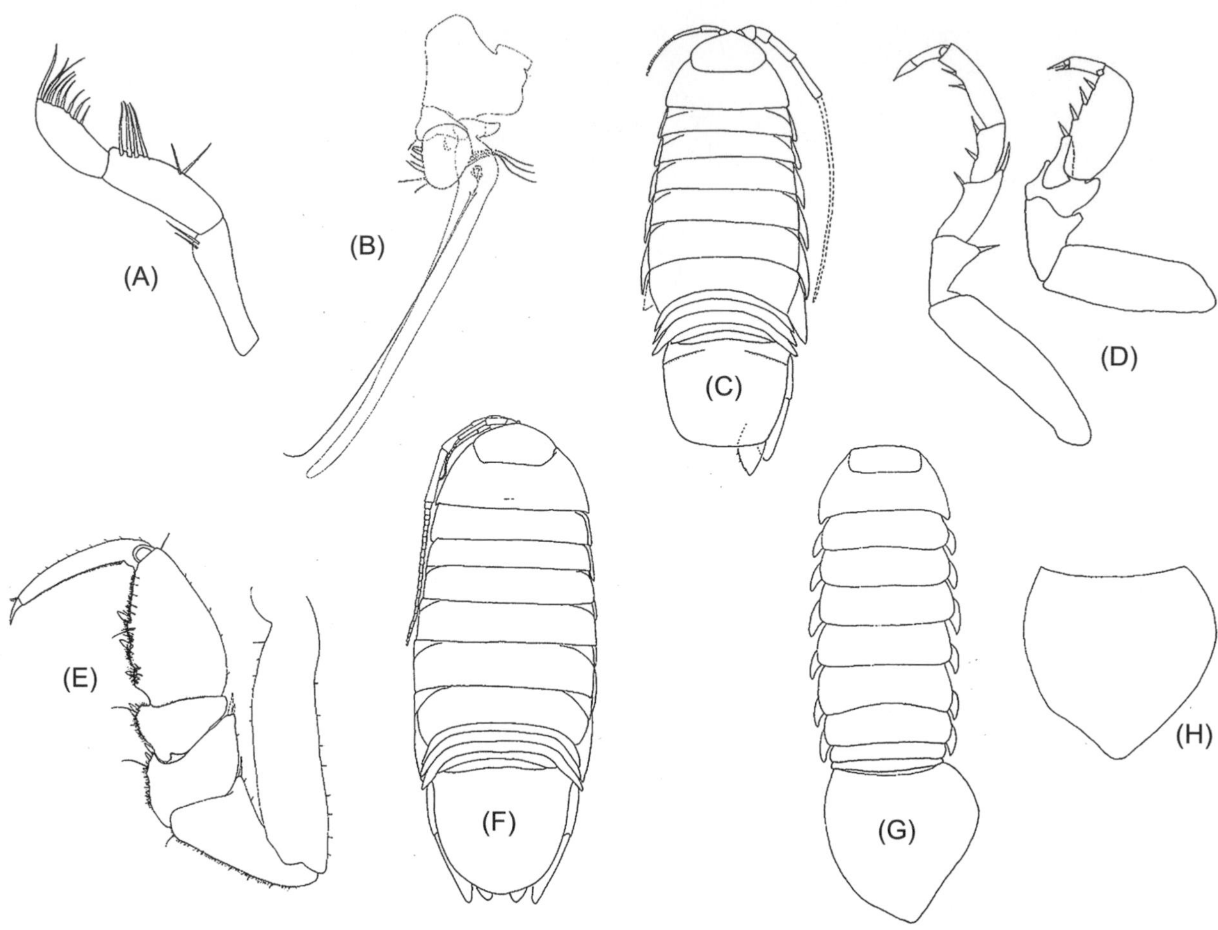

PLATE 16.185.09 (A) Mandibular palp of *Calasellus longus*; (B) pleopod 2 of *Calasellus longus*; (C) *Speocirolana endeca* habitus; note lack of eyes; (D) pereopods 1 and 2 of *Cirolanides texensis*; (E) pereopod of *Speocirolana pubens*; (F) *Speocirolana pubens* habitus; (G) *Sphaerolana karenae* habitus; (H) pleotelson of *Sphaerolana karenae*.

2(1) Maxilla 1, medial lobe with 5 plumose setae....3

2' Maxilla 1, medial lobe with 4 plumose setae .... *Lirceolus hardeni* Lewis & Bowman, 1996
[USA: Texas]

3(2) Body less than 5 mm long 4

3' Body more than 5 mm long .... *Lirceolus nidulus* Lewis, 2001
[USA: Texas]

4(3) Male pleopod 2 endopod tip with cannula extending beyond other processes 5

4' Male pleopod 2 endopod tip with cannula nestled among other processes .... *Lirceolus cocytus* Lewis, 2001
[USA: Texas. Mexico: Coahuila]

5(4) Male pleopod 2 endopod tip cannula not decurved, extending apically (Pl. 16.185.8 D) 6

5' Male pleopod 2 endopod tip with cannula decurved, beak-shaped (Pl. 16.185.8 D) .... *Lirceolus pilus* (Steeves, 1968)
[USA: Texas]

6(5) Male pleopod 2 endopod tip, cannula a blunt knob, slightly enlarged apically (Pl. 16.185.8 D) .... *Lirceolus bisetus* (Steeves, 1968)
[USA: Texas]

6' Male pleopod 2 endopod tip, cannula apically truncate cone.... *Lirceolus serratus* (Fleming, 1972)
[USA: Missouri]

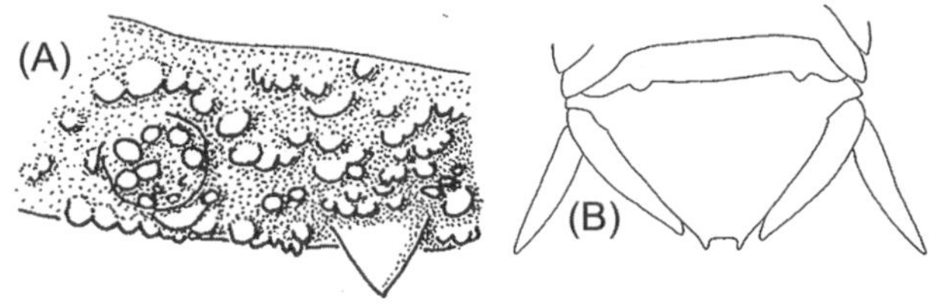

PLATE 16.185.10 (A) dorsal section of pleonite 1 in *Sphaeroma terebrans*; (B) pleotelson of *Thermosphaeroa cavicauda*.

Crustacea: Malacostraca: Isopoda: Asellota: Asellidae: *Salmasellus*: Species

1 Pereopod 1, palmar margin of propodus with 4 or fewer stout setae (Pl. 16.185.9 F) .... *Salmasellus howarthi* Lewis, 2001
[USA: Washington, Oregon]

1' Pereopod 1, palmar margin of propodus with 5 or more stout setae .... *Salmasellus steganothrix* Bowman, 1975
[Canada: Alberta; USA: Montana]

Crustacea: Malacostraca: Isopoda: Asellota: Asellidae: *Calasellus*: Species

1 Antenna 1 shorter than body; pleopod 2 endopod about 2× length of exopod .... *Calasellus californicus* (Miller, 1933)
[USA: California]

1' Antenna 1 longer than body; pleopod 2 endopod about 4× length of exopod (Pl. 16.185.9 A, B).......... *Calasellus longus* Bowman, 1981
[USA: California]

Crustacea: Malacostraca: Isopoda: Asellota: Stenasellidae: Genera

1 Protopods of male pleopod 1 and female pleopod 2 fused proximally .... ***Etlastenasellus* [p. 601]**

1' Protopods of male pleopod 1 and female pleopod 2 not fused .... ***Mexistenasellus* [p. 601]**

Crustacea: Malacostraca: Isopoda: Asellota: Stenasellidae: *Etlastenasellus*: Species

1 Uropod rami greater than 2× length of protopod .... *Etlastenasellus mixtecus* Argano, 1977
[Mexico: Oaxaca]

1' Uropod rami less than 2× length of protopod .... *Etlastenasellus confinus* Bowman, 2002
[Mexico: Oaxaca]

Crustacea: Malacostraca: Isopoda: Asellota: Stenasellidae: *Mexistenasellus*: Species

Based in part on Bowman, 2002.

1 Pleopod 4 exopod with distal setae 2

1' Pleopod 4 exopod without distal setae 5

2(1) Pleopod 4 exopod with diagonal, at least in medial half 3

2' Pleopod 4 exopod with transverse suture .... *Mexistenasellus magniezi* Argano, 1974
[Mexico: Veracruz]

3(2) Pleopod 4 exopod proximal article without marginal setae.......... 4

3' Pleopod 4 exopod proximal article with marginal setae .... *Mexistenasellus parzefalli* Magniez, 1972
[Mexico: San Luis Potosi, Tamaulipas]

4(3) Pleopod 4 exopod suture straight .... *Mexistenasellus nulemex* Bowman, 2002
[Mexico: Nuevo León]

4' Pleopod 4 exopod suture bent near mid-length.. *Mexistenasellus wilkensi* Magniez, 1972
[Mexico: San Luis Potosi]

5(1) Pleopod 4 exopod about 2× as long as wide, with strongly diagonal suture .... *Mexistenasellus coahuila* (Cole & Minckley, 1972)
[USA: Texas. Mexico: Coahuila]

5' Pleopod 4 exopod more than 4× as long as wide, with nearly transverse suture.................................. *Mexistenasellus colei* Bowman, 2002
[Mexico: Tamaulipas]

Crustacea: Malacostraca: Isopoda: Flabellifera: Families

1 Pleotelson longer than wide; eyeless, unpigmented, subterranean (Pl. 16.185.9 C) .... **Cirolanidae [p. 602]**

1' Pleotelson wider than long; eyes present, pigmented, epigean .... **Sphaeromatidae [p. 603]**

Crustacea: Malacostraca: Isopoda: Flabellifera: Cirolanidae: Genera

1 Pereopod 1 prehensile, pereopods 2-7 ambulatory (Pl. 16.185.9 D) 2

1' Pereopods 1-2 or 1-3 prehensile, pereopods 3/4-7 ambulatory (Pl. 16.185.9 E) 3

2(1) Exopod of pleopod 2, and endopods of pleopods 3-5 with 2 articles (Pl. 16.185.9 D) .... *Cirolanides texensis* Benedict, 1896
[USA: Texas. Mexico: Coahuila]

2' Exopod of pleopod 2, and endopods of pleopods 3-5 with one article... *Antrolana lira* Bowman, 1964
[USA: Virginia]

3(1) Pereopods 1-3 prehensile, pereopod 4-7 ambulatory..4

3' Pereopods 1-2 prehensile, pereopods 3-7 ambulatory .... *Mexilana saluposi* Bowman, 1975
[Mexico: San Luis Potosi]

4(3) Uropods not reduced, extending to or beyond the posterior margin of the pleotelson (Pl. 16. 185.9 F) .... ***Speocirolana* [p. 602]**

4' Uropods with rami reduced, minute (Pl. 16.185.9 G) .... ***Sphaerolana* [p. 603]**

Crustacea: Malacostraca: Isopoda: Flabellifera: Cirolanidae: *Speocirolana*: Species

Modified from Schotte (2002).

1 Pleotelson not acute 2

1' Pleotelson acute distally 12

2(1) Pleotelson rounded distally 3

2' Pleotelson truncated distally 6

3(2) Uropodal sympodite (protopod) not extending to telson distal margin 4

3' Uropodal sympodite extending beyond pleotelson distal margin 5

4(3) Male uropodal endopod clavate, longer than exopod .... *Speocirolana pubens* Bowman, 1982
[Mexico: San Luis Potosi]

4' Male uropodal endopod not clavate, subequal in length to exopod .... *Speocirolana prima* Schotte, 2002
[Mexico: Tamulipas]

5(3) Uropodal exopod acute, as long as endopod; second antenna extending no further than pereonite VI, and bearing 40 or less flagellomeres. .... *Speocirolana pelaezi* Bolivar & Pieltain, 1950
[Mexico: San Luis Potosi]

5' Uropodal exopod lanceolate, shorter than endopod; second antenna extending to pereonite VII, and bearing 50 or more flagellomeres .... *Speocirolana xilitla* Alvarez & Villalobos, 2008
[Mexico: San Luis Potosi]

6(2) Uropodal exopod not clavate 7

6' Uropodal exopod clavate, longer than endopod .... *Speocirolana fustiura* Botosaneanu & Iliffe, 1999
[Mexico: Nuevo León]

7(6) Pleotelson posterior margin width less than half basal width 8

7' Pleotelson posterior margin width greater than half basal width 10

8(7) Frontal lamina not narrow, not bent at one-third of length 9

8' Frontal lamina narrow, elongated, bent at nearly 90° at one-third of length .... *Speocirolana lapenita* Botosaneanu & Iliffe, 1999
[Mexico: Tamulipas]

9(8) Uropod rami extending beyond pleotelson posterior margin .... *Speocirolana guerrai* Contreras-Balderas & Purata-Velarde, 1982
[Mexico: Nuevo León]

9' Uropod exopod subequal in length to pleotelson, endopod distinctly shorter than pleotelson........ *Speocirolana zumbadora* Botosaneanu, Iliffe, & Hendrickson, 1998

10(7) Frontal lamina flat, pentagonal; second antennae not extending beyond pleotelson middle........ 11

10' Frontal lamina curved dorsally; second antennae nearly reaching pleotelson posterior margin .... *Speocirolana disparicornis* Botosaneanu & Iliffe, 1999
[Mexico: Tamulipas]

11(10) Frontal lamina length 1.5 times the width .... *Speocirolana endeca* Bowman, 1982
[Mexico: Tamulipas]

11' Frontal lamina length 2.5 times the width .... *Speocirolana bolivari* Rioja, 1953
[Mexico: San Luis Potosi]

12(1) Uropod endopod broadly rounded, one-half the length of the exopod .... *Speocirolana thermydronis* Cole & Minckley, 1966
[Mexico: Coahuila]

12' Uropod endopod subacute, more than one-half the length of the exopod .... *Speocirolana hardeni* Bowman, 1992
[USA: Texas]

Crustacea: Malacostraca: Isopoda: Flabellifera: Cirolanidae: *Sphaerolana*: Species

1 Pleotelson with apex broadly rounded 2

1' Pleotelson with apex tapered to a rounded point (Pl. 16.185.9 H) .... *Sphaerolana karenae* Rodríguez-Almaraz & Bowman, 1995
[Mexico: Nuevo León]

2(1) Pleotelson length width ratio 0.90–0.95 .... *Sphaerolana interstitialis* Cole & Minkley, 1970
[Mexico: Coahuila]

2' Pleotelson length width ratio 1.10–1.20 .... *Sphaerolana affinis* Cole & Minkley, 1970
[Mexico: Coahuila]

Crustacea: Malacostraca: Isopoda: Flabellifera: Sphaeromatidae: Genera

1 Pleonites smooth dorsally 2

1' Pleonites with coarse papillose dorsal surface (Pl. 16.185.10 A) .... *Sphaeroma terebrans* Bates 1866
[USA: Gulf of Mexico coastal states, brackish water]

2 Uropod lateral ramus prominent 3

2' Uropod lateral ramus reduced .... *Cassidinidea ovalis* (Say, 1818)
[USA: Atlantic coastal states, brackish water]

3 Uropod exopod rounded apically........ ***Gnorimosphaeroma* [p. 603]**

3' Uropod exopod acute apically (Pl. 16.185.10 B) .... ***Thermosphaeroma* [p. 604]**

Crustacea: Malacostraca: Isopoda: Flabellifera: Sphaeromatidae: *Gnorimosphaeroma*: Species

1 Antennal peduncles not in contact medially 2

1' Antennal peduncles in contact medially .... *Gnorimosphaeroma nobelei* Menzies, 1954
[USA: California, intertidal, but moving into shallow freshwater streams in rocky areas.]

2(2) Pleonites I-III all forming the lateral margin of the pleon 3

2' Pleonites I and II forming lateral margin of pleon, Pleonite III not reaching lateral margin .... *Gnorimosphaeroma insulare* (Van Name, 1940)

[USA: California]

3(2) Pereopod I ischium with sternal crest with seven to nine setae .... *Gnorimosphaeroma rayi* Hoestlandt, 1969

[Introduced. USA: California. Palaearctic]

3' Pereopod I ischium with sternal crest with rows of long setae, often reaching the dactyl .... *Gnorimospaheroma oregonense* (Dana, 1853)

[Canada: British Columbia. USA: Alaska, California, Oregon, Washington]

Crustacea: Malacostraca: Isopoda: Flabellifera: Sphaeromatidae: *Thermosphaeroma*: Species

1 Pleotelson apex entire 2

1' Pleotelson apex sulcate (Pl. 16.185.10 B) ....*Thermosphaeroma cavicauda* Bowman, 1985

[Mexico: Durango]

2(1) Frontal lamina not truncated 3

2' Frontal lamina anteriorly truncated ....*Thermosphaeroma mendozai* Shotte, 2000

[Mexico: Chihuahua]

3(2) Uropods with endopod and exopod subequal in length or exopod distinctly shorter than endopod 4

3' Uropods with exopod distinctly longer than endopods .... *Thermosphaeroma smithi* Bowman, 1981

[Mexico: Chihuahua]

4(3) Uropods with exopod distinctly shorter than endopod 5

4' Uropods with endopod and exopod subequal in length .... *Thermosphaeroma subequalum* Cole & Bane, 1978

[USA: Texas. Mexico: Chihuahua]

5(4) Pleotelson broadly rounded 6

5' Pleotelson narrowed apically, subtriangular 7

6(5) Pleotelson width approximately 1.5 times pleotelson length .... *Thermosphaeroma thermophilum* (Richardson, 1897)

[USA: New Mexico]

6' Pleotelson width subequal to pleotelson length .... *Thermosphaeroma macrura* Bowman, 1985

[Mexico: Chihuahua]

7(5) Pleotelson width approximately 1.5 times pleotelson length .... *Thermosphaeroma milleri* Bowman, 1981

[Mexico: Chihuahua]

7' Pleotelson width subequal to pleotelson length .... *Thermosphaeroma dugesi* (Dollfus, 1893)

[Mexico: Aguascalientes]

Crustacea: Malacostraca: Isopoda: Oniscoidea

1 Antennae with 6 fused articles that appear as a single section, eyes with 1 ocellus .... Trichoniscidae, one species: *Hyloniscus riparius* (Koch, 1838)

[Invasive. USA. Palaearctic; a riparian species that can tolerate prolonged submersion]

1' Antennae with 10 or more distinct articles, eyes large with 100+ ocelli .... Ligiidae, one genus: *Ligidium*

[Canada: southern and western forested regions. USA: Alaska, California, Oregon, Washington, west to New England. Genus in need of revision]

Crustacea: Malacostraca: Isopoda: Valvifera: Genera

1 Rostrum present .... *Saduria entomom* (Linnaeus, 1758)

[Canada: arctic coast. USA: Alaska]

1' Rostrum absent .... *Chiridotea almyra* Bowman, 1955

[Canada: Atlantic Coast. USA: Atlantic Coast, south to South Carolina. Deep coastal rivers]

REFERENCES

Bowman, T.E. 1982. Three new stenasellid isopods from Mexico (Crustacea: Asellota). Association of Mexican Cave Studies Bulletin. 8: 25–38.

Estes, J.A. & J.R. Holsinger. 1976. A second troglobitic species of the genus *Lirceus* (Isopoda, Asellidae) from southwestern Virginia. Proceedings of the Biological Society of Washington 89: 481–490.

Henry, J.P., J.J. Lewis & G. Magniez. 1986. Isopoda: Asellota: Aselloida, Gnathostenetroidoidea, Stenetrioidea). Pages 434–464 *in*: L. Botosaneanu (Ed.), Stygofauna Mundi: a faunistic, distributional and ecological synthesis of the world fauna inhabiting subterranean waters (including the marine interstitial). E. J. Brill, Leiden, Netherlands.

Hubricht, L. & J.G. Mackin. 1949. The freshwater isopods of the genus *Lirceus* (Asellota, Asellidae). American Midland Naturalist 42: 334–349.

Holsinger, J.R. & T.E. Bowman. 1973. A new troglobitic isopod of the genus *Lirceus* (Asellidae) from southwestern Virginia, with notes on its ecology and additional cave records for the genus in the Appalachians. International Journal of Speleology 5: 261–271.

Lewis, J.J. 2001. Three new species of subterranean asellids from western North America, with a synopsis of the species of the region (Crustacea: Isopoda: Asellidae). Speleology Monograph Series, Texas Memorial Museum 5: 1–15.

Lewis, J. J. 2009. Three new species of *Caecidotea*, with a synopsis of the asellids of Virginia (Crustacea: Isopoda: Asellidae). Martinsville, Virginia Museum of Natural History, Special Publication 16: 245–259.

Lewis, J.J. & T.E. Bowman. 1981. The subterranean asellids (*Caecidotea*) of Illinois (Crustacea: Isopoda: Asellidae). Smithsonian Contributions to Zoology 335.

Lewis, J.J. & T.E. Bowman. 1996. The subterranean asellids of Texas (Crustacea: Isopoda: Asellidae). Proceedings of the Biological Society of Washington, 109: 482–500.

Lewis, J.J. & T.E. Bowman. 2010. The subterranean asellids of Maryland: Description of *Caecidotea nordeni*, new species, and new records of *C. holsingeri* and *C. franzi* (Crustacea: Malacostraca: Isopoda. Journal of Cave & Karst Studies 72: 100–104.

Lewis, J.J., T.E. Bowman & D. Feller. 2011. A synopsis of the subterranean asellids of Maryland, U.S.A., with description of *Caecidotea alleghenyensis*, new species (Crustacea: Isopoda: Asellota). Zootaxa 2769: 54–64.

Schotte, M. 2002. *Speocirolana prima*, a new species from Tamaulipas, Mexico with a key to known species of the genus (Crustacea: Isopoda: Cirolanidae). Proceedings of the Biological Society of Washington 115: 628–635.

Williams, W. D. 1970. A revision of North American epigean species of *Asellus* (Crustacea: Isopoda). Smithsonian Contributions to Zoology 49: 1–80.

Order Decapoda

D. Christopher Rogers

Kansas Biological Survey, University of Kansas, Lawrence, KS, USA; The Biodiversity Institute, University of Kansas, Lawrence, KS, USA

INTRODUCTION

The freshwater and terrestrial decapods of the Nearctic are treated here because all terrestrial species are often found in and around freshwater, especially younger individuals. North America has more crayfish (crawdads, crawfish, mudbugs) than any other continent (Hobbs, 1972), with diversity being at its highest in southeastern USA.

LIMITATIONS

Specific limitations are described under each group. In general, sexually mature and/or "form I" males are required for species level identifications. In almost all crayfish keys, form I (fully mature) males are required for any meaningful identification at the species level. Females and juvenile or form II males cannot be identified to species with reasonable accuracy for the Cambaridae. Astacids and parastacids are not limited in this way.

However, the greatest limitation is that many native species of crayfish and shrimp are introduced to new areas (typically as escaped or released bait) or new invasive species of crayfish, shrimp, and crabs enter to the Nearctic Region. As a result, invasive crayfish species have driven some local crayfish species to extinction, either through competition, predation, or hybridization.

TERMINOLOGY AND MORPHOLOGY

Decapods are in part defined by their name: Decapoda, or "ten legs." Excellent illustrations of the general anatomy of shrimps, crayfishes, brachyuran crabs, and anomuran crabs can be seen in Volume I, Chapter 32 by Cumberlidge, Hobbs, and Lodge (Figs. 32.12–15) (Thorp & Rogers, 2015). The first or second pair of pereopods (walking limbs) may be chelate, or modified for filter feeding by possessing elongated setal fans.

The abdomen is posterior to the carapace and may be folded flat beneath the body (crabs) or projecting posteriorly (shrimp, crayfish).

The primary characters for the majority of species are on the male gonopods; these are the first pleopods which have been modified for sperm transfer. All references to gonopods in the key refer to form I gonopods. Plate 16.188.1 A–E show the difference in structure between form I and form II crayfish males. Form I males have much more detailed structures, and the central projection is sclerotized and brown in color. The position of the gonopods, their symmetry, and the distance anteriorly they extend are important to note before dissection.

Strangely, crayfish gonopod surfaces are referred to in terms of the appendage hanging perpendicular to the main cephalic/caudal axis of the crayfish body, rather than as "anterior/posterior" as in all other invertebrate groups. This is often a source of confusion. Thus, references to the cephalic surface of the gonopod equate to the dorsal surface when the appendage is lying in its normal position between the bases of the pereiopods and the caudal surface equates to the ventral surface. Thus: toward the attached end is proximal; toward the free end is distal; the side toward the head, cephalic; that toward the telson, caudal; that facing the corresponding pleopod of the pair, mesal; and that facing away from the midline of the body, lateral.

Crayfish gonopods generally possess several projections and processes (e.g., Pls. 16.186.14, 187.43). Most prominent is the central projection, which is the primary projecting terminus of the gonopod, and may be partially obscured by setal tufts. The mesal process is the secondary terminal projection, and originates just mesad to the central projection. Occasionally there is a cephalic process terminally or subterminally on the gonopod cephalic (dorsal) surface and a caudal process situated on the caudal surface. All of these structures may be variously curved, bent, or straight depending upon species. Often the angle of the bend or curve is diagnostic.

In crayfish, measurements of the length of the central projection of the gonopod and the total length of the gonopod require precision. It is strongly recommended that high quality calipers be used and in some cases that the structures be measured more than once and averaged. Anchor points for taking those measurements are shown in (Pl. 16.188.1 A–E). Typically, the crayfish's left gonopod is the gonopod examined (unless otherwise stated, all gonopod figures are of the left gonopod in medial view).

Decapods bear five pairs of pereopods (ambulatory limbs), with the first and sometimes the second pair modified as chelae (claws). Each pereopod is composed of several podomeres, which are (from basalmost to distalmost): coxa, basis, ischium, merus, carpus, propodus, and dactylus. The coxae and ischia may bear diagnostic structures in the form of bosses or hooks. A chela claw is composed of the propodus and the dactylus. The propodus is comprised of the broad palm and a distal extension called the pollex (or fixed "finger"). The dactylus is the "movable finger" which moves against the pollex. The opposing margins of the dactylus and pollex may have defining characters in terms of gape or tubercle arrangement.

The chelae are important in species diagnoses; however, crayfishes commonly lose and regenerate them. Chelae characteristics used in this key apply only to nonregenerated chelae. Regenerated chelae differ from nonregenerated in that they possess tubercles of the same size along the opposable margins of the dactyl and pollex, as well as by having a much shorter palm region (Pl. 16.188.1 F).

The carapace is the dorsal shield that covers the animal. Anteriorly, the carapace projects between the eyes. This projection is called the rostrum. The rostrum may have the distal portion constricted or otherwise offset from the basal portion. This offset distal portion is called the acumen. Sometimes the rostrum will have a longitudinal median carina.

The carapace has a shallow lateral cervical groove that arcs from behind the head anteriorly to the lower carapace margin. This groove may bear one or more branchostegial spines.

On the dorsal surface of the carapace, behind the medial arc of the cervical groove, are paired longitudinal, parallel lines (called the branchiocardic grooves) that arc towards each other medially. The region between them is the areola, which may be obliterated if the lines come together. Generally the areola is open and contains punctations, which are very small, shallow, usually circular depressions. The length and width of the areola, and sometimes the number of punctations are diagnostic characters.

MATERIAL PREPARATION AND PRESERVATION

Decapods are best preserved in ethyl alcohol. The gonopod is typically dissected out and examined separately. Generally, the gonopod may be temporarily mounted on an insect pin for easy manipulation under the microscope. The gonopod is then generally kept in a smaller vial within the jar holding the larger specimen.

KEYS TO DECAPODA

Crustacea: Malacostraca: Decapoda: Infraorders

1	Four or five pairs of well developed pereopods used for locomotion	2
1'	Three pairs of well developed walking pereopods, posterior most pair of pereopods (pereopod V) greatly reduced, at least 50% smaller than pereopod III; antennae always longer than eyestalks (Pl. 16.186.01)	**Anomura [p. 607]**
2(1)	Abdomen projecting posteriorly	3
2'	Abdomen folded beneath body, tightly pressed to thorax (Pl. 16.186.02)	**Brachyura [p. 607]**
3(2)	Abdomen compressed laterally	4
3'	Abdomen compressed dorsoventrally (Pl. 16.186.09)	**Astacidea [p. 611]**
4(3)	Abdominal pleuron I over lapping pleuron II	**Dendrobranchiata [p. 695]**
4'	Abdominal pleuron II overlapping pleura I and III (Pl. 16.190.1, 11)	**Caridea [p. 696]**

Crustacea: Malacostraca: Decapoda: Anomura

The "false crabs," includes hermit crabs and porcelain crabs.

1 Antennae long, filiform; body dorsoventrally flattened; chelae equal in size; abdomen folded beneath body, tightly pressed to thorax; free-living (Pl. 16.186.01) .... Porcellanidae, one species: *Petrolisthes armatus* (Gibbes, 1850)

[Invasive in estuarine, or in tidally influenced rivers. USA: Florida, Georgia, South Carolina, Gulf of Mexico. Mexico: Gulf of Mexico, Gulf of California, south. Neotropical]

1' Antennae geniculate; body not dorsoventrally flattened; chelae unequal in size; abdomen not folded beneath body, but projecting posteriorly; animal occupying empty gastropod shells .... Coenobitidae, one species: *Coenobita clypeatus* (Fabricius, 1787)

[Terrestrial, occasionally in or near freshwater. USA: Florida. Mexico: Gulf Coast]

Crustacea: Malacostraca: Decapoda: Brachyura: Families

1 Pereopod V ending in a dactyl similar to pereopods II-IV 2

1' Pereopod V dactyl paddle-shaped, used for swimming (Pl. 16.186.02) .... Portunidae, one genus; ***Callinectes* [p. 608]**

2(1) Chelae propodus not densely setose, scattered setae present or not 3

2' Chela propodus densely setose; setae obscuring chela surface (Pl. 16.186.05) .... Varunidae, one genus; ***Eriocheir* [p. 609]**

3 Maxilliped III palp articulating near merus distal margin middle or distolateral angle; palp may be partially or completely concealed by expanded merus 4

3' Maxilliped III palp articulating at or near the merus distomedial angle; palp never covered by expanded merus 7

4(3) Eyestalk bases separated by a distance greater than half the carapace width 5

4' Eye stalk bases separated by a distance less than half the carapace width .... Ocypodidae, one genus, ***Uca* [p. 609]**

5(4) Chelae equal or unequal; major chela propodus lacking posterior projections 6

5' Male chelae unequal, with major chela propodus greatly flattened, and projecting posteriorly, beyond its articulation point with the carpus .... Glyptograpsidae, one species: *Platychirograpsus spectabilis* De Man, 1896

[USA: invasive in Florida. Mexico: Chiapas, Nayarit, Tabasco, Veracruz]

6(5) Carapace quadrate to rectangular, lateral margins subparallel or posteriorly converging .... **Sesarmidae [p. 610]**

6' Carapace transversely oval, lateral margins arcuate .... **Gecarcinidae [p. 610]**

7(3) Carapace lateral margin lacking teeth (Pl. 16.186.07) .... Geothelphusidae, one species *Geothelphusa dehaani* (White, 1847)

[Invasive. USA: introduced to southern Nevada, possibly extirpated. Palaearctic]

7' Carapace with anterolateral teeth (Pl. 16.186.06) .... **Panopeidae [p. 610]**

PLATE 16.186.01 *Petrolisthes armatus*: "false crabs" of the infraorder Anomura; three pairs of well-developed walking legs (pereiopods 2-4; fifth pair reduced and beneath carapace).

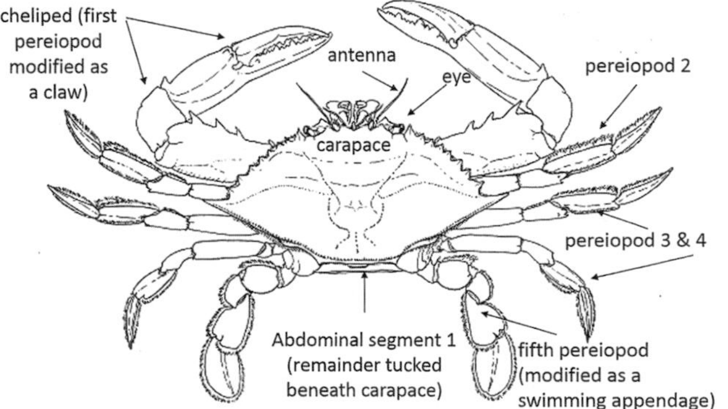

PLATE 16.186.02 Swimming crab *Callinectes*; paired pereiopods function as claws (#1), walking legs (#2–4), or swimming appendages (#5); abdomen is folded between carapace and is broader in females in order to carry eggs. Modified after several sources, including the crab figure from the United Nations Food and Agriculture Organization.

Crustacea: Malacostraca: Decapoda: Brachyura: Portunidae: *Callinectes*: Species

The following species are either known from estuaries at low salinities or may be found in low gradient, coastal freshwater (i.e., *Callinectes sapidus*).

1	Carapace frontal margin between orbit medial spines quadridentate, with margin between medial teeth sinuate (Pl. 16.186.03.3) or cleft (Pl. 16.186.03.1)	2
1'	Carapace frontal margin between orbit medial spines hexidentate (Pl. 16.186.03.2)	4
2(1)	Carapace frontal margin between medial spines sinuate (Pl. 16.186.03.3); other characters variable	3
2'	Carapace frontal margin between medial spines cleft (Pl. 16.186.03.1) carapace anteriolateral teeth directed laterally, not noticeably curving anteriorly; lateral spines generally more than twice as long as posterior margin of preceding tooth	*Callinectes sapidus* Rathbun, 1896
	[Canada: Nova Scotia. USA: Atlantic Coast, Gulf of Mexico, introduced to California and Washington. Mexico: Gulf of Mexico. Neotropics. Introduced to Palaearctic]	

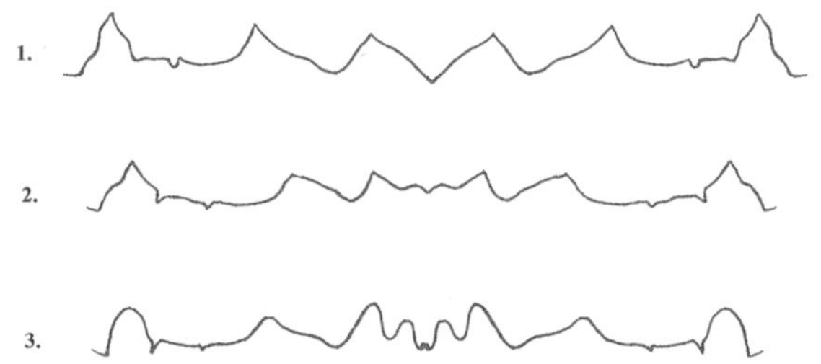

PLATE 16.186.03 Frontal margin of the carapace between orbit medial spines in *Callinectes* crabs are: (1) quadridentate (with margin between medial teeth sinuate or cleft); (2) hexidentate; or (3) sinuate.

PLATE 16.186.05 Drawing of the dorsal view of the Chinese mitten crab, *Eriocheir sinesis. From Hobbs in Thorp & Covich, 2010.*

PLATE 16.186.04 Photograph of a dorsal view of the invasive Chinese mitten crab, *Eriocheir sinensis. Courtesy of Matthew A. Hill & EcoAnalysts, Inc.*

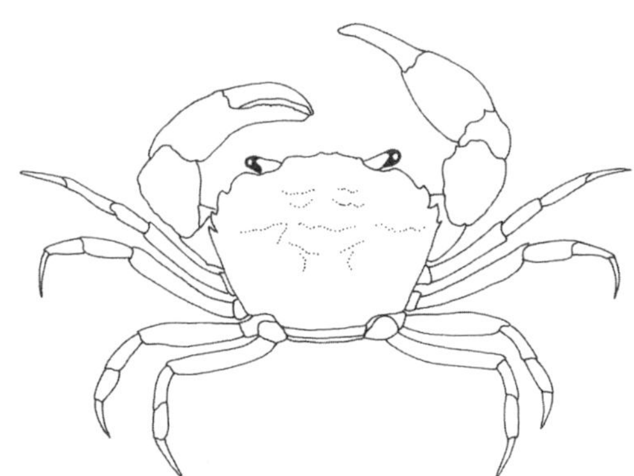

PLATE 16.186.06 Dorsal view of the crab *Rhithropanopeus harrisii.*

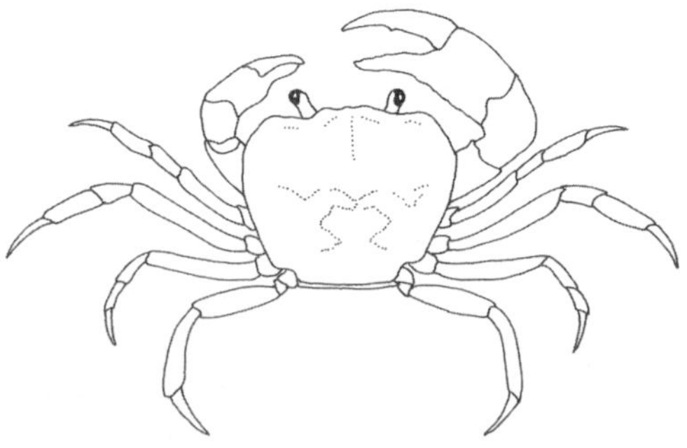

PLATE 16.186.07 Dorsal view of *Geothelphusa dehaani. From Hobbs in Thorp & Covich, 2010.*

3(2) Adult male gonopod (gonopod) with apicies curving medially .... *Callinectes ornatus* Ordway, 1863

[USA: Atlantic Coast from New Jersey to Florida. Neotropics]

3' Adult male gonopod (gonopod) with apicies curving laterally, and not reaching the suture between pereopods III and IV; carapace anteriolateral margin weakly arched .... *Callinectes larvatus* (Ordway, 1863)

[USA: Florida. Neotropics]

4(1) Front with submedian teeth smaller than lateral teeth 5

4' Front with submedian teeth extending nearly as far as lateral teeth .... *Callinectes bocourti* Milne-Edwards, 1879

[USA: Florida. Neotropics]

5(4) Carapace lateral spine more than twice the length of the preceding tooth's posterior margin .... *Callinectes arcuatus* Ordway, 1863

[USA: California. Mexico: Pacific Coast. Neotropics]

5' Carapace lateral spine less than twice the length of the preceding tooth's posterior margin .... *Callinectes exasperatus* (Gerstaecker, 1856)

[USA: Florida. Neotropics]

Crustacea: Malacostraca: Decapoda: Brachyura: Varunidae: *Eriocheir*: Species

1 Carapace lateral margin with four teeth behind exorbital angle; front quadridentate (Pls 16.186.04–05) .... Chinese mitten crab, *Eriochier sinensis* Milne-Edwards, 1853

[Invasive. USA: California. Native to Palaearctic]

1' Carapace lateral margin with two teeth behind exorbital angle; front with two teeth flanking a medial cleft .... Japanese mitten crab, *Eriochier japonica* (De Haan, 1835)

[Invasive. USA: Oregon, only one record (Jensen & Armstrong, 2004). Native to Palaearctic]

Crustacea: Malacostraca: Decapoda: Brachyura: Ocypodidae: *Uca*: Species

Following Rathbun, 1918.

1 Carapace convex, body shape transversely subcylindrical 2

1' Carapace flatter, not semicylindrical 3

2(1) Male major chela with propodus palm medial surface bearing an oblique tuberculate ridge .... *Uca musica* Rathbun, 1914

[River and stream mouths. Mexico: Pacific Coast]

2' Male major chela with propodus palm medial surface lacking an oblique tuberculate ridge .... *Uca subcylindricum* (Stimpson, 1859)

[Estuaries. USA: Texas. Mexico: Tamaulipas, Veracruz]

3(1) Carapace lateral margins parallel anteriorly, strongly angled medially, with posterior margins straight and converging; estuaries, salt marshes, tidally influenced freshwater 4

3' Carapace lateral margins convex, posterior portions converging posteriorly gradually 5

4(3) Carapace anterolateral angles acute, produced; major chela with propodus palm with dorsal surface rounded to lateral surface .... *Uca crenulata crenulata* (Lockington, 1877)

[USA: California. Mexico: Pacific Coast, Gulf of California]

4' Carapace anterolateral angles rectangular; major chela with propodus palm with dorsal surface at right angle to lateral surface .... *Uca crenulata coloradensis* (Rathbun, 1893)

[Mexico: Sonora]

5(3) Interorbital carapace front more than one-fifth the carapace width 6

5' Interorbital carapace front less than one-fifth the carapace width .... *Uca thayeri* Rathbun, 1900

[Estuaries, salt marshes, tidally influenced freshwater. USA: Florida]

6(5) Interorbital carapace front more than one-third the carapace width 7

6' Interorbital carapace front less than one-third the carapace width .... *Uca rapax* (Smith, 1870)

[River mouths, coastal freshwater. USA: Florida. Mexico: Tamaulipas, Veracruz. Neotropics]

7(6) Male major chela with propodus palm medial surface bearing an oblique tuberculate ridge .... *Uca minax* (Le Conte, 1855)

[Tidally influenced freshwater. USA: Atlantic Coast from Massachusetts to Florida, Gulf of Mexico from Texas to Florida]

7' Male major chela with propodus palm medial surface with a vestigial oblique tuberculate ridge lacking, or ridge lacking .... *Uca vocator vocator* (Herbst, 1804)

[USA: Texas? Mexico: Tamaulipas, Veracruz. Neotropics]

Crustacea: Malacostraca: Decapoda: Brachyura: Sesarmidae: Genera

After Abele (1992).

1 Chela propodus palmar dorsal surface bearing a longitudinal carina compose of a single row of acute tubercles, terminating at the base of the dactyl .... ***Sesarma* [p. 610]**

1' Chela propodus palmar dorsal surface lacking a carina, but bearing scattered granulae .... ***Armases* [p. 610]**

Crustacea: Malacostraca: Decapoda: Brachyura: Sesarmidae: *Sesarma*: Species

1 Carapace interorbital front cleft medially; carapace anterolateral angle with a single tooth behind orbital tooth .... *Sesarma sulatum* Smith, 1870

[Tidally influenced freshwater, brackish water. Mexico: Gulf of California. Neotropics]

1' Carapace interorbital front sinuate, not cleft; carapace anterolateral angle without an obvious tooth behind orbital angle, although an emargination is present .... *Sesarma reticulatum* (Say, 1817)

[Estuarine. Rare in brackish water. USA: Atlantic coast from Massachusetts south to Florida, Gulf of Mexico coast from Texas to Florida]

Crustacea: Malacostraca: Decapoda: Brachyura: Sesarmidae: *Armases*: Species

After Abele (1992).

1 Frontal region widening distally 2

1' Frontal region with lateral margins subparallel .... *Armases americanum* (de Saussure, 1858)

[Coastal streams. Mexico: Tamaulipas, Veracruz. Neotropics]

2(1) Gonopod apex subrectangular, directed distolaterally 3

2' Gonopod apex small, central, unarmed .... *Armases ricordi* (Milne-Edwards, 1853)

[Uplands near estuaries and coastal swamps. USA: Florida. Neotropics]

3(2) Pereopod IV with dactyl unarmed dorsally .... *Armases miersii* (Rathbun, 1897)

[River mouths. USA: Florida]

3' Pereopod IV with dactyl armed dorsally with short black spines .... *Armases cinereum* (Bosc, 1802)

[Uplands near estuaries and coastal swamps. USA: Atlantic Coast from Maryland south to Florida, Gulf of Mexico coast. Mexico: Gulf of Mexico coast. Neotropics]

Crustacea: Malacostraca: Decapoda: Brachyura: Gecarcinidae: Genera

This family is comprised of land crabs, however they are occasionally found near rivers and streams.

1 Carapace orbital width less than half the carapace width .... ***Gecarcinus* [p. 610]**

1' Carapace orbital width more than half the carapace width .... *Cardisoma guanhumi* Latreille, 1852

[USA: Florida, Texas. Mexico: Gulf of Mexico Coast. Neotropics]

Crustacea: Malacostraca: Decapoda: Brachyura: Gecarcinidae: *Gecarcinus*: Species

1 Pereopods II-V with dactyls armed with six spine rows; maxiliped III merus with margin entire .... *Gecarcinus ruricola* (Linnaeus, 1758)

[USA: Florida. Neotropics]

1' Pereopods II-V with dactyls armed with four spine rows; maxiliped III merus mediodistal margin with a shallow emargination .... *Gecarcinus lateralis* (Freminville, 1835)

[USA: Florida. Neotropics]

Crustacea: Malacostraca: Decapoda: Brachyura: Panopeidae: Genera

1 Carapace widest between tips of posterior lateral teeth; chelipeds with dactyl and pollex apicies variable 2

1' Carapace widest posterior to last pair of lateral teeth; chelipeds with dactyl and pollex apicies white .... *Eurytium limosum* (Say, 1818)

[Tidally influenced freshwater and brackish water streams. USA: Atlantic coast from New York south. Mexico: Gulf of Mexico Coast. Neotropics]

2(1) Carapace orbital width less than half the carapace width; chelipeds with dactyl and pollex apicies dark .... *Panopeus herbstii* (Milne-Edwards, 1834)

[Estuaries and brackish water. USA: Atlantic coast, from Massachusetts south. Mexico: Gulf of Mexico Coast. Neotropics]

2' Carapace orbital width more than half the carapace width; chelipeds with dactyl and pollex apicies light (Pl. 16.186.6).. *Rhithropanopeus harrisii* (Gould, 1841)

[Freshwater and brackish water. Canada: New Brunswick, Quebec. USA: Atlantic and Gulf of Mexico coasts. Introduced and invasive in Oregon, California, and inland lakes in Texas. Mexico: Gulf of Mexico coast. Neotropics]

REFERENCES

Abele, L.G. 1992. A review of the grapsid crab genus *Sesarma* (Crustacea: Decapoda: Grapsidae) in America with the description of a new genus. Smithsonian Contributions to Zoology 527: 1–60.

Jensen, G.C. & D.A. Armstrong. 2004. The occurrence of the Japanese mitten crab, *Eriochier japonica* (De Haan), on the west coast of North America. California Fish and Game, 90: 94–99.

Rathbun, M.J. 1918. The grapsoid crabs of America. United States national Museum Bulletin, 97: 1–461.

Thorp, J.H. & D.C. Rogers (eds.). 2015. Ecology and general biology. Volume I of Thorp and Covich Freshwater Invertebrates, Academic Press, Elsevier, Boston, MA. 1118 p.

Crustacea: Malacostraca: Decapoda: Astacidea: Families

D. Christopher Rogers

Kansas Biological Survey, University of Kansas, Lawrence, KS, USA; The Biodiversity Institute, University of Kansas, Lawrence, KS, USA

1 Last thoracic segment lacking a pleurobranch (a gill attached to the body) under the posteriolateral carapace margin at or above pereopod V; females with annulis ventralis and gonopod present or absent 2

1' Last thoracic segment bearing a small pleurobranch under the posteriolateral carapace margin; males with pleopods I and II never modified; females lacking first pair of pleopods and annulis ventralis (Pls. 16.186.08–10) .... Astacidae, one genus: ***Pacifastacus* [p. 611]**

2(1) Sternum narrow, cariniform; male pleopods I and II never modified; female lacking pleopod I and annulis ventralis (Pl. 16.186.12) .... Parastacidae, one species: *Cherax quadricarinatus* (Von Martens, 1868)

[Invasive species. Mexico: Morelos, Tamaulipas. Native to Australia. May appear in USA in future, as this species is common in the pet and aquaculture trades. At least five other *Cherax* species are common in the pet trade in North America]

2' Sternum triangular, broadest between posterior pereopod pairs; mature males with pleopods I and II modified; pleopod I modified as a gonopod modified for sperm transfer, and some ischia bearing hooks .... **Cambaridae [p. 612]**

Crustacea: Malacostraca: Decapoda: Astacidea: Astacidae: *Pacifastacus*: Species

1 Rostrum with two or more pairs of distolateral spines 2

1' Rostrum with one pair of distolateral spines 5

2(1) Chela bearing a dorsomedial and a dorsolateral, longitudinal tomentose area 3

2' Chela lacking tomentose areas 4

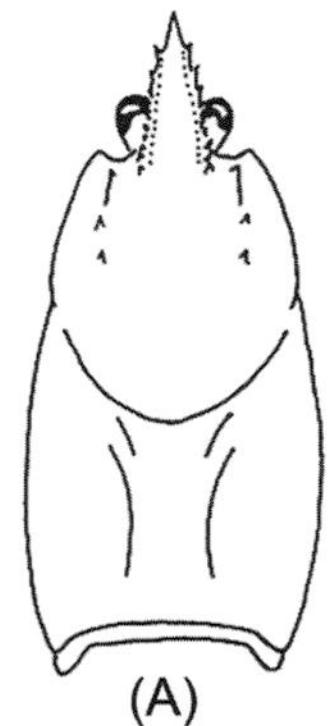

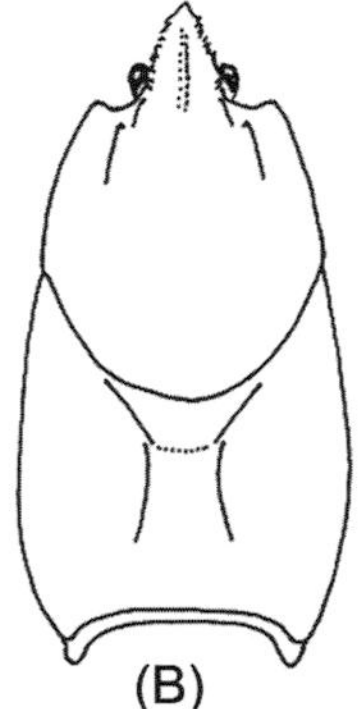

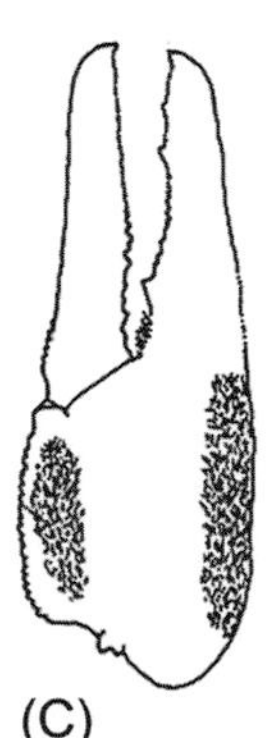

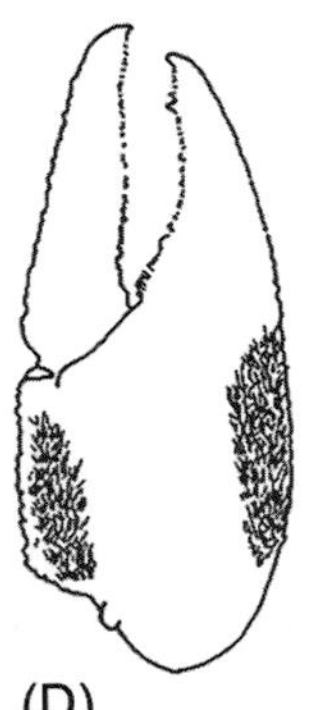

PLATE 16.186.08 Dorsal view of carapace and chela, respectively, in: (A, C) *Pacifastacus* (*Hobbsastacus*) *connectens,* and (B, D) *P.* (*Hobbsastacus*) *gambelii. Modified from Hobbs, 1989.*

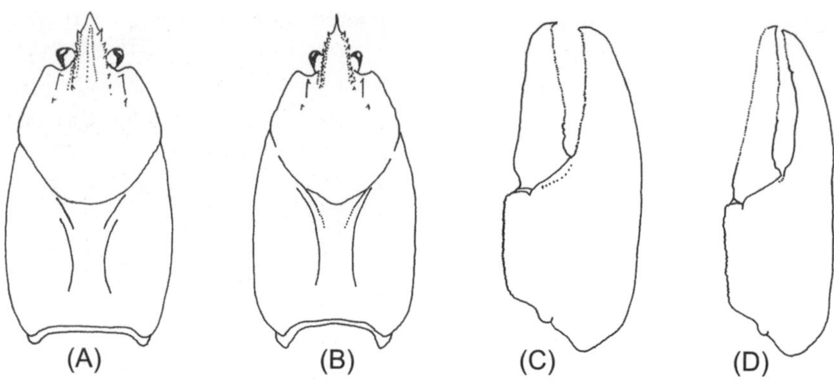

PLATE 16.186.09 (A, C) Dorsal view of carapace and chela, respectively, in: (A, C) *Pacifastacus (Hobbsastacus) fortis*, and (B, D) *P. (Hobbsastacus) nigrescens*. *Modified from Hobbs, 1989.*

3(2) Carapace postorbital ridge lacking a posterior tubercle or spine; rostrum frequently bearing a medial carina (Pl. 16.186.08 B, D) .... *Pacifastacus gambelii* (Girard, 1852)

[USA: California, Idaho, Montana, Nevada, Oregon, Utah, Washington, Wyoming]

3' Carapace postorbital ridge bearing a posterior tubercle or spine; rostrum lacking a medial carina (Pl. 16.186.08 A, C) .... *Pacifastacus connectens* (Faxon, 1914)

[USA: Idaho, Oregon]

4(2) Chela propodus greatest width subequal to dorsomedial margin (from distal margin of carpus to the base of the dactylus) length (Pl. 16.186.09 A, C) .... *Pacifastacus fortis* (Faxon, 1914)

[USA: California. Federally Endangered Species]

4' Chela propodus greatest width less than dorsomedial margin length (Pl. 16.186.09 B, D) .... *Pacifastacus nigrescens* (Faxon, 1914)

[USA: California. Probably extinct]

5(1) Carapace with postorbital ridges smooth anteriorly, tuberculate or smooth posteriorly; rostral acumen subequal or shorter than rostral width at lateral spines 6

5' Carapace with postorbital ridges spiniform anteriorly and posteriorly; rostral acumen longer than rostral width at lateral spines .... *Pacifastacus leniusculus leniusculus* (Dana, 1852)

[Canada: British Columbia. USA: Idaho, Nevada, Oregon, Washington. Invasive in California, Nevada]

6 Carapace with postorbital ridges smooth posteriorly (Pl. 16.186.10 B) .... *Pacifastacus leniusculus klamathensis* (Stimpson, 1857)

[Canada: British Columbia. USA: California, Idaho, Oregon, Washington]

6' Carapace with postorbital ridges tuberculate posteriorly (Pl. 16.186.10 A) .... *Pacifastacus leniusculus trowbridgii* (Stimpson, 1857)

[Canada: British Columbia. USA: California, Idaho, Nevada, Oregon, Washington]

Crustacea: Malacostraca: Decapoda: Astacidea: Cambaridae: Genera

The genera are poorly defined and need to be revised. This key requires form I male specimens. Refer to the Terminology and Morphology section above for information to distinguish between form I and form II males. Unless otherwise stated, all gonopod figures depict a medial view of the left form I gonopod.

1 Form I male pereopods II with ischia never bearing a prominent medial projection (hook), although other pereopods may have this character; subfamily Cambarinae 2

1' Form I male pereopods II and III with ischia each bearing a prominent medial projection (hook); subfamily Cambarellinae (Pls. 16.186.14, 16–19) .... ***Cambarellus* [p. 616]**

2(1) Second antenna with scattered or no setae, never tomentose 3

2' Second antennae mesal margin conspicuously and densely tomentose (Pl. 16.186.11 A) .... ***Barbicambarus* [p. 617]**

3(2) Maxilliped III not greatly enlarged, dactylus acute, at least three times longer than broad, ischium with a longitudinal row of spiniform tubercles 4

3' Maxilliped III greatly enlarged, dactylus rounded twice as long as broad, ischium lacking a longitudinal row of tubercles (Pl. 16.186.13 A) .... *Troglocambarus maclanei* Hobbs, 1942

[Troglobitic. USA: Florida]

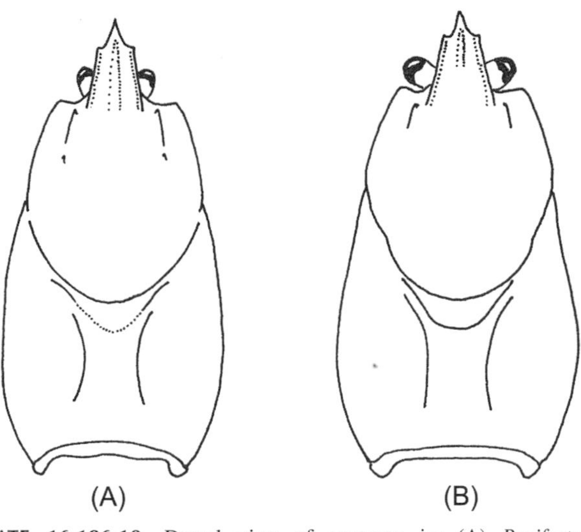

PLATE 16.186.10 Dorsal view of carapace in: (A) *Pacifastacus (Pacifastacus) leniusculus trowbridgii*, and (B) *P. (Pacifastacus) leniusculus klamathensis*. *After Hobbs, 1989.*

PLATE 16.186.12 *Cherax quadricarinatus*, the giant, invasive crayfish from Australia that has been imported to the Nearctic along with several other related species for aquaculture and the pet trade.

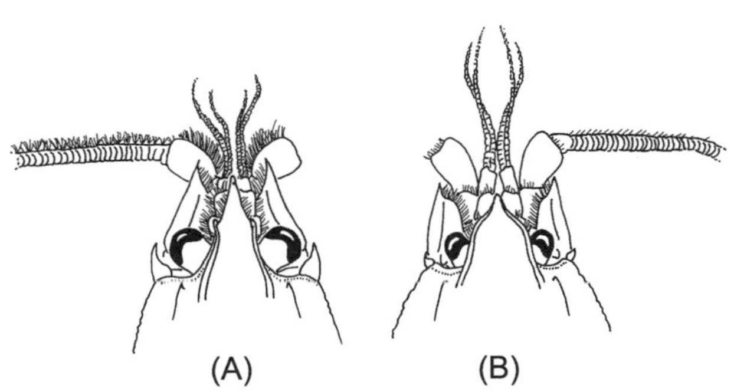

PLATE 16.186.11 Dorsal view of the cephalic region of: (A) *Barbicambarus cornutus*; (B) *Cambarus (Cambarus) b. bartonii*. Note the differences in degree of setation on the antennae; those of *Barbicambarus* are notably tomentose. *After Hobbs, 1972b.*

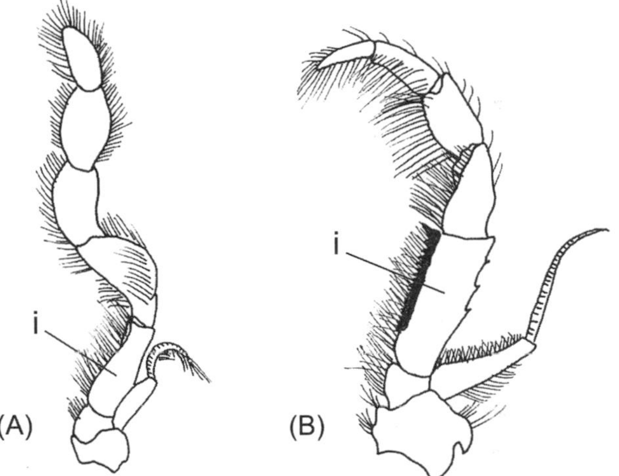

PLATE 16.186.13 Ventral view of left third maxilliped of: (A) the stygiobiont *Troglocambarus maclanei*, and (B) *Procambarus (Ortmannicus) pallidus*. *After Hobbs, 1972b.*

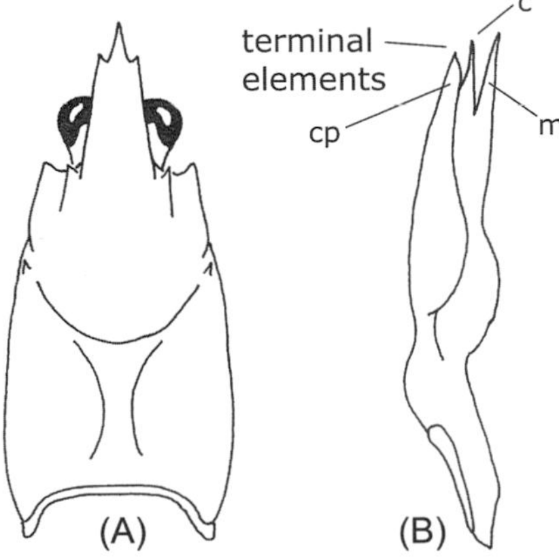

PLATE 16.186.14 *Cambarellus (Dirigicambarus) shufeldi*: (A) dorsal view of carapace; (B) mesial view of the left first pleopod. c = caudal process, cp = central projection, m = mesial process. *Modified from Hobbs, 1989.*

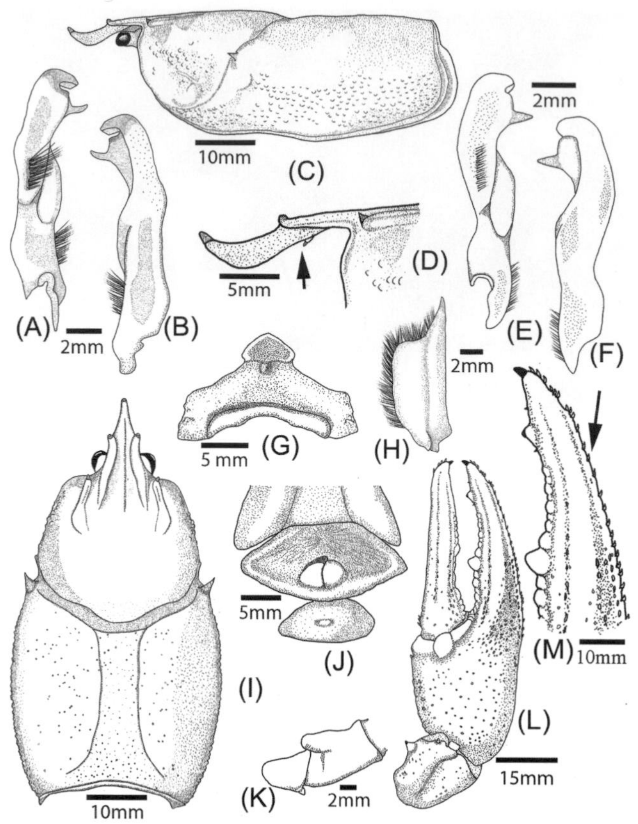

PLATE 16.186.15 Multiple anatomical components of *Barbicambarus simmonsi*. *Courtesy of Christopher Taylor from Taylor & Schuster, 2010.*

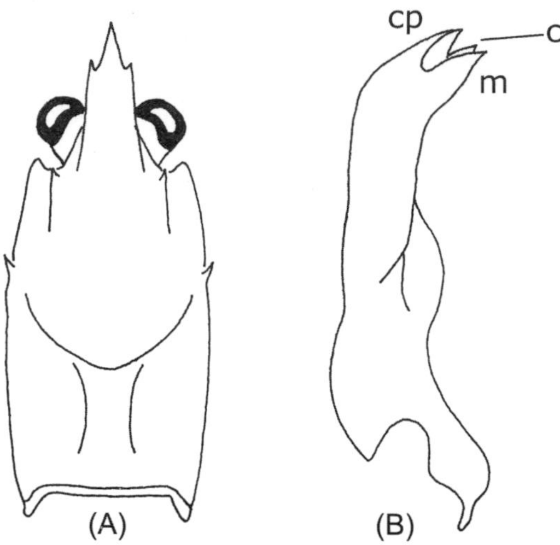

PLATE 16.186.16 *Cambarellus* (*Pandicambarus*) *diminutus*: (A) dorsal view of carapace; (B) mesial view of left first pleopod. c = caudal process, cp = central projection, m = mesial process. *Modified from Hobbs, 1989.*

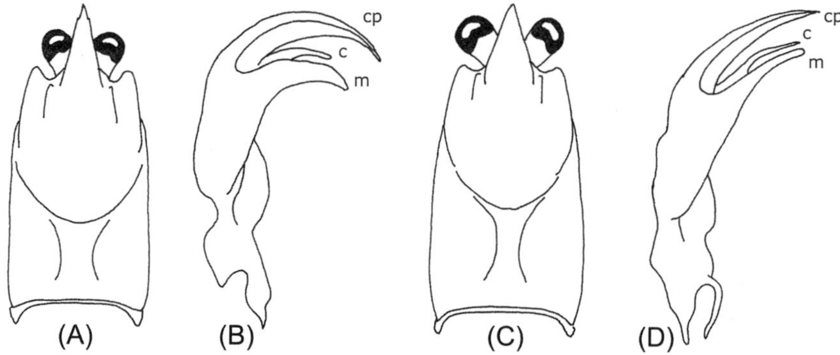

PLATE 16.186.17 Dorsal view of carapace and mesial view of left first pleopod, respectively, in: (A, B) *Cambarellus* (*Pandicambarus*) *ninae*; and (C, D) *C.* (*Pandicambarus*) *texanus*. c = caudal process, cp = central projection, m = mesial process. *Modified from Hobbs, 1989.*

4(3)	Gonopod apically biramal, with rami more than one-eighth the length of the pleopod	5
4'	Gonopod apically with more than two rami	***Procambarus*** (in part) **[p. 617]**
5(4)	Gonopod with mesal process greater than 50% as long as central projection; adult animal carries paired pleopods I parallel	6
5'	Gonopod with central projection long and slender; mesal process less than 50% as long as central projection; adult animal carries paired gonopod with distal portions crossed (Pls. 16.187.42–43, 188.60)	***Faxonella*** **[p. 642]**
6(5)	Pereopod IV coxa bearing a posteriomedial projection (boss)	7
6'	Pereopods with coxae lacking a posteriomedial projection	11
7(6)	Pleopods I with bases parallel	8

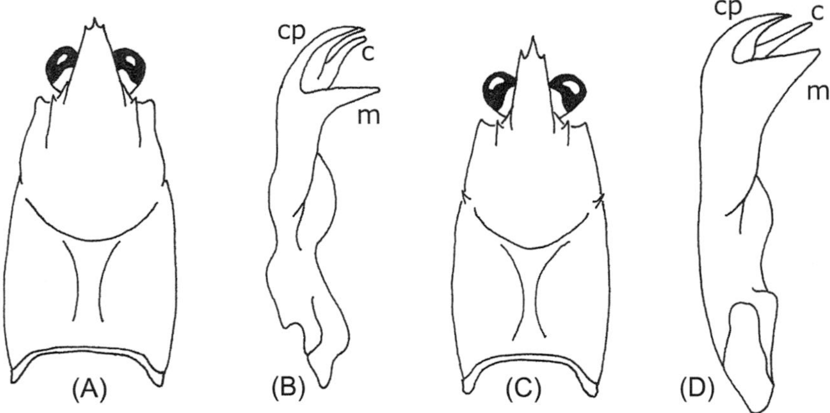

PLATE 16.186.18 Dorsal view of carapace and mesial view of left first pleopod, respectively, in: (A, B) *Cambarellus* (*Pandicambarus*) *puer*; and (C, D) *C.* (*Pandicambarus*) *lesliei*. *Modified from Hobbs, 1989.*

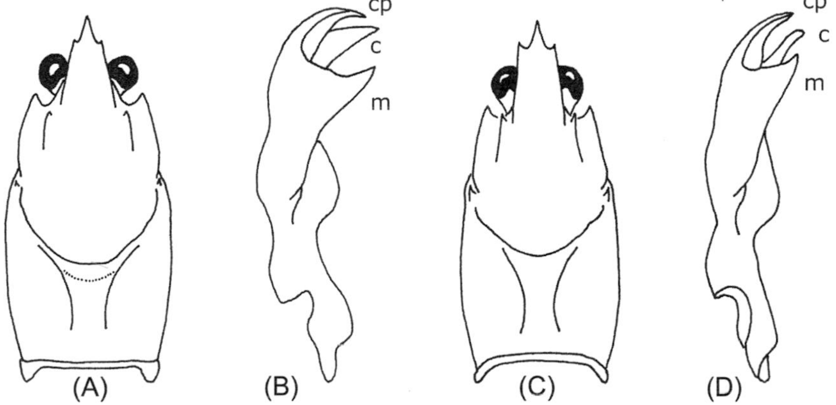

PLATE 16.186.19 Dorsal view of carapace and mesial view of left first pleopod, respectively, in: (A, B) *Cambarellus* (*Pandicambarus*) *blackii*; and (C, D) *C.* (*Pandicambarus*) *schmitti*. c = caudal process, cp = central projection, m = mesial process. *Modified from Hobbs, 1989.*

7’ Pleopods I with bases asymmetrically placed .... ***Procambarus*** (in part) **[p. 617]**

8(7) Pereopod III with ischium bearing a prominent medial projection (hook) 9

8’ Pereopod III and IV with each ischium bearing a prominent medial projection (hook) .... ***Orconectes*** (in part)

9(8) Chela dactyl ventral margin lacking a proximal, deep, angular emargination 10

9’ Chela dactyl ventral margin with a large, proximal, angular emargination (Pls. 16.188.60–76) .... ***Fallicambarus*** **[p. 642]**

10(9) Gonopod with terminal rami distally bent ventroposteriorly (caudodistally) or posteriorly (caudally) (Pls. 16.189.10, 31, 39) .... ***Cambarus*** **[p. 662]**

10’ Gonopod with terminal rami directed distally (may be slightly arcuate) .... ***Orconectes*** (in part) **[p. 643]**

11(6) Chela lateral portions with widely scattered small tubercles 12

11’ Chela surface densely covered with small, close set tubercles (Pls. 16.189.75–78) .... ***Hobbseus*** **[p. 667]**

12(11) Gonopod with proximomedial surface smooth, lacking an angular projection 13

12’ Gonopod with a proximomedial angular projection .... *Bouchardina robisoni* Hobbs, 1977

13(12) Cheliped with carpus as long as or longer than propodus medial margin; rostrum lacking marginal spines (Pls. 16.189.79–83) .... ***Distocambarus*** **[p. 691]**

13’ Cheliped with carpus shorter than propodus medial margin; if subequal or longer than propodus medial margin, then rostrum bearing marginal spines .... ***Orconectes*** (in part) **[p. 643]**

Crustacea: Malacostraca: Decapoda: Astacidea: Cambaridae: *Cambarellus*: Species

Form I males can only be identified with this key. After Hobbs (1972) and Hobbs & Lodge (2010).

1 Gonopod apical rami curved; pereopod IV coxa with or without an anteriomedial projection 2

1' Gonopod apical rami directed distally; pereopod IV coxa lacking an anteriomedial projection (Pl. 16.186.14) .... *Cambarellus* (*Dirigicambarus*) *shufeldi* (Faxon, 1884)

[USA: Mississippi River Drainage, Georgia, Louisiana, Texas]

2(1) Pereopod IV coxa with an anteriomedial (cephalomedial) projection (may be vestigial) subgenus *Pandicambarus* 3

2' Pereopod IV coxa without an anteriomedial projection subgenus *Cambarellus* 9

3(2) Male gonopod medial ramus with a longitudinal groove 4

3' Male gonopod medial ramus without a longitudinal groove 7

4(3) Male gonopod medial process in lateral view lobiform or digitiform, never obscuring the cephalic projection 5

4' Male gonopod medial process in lateral view broadly triangular (conical), mostly obscuring the cephalic projection (Pl. 16.186.16) .... *Cambarellus* (*Pandicambarus*) *diminutus* Hobbs, 1945

[USA: Alabama, Mississippi]

5(4) Male gonopod central projection one fourth the length of gonopod; central projection bent no more than 45° posteriorly from normal 6

5' Male gonopod central projection one third or more the length of gonopod; rami all bent at least 90° posteriorly from normal (Pl. 16.186.17 A, B) .... *Cambarellus* (*Pandicambarus*) *ninae* Hobbs, 1950

[USA: Arkansas, Texas]

6(5) Gonopod medial ramus in lateral view subtriangular, shorter than the anterior (cephalic) ramus (Pl. 16.186.19 C, D) .... *Cambarellus* (*Pandicambarus*) *schmitti* Hobbs, 1942

[USA: Florida]

6' Gonopod medial ramus in lateral view oblong, tapering, subequal in length to anterior (cephalic) (Pl. 16.186.18 A, B) .... *Cambarellus* (*Pandicambarus*) *puer* Hobbs, 1941

[USA: Arkansas, Illinois, Kentucky, Louisiana, Mississippi, Oklahoma, Tennessee, Texas]

7(3) Gonopod with central and cephalic projection subequal in length 8

7' Gonopod with central projection obviously longer than the other two rami (Pl. 16.186.17 C, D) .... *Cambarellus* (*Pandicambarus*) *texanus* Albaugh & Black, 1973

[USA: Texas]

8(7) Carapace with areola wider than rostrum at base of acumen (Pl. 16.186.19 A, B) .... *Cambarellus* (*Pandicambarus*) *blacki* Hobbs, 1980

[USA: Florida]

8' Carapace with areola as wide as rostrum at base of acumen (Pl. 16.186.18 C, D) .... *Cambarellus* (*Pandicambarus*) *lesliei* Fitzpatrick & Laning, 1976

[USA: Alabama, Mississippi]

9(2) Male gonopod medial projection with a longitudinal groove 10

9' Male gonopod medial projection without a longitudinal groove .... *Cambarellus alvarezi* Villalobosi, 1952

[Mexico: Nuevo León]

10(9) Rostral acumen at least twice as long as basal width 11

10' Rostral acumen length subequal to basal width 12

11(10) Rostral acumen twice as long as basal width .... *Cambarellus chapalanus* Faxon, 1898

[Mexico: Jalisco, Michoacán]

11' Rostral acumen at least three times as long as basal width .... *Cambarellus prolixus* Villalobos & Hobbs, 1981

[Mexico: Jalisco]

12(10) Chela with dactyl and pollex directed distally 13

12' Chela with dactyl and pollex gently curved medially; chela lateral margin strongly convex .... *Cambarellus zempoalensis* Villalobos, 1943

[Mexico: Morelos]

13(12) Areola length 3 times or more than width 14

13' Areola length 3 times width or less 15

14(13) Male gonopod medial projection with apex broad, rounded; chela palm greatest width one half medial margin length .... *Cambarellus montezumae* Saussure, 1857

[Mexico: Federal District, Guanajuato, Hidalgo, Jalisco, Mexico State, Michoacán]

14' Male gonopod medial projection with apex narrow, truncated; chela palm greatest width 1.5 times medial margin length .... *Cambarellus patzcuarensis* Villalobos, 1943

[Mexico: Michoacán]

15(13) Male gonopod central projection constituting approximately one-third the length of the pleopod 16

15' Male gonopod central projection constituting approximately one-fifth the length of the pleopod .... *Cambarellus chihuahuae* Hobbs, 1980

[Mexico: Chihuahua]

16(15) Rostrum width at acumen base half or more the width of rostral base; rostrum margins subparallel from middle to base .... *Cambarellus areolatus* Faxon, 1885

[Mexico: Coahuila]

16' Rostrum width at acumen base one-third or less the width of rostral base; rostrum margins converging anteriorly to acumen .... *Cambarellus occidentalis* Faxon, 1898

[Mexico: Michoacán, Sinaloa]

Crustacea: Malacostraca: Decapoda: Astacidea: Cambaridae: *Barbicambarus*: Species

1 Rostrum without a medial carina; cheliped merus lacking a prominent dorsodistal spine ....(Pl. 16.187.41 A) .... *Barbicambarus cornutus* (Faxon, 1884)

[USA: Kentucky, Tennessee]

1' Rostrum with a medial carina; cheliped merus with a prominent single, dorsodistal spine (Pl. 16.186.15) .... *Barbicambarus simmonsi* Taylor & Schuster, 2010

[USA: Tennessee]

Crustacea: Malacostraca: Decapoda: Astacidea: Cambaridae: *Procambarus*: Subgenera

James W. Fetzner Jr.
Section of Invertebrate Zoology, Carnegie Museum of Natural History, Pittsburgh, PA, USA

Only form I males can be identified with this key. These keys rely very heavily on the keys by Hobbs (1972a,b), Fitzpatrick (1978), Hobbs & Hobbs (1991), Hobbs & Robison (1988), Walls (2006), Walls & Black (2008), and Schuster et al. (2015). These keys include all currently accepted species and subspecies of *Procambarus* that fall within the geographic coverage of this book, totaling 133 of the 177 recognized extant taxa and including 10 of the 15 subgenera. The line drawings included below have been completely redrawn from their original sources.

1 Carapace with or without 1 cervical spine (Pl. 16.187.1 B) 2

1' Carapace with 2 or more cervical spines (Pl. 16.187.1 A *cs*) .... ***Pennides* [p. 642]**

2(1) Gonopod cephalic surface with or without prominent angular or subangular shoulder; if present, situated immediately proximal to base of terminal elements (Pl. 16.187.3 C–G), or, if situated more proximally, somewhat rounded and cephalic process never broadly rounded (Pls. 16.187.14 A, B, 39 F, 40 H) 3

2' Gonopod cephalic surface with prominent angular or subangular shoulder situated far proximal to base of terminal elements (Pls. 16.187.3 A *s*, B, 11, 16) .... ***Scapulicambarus* [p. 624]**

3(2) Pereopods III OR III and IV with ischia bearing well-developed hooks (Pl. 16.187.4 A, B, D) 4

3' Pereopods IV with ischia bearing hooks only (Pl. 16.187.4 C see also Pl. 16.187.05 D), occasionally with vestigial ones on pereopod III . ... *Procambarus* (*Acucauda*) *fitzpatricki* Hobbs, 1971

[USA: Mississippi]

4(3) Chela palm mesal surface with tubercles (Pl. 16.187.7 A, C–E), sometimes obscured by conspicuous brush of setae (Pl. 16.187.7 F) (tubercles almost obsolete in *P. youngi*); pereopods I-III always lacking conspicuous brush of plumose setae (Pl. 16.187.6 B) 5

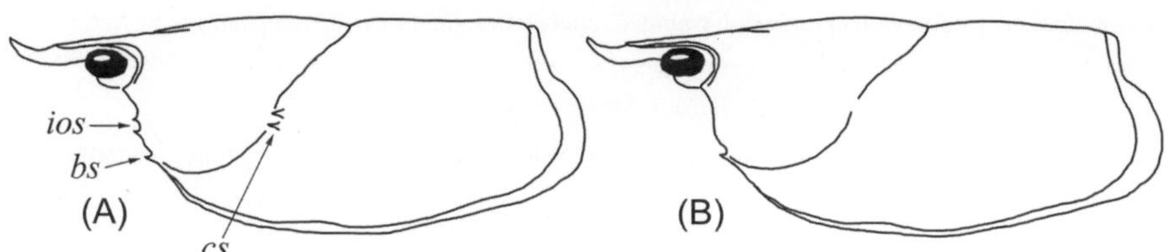

PLATE 16.187.01 Lateral view of carapaces showing: (A) presence and (B) absence of cervical spines (cs). ios, infraorbital spines; bs, branchiostegal spine.

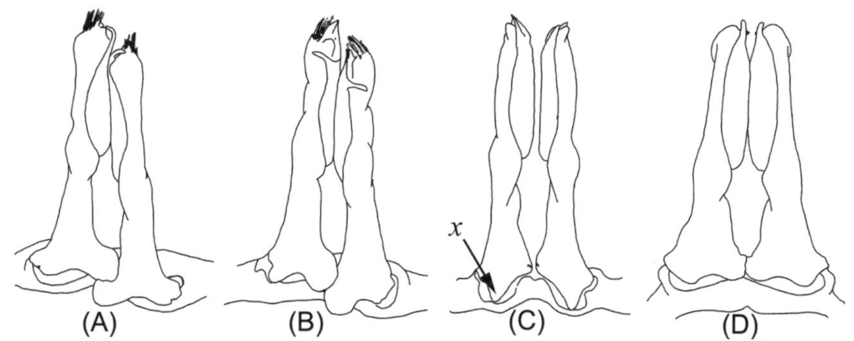

PLATE 16.187.02 Ventral view of first pleopods: (A & B) asymmetrical; (C & D) symmetrical; (A) *Procambarus (Ortmannicus) seminolae*; (B) *P. (Ortmannicus) acutissimus*; (C) *P. (Girardiella) h. hagenianus*; (D) *P. (Austrocambarus) llamasi*. x, proximomedian lobe.

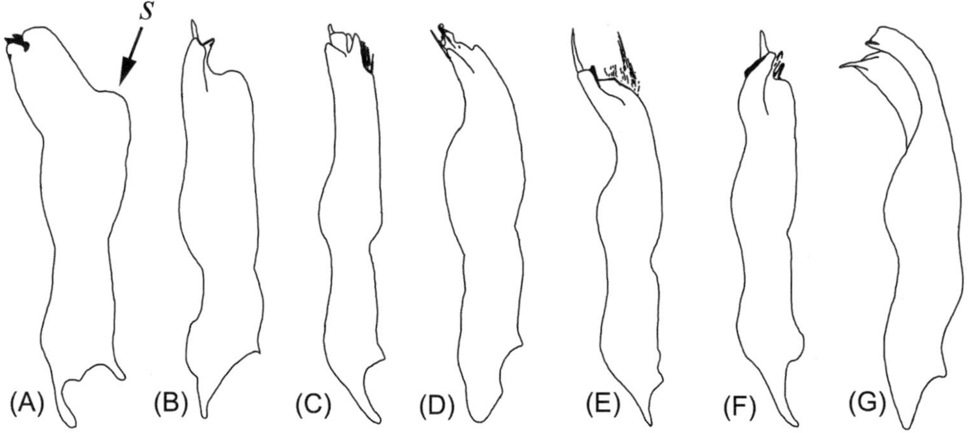

PLATE 16.187.03 Lateral view of left first pleopods: (A) *Procambarus* (*Scapulicambarus*) *okaloosae*; (B) *P.* (*Austrocambarus*) *rodriguezi*; (C) *P.* (*Acucauda*) *fitzpatricki*; (D) *P.* (*Procambarus*) *digueti*; (E) *P.* (*Leconticambarus*) *latipleurum*; (F) *P.* (*Girardiella*) *simulans*; (G) *P.* (*Tenuicambarus*) *tenuis*. (s, shoulder).

4'	Chela palm mesal surface without tubercles or setal brush; pereopods I-III with conspicuous brush of plumose setae extending from basis to at least merus proximal part (Pl. 16.187.6 A) .... ***Capillicambarus* [p. 625]**
5(4)	Chela subovate to cylindrical, mostly elongate, palm mesal margin lacking cristiform tubercle row (Pls. 16.187.7 B–F, 20 A–C) 6
5'	Chela strongly depressed, usually broadly triangular, palm mesal margin with a row of cristiform or subcristiform tubercles (Pl. 16.187.7 A) .... ***Hagenides* [p. 625]**
6(5)	Pereopods III with ischia bearing hooks only (Pl. 16.187.4 A) 7
6'	Pereopods III and IV with ischia bearing well-developed hooks (Pl. 16.187.4 B, D) 8
7(6)	Gonopod without subapical setae (Pls. 16.187.3 D, F, 9 A) .... ***Girardiella* [p. 629]**
7'	Gonopod with subapical setae (Pl. 16.187.3 F) .... ***Leconticambarus*** (in part) **[p. 627]** [USA: southeastern states]

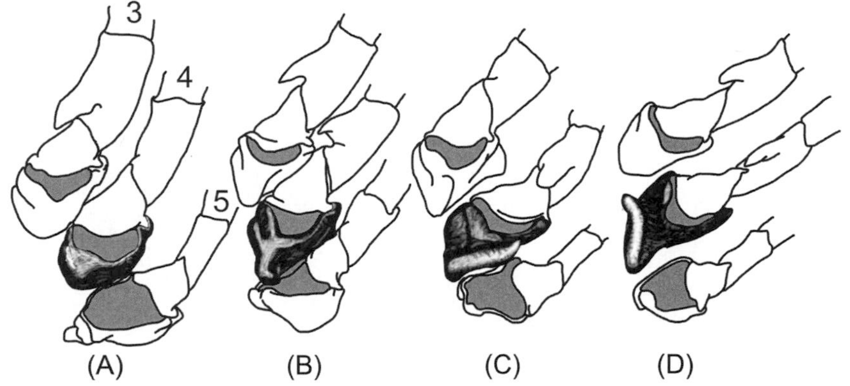

PLATE 16.187.04 Ventral view of basal podomeres of left third, fourth, and fifth pereiopods showing variations in coxae (blackened) of fourth: (A) without boss; (B–D) with boss. (A) *Procambarus digueti*; (B) *P.* (*Scapulicambarus*) *paeninsulanus*; (C) *P.* (*Villalobosus*) *riojai*; (D) *P.* (*Tenuicambarus*) *tenuis*.

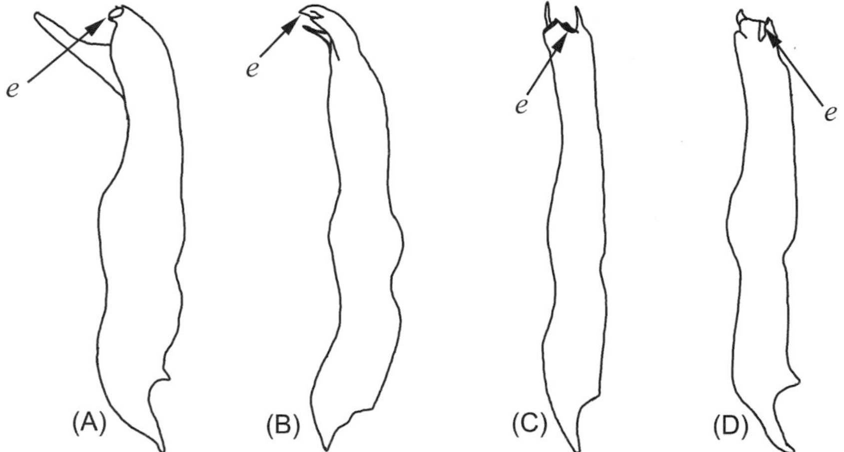

PLATE 16.187.05 Lateral view of left first pleopods: (A) *Procambarus* (*Paracambarus*) *paradoxus*; (B) *P.* (*Paracambarus*) *ortmannii*; (C) *P.* (*Villalobosus*) *riojai*; (D) *P.* (*Acucauda*) *fitzpatricki*. (e, central projection).

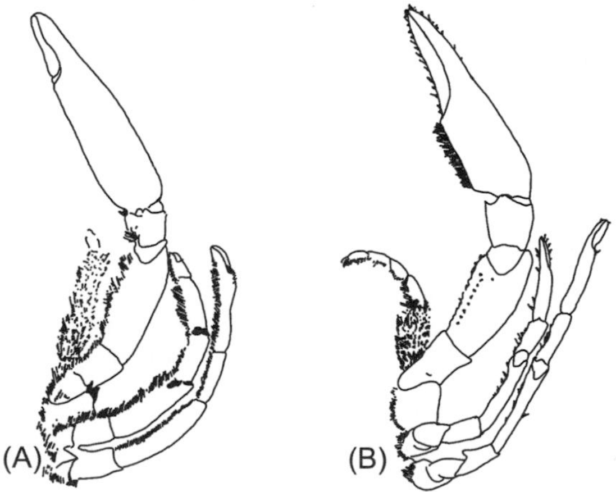

PLATE 16.187.06 Ventral view of left third maxillipeds and first three pereiopods: (A) *Procambarus* (*Capillicambarus*) *hinei;* (B) *P.* (*Leconticambarus*) *barbatus*.

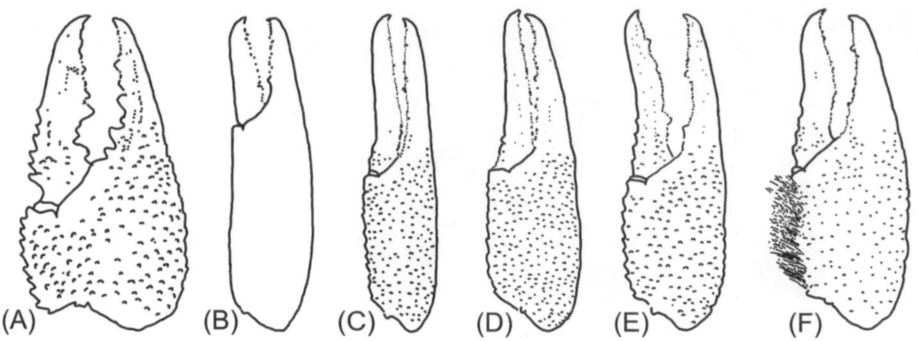

PLATE 16.187.07 Dorsal view of chelae: (A) *Procambarus* (*Hagenides*) *rogersi expletus*; (B) *P.* (*Ortmannicus*) *youngi*; (C) *P.* (*Ortmannicus*) *lecontei*; (D) *P.* (*Ortmannicus*) *hirsutus*; (E) *P.* (*Ortmannicus*) *seminolae*; (F) *P.* (*Leconticambarus*) *barbatus*.

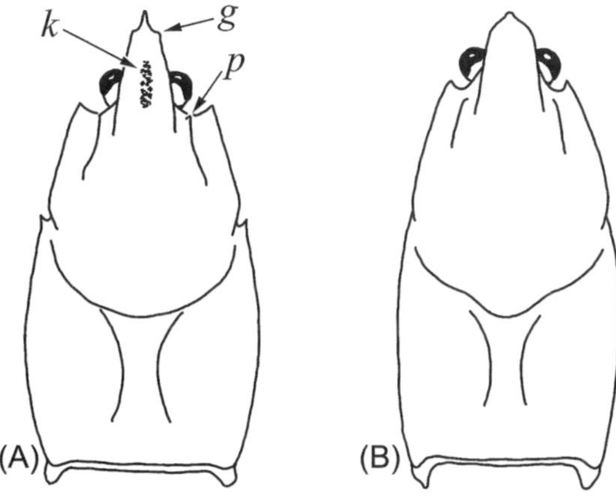

PLATE 16.187.08 Dorsal view of carapaces: (A) *Procambarus* (*Procambarus*) *digueti*; (B) *P.* (*Leconticambarus*) *rathbunae*. k, median carina; g, marginal spines; p, cephalic spines.

8(6) Pereopods III and IV with ischial hooks never bituberculate on both third and fourth pereopods; gonopod mesal process always present 9

8' Pereopods III and IV with ischial hooks bituberculate (Pl. 16.187.32 D; see also Pl. 16.187.32 A); or gonopod mesal process absent (Pl. 16.187.32 B) .... ***Lonnbergius*** **[p. 633]**

[USA: Florida]

9(8) Gonopod without subapical setae (Pl. 16.187.9 B, C) 10

9' Gonopod with subapical setae (Pls. 16.187.9 D–H, 10 C–F) 11

10(9) Gonopod central projection very conspicuous and extending distally far beyond cephalic and mesal processes; cephalic process situated distinctly mesal to central projection base (Pl. 16.187.9 B) .... *Procambarus* (*Tenuicambarus*) *tenuis* Hobbs, 1950

[USA: Arkansas, Oklahoma]

10' Gonopod central projection not conspicuously large, never extending beyond cephalic and mesal processes; cephalic process, if present, either cephalic or lateral to central projection (Pls. 16.187.9 C, 22 C, 39 F) .... ***Ortmannicus*** (in part) **[p. 634]**

11(9) Gonopod mesal process usually extending to or beyond apical plane perpendicular to axis of appendage shaft (Pl. 16.187.10 A); if not (Pl. 16.187.10 F), then chela palm mesal surface bearded (Pl. 16.187.7 F) .... ***Leconticambarus*** (in part) **[p. 627]**

11' Gonopod mesal process of first pleopod seldom extending to or beyond apical plane perpendicular to axis of appendage shaft (Pl. 16.187.10 B); if so, cephalic process situated caudomesal to central projection (Pl. 16.187.10 D) or cephalic process subapical setae situated lateral to base (Pl. 16.187.9 E, F); chela palm mesal surface never bearded (Pl. 16.187.7 B–E) .... ***Ortmannicus*** (in part) **[p. 634]**

Crustacea: Malacostraca: Decapoda: Astacidea: Cambaridae: *Procambarus*: *Pennides*: Species

1 Cheliped basis without a mesal spine (Pl. 16.187.11 B) 2

1' Cheliped basis with a mesal spine (Pl. 16.187.11 A; see also Pl. 16.187.12 A) .... *Procambarus* (*Pennides*) *versutus* (Hagen, 1870)

[USA: Alabama, Florida, Georgia]

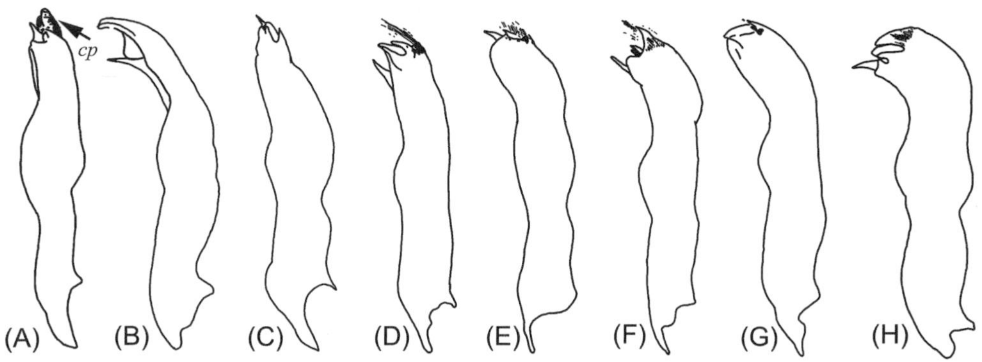

PLATE 16.187.09 Lateral view of left first pleopods: (A) *Procambarus (Mexicambarus) bouvieri*; (B) *P. (Tenuicambarus) tenuis*; (C) *P. (Ortmannicus) lewisi*; (D) *P. (Ortmannicus) acutus*; (E) *P. (Ortmannicus) fallax*; (F) *P. (Ortmannicus) pictus*; (G) *P. (Ortmannicus) planirostris*; (H) *P. (Ortmannicus) pearsei*. cp, cephalic process.

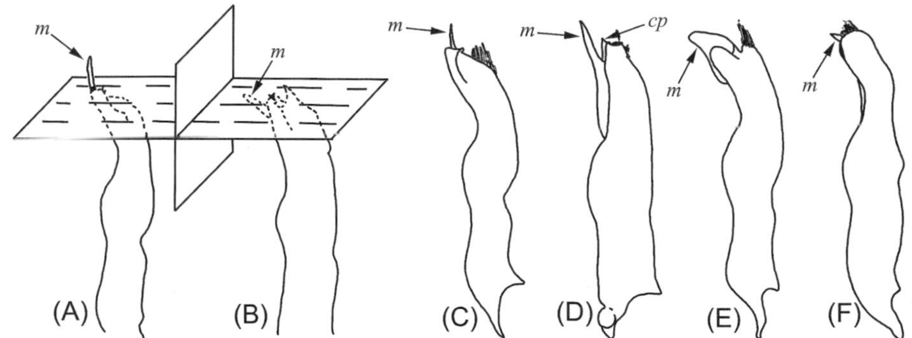

PLATE 16.187.10 Lateral view of left first pleopods: (A) *Procambarus (Leconticambarus) latipleurum*; (B) *P. (Ortmannicus) litosternum*; (C) *P. (Leconticambarus) barbatus*; (D) *P. (Ortmannicus) villalobosi*; (E) *P. (Leconticambarus) kilbyi*; (F) *P. (Leconticambarus) hubbelli*. cp, cephalic process; m, mesial process.

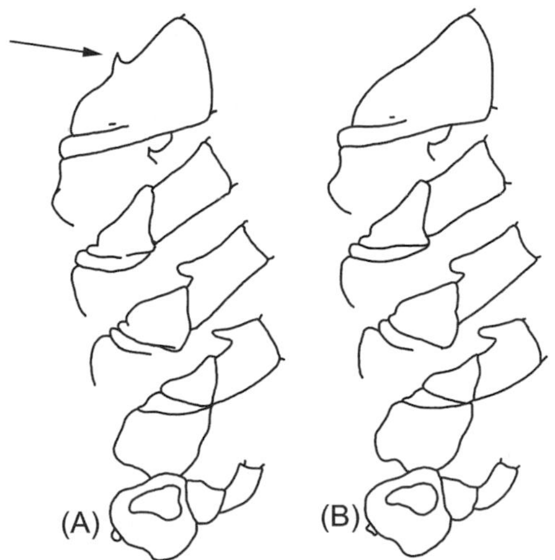

PLATE 16.187.11 Ventral view of basal portion of left pereopods: (A) *Procambarus (Pennides) versutus*, arrow indicating spine on basis of first pereopod (cheliped); (B) first pereopod lacking spine on basis.

2(1) Areola less than 6 times longer than wide; gonopod cephalic process erect and compressed 3

2' Areola greater than 7 times longer than wide; gonopod cephalic process a small, curved plate (Pl. 16.187.15 C) .... *Procambarus* (*Pennides*) *roberti* Villalobos & Hobbs, 1974

[Mexico: San Luis Potosí]

3(2) Gonopod cephalic surface without a distinct angular shoulder (Pls. 16.187.12 C–I, 14) 4

3' Gonopod cephalic surface with distinct angular shoulder (Pl. 16.187.12 B "*s*") .... *Procambarus* (*Pennides*) *lylei* Fitzpatrick & Hobbs, 1971

[USA: Mississippi]

4(3) Gonopod cephalic process well developed (Pls. 16.187.12 C–I, 14 A–B) 5

4' Gonopod cephalic process of first pleopod absent or rudimentary (Pl. 16.187.14 C–G) 15

5(4) Gonopod cephalic process situated cephalic, lateral, or cephalomesal to central projection, never entirely obscured by latter in lateral aspect (Pls. 16.187.12 E–I, 14 A–B) 6

5' Gonopod cephalic process situated entirely mesal to central projection and completely obscured by latter in lateral aspect (Pl. 16.187.12 C, D) .... *Procambarus* (*Pennides*) *suttkusi* Hobbs, 1953

[USA: Alabama, Florida]

6(5) Gonopod cephalic process subtruncate with a caudodistal acute angle (Pl. 16.187.12 E) 7

6' Gonopod cephalic process tapering from base (Pl. 16.187.12 F–I , 14 A–B) 8

7(6) Gonopod cephalic process very prominent, subrhomboid in mesal aspect and somewhat wrapped around central projection base (Pl. 16.187.15 D)........ *Procambarus* (*Pennides*) *vioscai paynei* Fitzpatrick, 1990

[USA: Alabama, Louisiana, Mississippi, Tennessee]

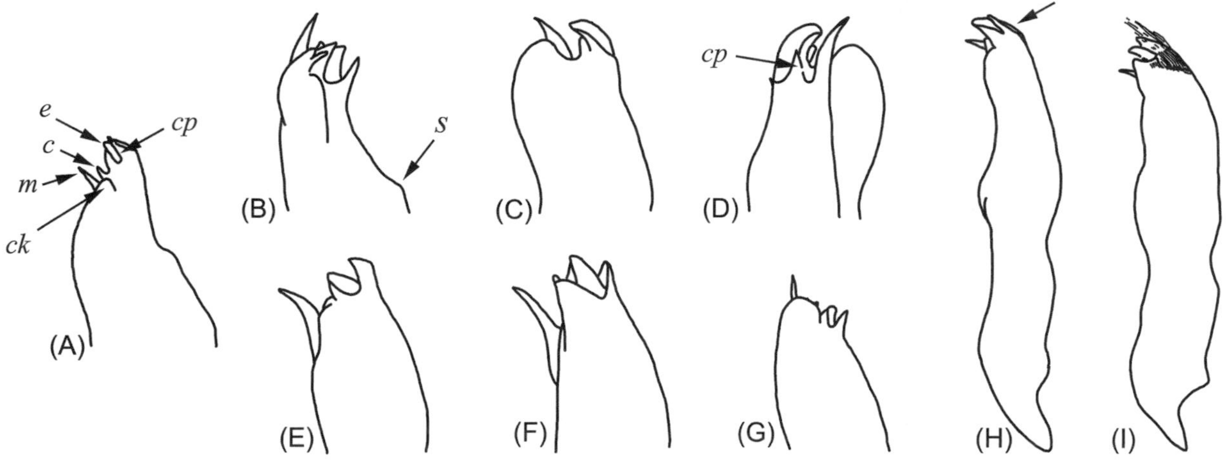

PLATE 16.187.12 (A–C, E, G) Lateral view of distal portion of left first pleopod, (D) mesial view of same, (H, I) lateral view of left first pleopods: (A) *Procambarus (Pennides) versutus*; (B) *P. (Pennides) lylei*; (C, D) *P. (Pennides) suttkusi*; (E) *P. (Pennides) vioscai vioscai*; (F) *P. (Pennides) penni*; (G) *P. (Pennides) elegans*; (H) *P. (Pennides) echinatus*; (I) *P. (Pennides) dupratzi*. c, caudal process; ck, caudal knob; cp, cephalic process; e, central projection; m, mesial process; s, shoulder on cephalic surface.

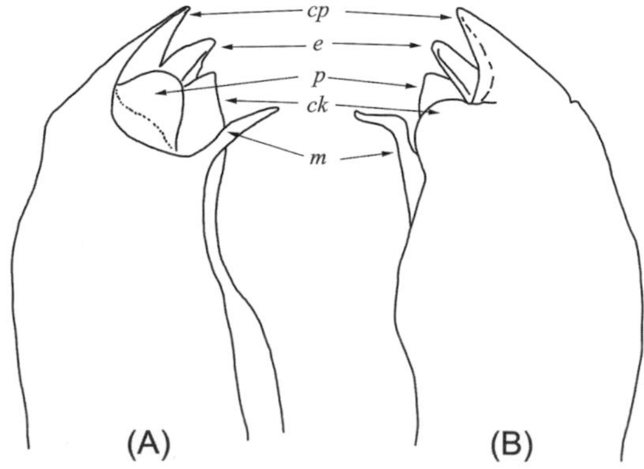

PLATE 16.187.13 Distal portion of the left form I male gonopod of *Procambarus (Pennides) pentastylus* showing five terminal elements: (A) mesial view; (B) lateral view cp, cephalic process; e, central projection; p, pentastyle (= adventitious process); ck, caudal knob; and m, mesial process.

7’ Gonopod cephalic process erect in basal half then abruptly bent caudally at about 90 degrees; caudal knob inflated, rounded; pentastyle (like Pl. 187.13 A–B “*p*”) a corneous ridge .... *Procambarus* (*Pennides*) *vioscai vioscai* Penn, 1946

[USA: Arkansas, Louisiana]

8(6) Gonopod cephalic process directed caudodistally (Pl. 16.187.12 H, I) 9

8’ Gonopod cephalic process directed distally (Pls. 16.187.12 F, G, 14 A–B) 11

9(8) Gonopod subapical setae abundant 10

9’ Gonopod subapical setae sparse; part of central projection evident in lateral aspect cephalodistal to cephalic process (Pl. 16.187.12 H), note arrow) .... *Procambarus* (*Pennides*) *echinatus* Hobbs, 1956

[USA: South Carolina]

10(9) Gonopod caudal knob small, not strongly inflated; pentastyle (like Pl. 16.187.13 A, B “*p*”) not visible; central projection extending distinctly beyond cephalic process (Pl. 16.187.12 I) .... *Procambarus* (*Pennides*) *dupratzi* Penn, 1953

[USA: Louisiana, Texas]

10’ Gonopod caudal knob large, inflated; pentastyle (Pl. 16.187.13 A, B “*p*”) an erect plate; central projection not projecting distinctly beyond cephalic process (Pl. 16.187.13) .... *Procambarus* (*Pennides*) *pentastylus* Walls & Black, 2008

[USA: Louisiana]

11(8) Gonopod cephalic and mesal processes in lateral aspect diverging at angle of at least 50° (Pl. 16.187.12 F) 12

11’ Gonopod cephalic and mesal processes in lateral aspect subparallel or diverging at angle of much less than 50° (Pls. 16.187.12 G, 14 A, B) 13

12(11) Gonopod caudal process directed caudally (Pl. 16.187.15 A “*c*”) .... *Procambarus* (*Pennides*) *clemmeri* Hobbs, 1975

[USA: Mississippi]

12’ Gonopod caudal process directed distally (Pl. 16.187.15 F “*c*”) .... *Procambarus* (*Pennides*) *penni* Hobbs, 1951

[USA: Louisiana, Mississippi]

13(11) Gonopod cephalic process reaching beyond caudal knob or caudal process (Pl. 16.187.14 A, B) 14

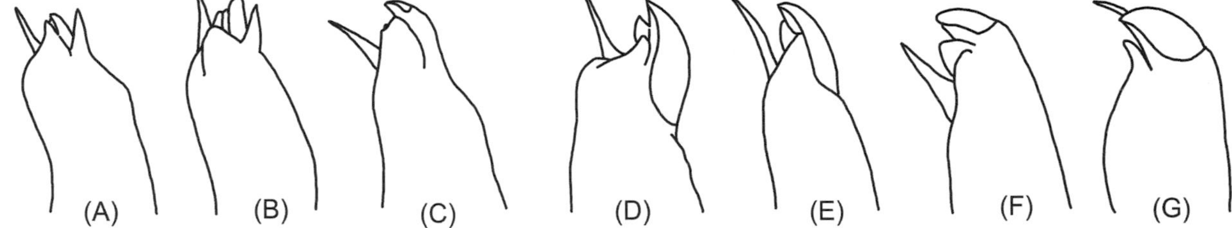

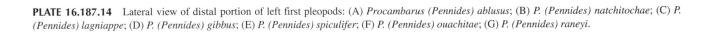

PLATE 16.187.14 Lateral view of distal portion of left first pleopods: (A) *Procambarus (Pennides) ablusus*; (B) *P. (Pennides) natchitochae*; (C) *P. (Pennides) lagniappe*; (D) *P. (Pennides) gibbus*; (E) *P. (Pennides) spiculifer*; (F) *P. (Pennides) ouachitae*; (G) *P. (Pennides) raneyi*.

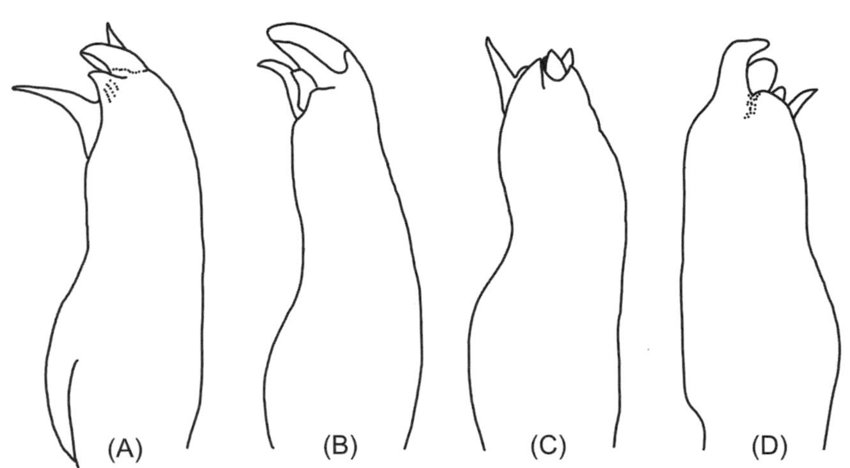

PLATE 16.187.15 Lateral view of distal portion of left first pleopods: (A) *Procambarus (Pennides) clemmeri*; (B) *P. (Pennides) petersi*; (C) *P. (Pennides) roberti*; (D) *P. (Pennides) vioscai paynei*.

13' Gonopod cephalic process not reaching beyond caudal knob or caudal process (Pl. 16.187.12 G) .... *Procambarus* (*Pennides*) *elegans* Hobbs, 1969

[USA: Louisiana]

14(13) Gonopod central projection cephalic margin rounded; caudal process small (Pl. 16.187.14 A) .... *Procambarus* (*Pennides*) *ablusus* Penn, 1963

[USA: Mississippi, Tennessee]

14' Gonopod central projection cephalic margin virtually straight; caudal process prominent (Pl. 16.187.14 B) .... *Procambarus* (*Pennides*) *natchitochae* Penn, 1953

[USA: Arkansas, Louisiana, Texas]

15(4) Gonopod distolateral surfaces without an excavation (Pl. 16.187.14 D–G) 16

15' Gonopod distolateral surface with a longitudinal excavation extending proximally from base of central projection (Pl. 16.187.14 C) .... *Procambarus* (*Pennides*) *lagniappe* Black, 1968

[USA: Mississippi]

16(15) Gonopod central projection arising from level distinctly proximal to base of caudal process (Pl. 16.187.14 D–E) 17

16' Gonopod central projection of first pleopod arising from level distinctly distal to base of caudal process (Pl. 16.187.14 F, G) 18

17(16) Gonopod caudodistal portion subtruncate with caudal element and central projection situated on cephalic 1/2 of tip (Pl. 16.187.14 D) .... *Procambarus* (*Pennides*) *gibbus* Hobbs, 1969

[USA: Georgia]

17' Gonopod caudodistal portion tapering with caudal element and central projection constituting almost entire tip (Pl. 16.187.14 E) .... *Procambarus* (*Pennides*) *spiculifer* (LeConte, 1856)

[Alabama, Florida, Georgia, Mississippi, South Carolina]

18(16) Gonopod caudal knob vestigial 19

18' Gonopod caudal knob conspicuous (Pl. 16.187.14 F) .... *Procambarus* (*Pennides*) *ouachitae* Penn, 1956

[USA: Arkansas, Mississippi]

19(18) Gonopod caudal process narrow and elongate (Pl. 16.187.14 G) .... *Procambarus* (*Pennides*) *raneyi* Hobbs, 1953

[USA: Georgia, South Carolina]

19' Gonopod caudal process small and sub-triangular (Pl. 16.187.15 B) .... *Procambarus* (*Pennides*) *petersi* Hobbs, 1981

[USA: Georgia]

Crustacea: Malacostraca: Decapoda: Astacidea: Cambaridae: *Procambarus*: *Scapulicambarus*: Species

1 Gonopod cephalic process acute (Pl. 16.187.16 A–C "*cp*") 2

1' Gonopod cephalic process lobiform, caudal margin with or without angle (Pl. 16.187.16 D–F "*cp*") 4

2(1) Gonopod not conspicuously setose, cephalic process blade-like and directed caudodistad; carapace dorsal surface with punctations normal 3

2' Gonopod conspicuously setose, cephalic process forming a gently curved lamelliform plate flanking the cephalic part of the central projection (Pl. 16.187.16 C); carapace dorsal surface with punctations deep .... *Procambarus* (*Scapulicambarus*) *strenthi* Hobbs, 1977

[Mexico: San Louis Potosí]

3(2) Gonopod in lateral aspect with distal portion tapering from level of shoulder distally (Pl. 16.187.16 A) .... *Procambarus* (*Scapulicambarus*) *howellae* Hobbs, 1952

[USA: Georgia]

3' Gonopod in lateral aspect with distal portion not markedly tapering from level of shoulder (Pl. 16.187.16 B; see also Pl. 16.187.4 B) .... *Procambarus* (*Scapulicambarus*) *paeninsulanus* (Faxon, 1914)

[USA: Alabama, Florida, Georgia]

4(1) Areola more than 12 times longer than broad (see also Pl. 16.187.16 E, F) 5

4' Areola less than 12 times longer than broad (see also Pl. 16.187.16 D) .... *Procambarus* (*Scapulicambarus*) *okaloosae* Hobbs, 1942

[USA: Alabama, Florida]

5(4) Gonopod cephalic process with caudal margin with distinct angle (Pl. 16.187.16 E) .... *Procambarus* (*Scapulicambarus*) *clarkii* (Girard, 1852)

[USA: Alabama, Arkansas, Florida, Illinois, Kentucky, Louisiana, Mississippi, Missouri, New Mexico, Texas. Introduced in California, Virginia. Mexico: Coahuila, Chihuahua]

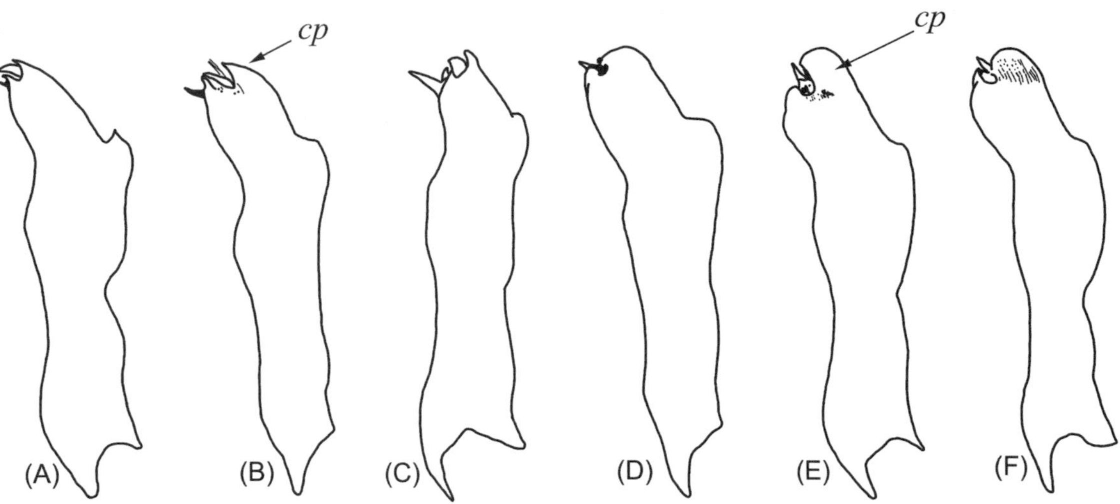

PLATE 16.187.16 Lateral view of left first pleopods: (A) *Procambarus (Scapulicambarus) howellae*; (B) *P. (Scapulicambarus) paeninsulanus*; (C) *P. (Scapulicambarus) strenthi*; (D) *P. (Scapulicambarus) okaloosae*; (E) *P. (Scapulicambarus) clarkii*; (F) *P. (Scapulicambarus) troglodytes*. cp, cephalic process.

5' Gonopod cephalic process entire margin rounded, lacking angle (Pl. 16.187.16 F) .... *Procambarus* (*Scapulicambarus*) *troglodytes* (LeConte, 1856)

[USA: Georgia, South Carolina]

Crustacea: Malacostraca: Decapoda: Astacidea: Cambaridae: *Procambarus*: *Capillicambarus*: Species

1 Gonopod distal third tapering (Pl. 16.187.17 B–C; see also Pl. 16.187.6 A) 2

1' Gonopod distal third almost uniformly broad, ending in 2 distinct tips, one of which is broadly truncate (Pl. 16.187.17 A) .... *Procambarus* (*Capillicambarus*) *incilis* Penn, 1962

[USA: Texas]

2(1) Gonopod apical third bent caudad at 35 degree angle, expanding distally in mesal view and ending abruptly in three small elements (Pl. 16.187.17 B); cephalic process both wider at base and at truncate tip, and lies in a more nearly longitudinal plane (with respect to axis of animal) .... *Procambarus* (*Capillicambarus*) *brazoriensis* Albaugh, 1975

[USA: Texas]

2' Gonopod apical third tapering to its apex distally in mesal view (Pl. 16.187.17 C); cephalic process nearly transverse in orientation .... *Procambarus* (*Capillicambarus*) *hinei* (Ortmann, 1905)

[USA: Louisiana, Texas]

Crustacea: Malacostraca: Decapoda: Astacidea: Cambaridae: *Procambarus*: *Hagenides*: Species

1 Gonopod central projection plate-like and directed laterally across pleopod cephalodistal surface (Pl. 16.187.18 A "*e*") 2

1' Gonopod central projection falciform and directed caudally, distally, or caudodistally (Pl. 16.187.18 B–G "*e*") 5

2(1) Gonopod caudal knob directed distolaterally at angle less than 90 degrees to principal axis of appendage shaft (Pl. 16.187.19 B–D "*ck*") 3

2' Gonopod caudal knob directed mesally at approximately right angle to principal axis of appendage shaft (Pl. 16.187.19 A; see also Pl. 16.187.18 A "*ck*") .... *Procambarus* (*Hagenides*) *rogersi rogersi* (Hobbs, 1938)

[USA: Florida]

3(2) Gonopod cephalic process not extending distally as far as central projection (Pl. 16.187.19 C–D "*cp*") 4

3' Gonopod cephalic process extending distally as far as central projection (Pl. 16.187.19 B; see also Pl. 16.187.7 A) .... *Procambarus* (*Hagenides*) *rogersi expletus* Hobbs & Hart, 1959

[USA: Florida]

4(3) Gonopod caudal knob in caudal aspect, distinctly digitiform, longer than broad (Pl. 16.187.19 C) .... *Procambarus* (*Hagenides*) *rogersi campestris* Hobbs, 1942

[USA: Florida]

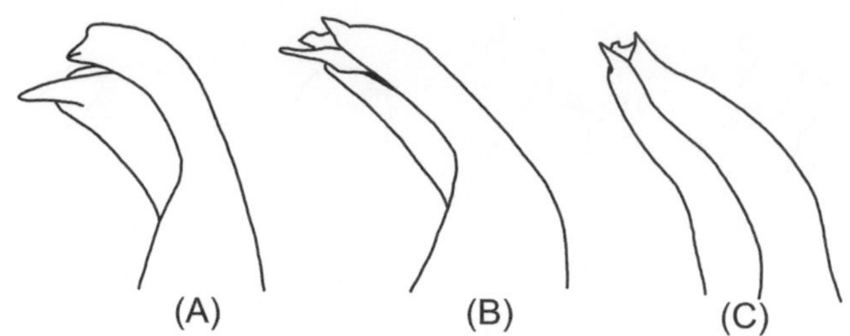

PLATE 16.187.17 Lateral view of distal portion of form I male first pleopod: (A) *Procambarus (Capillicambarus) incilis*; (B) *P. (Capillicambarus) brazoriensis*; (C) *P. (Capillicambarus) hinei*.

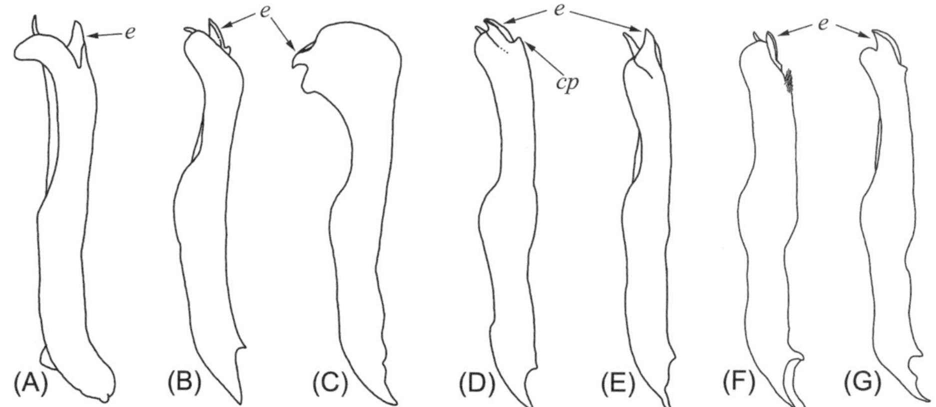

PLATE 16.187.18 Lateral view of left first pleopods: (A) *Procambarus (Hagenides) rogersi rogersi*; (B) *P. (Hagenides) geodytes*; (C) *P. (Hagenides) truculentus*; (D) *P. (Hagenides) advena*; (E) *P. (Hagenides) pygmaeus*; (F) *P. (Hagenides) caritus*; (G) *P. (Hagenides) talpoides*. cp, cephalic process; e, central projection.

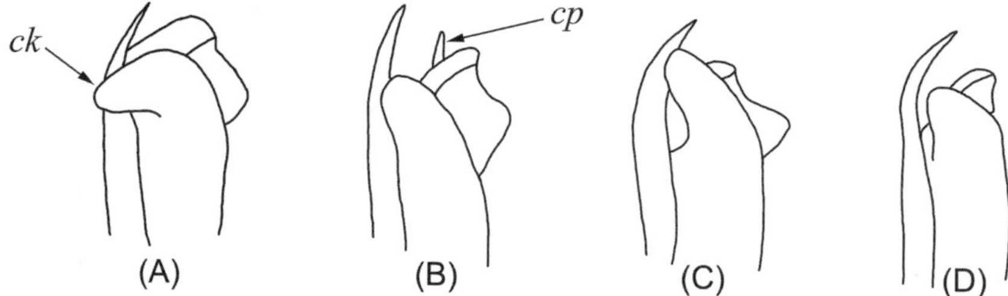

PLATE 16.187.19 Caudal view of distal portion of left first pleopods: (A) *Procambarus (Hagenides) rogersi rogersi*; (B) *P. (Hagenides) rogersi expletus*; (C) *P. (Hagenides) rogersi campestris*; (D) *P. (Hagenides) rogersi ochlocknensis*. ck, caudal knob; cp, cephalic process.

4' Gonopod caudal knob in caudal aspect, almost or quite as broad as long (Pl. 16.187.19 D) .... *Procambarus* (*Hagenides*) *rogersi ochlocknensis* Hobbs, 1942

[USA: Florida]

5(1) Pereopod III ischia only with hooks (Pl. 16.187.4 A) 6

5' Pereopod III and IV ischia with hooks (Pl. 16.187.4 B, D; see also Pl. 16.187.18 B) .... *Procambarus* (*Hagenides*) *geodytes* Hobbs, 1942

[USA: Florida]

6(5) Gonopod with terminal elements directed distally or caudodistally but never at more than 40 degree angle to appendage shaft 7

6' Gonopod with terminal elements directed caudally at approximately right angle to appendage shaft (Pl. 16.187.18 C) .... *Procambarus* (*Hagenides*) *truculentus* Hobbs, 1954

[USA: Georgia]

7(6) Gonopod with caudal element moderately to strongly inflated, and, in lateral view, base of falciform or dentiform central projection spanning scarcely more (usually less) than one-half the diameter of appendage distal portion; color variable but never green with scarlet markings 8

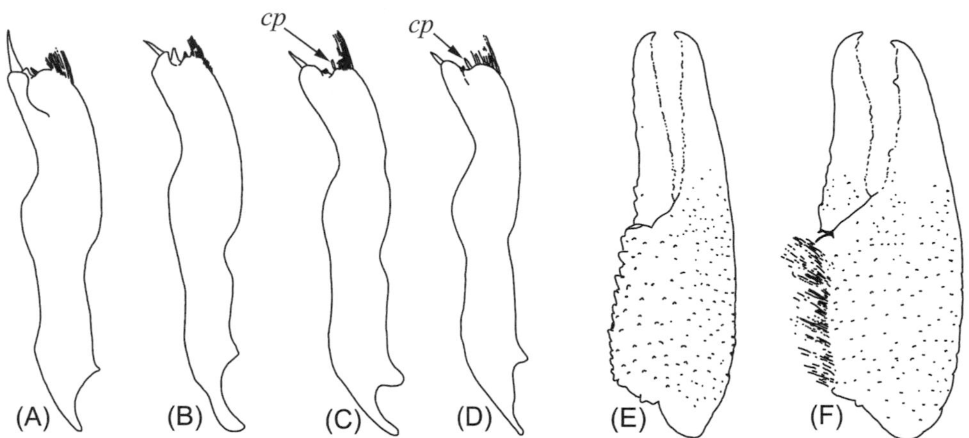

PLATE 16.187.20 (A–D) Lateral view of left first pleopods; (E, F) dorsal view of chelae: (A) *Procambarus (Leconticambarus) pubischelae*; (B) *P. (Leconticambarus) escambiensis*; (C) *P. (Leconticambarus) econfinae*; (D) *P. (Leconticambarus) apalachicolae*; (E) *P. (Leconticambarus) latipleurum*; (F) *P. (Leconticambarus) barbatus*. cp, cephalic process.

7' Gonopod with caudal element very weakly inflated, and, in lateral view, base of falciform central projection spanning about three-fourths the appendage distal diameter (Pl. 16.187.18 E); color forest green with scarlet markings .... *Procambarus* (*Hagenides*) *pygmaeus* Hobbs, 1942

[USA: Florida, Georgia]

8(7) Gonopod with cephalic process reduced or absent 9

8' Gonopod with well-developed cephalic process (Pl. 16.187.18 D) .... *Procambarus* (*Hagenides*) *advena* (LeConte, 1856)

[USA: Florida, Georgia]

9(8) Gonopod cephalic process vestigial, rarely absent; central projection falciform and strongly arched cephalically, appendage cephalocaudal diameter approximately half corresponding distal diameter (Pl. 16.187.18 G) .... *Procambarus* (*Hagenides*) *talpoides* Hobbs, 1981

[USA: Florida, Georgia]

9' Gonopod cephalic process absent; central projection dentiform and weakly arched cephalically, appendage cephalocaudal diameter little, if any, greater than one-fourth corresponding distal diameter (Pl. 16.187.18 F) .... *Procambarus* (*Hagenides*) *caritus* Hobbs, 1981

[USA: Georgia]

Crustacea: Malacostraca: Decapoda: Astacidea: Cambaridae: *Procambarus*: *Leconticambarus*: Species

1 Pigmented; eyes normally pigmented 2

1' Subterranean; albinistic; eyes with reduced pigment (see also Pl. 16.187.21 A) .... *Procambarus* (*Leconticambarus*) *milleri* Hobbs, 1971

[USA: Florida]

2(1) Pereopod III ischium only bearing a hook (Pl. 16.187.4 A) 3

2' Pereopods III and IV ischia bearing hooks (Pl. 16.187.4 B, D) 6

3(2) Gonopod mesal process slender and tapering to acute or subacute apex (Pls. 16.187.21 A–E, H "*m*", 20 A–D) 4

3' Gonopod mesal process massive and subspatulate (Pl. 16.187.21 G "*m*") .... *Procambarus* (*Leconticambarus*) *kilbyi* (Hobbs, 1940)

[USA: Florida]

4(3) Gonopod mesal process extending beyond cephalic process; subapical setae arranged in linear series on appendage cephalodistal margin (Pl. 16.187.21 C, D) 5

4' Gonopod with mesal process subequal to cephalic process; subapical setae clustered (Pl. 16.187.21 B) .... *Procambarus* (*Leconticambarus*) *hubbelli* (Hobbs, 1940)

[USA: Alabama, Florida]

5(4) Gonopod caudal process directed cephalodistally, not reaching apices of cephalic process and central projection (Pl. 16.187.21 C "*c*"; see also Pl. 16.187.8 B) .... *Procambarus* (*Leconticambarus*) *rathbunae* (Hobbs, 1940)

[USA: Florida]

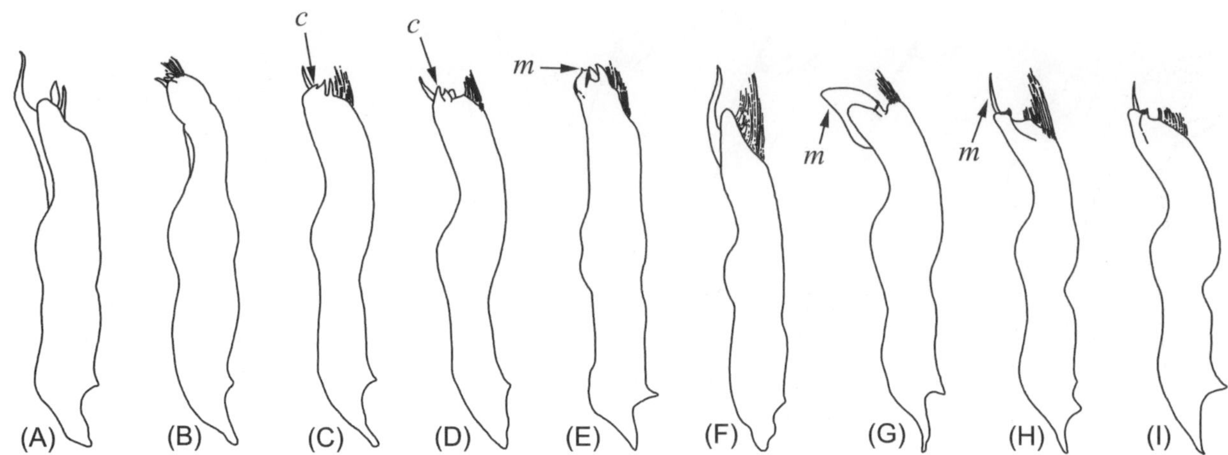

PLATE 16.187.21 Lateral view of left first pleopods: (A) *Procambarus (Leconticambarus) milleri*; (B) *P. (Leconticambarus) hubbelli*; (C) *P. (Leconticambarus) rathbunae*; (D) *P. (Leconticambarus) capillatus*; (E) *P. (Leconticambarus) shermani*; (F) *P. (Leconticambarus) alleni*; (G) *P. (Leconticambarus) kilbyi*; (H) *P. (Leconticambarus) latipleurum*; (I) *P. (Leconticambarus) barbatus*. c, caudal process; m, mesial process.

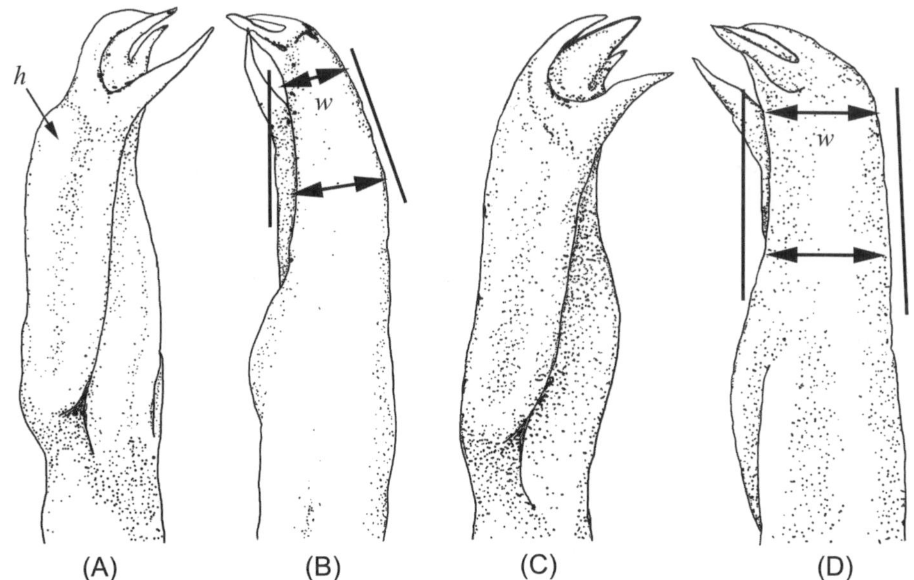

PLATE 16.187.22 Distal half of left first pleopod: mesial (A) and lateral (B) views of *Procambarus (Ortmannicus) zonangulus*; mesial (C) and lateral (D) view of *P. (Ortmannicus) nueces*. h, rounded hump; w, width of pleopod.

5' Gonopod caudal process directed distally and extending at least to apices of cephalic process and central projection (Pl. 16.187.21 D "*c*") .... *Procambarus* (*Leconticambarus*) *capillatus* Hobbs, 1971

[USA: Alabama, Florida]

6(2) Gonopod mesal process "*m*" extending well beyond cephalic process apex (Pls. 16.187.21 F–I "*m*", 20 A–D) 7

6' Gonopod mesal process subequal in length to cephalic process (Pl. 16.187.21 E "*m*") .... *Procambarus* (*Leconticambarus*) *shermani* Hobbs, 1942

[USA: Florida, Louisiana, Mississippi]

7(6) Gonopod mesal process straight or curved but never sinuous (Pls. 16.187.21 G–I, 20 A–D) 8

7' Gonopod mesal process sinuous (Pl. 16.187.21 F) .... *Procambarus* (*Leconticambarus*) *alleni* (Faxon, 1884)

[USA: Florida]

8(7) Gonopod mesal process acute to subspiculiform, apex directed caudo- or cephalodistally (Pls. 16.187.21 H, I "*m*", 20 A–D) 9

8' Gonopod mesal process massive, subspatulate, with apex directed almost caudally (Pl. 16.187.21 G "*m*") .... *Procambarus* (*Leconticambarus*) *kilbyi* (Hobbs, 1940)

[USA: Florida]

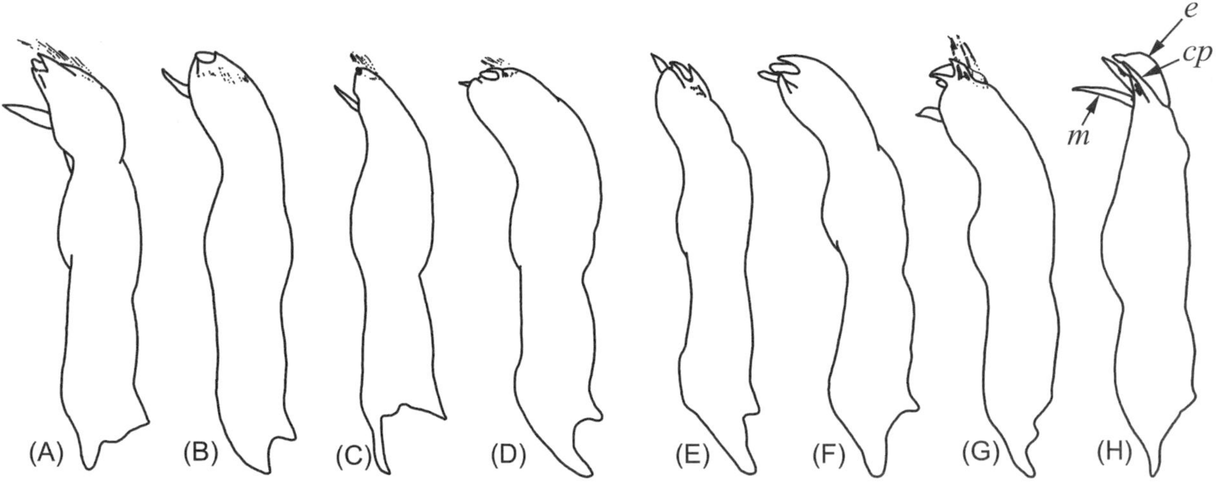

PLATE 16.187.23 Lateral view of left first pleopods: (A) *Procambarus (Ortmannicus) verrucosus*; (B) *P. (Ortmannicus) evermanni*; (C) *P. (Ortmannicus) caballeroi*; (D) *P. (Ortmannicus) plumimanus*; (E) *P. (Ortmannicus) jaculus*; (F) *P. (Ortmannicus) hybus*; (G) *P. (Ortmannicus) lepidodactylus*; (H) *P. (Ortmannicus) toltecae*. c, cephalic process; e, central projection; m, mesial process.

9(8) Gonopod mesal process directed distally or cephalodistally (Pls. 16.187.21 H, I "*m*", 20 A) 10

9' Gonopod mesal process directed caudodistally (Pl. 16.187.20 B–D) 13

10(9) Chela palm densely hirsute (Pl. 16.187.20 F) 11

10' Chela palm not densely hirsute (Pl. 16.187.20 E; see also Pl. 16.187.21 H) .... *Procambarus* (*Leconticambarus*) *latipleurum* Hobbs, 1942
[USA: Florida]

11(10) Gonopod cephalodistal margin forming round hump (Pl. 16.187.20 A) 12

11' Gonopod cephalodistal margin sloping steeply from base of cephalic process (Pl. 16.187.21 I; see also Pls. 16.187.6 B, 20 F) .... *Procambarus* (*Leconticambarus*) *barbatus* (Faxon, 1890)
[USA: Georgia, South Carolina]

12(11) Pereopods III and IV ischia each bearing a hook; pereopod IV coxa with a caudomesal boss .... *Procambarus* (*Leconticambarus*) *pubischelae pubischelae* Hobbs, 1942
[USA: Florida, Georgia]

12' Pereopod III ischium bearing a hook, pereopod IV ischium lacking hook and coxa lacking caudomesal boss .... *Procambarus* (*Leconticambarus*) *pubischelae deficiens* Hobbs, 1981
[USA: Georgia]

13(9) Chela palm not densely hirsute (like Pl. 16.187.20 E; see also 21) 14

13' Chela palm densely hirsute (like Pl. 16.187.20 F; also see 20 B) .... *Procambarus* (*Leconticambarus*) *escambiensis* Hobbs, 1942
[USA: Alabama, Florida]

14(13) Gonopod cephalic process curved, its apex directed cephalodistally (Pl. 16.187.20 C "*c*") .... *Procambarus* (*Leconticambarus*) *econfinae* Hobbs, 1942
[USA: Florida]

14' Gonopod cephalic process straight and directed caudodistally (Pl. 16.187.20 D "*c*") .... *Procambarus* (*Leconticambarus*) *apalachicolae* Hobbs, 1942
[USA: Florida]

Crustacea: Malacostraca: Decapoda: Astacidea: Cambaridae: *Procambarus*: *Girardiella*: Species

1 Uropod endopod distal margin without projecting spines (Pl. 16.187.25 A, see arrow); gonopod cephalic process distinctly U-shaped, but may be very small (e.g., in *P.* (*G.*) *ceruleus*, Pl. 16.187.28 A) (*gracilis* group) 2

1' Uropod endopod distal margin with two projecting spines (Pl. 16.187.25 B "*s*"); gonopod cephalic process forming a much reduced plate like structure when viewed from mesial or cephalic aspect (*hagenianus* group) 16

2(1) First pleopods asymmetrical (Pl. 16.187.2 A, B) 3

2' First pleopods symmetrical (Pl. 16.187.2 C, D) 4

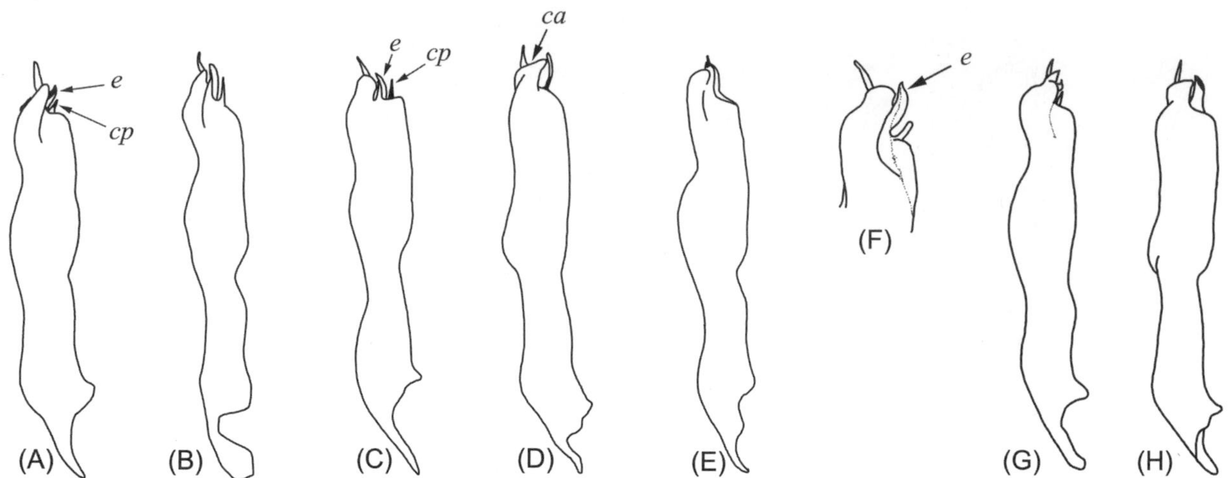

PLATE 16.187.24 (A–E) Lateral view of left first pleopods; (F) distal end of same: (A) *Procambarus (Girardiella) simulans*; (B) *P. (Girardiella) regiomontanus*; (C) *P. (Girardiella) gracilis*; (D) *P. (Girardiella) hagenianus hagenianus*; (E) *P. (Girardiella) tulanei*; (F) *P. (Girardiella) curdi*; (G) *P. (Girardiella) parasimulans*; (H) *P. (Girardiella) hagenianus vesticeps*. cp, cephalic process; (E) central projection.

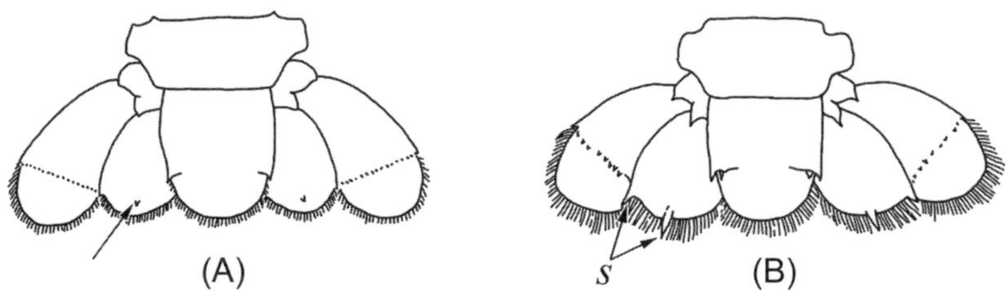

PLATE 16.187.25 Dorsal view of terminal portion of abdomens: (A) *Procambarus (Girardiella) gracilis*; (B) *P. (Girardiella) hagenianus hagenianus*. s, indicates spines projecting from distal margin of mesial ramus of uropod.

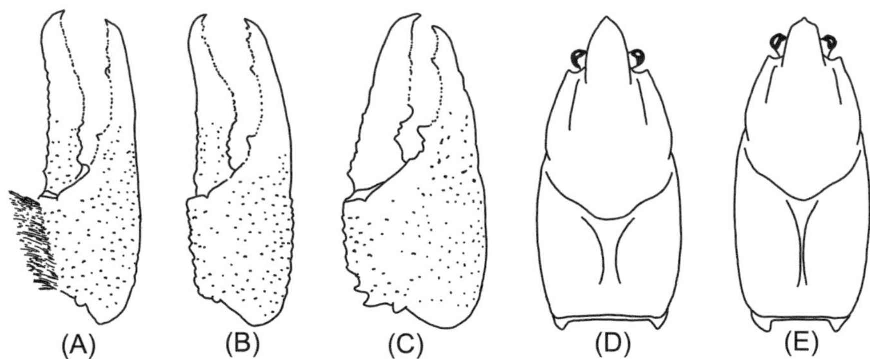

PLATE 16.187.26 (A–C) Dorsal view of chelae; (D, E) dorsal view of carapaces: (A) *Procambarus (Girardiella) tulanei*; (B) *P. (Girardiella) simulans*; (C) *P. (Girardiella) gracilis*; (D) *P. (Girardiella) tulanei*; (E) *P. (Girardiella) gracilis*.

3(2) Gonopod central projection strongly sloping laterally (Pl. 16.187.28 A); body color brownish with darker speckling .... *Procambarus* (*Girardiella*) *ceruleus* Fitzpatrick & Wicksten, 1998

[USA: Texas]

3' Gonopod central projection erect, not strongly sloping laterally (Pl. 16.187.28 D); body color red and blue without speckling .... *Procambarus* (*Girardiella*) *steigmani* Hobbs, 1991

[USA: Texas]

4(2) Chela dactyl lateral margin basal third with distinct excision (Pl. 16.187.26 C); chela palm dorsolateral surface punctate; color generally reddish without dark speckling 5

4' Chela dactyl lateral margin basal third without distinct excision (Pl. 16.187.27 A); chela palm dorsolateral surface tuberculate; color brownish with dark speckling 9

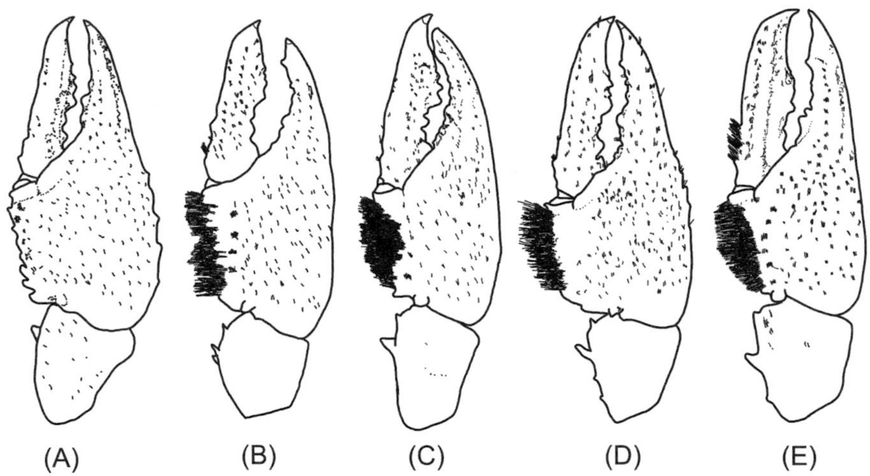

PLATE 16.187.27 Dorsal view of chelae: (A) *Procambarus (Girardiella) hagenianus hagenianus*; (B) *P. (Girardiella) barbiger*; (C) *P. (Girardiella) cometes*; (D) *P. (Girardiella) connus*; (E) *P. (Girardiella) pogum.*

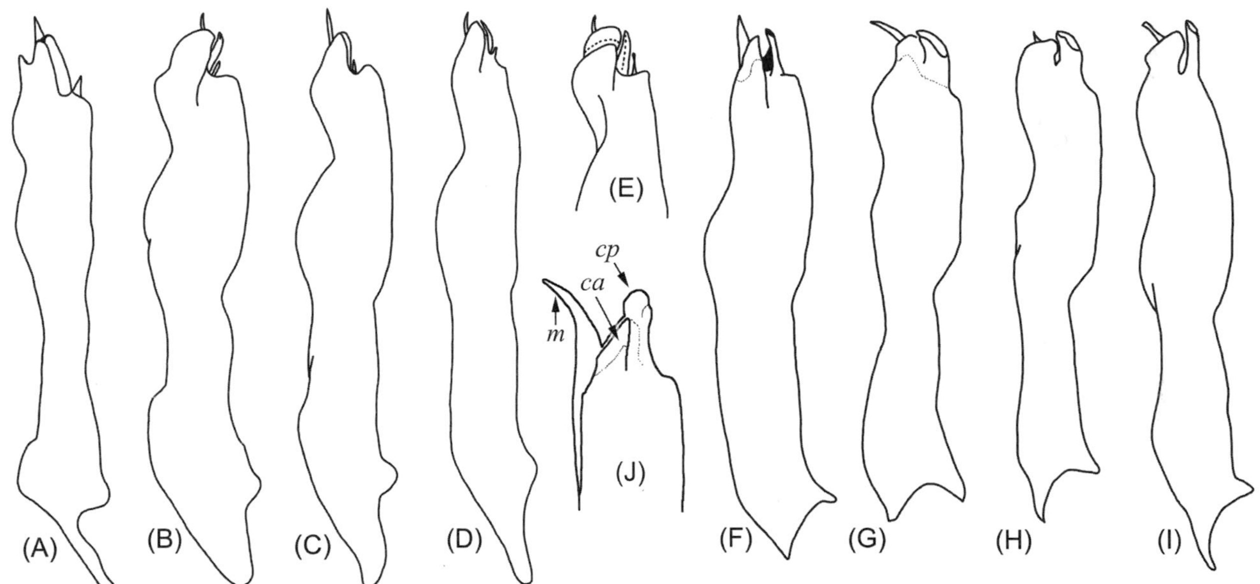

PLATE 16.187.28 (A–D) Lateral view of left first pleopods; (E, J) distal portion of same: (A) *Procambarus (Girardiella) ceruleus*; (B) *P. (Girardiella) kensleyi*; (C) *P. (Girardiella) nigrocinctus*; (D) *P. (Girardiella) steigmani*; (E) *P. (Girardiella) machardyi*; (F) *P. (Girardiella) barbiger*; (G) *P. (Girardiella) cometes*; (H) *P. (Girardiella) connus*; (I) *P. (Girardiella) pogum*; (J) *P. (Girardiella) holifieldi*. cp, central projection; ca, caudal process; m, mesial process.

5(4) Gonopod central projection not extending beyond caudal element 6

5' Gonopod central projection extending beyond caudal element (Pl. 16.187.24 C "*e*") .... *Procambarus* (*Girardiella*) *gracilis* (Bundy, 1876)

[USA: Arkansas, Iowa, Illinois, Kansas, Missouri, Oklahoma, Texas, Wisconsin]

6(4) Areola closed (Pl. 16.187.30 E), linear, or nearly so, 38 or more times longer than broad 7

6' Areola open, 16 to 25 times longer than broad (Pl. 16.187.26 D), with room for single row of punctations at middle .... *Procambarus* (*Girardiella*) *reimeri* Hobbs, 1979

[USA: Arkansas]

7(6) Chela palm mostly tuberculate, especially on dorsal and ventral surface mesal halves; gonopod caudoproximal ridge with setae splayed, directed distomesally and distolaterally 8

7' Chela palm punctate, mesal surface with tubercles; gonopod caudoproximal ridge with setae directed caudolaterally .... *Procambarus* (*Girardiella*) *liberorum* Fitzpatrick, 1978

[USA: Arkansas]

8(7') Rostral margins conspicuously thickened and tapering from base; ratio of rostrum length to width never more than 1.2 .... *Procambarus* (*Girardiella*) *regalis* Hobbs & Robison, 1988

[USA: Arkansas]

8' Rostral margins not thickened and subparallel to ill-defined base of acumen; ratio of rostrum length to width of greater than 1.2 .... *Procambarus* (*Girardiella*) *ferrugineus* Hobbs & Robison, 1988

[USA: Arkansas]

9(4) Chela palm mesal margin with obvious beard (Pls. 16.187.26 A, 27 B–E) 10

9' Chela palm mesal margin without beard (Pls. 16.187.26 B, C, 27 A) 11

10(9) Gonopod central projection not extending beyond caudal element (Pl. 16.187.28 C); rostral marginal spines present .... *Procambarus* (*Girardiella*) *nigrocinctus* Hobbs, 1990

[USA: Texas]

10' Gonopod central projection extending beyond caudal element (Pl. 16.187.24 E); no rostral marginal spines .... *Procambarus* (*Girardiella*) *tulanei* Penn, 1953

[USA: Arkansas, Louisiana]

11(9) Gonopod central projection erect or nearly so, not distinctly slanting distolaterally 12

11' Gonopod central projection slanting distolaterally (Pl. 16.187.24 F "*e*"), in caudal view extending lateral to caudal element .... *Procambarus* (*Girardiella*) *curdi* Reimer, 1975

[USA: Arkansas, Texas, Oklahoma]

12(11) Gonopod central projection not extending beyond caudal element; north of Rio Grande 13

12' Gonopod central projection extending distinctly beyond caudal element (Pl. 16.187.24 B); south of Rio Grande .... *Procambarus* (*Girardiella*) *regiomontanus* Villalobos, 1954

[Mexico: Nuevo Leon, Tamaulipas]

13(12) Areola wide, 14 or fewer times longer than broad; gonopod in mesal view, with inflated caudal knob distinctly angled 14

13' Areola narrow, more than 14 times longer than broad; gonopod in mesal view with inflated caudal knob gently rounded (Pl. 16.187.29 C, see also 24 A) .... *Procambarus* (*Girardiella*) *simulans* (Faxon, 1884)

[USA: Arkansas, Colorado, Kansas, Louisiana, New Mexico]

14(13) Gonopod caudal element in lateral and mesal views with lamellar crest broadly rounded, overlapping mesal process in mesal view 15

14' Gonopod caudal element in lateral and mesal views with lamellar crest acute, entirely cephalic to mesal process in mesal view (Pl. 16.187.29 D, see also 24 G) .... *Procambarus* (*Girardiella*) *parasimulans* Hobbs & Robison, 1982

[USA: Arkansas]

15(14) Gonopod central projection twisted, touching caudal element lamellar crest; in mesal view most of lamellar crest cephalic to mesal process (Pl. 16.187.28 B) .... *Procambarus* (*Girardiella*) *kensleyi* Hobbs, 1990

[USA: Texas]

15' Gonopod central projection distinctly spaced from caudal element lamellar crest; in mesal view lamellar crest bisected by mesal process, or nearly so (Pl. 16.187.28 E) .... *Procambarus* (*Girardiella*) *machardyi* Walls, 2006

[USA: Louisiana]

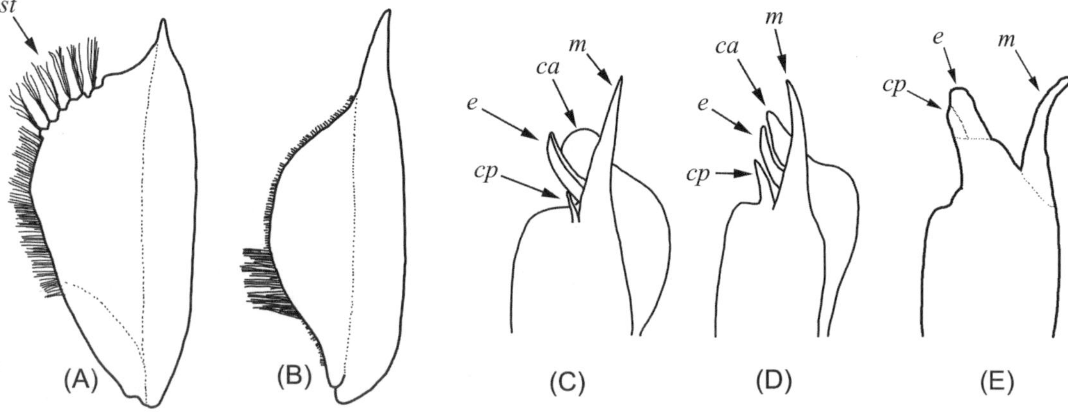

PLATE 16.187.29 (A, B) Antennal scale; (C–E) mesial view of distal end of first pleopod: (A) *Procambarus (Girardiella) hagenianus vesticeps*; (B) *P. (Girardiella) barbiger*; (C) *P. (Girardiella) simulans*; (D) *P. (Girardiella) parasimulans*. st, setal tufts; cp, cephalic process; e, central projection; ca, caudal process; m, mesial process.

16(1) Chela palm mesal margin with obvious setiferous beard (Pl. 16.187.27 B–E) 17

16' Chela palm mesal margin lacking setiferous beard (Pl. 16.187.27 A) 21

17(16) Chela dactyl lateral margin basal half with one or more tufts of long setae (Pl. 16.187.27 E) 18

17' Chela dactyl lateral margin lacking tufts of long setae 19

18(17) Chela dactyl setal tufts scanty (Pl. 16.187.27 D); gonopod central projection base enveloped by a tight fold (Pl. 16.187.28 H) .... *Procambarus* (*Girardiella*) *connus* Fitzpatrick, 1978

[USA: Mississippi]

18' Chela dactyl setal tufts well-developed (Pl. 16.187.27 E); gonopod central projection base lacking enveloping fold (Pl. 16.187.28 I) .... *Procambarus* (*Girardiella*) *pogum* Fitzpatrick, 1978

[USA: Mississippi]

19(17) Gonopod mesal process longer than central projection; gonopod central projection base enveloped by a tight fold (Pl. 16.187.29 C–D) 20

19' Gonopod mesal process subequal in length to central projection; gonopod central projection base lacking enveloping fold (Pl. 16.187.28 G) .... *Procambarus* (*Girardiella*) *cometes* Fitzpatrick, 1978

[USA: Mississippi]

20(19) Gonopod caudal process with distal margin subperpendicular to main pleopod axis (Pl. 16.187.24 H); antennal scale with lamellar portion setae arising in tufts from tubercular eminences (Pl. 16.187.29 A) .... *Procambarus* (*Girardiella*) *hagenianus vesticeps* Fitzpatrick, 1978

[USA: Mississippi. Intergrades with the nominate subspecies to the south]

20' Gonopod caudal process with distal margin not subperpendicular to main pleopod axis (Pl. 16.187.28 F); antennal scale with lamellar portion setae arising individually (Pl. 16.187.29 B) .... *Procambarus* (*Girardiella*) *barbiger* Fitzpatrick, 1978

[USA: Mississippi]

21(16) Gonopod caudal process in lateral view rhomboidal in outline, distally with an elongated straight edge (Pl. 16.187.24 D "*ca*") .... *Procambarus* (*Girardiella*) *hagenianus hagenianus* (Faxon, 1884)

21' Gonopod caudal process in lateral view triangular in outline, distally coming to a sharp point (Pl. 16.187.29 E "*ca*", see also Pl. 16.187.28 J) .... *Procambarus* (*Girardiella*) *holifieldi* Schuster et al., 2015

Crustacea: Malacostraca: Decapoda: Astacidea: Cambaridae: *Procambarus*: *Lonnbergius*: Species

Both species are subterranean.

1 Gonopod mesal process absent (Pl. 16.187.30 B); cephalic process well-developed; pereopods III and IV ischia never with clearly defined bituberculate hooks (Pl. 16.187.30 C) .... *Procambarus (Lonnbergius) morrisi* Hobbs & Franz, 1991

[USA: Florida]

1' Mesal process present but of variable length, ranging from falling short of, to almost reaching, distal margin of appendage shaft (Pl. 16.187.30 A "*m*"); cephalic process rudimentary or absent; pereopods III and IV ischia with hooks bituberculate (Pl. 16.187.27 D "*bt*") .... *Procambarus (Lonnbergius) archerontis* (Lönnberg, 1894)

[USA: Florida]

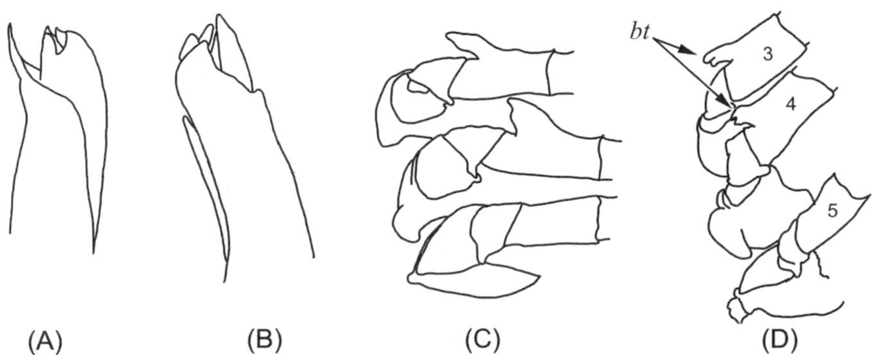

PLATE 16.187.30 (A, B) Lateral view of distal portion of left first pleopods; (C, D) ventral view of basal podomeres of left third, fourth and fifth pereiopods. (A, C), *Procambarus (Lonnbergius) archerontis*; (B, D), *P. (Lonnbergius) morrisi*. m, mesial process; cp, cephalic process; bt, bituberculate hooks.

Crustacea: Malacostraca: Decapoda: Astacidea: Cambaridae: *Procambarus*: *Ortmannicus*: Species

1 Albinistic, eyes without pigment or with small pigment spot, subterranean 2

1' Body pigmented, eyes well-developed and pigmented 12

2(1) Eyes with small pigment spot 3

2' Eyes without pigment 6

3(2) Gonopod cephalic process situated cephalic to and partly hooding central projection (Pl. 16.187.31 B–F) 4

3' Gonopod cephalic process situated lateral to central projection (Pl. 16.187.31 A) .... *Procambarus* (*Ortmannicus*) *orcinus* Hobbs & Means, 1972

[USA: Florida]

4(3) Gonopod with small caudal process (Pl. 16.187.31 C); pigment spot on eye black .... *Procambarus* (*Ortmannicus*) *lucifugus alachua* (Hobbs, 1940)

[USA: Florida. Intergrades with the nominate subspecies]

4' Gonopod lacking caudal process (Pl. 16.187.31 B, F); pigment spot on eye black or red 5

5(4) Rostral margins convex laterally; pigment spot on eye red; antennal scale less than twice as long as broad (see also Pl. 16.187.31 B) .... *Procambarus* (*Ortmannicus*) *erythrops* Relyea & Sutton, 1975

[USA: Florida]

5' Rostral margins subparallel basally and convergent distally; pigment spot on eye black; antennal scale at least twice as long as broad (see also Pl. 16.187.31 F) .... *Procambarus* (*Ortmannicus*) *leitheuseri* Franz & Hobbs, 1983

[USA: Florida]

6(2) Branchiostegite anteroventral area without distinct bulge (Pl. 16.187.32 B) 7

6' Branchiostegite anteroventral area with distinct bulge (Pl. 16.187.32 A, note arrow) .... *Procambarus* (*Ortmannicus*) *delicatus* Hobbs & Franz, 1986

[USA: Florida]

7(6) Maxilliped III not enlarged, ischium opposable margin smooth (Pl. 16.187.33 F), distolateral extremity without strong spine 8

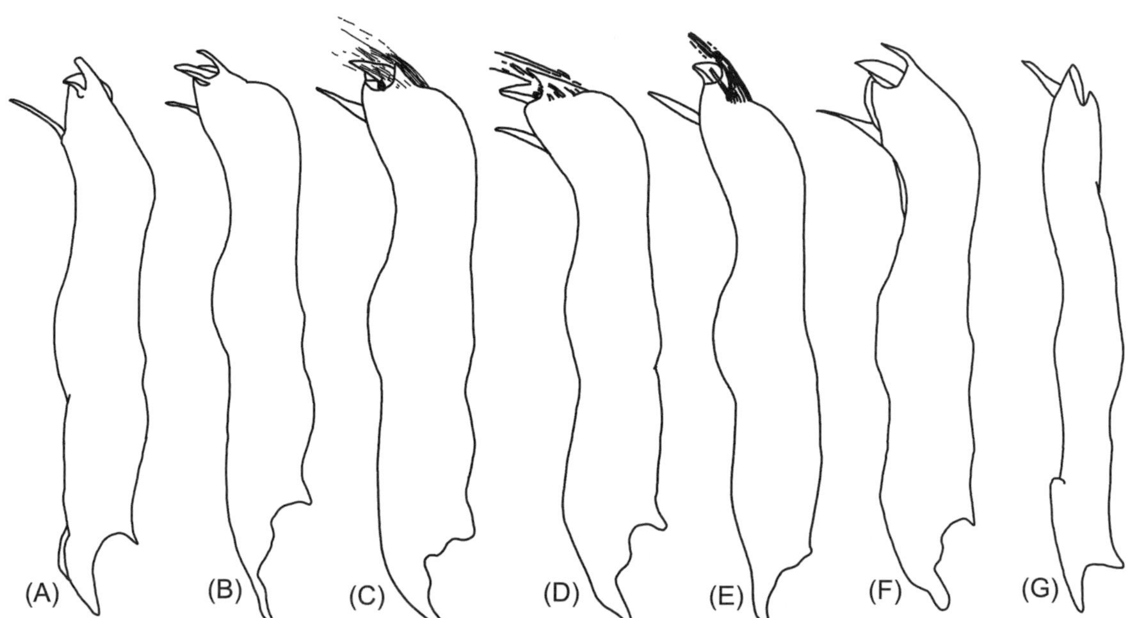

PLATE 16.187.31 Lateral view of left first pleopods: (A) *Procambarus (Ortmannicus) orcinus*; (B) *P. (Ortmannicus) erythrops*; (C) *P. (Ortmannicus) lucifugus alachua*; (D) *P. (Ortmannicus) lucifugus lucifugus*; (E) *P. (Ortmannicus) pallidus*; (F) *P. (Ortmannicus) leitheuseri*; (G) *P. (Ortmannicus) delicatus*.

7' Maxilliped III enlarged, ischium opposable margin with 7–9 denticles and as many as 8 small tubercles or spines laterally (Pl. 16.187.33 E, see also 33 A), distolateral extremity with spine .... *Procambarus* (*Ortmannicus*) *attiguus* Hobbs & Franz, 1992

[USA: Florida]

8(7) Gonopod caudal process consisting of two parts, a remnant or vestigial caudal knob and a prominent adventitious process (Pl. 16.187.33 C) 9

8' Gonopod caudal process well-developed, adventitious process absent (Pl. 16.187.33 B) .... *Procambarus* (*Ortmannicus*) *xilitlae* Hobbs & Grubbs, 1982

[Mexico: San Luis Potosí]

9(8) Gonopod cephalic process lying cephalic to central projection (like Pl. 16.187.33 C) 10

9' Gonopod cephalic process lying entirely or mostly lateral to central projection (like Pl. 16.187.33 D) 11

10(9) Areola less than 18 times as long as broad (see also Pl. 16.187.33 C) .... *Procambarus* (*Ortmannicus*) *franzi* Hobbs & Lee, 1976

[USA: Florida]

10' Areola more than 18 times as long as broad (see also Pl. 16.187.31 D) .... *Procambarus* (*Ortmannicus*) *lucifugus lucifugus* (Hobbs, 1940)

[USA: Florida. Intergrades with *P. lucifugus alachua*]

11(9) Areola less than 20 times as long as broad; postorbital ridge with spines or tubercles posteriorly (Pl. 16.187.32 C); gonopod cephalic process situated lateral to central projection (Pl. 16.187.33 D) .... *Procambarus* (*Ortmannicus*) *horsti* Hobbs & Means, 1972

[USA: Florida]

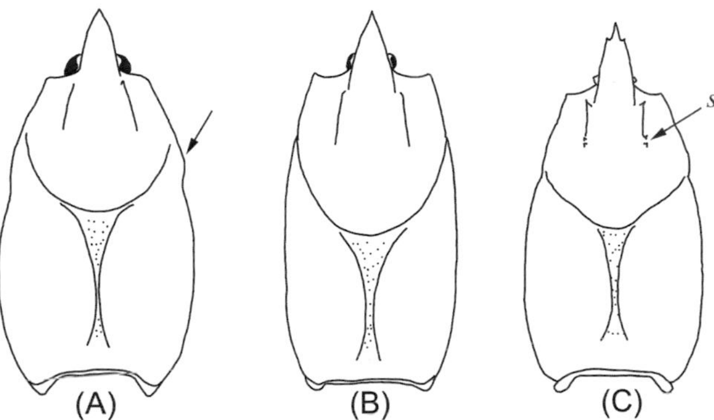

PLATE 16.187.32 Dorsal view of carapaces: (A) *Procambarus (Ortmannicus) delicatus*; (B) *P. (Ortmannicus) attiguus*; (C) *P. (Ortmannicus) horsti*. s, spines.

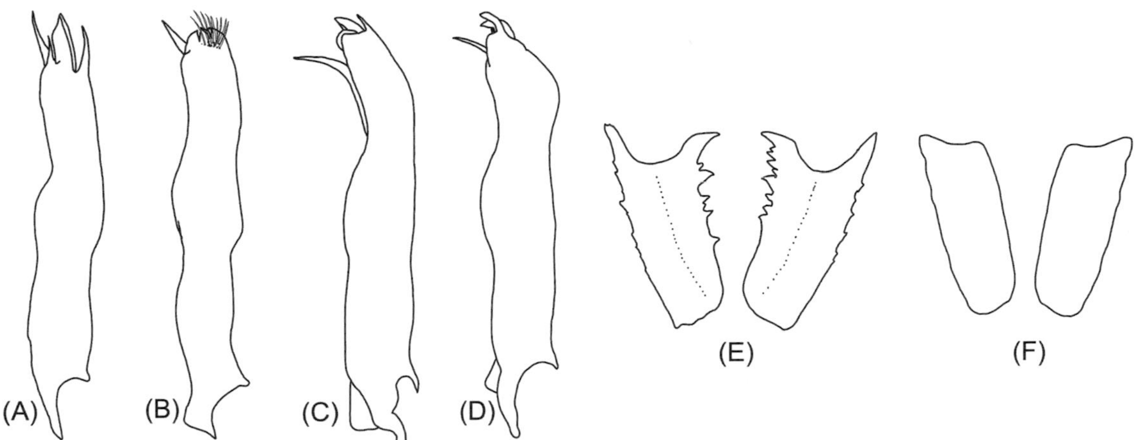

PLATE 16.187.33 (A–D) Lateral view of left first pleopods; (E, F) ventral view of ischium of third maxillipeds: (A) *Procambarus (Ortmannicus) attiguus*; (B) *P. (Ortmannicus) xilitlae*; (C) *P. (Ortmannicus) franzi*; (D) *P. (Ortmannicus) horsti*; (E) *P. (Ortmannicus) attiguus*; (F) *P. (Ortmannicus) pallidus*.

11' Areola more than 20 times as long as broad; postorbital ridge without spines or tubercles posteriorly; gonopod cephalic process situated cephalic to and partly hooding central projection (Pl. 16.187.31 E, see also 33 F) .... *Procambarus* (*Ortmannicus*) *pallidus* (Hobbs, 1940)

[USA: Florida]

12(1) Gonopod cephalodistal or laterodistal surface with subapical setae borne on knob (Pls. 16.187.34 A–H, 36 A) 13

12' Gonopod subapical setae, if present, never borne on distinct knob (Pls. 16.187.36 B–H, 23, 38, 40) 25

13(12) Gonopod cephalodistal surface bearing a setiferous knob usually laterally situated; if cephalic to cephalic process, then never widely separated from it (Pls. 16.187.34 B–H, 36 A) 14

13' Gonopod cephalodistal surface bearing a setiferous knob with broad gap between it and cephalic process (Pl. 16.187.34 A) .... *Procambarus* (*Ortmannicus*) *viaeviridis* (Faxon, 1914)

[USA: Alabama, Arkansas, Mississippi]

14(13) Gonopod cephalodistal surface bearing a setiferous knob situated cephalically or laterally but never so far caudally as at base of caudal process (Pls. 16.187.34 C–H, 36 A) 15

14' Gonopod cephalodistal surface bearing a setiferous knob situated at lateral base of caudal process (Pl. 16.187.34 B "*ck*") .... *Procambarus* (*Ortmannicus*) *hayi* (Faxon, 1884)

[USA: Alabama, Mississippi, Tennessee]

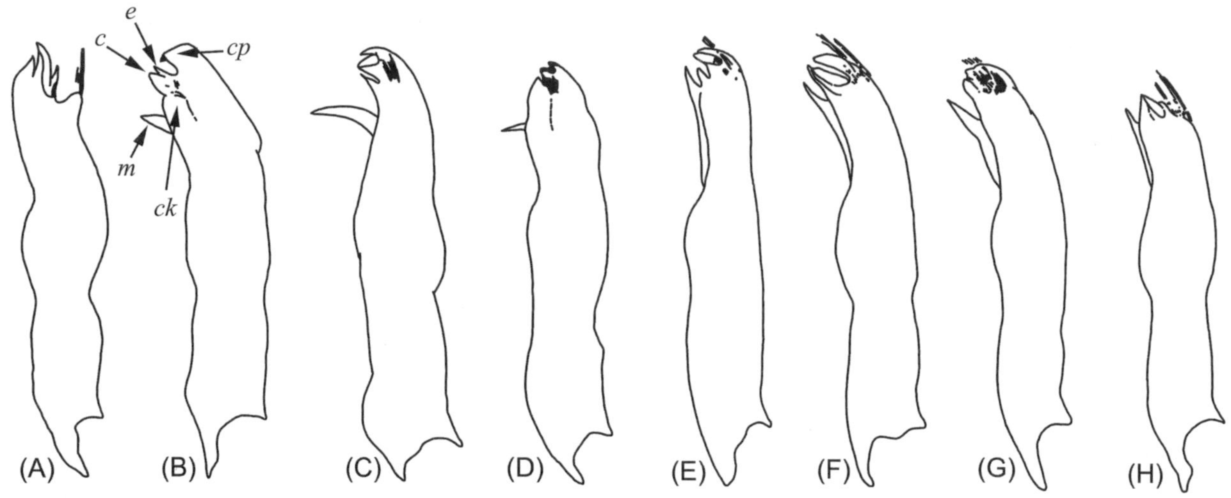

PLATE 16.187.34 Lateral view of left first pleopods: (A) *Procambarus (Ortmannicus) viaeviridis*; (B) P. *(Ortmannicus) hayi*; (C) *P. (Ortmannicus) lecontei*; (D) *P. (Ortmannicus) acutissimus*; (E) *P. (Ortmannicus) texanus*; (F) *P. (Ortmannicus) cuevachicae*; (G) *P. (Ortmannicus) lophotus*; (H) *P. (Ortmannicus) blandingii*. ck, caudal knob; cp, cephalic process; (E) central projection; m, mesial process; (C) caudal process.

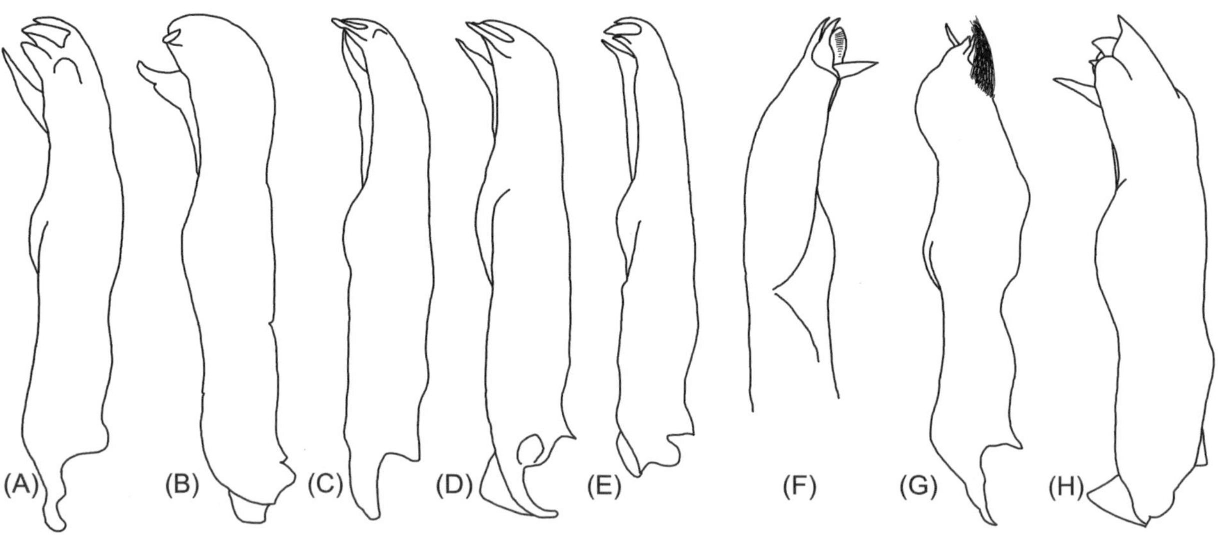

PLATE 16.187.35 Lateral view of left first pleopods: (A) *Procambarus (Ortmannicus) nechesae*; (B) *P. (Ortmannicus) geminus*; (C) *P. (Ortmannicus) zonangulus*; (D) *P. (Ortmannicus) nueces*; (E) *P. (Ortmannicus) luxus*; (F) *P. (Ortmannicus) hidalgoensis*; (G) *P. (Ortmannicus) marthae*; (H) *P. (Ortmannicus) braswelli.*

15(14) Gonopod mesal process directed at 90 degree angle to appendage main axis (Pl. 16.187.34 C–D), except in *P.* (*O.*) *nechesae*, where directed at angle of 45 degrees 16

15' Gonopod mesal process directed at angle less than 90 degrees to appendage main axis (Pls. 16.187.34 E–H, 36 A) 19

16(15) Gonopod terminal elements curved at angle of 80–90 degrees relative to main axis of appendage; pereopod IV basis without a tubercle opposing corresponding ischial hook; rostrum with tapering margins 17

16' Gonopod terminal elements curved at angle of 40–50 degrees relative to appendage main axis (Pl. 16.187.35 A); pereopod IV basis bearing a strong tubercle which opposes corresponding ischial hook; rostrum with strongly tapering margins .... *Procambarus* (*Ortmannicus*) *nechesae* Hobbs, 1990

[USA: Texas]

17(16) Gonopod cephalic surface lacking prominent rounded bulge proximal to the mesal process base (Pl. 16.187.34 C, D) 18

17' Gonopod cephalic surface bearing a prominent, rounded bulge proximal to the mesal process base (Pl. 16.187.35 B) .... *Procambarus* (*Ortmannicus*) *geminus* Hobbs, 1975

[USA: Arkansas, Louisiana]

18(17) Gonopod cephalic process and central projection directed caudally at 90 degree angle to appendage main axis (Pl. 16.187.34 C; see also Pl. 16.187.7 C) .... *Procambarus* (*Ortmannicus*) *lecontei* (Hagen, 1870)

[USA: Alabama, Mississippi]

18' Gonopod cephalic process and central projection directed caudodistally at angle much less than 90 degrees to appendage main axis (Pl. 16.187.34 D; see also Pl. 16.187.2 B) .... *Procambarus* (*Ortmannicus*) *acutissimus* (Girard, 1852)

[USA: Alabama, Mississippi]

19(15) Gonopod bases symmetrical (Pl. 16.187.2 C, D) 20

19' Gonopod bases asymmetrical (Pl. 16.187.2 A, B) 21

20(19) Gonopod conspicuously tapering distally (Pls. 16.187.35 C, 22 B "*w*"), cephalomesial margin with well-developed rounded hump projecting cephalomesially (Pl. 16.187.22 A "*h*"); rostrum with or without minute marginal tubercles .... *Procambarus* (*Ortmannicus*) *zonangulus* Hobbs & Hobbs, 1990

[USA: Texas. Widely introduced]

20' Gonopod not conspicuously tapering distally (Pls. 16.187.35 D, 22 D "*w*"), lacking any trace of shoulder on the cephalic or cephalomesial surface proximal to terminal elements (Pl. 16.187.22 C, note arrow); rostrum with marginal spines well-developed .... *Procambarus* (*Ortmannicus*) *nueces* Hobbs & Hobbs, 1995

[USA: Texas]

21(19) Gonopod with an exceedingly narrow gap between caudal process and central projection (Pls. 187.34 F–H, 36 A) 22

21' Gonopod with a prominent gap between caudal process and central projection (Pl. 16.187.34 E) .... *Procambarus* (*Ortmannicus*) *texanus* Hobbs, 1971

[USA: Texas]

22(21) Gonopod cephalic process, central projection and caudal process not strongly recurved, with apices of cephalic process and central projection directed at angle less than 90 degrees to appendage main axis (Pl. 16.187.34 F–H) 23

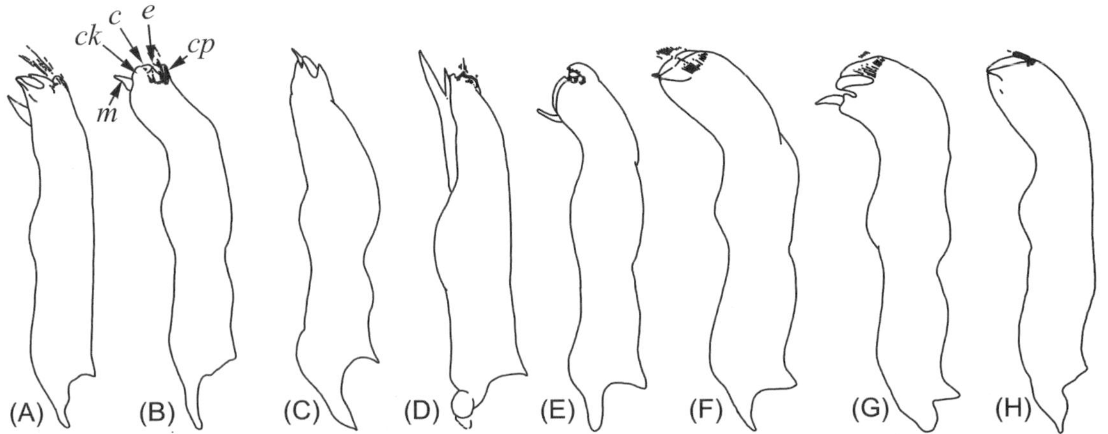

PLATE 16.187.36 Lateral view of left first pleopods: (A) *Procambarus (Ortmannicus) acutus*; (B) *P. (Ortmannicus) bivittatus*; (C) *P. (Ortmannicus) lewisi*; (D) *P. (Ortmannicus) villalobosi*; (E) *P. (Ortmannicus) gonopodocristatus*; (F) *P. (Ortmannicus) mancus*; (G) *P. (Ortmannicus) pearsei*; *P. (Ortmannicus) planirostris*. c, caudal process; ck, caudal knob; cp, cephalic process; e, central projection; m, mesial process.

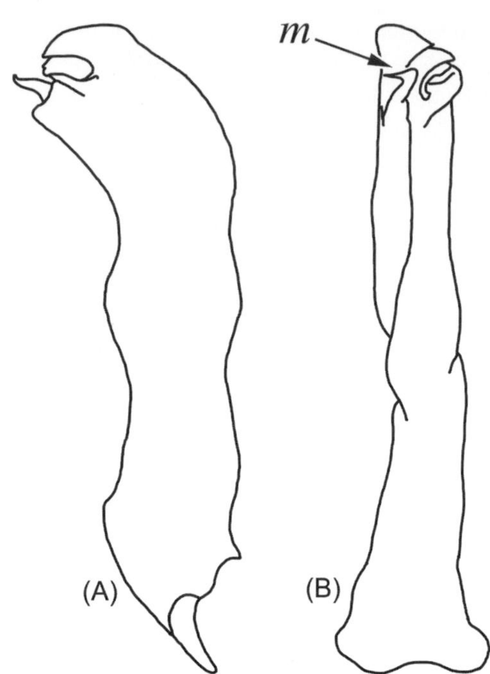

PLATE 16.187.37 *Procambarus (Ortmannicus) medialis*: (A) lateral view of first pleopod; (B) caudal view of same. m, mesial process.

22' Gonopod cephalic process, central projection and caudal process strongly recurved, with apices of cephalic process and central projection directed at angle of 90 degrees to appendage main axis (Pl. 16.187.37 E) .... *Procambarus* (*Ortmannicus*) *luxus* Johnson, 2011

[USA: Texas]

23(22) Gonopod caudal process tapering from base in lateral aspect (Pls. 16.187.34 H, 36 A) 24

23' Gonopod caudal process lanceolate in lateral aspect (Pl. 16.187.34 G) .... *Procambarus* (*Ortmannicus*) *lophotus* Hobbs & Walton, 1960

[USA: Alabama, Georgia, Tennessee]

24(23) Gonopod with cephalic process base bearing a setiferous knob; setae not obscuring part of central projection in lateral aspect (Pl. 16.187.34 H) .... *Procambarus* (*Ortmannicus*) *blandingii* (Harlan, 1830)

[USA: North Carolina, South Carolina]

24' Gonopod cephalic process with setiferous knob situated cephalolaterally; setae obscuring at least proximal 1/2 of central projection in lateral aspect (Pl. 16.187.36 A) .... *Procambarus* (*Ortmannicus*) *acutus* (Girard, 1852)

[USA: eastern states]

25(12) Gonopod with subapical setae absent (Pls. 16.187.36 C, 23 F) 26

25' Gonopod subapical setae present (Pl. 16.187.36 B, D, H) 27

26(25) Gonopod cephalic process directed distally (Pl. 16.187.36 C; see also Pl. 16.187.39 B) .... *Procambarus* (*Ortmannicus*) *lewisi* Hobbs & Walton, 1959

[USA: Alabama]

26' Gonopod cephalic process directed caudally (Pl. 16.187.23 F) .... *Procambarus* (*Ortmannicus*) *hybus* Hobbs & Walton, 1957

[USA: Alabama, Mississippi]

27(25) Gonopod caudal knob not distinct, OR if distinct, directed distally or caudodistally, shorter, never subequal in length to caudal process and central projection (Pls. 16.187.36 D–H, 23 A–E, G, H, 38, 40) 28

27' Gonopod caudal knob distinct and subequal in length as caudal process and central projection (Pl. 16.187.36 B; see also 39 A) .... *Procambarus* (*Ortmannicus*) *bivittatus* Hobbs, 1942

[USA: Florida, Louisiana]

28(27) Gonopod caudal process prominent, usually laterally compressed, arising from pleopod caudolateral surface; caudal knob never well-defined; cephalic process absent, or if present, arising from cephalic or cephalomesal side of central projection (Pls. 16.187.36 D–H, 23 A–E) 29

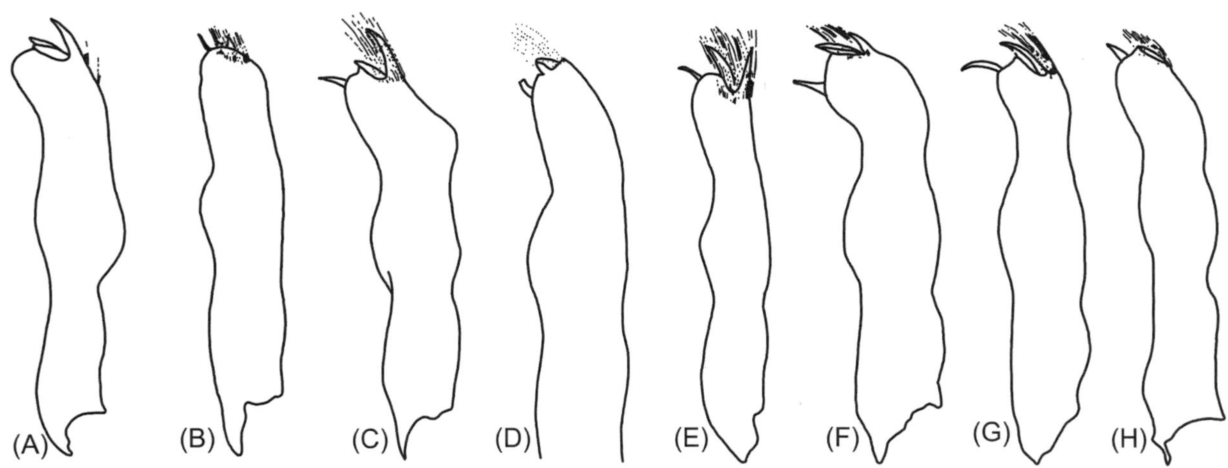

PLATE 16.187.38 Lateral view of left first pleopods: (A) *Procambarus (Ortmannicus) youngi*; (B) *P. (Ortmannicus) pycnogonopodus*; (C) *P. (Ortmannicus) hirsutus*; (D) *P. (Ortmannicus) angustatus*; (E) *P. (Ortmannicus) seminolae*; (F) *P. (Ortmannicus) lunzi*; (G) *P. (Ortmannicus) ancylus*; (H) *P. (Ortmannicus) fallax*.

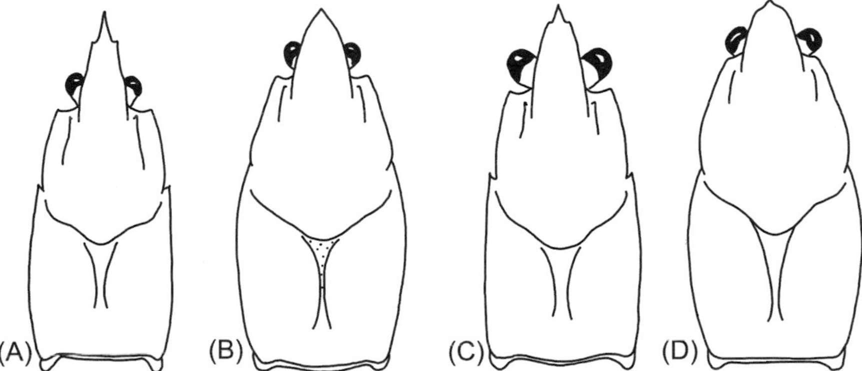

PLATE 16.187.39 Dorsal view of carapaces: (A) *Procambarus (Ortmannicus) bivittatus*; (B) *P. (Ortmannicus) lewisi*; (C) *P. (Ortmannicus) verrucosus*; (D) *P. (Ortmannicus) jaculus*.

28' Gonopod caudal process seldom prominent, sometimes absent, if present, arising distinctly mesal to caudal knob except in *P. lepidodactylus* and *P. toltecae* in which the cephalic process is situated lateral to central projection; cephalic process arising from central projection cephalic or lateral side (Pls. 16.187.23 G, H, 38, 40) 37

29(28) Gonopod cephalic process absent (Pl. 16.187.36 F) 30

29' Gonopod cephalic process present (Pls. 16.187.36 G, H, 23 A–E) 31

30(29) Gonopod mesal process and central projection directed strongly caudally (Pl. 16.187.36 F) .... *Procambarus (Ortmannicus) mancus* Hobbs & Walton, 1957

[USA: Mississippi]

30' Gonopod mesal process and central projection directed caudodistally (and somewhat laterally) and caudally, respectively (Pl. 16.187.35 G) .... *Procambarus (Ortmannicus) marthae* Hobbs, 1975

[USA: Alabama]

31(29) Gonopod central projection directly caudally at 90 degree angle to appendage main axis (Pl. 16.187.36 G, H) 32

31' Gonopod central projection never directed caudally so much as at 90degree angle to appendage main axis (Pl. 16.187.23 A–E) 34

32(31) Gonopod mesal process spiniform and directed caudodistad 33

32' Gonopod mesal process slender with distal fourth bent mesally at right angle to main shaft of the process (Pl. 16.187.22 B) .... *Procambarus (Ortmannicus) medialis* Hobbs, 1975

[USA: North Carolina]

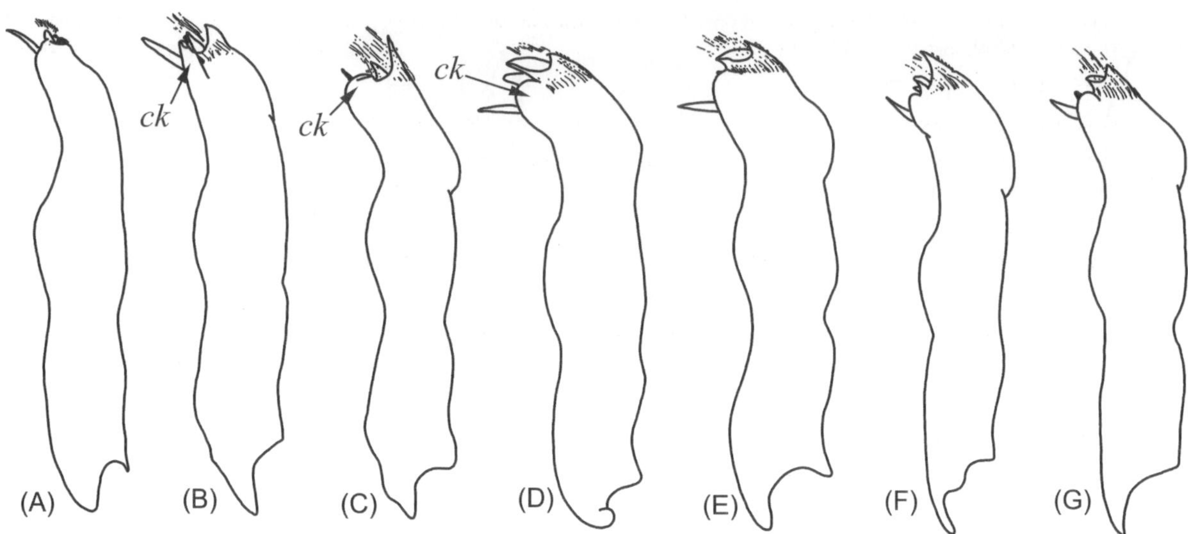

PLATE 16.187.40 Lateral view of left first pleopod: (A) *Procambarus (Ortmannicus) leonensis*; (B) *P. (Ortmannicus) litosternum*; (C) *P. (Ortmannicus) pubescens*; (D) *P. (Ortmannicus) epicyrtus*; (E) *P. (Ortmannicus) enoplosternum*; (F) *P. (Ortmannicus) pictus*; (G) *P. (Ortmannicus) chacei*. ck, caudal knob.

33(32) Gonopod with all terminal elements directed caudad at 90 degree angle to appendage main axis, and cephalic process extending as far caudad as caudal process (Pl. 16.187.36 G) .... *Procambarus* (*Ortmannicus*) *pearsei* (Creaser, 1934)

[USA: North Carolina, South Carolina]

33' Gonopod mesal process directed caudodistally and cephalic process never extending so far caudad as caudal process (Pl. 16.187.36 H) .... *Procambarus* (*Ortmannicus*) *planirostris* Penn, 1953

[USA: Louisiana, Mississippi]

34(31) Rostrum with marginal spines or tubercles, or at least angulate at base of acumen (Pl. 16.187.39 C) 35

34' Rostrum with acumen not distinctly delimited basally (Pl. 16.187.39 D) 36

35(34) Gonopod cephalodistal surface with long rounded hump extending proximally from cephalic process base; cephalic process cephalic area with subapical setae (Pl. 16.187.23 A; see also 39 C) .... *Procambarus* (*Ortmannicus*) *verrucosus* Hobbs, 1952

[USA: Alabama]

35' Gonopod cephalodistal surface not produced in long rounded hump; lateral bases of cephalic process and central projection with subapical setae (Pl. 16.187.23 B) .... *Procambarus* (*Ortmannicus*) *evermanni* (Faxon, 1890)

[USA: Florida, Mississippi]

36(34) Gonopod mesal process gently tapering and directed caudally; cephalic process situated cephalomesal to central projection; caudal process subtruncate (Pl. 16.187.22 D) .... *Procambarus* (*Ortmannicus*) *plumimanus* Hobbs & Walton, 1958

[USA: North Carolina]

36' Gonopod mesal process lanceolate, directed caudodistally; cephalic process situated cephalic to central projection; caudal process rounded apically (Pl. 16.187.22 E; see also 39 D) .... *Procambarus* (*Ortmannicus*) *jaculus* Hobbs & Walton, 1957

[USA: Louisiana, Mississippi]

37(28) Gonopod cephalic process situated cephalic to central projection (Pls. 16.187.38, 40) 38

37' Gonopod cephalic process situated distinctly lateral to central projection (Pl. 16.187.23 G, H) .... *Procambarus* (*Ortmannicus*) *lepidodactylus* Hobbs, 1947

[USA: North Carolina, South Carolina]

38(37) Gonopod caudal process absent or not evident in lateral aspect (Pl. 16.187.38 A–G) 39

38' Gonopod caudal process small to large, always evident in lateral aspect (Pls. 16.187.38 H, 40) 46

39(38) Gonopod cephalic and lateral surfaces with abundant subapical setae (Pl. 16.187.38 B–G); acumen, if distinct, much shorter than remainder of rostrum 40

39' Gonopod cephalic process with very few subapical setae, restricted to base of cephalodistal margin (Pl. 16.187.38 A); acumen as long as remainder of rostrum (see Pl. 16.187.7 B) .... *Procambarus* (*Ortmannicus*) *youngi* Hobbs, 1942

[USA: Florida]

40(39) Gonopod central projection always conspicuous, frequently as large as other terminal elements (Pl. 16.187.38 C–G) 41

40' Gonopod central projection minute, much less conspicuous than other terminal elements (Pl. 16.187.38 B) .... *Procambarus* (*Ortmannicus*) *pycnogonopodus* Hobbs, 1942

[USA: Florida]

41(40) Areola never more than 4 times as long as broad 42

41' Areola always more than 4 times as long as broad 43

42(41) Gonopod mesal process much longer than central projection (Pl. 16.187.38 C; see also 7 D) .... *Procambarus* (*Ortmannicus*) *hirsutus* Hobbs, 1958

[USA: South Carolina]

42' Gonopod mesal process shorter than central projection (Pl. 16.187.38 D) .... *Procambarus* (*Ortmannicus*) *angustatus* (LeConte, 1856)

[USA: "Georgia inferiore" (Known only from the single type specimen)]

43(41) Gonopod cephalic process directed caudodistally (Pls. 16.187.38 F, G, 40 A) 44

43' Gonopod cephalic process directed distally or cephalodistally (Pl. 16.187.23 E; see also 7 E) .... *Procambarus* (*Ortmannicus*) *seminolae* Hobbs, 1942

[USA: Florida, Georgia]

44(43) Gonopod central projection somewhat blade-like, curved, and as long as cephalic process (Pl. 16.187.38 F, G) 45

44' Gonopod central projection dentiform, almost straight, and distinctly shorter than cephalic process (Pl. 16.187.40 A) .... *Procambarus* (*Ortmannicus*) *leonensis* Hobbs, 1942

[USA: Florida]

45(44) Gonopod central projection laterodistal margin at base almost horizontal; central projection directed caudally (Pl. 16.187.38 F) .... *Procambarus* (*Ortmannicus*) *lunzi* Hobbs, 1940

[USA: Georgia, South Carolina]

45' Gonopod central projection laterodistal margin steeply oblique at base; central projection directed caudodistally (Pl. 16.187.38 G) .... *Procambarus* (*Ortmannicus*) *ancylus* Hobbs, 1958

[USA: North Carolina, South Carolina]

46(38) Gonopod caudal knob not well defined (Pls. 16.187.38 H "*ck*", 40 A); areola usually more than 5 times longer than broad 47

46' Gonopod caudal knob well defined (Pl. 16.187.40 B–G "*ck*"); areola usually less than 5 times longer than broad 48

47(46) Gonopod mesal process lanceolate; caudal process situated lateral to central projection (Pl. 16.187.38 H) .... *Procambarus* (*Ortmannicus*) *fallax* (Hagen, 1870)

[USA: Florida, Georgia]

47' Gonopod mesal process subspiculiform; caudal process situated caudal to central projection (Pl. 16.187.40 A) .... *Procambarus* (*Ortmannicus*) *leonensis* Hobbs, 1942

[USA: Florida]

48(46) Gonopod caudal knob cephalically with deep groove (Pl. 16.187.40 B) 49

48' Gonopod caudal knob inflated or truncate but never with a deep groove cephalically (Pl. 16.187.40 C–G) 50

49(48) Gonopod cephalic process spiniform, directed caudodistad (Pl. 16.187.40 F) .... *Procambarus* (*Ortmannicus*) *litosternum* Hobbs, 1947

[USA: Georgia]

49' Gonopod cephalic process distally directed, with caudal base transversely expanded to form cowl or hood around cephalic base of central projection (Pl. 16.187.35 H) .... *Procambarus* (*Ortmannicus*) *braswelli* Cooper, 1998

[USA: North Carolina, South Carolina]

50(48) Gonopod caudal knob inflated (Pl. 16.187.40 D–G) 51

50' Gonopod caudal knob truncate and somewhat compressed (Pl. 16.187.40 C) .... *Procambarus* (*Ortmannicus*) *pubescens* (Faxon, 1884)

[USA: Georgia, South Carolina]

51(50) Gonopod cephalic process much shorter than central projection (Pl. 16.187.40 D, E) 52

51' Gonopod cephalic process as long as central projection (Pl. 16.187.40 F, G) 53

52(51) Gonopod caudal process in lateral aspect projecting caudally between central projection and caudal knob, filling interval between them (Pl. 16.187.40 D) .... *Procambarus* (*Ortmannicus*) *epicyrtus* Hobbs, 1958

[USA: Georgia]

52' Gonopod caudal process in lateral aspect projecting caudally from level of caudal knob (Pl. 16.187.39 E) .... *Procambarus* (*Ortmannicus*) *enoplosternum* Hobbs, 1947

[USA: Georgia]

53(51) Gonopod caudal process in lateral aspect projecting caudally between central projection and caudal knob, filling interval between them (Pl. 16.187.40 F) .... *Procambarus* (*Ortmannicus*) *pictus* (Hobbs, 1940)

[USA: Florida]

53' Gonopod caudal process in lateral aspect projecting caudally from level of caudal knob (Pl. 16.187.40 G) .... *Procambarus* (*Ortmannicus*) *chacei* Hobbs, 1958

[USA: Georgia, South Carolina]

Crustacea: Malacostraca: Decapoda: Astacidea: Cambaridae: *Faxonella*: Species

D. Christopher Rogers

Kansas Biological Survey, University of Kansas, Lawrence, KS, USA; The Biodiversity Institute, University of Kansas, Lawrence, KS, USA

Only Form I males can be identified. Identification is based on gonopod, which must be viewed *in situ*, in caudal (posterior) view.

1 Gonopod extending to pereopod I coxa 2

1' Gonopod not reaching pereopod I coxa 3

2(1) Gonopod central projection arcing medially (Pls. 16.187.41 C, D, 43 C) .... *Faxonella creaseri* Walls, 1968

[USA: Louisiana (Arkansas? Mississippi?)]

2' Gonopod central projection bent medially at base, remaining portion straight (Pl. 16.187.42 C, D) .... *Faxonella blairi* Hayes & Reimer, 1977

[USA: Arkansas, Oklahoma]

3(1) Gonopod mesal process less than one-fourth the length of the central projection; mesal processes never overlapping each other (Pls. 16.187.42 A, B, 43 A) .... *Faxonella clypeata* (Hay, 1899)

[USA: Alabama, Florida, Georgia, Louisiana, Mississippi, Oklahoma, South Carolina, Texas (Arkansas?)]

3' Gonopod mesal process one-third or more the length of the central projection; mesal processes overlapping each other (Pls. 16.187.41 A, B, 43 B) .... *Faxonella beyeri* (Penn, 1950)

[USA: Louisiana, Texas]

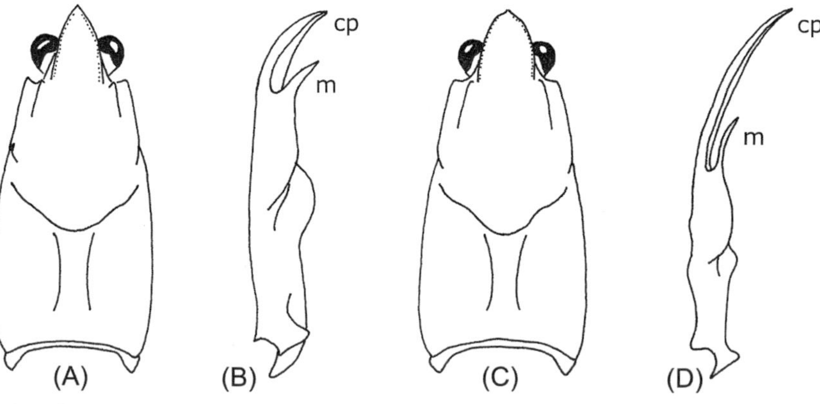

PLATE 16.187.41 Dorsal view of the carapace and mesial view of left first pleopod in: (A, B) *Faxonella beyeri*; and (C, D) *F. creaseri*. cp = central projection, m = mesial process. *Modified from Hobbs, 1989.*

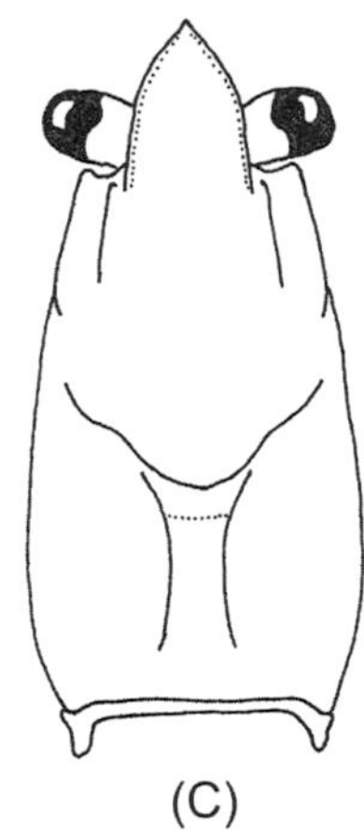

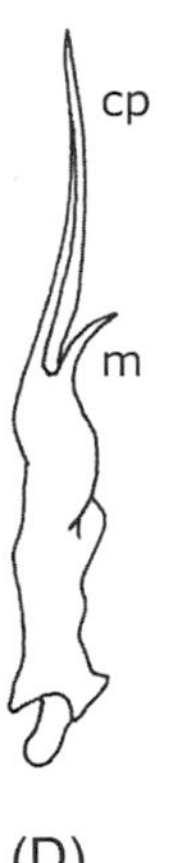

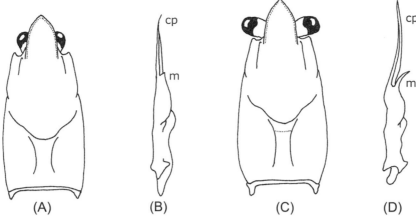

PLATE 16.187.42 Dorsal view of the carapace and mesial view of left first pleopod in: (A, B) *Faxonella clypeata*; and (C, D) *F. blairi*. cp = central projection, m = mesial process. *Modified from Hobbs, 1989.*

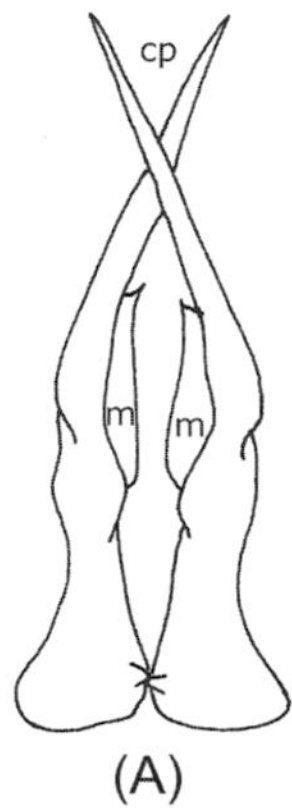

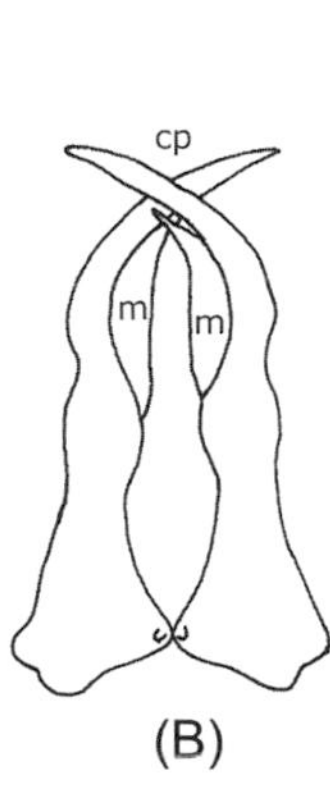

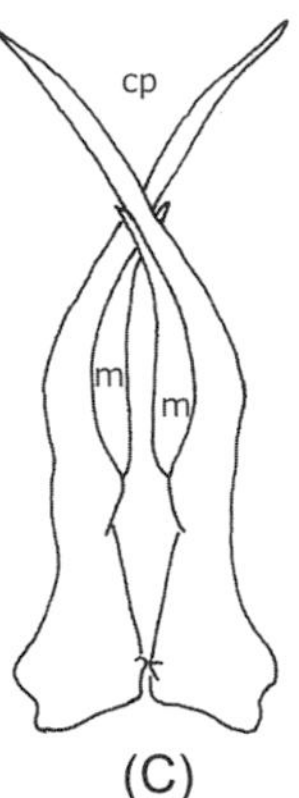

PLATE 16.187.43 Ventral views of first pleopods of: (A) *Faxonella clypeata*; (B) *F. beyeri*; and (C) *F. creaseri*. cp = central projection, m = mesial process. *After Hobbs, 1972b.*

Crustacea: Malacostraca: Decapoda: Astacidea: Cambaridae: *Orconectes*: Species

Christopher A. Taylor

Illinois Natural History Survey, Prairie Research Institute, University of Illinois at Urbana-Champaign, Champaign, IL, USA

This key is based on Form I males. Pl. 16.188.1(A, B) provides images of form I vs. form II gonopods. Pl. 16.188.1 (C, D, E) provides images of how to measure lengths of gonopod terminal elements.

1	Body lacking pigment, white in color; eyes unpigmented; subterranean	2
1'	Body with pigment, colors variable, ranging from brown to red, to cream, with or without dark stripes on chela tips and carapace; with pigmented eyes	10
2(1)	Gonopod central projection less than 10% of total gonopod length	3
2'	Gonopod central projection greater than 10% of total gonopod length (Pl. 16.188.2) [USA: Alabama]	*Orconectes sheltae* Cooper & Cooper, 1977
3(2)	Gonopod mesal process extending distinctly farther distally than central projection (Pl. 16.188.3 A)	4
3'	Gonopod mesal process extending only slightly, if at all, farther distally than central projection (Pl. 16.188.3 B)	5
4(3)	Gonopod mesal process recurved cephalically towards central projection (Pl. 16.188.4) [USA: Missouri]	*Orconectes stygocaneyi* Hobbs III, 2001

4' Gonopod mesal process directed distally and slightly caudally (Pl. 16.188.3 A) .... *Orconectes pellucidus* (Tellkampf, 1844)
[USA: Kentucky, Tennessee]

5(3) Not found in caves along the Cumberland Plateau in south central Wayne Co. Kentucky south to north central Fentress Co. Tennessee 6

5' Species diagnosed on basis of DNA in COI and 16S gene regions; found in caves along the Cumberland Plateau in south-central Wayne Co. Kentucky south to north-central Fentress Co. Tennessee .... *Orconectes barri* Buhay & Crandall, 2008
[USA: Kentucky, Tennessee]

6(5) Pereopods III and IV ischia bearing hooks (Pl. 16.188.5 A) 7

6' Pereopods III ischia bearing hooks only, hooks may be small or rudimentary (Pl. 16.188.5 B) 8

7(6) Rostrum without marginal spines or tubercles, areola constituting at least 43% of entire carapace length (Pl. 16.188.6 A) .... *Orconectes inermis testii* (Hay, 1891)
[USA: Indiana]

7' Rostrum with marginal spines or tubercles, areola constituting less than 43% of entire carapace length (Pl. 16.188.6 B) .... *Orconectes inermis inermis* Cope, 1872
[USA: Indiana, Kentucky]

8(6) Rostrum with marginal spines or tubercles; mesal process of gonopod slender (Pl. 16.188.6 B) 9

8' Rostrum lacking marginal spines or tubercles; gonopod mesal process broad, somewhat flattened (Pl. 16.188.6 A) .... *Orconectes incomptus* Hobbs & Barr, 1972
[USA: Tennessee]

9(8) Gonopod central projection cephalic base without rounded shoulder; caudal process usually present (Pl. 16.188.7 A) ....*Orconectes australis* (Rhoades, 1941)
[USA: Alabama, Kentucky, Tennessee]

9' Gonopod central projection with basal angular shoulder; caudal process absent (Pl. 16.188.7 B) .... *Orconectes packardi* Rhoades, 1944
[USA: Kentucky, Tennessee]

10(1) Chela dactyl and pollex longer, length of dactyl greater than or equal to length of palm (Pl. 16.188.8); acumen shorter than twice the width of the rostrum at marginal spines 11

10' Chela dactyl and pollex short, length of dactyl less than length of palm (Pl. 16.188.9); acumen very long, its length equal to or greater than twice the width of rostrum at marginal pines .... *Orconectes lancifer* (Hagen, 1870)
[USA: Alabama, Arkansas, Illinois, Kentucky, Louisiana, Missouri, Mississippi, Oklahoma, Tennessee, Texas]

11(10) Gonopods with elements greater than 11% of total gonopod length 12

11' Gonopod with extremely short elements, both central projection and mesal process less than 11% of total gonopod length (Pl. 16.188.10) .... *Orconectes pagei* Taylor & Sabaj, 1997
[USA: Tennessee]

12(11) Areola linear or obliterated along a portion of its length (Pl. 16.188.11 A) 13

12' Areola open, never linear or obliterated along a portion of its length (Pl. 16.188.11 B) 29

13(12) Gonopod central projection equal to or less than 26% of total gonopod length 14

13' Gonopod central projection greater than 26% of total gonopod length 21

14(13) Gonopod mesal process apex without flattened cephalic and caudal surfaces, or a groove on the cephalic surface 15

14' Gonopod mesal process apex with flattened cephalic and caudal surfaces and cephalic surface with a groove .... *Orconectes holti* Cooper & Hobbs, 1980
[USA: Alabama]

15(14) Gonopod mesal process equal to or longer than central projection 16

15' Gonopod mesal process shorter than central projection 18

16(15) Gonopod mesal process thin, gradually tapering to tip (Pl. 16.188.12 B) 17

16' Gonopod mesal process in lateral perspective wide, only narrowing at its blunted apex (Pl. 16.188.12 A) .... *Orconectes blacki* Walls, 1972
[USA: Louisiana]

17(16) Gonopod central projection apex angled less than 90° from gonopod main axis (Pl. 16.188.13 A) .... *Orconectes cyanodigitus* Johnson, 2010

[USA: Arkansas, Oklahoma, Texas]

17' Gonopod central projection apex angled 90° to gonopod main axis (Pl. 16.188.13 B) .... *Orconectes deanae* Reimer & Jester, 1975

[USA: New Mexico, Oklahoma]

18(15) Gonopod central projection gradually tapering to apex, only narrowing at apex 19

18' Gonopod central projection gradually tapering from base to acute apex (Pl. 16.188.12 B) .... *Orconectes difficilis* (Faxon, 1898)

[USA: Oklahoma]

19(18) Gonopod central projection caudodistal margin lacking a subapical notch or shallow depression 20

19' Gonopod central projection caudodistal margin with a subapical notch or shallow depression (Pl. 16.188.14) .... *Orconectes hartfieldi* Fitzpatrick & Suttkus, 1992

[USA: Mississippi]

20(19) Chela pollex flattened, with two rows of punctations, resulting in a distinct ridge .... *Orconectes hathawayi* Penn, 1952

[USA: Louisiana]

20' Chela pollex not flattened, with one distinct row of punctations, without a ridge .... *Orconectes perfectus* Walls, 1971

[USA: Alabama, Mississippi]

21(13) Gonopod central projection and/or mesal process with a portion of caudal edge straight or nearly straight (Pl. 16.188.15 A) 22

21' Gonopod central projection and/or mesal process with entire length curved (Pl. 16.188.15 B) 23

22(21) Chela palm and dactyl mesal margin with large, forward directed serrate tubercles (Pl. 16.188.16 A) .... *Orconectes mississippiensis* (Faxon, 1884)

[USA: Mississippi]

22' Chela palm and dactyl mesal margin with low, rounded tubercles (Pl. 16.188.16 B) .... *Orconectes validus* (Faxon, 1914) (in part)

[USA: Alabama]

23(21) Gonopod central projection greater than 32% of gonopod total length 24

23' Gonopod central projection 32% or less than total length of gonopod (Pl. 16.188.1 D) .... *Orconectes maleate* Walls, 1972

[USA: Louisiana]

24(23) Gonopod central projection constituting greater than 36% of total gonopod length OR carapace without semiprominent dark lateral longitudinal stripe 25

24' Gonopod central projection constituting less than 36% of gonopod total length; carapace with semiprominent dark lateral longitudinal stripe (Pl. 16.188.1 D) .... *Orconectes castaneus* Johnson, 2010

[USA: Texas]

25(24) Chela dactyl and pollex with apices obscurely orange to distinctly red, bordered proximally by cream band, gonopod processes with caudal offset 22–35% of main gonopod axis (Pl. 16.188.17 B) 26

25' Chela dactyl and pollex with apices red, bordered proximally by charcoal band, gonopod processes with caudal offset 20–24% of gonopod main axis (Pl. 16.188.17 A) .... *Orconectes palmeri longimanus* (Faxon, 1898)

[USA: Arkansas, Kansas, Louisiana, Oklahoma,Texas]

26(25) Distribution in Texas or in the Sabine River drainage of western Louisiana, dark lateral carapace stripe never present 27

26' Distribution not as above, dark lateral carapace stripe present or absent 28

27(26) Postorbital ridge bright red .... *Orconectes texanus* Johnson, 2010

[USA: Louisiana, Texas]

27' Postorbital ridge not bright red .... *Orconectes occidentalis* Johnson, 2010

[USA: Texas]

28(26) Gonopod central projection comprising more than 33% total gonopod length (Pl. 16.188.15 B) .... *Oronectes palmeri palmeri* (Faxon, 1884)

[USA: Arkansas, Kentucky, Louisiana, Missouri]

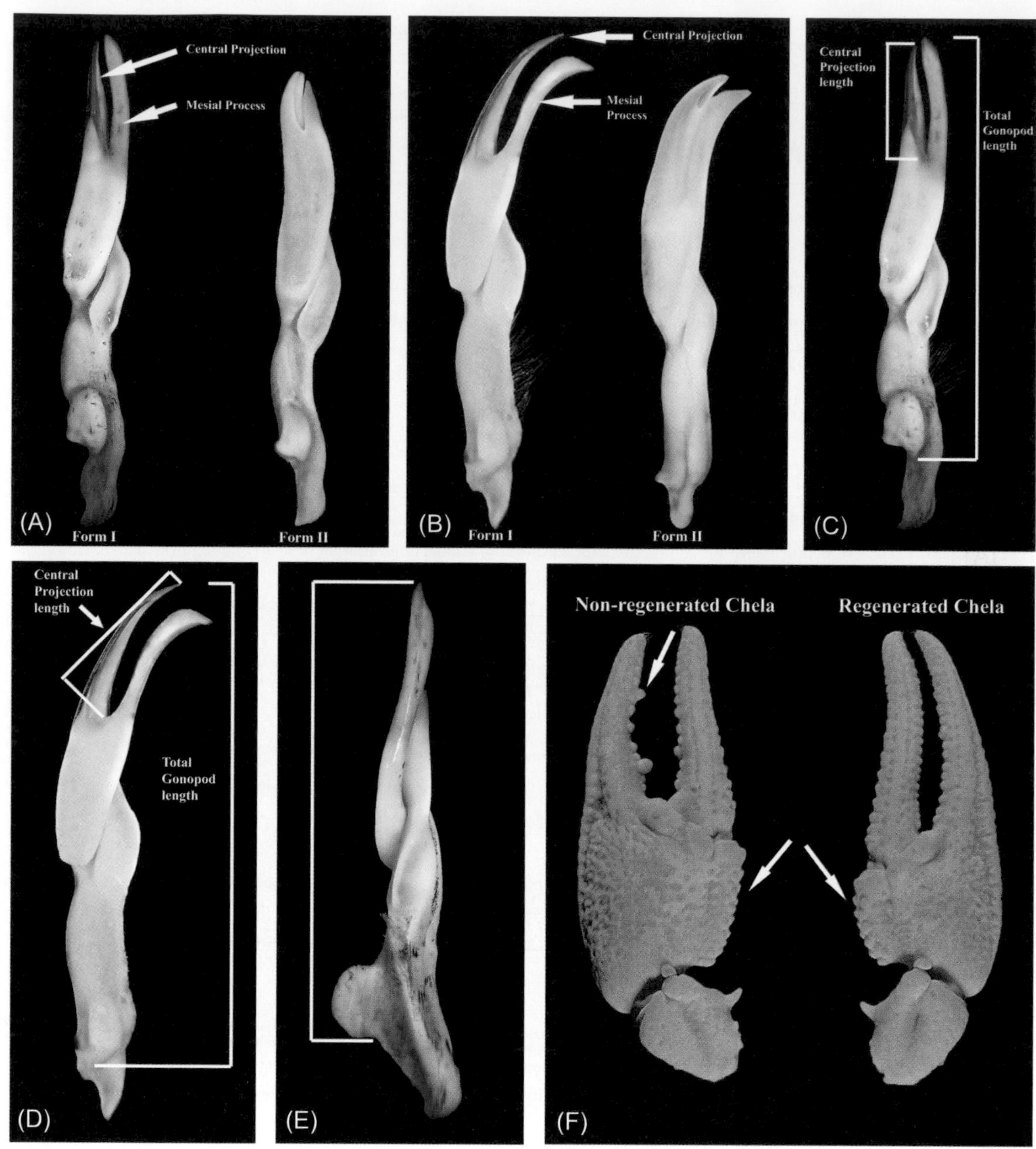

PLATE 16.188.01 (A–F) gonopods (first pleopod) of *Orconectes* species.

28' Gonopod central projection comprising less than 33% total gonopod length .... *Orconectes palmeri creolanus* (Creaser, 1933) [USA: Louisiana, Mississippi]

29(12) Areola open, from narrow to wide, with room for at least one punctation 30

29' Areola open but extremely narrow, lacking room for at least one punctation .... *Orconectes hobbsi* Penn, 1950 [USA: Louisiana, Mississippi]

30(29) Gonopod with two very short and curved elements, central projection laterally flattened and wide in lateral perspective, central projection length usually less than 20% of total gonopod length (Pl. 16.188.18 A) 31

30' Gonopod with two thin elements; elements either curved or straight, central projection length usually more than 20% of total gonopod length (Pl. 16.188.18 B) 33

31(30) Gonopod central projection tapering to a thin apex (Pl. 16.188.19 A) 32

31' Gonopod central projection apex truncated (Pl. 16.188.18 A) .... *Orconectes kentuckiensis* Rhoades, 1944 [USA: Illinois, Kentucky]

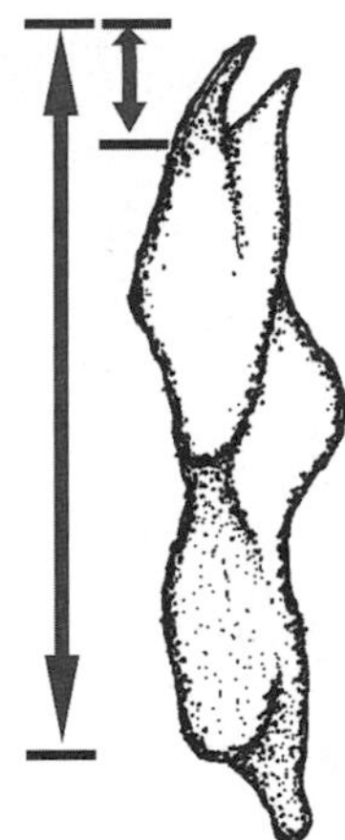

PLATE 16.188.02 Mesal view of form I gonopod (first pleopod).

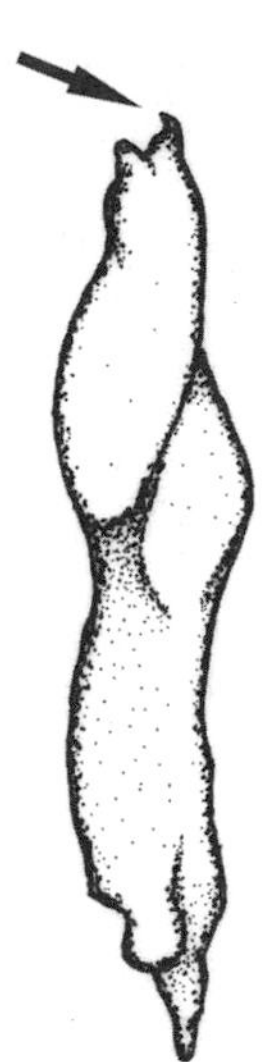

PLATE 16.188.04 Mesial view of form I gonopod of *Orconectes stygocaneyi.*

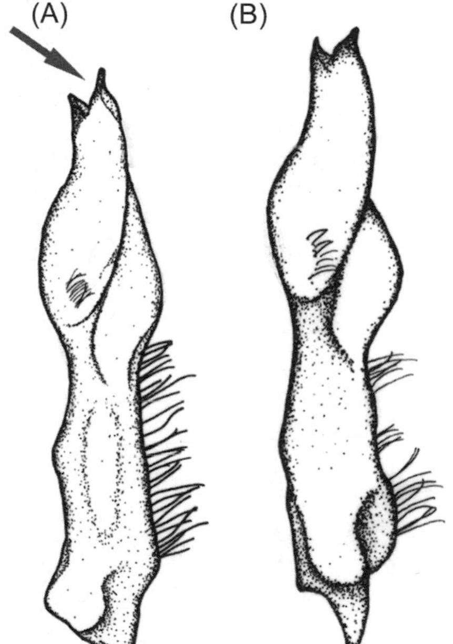

PLATE 16.188.03 Mesal view of form I gonopod showing longer mesial process extending (A) or not extending (B) past central projection.

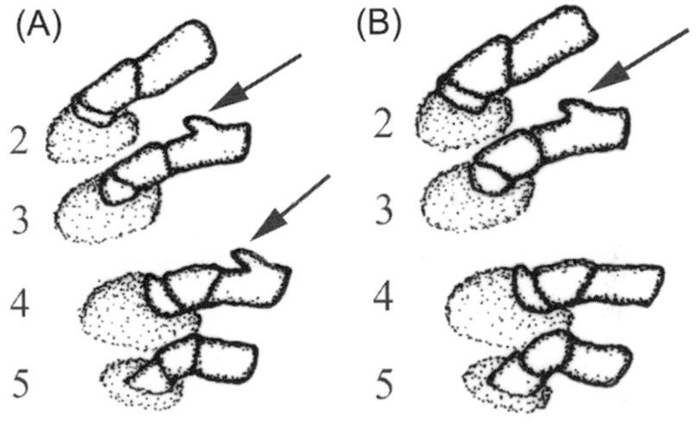

PLATE 16.188.05 (A) Second through fifth pairs of pereopods showing hooks on ishia of third and fourth pair, or (B) on third pair only.

32(31) Gonopod central projection and mesal process separated by a noticeable gap along their entire length, rostrum short and wide with subparallel margins (Pl. 16.188.19 A) .... *Orconectes sloanii* (Bundy, 1876)

[USA: Indiana, Ohio]

32' Gonopod central projection and mesal process not separated by a noticeable gap along their entire length, if present, gap only found at apices (Pl. 16.188.19 B), rostrum long and narrow with concave margins near their bases .... *Orconectes harrisonii* (Faxon, 1884)

[USA: Missouri]

33(30) Gonopod with at least one terminal element straight <u>or</u> both elements with apices divergent (Pl. 16.188.20 A) 34

33' Gonopod with both terminal elements curved caudistally for most of their length (Pl. 16.188.20 B) 55

34(33) Gonopod terminal elements short, forming a distinct V, apices divergent (Pl. 16.188.21) 35

34' Gonopod terminal elements not short and divergent 37

35(34) Carapace lacking small hepatic spines 36

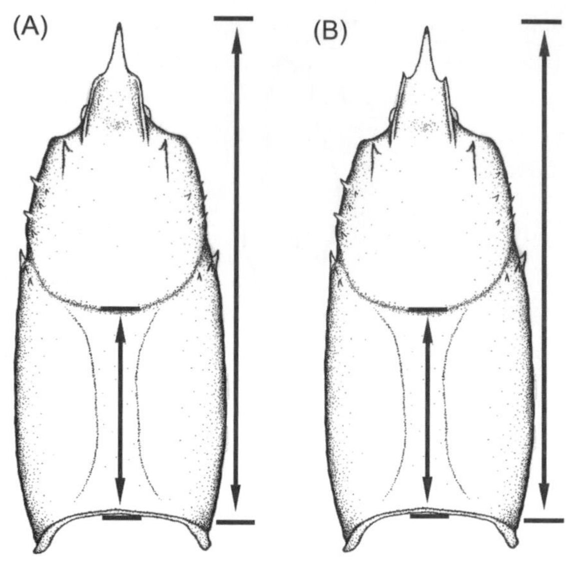

PLATE 16.188.06 Dorsal view of carapace of *Orconectes inermis testii* (A) and *Orconectes inermis inermis* (B).

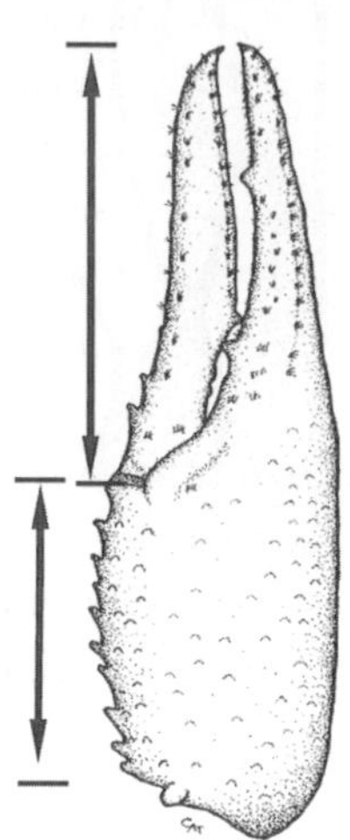

PLATE 16.188.08 Dorsal view of chela with a dactyl longer than palm region.

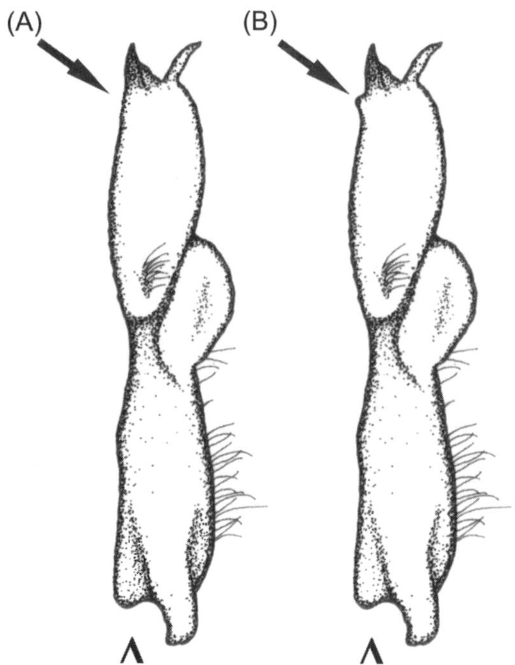

PLATE 16.188.07 Mesial view of form I gonopod without (A) or with (B) a shoulder at base of central projection.

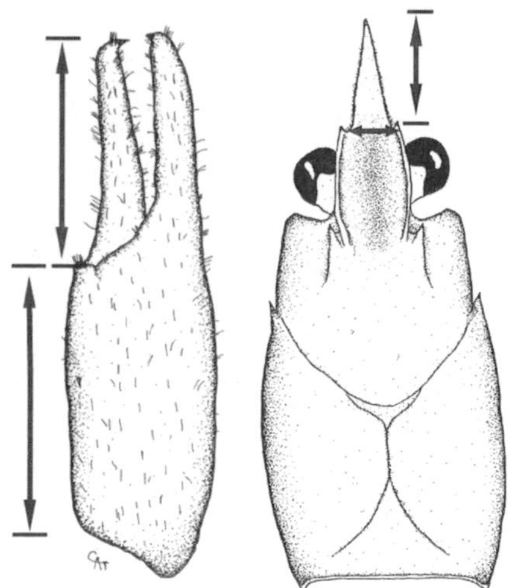

PLATE 16.188.09 Dorsal view of chela and carapace of *Orconectes lancifer*.

35' Carapace with several small hepatic spines (Pl. 16.188.22) .... *Orconectes limosus* (Rafinesque, 1817)

[USA: Atlantic Slope drainages, Maine to Virginia]

36(35) Found only in western tributaries of the Tennessee River in Hardin and McNairy counties, Tennessee .... *Orconectes wrighti* Hobbs, 1948

[USA: Tennessee]

36' Found only in Wabash and Ohio river tributaries in extreme southeastern Illinois and southwestern Indiana .... *Orconectes indianensis* (Hay, 1896)

[USA: Illinois, Indiana]

37(34) Gonopod terminal elements 33% or less of total gonopod length (Pl. 16.188.23 A) 38

37' Gonopod terminal elements greater than 33% total gonopod length (Pl. 16.188.23 B) 70

38(37) Gonopod central projection without a strong basal shoulder........ 39

38' Gonopod central projection with a strong basal shoulder (Pl. 16.188.24) .... *Orconectes obscurus* (Hagen, 1870)

[Canada: Ontario. USA: northeast. Introduced to parts of Maine and Massachusetts]

39(38) Gonopod elements apices curved caudistally (Pl. 16.188.25) 40

39' Gonopod elements apices not curved caudistally, at least one element with apex straight and in line with gonopod basal shaft 41

40(39) Rostrum narrow with strongly converging margins (Pl. 16.188.26 A) .... *Orconectes shoupi* Hobbs, 1948

[USA: Tennessee]

40' Rostrum broad, margins only moderately converging or subparallel (Pl. 16.188.26 B) .... *Orconectes marchandi* Hobbs, 1948

[USA: Arkansas, Missouri]

41(39) Mandible incisor region with edge straight (Pl. 16.188.27 A) 42

41' Mandible incisor region with edge dentated, with distinct teeth (Pl. 16.188.27 B) 43

42(41) Rostral margins nearly straight or subparallel basally, not curved (Pl. 16.188.28 A) .... *Orconectes margorectus* Taylor, 2002

[USA: Kentucky]

42' Rostral margins strongly curved and converging distally, much wider at base (Pl. 16.188.28 B) .... *Orconectes jeffersoni* Rhoades, 1944 (in part)

[USA: Kentucky]

43(41) Gonopod central projection apex not strongly curved and bent at 90° to main shaft and not overhanging mesal process 44

43' Gonopod central projection apex strongly curved, bent at 90° to main shaft and usually overhanging mesal process (Pl. 16.188.29) .... *Orconectes burri* Taylor & Sabaj, 1998

[USA: Kentucky, Tennessee]

44(43) Gonopod central projection length more than 25% of total gonopod length (Pl. 16.188.23 B) 45

44' Gonopod central projection length less than 25% of total gonopod length (Pl. 16.188.23 A) 47

45(44) Gonopod mesal process apex not thicker than central projection apex, both apices separated from one another by a space 46

45' Gonopod mesal process apex thicker than central projection apex, mesal process apex not separated from central projection by a space (Pl. 16.188.30) .... *Orconectes quadruncus* (Creaser, 1933)

[USA: Missouri]

46(45) Rostral margins strongly curved and tapering, much wider at base (Pl. 16.188.28 B) .... *Orconectes jeffersoni* Rhoades, 1944 (in part)

[USA: Kentucky]

46' Rostral margins nearly straight (Pl. 16.188.28 A) .... *Orconectes erichsonianus* (Faxon, 1898)

[USA: Alabama, Georgia, Tennessee, Virginia]

47(44) Gonopod mesal process lacking caudal spur 48

47' Gonopod mesal process with caudal spur (Pl. 16.188.31) .... *Orconectes stannardi* Page, 1985

[USA: Illinois]

48(47) Not found in Roanoke and Chowan river drainages of North Carolina 49

48' Found in Roanoke and Chowan river drainages of North Carolina .... *Orconectes virginiensis* Hobbs, 1951

49(48) Not found in the Spring and Eleven Point river drainages of Arkansas and Missouri 50

49' Found in the Spring and Eleven Point river drainages of Arkansas and Missouri .... *Orconectes eupunctus* Williams, 1952

50(49) Gonopod mesal processes not laterally divergent, central projection apex acute (Pl. 16.188.32 B) 51

50' Gonopod mesal processes laterally divergent, central projection apex blunt (Pl. 16.188.32 A) .... *Orconectes bisectus* Rhoades, 1944

[USA: Kentucky]

51(50) Rostrum with a medial carina (Pl. 16.188.33) 52

51' Rostrum without a medial carina 53

52(51) Gonopod mesal process curved cephalodistally, and with cephalic and caudal surfaces flattened (Pl. 16.188.34 A) .... *Orconectes tricuspis* Rhoades, 1944

[USA: Kentucky]

PHYLUM ARTHROPODA

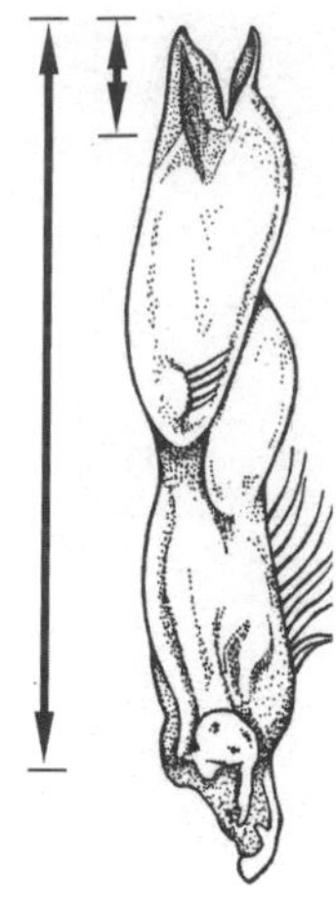

PLATE 16.188.10 Mesial view of form I gonopod of *Orconectes pagei*.

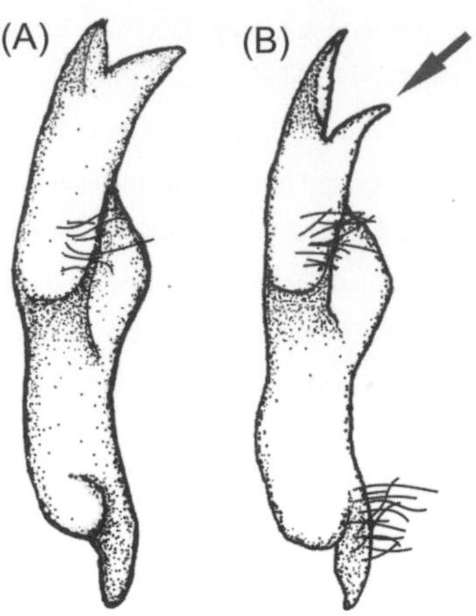

PLATE 16.188.12 Mesial view of form I gonopod with: (A) a wide mesial process, or (B) a thin mesial process tapering to a tip.

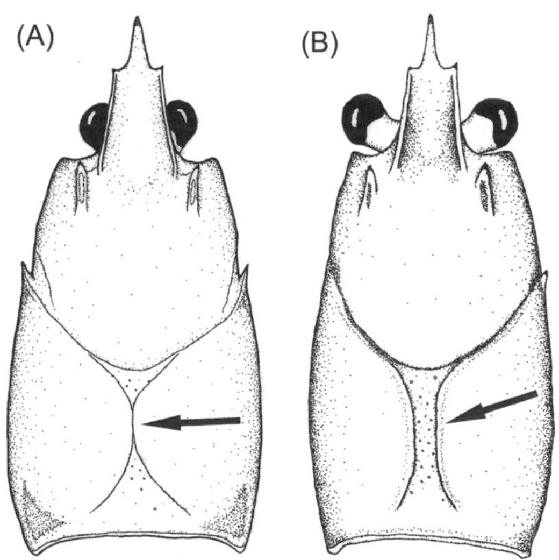

PLATE 16.188.11 Dorsal view of carapace with: (A) linear or obliterated areola, or (B) open areola.

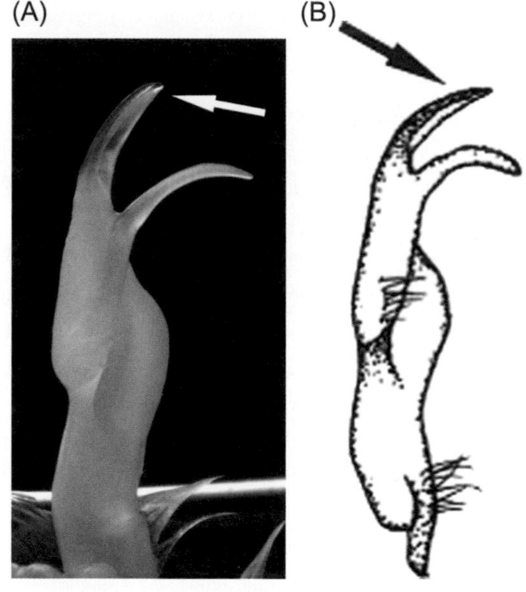

PLATE 16.188.13 Mesial view of form I gonopod with a central projection angled: (A) less than 90° to main axis; or (B) at 90° to main axis.

52' Gonopod mesal process straight, not curved cephalodistally or flattened (Pl. 16.188.34 B) .... *Orconectes propinquus* (Giard, 1852)

[Canada: Ontario, Quebec. USA: upper midwestern states from Iowa and Minnesota to Vermont]

53(51) Gonopod cephalic edge proximal to central projection slightly curved; chela palm mesal margin subequal to 50% of dactyl length (Pl. 16.188.35 C) 54

53' Gonopod cephalic edge proximal to central projection straight, chela palm mesal margin usually more than 50% of dactyl length .... *Orconectes illinoiensis* Brown, 1956

[USA: Illinois]

54(53) Gonopod mesal process flattened distally on cephalic and caudal surfaces (Pl. 16.188.36 A) .... *Orconectes sanbornii* (Faxon, 1884)

[USA: Kentucky, Ohio, West Virginia]

54' Gonopod mesal process not flattened (Pl. 16.188.36 B) .... *Orconectes rafinesquei* Rhoades, 1944

[USA: Kentucky]

55(33) Rostrum with median carina (Pl. 16.188.33) 56

55' Rostrum without median carina 58

56(55) Carapace not strongly laterally compressed, cervical spines present; gonopod central projection lacking a shoulder (Pl. 16.188.37 B) 57

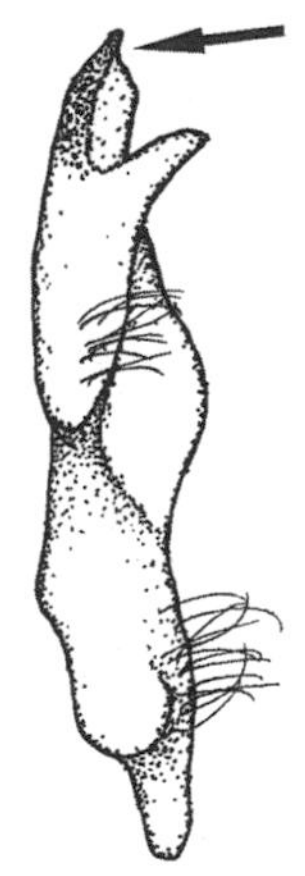

PLATE 16.188.14 Mesial view of form I gonopod of *Orconectes hartfieldi.*

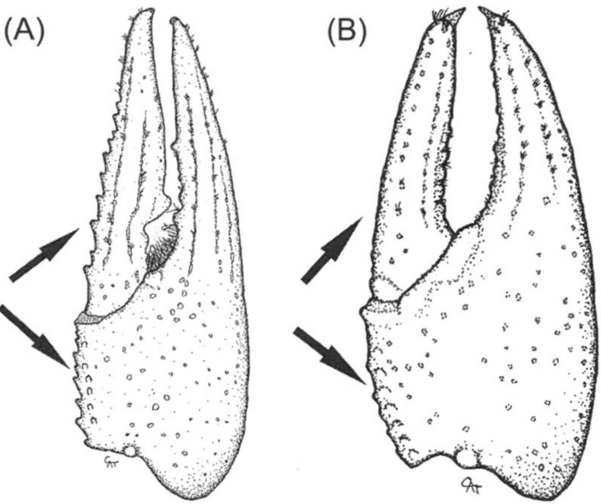

PLATE 16.188.16 Dorsal view of chela with: (A) serrate tubercles along mesial margin; or (B) low rounded tubercles along mesial margin.

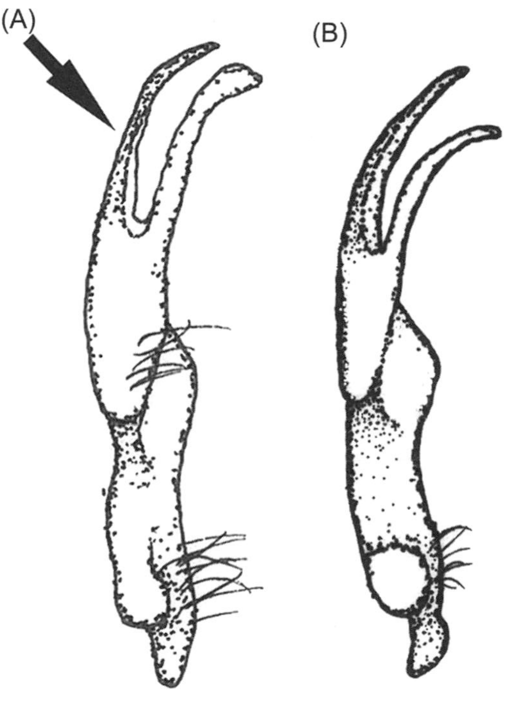

PLATE 16.188.15 (A) Mesial view of form I gonopod with a portion of central projection straight; (B) mesial view of form I gonopod with a central projection curved for its entire length.

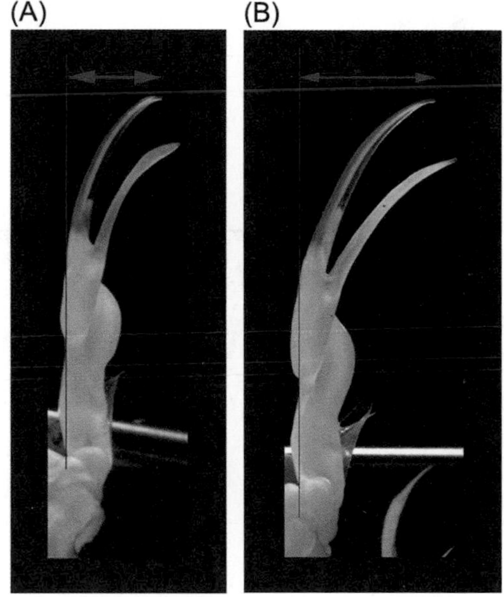

PLATE 16.188.17 Mesial view of form I gonopod with terminal elements offset: (A) 20–24% of main axis; or (B) 22–35% of main axis.

56' Carapace strongly laterally compressed, cervical spines absent; gonopod central projection basally usually with a weak shoulder (Pl. 16.188.37 A) .... *Orconectes compressus* (Faxon, 1884)

[USA: Alabama, Kentucky, Mississippi, Tennessee]

57(56) Gonopod central projection less than 28% of total gonopod length; central projection wide basally (Pl. 16.188.38) .... *Orconectes taylori* Schuster, 2008

[USA: Tennessee]

57' Gonopod central projection equal to or greater than 28% of total gonopod length; central projection not wide basally (Pl. 16.188.37 B) .... *Orconectes alabamensis* (Faxon, 1884)

[USA: Alabama, Tennessee]

58(55) Chela palm and dactyl mesal margin with tubercles low, rounded and not angled anteriorly (Pl. 16.188.16 B) 59

58' Chela palm and dactyl mesal margin with tubercles large, pointed and angled anteriorly (Pl. 16.188.16 A) 62

59(58) Not found in the Spring River drainage of Arkansas and Missouri 60

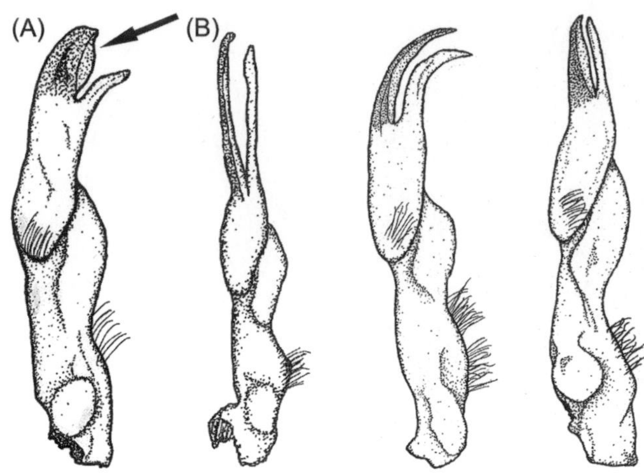

PLATE 16.188.18 Mesial view of form I gonopod with short central projection: (A) less than 20% of entire length of gonopod; or (B) more than 20% of entire length of gonopod.

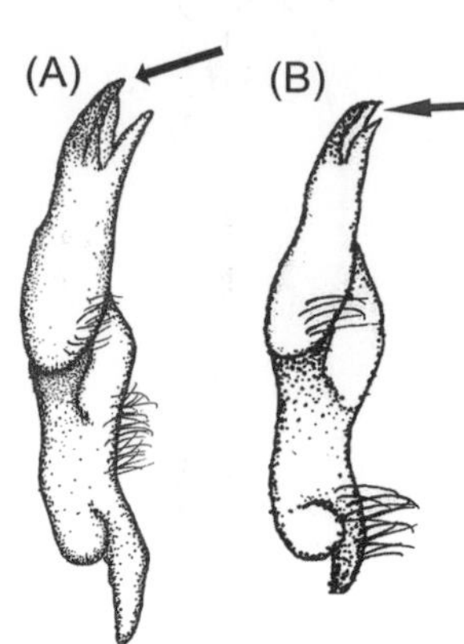

PLATE 16.188.19 Mesial view of form I gonopod with terminal elements: (A) separated by gap along their entire length; or (B) not separated by gap along their entire length.

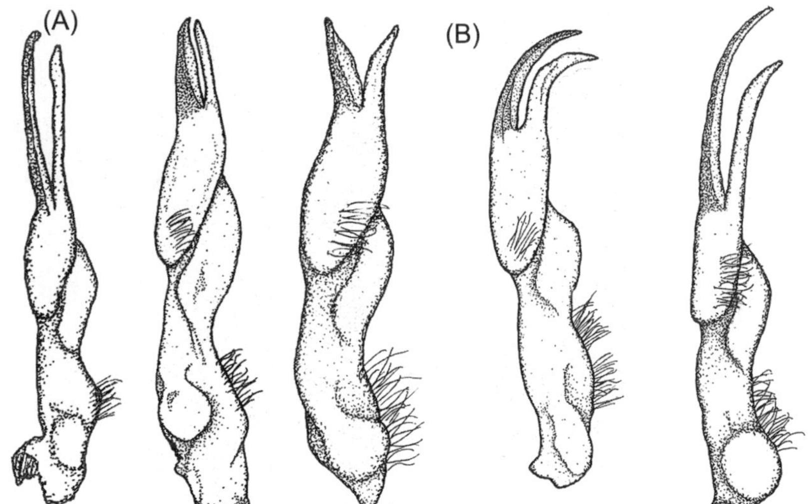

PLATE 16.188.20 Mesial view of form I gonopods with terminal elements: (A) straight or divergent; or (B) curved caudistally.

59’ Found in a restricted portion of the Spring River dr. in Sharp and Lawrence counties, Arkansas and Oregon Co., Missouri .... *Orconectes marchandi* Hobbs, 1948

60(59) Chelae with pollex and dactyl long, palm mesal margin length less than 60% of dactyl mesal margin length 61

60’ Chelae with pollex and dactyl short; palm mesal margin length at least 60% of dactyl mesal margin length (Pl. 16.188.39) .... *Orconectes cooperi* Cooper & Hobbs, 1980

[USA: Alabama]

61(60) Rostral margins terminate in low rounded tubercles, rostral edge from distal end of rostral margin to base of acumen not concave (Pl. 16.188.40) .... *Orconectes rhoadesi* Hobbs, 1949

[USA: Tennessee]

61’ Rostral margins terminate in spines, rostral edge from distal end of rostral margin to base of acumen concave (Pl. 16.188.26 B) .... *Orconectes validus* (Faxon, 1914) (in part)

[USA: Alabama, Tennessee]

62(58) Antennal scale widest at its midpoint, distal edge of scale angular and not perpendicular to lateral edge of scale (Pl. 16.188.41 B) 63

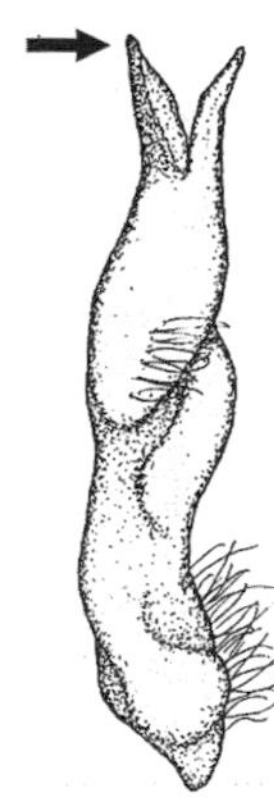

PLATE 16.188.21 Mesial view of form I gonopod with divergent terminal elements.

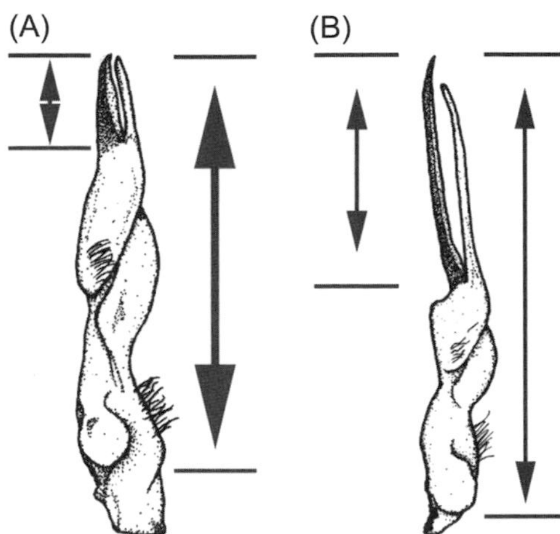

PLATE 16.188.23 Mesial view of form I gonopod with central projection: (A) less than 33% of entire length of gonopod; or (B) more than 33% of entire length of gonopod.

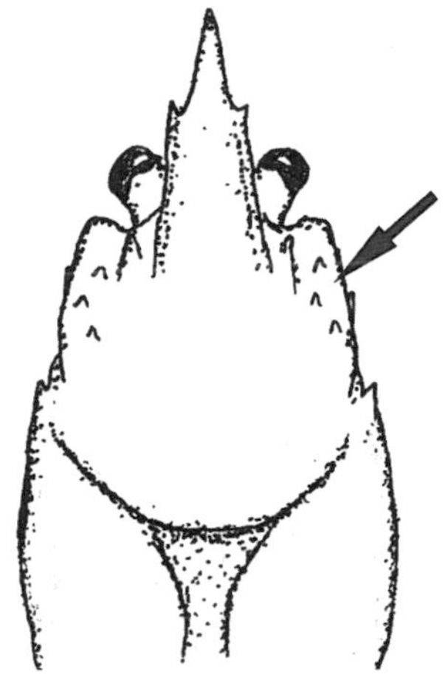

PLATE 16.188.22 Dorsal view of carapace of *Orconectes limosus* showing small hepatic spines.

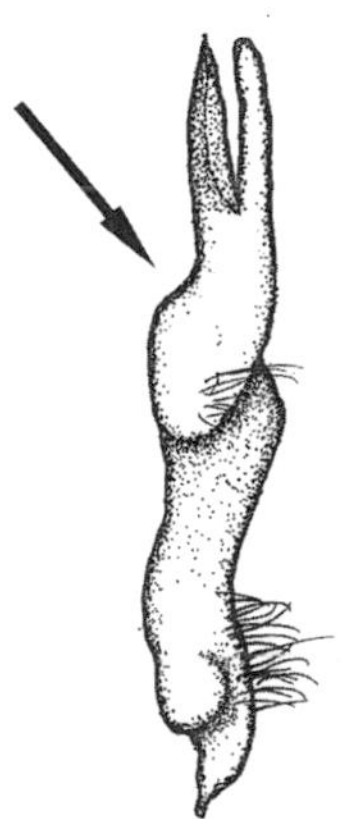

PLATE 16.188.24 Mesial view of form I gonopod of *Orconectes obscurus* showing large shoulder at base of central projection.

62' Antennal scale widest distal to its midpoint, distal margin of scale nearly perpendicular to lateral edge of scale (Pl. 16.188.41 A) .... *Orconectes meeki* (Faxon, 1898)

[USA: Arkansas, Missouri]

63(62) Gonopod elements longer, central projection more than 28% of total gonopod length 64

63' Gonopod elements short, central projection 28% or less than total gonopod length (Pl. 16.188.42) .... *Orconectes jonesi* Fitzpatrick, 1992/*Orconectes chickasawae* Cooper & Hobbs, 1980

[Both species, and *O. etnieri*, comprise a species complex and the taxonomic status of these three taxa are uncertain (Taylor et al., 2014) USA: Alabama, Tennessee]

64(63) Not found in drainages listed in 64' 65

64' Found in tributaries of the Tennessee River dr. in Hardin and McNairy counties, Tennessee and tributaries of the Forked Deer, Hatchie, and Loosahatchie river drs. in western Tennessee and northern Mississippi .... *Orconectes etnieri* Bouchard & Bouchard, 1976

[see notes for *O. jonesi* and *O. chickasawae* in couplet 63 above]

65(64) Chelae with pollex and dactyl short, not dark blue or black in life; chela palm mesal margin length greater than 30% dactyl mesal margin length 66

65' Chelae with pollex and dactyl very long and thin, fingers of chelae dark blue or black in life; chela palm mesal margin length less than 30% dactyl mesal margin length .... *Orconectes longidigitus* (Faxon, 1898)(in part)

[USA: Arkansas, Missouri]

66(65) Gonopod central projection greater than 38% of total gonopod length, central projection caudal surface basal 1/3 straight and in line with gonopod main axis (Pl. 16.188.43) 67

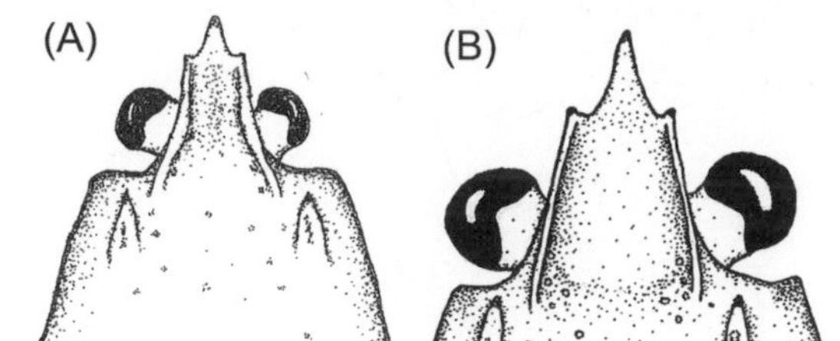

PLATE 16.188.26 Dorsal view of rostrum with: (A) strongly converging margins; or (B) moderately converging or subparallel margins.

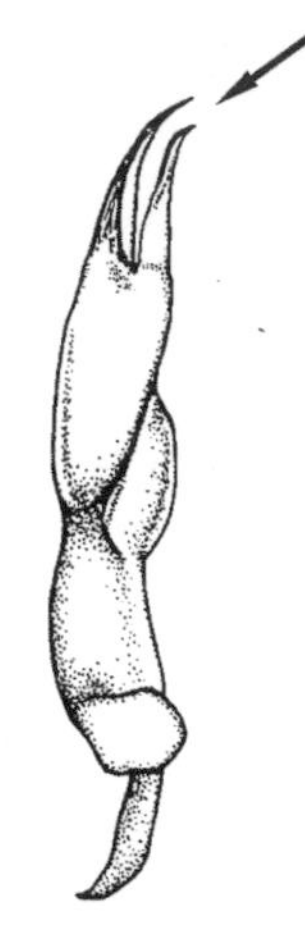

PLATE 16.188.25 Mesial view of form I gonopod with apcies of terminal elements curved caudistally.

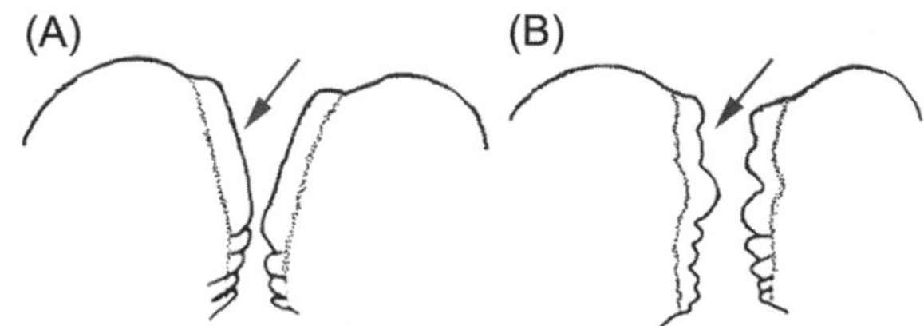

PLATE 16.188.27 Mandible with: (A) straight edged incisor region; or (B) toothed incisor region.

66' Gonopod central projection less than 38% of total gonopod length, central projection curved for most of its length (Pl. 16.188.15 B) 68

67(66) Gonopod central projection 45% or less than total gonopod length; chelae dorsal surface usually with scattered dark flecks; abdominal segments usually with paired dark blotches (Pl. 16.188.43) .... *Orconectes virilis* (Hagen, 1870) (including *O. causeyi* Jester, 1967)

[Canada: Hudson Bay drainages. USA: Missouri River system, upper Mississippi River system north of Ohio River, Atlantic Slope and St. Lawrence River system. Widely introduced and invasive across US. *Orconectes causeyi* is indistinguishable from *O. virilis* using morphology. Future analysis of molecular and morphological characters is needed to ascertain the true status of *O. causeyi*]

67' Gonopod central projection greater than 45% of total gonopod length; chelae dorsally without dark flecks; abdomen lacking paired dark blotches .... *Orconectes punctimanus* (Creaser, 1933)

[USA: Arkansas, Missouri]

68(66) Chela dactyl of adults without deep basal excision; adult rostral margins terminating in spines (Pl. 16.188.44 B) 69

68' Chela dactyl of adults with deep basal excision; adult rostral margins not terminating in large spines (Pls. 16.188.44 A, 40) .... *Orconectes immunis* (Hagen, 1870)

[Canada: Ontario. USA: Colorado to Maine, south to Kentucky]

69(68) Found in Massachusetts .... *Orconectes quinebaugensis* Mathews & Warren, 2008

69' Found in the southern Great Plains .... *Orconectes nais* (Faxon, 1885)

[USA: Arkansas, Colorado, Kansas, Louisiana, Missouri, New Mexico, Texas]

70(37) Chelae with pollex and dactyl short, not dark blue or black in life; chela palm mesal margin length greater than 30% dactyl mesal margin length 71

70' Chelae with pollex and dactyl very long and thin, dark blue or black in life; chela palm mesal margin length less than 30% dactyl mesal margin length .... *Orconectes longidigitus* (Faxon, 1898) (in part)

[USA: Arkansas, Missouri]

71(70) One or both mandibles with incisor region bearing a straight edge (Pls. 16.188.27 A, 45) 72

71' Both mandibles with incisor region not straight, bearing distinct teeth (Pl. 16.188.27 B) 78

72(71) Carapace with cervical spines (Pl. 16.188.46 A) 73

72' Cervical spines absent (Pl. 16.188.46 B) .... *Orconectes barrenensis* Rhoades, 1944 (in part)

[USA: Kentucky, Tennessee]

73(72) Rostrum with median carina (Pl. 16.188.33) 74

73' Rostrum without median carina 76

74(73) Chela palm tubercles scattered across dorsal surface (Pl. 16.188.47 B) 75

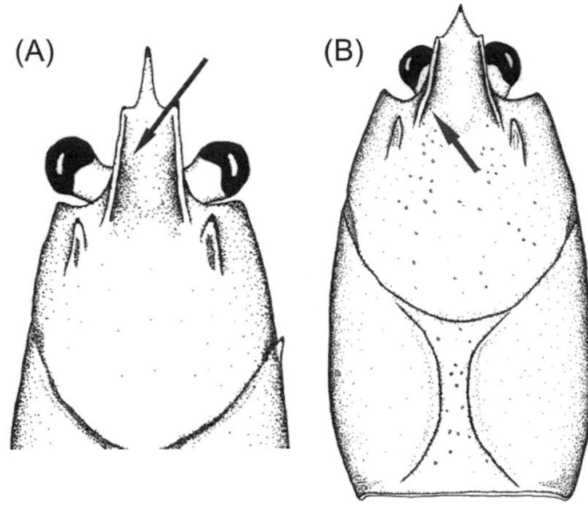

PLATE 16.188.28 Dorsal view of rostrum with: (A) straight or nearly straight margins; or (B) margins that strongly curve at their bases.

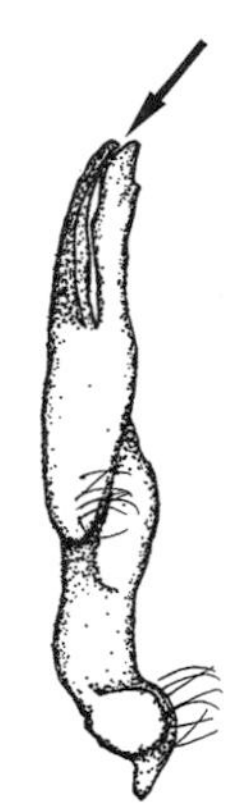

PLATE 16.188.30 Mesal view of form I gonopod of *Orconectes quadruncus* showing thick mesial process.

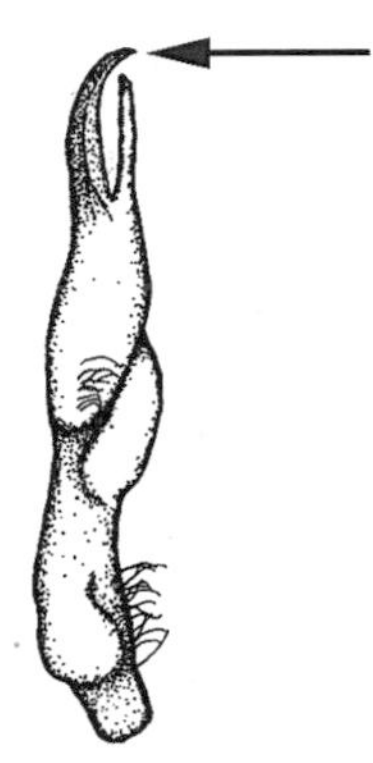

PLATE 16.188.29 Mesial view of form I gonopod of *Orconectes burri* showing curved central projection overhanging mesial process.

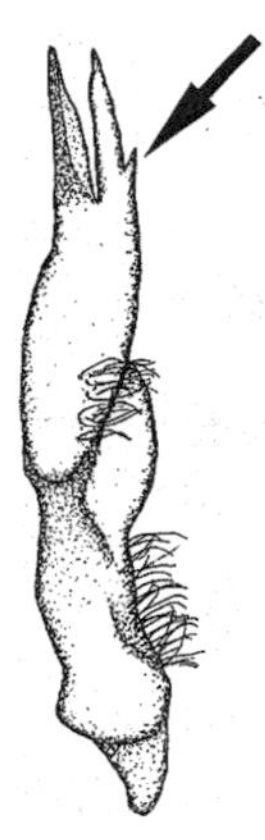

PLATE 16.188.31 Mesial view of form I gonopod of *Orconectes stannardi* showing mesial process with caudal spur.

74' Chela palm tubercles restricted to mesal margin in two well defined rows (Pl. 16.188.47 A) .... *Orconectes luteus* (Creaser, 1933) (in part)

[USA: Illinois, Iowa, Minnesota, Missouri]

75(74) Gonopod central projection straight (Pl. 16.188.48 A) .... *Orconectes theaphionensis* Simon, Timm, & Morris, 2005

[USA: Indiana]

75' Gonopod central projection with apex curved (Pl. 16.188.48 B) .... *Orconectes raymondi* Thoma & Stocker, 2009

[USA: Ohio]

76(73) Gonopod central projection less than 44% of total gonopod length; gonopods usually not reaching base of first pair of pereopods when abdomen flexed forward (Pl. 16.188.49 B, 49 C) 77

76' Gonopod central projection more than 44% of total gonopod length; gonopods usually reaching base of first pair of pereopods when abdomen flexed forward (Pls. 16.188.23 B, 49 A) .... *Orconectes juvenilis* (Hagen, 1870)

[USA: Indiana, Kentucky]

77(76) Gonopod central projection proximally with shoulder distinct and sharply angled; gonopods elements usually less than 40% of total gonopod length; carapace posteriolateral surface in life usually with large rust colored spot, (Pl. 16.188.50 A) .... *Orconectes rusticus* (Girard, 1852)

[Canada: introduced. USA: Indiana, Kentucky, Ohio. Widely introduced across midwestern and New England states]

77' Gonopod central projection proximally with shoulder not distinct, usually low and rounded; gonopods elements usually greater than 40% of total gonopod length; carapace posteriolateral surface in life lacking large rust-colored spot, (Pl. 16.188.50 B) .... *Orconectes luteus* (Creaser, 1933) (in part)

[USA: Illinois, Iowa, Minnesota, Missouri]

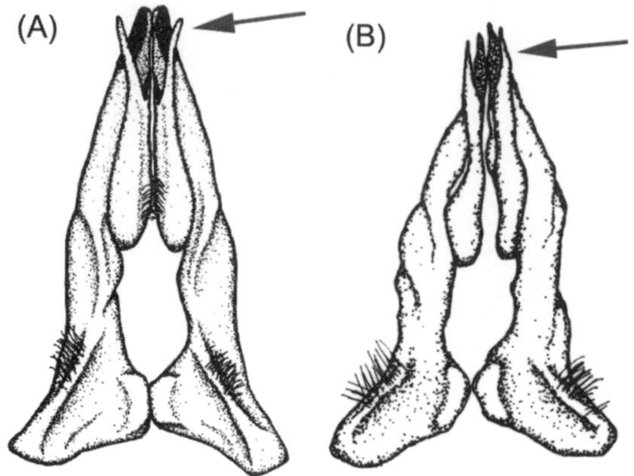

PLATE 16.188.32 (A) Ventral view of gonopods with strongly diverging mesial processes; (B) ventral view of gonopods without strongly diverging mesial processes.

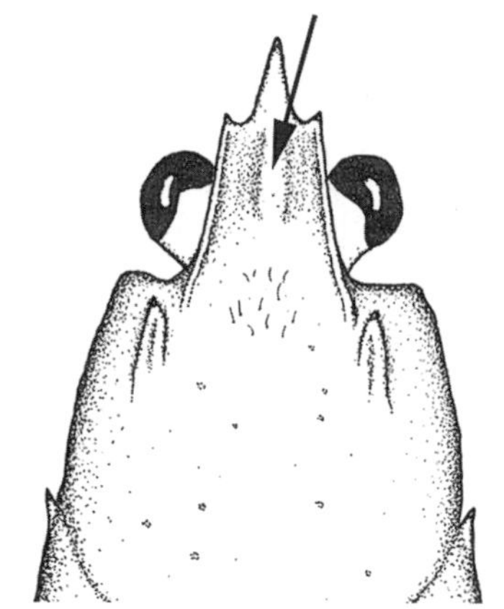

PLATE 16.188.33 Dorsal view of rostrum with a median carina.

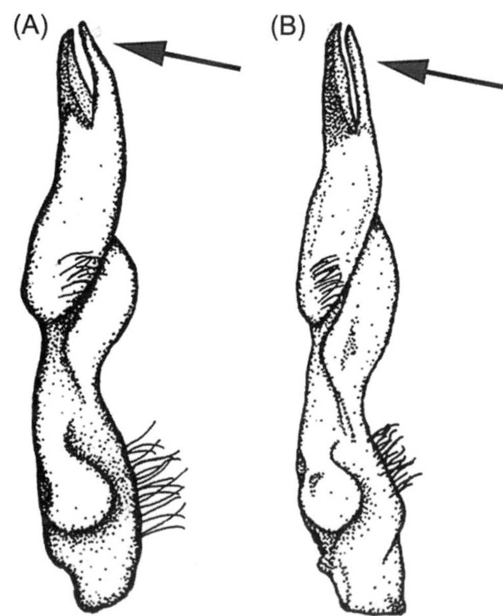

PLATE 16.188.34 (A) mesal view of form I gonopod with mesial process curved caudistally with caudal and cephalic surfaces flattened; (B) mesial view of form I gonopod with mesial process not curved caudistally with caudal and cephalic surfaces flattened.

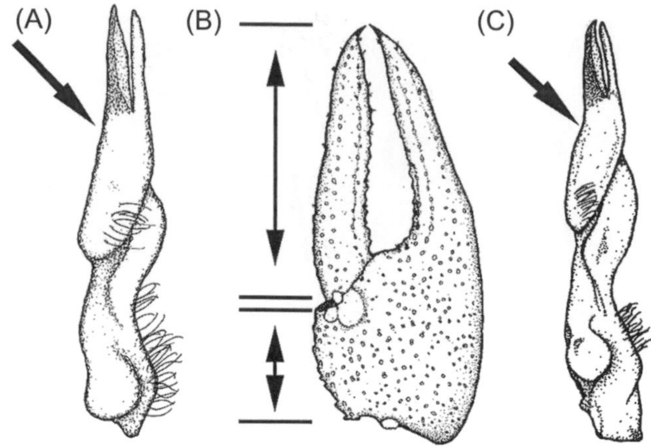

PLATE 16.188.35 (A) mesal view of form I gonopod with straight caudal edge proximal to central projection; (B) dorsal view of chela with palm more than 50% of dactyl length; (C) mesial view of form I gonopod with curved caudal edge proximal to central projection.

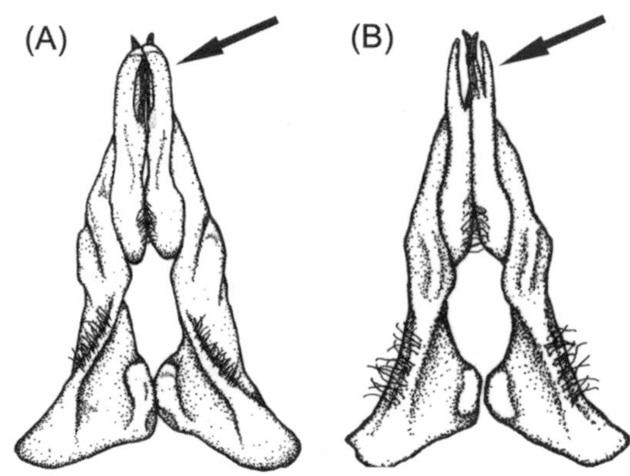

PLATE 16.188.36 Ventral view of form I gonopods with (A) or without (B) mesial processes with flattened caudal and cephalic surfaces.

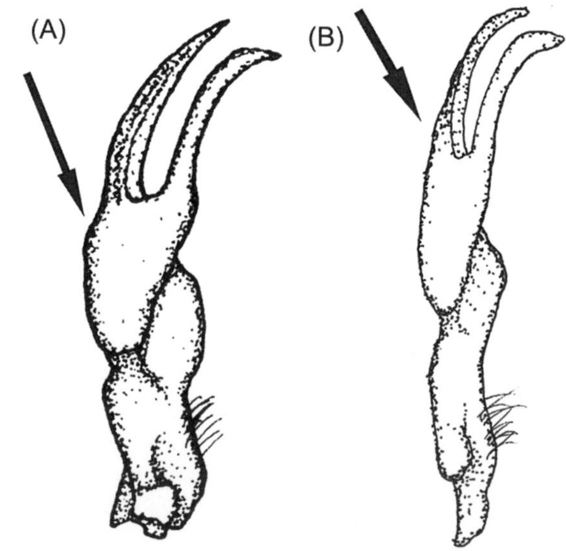

PLATE 16.188.37 Mesal view of form I gonopod with (A) or without (B) a weak shoulder at central projection base.

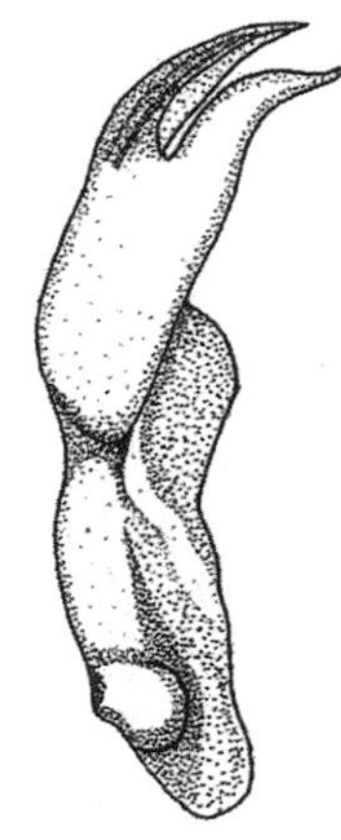

PLATE 16.188.38 Mesial view of form I gonopod of *Orconectes taylori* showing wide central projection less than 28% of total gonopod length.

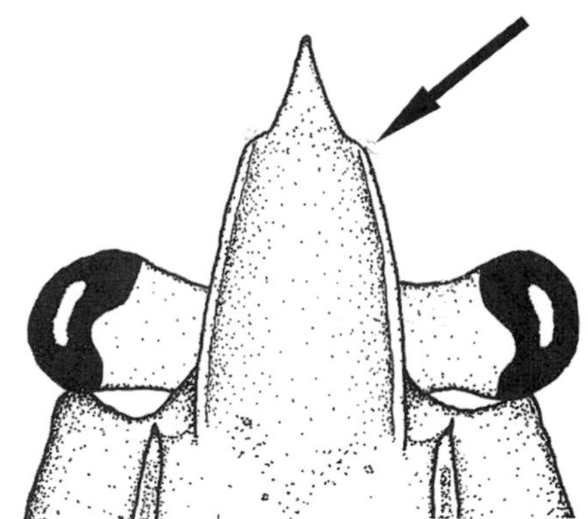

PLATE 16.188.40 Dorsal view of rostrum showing margins that terminate in low rounded tubercles.

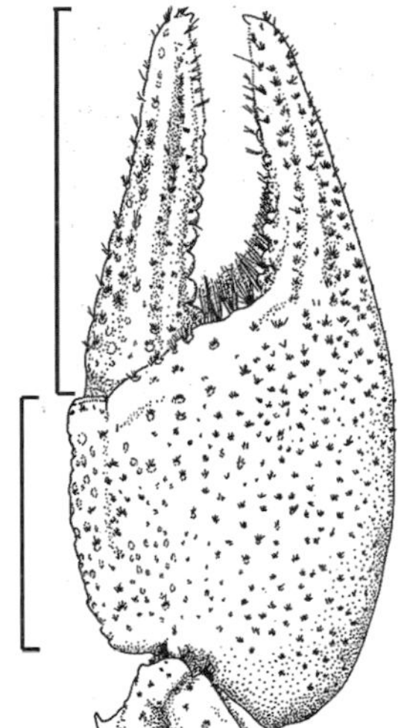

PLATE 16.188.39 Dorsal view of chela of *Orconectes cooperi* showing palm region more than 60% of dactyl length.

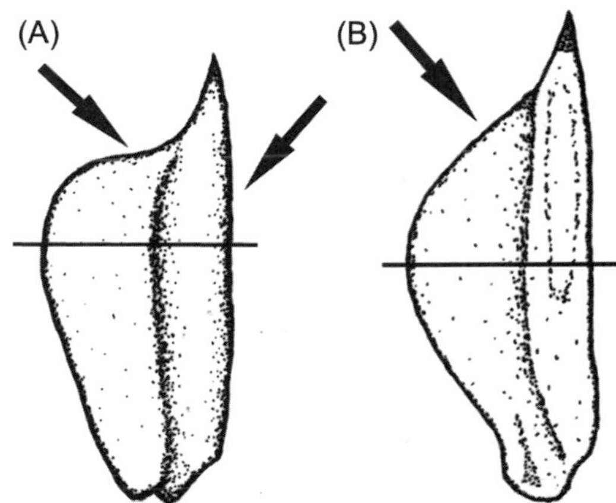

PLATE 16.188.41 (A) Dorsal view of antennal scale with widest point distal to its midpoint and with distal margin nearly perpendicular to lateral edge; (B) dorsal view of antennal scale with widest point distal at its midpoint and with distal margin not perpendicular to lateral edge.

78(71)	Rostrum with median carina (Pl. 16.188.33)	79
78'	Rostrum without median carina	87
79(78)	Cervical spines absent (Pl. 16.188.46 B)	80
79'	Cervical spines present (Pl. 16.188.46 A)	83
80(79)	Gonopod central projection 45% or less than total gonopod length	81
80'	Gonopod central projection greater than 45% of total gonopod length	82
81(80)	Found in upper and middle Green River drainage of Kentucky and Tennessee [USA: Kentucky, Tennessee]	*Orconectes barrenensis* Rhoades, 1944 (in part)
81'	Found in middle Tennessee River drainage of northern Alabama and southcentral Tennessee	*Orconectes mirus* (Ortmann, 1931)
82(80)	Gonopod central projection 55% or greater than total gonopod length [USA: Arkansas, Oklahoma]	*Orconectes leptogonopodus* Hobbs, 1948
82'	Gonopod central projection less than 55% of total gonopod length [USA: Arkansas]	*Orconectes acares* Fitzpatrick, 1965
83(79)	Chela carpus ventral surface with a distomedian spine (Pl. 16.188.51 A)	84
83'	Chela carpus ventral surface without a distomedian spine (Pl. 16.188.51 B) [USA: Tennessee]	*Orconectes forceps* (Faxon, 1884)

84(83) Gonopod central projection proximally with a strong angular shoulder (Pl. 16.188.52 A)........ 85

84' Gonopod central projection proximally with a weak or no shoulder (Pl. 16.188.52 B) 86

85(84) Gonopod central projection 42% or less than total gonopod length .... *Orconectes neglectus* (Faxon, 1885)

[USA: Arkansas, Colorado, Kansas, Missouri. Introduced in portions of Arkansas, Missouri, New York, Oregon]

85' Gonopod central projection more than 42% of total gonopod length .... *Orconectes cristavarius* Taylor, 2000 (in part)

[USA: Kentucky, Ohio, North Carolina, Tennessee, West Virginia, Virginia]

86(84) Chelae robust with short dactyl and pollex; dactyl length less than twice the palm mesal margin length (Pl. 16.188.53 A) .... *Orconectes durelli* Bouchard & Bouchard, 1995 (in part)

[USA: Kentucky, Tennessee]

86' Chelae not robust; dactyl more than twice the palm mesal margin length (Pl. 16.188.53 B) .... *Orconectes placidus* (Hagen, 1870) (in part)

[USA: Cumberland and Tennessee river drainages, Illinois]

87(78) Cervical spines absent (Pl. 16.188.46 B) 88

87' Cervical spines present (Pl. 16.188.46 A) 93

88(87) Gonopod central projection proximally with a weak or no shoulder (Pl. 16.188.52 B) 89

88' Gonopod central projection proximally with a strong angular shoulder (Pl. 16.188.52 A) 90

89(88) Chela palm with tubercles restricted to mesal margin in two well-defined rows (Pl. 16.188.47 A) .... *Orconectes williamsi* Fitzpatrick, 1966

[USA: Arkansas, Missouri]

89' Chela palm with tubercles scattered across dorsal surface (Pl. 16.188.47 B) .... *Orconectes menae* (Creaser, 1933)

[USA: Arkansas, Oklahoma]

90(88) Gonopod central projection 45% or more of total gonopod length; rostrum entire length not narrow 91

90' Gonopod central projection less than 45% of total gonopod length; rostrum entire length narrow (Pl. 16.188.54) .... *Orconectes nana* Williams, 1952

[USA: Arkansas, Oklahoma]

91(90) Chela carpus ventral surface with distomedian spine, rostrum not narrow 92

91' Chela carpus ventral surface without a strong distomedian spine, a low rounded tubercle may be present; rostrum distal 2/3rds narrow (Pls. 16.188.51 B, 54) .... *Orconectes macrus* Williams, 1952

[USA: Kansas, Missouri, Oklahoma]

92(91) Acumen length less than 55% of rostrum basal width; rostral margins terminate in weak spines or no spines at acumen base (Pl. 16.188.55 A) .... *Orconcetes medius* (Faxon, 1884) (in part)

[USA: Missouri]

92' Acumen length more than 55% of rostrum basal width; rostral margins terminate in large spines at acumen base (Pl. 16.188.55 B) .... *Orconectes saxatilis* Bouchard & Bouchard, 1976

[USA: Oklahoma]

93(87) Male third pereopods with ischia bearing hooks (Pl. 16.188.5 B) 94

93' Male third and fourth pereopods with ischia bearing hooks (Pl. 16.188.5 A) .... *Orconectes peruncus* (Creaser, 1931)

[USA: Missouri]

94(93) Chela carpus ventral surface without a strong distomedian spine, a low rounded tubercle may be present (Pl. 16.188.51 B) 95

94' Chela carpus ventral surface with distomedian spine (Pl. 16.188.51 A) 97

95(94) Gonopod central projection less than 45% of total gonopod length, gonopods usually not reaching first pereopods base when abdomen flexed forward (Pl. 16.188.49 C) 96

95' Gonopod central projection more than 45% of total gonopod length, gonopods reaching first pereopods base when abdomen flexed forward (Pl. 16.188.49 A) .... *Orconectes putnami* (Faxon, 1884)

[USA: Kentucky, Tennessee]

96(95) Chelae robust with dactyl and pollex short: dactyl length less than twice palm mesal margin length; carapace dorsal surfaces not covered with dark spots (Pl. 16.188.53 A) .... *Orconectes durelli* Bouchard & Bouchard, 1995 (in part)

[USA: Kentucky, Tennessee]

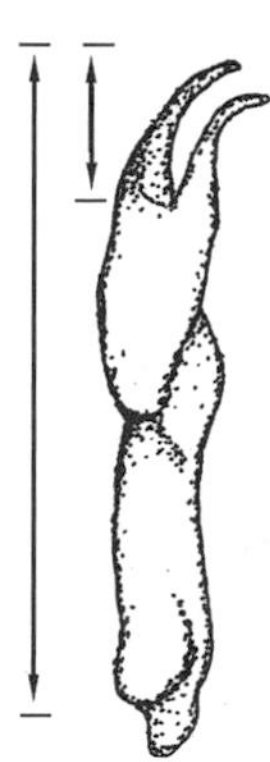

PLATE 16.188.42 Mesal view of form I gonopod with central projection 28% or less than total gonopod length.

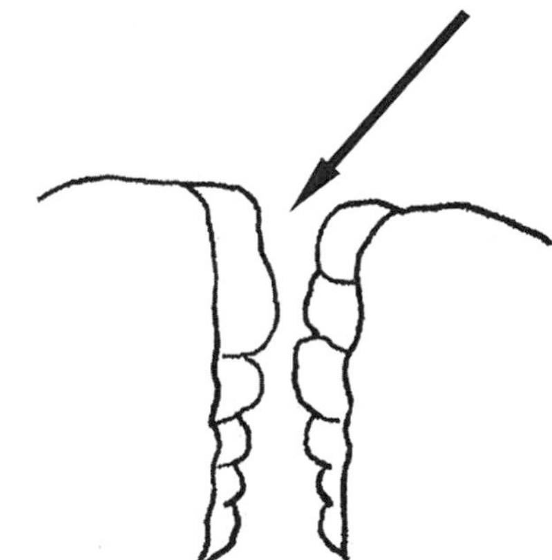

PLATE 16.188.45 Mandible with at least one straight edged incisor region.

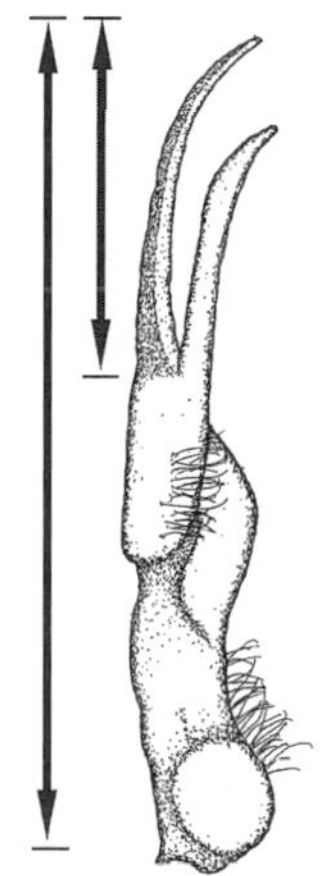

PLATE 16.188.43 Mesal view of form I gonopod with central projection 38% or more but less than 46% of total gonopod length.

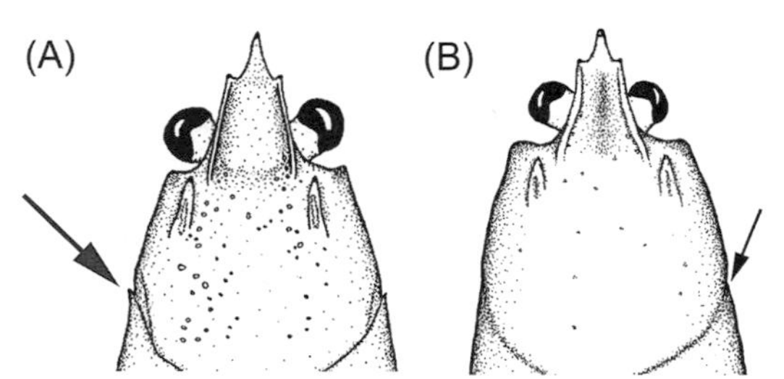

PLATE 16.188.46 Dorsal view of carapace with (A) or without (B) cervical spines.

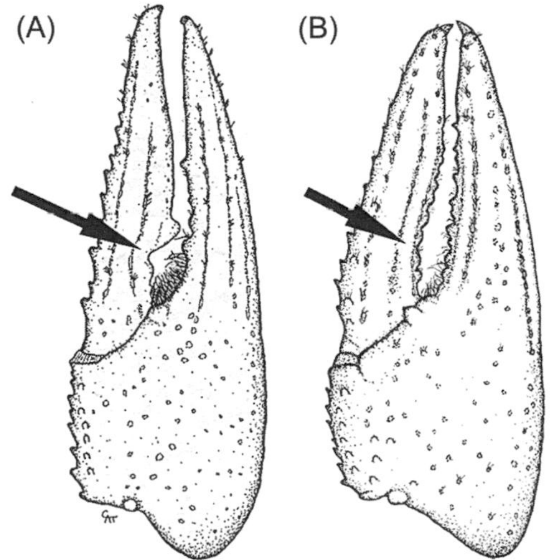

PLATE 16.188.44 Dorsal view of chela with (A) or without (B) a deep basal excision on dactyl.

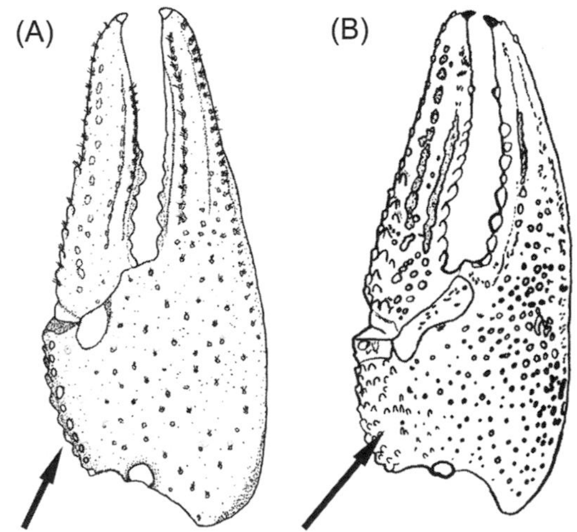

PLATE 16.188.47 Dorsal view of chela with tubercles: (A) along mesial margin of palm restricted to two rows; or (B) scattered across dorsal surface of palm region.

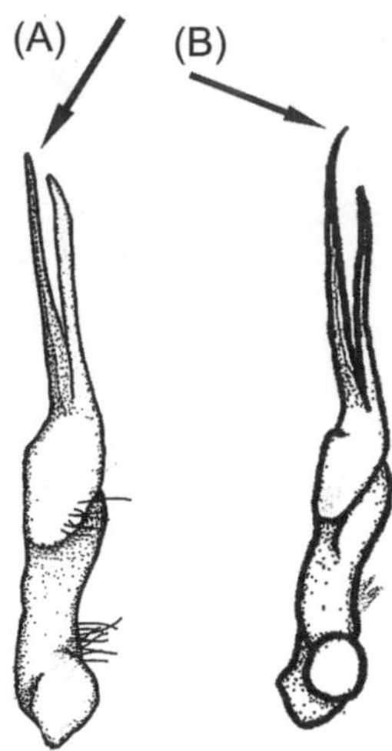

PLATE 16.188.48 Mesial view of form I gonopod with: (A) a straight central projection; or (B) a straight central projection with curved apex.

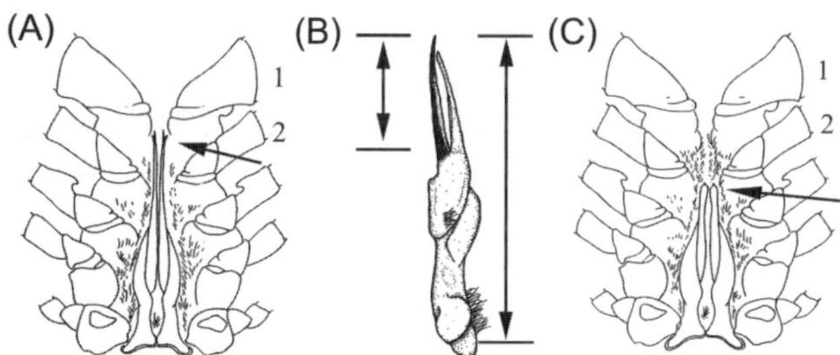

PLATE 16.188.49 (A) Ventral view of carapace of male with central projections of gonopods reaching base of first pair of pereopods when abdomen flexed forward; (B) mesal view of form I gonopod with central projection less than 44% of total gonopod length; (C) ventral view of carapace of male with central projections of gonopods not reaching base of first pair of pereopods when abdomen flexed forward.

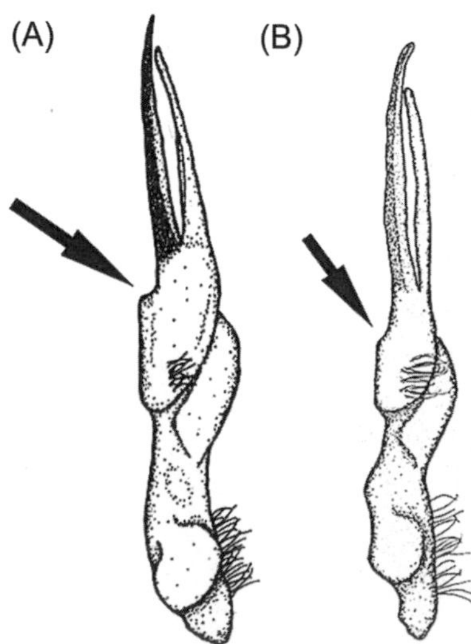

PLATE 16.188.50 Mesial view of form I gonopod with: (A) distinct, angled shoulder at central projection base; or (B) low, rounded shoulder at central projection base.

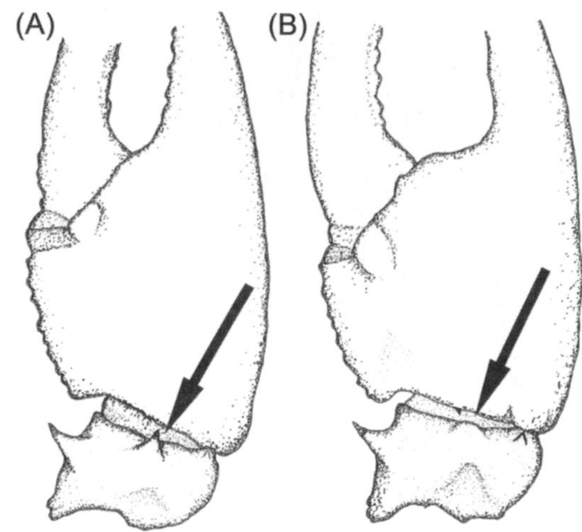

PLATE 16.188.51 Ventral view of carpus: (A) with a distomedian spine; or (B) without a distomedian spine.

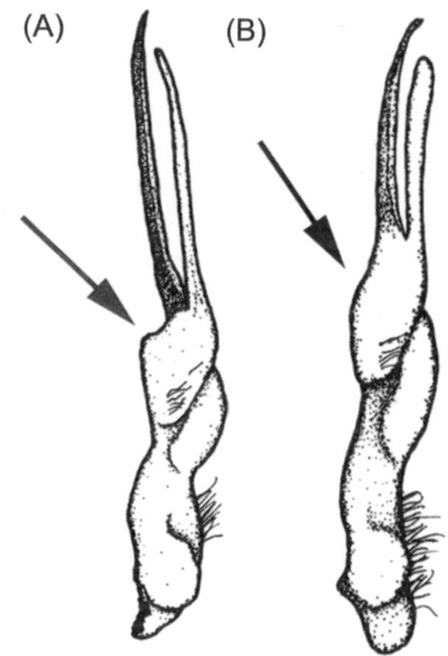

PLATE 16.188.52 Mesial view of form I gonopod with: (A) distinct, angled shoulder at central projection base; or (B) weak, rounded shoulder.

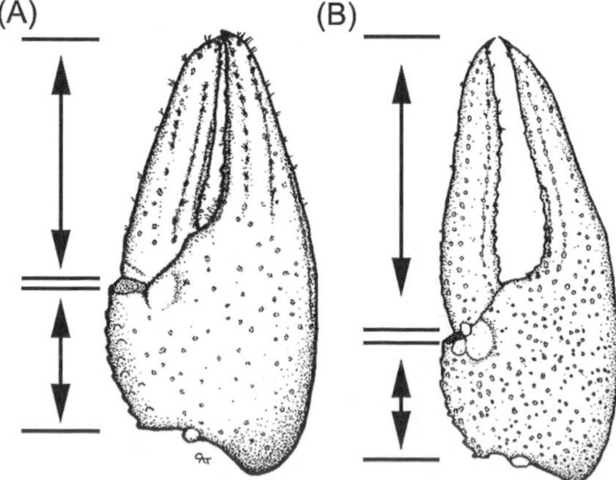

PLATE 16.188.53 Dorsal view of chela with dactyl: (A) less than twice the length of palm mesial margin; or (B) more than twice the length of palm mesial margin.

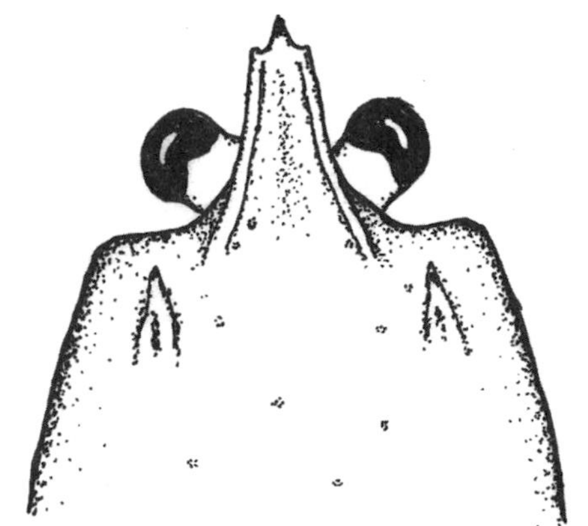

PLATE 16.188.54 Dorsal view of a narrow rostrum.

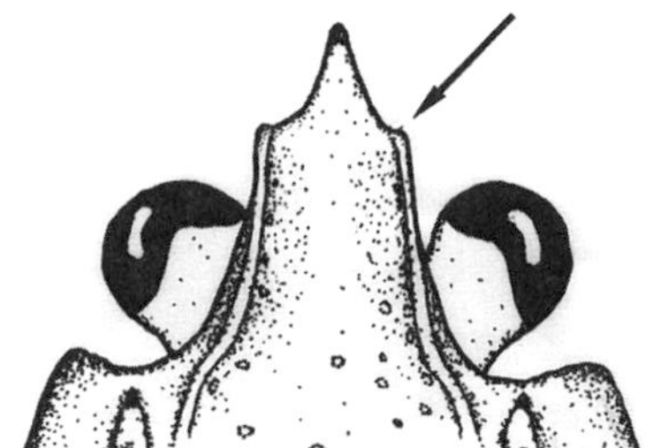

PLATE 16.188.57 Dorsal view of rostrum with margins that terminate in weak spines.

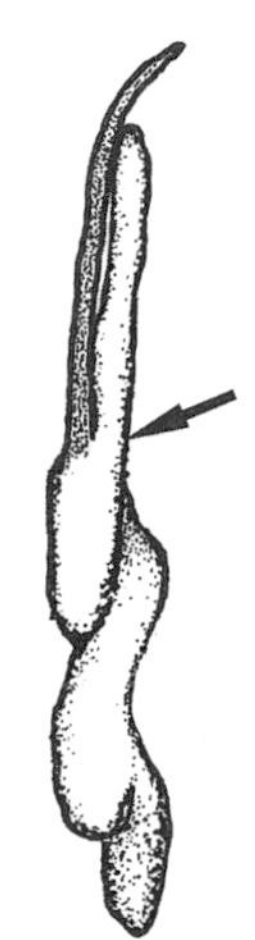

PLATE 16.188.58 Mesial view of form I gonopod of *Orconectes hylas* showing thick mesial process.

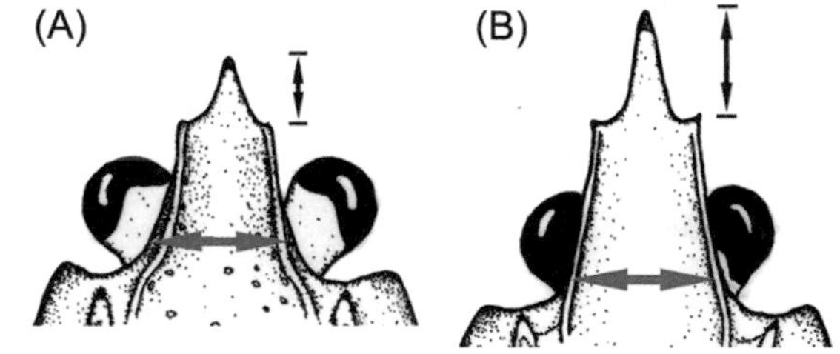

PLATE 16.188.55 Dorsal view of rostrum with acumen: (A) less than 55% of width of rostrum at its base; or (B) more than 55% of width of rostrum at its base.

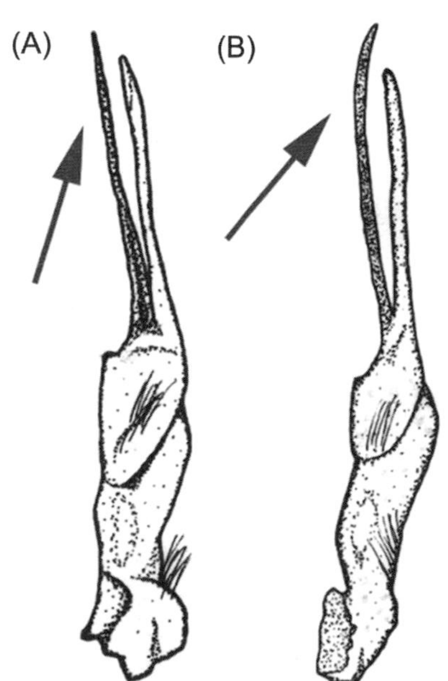

PLATE 16.188.56 Mesial view of form I gonopod with: (A) straight central projection; or (B) curved central projection.

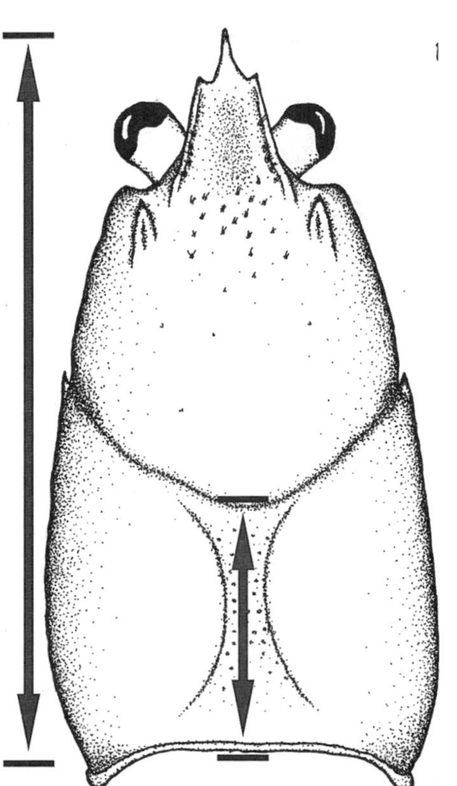

PLATE 16.188.59 Dorsal view of carapace with areola length 33% or less than length of carapace.

96' Chelae not robust, dactyl more than twice palm mesal margin length; chela usually with dactyl and pollex bases separated by a wide gap; body dorsal surfaces covered with dark spots (Pl. 16.188.53 B) .... *Orconectes pardalotus* Wetzel, Poly, & Fetzner, 2005

[USA: Illinois, Kentucky]

97(94) Gonopod central projection curved, either apically or through distal half (Pl. 16.188.56 B).. 98

97' Gonopod central projection straight with no curvature (Pl. 16.188.56 A) .... *Orconectes ronaldi* Taylor, 2000

[USA: Kentucky]

98(97) Gonopod central projection more than 45% of total gonopod length; gonopods reaching first pereopods pair base when abdomen flexed forward (Pl. 16.188.49 A) 99

98' Gonopod central projection 45% or less than total gonopod length; gonopods usually not reaching first pereopods pair base when abdomen flexed forward (Pl. 16.188.49 C) .... *Orconectes placidus* (Hagen, 1870) (in part)

[USA: Cumberland and Tennessee river drainages, Illinois]

99(98) Rostral margins terminate at base of acumen in large spines (Pl. 16.188.26 B) 100

99' Rostral margins terminate at acumen base in weak spines or no spines (Pl. 16.188.57) .... *Orconcetes medius* (Faxon, 1884) (in part)

[USA: Missouri]

100(99) Not found in Neuse and Tar-Pamlico river drainages of North Carolina 101

100' Found in Neuse and Tar-Pamlico river drainages of North Carolina .... *Orconectes carolinensis* Cooper & Cooper, 1995

101(100) Gonopod central projection greater than 49% of total gonopod length 102

101' Gonopod central projection 49% or less than total gonopod length .... *Orconectes cristavarius* Taylor, 2000 (in part)

[USA: Kentucky, Ohio, North Carolina, Tennessee, West Virginia, Virginia]

102(101) Gonopod mesal process not stouter basally, central projection stouter basally 103

102' Gonopod mesal process stouter basally, basal diameter equal to central projection basal diameter (Pl. 16.188.58) .... *Orconectes hylas* (Faxon, 1890)

[USA: Missouri. Introduced into other portions of Missouri]

103(102) Areola length 33% or less than carapace length (Pl. 16.188.59) .... *Orconectes spinosus* (Bundy, 1877)

[USA: Alabama, Georgia, Tennessee]

103' Areola length more than 33% carapace length .... *Orconectes ozarkae* Williams, 1952

[USA: Arkansas, Missouri]

ACKNOWLEDGMENTS

Chris Taylor is grateful to D. P. Johnson and G. A. Schuster for the use of their photographs, and is also grateful to D. P. Johnson for assistance in developing couplets 24-28 in the *Orconectes* keys.

Crustacea: Malacostraca: Decapoda: Astacidea: Cambaridae: *Fallicambarus*: Species

D. Christopher Rogers

Kansas Biological Survey, University of Kansas, Lawrence, KS, USA; The Biodiversity Institute, University of Kansas, Lawrence, KS, USA

Based on form I males. This key is modified from Hobbs (1972), Hobbs & Robison (1989), and Johnson (2011).

1 Gonopod mesal process with a distinct cepahlic process 2

1' Gonopod mesal process lacking a cephalic process (although a small protrusion may be present) 8

2(1) Cephalic process base subequal to or narrower than mesal process adjacent portion 3

2' Cephalic process base at least twice the width of mesal process adjacent portion (Pl. 16.188.60) .... *Fallicambarus devastator* Hobbs & Whiteman, 1987

[USA: Texas]

3(2) Gonopod with central projection not recurved across mesal process 4

3' Gonopod with central projection recurving strongly, arcing back across base of mesal process (Pl. 16.188.61) .... *Fallicambarus petilicarpus* Hobbs & Robison, 1989

[USA: Arkansas]

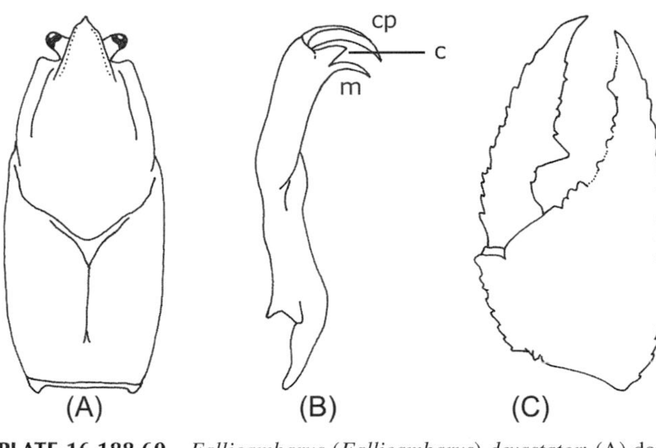

PLATE 16.188.60 *Fallicambarus* (*Fallicambarus*) *devastator*: (A) dorsal view of carapace; (B) mesial view of left first pleopod; (C) dorsal view of chela. c = cephalic process, cp = central projection, m = mesial process. *Modified from Hobbs, 1989.*

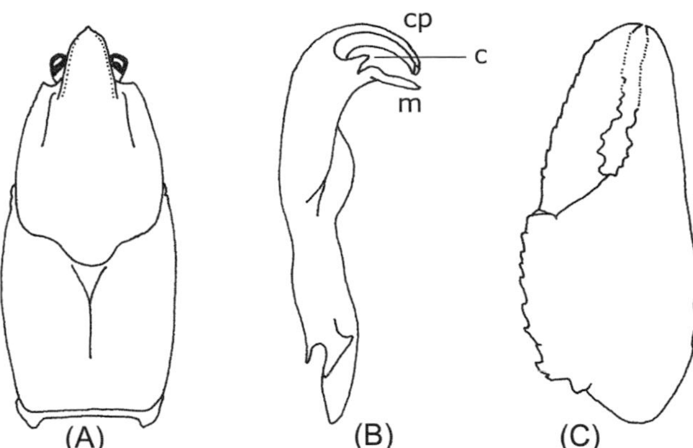

PLATE 16.188.62 *Fallicambarus* (*Fallicambarus*) *macneesei*: (A) dorsal view of carapace; (B) mesial view of left first pleopod; and (C) dorsal view of chela. c = cephalic process, cp = central projection, m = mesial process. *Modified from Hobbs, 1989.*

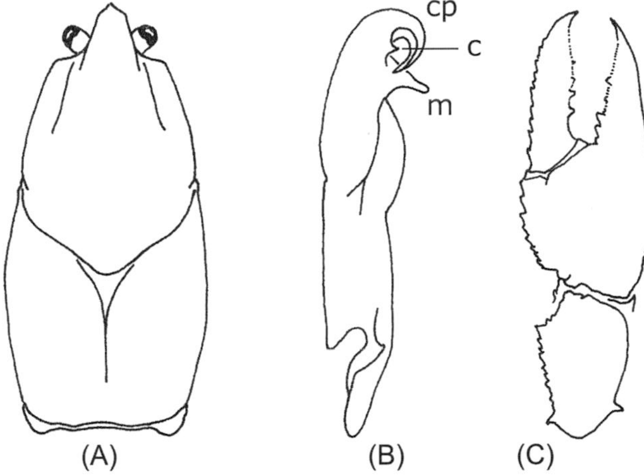

PLATE 16.188.61 *Fallicambarus* (*Fallicambarus*) *pelticarpus*: (A) dorsal view of carapace; (B) mesial view of left first pleopod; and (C) dorsal view of chela. c = cephalic process, cp = central projection, m = mesial process. *Modified from Hobbs & Robison, 1989.*

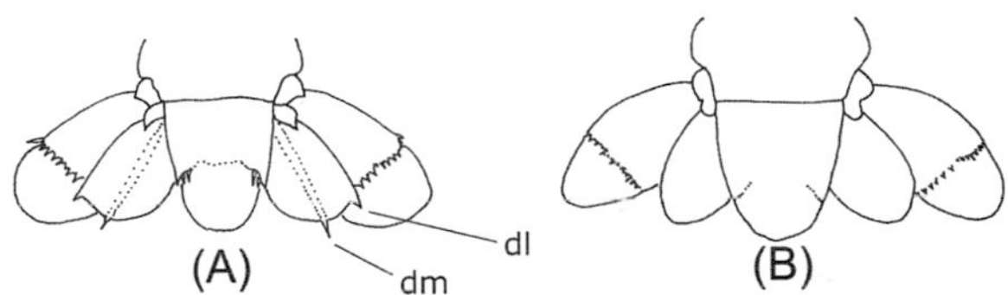

PLATE 16.188.63 (A, B) Dorsal view of telson and uropods of: (A) *Fallicambarus* (*Fallicambarus*) *macneesei*; and (B) *F.* (*Fallicambarus*) *jeanae*. Dl = distolateral spine on inner ramus of uropod, dm = distomesial spine. *Modified from Hobbs, 1981.*

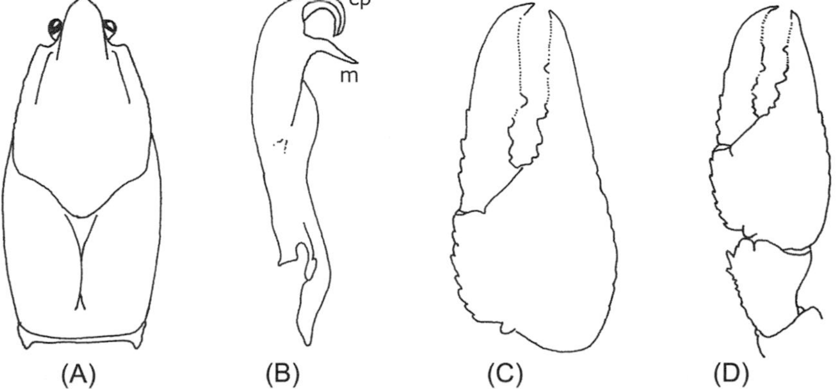

PLATE 16.188.64 *Fallicambarus* (*Falicambarus*) *dissitus*: (A) dorsal view of carapace; (B) mesial view of left first pleopod; and (C, D) dorsal view of chela. cp = central projection, m = mesial process. *Modified from Hobbs, 1989.*

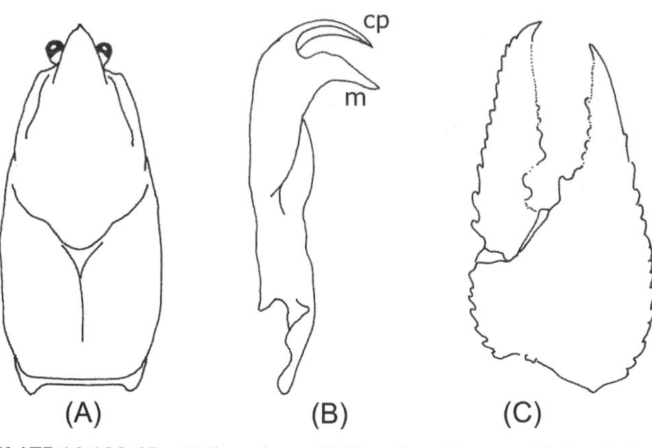

PLATE 16.188.65 *Fallicambarus* (*Fallicambarus*) *jeanae*: (A) dorsal view of carapace; (B) mesial view of left first pleopod; and (C) dorsal view of chela. cp=central projection, m=mesial process. *Modified from Hobbs, 1989.*

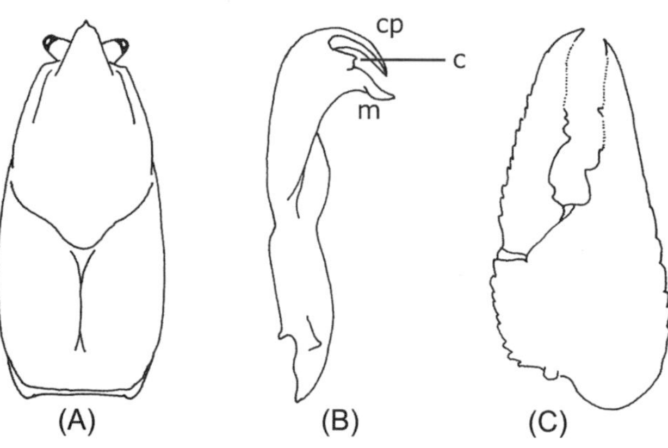

PLATE 16.188.66 *Fallicambarus* (*Fallicambarus*) *strawni*: (A) dorsal view of carapace; (B) mesial view of left first pleopod; and (C) dorsal view of chela. c=cephalic process, cp=central projection, m=mesial process. *Modified from Hobbs, 1989.*

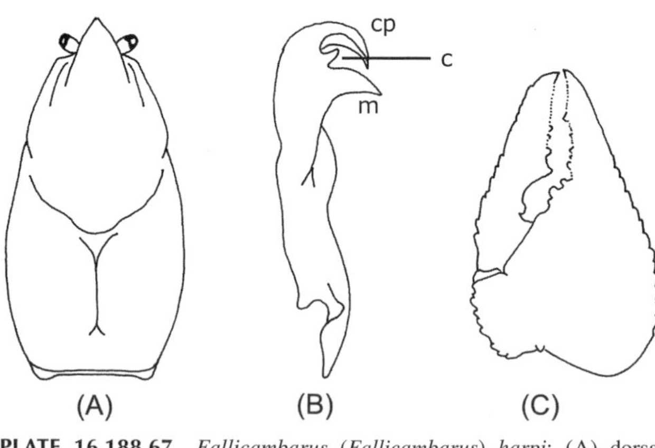

PLATE 16.188.67 *Fallicambarus* (*Fallicambarus*) *harpi*: (A) dorsal view of carapace; (B) mesial view of left first pleopod; and (C) dorsal view of chela. c=cephalic process, cp=central projection, m=mesial process. *Modified from Hobbs, 1989.*

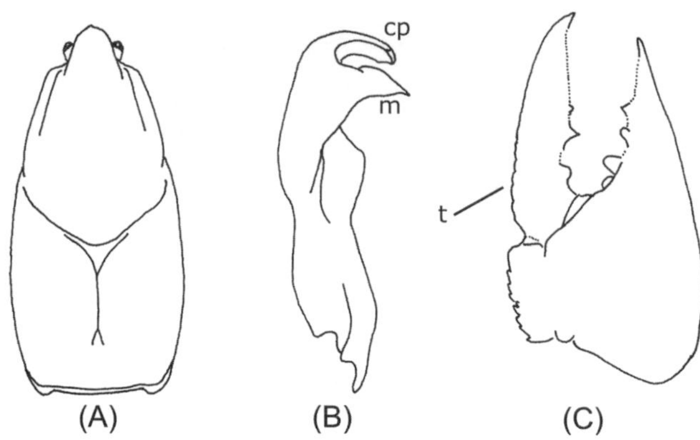

PLATE 16.188.68 *Fallicambarus* (*Creaserinus*) *caesius*: (A) dorsal view of carapace; (B) mesial view of left first pleopod; and (C) dorsal view of chela. c=cephalic process, cp=central projection, m=mesial process, t=tubercles. *Modified from Hobbs, 1989.*

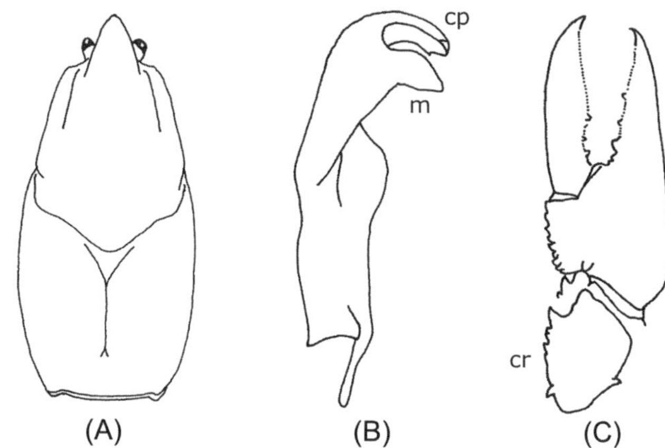

PLATE 16.188.69 *Fallicambarus* (*Creaserinus*) *gilpini*: (A) dorsal view of carapace; (B) mesial view of left first pleopod; and (C) dorsal view of chela. cp=central projection, m=mesial process. *Modified from Hobbs, 1989.*

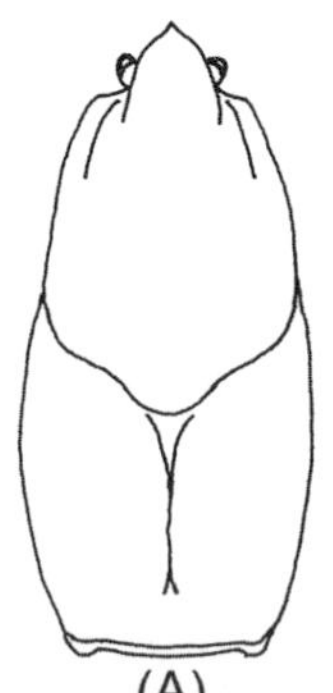

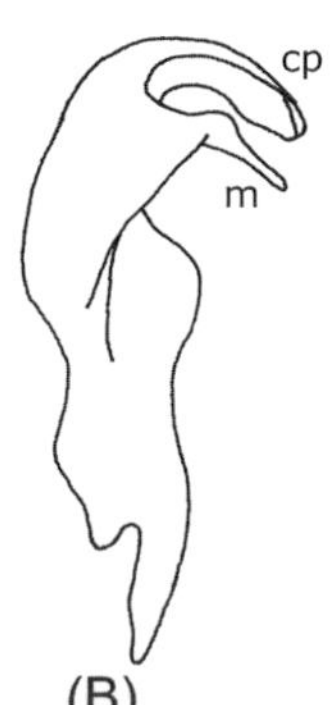

PLATE 16.188.70 *Fallicambarus* (*Creaserinus*) *byersi*: (A) dorsal view of carapace; (B) mesial view of left first pleopod; and (C) dorsal view of chela. cp=central projection, m=mesial process. *Modified from Hobbs, 1989.*

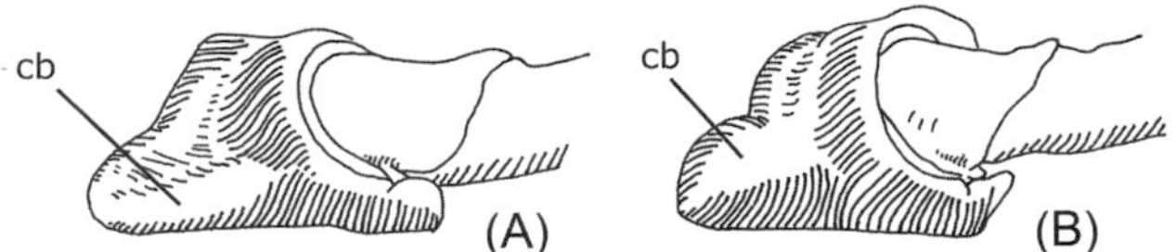

PLATE 16.188.71 Basal podomeres of fourth pereiopod; (A) *Fallicambarus* (*Creaserinus*) *gordoni;* and (B) *F.* (*Creaserinus*) *burrisi*. cb=coxal boss. *After Hobbs and Robinson, 1989.*

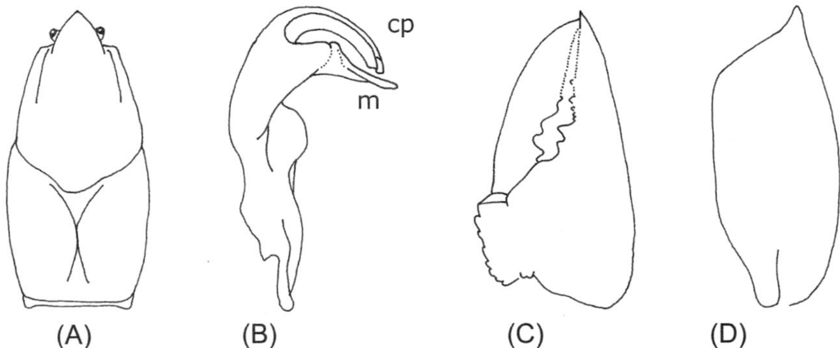

PLATE 16.188.72 *Fallicambarus* (*Creaserinus*) *gordoni*: (A) dorsal view of carapace; (B) mesial view of left first pleopod; and (C) dorsal view of chela; (D) antennal scale. cp=central projection, m=mesial process. *Modified from Hobbs, 1989.*

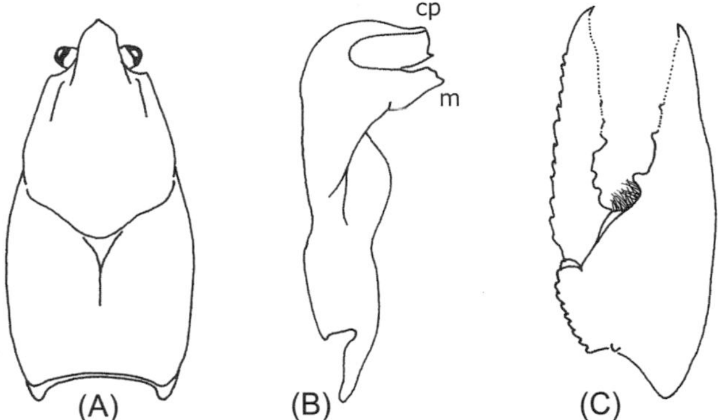

PLATE 16.188.73 *Fallicambarus* (*Creaserinus*) *hortoni*: (A) dorsal view of carapace; (B) mesial view of left first pleopod; and (C) dorsal view of chela. cp=central projection, m=mesial process. *Modified from Hobbs, 1989.*

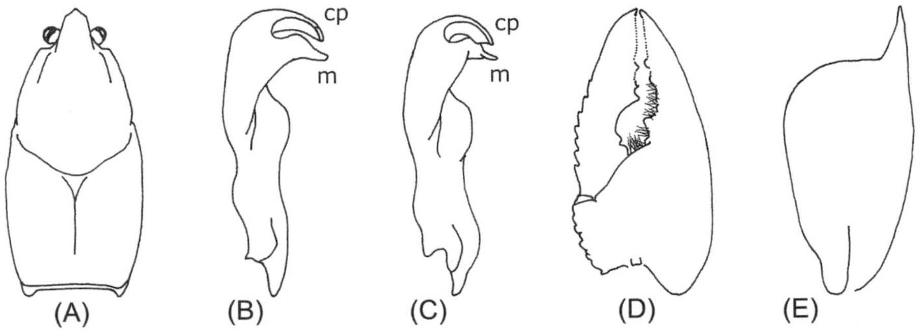

PLATE 16.188.74 *Fallicambarus* (*Creaserinus*) *fodiens*: (A) dorsal view of carapace; (B, C) mesial view of left first pleopod; (D) dorsal view of chela; and (E) antennal scale. cp=central projection, m=mesial process. *Modified from Hobbs, 1989, in part.*

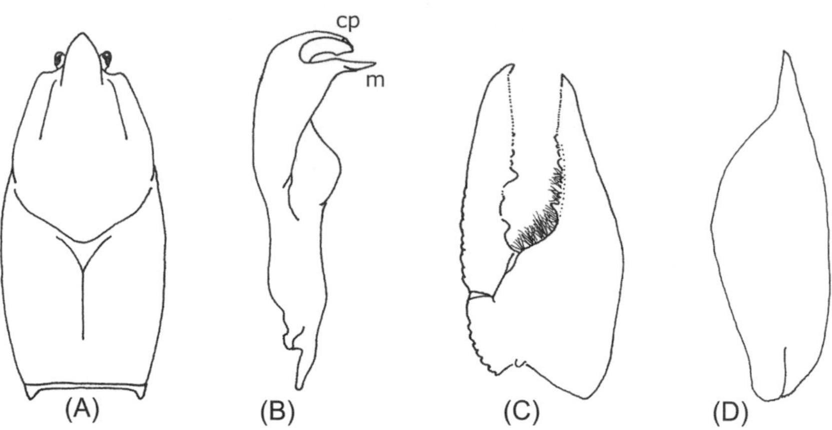

PLATE 16.188.75 *Fallicambarus* (*Creaserinus*) *danielae*: (A) dorsal view of carapace; (B) mesial view of left first pleopod; (C) dorsal view of chela; and (D) antennal scale. cp = central projection, m = mesial process. *Modified from Hobbs, 1989, in part.*

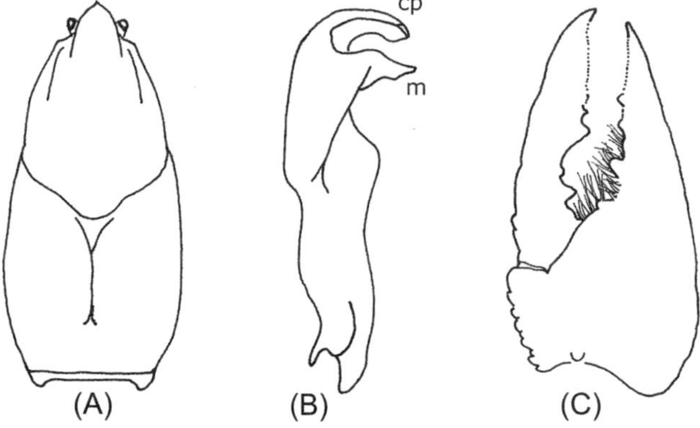

PLATE 16.188.76 *Fallicambarus* (*Creaserinus*) *oryktes*: (A) dorsal view of carapace; (B) mesial view of left first pleopod; and (C) dorsal view of chela. cp = central projection, m = mesial process. *Modified from Hobbs, 1989.*

4(3) Cheliped ischium with a basal protuberance (sufflamen) projecting against the basis (Pl. 16.188.62) 5

4' Cheliped ischium lacking a sufflamen 6

5(4) Uropod endopod bearing a distomedial spine that does not project beyond ramus distal margin (as shown for *F. jeanae* in Pl. 16.188.63 B) .... *Fallicambarus houstonensis* Johnson, 2008

[USA: Texas]

5' Uropod endopod bearing a distomedial spine projecting beyond ramus distal margin (Pl. 16.188.63 A) .... *Fallicambarus macneesei* (Black, 1967)

[USA: Louisiana, Texas]

6(4) Uropod endopod without a distinct distomedial spine 7

6' Uropod endopod bearing a distinct distomedial spine .... *Fallicambarus kountzeae* Johnson, 2008

[USA: Texas]

7(6) Uropod endopod bearing a distomedial premarginal spine; pleopod I cephalic process similar in curvature, and subparallel to mesal process basal portion (Pl. 16.188.66) .... *Fallicambarus strawni* (Reimer, 1996)

[USA: Arkansas]

7' Uropod endopod lacking a distomedial spine; pleopod I cephalic process arcing caudodistally (Pl. 16.188.67) .... *Fallicambarus harpi* Hobbs & Robison, 1985

[USA: Arkansas]

8(1) Gonopod with a basomedial spur (Pl. 16.188.65) 9

8' Gonopod without a basomedial spur 11

9(8) Uropod endopod bearing a distinct distomedial spine 10

9' Uropod endopod without a distinct distomedial spine (Pl. 16.188.63 B) .... *Falicambarus jeanae* Hobbs, 1973

[USA: Arkansas]

10(9) In situ both gonopods with central projections directed proximomesally and typically overlapping (Pl. 16.188.64) .... *Fallicambarus dissitus* (Penn, 1955)

[USA: Arkansas, Louisiana]

10' In situ both gonopods with central projections directed proximoventrally and never overlapping .... *Fallicambarus wallsi* Johnson, 2011

11(9) Cheliped merus ventral surface with tubercles in a single row 12

11' Cheliped merus ventral surface with tubercles in two rows 13

12(11) Chela dactyl mesal surface tuberculate proximally; uropod endopod distolateral spine absent (Pl. 16.188.68) .... *Fallicambarus caesius* Hobbs, 1975

[USA: Arkansas]

12' Chela dactyl mesal surface smooth; uropod endopod distolateral spine present (Pl. 16.188.69) .... *Fallicambarus gilpini* Hobbs & Robison, 1989

[USA: Arkansas]

13(11) Cheliped palm mesal surface glabrous, or with a very few scattered seate; pleopod I strongly reflexed 14

13' Cheliped palm mesal surface densely hirsute; pleopod I straight to slightly curved 16

14(13) Gonopod central projection and mesal process approximately one-third the length of gonopod shaft 15

14' Gonopod central projection and mesal process approximately one-fourth the length of gonopod shaft (Pl. 16.188.70) .... *Fallicambarus byersi* (Hobbs, 1941)

[USA: Alabama, Florida, Mississippi]

15(14) Antennal scale with apex produced, apically acute; pereopod IV coxa with boss produced as as longitudinal ridge (Pls. 16.188.71 A, 72) .... *Fallicambarus gordoni* Fitzpatrick, 1987

[USA: Mississippi]

15' Antennal scale with apex rounded to subtruncate; pereopod IV coxa with boss bulbiform (Pl. 16.188.71 B) .... *Fallicambarus burrisi* Fitzpatrick, 1987

[USA: Alabama, Mississippi]

16(13) Gonopod central projection straight, apex tapering or notched 17

16' Gonopod central projection with cephalic margin arcuate, apex projecting at right angle to gonopod shaft, with apex truncate (Pl. 16.188.73) .... *Fallicambarus hortoni* Hobbs & Fitzpatrick, 1970

[USA: Tennessee]

17(16) Antennal scale without angle between medial and distal margins; abdomen narrowly joined to thorax 18

17' Antennal scale with obtuse angle between medial and distal margins; abdomen broadly joined to thorax (Pl. 16.188.74) .... *Fallicambarus fodiens* (Cottle, 1863)

[Canada: Ontario. USA: Alabama, Arkansas, Illinios, Indiana, Kentucky, Michigan, Mississippi, Missouri, Ohio, Oklahoma, Tennessee, Wisconsin]

18(17) Gonopod mesal process extending beyond central projection (Pl. 16.188.75) .... *Fallicambarus danielae* Hobbs, 1975

[USA: Alabama, Mississippi]

18' Gonopod mesal process subequal in length to central projection (Pl. 16.188.76) .... *Fallicambarus oryktes* Penn & Marlow, 1959

[USA: Alabama, Louisiana, Mississippi]

Crustacea: Malacostraca: Decapoda: Astacidea: Cambaridae: *Cambarus*: Species

Roger F. Thoma
Midwest Biodiversity Institute, Hilliard, OH, USA

All gonopod characters require form I males.

1 Eyes not pigmented, highly reduced (Pl. 16.189.1 B) 2

1' Eyes pigmented, large (Pl. 16.189.1 A) 14

2(1) Chela palm with one row of tubercles 3

2' Chela palm with multiple rows of tubercles 6

3(2) Areola narrow, with no more than two rows of punctations; chelae studded with numerous setae (Pl. 16.189.2); gonopod curved more than 90° 4

3' Areola wide enough to allow three or more rows of punctations; chelae with few to no setae (Pl. 16.189.1 B), gonopod curved 90°, central projection shorter than mesal process and with subapical notch .... *Cambarus cryptodytes* Hobbs, 1941

[USA: Florida, Georgia]

4(3) Chela narrow, less than 30% of total length (Pl. 16.189.3 A) 5

4' Chela wider, greater than 30% of total length (Pl. 16.189.3 B); cervical spine present (Pl. 16.189.3 A); gonopod recurved more than 90°, central projection with subapical notch, slightly longer than mesal process .... *Cambarus setosus* (Faxon, 1889)

[USA: Missouri, Oklahoma]

5(4) Chela palm with one well-defined row of tubercles; gonopod central projection equal in length to mesal process and lacking a subapical notch .... *Cambarus zophonastes* Hobbs & Bedinger, 1964

[USA: Arkansas]

5' Chela palm with two rows of tubercles; gonopod central projection slightly shorter than mesal process, and with subapical notch .... *Cambarus tartarus* Hobbs & Cooper, 1972

[USA: Oklahoma]

6(2) Rostrum acuminate, without lateral spines (Pl. 16.189.5 A) 7

6' Rostrum with well defined angle at tip, lateral spines or tubercles present (Pl. 16.189.5 B) 8

7(6) Gonopod curved 90°, central projection slightly shorter than mesal process, and with subapical notch; posterior edge of annulus ventralis formed by two semi symmetrical, scleritized, elevated, ridges that meet at the midpoint (Pl. 16.189.6 A, C) .... *Cambarus hubrichti* Hobbs, 1952

[USA: Missouri]

7' Gonopod curved 45°, central projection lacking subapical notch and slightly shorter than mesal process; annulus ventralis posterior edge formed by two asymmetrical, scleritized, elevated, ridges; one bent at 90°angle at meeting point (Pl. 16.189.6 B, D) .... *Cambarus laconensis* Buhay & Crandall, 2009

[USA: Alabama]

8(6) Cervical spines absent (Pl. 16.189.7 B) 9

8' Cervical spines present (Pl. 16.189.7 A) 10

9(8) Chela palm approximately equal in length to dactyl (Pl. 16.189.8 A); gonopod curved 90°, central projection slightly shorter than or equal to mesal process and lacking subapical notch .... *Cambarus speleocoopi* Buhay & Crandall, 2009

[USA: Alabama]

9' Chela palm approximately one half length of dactyl (Pl. 16.189.8 B); gonopod curved more than 90°, central projection equal in length to mesal process and with subapical notch .... *Cambarus subterraneus* Hobbs III, 1993

[USA: Oklahoma]

10(8) Areola wide, with more than two punctations at narrowest part (Pl. 16.189.7 A) 11

10' Areola narrow, with two or one punctations at narrowest part (Pl. 16.189.5 B); gonopod recurved more than 90°, central projection slightly shorter than mesal process, and with subapical notch .... *Cambarus aculabrum* Hobbs & Brown, 1987

[USA: Arkansas]

11(10) Carapace hepatic region without numerous spines (Pl. 16.189.7 D) 12

11' Hepatic region with numerous spines (Pl. 16.189.7 C); gonopod curved slightly less than 90°, central projection extending slightly less than mesal process, and with subapical notch .... *Cambarus hamulatus* (Cope, 1881)

[USA: Alabama, Tennessee]

12(11) Chela width less than 25% of chela length (Pl. 16.189.9 B) 13

12' Chela width greater than 25% of chela length (Pl. 16.189.9 A); gonopod curved slightly more than 90°, central projection equal in length to mesal process, and with subapical notch .... *Cambarus jonesi* Hobbs & Barr, 1960

[USA: Alabama]

13(12) Gonopod curved slightly more than 90° from main shaft, central projection equal in length to mesal process, and with subapical notch; annulus ventralis posterior edge formed by two symmetrical, scleritized, elevated, ridges that meet at the midpoint (Pl. 16.189.10 A, B) .... *Cambarus veitchorum* Cooper & Cooper, 1997

[USA: Alabama]

13' Gonopod curved slightly less than 45° from main shaft, central projection equal in length to mesal process, and without subapical notch; annulus ventralis posterior edge lacking two asymmetrical, scleritized, elevated, ridges (Pl. 16.189.10 A, B) .... *Cambarus pecki* Hobbs, 1967

[USA: Louisiana, Mississippi]

14(1) Areola closed or very narrow with no more than two punctations present in narrowest part (Pl. 16.189.11) 15

14' Areola wide, with more than two rows of punctations (Pl. 16.189.11) 43

15(14) Areola closed (Pl. 16.189.11 A) 16

15' Areola narrowly open (Pl. 16.189.11 B) 31

16(15) Chelae palm mesal margin bearing one row of tubercles (Pl. 16.189.12 A) 17

16' Chelae palm mesal margin bearing two or more rows of tubercles or chela dorsal surface with numerous scattered tubercles (Pl. 16.189.12 B) 18

17(16) Color in life all red; antennal scale twice as long as wide (Pl. 16.189.13 A); gonopod curved 90°, central projection extending less than mesal process and with a subapical notch .... *Cambarus batchi* Schuster, 1976

[USA: Kentucky]

17' Color in life mostly blue with gold trim; antennal scale three times as long as wide (Pl. 16.189.13 B); gonopod curved slightly more than 90°, central projection extending slightly less than or equal to mesal process and without a subapical notch .... *Cambarus gentryi* Hobbs, 1970

[USA: Tennessee]

18(16) Chela palm mesal edge to dorsal midpoint or more with tubercles, only one distinct tubercle row at mesal most edge, all other tubercles scattered (Pl. 16.189.12 B) 19

18' Chela palm with mesal most margin bearing two complete tubercle rows, occasionally dorsal surface with a third row (Pl. 16.189.14) 22

19(18) Chela palm ventral surface with only one to three tubercles (Pl. 16.189.15 B) 20

19' Chela palm dorsal and ventral surface with numerous scattered tubercles (Pl. 16.189.15 A); gonopod curved 90°, central projection equal in length to mesal process, and with a subapical notch .... *Cambarus stockeri* Thoma, 2011

[USA: Georgia, Tennessee]

20(19) Uropod medial spine extending to distal margin (Pl. 16.189.16 B) 21

20' Uropod medial spine extending noticeably beyond distal margin (Pl. 16.189.16 A); gonopod curved 90°, central projection extending slightly less than mesal process, and without subapical notch .... *Cambarus acanthura* Hobbs, 1981

[USA: Alabama, Georgia, North Carolina, Tennessee]

21(20) Rostrum, in side view, strongly deflected beyond eyes (Pl. 16.189.17 A); gonopod recurved 90°, central projection extending less than mesal process, and with or without subapical notch .... *Cambarus polychromatus* Thoma, Jezerinac, & Simon, 2005

[USA: Alabama, Illinios, Indiana, Ohio, Tennessee, Kentucky]

21' Rostrum, in side view, straight throughout length (Pl. 16.189.17 B); gonopod recurved 90°, central projection extending less than mesal process, and without subapical notch .... *Cambarus thomai* Jezerinac, 1993

[USA: Kentucky, Ohio, Pennsylvania, Tennessee, West Virginia]

22(18) Suborbital angle obsolete, rounded (Pl. 16.189.18 C) 23

22' Suborbital angle acute or obtuse but clearly evident (Pl. 16.189.18 A, B) 24

23(22) Antennal scale widest at midpoint (Pl. 16.189.19) .... *Cambarus daughertyensis* Cooper & Skelton, 2003

[USA: Georgia]

23' Antennal scale widest distal to midpoint (Pl. 16.189.19) .... *Cambarus deweesae* Bouchard & Etnier, 1979

[USA: Kentucky, Tennessee]

24(22) Suborbital angle acute (Pl. 16.189.18 A) 25

24' Suborbital angle obtuse (Pl. 16.189.18 B) 28

25(24) Chelae pollex grasping surface without basal gap between tubercles (Pl. 16.189.20 B) 26

25' Chelae pollex grasping surface with basal gap between second and third tubercles (Pl. 16.189.20 A); gonopod curved 90°, central projection extending less than mesal process and with a subapical notch .... *Cambarus cymatilis* Hobbs, 1970

[USA: Georgia, Tennessee]

26(25) Carapace cephalic section less than 1.4 times areola length (Pl. 16.189.21) 27

26' Carapace cephalic section more than 1.4 times areola length (Pl. 16.189.21); gonopod curved 90°, central projection extending equal to mesal process, and with subapical notch .... *Cambarus diogenes* Girard, 1852

[Canada: Ontario. USA: east of Continental Divide]

27(26) Carapace width at suborbital angles approximately 54% of greatest carapace width; epistome truncate apically (Pl. 16.189.22 A); gonopod curved 90°, central projection equal in length to mesal process, and with a subapical notch .... *Cambarus miltus* Fitzpatrick, 1978

[USA: Alabama, Florida]

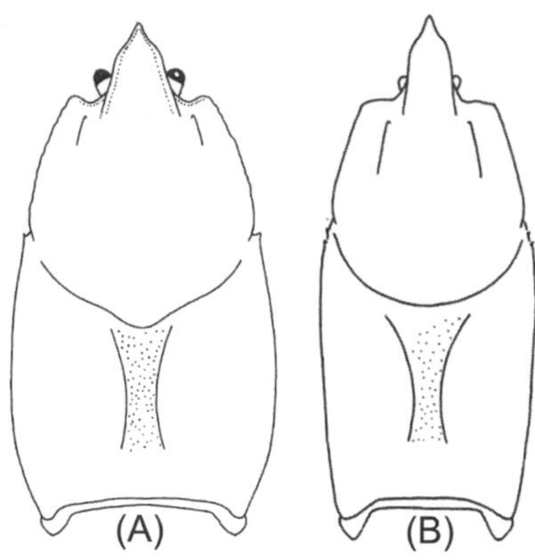

PLATE 16.189.01 Eye development: (A) eyes developed and pigmented; (B) eyes not developed or pigmented.

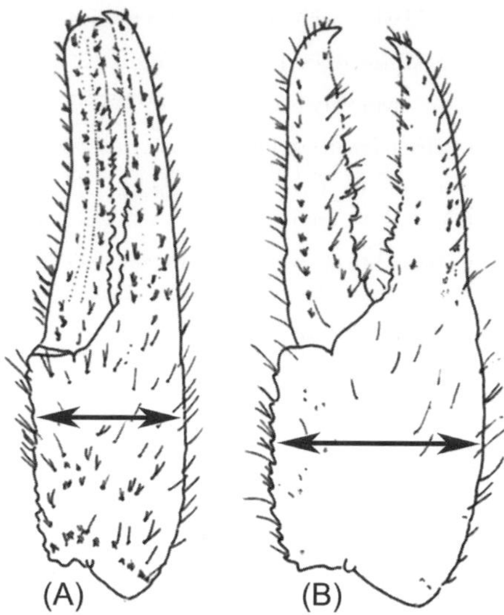

PLATE 16.189.03 Chela width: (A) narrow less than 30% length; (B) wider greater than 30% length.

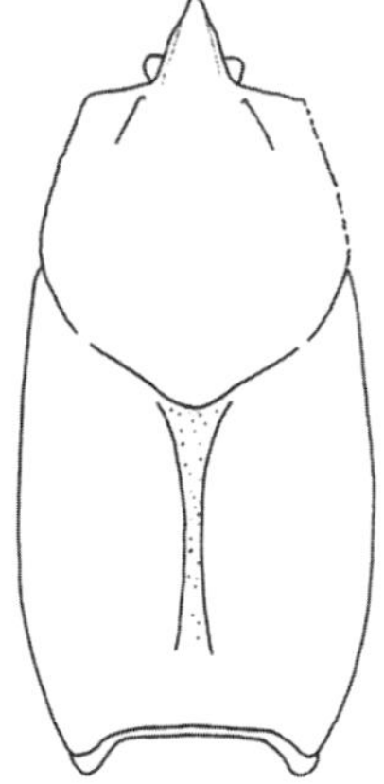

PLATE 16.189.02 Areola narrow, no more than two rows of punctations.

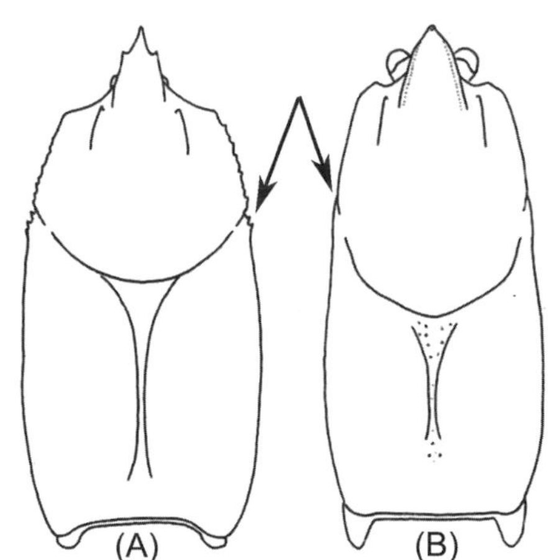

PLATE 16.189.04 Cervical spine present (A), cervical spine absent (B).

27' Carapace width at suborbital angles approximately 44% of greatest carapace width; epistome apically rounded (Pl. 16.189.22 B); gonopod curved 90°, central projection extending less than mesal process, and with a subapical notch .... *Cambarus ludovicianus* Faxon, 1884

[USA: Alabama, Arkansas, Louisiana, Tennessee, Texas]

28(24) Abdominal plura lateral margins rounded (Pl. 16.189.23 B) [Note: *Cambarus striatus* may key here if specimen has signs of a second row of palmer tubercles. It may be distinguished in that it has a developed obtuse orbital angle (Pl. 16.189.18 B) whereas *C. strigosus* and *C. truncatus* have the suborbital angle rounded or obsolete (Pl. 16.189.18 C) 29

28' Abdominal plura lateral margins slightly angular (Pl. 16.189.23 A); gonopod curved greater than 90°, central projection extending less than mesal process, and with a subapical notch .... *Cambarus harti* Hobbs, 1981

[USA: Georgia]

29(28) Uropod endopod with distomedian ridge terminating in spine, distolateral spine present (Pl. 16.189.24 A); gonopod curved slightly greater than 90°, central projection extending less than mesal process, and with a subapical notch, mesal process caudomesal surface with row of plumose setae .... *Cambarus strigosus* Hobbs, 1981

[USA: Georgia]

29' Uropod endopod almost always lacking a distomedian ridge and distolateral spines (Pl. 16.189.24 B); gonopod curved 90°, central projection extending less than mesal process, and with a subapical notch .... *Cambarus truncatus* Hobbs, 1981

[USA: Georgia]

30(15) Chela palm mesal surface with one complete tubercle row, if second row present, incomplete (Pl. 16.189.25 B) 31

30' Chela palm mesal surface with two complete tubercle rows (Pl. 16.189.25 A) 37

31(30) Chela palm mesal edge tubercles adpressed (Pl. 16.189.26 B) 32

31' Chela palm mesal edge tubercles cristaform or pronounced (Pl. 16.189.26 A) 33

32(31) Cervical groove lateral posterior edge with multiple strong tubercles (Pl. 16.189.27 A); gonopod curved 90°, central projection equal to mesal process, and with a subapical notch .... *Cambarus davidi* Cooper, 2000

[USA: North Carolina]

32' Cervical grove lateral posterior edge with weak or no tubercles (Pl. 16.189.27 B); gonopod curved 90°, central projection shorter than mesal process, and with a subapical notch .... *Cambarus ortmanni* Williamson, 1907

[USA: Indiana, Kentucky, Ohio]

33(31) Chela palm mesal edge tubercles cristaform (Pl. 16.189.26 A) 34

33' Chela palm mesal edge tubercles not cristaform (Pl. 16.189.20 A); gonopod highly variable, normally curved greater than 90°, central projection equal to or longer than mesal process, and with or without subapical notch .... *Cambarus striatus* Hay, 1902 (in part)

[USA: Alabama, Florida, Georgia, Kentucky, Mississippi, South Carolina, Tennessee]

34(33) Chelae lateral margin smooth (Pl. 16.189.28 B, C) 35

34' Chelae lateral margin costate (Pl. 16.189.28 A); gonopod curved 90°, central projection equal to or shorter than mesal process, and with a subapical notch .... *Cambarus dubius* Faxon, 1884 (in part)

[USA: Kentucky, Maryland, North Carolina, Pennsylvania, Tennessee, Virginia, West Virginia]

35(34) Chelae lacking well-developed setae 36

35' Chelae with well-developed prominent setae; gonopod recurved more than 90°, central projection sickle shaped, lacking subapical notch, longer than mesal process, mesal process with nipple .... *Cambarus causeyi* (Reimer, 1966)

[USA: Arkansas]

36(35) Chela palm basal tubercles acute (Pl. 16.189.28 C), color red; gonopod curved slightly greater than 90°, central projection equal in length to mesal process, and without a subapical notch .... *Cambarus carolinus* (Erichson, 1846)

[USA: North Carolina, South Carolina, Tennessee]

36' Chela palm basal tubercles rounded (Pl. 16.189.28 B), color blue; gonopod curved 90°, central projection slightly shorter than mesal process, and with subapical notch .... *Cambarus monongalensis* Ortmann, 1905

[USA: Pennsylvania, Virginia, West Virginia]

37(31) Epistomal zygoma moderately to strongly arched (Pl. 16.189.29 B) 38

37' Epistomal zygoma gently arched (Pl. 16.189.29 A); gonopod curved 90°, central projection shorter than mesal process, and without a subapical notch .... *Cambarus graysoni* Faxon, 1914

[USA: Alabama, Kentucky, Tennessee]

38(37) Antennal scale width greater than 35% of length (Pl. 16.189.30 A, B) 39

38' Antennal scale width less than 33% of length (Pl. 16.189.30 C); gonopod curved greater than 90°, central projection sickle-shaped, longer than mesal process, and without a subapical notch .... *Cambarus reduncus* Hobbs, 1956

[USA: North Carolina, South Carolina]

39(38) Gonopod central projection shorter than mesal process (Pl. 16.189.31 A, B); antennal scale width 40% or less of length 40

39' Gonopod central projection equal to or longer than mesal process (Pl. 16.189.31 C, D), antennal scale width 42% or more of length 41

40(38) Antennal scale mesal margin strongly declivous; rostrum width at orbits approximately 47% of length from post orbital ridge to rostral apex (Pl. 16.189.32 A); gonopod curved slightly more than 90°, central projection shorter than mesal process and with a slight subapical notch .... *Cambarus catagius* Hobbs & Perkins, 1967

[USA: North Carolina]

40' Antennal scale mesal margin not strongly declivous; rostrum width at orbits approximately 70% of length from post orbital ridge to rostral apex (Pl. 16.189.32 B); gonopod curved greater than 90°, central projection shorter than mesal process, and without a subapical notch .... *Cambarus reflexus* Hobbs, 1981

[USA: Georgia, South Carolina]

41(38) Suborbital angle obtuse to obsolete; areola more than 10 times longer than broad (Pl. 16.189.33 B, C) 42

41' Suborbital angle acute; areola less than 10 times longer than broad (Pl. 16.189.33 A); gonopod highly variable, normally curved more than 90°, central projection normally subequal in length to mesal process, and normally without a subapical notch .... *Cambarus latimanus* (Le Conte, 1856)

[USA: Alabama, Florida, Georgia, North Carolina, South Carolina, Tennessee]

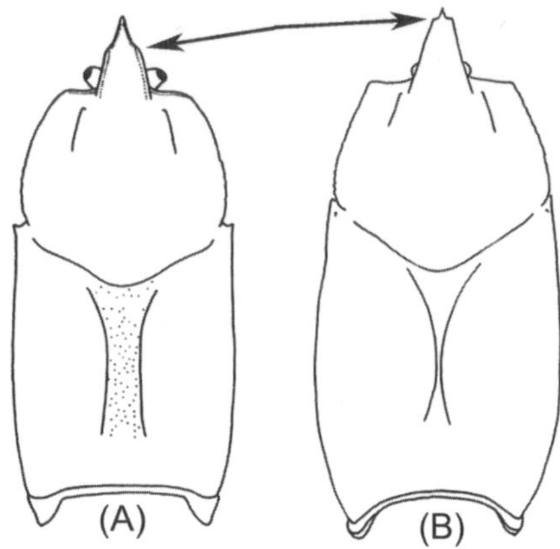

PLATE 16.189.05 Rostrum acuminate, without lateral spines (A), rostrum with angle near tip and lateral spines (B).

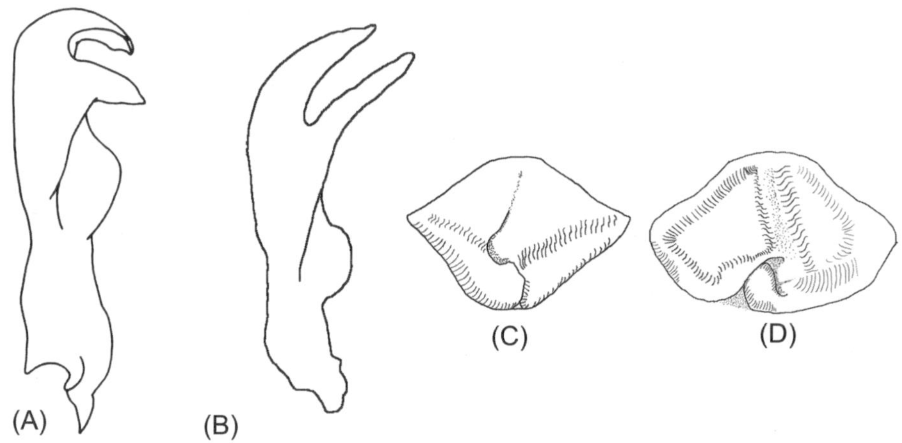

PLATE 16.189.06 *Cambarus hubrichti* first form male gonopod (A) and annulus ventralis (C), *C. laconensis* first form male gonopod (B) and annulus ventralis (D).

42(41) Antennal scale width 48% or more of length; suborbital angle obsolete (Pls. 16.189.30 A, 33 C); gonopod curved 90°, central projection shorter than mesal process, and without a subapical notch .... *Cambarus pyronotus* Bouchard, 1978

[USA: Florida]

42' Antennal scale width less than 48% of length; suborbital angle obtuse (Pls. 16.189.30 B, 33 B); gonopod highly variable, normally curved greater than 90°, central projection equal to or longer than mesal process, and with or without a subapical notch .... *Cambarus striatus* Hay, 1902 (in part)

[USA: Alabama, Florida, Georgia, Kentucky, Mississippi, South Carolina, Tennessee]

43(15) Rostrum lancelet-shaped, acumen margins lacking 45° angle and/or lateral spines or tubercles (Pl. 16.189.34 A) 44

43' Rostrum not lancelet-shaped, acumen margins angled at 45° or more toward apex, or with lateral spines or tubercles (Pl. 16.189.34 B–D) 76

44(43) Rostrum margins noticeably thicker before apex (Pl. 16.189.35) 45

44' Rostrum margins not thicker before apex, thickness uniform throughout length (Pl. 16.189.34 A) 48

45(44) Chelae margin with one complete row of tubercles, bearing numerous setae (Pl. 16.189.36) 46

45' Chelae margin with one complete row of tubercles and a second, dorsolateral, incomplete row, no setae present (Pl. 16.189.25 B); gonopod curved slightly more than 90°, central projection shorter than mesal process, and with a subapical notch .... *Cambarus tuckasegee* Cooper & Schofield, 2002

[USA: North Carolina]

46(45) Chela palm comprising less than 48% of total chelae length (Pl. 16.189.36 B) 47

46' Chela palm comprising more than 48% of total chelae length (Pl. 16.189.36 A); gonopod Recurved 90°, central projection equal in length to mesal process, and without a subapical notch .... *Cambarus brachydactylus* Hobbs, 1953

[USA: Tennessee]

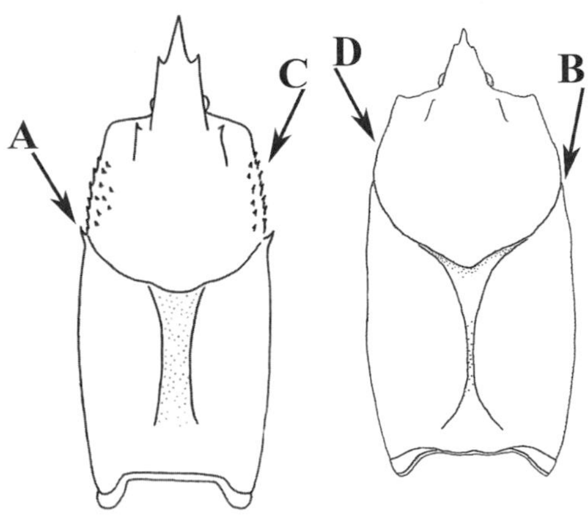

PLATE 16.189.07 Cervical spine present (A), cervical spine absent (B), hepatic spines present (C), hepatic spines absent (D).

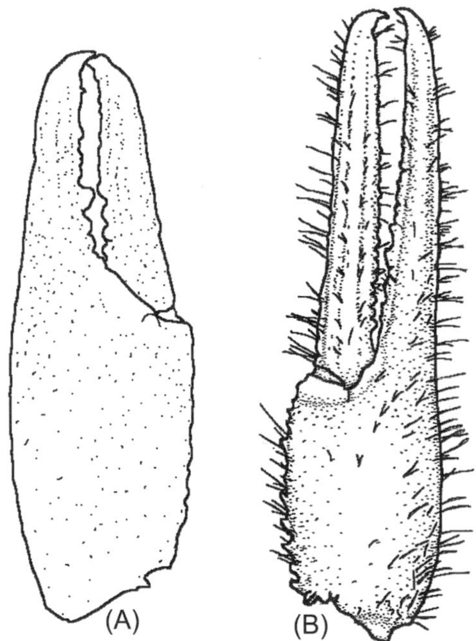

PLATE 16.189.08 Palm length approximately equal to dactyl length (A), palm length approximately half dactyl length (B).

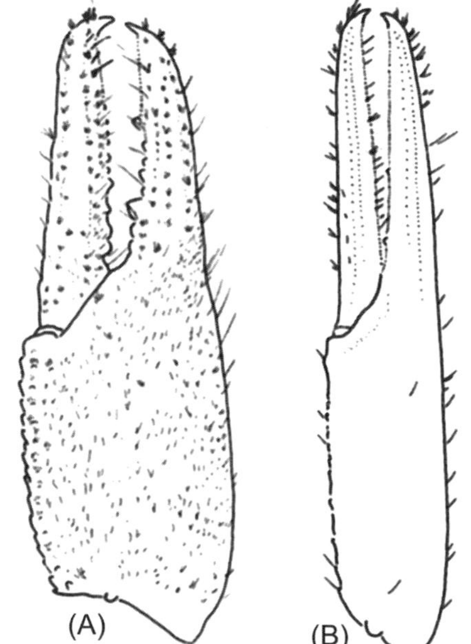

PLATE 16.189.09 Chela width greater than 25% of chela length (A), chela width less than 25% of chela length (B).

47(46) Gonopod central projection curved 90° from main axis, as long as mesal process, and without subapical notch, caudal process developed; annulus ventralis asymmetrical with sinuous sinus down middle line and a two-part, scleritized, raised posterior margin .... *Cambarus friaufi* Hobbs, 1953

[USA: Kentucky, Tennessee]

47' Gonopod central projection curved less than 45° from main axis, extending less than mesal process, and with a subapical notch, caudal process developed; annulus ventralis nearly symmetrical with linear sinus down middle of anterior 2/3 and a two-part, scleritized, raised anterior margin .... *Cambarus williami* Bouchard & Bouchard, 1995

[USA: Tennessee]

48(44) Chela margin with one complete tubercle row of and a second incomplete row, second row smaller than first (Pl. 16.189.25 B) 49

48' Chela margin with two complete tubercle rows (Pl. 16.189.25 A) 62

49(48) Cervical groove posterior margin bearing a cervical spine (Pl. 16.189.7 A); gonopod central projection with or without a subapical notch 50

49' Cervical groove posterior margin without a cervical spine (Pl. 16.189.7 B); gonopod central projection always with a subapical notch 53

50(49) Chela closing without a significant gap (Pl. 16.189.25 A) 51

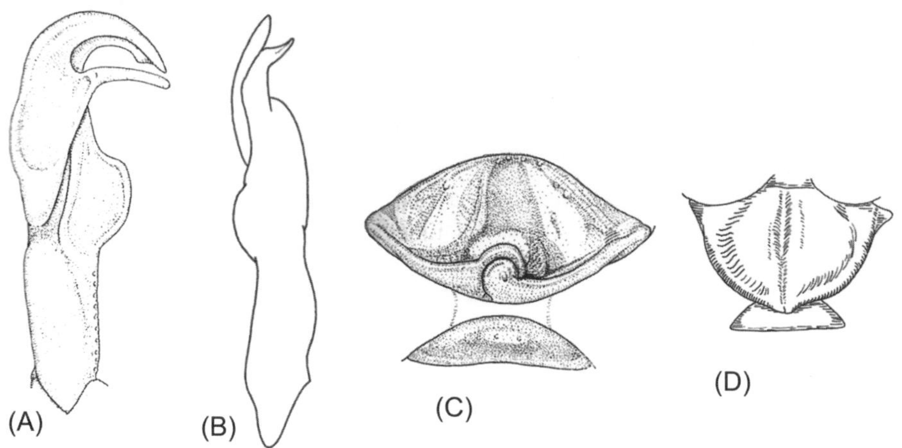

PLATE 16.189.10 *Cambarus veitchorum* first form male gonopod (A) and annulus ventralis (C), *C. pecki* first form male gonopod (B) and annulus ventralis (D).

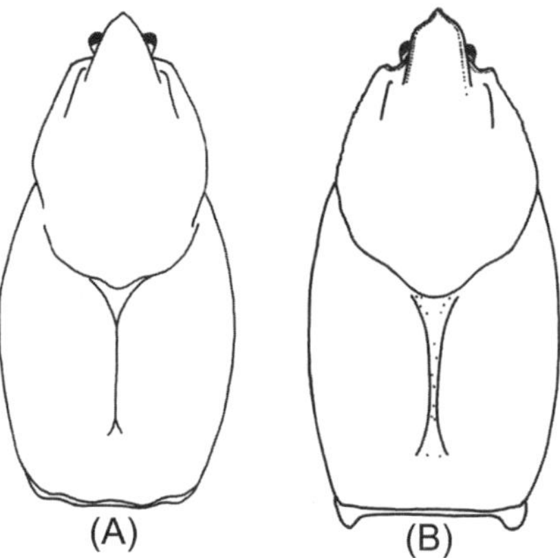

PLATE 16.189.11 Areola closed (A), areola narrowly open (B).

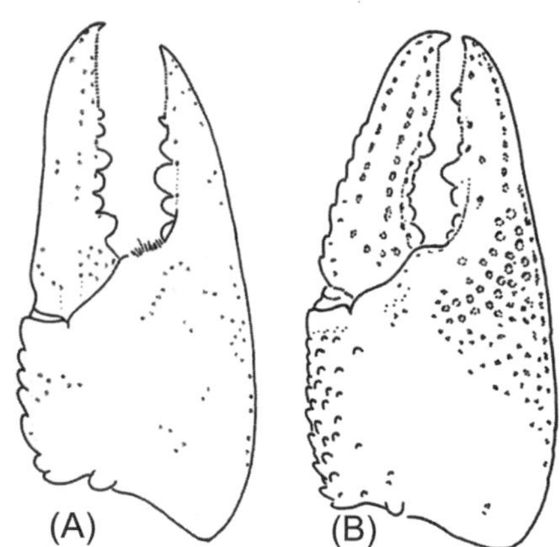

PLATE 16.189.12 One row of palmer tubercles (A), two plus rows of scattered palmer tubercles (B).

50' Chela closing with a wide gap, chela dorsal ridges very weak (Pl. 16.189.37); gonopod curved 90°, central projection without a subapical notch and equal in length to mesal process, both processes not extending beyond gonopod distal margin .... *Cambarus veteranus* Faxon, 1914

[USA: Kentucky, Virgina, West Virginia]

51(50) Areola less than six times as long as wide (Pl. 16.189.38 B, C) 52

51' Areola six or more times longer than wide (Pl. 16.189.38 A); gonopod curved slightly more than 90°, central projection slightly shorter than mesal process, and with a subapical notch .... *Cambarus tenebrosus* Hay, 1902

[USA: Alabama, Illinois, Indiana, Kentucky, Ohio, Tennessee]

52(51) Antennal scale width approximately 45.1% of length, broadest at midpoint; gonopod curved 90° to main axis (Pl. 16.189.39 A, C), central projection equal to or slightly shorter than mesal process, and without a subapical notch .... *Cambarus crinipes* Bouchard, 1973

[USA: Kentucky, Tennessee]

52' Antennal scale width approximately 40.0% of length, broadest distal to midpoint; gonopod scarcely curved in relation to main axis (Pl. 16.189.39 B, D), central projection shorter than mesal process, and with a subapical notch, caudal process developed .... *Cambarus bouchardi* Hobbs, 1970

[USA: Kentucky, Tennessee]

53(49) Chela closing with a wide gap, dorsal ridges very weak (Pl. 16.189.37) 54

53' Chela closing without a significant gap (Pl. 16.189.25 A) 59

54(53) Dactyl length on average, less than 2.3 times palm length (Pl. 16.189.40 B) 55

PHYLUM ARTHROPODA

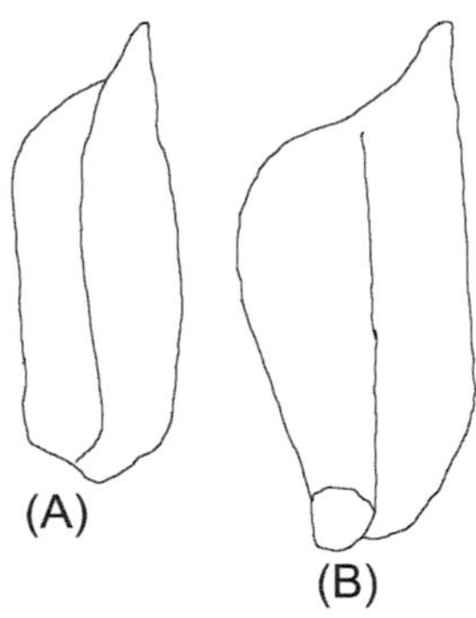

PLATE 16.189.13 *Cambarus batchi* antennal scale (A), *C. gentryi* antennal scale (B).

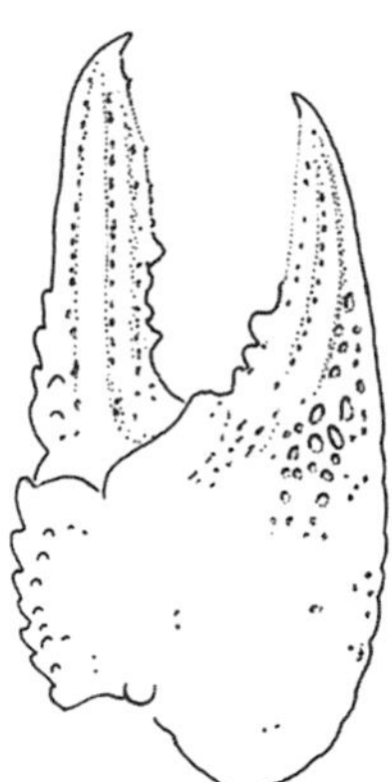

PLATE 16.189.14 Chela showing two rows of palmer tubercles.

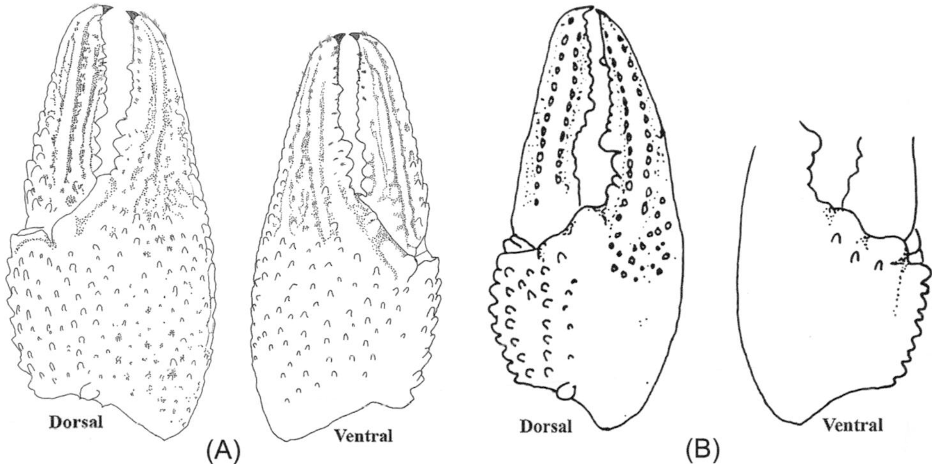

PLATE 16.189.15 Chela tuberculation both dorsal and ventral (A, B), and tuberculation limited to mesial half of dorsal chela and few in number ventrally.

54' Dactyl length on average, 2.3 or more times palm length (Pl. 16.189.40 A); gonopod curved 90°, central projection equal in length to mesal process, both processes extending slightly beyond gonopod margin .... *Cambarus chasmodactylus* James, 1966

[USA: North Carolina, Virginia, West Virginia]

55(54) Chela dactyl length more that 1.8 times palm mesal margin length (Pl. 16.189.41 B) 56

55' Chela dactyl length less than 1.8 times palm mesal margin length (Pl. 16.189.41 A); gonopod curved 90°, central projection projecting as long as mesal process, both processes extending slightly beyond gonopod margin .... *Cambarus longulus* Girard, 1852

[USA: North Carolina, Virgina, West Virginia]

56(55) Areola length, on average, almost always less than 38% of total carapace length (Pl. 16.189.42 B) 57

56' Areola length, on average, almost always 38% or more of total carapace length (Pl. 16.189.42 A); gonopod curved 90°, central projection equal in length to mesal process, both processes extending to gonopod main shaft margin .... *Cambarus manningi* Hobbs, 1981

[USA: Alabama, Georgia, Tennessee]

57(56) Suborbital angle acute (Pl. 16.189.33 A) 58

57' Suborbital angle obtuse to obsolete (Pl. 16.189.33 C); gonopod curved 90°, central projection equal in length to mesal process, both processes not extending beyond gonopod main shaft margin .... *Cambarus elkensis* Jezerinac & Stocker, 1993

[USA: West Virginia]

58(57) Abdominal segments II through V with pleura bearing a distinct posterioventral angle (Pl. 16.189.43 A) .... *Cambarus girardianus* Faxon, 1884

[USA: Alabama, Georgia, Tennessee]

58' Abdominal segments II through V with pleura posterioventrally rounded (Pl. 16.189.43 B) .... *Cambarus longirostrus* Faxon, 1885

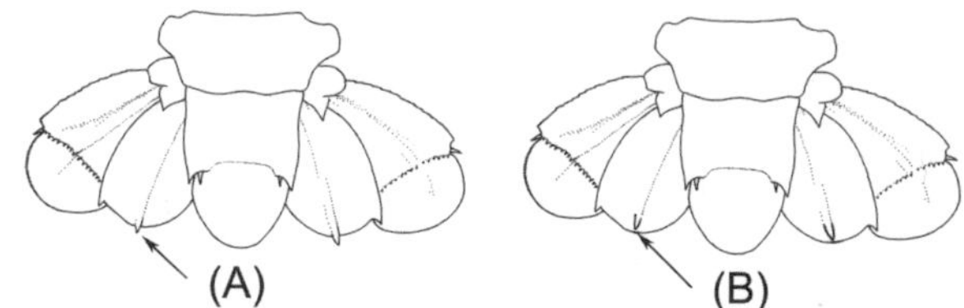

PLATE 16.189.16 (A) Median spine of uropod (*Cambarus acanthura*) extending beyond margin of uropod, (B) median spine of uropod that does not extend beyond margin of uropod.

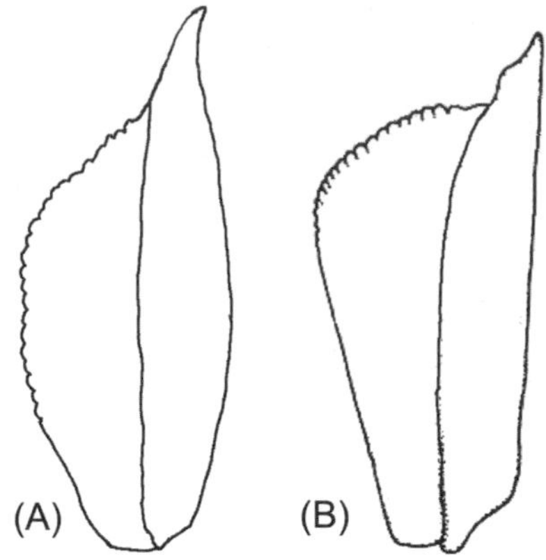

PLATE 16.189.19 Antennal scales showing position of widest point: (A) *Cambarus daughertyensis* widest at midpoint, (B) *C. deweesae* widest distal to midpoint.

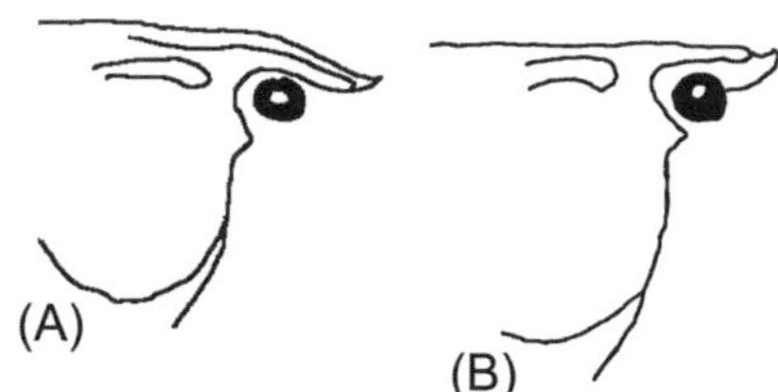

PLATE 16.189.17 (A) Rostrum of *Cambarus polychromatus* showing deflection anterior to eyes, (B) rostrum of *C. thomai* showing straight, non-deflected anterior tip.

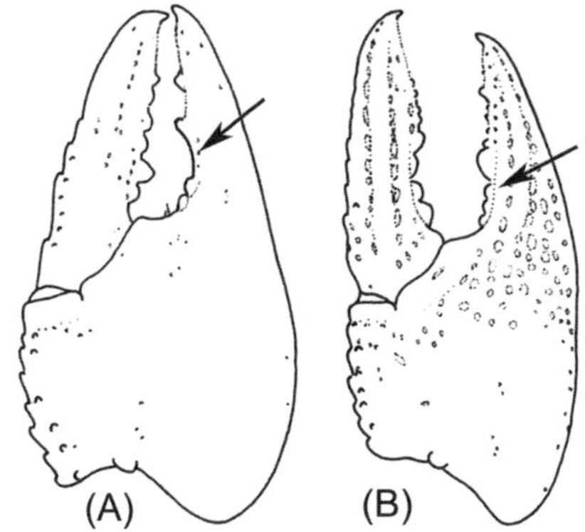

PLATE 16.189.20 (A) Chela of *Cambarus cymatilis* with arrow demarking gap in propodus tuberculation, (B) chela without gap in propodus tuberculation.

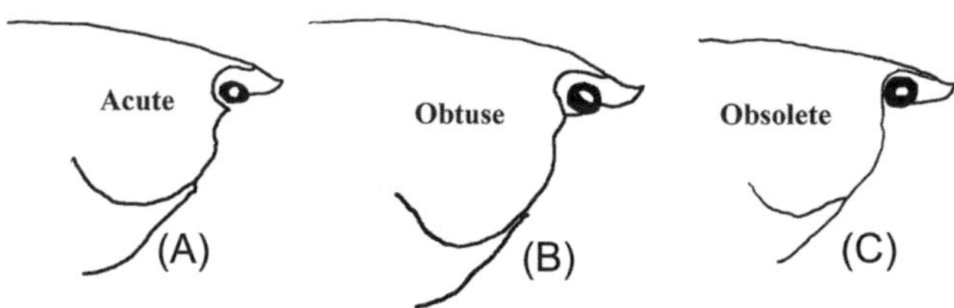

PLATE 16.189.18 Three generalized suborbital angles: (A) acute, (B) obtuse, and (C) obsolete.

[USA: Georgia, North Carolina, South Carolina, Tennessee, Virginia]

59(53) Chelae dactyl opposable margin with tubercles of variable sizes with some distal tubercles distinctly larger than proximal (Pl. 16.189.44 A) 60

59' Chelae dactyl opposable margin with tubercles subequal in size or generally decreasing in size distally (Pl. 16.189.44 B) 61

60(59) Chela palm length approximately 59% of dactyl length (Pl. 16.189.44 A); gonopod curved 90°, central projection shorter than mesal process, both processes extending beyond gonopod main shaft margin .... *Cambarus howardi* Hobbs & Hall, 1969

[USA: Alabama, Georgia, North Carolina, Tennessee, Virginia]

60' Chela palm length approximately 51% of dactyl length (Pl. 16.189.45) ; gonopod curved 90°, central projection distal end with a 90° mesally directed bend, and equal in length to mesal process, with both processes extending beyond margin of gonopod main shaft .... *Cambarus lenati* Cooper, 2000

[USA: North Carolina]

61(59) Suborbital angle acute, no tubercle or spine (Pl. 16.189.33 A); gonopod curved 45°, central projection shorter than mesal process, with both processes not extending beyond gonopod main shaft margin .... *Cambarus obeyensis* Hobbs & Shoup, 1947

[USA: Tennessee]

61' Suborbital angle usually obtuse to obsolete, with tubercle or small spine (Pl. 16.189.33 A, B); gonopod curved slightly more than 90°, central projection extending slightly less than mesal process, with both processes extending to gonopod main shaft margin .... *Cambarus brimleyorum* Cooper, 2006

[USA: North Carolina]

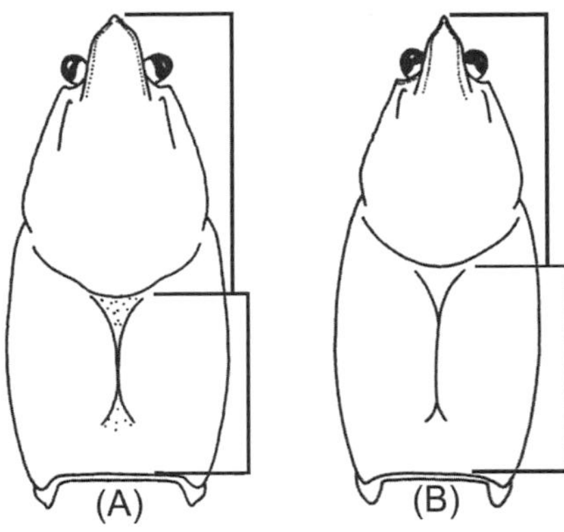

PLATE 16.189.21 (A) Carapace of *Cambarus diogenes* showing cephalic section 1.4 times as long as areola length, (B) carapace with cephalic section less than 1.4 times areola length.

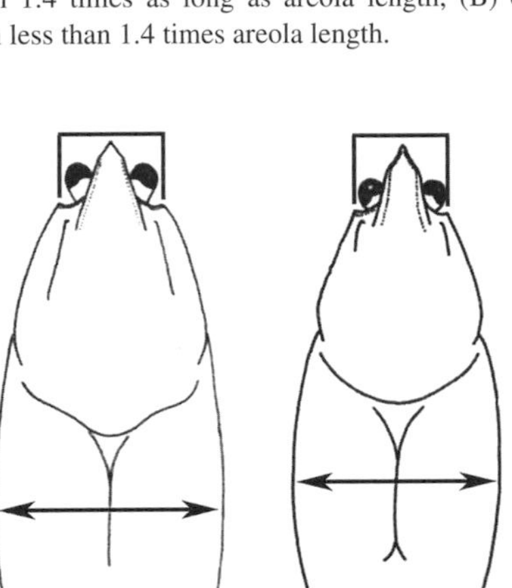

PLATE 16.189.22 Carapace width at suborbital angle in relation to greatest width: (A) *Cambarus miltus*, (B) *C. ludovicianus*.

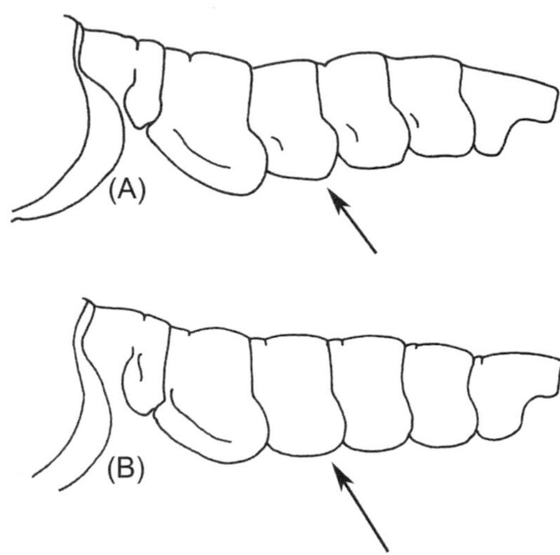

PLATE 16.189.23 Lateral view of abdominal plura: (A) slightly angled (*Cambarus harti*), (B) rounded.

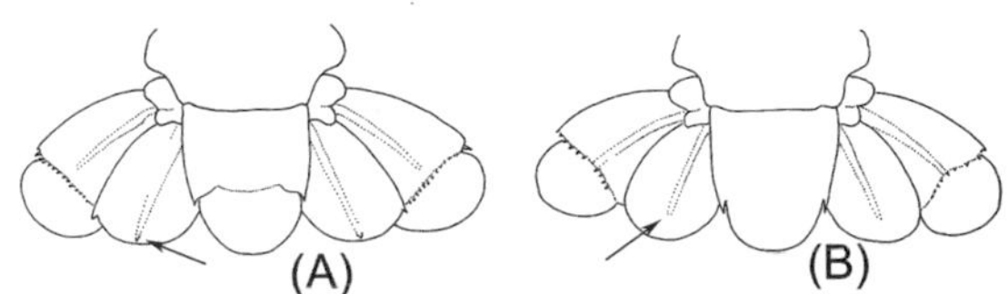

PLATE 16.189.24 Mesal ramus of uropod showing distomedian ridge: (A) terminating in spine (*Cambarus strigosus*), (B) lacking terminal spine (*C. truncates*).

62(48) Cervical groove posterior margin lacking a spine (Pl. 16.189.34 B, D); gonopod central projection with subapical notch 63

62' Cervical groove posterior margin with a spine (Pl. 16.189.34 A, C) 65

63(62) Chelae dactyl proximomedial margin with moderate tuberculation (Pl. 16.189.46 B); suborbital angle subacute to obtuse (Pl. 16.189.33 B, C) and may possess spine or tubercle; gonopod central projection longer than mesal process, with both processes extending beyond gonopod main shaft margin 64

63' Chelae dactyl proximomedial margin with strong tuberculation (Pl. 16.189.46 A); suborbital angle acute (Pl. 16.189.33 A); gonopod curved 90°, central projection equal in length to mesal process, with both processes extending to or just beyond gonopod main shaft margin .... *Cambarus hiwasseensis* Hobbs, 1981

[USA: Georgia, North Carolina, Tennessee]

64(63) Antennal scale widest distad of midpoint (Pl. 16.189.30 B); gonopod curved greater than 90° .... *Cambarus unestami* Hobbs & Hall, 1969

[USA: Alabama, Georgia]

64' Antennal scale widest at midpoint (Pl. 16.189.47); gonopod curved 90° .... *Cambarus smilax* Loughman et al., 2013

[USA: West Virginia]

65(62) Chela dorsal surface without or with very small setae (Pl. 16.189.44 B); gonopod always curved 90°, central projection with or without a subapical notch 66

65' Chela dorsal surface with easily discerned setae (Pl. 16.189.48); gonopod curved 45°, central projection always with a subapical notch 75

66(65) Chela dactyl and pollex not or only slightly gaping; dactyl opposable margin lacking a basal setal tuft of long setae (Pl. 16.189.46 A, B); gonopod with central projection bearing a subapical notch 67

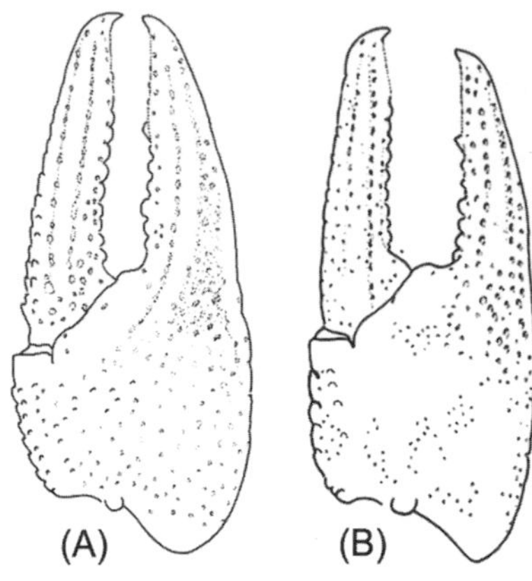

PLATE 16.189.25 Chela palm: (A) two complete rows of tubercles, (B) one complete row and one partial row of tubercles.

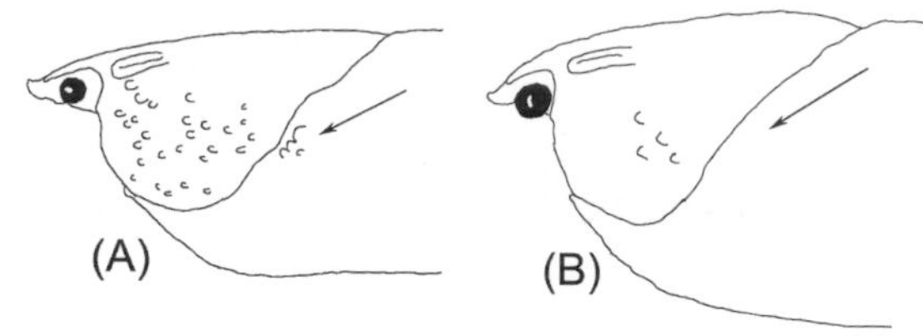

PLATE 16.189.27 Cervical grove lateral tubercles: (A) strong tubercles present (*Cambarus davidi*), (B) no tubercles present (*C. ortmanni*).

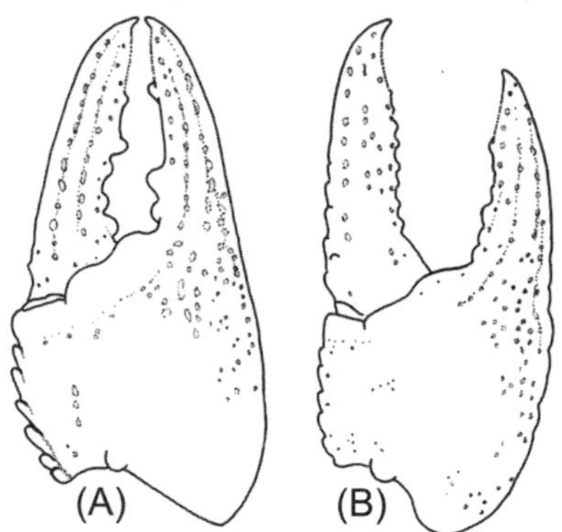

PL. 16.189.26 Chela palmer tubercles: (A) row of cristiform tubercles, (B) row of adpressed tubercles.

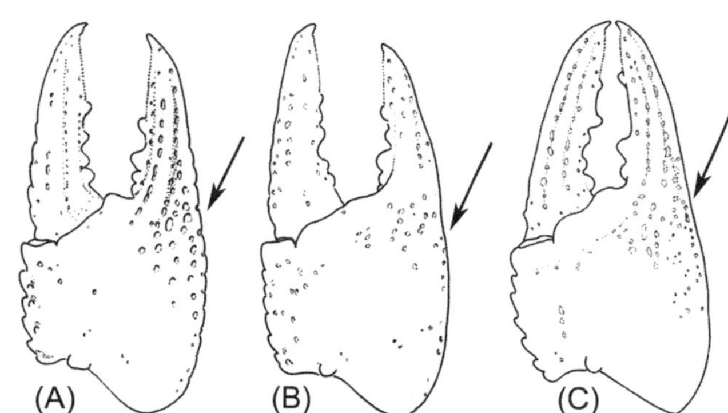

PLATE 16.189.28 Lateral margin of chela: (A) costate (*Cambarus dubius*), (B, C) smooth. Mesial palmer tubercles: (B) rounded (*C. monongalensis*), (C) forming sharp points (*C. carolinus*).

66' Chela dactyl and pollex widely gaping; dactyl opposable margin usually bearing a tuft of long setae (Pl. 16.189.49); gonopod with central projection lacking a subapical notch, and equal in length to mesal process, with both processes extending beyond gonopod main shaft margin .... *Cambarus speciosus* Hobbs, 1981

[USA: Georgia]

67(66) Hepatic region with strong spines (Pl. 16.189.50 A, B) 68

67' Hepatic region without spines (Pl. 16.189.27 A, B) 69

68(67) Cervical groove posterior margin with two to four spines (Pl. 16.189.50 A); gonopod with central projection shorter than mesal process, with both processes extending beyond gonopod main shaft margin .... *Cambarus hystricosus* Cooper & Cooper, 2003

[USA: North Carolina]

68' Cervical groove posterior margin with one spine (Pl. 16.189.50 B); gonopod central projection subequal in length to mesal process, with both processes extending beyond gonopod main shaft margin .... *Cambarus aldermanorum* Cooper & Price, 2010

[USA: South Carolina]

69(67) Areola length less than 4.2 times width (Pl. 16.189.34 A) 70

69' Areola length 4.2 times or more width (Pl. 16.189.51); gonopod central projection with subapical notch, and equal in length to mesal process, with both processes extending beyond gonopod main shaft margin .... *Cambarus hobbsorum* Cooper, 2001

[USA: North Carolina]

70(69) Epistomal zygoma moderately to strongly curved, narrower than space between renal pores (Pl. 16.189.52 B) 71

70' Epistomal zygoma weakly curved, wider than space between renal pores (Pl. 16.189.52 A); gonopod central projection equal in length to mesal process, with both processes extending beyond gonopod main shaft margin .... *Cambarus johni* Cooper, 2006

[USA: North Carolina]

71(70) Areola broad, length less than 4 times width (Pl. 16.189.53 A) 72

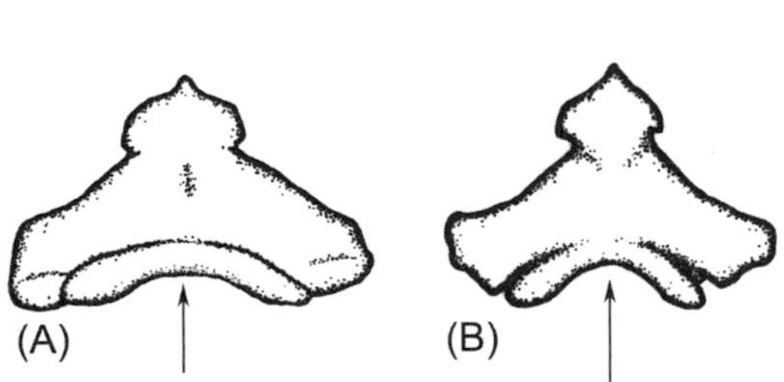

PLATE 16.189.29 Epistomal zygome: (A) gently arched (*Cambarus graysoni*), (B) moderately to strongly arched.

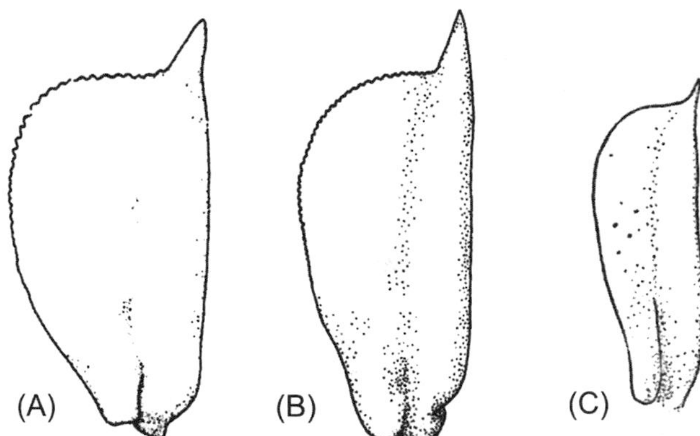

PLATE 16.189.30 Antennal scale: (C) width less than 33% length (*Cambarus reduncus*), (A, B) width greater than 33% length: (A) *C. pyronotus*, and (B) (*C. striatus*).

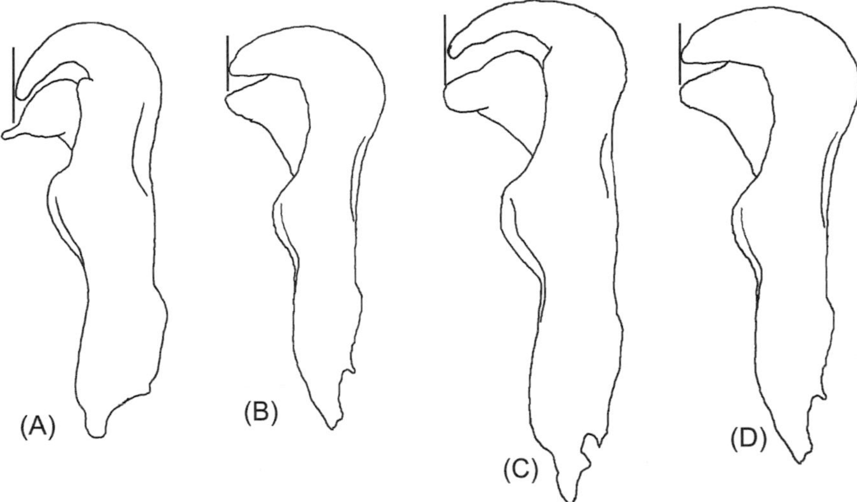

PLATE 16.189.31 Lateral view of male gonopods (A, C) first form, (B, D) second form; comparing projection of central projection to projection of mesial process: (A, B) projecting less than, (C, D) projecting further than mesial process.

71’ Areola narrower, more than 4 times width (Pl. 16.189.53 B) 73

72(71) Cervical and postorbital ridge spines strong (Pl. 16.189.53 A); chelae width less than 1.3 times length (Pl. 16.189.54 A); gonopod central projection projecting subequal in length or shorter than mesal process, with both processes extending to or falling short of gonopod main shaft margin .... *Cambarus scotti* Hobbs, 1981

[USA: Alabama, Georgia]

72’ Cervical and postorbital ridge spines weak; chelae width greater than 1.3 times length (Pl. 16.189.54 B, C); gonopod central projection equal in length to mesal process, with both processes extending beyond gonopod main shaft margin .... *Cambarus chaugaensis* Prins & Hobbs, 1969

[USA: Georgia, South Carolina]

73(71) Rostrum not strongly acuminate, slight angle distally just before acumen, suborbital angle acute (Pls. 16.189.33 A, 34 A) 74

73’ Rostrum strongly acuminate; suborbital angle obsolete (Pls. 16.189.33 C, 54 C); gonopod central projection equal in length to mesal process, with both processes extending beyond gonopod main shaft margin .... *Cambarus acuminatus* Faxon, 1884

[USA: Maryland, North Carolina, South Carolina, Virginia]

74(73) Rostrum narrow, margins convergent, not thickened to base of acumen (Pl. 16.189.34 A, 53 B); gonopod central projection equal in length to mesal process, with both processes extending beyond gonopod main shaft margin .... *Cambarus robustus* Girard, 1852

[Canada: Ontario. USA: Connecticut, Illinios, Indiana, Kentucky, Michigan, New York, North Carolina, Ohio, Pennsylvania, Tennessee, Virginia, West Virginia]

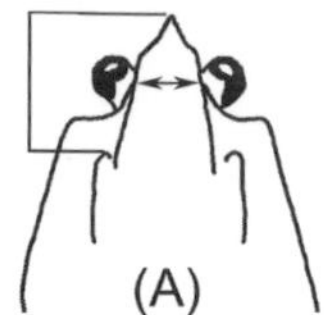

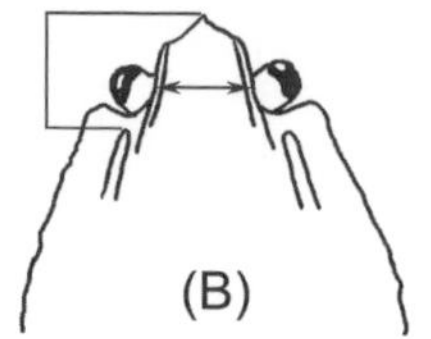

PLATE 16.189.32 Rostrum width in relation to length: (A) approximately 47% (*Cambarus catagius*), (B) approximately 70% (*C. refexus*).

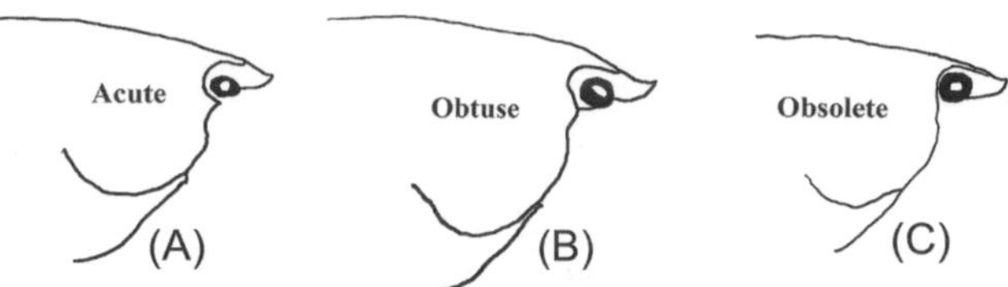

PLATE 16.189.33 Suborbital angle development: (A) acute (*Cambarus latimanus*), (B) obtuse (*C. striatus*), (C) obsolete (*C. pyronotus*).

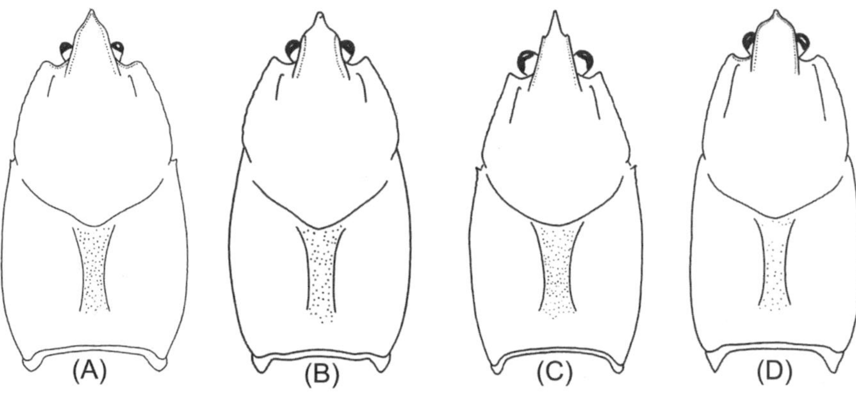

PLATE 16.189.34 Rostrum shape: (A) lancolete witout lateral spines or a 45° angle, (B, C, D) nonlancolete or with lateral spines or 45° angle.

74' Rostrum broad, margins parallel and thickened to base of acumen; gonopod central projection with a subapical notch, and equal in length to mesal process, with both processes extending beyond gonopod main shaft margin .... *Cambarus theepiensis* Loughman et al., 2013

[USA: Kentucky, Virginia, West Virginia]

75(65) Chela palm mesal margin with a row of ten tubercles, smaller tubercles irregularly arraigned dorsally (Pl. 16.189.47); gonopod curved 45°, central projection shorter than mesal process, with both processes not extending beyond margin of gonopod main shaft, a caudal knob is present .... *Cambarus pristinus* Hobbs, 1965

[USA: Tennessee]

75' Chela palm mesal margin with two rows of seven or eight tubercles, subequal in size or dorsal row larger (in adults) (Pl. 16.189.55); gonopod curved 90°, central projection subequal in length to mesal process, with both processes extending beyond gonopod main shaft margin .... *Cambarus reburrus* Prins, 1968

[USA: North Carolina]

76(43) Rostrum with lateral spines or tubercles; cervical groove posterior margin bearing a cervical spine (Pl. 16.189.34) 77

76' Rostrum with or without lateral spines; cervical groove posterior margin lacking a cervical spine (Pl. 16.189.34 B) 89

77(76) Chela palm dorsal surface without squamous tubercles 78

77' Chela palm dorsal surface studded with squamous tubercles (Pl. 16.189.56); gonopod curved more than 90°, central projection with subapical notch, and subequal in length to mesal process, with both processes extending beyond gonopod main shaft margin .... *Cambarus cracens* Bouchard & Hobbs, 1976

[USA: Alabama]

78(77) Chela palm with one tubercle row, occasionally a poorly defined, incomplete second row (Pl. 16.189.25 A); gonopod central projection with a subapical notch 79

78' Chela palm with two complete tubercle rows (Pl. 16.189.25 B); gonopod central projection with or without a subapical notch 81

79(78) Eyes large; chelae not narrow, dactyl and pollex opposable margins with tubercles of varying sizes 80

79' Eyes small; chelae elongate and narrow, dactyl and pollex opposable margins with tubercles on opposable margins of fingers of similar size (Pl. 16.189.57); gonopod curved 90°, central projection slightly shorter than mesal process, with both processes extending beyond gonopod main shaft margin .... *Cambarus nerterius* Hobbs, 1964

[USA: West Virginia]

PLATE 16.189.35 Rostral margins thickened (*Cambarus tuckasegee*).

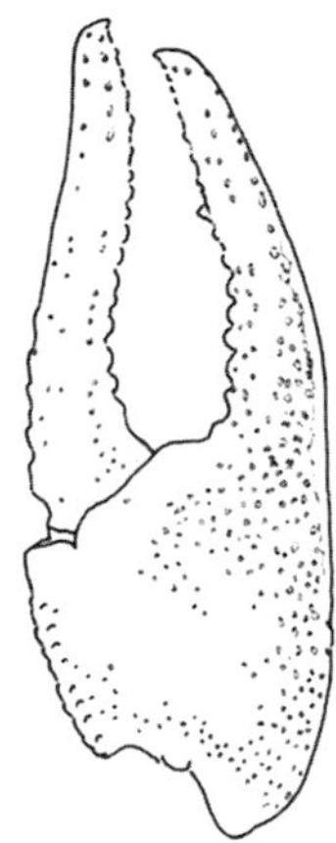

PLATE 16.189.37 Chela with gaping fingers (*Cambarus veteranus*).

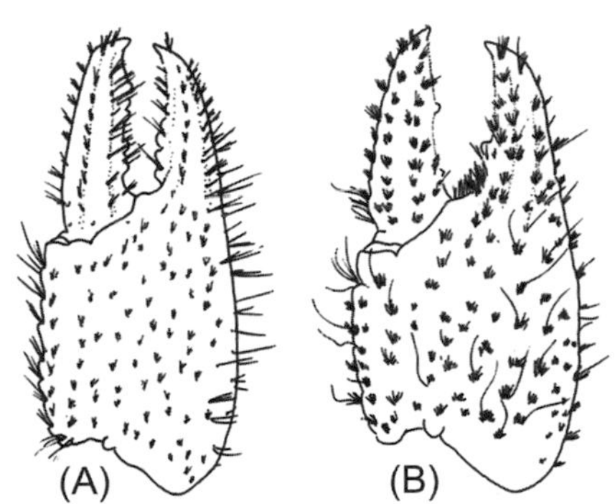

PLATE 16.189.36 Chela with single row of palmer tubercles and abundant setation. (A) *Cambarus brachydactylus*, (B) *C. friaufi*.

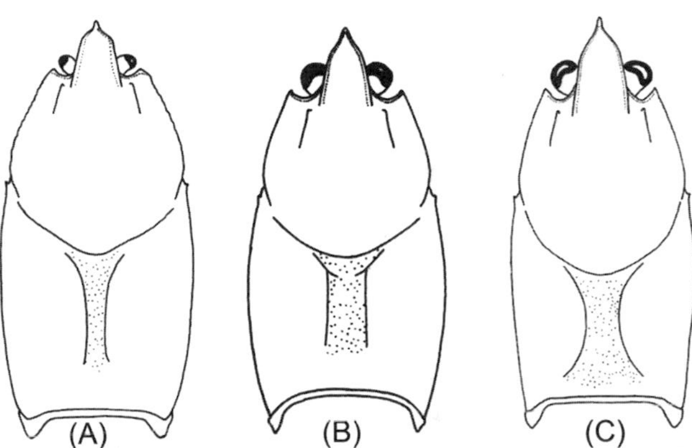

PLATE 16.189.38 Areola width: (A) six or more times as long as wide (*Cambarus tenebrosus*), (B) *C. crinipis* and (C) *C. Bouchardi* are less than six times as long as wide.

80(79) Rostrum margins convex and thin; areola wide; cervical spines strongly developed (Pl. 16.189.58 A); gonopod curved 90°, central projection equal in lengthto mesal process, with both processes extending beyond gonopod main shaft margin .... *Cambarus georgiae* Hobbs, 1981

[USA: Georgia, North Carolina]

80' Rostrum margins concave and thickened; areola width moderate; cervical spines missing in most populations (Pl. 16.189.58 B); gonopod curved slightly more than 90°, central projection slightly shorter than mesal process, with both processes extending beyond gonopod main shaft margin .... *Cambarus rusticiformis* Rhoades, 1944

[USA: Alabama, Illinios, Kentucky, Tennessee]

81(78) Chela with dactyl and pollex gaping; dactyl and pollex opposable margins with tubercles similar in size (Pl. 16.189.41 A, B); gonopod central projection with subapical notch 82

81' Chela with dactyl and pollex not gaping; dactyl and pollex opposable margins with tubercles of vaying sizes (Pl. 16.189.74); gonopod central projection with or without a subapical notch 83

82(81) Rostrum margins straight (Pl. 16.189.59 A); gonopod curved 90°, central projection equal in length to mesal process, with both processes extending to gonopod main shaft margin .... *Cambarus coosae* Hobbs, 1981

[USA: Alabama, Georgia, Tennessee]

82' Rostrum margins concave (Pl. 16.189.59 B); gonopod curved greater than 90°, central projection slightly longer than mesal process, with both processes extending beyond gonopod main shaft margin .... *Cambarus faciatus* Hobbs, 1981

[USA: Georgia]

83(81) Carapace without hepatic spines (Pl. 16.189.34 B, C) 84

83' Carapace with hepatic spines (Pl. 16.189.34 A, C); gonopod curved 90°, central projection with a subapical notch, equal in length to mesal process, with both processes extending beyond gonopod main shaft margin .... *Cambarus spicatus* Hobbs, 1956

[USA: North Carolina, South Carolina]

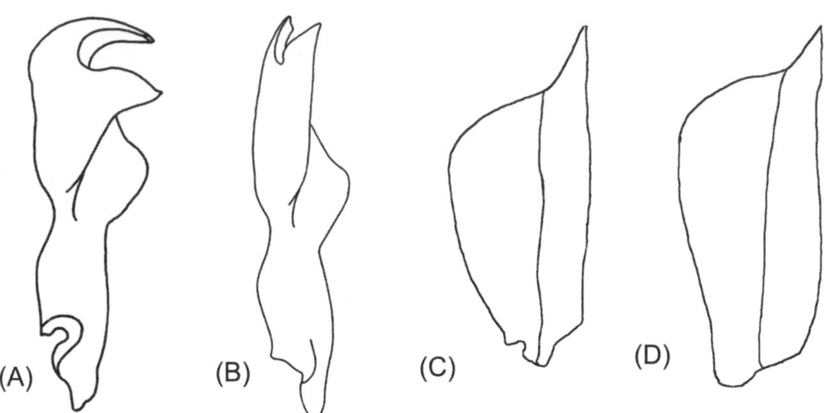

PLATE 16.189.39 Antennal scale (C & D) and gonopod shape (A & B): (A, C) *Cambarus crinipes*, (D, E) *C. bouchardi*.

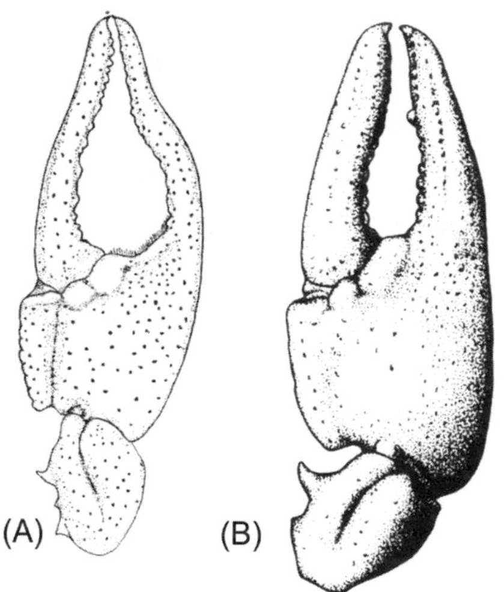

PLATE 16.189.40 Comparison of dactyl length to palm length: (A) *Cambarus chasmodactylus*, (B) *C. elkensis*.

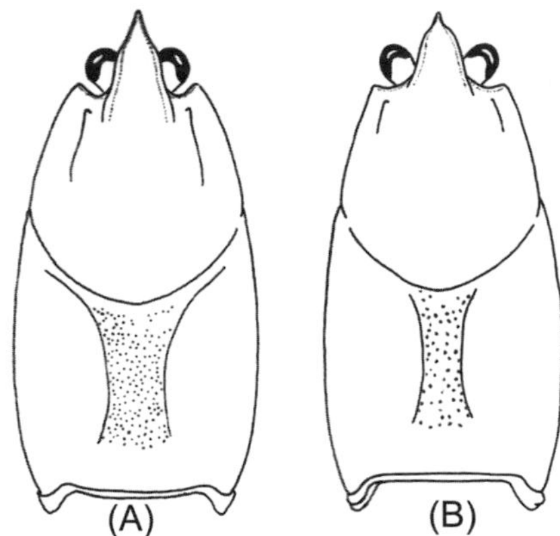

PLATE 16.189.42 Areola length comparison to total carapace length: (A) 38% or more (*Cambarus manningi*), (B) less than 38%.

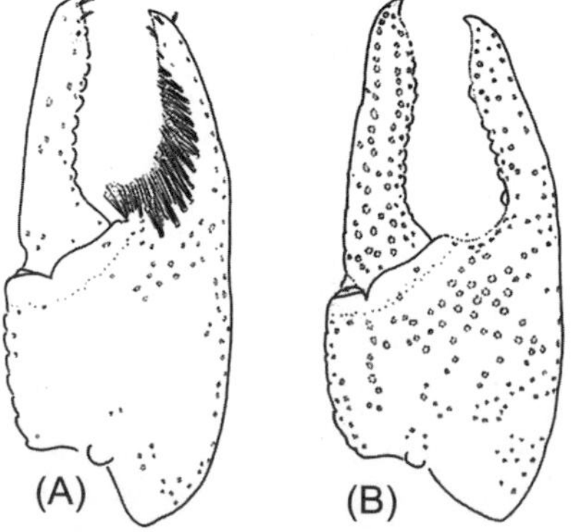

PLATE 16.189.41 Comparison of dactyl length to palm length: (A) less than 1.8 times (*Cambarus longulus*), (B) greater than 1.8 times.

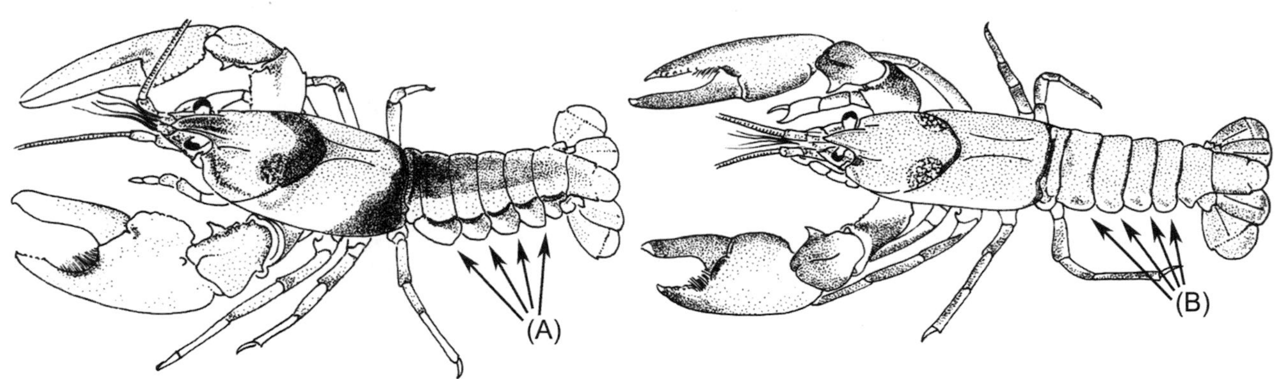

PLATE 16.189.43 Abdominal pleura (A) with distinct postventral angle (*Cambarus girardianus*), (B) with rounded postventral margin (*C. longulus*).

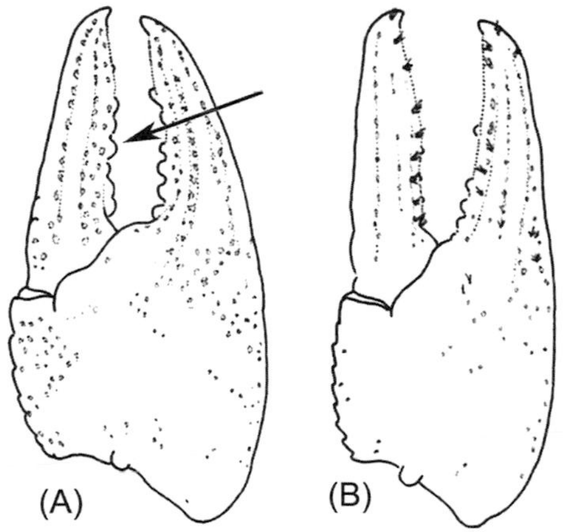

PLATE 16.189.44 Chela dactyl opposable margin tuberculation: (A) variable in size with some distal tubercles larger than proximal (*Cambarus howardi*), (B) tubercles of same size or decreasing in size from proximal to distal.

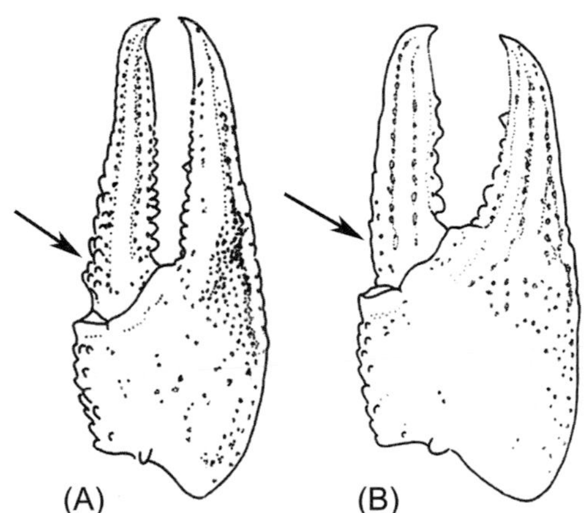

PLATE 16.189.46 Chela dactyl proximal mesial margin tuberculation: (A) strongly developed tubercles (*Cambarus hiwasseensis*), (B) moderately developed.

PLATE 16.189.45 *Cambarus lenati* chela: palm length approximately 51% of total length.

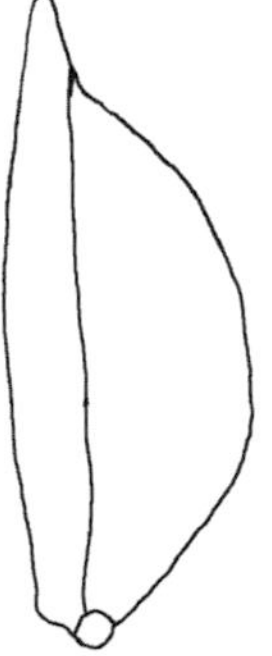

PLATE 16.189.47 Antennal scale of *Cambarus smilax*, widest at midpoint.

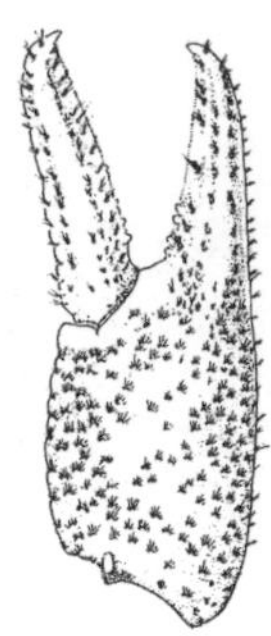

PLATE 16.189.48 Chela of *Cambarus pristinus* illustrating abundant setation.

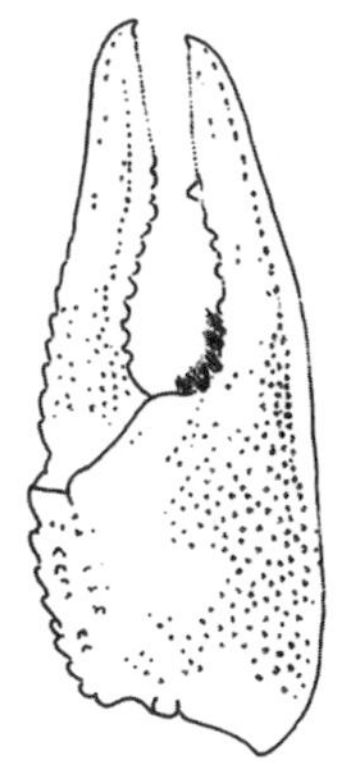

PLATE 16.189.49 Chela of *Cambarus speciosus* illustrating wide finger gap and tuft of setae.

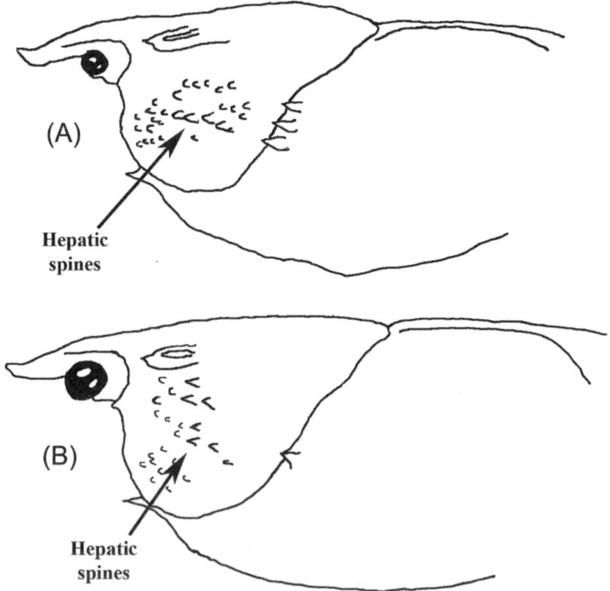

PLATE 16.189.50 Carapace lateral spination: (A) *Cambarus hystricosus* with strong hepatic spines and cervical groove with multiple spines, (B) *C. aldermanorum* with strong hepatic spines and cervical groove with one spine.

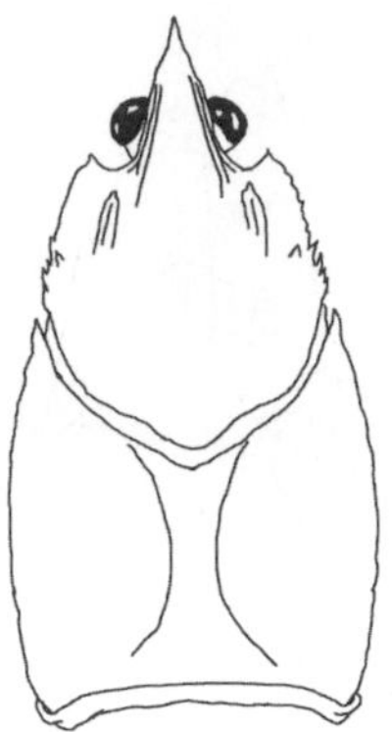

PLATE 16.189.51 *Cambarus hobbsorum* carapace. Areola length 4.2 times width.

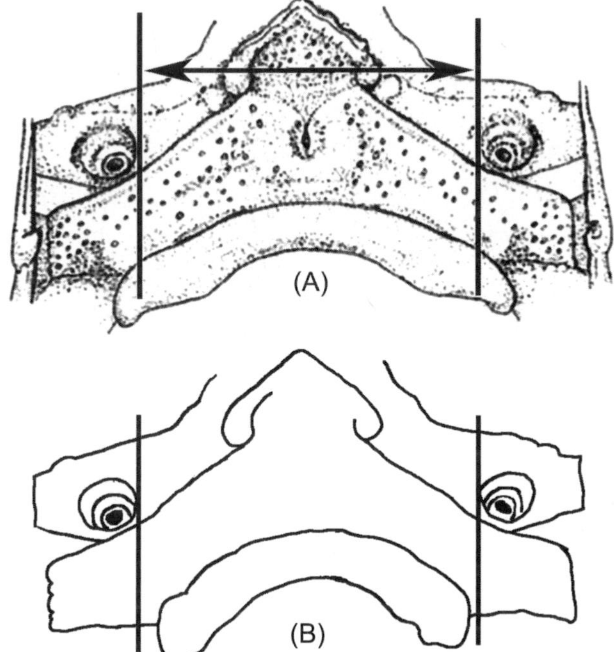

PLATE 16.189.52 Epistomal zygome development: (A) weakly curved and extending beyond inner edges of renal pores (*Cambarus johni*), (B) moderately curved and not extending beyond inner edges of renal pores.

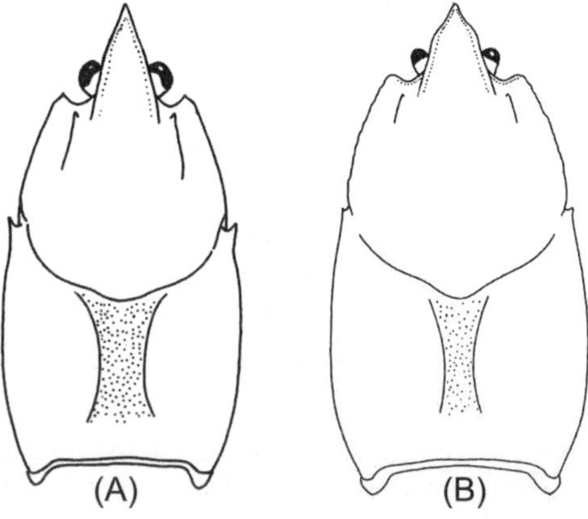

PLATE 16.189.53 Areola length width ratio: (A) broad, length less than four times width (*Cambarus scotti*), (B) narrower, length more than four times width.

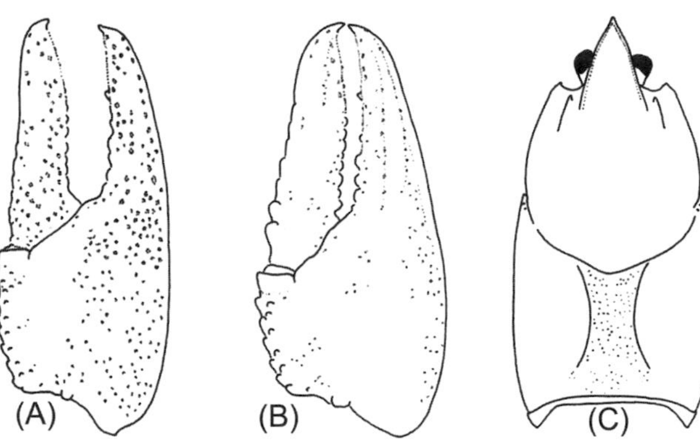

PLATE 16.189.54 (A) *Cambarus scotti* chela, width less than 1.3 times length, (B) *C. chaugaenis* chela, width greater than 1.3 times length, (C) carapace with weak cervical and postorbital ridge spines and strongly acuminate rostrum.

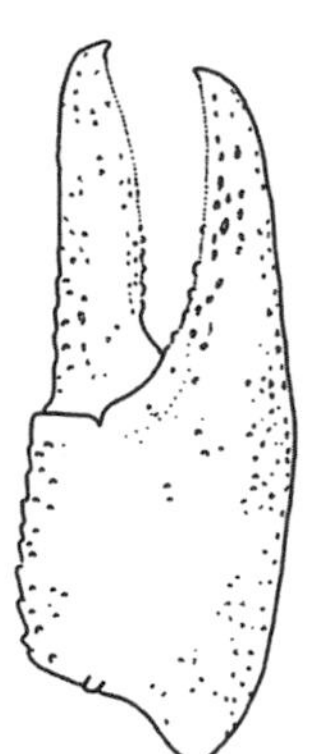

PLATE 16.189.56 *Cambarus cracens* chela showing position of squamous palmer tubercles.

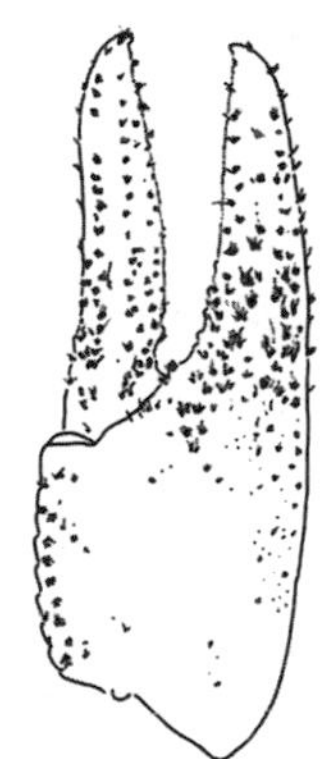

PLATE 16.189.55 *Cambarus reburrus* chela showing double row of palmer tubercles.

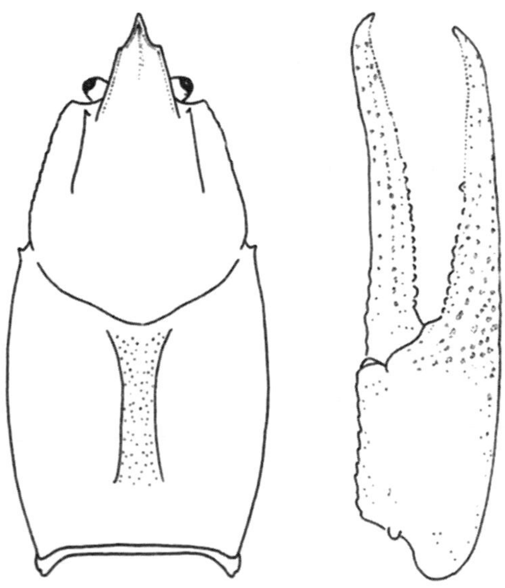

PLATE 16.189.57 *Cambarus nerterius* carapace with small eyes and elongate narrow chela.

84(83) Chela width approximately 46% or less of length (Pl. 16.189.60 A, B); gonopod curved 90° or greater, central projection with or without subapical notch 85

84' Chela width approximately 47% or more of length (Pl. 16.189.60 C, D); gonopod curved 90°, central projection with subapical notch 86

85(83) Rostrum margins thickened (Pl. 16.189.61 A); gonopod curved 90°, central projection sickle-shaped, lacking a subapical notch, and shorter than mesal process, with both processes extending beyond gonopod main shaft margin .... *Cambarus englishi* Hobbs & Hall, 1972

[USA: Alabama, Georgia]

85' Rostrum margins not thickened (Pl. 16.189.61 B); gonopod recurved greater than 90°, central projection with a subapical notch, and slightly shorter than mesal process, with both processes extending beyond gonopod shaft margin .... *Cambarus halli* Hobbs, 1968

[USA: Alabama, Georgia]

86(84) Suborbital angle obtuse, often with a short, acute, spine (Pl. 16.189.33 B) 87

86' Suborbital angle acute (Pl. 16.189.33 A) 88

87(86) Postorbital ridge with tubercle (Pl. 16.189.62 A); gonopod central projection shorter than mesal process, mesal processes extending beyond central projection and extending to gonopod shaft margin .... *Cambarus obstipus* Hall, 1959

[USA: Alabama]

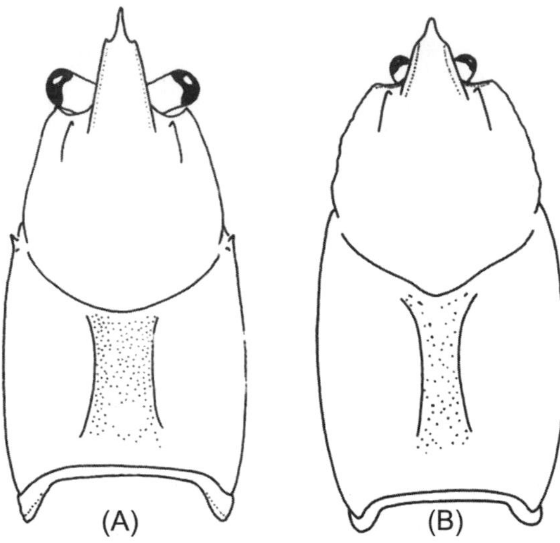

PLATE 16.189.58 Carapace dorsal view: (A) *Cambarus georgiae*, rostral margins convex and thin, areola wide, cervical spines strongly developed, (B) *C. rusticiformis*, rostral margins concave and thick, areola moderate width, cervical spines abscent.

87' Postorbital ridge with spines (Pl. 16.189.62 B); gonopod central projection longer than mesal process, with both processes extending beyond gonopod main shaft margin .... *Cambarus extraneus* Hagen, 1870

[USA: Georgia, Tennesse]

88(86) Rostral margins terminating in strong spines; carapace spines robust and sharp; cervical spine present (Pl. 16.189.63 A) .... *Cambarus cumberlandensis* Hobbs & Bouchard, 1973

[USA: Kentucky, Tennessee]

88' Rostral margins terminating in short spines or tubercles; carapace spines weakly developed; cervical spine absent (Pl. 16.189.63 B) .... *Cambarus buntingi* Bouchard, 1973

[USA: Kentucky, Tennessee]

89(76) Chela with dactyl and pollex not widely gaping, lacking a setal tuft; dactyl and pollex opposable margins with tubercles large and of varying sizes (Pl. 16.189.60 B, C); gonopod central projection with or without a subapical notch 90

89' Chela with dactyl and pollex gaping, frequently with a basal setal tuft; dactyl and pollex opposable margins with tubercles small and of similar size (Pl. 16.189.41 A, B); gonopod curved 90°, central projection with a subapical notch, and slightly longer than mesal process, with both processes extending beyond or just to gonopod main shaft margin .... *Cambarus coosawattae* Hobbs, 1981

[USA: Georgia]

90(89) Chela palm greater than 75% of dactyl length (Pl. 16.189.64 B) 91

90' Chela palm less than 75% of dactyl length (Pl. 16.189.64 A); gonopod curved greater than 90°, central projection with a subapical notch, and shorter than mesal process, with both processes extending beyond gonopod main shaft margin .... *Cambarus parrishi* Hobbs, 1981

[USA: Georgia, North Carolina]

91(90) Post orbital ridge and rostral margin spines robust, sharp, and easily discerned (Pl. 16.189.65 A); gonopod curved greater than 90°, central projection sickle-shaped, lacking a subapical notch, and slightly longer than mesal process, with both processes extending beyond gonopod main shaft margin .... *Cambarus maculatus* Hobbs & Pflieger, 1988

[USA: Missouri]

91' Post orbital ridge and rostral margin spines weakly developed and not readily discernable (Pl. 16.189.64 B); gonopod curved 90°, central projection with a subapical notch, subequal in length to mesal process, with both processes extending beyond gonopod main shaft margin .... *Cambarus hubbsi* Creaser, 1931

[USA: Arkansas, Missouri]

92(76) Chela palm mesal margin with two or more complete tubercle rows, tubercles easily discerned (Pl. 16.189.25 A); gonopod central projection always with a subapical notch 93

92' Chela palm mesal margin with one tubercle row, if second row present, then fewer tubercles than first row and tubercles not easily discerned (Pl. 16.189.25 B); gonopod central projection with or without a subapical notch 96

93(92) Rostral margins not thickened (Pl. 16.189.34 A, C) 94

93' Rostral margins thickened (Pl. 16.189.34 B); gonopod recurved 90°, central projection subequal in length to mesal process, with both processes extending just beyond or to gonopod main shaft margin .... *Cambarus sciotensis* Rhoades, 1944

[USA: Kentucky, Ohio Virginia, West Virginia]

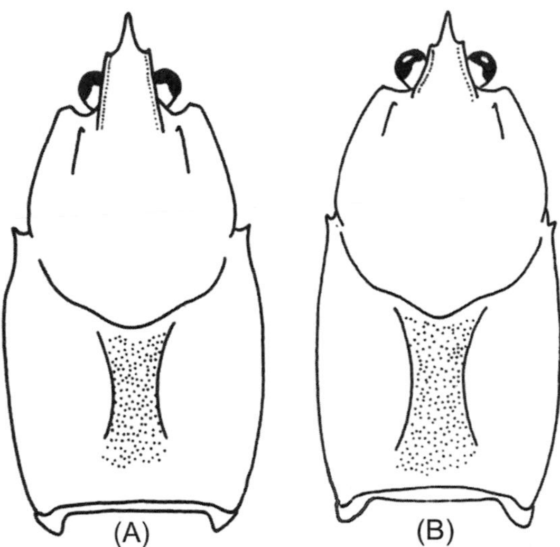

PLATE 16.189.59 Carapace dorsal view: (A) *Cambarus coosae*, rostral margins straight, (B) *C. faciatus*, rostral margins concave.

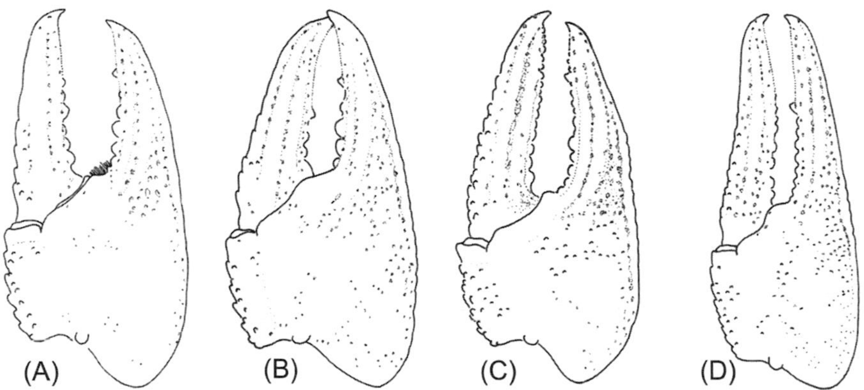

PLATE 16.189.60 Chelae: (A, B) width approximately 46% or less of length, (C, D) width approximately 47% or more of length.

94(93) Chela dactyl opposable surface with fourth tubercle from base equal in size or smaller than adjacent tubercles (Pl. 16.189.66 B) 95

94' Chela dactyl opposable surface with fourth tubercle from base enlarged (Pl. 16.189.66 A); gonopod normally curved greater than 90°, central projection shorter than mesal process, with both processes extending beyond gonopod main shaft margin .... *Cambarus sphenoides* Hobbs, 1968

[USA: Kentucky, Tennessee]

95(94) Rostrum with distal rounded margins leading to short acumen (Pl. 16.189.34 D); gonopod curved 90°, central projection shorter than the much enlarged mesal process, with both processes extending far beyond gonopod main shaft margin .... *Cambarus nodosus* Bouchard & Hobbs, 1976

[Distribution: Georgia, North Carolina, South Carolina, Tennessee]

95' Rostral distal margins forming distinct angles leading to long acumen (Pl. 16.189.34 B, C); gonopod curved 90°, central projection much shorter than mesal process, mesal processes extending beyond central projection and extending to gonopod main shaft margin .... *Cambarus obstipus* Hall, 1959

[USA: Alabama]

96(92) Rostral margins thickened, distal end truncated (Pl. 16.189.67 A); gonopod central projection always with a subapical notch (New river populations of *Cambarus sciotensis* may key here) 97

96' Rostral margins not thickened, distal end truncated or slightly curved to accumen (Pl. 16.189.67 B); gonopod central projection with or without a subapical notch 100

97(94) Chela palm mesal margin tubercles not noticeably raised or enlarged, basalmost tubercles not fused and forming ridge (Pl. 16.189.26 B) 98

97' Chela palm mesal margin tubercles noticeably raised, basalmost tubercles fused forming a cristiform ridge (Pl. 16.189.26 A); gonopod curved greater than 90°, central projection longer than mesal process, with both processes extending beyond gonopod main shaft margin .... *Cambarus jezerinaci* Thoma, 2000

[USA: Kentucky, Tennessee, Virginia]

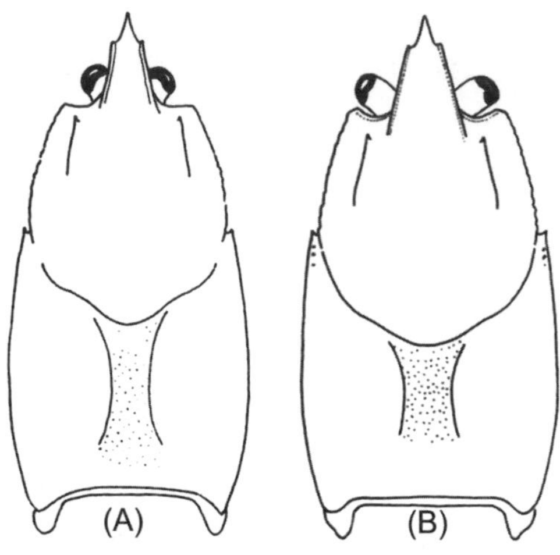

PLATE 16.189.61 Carapace dorsal view: (A) *Cambarus englishi*, rostral margins thickened, (B) *C. halli*, rostral margins not thickened.

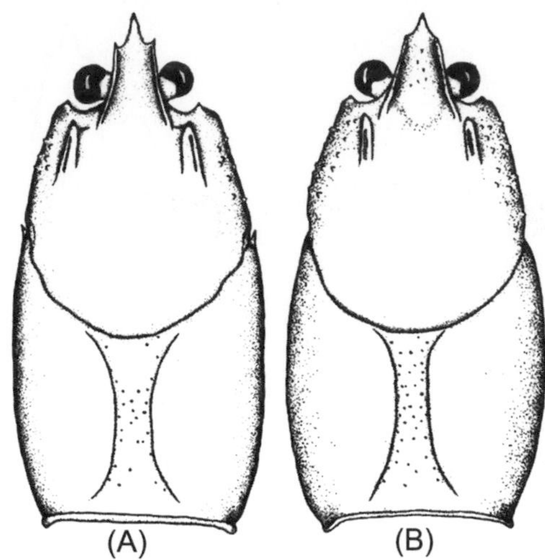

PLATE 16.189.63 Carapace dorsal view: (A) *Cambarus cumberlandensis*, rostral margins terminating in strong spines, cervical spine present, (B) *C. buntingi*, rostral margins terminating in short spines or tubercles, cervical spine abscent.

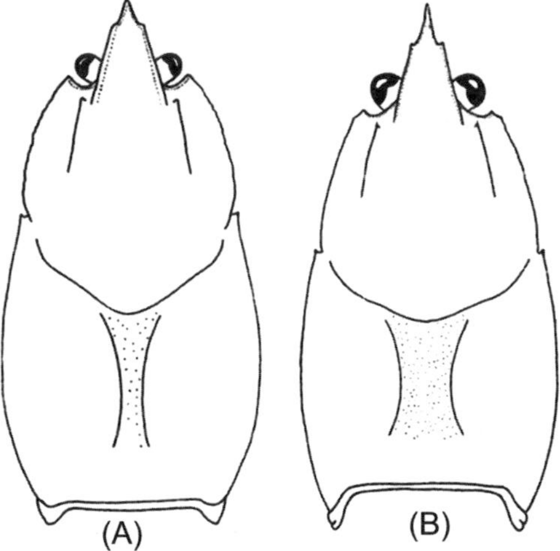

PLATE 16.189.62 Carapace dorsal view: (A) *Cambarus obstipus*, post orbital ridges with tubercles, (B) *C. extraneus*, post orbital ridges with spines.

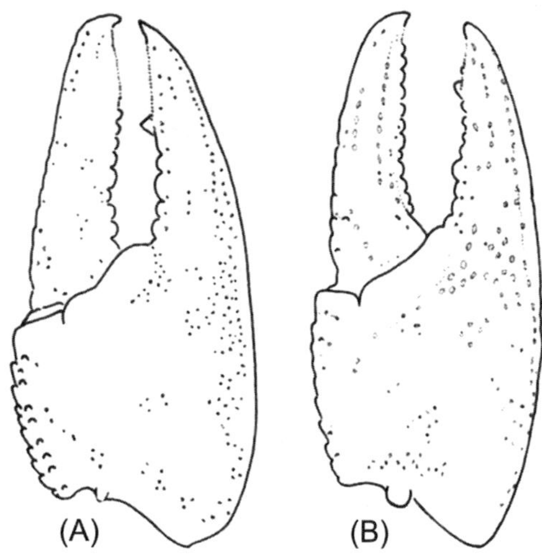

PLATE 16.189.64 Dorsal chelae: (A) *Cambarus parrishi*, palm less than 75% of length, (B) *C. hubbsi*, palm greater than 75% of length.

98(97) Chelae not inflated and rounded; pereopod I carpus margin with a single basomedial tubercle (Pl. 16.189.68 B); gonopod curved 90° 99

98' Chelae inflated, rounded; pereopod I carpus margin with two, basomedial, conjoined tubercles (Pl. 16.189.68 A); gonopod curved greater than 90°, central projection subequal in length or slightly shorter than mesal process, with both processes extending beyond gonopod main shaft margin .... *Cambarus angularis* Hobbs & Bouchard, 1994

[USA: Tennessee, Virginia]

99(98) Chela palm mesal margin with a partial second row of tubercles, especially in large males; areola four times longer than wide (Pl. 16.189.69 A); gonopod central projection subequal in length to mesal process, with both processes extending beyond gonopod main shaft margin .... *Cambarus eeseeohensis* Thoma, 2005

[USA: North Carolina]

99' Chela palm mesal margin without a partial second row of tubercles; areola five times longer than wide (Pl. 16.189.69 B); gonopod central projection subequal or shorter than mesal process, with both processes extending beyond gonopod main shaft margin .... *Cambarus carinirostris* Hay, 1914

[USA: Ohio, Pennsylvania, Virginia, West Virginia]

100(96) Chela palm mesal margin tubercles noticeably raised, basalmost tubercles fused forming a cristiform ridge (Pl. 16.189.26 A) 101

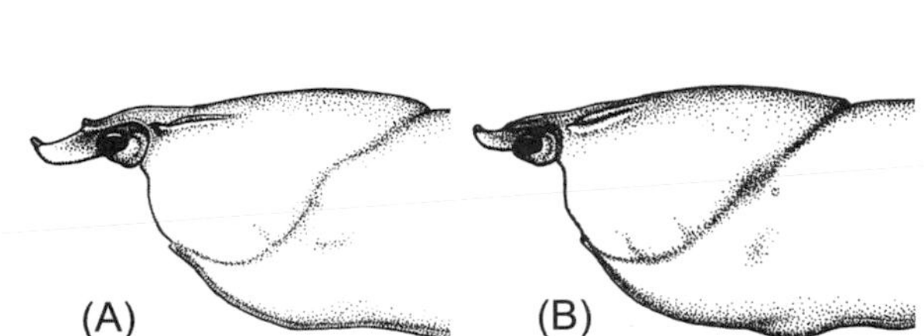

PLATE 16.189.65 Carapace lateral view: (A) *Cambarus maculates*, post orbital and rostral spines strongly developed, (B) *C. hubbsi*, post orbital and rostral spines weakly developed.

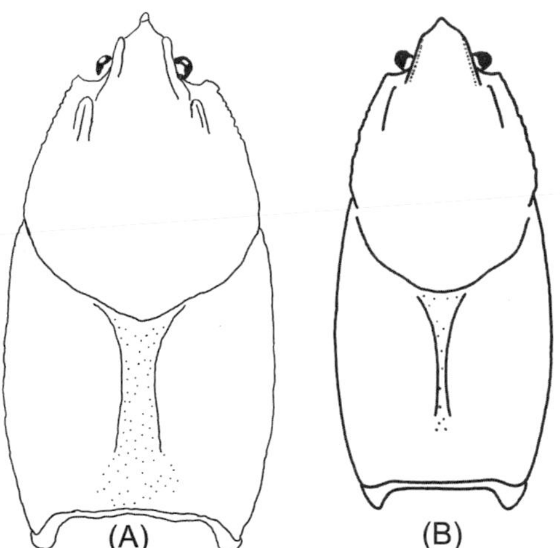

PLATE 16.189.67 Carapace dorsal view: (A) rostral margins thickened, distal end truncated, (B) rostral margins not thickened, distal end truncated or slightly curved to accumen.

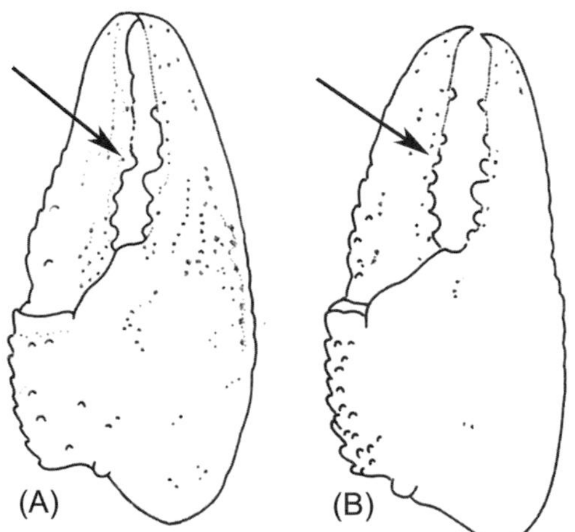

PLATE 16.189.66 Dorsal chelae: (A) *Cambarus sphenoides*, fourth tubercle from base of opposable dactyl enlarged, (B) fourth tubercle from base of opposable dactyl equal in size or smaller than adjacent.

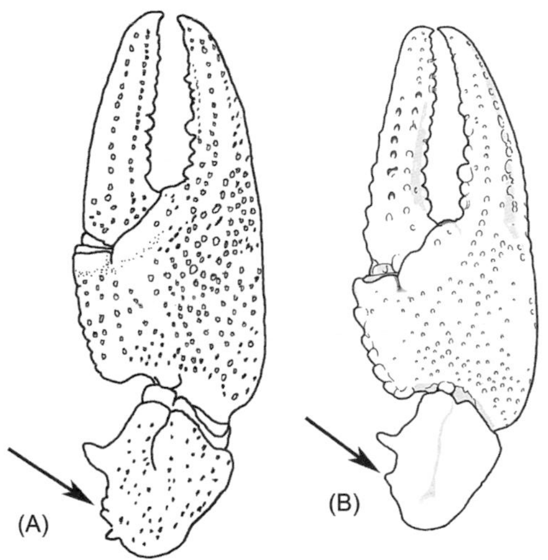

PLATE 16.189.68 Dorsal chelae: (A) *Cambarus angularis*, rounded appearance, carpus with two conjoined tubercles on basomesial margin, (B) not inflated and rounded in appearance, carpus with single tubercle on basomesial margin.

100' Chela palm mesal margin tubercles not noticeably raised or enlarged, basalmost tubercles not fused and forming ridge (Pl. 16.189.26 B); gonopod curved 90°, central projection with subapical notch, and subequal in length to mesal process, with both processes extending beyond gonopod main shaft margin [*Cambarus carinirostris* may key to this couplet. This species has no second row of tubercles as in *C. b. cavatus* and the fingers are not gaping as in *C. b. bartonii* from the northern portion of its range. Southern populations of *C. b. bartonii* (New River and south) have two enlarged tubercles (usually first and fourth) on the opposable margin of the pollex and *C. carinirostris* has none.] 106

101(100) Chelae not covered with easily discernable setae, or if setae present, not extending significantly onto palm and all setae of approximate equal size; gonopod central projection always with subapical notch 102

101' Chelae with easily discernable setae extending onto palm, some single setae longer than most others (Pl. 16.189.70); gonopod curved greater than 90°, central projection sickle-shaped, lacking a subapical notch, and shorter than mesal process, with both processes greatly enlarged and extending beyond gonopod main shaft margin ..... *Cambarus asperimanus* Faxon, 1914

[USA: Georgia, North Carolina, South Carolina, Tennessee]

102(101) Eye small in comparison to body size (Pl. 16.189.57) 103

102' Eyes normal size (Pl. 16.189.69 A, B) 104

103(102) Large cave dwelling species; chelae three times as long as wide, dactyl and pollex opposable margins with tubercles small and similar in size (Pl. 16.189.57); gonopod curved 90°, central projection equal to or slightly shorter than mesal process, with both processes extending beyond gonopod main shaft margin ..... *Cambarus neterius* Hobbs, 1964

[USA: West Virginia]

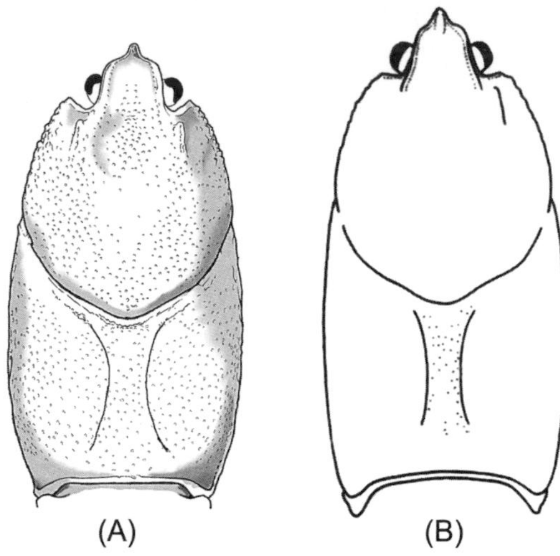

PLATE 16.189.69 Carapace dorsal view: (A) *Cambarus eeseeohensis*, areola four times longer than wide, (B) *C. carinirostris*, areola five times longer than wide.

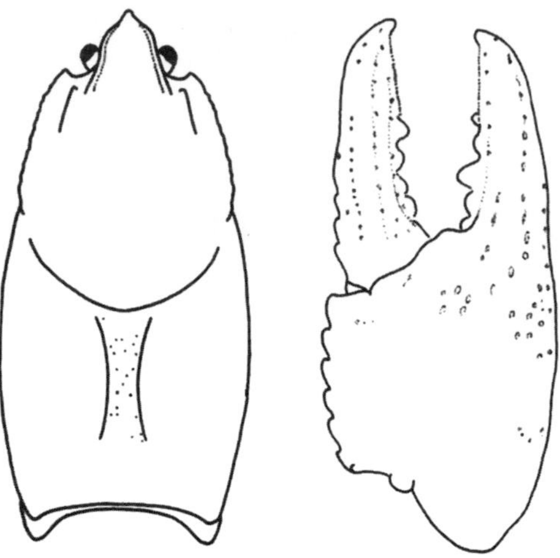

PLATE 16.189.71 *Cambarus parvoculus*, dorsal carapace and chela. Chela two times as long as wide, tubercles on opposable margins of fingers varying in size.

PLATE 16.189.70 Chela dorsal view, *Cambarus asperimanus*, with easily discernable hairs extending onto palm area, some single hairs longer than most others.

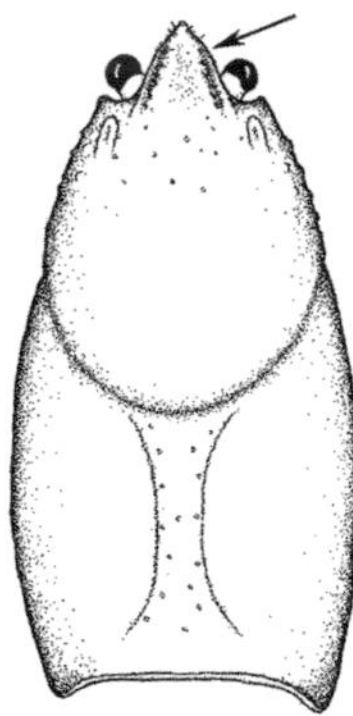

PLATE 16.189.72 *Cambarus clivosus*, rostrum short and triangular-shaped, small hairs lining inner edges of rostral margins.

103' Small species dwelling in springs and coldwater streams; chelae twice as long as wide, dactyl and pollex opposable margins with tubercles varying in size (Pl. 16.189.71); gonopod curved greater than 90°, central projection shorter than mesal process, with both processes extending beyond gonopod main shaft margin .... *Cambarus parvoculus* Hobbs & Shoup, 1947

[USA: Georgia, Kentucky, Tennessee]

104(102) Rostrum either spatulate or long with semiparallel sides, rostral margins lacking setae... 105

104' Rostrum short and triangular, rostral margins edged with small setae (Pl. 16.189.72); gonopod curved greater than 90°, central projection shorter than mesal process, with both processes extending beyond gonopod main shaft margin ...... *Cambarus clivosus* Taylor et al., 2006

[USA: Tennessee]

105(104) Suborbital angle obsolete (Pl. 16.189.33 C); gonopod curved greater than 90°, central projection slightly shorter than mesal process, with both processes extending beyond gonopod main shaft margin .... *Cambarus conasaugaensis* Hobbs & Hobbs, 1962

[USA: Georgia, Tennessee]

105' Suborbital angle acute (Pl. 16.189.33 A); gonopod curved 90°, central projection shorter than mesal process, mesal processes extending beyond or just to gonopod main shaft margin .... *Cambarus distans* Rhoades, 1944

[USA: Alabama, Georgia, Kentucky, Tennessee]

106(100) Chela palm mesal margin with one row of (~six) small, adpressed tubercles (Pl. 16.189.73)........... *Cambarus bartonii bartonii* (Fabricius, 1798)

[Canada: New Brunswick, Ontario, Quebec. USA: Alabama, Connecticut, Delaware, Georgia, Maine, Maryland, Massachuttesetes, New Jersey, New York, North Carolina, Pennsylvania, Rhode Island, South Carolina, Tennessee, Vermont, Virginia, West Virginia]

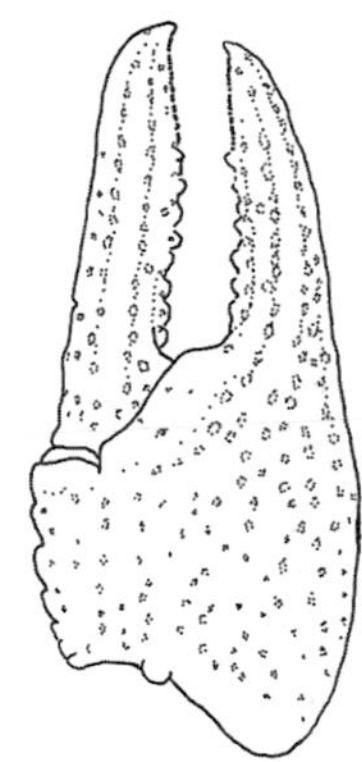

PLATE 16.189.73 *Cambarus bartonii bartonii*, chela with one row of six small, adpressed tubercles on mesial margin of chelae palm.

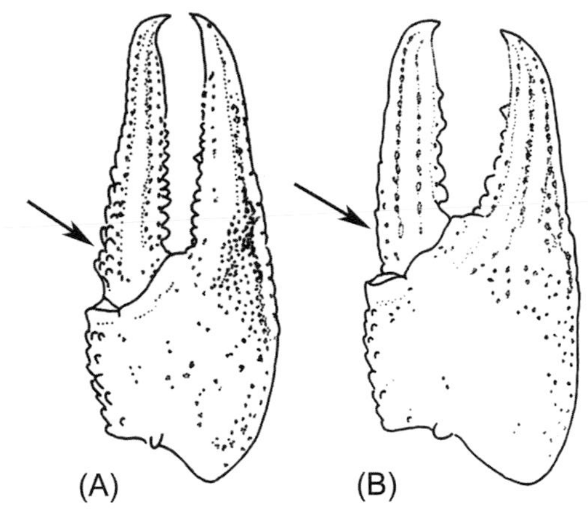

PLATE 16.189.74 *Cambarus* showing tubercles of various sizes.

106' Chela palm mesal margin with one row of (~six) tubercles subtended by second incomplete row (~four) tubercles increasing in size distally (Pl. 16.189.25 B) .... *Cambarus bartonii cavatus* Hay, 1902

[USA: Alabama, Georgia, Kentucky, Ohio, Tennessee, Virginia, West Virginia]

Crustacea: Malacostraca: Decapoda: Astacidea: Cambaridae: *Hobbseus*: Species

D. Christopher Rogers

Kansas Biological Survey, University of Kansas, Lawrence, KS, USA; The Biodiversity Institute, University of Kansas, Lawrence, KS, USA

Based on form I males. This key is based on Fitzpatrick & Payne (1968) and Hobbs (1972).

1 Gonopod central projection distally curved less than 90° from main gonopod axis (Pls. 16.189.75 B, 77 B) 2

1' Gonopod central projection distally curved 90° or more from main gonopod axis (Pl. 16.189.76 D) 3

2(1) Gonopod central projection distally curved less than 20° from main gonopod axis (Pl. 16.189.75 B) .... *Hobbseus orconectoides* Fitzopatrick & Payne, 1968

[USA: Mississippi]

2' Gonopod central projection distally curved approximately 45° from main gonopod axis (Pl. 16.189.76 B) .... *Hobbseus cristatus* (Hobbs, 1955)

[USA: Mississippi]

3(1) Gonopod mesal process curved 90° from main gonopod axis (Pls. 16.189.76 D, 77 D) 4

3' Gonopod mesal process curved more than 90° from main gonopod axis (Pl. 16.189.77 D) 5

4(3) Gonopod mesal process apical portion straight (Pl. 16.189.76 D) .... *Hobbseus attenuatus* Black, 1969

[USA: Mississippi]

4' Gonopod mesal process apical portion sinuate, with apex converging towards central projection (Pl. 16.189.77 D) .... *Hobbseus petilus* Fitzpatrick, 1977

[USA: Mississippi]

5(3) Gonopod central projection shorter than mesal process, mesal process with at least one-third of its length projecting beyond central projection apex (Pls. 16.189.77 B, 78 B) 6

5' Gonopod central projection subequal to or slightly shorter than mesal process (Pl. 16.189.78 D) .... *Hobbseus yalobushensis* Fitzpatrick & Busack, 1989

[USA: Mississippi]

6(5) Gonopod mesal process with at least one-third of its length projecting beyond central projection apex; central projection curved at approximately 110° from the main gonopod axis; pleopods symmetrical; rostrum margins convex (Pl. 16.189.77 B) .... *Hobbseus prominens* (Hobbs, 1966)

[USA: Alabama, Mississippi]

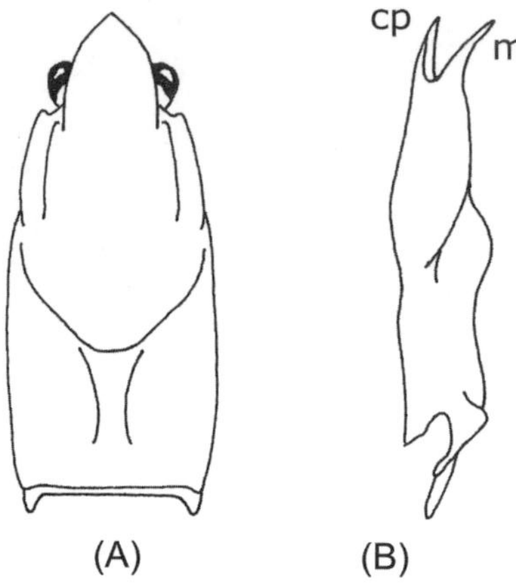

PLATE 16.189.75 *Hobbseus orconectoides*: (A) dorsal view of carapace; and (B) mesial view of left first pleopod. cp = central projection, m = mesial process. *Modified from Hobbs, 1989.*

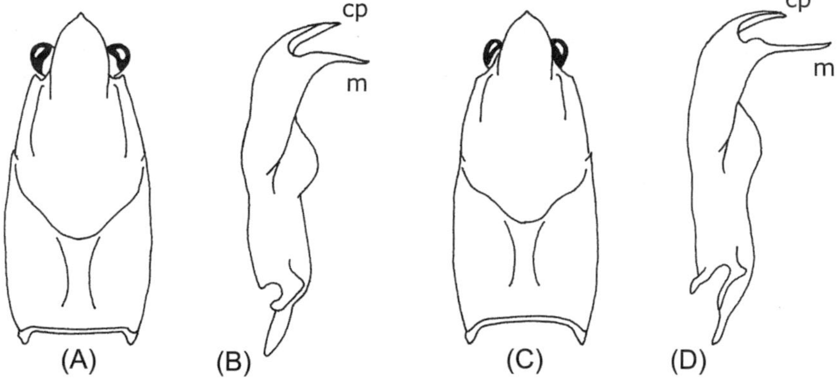

PLATE 16.189.76 Dorsal view of carapace (A, C) and mesial view of left first pleopod (B, D) in: (A, B) *Hobbseus cristatus*; and (C, D) *H. attenuatus*. cp = central projection, m = mesial process. *Modified from Hobbs, 1989.*

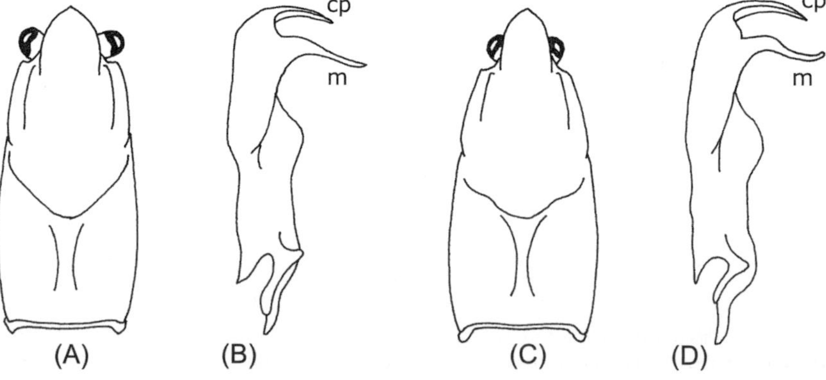

PLATE 16.189.77 Dorsal view of carapace (A, C) and mesial view of left first pleopod (B, D) in: (A, B) *Hobbseus prominens*; and (C, D) *H. petilus*. cp = central projection, m = mesial process. *Modified from Hobbs, 1989.*

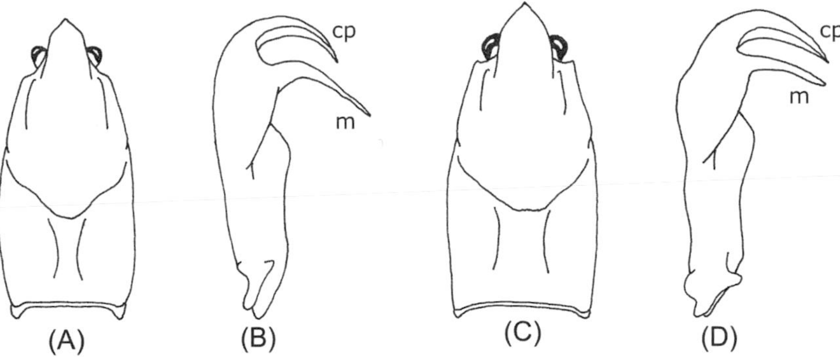

PLATE 16.189.78 (A, C) Dorsal view of carapace (A, C) and mesial view of left first pleopod (B, D) in: (A, B) *Hobbseus valleculus*; and (C, D) *H. yalobushensis*. cp = central projection, m = mesial process. *Modified from Hobbs, 1989.*

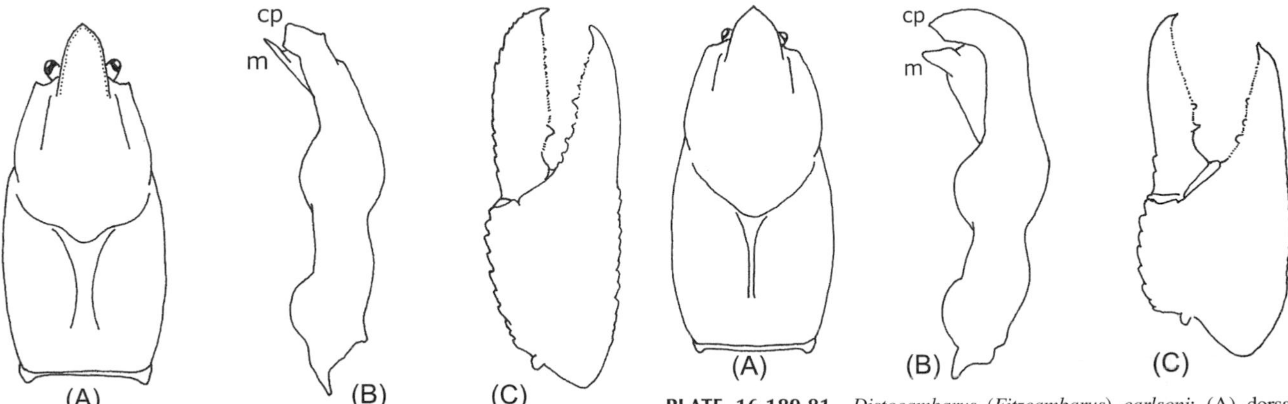

PLATE 16.189.79 *Distocambarus* (*Distocambarus*) *crockeri*: (A) dorsal view of carapace; (B) lateral view of left first pleopod; and (C) dorsal view of chela. cp = central projection, m = mesial process. *Modified from Hobbs, 1989.*

PLATE 16.189.81 *Distocambarus* (*Fitzcambarus*) *carlsoni*: (A) dorsal view of carapace; (B) lateral view of left first pleopod; and (C) dorsal view of chela. cp = central projection, m = mesial process. *Modified from Hobbs, 1989.*

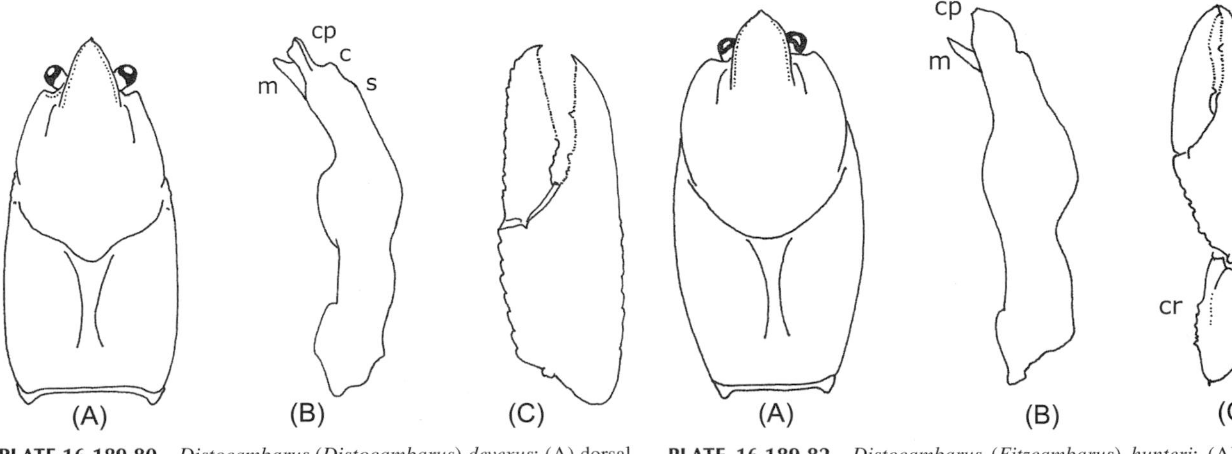

PLATE 16.189.80 *Distocambarus* (*Distocambarus*) *devexus*: (A) dorsal view of carapace; (B) lateral view of left first pleopod; and (C) dorsal view of chela. c = cephalic process, cp = central projection, m = mesial process, s = shoulder. *Modified from Hobbs, 1989.*

PLATE 16.189.82 *Distocambarus* (*Fitzcambarus*) *hunteri*: (A) dorsal view of carapace; (B) lateral view of left first pleopod; and (C) dorsal view of chela and carpus. cp = central projection, m = mesial process. *Modified from Hobbs, 1989.*

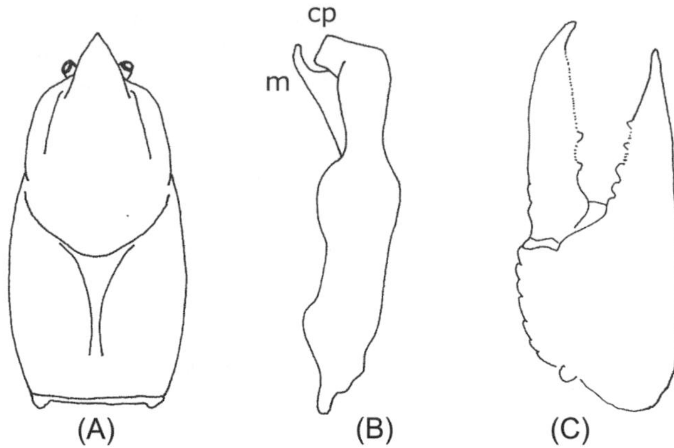

PLATE 16.189.83 *Distocambarus* (*Fitzcambarus*) *youngineri*: (A) dorsal view of carapace; (B) lateral view of left first pleopod; and (C) dorsal view of chela. cp = central projection, m = mesial process. *Modified from Hobbs, 1989.*

6' Gonopod mesal process with at least one-half of its length projecting beyond central projection apex; central projection curved at approximately 120° from the main gonopod axis; pleopods asymmetrical; rostral margins tapering distally (Pl. 16.189.78 B) .... *Hobbseus valleculus* (Fitzpatrick, 1967)

[USA: Mississippi]

Crustacea: Malacostraca: Decapoda: Astacidea: Cambaridae: *Distocambarus*: Species

D. Christopher Rogers

Kansas Biological Survey, The Biodiversity Institute, University of Kansas, Lawrence, KS, USA; The Biodiversity Institute, University of Kansas, Lawrence, KS, USA

Key based on form I males.

1 Chela palm medial margin longer than propodus width and as long or longer than carpus (subgenus *Distocambarus*) (Pl. 16.189.79 C) 2

1' Chela palm medial margin shorter than propodus width and shorter than carpus (subgenus *Fitzcambarus*) (Pl. 16.189.81 C) 3

2(1) Gonopod bearing a low, rounded cephalic process; central projection constricted basally (Pl. 16.189.80 B) .... *Distocambarus devexus* Hobbs, 1981

[USA: Georgia]

2' Gonopod lacking a cephalic process; central projection not constricted basally (Pl. 16.189.79 B) .... *Distocambarus crockeri* Hobbs & Carlson, 1983

[USA: South Carolina]

3(1) Gonopod central projection subquadrangular, with length subequal to or shorter than width 4

3' Gonopod central projection arcuate, length at least twice width; central projection with a shallow subapical notch (Pl. 16.189.81 B) .... *Distocambarus carlsoni* Hobbs, 1983

[USA: South Carolina]

4(3) Gonopod mesal process subequal in length to central projection; cephalic process barely discernible or absent; branchostegial spine present (Pl. 16.189.82 B) .... *Distocambarus hunteri* Fitzpatrick & Eversole, 1997

[USA: South Carolina]

4' Gonopod mesal process longer than central projection; cephalic process vestigial; branchostegial spine absent or obtuse (Pl. 16.186.01 B) .... *Distocambarus youngineri* Hobbs & Carlson, 1985

[USA: South Carolina]

REFERENCES

Cooper, M. R. & H. H. Hobbs, Jr. 1980. New and little known crayfishes of the *virilis* section of genus *Orconectes* (DecapodaL Cambaridae) from the southeastern United State. Smithsonian Contributions to Zoology 320:1 – 44.

Fitzpatrick, Jr., J. F. & J. F. Payne. 1968. A new genus and species of crawfish from the southeastern United States (Decapoda, Astacidae). Proceedings of the Biological Society of Washington, 81: 11–22.

Hobbs, H. H. Jr. 1972a. Biota of Freshwater Ecosystems, Identification Manual no. 9, Crayfishes (Astacidae) of North and Middle America. EPA project # 18050 ELD.

Hobbs, H.H. Jr. 1972b. Crayfishes (Astacidae) of North and Middle America. Biota of freshwater Ecosystems. U.S. Environmental Protection Agency, Water Pollution Control Research Service Identification Manual 9: 1–173.

Hobbs, H.H., Jr. 1981. The crayfishes of Georgia. Smithsonian Contributions to Zoology 318: 1–549.

Hobbs, H. H. Jr. 1989. An illustrated checklist of the American crayfishes (Decapoda: Astacidae, Cambaridae, and Parastacidae). Smithsonian Contributions to Zoology 480: 1–236.

Hobbs, H.H., Jr. & H.H. Hobbs, III. 1991. An illustrated key to the crayfishes of Florida (based on first form males). Florida Scientist 54: 13–24.

Hobbs, H.H., Jr. & H.W. Robison 1988. The crayfish subgenus *Girardiella*, Decapoda: Cambaridae, in Arkansas, with the descriptions of two new species and a key to the members of the *gracilis* group in the genus *Procambarus*. Proceedings of the Biological Society of Washington 101: 391–341.

Hobbs, H. H. Jr., & H. W. Robison. 1989. On the crayfish genus *Fallicambarus* (Decapoda: Cambaridae) in Arkansas, with notes on the *fodiens* complex and descriptions of two new species. Proceedings of the Biological Society of Washington 102: 651–697.

Hobbs, H. H. III, & D. M. Lodge. 2010. Decapoda. Pages 901 – 967 *in*: J. H. Thorp & A. P. Covich (eds.) Ecology and Classification of North American Freshwater Invertebrates, 3rd edition. Academic Press.

Johnson, D. P. 2011. *Fallicambarus* (*F.*) *wallsi* (Decapoda: Cambaridae), a new burrowing crayfish from eastern Texas. Zootaxa 2939: 59–68.

Pflieger, W. L. 1996. The Crayfishes of Missouri. Missouri Department of Conservation, Jefferson City, MO.

Schuster, G. A. 2008. *Orconectes* (*Trisellescens*) *taylori*, a new species of crayfish from western Tennessee (Decapoda: Cambaridae). Proceedings of Biological Society of Washington 121: 62–71.

Taylor, C. A. 2000. Systematic studies of the *Orconectes juvenilis* complex (Decapoda: Cambaridae), with descriptions of two new species. Journal of Crustacean Biology 20: 132–152.

Taylor, C. A., S. B. Adams, and G. A. Schuster. 2014. Systematics and biogeography of *Orconectes*, subgenus *Trisellescens*, in the southeastern United States, a test of morphology-based classification. Journal of Crustacean Biology 34: 1–14.

Walls, J.G. 2006. A new crayfish, *Procambarus* (*Girardiella*) *machardyi*, from northwestern Louisiana (Crustacea: Decapoda: Cambaridae). Proceedings of the Biological Society of Washington 119: 259–268.

Walls, J.G. & J.B. Black 2008. A new crayfish, *Procambarus* (*Pennides*) *pentastylus*, from southwestern Louisiana (Crustacea: Decapoda: Cambaridae), with a key to western species of the subgenus. Proceedings of the Biological Society of Washington 121: 49–61.

Walls, J. G. 2009. Crawfishes of Louisiana. Louisiana State University Press, Baton Rouge, Louisiana.

Dendrobranchiata

D. Christopher Rogers

Kansas Biological Survey, University of Kansas, Lawrence, KS, USA; The Biodiversity Institute, University of Kansas, Lawrence, KS, USA

Dendrobranchiate shrimp are primarily marine. However, two Nearctic species of this infraorder of Decapoda have juvenile stages that enter estuaries and tidally influenced freshwater. The juvenile forms are small, and spend little time in freshwater.

KEYS TO DENDROBRANCHIATA

1 Rostrum shorter than antennal scales, with ventral teeth present or not; antenna I with flagella shorter than peduncle; pereopod IV and V shorter than pereopod III .... *Farfantepenaeus brasiliensis* (Latreille, 1817)

[USA: coastal North Carolina south. Mexico: Gulf of Mexico states. Neotropics.]

1' Rostrum longer than antennal scales, ventral teeth lacking; antenna I with dorsal flagellum many times longer than peduncle; pereopods IV and V longer than pereopod III .... *Xiphopeneus kroyeri* (Heller, 1862)

[USA: coastal North Carolina south. Mexico: Gulf of Mexico states. Neotropics.]

Caridea

Fernando Alvarez

Colección Nacional de Crustáceos, Instituto de Biología, Universidad Nacional Autónoma de México, Ciudad de México, Mexico

INTRODUCTION

The freshwater shrimps from the Neartic region belong to two families in the infraorder Caridea: Atyidae and Palaemonidae. In contrast to the Palaemonidae that includes freshwater, estuarine, and marine species, the adult Atyidae are completely freshwater, although the juveniles and occasionally adults can venture into brackish and marine waters. The five species of Atyidae from the Neartic region in North America belong to three genera. *Palaemonias* Hay, 1901 is characterized by a long rostrum, the absence of eyes, spines on the anterior margin of the carapace, and long pereopodal exopods. *Potimirim* Holthuis, 1954 has a smooth carapace and is introduced along the southern Atlantic drainage of the United States (Abele, 1972; Gore et al., 1978; Beck, 1979). Finally, *Syncaris* Holmes, 1900 is limited to the west coast of the United States; members of this genus have a pair of spines on the anterior portion of the carapace. This list of atyid shrimps includes two species which seem to be extinct. The last known individuals of *Syncaris pasadenae* (Kingsley, 1896), the Pasadena freshwater shrimp, were collected in 1933 (Martin & Wicksten, 2004). *Potimirim potimirim* is probably extinct because none of these carideans have been collected since the 1970s when several records of reproductive individuals were reported from Florida canals (Fofonoff et al., 2003). The genera *Palaemonias* and *Syncaris* only occur in the United States, while *Potimirim* is distributed from Mexico to Brazil.

The palaemonid shrimps are represented by six species of *Macrobrachium* Bate, 1868 (Bowles et al., 2000), 15 species of *Palaemon* Weber, 1795 (Strenth, 1976, 1994; De Grave & Ashelby, 2013), and the monotypic *Calathaemon* Bruce & Short, 1993. *Macrobrachium* has a pantropical distribution with more than 240 species (De Grave et al., 2009), of which only a few, as presented here, have become established in the Neartic Region. Of the six species of *Macrobrachium*, in the United States, *M. acanthurus* (Wiegmann, 1836), *M. carcinus* (Linnaeus, 1758), and *M. ohione* (Smith, 1874) are considered native species. In contrast, *M. faustinum* (de Saussure, 1857), *M. heterochirus* (Wiegmann, 1836), and *M. olfersii* (Wiegmann, 1836) have been considered introduced species (Holthuis & Provenzano, 1970; Fofonoff et al., 2003); however, no detailed studies have been conducted, and these species are common in the Gulf of Mexico.

The genus *Palaemon* has a cosmopolitan distribution and includes 83 marine, estuarine, and freshwater species (De Grave & Fransen, 2011; De Grave & Ashelby, 2013). Of the freshwater species of *Palaemon* known worldwide, 12 species are native to the Nearctic and three are introduced. Many species were previously treated as *Palaemonetes* or *Exopalaemon*; however, these genera were based on unstable characters (De Grave & Ashelby, 2013). The genus *Calathaemon* is monotypic.

TERMIONOLOGY AND MORPHOLOGY

Atyid shrimps are characterized by modified chelae of the first two pairs of pereopods. These bear thick bands of setae along the distal margins of the dactyl and propodus, thereby composing a filtering or scraping apparatus when the chela is open. In addition, the carpus in both pereopods is modified to receive the propodus. The morphology of the first two pereopods is uniform across the family. The presence of spines on the carapace and the ornamentation of some appendages are of taxonomic value in the family. The morphology of the male's second pleopod is used to differentiate among the species of *Potimirim* (Villalobos, 1959).

Macrobrachium species differ in the shape, length, and dentition of the rostrum, as well as in the morphology of the second pereopod, or major cheliped. The number of teeth on dorsal and ventral margins of the rostrum, and the number of dorsal teeth behind the orbit are of taxonomic value. Regarding the second pereopod or major cheliped, the relative length of each article, the presence of spines or pubescence, the way the fingers of the chela close, the dentition of the cutting edges of the fingers, the shape of the chela, and the total length of the appendage relative to the total length of the organism are used to discriminate among species. Adult males should be used whenever possible to correctly identify the species. Moreover, juvenile *Macrobrachium* of different species may look very similar. The revision of the Palaemoninae by Holthuis (1952) remains an important reference on the genus.

The taxonomy of *Palaemon* is complex because of the reduced number of useful characters to discriminate among species, and because these characters have a limited extent of variation. The dentition of the rostrum and the spinulation of the appendix masculina, both with a small range of variation, are the main examples of taxonomically important characters for this genus. Typically, small morphological variations associated to disjunct or isolated distributions, have been used to describe and differentiate species. The key to species of *Palaemon* presented here is based on the one presented by Strenth (1976).

The following keys includes some drawings taken from Chapter 22 by Hobbs & Lodge in the third edition of *Ecology and Classification of North American Freshwater Invertebrates* (Thorp & Covich, 2010).

KEYS TO CARIDEA

Crustacea: Malacostraca: Caridea: Families

1 Pereopods I and II with chelae bearing elongated apical setal tufts which form a sieving or scraping apparatus .... **Atyidae [p. 697]**

1' Pereopods I and II with chelae lacking elongated, apical setal tufts .... **Palaemonidae [p. 697]**

Crustacea: Malacostraca: Caridea: Atyidae: Genera

1 Eyes normally developed and pigmented 2

1' Eyes reduced, unpigmented, cave adapted .... **_Palaemonias_ [p. 697]**

[USA: Alabama and Kentucky]

2(1) Carapace smooth, without spines .... *Potimirim potimirim* (F. Mueller, 1881)

[USA: Florida. Mexico: Tamaulipas. Neotropical]

2 Carapace with supraorbital and antennal spines .... **_Syncaris_ [p. 697]**

[USA: California]

Crustacea: Malacostraca: Caridea: Atyidae: *Palaemonias*: Species

Note: both species of *Palaemonias* are protected under the USA Endangered Species Act.

1 Rostrum with dorsal and ventral teeth, orbital margin devoid of spines, maxilliped III exopod reaching pereopod I propodus distal half; maximum length 23 mm (Pl. 16.190.06 B) .... *Palaemonias ganteri* Hay, 1901

[USA: Kentucky: Mammoth-Flint Ridge Cave System]

1' Rostrum without ventral teeth, orbital spine present, maxilliped III exopod reaching slightly beyond pereopod I carpus-propodus articulation, maximum length about 15 mm (Pl. 16.190.06 A) .... *Palaemonias alabamae* Smalley, 1961

[USA: Alabama: Madison County]

Crustacea: Malacostraca: Caridea: Atyidae: *Syncaris*: Species

Note: *Syncaris pacifica* is protected under the USA Endangered Species Act.

1 Rostrum lacking dorsal teeth; rostrum less than twice the length of first antenna peduncle; pereopods 3-4 without exopods (possibly extinct) (Pl. 16.190.05 B) .... *Syncaris pasadenae* (Kingsley, 1896)

[USA: California]

1' Rostrum with dorsal teeth; rostrum approximately twice as long as first antenna peduncle; pereopods 3-4 with exopods (Pl. 16.190.05 A) .... *Syncaris pacifica* (Holmes, 1895)

[USA: California]

Crustacea: Malacostraca: Caridea: Palaemonidae: Genera

1 Carapace with hepatic spine absent 2

1' Carapace hepatic spines present (Pls. 16.190.1–4, 7) .... **_Macrobrachium_ [p. 697]**

2(1) Carapace with anterior branchiostegal region not inflated, branchiostegal suture distinct (Pl. 16.190.10) .... **_Palaemon_ [p. 699]**

2' Carapace with anterior branchiostegal region inflated, branchiostegal suture absent .... *Calathaemon holthuisi* (Strenth, 1976)

[USA: Texas]

Crustacea: Malacostraca: Caridea: Palaemonidae: *Macrobrachium*: Species

Note: *Macrobrachium rosenbergii* De Man, 1879 is commercially raised in the Nearctic, but successful introduction to the wild has thus far not been reported.

1 Adult male second pereopod with carpus as long or longer than merus 2

1' Adult male second pereopod with carpus shorter than merus (Pls. 16.190.1, 2 A) .... *Macrobrachium carcinus* (Linnaeus, 1758)

[USA: Florida. Mexico: Tamaulipas. Neotropics.]

2(1) Chelipeds different in size and appearance, major chela globose, spinulated, pubescent 3

2' Chelipeds similar in appearance, may be subequal in size or one much larger than the other 4

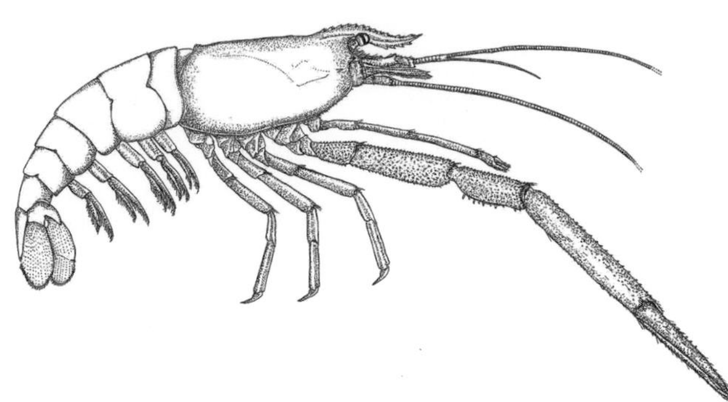

PLATE 16.190.01 *Macrobrachium carcinus.*

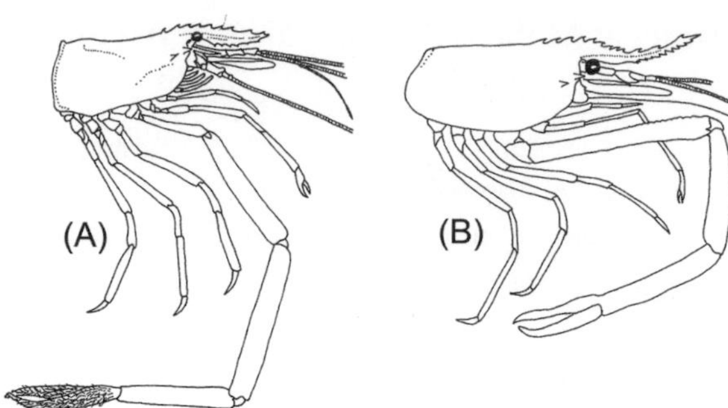

PLATE 16.190.03 Lateral view of the carapace of: (A) *Macrobrachium acanthurus*; (B) *M. rosenbergii*, a common mariculture species worldwide, including in parts of the USA. *From Hobbs, as modified from Holthuis, 1952.*

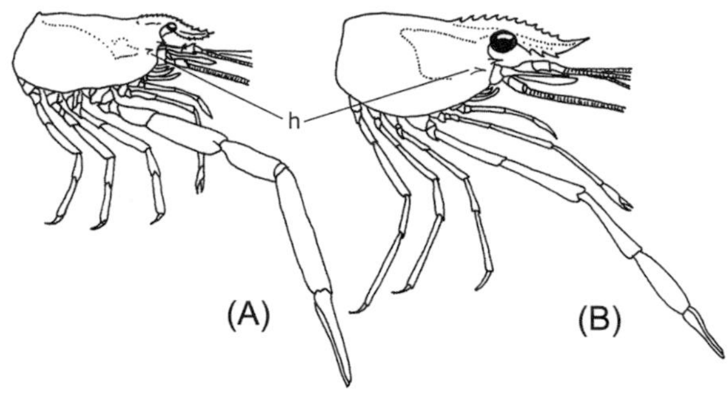

PLATE 16.190.02 Lateral view of anterior part of the body of: (A) *Macrobrachium carcinus* and (B) *M. ohione*. H=hepatic spine. *From Hobbs (in Hobbs & Lodge, 2010 for this and other noted shrimp figures) after Holthuis, 1952.*

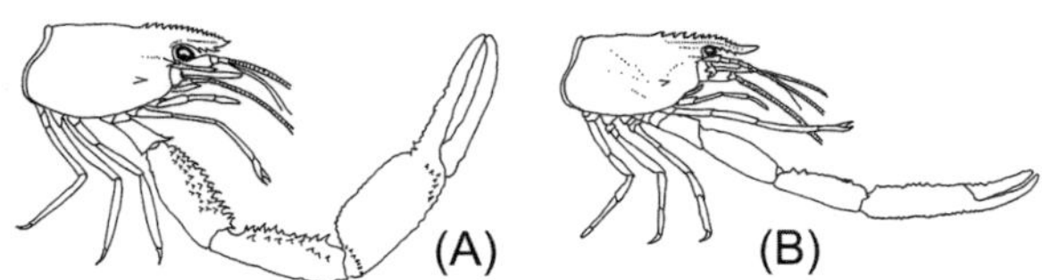

PLATE 16.190.04 Lateral view of the carapace of: (A) *Macrobrachium faustinum*; (B) *M. heterochirus. From Hobbs, as modified from Chace & Hobbs, 1969.*

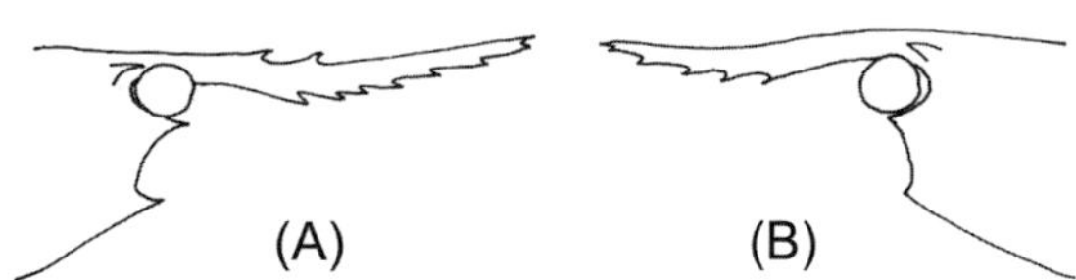

PLATE 16.190.05 Lateral view of: (A) *Syncaris pacifica*, an endangered species; and (B) *S. pasadenae* (probably extinct). *From Hobbs, after Martin & Wicksten, 2004.*

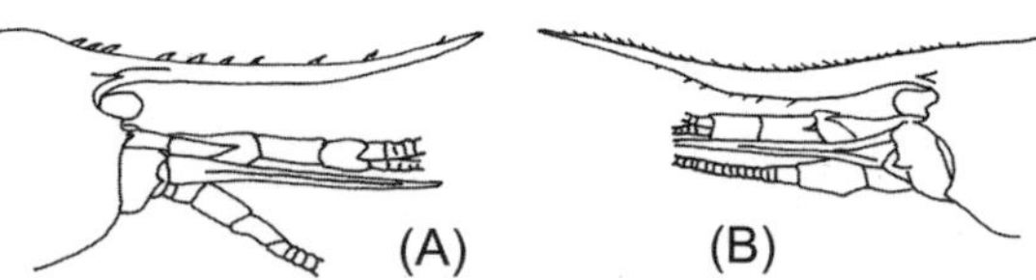

PLATE 16.190.06 Lateral view of the stygiobionts: (A) *Palaemonias alabamae*; and (B) *Palaemonias ganteri. After Hobbs et al., 1977.*

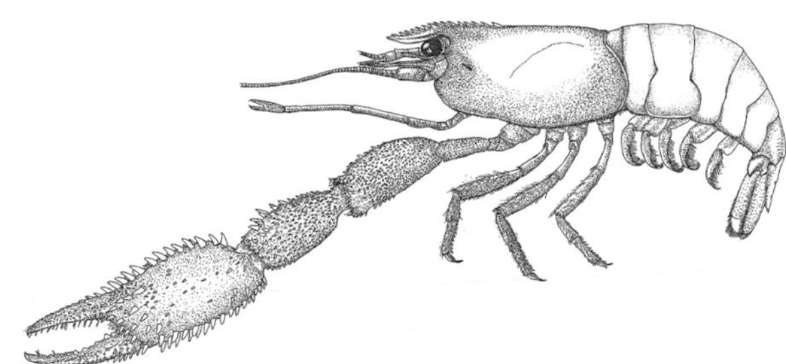

PLATE 16.190.07 Lateral view of *Macrobrachium olfersii.*

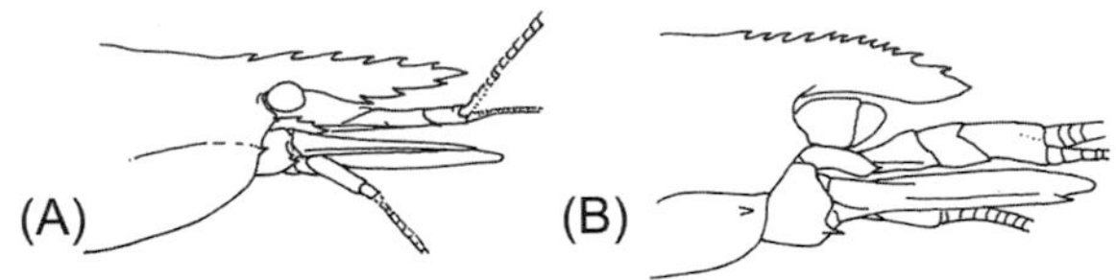

PLATE 16.190.08 Lateral view of anterior part of carapace: (A) *Palaemon cummingi* (after Chace, 1954); and (B) *Palaemon antrorum* (after Holthuis, 1949).

3(1) Major chela with merus 0.8 to 0.9 times the length of carpus, chela with palm of same length as dactylus (Pl. 16.190.04 A) .... *Macrobrachium faustinum* (de Saussure, 1857)

[USA: Florida. Neotropics.]

3' Major chela with merus 0.5 times the length of carpus, chela with palm longer than dactylus (Pl. 16.190.07) .... *Macrobrachium olfersii* (Wiegmann, 1836)

[USA: Florida, Louisiana, Mississippi, and Texas. Mexico: Tamaulipas. Neotropics.]

4(2) Rostrum longer than first antenna peduncle, second pereopods subequal in size 5

4' Rostrum shorter than first antenna peduncle, second pereopods unequal in size, merus and carpus of major cheliped of about same length, dactyl and manus one-third length of chela (Pl. 16.190.04 B) .... *Macrobrachium heterochirus* (Wiegmann, 1836)

[USA: Florida. Mexico: Tamaulipas. Neotropics.]

5(4) Rostrum with anterior portion lacking teeth; chelipeds not longer than total body length (Pl. 16.190.02 B) .... *Macrobrachium ohione* (Smith, 1874)

[USA: Mississippi and Gulf Coast drainages. Mexico: Tamaulipas. Neotropics.]

5' Rostrum anterior portion with teeth along dorsal and ventral margins, chelipeds longer than total body length (Pl. 16.190.3 A) .... *Macrobrachium acanthurus* (Wiegmann, 1836)

[USA: North Carolina to Florida, Gulf Coast drainages. Mexico: Tamaulipas. Neotropics.]

Crustacea: Malacostraca: Caridea: Palaemonidae: *Palaemon*: Species

1 Body, unpigmented, eyes reduced, subterranean 2

1' Shrimps distributed in epigean bodies of water, pigmented, eyes normally developed 3

2(1) Rostrum ventral margin with teeth; antenna I flagellum dorsal rami with free portion shorter than fused portion (Pl. 16.190.08 A) .... *Palaemon cummingi* (Chace, 1954)

[USA: Florida: Alachua County]

2' Rostrum ventral margin without teeth; antenna I flagellum dorsal rami with free portion longer than fused portion (Pl. 16.190.08 B) .... *Palaemon antrorum* (Benedict, 1896)

[USA: Texas: San Marcos]

3(1) Gastric spine absent 4

3' Gastric spine present .... *Palaemon macrodactylus* Rathbun, 1902

[USA: Invasive. California, Oregon, Washington. Palaearctic]

4(3) Rostrum not raised as a crest 6

4' Rostrum raised in proximal portion as a crest; gastric spine absent 5

5(4) Abdominal segments each with a dorsal, transverse carina .... *Palaemon carinicauda* Holthuis, 1950

[USA: California: introduced to the Sacramento River Delta, possibly extirpated. Palaearctic]

5' Abdominal segments lacking any such carinae .... *Palaemon modestus* (Heller, 1862)

[USA: California, Idaho, Oregon, Washington, invasive. Palaearctic]

6(4) Antenna I dorsal flagellum with fused portion of the two rami longer than free portion 7

6' Antenna I dorsal flagellum dorsal rami with free portion longer than or subequal to fused portion 13

7(6) Branchostegial spine at carapace margin 8

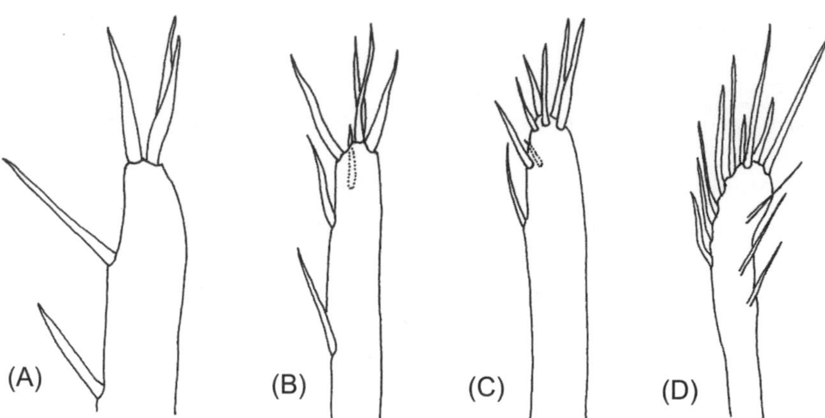

PLATE 16.190.09 Postaxial view of distal portion of appendix masculina demonstrating setal ornamentation: (A) *Palaemon kadiakensis*; (B) *P. paludosus*; (C) *P. cummingi*; and (D) *P. antrorum*. *After Villalobos & Hobbs, 1974.*

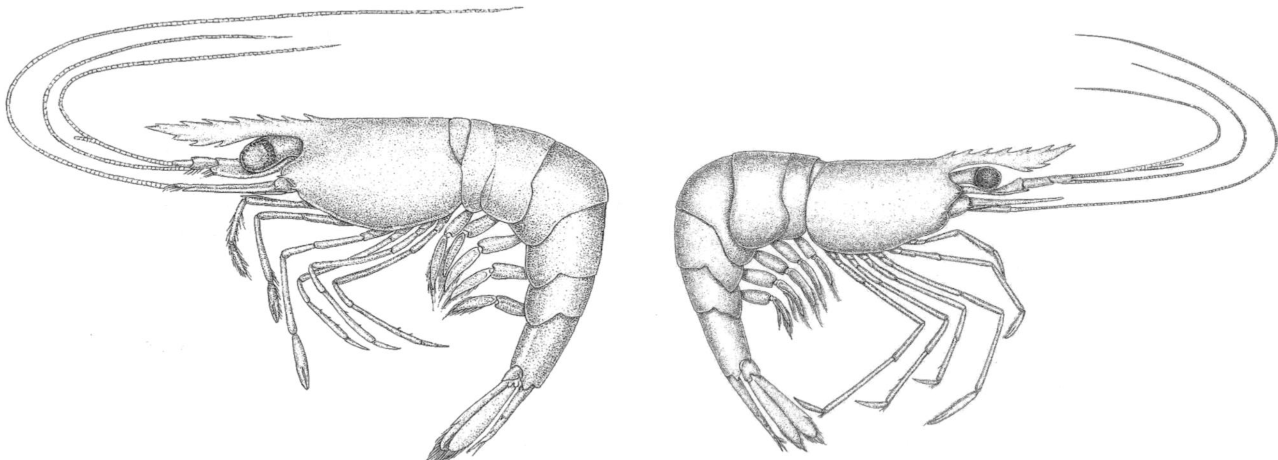

PLATE 16.190.10 Lateral view of *Palaemon paludosus.*

PLATE 16.190.11 Lateral view of *Palaemon lindsayi.*

7' Branchostegial spine set back from carapace margin, such that the apex of the spine does not attain the carapace margin; abdominal segment 6 longer than telson; appendix masculina with three apical setae (Pl. 16.190.09 A) .... *Palaemon kadiakensis* (Rathbun, 1902)

[Canada: Great Lakes region and Mississippi Drainage. USA: Mississippi and Gulf Coast drainages, introduced in western states. Mexico: Nuevo Leon, Tamulipas]

8(7) Rostrum subequal in length to scaphocerites 9

8' Rostrum reaching beyond apex of scaphocerites .... *Palaemon texanus* (Strenth, 1976)

[USA: Texas. Introduced in western states]

9(8) Appendix masculina apex with four setae (Pl. 16.190.09) 10

9' Appendix masculina apex with five or six setae 11

10(9) Rostrum with 3–4 ventral teeth (Pl. 16.190.10), branchiostegal spine as a continuation of branchiostegal groove, telson with posterior dorsal spine pair midway between anterior pair and posterior margin (Pl. 16.190.10) .... *Palaemon paludosus* (Gibbes, 1850)

[USA: Atlantic and Gulf Coastal drainages. Introduced into southern California and northern Mexico]

10' Rostrum with 2 ventral teeth, branchiostegal spine slightly below branchiostegal groove, telson with posterior dorsal spine pair next to posterior margin .... *Palaemon hobbsi* (Strenth, 1994)

[Mexico: Tamaulipas: Headwaters of Río Mante]

11(9) Appendix masculina apex with five setae 12

11' Appendix masculina apex with six setae; rostrum slightly curved, tip directed upwards; rostrum reaching antenna I peduncle distal margin .... *Palaemon suttkusi* (Smalley, 1964)

[Mexico: Coahuila: Cuatro Ciénegas Basin and Nadadores River]

12(11) Rostrum with six or seven dorsal teeth and two to four ventral teeth; telson posterior spine pair clearly separated form posterior margin (Pl. 16.190.11) .... *Palaemon lindsayi* (Villalobos & Hobbs, 1974)

[Mexico: San Luis Potosí: Media Luna Lagoon, Rio Verde]

12' Rostrum with six dorsal teeth and two ventral teeth, telson dorsal spines posterior on or near posterior margin .... *Palaemon mexicanus* (Strenth, 1976)

[Mexico: San Luis Potosí]

13(6) Rostrum with one tooth posterior to orbit; pereopod II with dactylus cutting edge bearing one or no teeth, propodus cutting edge bearing no teeth, carpus as long as or longer than chela palm 14

13' Rostrum with two teeth posterior to orbit; pereopod II with dactylus cutting edge bearing one tooth, propodus cutting edge bearing one tooth, carpus subequal or shorter than chela palm .... *Palaemon vulgaris* Say, 1818

[USA: Atlantic and Gulf Coast estuaries. Mexico: Tamaulipas estuaries]

14(13) Rostrum with dorsal teeth extending to apex .... *Palaemon mundusnovus* De Grave & Ashelby, 2013

[USA: Atlantic Coast estuaries. Neotropics]

14' Rostrum dorsal margin unarmed before apex .... *Palaemon pugio* (Holthuis, 1949)

[USA: Atlantic and Gulf Coast estuaries. Mexico: Tamualipas estuaries. Neotropics]

REFERENCES

Abele, L.E. 1972. Introductions of two freshwater decapod crustaceans (Hymenosomatidae and Atyidae) into Central and North America. Crustaceana 23: 209–218.

Beck, J.T. 1979. A third occurrence of the introduced atyid shrimp, *Potimirim potimirim*, in Florida. Florida Scientist 42: 256.

Bowles, D.E., K. Azis & C.L. Knight. 2000. *Macrobrachium* (Decapoda: Caridea: Palaemonidae) in the contiguous United States: a review of the species and an assessment of threats to their survival. Journal of Crustacean Biology 20:158–171.

Chace, F.A., Jr. 1954. Two new subterranean shrimps (Decapoda: Caridea) from Florida and the West Indies, with a revised key to the American species. Journal of the Washington Academy of Sciences 44: 318–324.

Chace, F.A., Jr. & H.H. Hobbs, Jr. 1969. The freshwater and terrestrial decapod crustaceans of the West Indies with special reference to Dominica. Bulletin of the U.S. National Museum 292: 1–258.

De Grave, S. & C.H.J.M. Fransen. 2011. *Carideorum catalogus*: the recent species of the dendrobranchiate, stenopodidean, procarididean and caridean shrimps (Crustacea: Decapoda). Zoologische Mededelingen 85: 195–588.

De Grave, S. & C. W. Ashelby. 2013. A re-appraisal of the systematic status of selected genera in Palaemoninae (Crustacea: Decapoda: Palaemonidae). Zootaxa 3734: 331–334.

De Grave, S., N.D. Pentcheff, S.T. Ahyong, T.Y. Chan, K.A. Crandall, P.C. Dworschak, D.L. Felder, R.M. Feldmann, C.H.J.M. Fransen, L.Y.D. Goulding, R. Lemaitre, M.E.Y. Low, J.W. Martin, P.K.L. Ng, C.E. Schweitzer, S.H. Tan, D. Tshudy & R. Wetzer. 2009. A classification of the living and fossil genera of decapod crustaceans. The Raffles Bulletin of Zoology Supplement 21: 1–109.

Fofonoff, P.W., G.M. Ruiz, B. Steves & J.T. Carlton. 2003. National exotic marine and estuarine species Information System. http://invasions.si.edu/nemesis/. Access date 12-September-2013.

Gore, R.H., G.R. Kulczycki & P.A. Hastings. 1978. A second occurrence of the Brazilian freshwater shrimp, *Potimirim potimirim*, along the central eastern Florida coast. Florida Scientist 41: 57–60.

Hobbs, H.H. & D.M. Lodge. 2010. Decapoda. Chapter 22 *in*: J.H. Thorp and A.P. Covich (eds.), Ecology and classification of North American freshwater invertebrates. Academic Press, Boston.

Hobbs, H.H. Jr., H.H. Hobbs, III & M.A.Daniel. 1977. A review of the troglobitic decapod crustaceans of the Americas. Smithsonian Contributions to Zoology 244: 1–183.

Holthuis, L.B. 1949. Note on the species of Palaemonetes (Crustacea, Decapoda) found in the United States of America. Proceedings of the Koninklijke Nederlandse Akademie van Wetenschappen 52: 87–95.

Holthuis, L.B. 1952. A general review of the Palaemonidae (Crustacea, Decapoda, Natantia) of the Americas. II. The subfamily Palaemoninae. Occasional Papers, Allan Hancock Foundation 22:1–396.

Holthuis, L.B. & A.J. Provenzano, Jr. 1970. New distribution records for species of *Macrobrachium* with notes on the distribution of the genus in Florida (Decapoda, Palaemonidae). Crustaceana 19: 211–213.

Martin, J.W. & M.K. Wicksten. 2004. Review and redescription of the freshwater atyid shrimp genus *Syncaris* Holmes, 1900, in California. Journal of Crustacean Biology 24: 447–462.

Strenth, N.E. 1976. A review of the systematics and zoogeography of the freshwater species of *Palaemonetes* Heller of North America (Crustacea: Decapoda). Smithsonian Contributions to Zoology 228: 1–27.

Strenth, N.E. 1994. A new species of *Palaemonetes* (Crustacea: Decapoda: Palaemonidae) from northeastern Mexico. Proceedings of the Biological Society of Washington 107: 291–295.

Taylor, C.A. & G.A. Schuster. 2010. Monotypic no more, a description of a new crayfish of the genus *Barbicambarus* Hobbs, 1969 (Decapoda: Cambaridae) from the Tennessee River drainage using morphology and molecules. Proceedings of the Biological Society of Washington 123: 324–334.

Villalobos, A. 1959. Contribución al conocimiento de los Atyidae de México. II (Crustacea, Decapoda). Estudio de algunas especies del género *Potimirim* (=*Ortmannia*), con descripción de una especie nueva de Brasil. Anales del Instituto de Biología, Universidad Nacional Autónoma de México 30: 269–330.

Villalobos, A. & H.H. Hobbs, Jr. 1974. Three new crustaceans from La Media Luna, San Luis Potosi, Mexico. Smithsonian Contributions in Zoology 174: 1–18.

Order Mysida

W. Wayne Price
Department of Biology, University of Tampa, Tampa, FL, USA

INTRODUCTION

Fewer than 25 species of mysidaceans or "opossum shrimp" are found in Nearctic freshwater habitats. Members of the Mysida are represented by two groups: glacial relict species and euryhaline estuarine fauna. The relict group is found in lakes of previously glaciated continental areas of North America and consists of two species. *Mysis diluviana* Audzijonte & Väinölä, 2005 (formally known as *Mysis relicta* Lovén, 1862)—the most studied of the two species—has a native distribution in northern North America but has been introduced widely into lakes to the south and west of its natural range to serve as a supplemental food source for fish (Audzijonte & Väinölä, 2005; Porter et al., 2008). Most Nearctic freshwater mysids are represented by estuarine species that can tolerate freshwater. These species occur in bays, but also in freshwater tributaries of estuarine systems, sometimes hundreds of kilometers inland (Porter et al., 2008). Most mysids are hyperbenthic, living on or just above the sediment surface, but a few species are pelagic. Most are omnivorous and have the potential to influence phytoplankton and zooplankton communities. Because they may occur in high densities in estuaries and lakes, mysids serve as a food source for higher trophic levels and are considered an important link between benthic and pelagic systems. Due to their sensitivity to toxicants at ecologically relevant concentrations, estuarine mysids are used in toxicity testing and environmental monitoring (Heard et al., 2006).

LIMITATIONS

Identification of Nearctic mysidaceans is based mainly on morphological variations in the telson, uropods, and antennal scales, characters that exhibit little sexual dimorphism. These and other basic morphological characters included in the key are presented in Pl. 16.191.1. Whenever possible, mature adult specimens should be examined because many characters used in the key are from adults as opposed to juveniles. This is especially important for meristic characters such as the number of spiniform setae on the lateral and posterior margins of telsons and uropodal endopods. Setae may be added with successive molts as the organism matures. In addition, the length:width ratio of the telson and antennal scales as well as the degree of male pleopod dimorphism may change with growth. Characters that separate immature from mature mysidaceans as well as females and males are presented in the next section.

Most Nearctic mysidaceans are relatively well-known and easily identified to species, but confusion concerning some groups from the Pacific Coast and subarctic and arctic continental regions have been resolved only in the last few decades. Revisions of existing genera and the establishment of new genera and species have taken place with reference to *Acanthomysis* (Holmquist, 1981), *Mysis* (Audzijonte & Väinölä, 2005, 2007), *Neomysis* (Holmquist, 1973), and *Orientomysis* (Fukuoka & Murano, 2005).

The *Mysis relicta* species group has a widespread distribution in boreal and subarctic lakes and estuarine/coastal regions of North America and Europe. Based on morphological and molecular evidence, Audzijonte & Väinölä (2005) separated this group into four species, including two, *M. diluviana* Audzijonte & Väinölä, 2005 and *M. segerstralei* Audzijonte & Väinölä 2005, from the Nearctic. In 2007, Audzijonte & Väinölä described *M. nordenkioldi*, a circumpolar coastal species found in freshwater lakes in Greenland and earlier considered conspecific with *M. litoralis*, a coastal species restricted to northwestern North America.

There have been four confirmed reports of eastern Asian and European species established in Nearctic waters in the last two decades. These species were introduced presumably through ballast water discharge from ships (Fukuoka & Murano, 2005; Kestrup & Ricciardi, 2008). A fifth nonindigenous species, *N. japonica*, has been reported (unconfirmed) from estuarine waters of San Francisco Bay (Modlin, 2007) and is similar morphologically to *N. kadiakensis*, a Nearctic endemic. However, these two species have rather different salinity requirements: *N. kadiakensis* is a coastal marine species and *N. japonica* is an oligohaline-freshwater species common in the mouths of rivers and brackish lakes. Both species are included in the Nearctic key.

TERMINOLOGY AND MORPHOLOGY

Nearctic adult mysidaceans range in size from less than 4 mm to nearly 30 mm. Mysidaceans have a carapace that covers most or all of the thoracic somites. The first antenna consists of a peduncle of three articles followed by two flagella and is often more robust in males than females. In addition, males have a setose conical process, the male lobe (appendix masculina), at the distal end of article 3 of the peduncle. This lobe is well-developed in most mature Nearctic mysidaceans, but is reduced in *Deltamysis*. The biramous second antenna has an endopod with a 3 article peduncle and a flagellum. The exopod

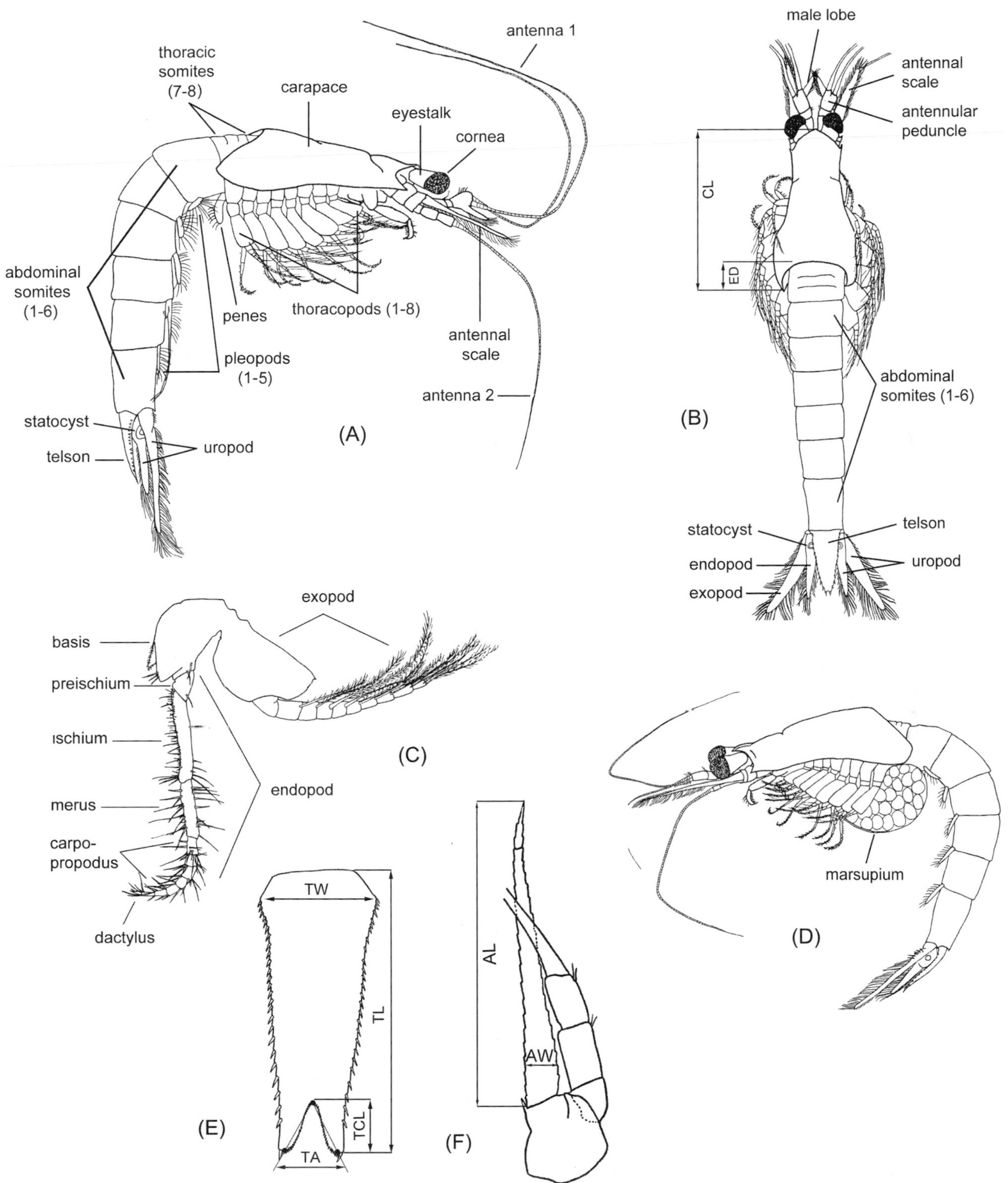

PLATE 16.191.01 Morphology of a typical mysid. *Neomysis mercedis*; (A) Male, lateral view; (B) Male, dorsal view; CL, carapace length; ED, emargination depth; (C) Thoracopod 4; (D) Ovigerous female, lateral view (after Daly & Holmquist, 1986); (E) *Mysis nordenskioldi*, telson; TA, telson cleft angle; TCL, telson cleft length; TL, telson length; TW, telson width at base (after Holmquist, 1959); (F) *Neomysis japonica*, antenna 2; AL, antennal scale length; AW, antennal scale greatest width (after Tattersall, 1951).

or antennal scale is in the form of a flat plate that is usually longer than the antennal peduncle in mysids (except *Deltamysis*).

Nearctic mysidaceans have eight pairs of biramous thoracic appendages (thoracopods) with the anterior ones modified for feeding and often termed maxillipeds or gnathopods. All exopods are well-developed and used for swimming. Mature females have a well developed marsupium or brood pouch, with or without eggs/embryos, that is composed of a number of paired plates (oostegites). These structures are attached to thoracopods and extend between them on the ventral side of the thorax. Mysids have two to three pairs of oostegites on the posterior thoracopods. Mature males have a pair of gonopods (penes) extending as lobes from the medial bases of the 8th thoracic appendages.

A pair of pleopods is present on each of the first five abdominal somites of mysidaceans. Although the pleopods in *Deltamysis* are uniramous simple plates in both sexes, the pleopods of most Nearctic mysids exhibit marked sexual dimorphism. Female pleopods are as in *Deltamysis*, but male pleopods are either all biramous with pleopods 2-5 well developed, or pleopods 1, 2, and 5 are uniramous plates and pleopod 4 is biramous with an elongated exopod. The modified pleopod 4 may be important in the taxonomic diagnosis of some genera (*Neomysis*, *Orientomysis*, *Taphromysis*) and their species.

Abdominal somite 6 bears the telson as well as the last abdominal appendages, the uropods. The telson is the terminal somite (rather than appendage) of the body, and thus the terms anterior-posterior rather than proximal-distal apply. The apex or posterior end of the telson may be entire, emarginate, or cleft; the cleft may be unarmed or armed partially or completely with spinules. The lateral margins of the telson may be unarmed or armed partially or completely with spiniform (spine-like) setae. Each uropod comprises a uniarticulate protopod and two branches, the exopod and endopod. The endopod bears a statocyst. Most mysids have a few to many spiniform setae on the medial margin of the endopod extending distally from the statocyst.

MATERIAL PREPARATION AND PRESERVATION

Mysidaceans can be preserved in the field in a 5% buffered formalin solution or 70–95% ethanol. If formalin is used, specimens should be transferred to 70% ethanol within 7 to 10 days. For long-term storage, 70% ethanol with 1% glycerine is a good medium because flexibility of specimens is maintained and desiccation is prevented.

For purposes of identifying mysidaceans, manipulate specimens in ethanol or water. Although dissection is often not necessary for larger specimens, that can be identified using a dissecting microscope, smaller specimens can require dissection, clearing, and viewing with a compound microscope. Examine smaller specimens <10 mm using a depression slide by placing the specimen in the depression with liquid, and covering with a cover slip. By rolling the specimen under the cover slip, you can view most desired orientations. If clearing is desired, place a few drops of an ethanol/glycerine mixture (8 parts of 70% ethanol:2 parts glycerine) in the depression, wait a few minutes for the alcohol to evaporate leaving mostly glycerine, and then place a cover slip over the depression. Any dissections should be performed in this mixture and body parts transferred with glycerine to the flat part of the depression slide or to another slide and covered with small round or square cover slips. To protect delicate or thick parts from too much compression, support the cover slip with small pieces of broken cover slips. Add pure glycerine as necessary. These "temporary" glycerine mounts partially clear specimens or parts and last for an extended period of time. If glycerine does not provide sufficient clearing, use lactic acid for a temporary mount. Place specimens in a 50% aqueous solution and then full strength lactic acid. However, do not store specimens or parts in acid as they will continue to clear and may disarticulate.

Staining specimens in methylene blue or chlorazol black E may facilitate the examination of certain external morphological structures, especially setae. Both stains are prepared as 1% solutions by weight in distilled water or 70% ethanol. Add a few drops of methylene blue to diluted or full strength lactic acid. To use chlorazol black E, add the stain to lactic acid or glycerine to make about a 5% solution. Since chlorazol black E is chitin specific, leaving specimens too long in the stain solution will turn them very dark.

Although specimens or parts can be mounted permanently, the mounted parts cannot be manipulated for 3-dimensional observation. I favor permanent storage in an ethanol-glycerine solution in glass or plastic vials with small specimens or parts placed in polyethylene embedding capsules with hinged caps, containers used in electron microscopy specimen preparation. Several capsules may be placed in one vial or larger container. For permanent slides, a variety of mounting media, such as glycerine jelly, lactophenol, and Canada balsam, may be used. Some of these substances must be sealed by ringing the cover slip with waterproof polyurethane, Canada balsam, or clear nail polish, and slides should be stored horizontally.

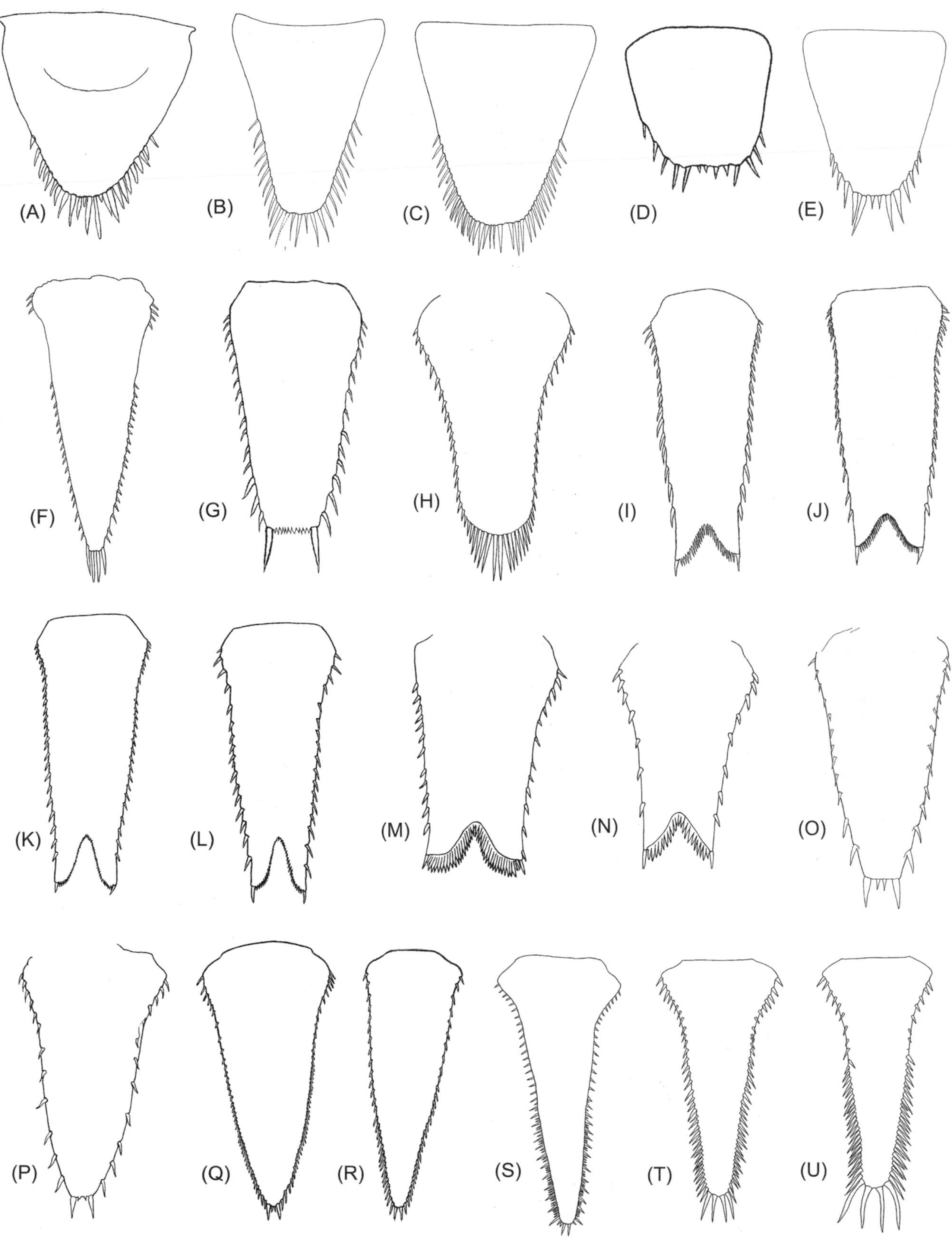

PLATE 16.191.02 Telson: (A) *Spelaeomysis villalobosi* (from García-Garza et al., 1996); (B) *S. quinterensis* (from Bowman, 1973); (C) *S. olivae* (from Bowman, 1973); (D) *Antromysis reddelli* (from Bowman, 1977); (E) *Deltamysis holmquistae* (after Bowman & Orsi, 1992); (F) *Hyperacanthomysis longirostris* (from Modlin & Orsi, 1997); (G) *Hemimysis anomala* (after Băcescu, 1954); (H) *Americamysis almyra* (from Brattegard, 1969); (I) *Mysis diluviana* (from Audzijonte & Väinölä, 2005); (J) *M. segerstralei* (from Audzijonte & Väinölä, 2005); (K) *M. nordenskioldi* (after Holmquist, 1959); (L) *M. litoralis* (after Holmquist, 1958); (M) *Taphromysis louisianae*; (N) *T. bowmani* (from Price, 1982); (O) *Neomysis intermedia*; (P) *N. mercedis* (after Holmquist, 1973); (Q) *N. japonica*; (R) *N. kadiakensis* (from Tattersall, 1951); (S) *Orientomysis hwanhaiensis* (from Modlin & Orsi, 2000); (T, U) *O. aspera*, male, female (from Fukuoka & Murano, 2005).

KEYS TO MYSIDA

Crustacea: Malacostraca: Mysida: Mysidae: Genera

1 Telson apex cleft (Pl. 16.191.2 I–N) 2

1' Telson apex entire, convex or truncate (Pl. 16.191.2 D–H, O–U) 3

2(1) Telson length more than two times basal width at broadest interval, lateral margins armed with 15–30 spiniform setae; uropodal endopod with 2–8 spiniform setae on medial margin distal to statocyst (Pl. 16.191.2 I–L) .... ***Mysis* [p. 706]**

2' Telson length less than two times basal width at broadest interval, lateral margins armed with 5–14 spiniform setae; uropodal endopod with 1 long spiniform seta on medial margin near distal edge of statocyst (Pl. 16.191.2 M, N) .... ***Taphromysis* [p. 709]**

3(1) Telson length no more than 1.3 times basal width at broadest interval, apex truncate, lateral margins with spiniform setae only on posterior 1/3 or less; uropodal endopod with no spiniform setae on medial margin near distal edge of statocyst (Pl. 16.191.2 D, E) 4

3' Telson length at least 1.6 times basal width at broadest interval, apex truncate or rounded, lateral margins with spiniform setae on anterior and posterior halves; uropodal endopod with 1 or more spiniform setae on medial margin distal to statocyst (Pls. 16.191.2 F–H, O–U, 3 L, N, 4 N, O) 5

4(3) Eyestalks without ommatidia or pigment; telson about as long as basal width at broadest interval, apex with 4 short apical spiniform setae flanked by 2 longer pairs (Pls. 16.191.2 D, 3 I) .... *Antromysis reddelli* Bowman, 1977

[Mexico: Oaxaca]

4' Eyes normal, cornea with pigmented ommatidia; telson length 1.2–1.3 times basal width at broadest interval, apex with 2 short apical spiniform setae flanked by 2 longer pairs (Pls. 16.191.2 E, 3 J) .... *Deltamysis holmquistae* Bowman & Orsi, 1992

[USA: California]

5(3) Eye (eyestalk and cornea), length less than 3 times basal width 6

5' Eye length at least 4 times basal width (Pl. 16.191.3 K) .... *Alienacanthomysis macropsis* (Tattersall, 1932)

[Canada: British Columbia. USA: Alaska, California, Oregon, Washington]

6(5) Telson, lateral margins with a continuous row of spiniform setae from base to apex (Pl. 16.191.2 G, H, O–U) 7

6' Telson, lateral margins with 2–4 spiniform setae at base, followed by unarmed space occupying 1/4 of margin, and 18–23 setae forming a continuous row along posterior 3/5 of margin (Pl. 16.191.2 F) .... *Hyperacanthomysis longirostris* (Ii, 1936)

[Invasive. USA: California. Palaearctic]

7(6) Antennal scale with rounded apex, length 6.5 times or less greatest width; uropodal endopod with 10 or fewer spiniform setae on medial margin distal to statocyst (Pls. 16.191.3 L–O, 4 N, O) 8

7' Antennal scale with pointed apex, length 7.5 times or more greatest width; uropodal endopod with 11 or more spiniform setae on medial margin distal to statocyst (Pl. 16.191.4 J, K) .... ***Neomysis* [p. 709]**

8(7) Telson length 1.8 times or less basal width at broadest interval; lateral margins armed with 21 or fewer spiniform setae, subequal in length or gradually increasing in length posteriorly (Pl. 16.191.2 G, H) 9

8' Telson length more than 2 times basal width at broadest interval; lateral margins armed with at least 30 spiniform setae, arranged in groups of larger and smaller ones in posterior half (Pl. 16.191.2 S–U) .... ***Orientomysis* [p. 709]**

9(8) Telson apex truncate; uropodal endopod with 6–9 spiniform setae on medial margin distal from statocyst; antennal scale length less than 4 times greatest width (Pls. 16.191.2 G, 3 L, M) .... *Hemimysis anomala* Sars, 1907

[Invasive. Canada: Ontario, Quebec; USA: Illinois, Michigan, New York, Wisconsin; Palaearctic]

9' Telson apex rounded; uropodal endopod with 1 spiniform seta on medial margin near distal edge of statocyst; antennal scale length 6–6.5 times greatest width (Pls. 16.191.2 H, 3 N, O) .... *Americamysis almyra* Bowman, 1964

[USA: Alabama, Florida, Louisiana, Maryland, Mississippi, South Carolina, Texas; Mexico: Campeche, Veracruz.]

Crustacea: Malacostraca: Mysida: Mysidae: *Mysis*: Species

1 Telson cleft broad and shallow, V-shaped with cleft angle of about 90° (65–122), cleft depth <0.15 telson length, telson apical lobes usually without (sometimes 1) lateral spiniform setae posterior to anterior margin of cleft (Pl. 16.191.2 I, J) 2

1' Telson cleft relatively narrow, V-shaped with cleft angle of 55° or less, cleft depth >0.15 telson length, telson apical lobes with 1–4 lateral spiniform setae posterior to anterior margin of cleft (Pl. 16.191.2 K, L) 3

2(1) Telson, lateral spiniform setae relatively robust, subequal in size to apical pair, lateral margin with 16–19 (21) spiniform setae; carapace posterior cleft deep, >0.2 carapace length (Pls. 16.191.2 I, 4 A) .... *Mysis diluviana* Audzijonte & Väinölä, 2005

[Native distribution: Canada: Alberta, Manitoba, Northwest Territories, Ontario, Quebec, Saskatchewan. USA: Michigan, Minnesota, New York, Wisconsin; Introduced distribution: Canada: British Columbia. USA: California, Colorado, Idaho, Montana, Nevada, North Dakota, Oregon, Washington, Wyoming]

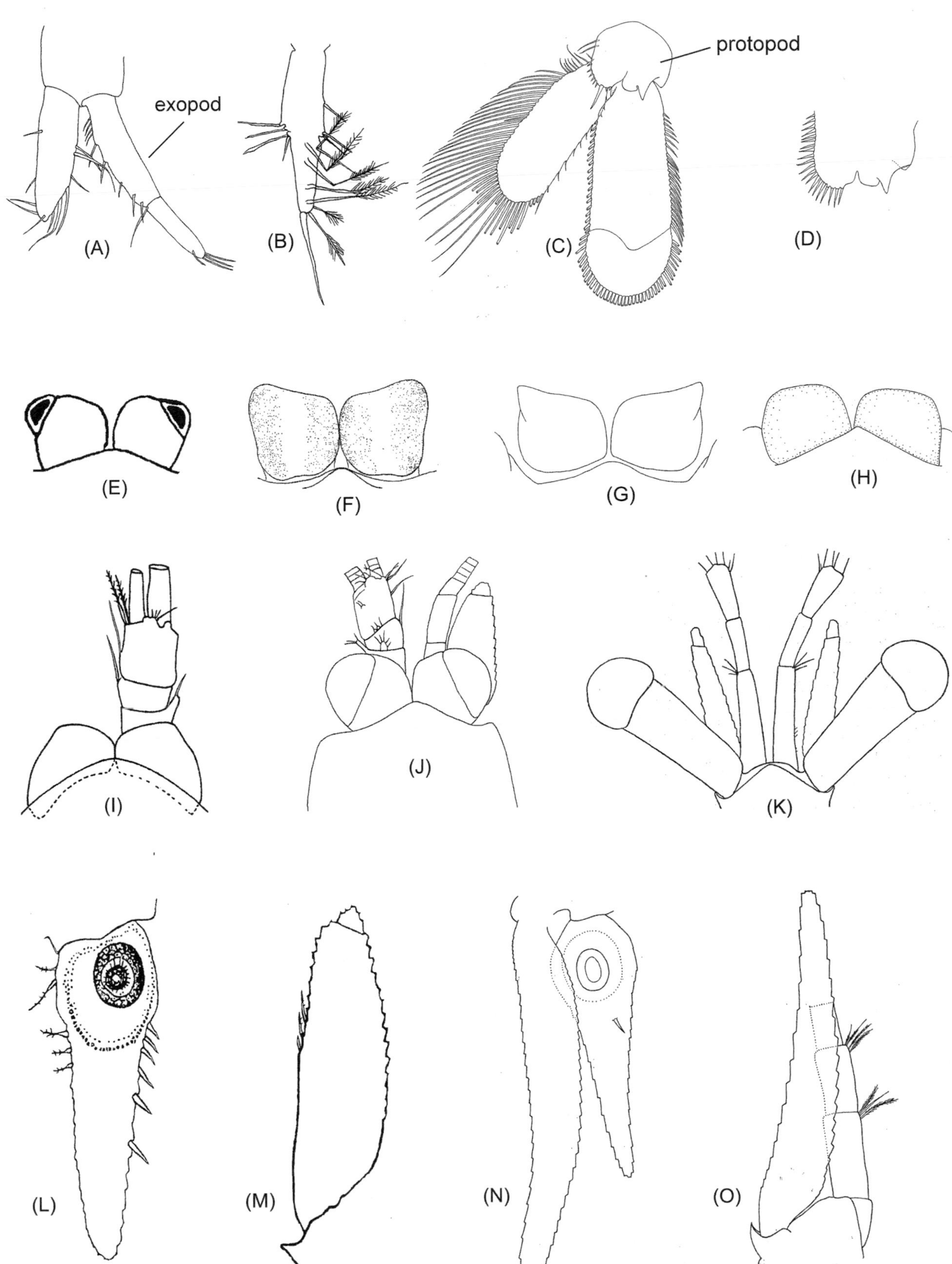

PLATE 16.191.03 (A) *Spelaeomysis villalobosi*, male pleopod 4 (from García-Garza et al., 1996); (B) *Taphromysis villalobosi*, female pleopod 5 (from Escobar-Briones & Soto, 1988); (C) *Spelaeomysis olivae*, uropod (from Bowman, 1973); (D) *S. quinterensis*, uropodal protopod (from Bowman, 1973); (E) *S. cardisomae*, eyes (from Bowman, 1973); (F) *S. quinterensis* eyes (after Villalobos, 1951); (G) *S. villalobosi*, eyes (after García-Garza et al., 1996); (H) *S. olivae*, eyes (after Bowman, 1973); (I) *Antromysis reddelli*, anterior body, dorsal (after Bowman, 1977); (J) *Deltamysis holmquistae*, anterior body, dorsal (after Bowman & Orsi, 1992); (K) *Alienacanthomysis macropsis*, anterior body, dorsal (from Tattersall, 1951); (L, M) *Hemimysis anomala*, uropodal endopod, antennal scale (after Băcescu, 1954); (N, O) *Americamysis almyra*, uropod, antenna 2 (from Brattegard, 1969).

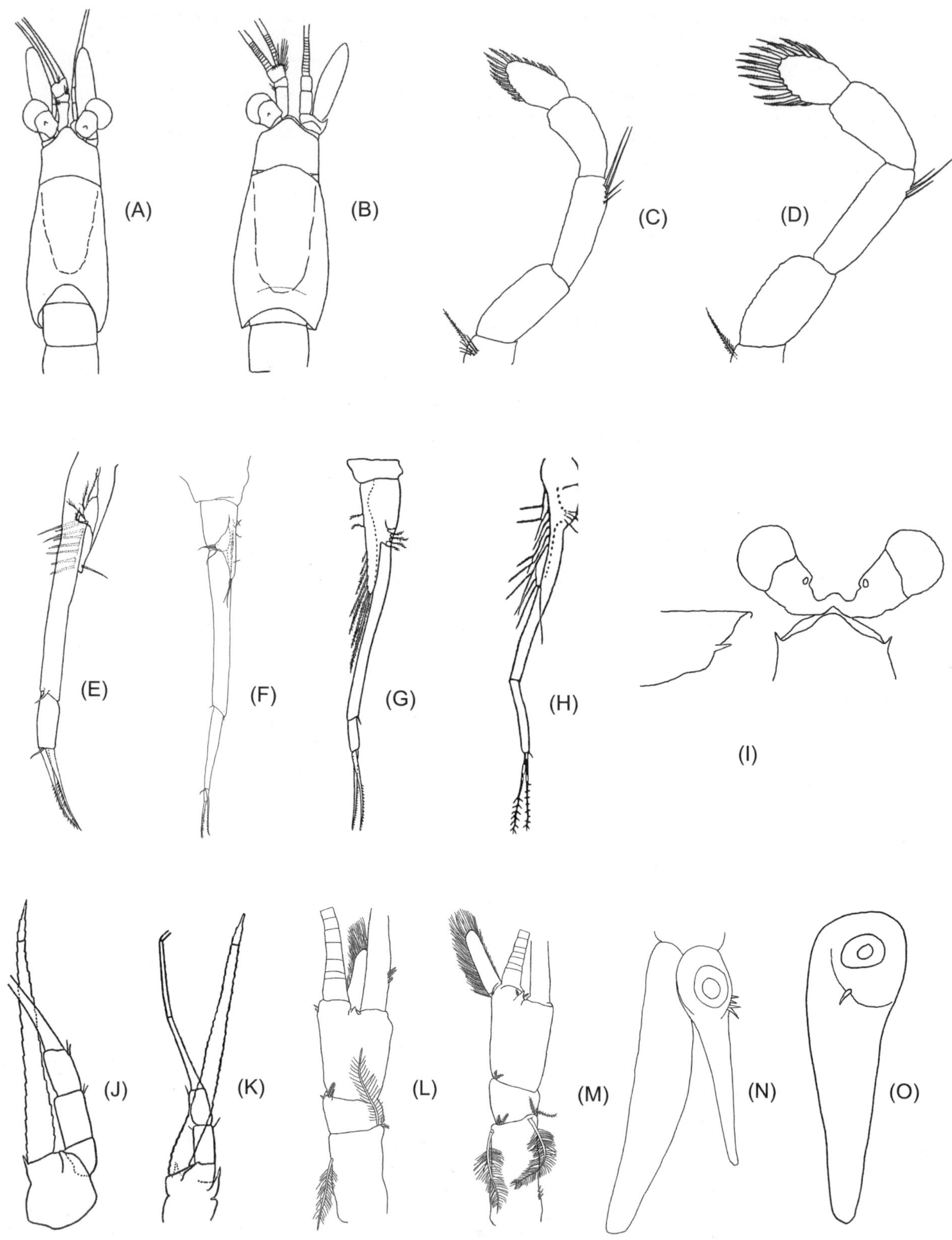

PLATE 16.191.04 (A) *Mysis diluviana*, anterior body, dorsal; (B) *M. segerstralei*, anterior body, dorsal (from Audzijonte & Väinölä, 2005); (C) *M. nordenskioldi*, thoracic endopod 2; (D) *M. litoralis*, thoracic endopod 2 (from Audzijonte & Väinölä, 2007); (E) *Neomysis mercedis,* male pleopod 4; (F) *N. intermedia*, male pleopod 4 (after Holmquist, 1973); (G) *N. japonica*, male pleopod 4 (from Tattersall, 1951); (H) N. *kadiakensis* male pleopod 4 (after Tattersall, 1951); (I) *Taphromysis louisianae*, anterior body, lateral, dorsal (from Price, 1982); (J) *Neomysis japonica*, antenna 2 (from Tattersall, 1951); (K) *N. kadiakensis*, antenna 2 (from Tattersall, 1951); (L) *N. japonica*, antenna 1 peduncle (original); (M) *N. kadiakensis*, antenna 1 peduncle (original); (N) *Orientomysis hwanhaiensis*, uropod (from Modlin & Orsi, 2000); (O) *Orientomysis aspera*, uropodal endopod (from Modlin, 2007).

2’ Telson, lateral spiniform setae smaller than apical pair, lateral margin with 22–26 (19–29) spiniform setae; carapace posterior cleft shallow, <0.14 carapace length (Pls. 16.191.2 J, 4 B) .... *Mysis segerstralei* Audzijonte & Väinölä, 2005

[Canada: Northwest Territories, Nunavut, Yukon. USA: Alaska. Greenland. Palaearctic]

3(1) Telson apical lobes with 2–4 lateral spiniform setae posterior to anterior margin of cleft; thoracic endopod 2 dactylus armed with serrated spiniform setae, leaving about 0.5 of mesial margin unarmed; thoracic endopod 2 merus with 2–3 long distolateral setae reaching nearly 0.5 length of carpo-propodus (Pls. 16.191.2 K, 4 C) .... *Mysis nordenskioldi* Audzijonte & Väinölä, 2007

[Canada: Nunavut, Quebec. USA: Alaska. Greenland. Palaearctic]

3’ Telson apical lobes with 1–2 (3) lateral spiniform setae posterior to anterior margin of cleft; thoracic endopod 2 dactylus armed with stout serrated spiniform setae, leaving only 0.2 of mesial margin unarmed; thoracic endopod 2 merus with 1 long distolateral seta reaching about 0.2 length of carpo-propodus (Pls. 16.191.2 L, 4 D) .... *Mysis litoralis* (Banner, 1948)

[Canada: British Columbia. USA: Alaska (?), Washington.]

Crustacea: Malacostraca: Mysida: Mysidae: *Taphromysis*: Species

1 Carapace, anterior margin with small lateral spine just below base of eyestalk; telsonal cleft armed with 30–40 spinules along entire margin (Pls. 16.191.2 M, 4 I) .... *Taphromysis louisianae* Banner, 1953

[USA: Alabama, Florida, Kentucky, Louisiana, Mississippi, Ohio, Oklahoma, Texas. Mexico: Tamaulipas.]

1’ Carapace, anterior margin without lateral spine just below base of eyestalk; telsonal cleft armed with 30 or fewer spinules along entire margin (Pl. 16.191.2 N) .... *Taphromysis bowmani* Băcescu, 1961

[USA: Florida, Mississippi, Texas]

Crustacea: Malacostraca: Mysida: Mysidae: *Neomysis*: Species

1 Telson, lateral margins with 16 or less, widely spaced spiniform setae, length slightly less than 2 times basal width at broadest interval (Pl. 16.191.2 O, P) 2

1’ Telson, lateral margins with 25 or more, closely spaced spiniform setae, length more than 2 times basal width at widest interval (Pl. 16.191.2 Q, R) 3

2(1) Telson, inner 2 apical spiniform setae 0.2–0.3 times as long as 2 outer apical setae, apex width 0.15–0.20 times basal width at broadest interval; male pleopod 4 slightly curved, extending almost to posterior end of last abdominal somite (Pls. 16.191.1 A, 2 P, 4 E) .... *Neomysis mercedis* Holmes, 1897

[Canada: British Colombia. USA: Alaska, California, Oregon, Washington]

2’ Telson, inner 2 apical spiniform setae at least 0.4 times as long as 2 outer apical setae, apex width 0.21–0.29 times basal width at broadest interval; male pleopod 4 almost straight, extending to the middle or posterior half of telson (Pls. 16.191.2 O, 4 F) .... *Neomysis intermedia* (Czerniasky, 1882)

[Canada: Northwest Territory. USA: Alaska. Palaearctic]

3(1) Antennal scale 8–10 times longer than greatest width; telson length 2.2 times or less basal width at broadest interval; male pleopod 4 exopod, distal article 1/6-1/9 length of proximal article; distal article less than 1/2 length of terminal pair of spiniform setae; antenna 1 peduncle, article 1 with a long plumose seta on distal 1/3 of mesial margin (Pls. 16.191.2 Q, 4 G, J, L) .... *Neomysis japonica* Nakazawa, 1910

[Invasive. USA: California? Palaearctic]

3’ Antennal scale 11–14 times longer than greatest width; telson length 2.3 times or more basal width at broadest interval; male pleopod 4 exopod, distal article 1/4-1/2 length of proximal article, distal article greater than 1/2 length of terminal pair of spiniform setae; antenna 1 peduncle, article 1 with a long plumose seta on or near distomedial corner (Pls. 16.191.2 R, 4 H, K, M) .... *Neomysis kadiakensis* Ortmann, 1908

[Canada: British Columbia. USA: Alaska, California, Washington]

Crustacea: Malacostraca: Mysida: Mysidae: *Orientomysis*: Species

[Modified from Fukuoka and Murano (2005)]

1 Telson, posterior margin armed with 2 pairs of large spiniform setae with 1–3 small spiniform setae between inner and outer pairs of large setae, large apical setae subequal with larger lateral spiniform setae; uropodal endopod with 4 spiniform setae on medial margin near distal edge of statocyst (Pls. 16.191.2 S, 4 N) .... *Orientomysis hwanhaiensis* (Ii, 1964)

[Invasive. USA: California. Palaearctic]

1’ Telson, posterior margin armed with 2 pairs of large subequal spiniform setae, all distinctly longer than larger lateral setae; uropodal endopod with 1–2 spiniform setae on medial margin near distal edge of statocyst (Pls. 16.191.2 T, U, 4 O) .... *Orientomysis aspera* (Ii, 1964)

[Invasive. USA: California. Palaearctic]

REFERENCES

Audzijonte, A. & R. Väinölä, 2005. Diversity and distributions of circumpolar fresh- and brackish-water *Mysis* (Crustacea: Mysida): descriptions of *M. relicta* Lovén, 1862, *M. salemaai* n. sp., *M. segerstralei* n. sp. and *M. diluviana* n. sp., based on molecular and morphological characters. Hydrobiologia 544: 89–141.

Audzijonte, A. & R. Väinölä. 2007. *Mysis nordenskioldi* n. sp. (Crustacea, Mysida), a circumpolar coastal mysid separated from the NE Pacific *M. litoralis* (Banner, 1948). Polar Biology 30: 1137–1157.

Băcescu, M. 1954. Mysidacea, Fauna Republicii Populare Romîne, Crustacea. Editura Academiei Republicii Populare Romîne 4: 1–126.

Bowman, T. 1973. Two new American species of *Spelaeomysis* (Crustacea: Mysidacea) from a Mexican cave and land crab burrows. Association for Mexican Cave Studies, Bulletin 5: 13–20.

Bowman, T. 1977. A review of the genus *Antromysis* (Crustacea: Mysidacea), including new species from Jamaica and Oaxaca, Mexico, and a redescription and new records for *A. cenotensis*. Association for Mexican Cave Studies, Bulletin 6: 27–38.

Bowman, T. & J. Orsi. 1992. *Deltamysis holmquistae*, a new genus and species of Mysidacea from the Sacramento-San Joaquin estuary of California (Mysidae: Mysinae: Heteromysini). Proceedings of the Biological Society of Washington 105: 733–742.

Brattegard, T. 1969. Marine biological investigations in the Bahamas. 10. Mysidacea from shallow water in the Bahamas and Southern Florida. Part I. Sarsia 39: 17–106.

Daly, K. & C. Holmquist. 1986. A key to the Mysidacea of the Pacific Northwest. Canadian Journal of Zoology 64: 1201–1210.

Escobar-Briones, E. & L. A. Soto. 1988. Mysidacea from Terminos Lagoon, Southern Gulf of Mexico, and description of a new species of *Taphromysis*. Journal of Crustacean Biology 8: 639–655.

Fukuoka, K. & M. Murano. 2005. A revision of East Asian *Acanthomysis* (Crustacea: Mysida: Mysidae) and redefinition of *Orientomysis*, with description of a new species. Journal of Natural History 39: 657–708.

García-Garza, M. E., G. A. Rodríguez-Almarez & T. E. Bowman. 1996. *Spelaeomysis villalobosi*, a new species of mysidacean from northeastern México (Crustacea: Mysidacea). Proceedings of the Biological Society of Washington 109: 97–102.

Heard, R. A., W. W. Price, D. M. Knott, R. A. King & D. M. Allen. 2006. A taxonomic guide to the mysids of the South Atlantic Bight. NOAA Professional Paper NMFS 4. 37 pp.

Holmquist, C. 1958. On a new species of the genus *Mysis*, with some notes on *Mysis oculata* (O. Fabricius). Meddelelser om Grönland 159: 1–17.

Holmquist, C. 1959. Problems In Marine-Glacial Relicts on Account of Investigations on the Genus *Mysis*. Lund, Berlingska Boktryckeriet, 270 pp.

Holmquist, C. 1973. Taxonomy, distribution and ecology of the three species *Neomysis intermedia* (Czerniavsky), *N. awatschensis* (Brandt) and *N. mercedis* Holmes (Crustacea, Mysidacea). Zoologische Jahrbücher Abteilung für Systematik, Ökologie, und Geographie der Tiere 100: 197–222.

Holmquist, C. 1981. The genus *Acanthomysis* Czerniavsky, 1882 (Crustacea, Mysidacea). Zoologische Jahrbücher Abteilung für Systematik, Ökologie, und Geographie der Tiere 108: 386–415.

Kestrup, A. & A. Ricciardi. 2008. Occurrence of the Ponto-Caspian mysid shrimp *Hemimysis anomala* (Crustacea, Mysida) in the St. Lawrence River. Aquatic Invasions 3: 461–464.

Modlin, R. 2007. Mysidacea. Pages 489–495 *in*: J. T. Carlton. (ed.) The Light and Smith Manual, Intertidal Invertebrates from Central California to Oregon. University of California Press, Berkeley, CA.

Modlin, R. & J. Orsi, 1997. *Acanthomysis bowmani*, a new species, and *A. aspera* Ii, Mysidacea newly reported from the Sacramento-San Joaquin Estuary, California (Crustacea: Mysidae). Proceedings of the Biological Society of Washington 110: 439–446.

Modlin, R. & J. Orsi, 2000. Range extension of *Acanthomysis hwanhaiensis* Ii, 1964, to the San Francisco estuary California, and notes on its description (Crustacea: Mysidacea). Proceedings of the Biological Society of Washington 113: 690–695.

Porter, M., K. Meland & W. Price. 2008. Global diversity of mysids (Crustacea-Mysida) in freshwater. Hydrobiologia 595: 213–218.

Price, W. W. 1982. A key to the Mysidacea of the Texas coast with notes on their ecology. Hydrobiology 93: 9–21.

Tattersall, W. 1951. A review of the Mysidacea of the United States National Museum. Smithsonian Institution, United States National Museum Bulletin 201: 1–292.

Villalobos, A. 1951. Un nuevo Misidáceo de las grutas de Quintero en el Estado de Tamaulipas. Anales del Instituto de Biología Universidad Nacional Autónoma de México 22: 191–218.

Order Stygiomysida

W. Wayne Price

Department of Biology, University of Tampa, Tampa, FL, USA

INTRODUCTION

Only four species of Stygiomysida (stygiomysids) occur in Nearctic freshwater habitats. Stygiomysids occur in subterranean habitats such as caves, groundwater, wells, and crab burrows.

LIMITATIONS

Identification of Nearctic stygiomysids is based mainly on morphological variations in the eyes, telson, and uropods, characters that exhibit little sexual dimorphism. These and other basic morphological characters included in the key are presented in Pl. 16.191.1. Whenever possible, mature adult specimens should be examined because many characters used in the key are from adults as opposed to juveniles. This is especially important for meristic characters such as the number of spiniform setae on the lateral and posterior margins of telsons and the uropods. Setae may be added with successive molts as the organism matures.

TERMINOLOGY AND MORPHOLOGY

The terminology and morphology is essentially the same as for the Mysida (see above). Stygiomysids have a carapace that covers most of the thoracic somites in *Spelaeomysis*, although it is somewhat shorter posteriorly in other members of the order. The stygiomysid and mysid first antenna consists of a peduncle of three articles followed by two flagella and is often more robust in males than females. In addition, males have a setose conical process, the male lobe (appendix masculina), at the distal end of article 3 of the peduncle. This lobe is reduced in stygiomysids. The biramous second antenna has an endopod with a 3-article peduncle and a flagellum. The stygiomysid and mysid exopod or antennal scale is in the form of a flat plate, that may be reduced in stygiomysids.

Mature female stygiomysids have 4–7 pairs of oostegites on their thoracopods. Unlike the Mysida, mature male stygiomysids do not have paired gonopods (penes) extending as tubular lobes from the medial bases of the 8th thoracic appendages (Wittman, 2013).

The pleopods are sexually undifferentiated (usually) and biramous with plate-like endopods and multiarticulated exopods. In addition, abdominal somites 3–5 have transverse lamellae extending from their posterior sternal margins.

Each uropod comprises a uniarticulate protopod and two branches, the exopod and endopod. In Nearctic stygiomysids, the protopod is produced on the ventral surface into a posteriorly directed prolongation, the exopod has a distal articulation (*Spelaeomysis*), and the endopod lacks a statocyst.

MATERIAL PREPARATION AND PRESERVATION

Stygiomysids are prepared and preserved as for mysids (above).

KEYS TO STYGIOMYSIDA

Crustacea: Malacostraca: Stygiomysida: Lepidomysidae: *Spelaeomysis*: Species

1 Eyes without ommatidia (Pl. 16.191.3 F–H) 2

1' Eyes with a few distolateral ommatidia (Pl. 16.191.3 E) .... *Spelaeomysis cardisomae* Bowman, 1973
[USA: Florida. Neotropics]

2(1) Eyestalks quadrate or produced distolaterally into rounded subtriangular lobes; telson length 1.1–1.3 times basal width at broadest interval, lateral margins straight or slightly convex (Pls. 16.191.2 B, C, 3 F, H) 3

2' Eyestalks produced distolaterally into pointed subtriangular lobes; telson as long as basal width at broadest interval, lateral margins evenly convex (Pls. 16.191.2 A, 3 G) .... *Spelaeomysis villalobosi* García-Garza, Rodríguez-Almaraz & Bowman, 1996
[Mexico: Nuevo León]

3(2) Eyestalks produced distolaterally into rounded subtriangular lobes; telson with about 25 marginal spiniform setae; uropodal protopod medial lobe with no long apical spiniform seta (Pls. 16.191.2 B, 3 D, F) .... *Spelaeomysis quinterensis* (Villalobos, 1951)
[Mexico: Tamaulipas, San Luis Potosí]

3' Eyestalks quadrate; telson with about 40 marginal spiniform setae; uropodal protopod medial lobe with a long apical spiniform seta (Pls. 16.191.2 C, 3 C, H) .... *Spelaeomysis olivae* Bowman, 1973
[Mexico: Oaxaca]

REFERENCE

Wittmann, K. J. 2013. Comparative morphology of the external male genitalia in Lophogastrida, Stygiomysida, and Mysida (Crustacea, Eumalacostraca). Zoomorphology 132: 389–401.

PHYLUM ARTHROPODA

Taxonomy Index

'*Note:* Page numbers followed by "f" indicate figures, "t" indicate tables.'

B

C

D

F

G

H

P

Q

R

S

U

V